Microbiology

Third Edition

Lansing M. Prescott
Augustana College

John P. Harley
Eastern Kentucky University

Donald A. Klein
Colorado State University

WCB **Wm. C. Brown Publishers**

Dubuque, IA   Bogota   Boston   Buenos Aires   Caracas   Chicago
Guilford, CT   London   Madrid   Mexico City   Sydney   Toronto

Book Team

Editor *Elizabeth M. Sievers*
Developmental Editor *Terry Stanton*
Production Editor *Julie L. Wilde*
Cover Designer *Lu Ann Schrandt*
Art Editor *Jodi K. Banowetz*
Photo Editor *Lori Hancock*
Permissions Coordinator *LouAnn K. Wilson*

Wm. C. Brown Publishers

President and Chief Executive Officer *Beverly Kolz*
Vice President, Publisher *Kevin Kane*
Vice President, Director of Sales and Marketing *Virginia S. Moffat*
Vice President, Director of Production *Colleen A. Yonda*
National Sales Manager *Douglas J. DiNardo*
Marketing Manager *Patrick E. Reidy*
Advertising Manager *Janelle Keeffer*
Production Editorial Manager *Renée Menne*
Publishing Services Manager *Karen J. Slaght*
Royalty/Permissions Manager *Connie Allendorf*

A Times Mirror Company

Copyedited by Beatrice Sussman

Cover—Main Text: From *The Incredible Machine,* photo by Lennart Nilsson
Volume I: Esther R. Angert and Norman R. Pace, Indiana University
Volume II: © A.B. Dowsett/SPL/Photo Researchers, Inc.

The credits section for this book begins on page C1 and is considered
an extension of the copyright page.

Copyright © 1996 Times Mirror Higher Education Group, Inc.
All rights reserved

Library of Congress Catalog Card Number: 95–80772

ISBN 0–697–29390–4 Main Text
ISBN 0–697–21865–1 Vol. I
ISBN 0–697–21866–X Vol. II

No part of this publication may be reproduced, stored in a retrieval
system, or transmitted, in any form or by any means, electronic,
mechanical, photocopying, recording, or otherwise, without the
prior written permission of the publisher.

Printed in the United States of America by Times Mirror Higher Education Group, Inc.,
2460 Kerper Boulevard, Dubuque, IA 52001

10 9 8 7 6 5 4 3 2 1

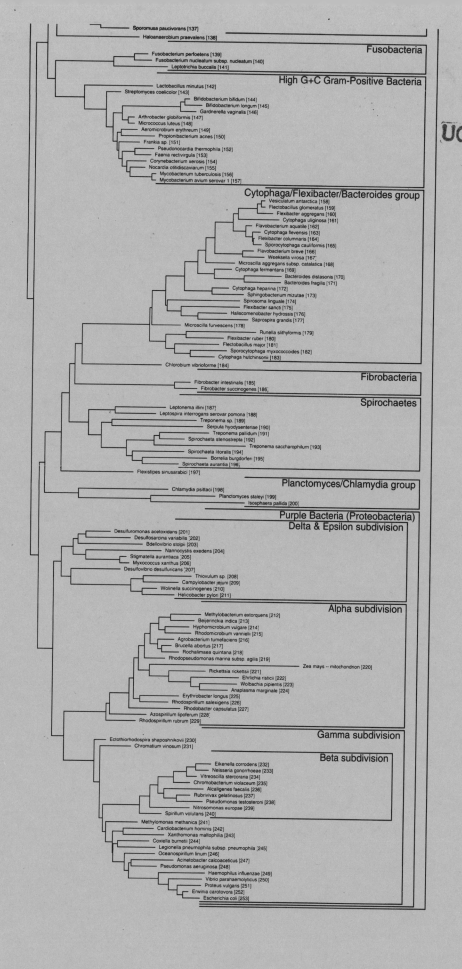

Sporomusa paucivorans [137]
Haloanaerobium praevalens [138]

Fusobacteria

Fusobacterium perfoetens [139]
Fusobacterium nucleatum subsp. nucleatum [140]
Leptotrichia buccalis [141]

High G+C Gram-Positive Bacteria

Lactobacillus minutus [142]
Streptomyces coelicolor [143]
Bifidobacterium bifidum [144]
Bifidobacterium longum [145]
Gardnerella vaginalis [146]
Arthrobacter globiformis [147]
Micrococcus luteus [148]
Aeromicrobium erythreum [149]
Propionibacterium acnes [150]
Frankia sp. [151]
Pseudonocardia thermophila [152]
Faenia rectivirgula [153]
Corynebacterium xerosis [154]
Nocardia otitidiscaviarum [155]
Mycobacterium tuberculosis [156]
Mycobacterium avium serovar 1 [157]

Cytophaga/Flexibacter/Bacteroides group

Vesiculatum antarctica [158]
Flectobacillus glomeratus [159]
Flexibacter aggregans [160]
Cytophaga uliginosa [161]
Flavobacterium aquatile [162]
Cytophaga flevensis [163]
Flexibacter columnaris [164]
Sporocytophaga cauliformis [165]
Flavobacterium breve [166]
Weeksella virosa [167]
Microscilla aggregans subsp. catalatica [168]
Cytophaga fermentans [169]
Bacteroides distasonis [170]
Bacteroides fragilis [171]
Cytophaga heparina [172]
Sphingobacterium mizutae [173]
Sprosoma linguale [174]
Flexibacter sancti [175]
Haliscomenobacter hydrossis [176]
Saprospira grandis [177]
Microscilla furvescens [178]
Runella slithyformis [179]
Flexibacter ruber [180]
Flectobacillus major [181]
Sporocytophaga myxococcoides [182]
Cytophaga hutchinsonii [183]
Chlorobium vibrioforme [184]

Fibrobacteria

Fibrobacter intestinalis [185]
Fibrobacter succinogenes [186]

Spirochaetes

Leptonema illini [187]
Leptospira interrogans serovar pomona [188]
Treponema sp. [189]
Serpula hyodysenteriae [190]
Treponema pallidum [191]
Spirochaeta stenostrepta [192]
Treponema saccharophilum [193]
Spirochaeta litoralis [194]
Borrelia burgdorferi [195]
Spirochaeta aurantia [196]
Flexistipes sinusarabici [197]

Planctomyces/Chlamydia group

Chlamydia psittaci [198]
Planctomyces staleyi [199]
Isosphaera pallida [200]

Purple Bacteria (Proteobacteria)
Delta & Epsilon subdivision

Desulfuromonas acetoxidans [201]
Desulfosarcina variabilis [202]
Bdellovibrio stolpii [203]
Nannocystis exedens [204]
Stigmatella aurantiaca [205]
Myxococcus xanthus [206]
Desulfovibrio desulfuricans [207]
Thiovulum sp. [208]
Campylobacter jejuni [209]
Wolinella succinogenes [210]
Helicobacter pylori [211]

Alpha subdivision

Methylobacterium extorquens [212]
Beijerinckia indica [213]
Hyphomicrobium vulgare [214]
Rhodomicrobium vannielii [215]
Agrobacterium tumefaciens [216]
Brucella abortus [217]
Rochalimaea quintana [218]
Rhodopseudomonas marina subsp. agilis [219]
Zea mays -- mitochondrion [220]
Rickettsia rickettsii [221]
Ehrlichia risticii [222]
Wolbachia pipientis [223]
Anaplasma marginale [224]
Erythrobacter longus [225]
Rhodospirillum salexigens [226]
Rhodobacter capsulatus [227]
Azospirillum lipoferum [228]
Rhodospirillum rubrum [229]

Gamma subdivision

Ectothiorhodospira shaposhnikovii [230]
Chromatium vinosum [231]

Beta subdivision

Eikenella corrodens [232]
Neisseria gonorrhoeae [233]
Vitreoscilla stercoraria [234]
Chromobacterium violaceum [235]
Alcaligenes faecalis [236]
Rubrivivax gelatinosus [237]
Pseudomonas testosteroni [238]
Nitrosomonas europae [239]
Spirillum volutans [240]
Methylomonas methanica [241]
Cardiobacterium hominis [242]
Xanthomonas maltophilia [243]
Coxiella burnetii [244]
Legionella pneumophila subsp. pneumophila [245]
Oceanospirillum linum [246]
Acinetobacter calcoaceticus [247]
Pseudomonas aeruginosa [248]
Haemophilus influenzae [249]
Vibrio parahaemolyticus [250]
Proteus vulgaris [251]
Erwinia carotovora [252]
Escherichia coli [253]

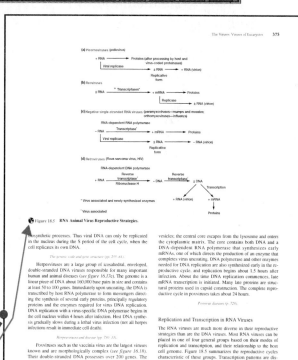

What do we REALLY need to KNOW?

Wm. C. Brown Publishers has THE ANSWER—microbiology aids that are very handy to have come study time.

When it's time to study, **TIME is of the essence. You need to figure out what you know—and don't know—and focus on the areas where you need help.**

Now zeroing in on what to study is easy with this computerized study guide.

BRIEF CONTENTS

EXPANDED CONTENTS

PART TWO

Microbial Nutrition, Growth, and Control

PART THREE
Microbial Metabolism

PART FOUR
Microbial Genetics

13   Microbial Genetics: General Principles   254

14   Microbial Genetics: Recombination and Plasmids   277

15   Recombinant DNA Technology   305

PART FIVE
The Viruses

PART SIX
The Diversity of the Microbial World

PART SEVEN
The Nature of Symbiotic Associations

PART EIGHT
The Fundamentals of Immunology

30   The Immune Response: Antigens and Antibodies   586

31   The Immune Response: Chemical Mediators, B- and T-Cell Biology, and Immune Disorders   606

32   The Immune Response: Antigen–Antibody Reactions   632

PART NINE
Microbial Diseases and Their Control

PART TEN
Microorganisms and the Environment

PART ELEVEN
Food and Industrial Microbiology

44   Industrial Microbiology and Biotechnology   905

PREFACE

Books are the carriers of civilization. Without books, history is silent, literature dumb, science crippled, thought and speculation at a standstill. They are engines of change, windows on the world, lighthouses erected in a sea of time.

—Barbara Tuchman

Microbiology is an exceptionally broad discipline encompassing specialties as diverse as biochemistry, cell biology, genetics, taxonomy, pathogenic bacteriology, food and industrial microbiology, and ecology. A microbiologist must be acquainted with many biological disciplines and with all major groups of microorganisms: viruses, bacteria, fungi, algae, and protozoa. The key is balance. Students new to the subject need an introduction to the whole before concentrating on those parts of greatest concern to them. This text provides a balanced introduction to all major areas of microbiology for a variety of students. Because of this balance, the book is suitable for courses with orientations ranging from basic microbiology to medical and applied microbiology. Students preparing for careers in medicine, dentistry, nursing, and allied health professions will find the text just as useful as those aiming for careers in research, teaching, and industry. Two quarters/semesters each of biology and chemistry are assumed, and an overview of relevant chemistry is also provided in appendix I.

Organization and Approach

The book is organized flexibly so that chapters and topics may be arranged in almost any order. Each chapter has been made as self-contained as possible to promote this flexibility. Some core topics are essential to microbiology and have been given more extensive treatment. The chapters that cover these topics (chapters 2–18 and 29–32) are somewhat longer than others and may be combined with a selection of noncore chapters to achieve the desired orientation, whether basic, applied, or medical. The first nine parts of the text also are available as two separate paperbound volumes. Volume one covers the general aspects of microbiology (chapters 1–27); and volume two, immunology and medical microbiology (chapters 28–39). This allows use of separate volumes in courses with a specific focus.

The book is divided into 11 parts. The first 5 introduce the foundations of microbiology: the development of microbiology, the structure of microorganisms, microbial growth and its

control, metabolism, microbial genetics, and the nature of viruses. In this third edition the microbial genetics unit directly follows microbial growth and metabolism (Parts Two and Three) and discusses genetics before the introduction to viruses (Part Five). Part Six is a survey of the microbial world. The bacterial survey closely follows the organization of *Bergey's Manual of Systematic Bacteriology*. Because of the importance and uniqueness of archaeobacteria, the third edition contains a separate chapter on them. Although principal attention is devoted to bacteria, eucaryotic microorganisms receive more than usual coverage. Fungi, algae, and protozoa are important in their own right. The introduction to their biology in chapters 25–27 is essential to understanding topics as diverse as clinical microbiology and microbial ecology. Part Seven discusses symbiotic associations and parasitism in depth, providing a good foundation for the later survey of specific diseases. Three chapters in Part Eight describe in detail all major aspects of the immune response. Part Nine begins with an introduction to antimicrobial chemotherapy, epidemiology, and clinical microbiology. This is followed by a survey of the major human microbial diseases. The disease survey is primarily organized taxonomically on the chapter level; within each chapter diseases are covered according to mode of transmission. This approach provides flexibility and allows the student easy access to information about any disease of interest. The survey is not a simple cataloguing of diseases; diseases are included because of their medical importance and their ability to illuminate the basic principles of disease and resistance. Part Ten focuses on the relationship of microorganisms to aquatic and terrestrial environments. Chapter 40 presents the general principles underlying microbial ecology and environmental microbiology so that the chapters on aquatic and terrestrial habitats can be used without excessive redundancy. Part Eleven concludes the text with an introduction to food and industrial microbiology. Five appendices aid the student with a review of some basic chemical concepts and with extra information about important topics not completely covered in the text.

Many substantial changes and improvements have been made in the third edition, including the following:

1. The organization of the book has been modified slightly to make it more logical. The chapter on the control of microorganisms by physical and chemical agents has been moved to the section on growth and nutrition (Part Two). The antimicrobial therapy chapter now is in the section on microbial diseases (Part Nine). Parts Two and Nine have been renamed to reflect this change in organization.

2. Reference sections have been reorganized to group references under topical heads (these usually correspond to the major sections in each chapter). This arrangement provides ease of access for students interested in particular topics. New references have been added through the middle of 1995 and some older references deleted.

3. The appendices have been revised. A new appendix that visually compares the relative sizes of biomolecules, viruses, bacteria, and eucaryotic cells has been added.

4. Many new topics and sections have been added and others have been substantially revised. Some examples are the following:

    a. Chapter 3—The molecular mechanism of chemotaxis is described, and molecular chaperones and heat-shock proteins are discussed. A box on *Epulopiscium fishelsoni* has been added.

    b. Chapter 12—There is a new section on regulation by sigma factors and updated material on regulation of the bacterial cell cycle.

    c. Chapter 19—The chapter on microbial taxonomy has been reorganized to make it more logical. Phylogenetic diagrams are updated. Six-and eight-kingdom systems are described.

    d. Chapters 20–24—The taxonomy of bacteria has been updated to correspond more closely to that in the ninth edition of *Bergey's Manual of Determinative Bacteriology* and reflect recent developments in bacterial classification.

    e. Chapter 26—There is new material on toxic algal blooms.

    f. Chapter 28—A section covering the symbiosis between hydrothermal vent tube worms and bacteria has been added. There is a new box on probiotics for humans and animals.

    g. Chapter 29—There is new material on the causes of inflammation.

    h. Chapter 31—The section on cytokines has been expanded to incorporate recent research findings. There is new material on T cells and their activation. Sections on superantigens and cell-associated differentiation antigens have been added.

    i. Chapter 35—The relationship between virulence and the mode of transmission is discussed.

    j. Chapter 36—New material or substantial updates have been added for hantaviruses, AIDS, herpesvirus 6, parvovirus, rabies virus, and the hepatitis viruses.

    k. Chapter 37—There is new or expanded coverage of invasive streptococci and necrotizing fasciitis, multi-drug-resistant tuberculosis, *Helicobacter pylori,* shigellosis, diarrheagenic *Escherichia coli* infections, septic shock, and the medical aspects of biofilms.

    l. Chapter 38—A section on ehrlichosis has been included.

    m. Chapter 41—There is new material on biofilms in microbial ecology, confocal scanning microscopy, the sea as a source of biomedicines, and *Cryptosporidium* as a waterborne disease agent.

    n. Chapter 42—The sections on the *Rhizobium* infection of legumes and microbial effects on atmospheric gases have been significantly updated.

    o. Chapter 44—The material on bioremediation and the use of *Bacillus thuringiensis* as an insecticide has been revised and updated.

This text is designed to be an effective teaching tool. A text is only as easy for a student to use as it is easy to read. Readability has been enhanced by using a relatively simple, direct writing style, many section headings, and an organized outline format within each chapter. The level of difficulty has been carefully set with the target audience in mind. During preparation of the third edition, every sentence was carefully checked for clarity and revised when necessary. The American Society for Microbiology's *ASM Style Manual* conventions for nomenclature and abbreviations have been followed as consistently as possible.

The many new terms encountered in studying microbiology are a major stumbling block for students. This text lessens the problem by addressing and reinforcing a student's vocabulary development in three ways: (1) no new term is used without being clearly defined (often derivations also are given)—a student does not have to be familiar with the terminology of microbiology to use this text; (2) the most important terms are printed in boldface when first used; and (3) an extensive, up-to-date, page-referenced glossary is included at the end of the text.

Because illustrations are critical to a student's learning and enjoyment of microbiology, all illustrations are full-color, and as many color photographs as possible have been used. Color not only enhances the text's attractiveness but also increases each figure's teaching effectiveness. Considerable effort has gone into making the art as attractive and useful as possible. About a third of the art in this edition is either new or revised, and all line art has been produced under the direct supervision of the authors and designed to illustrate and reinforce specific points in the text. Consequently, every illustration is directly related to the narrative and specifically cited where appropriate. Great care has been taken to position illustrations as close as possible to the places where they are cited. Illustrations and captions have been reviewed for accuracy and clarity.

Themes in the Book

At least seven themes permeate the text, though a particular one may be more obvious at some points than are others. These themes or emphases are the following:

1. The development of microbiology as a science

2. The nature and importance of the techniques used to isolate, culture, observe, and identify microorganisms

3. The control of microorganisms and reduction of their detrimental effects

4. The importance of molecular biology for microbiology

5. The medical significance of microbiology

6. The ways in which microorganisms interact with their environments and the consequences of these interactions

7. The influences that microorganisms and microbiological applications have on everyday life

These themes help unify the text and enhance continuity. The student should get a feeling for what microbiologists do and for how this affects society.

Aids to the Student

It is hard to overemphasize the importance of pedagogical aids for the student. Accuracy is most important, but if a text is not clear, readable, and attractive, up-to-dateness and accuracy are wasted because students will not read it. Students must be able to understand the material being presented, effectively use the text as a learning tool, and enjoy reading the book.

To be an effective teaching tool, a text must present the science of microbiology in a way that can be clearly taught and easily learned. Therefore many aids are included to make the task of learning more efficient and enjoyable. Following the preface a special section addressed to the student user reviews the principles of effective learning, including the SQ4R (survey, question, read, revise, record, and review) study technique. Each chapter contains the following:

1. *Opening quote(s):* These are designed to perk student interest and provide perspective on the chapter's contents.

2. *Chapter outline:* The chapter outline, with page numbers, includes all major headings in the chapter. This helps the reader locate particular topics of interest.

3. *Chapter concepts:* Several statements briefly summarize the most important concepts the student should master.

4. *Chapter preface:* One or two short paragraphs preview the chapter's contents and relate it to the rest of the text. The preface is not a summary but allows the student to put the chapter in perspective at the start.

5. *Boldfaced terms:* Important terms are emphasized and clearly defined when they are first used.

6. *Chapter summaries:* A series of brief, numbered statements is designed to serve more as a study guide than as a complete, detailed summary of the chapter.

7. *Key terms:* A list of all boldfaced terms is provided at the end of the chapter to emphasize the most significant facts and concepts. Each term is page-referenced to the page on which the term is first introduced in the chapter.

8. *Review questions and activities:* Two kinds of review questions appear in each chapter. A small box with one to five brief review questions is located at the end of most major sections. These questions help the student master the

section's factual material and major concepts before continuing with the chapter. An example of these in-chapter review questions from page 49 has been reproduced here.

---

1. Briefly describe the nature and function of the cytoplasmic matrix and the ribosome. What is a protoplast?

2. What kinds of inclusion bodies do procaryotes have? What are their functions?

3. What is a gas vacuole? Relate its structure to its function.

4. What are molecular chaperones and heat-shock proteins? Describe their functions.

The "Questions for Thought and Review" section at the end of the chapter contains factual questions and some synthetic questions to aid the student in reviewing, integrating, and applying the material in the chapter.

9. *Multimedia-supported illustrations:* Throughout the text the reader will find illustrations of microbiological concepts and processes that can be supplemented with full-color video, animations, or interactive screens from *Microbes in Motion* (0-697-24596-9), the interactive CD-ROM available from Wm. C. Brown Publishers. The reader will be able to easily recognize these figures, as the figure legends are preceded by a CD icon. Figure 3.37, page 62, is one example of such an illustration.

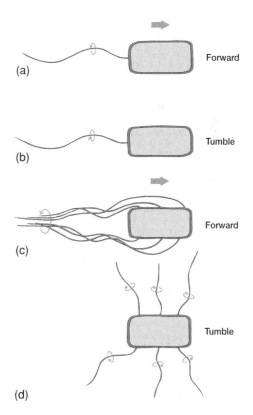

Figure 3.37   Flagellar Motility. The relationship of flagellar rotation to bacterial movement. Parts (*a*) and (*b*) describe the motion of monotrichous, polar bacteria. Parts (*c*) and (*d*) illustrate the movements of peritrichous organisms.

10. *Correlation guide to microbes in motion:* To facilitate finding the corresponding information on the *Microbes in Motion* CD-ROM for the multimedia-supported illustrations described above, a correlation guide is provided following the preface. The *Microbes in Motion* CD-ROM is organized into 15 topical "books," the books are divided up into "chapters," and the chapters have numbered "pages." For each multimedia-supported illustration the correlation guide directs the reader to the book, chapter, and page on the CD-ROM where corresponding material can be found. The correlation guide entry is shown here for the multimedia-supported illustration referred to above.

> Fig. 3.7   Bacterial Structure & Function Book
>   Cell Membrane Chapter
>   pp. 8 & 10 (animation)

11. *Additional readings:* References are provided for further study. Most are reviews, monographs, and *Scientific American* articles rather than original research papers. Publications cited in these reviews introduce sufficiently interested students to the research literature. References through the middle of 1995 have been included. The reference sections also are organized into topical groups that correspond to the major sections in each chapter. This arrangement provides ease of access for students interested in particular topics.

12. *Cross-reference notes:* These notes refer the student to major topics that are difficult and may need review in order to understand the current material. They also point the student either forward or backward to a related item of unusual interest or importance. An example of the usage of the cross-reference note from page 44 has been reproduced here.

> *Osmosis (p. 57).*
> *Transport of substances across membranes (pp. 102–106).*

13. *Boxed readings:* Most chapters also contain one or more boxes, which describe items of interest that are not essential to the primary thrust of the chapter. Topics include currently exciting research areas, the practical impact of microbial activities, items of medical significance, historical anecdotes, and descriptions of extraordinary microorganisms. An example of a boxed reading from page 41 has been reduced and reproduced here.

Besides the chapter aids the text also contains a glossary, an index, and five appendices. The *glossary* defines the most important terms from each chapter and includes page references. Where desirable, phonetic pronunciations also are given. Most of the glossary definitions have not been taken directly from the text but have been rewritten to give the student further understanding of the item. To improve ease of use, the third edition has an extensive, detailed *index*. It has been carefully designed to make text material more accessible. The *appendices* aid the student with extra review of chemical principles and metabolic pathways, provide further details about the taxonomy of bacteria and viruses, and compare the sizes of various molecules and cells.

Supplementary Materials

Many supplementary materials are available to help instructors with their presentations and general course management.

1. An *Instructor's Manual,* written by Ralph Rascati, Kennesaw College, includes a list of transparencies, a list of two slide sets, and a correlation guide that ties *Microbiology,* third edition, to the Microbes in Motion CD-rom. In addition, the number of test questions has been increased to about 60 for each chapter. This extensive battery of more than 2,600 test items is a powerful instructional tool.

2. The *WCB Microtest* is a computerized testing and classroom management service that is offered free on request to adopters of this text. The service provides a database of objective questions for preparing exams and a grade-recording program. The software requires no programming experience and is available in DOS (3.5 inch disk only), WINDOWS, and MAC.

3. A set of 150 full-color acetate *transparencies* is available and may be used to supplement classroom lectures.

4. A *Student Study Art Notebook* is available as a free ancillary with all new textbook orders. The Student Study Art Notebook will come packaged with *Microbiology,* third edition by Prescott, Harley, and Klein and contains the 150 illustrations found in the transparency set. With this notebook, students no longer have to worry about trying to sketch out the images that appear on the overhead while trying to take notes and listen to the professor's lecture. The art is reproduced in a handy notebook that the student can use to take notes on the figures.

5. The *WCB Electronic Image Bank* contains virtually all line art illustrations from *Microbiology,* third edition, on CD-ROM with easy to use presentation software. The

Box 3.1

A Monstrous Microbe

iologists often have distinguished between procaryotes and eucaryotes based in part on cell size. Generally, procaryotic cells are supposed to be smaller than eucaryotic cells. Procaryotes grow extremely rapidly compared to most eucaryotes and lack the complex vesicular transport systems of eucaryotic cells (*see chapter 4*). It has been assumed that they must be small because of the slowness of nutrient diffusion and the need for a large surface-to-volume ratio. Thus when Fishelson, Montgomery, and Myrberg discovered a large, cigar-shaped microorganism in the intestinal tract of the Red Sea brown surgeonfish, *Acanthurus nigrofuscus*, they suggested in their 1985 publication that it was a protist. The microorganism seemed too large to be anything else. In 1993 Esther Angert, Kendall Clemens, and Norman Pace used rRNA sequence comparisons (*pp. 00–00*) to identify the microorganism, now called *Epulopiscium fishelsoni*, as a procaryote related to the gram-positive genus *Clostridium*.

E. fishelsoni [Latin, *epulum*, a feast or banquet, and *piscium*, fish] can reach a size of 80 μm by 600 μm, and normally ranges from 200 to 500 μm in length (Box Figure 3.1). It is about a million times larger in volume than *Escherichia coli*. Despite its huge size the organism must have procaryotic cell structure. It is motile and swims at about two body lengths a second (approximately 2.4 cm/min) using the bacterial-type flagella that cover its surface. The cytoplasm contains large nucleoids and many ribosomes, as would be required for such a large cell. *Epulopiscium* appears to overcome the size limits set by diffusion by having an outer layer consisting of a highly convoluted plasma membrane. This increases the cell's surface area and aids in nutrient transport. The bacterium reproduces when between

one and about seven daughter cells develop asexually within the parent and then escape through a centrally located slit in the parental cell envelope.

Epulopiscium also has been found in surgeonfish from the Great Barrier Reef in Australia. Presumably it is a symbiont of the surgeonfish, and its reproduction and behavior are closely correlated with its host's feeding habits. During the day when the surgeonfish feeds, the bacteria are in the middle intestine where they reside in a bolus of food and grow larger. At night when the fish is resting, the bacteria move to the back of the intestine where they reside in a bolus of food and divide repeatedly. Because there is little growth between divisions, the bacteria become smaller. In the morning *Epulopiscium* leaves the bolus of food and moves forward again.

It appears that *Epulopiscium* is transmitted between hosts through fecal contamination of the fish's food. The bacterium can be eliminated by starving the surgeonfish for a few days. If juvenile fish that lack the bacterium are placed with infected hosts, they are reinoculated. Interestingly this does not work with uninfected adult surgeonfish.

The discovery of this procaryote greatly weakens the distinction between procaryotes and eucaryotes based on cell size. *Epulopiscium* is certainly larger than a normal eucaryotic cell. In addition, some eucaryotic cells have been discovered that are smaller than previously thought possible. The best example is *Nanochlorum eukaryotum*. *Nanochlorum* is only about 1 to 2 μm in diameter, yet is truly eucaryotic and has a nucleus, a chloroplast, and a mitochondrion. Our understanding of the factors limiting procaryotic cell size must be reevaluated. It is no longer safe to assume that large cells are eucaryotic and small cells are procaryotic.

Box Figure 3.1   *Epulopiscium fishelsoni*. This photograph, taken with pseudo dark field illumination, shows *E. fishelsoni* at the top of the figure dwarfing paramecia at the bottom (×200).

Interactive Slide Show program allows you to present the ready to play slide show offered on CD-ROM, and the *Slide Editor* program enables you to maneuver among the images to create your own multimedia presentation.

6. A set of *projection slides* provides clinical examples of diseases to supplement the illustrations in the text. A second set of *projection slides* provides examples of microbial specimens.

7. A *Student Study Guide* by Ralph Rascati contains chapter objectives, focus questions, mastery tests, and other activities to aid student comprehension.

8. *QuickStudy,* written by David Mardon of Eastern Kentucky University, is a *computerized study guide* available in IBM and Macintosh formats for student purchase. This program contains approximately 1,700 page-referenced questions that review the material covered in *Microbiology.* If incorrect answers are chosen, this study guide provides the student with a page-referenced study plan for further review. See page iv preceding the preface for further information about this study aid.

9. *Microbes in Motion* is an interactive CD-ROM for Windows that brings microbiology to life. This easy to use tutorial can go from the classroom to the resource center to your students' own personal computers. *Microbes in Motion* brings discovery back into the learning and education process through interactive screens, animations, video, audio, and hyperlinking questions. The applications of this CD-ROM are only as limited as your good ideas. See page iii preceding the preface for further information regarding this valuable study/teaching aid.

10. A third edition of the *laboratory manual, Laboratory Exercises in Microbiology* by John P. Harley and Lansing M. Prescott, has been prepared to accompany the text. This manual is directly correlated with the text and designed to be used with it (although it may be used easily with other microbiology textbooks). Like the text the laboratory manual provides a balanced introduction to laboratory techniques and principles that are important in each area of microbiology. The class-tested exercises are modular and short so that an instructor can easily choose only those exercises that fit his or her course. The third edition of the manual contains recipes for all reagents and media. Each exercise in this manual can also be customized by combining the exercises with one's own materials. The local Wm. C. Brown Publishers representative should be contacted for more details on this custom publishing service. A *lab resource guide* has been prepared by David Mardon of Eastern Kentucky University. It has complete answers to all laboratory report questions.

11. A set of 305, 3″ × 5″ *Microbiology Study Cards* prepared by Kent M. Van De Graaff, F. Brent Johnson, Brigham Young University, and Christopher H. Creek features complete descriptions of terms, clearly labeled drawings, clinical information on diseases, and much more.

Acknowledgments

The authors wish to thank the reviewers, who provided detailed criticism and analysis. Their suggestions greatly improved the final product.

Reviewers of the First and Second Editions

Richard J. Alperin
Community College of Philadelphia

Susan T. Bagley
Michigan Technological University

Dwight Baker
Yale University

R. A. Bender
University of Michigan

Hans P. Blaschek
University of Illinois

Dennis Bryant
University of Illinois

Douglas E. Caldwell
University of Saskatchewan

Arnold L. Demain
Massachusetts Institute of Technology

A. S. Dhaliwal
Loyola University of Chicago

Donald P. Durand
Iowa State University

John Hare
Linfield College

Robert B. Helling
University of Michigan–Ann Arbor

Barbara Bruff Hemmingsen
San Diego State University

R. D. Hinsdill
University of Wisconsin–Madison

John G. Holt
Michigan State University

Robert L. Jones
Colorado State University

Martha M. Kory
University of Akron

Robert I. Krasner
Providence College

Ron W. Leavitt
Brigham Young University

David Mardon
Eastern Kentucky University

Glendon R. Miller
Wichita State University

Richard L. Myers
Southwest Missouri State University

G. A. O'Donovan
North Texas State University

Pattle P. T. Pun
Wheaton College

Ralph J. Rascati
Kennesaw State College

Albert D. Robinson
SUNY–Potsdam

Ronald Wayne Roncadori
University of Georgia–Athens

Ivan Roth
University of Georgia–Athens

Thomas Santoro
SUNY–New Paltz

Ann C. Smith
University of Maryland, College Park

David W. Smith
University of Delaware

Paul Smith
University of South Dakota

James F. Steenbergen
San Diego State University

Henry O. Stone, Jr.
East Carolina University

James E. Struble
North Dakota State University

Kathleen Talaro
Pasadena City College

Thomas M. Terry
The University of Connecticut

Michael J. Timmons
Moraine Valley Community College

John Tudor
St. Joseph's University

Robert Twarog
University of North Carolina

Blake Whitaker
Bates College

Oscar Will
Augustana College

Calvin Young
California State University–Fullerton

Reviewers of the Third Edition

Allen C. Rogerson
St. Lawrence University

James E. Miller
Delaware Valley College

Richard A. Patrick, Ph.D.
Summit Editorial Group

Gary Armour
MacMurray College

Penelope J. Padgett
Shippensburg University

Bobbie Pettriess
Wichita State University

Ronald E. Hurlbert
Washington State University

B. T. Lingappa, Ph.D.
College of the Holy Cross

Valdis A. Dzelzkalns
Case Western Reserve University

Elizabeth D. Wolfinger
Meredith College

Jackie S. Reynolds
Richland College

David W. Smith
University of Delaware

James L. Botsford
New Mexico State University

Alfred E. Brown
Auburn University

John W. Fitzgerald
University of Georgia

David A. Mullin
Tulane University

William H. Coleman
University of Hartford

Bernard Frye
University of Texas at Arlington

Dr. David C. Reff
Middle Georgia College

Robert J. Kearns
University of Dayton

Katharine B. Gregg
West Virginia Wesleyan College

Lynn O. Lewis
Mary Washington College

Dr. Deborah Rochefort
Shepherd College

Patrick M. Weir
Felician College

Thomas Punnett
Temple University

Donald C. Cox
Miami University

Russell C. Baskett
Germanna Community College

Van H. Grosse
Columbus College–Georgia

Ann C. Smith
University of Maryland

Michael J. Lemke
Kent State University

Henry Keil
Brunel University

Jill M. Williams
University of Glamorgan

Joan Henson
Montana State University

Garriet W. Smith
University of South Carolina at Aiken

George N. Bennett
Rice University

Merrill Emmett
University of Colorado at Denver

Eileen Gregory
Rollins College

Richard J. Ellis
Bucknell University

Phillip Cunningham
Wayne State University

B. G. Foster
Texas A&M University

Thomas M. Terry
University of Connecticut

Mary Burke
Oregon State University

Heman Witmer
University of Illinois at Chicago

Robert Zdor
Andrews University

Jackie S. Reynolds
Richland College

Dr. K. J. Reddy
SUNY–Binghamton

William G. Hixon
St. Ambrose University

Prakash H. Bhuta
Eastern Washington University

Richard P. Cunningham
SUNY at Albany

Linda E. Fisher
University of Michigan–Dearborn

James Daly
Purchase College, SUNY

Robert Gunsalus
UCLA

Thomas M. Walker
University of Central Arkansas

Barbara B. Hemmingsen
San Diego State University

Harold F. Foerster
Sam Houston State University

Jo Anne Quinlivan, Ph.D.
Holy Names College

Frank B. Dazzo
Michigan State University

David P. Clark
Southern Illinois University

Michael Shiaris
University of Massachusetts at Boston

Laurie A. Achenbach
Southern Illinois University

Michael J. San Francisco
Texas Tech University

Mary L. Taylor
Portland State University

The publication of a textbook requires the effort of many people besides the authors. We would like to express special appreciation to the editorial and production staffs of Wm. C. Brown Publishers for their excellent work. In particular, we would like to thank our project editor, Elizabeth Sievers, for her guidance and support, and our production editor, Julie Wilde, for her effort and attention to detail. Our copy editor, Beatrice Sussman, contributed greatly to the third edition's clarity, consistency, and readability. We would like to offer our special thanks to Raymond B. Otero, Eastern Kentucky University, for the many excellent clinical photographs he contributed to this textbook.

Finally, but most important, we wish to extend appreciation to our families for their patience and encouragement, especially to our wives, Linda Prescott, Jane Harley, and Sandra Klein. To them, we dedicate this book.

Lansing M. Prescott
John P. Harley
Donald A. Klein

Correlation Guide for
Microbes in Motion to *Microbiology*

This correlation guide was compiled in order to assist you in integrating the information from *Microbes in Motion,* the tutorial CD-ROM developed and published by Wm. C. Brown Publishers, into the textual material presented in this third edition of *Microbiology* by Prescott, Harley, and Klein.

Microbes in Motion is an interactive tutorial that can be used by professors and students alike. This highly visual tutorial brings the study of microbiology to life through use of animations, video, audio, music, interactive screens, and quizzes. Your understanding of microbiological concepts will be enhanced using this CD-ROM

tutorial. In some cases the discussion on *Microbes in Motion* will be different in level, emphasis, or presentation than those found in this textbook, but this can actually enhance understanding.

You will encounter a CD-ROM icon ✪ in figure legends throughout the text. When you see a CD-ROM icon be sure to refer to this correlation guide for corresponding information on *Microbes in Motion*. In some cases, there will be a reference to supporting information on *Microbes in Motion* in this correlation guide but no icon in the text. This CD material usually is supporting textual material in the third edition of *Microbiology* and an icon could not be inserted into the text narrative. Be sure to check this correlation table prior to reading a chapter in case there is supporting information on *Microbes in Motion.*

Chapter 2

Fig. 2.13	Bacterial Structure & Function Book Bacteria Groups/Overview
Fig. 2.14	Bacterial Structure & Function Book Bacteria Groups/Categories and Classification

Chapter 3

Fig. 3.5	Bacterial Structure & Function Book Cell Membrane Chapter pp. 5 & 6 (animation)
Fig. 3.7	Bacterial Structure & Function Book Cell Membrane Chapter pp. 8 & 10 (animation)
Fig. 3.18	Bacterial Structure & Function Book Gram Positive Cell Chapter pp. 1–5 (enhanced still images) Gram Negative Cell Chapter pp. 1–5 (enhanced still images)
Fig. 3.22	Bacterial Structure & Function Book Cell Wall Chapter/Peptidoglycan pp. 2 & 3 Antimicrobial Action Book Cell Wall Inhibitors Chapter/Mechanism of Action p. 7 (Peptidoglycan Synthesis Animation)
Fig. 3.24	Bacterial Structure & Function Book Gram Positive Cell Chapter/Teichoic Acids pp. 4 & 5
Fig. 3.26	Bacterial Structure & Function Book Gram Negative Cell Chapter pp. 1–4
Fig. 3.28	Antimicrobial Action Book Cell Wall Inhibitors Chapter/Mechanism of Action p. 7 (animation)
Fig. 3.29	Bacterial Structure & Function Book External Structures Chapter/Glycocalyx pp. 8 & 9
Fig. 3.37	Bacterial Structure & Function Book External Structures Chapter/Flagella p. 24 (video)

Chapter 4

Fig. 4.7	Fungal Structure & Function Book General Eucaryotic Structures Chapter/Internal Structures pp. 8–10

Chapter 38

Chapter 39

TO THE STUDENT

One of the most important factors contributing to success in college, and in microbiology courses, is the use of good study techniques. This textbook is organized to help you to study more efficiently. But even a text with many learning aids is not effective unless used properly. Thus this section briefly outlines some practical study skills that will help ensure success in microbiology and make your use of this textbook more productive. Many of you already have the study skills mentioned here and will not need to spend time reviewing familiar material. These suggestions are made in the hope that they may be useful to those who are unaware of approaches like the SQ4R technique for studying textbooks.

Time Management and Study Environment

Many students find it difficult to study effectively because of a lack of time management and a proper place to study. Often a student will do poorly in courses because not enough time has been spent studying outside class. For best results you should plan to spend at least an average of four to eight hours a week outside class working on each course. There is sufficient time in the week for this, but it does require time management. If you spend a few minutes early in the morning planning how the day is to be used and allow adequate time for studying, much more will be accomplished. Students who make efficient use of every moment find that they have plenty of time for recreation.

A second important factor is a proper place to study so that you can concentrate and efficiently use your study time. Try to find a quiet location with a desk and adequate lighting. If possible, always study in the same place and use it only for studying. In this way you will be mentally prepared to study when you are at your desk. This location may be in the dorm, the library, a special study room, or somewhere else. Wherever it is, your study area should be free from distractions—including friends who drop by to socialize. Much more will be accomplished if you really study during your designated study times.

Making the Most of Lectures

Attendance at lectures is essential for success. Students who chronically miss classes usually do not do well. To gain the most from lectures, it is best to read any relevant text material beforehand. Be prepared to concentrate during lectures; do not simply sit back passively and listen to the instructor. During the lecture record your notes in a legible way so that you can understand them later. It is most efficient to employ an outline or simple paragraph format. The use of abbreviations or some type of shorthand notation often is effective. During lecture concentrate on what is being said and be sure to capture all of the main ideas, concepts, and definitions of important terms. Do not take sketchy notes assuming that you will remember things because they are easy or obvious; you won't. Diagrams, lists, and terms written on the board are almost always important, as is anything the instructor clearly emphasizes by tone of voice. Feel free to ask questions during class when you don't understand something or wish the instructor to pursue a point further. Remember that if you don't understand, it is very likely that others in the class don't either but simply aren't willing to show their confusion. As soon as possible after a lecture, carefully review your notes to be certain that they are complete and understandable. Refer to the textbook when uncertain about something in your notes; it will be invaluable in clearing up questions and amplifying major points. When studying your notes for tests, it is a good idea to emphasize the most important points with a felt-tip marker just as you would when reading the textbook.

Studying the Textbook

Your textbook is one of the most important learning tools in any course and should be very carefully and conscientiously used. Many years ago Francis P. Robinson developed a very effective study technique called SQ3R (survey, question, read, recite, and review). More recently L. L. Thistlethwaite and N. K. Snouffer have slightly modified it to yield the SQ4R approach (survey, question, read, revise, record, and review). This latter approach is summarized below.

1. *Survey.* Briefly scan the chapter to become familiar with its general content. Quickly read the title, introduction, summary, and main headings. Record the major ideas and points that you think the chapter will make. If there are a list of chapter concepts and a chapter outline, pay close attention to these. This survey should give you a feel for the topic and how the chapter is approaching it.

2. *Question.* As you reach each main heading or subheading, try to compose an important question or two that you believe the section will answer. This preview question will help focus your reading of the section. It is also a good idea to keep asking yourself questions as you read. This habit facilitates active reading and learning.

3. *Read.* Carefully read the section. Read to understand concepts and major points, and try to find the answer to your preview question(s). You may want to highlight very important terms or explanations of concepts, but do not indiscriminately highlight everything. Be sure to pay close attention to any terms printed in color or boldface since the author(s) considered these to be important.

4. *Revise.* After reading the section, revise your question(s) to more accurately reflect the section's contents. These questions should be concept type questions that force you to bring together a number of details. They can be written in the margins of your text.

5. *Record.* Underline the information in the text that answers your questions, if you have not already done so. You may wish to write down the answers in note form as well. This process will give you good material to use in preparing for exams.

6. *Review.* Review the information by trying to answer your questions without looking at the text. If the text has a list of key words and a set of study questions, be sure to use these in your review. You will retain much more if you review the material several times.

Preparing for Examinations

It is extremely important to prepare for examinations properly so that you will not be rushed and tired on examination day. All textbook reading and lecture note revision should be completed well ahead of time so that the last few days can be spent in mastering the material, not in trying to understand the basic concepts. Cramming at the last moment for an exam is no substitute for daily preparation and review. By managing time carefully and keeping up with your studies, you will have plenty of time to review thoroughly and clear up any questions. This will allow you to get sufficient rest before the test and to feel confident in your preparation. Because both physical condition and general attitude are important factors in test performance, you will automatically do better. Proper reviewing techniques also aid retention of the material.

Further Reading

Grassick, P. 1983. *Making the grade: How to score high on all scholastic tests.* New York: Arco.

Shaw, H. 1976. *30 Ways to improve your grades.* New York: McGraw-Hill.

Shepherd, J. F. 1988. *RSVP: The Houghton Mifflin reading, study, and vocabulary program,* 3d ed. Boston: Houghton Mifflin.

Thistlethwaite, L. L., and Snouffer, N. K. 1976. *College reading power,* 3d ed. Dubuque, Iowa: Kendall/Hunt.

PART ONE
Introduction to Microbiology

The ultrastructure of *Salmonella*, a bacterium that
causes human diseases such as typhoid fever and
food poisoning. A procaryotic cell such as this does
not contain a membrane-delimited nucleus like
eucaryotic cells but has its DNA located in a more
diffuse nucleoid (pictured in yellow). The
cytoplasm (green) is enclosed in a plasma
membrane (violet) and a complex cell wall (brown).
The cell wall contains lipopolysaccharides, which
are toxic for humans. The bacterium moves with
whiplike flagella that project far from its surface. It
bristles with shorter pili that help attach it to
objects. A polysaccharide capsule (green) outside
the cell wall provides protection.

CHAPTER 1

The History and Scope of Microbiology

Dans les champs de l'observation, le hasard ne favorise que les esprits prepares.

—Louis Pasteur

(In the field of observation, chance favors only prepared minds.)

Outline

Concepts

1. Microbiology is the study of organisms that are usually too small to be seen by the unaided eye; it employs techniques—such as sterilization and the use of culture media—that are required to isolate and grow these microorganisms.

2. Microorganisms are not spontaneously generated from inanimate matter but arise from other microorganisms.

3. Many diseases result from viral, bacterial, fungal, or protozoan infections. Koch's postulates may be used to establish a causal link between the suspected microorganism and a disease.

4. The development of microbiology as a scientific discipline has depended on the availability of the microscope and the ability to isolate and grow pure cultures of microorganisms.

5. Microorganisms are responsible for many of the changes observed in organic and inorganic matter (e.g., fermentation and the carbon, nitrogen, and sulfur cycles that occur in nature).

6. Microorganisms have two fundamentally different types of cells—procaryotic and eucaryotic—and are distributed among several kingdoms.

7. Microbiology is a large discipline and has a great impact on other areas of biology and general human welfare.

One can't overemphasize the importance of microbiology. Society benefits from microorganisms in many ways. They are necessary for the production of bread, cheese, beer, antibiotics, vaccines, vitamins, enzymes, and many other important products. Indeed, modern biotechnology rests upon a microbiological foundation. Microorganisms are indispensable components of our ecosystem. They make possible the cycles of carbon, oxygen, nitrogen, and sulfur that take place in terrestrial and aquatic systems. They also are a source of nutrients at the base of all ecological food chains and webs.

Of course microorganisms also have harmed humans and disrupted society over the millenia. Microbial diseases undoubtedly played a major role in historical events such as the decline of the Roman Empire and the conquest of the New World. In 1347 plague or black death (*see chapter 37*) struck Europe with brutal force. By 1351, only four years later, the plague had killed 1/3 of the population (about 25 million people). Over the next 80 years, the disease struck again and again, eventually wiping out 75% of the European population. Some historians believe that this disaster changed European culture and prepared the way for the Renaissance. Today the struggle by microbiologists and others against killers like AIDS and malaria continues.

The biology of AIDS and its impact (pp. 722–30).

In this introductory chapter the historical development of the science of microbiology is described, and its relationship to medicine and other areas of biology is considered. The nature of the microbial world is then surveyed to provide a general idea of the organisms and agents that microbiologists study. Finally, the scope and relevance of modern microbiology are discussed.

Microbiology often has been defined as the study of organisms and agents too small to be seen clearly by the unaided eye—that is, the study of **microorganisms.** Because objects less than about one millimeter in diameter cannot be seen clearly and must be examined with a microscope, microbiology is concerned primarily with organisms and agents this small and smaller. Its subjects are viruses, bacteria, many algae and fungi, and protozoa (*see table 29.1*). Yet other members of these groups, particularly some of the algae and fungi, are larger and quite visible. For example, bread molds and filamentous algae are studied by microbiologists, yet are visible to the naked eye. A bacterium that is visible without a microscope, *Epulopiscium,* also has been discovered (*see p. 41*). The difficulty in setting the boundaries of microbiology led Roger Stanier to suggest that the field be defined not only in terms of the size of its subjects but also in terms of its techniques. A microbiologist usually first isolates a specific microorganism from a population and then cultures it. Thus microbiology employs techniques—such as sterilization and the use of culture media—that are necessary for successful isolation and growth of microorganisms.

The development of microbiology as a science is described in the following sections. Table 1.1 presents a summary of some of the major events in this process and their relationship to other historical landmarks.

The Discovery of Microorganisms

Even before microorganisms were seen, some investigators suspected their existence and responsibility for disease. Among others, the Roman philosopher Lucretius (about 98–55 B.C.) and the physician Girolamo Fracastoro (1478–1553) suggested that disease was caused by invisible living creatures. The earliest microscopic observations appear to have been made between 1625 and 1630 on bees and weevils by the Italian Francesco Stelluti, using a microscope probably supplied by Galileo. However, the first person to observe and describe microorganisms accurately was the amateur microscopist Antony van Leeuwenhoek (1632–1723) of Delft, Holland (figure 1.1*a*). Leeuwenhoek earned his living as a draper and haberdasher (a dealer in men's clothing and accessories), but spent much of his spare time constructing simple microscopes composed of double convex glass lenses held between two silver plates (figure 1.1*b*). His microscopes could magnify around 50 to 300 times, and he may have illuminated his liquid specimens by placing them between two pieces of glass and shining light on them at a 45° angle to the specimen plane. This would have provided a form of dark-field illumination (*see chapter 2*) and made bacteria clearly visible. Beginning in 1673 Leeuwenhoek sent detailed letters describing his discoveries to the Royal Society of London. It is clear from his descriptions that he saw both bacteria and protozoa.

The Spontaneous Generation Conflict

From earliest times, people had believed in **spontaneous generation**—that living organisms could develop from nonliving matter. Even the great Aristotle (384–322 B.C.) thought some of the simpler invertebrates could arise by spontaneous generation. This view finally was challenged by the Italian physician Francesco Redi (1626–1697), who carried out a series of experiments on decaying meat and its ability to produce maggots spontaneously. Redi placed meat in three containers. One was uncovered, a second was covered with paper, and the third was covered with a fine gauze that would exclude flies. Flies laid their eggs on the uncovered meat and maggots developed. The other two pieces of meat did not produce maggots spontaneously. However, flies were attracted to the gauze-covered container and laid their eggs on the gauze; these eggs produced maggots. Thus the generation of maggots by decaying meat resulted from the presence of fly eggs, and meat did not spontaneously generate maggots as previously believed. Similar experiments by others helped discredit the theory for larger organisms.

Leeuwenhoek's discovery of microorganisms renewed the controversy. Some proposed that microorganisms arose by

TABLE 1.1   Some Important Events in the Development of Microbiology

Date	Microbiological History	Other Historical Events
1546	Fracastoro suggests that invisible organisms cause disease	Publication of Copernicus's work on the heliocentric solar system (1543)
		Shakespeare's *Hamlet* (1600–1601)
1676	Leeuwenhoek discovers "animalcules"	J. S. Bach and Handel born (1685)
1688	Redi publishes work on spontaneous generation of maggots	Isaac Newton publishes the *Principia* (1687)
		Linnaeus's *Systema Naturae* (1735)
		Mozart born (1756)
1786	Müller produces first classification of bacteria	French Revolution (1789)
1798	Jenner introduces cowpox vaccination for smallpox	
1799	Spallanzani attacks spontaneous generation	Beethoven's first symphony (1800)
		The battle of Waterloo and the defeat of Napoleon (1815)
		Faraday demonstrates the principle of an electric motor (1821)
1838–1839	Schwann and Schleiden, the Cell Theory	England issues first postage stamp (1840)
1835–1844	Bassi discovers that silkworm disease is caused by a fungus and proposes that many diseases are microbial in origin	Marx's *Communist Manifesto* (1848)
1847–1850	Semmelweis shows that childbed fever is transmitted by physicians and introduces the use of antiseptics to prevent the disease	Velocity of light first measured by Fizeau (1849)
		Clausius states the first and second laws of thermodynamics (1850)
1849	Snow studies the epidemiology of a cholera epidemic in London	Graham distinguishes between colloids and crystalloids
		Melville's *Moby Dick* (1851)
		Otis installs first safe elevator (1854)
		Bunsen introduces the use of the gas burner (1855)
		Mendel begins his genetics experiments (1856)
1857	Pasteur shows that lactic acid fermentation is due to a microorganism	
1858	Virchow states that all cells come from cells	Darwin's *On the Origin of Species* (1859)
1861	Pasteur shows that microorganisms do not arise by spontaneous generation	American Civil War (1861–1865)
		Cross-Atlantic cable laid (1865)
1867	Lister publishes his work on antiseptic surgery	Dostoevski's *Crime and Punishment* (1866)
1869	Miescher discovers nucleic acids	Franco-German War (1870–1871)
1876–1877	Koch demonstrates that anthrax is caused by *Bacillus anthracis*	Bell invents telephone (1876)
		Edison's first light bulb (1879)
1881	Koch cultures bacteria on gelatin	Ives produces first color photograph (1881)
	Pasteur develops anthrax vaccine	
1882	Koch discovers tubercle bacillus	First central electric power station constructed by Edison (1882)
1884	Koch's postulates first published	Mark Twain's *The Adventures of Huckleberry Finn* (1884)
	Metchnikoff describes phagocytosis	
	Autoclave developed	
	Gram stain developed	
1885	Pasteur develops rabies vaccine	First motor vehicles developed by Daimler (1885–1886)
1887	Petri dish (plate) developed by Richard Petri	
1887–1890	Winogradsky studies sulfur and nitrifying bacteria	Hertz discovers radio waves (1888)
1888	Beijerinck isolates root nodule bacteria	Eastman makes box camera (1888)
1890	Von Behring prepares antitoxins for diphtheria and tetanus	
1892	Ivanowsky provides evidence for virus causation of tobacco mosaic disease	First zipper patented (1895)
1895	Bordet discovers complement	Röntgen discovers X rays (1895)
1897	Buchner prepares yeast extract that ferments	Thomson discovers the electron (1897)
	Ross shows that malaria parasite is carried by the mosquito	Spanish-American War (1898)
1899	Beijerinck proves that a virus particle causes the tobacco mosaic disease	

TABLE 1.1   Continued

Date	Microbiological History	Other Historical Events
1900	Reed proves that yellow fever is transmitted by the mosquito	Planck develops the quantum theory (1900)
1902	Landsteiner discovers blood groups	First electric typewriter (1901)
1903	Wright and others discover antibodies in the blood of immunized animals	First powered aircraft (1903)
		Einstein's special theory of relativity (1905)
1906	Schaudinn and Hoffmann show *Treponema pallidum* causes syphilis	First model T Ford (1908)
	Wassermann develops complement fixation test for syphilis	
	Ricketts shows that Rocky Mountain spotted fever is transmitted by ticks	Peary and Hensen reach North Pole (1909)
1910	Ehrlich develops chemotherapeutic agent for syphilis	Rutherford presents his theory of the atom (1911)
		Picasso and cubism (1912)
		World War I begins (1914)
1915–1917	D'Herelle and Twort discover bacterial viruses	Einstein's general theory of relativity (1916)
		Russian Revolution (1917)
1921	Fleming discovers lysozyme	
1923	First edition of *Bergey's Manual*	Lindberg's transatlantic flight (1927)
1928	Griffith discovers bacterial transformation	
1929	Fleming discovers penicillin	Stock market crash (1929)
1933	Ruska develops first transmission electron microscope	Hitler becomes chancellor of Germany (1933)
1935	Stanley crystallizes the tobacco mosaic virus	
	Domagk discovers sulfa drugs	
1937	Chatton divides living organisms into procaryotes and eucaryotes	Krebs discovers the citric acid cycle (1937)
		World War II begins (1939)
1941	Beadle and Tatum, one-gene-one-enzyme hypothesis	
1944	Avery shows that DNA carries information during transformation	The insecticide DDT introduced (1944)
	Waksman discovers streptomycin	
		Atomic bombs dropped on Hiroshima and Nagasaki (1945)
1946	Lederberg and Tatum describe bacterial conjugation	United Nations formed (1945)
		First electronic computer (1946)
1950	Lwoff induces lysogenic bacteriophages	Korean War begins (1950)
1952	Hershey and Chase show that bacteriophages inject DNA into host cells	First hydrogen bomb exploded (1952) Stalin dies (1952)
	Zinder and Lederberg discover generalized transduction	First commercial transistorized product (1952)
1953	Phase-contrast microscope developed	U.S. Supreme Court rules against segregated schools (1954)
	Medawar discovers immune tolerance	
	Watson and Crick propose the double helix structure for DNA	
1955	Jacob and Wollman discover the F factor is a plasmid	Montgomery bus boycott (1955)
	Jerne and Burnet propose the clonal selection theory	Sputnik launched by Soviet Union (1957)
1959	Yalow develops the radioimmunoassay technique	Birth control pill (1960)
1961	Jacob and Monod propose the operon model of gene regulation	First humans in space (1961)
1961–1966	Nirenberg, Khorana, and others elucidate the genetic code	Cuban missile crisis (1962)
		Nuclear test ban treaty (1963)
1962	Porter proposes the basic structure for immunoglobulin G	Civil Rights March on Washington (1963)
	First quinolone antimicrobial (nalidixic acid) synthesized	President Kennedy assassinated (1963)
		Bombing of North Vietnam begun and first regular U.S. combat troops sent (1965)
		Arab-Israeli War (1967)
		Martin Luther King assassination (1968)
		Neil Armstrong walks on the moon (1969)
1970	Discovery of restriction endonucleases by Arber and Smith	
	Discovery of reverse transcriptase in retroviruses by Temin and Baltimore	

TABLE 1.1   Continued

Date	Microbiological History	Other Historical Events
1973	Ames develops a bacterial assay for the detection of mutagens and carcinogens	Salt I Treaty (1972) Vietnam War ends (1973)
1975	Kohler and Milstein develop technique for the production of monoclonal antibodies	President Nixon resigns because of Watergate cover-up (1975)
	Lyme disease discovered	
1977	Recognition of archaeobacteria as a distinct microbial group	Panama Canal Treaty (1977)
1979	Insulin synthesized using recombinant DNA techniques	Hostages seized in Iran (1978)
		Three Mile Island disaster (1979)
1980	Development of the scanning tunneling microscope	Home computers marketed (1980)
1982	Recombinant hepatitis B vaccine developed	AIDS first recognized (1981)
1982–1983	Discovery of catalytic RNA by Cech and Altman	First artificial heart implanted (1982)
1983–1984	The human immunodeficiency virus isolated and identified by Gallo and Montagnier	Meter redefined in terms of distance light travels (1983)
	The polymerase chain reaction developed by Mullis	
1986	First vaccine (hepatitis B vaccine) produced by genetic engineering approved for human use	Gorbachev becomes Communist party general secretary (1985)
		Chernobyl reactor disaster (1986)
1990	First human gene-therapy testing begun	Berlin Wall falls (1989)
		Persian Gulf War with Iraq begins (1990)
1992	First human trials of antisense therapy	Failed Soviet coup; Boris Yeltsin comes to power (1991)
1995	Chickenpox vaccine approved for U.S. use	
	Haemophilus influenzae genome sequenced	

(a)

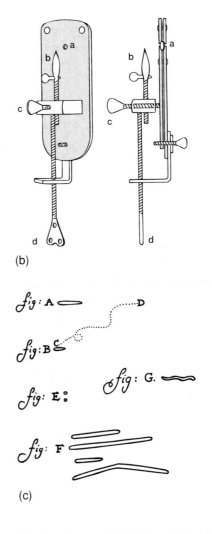

(b)

(c)

Figure 1.1   **Antony van Leeuwenhoek.** Leeuwenhoek (1632–1723) and his microscopes. (*a*) Leeuwenhoek holding a microscope. (*b*) A drawing of one of the microscopes showing the lens, *a;* mounting pin, *b;* and focusing screws, *c* and *d.* (*c*) Leeuwenhoek's drawings of bacteria from the human mouth. (*b*) *Source: C. E. Dobell, Antony van Leeuwenhoek and His Little Animals (1932), Russell and Russell, 1958.*

spontaneous generation even though larger organisms did not. They pointed out that boiled extracts of hay or meat would give rise to microorganisms after sitting for a while. In 1748 the English priest John Needham (1713–1781) reported the results of his experiments on spontaneous generation. Needham boiled mutton broth and then tightly stoppered the flasks. Eventually many of the flasks became cloudy and contained microorganisms. He thought organic matter contained a vital force that could confer the properties of life on nonliving matter. A few years later the Italian priest and naturalist Lazzaro Spallanzani (1729–1799) improved on Needham's experimental design by first sealing glass flasks that contained water and seeds. If the sealed flasks were placed in boiling water for 3/4 of an hour, no growth took place as long as the flasks remained sealed. He proposed that air carried germs to the culture medium, but also commented that the external air might be required for growth of animals already in the medium. The supporters of spontaneous generation maintained that heating the air in sealed flasks destroyed its ability to support life.

Several investigators attempted to counter such arguments. Theodore Schwann (1810–1882) allowed air to enter a flask containing a sterile nutrient solution after the air had passed through a red-hot tube. The flask remained sterile. Subsequently Georg Friedrich Schroder and Theodor von Dusch allowed air to enter a flask of heat-sterilized medium after it had passed through sterile cotton wool. No growth occurred in the medium even though the air had not been heated. Despite these experiments the French naturalist Felix Pouchet claimed in 1859 to have carried out experiments conclusively proving that microbial growth could occur without air contamination. This claim provoked Louis Pasteur (1822–1895) to settle the matter once and for all. Pasteur (figure 1.2) first filtered air through cotton and found that objects resembling plant spores had been trapped. If a piece of the cotton was placed in sterile medium after air had been filtered through it, microbial growth appeared. Next he placed nutrient solutions in flasks, heated their necks in a flame, and drew them out into a variety of curves, while keeping the ends of the necks open to the atmosphere (figure 1.3). Pasteur then boiled the solutions for a few minutes and allowed them to cool. No growth took place even though the contents of the flasks were exposed to the air. Pasteur pointed out that no growth occurred because dust and germs had been trapped on the walls of the curved necks. If the necks were broken, growth commenced immediately. Pasteur had not only resolved the controversy by 1861 but also had shown how to keep solutions sterile.

Figure 1.2   Louis Pasteur. Pasteur (1822–1895) working in his laboratory.

Figure 1.3   The Spontaneous Generation Experiment. Pasteur's swan neck flasks used in his experiments on the spontaneous generation of microorganisms. *Source:* Annales Sciences Naturelle, *4th Series, Vol. 16, pp. 1–98, Pasteur, L., 1861, "Mémoire sur les Corpuscules Organisés Qui Existent Dans L'Atmosphère: Examen de la Doctrine des Générations Spontanées."*

The English physicist John Tyndall (1820–1893) dealt a final blow to spontaneous generation in 1877 by demonstrating that dust did indeed carry germs and that if dust were absent, broth remained sterile even if directly exposed to air. During the course of his studies, Tyndall provided evidence for the existence of exceptionally heat-resistant forms of bacteria. Working independently, the German botanist Ferdinand Cohn (1828–1898) discovered the existence of heat-resistant bacterial endospores (*see chapter 3*).

1. Describe the field of microbiology in terms of the size of its subject material and the nature of its techniques.
2. How did Pasteur and Tyndall finally settle the spontaneous generation controversy?

The Recognition of the Microbial Role in Disease

Although Fracastoro and a few others had suggested that invisible organisms produced disease, most believed that disease was due to causes such as supernatural forces, poisonous vapors called miasmas, and imbalances between the four humors thought to be present in the body. The idea that an imbalance between the four humors (blood, phlegm, yellow bile [choler], and black bile [melancholy]) led to disease had been widely accepted since the time of the Greek physician Galen (129–199). Support for the germ theory of disease began to accumulate in the early nineteenth century. Agostino Bassi (1773–1856) first showed a microorganism could cause disease when he demonstrated in 1835 that a silkworm disease was due to a fungal infection. He also suggested that many diseases were due to microbial infections. In 1845 M. J. Berkeley proved that the great Potato Blight of Ireland was caused by a fungus. Following his successes with the study of fermentation, Pasteur was asked by the French government to investigate the pébrine disease of silkworms that was disrupting the silk industry. After several years of work, he showed that the disease was due to a protozoan parasite. The disease was controlled by raising caterpillars from eggs produced by healthy moths.

Indirect evidence that microorganisms were agents of human disease came from the work of the English surgeon Joseph Lister (1827–1912) on the prevention of wound infections. Lister (figure 1.4), impressed with Pasteur's studies on the involvement of microorganisms in fermentation and putrefaction, developed a system of antiseptic surgery designed to prevent microorganisms from entering wounds. Instruments were heat sterilized, and phenol was used on surgical dressings and at times sprayed over the surgical area. The approach was remarkably successful and transformed surgery after Lister published his findings in 1867. It also provided strong indirect evidence for the role of microorganisms in disease because phenol, which killed bacteria, also prevented wound infections.

Figure 1.4 Joseph Lister. Lister (1827–1912) performing surgery using antiseptic techniques.

Figure 1.5 Robert Koch. Koch (1843–1910) examining a specimen in his laboratory.

The first direct demonstration of the role of bacteria in causing disease came from the study of anthrax (*see chapter 37*) by the German physician Robert Koch (1843–1910). Koch (figure 1.5) used the criteria proposed by his former teacher, Jacob Henle (1809–1885), to establish the relationship between *Bacillus anthracis* and anthrax, and published his findings in 1876 (Box 1.1 briefly discusses the scientific method). Koch injected healthy mice with material from diseased animals, and the mice became ill. After transferring anthrax by inoculation through a series of 20 mice, he incubated a piece of spleen containing the anthrax bacillus in beef serum. The bacilli grew, reproduced, and produced spores. When the isolated bacilli or spores were injected into mice, anthrax developed. His criteria for proving the causal relationship between a microorganism and a specific disease are known as **Koch's postulates** and can be summarized as follows:

1. The microorganism must be present in every case of the disease but absent from healthy organisms.
2. The suspected microorganism must be isolated and grown in a pure culture.
3. The same disease must result when the isolated microorganism is inoculated into a healthy host.
4. The same microorganism must be isolated again from the diseased host.

Although Koch used the general approach described in the postulates during his anthrax studies, he did not outline them fully until his 1884 publication on the cause of tuberculosis.

Koch's proof that *Bacillus anthracis* caused anthrax was independently confirmed by Pasteur and his coworkers. They discovered that after burial of dead animals, anthrax spores survived and were brought to the surface by earthworms. Healthy animals then ingested the spores and became ill.

During Koch's studies on bacterial diseases, it became necessary to isolate suspected bacterial pathogens. At first he cultured bacteria on the sterile surfaces of cut, boiled potatoes. This was unsatisfactory because bacteria would not always grow well on potatoes. He then tried to solidify regular liquid medium by adding gelatin. Separate bacterial colonies developed after the surface had been streaked with a bacterial sample. The sample could also be mixed with liquefied gelatin medium. When the gelatin medium hardened, individual bacteria produced separate colonies. Despite its advantages gelatin was not an ideal solidifying agent because it was digested by many bacteria and melted when the temperature rose above 28°C. A better alternative was provided by Fannie Eilshemius Hesse, the wife of Walther Hesse, one of Koch's assistants (figure 1.6). She suggested the use of agar as a solidifying agent—she had been using it successfully to make jellies for some time. Agar was not attacked by most bacteria and did not melt until reaching a temperature of 100°C. One of Koch's assistants, Richard Petri, developed the petri dish (plate), a container for solid culture media. These developments made possible the isolation of pure cultures that contained only one type of bacterium, and directly stimulated progress in all areas of bacteriology.

Isolation of bacteria and pure culture techniques (pp. 108–10).

Koch also developed media suitable for growing bacteria isolated from the body. Because of their similarity to body fluids, meat extracts and protein digests were used as nutrient sources. The result was the development of nutrient broth and nutrient agar, media that are still in wide use today.

Figure 1.6   Fannie Eilshemius (1850–1934) and Walther Hesse (1846–1911). Fannie Hesse first proposed using agar in culture media.

By 1882 Koch had used these techniques to isolate the bacillus that caused tuberculosis. There followed a golden age of about 30 to 40 years in which most of the major bacterial pathogens were isolated (table 1.2).

The discovery of viruses and their role in disease was made possible when Charles Chamberland (1851–1908), one of Pasteur's associates, constructed a porcelain bacterial filter in 1884. The first viral pathogen to be studied was the tobacco mosaic disease virus (*see chapter 16*).

The development of virology (pp. 333–34).

In this period progress also was made in determining how animals resisted disease and in developing techniques for protecting humans and livestock against pathogens. During studies on chicken cholera, Pasteur and Roux discovered that incubating their cultures for long intervals between transfers would attenuate the bacteria, which meant they had lost their ability to cause the disease. If the chickens were injected with these attenuated cultures, they remained healthy but developed the ability to resist the disease. He called the attenuated culture a vaccine [Latin *vacca,* cow] in honor of Edward Jenner because, many years earlier, Jenner had used vaccination with material from cowpox lesions to protect people against smallpox (*see chapter 16*). Shortly after this, Pasteur and Chamberland developed an attenuated anthrax vaccine in two ways: by treating cultures with potassium bichromate and by incubating the bacteria at 42 to 43°C.

The development of chemotherapy (pp. 657–58).

Pasteur next prepared rabies vaccine by a different approach. The pathogen was attenuated by growing it in an abnormal host, the rabbit. After infected rabbits had died, their brains and spinal cords were removed and dried. During the course of these studies, Joseph Meister, a nine-year-old boy who had been bitten by a rabid dog, was brought to Pasteur. Since the boy's death was certain in

TABLE 1.2   The Discovery of Some of the Major Pathogens

Disease	Causative Agent[a]	Discoverer	Date
Anthrax	*Bacillus anthracis*	Koch	1876
Gonorrhea	*Neisseria gonorrhoeae*	Neisser	1879
Typhoid fever	*Salmonella typhi*	Gaffky	1884
Malaria	*Plasmodium* spp.	Laveran	1880
Tuberculosis	*Mycobacterium tuberculosis*	Koch	1882
Cholera	*Vibrio cholerae*	Koch	1883
Diphtheria	*Corynebacterium diphtheriae*	Klebs and Loeffler	1883–1884
Tetanus	*Clostridium tetani*	Nicolaier and Kitasato	1885–1889
Diarrhea	*Escherichia coli*	Escherich	1885
Pneumonia	*Streptococcus pneumoniae*	Fraenkel	1886
Meningitis	*Neisseria meningitidis*	Weichselbaum	1887
Undulant fever	*Brucella* spp.	Bruce	1887
Gas gangrene	*Clostridium perfringens*	Welch and Nuttal	1892
Plague	*Yersinia pestis*	Kitasato and Yersin	1894
Botulism	*Clostridium botulinum*	Van Ermengem	1896
Dysentery	*Shigella dysenteriae*	Shiga	1898
Syphilis	*Treponema pallidum*	Schaudin and Hoffmann	1905
Whooping cough	*Bordetella pertussis*	Bordet and Gengou	1906
Rocky Mountain spotted fever	*Rickettsia rickettsii*	Ricketts	1909
Tularemia	*Francisella tularensis*	McCoy and Chapin	1912

[a]Modern name used.

the absence of treatment, Pasteur agreed to try vaccination. Joseph was injected 13 times over the next 10 days with increasingly virulent preparations of the attenuated virus. He survived.

In gratitude for Pasteur's development of vaccines, people from around the world contributed to the construction of the Pasteur Institute in Paris, France. One of the initial tasks of the Institute was vaccine production.

After the discovery that the diphtheria bacillus produced a toxin, Emil von Behring (1854–1917) and Shibasaburo Kitasato (1852–1931) injected inactivated toxin into rabbits, inducing them to produce an antitoxin, a substance in the blood that would inactivate the toxin and protect against the disease. A tetanus antitoxin was then prepared and both antitoxins were used in the treatment of people.

The antitoxin work provided evidence that immunity could result from soluble substances in the blood, now known to be antibodies (humoral immunity). It became clear that blood cells were also important in immunity (cellular immunity) when Elie Metchnikoff (1845–1916) discovered that some blood leukocytes could engulf disease-causing bacteria (figure 1.7). He called these cells phagocytes and the process phagocytosis [Greek *phagein*, eating].

1. Discuss the contributions of Lister, Pasteur, and Koch to the germ theory of disease and to the treatment or prevention of diseases.
2. What other contributions did Koch make to microbiology?
3. Describe Koch's postulates in detail.
4. How did von Behring and Metchnikoff contribute to the development of immunology?

Figure 1.7   Elie Metchnikoff. Metchnikoff (1845–1916) shown here at work in his laboratory.

Box 1.1

The Scientific Method

Although biologists employ a variety of approaches in conducting research, microbiologists and other experimentally oriented biologists often use the general approach known as the scientific method. They first gather observations of the process to be studied and then develop a tentative **hypothesis**—an educated guess—to explain the observations (Box Figure 1.1). This step often is inductive and creative because there is no detailed, automatic technique for generating hypotheses. Next they decide what information is required to test the hypothesis and collect this information through observation or carefully designed experiments. After the information has been collected, they decide whether the hypothesis has been supported or falsified. If it has failed to pass the test, the hypothesis is rejected, and a new explanation or hypothesis is constructed. If the hypothesis passes the test, it is subjected to more severe testing. The procedure often is made more efficient by constructing and testing alternative hypotheses and then refining the hypothesis that survives testing. This general approach is often called the hypothetico-deductive method. One deduces predictions from the currently accepted hypothesis and tests them. In deduction the conclusion about specific cases follows logically from a general premise ("if . . . , then . . ." reasoning). Induction is the opposite. A general conclusion is reached after considering many specific examples. Both types of reasoning are used by scientists.

When carrying out an experiment, it is essential to use a control group as well as an experimental group. The control group is treated precisely the same as the experimental group except that the experimental manipulation is not performed on it. In this way one can be sure that any changes in the experimental group are due to the experimental manipulation rather than to some other factor not taken into account.

If a hypothesis continues to survive testing, it may be accepted as a valid theory. A **theory** is a set of propositions and concepts that provides a reliable, systematic, and rigorous account of an aspect of nature. It is important to note that hypotheses and theories are never absolutely proven. Scientists simply gain more and more confidence in their accuracy as they continue to survive testing, fit with new observations and experiments, and satisfactorily explain the observed phenomena.

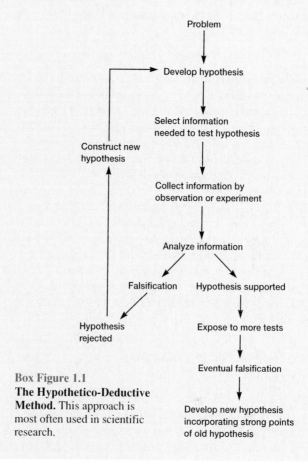

Box Figure 1.1
The Hypothetico-Deductive Method. This approach is most often used in scientific research.

The Discovery of Microbial Effects on Organic and Inorganic Matter

Although Theodore Schwann and others had proposed in 1837 that yeast cells were responsible for the conversion of sugars to alcohol, a process they called alcoholic fermentation, the leading chemists of the time believed microorganisms were not involved. They were convinced that fermentation was due to a chemical instability that degraded the sugars to alcohol. Pasteur did not agree. It appears that early in his career Pasteur became interested in fermentation because of his research on the optical activity of molecules. He believed that fermentations were carried out by living organisms and produced asymmetric products such as amyl alcohol that had optical activity. There was an intimate connection between molecular asymmetry, optical activity, and life. Then in 1856 M. Bigo, an industrialist in Lille, France, where Pasteur worked, requested Pasteur's assistance. His business produced ethanol from the fermentation of beet sugars, and the alcohol yields had recently declined and the product had become sour. Pasteur discovered that the fermentation was failing because the yeast normally responsible for alcohol formation had been replaced by microorganisms producing lactic acid rather than ethanol. In solving this practical problem, Pasteur demonstrated that all fermentations were due to the activities of specific yeasts and bacteria, and he published several papers on fermentation between 1857 and 1860. His success led to a study of wine diseases and the development of pasteurization (*see chapter 7*) to preserve wine during storage. Pasteur's studies on fermentation continued for almost 20 years. One of his most important discoveries was that some fermentative microorganisms were anaerobic and could live only in the absence of oxygen, whereas others were able to live either aerobically or anaerobically.

Fermentation (pp. 175–76).
The effect of oxygen on microorganisms (pp. 130–31).

A few of the early microbiologists chose to investigate the ecological role of microorganisms. In particular they studied microbial involvement in the carbon, nitrogen, and sulfur cycles taking place in soil and aquatic habitats. Two of the pioneers in this endeavor were Sergei N. Winogradsky (1856–1953) and Martinus W. Beijerinck (1851–1931).

Biogeochemical cycles (pp. 814–19).

The Russian microbiologist Sergei N. Winogradsky (figure 1.8) made many contributions to soil microbiology. He discovered that soil bacteria could oxidize iron, sulfur, and ammonia to obtain energy, and that many bacteria could incorporate CO_2 into organic matter much like photosynthetic organisms do. Winogradsky also isolated anaerobic nitrogen-fixing soil bacteria and studied the decomposition of cellulose.

Martinus W. Beijerinck (figure 1.9) was one of the great general microbiologists who made fundamental contributions to microbial ecology and many other fields. He isolated the aerobic nitrogen-fixing bacterium *Azotobacter;* a root nodule bacterium also capable of fixing nitrogen (later named *Rhizobium*); and sulfate-reducing bacteria. Beijerinck and Winogradsky developed the enrichment-culture technique and the use of selective media (*see chapter 5*), which have been of such great importance in microbiology.

The Development of Microbiology in This Century

During the early part of the twentieth century, microbiology developed somewhat independently of other biological disciplines, at least partially because of a difference in focus. Many biologists were interested in such topics as cell structure and function, the ecology of plants and animals, the reproduction and development of organisms, the nature of heredity, and the mechanism of evolution. In contrast, microbiologists were more concerned with the agents of infectious disease, the immune response, the search for new chemotherapeutic agents, and bacterial metabolism.

Microbiology established a closer relationship with other biological disciplines during the 1940s because of its association with genetics and biochemistry. Microorganisms are extremely useful experimental subjects because they are relatively simple, grow rapidly, and can be cultured in large quantities. George W. Beadle and Edward L. Tatum studied the relationship between genes and enzymes in 1941 using mutants of the bread mold *Neurospora*. Salvadore Luria and Max Delbrück (1943) used bacterial mutants to show that gene mutations were

Figure 1.8   **Sergei N. Winogradsky** (1856–1953).

Figure 1.9   **Martinus W. Beijerinck** (1851–1931).

truly spontaneous and not directed by the environment. Oswald T. Avery, Colin M. MacLeod, and Maclyn McCarty provided strong evidence in 1944 that DNA was the genetic material and carried genetic information during transformation (*see chapter 13*). The interactions between microbiology, genetics, and biochemistry soon led to the development of modern molecularly oriented genetics.

More recently microbiology has been a major contributor to the rise of molecular biology, the branch of biology dealing with the physical and chemical aspects of living matter and its function. Microbiologists have been deeply involved in studies on the genetic code and the mechanisms of DNA, RNA, and protein synthesis. Microorganisms were used in many of the early studies on the regulation of gene expression and the control of enzyme activity (*see chapter 12*). In the 1970s new discoveries in microbiology led to the development of recombinant DNA technology and genetic engineering.

The mechanisms of DNA, RNA, and protein synthesis (chapter 11).
Recombinant DNA and genetic engineering (chapter 15).

One indication of the importance of microbiology in the twentieth century is the Nobel Prize given for work in physiology or medicine. About 1/3 of these have been awarded to scientists working on microbiological problems (table 1.3).

---

1. Briefly summarize the contributions of Pasteur, Winogradsky, and Beijerinck to the study of microbial effects on organic and inorganic matter.
2. What is molecular biology, and why has microbiology been so important to its development?

The Composition of the Microbial World

Although the kingdoms of organisms and the differences between procaryotic and eucaryotic cells are discussed in much more detail later, a brief introduction to the organisms a microbiologist studies is given here.

Comparison of procaryotic and eucaryotic cells (pp. 91–94).

Two fundamentally different types of cells exist. **Procaryotic cells** [Greek *pro*, before, and *karyon*, nut or kernel; organism with a primordial nucleus] have a much simpler morphology than eucaryotic cells and lack a true membrane-delimited nucleus. All bacteria are procaryotic. In contrast, **eucaryotic cells** [Greek *eu*, true, and *karyon*, nut or kernel] have a membrane-enclosed nucleus; they are more complex morphologically and are usually larger than procaryotes. Algae, fungi, protozoa, higher plants, and animals are eucaryotic. Procaryotic and eucaryotic cells differ in many other ways as well (*see chapter 4*).

The early description of organisms as either plants or animals clearly is too simplified, and for many years biologists have divided organisms into five kingdoms (*see chapter 19*).

1. Procaryotic organisms are found in the kingdom *Monera* or *Procaryotae.*
2. The kingdom *Protista* contains either unicellular or colonial eucaryotic organisms that lack true tissues. Protozoa, the lower fungi, and most of the smaller algae are placed in this kingdom.
3. Members of the kingdom *Fungi* are eucaryotic organisms with absorptive nutrition and often are multinucleate.
4. The kingdom *Animalia* contains multicellular animals with ingestive nutrition.
5. Multicellular plants with walled eucaryotic cells and photosynthesis are located in the kingdom *Plantae.*

Microbiologists study primarily members of the first three kingdoms—the *Monera* or *Procaryotae, Protista,* and *Fungi.* Although they are not included in the five kingdoms, viruses are also studied by microbiologists.

Fungi (chapter 25).
Algae (chapter 26).
Protozoa (chapter 27).
Introduction to the viruses (chapters 16–18).

In the last few decades great progress has been made in three areas that profoundly affect microbial classification. First, much has been learned about the detailed structure of microbial cells from the use of electron microscopy. Second, microbiologists have determined the biochemical and physiological characteristics of many different microorganisms. Third, the sequences of ribosomal RNAs and proteins from a wide variety of organisms have been compared. It is now clear that there are two quite different groups of procaryotic organisms: eubacteria and archaeobacteria. Furthermore, the protists are so diverse that it may be neceessary to divide the kingdom *Protista* into three or more kingdoms. Thus many taxonomists have concluded that the five-kingdom system is too simple. One recent classification system divides organisms into two empires and eight kingdoms (*see pp. 410–11*). Other taxonomists have proposed more radical solutions. The differences between eubacteria, archaeobacteria, and eucaryotes seem so great some microbiologists have proposed that organisms should be divided among three domains or empires: bacteria (the true bacteria or eubacteria), archaea (the archaeobacteria), and eucarya (all eucaryotic organisms). This proposal and the results leading to it are discussed in chapter 19.

---

1. Describe and contrast procaryotic and eucaryotic cells.
2. Briefly describe the five-kingdom system and give the major characteristics of each kingdom. How might the system change in the future?

The Scope and Relevance of Microbiology

Microorganisms are exceptionally diverse, are found almost everywhere, and affect human society in countless ways. Thus modern microbiology is a large discipline with many different specialties; it has a great impact on medicine, agricultural and food sciences, ecology, genetics, biochemistry, and many other fields.

Microbiology has both basic and applied aspects. Many microbiologists are interested primarily in the biology of the

TABLE 1.3   Nobel Prizes Awarded for Research in Microbiology

Date	Scientist[a]	Research	Date	Scientist[a]	Research
1901	E. von Behring (GR)	Diphtheria antitoxin	1977	R. Yalow (US)	Development of the radioimmunoassay technique
1902	R. Ross (GB)	Cause and transmission of malaria			
1905	R. Koch (GR)	Tuberculosis research	1978	H. O. Smith (US)	Discovery of restriction
1907	C. Laveran (F)	Role of protozoa in disease		D. Nathans (US)	enzymes and their application
1908	P. Ehrlich (GR)	Work on immunity		W. Arber (SW)	to the problems of molecular genetics
	E. Metchnikoff (R)				
1913	C. Richet (F)	Work on anaphylaxis	1980	B. Benacerraf (US)	Discovery of the
1919	J. Bordet (B)	Discoveries about immunity		G. Snell (US)	histocompatibility antigens
1928	C. Nicolle (F)	Work on typhus fever		J. Dausset (F)	
1930	K. Landsteiner (US)	Discovery of human blood groups		P. Berg (US)	Development of recombinant
1939	G. Domagk (GR)	Antibacterial effect of prontosil		W. Gilbert (US) &	DNA technology (Berg);
1945	A. Fleming (GB)	Discovery of penicillin and its therapeutic value		F. Sanger (GB)	development of DNA sequencing techniques (Chemistry Prize)
	E. B. Chain (GB)		1982	A. Klug (GB)	Development of
	H. W. Florey (AU)				crystallographic electron
1951	M. Theiler (SA)	Development of yellow fever vaccine			microscopy and the elucidation of the structure
1952	S. A. Waksman (US)	Discovery of streptomycin			of viruses and other
1954	J. F. Enders (US)	Cultivation of poliovirus in tissue culture			nucleic-acid–protein complexes (Chemistry Prize)
	T. H. Weller (US)		1984	C. Milstein (GB)	Development of the technique
	F. Robbins (US)			G. J. F. Kohler (GR)	for formation of monoclonal
1957	D. Bovet (I)	Discovery of the first antihistamine		N. K. Jerne (D)	antibodies (Milstein & Kohler); theoretical work in
1958	G. W. Beadle (US)	Microbial genetics			immunology (Jerne)
	E. L. Tatum (US)		1986	E. Ruska (GR)	Development of the
	J. Lederberg (US)				transmission electron
1959	S. Ochoa (US)	Discovery of enzymes catalyzing nucleic acid synthesis			microscope (Physics Prize)
	A. Kornberg (US)		1987	S. Tonegawa (J)	The genetic principle for
1960	F. M. Burnet (AU)	Discovery of acquired immune			generation of antibody
	P. B. Medawar (GB)	tolerance to tissue transplants			diversity
1962	F. H. C. Crick (GB)	Discoveries concerning the	1988	J. Deisenhofer, R. Huber,	Crystallization and study of the
	J. D. Watson (US)	structure of DNA		and H. Michel (GR)	photosynthetic reaction
	M. Wilkins (GB)				center from a bacterial membrane
1965	F. Jacob (F)	Discoveries about the		G. Elion (US)	Development of drugs for the
	A. Lwoff (F)	regulation of genes		G. Hitchings (US)	treatment of cancer, malaria,
	J. Monod (F)				and viral infections
1966	F. P. Rous (US)	Discovery of cancer viruses	1989	J. M. Bishop (US)	Discovery of oncogenes
1968	R. W. Holley (US)	Deciphering of the genetic		H. E. Varmus (US)	
	H. G. Khorana (US)	code		S. Altman (US)	Discovery of catalytic RNA
	M. W. Nirenberg (US)			T. R. Cech (US)	
1969	M. Delbrück (US)	Discoveries concerning viruses	1993	K. B. Mullis (US)	Invention of the polymerase
	A. D. Hershey (US)	and viral infection of cells			chain reaction
	S. E. Luria (US)			M. Smith (US)	Development of site-directed
1972	G. Edelman (US)	Research on the structure of			mutagenesis
	R. Porter (GB)	antibodies		R. J. Roberts (US)	Discovery of split genes
1975	H. Temin (US)	Discovery of RNA-dependent		P. A. Sharp (US)	
	D. Baltimore (US)	DNA synthesis by RNA			
	R. Dulbecco (US)	tumor viruses; reproduction of DNA tumor viruses			
1976	B. Blumberg (US)	Mechanism for the origin and			
	D. C. Gajdusek (US)	dissemination of hepatitis B virus; research on slow virus infections			

[a]The Nobel laureates were citizens of the following countries: Australia (AU), Belgium (B), Denmark (D), France (F), Germany (GR), Great Britain (GB), Italy (I), Japan (J), Russia (R), South Africa (SA), Switzerland (SW), and the United States (US).

microorganisms themselves. They may focus on a specific group of microorganisms and be called virologists (viruses), bacteriologists (bacteria), phycologists or algologists (algae), mycologists (fungi), or protozoologists (protozoa). Others are interested in microbial morphology or particular functional processes and work in fields such as microbial cytology,

microbial physiology, microbial ecology, microbial genetics and molecular biology, and microbial taxonomy. Of course a person can be thought of in both ways (e.g., as a bacteriologist who works on taxonomic problems). Many microbiologists have a more applied orientation and work on practical problems in fields such as medical microbiology, food and dairy microbiology, and public health microbiology (basic research is also conducted in these fields). Because the various fields of microbiology are interrelated, an applied microbiologist must be familiar with basic microbiology. For example, a medical microbiologist must have a good understanding of microbial taxonomy, genetics, immunology, and physiology to identify and properly respond to the pathogen of concern.

What are some of the current occupations of professional microbiologists? One of the most active and important is medical microbiology, which deals with the diseases of humans and animals. Medical microbiologists identify the agent causing an infectious disease and plan measures to eliminate it. Frequently they are involved in tracking down new, unidentified pathogens such as the bacterium that causes Legionnaires' disease, the hantavirus, and the virus responsible for AIDS. These microbiologists also study the ways in which microorganisms cause disease.

Legionnaires' disease (pp. 744–45).
Hantavirus pulmonary syndrome (p. 721).
AIDS (pp. 722–30).

Public health microbiology is closely related to medical microbiology. Public health microbiologists try to control the spread of communicable diseases. They often monitor community food establishments and water supplies in an attempt to keep them safe and free from infectious disease agents.

Immunology is concerned with how the immune system protects the body from pathogens and the response of infectious agents. It is one of the fastest growing areas in science; for example, techniques for the production and use of monoclonal antibodies have developed extremely rapidly. Immunology also deals with practical health problems such as the nature and treatment of allergies and autoimmune diseases like rheumatoid arthritis.

Monoclonal antibodies and their uses (chapter 30 and Box 34.2).

Many important areas of microbiology do not deal directly with human health and disease but certainly contribute to human welfare. Agricultural microbiology is concerned with the impact of microorganisms on agriculture. Agricultural microbiologists try to combat plant diseases that attack important food crops, work on methods to increase soil fertility and crop yields, and study the role of microorganisms living in the digestive tracts of ruminants such as cattle. Currently there is great interest in using bacterial and viral insect pathogens as substitutes for chemical pesticides.

The field of microbial ecology is closely related to agricultural microbiology. Microbial ecologists study the relationships between microorganisms and their habitats. They are concerned with the contribution of microorganisms to the carbon, nitrogen, and sulfur cycles in soil and in fresh water. The study of pollution effects on microorganisms also is important because of the impact these organisms have on the environment. Microbial ecologists are now using microorganisms in bioremediation to reduce pollution effects.

Scientists working in food and dairy microbiology try to prevent microbial spoilage of food and the transmission of foodborne diseases such as botulism and salmonellosis (*see chapter 37*). They also use microorganisms to make foods such as cheeses, yogurts, pickles, and beer. In the future microorganisms themselves will become a more important nutrient source for livestock and humans.

In industrial microbiology microorganisms are used to make products such as antibiotics, vaccines, steroids, alcohols and other solvents, vitamins, amino acids, and enzymes. Microorganisms can even leach valuable minerals from low-grade ores.

Research on the biology of microorganisms occupies the time of many microbiologists and also has practical applications. Those working in microbial physiology and biochemistry study the synthesis of antibiotics and toxins, microbial energy production, the ways in which microorganisms survive harsh environmental conditions, microbial nitrogen fixation, the effects of chemical and physical agents on microbial growth and survival, and many other topics.

Microbial genetics and molecular biology focus on the nature of genetic information and how it regulates the development and function of cells and organisms. The use of microorganisms has been very helpful in understanding gene function. Microbial geneticists play an important role in applied microbiology by producing new microbial strains that are more efficient in synthesizing useful products. Genetic techniques are used to test substances for their ability to cause cancer. More recently the field of genetic engineering (*see chapter 15*) has arisen from work in microbial genetics and molecular biology and will contribute substantially to microbiology, biology as a whole, and medicine. Engineered microorganisms are used to make hormones, antibiotics, vaccines, and other products. New genes can be inserted into plants and animals; for example, it may be possible to give corn and wheat nitrogen-fixation genes so they will not require nitrogen fertilizers.

The future of microbiology is bright. With the advent of recombinant DNA technology and genetic engineering, microbiology will expand and change even faster in the future. The need for microbiologists to work on problems in medicine, environmental protection, food production and preservation, and industrial development will undoubtedly grow. The microbiologist René Dubos has summarized well the excitement and promise of microbiology:

> How extraordinary that, all over the world, microbiologists are now involved in activities as different as the study of gene structure, the control of disease, and the industrial processes based on the phenomenal ability of microorganisms to decompose and synthesize complex organic molecules. Microbiology is one of the most rewarding of professions because it gives its practitioners the opportunity to be in contact with all the other natural sciences and thus to contribute in many different ways to the betterment of human life.

Summary

1. Microbiology may be defined in terms of the size of the organisms studied and the techniques employed.

2. Antony van Leeuwenhoek was the first person to observe microorganisms.

3. Experiments by Redi and others disproved the theory of spontaneous generation in regard to larger organisms.

4. The spontaneous generation of microorganisms was disproved by Spallanzani, Pasteur, Tyndall, and others.

5. Support for the germ theory of disease came from the work of Bassi, Pasteur, Koch, and others. Lister provided indirect evidence with his development of antiseptic surgery.

6. Koch's postulates are used to prove a direct relationship between a suspected pathogen and a disease.

7. Koch also developed the techniques required to grow bacteria on solid media and to isolate pure cultures of pathogens.

8. Vaccines against anthrax and rabies were made by Pasteur; von Behring and Kitasato prepared antitoxins for diphtheria and tetanus.

9. Metchnikoff discovered some blood leukocytes could phagocytize and destroy bacterial pathogens.

10. Pasteur showed that fermentations were caused by microorganisms and that some microorganisms could live in the absence of oxygen.

11. The role of microorganisms in carbon, nitrogen, and sulfur cycles was first studied by Winogradsky and Beijerinck.

12. In the twentieth century microbiology has contributed greatly to the fields of biochemistry and genetics. It also has helped stimulate the rise of molecular biology.

13. Procaryotic cells differ from eucaryotic cells in that the former lack a membrane-delimited nucleus and in other ways as well.

14. There are at least five different kingdoms of living organisms. Microbiologists study primarily members of the kingdoms *Monera* or *Procaryotae, Protista,* and *Fungi.* Some have proposed that the archaeobacteria are so different that organisms should be divided into three empires: bacteria, archaea, and eucarya.

15. There is a wide variety of fields in microbiology, and many have a great impact on society. These include the more applied disciplines such as medical, public health, industrial, food, and dairy microbiology. Microbial ecology, physiology, biochemistry, and genetics are examples of basic microbiological research fields.

Key Terms

eucaryotic cell   *13*

hypothesis   *11*

Koch's postulates   *9*

microbiology   *3*

microorganism   *3*

procaryotic cell   *13*

spontaneous generation   *3*

theory   *11*

Questions for Thought and Review

1. Why was the belief in spontaneous generation an obstacle to the development of microbiology as a scientific discipline?

2. Describe the major contributions of the following people to the development of microbiology: Leeuwenhoek, Spallanzani, Fracastoro, Pasteur, Tyndall, Cohn, Bassi, Lister, Koch, Chamberland, von Behring, Metchnikoff, Winogradsky, and Beijerinck.

3. Would microbiology have developed more slowly if Fannie Hesse had not suggested the use of agar? Give your reasoning. What is a pure culture?

4. Why do you think viruses are not included in the five-kingdom system?

5. Why are microorganisms so useful to biologists as experimental models?

6. What do you think were the most important discoveries in the development of microbiology? Why?

7. List all the activities or businesses you can think of in your community that are directly dependent on microbiology.

8. Describe in your own words the scientific method. How does a theory differ from a hypothesis? Why is it important to have a control group?

Additional Reading

General

Baker, J. J. W., and Allen, G. E. 1968. *Hypothesis, prediction, and implication in biology.* Reading, Mass.: Addison-Wesley.

Brock, T. D. 1961. *Milestones in microbiology.* Englewood Cliffs, N.J.: Prentice-Hall.

Bulloch, W. 1979. *The history of bacteriology.* New York: Dover.

Chung, K.-T.; Stevens, Jr., S.E.; and Ferris, D. H. 1995. A chronology of events and pioneers of microbiology. *SIM News* 45(1):3–13.

Clark, P. F. 1961. *Pioneer microbiologists of America.* Madison: Univ. of Wisconsin Press.

Collard, P. 1976. *The development of microbiology.* New York: Cambridge University Press.

de Kruif, P. 1937. *Microbe hunters.* New York: Harcourt, Brace.

Gabriel, M. L., and Fogel, S., eds. 1955. *Great experiments in biology.* Englewood Cliffs, N.J.: Prentice-Hall.

Geison, G. L. 1995. *The private science of Louis Pasteur.* Princeton, N. J.: Princeton University Press.

Hellemans, A., and Bunch, B. 1988. *The timetables of science.* New York: Simon and Schuster.

Hill, L. 1985. Biology, philosophy, and scientific method. *J. Biol. Educ.* 19(3):227–31.

Lechevalier, H. A., and Solotorovsky, M. 1965. *Three centuries of microbiology.* New York: McGraw-Hill.

McNeill, W. H. 1976. *Plagues and peoples.* Garden City, N.Y.: Anchor Press/Doubleday.

Singer, C. 1959. *A history of biology,* 3d ed. New York: Abelard-Schuman.

Singleton, P., and Sainsbury, D. 1995. *Dictionary of microbiology and molecular biology,* 3d ed. New York: John Wiley and Sons.

Stanier, R. Y. 1978. What is microbiology? In *Essays in microbiology,* J. R. Norris and M. H. Richmond, editors, 1/1–1/32. New York: John Wiley and Sons.

Wainwright, M., and Lederberg, J. L. 1992. History of microbiology. In *Encyclopedia of microbiology,* vol. 2, J. Lederberg, editor, 419–37. San Diego: Academic Press.

The Discovery of Microorganisms

Dobell, C. 1960. *Antony van Leeuwenhoek and his "little animals."* New York: Dover.

The Spontaneous Generation Conflict

Dubos, R. J. 1950. *Louis Pasteur: Free lance of science.* Boston: Little, Brown.

Vallery-Radot, R. 1923. *The life of Pasteur.* New York: Doubleday.

The Recognition of the Microbial Role in Disease

Brock, T. D. 1988. *Robert Koch: A life in medicine and bacteriology.* Madison, Wis.: Science Tech Publishers.

Hesse, W. 1992. Walther and Angelina Hesse—early contributors to bacteriology. *ASM News* 58(8):425–28.

Hitchens, A. P., and Leikind, M. C. 1939. The introduction of agar-agar into bacteriology. *J. Bacteriol.* 37(5):485–93.

Silverstein, A. M. 1989. *A history of immunology.* San Diego: Academic Press.

CHAPTER 2
The Study of Microbial Structure: Microscopy and Specimen Preparation

There are more animals living in the scum on the teeth in a man's mouth than there are men in a whole kingdom.
—Antony van Leeuwenhoek

Concepts

1. Light microscopes use glass lenses to bend and focus light rays and produce enlarged images of small objects. The resolution of a light microscope is determined by the numerical aperture of its lens system and by the wavelength of the light it employs; maximum resolution is about 0.2 µm.

2. The most common types of light microscopes are the bright-field, dark-field, phase-contrast, and fluorescence microscopes. Each yields a distinctive image and may be used to observe different aspects of microbial morphology.

3. Because most microorganisms are colorless and therefore not easily seen in the bright-field microscope, they are usually fixed and stained before observation. Either simple or differential staining can be used to enhance contrast. Specific bacterial structures such as capsules, endospores, and flagella also can be selectively stained.

4. The transmission electron microscope achieves great resolution (about 0.5 nm) by using electron beams of very short wavelength rather than visible light. Although one can prepare microorganisms for observation in other ways, one normally views thin sections of plastic-embedded specimens treated with heavy metals to improve contrast.

5. External features can be observed in great detail with the scanning electron microscope, which generates an image by scanning a fine electron beam over the surface of specimens rather than projecting electrons through them.

Microbiology usually is concerned with organisms so small they cannot be seen distinctly with the unaided eye. Because of the nature of this discipline, the microscope is of crucial importance; much of what is known about microorganisms has been discovered with microscopes. Thus it is important to understand how the microscope works and the way in which specimens are prepared for examination.

The chapter begins with a detailed treatment of the standard bright-field microscope and then describes other common types of light microscopes. Next preparation and staining of specimens for examination with the light microscope are discussed. The chapter closes with a description of transmission and scanning electron microscopes, both of which are used extensively in current microbiological research.

Lenses and the Bending of Light

To understand how a light microscope operates, one must know something about the way in which lenses bend and focus light to form images. When a ray of light passes from one medium to another, **refraction** occurs—that is, the ray is bent at the interface. The **refractive index** is a measure of how greatly a substance slows the velocity of light, and the direction and magnitude of bending is determined by the refractive indexes of the two media forming the interface. When light passes from air into glass, a medium with a greater refractive index, it is slowed and bent toward the normal, a line perpendicular to the surface (figure 2.1). As light leaves glass and returns to air, a medium with a lower refractive index, it accelerates and is bent away from the normal. Thus a prism bends light because glass has a different refractive index from air, and the light strikes its surface at an angle.

Lenses act like a collection of prisms operating as a unit. When the light source is distant so that parallel rays of light strike the lens, a convex lens will focus these rays at a specific point, the **focal point** (F in figure 2.2). The distance between the center of the lens and the focal point is called the **focal length** (f in figure 2.2).

Our eyes cannot focus on objects nearer than about 25 cm or 10 inches (table 2.1). This limitation may be overcome by using a convex lens as a simple magnifier (or microscope) and holding it close to an object. A magnifying glass provides a clear image at much closer range, and the object appears larger. Lens strength is related to focal length; a lens with a short focal length will magnify an object more than a weaker lens having a longer focal length.

1. Define refraction, refractive index, focal point, and focal length.
2. Describe the path of a light ray through a prism or lens.
3. How is lens strength related to focal length?

Figure 2.1   The Bending of Light by a Prism. Normals (lines perpendicular to the surface of the prism) are indicated by dashed lines. As light enters the glass, it is bent toward the first normal (angle θ_2 is less than θ_1). When light leaves the glass and returns to air, it is bent away from the second normal (θ_4 is greater than θ_3). As a result the prism bends light passing through it.

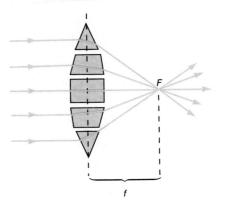

Figure 2.2   Lens Function. A lens functions somewhat like a collection of prisms. Light rays from a distant source are focused at the focal point F. The focal point lies a distance f, the focal length, from the lens center.

TABLE 2.1	Common Units of Measurement	
Unit	**Abbreviation**	**Value**
1 centimeter	cm	10^{-2} meter or 0.394 inches
1 millimeter	mm	10^{-3} meter
1 micrometer	μm	10^{-6} meter
1 nanometer	nm	10^{-9} meter
1 Angstrom	Å	10^{-10} meter
1 picometer	pm	10^{-12} meter

The Light Microscope

Microbiologists currently employ a variety of light microscopes in their work; bright-field, dark-field, phase-contrast, and fluorescence microscopes are most commonly used (*see confocal scanning laser microscopy, Box 41.1*). Modern microscopes are all compound microscopes. That is, the magnified image formed by the objective lens is further enlarged by one or more additional lenses.

The Bright-Field Microscope

The ordinary microscope is called a **bright-field microscope** because it forms a dark image against a brighter background.

Figure 2.3   A Bright-Field Microscope. The parts of a modern bright-field microscope. The microscope pictured is somewhat more sophisticated than those found in many student laboratories. For example, it is binocular (has two eyepieces) and has a mechanical stage, an adjustable substage condenser, and a built-in illuminator.

The microscope consists of a sturdy metal body or stand composed of a base and an arm to which the remaining parts are attached (figure 2.3). A light source, either a mirror or an electric illuminator, is located in the base. Two focusing knobs, the **fine** and **coarse adjustment knobs,** are located on the arm and can move either the stage or the nosepiece to focus the image.

The **stage** is positioned about halfway up the arm and holds microscope slides by either simple slide clips or a mechanical stage clip. A mechanical stage allows the operator to move a slide around smoothly during viewing by use of stage control knobs. The **substage condenser** is mounted within or beneath the stage and focuses a cone of light on the slide. Its position often is fixed in simpler microscopes but can be adjusted vertically in more advanced models.

The curved upper part of the arm holds the body assembly, to which a **nosepiece** and one or more **eyepieces** or **oculars** are attached. More advanced microscopes have eyepieces for both eyes and are called binocular microscopes. The body assembly itself contains a series of mirrors and prisms so that the barrel holding the eyepiece may be tilted for ease in viewing (figure 2.4). The nosepiece holds three to five **objectives** with lenses of differing magnifying power and can be rotated to position any objective beneath the body assembly. Ideally a microscope should be **parfocal**—that is, the image should remain in focus when objectives are changed.

The path of light through a bright-field microscope is shown in figure 2.4. The objective lens forms an enlarged real image within the microscope, and the eyepiece lens further magnifies this primary image. When one looks into a microscope, the enlarged specimen image, called the virtual image,

Figure 2.4   A Microscope's Light Path. The light path in an advanced bright-field microscope and the location of the virtual image are shown. (See also figure 2.22.)

appears to lie just beyond the stage about 25 cm away. The total magnification is calculated by multiplying the objective and eyepiece magnifications together. For example, if a 45× objective is used with a 10× eyepiece, the overall magnification of the specimen will be 450×.

Microscope Resolution

The most important part of the microscope is the objective, which must produce a clear image, not just a magnified one. Thus resolution is extremely important. **Resolution** is the ability of a lens to separate or distinguish between small objects that are close together. Much of the optical theory underlying microscope design was developed by the German physicist Ernst Abbé in the 1870s. The minimum distance (d) between two objects that reveals them as separate entities is given by the Abbé equation, in which lambda (λ) is the wavelength of light used to illuminate the specimen and $n \sin \theta$ is the numerical aperture (NA).

$$d = \frac{0.5\lambda}{n \sin \theta}$$

As d becomes smaller, the resolution increases, and finer detail can be discerned in a specimen.

The preceding equation indicates that a major factor in resolution is the wavelength of light used. The wavelength must be shorter than the distance between two objects or they will not be seen clearly. Thus the greatest resolution is obtained with light of the shortest wavelength, light at the blue end of the visible spectrum (in the range of 450 to 500 nm).

The electromagnetic spectrum of radiation (p. 132).

The **numerical aperture** ($n \sin \theta$) is more difficult to understand. Theta is defined as ½ the angle of the cone of light entering an objective (figure 2.5). Light that strikes the microorganism after passing through a condenser is cone-shaped. When this cone has a narrow angle and tapers to a sharp point, it does not spread out much after leaving the slide and therefore does not adequately separate images of closely packed objects. The resolution is low. If the cone of light has a very wide angle and spreads out rapidly after passing through a specimen, closely packed objects appear widely separated and are resolved. The angle of the cone of light that can enter a lens depends on the refractive index (n) of the medium in which the lens works, as well as upon the objective itself. The refractive index for air is 1.00. Since $\sin \theta$ cannot be greater than 1 (the maximum θ is 90° and $\sin 90°$ is 1.00), no lens working in air can have a numerical aperture greater than 1.00. The only practical way to raise the numerical aperture above 1.00, and therefore achieve higher resolution, is to increase the refractive index with **immersion oil,** a colorless liquid with the same refractive index as glass

(table 2.2). If air is replaced with immersion oil, many light rays that did not enter the objective due to reflection and refraction at the surfaces of the objective lens and slide will now do so (figure 2.6). An increase in numerical aperture and resolution results.

The resolution of a microscope depends upon the numerical aperture of its condenser as well as that of the objective. This is evident from the equation describing the resolution of the complete microscope.

$$d_{\text{microscope}} = \frac{\lambda}{(\text{NA}_{\text{objective}} + \text{NA}_{\text{condenser}})}$$

Most microscopes have a condenser with a numerical aperture between 1.2 and 1.4. However, the condenser numerical aperture will not be much above about 0.9 unless the top of the condenser is oiled to the bottom of the slide. During routine microscope

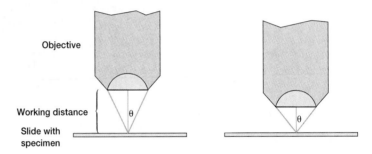

Figure 2.5   Numerical Aperture in Microscopy. The angular aperture θ is ½ the angle of the cone of light that enters a lens from a specimen, and the numerical aperture is $n \sin \theta$. In the right-hand illustration the lens has larger angular and numerical apertures; its resolution is greater and its working distance smaller.

Figure 2.6   The Oil Immersion Objective. An oil immersion objective operating in air and with immersion oil.

TABLE 2.2   The Properties of Microscope Objectives

| | Objective | | | |
Property	Scanning	Low Power	High Power	Oil Immersion
Magnification	4×	10×	40–45×	90–100×
Numerical aperture	0.10	0.25	0.55–0.65	1.25–1.4
Approximate focal length (*f*)	40 mm	16 mm	4 mm	1.8–2.0 mm
Working distance	17–20 mm	4–8 mm	0.5–0.7 mm	0.1 mm
Approximate resolving power with light of 450 nm (blue light)	2.3 μm	0.9 μm	0.35 μm	0.18 μm

operation, the condenser usually is not oiled and this limits the overall resolution, even with an oil immersion objective.

The limits set on the resolution of a light microscope can be calculated using the Abbé equation. The maximum theoretical resolving power of a microscope with an oil immersion objective (numerical aperture of 1.25) and blue-green light is approximately 0.2 μm.

$$d = \frac{(0.5)(530 \text{ nm})}{1.25} = 212 \text{ nm or } 0.2 \text{ μm}$$

At best, a bright-field microscope can distinguish between two dots around 0.2 μm apart (the same size as a very small bacterium).

Normally a microscope is equipped with three or four objectives ranging in magnifying power from 4× to 100× (table 2.2). The **working distance** of an objective is the distance between the front surface of the lens and the surface of the cover glass (if one is used) or the specimen when it is in sharp focus. Objectives with large numerical apertures and great resolving power have short working distances.

The largest useful magnification increases the size of the smallest resolvable object enough to be visible. Our eye can just detect a speck 0.2 mm in diameter, and consequently the useful limit of magnification is about 1,000 times the numerical aperture of the objective lens. Most standard microscopes come with 10× eyepieces and have an upper limit of about 1,000× with oil immersion. A 15× eyepiece may be used with good objectives to achieve a useful magnification of 1,500×. Any further magnification increase does not enable a person to see more detail. A light microscope can be built to yield a final magnification of 10,000×, but it would simply be magnifying a blur. Only the electron microscope provides sufficient resolution to make higher magnifications useful.

Proper specimen illumination also is extremely important in determining resolution. A microscope equipped with a concave mirror between the light source and the specimen illuminates the slide with a fairly narrow cone of light and has a small numerical aperture. Resolution can be improved with a substage condenser, a large light-gathering lens used to project a wide cone of light through the slide and into the objective lens, thus increasing the numerical aperture.

The Dark-Field Microscope

Living, unstained cells and organisms can be observed by simply changing the way in which they are illuminated. A hollow cone of light is focused on the specimen in such a way that unreflected and unrefracted rays do not enter the objective. Only light that has been reflected or refracted by the specimen forms an image (figure 2.7). The field surrounding a specimen appears black, while the object itself is brightly illuminated (figure 2.8a,b); because the background is dark, this type of microscopy is called **dark-field microscopy.** Considerable internal structure is often visible in larger eucaryotic microorganisms (figure 2.8b). The dark-field microscope is used to identify bacteria like the thin and distinctively shaped *Treponema pallidum* (figure 2.8a), the causative agent of syphilis.

(a)

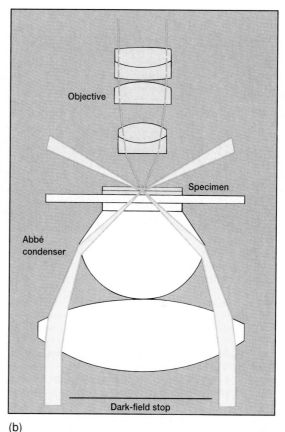

Objective

Specimen

Abbé condenser

Dark-field stop

(b)

Figure 2.7 Dark-Field Microscopy. The simplest way to convert a microscope to dark-field microscopy is to place (*a*) a dark-field stop underneath (*b*) the condenser lens system. The condenser then produces a hollow cone of light so that the only light entering the objective comes from the specimen.

The Phase-Contrast Microscope

Unpigmented living cells are not clearly visible in the bright-field microscope because there is little difference in contrast between the cells and water. Thus microorganisms often must be fixed and stained before observation to increase contrast and create variations in color between cell structures. A **phase-contrast**

Figure 2.8   Examples of Dark-Field and Phase-Contrast Microscopy. (*a*) *Treponema pallidum*, the spirochete that causes syphilis; dark-field microscopy (×500). (*b*) *Volvox* and *Spirogyra;* dark-field microscopy (×175). Note daughter colonies within the mature *Volvox* colony (center) and the spiral chloroplasts of *Spirogyra* (left and right). (*c*) *Spirillum volutans,* a very large bacterium with flagellar bundles; phase-contrast microscopy (×210). (*d*) *Clostridium botulinum,* the bacterium responsible for botulism, with subterminal oval endospores; phase-contrast microscopy (×600). (*e*) *Paramecium* stained to show a large central macronucleus with a small spherical micronucleus at its side; phase-contrast microscopy (×100).

microscope converts slight differences in refractive index and cell density into easily detected variations in light intensity and is an excellent way to observe living cells (figure 2.8*c–e*).

The condenser of a phase-contrast microscope has an annular stop, an opaque disk with a thin transparent ring, which produces a hollow cone of light (figure 2.9). As this cone passes through a cell, some light rays are bent due to variations in density and refractive index within the specimen and are retarded by about ¼ wavelength. The deviated light is focused to form an image of the object. Undeviated light rays strike a phase ring in the phase plate, a special optical disk located in the objective, while the deviated rays miss the ring and pass through the rest of the plate. If the phase ring is constructed in such a way that the undeviated light passing through it is

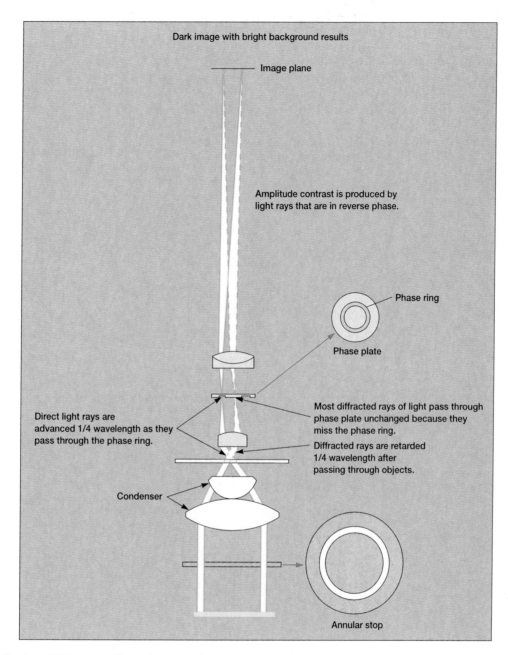

Dark image with bright background results

Image plane

Amplitude contrast is produced by light rays that are in reverse phase.

Phase ring

Phase plate

Direct light rays are advanced 1/4 wavelength as they pass through the phase ring.

Most diffracted rays of light pass through phase plate unchanged because they miss the phase ring.

Diffracted rays are retarded 1/4 wavelength after passing through objects.

Condenser

Annular stop

Figure 2.9   Phase-Contrast Microscopy. The optics of a dark-phase-contrast microscope.

advanced by ¼ wavelength, the deviated and undeviated waves will be about ½ wavelength out of phase and will cancel each other when they come together to form an image (figure 2.10). The background, formed by undeviated light, is bright, while the unstained object appears dark and well-defined. This type of microscopy is called **dark-phase-contrast microscopy.** Color filters often are used to improve the image (figure 2.8*c,d*).

Phase-contrast microscopy is especially useful for the detection of bacterial components such as endospores and inclusion bodies containing poly-β-hydroxybutyrate, polymetaphos-

phate, sulfur, or other substances (*see chapter 3*). These are clearly visible (figure 2.8*d*) because they have refractive indexes markedly different from that of water. Phase-contrast microscopes also are widely used in studying eucaryotic cells.

The Fluorescence Microscope

The microscopes thus far considered produce an image from light that passes through a specimen. An object also can be seen because it actually emits light, and this is the basis of fluorescence microscopy. When some molecules absorb radiant energy, they become excited and later release much of their trapped energy as light. Any light emitted by an excited molecule will have a longer wavelength (or be of lower energy) than the radiation originally absorbed. **Fluorescent light** is emitted very quickly by the excited molecule as it gives up its trapped energy and returns to a more stable state.

The **fluorescence microscope** (figure 2.11) exposes a specimen to ultraviolet, violet, or blue light and forms an image of the object with the resulting fluorescent light. A mercury vapor arc lamp or other source produces an intense beam, and heat transfer is limited by a special infrared filter. The light passes through an exciter filter that transmits only the desired wavelength. A dark-field condenser provides a black background against which the fluorescent objects glow. Usually the specimens have been stained with dye molecules, called **fluorochromes,** that fluoresce brightly upon exposure to light of a specific wavelength, but some microorganisms are

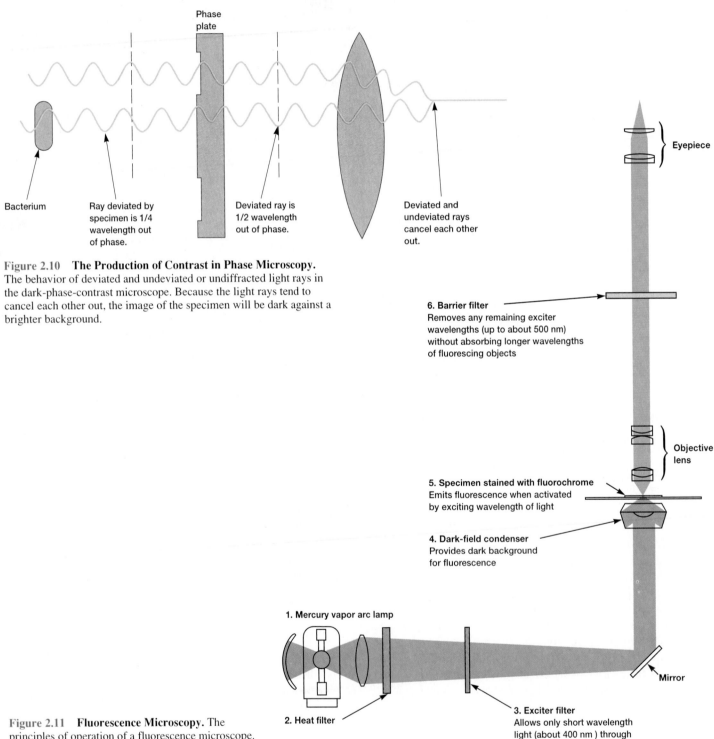

Figure 2.10   The Production of Contrast in Phase Microscopy. The behavior of deviated and undeviated or undiffracted light rays in the dark-phase-contrast microscope. Because the light rays tend to cancel each other out, the image of the specimen will be dark against a brighter background.

Bacterium

Ray deviated by specimen is 1/4 wavelength out of phase.

Deviated ray is 1/2 wavelength out of phase.

Deviated and undeviated rays cancel each other out.

Phase plate

Eyepiece

6. Barrier filter
Removes any remaining exciter wavelengths (up to about 500 nm) without absorbing longer wavelengths of fluorescing objects

Objective lens

5. Specimen stained with fluorochrome
Emits fluorescence when activated by exciting wavelength of light

4. Dark-field condenser
Provides dark background for fluorescence

1. Mercury vapor arc lamp

Mirror

2. Heat filter

3. Exciter filter
Allows only short wavelength light (about 400 nm) through

Figure 2.11   Fluorescence Microscopy. The principles of operation of a fluorescence microscope.

autofluorescing. The microscope forms an image of the fluorochrome-labeled microorganisms from the light emitted when they fluoresce (figure 2.12). A barrier filter positioned after the objective lenses removes any remaining ultraviolet light, which could damage the viewer's eyes, or blue and violet light, which would reduce the image's contrast.

The fluorescence microscope has become an essential tool in medical microbiology and microbial ecology. Bacterial pathogens (e.g., *Mycobacterium tuberculosis,* the cause of tuberculosis) can be identified after staining them with fluorochromes or specifically labeling them with fluorescent antibodies using immunofluorescence procedures. In ecological studies the fluorescence microscope is used to observe microorganisms after staining them with the fluorochrome acridine orange. The stained organisms will fluoresce orange or green and can be detected even in the midst of other particulate material. Thus the microorganisms can be viewed and directly counted in a relatively undisturbed ecological niche.

Immunofluorescence and diagnostic microbiology (pp. 649–50, 682).

1. List the parts of a light microscope and their functions.
2. Define resolution, numerical aperture, working distance, and fluorochrome.
3. How does resolution depend upon the wavelength of light, refractive index, and the numerical aperture? What are the functions of immersion oil and the substage condenser?
4. Briefly describe how dark-field, phase-contrast, and fluorescence microscopes work and the kind of image provided by each. Give a specific use for each type.

Preparation and Staining of Specimens

Although living microorganisms can be directly examined with the light microscope, they often must be fixed and stained to increase visibility, accentuate specific morphological features, and preserve them for future study.

Fixation

The stained cells seen in a microscope should resemble living cells as closely as possible. **Fixation** is the process by which the internal and external structures of cells and microorganisms are preserved and fixed in position. It inactivates enzymes that might disrupt cell morphology and toughens cell structures so that they do not change during staining and observation. A microorganism usually is killed and attached firmly to the microscope slide during fixation.

There are two fundamentally different types of fixation. (1) Bacteriologists heat-fix bacterial smears by gently flame heating an air-dried film of bacteria. This adequately preserves overall morphology but not structures within cells. (2) Chemical fixation must be used to protect fine cellular substructure and the morphology of larger, more delicate microorganisms. Chemical fixatives penetrate cells and react with cellular components, usually proteins and lipids, to render them inactive, insoluble, and immobile. Common fixative mixtures contain such components as ethanol, acetic acid, mercuric chloride, formaldehyde, and glutaraldehyde.

Dyes and Simple Staining

The many types of dyes used to stain microorganisms have two features in common. (1) They have **chromophore groups,** groups with double bonds that give the dye its color.

(a) (b) (c)

Figure 2.12 **Examples of Fluorescence Microscopy.** (*a*) *Escherichia coli* stained with fluorescent antibodies (×600). The green material is debris. (*b*) *Paramecium tetraurelia* conjugating; acridine-orange fluorescence (×125). (*c*) The flagellate protozoan *Crithidia luciliae* stained with fluorescent antibodies to show the kinetoplast (×1,000).

(2) They can bind with cells by ionic, covalent, or hydrophobic bonding. For example, a positively charged dye binds to negatively charged structures on the cell.

Ionizable dyes may be divided into two general classes based on the nature of their charged group.

1. **Basic dyes**—methylene blue, basic fuchsin, crystal violet, safranin, malachite green—are cationic or have positively charged groups (usually some form of pentavalent nitrogen) and are generally sold as chloride salts. Basic dyes bind to negatively charged molecules like nucleic acids and many proteins. Because the surfaces of bacterial cells also are negatively charged, basic dyes are most often used in bacteriology.

2. **Acid dyes**—eosin, rose bengal, and acid fuchsin—are anionic or possess negatively charged groups such as carboxyls (— COOH) and phenolic hydroxyls (— OH). Acid dyes, because of their negative charge, bind to positively charged cell structures. The pH may alter staining effectiveness since the nature and degree of the charge on cell components change with pH.

Thus anionic dyes stain best under acidic conditions when proteins and many other molecules carry a positive charge; basic dyes are most effective at higher pHs.

Although ionic interactions are probably the most common means of attachment, dyes also bind through covalent bonds or because of their solubility characteristics. For instance, DNA can be stained by the Feulgen procedure in which Schiff's reagent is covalently attached to its deoxyribose sugars after hydrochloric acid treatment. Sudan III (Sudan Black) selectively stains lipids because it is lipid soluble but will not dissolve in aqueous portions of the cell.

Microorganisms often can be stained very satisfactorily by **simple staining,** in which a single staining agent is used. Simple staining's value lies in its simplicity and ease of use. One covers the fixed smear with stain for the proper length of time, washes the excess stain off with water, and blots the slide dry. Basic dyes like crystal violet, methylene blue, and carbolfuchsin are frequently used to determine the size, shape, and arrangement of bacteria.

Differential Staining

Differential staining procedures divide bacteria into separate groups based on staining properties. The **Gram stain,** developed in 1884 by the Danish physician Christian Gram, is the most widely employed staining method in bacteriology. It is a differential staining procedure because it divides bacteria into two classes—gram negative and gram positive.

Gram-positive and gram-negative bacteria (pp. 51–56, 400).

In the first step of the Gram-staining procedure (figure 2.13), the smear is stained with the basic dye crystal violet, the primary stain. It is followed by treatment with an iodine solution functioning as a **mordant.** That is, the iodine increases the interaction between the cell and the dye so that the cell is stained

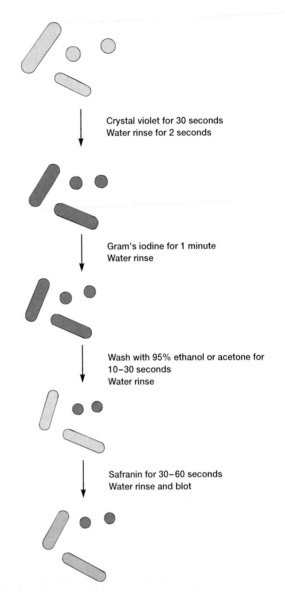

Crystal violet for 30 seconds
Water rinse for 2 seconds

Gram's iodine for 1 minute
Water rinse

Wash with 95% ethanol or acetone for 10–30 seconds
Water rinse

Safranin for 30–60 seconds
Water rinse and blot

Figure 2.13   The Gram-Staining Procedure. Note that decolorization with ethanol or acetone removes crystal violet from gram-negative cells but not from gram-positive cells. The gram-negative cells then turn pink to red when counterstained with safranin.

more strongly. The smear is next decolorized by washing with ethanol or acetone. This step generates the differential aspect of the Gram stain; gram-positive bacteria retain the crystal violet, whereas gram-negative bacteria lose their crystal violet and become colorless. Finally, the smear is counterstained with a basic dye different in color from crystal violet. Safranin, the most common counterstain, colors gram-negative bacteria pink to red and leaves gram-positive bacteria dark purple (figure 2.14).

Cell wall structure and the mechanism of the Gram stain (pp. 56–57).

Acid-fast staining is another important differential staining procedure. A few species, particularly those in the genus *Mycobacterium* (*see chapter 21*) do not bind simple stains readily and must be stained by a harsher treatment: heating with a mixture of basic fuchsin and phenol (the

(a)

(b)

(c)

(d)

Figure 2.14 **Examples of Gram Staining.** (*a*) Gram-positive *Clostridium perfringens* (×800). Some rods have stained pink rather than purple, as often happens when gram-positive cells age. (*b*) *Staphylococcus aureus.* Gram stain, bright-field microscopy (×1,000). The gram-positive cocci associate in grapelike clusters. (*c*) *Escherichia coli,* Gram stain (×500). (*d*) *Neisseria gonorrhoeae.* The diplococci are often within white blood cells from pus (×1,000).

Ziehl-Neelsen method). Once basic fuchsin has penetrated with the aid of heat and phenol, acid-fast cells are not easily decolorized by an acid-alcohol wash and hence remain red. This is due to the quite high lipid content of acid-fast cell walls; in particular, mycolic acid—a group of branched chain hydroxy lipids—appears responsible for acid-fastness. Non-acid-fast bacteria are decolorized by acid-alcohol and thus are stained blue by methylene blue counterstain. This method is used to identify *Mycobacterium tuberculosis* and *M. leprae* (figure 2.15), the pathogens responsible for tuberculosis and leprosy, respectively.

Staining Specific Structures

Many special staining procedures have been developed over the years to study specific bacterial structures with the light microscope. One of the simplest is **negative staining,** a technique that reveals the presence of the diffuse capsules surrounding many bacteria. Bacteria are mixed with India ink or Nigrosin dye and spread out in a thin film on a slide. After air-drying, bacteria appear as lighter bodies in the midst of a blue-black background because ink and dye particles cannot penetrate either the bacterial cell or its capsule. The extent of the

Figure 2.15 **Acid-Fast Staining.** *Mycobacterium leprae.* Acid-fast stain (×380). Note the masses of red bacteria within host cells.

light region is determined by the size of the capsule and of the cell itself. There is little distortion of bacterial shape, and the cell can be counterstained for even greater visibility (figure 2.16).

Capsules and slime layers (pp. 58–59).

Figure 2.16   Negative Staining. *Klebsiella pneumoniae* negatively stained with India ink to show its capsules (×900).

Figure 2.17   Spore Staining. *Bacillus cereus* stained with the Schaeffer-Fulton procedure. Note the central, elliptical blue to green spores within the red to purple cells (×1,000).

washed free of dye with water and is counterstained with safranin. This technique yields a green endospore resting in a pink to red cell.

Bacterial endospore structure (pp. 67–69).

Bacterial flagella are fine, threadlike organelles of locomotion that are so slender (about 10 to 30 nm in diameter) they can only be seen directly using the electron microscope. To observe them with the light microscope, the thickness of flagella is increased by coating them with mordants like tannic acid and potassium alum, and they are stained with pararosaniline (Leifson method) or basic fuchsin (Gray method). **Flagella staining** procedures provide taxonomically valuable information about the presence and location of flagella (figure 2.18; *see also figure 3.33*).

The bacterial flagellum (pp. 60–63).

Figure 2.18   Example of Flagella Staining. *Spirillum volutans* with bipolar tufts of flagella (×400). (*See also figure 3.33.*)

1. Define fixation, dye, chromophore, basic dye, acid dye, simple staining, differential staining, mordant, negative staining, and acid-fast staining.
2. Describe the Gram-stain procedure and how it works. How would you visualize capsules, endospores, and flagella?

Bacteria in the genera *Bacillus* and *Clostridium* (*see chapter 21*) form an exceptionally resistant structure capable of surviving for long periods in an unfavorable environment. This dormant structure is called an endospore since it develops within the cell. Endospore morphology and location vary with species and often are valuable in identification; endospores may be spherical to elliptical and either smaller or larger than the diameter of the parent bacterium. They can be observed with the phase-contrast microscope or negative staining. Endospores are not stained well by most dyes, but once stained, they strongly resist decolorization. This property is the basis of most **spore staining** methods (figure 2.17). In the Schaeffer-Fulton procedure, endospores are first stained by heating bacteria with malachite green, then the rest of the cell is

Electron Microscopy

For centuries the light microscope has been the most important instrument for studying microorganisms. More recently the electron microscope has transformed microbiology and added immeasurably to our knowledge. The nature of the electron microscope and the ways in which specimens are prepared for observation are reviewed briefly in this section.

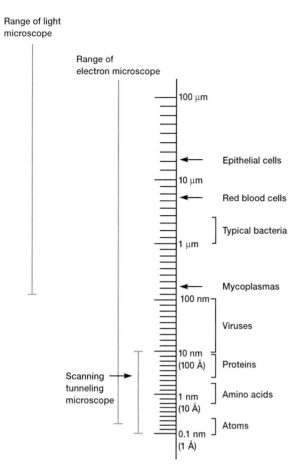

Figure 2.19   The Limits of Microscopic Resolution. Dimensions are indicated with a logarithmic scale (each major division represents a tenfold change in size). To the right side of the scale are the approximate sizes of cells, bacteria, viruses, molecules, and atoms.

The Transmission Electron Microscope

The very best light microscope has a resolution limit of about 0.2 μm. Because bacteria usually are around 1 μm in diameter, only their general shape and major morphological features are visible in the light microscope. The detailed internal structure of larger microorganisms also cannot be effectively studied by light microscopy. These limitations arise from the nature of visible light waves, not from any inadequacy of the light microscope itself.

Recall that the resolution of a light microscope increases with a decrease in the wavelength of the light it uses for illumination. Electron beams behave like radiation and can be focused much as light is in a light microscope. If electrons illuminate the specimen, the microscope's resolution is enormously increased because the wavelength of the radiation is around 0.005 nm, approximately 100,000 times shorter than that of visible light. The transmission electron microscope has a practical resolution roughly 1,000 times better than the light microscope; with many electron microscopes, points closer than 5 Å or 0.5 nm can be

(a)

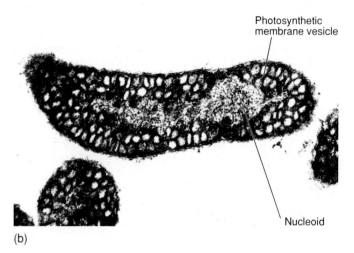

(b)

Figure 2.20   Light and Electron Microscopy. A comparison of light and electron microscopic resolution. (*a*) *Rhodospirillum rubrum* in phase-contrast light microscope (×600). (*b*) A thin section of *R. rubrum* in transmission electron microscope (×100,000).

distinguished, and the useful magnification is well over 100,000× (figure 2.19). The value of the electron microscope is evident on comparison of the photographs in figure 2.20; microbial morphology can now be studied in great detail.

A modern **transmission electron microscope (TEM)** is complex and sophisticated (figure 2.21), but the basic principles behind its operation can be understood readily. A heated tungsten filament in the electron gun generates a beam of electrons that is then focused on the specimen by the condenser (figure 2.22). Since electrons cannot pass through a glass lens, doughnut-shaped electromagnets called magnetic lenses are used to focus the beam. The column containing the lenses and

Figure 2.21 A Modern Transmission Electron Microscope. The electron gun is at the top of the central column, and the magnetic lenses are within the column. The image on the fluorescent screen may be viewed through a magnifier positioned over the viewing window. The camera is in a compartment below the screen.

specimen must be under high vacuum to obtain a clear image because electrons are deflected by collisions with air molecules. The specimen scatters electrons passing through it, and the beam is focused by magnetic lenses to form an enlarged, visible image of the specimen on a fluorescent screen. A denser region in the specimen scatters more electrons and therefore appears darker in the image since fewer electrons strike that area of the screen. In contrast, electron-transparent regions are brighter. The screen can also be moved aside and the image captured on photographic film as a permanent record.

Specimen Preparation

Table 2.3 compares some of the important features of light and electron microscopes. The distinctive features of the TEM place harsh restrictions on the nature of samples that can be viewed and the means by which those samples must be prepared. Since electrons are quite easily absorbed and scattered by solid matter, only extremely thin slices of a microbial specimen can be viewed in the average TEM. The specimen must be around 20 to 100 nm thick, about $\frac{1}{50}$ to $\frac{1}{10}$ the diameter of a typical bacterium, and able to maintain its structure when bombarded with electrons under a high vacuum! Such a thin slice cannot be cut unless the specimen has support of some kind; the necessary support is provided by plastic. After fixation with chemicals

Figure 2.22 Transmission Electron Microscope Operation. An overview of TEM operation and a comparison of the operation of light and transmission electron microscopes.

TABLE 2.3   Characteristics of Light and Transmission Electron Microscopes

Feature	Light Microscope	Electron Microscope
Highest practical magnification	About 1,000–1,500	Over 100,000
Best resolution[a]	0.2 μm	0.5 nm
Radiation source	Visible light	Electron beam
Medium of travel	Air	High vacuum
Type of lens	Glass	Electromagnet
Source of contrast	Differential light absorption	Scattering of electrons
Focusing mechanism	Adjust lens position mechanically	Adjust current to the magnetic lens
Method of changing magnification	Switch the objective lens or eyepiece	Adjust current to the magnetic lens
Specimen mount	Glass slide	Metal grid (usually copper)

[a]The resolution limit of a human eye is about 0.2 mm.

like glutaraldehyde or osmium tetroxide to stabilize cell structure, the specimen is dehydrated with organic solvents (e.g., acetone or ethanol). Complete dehydration is essential because most plastics used for embedding are not water soluble. Next the specimen is soaked in unpolymerized, liquid epoxy plastic until it is completely permeated, and then the plastic is hardened to form a solid block. Thin sections are cut from this block with a glass or diamond knife using a special instrument called an ultramicrotome.

Cells usually must be stained before they can be seen clearly in the bright-field microscope; the same is true for observations with the TEM. The probability of electron scattering is determined by the density (atomic number) of the specimen atoms. Biological molecules are composed primarily of atoms with low atomic numbers (H, C, N, and O), and electron scattering is fairly constant throughout the unstained cell. Therefore specimens are prepared for observation by soaking thin sections with solutions of heavy metal salts like lead citrate and uranyl acetate. The lead and uranium ions bind to cell structures and make them more electron opaque, thus increasing contrast in the material. Heavy osmium atoms from the osmium tetroxide fixative also "stain" cells and increase their contrast. The stained thin sections are then mounted on tiny copper grids and viewed.

Although the above procedure for preparing thin sections normally is used to reveal the internal structure of the smallest cell, there are other ways in which microorganisms and smaller objects can be readied for viewing. One very useful technique is negative staining. The specimen is spread out in a thin film with either phosphotungstic acid or uranyl acetate. Just as in negative staining for light microscopy, heavy metals do not penetrate the specimen but render the background dark, whereas the specimen appears bright in photographs. Negative staining is an excellent way to study the structure of viruses, bacterial gas vacuoles, and other similar material. A microorganism also can be viewed after **shadowing** with metal. It is coated with a thin film of platinum or other heavy metal by evaporation at an angle of about 45° from horizontal so that the metal strikes the microorganism on only one side. The area coated with metal scatters electrons and appears light in photographs, whereas the uncoated side and the shadow region created by the object is dark

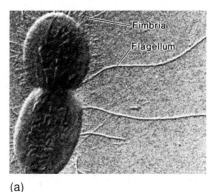

(a)

(b)

Figure 2.23   Specimen Shadowing for the TEM. Examples of specimens viewed in the TEM after shadowing with uranium metal. (*a*) *Proteus mirabilis* (×42,750); note flagella and fimbriae. (*b*) T4 coliphage (×72,000).

(figure 2.23). The specimen looks much as it would if light were shining on it to cast a shadow. This technique is particularly useful in studying virus morphology, bacterial flagella, and plasmids (*see chapter 14*).

The TEM will also disclose the shape of organelles within microorganisms if specimens are prepared by the **freeze-etching** procedure. Cells are rapidly frozen in liquid nitrogen and then warmed to –100°C in a vacuum chamber. Next a knife that has been precooled with liquid nitrogen (–196°C) fractures the frozen cells, which are very brittle and break along lines of

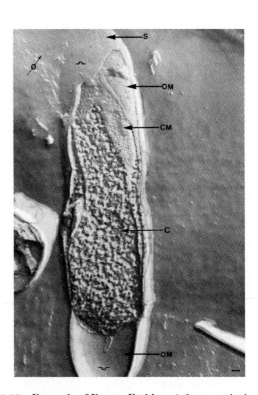

Figure 2.24   **The Freeze-Etching Technique.** In steps (*a*) and (*b*), a frozen eucaryotic cell is fractured with a cold knife. Etching by sublimation is depicted in *c*. Shadowing with platinum plus carbon and replica formation are shown in (*d*) and (*e*). See text for details.

greatest weakness, usually down the middle of internal membranes (figure 2.24). The specimen is left in the high vacuum for a minute or more so that some of the ice can sublimate away and uncover more structural detail (sometimes this etching step is eliminated). Finally, the exposed surfaces are shadowed and coated with layers of platinum and carbon to form a replica of the surface. After the specimen has been removed chemically, this replica is studied in the TEM and provides a detailed, three-dimensional view of intracellular structure (figure 2.25). An advantage of freeze-etching is that it minimizes the danger of artifacts because the cells are frozen quickly rather than being subjected to chemical fixation, dehydration, and plastic embedding.

The Scanning Electron Microscope

The previously described microscopes form an image from radiation that has passed through a specimen. More recently the **scanning electron microscope (SEM)** has been used to examine the surfaces of microorganisms in great detail; many instruments have a resolution of 7 nm or less. The SEM differs from other electron microscopes in producing an image from electrons emitted by an object's surface rather than from transmitted electrons (Box 2.1).

Figure 2.25   **Example of Freeze-Etching.** A freeze-etched preparation of the bacterium *Thiobacillus kabobis*. Note the differences in structure between the outer surface, S; the outer membrane of the cell wall, OM; the cytoplasmic membrane, CM; and the cytoplasm, C. Bar = 0.1 μm.

Box 2.1

A Microscope That Sees Atoms

lthough light and electron microscopes have become quite sophisticated and reached an advanced state of development, powerful new microscopes are still being created. A new class of microscopes, called **scanning probe microscopes,** measure surface features by moving a sharp probe over the object's surface. The scanning tunneling microscope, invented in 1980, is an excellent example of a scanning probe microscope. It can achieve magnifications of 100 million and allow scientists to view atoms on the surface of a solid. The electrons surrounding surface atoms tunnel or project out from the surface boundary a very short distance. The scanning tunneling microscope has a needlelike probe with a point so sharp that often there is only one atom at its tip. The probe is lowered toward the specimen surface until its electron cloud just touches that of the surface atoms. If a small voltage is applied between the tip and specimen, electrons flow through a narrow channel in the electron clouds. This tunneling current, as it is called, is extraordinarily sensitive to distance and will decrease about a thousandfold if the probe is moved away from the surface by a distance equivalent to the diameter of an atom.

The arrangement of atoms on the specimen surface is determined by moving the probe tip back and forth over the surface while keeping it at a constant height by adjusting the probe distance to maintain a steady tunneling current. As the tip moves up and down while following the surface contours, its motion is recorded and analyzed by a computer to create an accurate three-dimensional image of the surface atoms. The surface map can be displayed on a computer screen or plotted on paper. The resolution is so great that individual atoms are observed easily. The microscope's inventors,

Gerd Binnig and Heinrich Rohrer, shared the 1986 Nobel Prize in Physics for their work. Interestingly the other recipient of the prize was Ernst Ruska, the designer of the first transmission electron microscope.

The scanning tunneling microscope will likely have a major impact in biology. Recently it has been used to directly view DNA (Box Figure 2.1*b*). Since the microscope can examine objects when they are immersed in water, it may be particularly useful in studying biological molecules.

More recently a second type of scanning probe microscope has been developed. The atomic force microscope moves a sharp probe over the specimen surface while keeping the distance between the probe tip and the surface constant. It does this by exerting a very small amount of force on the tip, just enough to maintain a constant distance but not enough force to damage the surface. The vertical motion of the tip usually is followed by measuring the deflection of a laser beam that strikes the lever holding the probe. Unlike the scanning tunneling microscope, the atomic force microscope can be used to study surfaces that do not conduct electricity well.

(a)   (b)

Box Figure 2.1   Examples of Scanning Tunneling Microscopy. (*a*) Silicon (Si) atoms viewed through a scanning tunneling microscope. (*b*) A scanning tunneling micrograph of the DNA double helix with approximately three turns shown (false color; ×2,000,000).

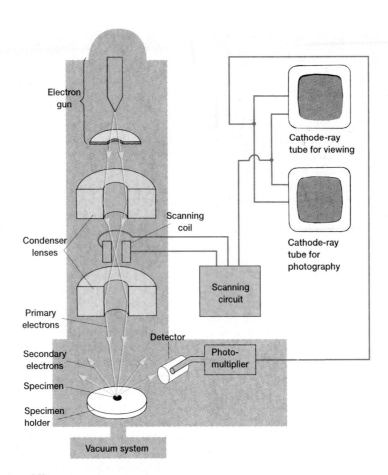

Figure 2.26   **The Scanning Electron Microscope.**

Specimen preparation is easy, and in some cases air-dried material can be examined directly. Most often, however, microorganisms must first be fixed, dehydrated, and dried to preserve surface structure and prevent collapse of the cells when they are exposed to the SEM's high vacuum. Before viewing, dried samples are mounted and coated with a thin layer of metal to prevent the buildup of an electrical charge on the surface and to give a better image.

The SEM scans a narrow, tapered electron beam back and forth over the specimen (figure 2.26). When the beam strikes a particular area, surface atoms discharge a tiny shower of electrons called secondary electrons, and these are trapped by a special detector. Secondary electrons entering the detector strike a scintillator causing it to emit light flashes that a photomultiplier converts to an electrical current and amplifies. The signal is sent to a cathode-ray tube and produces an image like a television picture, which can be viewed or photographed.

The number of secondary electrons reaching the detector depends on the nature of the specimen's surface. When the electron beam strikes a raised area, a large number of secondary electrons enter the detector; in contrast, fewer electrons escape a depression in the surface and reach the detector. Thus raised areas appear lighter on the screen and depressions are darker. A realistic three-dimensional image of the microorganism's surface with great depth of focus results (figure 2.27). The actual in situ location of microorganisms in ecological niches such as the human skin and the lining of the gut also can be examined.

1. Why does the transmission electron microscope have much greater resolution than the light microscope? Describe in general terms how the TEM functions.

2. Describe how specimens are prepared for viewing in the TEM. How are sections usually stained to increase contrast? What is negative staining, shadowing, and freeze-etching?

3. How does the scanning electron microscope operate and in what way does its function differ from that of the TEM? What aspect of morphology is the SEM used to study?

(a)   (b)

Figure 2.27   Scanning Electron Micrographs of Bacteria. (*a*) *Staphylococcus aureus* (×32,000). (*b*) *Cristispira,* a spirochete from the crystalline style of the oyster, *Ostrea virginica.* The axial fibrils or periplasmic flagella are visible around the protoplasmic cylinder (×6,000).

Summary

1. A light ray moving from air to glass, or vice versa, is bent in a process known as refraction. Lenses focus light rays at a focal point and magnify images.

2. In a compound microscope like the bright-field microscope, the primary image is formed by an objective lens and enlarged by the eyepiece or ocular lens to yield a virtual image.

3. A substage condenser focuses a cone of light on the specimen.

4. Microscope resolution increases as the wavelength of radiation used to illuminate the specimen decreases. The maximum resolution of a light microscope is about 0.2 μm.

5. The dark-field microscope uses only refracted light to form an image, and objects glow against a black background.

6. The phase-contrast microscope converts variations in the refractive index and density of cells into changes in light intensity and thus makes colorless, unstained cells visible.

7. The fluorescence microscope illuminates a fluorochrome-labeled specimen and forms an image from its fluorescence.

8. Specimens usually must be fixed and stained before viewing them in the bright-field microscope.

9. Most dyes are either positively charged basic dyes or negative acid dyes and bind to ionized parts of cells.

10. In simple staining a single dye mixture is used to stain microorganisms.

11. Differential staining procedures like the Gram stain and acid-fast stain distinguish between microbial groups by staining them differently.

12. Some staining techniques are specific for particular structures like bacterial capsules, flagella, and endospores.

13. The transmission electron microscope uses magnetic lenses to form an image from electrons that have passed through a very thin section of a specimen. Resolution is great because the wavelength of electrons is very short.

14. Thin section contrast can be increased by treatment with solutions of heavy metals like osmium tetroxide, uranium, and lead.

15. Specimens are also prepared for the TEM by negative staining, shadowing with metal, or freeze-etching.

16. The scanning electron microscope is used to study external surface features of microorganisms.

Key Terms

acid dyes   *26*

acid-fast staining   *26*

basic dyes   *26*

bright-field microscope   *18*

chromophore groups   *25*

dark-field microscopy   *21*

dark-phase-contrast microscopy   *23*

differential staining procedures   *26*

eyepieces   *19*

fine and coarse adjustment knobs   *19*

fixation   *25*

flagella staining   *28*

fluorescence microscope   *24*

fluorescent light   *24*

fluorochromes   *24*

focal length   *18*

focal point   *18*

freeze-etching   *31*

Gram stain   *26*

immersion oil   *20*

mordant   *26*

negative staining   *27*

nosepiece   *19*

numerical aperture   *20*

objectives   *19*

oculars   *19*

parfocal   *19*

phase-contrast microscope   *21*

refraction   *18*

refractive index   *18*

resolution   *20*

scanning electron microscope (SEM)   *32*

scanning probe microscope   *33*

shadowing   *31*

simple staining   *26*

spore staining   *28*

stage   *19*

substage condenser   *19*

transmission electron microscope (TEM)   *29*

working distance   *21*

Questions for Thought and Review

1. How are real and virtual images produced in a light microscope? Which one is a person actually seeing?

2. If a specimen is viewed using a 43× objective in a microscope with a 15× eyepiece, how many times has the image been magnified?

3. Why don't most light microscopes use 30× eyepieces for greater magnification?

4. Describe the two general types of fixation. Which would you normally use for bacteria? For protozoa?

5. Why would one expect basic dyes to be more effective under alkaline conditions?

6. What step in the Gram-stain procedure could be dropped without losing the ability to distinguish between gram-positive and gram-negative bacteria? Why?

7. Why must the TEM use a high vacuum and very thin sections?

8. Material is often embedded in paraffin before sectioning for light microscopy. Why can't this approach be used when preparing a specimen for the TEM?

9. Under what circumstances would it be desirable to prepare specimens for the TEM by use of negative staining? Shadowing? Freeze-etching?

10. Compare the microscopes described in this chapter—bright-field, dark-field, phase-contrast, fluorescence, TEM, and SEM—in terms of the images they provide and the purposes for which they are most often used.

11. Describe briefly how the scanning probe microscope operates. For what is it used? Distinguish between the two types of scanning probe microscopes with respect to their mechanism of operation.

Additional Reading

General

Boatman, E. S.; Berns, M. W.; Walter, R. J.; and Foster, J. S. 1987. Today's microscopy. *BioScience* 37(6):384–94.

Clark, G. L. 1961. *The encyclopedia of microscopy.* New York: Van Nostrand Reinhold.

Gerhard, P.; Murray, R. G. E.; Wood, W. A.; and Krieg, N. R., editors, 1994. *Methods for general and molecular bacteriology.* Washington, D.C.: American Society for Microbiology.

The Light Microscope

Cosslett, V. E. 1966. *Modern microscopy or seeing the very small.* Ithaca, N.Y.: Cornell University Press.

Preparation and Staining of Specimens

Clark, G. L., ed. 1973. *Staining procedures used by the Biological Stain Commission,* 3d ed. Baltimore: Williams & Wilkins.

Gray, Peter. 1964. *Handbook of basic microtechnique,* 3d ed. New York: McGraw-Hill.

Lillie, R. D. 1969. *H. J. Conn's biological stains,* 8th ed. Baltimore: Williams & Wilkins.

Scherrer, Rene. 1984. Gram's staining reaction, Gram types and cell walls of bacteria. *Trends Biochem. Sci.* 9:242–45.

Electron Microscopy

Binnig, G., and Rohrer, H. 1985. The scanning tunneling microscope. *Sci. Am.* 253(2):50–56.

Kopp, Friedrich. 1981. Electron microscopy. *Carolina Biology Reader,* no. 105. Burlington, N.C.: Carolina Biological Supply Co.

Meek, G. A. 1976. *Practical electron microscopy for biologists,* 2d ed. New York: John Wiley and Sons.

Postek, M. T.; Howard, K. S.; Johnson, A. H.; and McMichael, K. L. 1980. *Scanning electron microscopy: A student's handbook.* Burlington, Vt.: Ladd Research Industries.

Wickramasinghe, H. K. 1989. Scanned-probe microscopes. *Sci. Am.* 261(4):98–105.

Wischnitzer, S. 1981. *Introduction to electron microscopy,* 3d ed. New York: Pergamon Press.

CHAPTER 3
Procaryotic Cell Structure and Function

The era in which workers tended to look at bacteria as very small bags of enzymes has long passed.

—Howard J. Rogers

Concepts

1. Bacteria are small and simple in structure when compared with eucaryotes, yet they often have characteristic shapes and sizes.
2. Although they have a plasma membrane, which is required by all living cells, bacteria generally lack extensive, complex, internal membrane systems.
3. The cytoplasmic matrix typically contains several constituents that are not membrane-enclosed: inclusion bodies, ribosomes, and the nucleoid with its genetic material.
4. The procaryotic cell wall almost always has peptidoglycan and is chemically and morphologically complex. Most bacteria can be divided into gram-positive and gram-negative groups based on their cell wall structure and response to the Gram stain.
5. Components like capsules and fimbriae are located outside the cell wall. One of these is the flagellum, which many bacteria use like a propeller to swim toward attractants and away from repellents.
6. Some bacteria form resistant endospores to survive harsh environmental conditions in a dormant state.

Even a superficial examination of the microbial world shows that bacteria are one of the most important groups by any criterion: numbers of organisms, general ecological importance, or practical importance for humans. Indeed, much of our understanding of phenomena in biochemistry and molecular biology comes from research on bacteria. Although considerable space is devoted to eucaryotic microorganisms, the major focus is on procaryotes or bacteria. Therefore the unit on microbial morphology begins with the structure of procaryotes.

Eucaryotes, procaryotes, and the composition of the microbial world (pp. 13; 91–94).

(a)

(b)

An Overview of Procaryotic Cell Structure

Because much of this chapter is devoted to a discussion of individual cell components, a preliminary overview of the procaryotic cell as a whole is in order.

Size, Shape, and Arrangement

One might expect that small, relatively simple organisms like bacteria would be uniform in shape and size. Although it is true that many bacteria are similar in morphology, there is a remarkable amount of variation (figures 3.1 and 3.2; *see also figures 2.8 and 2.14*). Major morphological patterns are described here, and interesting variants are mentioned in the bacterial survey (*see chapters 20–24*).

Most commonly encountered bacteria have one of two shapes. **Cocci** (s., **coccus**) are roughly spherical cells. They can exist as individual cells, but also are associated in characteristic arrangements that are frequently useful in bacterial identification. **Diplococci** (s., **diplococcus**) arise when cocci divide and remain together to form pairs (*Neisseria; see figure 2.14d*). Long chains of cocci result when cells adhere after repeated divisions in one plane; this pattern is seen in the genera *Streptococcus, Enterococcus,* and *Lactococcus* (figure 3.1*b*). *Staphylococcus* divides in random planes to generate irregular grapelike clumps (figure 3.1*a*). Divisions in two or three planes can produce symmetrical clusters of cocci. Members of the genus *Micrococcus* often divide in two planes to form square groups of four cells called tetrads. In the genus *Sarcina*, cocci divide in three planes producing cubical packets of eight cells.

(c)

(d)

Figure 3.1 Representative Bacteria. Stained bacterial cultures as seen in the light microscope. (*a*) *Staphylococcus aureus.* Note the gram-positive spheres in irregular clusters. Gram stain (×1,000). (*b*) *Enterococcus faecalis.* Note the chains of cocci; phase contrast (×200). (*c*) *Bacillus megaterium,* a rod-shaped bacterium in chains. Gram stain (×600). (*d*) *Rhodospirillum rubrum.* Phase contrast (×500). (*e*) *Vibrio cholerae.* Curved rods with polar flagella (×1,000).

(e)

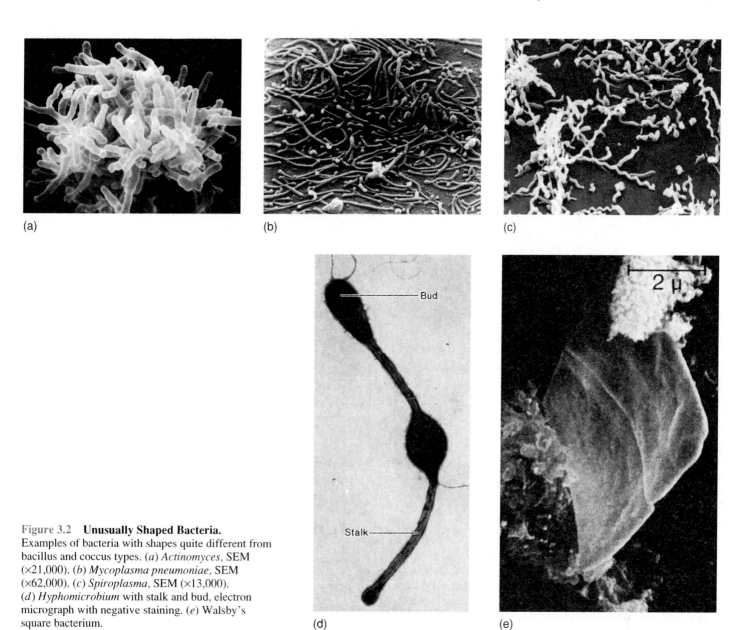

Figure 3.2   Unusually Shaped Bacteria.
Examples of bacteria with shapes quite different from bacillus and coccus types. (*a*) *Actinomyces,* SEM (×21,000). (*b*) *Mycoplasma pneumoniae,* SEM (×62,000). (*c*) *Spiroplasma,* SEM (×13,000). (*d*) *Hyphomicrobium* with stalk and bud, electron micrograph with negative staining. (*e*) Walsby's square bacterium.

The other common bacterial shape is that of a **rod,** often called a **bacillus** (pl., **bacilli**). *Bacillus megaterium* is a typical example of a bacterium with a rod shape (figure 3.1*c; see also figure 2.14*a,c). Bacilli differ considerably in their length-to-width ratio, the coccobacilli being so short and wide that they resemble cocci. The shape of the rod's end often varies between species and may be flat, rounded, cigar-shaped, or bifurcated. Although many rods do occur singly, they may remain together after division to form pairs or chains (e.g., *Bacillus megaterium* is found in long chains). A few rod-shaped bacteria, the **vibrios,** are curved to form distinctive commas or incomplete spirals (figure 3.1*e*).

Bacteria can assume a great variety of shapes, although they often are simple spheres or rods. Actinomycetes characteristically form long multinucleate filaments or hyphae that may

branch to produce a network called a **mycelium** (figure 3.2*a*). Many bacteria are shaped like long rods twisted into spirals or helices; they are called **spirilla** if rigid and **spirochetes** when flexible (figures 3.1*d*, 3.2*c; see also figure 2.8*a,c*). The oval- to pear-shaped *Hyphomicrobium* (figure 3.2*d*) produces a bud at the end of a long hypha. A few bacteria actually are flat. For example, Anthony E. Walsby has discovered square bacteria living in salt ponds (figure 3.2*e*). These bacteria are shaped like flat, square-to-rectangular boxes about 2 μm by 2 to 4 μm, and only 0.25 μm thick. Finally, some bacteria are variable in shape and lack a single, characteristic form (figure 3.2*b*). These are called **pleomorphic** even though they may, like *Corynebacterium,* have a generally rodlike form.

Bacteria vary in size as much as in shape (figure 3.3). The smallest (e.g., some members of the genus *Mycoplasma*) are

	Approximate diameter or width × length in nm	
Oscillatoria Red blood cell	7,000	5.4 times the diameter of *E. coli* below
E. coli	1,300 × 4,000	
Rickettsias	475	
Poxviruses	230 × 320	
Influenza virus	85	
T2 *E. coli* bacteriophage	65 × 95	
Tobacco mosaic virus	15 × 300	
Poliomyelitis virus	27	

Figure 3.3   Sizes of Bacteria and Viruses. The sizes of selected bacteria relative to the red blood cell and viruses.

about 100 to 200 nm in diameter, approximately the size of the largest viruses (the poxviruses). *Escherichia coli,* a bacillus of about average size, is 1.1 to 1.5 μm wide by 2.0 to 6.0 μm long. A few become fairly large; some spirochetes occasionally reach 500 μm in length, and the cyanobacterium *Oscillatoria* is about 7 μm in diameter (the same diameter as a red blood cell). Recently a huge bacterium has been discovered in the intestine of the brown surgeonfish, *Acanthurus nigrofuscus. Epulopiscium fishelsoni* grows as large as 600 μm by 80 μm, a little smaller than a printed hyphen (Box 3.1). It is now clear that a few bacteria are much larger than the average eucaryotic cell.

Comparative sizes of molecules, viruses, and procaryotic and eucaryotic cells (appendix II).

Procaryotic Cell Organization

A variety of structures is found in procaryotic cells. Their major functions are summarized in table 3.1, and figure 3.4 illustrates many of them. Not all structures are found in every genus. Furthermore, gram-negative and gram-positive cells differ, particularly with respect to their cell walls. Despite these variations procaryotes are consistent in their fundamental structure and most important components.

Procaryotic cells almost always are bounded by a chemically complex cell wall. Inside this wall, and separated from it by a periplasmic space, lies the plasma membrane. This membrane

TABLE 3.1 Functions of Procaryotic Structures	
Plasma membrane	Selectively permeable barrier, mechanical boundary of cell, nutrient and waste transport, location of many metabolic processes (respiration, photosynthesis), detection of environmental cues for chemotaxis
Gas vacuole	Buoyancy for floating in aquatic environments
Ribosomes	Protein synthesis
Inclusion bodies	Storage of carbon, phosphate, and other substances
Nucleoid	Localization of genetic material (DNA)
Periplasmic space	Contains hydrolytic enzymes and binding proteins for nutrient processing and uptake
Cell wall	Gives bacteria shape and protection from lysis in dilute solutions
Capsules and slime layers	Resistance to phagocytosis, adherence to surfaces
Fimbriae and pili	Attachment to surfaces, bacterial mating
Flagella	Movement
Endospore	Survival under harsh environmental conditions

Box 3.1

A Monstrous Microbe

Biologists often have distinguished between procaryotes and eucaryotes based in part on cell size. Generally, procaryotic cells are supposed to be smaller than eucaryotic cells. Procaryotes grow extremely rapidly compared to most eucaryotes and lack the complex vesicular transport systems of eucaryotic cells (*see chapter 4*). It has been assumed that they must be small because of the slowness of nutrient diffusion and the need for a large surface-to-volume ratio. Thus when Fishelson, Montgomery, and Myrberg discovered a large, cigar-shaped microorganism in the intestinal tract of the Red Sea brown surgeonfish, *Acanthurus nigrofuscus,* they suggested in their 1985 publication that it was a protist. The microorganism seemed too large to be anything else. In 1993 Esther Angert, Kendall Clemens, and Norman Pace used rRNA sequence comparisons (*pp. 396–97*) to identify the microorganism, now called *Epulopiscium fishelsoni,* as a procaryote related to the gram-positive genus *Clostridium.*

E. fishelsoni [Latin, *epulum,* a feast or banquet, and *piscium,* fish] can reach a size of 80 μm by 600 μm, and normally ranges from 200 to 500 μm in length (Box Figure 3.1). It is about a million times larger in volume than *Escherichia coli.* Despite its huge size the organism does have procaryotic cell structure. It is motile and swims at about two body lengths a second (approximately 2.4 cm/min) using the bacterial-type flagella that cover its surface. The cytoplasm contains large nucleoids and many ribosomes, as would be required for such a large cell. *Epulopiscium* appears to overcome the size limits set by diffusion by having an outer layer consisting of a highly convoluted plasma membrane. This increases the cell's surface area and aids in nutrient transport. The bacterium reproduces when between

Box Figure 3.1 *Epulopiscium fishelsoni.* This photograph, taken with pseudo dark field illumination, shows *E. fishelsoni* at the top of the figure dwarfing paramecia at the bottom (×200).

one and about seven daughter cells develop asexually within the parent and then escape through a centrally located slit in the parental cell envelope.

Epulopiscium also has been found in surgeonfish from the Great Barrier Reef in Australia. Presumably it is a symbiont of the surgeonfish, and its reproduction and behavior are closely correlated with its host's feeding habits. During the day when the surgeonfish feeds, the bacteria are in the middle intestine with the partially digested food and grow larger. At night when the fish is resting, the bacteria move to the back of the intestine where they reside in a bolus of food and divide repeatedly. Because there is little growth between divisions, the bacteria become smaller. In the morning *Epulopiscium* leaves the bolus of food and moves forward again.

It appears that *Epulopiscium* is transmitted between hosts through fecal contamination of the fish's food. The bacterium can be eliminated by starving the surgeonfish for a few days. If juvenile fish that lack the bacterium are placed with infected hosts, they are reinoculated. Interestingly this does not work with uninfected adult surgeonfish.

The discovery of this procaryote greatly weakens the distinction between procaryotes and eucaryotes based on cell size. *Epulopiscium* is certainly larger than a normal eucaryotic cell. In addition, some eucaryotic cells have been discovered that are smaller than previously thought possible. The best example is *Nanochlorum eukaryotum. Nanochlorum* is only about 1 to 2 μm in diameter, yet is truly eucaryotic and has a nucleus, a chloroplast, and a mitochondrion. Our understanding of the factors limiting procaryotic cell size must be reevaluated. It is no longer safe to assume that large cells are eucaryotic and small cells are procaryotic.

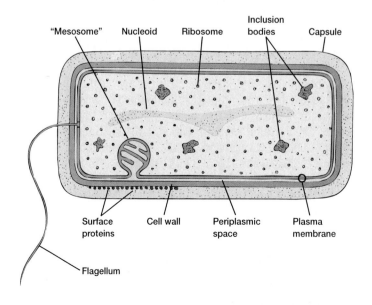

Figure 3.4   Morphology of a Gram-Positive Bacterial Cell. The majority of the structures shown here are found in all gram-positive cells. Only a small stretch of surface proteins has been included to simplify the drawing; when present, these proteins cover the surface. Gram-negative bacteria are similar in morphology.

can be invaginated to form simple internal membranous structures. Since the procaryotic cell does not contain internal membrane-bound organelles, its interior appears morphologically simple. The genetic material is localized in a discrete region, the nucleoid, and is not separated from the surrounding cytoplasm by membranes. Ribosomes and larger masses called inclusion bodies are scattered about in the cytoplasmic matrix. Both grampositive and gram-negative cells can use flagella for locomotion. In addition, many cells are surrounded by a capsule or slime layer external to the cell wall.

Procaryotic cells are morphologically much simpler than eucaryotic cells. These two cell types are compared following the review of eucaryotic cell structure (*see pp. 91–94*).

1. What characteristic shapes can bacteria assume? Describe the ways in which bacterial cells cluster together.
2. Draw a bacterial cell and label all important structures.

Procaryotic Cell Membranes

Membranes are an absolute requirement for all living organisms. Cells must interact in a selective fashion with their environment, whether it is the internal environment of a multicellular organism or a less protected and more variable external environment. Cells must not only be able to acquire nutrients and eliminate wastes, but they also have to maintain their interior in a constant, highly organized state in the face of external changes. The **plasma membrane** encompasses the cytoplasm of both procaryotic and eucaryotic cells. This membrane is the chief point of contact with the cell's environment and thus is responsible for much of its relationship with the outside world. To understand membrane function, it is necessary to become familiar with membrane structure, and particularly with plasma membrane structure.

The Plasma Membrane

Membranes contain both proteins and lipids, although the exact proportions of protein and lipid vary widely. Most membrane-associated lipids are structurally asymmetric with polar and nonpolar ends (figure 3.5) and are called amphipathic. The polar ends interact with water and are **hydrophilic;** the nonpolar **hydrophobic** ends are insoluble in water and tend to associate with one another. This property of lipids enables them to form a bilayer in membranes. The outer surfaces are hydrophilic, whereas hydrophobic ends are buried in the interior away from the surrounding water. Many of these amphipathic lipids are phospholipids (figure 3.5). Bacterial membranes usually differ from eucaryotic membranes in lacking sterols such as cholesterol (figure 3.6a). However, many bacterial membranes do

Figure 3.5 The Structure of a Polar Membrane Lipid. Phosphatidylethanolamine, an amphipathic phospholipid often found in bacterial membranes. The R groups are long, nonpolar fatty acid chains.

(a)   Cholesterol (a steroid)

(b)   A bacteriohopanetetrol (a hopanoid)

Figure 3.6   Membrane Steroids and Hopanoids. Common examples.

contain pentacyclic sterol-like molecules called hopanoids (figure 3.6b), and huge quantities of hopanoids are present in our ecosystem (Box 3.2). Hopanoids are synthesized from the same precursors as steroids. Like steroids in eucaryotes, they probably stabilize the bacterial membrane. The membrane lipid is organized in two layers, or sheets, of molecules arranged end-to-end (figure 3.6).

Many archaeobacterial membranes differ from other bacterial membranes in having a monolayer with lipid molecules spanning the whole membrane.

Archaeobacteria (chapter 23).

Box 3.2

Bacteria and Fossil Fuels

For many years there has been great interest in the origin of fossil fuels such as coal and petroleum. Fossil fuel formation begins when organic matter is buried before it can be oxidized to carbon dioxide by microorganisms. When organic matter is buried deeply and subjected to increasing temperature under anaerobic conditions, petroleum and coal are often formed. The quantities involved in these processes are enormous. It has been estimated that the earth contains about 10^{16} tons of carbon in its sediments.

There is increasing evidence that much of the organic material in sediments is bacterial in origin. About 90% of this material is in the form of insoluble kerogen, an organic precursor of petroleum. Recently the hopanoid bacteriohopanetetrol (figure 3.6*b*) was isolated from kerogen, and evidence is accumulating that kerogen arises from bacterial activity. We may owe our supply of fossil fuels largely to bacteria that serve as the final degraders of the organic material in dead organisms.

It has been estimated that the total mass of hopanoids in sediments is around 10^{11-12} tons, about as much as the total mass of organic carbon in all living organisms (10^{12} tons). Hopanoids may be the most abundant biomolecules on our planet.

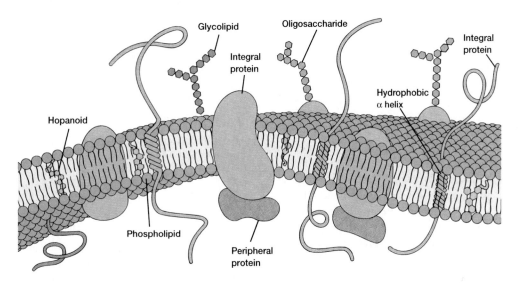

Figure 3.7   Plasma Membrane Structure. This diagram of the fluid mosaic model of bacterial membrane structure shows the integral proteins (blue) floating in a lipid bilayer. Peripheral proteins (purple) are associated loosely with the membrane surface. Small spheres represent the hydrophilic ends of membrane phospholipids and wiggly tails, the hydrophobic fatty acid chains. Other membrane lipids such as hopanoids (gold) may be present. For the sake of clarity, phospholipids are shown in proportionately much larger size than in real membranes.

Cell membranes are very thin structures, about 5 to 10 nm thick, and can only be seen with the electron microscope. The freeze-etching technique has been used to cleave membranes down the center of the lipid bilayer, splitting it in half and exposing the interior. In this way it has been discovered that many membranes, including the plasma membrane, have a complex internal structure. The small globular particles seen in these membranes are thought to be membrane proteins that lie within the membrane lipid bilayer (*see figure 2.25*).

Freeze-etching technique (pp. 31–32).

The most widely accepted current model for membrane structure is the **fluid mosaic model** of S. Jonathan Singer and Garth Nicholson (figure 3.7). They distinguish between two types of membrane proteins. **Peripheral proteins** are loosely connected to the membrane and can be easily removed. They are soluble in aqueous solutions and make up about 20 to 30% of total membrane protein. About 70 to 80% of membrane proteins are **integral proteins.** These are not easily extracted from membranes and are insoluble in aqueous solutions when freed of lipids.

Protein and lipid chemistry (appendix I).

Integral proteins, like membrane lipids, are amphipathic; their hydrophobic regions are buried in the lipid while the hydrophilic portions project from the membrane surface (figure 3.7). Some of these proteins even extend all the way through the lipid layer. Integral proteins can diffuse laterally around the surface to new locations, but do not flip-flop or rotate through the lipid layer. Often carbohydrates are attached to the outer surface of plasma membrane proteins and seem to have important functions.

The emerging picture of the cell membrane is one of a highly organized and asymmetric system, which also is flexible and dynamic. Although membranes apparently have a common basic design, there are wide variations in both their structure and functional capacities.

The plasma membranes of bacterial cells must fill an incredible variety of roles successfully. Many major plasma membrane functions are noted here even though they are discussed individually at later points in the text. The plasma membrane retains the cytoplasm, particularly in cells without cell walls, and separates it from the surroundings. The plasma membrane also serves as a selectively permeable barrier: it allows particular ions and molecules to pass, either into or out of the cell, while preventing the movement of others. Thus the membrane prevents the loss of essential components through leakage while allowing the movement of other molecules. Because many substances cannot cross the plasma membrane without assistance, it must aid such movement when necessary. Transport systems can be used for such tasks as nutrient uptake, waste excretion, and protein secretion. The bacterial plasma membrane also is the location of a variety of crucial metabolic processes: respiration, photosynthesis, and the synthesis of lipids and cell wall constituents. Finally, the membrane contains special receptor molecules that help bacteria detect and respond to chemicals in their surroundings. Clearly the plasma membrane is essential to the survival of microorganisms.

Osmosis (p. 57).
Transport of substances across membranes (pp. 102–106).

Internal Membrane Systems

Although bacterial cytoplasm does not contain complex membranous organelles like mitochondria or chloroplasts, membranous structures of several kinds can be observed. A common structure is the **mesosome.** Mesosomes are invaginations of the plasma membrane in the shape of vesicles, tubules, or lamellae (figures 3.4, 3.8, and 3.11). They are seen in both gram-positive and gram-negative bacteria, although they are generally more prominent in the former.

Despite years of research on mesosomes, their exact functions are still unknown. They often are found next to septa or cross-walls in dividing bacteria and sometimes seem attached to the bacterial chromosome. Thus they may be involved in cell wall formation during division or play a role in chromosome replication and distribution to daughter cells. Mesosomes also may be involved in secretory processes.

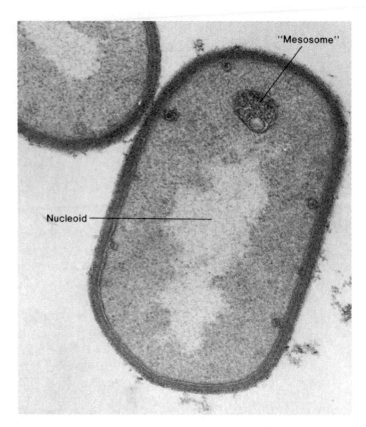

Figure 3.8 Mesosome Structure. *Bacillus fastidiosus* (×91,000). A large mesosome lies adjacent to the nucleoid.

Currently many bacteriologists believe that mesosomes are artifacts generated during the chemical fixation of bacteria for electron microscopy. Possibly they represent parts of the plasma membrane that are chemically different and more disrupted by fixatives. However, mesosomes are seen occasionally in freeze-etched bacteria and therefore sometimes may be present in living cells. Further research should resolve this controversy.

Many bacteria have internal membrane systems quite different from the mesosome (figure 3.9). Plasma membrane infoldings can become extensive and complex in photosynthetic bacteria such as the cyanobacteria and purple bacteria or in bacteria with very high respiratory activity like the nitrifying bacteria (*see chapter 22*). They may be aggregates of spherical vesicles, flattened vesicles, or tubular membranes. Their function may be to provide a larger membrane surface for greater metabolic activity.

---

1. Describe with a labeled diagram and in words the fluid mosaic model for cell membranes.
2. List the functions of the plasma membrane.
3. Discuss the nature, structure, and possible functions of the mesosome.

(a)

(b)

Figure 3.9 **Internal Bacterial Membranes.** Membranes of nitrifying and photosynthetic bacteria. (*a*) *Nitrocystis oceanus* with parallel membranes traversing the whole cell. Note nucleoplasm (n) with fibrillar structure. (*b*) *Ectothiorhodospira mobilis* with an extensive intracytoplasmic membrane system (×60,000).

The Cytoplasmic Matrix

Procaryotic cytoplasm, unlike that of eucaryotes, lacks unit membrane-bound organelles. The **cytoplasmic matrix** is the substance lying between the plasma membrane and the nucleoid (*pp. 49–51*). The matrix is largely water (about 70% of bacterial mass is water). It is featureless in electron micrographs but often is packed with ribosomes and highly organized (figure 3.10). The plasma membrane and everything within is called the **protoplast;** thus the cytoplasmic matrix is a major part of the protoplast.

Inclusion Bodies

A variety of **inclusion bodies,** granules of organic or inorganic material that often are clearly visible in a light microscope, is present in the cytoplasmic matrix. Some inclusion bodies are not bounded by a membrane and lie free in the cytoplasm—for example, polyphosphate granules, cyanophycin granules, and some glycogen granules. Other inclusion bodies are enclosed by a single-layered, nonunit membrane about 2.0 to 4.0 nm thick. Examples of nonunit membrane-enclosed inclusion bodies are poly-β-hydroxybutyrate granules, some glycogen and sulfur granules, carboxysomes, and gas vacuoles. Inclusion body membranes vary in composition. Some are protein in nature, whereas others contain lipid. A brief description of several important inclusion bodies follows.

Organic inclusion bodies usually contain either glycogen or poly-β-hydroxybutyrate. **Glycogen** is a polymer of glucose units composed of long chains formed by $\alpha(1\rightarrow4)$ glycosidic bonds and branching chains connected to them by $\alpha(1\rightarrow6)$ glycosidic bonds (*see appendix I*). **Poly-β-hydroxybutyrate (PHB)** contains β-hydroxybutyrate molecules joined by ester bonds between the carboxyl and hydroxyl groups of adjacent molecules. Usually only one of these polymers is found in a species, but purple photosynthetic bacteria have both. Poly-β-hydroxybutyrate accumulates in distinct bodies, around 0.2 to 0.7 μm in diameter, that are readily stained with Sudan black for light microscopy and are clearly visible in the electron microscope (figure 3.11). Glycogen is dispersed more evenly throughout the matrix as small granules (about 20 to 100 nm in diameter) and often can be seen only with the electron microscope. If cells contain a large amount of glycogen, staining with an iodine solution will turn them reddish-brown. Glycogen and PHB inclusion bodies are carbon storage reservoirs providing material for energy and biosynthesis. Many bacteria also store carbon as lipid droplets.

Cyanobacteria have two distinctive organic inclusion bodies. **Cyanophycin granules** (figure 3.15*a*) are composed of large polypeptides containing approximately equal amounts of the amino acids arginine and aspartic acid. The granules often are large enough to be visible in the light microscope and store extra nitrogen for the bacteria. **Carboxysomes** are present in many cyanobacteria, nitrifying bacteria, and thiobacilli. They are polyhedral, about 100 nm in diameter, and contain the enzyme ribulose-1,5-bisphosphate carboxylase (*see p. 194*) in a paracrystalline arrangement. They serve as a reserve of this enzyme and may be a site of CO_2 fixation.

A most remarkable organic inclusion body, the **gas vacuole,** is present in many cyanobacteria (*see chapter 22*), purple and green photosynthetic bacteria, and a few other aquatic forms such as *Halobacterium* and *Thiothrix*. These bacteria float at or near the surface, because gas vacuoles give them buoyancy. This is vividly demonstrated by a simple but dramatic experiment. Cyanobacteria held in a full, tightly stoppered bottle will float (figure 3.12*a*), but if the stopper is struck

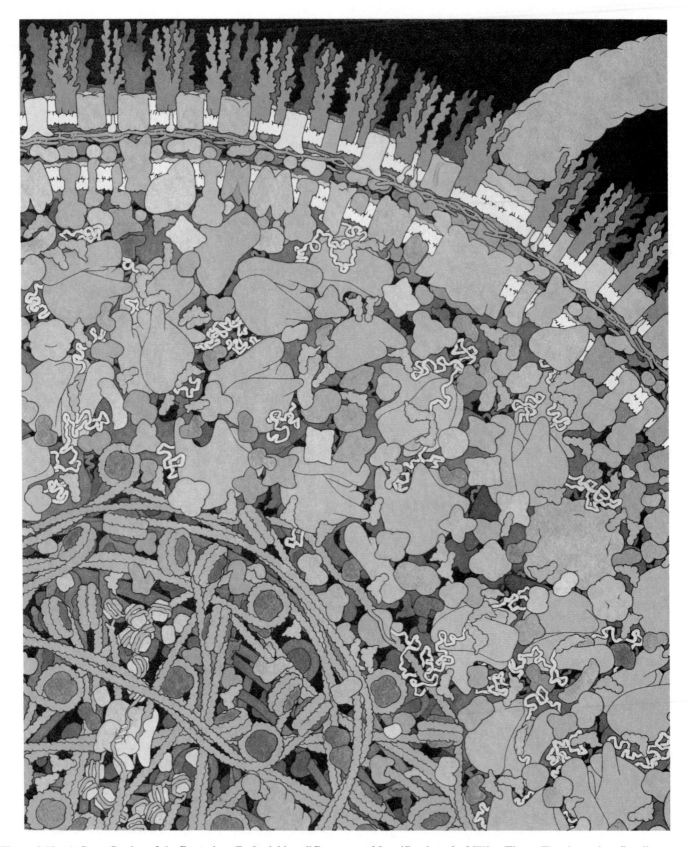

Figure 3.10   A Cross Section of the Bacterium *Escherichia coli* Drawn at a Magnification of a Million Times. The glycocalyx, flagellum, gram-negative cell wall, and plasma membrane are at the top. Ribosomes synthesizing proteins fill the underlying cytoplasmic matrix. At the bottom is the nucleoid with its dense tangle of DNA and associated proteins.

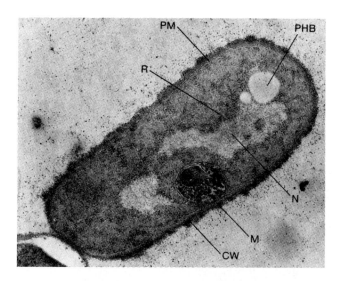

Figure 3.11   **The Structure of a Typical Gram-Positive Cell.**
Electron micrograph of *Bacillus megaterium* (×30,500). Note the thick
cell wall, CW; "mesosome," M; nucleoid, N; poly-β-hydroxybutyrate
inclusion body, PHB; plasma membrane, PM; and ribosomes, R.

(a)   (b)

Figure 3.12   **Gas Vacuoles and Buoyancy.** A hammer and stopper
experiment with the cyanobacterium *Microcystis aeruginosa*. (*a*) The
experimental bottle has been filled and tightly corked while the control
bottle is left open to the atmosphere. (*b*) After the stopper is struck with
a hammer, the cyanobacteria settle to the bottom. The undisturbed
control organisms float.

(a)

(b)

Figure 3.13   **Gas Vacuole Collapse.** Filaments of the
cyanobacterium *Anabaena flos-aquae*. (*a*) Before application of
pressure. (*b*) After application of pressure, the vacuoles have collapsed
and are no longer visible.

with a hammer, the bacteria sink to the bottom (figure 3.12*b*).
Examination of the bacteria at the beginning and end of the experiment shows that the sudden pressure increase has collapsed
the gas vacuoles and destroyed the microorganisms' buoyancy
(figure 3.13).

Gas vacuoles are aggregates of enormous numbers of
small, hollow, cylindrical structures called **gas vesicles** (figure
3.14). Gas vesicle walls do not contain lipid and are composed
entirely of a single small protein. These protein subunits assemble to form a rigid enclosed cylinder that is hollow and impermeable to water but freely permeable to atmospheric gases.
Bacteria with gas vacuoles can regulate their buoyancy to float

at the depth necessary for proper light intensity, oxygen concentration, and nutrient levels. They descend by simply collapsing
vesicles and float upward when new ones are constructed.

Two major types of inorganic inclusion bodies are seen.
Many bacteria store phosphate as **polyphosphate granules** or
volutin granules (figure 3.15*a*). Polyphosphate is a linear polymer of orthophosphates joined by ester bonds. Thus volutin
granules function as storage reservoirs for phosphate, an important component of cell constituents such as nucleic acids. In
some cells they act as an energy reserve, and polyphosphate can
serve as an energy source in reactions. These granules are sometimes called **metachromatic granules** because they show the
metachromatic effect; that is, they appear red or a different shade
of blue when stained with the blue dyes methylene blue or tolui-
dine blue. Some bacteria also store sulfur temporarily as sulfur
granules, a second type of inorganic inclusion body (figure
3.15*b*). For example, purple photosynthetic bacteria can use hydrogen sulfide as a photosynthetic electron donor (*see chapter 9*)
and accumulate the resulting sulfur in either the periplasmic
space or in special cytoplasmic globules.

Figure 3.14 Gas Vesicles. A freeze-fracture preparation of the cyanobacterium *Anabaena flos-quae* (×89,000). Clusters of the cigar-shaped vesicles form gas vacuoles. Both longitudinal and cross-sectional views of gas vesicles can be seen.

Ribosomes

As mentioned earlier, the cytoplasmic matrix often is packed with **ribosomes;** they also are loosely attached to the plasma membrane. Ribosomes look like small, featureless particles at low magnification in electron micrographs (figure 3.11) but are actually very complex objects made of both protein and ribonucleic acid (RNA). They are the site of protein synthesis; matrix ribosomes synthesize proteins destined to remain within the cell, whereas the plasma membrane ribosomes make proteins for transport to the outside. Protein synthesis, including a detailed treatment of ribosomes, is discussed at considerable length in chapter 11. At present, note that procaryotic ribosomes are smaller than eucaryotic ribosomes. They commonly are called 70S ribosomes, have dimensions of about 14 to 15 nm by 20 nm, a molecular weight of approximately 2.7 million, and are constructed of a 50S and a 30S subunit. The S in 70S and similar values stands for **Svedberg unit.** This is the unit of the sedimentation coefficient, a measure of the sedimentation velocity in a centrifuge; the faster a particle travels when centrifuged, the greater its Svedberg

Thylakoids

(a) (b)

Figure 3.15 Inclusion Bodies in Bacteria. (*a*) Ultrastructure of the cyanobacterium *Anacystis nidulans.* The bacterium is dividing, and a septum is partially formed, LI and LII. Several structural features can be seen, including cell wall layers, LIII and LIV; the plasma membrane, pm; polyphosphate granules, pp; a polyhedral body, pb; and cyanophycin material, c. Thylakoids run along the length of the cell. Bar = 0.1 μm. (*b*) *Chromatium vinosum,* a purple sulfur bacterium, with intracellular sulfur granules, light field (×2,000).

value or sedimentation coefficient. The sedimentation coefficient is a function of a particle's molecular weight, volume, and shape (*see figure 16.7*). Heavier and more compact particles normally have larger Svedberg numbers or sediment faster. It must be emphasized that Svedberg values are not directly proportional to molecular weights: the weight of a 70S ribosome equals the sum of the 50S and 30S subunit weights even though the sum of 50 and 30 is 80, not 70. Ribosomes in the cytoplasmic matrix of eucaryotic cells are 80S ribosomes and about 22 nm in diameter. Despite their overall difference in size, both types of ribosomes are similarly composed of a large and a small subunit.

Molecular Chaperones

For many years it was believed that polypeptides would spontaneously fold into their final native shape, either as they were synthesized by ribosomes or shortly after completion of protein synthesis. Although the amino acid sequence of a polypeptide does determine its final conformation, it is now clear that special helper proteins aid the newly formed or nascent polypeptide in folding to its proper shape. These proteins, called **molecular chaperones** or chaperones, recognize only unfolded polypeptides or partly denatured proteins and do not bind to native proteins. Their role is essential because the cytoplasmic matrix is filled with nascent polypeptide chains and proteins. Under such conditions it is quite likely that new polypeptide chains often will fold improperly and aggregate to form nonfunctional complexes. Molecular chaperones suppress incorrect folding and may reverse any incorrect folding that has already taken place. They are so important that chaperones are present in all cells, procaryotic and eucaryotic.

Several chaperones and cooperating proteins aid proper protein folding in bacteria. The process has been most studied in *Escherichia coli* and involves at least four chaperones—DnaK, DnaJ, GroEL, and GroES—and the stress protein GrpE. After a sufficient length of nascent polypeptide extends from the ribosome, DnaJ binds to the unfolded chain (figure 3.16). DnaK, which is complexed with ATP, then attaches to the polypeptide. These two chaperones prevent the polypeptide from folding improperly as it is synthesized. The ATP is hydrolyzed to ADP after DnaK binding, and this increases the affinity of DnaK for the unfolded polypeptide. When the polypeptide has been synthesized, the GrpE protein binds to the chaperone-polypeptide complex and causes DnaK to release ADP. Then ATP binds to DnaK and both DnaK and DnaJ dissociate from the polypeptide. The polypeptide has been folding during this sequence of events and may have reached its final native conformation. If it is still only partially folded, it can bind DnaJ and DnaK again and repeat the process. Often DnaK and DnaJ will transfer the polypeptide to the chaperones GroEL and GroES, where the final folding takes place. GroEL is a large, hollow barrel-shaped complex of 14 subunits arranged in two stacked rings. GroES exists as a single ring of seven subunits and can bind to one or both ends of the GroEL cylinder. As with DnaK, ATP binding to GroEL and ATP hydrolysis change the chaperone's affinity for the folding polypeptide and regulate polypeptide binding and release (polypeptide release is ATP-dependent). GroES binds to GroEL and assists in its binding and release of the refolding polypeptide.

Chaperones were first discovered because they dramatically increased in concentration when cells were exposed to high temperatures, metabolic poisons, and other stressful conditions. Thus many chaperones often are called **heat-shock proteins** or stress proteins. When an *E. coli* culture is switched from 30 to 42°C, the concentrations of some 20 different heat-shock proteins increase greatly within about 5 minutes. If the cells are exposed to a lethal temperature, the heat-shock proteins are still synthesized but most proteins are not. Thus chaperones protect the cell from thermal damage and other stresses as well as promote the proper folding of new polypeptides. For example, DnaK protects *E. coli* RNA polymerase from thermal inactivation in vitro. In addition, DnaK reactivates thermally inactivated RNA polymerase, especially if ATP, DnaJ, and GrpE are present. GroEL and GroES also protect intracellular proteins from aggregation. As one would expect, large quantities of chaperones are present in hyperthermophiles such as *Pyrodictium occultum,* a bacterium that will grow at temperatures as high as 110°C. *Pyrodictium* has a chaperone similar to the GroEL of *E. coli.* The chaperone hydrolyzes ATP most rapidly at 100°C and makes up almost 3/4 of the cell's soluble protein when *P. occultum* grows at 108°C.

Thermophilic and hyperthermophilic bacteria (pp. 129–130).

Chaperones have other functions as well. They are particularly important in the transport of proteins across membranes. For example, in *E. coli* the chaperone SecB binds to the partially unfolded forms of many proteins and keeps them in an export-competent state until they are translocated across the plasma membrane. DnaK, DnaJ, and GroEL/GroES also can aid in protein translocation across membranes. Proteins destined for the periplasm or outer membrane (pp. 54–56) are synthesized with the proper amino-terminal signal sequence. The signal sequence is a short stretch of amino acids that help direct the completed polypeptide to its final destination. Polypeptides associate with SecB and the chaperone then attaches to the membrane translocase. The polypeptides are transported through the membrane as ATP is hydrolyzed. When they enter the periplasm, the signal peptidase enzyme removes the signal sequence and the protein moves to its final location.

---

1. Briefly describe the nature and function of the cytoplasmic matrix and the ribosome. What is a protoplast?
2. What kinds of inclusion bodies do procaryotes have? What are their functions?
3. What is a gas vacuole? Relate its structure to its function.
4. What are molecular chaperones and heat-shock proteins? Describe their functions.

The Nucleoid

Probably the most striking difference between procaryotes and eucaryotes is the way in which their genetic material is packaged. Eucaryotic cells have two or more chromosomes contained within a membrane-delimited organelle, the nucleus. In contrast, procaryotes lack a membrane-delimited nucleus. The procaryotic chromosome, almost always a single circle of double-stranded **deoxyribonucleic acid (DNA),** is located in an irregularly

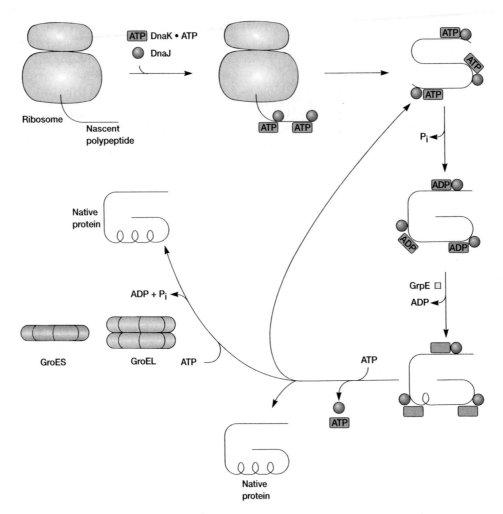

Figure 3.16 Chaperones and Polypeptide Folding. The involvement of bacterial chaperones in the proper folding of a newly synthesized polypeptide chain is depicted in this diagram. Three possible outcomes of a chaperone reaction cycle are shown. A native protein may result, the partially folded polypeptide may bind again to DnaK and DnaJ, or the polypeptide may be transferred to GroEL and GroES. See text for details.

shaped region called the **nucleoid** (other names are also used: the nuclear body, chromatin body, nuclear region). Although nucleoid appearance varies with the method of fixation and staining, fibers often are seen in electron micrographs (figures 3.11 and 3.17) and are probably DNA. The nucleoid also is visible in the light microscope after staining with the Feulgen stain, which specifically reacts with DNA. A cell can have more than one nucleoid when cell division occurs after the genetic material has been duplicated (figure 3.17a). In actively growing bacteria, the nucleoid has projections that extend into the cytoplasmic matrix (figure 3.17b,c). Presumably these projections contain DNA that is being actively transcribed to produce mRNA.

Careful electron microscopic studies often have shown the nucleoid in contact with either the mesosome or the plasma membrane. Membranes also are found attached to isolated nucleoids. Thus there is evidence that bacterial DNA is attached to cell membranes, and membranes may be involved in the separation of DNA into daughter cells during division.

Nucleoids have been isolated intact and free from membranes. Chemical analysis reveals that they are composed of about 60% DNA, some RNA, and a small amount of protein. In *Escherichia coli,* a rod-shaped cell about 2 to 6 μm long, the closed DNA circle measures approximately 1,400 μm. Obviously it must be very efficiently packaged to fit within the nucleoid. The DNA is looped and coiled extensively, probably with the aid of nucleoid proteins (these proteins differ from the histone proteins present in eucaryotic nuclei).

Procaryotic DNA and its function (chapter 11).

(a)

(b)

Figure 3.17 **The Bacterial Nucleoid.** (*a*) Nucleoids in growing *Bacillus* cells stained using HCl-Giemsa stain and viewed with a light microscope (bar = 5 μm). (*b*) A section of actively growing *E. coli* immunostained specifically for DNA and examined in the transmission electron microscope. Coupled transcription and translation occur in parts of the nucleoid that extend out into the cytoplasm. (*c*) A model of two nucleoids in an actively growing *E. coli* cell. Note that a metabolically active nucleoid is not compact and spherical but has projections that extend into the cytoplasmic matrix.

(c)

Many bacteria possess **plasmids** in addition to their chromosome. These are circular, double-stranded DNA molecules that can exist and replicate independently of the chromosome or may be integrated with it; in either case they are inherited or passed on to the progeny. Plasmids are not required for host growth and reproduction, although they may carry genes that give their bacterial host a selective advantage. Plasmid genes can render bacteria drug-resistant, give them new metabolic abilities, make them pathogenic, or endow them with a number of other properties.

Plasmids (pp. 280–83).

1. Characterize the nucleoid with respect to its structure and function.
2. What is a plasmid?

The Procaryotic Cell Wall

The cell wall is one of the most important parts of a procaryotic cell for several reasons. Except for the mycoplasmas (*see chapter 20*) and some archaeobacteria (*see chapter 23*), most bacteria have strong walls that give them shape and protect them from osmotic lysis (*p. 57*). The cell walls of many pathogens have components that contribute to their pathogenicity. The wall can protect a cell from toxic substances and is the site of action of several antibiotics.

After Christian Gram developed the Gram stain in 1884, it soon became evident that bacteria could be divided into two major groups based on their response to the Gram-stain procedure. Gram-positive bacteria stained purple, whereas gram-negative bacteria were colored pink or red by the technique. The true structural difference between these two groups became clear with the advent of the transmission electron microscope. The gram-positive cell wall consists of a single 20 to 80 nm thick homogeneous **peptidoglycan** or **murein** layer lying outside the plasma membrane (figure 3.18). In contrast, the gram-negative cell wall is quite complex. It has a 1 to 3 nm peptidoglycan layer surrounded by a 7 to 8 nm thick **outer membrane.** Microbiologists often call all the structures outside the plasma membrane the **envelope.** This includes the wall and structures like capsules (*p. 58*) when present.

Gram-stain procedure (p. 26).

Frequently a space is seen between the plasma membrane and the outer membrane in electron micrographs of gram-negative bacteria, and sometimes a similar but smaller gap may be observed between the plasma membrane and wall in gram-positive bacteria. This space is called the **periplasmic space.** Recent evidence indicates that the periplasmic space may be filled with a loose network of peptidoglycan. Possibly it is more a gel than a fluid-filled space. The substance that occupies the periplasmic space is the **periplasm.** Size estimates of the periplasmic space in gram-negative bacteria range from 1 nm to as great as 71 nm. Some recent studies

The gram-positive cell wall

Peptidoglycan
Plasma membrane

The gram-negative cell wall

Outer membrane
Peptidoglycan
Plasma membrane

Periplasmic
space

🔵 Figure 3.18   **Gram-Positive and Gram-Negative Cell Walls.** The gram-positive envelope is from *Bacillus licheniformis* (left), and the gram-negative micrograph is of *Aquaspirillum serpens* (right). M; peptidoglycan or murein layer; OM, outer membrane; PM, plasma membrane; P, periplasmic space; W, gram-positive peptidoglycan wall. Bar = 100 nm.

indicate that it may constitute about 20 to 40% of the total cell volume (around 30 to 70 nm), but more research is required to establish an accurate value. When cell walls are disrupted carefully or removed without disturbing the underlying plasma membrane, periplasmic enzymes and other proteins are released and may be easily studied. The periplasmic space of gram-negative bacteria contains many proteins that participate in nutrient acquisition—for example, hydrolytic enzymes attacking nucleic acids and phosphorylated molecules, and binding proteins involved in transport of materials into the cell. Denitrifying and chemolithoautotrophic bacteria (*see chapter 9*) often have electron transport proteins in their periplasm. The periplasmic space also contains enzymes involved in peptidoglycan synthesis and the modification of toxic compounds that could harm the cell. Gram-positive bacteria may not have a visible periplasmic space and do not appear to have as many periplasmic proteins; rather, they secrete several enzymes that ordinarily would be periplasmic in gram-negative bacteria. Such secreted enzymes are often called **exoenzymes.**

The recently discovered archaeobacteria differ from other bacteria in many respects (*see chapter 23*). Although they may be either gram positive or gram negative, their cell walls are distinctive in structure and chemical composition. The walls lack peptidoglycan and are composed of proteins, glycoproteins, or polysaccharides.

Following this overview of the envelope, peptidoglycan structure and the organization of gram-positive and gram-negative cell walls are discussed in more detail.

Peptidoglycan Structure

Peptidoglycan or murein is an enormous polymer composed of many identical subunits. The polymer contains two sugar derivatives, N-acetylglucosamine and N-acetylmuramic acid (the lactyl ether of N-acetylglucosamine), and several different amino acids, three of which—D-glutamic acid, D-alanine, and *meso*-diaminopimelic acid—are not found in proteins. The peptidoglycan subunit present in most gram-negative bacteria and many gram-positive ones is shown in figure 3.19. The backbone of this polymer is composed of alternating N-acetylglucosamine and N-acetylmuramic acid residues. A peptide chain of four alternating D- and L-amino acids is connected to the carboxyl group of N-acetylmuramic acid. Many bacteria substitute another diaminoacid, usually L-lysine, in the third position for *meso*-diaminopimelic acid (figure 3.20).

A review of the chemistry of biological molecules (appendix I).

Peptidoglycan structural variations (pp. 439–40).

Chains of linked peptidoglycan subunits are joined by cross-links between the peptides. Often the carboxyl group of the terminal D-alanine is connected directly to the amino group of diaminopimelic acid, but a **peptide interbridge** may be used instead (figure 3.21). Most gram-negative cell wall peptidoglycan lacks the peptide interbridge. This cross-linking results in an enormous peptidoglycan sac that is actually one dense, interconnected network (figure 3.22). These sacs have been isolated

Figure 3.19   Peptidoglycan Subunit Composition. The peptidoglycan subunit of *Escherichia coli,* most other gram-negative bacteria, and many gram-positive bacteria. NAG is N-acetylglucosamine. NAM is N-acetylmuramic acid (NAG with lactic acid attached by an ether linkage). The tetrapeptide side chain is composed of alternating D- and L-amino acids since *meso*-diaminopimelic acid is connected through its L-carbon. NAM and the tetrapeptide chain attached to it are shown in different shades of color for clarity.

Figure 3.20   Diaminoacids Present in Peptidoglycan. (*a*) L-lysine. (*b*) *Meso*-diaminopimelic acid.

Figure 3.21   Peptidoglycan Cross-Links. (*a*) *E. coli* peptidoglycan with direct cross-linking, typical of many gram-negative bacteria. (*b*) *Staphylococcus aureus* peptidoglycan. *S. aureus* is a gram-positive bacterium. NAM is N-acetylmuramic acid. NAG is N-acetylglucosamine. Gly is glycine. Although the polysaccharide chains are drawn opposite each other for the sake of clarity, two chains lying side-by-side may be linked together (see figure 3.22).

N-Acetylmuramic acid

N-Acetylglucosamine

Peptide chain

Pentaglycine interbridge

(a)

(b)

🌀 **Figure 3.22 Peptidoglycan Structure.** A peptidoglycan segment showing the polysaccharide chains, tetrapeptide side chains, and peptide interbridges. (*a*) A schematic diagram. (*b*) A space-filling model of gram-negative murein with four repeating peptidoglycan subunits in the plane of the paper. Two chains are arranged vertical to this direction.

Figure 3.23   Isolated Gram-Positive Cell Wall. The peptidoglycan wall from *Bacillus megaterium*, a gram-positive bacterium. The latex balls have a diameter of 0.25 µm.

acids such as D-alanine or sugars like glucose are attached to the glycerol and ribitol groups. The teichoic acids are connected to either the peptidoglycan itself by a covalent bond with the six hydroxyl of N-acetylmuramic acid or to plasma membrane lipids; in the latter case they are called lipoteichoic acids. Teichoic acids appear to extend to the surface of the peptidoglycan, and, because they are negatively charged, help give the gram-positive cell wall its negative charge. The functions of these molecules are still unclear, but they may be important in maintaining the structure of the wall. Teichoic acids are not present in gram-negative bacteria.

Gram-Negative Cell Walls

Even a brief inspection of figure 3.18 shows that gram-negative cell walls are much more complex than gram-positive walls. The thin peptidoglycan layer next to the plasma membrane may constitute not more than 5 to 10% of the wall weight. In *E. coli* it is about 1 nm thick and contains only one or two layers or sheets of peptidoglycan. As mentioned earlier, the peptidoglycan may be in the form of a gel rather than a compact layer.

The outer membrane lies outside the thin peptidoglycan layer (figure 3.26). The most abundant membrane protein is Braun's lipoprotein, a small lipoprotein covalently joined to the underlying peptidoglycan and embedded in the outer membrane by its hydrophobic end. The outer membrane and peptidoglycan are so firmly linked by this lipoprotein that they can be isolated as one unit.

from gram-positive bacteria and are strong enough to retain their shape and integrity (figure 3.23), yet they are elastic and somewhat stretchable, unlike cellulose. They also must be porous, as molecules can penetrate them.

Gram-Positive Cell Walls

Normally the thick, homogeneous cell wall of gram-positive bacteria is composed primarily of peptidoglycan, which often contains a peptide interbridge (figures 3.23 and 3.24). However gram-positive cell walls usually also contain large amounts of **teichoic acids,** polymers of glycerol or ribitol joined by phosphate groups (figures 3.24 and 3.25). Amino

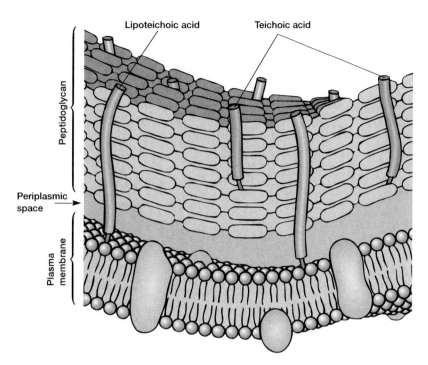

Lipoteichoic acid   Teichoic acid

Peptidoglycan

Periplasmic
space

Plasma
membrane

Figure 3.24   **The Gram-Positive Envelope.**

Figure 3.25   **Teichoic Acid Structure.** The segment of a teichoic
acid made of phosphate, glycerol, and a side chain, R. R may represent
D-alanine, glucose, or other molecules.

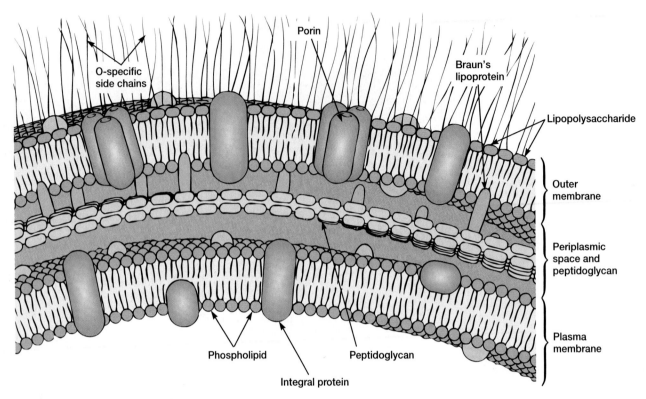

🔆 Figure 3.26   **The Gram-Negative Envelope.**

Possibly the most unusual constituents of the outer membrane are its **lipopolysaccharides (LPSs).** These large, complex molecules contain both lipid and carbohydrate, and consist of three parts: (1) lipid A, (2) the core polysaccharide, and (3) the O side chain. The LPS from *Salmonella typhimurium* has been studied most, and its general structure is described here (figure 3.27). The **lipid A** region contains two glucosamine sugar derivatives, each with three fatty acids and phosphate or pyrophosphate attached. It is buried in the outer membrane and the remainder of the LPS molecule projects from the surface. The **core polysaccharide** is joined to lipid A. In *Salmonella* it is constructed of 10 sugars, many of them unusual in structure. The **O side chain** or **O antigen** is a short polysaccharide chain extending outward from the core. It has several peculiar sugars and varies in composition between bacterial strains. Although O side chains are readily recognized by host antibodies, gram-negative bacteria may thwart host defenses by rapidly changing the nature of their O side chains to avoid detection. Antibody interaction with the LPS before reaching the outer membrane proper may also protect the cell wall from direct attack.

Antibodies and antigens (chapters 30 and 32).

The LPS is important for several reasons other than the avoidance of host defenses. Since the core polysaccharide usually contains charged sugars and phosphate (figure 3.27), LPS contributes to the negative charge on the bacterial surface. Lipid A is a major constituent of the outer membrane, and the LPS helps stabilize membrane structure. Furthermore, lipid A often is toxic; as a result the LPS can act as an endotoxin (*see chapter 29*) and cause some of the symptoms that arise in gram-negative bacterial infections.

A most important outer membrane function is to serve as a protective barrier. It prevents or slows the entry of bile salts, antibiotics, and other toxic substances that might kill or injure the bacterium. Even so, the outer membrane is more permeable than the plasma membrane and permits the passage of small molecules like glucose and other monosaccharides. This is due to the presence of special **porin proteins.** Three porin molecules cluster together and span the outer membrane to form a narrow channel through which molecules smaller than about 600 to 700 daltons can pass. Larger molecules such as vitamin B_{12} must be transported across the outer membrane by specific carriers. The outer membrane also prevents the loss of constituents like periplasmic enzymes.

The Mechanism of Gram Staining

Although several explanations have been given for the Gram-stain reaction results, it seems likely that the difference between gram-positive and gram-negative bacteria is due to the physical nature of their cell walls. If the cell wall is removed from gram-positive bacteria, they become gram negative. The peptidoglycan itself is not stained; instead it seems to act as a permeability barrier preventing loss of crystal violet. During the procedure the bacteria are first stained with crystal violet and next treated with iodine to promote dye retention. When

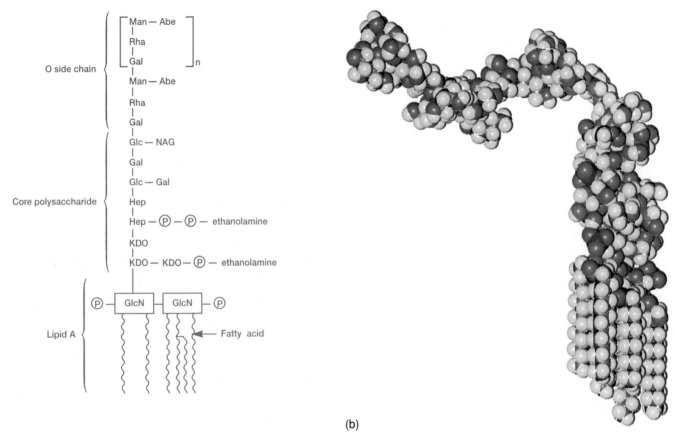

(a)

(b)

Figure 3.27 **Lipopolysaccharide Structure.** (*a*) The lipopolysaccharide from *Salmonella*. This slightly simplified diagram illustrates one form of the LPS. Abbreviations: Abe, abequose; Gal, galactose; Glc, glucose; GlcN, glucosamine; Hep, heptulose; KDO, 2-keto-3-deoxyoctonate; Man, mannose; NAG, N-acetylglucosamine; P, phosphate; Rha, L-rhamnose. (*b*) Molecular model of an *Escherichia coli* lipopolysaccharide. The lipid A and core polysaccharide are straight; the O side chain is bent at an angle in this model.

gram-positive bacteria then are decolorized with ethanol, the alcohol is thought to shrink the pores of the thick peptidoglycan. Thus the dye-iodine complex is retained during the short decolorization step and the bacteria remain purple. In contrast, gram-negative peptidoglycan is very thin, not as highly cross-linked, and has larger pores. Alcohol treatment also may extract enough lipid from the gram-negative wall to increase its porosity further. For these reasons, alcohol more readily removes the purple crystal violet-iodine complex from gram-negative bacteria.

The Cell Wall and Osmotic Protection

The cell wall usually is required to protect bacteria against destruction by osmotic pressure. Solutes are much more concentrated in bacterial cytoplasm than in most microbial habitats, which are hypotonic. During **osmosis,** water moves across selectively permeable membranes such as the plasma membrane from dilute solutions (higher water concentration) to more concentrated solutions (lower water concentration). Thus water normally enters bacterial cells and the osmotic pressure may reach 20 atmospheres or 300 pounds/square inch. The plasma membrane cannot withstand such pressures and the cell will swell and be physically disrupted and destroyed, a process called

lysis, without the wall that resists cell swelling and protects it. Solutes are more concentrated in hypertonic habitats than in the cell. Thus water flows outward, and the cytoplasm shrivels up and pulls away from the cell wall. This phenomenon is known as **plasmolysis** and is useful in food preservation because many microorganisms cannot grow in dried foods and jellies as they cannot avoid plasmolysis (*see pp. 124–25, chapter 43*).

The importance of the cell wall in protecting bacteria against osmotic lysis is demonstrated by treatment with lysozyme or penicillin. The enzyme **lysozyme** attacks peptidoglycan by hydrolyzing the bond that connects N-acetylmuramic acid with carbon four of N-acetylglucosamine. **Penicillin** inhibits peptidoglycan synthesis (*see chapter 33*). If bacteria are incubated with penicillin in an isotonic solution, gram-positive bacteria are converted to protoplasts that continue to grow normally when isotonicity is maintained even though they completely lack a wall. Gram-negative cells retain their outer membrane after penicillin treatment and are classified as **spheroplasts** because some of their cell wall remains. Protoplasts and spheroplasts are osmotically sensitive. If they are transferred to a dilute solution, they will lyse due to uncontrolled water influx (figure 3.28).

Although most bacteria require an intact cell wall for survival, some have none at all. For example, the mycoplasmas

Figure 3.28   Protoplast Formation. Protoplast formation induced by incubation with penicillin in an isotonic medium. Transfer to dilute medium will result in lysis.

(a)

lack a cell wall, yet often can grow in dilute media or terrestrial environments because their plasma membranes are stronger than normal. The precise reason for this is not known, although the presence of sterols in the membranes of many species may provide added strength. Without a rigid cell wall, mycoplasmas tend to be pleomorphic or variable in shape.

The **L forms** (named after the Lister Institute in London where they were discovered) also lack cell walls. The loss may be complete or partial (some have a defective wall), and the parent organism may be either gram positive or gram negative. They are pleomorphic like mycoplasmas and continue to reproduce. These organisms can arise through spontaneous mutations or from treatments such as growth in isotonic or hypertonic media containing penicillin. If all traces of the peptidoglycan disappear, bacteria cannot resynthesize it because preexisting wall is necessary to construct new peptidoglycan. In this case the L form may be stable—that is, it may continue to grow and reproduce after the penicillin treatment has ceased. Other L forms sometimes synthesize a wall again. L forms are not closely related to mycoplasmas and should not be confused with them.

1. Describe in some detail the composition and structure of peptidoglycan, gram-positive cell walls, and gram-negative cell walls. Include labeled diagrams in the answer.
2. Define or describe the following: outer membrane, periplasmic space, periplasm, envelope, teichoic acid, lipopolysaccharide, and porin protein.
3. Explain the role of the cell wall in protecting against lysis and how this role may be experimentally demonstrated. What are protoplasts, spheroplasts, and L forms?

Components External to the Cell Wall

Bacteria have a variety of structures outside the cell wall that function in protection, attachment to objects, or cell movement. Several of these are discussed.

Capsules, Slime Layers, and S-Layers

Some bacteria have a layer of material lying outside the cell wall. When the layer is well organized and not easily washed off, it is called a **capsule**. A **slime layer** is a zone of diffuse, unorganized material that is removed easily. A **glycocalyx**

(b)

Figure 3.29   Bacterial Capsules. (a) *Klebsiella pneumoniae* with its capsule stained for observation in the light microscope (×1,500). (b) *Bacteroides* glycocalyx (gly), TEM (×71,250).

(figure 3.29b) is a network of polysaccharides extending from the surface of bacteria and other cells (in this sense it could encompass both capsules and slime layers). Capsules and slime layers usually are composed of polysaccharides, but they may be constructed of other materials. For example, *Bacillus anthracis* has a capsule of poly-D-glutamic acid. Capsules are clearly visible in the light microscope when negative stains or special capsule stains are employed (figure 3.29a); they also can be studied with the electron microscope (figure 3.29b).

Although capsules are not required for bacterial growth and reproduction in laboratory cultures, they do confer several advantages when bacteria grow in their normal habitats. They help bacteria resist phagocytosis by host phagocytic cells. *Streptococcus pneumoniae* provides a classic example. When it

Figure 3.30 **Bacterial Glycocalyx.** Bacteria connected to each other and to the intestinal wall, by their glycocalyxes, the extensive networks of fibers extending from the cells (×17,500).

lacks a capsule, it is destroyed easily and does not cause disease, whereas the capsulated variant quickly kills mice. Capsules contain a great deal of water and can protect bacteria against desiccation. They exclude bacterial viruses and most hydrophobic toxic materials such as detergents. The glycocalyx also aids bacterial attachment to surfaces of solid objects in aquatic environments or to tissue surfaces in plant and animal hosts (figure 3.30).

The relationship of surface polysaccharides to phagocytosis and host colonization (chapter 29).

Many gram-positive and gram-negative bacteria have a regularly structured layer called an **S-layer** on their surface. S-layers also are common among archaeobacteria, where they may be the only wall structure outside the plasma membrane. The S-layer has a pattern something like floor tiles and is composed of protein or glycoprotein (figure 3.31). In gram-negative bacteria the S-layer adheres directly to the outer membrane; it is associated with the peptidoglycan surface in gram-positive bacteria. It may protect the cell against ion and pH fluctuations, osmotic stress, enzymes, or the predacious bacterium *Bdellovibrio* (*see chapter 20*). The S-layer also helps maintain the shape and envelope rigidity of at least some bacterial cells. It can promote cell adhesion to surfaces. Finally, the layer seems to protect some pathogens against complement attack and phagocytosis, thus contributing to their virulence.

Pili and Fimbriae

Many gram-negative bacteria have short, fine, hairlike appendages that are thinner than flagella and not involved in motility. These are usually called **fimbriae** (s., **fimbria**). They also are sometimes referred to as pili. Although a cell may be covered with up to 1,000 fimbriae, they are only visible in an electron microscope due to their small size (figure 3.32). They

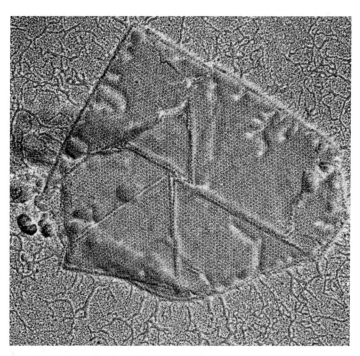

Figure 3.31 **The S-Layer.** An electron micrograph of the S-layer of *Deinococcus radiodurans* after shadowing.

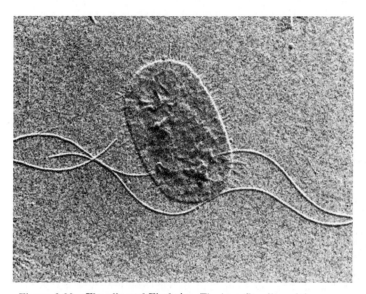

Figure 3.32 **Flagella and Fimbriae.** The long flagella and the numerous shorter fimbriae are very evident in this electron micrograph of *Proteus vulgaris* (×39,000).

seem to be slender tubes composed of helically arranged protein subunits and are about 3 to 10 nm in diameter and up to several μm long. At least some types of fimbriae attach bacteria to solid surfaces such as rocks in streams and host tissues.

Sex pili (s., **pilus**) are similar appendages, about 1 to 10 per cell, that differ from fimbriae in the following ways. Pili often are larger than fimbriae (around 9 to 10 nm in diameter). They are genetically determined by sex factors or conjugative plasmids and are required for bacterial mating (*see chapter 14*). Some bacterial viruses attach specifically to receptors on sex pili at the start of their reproductive cycle.

Flagella and Motility

Most motile bacteria move by use of **flagella** (s., **flagellum**), threadlike locomotor appendages extending outward from the plasma membrane and cell wall. They are slender, rigid structures, about 20 nm across and up to 15 or 20 μm long. Flagella are so thin they cannot be observed directly with a bright-field microscope, but must be stained with special techniques designed to increase their thickness (*see chapter 2*). The detailed structure of a flagellum can only be seen in the electron microscope (figure 3.32).

Bacterial species often differ distinctively in their patterns of flagella distribution. **Monotrichous** bacteria (*trichous* means hair) have one flagellum; if it is located at an end, it is said to be a **polar flagellum** (figure 3.33*a*). **Amphitrichous** bacteria (*amphi* means "on both sides") have a single flagellum at each pole. In contrast, **lophotrichous** bacteria (*lopho* means tuft) have a cluster of flagella at one or both ends (figure 3.33*b*). Flagella are spread fairly evenly over the whole surface of **peritrichous** (*peri* means "around") bacteria (figure 3.33*c*). Flagellation patterns are very useful in identifying bacteria.

Flagellar Ultrastructure

Transmission electron microscope studies have shown that the bacterial flagellum is composed of three parts. (1) The longest and most obvious portion is the **filament,** which extends from the cell surface to the tip. (2) A **basal body** is embedded in the cell; and (3) a short, curved segment, the **hook,** links the filament to its basal body and acts as a flexible coupling. The filament is a hollow, rigid cylinder constructed of a single protein called **flagellin,** which ranges in molecular weight from 30,000 to 60,000.

The hook and basal body are quite different from the filament (figure 3.34). Slightly wider than the filament, the hook is made of different protein subunits. The basal body is the most complex part of a flagellum (figures 3.34 and 3.35). In *E. coli* and most gram-negative bacteria, the body has four rings connected to a central rod. The outer L and P rings associate with the lipopolysaccharide and peptidoglycan layers, respectively. The inner M ring contacts the plasma membrane. Gram-positive bacteria have only two basal body rings, an inner ring connected to the plasma membrane and an outer one probably attached to the peptidoglycan.

Flagellar Synthesis

The synthesis of flagella is a complex process involving at least 20 to 30 genes. Besides the gene for flagellin, 10 or more genes code for hook and basal body proteins; other genes are concerned with the control of flagellar construction or function. It is not known how the cell regulates or determines the exact location of flagella.

Bacteria can be deflagellated, and the regeneration of the flagellar filament can then be studied. It is believed that flagellin subunits are transported through the filament's hollow

(a)

(b)

(c)

Figure 3.33 Flagellar Distribution. Examples of various patterns of flagellation as seen in the light microscope. (*a*) Monotrichous polar (*Pseudomonas*). (*b*) Lophotrichous (*Spirillum*). (*c*) Peritrichous (*Proteus vulgaris,* ×600). Bars = 5 μm.

Figure 3.34   **The Ultrastructure of Gram-Negative Flagella.** (*a*) Negatively stained flagella from *Escherichia coli* (×66,000). Arrows indicate the location of curved hooks and basal bodies. (*b*) An enlarged view of the basal body of an *E. coli* flagellum (×485,000). All four rings (L, P, S, and M) can be clearly seen. The uppermost arrow is at the junction of the hook and filament. Bar = 30 nm.

(a)   (b)

(a)   (b)

Figure 3.35   **The Ultrastructure of Bacterial Flagella.** Flagellar basal bodies and hooks in (*a*) gram-negative and (*b*) gram-positive bacteria.

internal core. When they reach the tip, the subunits spontaneously aggregate so that the filament grows at its tip rather than at the base (figure 3.36). Filament synthesis is an excellent example of **self-assembly.** Many structures form spontaneously through the association of their component parts without the aid of any special enzymes or other factors. The information required for filament construction is present in the structure of the flagellin subunit itself.

The Mechanism of Flagellar Movement

Procaryotic flagella operate differently from eucaryotic flagella. The filament is in the shape of a rigid helix, and the bacterium moves when this helix rotates. Considerable evidence shows that flagella act just like propellers on a boat. Bacterial mutants with straight flagella or abnormally long hook regions (polyhook mutants) cannot swim. When bacteria are tethered to a glass slide using antibodies to filament or hook proteins, the cell body rotates

Figure 3.36   Growth of Flagellar Filaments. Flagellin subunits travel through the flagellar core and attach to the growing tip.

rapidly about the stationary flagellum. If polystyrene-latex beads are attached to flagella, the beads spin about the flagellar axis due to flagellar rotation. The flagellar motor can rotate very rapidly. The *E. coli* motor rotates 270 revolutions per second; *Vibrio alginolyticus* averages 1,100 r.p.s.

Eucaryotic flagella and motility (pp. 90–91).

The direction of flagellar rotation determines the nature of bacterial movement. Monotrichous, polar flagella rotate counterclockwise (when viewed from outside the cell) during normal forward movement, whereas the cell itself rotates slowly clockwise. The rotating helical flagellar filament thrusts the cell forward with the flagellum trailing behind (figure 3.37). Monotrichous bacteria stop and tumble randomly by reversing the direction of flagellar rotation. Peritrichously flagellated bacteria operate in a somewhat similar way. To move forward, the flagella rotate counterclockwise. As they do so, they bend at their hooks to form a rotating bundle that propels them forward. Clockwise rotation of the flagella disrupts the bundle and the cell tumbles.

Because bacteria swim through rotation of their rigid flagella, there must be some sort of motor at the base. According to one hypothesis, a flagellum rotates because of interactions between its S ring and M ring. A rod or shaft extends from the hook and ends in the M ring, which can rotate freely in the plasma membrane (figure 3.35). It is believed that the S ring is

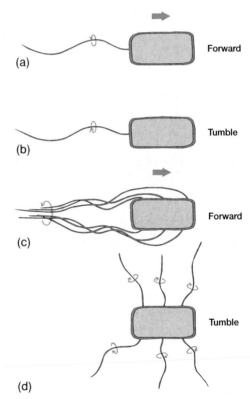

Figure 3.37   Flagellar Motility. The relationship of flagellar rotation to bacterial movement. Parts (*a*) and (*b*) describe the motion of monotrichous, polar bacteria. Parts (*c*) and (*d*) illustrate the movements of peritrichous organisms.

attached to the cell wall in gram-positive cells and does not rotate. Thus a turning motion will result if the two rings interact in some way. The P and L rings of gram-negative bacteria would act as bearings for the rotating rod. In contrast, there is some evidence that the basal body is a passive structure and rotates within a membrane-embedded protein complex much like the rotor of an electrical motor turns in the center of a ring of electromagnets (the stator).

The exact mechanism that drives basal body rotation still is not clear. There is evidence that the flow of protons past the two rings or between the basal body and surrounding membrane proteins leads to rotation. It does not appear that ATP directly provides the energy for flagellar rotation in bacteria.

The flagellum is a very effective swimming device. From the bacterium's point of view, swimming is quite a task because the surrounding water seems as thick and viscous as molasses. The cell must bore through the water with its helical or corkscrew-shaped flagella, and if flagellar activity ceases, it stops almost instantly. Despite such environmental resistance to movement, bacteria can swim from 20 to almost 90 µm/second. This is equivalent to traveling from 2 to over 100 cell lengths per second. In contrast, an exceptionally fast 6 ft human might be able to run around 5 body lengths per second.

Bacteria can move by mechanisms other than flagellar rotation. Spirochetes are helical bacteria that travel through viscous substances such as mucus or mud by flexing and spinning movements caused by a special **axial filament** (*see chapter 20*). A very different type of motility, **gliding motility,** is employed by many bacteria: cyanobacteria (*see chapter 22*), members of the orders *Myxobacteriales* (*see chapter 22*) and *Cytophagales* (*see chapter 22*), and some mycoplasmas (*see chapter 20*). Although there are no visible structures associated with gliding motility, these bacteria can coast along solid surfaces at rates up to 3 µm/second.

The mechanism of gliding motility (Box 22.1).

1. Briefly describe capsules, slime layers, glycocalyxes, and S-layers. What are their functions?
2. Distinguish between fimbriae and sex pili, and give the function of each.
3. Be able to discuss the following: flagella distribution patterns, flagella structure and synthesis, and the way in which flagella operate to move a bacterium.

Chemotaxis

Bacteria do not always swim aimlessly but are attracted by such nutrients as sugars and amino acids and are repelled by many harmful substances and bacterial waste products. (Bacteria also can respond to other environmental cues such as temperature, light, and gravity; Box 3.3.) Movement toward chemical attractants and away from repellents is known as **chemotaxis.** Such behavior is of obvious advantage to bacteria.

Chemotaxis may be demonstrated by observing bacteria in the chemical gradient produced when a thin capillary tube is filled with an attractant and lowered into a bacterial suspension. As the attractant diffuses from the end of the capillary, bacteria collect and swim up the tube. The number of bacteria within the capillary after a short length of time reflects the strength of attraction and rate of chemotaxis. Positive and negative chemotaxis also can be studied with petri dish cultures (figure 3.38). If bacteria are placed in the center of a dish of agar containing an attractant, the bacteria will exhaust the local supply and then swim outward following the attractant gradient they have created. The result is an expanding ring of bacteria. When a disk of repellent is placed in a petri dish of semisolid agar and bacteria, the bacteria will swim away from the repellent, creating a clear zone around the disk (figure 3.39).

Bacteria can respond to very low levels of attractants (about 10^{-8} M for some sugars), the magnitude of their response increasing with attractant concentration. Usually they sense repellents only at higher concentrations. If an attractant and a repellent are present together, the bacterium will compare both signals and respond to the chemical with the most effective concentration.

Attractants and repellents are detected by **chemoreceptors,** special proteins that bind chemicals and transmit signals to the other components of the chemosensing system. About 20 attractant chemoreceptors and 10 chemoreceptors for repellents have been discovered thus far. These chemoreceptor proteins may be located in the periplasmic space or the plasma membrane. Some receptors participate in the initial stages of sugar transport into the cell.

The chemotactic behavior of bacteria has been studied using the tracking microscope, a microscope with a moving stage that automatically keeps an individual bacterium in view. In the absence of a chemical gradient, *E. coli* and other bacteria move randomly. A bacterium travels in a straight or slightly curved line, a **run,** for a few seconds; then it will stop and **tumble** or **twiddle** about. The tumble is followed by a run in a different direction (figure 3.40). When the bacterium is exposed to an attractant gradient, it tumbles less frequently (or has longer runs) when traveling up the gradient, but tumbles at normal frequency if moving down the gradient. Consequently the bacterium moves up the gradient. Behavior is shaped by temporal changes in chemical concentration: the bacterium compares its current environment with that experienced a few moments previously; if the attractant concentration is higher, tumbling is suppressed and the run is longer. The opposite response occurs with a repellent gradient. Tumbling frequency decreases (the run lengthens) when the bacterium moves down the gradient away from the repellent.

Box 3.3

Living Magnets

Bacteria can respond to environmental factors other than chemicals. A fascinating example is that of the aquatic **magnetotactic bacteria** that orient themselves in the earth's magnetic field. Most of these bacteria have intracellular chains of magnetite (Fe_3O_4) particles or **magnetosomes,** around 40 to 100 nm in diameter and bounded by a membrane (Box Figure 3.3). Some species from sulfidic habitats have magnetosomes containing greigite (Fe_3S_4) and

pyrite (FeS_2). Since each iron particle is a tiny magnet, the bacteria use their magnetosome chain to determine northward and downward directions, and swim down to nutrient-rich sediments or locate the optimum depth in freshwater and marine habitats. Magnetosomes also are present in the heads of birds, tuna, dolphins, green turtles, and other animals, presumably to aid navigation. Animals and bacteria share more in common behaviorally than previously imagined.

(a)

(b)

(c)

Box Figure 3.3 **Magnetotactic Bacteria.** (*a*) Transmission electron micrograph of the magnetotactic bacterium *Aquaspirillum magnetotacticum* (×123,000). Note the long chain of electron-dense magnetite particles, MP. Other structures; OM, outer membrane, P, periplasmic space; CM, cytoplasmic membrane. (*b*) Isolated magnetosomes (×140,000). (*c*) Bacteria migrating in waves when exposed to a magnetic field.

Figure 3.38 Positive Bacterial Chemotaxis. Chemotaxis can be demonstrated on an agar plate that contains various nutrients. Positive chemotaxis by *Escherichia coli* on the left. The outer ring is composed of bacteria consuming serine. The second ring was formed by *E. coli* consuming aspartate, a less powerful attractant. The upper right colony is composed of motile, but nonchemotactic mutants. The bottom right colony is formed by nonmotile bacteria.

Figure 3.39 Negative Bacterial Chemotaxis. Negative chemotaxis by *E. coli* in response to the repellent acetate. The bright disks are plugs of concentrated agar containing acetate that have been placed in dilute agar inoculated with *E. coli*. Acetate concentration increases from zero at the top right to 3 M at top left. Note the increasing size of bacteria-free zones with increasing acetate. The bacteria have migrated for 30 minutes.

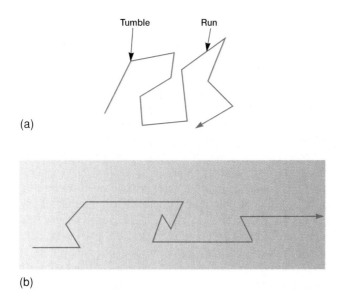

Figure 3.40 Directed Movement in Bacteria. (*a*) Random movement of a bacterium in the absence of a concentration gradient. Tumbling frequency is fairly constant. (*b*) Movement in an attractant gradient. Tumbling frequency is reduced when the bacterium is moving up the gradient. Therefore runs in the direction of increasing attractant are longer.

Much work has been done on the mechanism of chemotaxis in *Escherichia coli*. Recall that forward swimming is due to counterclockwise rotation of the flagellum, whereas tumbling results from clockwise rotation. The bacteria must be able to respond to gradients in such a way that they collect in nutrient-rich regions and at the proper oxygen level while avoiding toxic materials. *E. coli* has four different chemoreceptors that recognize serine, aspartate and maltose, ribose and galactose, and dipeptides, respectively. These chemoreceptors often are called **methyl-accepting chemotaxis proteins (MCPs).** They seem to be localized in patches, often at the end of a rod-shaped cell like *E. coli*. The MCPs do not directly influence flagellar rotation but act through a series of proteins (figure 3.41). The whole process is so efficient that a stimulus can trigger a motor response in less than 200 milliseconds.

The chemoreceptors are buried in the plasma membrane with major parts exposed on both sides. The periplasmic side of each MCP has a binding site for one or more attractant molecules and may also bind repellents. Although attractants often bind directly to the MCP, in some cases they may attach to special periplasmic binding proteins, which then interact with the MCPs. The cytoplasmic side of an MCP interacts with two proteins. The CheW protein binds to MCPs and helps attach the

Figure 3.41   The Mechanism of Chemotaxis in *Escherichia coli*. The chemotaxis system is designed to control counterclockwise (ccw) and clockwise (cw) flagellar rotation so that *E. coli* moves up an attractant gradient by a sequence of runs and tumbles. See the text for a description of the process.

CheA protein. The full complex is composed of an MCP dimer, two CheW monomers, and a CheA dimer. When the MCP is not bound to an attractant, it stimulates CheA to phosphorylate itself using ATP, a process called autophosphorylation. CheA autophosphorylation is inhibited when the attractant is bound to its MCP. Phosphorylated CheA can donate its phosphate to one of two receptor proteins, CheY or CheB. If CheY is phosphorylated by CheA, it moves to the flagellum and interacts with the switch proteins at its base. This causes the flagellum to rotate clockwise. Thus a decrease in attractant level promotes clockwise rotation and tumbling. The phosphate is removed from CheY in about 10 seconds in a process aided by the protein CheZ. The short lifetime of phosphorylated CheY means that the bacterium is very responsive to changes in attractant concentration. It will not be stuck in the tumble mode for too long a time when the attractant level changes. When no attractants or repellents are present, the system maintains intermediate concentrations of CheA phosphate and CheY phosphate. This produces a normal run-tumble swimming pattern.

The *E. coli* cell must ignore past stimulus responses so that it can compare the most recent attractant or repellent concentration with the immediately previous one and respond to any changes. This means that it must be able to adapt to a concentration change in order to detect still further changes. That is, it must have a short-term memory with a retention time of only seconds. This adaptation is accomplished by methylation of the MCP receptors. The cytoplasmic portion or domain of MCP molecules usually has about four or five methylation sites containing special glutamic acid residues. Methyls can be added to

these glutamic acid carboxyl groups using S-adenosylmethionine as the methylating agent. The reaction is catalyzed by the CheR protein and occurs at a fairly steady rate regardless of the attractant level. Methyl groups are hydrolytically removed from MCPs by the phosphorylated CheB protein, a methylesterase. These enzymes are part of a feedback circuit that stops motor responses a short time after they have commenced. The attractant-MCP complex is a good substrate for CheR and a poor substrate for CheB. When an attractant binds to the MCP, the levels of both CheY phosphate and CheB phosphate drop because the autophosphorylation of CheA is inhibited. This not only causes counterclockwise rotation and a run but also lowers methylesterase activity so that MCP methylation increases. Increased methylation changes the conformation of the MCP so that it again supports an intermediate level of CheA autophosphorylation. CheY phosphate and CheB phosphate return to intermediate levels and restore the normal run-tumble behavior. Removal of the attractant causes the overmethylated MCP to stimulate CheA autophosphorylation and the levels of CheY phosphate and CheB phosphate increase. This induces tumbling and simultaneously promotes MCP demethylation so that the system returns to an intermediate level of CheA autophosphorylation.

The chemotactic response is a very complex one involving many different proteins and two forms of covalent protein regulation (*see chapter 12*). The actual response arises from a combination of (1) the control of CheA phosphorylation by attractant and repellent levels; (2) the clockwise rotation promoted by phosphorylated CheY; and (3) a feedback regulatory circuit involving CheR, phosphorylated CheB, and variations in MCP methylation.

1. Define chemotaxis, run, and tumble or twiddle.

2. Explain in a general way how bacteria are attracted to substances like nutrients while being repelled by toxic materials.

3. Describe the molecular mechanism by which molecules attract *E. coli.*

The Bacterial Endospore

A number of gram-positive bacteria can form a special resistant, dormant structure called an **endospore.** Endospores develop within vegetative bacterial cells of several genera: *Bacillus* and *Clostridium* (rods), *Sporosarcina* (cocci), and others. These structures are extraordinarily resistant to environmental stresses such as heat, ultraviolet radiation, chemical disinfectants, and desiccation. In fact, some endospores have remained viable for well over 500 years, and actinomycete spores (which are not true endospores) have been recovered alive after burial in the mud for 7,500 years! Because of their resistance and the fact that several species of endospore-forming bacteria are dangerous pathogens, endospores are of great practical importance in food, industrial, and medical microbiology. This is because it is essential to be able to sterilize solutions and solid objects. Endospores often survive boiling for an hour or more; therefore autoclaves (*see chapter 7*) must be used to sterilize many materials. Endospores are also of considerable theoretical interest. Because bacteria manufacture these intricate entities in a very organized fashion over a period of a few hours, spore formation is well suited for research on the construction of complex biological structures. In the environment, endospores aid in survival when moisture or nutrients are scarce.

Resistance of endospores to high temperature (chapter 7).

Endospores can be examined with both light and electron microscopes. Because spores are impermeable to most stains, they often are seen as colorless areas in bacteria treated with methylene blue and other simple stains; special spore stains are used to make them clearly visible (*see chapter 2*). Spore position in the mother cell or **sporangium** frequently differs among species, making it of considerable value in identification. Spores may be centrally located, close to one end (subterminal), or definitely terminal (figure 3.42). Sometimes a spore is so large that it swells the sporangium.

Electron micrographs show that endospore structure is complex (figure 3.43). The spore often is surrounded by a thin, delicate covering called the **exosporium.** A **spore coat** lies beneath the exosporium, is composed of several protein layers, and may be fairly thick. It is impermeable and responsible for the spore's resistance to chemicals. The **cortex,** which may occupy as much as half the spore volume, rests beneath the spore coat. It is made of a peptidoglycan that is less cross-linked than that in vegetative cells. The **spore cell wall**

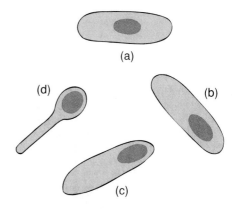

Figure 3.42   **Examples of Endospore Location and Size.** (*a*) Central spore. (*b*) Subterminal spore. (*c*) Terminal spore. (*d*) Terminal spore with swollen sporangium.

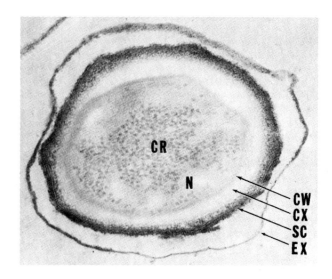

Figure 3.43   **Endospore Structure.** *Bacillus anthracis* endospore (×151,000). Note the following structures: exosporium, EX; spore coat, SC; cortex, CX; core wall, CW; and the protoplast or core with its nucleoid, N, and ribosomes, CR.

Figure 3.44   **Dipicolinic Acid.**

(or core wall) is inside the cortex and surrounds the protoplast or **core.** The core has the normal cell structures such as ribosomes and a nucleoid.

It is still not known precisely why the endospore is so resistant to heat and other lethal agents. As much as 15% of the spore's dry weight consists of **dipicolinic acid** complexed with calcium ions (figure 3.44). It has long been thought that dipicolinic acid was directly involved in spore heat resistance, but heat-resistant mutants lacking dipicolinic acid now have been

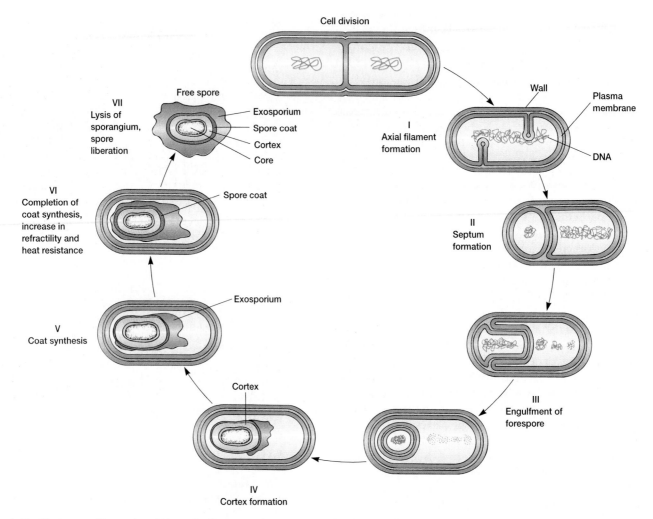

Figure 3.45   Endospore Formation. Life cycle of a sporulating bacterium. Stages are indicated by Roman numerals.

isolated. It may be that calcium-dipicolinate often stabilizes spore nucleic acids. Dehydration of the protoplast appears to be very important in heat resistance. The cortex may osmotically remove water from the protoplast, thereby protecting it from both heat and radiation damage. In summary, endospore resistance probably is due to several factors: calcium-dipicolinate stabilization of components like DNA, protoplast dehydration, the greater stability of cell proteins in bacteria adapted to growth at high temperatures, and others.

Spore formation, **sporogenesis** or **sporulation,** normally commences when growth ceases due to lack of nutrients. It is a complex process and may be divided into seven stages (figure 3.45). An axial filament of nuclear material forms (stage I), followed by an inward folding of the cell membrane to enclose part of the DNA and produce the forespore septum (stage II). The membrane continues to grow and engulfs the immature spore in a second membrane (stage III). Next, cortex is laid down in the space between the two membranes, and both cal-

Figure 3.46   Stages of Sporulation in *Bacillus megaterium*. The circled numbers refer to the hours from the end of the logarithmic phase of growth. (*a*) 0.25 h: a typical vegetative cell. (*b*) 4 h: stage II cell, septation. (*c*) 5.5 h: stage III cell, engulfment. (*d*) 6.5 h: stage IV cell, cortex formation. (*e*) 8 h: stage V cell, coat formation. (*f*) 10.5 h: stage VI cell, mature spore in sporangium. C, cortex; IFM and OFM, inner and outer forespore membranes; M, mesosome; N, nucleoid; S, septum; SC, spore coats. Bars = 0.5 μm.

cium and dipicolinic acid are accumulated (stage IV). Protein coats then are formed around the cortex (stage V), and maturation of the spore occurs (stage VI). Finally, lytic enzymes destroy the sporangium releasing the spore (stage VII). Figure 3.46 shows this process taking place in *Bacillus megaterium.* Sporulation requires only about 10 hours.

The transformation of dormant spores into active vegetative cells seems almost as complex a process as sporogenesis. It occurs in three stages: (1) activation, (2) germination, and (3) outgrowth. Often an endospore will not germinate successfully, even in a nutrient-rich medium, unless it has been activated. Activation is a reversible process that prepares spores for germination and usually results from treatments like heating. It is followed by **germination,** the breaking of the spore's dormant state. This process is characterized by spore swelling, rupture or

absorption of the spore coat, loss of resistance to heat and other stresses, loss of refractility, release of spore components, and increase in metabolic activity. Many normal metabolites or nutrients (e.g., amino acids and sugars) can trigger germination after activation. Germination is followed by the third stage, outgrowth. The spore protoplast makes new components, emerges from the remains of the spore coat, and develops again into an active bacterium (figure 3.47).

1. Describe the structure of the bacterial endospore using a labeled diagram.

2. Briefly describe endospore formation and germination. What is the importance of the endospore? What might account for its heat resistance?

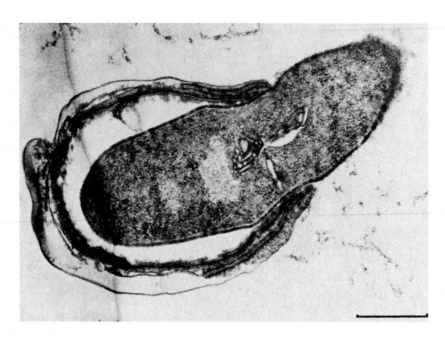

Figure 3.47 Endospore Germination. *Clostridium pectinovorum* emerging from the spore during germination. Bar = 0.5 μm.

Summary

1. Bacteria may be spherical (cocci), rod-shaped (bacilli), or spiral; resemble fungi; form buds and stalks; or even have no characteristic shape at all (pleomorphic).

2. Bacterial cells can remain together after division to form pairs, chains, and clusters of various sizes and shapes.

3. All bacteria are procaryotes and much simpler structurally than eucaryotes. Table 3.1 summarizes the major functions of bacterial cell structures.

4. The plasma membrane and most other membranes are composed of a lipid bilayer in which integral proteins are buried. Peripheral proteins are more loosely attached to membranes.

5. The plasma membrane may invaginate to form some simple structures such as membrane systems containing photosynthetic and respiratory assemblies, and possibly mesosomes.

6. The cytoplasmic matrix contains inclusion bodies and ribosomes.

7. Molecular chaperones help proteins fold properly, protect cells against environmental stresses, and transport proteins across membranes.

8. Procaryotic genetic material is located in an area called the nucleoid and is not enclosed by a membrane.

9. Most bacteria have a cell wall outside the plasma membrane to give them shape and protect them from osmotic lysis.

10. Bacterial walls are chemically complex and usually contain peptidoglycan or murein.

11. Bacteria often are classified as either gram positive or gram negative based on differences in cell wall structure and their response to Gram staining.

12. Gram-positive walls have thick, homogeneous layers of peptidoglycan and teichoic acids. Gram-negative bacteria have a thin peptidoglycan layer surrounded by a complex outer membrane containing lipopolysaccharides (LPSs) and other components.

13. Bacteria such as mycoplasmas and L forms lack a cell wall.

14. Structures such as capsules, fimbriae, and sex pili are found outside the cell wall.

15. Many bacteria are motile, usually by means of threadlike locomotory organelles called flagella.

16. Bacterial species differ in the number and distribution of their flagella.

17. The flagellar filament is a rigid helix and rotates like a propeller to push the bacterium through the water.

18. Motile bacteria can respond to gradients of attractants and repellents, a phenomenon known as chemotaxis.

19. Some bacteria survive adverse environmental conditions by forming endospores, dormant structures resistant to heat, desiccation, and many chemicals.

Key Terms

amphitrichous   *60*
axial filament   *63*
bacillus   *39*
basal body   *60*
capsule   *58*
carboxysomes   *45*
chemoreceptors   *63*
chemotaxis   *63*
coccus   *38*
core   *67*
core polysaccharide   *56*
cortex   *67*
cyanophycin granules   *45*
cytoplasmic matrix   *45*
deoxyribonucleic acid (DNA)   *49*
dipicolinic acid   *67*
diplococcus   *38*
endospore   *67*
envelope   *51*
exoenzyme   *52*
exosporium   *67*
filament   *60*
fimbriae   *59*
flagellin   *60*
flagellum   *60*
fluid mosaic model   *43*
gas vacuole   *45*
gas vesicles   *47*
germination   *69*
gliding motility   *63*
glycocalyx   *58*
glycogen   *45*

heat-shock proteins   *49*
hook   *60*
hydrophilic   *42*
hydrophobic   *42*
inclusion body   *45*
integral proteins   *43*
L form   *58*
lipid A   *56*
lipopolysaccharides (LPSs)   *56*
lophotrichous   *60*
lysis   *57*
lysozyme   *57*
magnetosomes   *64*
magnetotactic bacteria   *64*
mesosome   *44*
metachromatic granule   *47*
methyl-accepting chemotaxis proteins
(MCPs)   *65*
molecular chaperones   *49*
monotrichous   *60*
murein   *51*
mycelium   *39*
nucleoid   *50*
O antigen   *56*
O side chain   *56*
osmosis   *57*
outer membrane   *51*
penicillin   *57*
peptide interbridge   *52*
peptidoglycan   *51*
peripheral proteins   *43*
periplasm   *51*

periplasmic space   *51*
peritrichous   *60*
plasma membrane   *42*
plasmid   *51*
plasmolysis   *57*
pleomorphic   *39*
polar flagellum   *60*
poly-β-hydroxybutyrate (PHB)   *45*
polyphosphate granules   *47*
porin protein   *56*
protoplast   *45*
ribosome   *48*
rod   *39*
run   *63*
S-layer   *59*
self-assembly   *61*
sex pili   *59*
slime layer   *58*
spheroplast   *57*
spirilla   *39*
spirochete   *39*
sporangium   *67*
spore cell wall   *67*
spore coat   *67*
sporogenesis   *68*
sporulation   *68*
Svedberg unit   *48*
teichoic acid   *54*
tumble   *63*
twiddle   *63*
vibrio   *39*
volutin granule   *47*

Questions for Thought and Review

1. List all the major procaryotic structures discussed in this chapter and provide a brief description of the functions of each.

2. Why are procaryotic cells usually smaller than eucaryotic cells?

3. Some microbiologists believe that the plasma membrane is involved in DNA synthesis during bacterial reproduction. How might one prove that this is the case?

4. Discuss a possible mechanism of Gram staining in terms of differences in structure and chemistry between the walls of gram-positive and gram-negative bacteria.

5. What is self-assembly and why is it so important to cells?

6. How might one go about showing that a bacterium could form true endospores?

7. What is adaptation in chemotaxis and how does it occur?

Additional Reading

General

Balows, A.; Truper, H. G.; Dworkin, M.; Harder, W.; and Schleifer, K.-H. 1992. *The prokaryotes,* 2d ed. New York: Springer-Verlag.

Beveridge, T. J. 1989. The structure of bacteria. In *Bacteria in nature,* vol. 3, J. S. Poindexter and E. R. Leadbetter, editors, pp. 1–65. New York: Plenum.

Beveridge, T. J. 1995. The periplasmic space and the periplasm in gram-positive and gram-negative bacteria. *ASM News* 61(3):125–30.

Chung, K.-T.; Stevens, Jr., S. E.; and Ferris, D. H. 1995. A chronology of events and pioneers of microbiology. *SIM News* 45(1):3–13.

Gest, H., and Mandelstam, J. 1987. Longevity of microorganisms in natural environments. *Microbiol. Sci.* 4(3):69–71.

Goodsell, D. S. 1991. Inside a living cell. In *Trends Biochem. Sci.,* 16:203–6.

Henning, U. 1975. Determination of cell shape in bacteria. *Ann. Rev. Microbiol.* 29:45–60.

Koch, A. L. 1990. Growth and form of the bacterial cell wall. *Am. Sci.* 78:327–41.

Lederberg, J. 1992. *Encyclopedia of microbiology.* San Diego: Academic Press.

Mayer, F. 1986. *Cytology and morphogenesis of bacteria.* Berlin: Gebruder Borntraeger.

Neidhardt, F. C.; Ingraham, J. L.; and Schaechter, M. 1990. *Physiology of the bacterial cell: A molecular approach.* Sunderland, Mass.: Sinauer Associates.

Rogers, H. J. 1983. *Bacterial cell structure.* Washington: American Society for Microbiology.

Procaryotic Cell Membranes

Drews, G. 1992. Intracytoplasmic membranes in bacterial cells: Organization, function and biosynthesis. In *Prokaryotic structure and function,* S. Mohan, C. Dow, and J. A. Coles, editors, 249–74. New York: Cambridge University Press.

Ourisson, G.; Albrecht, P.; and Rohmer, M. 1984. The microbial origin of fossil fuels. *Sci. Am.* 251(2):44–51.

Salton, M. R. J., and Owen, P. 1976. Bacterial membrane structure. *Ann. Rev. Microbiol.* 30:451–82.

The Cytoplasmic Matrix

Dawes, E. A. 1992. Storage polymers in prokaryotes. In *Prokaryotic structure and function,* S. Mohan, C. Dow, and J. A. Coles, editors, 81–122. New York: Cambridge University Press.

Georgopoulos, C., and Welch, W. J. 1993. Role of the major heat shock proteins as molecular chaperones. *Ann. Rev. Cell Biol.* 9:601–34.

Hartl, F.-U.; Hlodan, R.; and Langer, T. 1994. Molecular chaperones in protein folding: The art of avoiding sticky situations. *Trends Biochem. Sci.* 19:20–25.

Hendrick, J. P., and Hartl, F.-U. 1993. Molecular chaperone functions of heat-shock proteins. *Ann Rev. Biochem.* 62:349–84.

Walsby, A. E. 1977. The gas vacuoles of blue-green algae. *Sci. Am.* 237(2):90–97.

Welch, W. J. 1993. How cells respond to stress. *Sci. Am.* 268(5):56–64.

Wittmann, H. G. 1983. Architecture of prokaryotic ribosomes. *Ann. Rev. Biochem.* 52:35–65.

Yura, T.; Nagai, H.; and Mori, H. 1993. Regulation of the heat-shock response in bacteria. *Ann. Rev. Microbiol.* 47:321–50.

The Nucleoid

Brock, T. D. 1988. The bacterial nucleus: A history. *Microbiol. Rev.* 52:397–411.

Robinow, C., and Kellenberger, E. 1994. The bacterial nucleoid revisited. *Microbiol. Rev.* 58(2): 211–32.

Schmidt, M. B. 1988. Structure and function of the bacterial chromosome. *Trends Biochem. Sci.* 13(4):131–35.

The Procaryotic Cell Wall

Ferguson, S. J. 1992. The periplasm. In *Prokaryotic structure and function,* S. Mohan, C. Dow, and J. A. Coles, editors, 311–40. New York: Cambridge University Press.

Hancock, R. E. W. 1991. Bacterial outer membranes: Evolving concepts. *ASM News* 57(4):175–82.

Osborne, M. J., and Wu, H. C. P. 1980. Proteins of the outer membrane of gram-negative bacteria. *Ann. Rev. Microbiol.* 34:369–422.

Rietschel, E. T., et al. 1994. Bacterial endotoxin: Molecular relationships of structure to activity and function. *The FASEB Journal* 8:217–25.

Scherrer, R. 1984. Gram's staining reaction, Gram types and cell walls of bacteria. *Trends Biochem. Sci.* 9:242–45.

Sharon, N. 1969. The bacterial cell wall. *Sci. Am.* 221(5):92–98.

Components External to the Cell Wall

Bayer, M. E., and Bayer, M. H. 1994. Biophysical and structural aspects of the bacterial capsule. *ASM News* 60(4):192–98.

Beveridge, T. J., and Graham, L. L. 1991. Surface layers of bacteria. *Microbiol. Rev.* 55(4):684–705.

Costerton, J. W.; Geesey, G. G.; and Cheng, K.-J. 1978. How bacteria stick. *Sci. Am.* 238(1):86–95.

Ferris, F. G., and Beveridge, T. J. 1985. Functions of bacterial cell surface structures. *BioScience* 35(3): 172–77.

Hultgren, S. J.; Abraham, S.; Caparon, M.; Falk, P.; St. Geme, III, J. W.; and Normark, S. 1993. Pilus and nonpilus bacterial adhesins: Assembly and function in cell recognition. *Cell* 73:887–901.

Messner, P., and Sleytr, U. B. 1992. Crystalline bacterial cell-surface layers. In *Advances in microbial physiology,* vol. 33, A. H. Rose, editor, 213–75. New York: Academic Press.

Troy, F. A. 1979. The chemistry and biosynthesis of selected bacterial capsular polymers. *Ann. Rev. Microbiol.* 33:519–60.

Chemotaxis

Adler, J. 1976. The sensing of chemicals by bacteria. *Sci. Am.* 234(4):40–47.

Bazylinski, D. A. 1995. Structure and function of the bacterial magnetosome. *ASM News* 61(7):337-43.

Berg, H. C. 1975. How bacteria swim. *Sci. Am.* 233(2):36–44.

Blakemore, R. P. 1982. Magnetotactic bacteria. *Ann. Rev. Microbiol.* 36:217–38.

Doetsch, R. N., and Sjoblad, R. D. 1980. Flagellar structure and function in eubacteria. *Ann. Rev. Microbiol.* 34:69–108.

Manson, M. D. 1992. Bacterial motility and chemotaxis. In *Advances in microbial physiology,* vol. 33, A. H. Rose, editor, 277–346. New York: Academic Press.

Parkinson, J. S. 1993. Signal transduction schemes of bacteria. *Cell* 73:857–71.

Stolz, J. F. 1993. Magnetosomes. *J. Gen. Microbiol.* 139:1663–70.

The Bacterial Endospore

Aronson, A. I., and Fitz-James, P. 1976. Structure and morphogenesis of the bacterial spore coat. *Bacteriol. Rev.* 40(2):360–402.

Errington, J. 1993. *Bacillus subtilis* sporulation: Regulation of gene expression and control of morphogenesis. *Microbiol. Rev.* 57(1):1–33.

Slepecky, R. A. 1978. Resistant forms. In *Essays in microbiology,* J. R. Norris and M. H. Richmond, editors, 14/1–14/31. New York: John Wiley and Sons.

CHAPTER 4
Eucaryotic Cell Structure and Function

The key to every biological

problem must finally be

sought in the cell.

—E. B. Wilson

Concepts

1. Eucaryotic cells differ most obviously from procaryotic cells in having a variety of complex membranous organelles in the cytoplasmic matrix and their genetic material within membrane-delimited nuclei. Each organelle has a distinctive structure directly related to specific functions.

2. A cytoskeleton composed of microtubules, microfilaments, and intermediate filaments helps give eucaryotic cells shape; microtubules and microfilaments are also involved in cell movements and intracellular transport.

3. In eucaryotes, genetic material is distributed between cells by the highly organized, complex processes called mitosis and meiosis.

4. Despite great differences between eucaryotes and procaryotes with respect to such things as morphology, they are similar on the biochemical level.

In chapter 3 considerable attention is devoted to procaryotic cell structure and function because bacteria are immensely important in microbiology and have occupied a large portion of microbiologists' attention in the past. Nevertheless, eucaryotic algae, fungi, and protozoa also are microorganisms and have been extensively studied. These organisms often are extraordinarily complex, interesting in their own right, and prominent members of the ecosystem (figure 4.1). In addition, fungi (and to some extent, algae) are exceptionally useful in industrial microbiology. Many fungi and protozoa are also major human pathogens; one only need think of either malaria or African sleeping sickness (figure 4.1c) to appreciate the significance of eucaryotes in pathogenic microbiology. So, although this text emphasizes bacteria, eucaryotic microorganisms are discussed at many points.

Chapter 4 focuses on eucaryotic cell structure and its relationship to cell function. Because many valuable studies on eucaryotic cell ultrastructure have used organisms other than microorganisms, some work on nonmicrobial cells is presented. At the end of the chapter, procaryotic and eucaryotic cells are compared in some depth.

An Overview of Eucaryotic Cell Structure

The most obvious difference between eucaryotic and procaryotic cells is in their use of membranes. Eucaryotic cells have membrane-delimited nuclei, and membranes also play a prominent part in the structure of many other organelles (figures 4.2 and 4.3). **Organelles** are intracellular structures that perform

(a)   (b)   (c)   (d)   (e)   (f)   (g)

Figure 4.1   Representative Examples of Eucaryotic Microorganisms.
(*a*) *Paramecium* as seen with interference-contrast microscopy (×115). (*b*) Mixed diatom frustules (×100). (*c*) The protozoan *Trypanosoma brucei gambiense,* blood smear (×1,000), one cause of African sleeping sickness. (*d*) *Penicillium* colonies, and (*e*) a microscopic view of the mold's hyphae and conidia (×220). (*f*) *Stentor.* The ciliated protozoa are extended and actively feeding, dark-field microscopy (×100).
(*g*) *Amanita muscaria,* a large poisonous mushroom (×5).

(a)

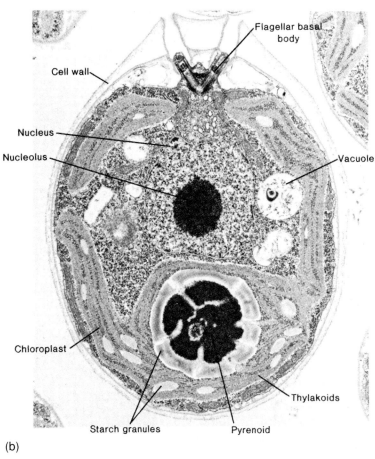

(b)

Figure 4.2   Eucaryotic Cell Ultrastructure. (*a*) A lymphoblast in the rat lymph node (×17,500). (*b*) The alga *Chlamydomonas reinhardtii*, a deflagellated cell. Note the large chloroplast with its pyrenoid body (×30,000).

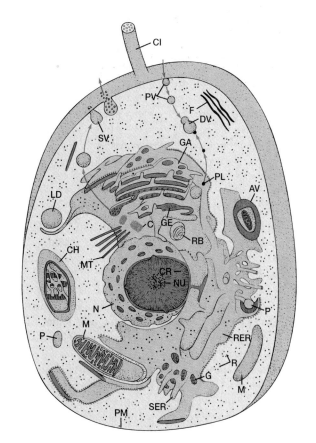

Figure 4.3   Eucaryotic Cell Ultrastructure. This is a schematic, three-dimensional diagram of a cell with the most important organelles identified in the illustration. AV, autophagic vacuole; C, centriole; CH, chloroplast; CI, cilium; CR, chromatin; DV, digestion vacuole; F, microfilaments; G, glycogen; GA, Golgi apparatus; GE, GERL; LD, lipid droplet; M, mitochondrion; MT, microtubules; N, nucleus; NU, nucleolus; P, peroxisome; PL, primary lysosome; PM, plasma membrane; PV, pinocytotic vesicle; R, ribosomes and polysomes; RB, residual body; RER, rough endoplasmic reticulum; SER, smooth endoplasmic reticulum; SV, secretion vacuole.

specific functions in cells. The name organelle (little organ) was coined because biologists saw a parallel between the relationship of organelles to a cell and that of organs to the whole body. A comparison of figures 4.2 and 4.3 with figure 3.11 (*p. 47*) shows how much more structurally complex the eucaryotic cell is. This complexity is due chiefly to the use of internal membranes for several purposes. The partitioning of the eucaryotic cell interior by membranes makes possible the placement of different biochemical and physiological functions in separate compartments so that they can more easily take place simultaneously under independent control and proper coordination. Large membrane surfaces make possible greater respiratory and photosynthetic activity because these processes are located exclusively in membranes. The intracytoplasmic membrane complex also serves as a transport system to move materials between different cell locations. Thus abundant membrane systems probably are necessary in eucaryotic cells because of their large volume and the need for adequate regulation, metabolic activity, and transport.

Figures 4.2 and 4.3 provide a generalized view of eucaryotic cell structure and illustrate most of the organelles to be discussed. Table 4.1 briefly summarizes the functions of the major eucaryotic organelles. Those organelles lying inside the plasma membrane are first described, and then components outside the membrane are discussed.

The Cytoplasmic Matrix, Microfilaments, Intermediate Filaments, and Microtubules

When a eucaryotic cell is examined at low power with the electron microscope, its larger organelles are seen to lie in an apparently featureless, homogeneous substance called the **cytoplasmic matrix.** The matrix, although superficially uninteresting, is actually one of the most important and complex parts of the cell. It is the "environment" of the organelles and the location of many important biochemical processes. Several physical changes seen in cells—viscosity changes, cytoplasmic streaming, and others—also are due to matrix activity.

Water constitutes about 70 to 85% by weight of a eucaryotic cell. Thus a large part of the cytoplasmic matrix is water. Cellular water can exist in two different forms. Some of it is bulk or free water; this is normal, osmotically active water.

Osmosis, water activity, and growth (pp. 57, 124–25).

Water also can exist as bound water or water of hydration. This water is bound to the surface of proteins and other macromolecules and is osmotically inactive and more ordered than bulk water. There is some evidence that much of metabolism occurs in bound water. The protein content of cells is so high that the cytoplasmic matrix often may be semicrystalline. Usually matrix pH is around neutrality, about pH 6.8 to 7.1, but can vary widely. For example, protozoan digestive vacuoles may reach pHs as low as 3 to 4.

Probably all eucaryotic cells have **microfilaments,** minute protein filaments, 4 to 7 nm in diameter, which may be either scattered within the cytoplasmic matrix or organized into networks and parallel arrays. Microfilaments play a major role in

TABLE 4.1	Functions of Eucaryotic Organelles
Plasma membrane	Mechanical cell boundary, selectively permeable barrier with transport systems, mediates cell-cell interactions and adhesion to surfaces, secretion
Cytoplasmic matrix	Environment for other organelles, location of many metabolic processes
Microfilaments, intermediate filaments, and microtubules	Cell structure and movements, form the cytoskeleton
Endoplasmic reticulum	Transport of materials, protein and lipid synthesis
Ribosomes	Protein synthesis
Golgi apparatus	Packaging and secretion of materials for various purposes, lysosome formation
Lysosomes	Intracellular digestion
Mitochondria	Energy production through use of the tricarboxylic acid cycle, electron transport, oxidative phosphorylation, and other pathways
Chloroplasts	Photosynthesis—trapping light energy and formation of carbohydrate from CO_2 and water
Nucleus	Repository for genetic information, control center for cell
Nucleolus	Ribosomal RNA synthesis, ribosome construction
Cell wall and pellicle	Strengthen and give shape to the cell
Cilia and flagella	Cell movement

cell motion and shape changes. Some examples of cellular movements associated with microfilament activity are the motion of pigment granules, amoeboid movement, and protoplasmic streaming in slime molds (*see chapter 25*).

The participation of microfilaments in cell movement is suggested by electron microscopic studies showing that they frequently are found at locations appropriate for such a role. For example, they are concentrated at the interface between stationary and flowing cytoplasm in plant cells and slime molds. Experiments using the drug cytochalasin B have provided additional evidence. Cytochalasin B disrupts microfilament structure and often simultaneously inhibits cell movements. However, because the drug has additional effects in cells, a direct cause-and-effect interpretation of these experiments is sometimes difficult.

Microfilament protein has been isolated and analyzed chemically. It is an actin, very similar to the actin contractile protein of muscle tissue. This is further indirect evidence for microfilament involvement in cell movement.

A second type of small filamentous organelle in the cytoplasmic matrix is shaped like a thin cylinder about 25 nm in diameter. Because of its tubular nature this organelle is called a **microtubule.** Microtubules are complex structures constructed of two slightly different spherical protein subunits named tubulins, each of which is approximately 4 to 5 nm in diameter. These subunits are assembled in a helical arrangement to form a cylinder with an average of 13 subunits in one turn or circumference (figure 4.4).

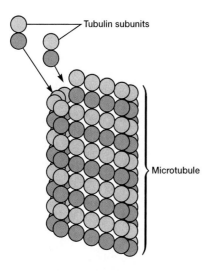

Figure 4.4   Microtubule Structure. The hollow cylinder, about 25 nm in diameter, is made of two kinds of protein subunits, α-tubulin and β-tubulin.

Microtubules serve at least three purposes: (1) they help maintain cell shape, (2) are involved with microfilaments in cell movements, and (3) participate in the intracellular transport of substances. Evidence for a structural role comes from their intracellular distribution and studies on the effects of the drug colchicine. Long, thin cell structures requiring support such as the axopodia (long, slender, rigid pseudopodia) of protozoa contain microtubules (figure 4.5). When migrating embryonic nerve and heart cells are exposed to colchicine, they simultaneously lose their microtubules and their characteristic shapes. The shapeless cells seem to wander aimlessly as if incapable of directed movement without their normal form. Their microfilaments are still intact, but due to the disruption of their microtubules by colchicine, they no longer behave normally.

Microtubules also are present in structures that participate in cell or organelle movements—the mitotic spindle, cilia, and flagella. For example, the mitotic spindle is constructed of microtubules; when a dividing cell is treated with colchicine, the spindle is disrupted and chromosome separation blocked. Microtubules also are essential to the movement of eucaryotic cilia and flagella.

Other kinds of filamentous components also are present in the matrix, the most important of which are the **intermediate filaments** (about 8 to 10 nm in diameter). The microfilaments, microtubules, and intermediate filaments are major components of a vast, intricate network of interconnected filaments called the **cytoskeleton** (figure 4.6). As mentioned previously, the cytoskeleton plays a role in both cell shape and movement. Procaryotes lack an organized cytoskeleton and may not possess actinlike proteins.

1. What is an organelle?
2. Define cytoplasmic matrix, bulk or free water, bound water, microfilament, microtubule, and tubulin. Discuss the roles of microfilaments, intermediate filaments, and microtubules.
3. Describe the cytoskeleton. What are its functions?

(a)

(b)

Figure 4.5   Cytoplasmic Microtubules. Electron micrographs of pseudopodia with microtubules. (*a*) Microtubules in a pseudopodium from the protozoan *Reticulomyxa* (×65,000). (*b*) A transverse section of a heliozoan axopodium (×48,000). Note the parallel array of microtubules organized in a spiral pattern.

(a)   (b)

Figure 4.6   The Eucaryotic Cytoskeleton. (*a*) Antibody-stained microfilament system in a mammal cell (×400). (*b*) Antibody-stained microtubule system in a mammal cell (×1,000).

The Endoplasmic Reticulum

Besides the cytoskeleton, the cytoplasmic matrix is permeated with an irregular network of branching and fusing membranous tubules, around 40 to 70 nm in diameter, and many flattened sacs called **cisternae** (s., **cisterna**). This network of tubules and cisternae is the **endoplasmic reticulum (ER)** (figures 4.2*a* and 4.7). The nature of the ER varies with the functional and

Figure 4.7   The Endoplasmic Reticulum. A transmission electron micrograph of the human corpus luteum showing structural variations in eucaryotic endoplasmic reticulum. Note the presence of both rough endoplasmic reticulum lined with ribosomes and smooth endoplasmic reticulum without ribosomes (×26,500).

Rough endoplasmic reticulum

Smooth endoplasmic reticulum

(a)

Figure 4.8   Golgi Apparatus Structure. Golgi apparatus of *Euglena gracilis*. Cisternal stacks are shown in the electron micrograph (×165,000) in (*a*) and diagrammatically in (*b*).

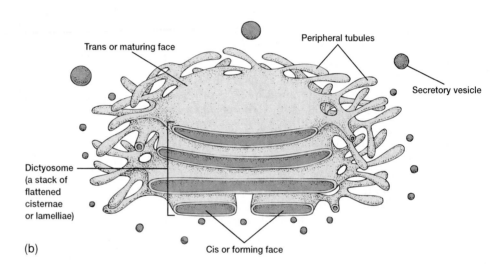

Trans or maturing face

Peripheral tubules

Secretory vesicle

Dictyosome (a stack of flattened cisternae or lamelliae)

(b)

Cis or forming face

physiological status of the cell. In cells synthesizing a great deal of protein for purposes such as secretion, a large part of the ER is studded on its outer surface with ribosomes and is called **rough** or **granular endoplasmic reticulum (RER or GER).** Other cells, such as those producing large quantities of lipids, have ER that lacks ribosomes. This is **smooth** or **agranular ER (SER or AER).**

The endoplasmic reticulum has many important functions. It transports proteins, lipids, and probably other materials through the cell. Lipids and proteins are synthesized by ER-associated enzymes and ribosomes. Polypeptide chains synthesized on RER-bound ribosomes may be inserted either into the ER membrane or into its lumen for transport elsewhere. The ER is also a major site of cell membrane synthesis.

New endoplasmic reticulum is produced through expansion of the old. Many biologists think the RER synthesizes new ER proteins and lipids. "Older" RER then loses its connected

ribosomes and is modified to become SER. Not everyone agrees with this interpretation, and other mechanisms of growth of ER are possible.

The Golgi Apparatus

The **Golgi apparatus** is a membranous organelle composed of flattened, saclike cisternae stacked on each other (figure 4.8). These membranes, like the smooth ER, lack bound ribosomes. There are usually around 4 to 8 cisternae or sacs in a stack, although there may be many more. Each sac is 15 to 20 nm thick and separated from other cisternae by 20 to 30 nm. A complex network of tubules and vesicles (20 to 100 nm in diameter) is located at the edges of the cisternae. The stack of cisternae has a definite polarity because there are two ends or faces that are quite different from one another. The sacs on the cis or forming face often are associated with the ER and differ from the sacs on the trans or maturing face in thickness, enzyme content, and degree of vesicle formation. It appears that material is transported from cis to trans cisternae by vesicles that bud off the cisternal edges and move to the next sac.

The Golgi apparatus is present in most eucaryotic cells, but many fungi and ciliate protozoa may lack a well-formed structure. Sometimes it consists of a single stack of cisternae; however, many cells may contain up to 20, and sometimes more, separate stacks. These stacks of cisternae, often called **dictyosomes,** can be clustered in one region or scattered about the cell.

The Golgi apparatus packages materials and prepares them for secretion, the exact nature of its role varying with the organism. The surface scales of some flagellated algae and radiolarian protozoa appear to be constructed within the Golgi apparatus and then transported to the surface in vesicles. It often participates in the development of cell membranes and in the packaging of cell products. The growth of some fungal hyphae occurs when Golgi vesicles contribute their contents to the wall at the hyphal tip.

In all these processes, materials move from the ER to the Golgi apparatus. Most often vesicles bud off the ER, travel to the Golgi apparatus, and fuse with the cis cisternae. Thus the Golgi apparatus is closely related to the ER in both a structural and a functional sense. Most proteins entering the Golgi apparatus from the ER are glycoproteins containing short carbohydrate chains. The Golgi apparatus frequently modifies proteins destined for different fates by adding specific groups and then sends the proteins on their way to the proper location (e.g., lysosomal proteins have phosphates added to their mannose sugars).

Lysosomes and Endocytosis

A very important function of the Golgi apparatus and endoplasmic reticulum is the synthesis of another organelle, the **lysosome.** This organelle (or a structure very much like it) is found in a variety of microorganisms—protozoa, some algae, and fungi—as well as in plants and animals. Lysosomes are roughly spherical and enclosed in a single membrane; they average about 500 nm in diameter, but range from 50 nm to several μm in size. They are involved in intracellular digestion and contain the enzymes needed to digest all types of macromolecules. These enzymes, called hydrolases, catalyze the hydrolysis of molecules and function best under slightly acid conditions (usually around pH 3.5 to 5.0). Lysosomes maintain an acidic environment by pumping protons into their interior. Digestive enzymes are manufactured by the RER and packaged to form lysosomes by the Golgi apparatus. A segment of smooth ER near the Golgi apparatus also may bud off lysosomes.

Lysosomes are particularly important in those cells that obtain nutrients through **endocytosis.** In this process a cell takes up solutes or particles by enclosing them in vacuoles and vesicles pinched off from its plasma membrane. Vacuoles and vesicles are membrane-delimited cavities that contain fluid, and often solid material. Larger cavities will be called vacuoles, and smaller cavities, vesicles. There are two major forms of endocytosis: phagocytosis and pinocytosis. During **phagocytosis** large particles and even other microorganisms are enclosed in a phagocytic vacuole or phagosome and engulfed (figure 4.9*a*). In

pinocytosis small amounts of the surrounding liquid with its solute molecules are pinched off as tiny pinocytotic vesicles (also called pinocytic vesicles) or pinosomes. Often phagosomes and pinosomes are collectively called **endosomes** because they are formed by endocytosis. The type of pinocytosis, receptor-mediated endocytosis, that produces coated vesicles (*see p. 373*) is important in the entry of animal viruses into host cells.

Material in endosomes is digested with the aid of lysosomes. Newly formed lysosomes, or **primary lysosomes,** fuse with phagocytic vacuoles to yield **secondary lysosomes,** lysosomes with material being digested (figure 4.9). These phagocytic vacuoles or secondary lysosomes often are called food vacuoles. Digested nutrients then leave the secondary lysosome and enter the cytoplasm. When the lysosome has accumulated large quantities of indigestible material, it is known as a **residual body.**

Lysosomes join with phagosomes for defensive purposes as well as to acquire nutrients. Invading bacteria, ingested by a phagocytic cell, usually are destroyed when lysosomes fuse with the phagosome. This is commonly seen in leukocytes (white blood cells) of vertebrates.

Phagocytosis and resistance to pathogens (pp. 582–83).

Cells can selectively digest portions of their own cytoplasm in a type of secondary lysosome called an **autophagic vacuole** (figure 4.9*a*). It is thought that these arise by lysosomal engulfment of a piece of cytoplasm (figure 4.10), or when the ER pinches off cytoplasm to form a vesicle that subsequently fuses with lysosomes. Autophagy probably plays a role in the normal turnover or recycling of cell constituents. A cell also can survive a period of starvation by selectively digesting portions of itself to remain alive. Following cell death, lysosomes aid in digestion and removal of cell debris.

A most remarkable thing about lysosomes is that they accomplish all these tasks without releasing their digestive enzymes into the cytoplasmic matrix, a catastrophe that would destroy the cell. The lysosomal membrane retains digestive enzymes and other macromolecules while allowing small digestion products to leave.

The intricate complex of membranous organelles composed of the Golgi apparatus, lysosomes, endosomes, and associated structures seems to operate as a coordinated whole whose main function is the import and export of materials (figure 4.10). Christian de Duve has suggested that this complex be called the vacuome in recognition of its functional unity. The ER manufactures secretory proteins and membrane, and contributes these to the Golgi apparatus. The Golgi apparatus then forms secretory vesicles that fuse with the plasma membrane and release material to the outside. It also produces lysosomes that fuse with endosomes to digest material acquired through phagocytosis and pinocytosis. Membrane movement in the region of the vacuome lying between the Golgi apparatus and the plasma membrane is two-way. Empty vesicles often are recycled and returned to the Golgi apparatus and plasma membrane rather than being destroyed. These exchanges in the vacuome occur without membrane rupture so that vesicle contents never escape directly into the cytoplasmic matrix.

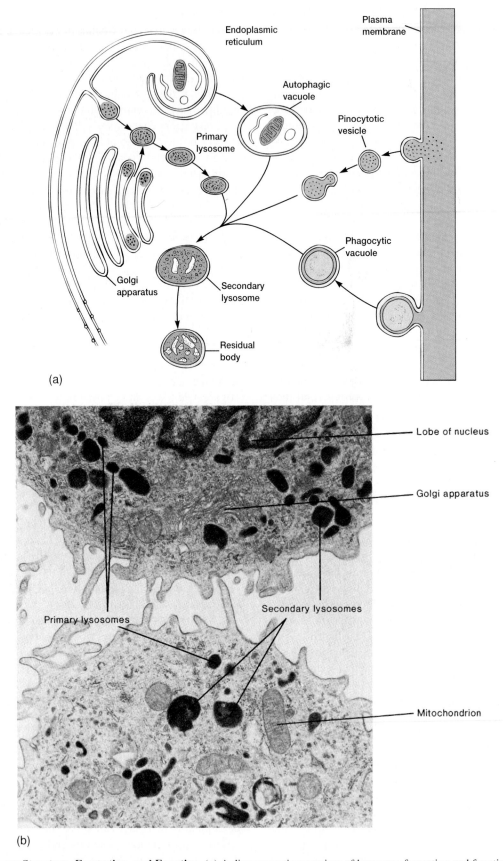

(a)

(b)

Figure 4.9   Lysosome Structure, Formation, and Function. (*a*) A diagrammatic overview of lysosome formation and function. (*b*) Lysosomes in macrophages from the lung. Secondary lysosomes contain partially digested material and are formed by fusion of primary lysosomes and phagocytic vacuoles (×14,137).

Figure 4.10　Membrane Flow in the Vacuome.
The flow of material and membranes between
organelles in a eucaryotic cell.
(*1*) Vesicles shuttling between the ER and Golgi
　apparatus.
(*2*) The Golgi-plasma membrane shuttle for secretion of
　materials.
(*3*) The Golgi-lysosome shuttle.
(*4*) The movement of material and membranes during
　endocytosis.
(*5*) Pathways of plasma membrane recovery from
　endosomes, lysosomes, and through the Golgi
　apparatus.
(*6*) Movement of vesicles from endosomes to
　lysosomes.
(*7*) Autophagy by a lysosome.

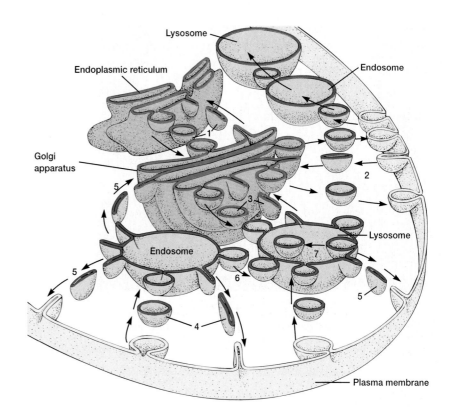

1. How do the rough and smooth endoplasmic reticulum differ from
   one another in terms of structure and function? List the processes
   in which the ER is involved.

2. Describe the structure of a Golgi apparatus in words and with a
   diagram. How do the cis and trans faces of the Golgi apparatus
   differ? List the major Golgi apparatus functions discussed in the
   text.

3. How are lysosomes formed? Describe the various forms of
   lysosomes and the way in which they participate in intracellular
   digestion. What is an autophagic vacuole? Define endocytosis,
   pinocytosis, and phagocytosis.

Eucaryotic Ribosomes

The eucaryotic ribosome can either be associated with the endoplasmic reticulum or be free in the cytoplasmic matrix and is
larger than the bacterial 70S ribosome. It is a dimer of a 60S
and a 40S subunit, about 22 nm in diameter, and has a sedimentation coefficient of 80S and a molecular weight of 4 million.
When bound to the endoplasmic reticulum to form rough ER, it
is attached through its 60S subunit.

　Both free and RER-bound ribosomes synthesize proteins. As
mentioned earlier, proteins made on the ribosomes of the RER either enter its lumen for transport, and often for secretion, or are
inserted into the ER membrane as integral membrane proteins.
Free ribosomes are the sites of synthesis for nonsecretory and
nonmembrane proteins. Some proteins synthesized by free ribosomes are inserted into organelles such as the nucleus, mitochondrion, and chloroplast. As discussed in chapter 3 (*see pp. 49–50*),
molecular chaperones aid the proper folding of proteins after synthesis. They also assist the transport of proteins into eucaryotic
organelles such as mitochondria. Several ribosomes usually attach to a single messenger RNA and simultaneously translate its
message into protein. These complexes of messenger RNA and
ribosomes are called **polyribosomes** or **polysomes.** Ribosomal
participation in protein synthesis is dealt with later.

The role of ribosomes in protein synthesis (pp. 226–33).

1. Describe the structure of the eucaryotic 80S ribosome and contrast
   it with the procaryotic ribosome.

2. How do free ribosomes and those bound to the ER differ in
   function?

Mitochondria

Found in most eucaryotic cells, **mitochondria** (s., **mitochondrion**) frequently are called the "powerhouses" of the cell. Tricarboxylic acid cycle activity and the generation of ATP by electron transport and oxidative phosphorylation take place here. In
the transmission electron microscope, mitochondria usually are
cylindrical structures and measure approximately 0.3 to 1.0 μm

Figure 4.11   Mitochondria in Living Cells. A mammalian cell stained with the fluorescent dye Rhodamine 123. Note the long, threadlike mitochondria that twist and turn through the cytoplasmic matrix.

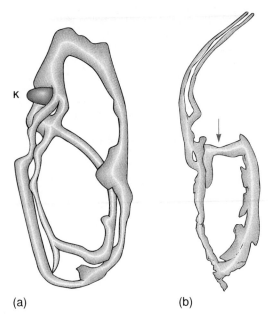

(a)   (b)

Figure 4.12   Trypanosome Mitochondria. The giant mitochondria from trypanosome protozoa. (*a*) *Crithidia fasciculata* mitochondrion with kinetoplast, K. The kinetoplast contains DNA that codes for mitochondrial RNA and protein. (*b*) *Trypanosoma cruzi* mitochondrion with arrow indicating position of kinetoplast.

by 5 to 10 µm. (In other words, they are about the same size as bacterial cells.) However, in living cells stained with a fluorescent dye, mitochondria often appear to be very long, threadlike objects (figure 4.11). Although cells can possess as many as 1,000 or more mitochondria, at least a few cells (some yeasts, unicellular algae, and trypanosome protozoa) have a single giant tubular mitochondrion twisted into a continuous network permeating the cytoplasm (figure 4.12).

The tricarboxylic acid cycle, electron transport, and oxidative phosphorylation (pp. 169–74).

The mitochondrion is bounded by two membranes, an outer mitochondrial membrane separated from an inner mitochondrial membrane by a 6 to 8 nm intermembrane space (figure 4.13). Special infoldings of the inner membrane, called **cristae** (s., **crista**), greatly increase its surface area. Their shape differs in mitochondria from various species. Fungi often have platelike cristae, whereas euglenoid flagellates may have cristae shaped like disks. Tubular cristae are found in some fungi and most protozoa; however, amoebae can possess mitochondria with cristae in the shape of vesicles (figure 4.14). The inner membrane encloses the mitochondrial matrix, a dense matrix containing ribosomes, DNA, and often large calcium phosphate granules. Mitochondrial ribosomes are smaller than cytoplasmic ribosomes and resemble those of bacteria in several ways, including their size and subunit composition. Mitochondrial DNA is a closed circle like bacterial DNA.

Each mitochondrial compartment is different from the others in chemical and enzymatic composition. The outer and inner mitochondrial membranes, for example, possess different

lipids. Enzymes and electron carriers involved in electron transport and oxidative phosphorylation (the formation of ATP as a consequence of electron transport) are located only in the inner membrane. The enzymes of the tricarboxylic acid cycle and the ß-oxidation pathway for fatty acids (*see chapter 9*) are located in the matrix.

The inner membrane of the mitochondrion has another distinctive structural feature related to its function. Many small spheres, about 8.5 nm diameter, are attached by stalks to its inner surface. The spheres are called **F₁ particles** and synthesize ATP during cellular respiration (*see pp. 172–74*).

The mitochondrion uses its DNA and ribosomes to synthesize some of its own proteins. Most mitochondrial proteins, however, are manufactured under the direction of the nucleus. Mitochondria reproduce by binary fission. Chloroplasts show similar partial independence and reproduction by binary fission. Because both organelles resemble bacteria to some extent, it has been suggested that these organelles arose from symbiotic associations between bacteria and larger cells (Box 4.1).

Chloroplasts

Plastids are cytoplasmic organelles of algae and higher plants that often possess pigments such as chlorophylls and carotenoids, and are the sites of synthesis and storage of food reserves. The most important type of plastid is the chloroplast. **Chloroplasts** contain chlorophyll and use light energy

Figure 4.13   Mitochondrial Structure. (*a*) A diagram of mitochondrial structure. The insert shows F_1F_0 complexes lining the inner surface of the cristae. (*b*) Scanning electron micrograph (×70,000) of a freeze-fractured mitochondrion showing the cristae (arrows). The outer and inner mitochondrial membranes also are evident. (*c*) Transmission electron micrograph of a mitochondrion from a bat pancreas (×85,000). Note outer and inner mitochondrial membranes, cristae, and inclusions in the matrix. The mitochondrion is surrounded by rough endoplasmic reticulum.

(a)

(b)

(c)

Figure 4.14   Mitochondrial Cristae. Mitochondria with a variety of cristae shapes. (*a*) Mitochondria from the protostelid *Schizoplasmodiopsis micropunctata*. Note the tubular cristae (×49,500). (*b*) The protozoan *Actinosphaerium* with vesicular cristae (×75,000).

(a)

(b)

The Origin of the Eucaryotic Cell

The profound differences between eucaryotic and procaryotic cells have stimulated much discussion about how the more complex eucaryotic cell arose. Some biologists believe the original "protoeucaryote" was a large aerobic bacterium that formed mitochondria, chloroplasts, and nuclei when its plasma membrane invaginated and enclosed genetic material in a double membrane. The organelles could then evolve independently. It also is possible that a large blue-green bacterium lost its cell wall and became phagocytic. Subsequently, primitive chloroplasts, mitochondria, and nuclei would be formed by the fusion of thylakoids and endoplasmic reticulum cisternae to enclose specific areas of cytoplasm.

By far the most popular theory for the origin of eucaryotic cells is the **endosymbiotic theory.** In brief, it is supposed that the ancestral procaryotic cell lost its cell wall and gained the ability to obtain nutrients by phagocytosing other bacteria. When photosynthetic cyanobacteria arose, the environment slowly became aerobic. If an anaerobic, amoeboid, phagocytic procaryote—possibly already possessing a developed nucleus—engulfed an aerobic bacterial cell and established a permanent symbiotic relationship with it, the host would be better adapted to its increasingly aerobic environment. The endosymbiotic aerobic bacterium eventually would develop into the mitochondrion. Similarly, symbiotic associations with cyanobacteria could lead to the formation of chloroplasts and photosynthetic eucaryotes. Cilia and flagella might have arisen from the attachment of spirochete bacteria (*see chapter 20*) to the surface of eucaryotic cells, much as spirochetes attach themselves to the surface of the motile protozoan *Myxotricha paradoxa* that grows in the digestive tract of termites.

There is evidence to support the endosymbiotic theory. Both mitochondria and chloroplasts resemble bacteria in size and appearance, contain DNA in the form of a closed circle like that of bacteria, and reproduce semiautonomously. Mitochondrial and chloroplast ribosomes resemble procaryotic ribosomes more closely than those in the eucaryotic cytoplasmic matrix. The sequences of the chloroplast and mitochondrial genes for ribosomal RNA and transfer RNA are more similar to bacterial gene sequences than to those of eucaryotic rRNA and tRNA nuclear genes. Finally, there are symbiotic associations that appear to be bacterial endosymbioses in which distinctive procaryotic characteristics are being lost. For example, the protozoan flagellate *Cyanophora paradoxa* has photosynthetic organelles called cyanelles with a structure similar to that of cyanobacteria and the remains of peptidoglycan in their walls. Their DNA is much smaller than that of cyanobacteria and resembles chloroplast DNA. Despite such evidence, the endosymbiotic theory still is somewhat speculative and the center of much continuing research and discussion.

to convert CO_2 and water to carbohydrates and O_2. That is, they are the site of photosynthesis.

Although chloroplasts are quite variable in size and shape, they share many structural features. Most often they are oval with dimensions of 2 to 4 μm by 5 to 10 μm, but some algae possess one huge chloroplast that fills much of the cell. Like mitochondria, chloroplasts are encompassed by two membranes (figure 4.15). A matrix, the **stroma,** lies within the inner membrane. It contains DNA, ribosomes, lipid droplets, starch granules, and a complex internal membrane system whose most prominent components are flattened, membrane-delimited sacs, the **thylakoids.** Clusters of two or more thylakoids are dispersed within the stroma of most algal chloroplasts (figures 4.2, 4.15, and 4.16). In some groups of algae, several disklike thylakoids are stacked on each other like coins to form **grana** (s., **granum**).

Photosynthetic reactions are separated structurally in the chloroplast just as electron transport and the tricarboxylic acid cycle are in the mitochondrion. The formation of carbohydrate from CO_2 and water, the dark reaction, takes place in the stroma. The trapping of light energy to generate ATP, NADPH, and O_2, the light reaction, is located in the thylakoid membranes, where chlorophyll and electron transport components are also found.

Photosynthesis (pp. 181–88).

The chloroplasts of many algae contain a **pyrenoid** (figures 4.2 and 4.16), a dense region of protein surrounded by starch or another polysaccharide. Pyrenoids participate in polysaccharide synthesis.

1. Describe in detail the structure of mitochondria and chloroplasts. Where are the different components of these organelles' energy trapping systems located?
2. Define F_1 particle, plastid, dark reaction, light reaction, and pyrenoid.
3. What is the role of mitochondrial DNA?

The Nucleus and Cell Division

The cell **nucleus** is by far the most visually prominent organelle. It was discovered early in the study of cell structure and was shown by Robert Brown in 1831 to be a constant feature of eucaryotic cells. The nucleus is the repository for the cell's genetic information and is its control center.

Nuclear Structure

Nuclei are membrane-delimited spherical bodies about 5 to 7 μm in diameter (figures 4.2, and 4.16). Dense fibrous material called **chromatin** can be seen within the nucleoplasm of the nucleus of a stained cell. This is the DNA-containing part of the nucleus. In nondividing cells, chromatin exists in a dispersed condition, but condenses during mitosis to become visible as **chromosomes.** Some nuclear chromatin, the euchromatin, is loosely organized and contains those genes that are expressing themselves actively. Heterochromatin is coiled

(a)

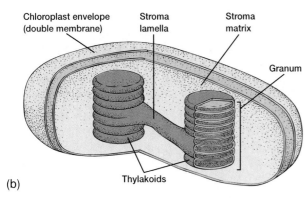

(b)

Figure 4.15   Chloroplast Structure. (*a*) The chloroplast (Chl), of the euglenoid flagellate *Colacium cyclopicolum.* The chloroplast is bounded by a double membrane and has its thylakoids in groups of three or more. A paramylon granule (P), lipid droplets (L), and the pellicular strips (Pe), can be seen (×40,000). (*b*) A diagram of chloroplast structure.

more tightly, appears darker in the electron microscope, and is not genetically active most of the time.

Organization of DNA in eucaryotic nuclei (pp. 214–16).

The nucleus is bounded by the **nuclear envelope** (figures 4.2 and 4.16), a complex structure consisting of inner and outer membranes separated by a 15 to 75 nm perinuclear space. The envelope is continuous with the ER at several points and its outer membrane is covered with ribosomes. A network of intermediate filaments, called the nuclear lamina, lies against the inner surface of the envelope and supports it. Chromatin usually is associated with the inner membrane.

Figure 4.16   Algal Cell Structure. The nucleus and other organelles of the green alga *Dunaliella primolecta.* Note the following nuclear structures: heterochromatin, chr; nucleolus, nu; nuclear envelope, ne; and nuclear pores, np. Many other important organelles may be seen: dictyosome, D; chloroplast, chl; lipid inclusion, li´ mitochondrion, m; pyrenoid, p; plasma membrane, pm; starch granule, st; thylakoids, th; thylakoids penetrating pyrenoid, th´ and vacuole, v (×30,400). Bar = 0.5 μm.

Many **nuclear pores** penetrate the envelope (figures 4.16 and 4.17), each pore formed by a fusion of the outer and inner membranes. Pores are about 70 nm in diameter and collectively occupy about 10 to 25% of the nuclear surface. A complex ring-like arrangement of granular and fibrous material called the annulus is located at the edge of each pore.

The nuclear pores serve as a transport route between the nucleus and surrounding cytoplasm. Particles have been observed moving into the nucleus through the pores. Although the function of the annulus is not understood, it may either regulate or aid the movement of material through the pores. Substances also move directly through the nuclear envelope by unknown mechanisms.

The Nucleolus

Often the most noticeable structure within the nucleus is the **nucleolus** (figures 4.2 and 4.16). A nucleus may contain from one to many nucleoli. Although the nucleolus is not membrane-enclosed, it is a complex organelle with separate granular and fibrillar regions. It is present in nondividing cells, but frequently disappears during mitosis. After mitosis the nucleolus reforms around the nucleolar organizer, a particular part of a specific chromosome.

The nucleolus plays a major role in ribosome synthesis. The nucleolar organizer DNA directs the production of **ribosomal RNA (rRNA).** This RNA is synthesized in a single long piece that then is cut to form the final rRNA molecules. The processed rRNAs next combine with ribosomal proteins (which have been synthesized in the cytoplasmic matrix) to form partially completed ribosomal subunits. The granules

seen in the nucleolus are probably these subunits. Immature ribosomal subunits then leave the nucleus, presumably by way of the nuclear envelope pores and mature in the cytoplasm.

RNA splicing (p. 224).

Mitosis and Meiosis

When a eucaryotic microorganism reproduces itself, its genetic material must be duplicated and then separated so that each new nucleus possesses a complete set of chromosomes. This process of nuclear division and chromosome distribution in eucaryotic cells is called **mitosis.** Mitosis actually occupies only a small portion of a microorganism's life as can be seen by examining the **cell cycle** (figure 4.18). The cell cycle is the total sequence of events in the growth-division cycle between the end of one division and the end of the next. Cell growth takes place in the **interphase,** that portion of the cycle between periods of mitosis. Interphase is composed of three parts. The G_1 period (gap 1 period) is a time of active synthesis of RNA, ribosomes, and other cytoplasmic constituents accompanied by considerable cell growth. This is followed by the S period (synthesis period) in which DNA is replicated and doubles in quantity. Finally, there is a second gap, the G_2 period, when the cell prepares for mitosis, the M period, by activities such as the synthesis of special division proteins. The total length of the cycle differs considerably between microorganisms, usually due to variations in the length of G_1.

Mitotic events are summarized in figure 4.19. During mitosis, the genetic material duplicated during the S period is distributed equally to the two new nuclei so that each has a full set of genes. Mitosis in eucaryotic microorganisms can differ from that pictured in figure 4.19. For example, the nuclear envelope

Figure 4.17   The Nucleus. A freeze-etch preparation of the conidium of the fungus *Geotrichum candidum* (×44,600). Note the large convex nuclear surface with nuclear pores scattered over it.

Figure 4.18   The Eucaryotic Cell Cycle. The length of the M period has been increased disproportionately in order to show the phases of mitosis. G_1 period: synthesis of mRNA, tRNA, ribosomes, and cytoplasmic constituents. Nucleolus grows rapidly. S period: rapid synthesis and doubling of nuclear DNA and histones. G_2 period: preparation for mitosis and cell division. M period: mitosis (prophase, metaphase, anaphase, telophase) and cytokinesis.

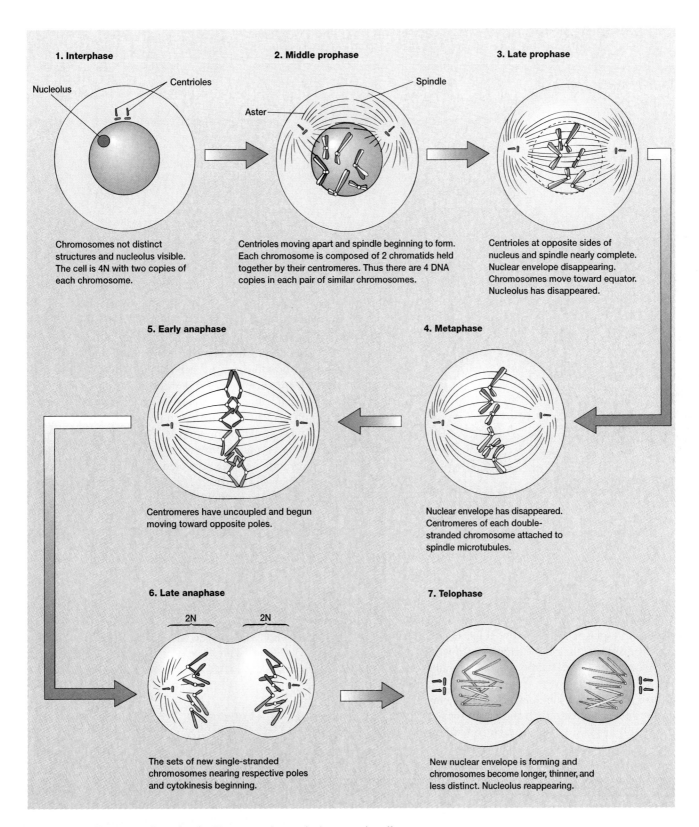

1. Interphase

Chromosomes not distinct structures and nucleolus visible. The cell is 4N with two copies of each chromosome.

2. Middle prophase

Centrioles moving apart and spindle beginning to form. Each chromosome is composed of 2 chromatids held together by their centromeres. Thus there are 4 DNA copies in each pair of similar chromosomes.

3. Late prophase

Centrioles at opposite sides of nucleus and spindle nearly complete. Nuclear envelope disappearing. Chromosomes move toward equator. Nucleolus has disappeared.

5. Early anaphase

Centromeres have uncoupled and begun moving toward opposite poles.

4. Metaphase

Nuclear envelope has disappeared. Centromeres of each double-stranded chromosome attached to spindle microtubules.

6. Late anaphase

The sets of new single-stranded chromosomes nearing respective poles and cytokinesis beginning.

7. Telophase

New nuclear envelope is forming and chromosomes become longer, thinner, and less distinct. Nucleolus reappearing.

Figure 4.19   Mitosis and Cytokinesis. The process in a typical eucaryotic cell.

Figure 4.20 **Mitosis with an Intact Nuclear Envelope.** Mitosis in the slime mold *Physarum flavicomum.* The nuclear envelope, NE, remains intact, and the spindle is intranuclear. The process is at metaphase with the chromosomes, Chr, aligned in the center and attached to spindle fibers, SF (×15,000).

does not disappear in many fungi and some protozoa and algae (figure 4.20). Frequently **cytokinesis,** the division of the parental cell's cytoplasm to form new cells, begins during anaphase and finishes by the end of telophase. However, mitosis can take place without cytokinesis to generate multinucleate or coenocytic cells.

In mitosis the original number of chromosomes is the same after division and a diploid organism will remain **diploid** or 2N (i.e., it still has two copies of each chromosome). Frequently a microorganism reduces its chromosome number by half, from the diploid state to the **haploid** or 1N (a single copy of each chromosome). Haploid cells may immediately act as gametes and fuse to reform diploid organisms or may form gametes only after a considerable delay (figure 4.21). The process by which the number of chromosomes is reduced in half with each daughter cell receiving one complete set of chromosomes is called **meiosis.**

Life cycles of eucaryotic microorganisms (chapters 25–27).

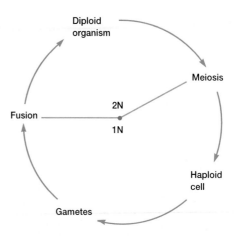

Figure 4.21   Generalized Eucaryotic Life Cycle.

Meiosis is quite complex and involves two stages (figure 4.22). The first stage differs markedly from mitosis. During prophase, homologous chromosomes come together and lie side-by-side, a process known as synapsis. Then the double-stranded chromosomes from each homologous pair move to opposite poles in anaphase. In contrast, during mitotic anaphase the two strands of each chromosome separate and move to opposite poles. Consequently the number of chromosomes is halved in meiosis but not in mitosis. The second stage of meiosis is similar to mitosis in terms of mechanics, and single-stranded chromosomes are separated. After completion of meiosis I and meiosis II, the original diploid cell has been transformed into four haploid cells.

1. Describe the structure of the nucleus. What are euchromatin and heterochromatin? What is the role of the pores in the nuclear envelope?
2. Briefly discuss the structure and function of the nucleolus. What is the nucleolar organizer?
3. Describe the eucaryotic cell cycle, its periods, and the process of mitosis. What is meiosis, how does it take place, and what is its role in the microbial life cycle?

External Cell Coverings

Eucaryotic microorganisms differ greatly from procaryotes in the supporting or protective structures they have external to the plasma membrane. In contrast with most bacteria, many eucaryotes lack an external cell wall. The amoeba is an excellent example. Eucaryotic cell membranes, unlike most procaryotic membranes, contain sterols such as cholesterol, and this may make them mechanically stronger, thus reducing the need for external support. (However, as mentioned on page 42, many procaryotic membranes are strengthened by hopanoids.) Of course many eucaryotes do have a rigid external **cell wall.** Algal

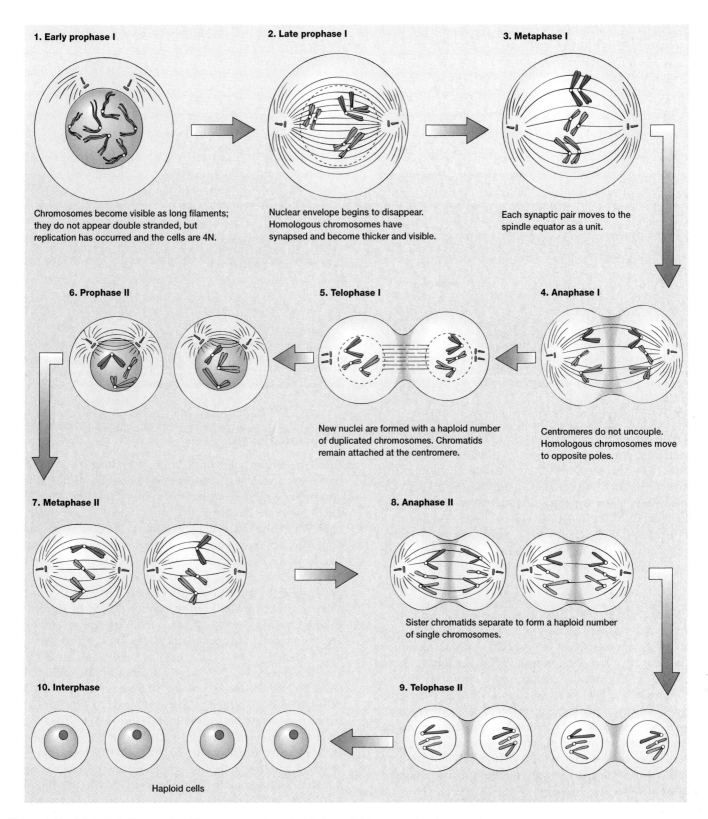

Figure 4.22 **Meiosis in Eucaryotes.** There are a number of variations of this process in microorganisms.

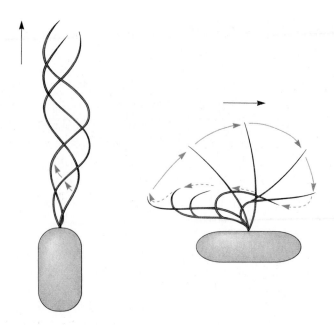

Figure 4.23   Patterns of Flagellar Movement. Flagellar movement (left illustration) often takes the form of waves that move either from the base of the flagellum to its tip or in the opposite direction. The motion of these waves propels the organism along. The beat of a cilium (right illustration) may be divided into two phases. In the effective stroke, the cilium remains fairly stiff as it swings through the water. This is followed by a recovery stroke in which the cilium bends and returns to its initial position. The black arrows indicate the direction of water movement in these examples.

cell walls usually have a layered appearance and contain large quantities of polysaccharides such as cellulose and pectin. In addition, inorganic substances like silica (in diatoms) or calcium carbonate (some red algae) may be present. Fungal cell walls normally are rigid. Their exact composition varies with the organism; but usually, cellulose, chitin, or glucan (a glucose polymer different from cellulose) are present. Despite their nature the rigid materials in eucaryotic walls are chemically simpler than procaryotic peptidoglycan.

Bacterial cell wall structure and chemistry (pp. 51–56).

Many protozoa and some algae have a different external structure, the **pellicle** (figure 4.15*a*). This is a relatively rigid layer of components just beneath the plasma membrane (sometimes the plasma membrane is also considered part of the pellicle). The pellicle may be fairly simple in structure. For example, *Euglena spirogyra* has a series of overlapping strips with a ridge at the edge of each strip fitting into a groove on the adjacent one. In contrast, ciliate protozoan pellicles are exceptionally complex with two membranes and a variety of associated structures. Although pellicles are not as strong and rigid as cell walls, they do give their possessors a characteristic shape.

Cilia and Flagella

Cilia (s., **cilium**) and **flagella** (s., **flagellum**) are the most prominent organelles associated with motility. Although both are whiplike and beat to move the microorganism along, they differ from one another in two ways. First, cilia are typically only 5 to 20 μm in length, whereas flagella are 100 to 200 μm long. Second, their patterns of movement are usually distinctive (figure 4.23). Flagella move in an undulating fashion and generate planar or helical waves originating at either the base or the tip. If the wave moves from base to tip, the cell is pushed along; a beat traveling from the tip toward the base pulls the cell through the water. Sometimes the flagellum will have lateral hairs called flimmer filaments (thicker, stiffer hairs are called mastigonemes). These filaments change flagellar action so that a wave moving down the filament toward the tip pulls the cell along instead of pushing it. Such a flagellum often is called a tinsel flagellum, whereas the naked flagellum is referred to as a whiplash flagellum (figure 4.24). Cilia, on the other hand, normally have a beat with two distinctive phases. In the effective stroke, the cilium strokes through the surrounding fluid like an oar, thereby propelling the organism along in the water. The cilium next bends along its length while it is

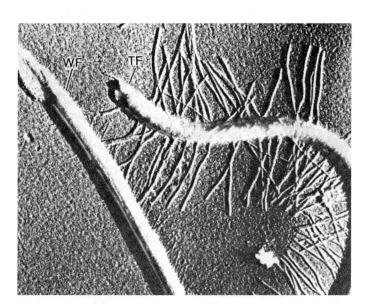

Figure 4.24   Whiplash and Tinsel Flagella. Transmission electron micrograph of a shadowed whiplash flagellum, WF, and a tinsel flagellum, TF, with mastigonemes.

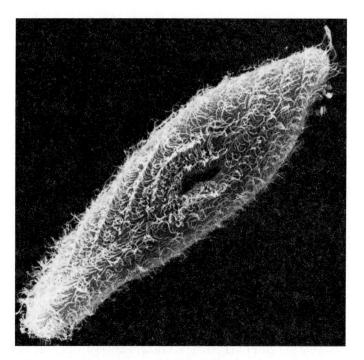

Figure 4.25   Coordination of Ciliary Activity. A scanning electron micrograph of *Paramecium* showing cilia (×1,500). The ciliary beat is coordinated and moves in waves across the protozoan's surface, as can be seen in the photograph.

pulled forward during the recovery stroke in preparation for another effective stroke. A ciliated microorganism actually coordinates the beats so that some of its cilia are in the recovery phase while others are carrying out their effective stroke (figure 4.25). This coordination allows the organism to move smoothly through the water.

Despite their differences, cilia and flagella are very similar in ultrastructure. They are membrane-bound cylinders about 0.2 μm in diameter. Located in the matrix of the organelle is a complex, the **axoneme,** consisting of nine pairs of microtubule doublets arranged in a circle around two central tubules (figures 4.26 and 4.27). This is called the 9 + 2 pattern of microtubules. Each doublet also has pairs of arms projecting from subtubule A (the complete microtubule) toward a neighboring doublet. A radial spoke extends from subtubule A toward the internal pair of microtubules with their central sheath. These microtubules are similar to those found in the cytoplasm. Each is constructed of two types of tubulin subunits, α- and β-tubulins, that resemble the contractile protein actin in their composition.

Bacterial flagella and motility (pp. 60–63).

A **basal body** lies in the cytoplasm at the base of each cilium or flagellum (figure 4.26). It is a short cylinder with nine microtubule triplets around its periphery (a 9 + 0 pattern) and is separated from the rest of the organelle by a basal plate. The basal body directs the construction of these organelles. Cilia and flagella appear to grow through the addition of preformed microtubule subunits at their tips.

Cilia and flagella bend because adjacent microtubule doublets slide along one another while maintaining their individual lengths. The doublet arms (figure 4.27), about 15 nm long, are made of the protein **dynein.** ATP powers the movement of cilia

and flagella, and isolated dynein hydrolyzes ATP. It appears that dynein arms interact with the B subtubules of adjacent doublets to cause the sliding. The radial spokes also participate in this sliding motion.

Cilia and flagella beat at a rate of about 10 to 40 strokes or waves per second and propel microorganisms rapidly. The record holder is the flagellate *Monas stigmatica,* which swims at a rate of 260 μm/second (approximately 40 cell lengths per second); the common euglenoid flagellate, *Euglena gracilis,* travels at around 170 μm or 3 cell lengths per second. The ciliate protozoan *Paramecium caudatum* swims at about 2,700 μm/second (12 lengths per second). Such speeds are equivalent to or much faster than those seen in higher animals.

1. How do eucaryotic microorganisms differ from procaryotes with respect to supporting or protective structures external to the plasma membrane? Describe the pellicle and indicate which microorganisms have one.
2. Prepare and label a diagram showing the detailed structure of a cilium or flagellum. How do cilia and flagella move, and what is dynein's role in the process?

Comparison of Procaryotic and Eucaryotic Cells

A comparison of the cells in figure 4.28 demonstrates that there are many fundamental differences between eucaryotic and procaryotic cells. **Eucaryotic cells** have a membrane-enclosed

(a)   (b)

Figure 4.26   Cilia Ultrastructure. In micrograph (*a*) one can see the basal body, b, and the cilium, cl, itself with its central and peripheral microtubules (×100,000). The basal body is cross-sectioned at one end to show typical microtubule triplets and an absence of central microtubules. (*b*) Cross section of cilia. Note the two central microtubules surrounded by nine microbutule doublets (×160,000).

Figure 4.27   Diagram of Cilia and Flagella Structure. Two doublets have been removed for sake of visibility.

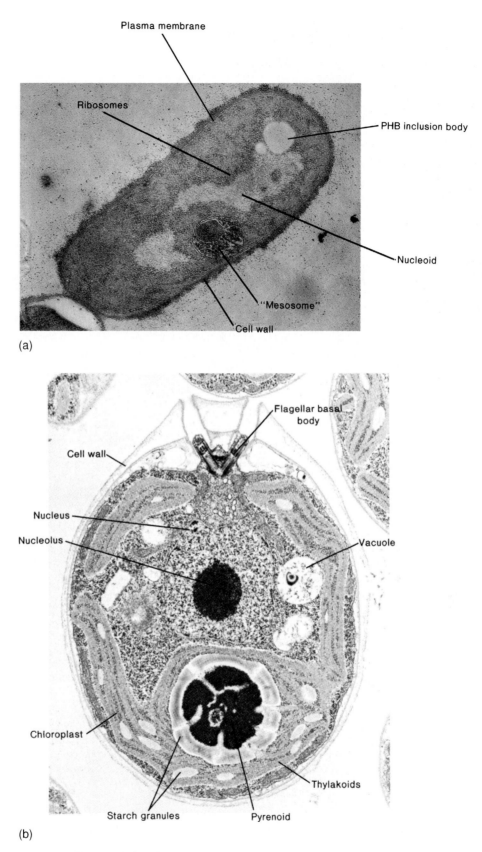

Figure 4.28   Comparison of Procaryotic and Eucaryotic Cell Structure. (*a*) The procaryote *Bacillus megaterium* (×30,500).
(*b*) The eucaryotic alga *Chlamydomonas reinhardtii* (×30,000).

TABLE 4.2   Comparison of Procaryotic and Eucaryotic Cells

Property	Procaryotes	Eucaryotes
Organization of Genetic Material		
True membrane-bound nucleus	Absent	Present
DNA complexed with histones	No	Yes
Number of chromosomes	One[a]	More than one
Introns in genes	Rare	Common
Nucleolus	Absent	Present
Mitosis occurs	No	Yes
Genetic Recombination	Partial, unidirectional transfer of DNA	Meiosis and fusion of gametes
Mitochondria	Absent	Present
Chloroplasts	Absent	Present
Plasma Membrane with Sterols	Usually no[b]	Yes
Flagella	Submicrosopic in size; composed of one fiber	Microscopic in size; membrane bound; usually 20 microtubules in 9 + 2 pattern
Endoplasmic Reticulum	Absent	Present
Golgi Apparatus	Absent	Present
Cell Walls	Usually chemically complex with peptidoglycan[c]	Chemically simpler and lacking peptidoglycan
Differences in Simpler Organelles		
Ribosomes	70S	80S (except in mitochondria and chloroplasts)
Lysosomes and peroxisomes	Absent	Present
Microtubules	Absent or rare	Present
Cytoskeleton	Absent	Present
Differentiation	Rudimentary	Tissues and organs

[a]Plasmids may provide additional genetic information.

[b]Only the mycoplasmas and methanotrophs (methane utilizers) contain sterols. The mycoplasmas cannot synthesize sterols and require them preformed. Many procaryotes contain hopanoids.

[c]The mycoplasmas and archaeobacteria do not have peptidoglycan cell walls.

nucleus. In contrast, **procaryotic cells** lack a true, membrane-delimited nucleus. Bacteria are procaryotes; all other organisms—algae, fungi, protozoa, higher plants, and animals—are eucaryotic. Procaryotes normally are smaller than eucaryotic cells, often about the size of eucaryotic mitochondria and chloroplasts.

The presence of the eucaryotic nucleus is the most obvious difference between these two cell types, but several other major distinctions should be noted. It is clear from table 4.2 that procaryotic cells are much simpler structurally. In particular, an extensive and diverse collection of membrane-delimited organelles is missing. Furthermore, procaryotes are simpler functionally in several ways. They lack mitosis and meiosis, and have a simpler genetic organization. Many complex eucaryotic

processes are absent in procaryotes: phagocytosis and pinocytosis, intracellular digestion, directed cytoplasmic streaming, ameboid movement, and others.

Despite the many significant differences between these two basic cell forms, they are remarkably similar on the biochemical level as discussed in succeeding chapters. Procaryotes and eucaryotes are composed of similar chemical constituents. With a few exceptions the genetic code is the same in both, as is the way in which the genetic information in DNA is expressed. The principles underlying metabolic processes and most of the more important metabolic pathways are identical. Thus beneath the profound structural and functional differences between procaryotes and eucaryotes, there is an even more fundamental unity: a molecular unity that is basic to all life processes.

Summary

1. The eucaryotic cell has a true, membrane-delimited nucleus and many membranous organelles.

2. The cytoplasmic matrix contains microfilaments, intermediate filaments, and microtubules, small organelles partly responsible for cell structure and movement. These and other types of filaments are organized into a cytoskeleton.

3. The matrix is permeated by an irregular network of tubules and flattened sacs or cisternae known as the endoplasmic reticulum (ER). The ER may have attached ribosomes and be active in protein synthesis (rough or granular endoplasmic reticulum) or lack ribosomes (smooth or agranular ER).

4. The ER can donate materials to the Golgi apparatus, an organelle composed of one or more stacks of cisternae. This organelle prepares and packages cell products for secretion.

5. The Golgi apparatus also forms lysosomes. These organelles contain digestive enzymes and aid in intracellular digestion of materials, including those taken up by endocytosis.

6. Eucaryotic ribosomes found free in the cytoplasmic matrix or bound to the ER are 80S ribosomes. Several may be attached to the same messenger RNA forming polyribosomes or polysomes.

7. Mitochondria are organelles bounded by two membranes, with the inner membrane folded into cristae, and are responsible for energy generation by the tricarboxylic acid cycle, electron transport, and oxidative phosphorylation.

8. Chloroplasts are the site of photosynthesis. The trapping of light energy takes place in the thylakoid membranes, whereas CO_2 incorporation is located in the stroma.

9. The nucleus is a large organelle containing the cell's chromosomes. It is bounded by a complex, double-membrane envelope perforated by pores through which materials can move.

10. The nucleolus lies within the nucleus and participates in the synthesis of ribosomal RNA and ribosomal subunits.

11. Eucaryotic chromosomes are distributed to daughter cells during regular cell division by mitosis. Meiosis is used to halve the chromosome number during sexual reproduction.

12. When a cell wall is present, it is constructed from polysaccharides, like cellulose, that are chemically simpler than procaryotic peptidoglycan. Many protozoa have a pellicle rather than a cell wall.

13. Many eucaryotic cells are motile because of cilia and flagella, membrane-delimited organelles with nine microtubule doublets surrounding two central microtubules. The doublets slide along each other to bend the cilium or flagellum.

14. Despite the fact that eucaryotes and procaryotes differ in many ways (table 4.2), they are quite similar metabolically.

Key Terms

autophagic vacuole *79*

axoneme *91*

basal body *91*

cell cycle *86*

cell wall *88*

chloroplast *82*

chromatin *84*

chromosome *84*

cilia *90*

cisternae *77*

cristae *82*

cytokinesis *88*

cytoplasmic matrix *76*

cytoskeleton *77*

dictyosome *79*

diploid *88*

dynein *91*

endocytosis *79*

endoplasmic reticulum (ER) *77*

endosome *79*

endosymbiotic theory *84*

eucaryotic cells *91*

F_1 particle *82*

flagella *90*

Golgi apparatus *78*

grana *84*

haploid *88*

intermediate filament *77*

interphase *86*

lysosome *79*

meiosis *88*

microfilament *76*

microtubule *76*

mitochondrion *81*

mitosis *86*

nuclear envelope *85*

nuclear pores *85*

nucleolus *85*

nucleus *84*

organelle *74*

pellicle *90*

phagocytosis *79*

pinocytosis *79*

plastid *82*

polyribosomes *81*

polysomes *81*

primary lysosomes *79*

procaryotic cells *94*

pyrenoid *84*

residual body *79*

ribosomal RNA (rRNA) *86*

rough or granular ER (RER or GER) *78*

secondary lysosomes *79*

smooth or agranular ER (SER or AER) *78*

stroma *84*

thylakoid *84*

Questions for Thought and Review

1. Describe the structure and function of every eucaryotic organelle discussed in the chapter.

2. Discuss the statement: "The most obvious difference between eucaryotic and procaryotic cells is in their use of membranes." What general roles do membranes play in eucaryotic cells?

3. Describe how the Golgi apparatus distributes proteins it receives from the ER to different organelles.

4. Briefly discuss how the complex of membranous organelles that de Duve calls the "vacuome" functions as a coordinated whole. What is its function?

5. Describe and contrast the ways in which flagella and cilia propel microorganisms through the water.

6. Outline the major differences between procaryotes and eucaryotes. In what way are they similar?

Additional Reading

General

Alberts, B.; Bray, D.; Lewis, J.; Raff, M.; Roberts, K.; and Watson, J. D. 1994. *Molecular biology of the cell,* 3d ed. New York: Garland Publishing.

Becker, W. M., and Deamer, D. W. 1991. *The world of the cell,* 2d ed. Redwood City, Calif.: Benjamin/Cummings.

de Duve, C. 1985. *A guided tour of the living cell.* New York: Scientific American Books.

Gray, M. W. 1983. The bacterial ancestry of plastids and mitochondria. *BioScience* 33(11):693–99.

Lodish, H.; Baltimore, D.; Berk, A.; Zipursky, S.L.; Matsudaira, P.; and Darnell, J. 1995. *Molecular cell biology,* 3d ed. New York: Scientific American Books.

Margulis, L. 1971. Symbiosis and evolution. *Sci. Am.* 225(2):49–57.

Wolfe, S. L. 1993. *Molecular and cellular biology.* Belmont, Calif.: Wadsworth.

The Cytoplasmic Matrix

Bretscher, A.; Drees, B.; Harsay, E.; Schott, D.; and Wang, T. 1994. What are the basic functions of microfilaments? Insights from studies in budding yeast. *J. Cell Biology* 126(4):821–25.

Porter, K. R., and Tucker, J. B. 1981. The ground substance of the living cell. *Sci. Am.* 244(3):57–67.

Pumplin, D. W., and Bloch, R. J. 1993. The membrane skeleton. *Trends Cell Biol.* 3:113–17.

Stossel, T. P. 1994. The machinery of cell crawling. *Sci. Am.* 271(3):54–63.

The Golgi Apparatus

Rothman, J. E. 1985. The compartmental organization of the Golgi apparatus. *Sci. Am.* 253(3):74–89.

Lysosomes and Endocytosis

Dautry-Varsat, A., and Lodish, H. F. 1984. How receptors bring proteins and particles into cells. *Sci. Am.* 250(5):52–8.

Helenius, A.; Mellman, I.; Wall, D.; and Hubbard, A. 1983. Endosomes. *Trends Biochem. Sci.* 8(7):245–50.

Holtzman, E. 1989. *Lysosomes.* New York: Academic Press.

Eucaryotic Ribosomes

Craig, E. A.; Gambill, B. D.; and Nelson, R. J. 1993. Heat shock proteins: Molecular chaperones of protein biosynthesis. *Microbiol. Rev.* 57(2):402–14.

Lake, J. A. 1985. Evolving ribosome structure: Domains in archaebacteria, eubacteria, eocytes, and eukaryotes. *Ann. Rev. Biochem.* 54:507–30.

Welch, W. J. 1993. How cells respond to stress. *Sci. Am.* 268(5):56–64.

The Nucleus and Cell Division

Glover, D. M.; Gonzalez, C.; and Raff, J. W. 1993. The centrosome. *Sci. Am.* 268(6):62–8.

Heywood, P., and Magee, P. T. 1976. Meiosis in protists. *Bacteriol. Rev.* 40:190–240.

McIntosh, J. R., and McDonald, K. L. 1989. The mitotic spindle. *Sci. Am.* 261(4):48–56.

Murray, A., and Hunt, T. 1993. *The cell cycle: An introduction.* New York: W. H. Freeman.

Newport, J. W., and Forbes, D. J. 1987. The nucleus: Structure, function, and dynamics. *Ann. Rev. Biochem.* 56:535–65.

Spector, D. L. 1993. Macromolecular domains within the cell nucleus. *Ann. Rev. Cell Biol.* 9:265–315.

Cilia and Flagella

Satir, P. 1983. Cilia and related organelles. *Carolina Biology Reader,* no. 123. Burlington, N.C.: Carolina Biological Supply Co.

PART TWO
Microbial Nutrition, Growth, and Control

Culture media are used to grow bacteria for many
purposes. In this picture, agar plates are being
inoculated to cultivate bacteria in a search for new
antibiotics. After inoculation, pellets of experimental
drugs are placed on the agar and the cultures are
incubated. The bacteria grow and form an opaque
lawn of cells over most of the plate. An effective
antibiotic kills the bacteria or inhibits their growth
and produces a wide, clear zone around the pellet.

CHAPTER 5
Microbial Nutrition

The whole of nature, as has been said, is a conjugation of the verb to eat, in the active and passive.

—William Ralph Inge

Concepts

1. Microorganisms require about 10 elements in large quantities, in part because they are used to construct carbohydrates, lipids, proteins, and nucleic acids. Several other elements are needed in very small amounts and are parts of enzymes and cofactors.

2. All microorganisms can be placed in one of a few nutritional categories on the basis of their requirements for carbon, energy, and hydrogen atoms or electrons.

3. Nutrient molecules frequently cannot cross selectively permeable plasma membranes through passive diffusion. They must be transported by one of three major mechanisms involving the use of membrane carrier proteins. Eucaryotic microorganisms also employ endocytosis for nutrient uptake.

4. Culture media are needed to grow microorganisms in the laboratory and to carry out specialized procedures like microbial identification, water and food analysis, and the isolation of particular microorganisms. A wide variety of media is available for these and other purposes.

5. Pure cultures can be obtained through the use of spread plates, streak plates, or pour plates and are required for the careful study of an individual microbial species.

To obtain energy and construct new cellular components, organisms must have a supply of raw materials or nutrients. **Nutrients** are substances used in biosynthesis and energy production and therefore are required for microbial growth. This chapter describes the nutritional requirements of microorganisms, how nutrients are acquired, and the cultivation of microorganisms.

Environmental factors such as temperature, oxygen levels, and the osmotic concentration of the medium are critical in the successful cultivation of microorganisms. These topics are discussed in chapter 6 after an introduction to microbial growth.

The Common Nutrient Requirements

Analysis of microbial cell composition shows that over 95% of cell dry weight is made up of a few major elements: carbon, oxygen, hydrogen, nitrogen, sulfur, phosphorus, potassium, calcium, magnesium, and iron. These are called **macroelements** or macronutrients because they are required by microorganisms in relatively large amounts. The first six (C, O, H, N, S, and P) are components of carbohydrates, lipids, proteins, and nucleic acids. The remaining four macroelements exist in the cell as cations and play a variety of roles. For example, potassium (K^+) is required for activity by a number of enzymes, including some of those involved in protein synthesis. Calcium (Ca^{2+}), among other functions, contributes to the heat resistance of bacterial endospores. Magnesium (Mg^{2+}) serves as a cofactor for many enzymes, complexes with ATP, and stabilizes ribosomes and cell membranes. Iron (Fe^{2+} and Fe^{3+}) is a part of cytochromes and a cofactor for enzymes and electron-carrying proteins.

All microorganisms require several **trace elements** (also called microelements or micronutrients) besides macroelements. The trace elements—manganese, zinc, cobalt, molybdenum, nickel, and copper—are needed by most cells. However, cells require such small amounts that contaminants in water, glassware, and regular media components often are adequate for growth. Therefore it is very difficult to demonstrate a trace element requirement. Trace elements are normally a part of enzymes and cofactors, and they aid in the catalysis of reactions and maintenance of protein structure. For example, zinc (Zn^{2+}) is present at the active site of some enzymes but is also involved in the association of regulatory and catalytic subunits in *E. coli* aspartate carbamoyltransferase (*see chapter 12*). Manganese (Mn^{2+}) aids many enzymes catalyzing the transfer of phosphate groups. Molybdenum (Mo^{2+}) is required for nitrogen fixation, and cobalt (Co^{2+}) is a component of vitamin B_{12}.

Electron carriers and enzymes (pp. 156–162).

Elements can be categorized somewhat differently with respect to nutritional requirements. The major elements (C, O, H, N, S, P) are needed in gram quantities in a liter of culture medium. The minor elements (K, Ca, Mg, Fe) often are required in milligram quantities. Trace elements (Mn, Zn, Co, Mo, Ni, Cu) must be available in microgram amounts.

Besides the common macroelements and trace elements, microorganisms may have particular requirements that reflect the special nature of their morphology or environment. Diatoms (*see figure 26.6 c,d*) need silicic acid (H_4SiO_4) to construct their beautiful cell walls of silica [$(SiO_2)_n$]. Although most bacteria do not require large amounts of sodium, many bacteria growing in saline lakes and oceans (*see pp. 125, 485–86*) depend on the presence of high concentrations of sodium ion (Na^+).

Requirements for Carbon, Hydrogen, and Oxygen

The requirements for carbon, hydrogen, and oxygen often are satisfied together. Carbon is required for the skeleton or backbone of all organic molecules, and molecules serving as carbon sources usually also contribute both oxygen and hydrogen atoms. One carbon source for which this is not true is carbon dioxide (CO_2) because it is oxidized and lacks hydrogen. Probably all microorganisms can fix CO_2—that is, reduce it and incorporate it into organic molecules. However, by definition, only **autotrophs** can use CO_2 as their sole or principal source of carbon. Many microorganisms are autotrophic; many of these are photosynthetic, but some autotrophs oxidize inorganic molecules to obtain energy.

Photosynthetic carbon dioxide fixation (pp. 193–96).

The reduction of CO_2 is a very energy expensive process. Thus many microorganisms cannot use CO_2 as their sole carbon source but must rely on the presence of more reduced, complex molecules for a supply of carbon. Organisms that use reduced, preformed organic molecules as carbon sources are **heterotrophs** (these preformed molecules normally come from other organisms). Most heterotrophs use organic nutrients as a source of both carbon and energy. For example, the glycolytic pathway traps energy as ATP and NADH and also produces carbon skeletons for use in biosynthesis.

The glycolytic pathway (pp. 166–67).

A most remarkable nutritional characteristic of microorganisms is their extraordinary flexibility with respect to carbon sources. There is no naturally occurring organic molecule that cannot be used by some microorganism. Actinomycetes can degrade amyl alcohol, paraffin, and even rubber. Some bacteria seem able to employ almost anything as a carbon source; for example, *Pseudomonas cepacia* can use over 100 different carbon compounds. Unfortunately many man-made substances like plastics and DDT are degraded slowly or not at all. In contrast to bacterial omnivores, some bacteria are exceedingly fastidious and catabolize few carbon compounds. Methylotrophic bacteria metabolize only methane, methanol, carbon monoxide, formic acid, and a few related one-carbon molecules. Parasitic members of the genus *Leptospira* use only long-chain fatty acids as their major source of carbon and energy.

The nutritional requirements of microorganisms vary enormously among different species. In addition, these requirements can change within a species because of mutations. A microorganism requiring the same nutrients as most naturally occurring members of its species is a **prototroph.** A prototrophic microorganism may mutate so that it cannot synthesize a molecule essential for growth and reproduction. It will then require that molecule, or a compound that can be converted to it, as a nutrient. A mutated microorganism that lacks the ability to synthesize an essential nutrient and therefore must obtain it or a precursor from the surroundings is an **auxotroph.** The requirement for a specific amino acid is a common form of auxotrophy. Many microorganisms can synthesize all the common amino acids needed for growth. Occasionally a mutation will block the synthesis of an essential amino acid and the microorganism becomes auxotrophic for it—that is, the amino acid must be available for growth to take place. The production of auxotrophs is useful in the study of microbial genetics.

Genetic mutations (pp. 261–70).

Nutritional Types of Microorganisms

All organisms also require sources of energy, hydrogen, and electrons for growth to take place. Microorganisms can be grouped into nutritional classes based on how they satisfy these requirements (tables 5.1 and 5.2). There are only two sources of energy available to organisms: (1) light energy trapped during photosynthesis, and (2) the energy derived from oxidizing organic or inorganic molecules. **Phototrophs** use light as their energy source; **chemotrophs** obtain energy from the oxidation of chemical compounds (either organic or inorganic). Microorganisms also have only two sources for hydrogen atoms or electrons. **Lithotrophs** (that is, "rock-eaters") use reduced inorganic substances as their electron source, whereas **organotrophs** extract electrons or hydrogen from organic compounds.

Photosynthesis light reactions (pp. 181–88).
Oxidation of organic and inorganic molecules (pp. 166–81).

Despite the great metabolic diversity seen in microorganisms, most may be placed in one of four nutritional classes based on their primary sources of energy, hydrogen and/or electrons, and carbon (table 5.2). The large majority of microorganisms thus far studied are either photolithotrophic autotrophs or chemoorganotrophic heterotrophs. **Photolithotrophic autotrophs** (often called **photoautotrophs**) use light energy and CO_2 as a carbon source. Eucaryotic algae and blue-green bacteria (cyanobacteria) employ water as the electron donor and release oxygen. Purple and green sulfur bacteria cannot oxidize water but extract electrons from inorganic donors like hydrogen, hydrogen sulfide, and elemental sulfur. **Chemoorganotrophic heterotrophs** (often called **chemoheterotrophs,** or even **heterotrophs**) use organic compounds as sources of energy, hydrogen, electrons, and carbon for biosynthesis. Frequently the same organic nutrient will satisfy all these requirements. It should be noted that essentially all pathogenic microorganisms are chemoheterotrophs. The other two classes have fewer microorganisms

---

TABLE 5.1   Sources of Carbon, Energy, and Hydrogen/Electrons

Carbon Sources

Autotrophs	CO_2 sole or principal biosynthetic carbon source (*pp. 193–96*)[a]
Heterotrophs	Reduced, preformed, organic molecules from other organisms (*chapters 9 and 10*)

Energy Sources

Phototrophs	Light (*pp. 181–88*)
Chemotrophs	Oxidation of organic or inorganic compounds (*chapter 9*)

Hydrogen or Electron Sources

Lithotrophs	Reduced inorganic molecules (*pp. 180–81*)
Organotrophs	Organic molecules (*chapter 9*)

[a]For each category, the location of material describing the participating metabolic pathways is given within the parentheses.

TABLE 5.2   Major Nutritional Types of Microorganisms

Major Nutritional Types[a]	Sources of Energy, Hydrogen/Electrons, and Carbon	Representative Microorganisms
Photolithotrophic autotrophy	Light energy Inorganic hydrogen/electron (H/e⁻) donor CO_2 carbon source	Algae Purple and green sulfur bacteria Blue-green bacteria (cyanobacteria)
Photoorganotrophic heterotrophy	Light energy Organic H/e⁻ donor Organic carbon source (CO_2 may also be used)	Purple nonsulfur bacteria Green nonsulfur bacteria
Chemolithotrophic autotrophy	Chemical energy source (inorganic) Inorganic H/e⁻ donor CO_2 carbon source	Sulfur-oxidizing bacteria Hydrogen bacteria Nitrifying bacteria Iron bacteria
Chemoorganotrophic heterotrophy	Chemical energy source (organic) Organic H/e⁻ donor Organic carbon source	Protozoa Fungi Most nonphotosynthetic bacteria (including most pathogens)

[a]Bacteria in other nutritional categories have been found. The categories are defined in terms of energy, electron, and carbon sources.

but often are very important ecologically. Some purple and green bacteria are photosynthetic and use organic matter as their electron donor and carbon source. These **photoorganotrophic heterotrophs** are common inhabitants of polluted lakes and streams. Some of these bacteria also can grow as photoautotrophs with molecular hydrogen as an electron donor. The fourth group, the **chemolithotrophic autotrophs,** oxidizes reduced inorganic compounds such as iron, nitrogen, or sulfur molecules to derive both energy and electrons for biosynthesis. Carbon dioxide is the carbon source. A few chemolithotrophs can derive their carbon from organic sources and thus are heterotrophic. Bacteria relying on inorganic energy sources and organic (or sometimes CO_2) carbon sources may be called **mixotrophic** (they are combining autotrophic and heterotrophic metabolic processes). Chemolithotrophs contribute greatly to the chemical transformations of elements (e.g., the conversion of ammonia to nitrate or sulfur to sulfate) that continually occur in the ecosystem.

Photosynthetic and chemolithotrophic bacteria (chapter 22).

Although a particular species usually belongs in only one of the four nutritional classes, some show great metabolic flexibility and alter their metabolic patterns in response to environmental changes. For example, many purple nonsulfur bacteria (*see chapter 22*) act as photoorganotrophic heterotrophs in the absence of oxygen but oxidize organic molecules and function chemotrophically at normal oxygen levels. When oxygen is low, photosynthesis and oxidative metabolism may function simultaneously. This sort of flexibility seems complex and confusing, yet it gives its possessor a definite advantage if environmental conditions frequently change.

---

1. What are nutrients, and on what basis are they divided into macroelements and trace elements? Describe some ways in which macronutrients and trace elements are used by an organism.
2. Define autotroph, heterotroph, prototroph, and auxotroph.
3. Discuss the ways in which microorganisms are classified based on their requirements for energy, hydrogen, and electrons. Describe the nutritional requirements of the four major nutritional groups and give some microbial examples of each. What is a mixotroph?

Requirements for Nitrogen, Phosphorus, and Sulfur

To grow, a microorganism must be able to incorporate large quantities of nitrogen, phosphorus, and sulfur. Although these elements may be acquired from the same nutrients that supply carbon, microorganisms usually employ inorganic sources as well.

Biochemical mechanisms for the incorporation of nitrogen, phosphorus, and sulfur (pp. 197–201).

Nitrogen is needed for the synthesis of amino acids, purines, pyrimidines, some carbohydrates and lipids, enzyme cofactors, and other substances. Many microorganisms can use the nitrogen in amino acids, and ammonia often is directly incorporated

through the action of such enzymes as glutamate dehydrogenase or glutamine synthetase and glutamate synthase (*see chapter 10*). Most phototrophs and many nonphotosynthetic microorganisms reduce nitrate to ammonia and incorporate the ammonia in assimilatory nitrate reduction (*pp. 198–99*). A variety of bacteria (e.g., many cyanobacteria and the symbiotic bacterium *Rhizobium*) can reduce and assimilate atmospheric nitrogen using the nitrogenase system (*see chapter 10*).

Phosphorus is present in nucleic acids, phospholipids, nucleotides like ATP, several cofactors, some proteins, and other cell components. Almost all microorganisms use inorganic phosphate as their phosphorus source and incorporate it directly. Some microorganisms (e.g., *E. coli*) actively acquire phosphate from their surroundings. Low phosphate levels actually limit microbial growth in many aquatic environments.

Sulfur is needed for the synthesis of substances like the amino acids cysteine and methionine, some carbohydrates, biotin, and thiamine. Most microorganisms use sulfate as a source of sulfur and reduce it by assimilatory sulfate reduction (*see chapter 10*); a few require a reduced form of sulfur such as cysteine.

Growth Factors

Microorganisms, especially many photolithotrophic autotrophs, often grow and reproduce when minerals and sources of energy, carbon, nitrogen, phosphorus, and sulfur are supplied. These organisms have the enzymes and pathways necessary to synthesize all cell components required for their well-being. Many microorganisms, on the other hand, lack one or more essential enzymes. Therefore they cannot manufacture all indispensable constituents but must obtain them or their precursors from the environment. Organic compounds required because they are essential cell components or precursors of such components and cannot be synthesized by the organism are called **growth factors.** There are three major classes of growth factors: (1) amino acids, (2) purines and pyrimidines, and (3) vitamins. Amino acids are needed for protein synthesis, purines and pyrimidines for nucleic acid synthesis. **Vitamins** are small organic molecules that usually make up all or part of enzyme cofactors (*see chapter 8*), and only very small amounts sustain growth. The functions of selected vitamins, and examples of microorganisms requiring them, are given in table 5.3. Some microorganisms require many vitamins; for example, *Enterococcus faecalis* (a lactic acid bacterium) needs eight different vitamins for growth. Other growth factors are also seen; heme (from hemoglobin or cytochromes) is required by *Haemophilus influenzae,* and some mycoplasmas need cholesterol.

Knowledge of the specific growth factor requirements of many microorganisms makes possible quantitative growth-response assays for a variety of substances. For example, species from the bacterial genera *Lactobacillus* and *Streptococcus* can be used in microbiological assays of most vitamins and amino acids. The appropriate bacterium is grown in a series of culture vessels, each containing medium with an excess amount of all required components except the growth factor to be assayed. A

TABLE 5.3   Functions of Some Common Vitamins in Microorganisms

Vitamin	Functions	Examples of Microorganisms Requiring Vitamin[a]
Biotin	Carboxylation (CO_2 fixation) One-carbon metabolism	*Leuconostoc mesenteroides* (B) *Saccharomyces cerevisiae* (F) *Ochromonas malhamensis* (A) *Acanthamoeba castellanii* (P)
Cyanocobalamin (B_{12})	Molecular rearrangements One-carbon metabolism—carries methyl groups	*Lactobacillus* spp. (B) *Euglena gracilis* (A) Diatoms and many other algae (A) *Acanthamoeba castellanii* (P)
Folic acid	One-carbon metabolism	*Enterococcus faecalis* (B) *Tetrahymena pyriformis* (P)
Lipoic acid	Transfer of acyl groups	*Lactobacillus casei* (B) *Tetrahymena* spp. (P)
Pantothenic acid	Precursor of coenzyme A—carries acyl groups (pyruvate oxidation, fatty acid metabolism)	*Proteus morganii* (B) *Hanseniaspora* spp. (F) *Paramecium* spp. (P)
Pyridoxine (B_6)	Amino acid metabolism (e.g., transamination)	*Lactobacillus* spp. (B) *Tetrahymena pyriformis* (P)
Niacin (nicotinic acid)	Precursor of NAD and NADP—carry electrons and hydrogen atoms	*Brucella abortus, Haemophilus influenzae* (B) *Blastocladia pringsheimi* (F) *Crithidia fasciculata* (P)
Riboflavin (B_2)	Precursor of FAD and FMN—carry electrons or hydrogen atoms	*Caulobacter vibrioides* (B) *Dictyostelium* spp. (F) *Tetrahymena pyriformis* (P)
Thiamine (B_1)	Aldehyde group transfer (pyruvate decarboxylation, α-keto acid oxidation)	*Bacillus anthracis* (B) *Phycomyces blakesleeanus* (F) *Ochromonas malhamensis* (A) *Colpidium campylum* (P)

[a]The representative microorganisms are members of the following groups: bacteria (*B*), fungi (*F*), algae (*A*), and protozoa (*P*).

different amount of growth factor is added to each vessel. The standard curve is prepared by plotting the growth factor quantity or concentration against the total extent of bacterial growth. Ideally the amount of growth resulting is directly proportional to the quantity of growth factor present; if the growth factor concentration doubles, the final extent of bacterial growth doubles. The quantity of the growth factor in a test sample is determined by comparing the extent of growth caused by the unknown sample with that resulting from the standards. Microbiological assays are specific, sensitive, and simple. They still are used in the assay of substances like vitamin B_{12} and biotin, despite advances in chemical assay techniques.

1. Briefly summarize the ways in which microorganisms obtain nitrogen, phosphorus, and sulfur from their environment.
2. What are growth factors? What are vitamins? How can microorganisms be used to determine the quantity of a specific substance in a sample?

Uptake of Nutrients by the Cell

The first step in nutrient use is uptake of the required nutrients by the microbial cell. Uptake mechanisms must be specific—that is, the necessary substances, and not others, must be acquired. It

does a cell no good to take in a substance that it cannot use. Since microorganisms often live in nutrient-poor habitats, they must be able to transport nutrients from dilute solutions into the cell against a concentration gradient. Finally, nutrient molecules must pass through a selectively permeable plasma membrane that will not permit the free passage of most substances. In view of the enormous variety of nutrients and the complexity of the task, it is not surprising that microorganisms make use of several different transport mechanisms. The most important of these are facilitated diffusion, active transport, and group translocation. Eucaryotic microorganisms do not appear to employ group translocation but take up nutrients by the process of endocytosis (*see chapter 4*).

Plasma membrane structure and properties (pp. 42–44).

Facilitated Diffusion

A few substances, such as glycerol, can cross the plasma membrane by **passive diffusion.** Passive diffusion, often simply called diffusion, is the process in which molecules move from a region of higher concentration to one of lower concentration because of random thermal agitation. The rate of passive diffusion is dependent on the size of the concentration gradient between a cell's exterior and its interior (figure 5.1). A fairly large concentration gradient is required for adequate nutrient uptake by

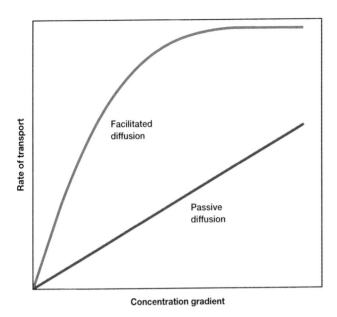

Figure 5.1   Passive and Facilitated Diffusion. The dependence of diffusion rate on the size of the solute's concentration gradient. Note the saturation effect or plateau above a specific gradient value when a facilitated diffusion carrier is operating.

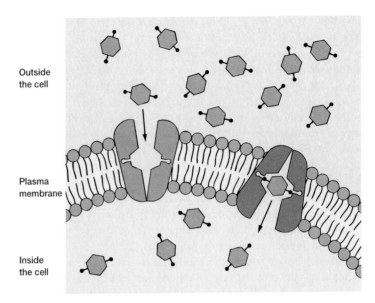

Figure 5.2   A Model of Facilitated Diffusion. The membrane carrier can change conformation after binding an external molecule and subsequently release the molecule on the cell interior. It then returns to the outward oriented position and is ready to bind another solute molecule. Because there is no energy input, molecules will continue to enter only as long as their concentration is greater on the outside.

passive diffusion (i.e., the external nutrient concentration must be high), and the rate of uptake decreases as more nutrient is acquired unless it is used immediately. Thus passive diffusion is an inefficient process and is not employed extensively by microorganisms. Although glycerol can enter cells by passive diffusion, another transport mechanism also operates. Very small molecules such as H_2O, O_2, and CO_2 often move across membranes by passive diffusion.

The rate of diffusion across selectively permeable membranes is greatly increased by using carrier proteins, sometimes called **permeases,** which are embedded in the plasma membrane. Because a carrier aids the diffusion process, it is called **facilitated diffusion.** The rate of facilitated diffusion increases with the concentration gradient much more rapidly and at lower concentrations of the diffusing molecule than that of passive diffusion (figure 5.1). Note that the diffusion rate levels off or reaches a plateau above a specific gradient value because the carrier is saturated—that is, the carrier protein is binding and transporting as many solute molecules as possible. The resulting curve resembles an enzyme-substrate curve (*see chapter 8*) and is different from the linear response seen with passive diffusion. Carrier proteins also resemble enzymes in their specificity for the substance to be transported; each carrier is selective and will transport only closely related solutes. Although a carrier protein is involved, facilitated diffusion is truly diffusion. A concentration gradient spanning the membrane drives the movement of molecules, and no extra energy input is required. If the concentration gradient disappears, net inward movement ceases.

Although much work has been done on the mechanism of facilitated diffusion, the process is not yet understood completely. It appears that the carrier protein complex spans the

membrane (figure 5.2). After the solute molecule binds to the outside, the carrier may change conformation and release the molecule on the cell interior. The carrier would subsequently change back to its original shape and be ready to pick up another molecule. The net effect is that a lipid-insoluble molecule can enter the cell in response to its concentration gradient. Remember that the mechanism is reversible; if the solute's concentration is greater inside the cell, it will move outward. Because the cell metabolizes nutrients upon entry, influx is favored.

Facilitated diffusion does not seem to be important in procaryotes. Glycerol is transported by facilitated diffusion in *E. coli, Salmonella typhimurium, Pseudomonas, Bacillus,* and many other bacteria. The process is much more prominent in eucaryotic cells where it is used to transport a variety of sugars and amino acids.

Active Transport

Although facilitated diffusion carriers can efficiently move molecules to the interior when the solute concentration is higher on the outside, they cannot take up solutes that are already more concentrated within the cell (i.e., against a concentration gradient). Microorganisms often live in habitats characterized by very dilute nutrient sources, and, to flourish, they must be able to transport and concentrate these nutrients. Thus facilitated diffusion mechanisms are not always adequate, and other approaches must be used. The two most important transport processes in such situations are active transport and group translocation.

Active transport is the transport of solute molecules to higher concentrations, or against a concentration gradient, with the use of metabolic energy input. Because active transport involves protein carrier activity, it resembles facilitated diffusion in some ways. The carrier proteins bind particular solutes with great specificity for the molecules transported. Similar solute molecules can compete for the same carrier protein in both facilitated diffusion and active transport. Active transport is also characterized by the carrier saturation effect at high solute concentrations (figure 5.1). Nevertheless, active transport differs from facilitated diffusion in its use of metabolic energy and in its ability to concentrate substances. Metabolic inhibitors that block energy production will inhibit active transport but will not affect facilitated diffusion (at least for a short time).

Binding protein transport systems employ special substrate binding proteins located in the periplasmic space of gram-negative bacteria (*see figure 3.26*). These periplasmic proteins, which also participate in chemotaxis (*see pp. 63–66*), bind the molecule to be transported and then interact with the membrane transport proteins to move the solute molecule inside the cell. The membrane-bound complex seems to usually have several subunits and form a pore in the membrane. The energy source is ATP, although other high-energy phosphate compounds may drive uptake in some transport systems. *E. coli* transports a variety of sugars (arabinose, maltose, galactose, ribose) and amino acids (glutamate, histidine, leucine) by this mechanism. ATP-driven transport is present in gram-positive bacteria but is not as well understood as in gram-negative organisms.

Bacteria also use protonmotive force (usually in the form of a proton gradient generated during electron transport) to drive active transport. The membrane transport proteins responsible for this process lack special periplasmic solute-binding proteins. The lactose permease of *E. coli* is a well-studied example. The permease is a single protein having a molecular weight of about 30,000. It transports a lactose molecule inward as a proton simultaneously enters the cell (a higher concentration of protons is maintained outside the membrane by electron transport chain activity). Such linked transport of two substances in the same direction is called **symport.** Here, energy stored as a proton gradient drives solute transport. Although the mechanism of transport is not completely understood, it is thought that binding of a proton to the transport protein changes its shape and affinity for the solute to be transported. *E. coli* also uses proton symport to take up amino acids and organic acids like succinate and malate.

The chemiosmotic hypothesis and protonmotive force (p. 172).

Protonmotive force can power active transport indirectly, often through the formation of a sodium ion gradient. For example, an *E. coli* sodium transport system pumps sodium outward in response to the inward movement of protons (figure 5.3). Such linked transport in which the transported substances move in opposite directions is termed **antiport.** The sodium gradient generated by this proton antiport system then drives the uptake of sugars and amino acids. A sodium ion could attach to a carrier protein, causing it to change shape. The carrier would then bind the sugar or amino acid tightly and orient its binding sites toward the cell interior. Because of the low intracellular sodium concentration, the sodium ion would dissociate from the carrier, and the other molecule would follow. *E. coli*

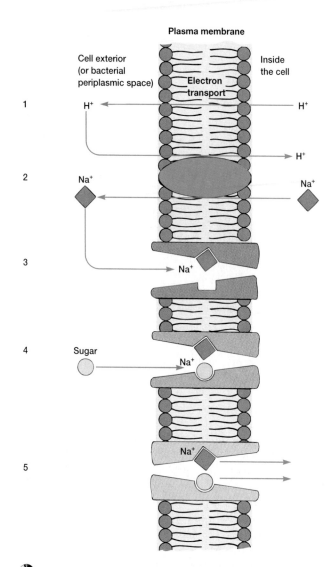

Figure 5.3 Active Transport Mechanism. The use of proton and sodium gradients in active transport. (*1*) Electron transport energy is used to pump protons to the outside of the plasma membrane. (*2*) The proton gradient drives sodium ion expulsion by an antiport mechanism. (*3*) Sodium binds to the carrier protein complex. (*4*) The shape of the solute binding site changes, and it binds the solute (e.g., a sugar or amino acid). (*5*) The carrier's conformation then alters so that sodium is released on the inside of the membrane. This is followed by solute dissociation from the carrier.

transport proteins carry the sugar melibiose and the amino acid glutamate when sodium simultaneously moves inward. Sodium symport or cotransport also is an important process in eucaryotic cells where it is used in sugar and amino acid uptake. ATP, rather than protonmotive force, usually drives sodium transport in eucaryotic cells.

It seems reasonable that a microorganism would have only one transport system for each nutrient, but often this is not so, as can be seen with *E. coli*. This bacterium has at least five transport systems for the sugar galactose, three systems each for the amino acids glutamate and leucine, and two potassium transport complexes. When there are several transport

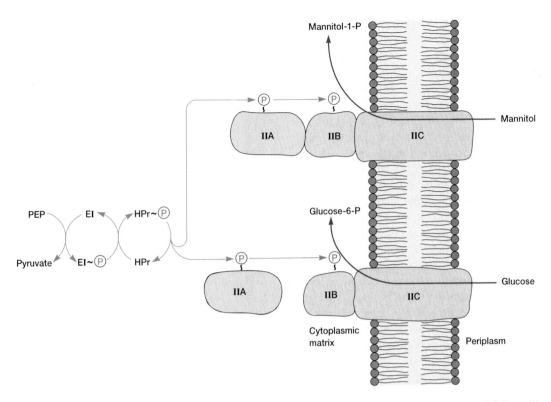

Figure 5.4   Bacterial PTS Transport. Two examples of the phosphoenolpyruvate: sugar phosphotransferase system (PTS) are illustrated. The following components are involved in the system: phosphoenolpyruvate, PEP; enzyme I, EI; the low molecular weight heat-stable protein, HPr; and enzyme II, EII. The high-energy phosphate is transferred from HPr to the soluble EIIA. EIIA is attached to EIIB in the mannitol transport system and is separate from EIIB in the glucose system. In either case the phosphate moves from EIIA to EIIB, and then is transferred to the sugar during transport through the membrane. Other relationships between the EII components are possible. For example, IIA and IIB may form a soluble protein separate from the membrane complex; the phosphate still moves from IIA to IIB and then to the membrane domain(s).

systems for the same substance, the systems differ in such properties as their energy source, their affinity for the solute transported, and the nature of their regulation. Presumably this diversity gives its possessor an added competitive advantage in a variable environment.

Group Translocation

In active transport, solute molecules move across a membrane without modification. Many procaryotes also take up molecules by **group translocation,** a process in which a molecule is transported into the cell while being chemically altered. The best-known group translocation system is the **phosphoenolpyruvate: sugar phosphotransferase system (PTS).** It transports a variety of sugars into procaryotic cells while phosphorylating them using phosphoenolpyruvate (PEP) as the phosphate donor.

$$PEP + sugar\ (outside) \longrightarrow pyruvate + sugar - P\ (inside)$$

The PTS is quite complex. In *E. coli* and *Salmonella typhimurium,* it consists of two enzymes and a low molecular weight heat-stable protein (HPr). HPr and enzyme I (EI) are cytoplasmic. Enzyme II (EII) is more variable in structure and often composed of three subunits or domains. EIIA (formerly called EIII) is cytoplasmic and soluble. EIIB also is hydrophilic

but frequently is attached to EIIC, a hydrophobic protein that is embedded in the membrane. A high-energy phosphate is transferred from PEP to enzyme II with the aid of enzyme I and HPr (figure 5.4). Then, as a sugar molecule is carried across the membrane by enzyme II, it is phosphorylated. Enzyme II transports only specific sugars and varies with PTS, whereas enzyme I and HPr are common to all PTSs.

PTSs are widely distributed in procaryotes. Except for some species of *Bacillus* that have both the Embden-Meyerhof pathway and phosphotransferase systems, aerobic bacteria seem to lack PTSs. Members of the genera *Escherichia, Salmonella, Staphylococcus,* and other facultatively anaerobic bacteria have phosphotransferase systems; some obligately anaerobic bacteria (e.g., *Clostridium*) also have PTSs. Many carbohydrates are transported by these systems. *E. coli* takes up glucose, fructose, mannitol, sucrose, N-acetylglucosamine, cellobiose, and other carbohydrates by group translocation. Besides their role in transport, PTS proteins can act as chemoreceptors for chemotaxis.

Iron Uptake

Almost all microorganisms require iron for use in cytochromes and many enzymes. Iron uptake is made difficult by the great insolubility of ferric iron (Fe^{3+}) and its derivatives, which leaves little free iron available for transport. Many bacteria and fungi have overcome this difficulty by secreting siderophores [Greek for iron bearers]. **Siderophores** are low molecular weight molecules that are able to complex with ferric iron and supply it to the cell. These iron-transport molecules are normally either hydroxamates or phenolates-catecholates. Ferrichrome is a hydroxamate produced by many fungi; enterobactin is the catecholate formed by *E. coli* (figure 5.5*a,b*). It appears that three siderophore groups complex with iron orbitals to form a six-coordinate, octahedral complex (figure 5.5*c*).

Microorganisms secrete siderophores when little iron is available in the medium. Once the iron-siderophore complex has reached the cell surface, it binds to a siderophore-receptor protein. Then the iron is either released to enter the cell directly or the whole iron-siderophore complex is transported inside. In *E. coli* the siderophore receptor is in the outer membrane of the cell envelope; when the iron reaches the periplasmic space, it moves through the plasma membrane with the aid of other proteins. After the iron has entered the cell, it is reduced to the ferrous form (Fe^{2+}). Iron is so crucial to microorganisms that they may use more than one route of iron uptake to ensure an adequate supply.

---

1. Describe facilitated diffusion, active transport, and group translocation in terms of their distinctive characteristics and mechanisms.
2. How do binding protein transport systems and membrane-bound transport systems differ with respect to energy sources? What are symport and antiport processes?
3. How are siderophores involved in iron transport?

Culture Media

Much of microbiology depends on the ability to grow and maintain microorganisms in the laboratory, and this is possible only if suitable culture media are available. In addition, specialized media are essential in the isolation and identification of microorganisms, the testing of antibiotic sensitivities, water and food analysis, industrial microbiology, and other activities. Although all microorganisms need sources of energy, carbon, nitrogen, phosphorus, sulfur, and various minerals, the precise composition of a satisfactory medium will depend on the species one is trying to cultivate because nutritional requirements vary so greatly. Knowledge of a microorganism's normal habitat often is useful in selecting an appropriate culture medium because its nutrient requirements reflect its natural surroundings. Frequently a medium is used to select and grow specific microorganisms or to help identify a particular species. In such cases the function of the medium also will determine its composition.

Synthetic or Defined Media

Some microorganisms, particularly photolithotrophic autotrophs such as cyanobacteria and eucaryotic algae, can be grown on relatively simple media containing CO_2 as a carbon source (often

Figure 5.5 Siderophore Ferric Iron Complexes. (*a*) Ferrichrome is a cyclic hydroxamate [—CO—N(O⁻)—] molecule formed by many fungi. (*b*) *E. coli* produces the cyclic catecholate derivative, enterobactin. (*c*) Ferric iron probably complexes with three siderophore groups to form a six-coordinate, octahedral complex as shown in this illustration of the enterobactin-iron complex.

added as sodium carbonate or bicarbonate), nitrate or ammonia as a nitrogen source, sulfate, phosphate, and a variety of minerals (table 5.4). Such a medium in which all components are known is a **defined medium** or **synthetic medium.** Many chemoorganotrophic heterotrophs also can be grown in defined media with glucose as a carbon source and an ammonium salt as a nitrogen source. Not all defined media are as simple as the examples in table 5.4 but may be constructed from dozens of components. Defined media are used widely in research, as it is often desirable to know what the experimental microorganism is metabolizing.

TABLE 5.4	Examples of Defined Media
BG–11 Medium for Cyanobacteria	**Amount (g/liter)**
$NaNO_3$	1.5
$K_2HPO_4 \cdot 3H_2O$	0.04
$MgSO_4 \cdot 7H_2O$	0.075
$CaCl_2 \cdot 2H_2O$	0.036
Citric acid	0.006
Ferric ammonium citrate	0.006
EDTA (Na_2Mg salt)	0.001
Na_2CO_3	0.02
Trace metal solution[a]	1.0 ml/liter
Final pH 7.4	
Medium for *Escherichia coli*	**Amount (g/liter)**
Glucose	1.0
Na_2HPO_4	16.4
KH_2PO_4	1.5
$(NH_4)_2SO_4$	2.0
$MgSO_4 \cdot 7H_2O$	200.0 mg
$CaCl_2$	10.0 mg
$FeSO_4 \cdot 7H_2O$	0.5 mg
Final pH 6.8–7.0	

Sources: Data from Rippka, et al. *Journal of General Microbiology*, 111:1–61, 1979; and S. S. Cohen, and R. Arbogast, *Journal of Experimental Medicine*, 91:619, 1950.

[a]The trace metal solution contains H_3BO_3, $MnCl_2 \cdot 4H_2O$, $ZnSO_4 \cdot 7H_2O$, $Na_2Mo_4 \cdot 2H_2O$, $CuSO_4 \cdot 5H_2O$, and $Co(NO_3)_2 \cdot 6H_2O$.

TABLE 5.5	Some Common Complex Media
Nutrient Broth	**Amount (g/liter)**
Peptone (gelatin hydrolysate)	5
Beef extract	3
Tryptic Soy Broth	
Tryptone (pancreatic digest of casein)	17
Peptone (soybean digest)	3
Glucose	2.5
Sodium chloride	5
Dipotassium phosphate	2.5
MacConkey Agar	**Amount (g/liter)**
Pancreatic digest of gelatin	17.0
Pancreatic digest of casein	1.5
Peptic digest of animal tissue	1.5
Lactose	10.0
Bile salts	1.5
Sodium chloride	5.0
Neutral red	0.03
Crystal violet	0.001
Agar	13.5

Complex Media

Media that contain some ingredients of unknown chemical composition are **complex media.** Such media are very useful, as a single complex medium may be sufficiently rich and complete to meet the nutritional requirements of many different microorganisms. In addition, complex media often are needed because the nutritional requirements of a particular microorganism are unknown, and thus a defined medium cannot be constructed. This is the situation with many fastidious bacteria, some of which may even require a medium containing blood or serum.

Complex media contain undefined components like peptones, meat extract, and yeast extract. **Peptones** are protein hydrolysates prepared by partial proteolytic digestion of meat, casein, soya meal, gelatin, and other protein sources. They serve as sources of carbon, energy, and nitrogen. Beef extract and yeast extract are aqueous extracts of lean beef and brewer's yeast, respectively. Beef extract contains amino acids, peptides, nucleotides, organic acids, vitamins, and minerals. Yeast extract is an excellent source of B vitamins as well as nitrogen and carbon compounds. Three commonly used complex media are (1) nutrient broth, (2) tryptic soy broth, and (3) MacConkey agar (table 5.5).

If a solid medium is needed for surface cultivation of microorganisms, liquid media can be solidified with the addition of 1.0% to 2.0% agar, most commonly 1.5%. **Agar** is a sulfated polymer composed mainly of D-galactose, 3,6-anhydro-L-galactose, and D-glucuronic acid. It usually is extracted from red algae (*see figure 26.8*). Agar is well suited as a solidifying agent because after it has been melted in boiling water, it can be cooled to about 40 to 42°C before hardening and will not melt again until the temperature rises to about 80 to 90°C. Agar is also an excellent hardening agent because most microorganisms cannot degrade it (Box 5.1).

Other solidifying agents are sometimes employed. For example, silica gel is used to grow autotrophic bacteria on solid media in the absence of organic substances and to determine carbon sources for heterotrophic bacteria by supplementing the medium with various organic compounds.

Types of Media

Media like tryptic soy broth and tryptic soy agar are called general purpose media because they support the growth of many microorganisms. Blood and other special nutrients may be added to general purpose media to encourage the growth of fastidious heterotrophs. These specially fortified media (e.g., blood agar) are called enriched media.

Selective media favor the growth of particular microorganisms. Bile salts or dyes like basic fuchsin and crystal violet favor the growth of gram-negative bacteria by inhibiting the growth of gram-positive bacteria without affecting gram-negative organisms. Endo agar, eosin methylene blue agar, and MacConkey agar (table 5.5), three media widely used for the detection of *E. coli* and related bacteria in water supplies and elsewhere, contain dyes that suppress gram-positive bacterial growth. MacConkey agar also contains bile salts. Bacteria also may be selected by incubation with nutrients that they specifically can use. A medium containing only cellulose as a carbon and energy source is quite effective in the isolation of

---

Box 5.1

The Discovery of Agar as a Solidifying Agent and the Isolation of Pure Cultures

The earliest culture media were liquid, which made the isolation of bacteria to prepare pure cultures extremely difficult. In practice, a mixture of bacteria was diluted successively until only one organism, as an average, was present in a culture vessel. If everything went well, the individual bacterium thus isolated would reproduce to give a pure culture. This approach was tedious, gave variable results, and was plagued by contamination problems. Progress in isolating pathogenic bacteria understandably was slow.

The development of techniques for growing microorganisms on solid media and efficiently obtaining pure cultures was due to the efforts of the German bacteriologist Robert Koch and his associates. In 1881 Koch published an article describing the use of boiled potatoes, sliced with a flame-sterilized knife, in culturing bacteria. The surface of a sterile slice of potato was inoculated with bacteria from a needle tip, and then the bacteria were streaked out over the surface so that a few individual cells would be separated from the remainder. The slices were incubated beneath bell jars to prevent airborne contamination, and the isolated cells developed into pure colonies. Unfortunately many bacteria would not grow well on potato slices.

At about the same time, Frederick Loeffler, an associate of Koch's, developed a meat extract peptone medium for cultivating pathogenic bacteria. Koch decided to try solidifying this medium. Koch was an amateur photographer—he was the first to take photomicrographs of bacteria—and was experienced in preparing his own photographic plates from silver salts and gelatin. Precisely the same approach was employed for preparing solid media. He spread a mixture of Loeffler's medium and gelatin over a glass plate, allowed it to harden, and inoculated the surface in the same way he had inoculated his sliced potatoes. The new solid medium worked well, but it could not be incubated at 37°C (the best temperature for most human bacterial pathogens) because the gelatin would melt. Furthermore, some bacteria digested the gelatin.

About a year later, in 1882, agar was first used as a solidifying agent. It had long been added to jelly to provide the appropriate solid consistency, and anyone who had made jelly or jam would be well acquainted with it. Not surprisingly, Fannie Eilshemius Hesse, the New Jersey–born wife of Walther Hesse, one of Koch's assistants, suggested its use when she heard of the difficulties with gelatin. Agar-solidified medium was an instant success and continues to be essential in all areas of microbiology.

---

cellulose-digesting bacteria. The possibilities for selection are endless, and there are dozens of special selective media in use.

Differential media are media that distinguish between different groups of bacteria and even permit tentative identification of microorganisms based on their biological characteristics. Blood agar is both a differential medium and an enriched one. It distinguishes between hemolytic and nonhemolytic bacteria. Hemolytic bacteria (e.g., many streptococci and staphylococci isolated from throats) produce clear zones around their colonies because of red blood cell destruction. MacConkey agar is both differential and selective. Since it contains lactose and neutral red dye, lactose-fermenting colonies appear pink to red in color and are easily distinguished from colonies of nonfermenters.

---

1. Describe the following kinds of media and their uses: defined or synthetic media, complex media, general purpose media, enriched media, selective media, and differential media. Give an example of each kind.
2. What are peptones, yeast extract, beef extract, and agar? Why are they found in media?

Isolation of Pure Cultures

In natural habitats microorganisms usually grow in complex, mixed populations containing several species. This presents a problem for the microbiologist because a single type of microorganism cannot be studied adequately in a mixed culture. One needs a **pure culture,** a population of cells arising from a single cell, to characterize an individual species. Pure cultures are so important that the development of pure culture techniques by the German bacteriologist Robert Koch transformed microbiology. Within about 20 years most pathogens responsible for the major human bacterial diseases had been isolated. There are several ways to prepare pure cultures; a few of the more common approaches are reviewed.

A brief survey of some major milestones in microbiology (chapter 1).

The Spread Plate and Streak Plate

If a mixture of cells is spread out on an agar surface so that every cell grows into a completely separate **colony,** a macroscopically visible growth or cluster of microorganisms on a solid medium, each colony represents a pure culture. The **spread plate** is an easy, direct way of achieving this result. A small volume of dilute microbial mixture containing around 100 to 200 cells or less is transferred to the center of an agar plate and spread evenly over the surface with a sterile bent-glass rod (figure 5.6). The dispersed cells develop into isolated colonies. Because the number of colonies should equal the number of viable organisms in the sample, spread plates can be used to count the microbial population.

Figure 5.6   Spread-Plate Technique. The preparation of a spread plate. (*1*) Pipette a small sample onto the center of an agar medium plate. (*2*) Dip a glass spreader into a beaker of ethanol. (*3*) Briefly flame the ethanol-soaked spreader and allow it to cool. (*4*) Spread the sample evenly over the agar surface with the sterilized spreader. Incubate.

Pure colonies also can be obtained from **streak plates.** The microbial mixture is transferred to the edge of an agar plate with an inoculating loop or swab and then streaked out over the surface in one of several patterns (figure 5.7). At some point in the process, single cells drop from the loop as it is rubbed along the agar surface and develop into separate colonies (figure 5.8). In both spread-plate and streak-plate techniques, successful isolation depends on spatial separation of single cells.

The Pour Plate

Extensively used with bacteria and fungi, a **pour plate** also yields isolated colonies. The original sample is diluted several times to reduce the microbial population sufficiently to obtain separate colonies when plating (figure 5.9). Then small volumes of several diluted samples are mixed with liquid agar that has been cooled to about 45°C, and the mixtures are poured immediately into sterile culture dishes. Most bacteria and fungi

Figure 5.7   Streak-Plate Technique. Preparation of streak plates. The upper illustration shows a petri dish of agar being streaked with an inoculating loop. A commonly used streaking pattern is pictured at the bottom.

Figure 5.8   Bacterial Colonies on Agar. Colonies growing on a streak plate. A blood-agar plate has been inoculated with *Staphylococcus aureus.* After incubation, large, golden colonies have formed on the agar.

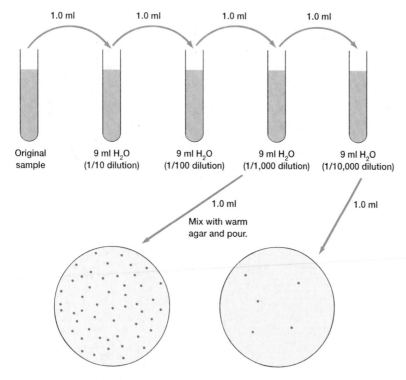

Figure 5.9   The Pour-Plate Technique. The original sample is diluted several times to thin out the population sufficiently. The most diluted samples are then mixed with warm agar and poured into petri dishes. Isolated cells grow into colonies and can be used to establish pure cultures. The surface colonies are circular; subsurface colonies would be lenticular or lens shaped.

are not killed by a brief exposure to the warm agar. After the agar has hardened, each cell is fixed in place and forms an individual colony. The total number of colonies equals the number of viable microorganisms in the diluted sample. Colonies growing on the surface also can be used to inoculate fresh medium and prepare pure cultures.

The preceding techniques require the use of special culture dishes named **petri dishes** or plates after their inventor Julius Richard Petri, a member of Robert Koch's laboratory; Petri developed these dishes around 1887 and they immediately replaced agar-coated glass plates. They consist of two round halves, the top half overlapping the bottom (figure 5.7). Petri dishes are very easy to use, may be stacked on each other to save space, and are one of the most common items in microbiology laboratories.

The plating methods just described are even more effective in producing pure cultures when used with selective or differential media. A good example is the isolation of bacteria that degrade the herbicide 2,4-dichlorophenoxyacetic acid (2,4-D). Bacteria able to metabolize 2,4-D can be obtained with a liquid medium containing 2,4-D as its sole carbon source and the required nitrogen, phosphorus, sulfur, and mineral components. When this medium is inoculated with soil, only bacteria able to use 2,4-D will grow. After incubation, a sample of the original culture is transferred to a fresh flask of selective medium for further enrichment of 2,4-D metabolizing bacteria. A mixed population of 2,4-D degrading bacteria will arise after several such transfers. Pure cultures can be obtained by plating this mixture on agar containing 2,4-D as the sole carbon source. Only bacteria able to grow on 2,4-D form visible colonies and can be subcultured. This same general approach is used to isolate and purify a variety of bacteria by selecting for specific physiological characteristics.

Colony Morphology and Growth

Colony development on agar surfaces aids the microbiologist in identifying bacteria because individual species often form colonies of characteristic size and appearance (figure 5.10). When a mixed population has been plated properly, it sometimes is possible to identify the desired colony based on its overall appearance and use it to obtain a pure culture. The structure of bacterial colonies also has been examined with the scanning electron microscope. The microscopic structure of colonies is often as variable as their visible appearance (figure 5.11).

In nature, bacteria and many other microorganisms often grow as colonies on surfaces. Therefore an understanding of colony growth is essential to microbial ecologists, and the growth of colonies on agar has been frequently studied. Generally the most rapid cell growth

Form

Punctiform   Circular   Filamentous   Irregular   Rhizoid   Spindle

Elevation

Flat   Raised   Convex   Pulvinate   Umbonate

Margin

Entire   Undulate   Lobate   Erose   Filamentous   Curled

(a)

(b)

Figure 5.10   Bacterial Colony Morphology. (*a*) Variations in bacterial colony morphology seen with the naked eye. The general form of the colony and the shape of the edge or margin can be determined by looking down at the top of the colony. The nature of colony elevation is apparent when viewed from the side as the plate is held at eye level. (*b*) Colony morphology can vary dramatically with the medium on which the bacteria are growing. These beautiful snowflakelike colonies were formed by *Bacillus subtilis* growing on nutrient-poor agar. The bacteria apparently behave cooperatively when confronted with poor growth conditions, and often the result is an intricate structure that resembles the fractal patterns seen in nonliving systems.

(a)   (b)   (c)

(d)   (e)   (f)

Figure 5.11   Scanning Electron Micrographs of Bacterial Colonies. (*a*) *Micrococcus* on agar ($\times$31,000). (*b*) *Streptococcus pneumoniae* ($\times$100). (*c*) *Clostridium* ($\times$12,000). (*d*) *Mycoplasma pneumoniae* ($\times$26,000). (*e*) *Escherichia coli* ($\times$14,000). (*f*) *Legionella pneumophila* ($\times$3,600).

occurs at the colony edge. Growth is much slower in the center, and cell autolysis takes place in the older central portions of some colonies (figure 5.11*b*). These differences in growth appear due to gradients of oxygen, nutrients, and toxic products within the colony. At the colony edge, oxygen and nutrients are plentiful. The colony center, of course, is much thicker than the edge. Consequently oxygen and nutrients do not diffuse readily into the center, toxic metabolic products cannot be quickly eliminated, and growth in the colony center is slowed or stopped. Because of these environmental variations within a colony, cells on the periphery can be growing at maximum rates while cells in the center are dying.

---

1. What are pure cultures, and why are they important? How are spread plates, streak plates, and pour plates prepared?
2. In what way does microbial growth vary within a colony? What factors might cause these variations in growth?

Summary

1. Microorganisms require nutrients, materials that are used in biosynthesis and energy production.

2. Macronutrients or macroelements (C, O, H, N, S, P, K, Ca, Mg, and Fe) are needed in large quantities; trace elements (e.g., Mn, Zn, Co, Mo, Ni, and Cu) are used in very small amounts.

3. Autotrophs use CO_2 as their primary or sole carbon source; heterotrophs employ organic molecules.

4. Microorganisms can be classified based on their energy and electron sources. Phototrophs use light energy, and chemotrophs obtain energy from the oxidation of chemical compounds. Electrons are extracted from reduced inorganic substances by lithotrophs and from organic compounds by organotrophs.

5. Nitrogen, phosphorus, and sulfur may be obtained from the same organic molecules that supply carbon, from the direct incorporation of ammonia and phosphate, and by the reduction and assimilation of oxidized inorganic molecules.

6. Probably most microorganisms need growth factors. Growth factor requirements make microbiological assays possible.

7. Although some nutrients can enter cells through passive diffusion, a membrane carrier protein is usually required.

8. In facilitated diffusion the transport protein simply carries a molecule across the membrane in the direction of decreasing concentration, and no energy source is required.

9. Active transport systems use metabolic energy and membrane carrier proteins to concentrate substances actively by transporting them against a gradient. Either ATP or protonmotive force may be used as an energy source, and sodium gradients also can drive solute uptake.

10. Bacteria also transport molecules while modifying them, a process known as group translocation. For example, many sugars are transported and phosphorylated simultaneously.

11. Iron is accumulated by the secretion of siderophores, small molecules able to complex with ferric iron. When the iron-siderophore complex reaches the cell surface, it is taken inside and the iron is reduced to the ferrous form.

12. Culture media can be constructed completely from chemically defined components (defined media or synthetic media) or may contain constituents like peptones and yeast extract whose precise composition is unknown (complex media).

13. Culture medium can be solidified with the addition of agar, a complex polysaccharide from red algae.

14. Culture media are classified based on function and composition as general purpose media, enriched media, selective media, and differential media.

15. The pure cultures are obtained by isolating individual cells with any of three plating techniques: the spread-plate, streak-plate, and pour-plate methods.

16. Microorganisms growing on solid surfaces tend to form colonies with distinctive morphology. Colonies usually grow most rapidly at the edge where there is an ample supply of oxygen and nutrients.

Key Terms

active transport   *103*

agar   *107*

antiport   *104*

autotrophs   *99*

auxotrophs   *100*

chemoheterotrophs   *100*

chemolithotrophic autotrophs   *101*

chemoorganotrophic heterotrophs   *100*

chemotrophs   *100*

colony   *108*

complex medium   *107*

defined medium   *106*

differential media   *108*

facilitated diffusion   *103*

group translocation   *105*

growth factors   *101*

heterotrophs   *99*

lithotrophs   *100*

macroelements   *99*

mixotrophic   *101*

nutrient   *99*

organotrophs   *100*

passive diffusion   *102*

peptones   *107*

permease   *103*

petri dish   *110*

phosphoenolpyruvate: sugar phosphotransferase system (PTS)   *105*

photoautotrophs   *100*

photolithotrophic autotrophs   *100*

photoorganotrophic heterotrophs   *101*

phototrophs   *100*

pour plate   *109*

prototrophs   *100*

pure culture   *108*

selective media   *107*

siderophores   *106*

spread plate   *108*

streak plate   *108*

symport   *104*

synthetic medium   *106*

trace elements   *99*

vitamins   *101*

Questions for Thought and Review

1. Why is it so difficult to demonstrate the micronutrient requirements of microorganisms?

2. Where might one expect to find microorganisms that are unable to catabolize a variety of nutrients?

3. List some of the most important uses of the nitrogen, phosphorus, and sulfur that microorganisms obtain from their surroundings.

4. Why are amino acids, purines, and pyrimidines often growth factors, whereas glucose is usually not?

5. Why do microorganisms normally take up nutrients using transport proteins or permeases? What advantage does a microorganism gain by employing active transport rather than facilitated diffusion?

6. If you wished to obtain a pure culture of bacteria that could degrade benzene and use it as a carbon and energy source, how would you proceed?

7. Describe the nutritional requirements of a chemolithotrophic heterotroph. Where might you search for such a bacterium?

Additional Reading

General

Conn, H. J., editor. 1957. *Manual of microbiological methods.* New York: McGraw-Hill.

Gottschall, J. C.; Harder, W.; and Prins, R. A. 1992. Principles of enrichment, isolation, cultivation, and preservation of bacteria. In *The prokaryotes,* 2d ed., A. Balows et al., editors, 149–96. New York: Springer-Verlag.

Holt, J. G., and Krieg, N. R. 1994. Enrichment and isolation. In *Methods for general and molecular bacteriology,* 2d ed., P. Gerhardt, editor, 179–215. Washington, D.C.: American Society for Microbiology.

Neidhardt, F. C.; Ingraham, J. L.; and Schaechter, M. 1990. *Physiology of the bacterial cell: A molecular approach.* Sunderland, Mass.: Sinauer.

Whittenbury, R. 1978. Bacterial nutrition. In *Essays in microbiology,* J. R. Norris and M. H. Richmond, editors, 16/1–16/32. New York: John Wiley and Sons.

Requirements for Carbon, Hydrogen, and Oxygen

Kelly, D. P. 1992. The chemolithotrophic prokaryotes. In *The prokaryotes,* 2d ed., A. Balows et al., editors, 331–43. New York: Springer-Verlag.

Whittenbury, R., and Kelly, D. P. 1977. Autotrophy: A conceptual phoenix. In *Microbial energetics,* B. A. Haddock and W. A. Hamilton, editors, 121–49. New York: Cambridge University Press.

Uptake of Nutrients by the Cell

Ames, G. F.-L.; Mimura, C. S.; Holbrook, S. R.; and Shyamala, V. 1992. Traffic ATPases: A superfamily of transport proteins operating from *E. coli* to humans. *Adv. Enzymol.* 65:1–47.

Braun, V. 1985. The unusual features of the iron transport systems of *Escherichia coli. Trends Biochem. Sci.* 10(2):75–78.

Dills, S. S.; Apperson, A.; Schmidt, M. R.; and Saier, M. H., Jr. 1980. Carbohydrate transport in bacteria. *Microbiol. Rev.* 44(3):385–418.

Doige, C. A., and Ames, G. F.-L. 1993. ATP-dependent transport systems in bacteria and humans: Relevance to cystic fibrosis and multidrug resistance. *Ann. Rev. Microbiol.* 47:291–319.

Harder, W., and Dijkhuizen, L. 1983. Physiological responses to nutrient limitation. *Ann. Rev. Microbiol.* 37:1–23.

Maloney, P. C.; Ambudkar, S. V.; Anantharam, V.; Sonna, L. A.; and Varadhachary, A. 1990. Anion-exchange mechanisms in bacteria. *Microbiol. Rev.* 54(1):1–17.

Meadow, N. D.; Fox, D. K.; and Roseman, S. 1990. The bacterial phosphoenolpyruvate: glycose phosphotransferase system. *Ann. Rev. Biochem.* 59:497–542.

Neilands, J. B. 1991. Microbial iron compounds. *Ann. Rev. Biochem.* 50:715–31.

Postma, P. W.; Lengeler, J. W.; and Jacobson, G. R. 1993. Phosphoenolpyruvate: carbohydrate phosphotransferase systems of bacteria. *Microbiol. Rev.* 57(3):543–94.

Culture Media

Atlas, R. M. 1993. *Handbook of microbiological media.* Boca Raton, Fla.: CRC Press.

Bridson, E. Y. 1990. Media in microbiology. *Rev. Med. Microbiol.* 1:1–9.

Cote, R. J., and Gherna, R. L. 1994. Nutrition and media. In *Methods for general and molecular bacteriology,* 2d ed., P. Gerhardt, editor, 155–78. Washington, D.C.: American Society for Microbiology.

Difco Laboratories. 1984. *Difco manual of dehydrated culture media and reagents for microbiology.* 10th ed. Detroit: Difco.

Power, D. A., editor. 1988. *Manual of BBL products and laboratory procedures,* 6th ed. Cockeysville, Md.: Becton, Dickinson and Company.

Isolation of Pure Cultures

Schindler, J. 1993. Dynamics of *Bacillus* colony growth. *Trends Microbiol.* 1(9):333–38.

Shapiro, J. A. 1988. Bacteria as multicellular organisms. *Sci. Am.* 258(6):82–89.

CHAPTER 6
Microbial Growth

The paramount evolutionary

accomplishment of bacteria as

a group is rapid, efficient cell

growth in many

environments.

—J. L. Ingraham, O. Maaloe,
and F. C. Neidhardt

Concepts

1. Growth is defined as an increase in cellular constituents and may result in an increase in a microorganism's size, population number, or both.

2. When microorganisms are grown in a closed system, population growth remains exponential for only a few generations and then enters a stationary phase due to factors such as nutrient limitation and waste accumulation. In an open system with continual nutrient addition and waste removal, the exponential phase can be maintained for long periods.

3. A wide variety of techniques can be used to study microbial growth by following changes in the total cell number, the population of viable microorganisms, or the cell mass.

4. Water availability, pH, temperature, oxygen concentration, pressure, radiation, and a number of other environmental factors influence microbial growth. Yet many microorganisms, and particularly bacteria, have managed to adapt and flourish under environmental extremes that would destroy most higher organisms.

Chapter 5 emphasizes that microorganisms need access to a source of energy and the raw materials essential for the construction of cellular components. All organisms must have carbon, hydrogen, oxygen, nitrogen, sulfur, phosphorus, and a variety of minerals; many also require one or more special growth factors. The cell takes up these substances by membrane transport processes, the most important of which are facilitated diffusion, active transport, and group translocation. Eucaryotic cells also employ endocytosis.

Chapter 6 concentrates more directly on the growth that takes place in the presence of an adequate nutrient supply. The nature of growth and the ways in which it can be measured are described first, followed by consideration of some specific aspects of culture growth. An account of the influence of environmental factors on microbial growth completes the chapter.

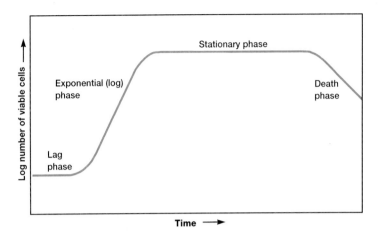

Figure 6.1   Microbial Growth Curve in a Closed System. The four phases of the growth curve are identified on the curve and discussed in the text.

Growth may be defined as an increase in cellular constituents. If the microorganism is **coenocytic**—that is, a multinucleate organism in which nuclear divisions are not accompanied by cell divisions—growth results in an increase in cell size but not cell number. Growth leads to a rise in cell number when microorganisms reproduce by processes like budding or binary fission. In the latter, individual cells enlarge and divide to yield two progeny of approximately equal size. It is usually not convenient to investigate the growth and reproduction of individual microorganisms because of their small size. Therefore, when studying growth, microbiologists normally follow changes in the total population number.

The cell cycle (pp. 86; 250–51).

The Growth Curve

Population growth is studied by analyzing the growth curve of a microbial culture. When microorganisms are cultivated in liquid medium, they usually are grown in a **batch culture** or closed system—that is, they are incubated in a closed culture vessel with a single batch of medium. Because no fresh medium is provided during incubation, nutrient concentrations decline and concentrations of wastes increase. The growth of microorganisms reproducing by binary fission can be plotted as the logarithm of cell number versus the incubation time. The resulting curve has four distinct phases (figure 6.1).

Lag Phase

When microorganisms are introduced into fresh culture medium, usually no immediate increase in cell number or mass occurs, and therefore this period is called the **lag phase.** Although cell division does not take place right away and there is no net increase in mass, the cell is synthesizing new components. A lag phase prior to the start of cell division can be necessary for a variety of reasons. The cells may be old and depleted of ATP, essential cofactors, and ribosomes; these must be synthesized

before growth can begin. The medium may be different from the one the microorganism was growing in previously. Here new enzymes would be needed to use different nutrients. Possibly the microorganisms have been injured and require time to recover. Whatever the causes, eventually the cells retool, replicate their DNA, begin to increase in mass, and finally divide.

The lag phase varies considerably in length with the condition of the microorganisms and the nature of the medium. This phase may be quite long if the inoculum is from an old culture or one that has been refrigerated. Inoculation of a culture into a chemically different medium also results in a long lag phase. On the other hand, when a young, vigorously growing exponential phase culture is transferred to fresh medium of the same composition, the lag phase will be short or absent.

Exponential Phase

During the **exponential** or **log phase,** microorganisms are growing and dividing at the maximal rate possible given their genetic potential, the nature of the medium, and the conditions under which they are growing. Their rate of growth is constant during the exponential phase, the microorganisms dividing and doubling in number at regular intervals. Because each individual divides at a slightly different moment, the growth curve rises smoothly rather than in discrete jumps (figure 6.1). The population is most uniform in terms of chemical and physiological properties during this phase; therefore exponential phase cultures are usually used in biochemical and physiological studies.

Stationary Phase

Eventually population growth ceases and the growth curve becomes horizontal (figure 6.1). This **stationary phase** usually is attained by bacteria at a population level of around 10^9 cells per ml. Other microorganisms normally do not reach such high population densities, protozoan and algal cultures often having maximum concentrations of about 10^6 cells per ml. Of course final population size depends on nutrient availability and other factors, as well as the type of microorganism being cultured. In the stationary phase the total number of viable microorganisms remains constant. This may result from a balance between cell division and cell death, or the population may simply cease to divide though remaining metabolically active.

Microbial populations enter the stationary phase for several reasons. One obvious factor is nutrient limitation; if an essential nutrient is severely depleted, population growth will slow. Aerobic organisms often are limited by O_2 availability. Oxygen is not very soluble and may be depleted so quickly that only the surface of a culture will have an O_2 concentration adequate for growth. The cells beneath the surface will not be able to grow unless the culture is shaken or aerated in another way. Population growth also may cease due to the accumulation of toxic waste products. This factor seems to limit the growth of many anaerobic cultures (cultures growing in the absence of O_2). For example, streptococci can produce so much lactic acid and other organic acids from sugar fermentation that their medium becomes acidic and growth is inhibited. Streptococcal cultures also can enter the stationary phase due to depletion of their sugar supply. Thus entrance into the stationary phase may result from several factors operating in concert.

Death Phase

Detrimental environmental changes like nutrient deprivation and the buildup of toxic wastes lead to the decline in the number of viable cells characteristic of the **death phase.** The death of a microbial population, like its growth during the exponential phase, is usually logarithmic (that is, a constant proportion of cells dies every hour). This pattern holds even when the total cell number remains constant because the cells simply fail to lyse after dying. Often the only way of deciding whether a bacterial cell is viable is by incubating it in fresh medium; if it does not grow and reproduce, it is assumed to be dead.

Although most of a microbial population usually dies in a logarithmic fashion, the death rate may decrease after the population has been drastically reduced. This is due to the extended survival of particularly resistant cells.

The Mathematics of Growth

Knowledge of microbial growth rates during the exponential phase is indispensable to microbiologists. Growth rate studies contribute to basic physiological and ecological research and the solution of applied problems in industry. Therefore the quantitative aspects of exponential phase growth will be discussed.

During the exponential phase each microorganism is dividing at constant intervals. Thus the population will double in number during a specific length of time called the **generation time** or **doubling time.** This situation can be illustrated with a simple example. Suppose that a culture tube is inoculated with

TABLE 6.1   An Example of Exponential Growth

Time[a]	Division Number	2^n	Population ($N_0 \times 2^n$)	$\log_{10} N_t$
0	0	$2^0 = 1$	1	0.000
20	1	$2^1 = 2$	2	0.301
40	2	$2^2 = 4$	4	0.602
60	3	$2^3 = 8$	8	0.903
80	4	$2^4 = 16$	16	1.204
100	5	$2^5 = 32$	32	1.505
120	6	$2^6 = 64$	64	1.806

[a]The hypothetical culture begins with one cell having a 20-minute generation time.

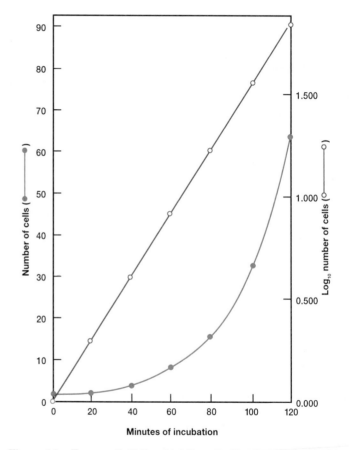

Figure 6.2   Exponential Microbial Growth. The data from table 6.1 for six generations of growth are plotted directly (●—●) and in the logarithmic form (○—○). The growth curve is exponential as shown by the linearity of the log plot.

one cell that divides every 20 minutes (table 6.1). The population will be 2 cells after 20 minutes, 4 cells after 40 minutes, and so forth. Because the population is doubling every generation, the increase in population is always 2^n where n is the number of generations. The resulting population increase is exponential or logarithmic (figure 6.2).

These observations can be expressed as equations for the generation time.

Let N_0 = the initial population number

N_t = the population at time t

n = the number of generations in time t

Then inspection of the results in table 6.1 will show that

$$N_t = N_0 \times 2^n.$$

Solving for n, the number of generations, where all logarithms are to the base 10,

$$\log N_t = \log N_0 + n \cdot \log 2, \text{ and}$$

$$n = \frac{\log N_t - \log N_0}{\log 2} = \frac{\log N_t - \log N_0}{0.301}.$$

The rate of growth in a batch culture can be expressed in terms of the **mean growth rate constant (k).** This is the number of generations per unit time, often expressed as the generations per hour.

$$k = \frac{n}{t} = \frac{\log N_t - \log N_0}{0.301 t}$$

The time it takes a population to double in size—that is, the **mean generation time** or mean doubling time (g), can now be calculated. If the population doubles ($t = g$), then

$$N_t = 2 N_0.$$

Substitute $2N_0$ into the mean growth rate equation and solve for k.

$$k = \frac{\log (2N_0) - \log N_0}{0.301 g} = \frac{\log 2 + \log N_0 - \log N_0}{0.301 g}$$

$$k = \frac{1}{g}$$

The mean generation time is the reciprocal of the mean growth rate constant.

$$g = \frac{1}{k}$$

The mean generation time (g) can be determined directly from a semilogarithmic plot of the growth data (figure 6.3) and the growth rate constant calculated from the g value. The generation time also may be calculated directly from the previous equations. For example, suppose that a bacterial population increases from 10^3 cells to 10^9 cells in 10 hours.

$$k = \frac{\log 10^9 - \log 10^3}{(0.301)(10 \text{ hr})} = \frac{9 - 3}{3.01 \text{ hr}} = 2.0 \text{ generations/hr}$$

$$g = \frac{1}{2.0 \text{ gen./hr}} = 0.5 \text{ hr/gen. or } 30 \text{ min/gen.}$$

Generation times vary markedly with the species of microorganism and environmental conditions. They range from

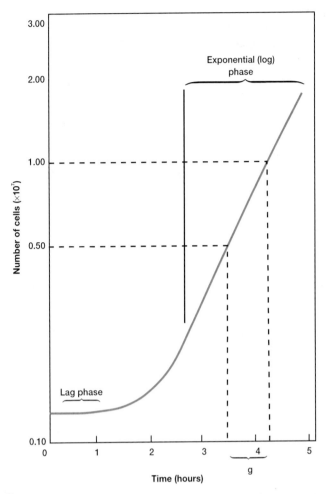

Figure 6.3 Generation Time Determination. The generation time can be determined from a microbial growth curve. The population data are plotted on semilogarithmic graph paper with the logarithmic axis used for the number of cells. The time to double the population number is then read directly from the plot. The log of the population number can also be plotted against time on regular graph paper.

less than 10 minutes (0.17 hours) for a few bacteria to several days with some eucaryotic microorganisms (table 6.2). Generation times in nature are usually much longer than in culture.

1. Define growth. Describe the four phases of the growth curve in a closed system and discuss the causes of each.

2. What are the generation or doubling time and the mean growth rate constant? How can they be determined from growth data?

Measurement of Microbial Growth

There are many ways to measure microbial growth to determine growth rates and generation times. Either population mass or number may be followed because growth leads to

TABLE 6.2	Generation Times for Selected Microorganisms	
Microorganism	**Temperature (°C)**	**Generation Time (Hours)**
Bacteria		
Beneckea natriegens	37	0.16
Escherichia coli	40	0.35
Bacillus subtilis	40	0.43
Clostridium botulinum	37	0.58
Mycobacterium tuberculosis	37	≈ 12
Anacystis nidulans	41	2.0
Anabaena cylindrica	25	10.6
Rhodospirillum rubrum	25	4.6–5.3
Algae		
Chlorella pyrenoidosa	25	7.75
Scenedesmus quadricauda	25	5.9
Asterionella formosa	20	9.6
Skeletonema costatum	18	13.1
Ceratium tripos	20	82.8
Euglena gracilis	25	10.9
Protozoa		
Acanthamoeba castellanii	30	11–12
Paramecium caudatum	26	10.4
Tetrahymena geleii	24	2.2–4.2
Leishmania donovani	26	10–12
Giardia lamblia	37	18
Fungi		
Saccharomyces cerevisiae	30	2
Monilinia fructicola	25	30

(a)

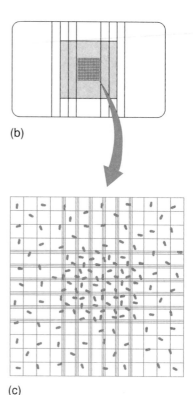

(b)

(c)

Figure 6.4 The Petroff-Hausser Counting Chamber. (*a*) Side view of the chamber showing the cover glass and the space beneath it that holds a bacterial suspension. (*b*) A top view of the chamber. The grid is located in the center of the slide. (*c*) An enlarged view of the grid. The bacteria in several of the central squares are counted, usually at 400× to 500× magnification. The average number of bacteria in these squares is used to calculate the concentration of cells in the original sample. Since there are 25 squares covering an area of 1 mm^2, the total number of bacteria in 1 mm^2 of the chamber is (number/square)(25 squares). The chamber is 0.02 mm deep and therefore,

$$\text{bacteria/mm}^3 = \text{(bacteria/square)(25 squares)(50)}.$$

The number of bacteria per cm^3 is 10^3 times this value. For example, suppose the average count per square is 28 bacteria:

$$\text{bacteria/cm}^3 = \text{(28 bacteria) (25 squares)(50)}(10^3) = 3.5 \times 10^7.$$

increases in both. The most commonly employed techniques for growth measurement are examined briefly and the advantages and disadvantages of each noted.

Measurement of Cell Numbers

The most obvious way to determine microbial numbers is through direct counting. Using a counting chamber is easy, inexpensive, and relatively quick; it also gives information about the size and morphology of microorganisms. Petroff-Hausser counting chambers can be used for counting bacteria; hemocytometers can be used for the larger eucaryotic microorganisms. These specially designed slides have chambers of known depth with an etched grid on the chamber bottom (figure 6.4). The number of microorganisms in a sample can be calculated by taking into account the chamber's volume and any sample dilutions required. There are some disadvantages to the technique. The microbial population must be fairly large for accuracy because such a small volume is sampled. It is also difficult or impossible to distinguish between living and dead cells in counting chambers.

Larger microorganisms such as protozoa, algae, and nonfilamentous yeasts can be directly counted with electronic coun-

ters such as the Coulter Counter. The microbial suspension is forced through a small hole or orifice. An electrical current flows through the hole, and electrodes placed on both sides of the orifice measure its electrical resistance. Every time a microbial cell passes through the orifice, electrical resistance increases (or the conductivity drops) and the cell is counted. The Coulter Counter gives accurate results with larger cells and is

(a)

(b)

Figure 6.5 Equipment for Counting Bacterial Colonies. (*a*) The Quebec colony counter. The counter illuminates the petri plate uniformly from the side, and the plate is magnified for easier counting of small colonies. In this sophisticated model an electric probe is touched to each colony to record the count. (*b*) An automated counter. The camera forms an enlarged image of the plate; all objects of the desired size range in any part of the image are counted by the instrument. The counter may be connected to a separate computer.

extensively used in hospital laboratories to count red and white blood cells. It is not as useful in counting bacteria because of interference by small debris particles, the formation of filaments, and other problems.

Counting chambers and electronic counters yield counts of all cells, whether alive or dead. There are also several viable counting techniques, procedures specific for cells able to grow and reproduce. In most viable counting procedures, a diluted sample of bacteria or other microorganisms is dispersed over a solid surface. Each microorganism or group of microorganisms develops into a distinct colony. The original number of viable microorganisms in the sample can be calculated from the number of colonies formed and the sample dilution. For example, if 1.0 ml of a 1×10^{-6} dilution yielded 150 colonies, the original sample contained around 1.5×10^8 cells per ml. Usually the count is made more accurate by use of a special colony counter (figure 6.5). In this way the spread-plate and pour-plate techniques may be used to find the number of microorganisms in a sample.

Spread-plate and pour-plate techniques (pp. 108–10).

Plating techniques are simple, sensitive, and widely used to count bacteria and other microorganisms in samples of food, water, and soil. Several problems, however, can lead to inaccurate counts. Low counts will result if clumps of cells are not broken up and the microorganisms well dispersed. Because it is not possible to be absolutely certain that each colony arose from an individual cell, the results are often expressed in terms of **colony forming units (CFU)** rather than

the number of microorganisms. The samples should yield between 25 and 250 colonies for best results. Of course the counts will also be low if the agar medium employed cannot support growth of all the viable microorganisms present. The hot agar used in the pour-plate technique may injure or kill sensitive cells; thus spread plates sometimes give higher counts than pour plates.

Analysis of water purity (pp. 850–53).

Microbial numbers are frequently determined from counts of colonies growing on special membrane filters having pores small enough to trap bacteria. In the membrane filter technique, a sample is drawn through a special **membrane filter** (figures 6.6 and 6.7). The filter is then placed on an agar medium or on a pad soaked with liquid media and incubated until each cell forms a separate colony. A colony count gives the number of microorganisms in the filtered sample, and special media can be used to select for specific microorganisms (figure 6.8).

Membrane filters also are used to count bacteria directly. The sample is first filtered through a polycarbonate membrane filter that has been stained black to provide a good background for observing fluorescent objects. The bacteria then are stained with a fluorescent dye such as acridine orange and observed microscopically. Acridine orange–stained microorganisms glow orange or green and are easily counted with an epifluorescence microscope (*see chapter 2*). Usually the counts obtained with this approach are much higher than those from culture techniques. Commercial kits that use fluorescent reagents to stain live and

Figure 6.6   A Millipore Membrane Filter System.
(*a*) Parts of the filter assembly. Some of the more
important components are the following: (1) funnel,
(2) membrane filter, (3) filter holder base and
support, and (4) receiver flask. (*b*) A complete
filtering system. The sample is poured into the
funnel and sucked through a membrane filter with
the use of vacuum.

(a)   (b)

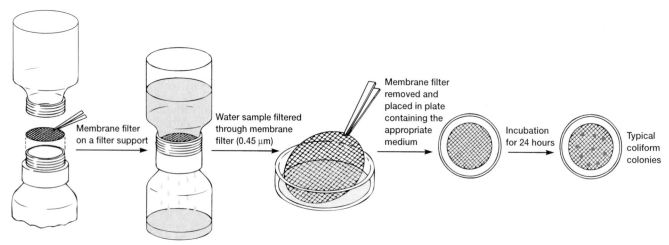

Figure 6.7   The Membrane Filtration Procedure.

(a)   (b)   (c)   (d)

Figure 6.8   Colonies on Membrane Filters. Membrane-filtered samples grown on a variety of media. (*a*) Standard nutrient media for a total
bacterial count. An indicator colors colonies red for easy counting. (*b*) Fecal coliform medium for detecting fecal coliforms that form blue colonies.
(*c*) m-Endo agar for detecting *E. coli* and other coliforms that produce colonies with a green sheen. (*d*) Wort agar for the culture of yeasts and molds.

dead cells differently are now available. This makes it possible to directly count the number of live and dead microorganisms in a sample.

Measurement of Cell Mass

Increases in the total cell mass, as well as in cell numbers, accompany population growth. Therefore techniques for measuring changes in cell mass can be used in following growth. The most direct approach is the determination of microbial dry weight. Cells growing in liquid medium are collected by centrifugation, washed, dried in an oven, and weighed. This is an especially useful technique for measuring the growth of fungi. It is time consuming, however, and not very sensitive. Because bacteria weigh so little, it may be necessary to centrifuge several hundred milliliters of culture to collect a sufficient quantity.

More rapid, sensitive techniques depend on the fact that microbial cells scatter light striking them. Because microbial cells in a population are of roughly constant size, the amount of scattering is proportional to the concentration of cells present. When the concentration of bacteria reaches about 10 million cells (10^7) per ml, the medium appears slightly cloudy or turbid. Further increases in concentration result in greater turbidity and less light is transmitted through the medium. The extent of light scattering can be measured by a spectrophotometer and is almost linearly related to bacterial concentration at low absorbance levels (figure 6.9). Thus population growth can be easily measured spectrophotometrically as long as the population is high enough to give detectable turbidity.

If the amount of a substance in each cell is constant, the total quantity of that cell constituent is directly related to the total microbial cell mass. For example, a sample of washed cells collected from a known volume of medium can be analyzed for total protein or nitrogen. An increase in the microbial population will be reflected in higher total protein levels. Similarly, chlorophyll determinations can be used to measure algal populations, and the quantity of ATP can be used to estimate the amount of living microbial mass.

---

1. Briefly describe each technique by which microbial population numbers may be determined and give its advantages and disadvantages.
2. Why are plate count results often expressed as colony forming units?

Growth Yields and the Effects of a Limiting Nutrient

When microbial growth is limited by the low concentration of a required nutrient, the final net growth or yield of cells increases with the initial amount of the limiting nutrient present (figure 6.10*a*). This is the basis of microbiological assays for vitamins and other growth factors. The rate of growth also increases with nutrient concentration (figure 6.10*b*), but in a hyperbolic manner. The shape of the curve seems to reflect the

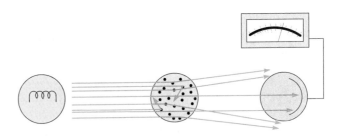

Figure 6.9   Turbidity and Microbial Mass Measurement. Determination of microbial mass by measurement of light absorption. As the population and turbidity increase, more light is scattered and the absorbance reading given by the spectrophotometer increases. The spectrophotometer meter has two scales. The bottom scale displays absorbance and the top scale, percent transmittance. Absorbance increases as percent transmittance decreases.

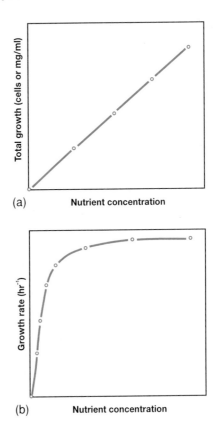

Figure 6.10   Nutrient Concentration and Growth. (*a*) The effect of changes in limiting nutrient concentration on total microbial yield. (*b*) The effect on growth rate.

rate of nutrient uptake by microbial transport proteins. At sufficiently high nutrient levels the transport systems are saturated, and the growth rate does not rise further with increasing nutrient concentration.

Microbiological assays (pp. 101–2).
Nutrient transport systems (pp. 102–6).

The amount of microbial mass produced from a nutrient can be expressed quantitatively as the **growth yield (Y)**.

$$Y = \frac{\text{mass of microorganisms formed}}{\text{mass of substrate consumed}}$$

The Y value is often expressed in terms of grams of cells formed per gram of substrate used or as the molar growth yield (grams of cells per mole of nutrient consumed). It is an index of the efficiency of conversion of nutrients into cell material. Aerobic microorganisms in rich media may assimilate from 20 to 50% of the sugar carbon taken up. In dilute media some bacteria seem able to increase their efficiency and assimilate up to 80% of the sugar acquired.

Growth yields of fermenting bacteria have been studied intensively. When growing in nutritionally rich media, bacteria may use carbohydrate fermentation only to generate energy and employ other substances like amino acids as raw material for biosynthesis. The efficiency of ATP use during fermentation is reflected in the Y_{ATP} value.

$$Y_{ATP} = \frac{\text{grams of cells formed}}{\text{mole ATP produced}}$$

Although Y_{ATP} values may vary at very low sugar concentrations, the Y_{ATP} for most bacteria is about 10.5 g/mole at higher sugar levels. Thus, it appears that bacteria generally use energy in biosynthesis with a fairly constant efficiency. Furthermore, the measured Y_{ATP} value is much lower than the theoretical maximum of 32, which is the Y_{ATP} value expected if all ATP is used in biosynthesis. Much of the ATP is used in transport, mechanical work, and probably other ways.

The Continuous Culture of Microorganisms

Up to this point the focus has been on closed systems called batch cultures in which nutrient supplies are not renewed nor wastes removed. Exponential growth lasts for only a few generations and soon the stationary phase is reached. However, it is possible to grow microorganisms in an open system, a system with constant environmental conditions maintained through continual provision of nutrients and removal of wastes. These conditions are met in the laboratory by a **continuous culture system.** A microbial population can be maintained in the exponential growth phase and at a constant biomass concentration for extended periods in a continuous culture system.

The Chemostat

Two major types of continuous culture systems commonly are used: (1) chemostats and (2) turbidostats. A **chemostat** is constructed so that sterile medium is fed into the culture vessel at the same rate as the media containing microorganisms is removed (figure 6.11). The culture medium for a chemostat possesses an essential nutrient (e.g., an amino acid) in limiting quantities. Because of the presence of a limiting nutrient, the growth rate is determined by the rate at which new medium is fed into the growth chamber, and the final cell density depends on the concentration of the limiting nutrient. The rate of nutrient exchange is expressed as the dilution rate (D), the rate at which medium flows through the culture vessel relative to the vessel volume, where f is the flow rate (ml/hr) and V is the vessel volume (ml).

$$D = f/V$$

For example, if f is 30 ml/hr and V is 100 ml, the dilution rate is 0.30 hr^{-1}.

Both the microbial population level and the generation time are related to the dilution rate (figure 6.12). The microbial population density remains unchanged over a wide range of dilution rates. The generation time decreases (i.e., the growth rate rises) as the dilution rate increases. The limiting nutrient will be almost completely depleted under these balanced conditions. If the dilution rate rises too high, the microorganisms can actually be washed out of the culture vessel before reproducing because the dilution rate is greater than the maximum growth rate. The limiting nutrient concentration rises at higher dilution rates because fewer microorganisms are present to use it.

At very low dilution rates, an increase in D causes a rise in both cell density and the growth rate. This is because of the effect of nutrient concentration on the growth rate, sometimes called the Monod relationship (figure 6.10*b*). Only a limited supply of nutrient is available at low dilution rates. Much of the available energy must be used for cell maintenance, not for growth and reproduction. As the dilution rate increases, the amount of nutrients and the resulting cell density rise because energy is available for both maintenance and growth. The growth rate increases when the total available energy exceeds the **maintenance energy.**

The Turbidostat

The second type of continuous culture system, the **turbidostat,** has a photocell that measures the absorbance or turbidity of the culture in the growth vessel. The flow rate of media through the vessel is automatically regulated to maintain a predetermined turbidity or cell density. The turbidostat differs from the chemostat in several ways. The dilution rate in a turbidostat varies rather than remaining constant, and its culture medium lacks a limiting nutrient. The turbidostat operates best at high dilution rates; the chemostat is most stable and effective at lower dilution rates.

(a)   (b)

Figure 6.11   A Continuous Culture System: The Chemostat. (*a*) Schematic diagram of the system. The fresh medium contains a limiting amount of an essential nutrient. Growth rate is determined by the rate of flow of medium through the culture vessel. (*b*) A commercial fermenter that may be employed as a chemostat. The fermenter uses 0.75 to 1.5 liter culture vessels, and factors such as temperature, pH, and oxygen concentration can be regulated during growth.

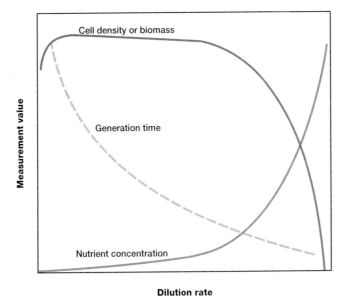

Figure 6.12   Chemostat Dilution Rate and Microbial Growth. The effects of changing the dilution rate in a chemostat.

Continuous culture systems are very useful because they provide a constant supply of cells in exponential phase and growing at a known rate. They make possible the study of microbial growth at very low nutrient levels, concentrations close to those present in natural environments. These systems are essential for research in many areas—for example, in studies on interactions between microbial species under environmental conditions resembling those in a freshwater lake or pond.

Balanced and Unbalanced Growth

Exponential growth, whether in batch or continuous cultures, is **balanced growth.** That is, all cellular constituents are manufactured at constant rates relative to each other. If nutrient levels or other environmental conditions change, **unbalanced growth** results because the rates of synthesis of cell components vary relative to one another until a new balanced state is reached. This response is readily observed in a shift-up experiment in which bacteria are transferred from a nutritionally poor medium to a richer one. The cells first construct new ribosomes to enhance their capacity for protein synthesis. This is followed by increases in protein and DNA synthesis. Finally, the expected rise in reproductive rate takes place.

Protein and DNA synthesis (chapter 11).

Unbalanced growth also results when a bacterial population is shifted down from a rich medium to a poor one. The organisms may previously have been able to obtain many cell components directly from the medium. When shifted to a nutritionally inadequate medium, they need time to make the enzymes required for the biosynthesis of unavailable nutrients. Consequently cell division and DNA replication continue after the shift-down, but net protein and RNA synthesis slow. The cells become smaller and reorganize themselves metabolically until they are able to grow again. Then balanced growth is resumed.

Regulation of nucleic acid synthesis (pp. 242–50).

These shift-up and shift-down experiments demonstrate that microbial growth is under precise, coordinated control and responds quickly to changes in environmental conditions.

---

1. What effect does increasing a limiting nutrient have on the yield of cells and the growth rate? Explain what a growth yield value is and what information it provides.
2. How does an open system differ from a closed culture system or batch culture? Describe how the two different kinds of continuous culture systems, the chemostat and turbidostat, operate. What is the dilution rate? What is maintenance energy?
3. Define balanced growth, unbalanced growth, shift-up experiment, and shift-down experiment.

The Influence of Environmental Factors on Growth

The growth of microorganisms is greatly affected by the chemical and physical nature of their surroundings. An understanding of environmental influences aids in the control of microbial growth and the study of the ecological distribution of microorganisms.

Solutes and Water Activity

Because a selectively permeable plasma membrane separates microorganisms from their environment, they can be affected by changes in the osmotic concentration of their surroundings. If a microorganism is placed in a hypotonic solution, water will enter the cell and cause it to burst unless something is done to prevent the influx. Most bacteria, algae, and fungi have rigid cell walls that maintain the shape and integrity of the cell. Indeed, many microorganisms keep the osmotic concentration of their protoplasm above that of the habitat by the use of compatible solutes, so that the plasma membrane is always pressed firmly against their cell wall. **Compatible solutes** are solutes that are compatible with metabolism and growth when at high intracellular concentrations. Most bacteria increase their internal osmotic concentration through the synthesis or uptake of choline, betaine, proline, glutamic acid, and other amino acids; elevated levels of potassium ions are also involved to some extent. Algae and fungi employ sucrose and polyols—for example, arabitol, glycerol, and mannitol—for the same purpose. Polyols and amino acids are ideal solutes for this function because they normally do not disrupt enzyme structure and function. A few bacteria like *Halobacterium salinarium* raise their osmotic concentration with potassium ions (sodium ions are also elevated but not as much as potassium). *Halobacterium*'s enzymes have been altered so that they actually require high salt concentrations for normal activity (*see chapter 23*). Since protozoa do not have a cell wall, they must use contractile vacuoles (*see figure 27.3*) to eliminate excess water when living in hypotonic environments.

Osmosis and the protective function of the cell wall (pp. 57–58).

When microorganisms with rigid cell walls are placed in a hypertonic environment, water leaves and the plasma membrane shrinks away from the wall, a process known as plasmolysis. This dehydrates the cell and may damage the plasma membrane; the cell usually becomes metabolically inactive and ceases to grow.

The amount of water available to microorganisms can be reduced by interaction with solute molecules (the osmotic effect) or by adsorption to the surfaces of solids (the matric effect). Because the osmotic concentration of a habitat has such profound effects on microorganisms, it is useful to be able to express quantitatively the degree of water availability. Microbiologists generally use **water activity (a_w)** for this purpose. The water activity of a solution is 1/100 the relative humidity of the solution (when expressed as a percent). It is also equivalent to the ratio of the solution's vapor pressure (P_{soln}) to that of pure water (P_{water}).

$$a_w = \frac{P_{soln}}{P_{water}}$$

The water activity of a solution or solid can be determined by sealing it in a chamber and measuring the relative humidity after the system has come to equilibrium. Suppose after a sample is treated in this way, the air above it is 95% saturated—that is, the air contains 95% of the moisture it would have when equilibrated at the same temperature with a sample of pure water. The relative humidity would be 95% and the sample's water activity, 0.95. Water activity is inversely related to osmotic pressure; if a solution has high osmotic pressure, its a_w is low.

Microorganisms differ greatly in their ability to adapt to habitats with low water activity (table 6.3). A microorganism must expend extra effort to grow in a habitat with a low a_w value because it must maintain a high internal solute concentra-

TABLE 6.3   Approximate Lower a_w Limits for Microbial Growth

Water Activity	Environment		Bacteria	Fungi	Algae
1.00—Pure water	Blood Plant wilt Seawater	{ Vegetables, meat, fruit	Most gram-negative nonhalophiles		
0.95	Bread		Most gram-positive rods	*Basidiomycetes*	Most algae
0.90	Ham		Most cocci, *Bacillus*	*Fusarium* *Mucor, Rhizopus* Ascomycetous yeasts	
0.85	Salami		*Staphylococcus*	*Saccharomyces rouxii* (in salt)	
0.80	Preserves			*Penicillium*	
0.75	Salt lakes Salted fish		*Halobacterium* *Actinospora*	*Aspergillus*	*Dunaliella*
0.70				*Aspergillus*	
0.60	Cereals, candy, dried fruit Chocolate Honey Dried milk			*Saccharomyces rouxii* (in sugars) *Xeromyces bisporus*	
0.55—DNA disordered					

Adapted from A. D. Brown, "Microbial Water Stress," in *Bacteriological Reviews*, 40(4):803–846 1976. Copyright © 1976 by the American Society for Microbiology. Reprinted by permission.

tion to retain water. Some microorganisms can do this and are **osmotolerant;** they will grow over wide ranges of water activity or osmotic concentration. For example, *Staphylococcus aureus* can be cultured in media containing any sodium chloride concentration up to about 3 M. The yeast *Saccharomyces rouxii* will grow in sugar solutions with a_w values as low as 0.6. The alga *Dunaliella viridis* tolerates sodium chloride concentrations from 1.7 M to a saturated solution.

Although a few microorganisms are truly osmotolerant, most only grow well at water activities around 0.98 (the approximate a_w for seawater) or higher. This is why drying food or adding large quantities of salt and sugar is so effective in preventing food spoilage. As table 6.3 shows, many fungi are osmotolerant and thus particularly important in the spoilage of salted or dried foods.

Food spoilage (pp. 889–92).

Halophiles have adapted so completely to saline conditions that they require high levels of sodium chloride to grow, concentrations between about 2.8 M and saturation (about 6.2 M) for extreme halophilic bacteria. The genus *Halobacterium* can be isolated from the Dead Sea (a salt lake between Israel and Jordan and the lowest lake in the world), the Great Salt Lake in Utah, and other aquatic habitats with salt concentrations approaching saturation. *Halobacterium* and other extremely halophilic bacteria have significantly modified the structure of their proteins and membranes rather than simply increasing the intracellular concentrations of solutes, the approach used by most osmotolerant microorganisms. Halophiles are archaeobacteria

and thus very different from other bacteria. These extreme halophiles accumulate enormous quantities of potassium in order to remain hypertonic to their environment; the internal potassium concentration may reach 4 to 7 M. The enzymes, ribosomes, and transport proteins of these bacteria require high levels of potassium for stability and activity. In addition, the plasma membrane and cell wall of *Halobacterium* are stabilized by high concentrations of sodium ion. If the sodium concentration decreases too much, the wall and plasma membrane literally disintegrate. Extreme halophilic bacteria have successfully adapted to environmental conditions that would destroy most organisms. In the process they have become so specialized that they have lost ecological flexibility and can prosper only in a few extreme habitats.

Archaeobacteria (chapter 23).

pH

pH is a measure of the hydrogen ion activity of a solution and is defined as the negative logarithm of the hydrogen ion concentration (expressed in terms of molarity).

$$pH = - \log [H^+] = \log(1/[H^+])$$

The pH scale extends from pH 0.0 (1.0 M H^+) to pH 14.0 (1.0×10^{-14} M H^+), and each pH unit represents a tenfold change in hydrogen ion concentration. Figure 6.13 shows that the habitats in which microorganisms grow vary widely—from pH 1 to 2 at the acid end to alkaline lakes and soil that may have pH values between 9 and 10.

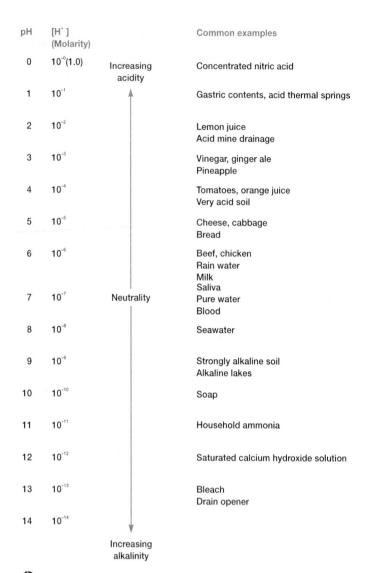

pH	[H⁺] (Molarity)		Common examples

pH	[H⁺] (Molarity)		Common examples
0	10^{-0}(1.0)	Increasing acidity	Concentrated nitric acid
1	10^{-1}		Gastric contents, acid thermal springs
2	10^{-2}		Lemon juice / Acid mine drainage
3	10^{-3}		Vinegar, ginger ale / Pineapple
4	10^{-4}		Tomatoes, orange juice / Very acid soil
5	10^{-5}		Cheese, cabbage / Bread
6	10^{-6}		Beef, chicken / Rain water / Milk / Saliva
7	10^{-7}	Neutrality	Pure water / Blood
8	10^{-8}		Seawater
9	10^{-9}		Strongly alkaline soil / Alkaline lakes
10	10^{-10}		Soap
11	10^{-11}		Household ammonia
12	10^{-12}		Saturated calcium hydroxide solution
13	10^{-13}		Bleach / Drain opener
14	10^{-14}	Increasing alkalinity	

Figure 6.13   The pH Scale. The pH scale and examples of substances with different pH values.

It is not surprising that pH dramatically affects microbial growth. Each species has a definite pH growth range and pH growth optimum (table 6.4). **Acidophiles** have their growth optimum between pH 0 and 5.5; **neutrophiles,** between pH 5.5 and 8.0; and **alkalophiles** prefer the pH range of 8.5 to 11.5. Extreme alkalophiles have growth optima at pH 10 or higher. In general, different microbial groups have characteristic pH preferences. Most bacteria and protozoa are neutrophiles. Most fungi prefer slightly acid surroundings, about pH 4 to 6; algae also seem to favor slight acidity. There are many exceptions to these generalizations. For example, the alga *Cyanidium caldarium* and the bacterium *Sulfolobus acidocaldarius* are common inhabitants of acidic hot springs; both grow well around pH 1 to 3 and at high temperatures.

Although microorganisms will often grow over wide ranges of pH, there are limits to their tolerance. Drastic variations in pH can harm microorganisms by disrupting the plasma membrane or inhibiting the activity of enzymes and membrane transport proteins. Bacterial death occurs if the internal pH drops much below 5.0 to 5.5. Changes in the external pH also might alter the ionization of nutrient molecules and thus reduce their availability to the organism.

Despite wide variations in habitat pH, the internal pH of most microorganisms is close to neutrality. Several mechanisms have been proposed to account for this. The plasma membrane may be relatively impermeable to protons. Neutrophiles appear to exchange potassium for protons using an antiport transport system (*p. 104*). Extreme alkalophiles maintain their internal pH closer to neutrality by exchanging internal sodium ions for external protons. Internal buffering also may contribute to pH homeostasis.

Microorganisms often must adapt to environmental pH changes to survive. In bacteria, potassium/proton and sodium/proton antiport systems probably correct small variations in pH. If the pH becomes too acidic, other mechanisms come into play. When the pH drops below about 5.5 to 6.0, *Salmonella typhimurium* and *E. coli* synthesize an array of new proteins as part of what has been called their acidic tolerance response. A proton-translocating ATPase contributes to this protective response, either by making more ATP or by pumping protons out of the cell. If the external pH decreases to 4.5 or lower, chaperones such as acid shock proteins and heat shock proteins are synthesized (*see pp. 49–50*). Presumably these prevent the acid denaturation of proteins and aid in the refolding of denatured proteins.

Microorganisms frequently change the pH of their own habitat by producing acidic or basic metabolic waste products. Fermentative microorganisms form organic acids from carbohydrates, whereas chemolithotrophs like *Thiobacillus* oxidize reduced sulfur components to sulfuric acid. Other microorganisms make their environment more alkaline by generating ammonia through amino acid degradation.

Microbial fermentations (pp. 175–76).
Sulfur-oxidizing bacteria (pp. 464–66).

Buffers often are included in media to prevent growth inhibition by large pH changes. Phosphate is a commonly used buffer and a good example of buffering by a weak acid ($H_2PO_4^-$) and its salt (HPO_4^{2-}).

$$H^+ + HPO_4^{2-} \longrightarrow H_2PO_4^-$$

$$OH^- + H_2PO_4^- \longrightarrow HPO_4^{2-} + HOH$$

If protons are added to the mixture, they combine with the salt form to yield a weak acid. An increase in alkalinity is resisted because the weak acid will neutralize hydroxyl ions through proton donation to give water. Peptides and amino acids in complex media also have a strong buffering effect.

TABLE **6.4** The Effect of pH on Microbial Growth			
Microorganism	Lower Limit	pH Optimum	Upper Limit
Bacteria			
Picrophilus oshimae	≈0	0.7	ND[a]
Thiobacillus thiooxidans	0.5	2.0–3.5	6.0
Sulfolobus acidocaldarius	1.0	2.5	4.0
Lactobacillus acidophilus	4.0–4.6	5.8–6.6	6.8
Staphylococcus aureus	4.2	7.0–7.5	9.3
Proteus vulgaris	4.4	6.0–7.0	8.4
Escherichia coli	4.4	6.0–7.0	9.0
Clostridium sporogenes	5.5–5.8	6.0–7.6	8.5–9.0
Pseudomonas aeruginosa	5.6	6.6–7.0	8.0
Chlorobium limicola	6.0	6.8	7.0
Nitrosomonas spp.	7.0–7.6	8.0–8.8	9.4
Bacillus pasteurii	8.5	ND	ND
Nostoc muscorum	ND	6.9–7.7	ND
Anabaena variabilis	ND	6.9–9	ND
Microcystis aeruginosa	ND	10	ND
Bacillus alcalophilus	8.5	10.6	11.5
Algae			
Dunaliella acidophila	ND	1.0	ND
Cyanidium caldarium	0	2	5
Euglena gracilis	3.9	6.6	9.9
Fungi			
Acontium velatum	0	ND	ND
Physarum polycephalum	ND	5	6.8
Protozoa			
Paramecium bursaria	4.9	6.7–6.8	8.0
Tetrahymena pyriformis	4.0	7.3	9.0
Acanthamoeba castellanii	3.2	5–6	8.7

[a]ND, no data.

Temperature

Environmental temperature profoundly affects microorganisms, like all other organisms. Indeed, microorganisms are particularly susceptible because they are usually unicellular and also poikilothermic—their temperature varies with that of the external environment. For these reasons, microbial cell temperature directly reflects that of the cell's surroundings. A most important factor influencing the effect of temperature on growth is the temperature sensitivity of enzyme-catalyzed reactions. At low temperatures a temperature rise increases the growth rate because the velocity of an enzyme-catalyzed reaction, like that of any chemical reaction, will roughly double for every 10° C rise in temperature. Because the rate of each reaction increases, metabolism as a whole is more active at higher temperatures, and the microorganism grows faster. Beyond a certain point further increases actually slow growth, and sufficiently high temperatures are lethal. High temperatures damage microorganisms by denaturing enzymes, transport carriers, and other proteins. Microbial membranes are also disrupted by heat; the lipid bilayer simply melts apart. Thus, although functional enzymes operate more rapidly at higher temperatures, the microorganism may be damaged to such an extent that growth is inhibited because the damage cannot be adequately repaired.

The temperature dependence of enzyme activity (pp. 161–62).

Because of these opposing temperature influences, microbial growth has a fairly characteristic temperature dependence with distinct **cardinal temperatures**—minimum, optimum, and maximum growth temperatures (figure 6.14). Although the shape of the temperature dependence curve can vary, the temperature optimum is always closer to the maximum than to the minimum. The cardinal temperatures for a particular species are not rigidly fixed but often depend to some extent on other environmental factors such as pH and the available nutrients. For example, *Crithidia fasciculata,* a flagellated protozoan living in

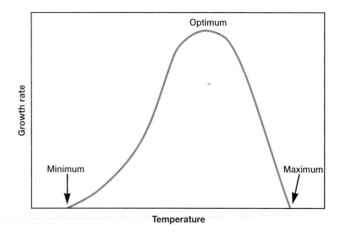

Figure 6.14 Temperature and Growth. The effect of temperature on growth rate.

	Cardinal Temperatures (°C)		
TABLE 6.5 Temperature Ranges for Microbial Growth			
Microorganism	**Minimum**	**Optimum**	**Maximum**
Nonphotosynthetic Bacteria			
Bacillus psychrophilus	–10	23–24	28–30
Micrococcus cryophilus	–4	10	24
Pseudomonas fluorescens	4	25–30	40
Staphylococcus aureus	6.5	30–37	46
Enterococcus faecalis	0	37	44
Escherichia coli	10	37	45
Neisseria gonorrhoeae	30	35–36	38
Thermoplasma acidophilum	45	59	62
Bacillus stearothermophilus	30	60–65	75
Thermus aquaticus	40	70–72	79
Sulfolobus acidocaldarius	60	80	85
Pyrococcus abyssi	67	96	102
Pyrodictium occultum	82	105	110
Photosynthetic Bacteria			
Rhodospirillum rubrum	ND[a]	30–35	ND
Nostoc muscorum	ND	32.5	ND
Anabaena variabilis	ND	35	ND
Anacystis nidulans	ND	41	ND
Oscillatoria tenuis	ND	ND	45–47
Synechococcus lividus	ND	ND	74
Synechococcus eximius	70	79	84
Eucaryotic Algae			
Chlamydomonas nivalis	–36	0	4
Fragilaria sublinearis	–2	5–6	8–9
Chlorella pyrenoidosa	ND	25–26	29
Euglena gracilis	ND	23	ND
Chilomonas paramecium	10	28	35
Skeletonema costatum	6	16–26	>28
Cyanidium caldarium	30–34	45–50	56
Fungi			
Candida scottii	0	4–15	15
Saccharomyces cerevisiae	1–3	28	40
Mucor pusillus	21–23	45–50	50–58
Protozoa			
Amoeba proteus	4–6	22	35
Naegleria fowleri	20–25	35	40
Trichomonas vaginalis	25	32–39	42
Paramecium caudatum		25	28–30
Tetrahymena pyriformis	6–7	20–25	33
Blepharisma intermedium	24–25	28	38
Cyclidium citrullus	18	43	47

[a]ND, no data.

the gut of mosquitos, will grow in a simple medium at 22 to 27°C. However, it cannot be cultured at 33 to 34°C without the addition of extra metals, amino acids, vitamins, and lipids.

The cardinal temperatures vary greatly between microorganisms (table 6.5). Optima normally range from 0°C to as high as 75°C, whereas microbial growth occurs at temperatures extending from –20°C to over 100°C. The growth temperature range for a particular microorganism usually spans about 30 degrees. Some species (e.g., *Neisseria gonorrhoeae*) have a small range and are called **stenothermal;** others, like *Enterococcus faecalis,* will grow over a wide range of temperatures and are **eurythermal.** The major microbial groups differ from one another regarding their maximum growth temperature. The upper limit for protozoa is around 50°C. Some algae and fungi can grow at temperatures as high as 55 to 60°C. Bacteria have been found growing at or close to 100°C, the boiling point of water (*see figure 23.12*), and recently strains growing at even higher temperatures have been discovered (Box 6.1). Clearly, procaryotic organisms can grow at much higher temperatures than eucaryotes. It has been suggested that eucaryotes are not able to manufacture organellar membranes that are stable and functional at temperatures above 60°C. The structure of water also limits the temperature range for growth. It is unlikely that microorganisms can grow in solid ice or steam—almost certainly, liquid water is necessary for microbial growth.

Microorganisms such as those in table 6.5 can be placed in one of five classes based on their temperature ranges for growth (figure 6.15).

1. **Psychrophiles** grow well at 0°C and have an optimum growth temperature of 15°C or lower; the maximum is around 20°C. They are readily isolated from Arctic and Antarctic habitats; because 90% of the ocean is 5°C or colder, it constitutes an enormous habitat for psychrophiles. The psychrophilic alga *Chlamydomonas nivalis* can actually turn a snowfield or glacier pink with its bright red spores. Most psychrophilic bacteria are members of the genera *Pseudomonas, Flavobacterium, Achromobacter,* and *Alcaligenes.* Psychrophilic microorganisms have adapted to their environment in several ways. Their enzymes, transport systems, and protein synthetic mechanisms function well at low temperatures. The cell membranes of psychrophilic microorganisms have high levels of unsaturated fatty acids and remain semifluid when cold. Indeed, many psychrophiles begin to leak cellular constituents at temperatures higher than 20°C because of cell membrane disruption.

Box 6.1

Life above 100°C

U ntil recently the highest reported temperature for bacterial growth was 105°C. It seemed that the upper temperature limit for life was about 100°C, the boiling point of water. Now thermophilic bacteria have been reported growing in sulfide chimneys or "black smokers," located along rifts and ridges on the ocean floor, that spew sulfide-rich superheated vent water with temperatures above 350°C (Box Figure 6.1). Evidence has been presented that these bacteria can grow and reproduce at 115°C. It is possible that bacteria may grow at higher temperatures. The pressure present in their habitat is sufficient to keep water liquid (at 265 atm; seawater doesn't boil until 460°C).

The implications of this discovery are many. The proteins, membranes, and nucleic acids of these bacteria are remarkably temperature stable and provide ideal subjects for studying the ways in which macromolecules and membranes are stabilized. In the future it may be possible to design enzymes that can operate at very high temperatures. Some thermostable enzymes from these bacteria may have important industrial uses.

Box figure 6.1

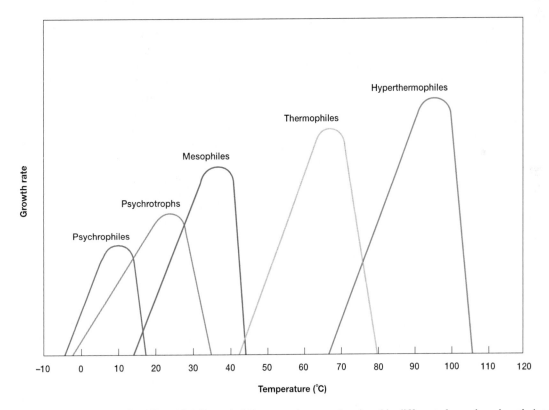

Figure 6.15   Temperature Ranges for Microbial Growth. Microorganisms can be placed in different classes based on their temperature ranges for growth. They are ranked in order of increasing growth temperature range as psychrophiles, psychrotrophs, mesophiles, thermophiles, and hyperthermophiles. Representative ranges and optima for these five types are illustrated here.

2. Many species can grow at 0°C even though they have optima between 20 and 30°C, and maxima at about 35°C. These are called **psychrotrophs** or **facultative psychrophiles.** Psychrotrophic bacteria and fungi are major factors in the spoilage of refrigerated foods (*see chapter 43*).

3. **Mesophiles** are microorganisms with growth optima around 20 to 45°C and a temperature minimum of 15 to 20°C. Their maximum is about 45°C or lower. Most microorganisms probably fall within this category. Almost all human pathogens are mesophiles, as might be expected since their environment is a fairly constant 37°C.

4. Some microorganisms are **thermophiles;** they can grow at temperatures of 55°C or higher. Their growth minimum is usually around 45°C and they often have optima between 55 and 65°C. The vast majority are bacteria although a few algae and fungi are thermophilic (table 6.5). These organisms flourish in many habitats including composts, self-heating hay stacks, hot water lines, and hot springs. Thermophiles differ from mesophiles in having much more heat-stable enzymes and protein synthesis systems able to function at high temperatures. Their membrane lipids are also more saturated than those of mesophiles and have higher melting points; therefore thermophile membranes remain intact at higher temperatures.

5. As mentioned previously, a few thermophiles can grow at 90°C or above and some have maxima above 100°C. Bacteria that have growth optima between 80°C and about 110°C are called **hyperthermophiles.** They usually do not grow well below 55°C. *Pyrococcus abyssi* and *Pyrodictium occultum* are examples of marine hyperthermophiles found in hot areas of the seafloor.

---

1. How do microorganisms adapt to hypotonic and hypertonic environments? What is plasmolysis? Define water activity and briefly describe how it can be determined. Why is it difficult for microorganisms to grow at low a_w values? What are halophiles and why does *Halobacterium* require sodium and potassium ions?

2. Define pH, acidophile, neutrophile, and alkalophile. How can microorganisms change the pH of their environment, and how does the microbiologist minimize this effect?

3. What are cardinal temperatures? Why does the growth rate rise with increasing temperature and then fall again at higher temperatures? Define stenothermal, eurythermal, psychrophile, psychrotroph, mesophile, thermophile, and hyperthermophile.

Oxygen Concentration

An organism able to grow in the presence of atmospheric O_2 is an **aerobe,** whereas one that can grow in its absence is an **anaerobe.** Almost all higher organisms are completely dependent

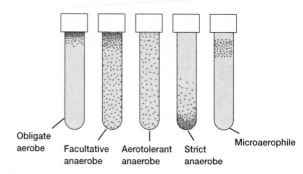

Obligate aerobe   Facultative anaerobe   Aerotolerant anaerobe   Strict anaerobe   Microaerophile

Figure 6.16   Oxygen and Bacterial Growth. An illustration of the growth of bacteria with varying responses to oxygen. Each dot represents an individual bacterial colony within the agar or on its surface. The surface, which is directly exposed to atmospheric oxygen, will be aerobic. The oxygen content of the medium decreases with depth until the medium becomes anaerobic toward the bottom of the tube.

on atmospheric O_2 for growth—that is, they are **obligate aerobes.** Oxygen serves as the terminal electron acceptor for the electron-transport chain in aerobic respiration. In addition, aerobic eucaryotes employ O_2 in the synthesis of sterols and unsaturated fatty acids. **Facultative anaerobes** do not require O_2 for growth but do grow better in its presence. **Aerotolerant anaerobes** such as *Enterococcus faecalis* simply ignore O_2 and grow equally well whether it is present or not. In contrast, **strict** or **obligate anaerobes** (e.g., *Bacteroides, Fusobacterium, Clostridium pasteurianum, Methanococcus*) do not tolerate O_2 at all and die in its presence. Aerotolerant and strict anaerobes cannot generate energy through respiration and must employ fermentation or anaerobic respiration pathways for this purpose. Finally, there are a few aerobes such as *Campylobacter,* called **microaerophiles,** that are damaged by the normal atmospheric level of O_2 (20%) and require O_2 levels below the range of 2 to 10% for growth. The nature of bacterial O_2 responses can be readily determined by growing the bacteria in culture tubes filled with a solid culture medium or a special medium like thioglycollate broth, which contains a reducing agent to lower O_2 levels (figure 6.16).

Electron transport and aerobic respiration (pp. 171–75).

Fermentation (pp. 175–76).

Anaerobic respiration (pp. 176–78).

A microbial group may show more than one type of relationship to O_2. All five types are found among the bacteria and protozoa. Fungi are normally aerobic, but a number of species—particularly among the yeasts—are facultative anaerobes. Algae are almost always obligate aerobes.

Although strict anaerobes are killed by O_2, they may be recovered from habitats that appear to be aerobic. In such cases they associate with facultative anaerobes that use up the available O_2 and thus make the growth of strict anaerobes possible.

For example, the strict anaerobe *Bacteroides gingivalis* lives in the mouth where it grows in the anaerobic crevices around the teeth.

These different relationships with O_2 appear due to several factors, including the inactivation of proteins and the effect of toxic O_2 derivatives. Enzymes can be inactivated when sensitive groups like sulfhydryls are oxidized. A notable example is the nitrogen-fixation enzyme nitrogenase (*see chapter 10*), which is very oxygen sensitive.

Oxygen accepts electrons and is readily reduced because its two outer orbital electrons are unpaired. Flavoproteins (*see chapter 8*), several other cell constituents, and radiation (*see pp. 132–33*) promote oxygen reduction. The result is usually some combination of the reduction products **superoxide radical, hydrogen peroxide,** and **hydroxyl radical.**

$$O_2 + e^- \rightarrow O_2^{-} \text{ (superoxide radical)}$$

$$O_2^{-} + e^- + 2H^+ \longrightarrow H_2O_2 \text{ (hydrogen peroxide)}$$

$$H_2O_2 + e^- + H^+ \longrightarrow H_2O + OH \cdot \text{ (hydroxyl radical)}$$

These products of oxygen reduction are extremely toxic because they are powerful oxidizing agents and destroy cellular constituents very rapidly. A microorganism must be able to protect itself against such O_2 products or it will be killed. Neutrophils and macrophages use these toxic oxygen products to destroy invading pathogens.

Oxygen-dependent killing of pathogens (p. 582).

Many microorganisms possess enzymes that afford protection against toxic O_2 products. Obligate aerobes and facultative anaerobes usually contain the enzymes **superoxide dismutase** and **catalase,** which catalyze the destruction of superoxide radical and hydrogen peroxide, respectively.

$$2O_2^{-} + 2H^+ \xrightarrow{\text{superoxide dismutase}} O_2 + H_2O_2$$

$$2H_2O_2 \xrightarrow{\text{catalase}} 2H_2O + O_2$$

Aerotolerant microorganisms may lack catalase but almost always have superoxide dismutase. The aerotolerant *Lactobacillus plantarum* uses manganous ions instead of superoxide dismutase to destroy the superoxide radical. All strict anaerobes lack both enzymes or have them in very low concentrations and therefore cannot tolerate O_2.

Because aerobes need O_2 and anaerobes are killed by it, radically different approaches must be used when growing the two types of microorganisms. When large volumes of aerobic microorganisms are cultured, either the culture vessel is shaken to aerate the medium or sterile air must be pumped through the culture vessel. Precisely the opposite problem arises with anaerobes; all O_2 must be excluded. This can be accomplished in several ways. (1) Special anaerobic media containing reducing agents such as thioglycollate or cysteine may be used. The reducing agents will eliminate any dissolved O_2 present within the

Figure 6.17   An Anaerobic Work Chamber and Incubator. This anaerobic system contains an oxygen-free work area and an incubator. The interchange compartment on the right of the work area allows materials to be transferred inside without exposing the interior to oxygen. The anaerobic atmosphere is maintained largely with a vacuum pump and nitrogen purges. The remaining oxygen is removed by a palladium catalyst and hydrogen. The oxygen reacts with hydrogen to form water, which is absorbed by desiccant.

medium so that anaerobes can grow beneath its surface. (2) Oxygen also may be eliminated from an anaerobic system by removing air with a vacuum pump and flushing out residual O_2 with nitrogen gas (figure 6.17). Often CO_2 as well as nitrogen is added to the chamber since many anaerobes require a small amount of CO_2 for best growth. (3) One of the most popular ways of culturing small numbers of anaerobes is by use of a Gas-Pak jar (figure 6.18). In this procedure the environment is made anaerobic by using hydrogen and a palladium catalyst to remove O_2 through the formation of water. The reducing agents in anaerobic agar also remove oxygen, as mentioned previously. (4) Plastic bags or pouches make convenient containers when only a few samples are to be incubated anaerobically. These have a catalyst and calcium carbonate to produce an anaerobic, carbon-dioxide-rich atmosphere. A special solution is added to the pouch's reagent compartment; petri dishes or other containers are placed in the pouch; it then is clamped shut and placed in an incubator. A laboratory may make use of all these techniques since each is best suited for different purposes.

Summary

1. Growth is an increase in cellular constituents and results in an increase in cell size, cell number, or both.

2. When microorganisms are grown in a closed system or batch culture, the resulting growth curve usually has four phases: the lag, exponential or log, stationary, and death phases.

3. In the exponential phase, the population number doubles at a constant interval called the doubling or generation time. The mean growth rate constant (k) is the reciprocal of the generation time.

4. Microbial populations can be counted directly with counting chambers, electronic counters, or fluorescence microscopy.

5. Viable counting techniques such as the spread plate, the pour plate, or the membrane filter can be employed.

6. Population changes also can be followed by determining variations in microbial mass through the measurement of dry weight, turbidity, or the amount of a cell component.

7. The amount of microbial mass produced from a nutrient can be expressed in terms of the growth yield (Y). Only part of the ATP generated is used in biosynthesis; the majority is used in transport, movement, and other processes.

8. Microorganisms can be grown in an open system in which nutrients are constantly provided and wastes removed.

9. A continuous culture system is an open system that can maintain a microbial population in the log phase. There are two types of these systems: chemostats and turbidostats.

10. Exponential growth is balanced growth, cell components are synthesized at constant rates relative to one another. Changes in culture conditions (e.g., in shift-up and shift-down experiments) lead to unbalanced growth. A portion of the available nutrients is used to generate maintenance energy.

11. Most bacteria, algae, and fungi have rigid cell walls and are hypertonic to the habitat because of solutes such as amino acids, polyols, and potassium ions.

12. The amount of water actually available to microorganisms is expressed in terms of the water activity (a_w).

13. Although most microorganisms will not grow well at water activities below 0.98 due to plasmolysis and associated effects, osmotolerant organisms survive and even flourish at low a_w values. Halophiles actually require high sodium chloride concentrations for growth.

14. Each species of microorganism has an optimum pH for growth and can be classified as an acidophile, neutrophile, or alkalophile.

15. Microorganisms can alter the pH of their surroundings, and most culture media must be buffered to stabilize the pH.

16. Microorganisms have distinct temperature ranges for growth with minima, maxima, and optima—the cardinal temperatures. These ranges are determined by the effects of temperature on the rates of catalysis, protein denaturation, and membrane disruption.

17. There are four major classes of microorganisms with respect to temperature preferences: (1) psychrophiles, (2) facultative psychrophiles or psychrotrophs, (3) mesophiles, and (4) thermophiles. Hyperthermophiles grow at even higher temperatures.

18. Microorganisms can be placed into at least five different categories based on their response to the presence of O_2: obligate aerobes, facultative anaerobes, aerotolerant anaerobes, strict or obligate anaerobes, and microaerophiles.

19. Oxygen is toxic because of the production of hydrogen peroxide, superoxide radical, and hydroxyl radical. These are destroyed by the enzymes superoxide dismutase and catalase.

20. Most deep-sea microorganisms are barotolerant, but some are barophilic and require high pressure for optimal growth.

21. High-energy or short-wavelength radiation harms organisms in several ways. Ionizing radiation—X rays and gamma rays—ionizes molecules and destroys DNA and other cell components.

22. Ultraviolet (UV) radiation induces the formation of thymine dimers and strand breaks in DNA. Such damage can be repaired by photoreactivation or dark reactivation mechanisms.

23. Visible light can provide energy for the formation of reactive singlet oxygen, which will destroy cells.

Key Terms

acidophile *126*

aerobe *130*

aerotolerant anaerobe *130*

alkalophile *126*

anaerobe *130*

balanced growth *123*

barophilic *132*

barotolerant *132*

batch culture *115*

cardinal temperatures *127*

catalase *131*

chemostat *122*

coenocytic *115*

colony forming units (CFU) *119*

compatible solutes *124*

continuous culture system *122*

dark reactivation *133*

death phase *116*

doubling time *116*

eurythermal *128*

exponential phase *115*

facultative anaerobe *130*

facultative psychrophiles *130*

generation time *116*

growth *115*

growth yield (Y) *122*

halophile *125*

hydrogen peroxide *131*

hydroxyl radical *131*

hyperthermophile *130*

ionizing radiation *133*

lag phase *115*

log phase *115*

maintenance energy *122*

mean generation time *117*

mean growth rate constant (k) *117*

membrane filter *119*

mesophile *130*

microaerophile *130*

neutrophile *126*

obligate aerobe *130*

obligate anaerobe *130*

osmotolerant *125*

photoreactivation *133*

psychrophile *128*

psychrotroph *130*

singlet oxygen *133*

stationary phase *115*

stenothermal *128*

strict anaerobe *130*

superoxide dismutase *131*

superoxide radical *131*

thermophile *130*

turbidostat *122*

ultraviolet (UV) radiation *133*

unbalanced growth *123*

water activity (a_w) *124*

Questions for Thought and Review

1. Discuss the reasons why a culture might have a long lag phase after inoculation.

2. Why can't one always tell when a culture enters the death phase by the use of total cell counts?

3. Calculate the mean growth rate and generation time of a culture that increases from 5×10^2 to 1×10^8 cells in 12 hours.

4. If the generation time is 90 minutes and the initial population contains 10^3 cells, how many bacteria will there be after 8 hours of exponential growth?

5. What does the fact that the Y_{ATP} of most fermenting bacteria is about 10.5 at high sugar levels mean in terms of the efficiency of biosynthesis and the uses of ATP?

6. Why are continuous culture systems so useful to microbiologists?

7. How do bacterial populations respond in shift-up and shift-down experiments? Account for their behavior in molecular terms.

8. Does the internal pH remain constant despite changes in the external pH? How might this be achieved? Explain how extreme pH values might harm microorganisms.

9. What metabolic and structural adaptations for extreme temperatures have psychrophiles and thermophiles made?

10. Why are generation times in nature usually much longer than in culture?

Additional Reading

General

Atlas, R. M., and Bartha, R. 1993. *Microbial ecology: Fundamentals and applications,* 3d ed. Menlo Park, Calif.: Benjamin/Cummings.

Caldwell, D. R. 1995. *Microbial physiology and metabolism.* Dubuque, Iowa: Wm. C. Brown.

Gerhardt, P.; Murray, R. G. E.; Wood, W. A.; and Krieg, N. R., editors. 1994. *Methods for general and molecular bacteriology,* chaps. 6–12. Washington, D.C.: American Society for Microbiology.

Kogut, M. 1980. Are there strategies of microbial adaptation to extreme environments? *Trends Biochem. Sci.* 5(1):15–18.

Kogut, M. 1980. Microbial strategies of adaptability. *Trends Biochem. Sci.* 5(2):47–50.

Kushner, D. J., editor. 1978. *Microbial life in extreme environments.* New York: Academic Press.

Moat, A. G., and Foster, J. W. 1995. *Microbial physiology,* 3d ed. New York: John Wiley and Sons.

Neidhardt, F. C.; Ingraham, J. L.; and Schaechter, M. 1990. *Physiology of the bacterial cell: A molecular approach.* Sunderland, Mass.: Sinauer Associates.

Schlegel, H. G., and Jannasch, H. W. 1992. Prokaryotes and their habitats. In *The prokaryotes,* 2d ed. A. Balows et al., editors, 75–125. New York: Springer-Verlag.

The Growth Curve

Kolter, R.; Siegele, D. A.; and Tormo, A. 1993. The stationary phase of the bacterial life cycle. *Ann. Rev. Microbiol.* 47:855–74.

Prosser, J. I., and Tough, A. J. 1991. Growth mechanisms and growth kinetics of filamentous microorganisms. *Critical reviews in biotechnology* 10(4):253–74.

Tempest, D. W. 1978. Dynamics of microbial growth. In *Essays in microbiology,* J. R. Norris and M. H. Richmond, editors, 7/1–7/32. New York: John Wiley and Sons.

Growth Yields and the Effects of a Limiting Nutrient

Harrison, D. E. F. 1978. Efficiency of microbial growth. In *Companion to microbiology,* A. T. Bull and P. M. Meadow, editors, 155–79. New York: Longman.

Payne, W. J. 1978. Growth yield and efficiency in chemosynthetic microorganisms. *Ann. Rev. Microbiol.* 32:155–83.

Russell, J. B., and Cook, G. M. 1995. Energetics of bacterial growth: Balance of anabolic and catabolic reactions. *Microbiol. Rev.* 59(1):48–62.

Stouthamer, A. H. 1977. Energetic aspects of the growth of micro-organisms. In *Microbial energetics,* B. A. Haddock and W. A. Hamilton, editors, 285–315. New York: Cambridge University Press.

The Influence of Environmental Factors on Growth

Adams, M. W. W. 1993. Enzymes and proteins from organisms that grow near and above 100° C. *Ann. Rev. Microbiol.* 47:627–58.

Blomberg, A., and Adler, L. 1992. Physiology of osmotolerance in fungi. In *Advances in microbial physiology,* vol. 33, A. H. Rose., editor, 145–212. San Diego: Academic Press.

Booth, I. R. 1985. Regulation of cytoplasmic pH in bacteria. *Microbiol. Rev.* 49(4):359–78.

Brock, T. D. 1985. Life at high temperatures. *Science* 230:132–38.

Brown, A. D. 1976. Microbial water stress. *Bacteriol. Rev.* 40(4):803–46.

Csonka, L. N. 1989. Physiological and genetic responses of bacteria to osmotic stress. *Microbiol. Rev.* 53(1):121–47.

Fridovich, I. 1977. Oxygen is toxic! *BioScience* 27(7):462–66.

Gaill, F. 1993. Aspects of life development at deep sea hydrothermal vents. *FASEB J.* 7:558–65.

Gottschal, J. C., and Prins, R. A. 1991. Thermophiles: A life at elevated temperatures. *Trends Ecol. & Evol.* 6(5):157–62.

Inlag, J. A., and Linn, S. 1988. DNA damage and oxygen radical toxicity. *Science* 240:1302–9.

Jannasch, H. W., and Taylor, C. D. 1984. Deep-sea microbiology. *Ann. Rev. Microbiol.* 38:487–514.

Jannasch, H. W., and Wirsen, C. O. 1977. Microbial life in the deep sea. *Sci. Am.* 236(6):42–52.

Krieg, N. R., and Hoffman, P. S. 1986. Microaerophily and oxygen toxicity. *Ann. Rev. Microbiol.* 40:107–30.

Krulwich, T. A., and Guffanti, A. A. 1989. Alkalophilic bacteria. *Ann. Rev. Microbiol.* 43:435–63.

Le Rudulier, D.; Strom, A. R.; Dandekar, A. M.; Smith, L. T.; and Valentine, R. C. 1984. Molecular biology of osmoregulation. *Science* 224:1064–68.

Morita, R. Y. 1975. Psychrophilic bacteria. *Bacteriol. Rev.* 39(2):144–67.

Potts, M. 1994. Desiccation tolerance of prokaryotes. *Microbiol. Rev.* 58(4):755–805.

Russell, N. J. 1984. Mechanisms of thermal adaptation in bacteria: Blueprints for survival. *Trends Biochem. Sci.* 9(3):108–12.

Stetter, K. O. 1995. Microbial life in hyperthermal environments. *ASM News* 61(6):285–90.

Yancey, P. H.; Clark, M. E.; Hand, S. C.; Bowlus, R. D.; and Somero, G. N. 1982. Living with water stress: Evolution of osmolyte systems. *Science* 217:1214–22.

CHAPTER 7

Control of Microorganisms by Physical and Chemical Agents

Were we but able to explain

The fiefdom of the microbe—

Why one man is his serf,

Another is his lord

When all are his

domain. . . .

— Anonymous

We all labour against our

own cure, for death is the cure

of all diseases.

—Sir Thomas Browne

Outline

Concepts

1. Microbial population death is exponential, and the effectiveness of an agent is not fixed but influenced by many environmental factors.

2. Solid objects can be sterilized by physical agents such as heat and radiation; liquids and gases are sterilized by heat, radiation, and filtration through the proper filter.

3. Most chemical agents do not readily destroy bacterial endospores and therefore cannot sterilize objects; they are used as disinfectants, sanitizers, and antiseptics. Objects can be sterilized by gases like ethylene oxide that destroy endospores.

4. A knowledge of methods used for microbial control is essential for personal and public safety.

The chapters in Part II are concerned with the nutrition, growth, and control of microorganisms. This chapter specifically addresses the subject of their control and destruction, a topic of immense practical importance. Although many microorganisms are beneficial and necessary for human well-being, microbial activities may have undesirable consequences, such as food spoilage and disease. Therefore it is essential to be able to kill microorganisms or inhibit their growth to minimize their destructive effects.

This chapter focuses on the control of microorganisms by physical and chemical agents. Chapter 33 introduces the use of antimicrobial chemotherapy to control microbial disease.

From the beginning of recorded history, people have practiced disinfection and sterilization, even though the existence of microorganisms was long unsuspected. The Egyptians used fire to sterilize infectious material and disinfectants to embalm bodies, and the Greeks burned sulfur to fumigate buildings. Mosaic law commanded the Hebrews to burn any clothing suspected of being contaminated with the leprosy bacterium. Today the ability to destroy microorganisms is no less important: it makes possible the aseptic techniques used in microbiological research, the preservation of food, and the prevention of disease. The techniques described in this chapter are also essential to personal safety in both the laboratory and hospital (Box 7.1).

There are several ways to control microbial growth that have not been included in this chapter, but they should be considered for a more complete picture of how microorganisms are controlled. Chapter 6 describes the effects of osmotic activity, pH, temperature, O_2, and radiation on microbial growth and survival (*see pp. 124–33*). Chapter 43 discusses the use of physical and chemical agents in food preservation (*see pp. 883–85*).

Definition of Frequently Used Terms

Terminology is especially important when the control of microorganisms is discussed because words like disinfectant and antiseptic often are used loosely. The situation is even more confusing because a particular treatment can either inhibit growth or kill depending on the conditions.

The ability to control microbial populations on inanimate objects, like eating utensils and surgical instruments, is of considerable practical importance. Sometimes it is necessary to eliminate all microorganisms from an object, whereas only partial destruction of the microbial population may be required in other situations. **Sterilization** [Latin *sterilis*, unable to produce offspring or barren] is the process by which all living cells, viable spores, viruses, and viroids (*see chapter 18*) are either destroyed or removed from an object or habitat. A sterile object is totally free of viable microorganisms, spores, and other infectious agents. When sterilization is achieved by a chemical agent, the chemical is called a sterilant. In contrast, **disinfection** is the killing, inhibition, or removal of microorganisms that may cause disease. **Disinfectants** are agents, usually chemical, used to carry out disinfection and are normally used only on inanimate objects. A disinfectant does not necessarily sterilize an object because viable spores and a few microorganisms may remain. **Sanitization** is closely related to disinfection. In

---

sanitization, the microbial population is reduced to levels that are considered safe by public health standards. The inanimate object is usually cleaned as well as partially disinfected. For example, sanitizers are used to clean eating utensils in restaurants.

It is frequently necessary to control microorganisms on living tissue with chemical agents. **Antisepsis** [Greek *anti,* against, and *sepsis,* putrefaction] is the prevention of infection or sepsis and is accomplished with **antiseptics.** These are chemical agents applied to tissue to prevent infection by killing or inhibiting pathogen growth. Because they must not destroy too much host tissue, antiseptics are generally not as toxic as disinfectants.

A suffix can be employed to denote the type of antimicrobial agent. Substances that kill organisms often have the suffix -cide [Latin *cida,* to kill]: a **germicide** kills pathogens (and many nonpathogens) but not necessarily endospores. A disinfectant or antiseptic can be particularly effective against a specific group, in which case it may be called a **bactericide, fungicide, algicide,** or **viricide.** Other chemicals do not kill, but they do prevent growth. Their names end in -static [Greek *statikos,* causing to stand or stopping]—for example, **bacteriostatic** and **fungistatic.**

Although these agents have been described in terms of their effects on pathogens, it should be noted that they also kill or inhibit the growth of nonpathogens as well. Their ability to reduce the total microbial population, not just to affect pathogen levels, is quite important in many situations.

---

1. Define the following terms: sterilization, sterilant, disinfection, disinfectant, sanitization, antisepsis, antiseptic, germicide, bactericide, bacteriostatic.

The Pattern of Microbial Death

A microbial population is not killed instantly when exposed to a lethal agent. Population death, like population growth, is generally exponential or logarithmic—that is, the population will be reduced by the same fraction at constant intervals (table 7.1). If the logarithm of the population number remaining is plotted against the time of exposure of the microorganism to the agent,

a straight line plot will result (compare figure 7.1 with figure 6.2). When the population has been greatly reduced, the rate of killing may slow due to the survival of a more resistant strain of the microorganism.

To study the effectiveness of a lethal agent, one must be able to decide when microorganisms are dead, a task by no means as easy as with macroorganisms. It is hardly possible to take a bacterium's pulse. A bacterium is defined as dead if it does not grow when inoculated into culture medium that would normally support its growth (*however, see Box 40.2*). In like manner an inactive virus cannot infect a suitable host.

---

1. Describe the pattern of microbial death and how one decides whether microorganisms are actually dead.

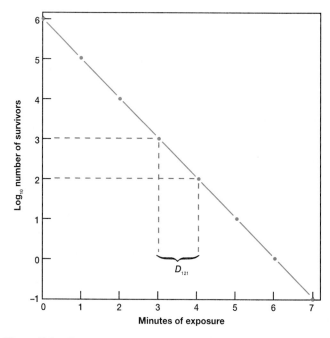

Figure 7.1   The Pattern of Microbial Death. An exponential plot of the survivors versus the minutes of exposure to heating at 121°C. In this example the D_{121} value is 1 minute. The data are from table 7.1.

TABLE 7.1	A Theoretical Microbial Heat-Killing Experiment			
Minute	Microbial Number at Start of Minute[a]	Microorganisms Killed in 1 Minute (90% of total)[a]	Microorganisms at End of 1 Minute	Log$_{10}$ of Survivors
1	10^6	9×10^5	10^5	5
2	10^5	9×10^4	10^4	4
3	10^4	9×10^3	10^3	3
4	10^3	9×10^2	10^2	2
5	10^2	9×10^1	10	1
6	10^1	9	1	0
7	1	0.9	0.1	−1

[a]Assume that the initial sample contains 10^6 vegetative microorganisms per ml and that 90% of the organisms are killed during each minute of exposure. The temperature is 121° C.

Conditions Influencing the Effectiveness of Antimicrobial Agent Activity

Destruction of microorganisms and inhibition of microbial growth are not simple matters because the efficiency of an **antimicrobial agent** (an agent that kills microorganisms or inhibits their growth) is affected by at least six factors.

1. Population size. Because an equal fraction of a microbial population is killed during each interval, a larger population requires a longer time to die than a smaller one. This can be seen in the theoretical heat-killing experiment shown in table 7.1 and figure 7.1. The same principle applies to chemical antimicrobial agents.

2. Population composition. The effectiveness of an agent varies greatly with the nature of the organisms being treated because microorganisms differ markedly in susceptibility. Bacterial endospores are much more resistant to most antimicrobial agents than are vegetative forms, and younger cells are usually more readily destroyed than mature organisms. Some species are able to withstand adverse conditions better than others. *Mycobacterium tuberculosis,* which causes tuberculosis, is much more resistant to antimicrobial agents than most other bacteria.

3. Concentration or intensity of an antimicrobial agent. Often, but not always, the more concentrated a chemical agent or intense a physical agent, the more rapidly microorganisms are destroyed. Dependence of effectiveness on concentration or intensity is generally not linear. Over a short range a small increase in concentration leads to an exponential rise in effectiveness; beyond a certain point, further increases may not raise the killing rate much at all. Sometimes an agent is more effective at lower concentrations. For example, 70% ethanol is more effective than 95% ethanol.

4. Duration of exposure. The longer a population is exposed to a microbicidal agent, the more organisms are killed (figure 7.1). To achieve sterilization, an exposure duration sufficient to reduce the probability of survival to 10^{-6} or less should be used.

5. Temperature. An increase in the temperature at which a chemical acts often enhances its activity. Frequently a lower concentration of disinfectant or sterilizing agent can be used at a higher temperature.

6. Local environment. The population to be controlled is not isolated but surrounded by environmental factors that may either offer protection or aid in its destruction. For example, because heat kills more readily at an acid pH, acid foods and beverages such as fruits and tomatoes are much easier to pasteurize than more neutral foods like milk. A second important environmental factor is organic matter that can protect microorganisms against heating and chemical disinfectants. It may be necessary to clean an object before it is disinfected or sterilized. Syringes and medical or dental equipment should be cleaned before

sterilization because the presence of too much organic matter could protect pathogens and increase the risk of infection. The same care must be taken when pathogens are destroyed during the preparation of drinking water. When a city's water supply has a high content of organic material, more chlorine must be added to disinfect it.

---

1. Briefly explain how the effectiveness of antimicrobial agents varies with population size, population composition, concentration or intensity of the agent, treatment duration, temperature, and local environmental conditions.

Table 7.2	**Approximate Conditions for Moist Heat Killing**	
Organism	Vegetative Cells	Spores
Yeasts	5 minutes at 50–60°C	5 minutes at 70–80°C
Molds	30 minutes at 62°C	30 minutes at 80°C
Bacteria[a]	10 minutes at 60–70°C	2 to over 800 minutes at 100°C
		0.5–12 minutes at 121°C
Viruses	30 minutes at 60°C	

[a]Conditions for mesophilic bacteria.

The Use of Physical Methods in Control

Heat and other physical agents are normally used to sterilize objects, as can be seen from the continual operation of the autoclave in every microbiology laboratory. The four most frequently employed physical agents are heat, filtration, ultraviolet radiation, and ionizing radiation.

Heat

Fire and boiling water have been used for sterilization and disinfection since the time of the Greeks, and heating is still one of the most popular ways to destroy microorganisms. Either moist or dry heat may be applied.

Moist heat readily kills viruses, bacteria, and fungi (table 7.2). Exposure to boiling water for 10 minutes is sufficient to destroy vegetative cells and eucaryotic spores. Unfortunately the temperature of boiling water (100°C or 212°F) is not high enough to destroy bacterial endospores that may survive hours of boiling. Therefore boiling can be used for disinfection of drinking water and objects not harmed by water, but boiling does not sterilize.

Because heat is so useful in controlling microorganisms, it is essential to have a precise measure of the heat-killing efficiency. Initially effectiveness was expressed in terms of thermal death point (TDP), the lowest temperature at which a microbial suspension is killed in 10 minutes. Because TDP implies that a certain temperature is immediately lethal despite the conditions,

thermal death time (TDT) is now more commonly used. This is the shortest time needed to kill all organisms in a microbial suspension at a specific temperature and under defined conditions. However, such destruction is logarithmic, and it is theoretically not possible to "completely destroy" microorganisms in a sample, even with extended heating. Therefore an even more precise figure, the **decimal reduction time (D)** or **D value** has gained wide acceptance. The decimal reduction time is the time required to kill 90% of the microorganisms or spores in a sample at a specified temperature. In a semilogarithmic plot of the population remaining versus the time of heating (figure 7.1), the D value is the time required for the line to drop by one log cycle or tenfold. The D value is usually written with a subscript, indicating the temperature for which it applies. D values are used to estimate the relative resistance of a microorganism to different temperatures through calculation of the **z value.** The z value is the increase in temperature required to reduce D to 1/10 its value or to reduce it by one log cycle when log D is plotted against temperature (figure 7.2). Another way to describe heating effectiveness is with the F value. The **F value** is the time in minutes at a specific temperature (usually 250°F or 121.1°C) needed to kill a population of cells or spores.

The food processing industry makes extensive use of D and z values. After a food has been canned, it must be heated to eliminate the risk of botulism arising from the presence of *Clostridium botulinum* spores. Heat treatment is carried out long enough to reduce a population of 10^{12} *C. botulinum* spores to 10^0 (one spore); thus there is a very small chance of any can having a viable spore. The D value for these spores at 121°C is 0.204 minutes. Therefore it would take 12D or 2.5 minutes to reduce 10^{12} spores to one spore by heating at 121°C. The z value for *C. botulinum* spores is 10°C— that is, it takes a 10°C change in temperature to alter the D value tenfold. If the cans were to be processed at 111°C rather than at 121°C, the D value would increase by tenfold to 2.04 minutes and the 12D value to 24.5 minutes. D values and z values for some common food-borne pathogens are given in table 7.3. Three D values are included for *Staphylococcus aureus* to illustrate the variation of killing rate with environment and the protective effect of organic material.

Food processing (pp. 883–85).

Botulism (p. 767).

Moist heat sterilization must be carried out at temperatures above 100°C in order to destroy bacterial endospores, and this requires the use of saturated steam under pressure. Steam sterilization is carried out with an **autoclave** (figure 7.3), a device somewhat like a fancy pressure cooker. The development of the autoclave by Chamberland in 1884 tremendously stimulated the growth of microbiology. Water is boiled to produce steam, which is released through the jacket and into the autoclave's chamber. The air initially present in the chamber is forced out until the chamber is filled with saturated steam and the outlets are closed. Hot, saturated steam continues to enter until the chamber reaches the desired temperature and pressure, usually 121°C and 15 pounds of pressure. At this temperature saturated

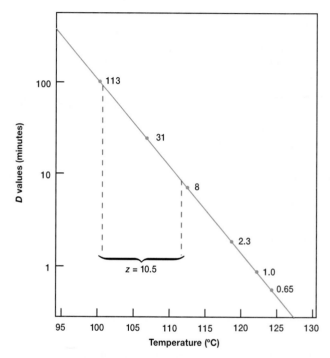

Figure 7.2   z Value Calculation. The z value used in calculation of time-temperature relationships for survival of a test microorganism, based on D value responses at various temperatures. The z value is the increase in temperature needed to reduce the decimal reduction time (D) to 10% of the original value. For this test microorganism the z value is 10.5°. The D values are plotted on a logarithmic scale.

TABLE 7.3 D Values and z Values for Some Food-Borne Pathogens

Organism	Substrate	D Value (°C) in Minutes	z Value (°C)
Clostridium botulinum	Phosphate buffer	$D_{121} = 0.204$	10
Clostridium perfringens (heat-resistant strain)	Culture media	$D_{90} = 3$–5	6–8
Salmonella	Chicken à la king	$D_{60} = 0.39$–0.40	4.9–5.1
Staphylococcus aureus	Chicken à la king	$D_{60} = 5.17$–5.37	5.2–5.8
	Turkey stuffing	$D_{60} = 15.4$	6.8
	0.5% NaCl	$D_{60} = 2.0$–2.5	5.6

Values taken from F. L. Bryan, 1979, "Processes That Affect Survival and Growth of Microorganisms," *Time-Temperature Control of Foodborne Pathogens*, 1979. Atlanta: Centers for Disease Control and Prevention, Atlanta, GA.

Figure 7.3 The Autoclave or Steam Sterilizer. (*a*) A modern, automatically controlled autoclave or sterilizer. (*b*) Longitudinal cross section of a typical autoclave showing some of its parts and the pathway of steam. (*b*) *From John J. Perkins,* Principles and Methods of Sterilization in Health Science, *2nd edition, 1969. Courtesy of Charles C Thomas, Publisher, Springfield, Illinois.*

steam destroys all vegetative cells and endospores in a small volume of liquid within 10 to 12 minutes. Treatment is continued for about 15 minutes to provide a margin of safety.

Autoclaving must be carried out properly or the processed materials will not be sterile. If all air has not been flushed out of the chamber, it will not reach 121°C even though it may reach a pressure of 15 pounds. The chamber should not be packed too tightly because the steam needs to circulate freely and contact everything in the autoclave. Bacterial endospores will be killed only if they are kept at 121°C for 10 to 12 minutes. When a large volume of liquid must be sterilized, an extended sterilization time will be needed because it will take longer for the center of the liquid to reach 121°C; 5 liters of liquid may require about 70 minutes. In view of these potential

difficulties, a biological indicator is often autoclaved along with other material. This indicator commonly consists of a culture tube containing a sterile ampule of medium and a paper strip covered with spores of *Bacillus stearothermophilus* or *Clostridium* PA3679. After autoclaving, the ampule is aseptically broken and the culture incubated for several days. If the test bacterium does not grow in the medium, the sterilization run has been successful. Sometimes either special tape that spells out the word *sterile* or a paper indicator strip that changes color upon sufficient heating is autoclaved with a load of material. If the word appears on the tape or if the color changes after autoclaving, the material is supposed to be sterile. These approaches are convenient and save time but are not as reliable as the use of bacterial endospores.

Moist heat is thought to kill so effectively by degrading nucleic acids and by denaturing enzymes and other essential proteins. It also may disrupt cell membranes.

Many substances, such as milk, are treated with controlled heating at temperatures well below boiling, a process known as **pasteurization** in honor of its developer Louis Pasteur. In the 1860s the French wine industry was plagued by the problem of wine spoilage, which made wine storage and shipping difficult. Pasteur examined spoiled wine under the microscope and detected microorganisms that looked like the bacteria responsible for lactic acid and acetic acid fermentations. He then discovered that a brief heating at 55 to 60°C would destroy these microorganisms and preserve wine for long periods. In 1886 the German chemists V. H. and F. Soxhlet adapted the technique for preserving milk and reducing milk-transmissible diseases. Milk pasteurization was introduced into the United States in 1889. Milk, beer, and many other beverages are now pasteurized. Pasteurization does not sterilize a beverage, but it does kill any pathogens present and drastically slows spoilage by reducing the level of nonpathogenic spoilage microorganisms.

Milk can be pasteurized in two ways. In the older method the milk is held at 63°C for 30 minutes. Large quantities of milk are now usually subjected to **flash pasteurization** or high-temperature short-term (HTST) pasteurization, which consists of quick heating to about 72°C for 15 seconds, then rapid cooling. The dairy industry also sometimes uses **ultrahigh temperature (UHT) sterilization.** Milk and milk products are heated at 140 to 150°C for 1 to 3 seconds. UHT-processed milk does not require refrigeration and can be stored at room temperature for about 2 months without flavor changes. The small coffee creamer portions provided by restaurants often are prepared using UHT sterilization.

Pasteurization and the dairy industry (pp. 883–84).

Sometimes a heat-sensitive material can be heat sterilized by **tyndallization** or **fractional steam sterilization.** The container with the material to be sterilized is heated at 90 to 100°C for 30 minutes on each of three consecutive days and incubated at 37°C in between. The first heating will destroy all organisms but bacterial endospores. Most of these germinate during the subsequent 37°C incubation and are killed by the second heating period. Any remaining spores are destroyed by the second incubation and third heat treatment.

Many objects are best sterilized in the absence of water by **dry heat sterilization.** The items to be sterilized are placed in an oven at 160 to 170°C for 2 to 3 hours. Microbial death apparently results from the oxidation of cell constituents and denaturation of proteins. Although dry air heat is less effective than moist heat—*Clostridium botulinum* spores are killed in 5 minutes at 121°C by moist heat but only after 2 hours at 160°C with dry heat—it has some definite advantages. Dry heat does not corrode glassware and metal instruments as moist heat does, and it can be used to sterilize powders, oils, and similar items. Most laboratories sterilize glass petri dishes and pipettes with dry heat. Despite these advantages, dry heat sterilization is slow and not suitable for heat-sensitive materials like many plastic and rubber items.

Filtration

Filtration is an excellent way to reduce the microbial population in solutions of heat-sensitive material, and sometimes it can be used to sterilize solutions. Rather than directly destroying contaminating microorganisms, the filter simply removes them. There are two types of filters. **Depth filters** consist of fibrous or granular materials that have been bonded into a thick layer filled with twisting channels of small diameter. The solution containing microorganisms is sucked through this layer under vacuum, and microbial cells are removed by physical screening or entrapment and also by adsorption to the surface of the filter material. Depth filters are made of diatomaceous earth (Berkefield filters), unglazed porcelain (Chamberlain filters), asbestos, or other similar materials.

Recently **membrane filters** have replaced depth filters for many purposes. These circular filters are porous membranes, a little over 0.1 mm thick, made of cellulose acetate, cellulose nitrate, polycarbonate, polyvinylidene fluoride, or other synthetic materials. Although a wide variety of pore sizes are available, membranes with pores about 0.2 μm in diameter are used to remove vegetative cells from solutions ranging in volume from 1 ml to many liters. The membranes are held in special holders (figure 7.4) and often preceded by depth filters made of glass fibers to remove larger particles that might clog the membrane filter. The solution is forced through the filter with a vacuum or with pressure from a syringe, peristaltic pump, or nitrogen gas bottle, and collected in previously sterilized containers. Membrane filters remove microorganisms by screening them out much as a sieve separates large sand particles from small ones (figure 7.5). These filters are used to sterilize pharmaceuticals, culture media, oils, antibiotics, and other heat-sensitive solutions.

The use of membrane filters in microbial counting (pp. 119–21).

Air also can be sterilized by filtration. Two common examples are surgical masks and cotton plugs on culture vessels that let air in but keep microorganisms out. **Laminar flow biological safety cabinets** employing **high-efficiency particulate air (HEPA) filters,** which remove 99.97% of 0.3 μm particles, are one of the most important air filtration systems. Laminar flow biological safety cabinets force air through HEPA filters, then project a vertical curtain of sterile air across the cabinet opening. This protects a worker from microorganisms being handled within the cabinet and prevents room contamination (figure 7.6). A person uses these cabinets when working with dangerous agents such as *Mycobacterium tuberculosis,* tumor viruses, and recombinant DNA. They are also employed in research labs and industries, such as the pharmaceutical industry, when a sterile working surface is needed for conducting assays, preparing media, examining tissue cultures, and the like.

1/4" stepped
hose connector

Vent

Durapore filters

Filter support

Filling bell

Bell cap

Cross section
Millipak-40 filter unit

(a)

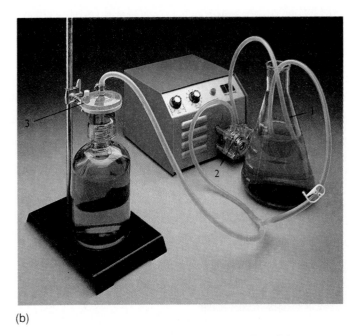

(b)

Figure 7.4   Membrane Filter Sterilization. A membrane filter outfit for sterilizing medium volumes of solution. (*a*) Cross section of the membrane filtering unit. Several membranes are used to increase capacity. (*b*) A complete filtering setup. The solution to be sterilized is kept in the Erlenmeyer flask, *1*, and forced through the filter by a peristaltic pump, *2*. The solution is sterilized by flowing through a membrane filter unit, *3*, and into a sterile container. A wide variety of other kinds of filtering outfits are also available.

(a)

(b)

Figure 7.5   Membrane Filter Types. (*a*) *Bacillus megaterium* on an Ultipor nylon membrane with a bacterial removal rating of 0.2 μm (×2,000). (*b*) *Enterococcus faecalis* resting on a polycarbonate membrane filter with 0.4 μm holes (×5,900).

Radiation

The types of radiation and the ways in which radiation damages or destroys microorganisms have already been discussed. The practical uses of ultraviolet and ionizing radiation in sterilizing objects are briefly described next.

Radiation and its effects on microorganisms (pp. 132–133).

Ultraviolet (UV) radiation around 260 nm (*see figure 6.19*) is quite lethal but does not penetrate glass, dirt films, water, and other substances very effectively. Because of this

disadvantage, UV radiation is used as a sterilizing agent only in a few specific situations. UV lamps are sometimes placed on the ceilings of rooms or in biological safety cabinets to sterilize the air and any exposed surfaces. Because UV radiation burns the skin and damages eyes, people working in such areas must be certain the UV lamps are off when the areas are in use. Commercial UV units are available for water treatment. Pathogens and other microorganisms are destroyed when a thin layer of water is passed under the lamps.

Ionizing radiation is an excellent sterilizing agent and penetrates deep into objects. Gamma radiation from a cobalt 60

(a)

(b)

Exhaust HEPA filter
Motor/blower
Supply HEPA filter
Special light and electrical compartment
Safety glass viewscreen
High-velocity air barrier
Optional support stand

Figure 7.6   A Laminar Flow Biological Safety Cabinet. (*a*) A technician pipetting potentially hazardous material in a safety cabinet. (*b*) A schematic diagram showing the airflow pattern.

source is used in the cold sterilization of antibiotics, hormones, sutures, and plastic disposable supplies such as syringes. Gamma radiation has also been used to sterilize and "pasteurize" meat and other food. Both the Food and Drug Administration and the World Health Organization have approved food irradiation and declared it safe. A commercial irradiation plant operates near Tampa, Florida. However, this process has not yet been widely employed in the United States because of the cost and concerns about the effects of gamma radiation on food. In some countries ionizing radiation is used to destroy pathogens in poultry and extend the shelf life of seafoods and fruits. It will probably be more extensively employed in the future.

1. Define thermal death point (TDP), thermal death time (TDT), decimal reduction time (*D*) or *D* value, *z* value, and the *F* value.

2. Describe how an autoclave works. What conditions are required for sterilization by moist heat, and what three things must one do when operating an autoclave to help ensure success?

3. How are pasteurization, flash pasteurization, ultrahigh temperature sterilization, fractional steam sterilization (tyndallization), and dry heat sterilization carried out? Give some practical applications for each of these procedures.

4. What are depth filters and membrane filters, and how are they used to sterilize liquids? Describe the operation of a biological safety cabinet.

5. Give the advantages and disadvantages of ultraviolet light and ionizing radiation as sterilizing agents. Provide a few examples of how each is used for this purpose.

The Use of Chemical Agents in Control

Although objects are sometimes disinfected with physical agents, chemicals are more often employed in disinfection and antisepsis. Many factors influence the effectiveness of chemical disinfectants and antiseptics as previously discussed. Factors such as the kinds of microorganisms potentially present, the concentration and nature of the disinfectant to be used, and the length of treatment should be considered. Dirty surfaces must be cleaned before a disinfectant or antiseptic is applied. The proper use of chemical agents is essential to laboratory and hospital safety (*Box 7.2; see also Box 34.1*). It should be noted that chemicals also are employed to prevent microbial growth in food. This is discussed in the chapter on food microbiology (*see pp. 884–85*).

Many different chemicals are available for use as disinfectants, and each has its own advantages and disadvantages. In selecting an agent, it is important to keep in mind the characteristics of a desirable disinfectant. Ideally the disinfectant must be effective against a wide variety of infectious agents (gram-positive and gram-negative bacteria, acid-fast bacteria, bacterial endospores, fungi, and viruses) at high dilutions and in the presence of organic matter. Although the chemical must be toxic for infectious agents, it should not be toxic to people or corrosive for common materials. It should be stable upon storage, odorless or with a pleasant odor, soluble in water and lipids for penetration into microorganisms, and have a low surface tension so that it can enter cracks in surfaces. If possible the disinfectant should be relatively inexpensive.

The properties and uses of several groups of common disinfectants and antiseptics are surveyed next. Many of their characteristics are summarized in tables 7.4 and 7.5. Structures of some common agents are given in figure 7.7.

Phenolics

Phenol was the first widely used antiseptic and disinfectant. In 1867 Joseph Lister employed it to reduce the risk of infection during operations. Today phenol and phenolics (phenol derivatives) such as cresols, xylenols, and orthophenylphenol are used as disinfectants in laboratories and hospitals. The commercial disinfectant Lysol is made of a mixture of phenolics. Phenolics act by denaturing proteins and disrupting cell membranes. They have some real advantages as disinfectants: phenolics are tuberculocidal, effective in the presence of organic material, and remain active on surfaces long after application. However, they do have a disagreeable odor and can cause skin irritation.

Box 7.2

Universal Precautions for Microbiology Laboratories

Blood and other body fluids from all patients should be considered infective.

1. All specimens of blood and body fluids should be put in a well-constructed container with a secure lid to prevent leaking during transport. Care should be taken when collecting each specimen to avoid contaminating the outside of the container and of the laboratory form accompanying the specimen.

2. All persons processing blood and body-fluid specimens should wear gloves. Masks and protective eyewear should be worn if mucous membrane contact with blood or body fluids is anticipated. Gloves should be changed and hands washed after completion of specimen processing.

3. For routine procedures, such as histologic and pathological studies or microbiologic culturing, a biological safety cabinet is not necessary. However, biological safety cabinets should be used whenever procedures are conducted that have a high potential for generating droplets. These include activities such as blending, sonicating, and vigorous mixing.

4. Mechanical pipetting devices should be used for manipulating all liquids in the laboratory. Mouth pipetting must not be done.

5. Use of needles and syringes should be limited to situations in which there is no alternative, and the recommendations for preventing injuries with needles outlined under universal precautions (*see Box 34.1, p. 679*) should be followed.

6. Laboratory work surfaces should be decontaminated with an appropriate chemical germicide after a spill of blood or other body fluids and when work activities are completed.

7. Contaminated materials used in laboratory tests should be decontaminated before reprocessing or be placed in bags and disposed of in accordance with institutional policies for disposal of infective waste.

8. Scientific equipment that has been contaminated with blood or other body fluids should be decontaminated and cleaned before being repaired in the laboratory or transported to the manufacturer.

9. All persons should wash their hands after completing laboratory activities and should remove protective clothing before leaving the laboratory.

10. There should be no eating, drinking, or smoking in the work area.

Source: Adapted from *Morbidity and Mortality Weekly Report,* 36 (Suppl. 2S) 5S–10S, 1987, the Centers for Disease Control and Prevention Guidelines.

TABLE 7.4   Activity Levels of Selected Germicides

Class	Use Concentration of Active Ingredient	Activity Level[a]
Gas		
Ethylene oxide	450–500 mg/liter [b]	High
Liquid		
Glutaraldehyde, aqueous	2%	High to intermediate
Formaldehyde + alcohol	8 + 70%	High
Stabilized hydrogen peroxide	6–30%	High to intermediate
Formaldehyde, aqueous	6–8%	High to intermediate
Iodophors	750–5,000 mg/liter[c]	High to intermediate
Iodophors	75–150 mg/liter[c]	Intermediate to low
Iodine + alcohol	0.5 + 70%	Intermediate
Chlorine compounds	0.1–0.5%[d]	Intermediate
Phenolic compounds, aqueous	0.5–3%	Intermediate to low
Iodine, aqueous	1%	Intermediate
Alcohols (ethyl, isopropyl)	70%	Intermediate
Quaternary ammonium compounds	0.1–0.2% aqueous	Low
Chlorhexidine	0.75–4%	Low
Hexachlorophene	1–3%	Low
Mercurial compounds	0.1–0.2%	Low

From Seymour S. Block, *Disinfection, Sterilization and Preservation.* Copyright © 1983 Lea & Febiger, Malvern, Pa. 1983. Reprinted by permission.

[a]High-level disinfectants destroy vegetative bacterial cells including *M. tuberculosis,* bacterial endospores, fungi, and viruses. Intermediate-level disinfectants destroy all of the above except endospores. Low-level agents kill bacterial vegetative cells except for *M. tuberculosis,* fungi, and medium-sized lipid-containing viruses (but not bacterial endospores or small, nonlipid viruses).

[b]In autoclave-type equipment at 55 to 60°C.

[c]Available iodine.

[d]Free chlorine.

TABLE 7.5   Relative Efficacy of Commonly Used Disinfectants and Antiseptics

Class	Disinfectant	Antiseptic	Comment
Gas			
Ethylene oxide	3–4[a]	0	Sporicidal; toxic; good penetration; requires relative humidity of 30% or more; microbicidal activity varies with apparatus used; absorbed by porous material; dry spores highly resistant; moisture must be present, and presoaking is most desirable
Liquid			
Glutaraldehyde, aqueous	3	0	Sporicidal; active solution unstable; toxic
Stabilized hydrogen peroxide	3	0	Sporicidal; use solution stable up to 6 weeks; toxic orally and to eyes; mildly skin toxic; little inactivated by organic matter
Formaldehyde + alcohol	3	0	Sporicidal; noxious fumes; toxic; volatile
Formaldehyde, aqueous	1–2	0	Sporicidal; noxious fumes; toxic
Phenolic compounds	3	0	Stable; corrosive; little inactivation by organic matter; irritates skin
Chlorine compounds	1–2	0	Fast action; inactivation by organic matter; corrosive; irritates skin
Alcohol	1	3	Rapidly microbicidal except for bacterial spores and some viruses; volatile; flammable; dries and irritates skin
Iodine + alcohol	0	4	Corrosive; very rapidly microbicidal; causes staining; irritates skin; flammable
Iodophors	1–2	3	Somewhat unstable; relatively bland; staining temporary; corrosive
Iodine, aqueous	0	2	Rapidly microbicidal; corrosive; stains fabrics; stains and irritates skin
Quaternary ammonium compounds	1	0	Bland; inactivated by soap and anionics; compounds absorbed by fabrics; old or dilute solution can support growth of gram-negative bacteria
Hexachlorophene	0	2	Bland; insoluble in water, soluble in alcohol; not inactivated by soap; weakly bactericidal
Chlorhexidine	0	3	Bland; soluble in water and alcohol; weakly bactericidal
Mercurial compounds	0	±	Bland; much inactivated by organic matter; weakly bactericidal

From Seymour S. Block, *Disinfection, Sterilization and Preservation.* Copyright © 1983 Lea & Febiger, Malvern, Pa. 1983. Reprinted by permission.

[a]Subjective ratings of practical usefulness in a hospital environment—4 is maximal usefulness; 0 is little or no usefulness; ± signifies that the substance is sometimes useful but not always.

Hexachlorophene (figure 7.7) has been one of the most popular antiseptics because it persists on the skin once applied and reduces skin bacteria for long periods. However, it can cause brain damage and is now used in hospital nurseries only in response to a staphylococcal outbreak.

Alcohols

Alcohols are among the most widely used disinfectants and antiseptics. They are bactericidal and fungicidal but not sporicidal; some lipid-containing viruses are also destroyed. The two most popular alcohol germicides are ethanol and isopropanol, usually used in about 70 to 80% concentration. They act by denaturing proteins and possibly by dissolving membrane lipids. A 10 to 15 minute soaking is sufficient to disinfect thermometers and small instruments.

Halogens

A halogen is any of the five elements (fluorine, chlorine, bromine, iodine, and astatine) in group VIIA of the periodic table. They exist as diatomic molecules in the free state and form saltlike compounds with sodium and most other metals. The halogens iodine and chlorine are important antimicrobial agents. Iodine is used as a skin antiseptic and kills by oxidizing cell constituents and iodinating cell proteins. At higher concentrations, it may even kill some spores. Iodine often has been applied as tincture of iodine, 2% or more iodine in a water-ethanol solution of potassium iodide. Although it is an effective antiseptic, the skin may be damaged, a stain is left, and iodine allergies can result. More recently iodine has been complexed with an organic carrier to form an **iodophor.** Iodophors are water soluble, stable, and nonstaining, and release iodine slowly to minimize skin burns and irritation. They are used in hospitals for preoperative skin degerming and in hospitals and laboratories for disinfecting. Some popular brands are Wescodyne for skin and laboratory disinfection and Betadine for wounds.

Chlorine is the usual disinfectant for municipal water supplies and swimming pools and is also employed in the dairy and food industries. It may be applied as chlorine gas, sodium hypochlorite, or calcium hypochlorite, all of which yield hypochlorous acid (HClO) and then atomic oxygen. The result is oxidation of cellular materials and destruction of vegetative bacteria and fungi, although not spores.

Phenolics

Phenol   Orthocresol   Orthophenylphenol   Hexachlorophene

Alcohols

CH_3-CH_2-OH
Ethanol

$CH_3-CH-CH_3$ (OH)
Isopropanol

Halogenated compound

Halazone

Aldehydes

Formaldehyde   Glutaraldehyde

Quaternary ammonium compounds

Cetylpyridinium chloride

Benzalkonium chloride

Gases

Ethylene oxide   Betapropiolactone

Figure 7.7 **Disinfectants and Antiseptics.** The structures of some frequently used disinfectants and antiseptics.

$$Cl_2 + H_2O \longrightarrow HCl + HClO$$

$$Ca(OCl)_2 + 2H_2O \longrightarrow Ca(OH)_2 + 2HClO$$

$$HClO \longrightarrow HCl + O$$

Death of almost all microorganisms usually occurs within 30 minutes. Since organic material interferes with chlorine action by reacting with chlorine and its products, an excess of chlorine is added to ensure microbial destruction.

Municipal water purification (pp. 848–50).

Chlorine is also an excellent disinfectant for individual use because it is effective, inexpensive, and easy to employ. Small quantities of drinking water can be disinfected with halazone tablets. Halazone (parasulfone dichloramidobenzoic acid) slowly releases chloride when added to water and disinfects it in about a half hour. It is frequently used by campers lacking access to uncontaminated drinking water.

Chlorine solutions make very effective laboratory and household disinfectants. An excellent disinfectant-detergent combination can be prepared if a 1/100 dilution of household bleach (e.g., 1.3 fl oz of Clorox or Purex bleach in 1 gal or 10 ml/liter) is combined with sufficient nonionic detergent (about 1 oz/gal or 7.8 ml/liter) to give a 0.8% detergent concentration. This mixture will remove both dirt and bacteria.

Heavy Metals

For many years the ions of heavy metals such as mercury, silver, arsenic, zinc, and copper were used as germicides. More recently these have been superseded by other less toxic and more effective germicides (many heavy metals are more bacteriostatic than bactericidal). There are a few exceptions. A 1% solution of silver nitrate is often added to the eyes of infants to prevent ophthalmic gonorrhea (in many hospitals, erythromycin is used instead of silver nitrate because it is effective against *Chlamydia* as well as *Neisseria*). Silver sulfadiazine is used on burns. Copper sulfate is an effective algicide in lakes and swimming pools.

Heavy metals combine with proteins, often with their sulfhydryl groups, and inactivate them. They may also precipitate cell proteins.

Quaternary Ammonium Compounds

Detergents [Latin *detergere,* to wipe off or away] are organic molecules that serve as wetting agents and emulsifiers because they have both polar hydrophilic and nonpolar hydrophobic ends. Due to their amphipathic nature (*see chapter 3*), detergents solubilize otherwise insoluble residues and are very effective cleansing agents. They are different than soaps, which are derived from fats.

Although anionic detergents have some antimicrobial properties, only cationic detergents are effective disinfectants. The most popular of these disinfectants are quaternary ammonium compounds characterized by a positively charged quaternary nitrogen and a long hydrophobic aliphatic chain (figure 7.7). They disrupt microbial membranes and may also denature proteins.

Cationic detergents like benzalkonium chloride and cetylpyridinium chloride kill most bacteria but not *M. tuberculosis* or endospores. They do have the advantages of being stable, nontoxic, and bland but they are inactivated by hard water and soap. Cationic detergents are often used as disinfectants for food utensils and small instruments and as skin antiseptics. Several brands are on the market. Zephiran contains benzalkonium chloride and Ceepryn, cetylpyridinium chloride.

Aldehydes

Both of the commonly used aldehydes, formaldehyde and glutaraldehyde, are highly reactive molecules that combine with proteins and inactivate them (figure 7.7). They are sporicidal and can be used as chemical sterilants. Formaldehyde is usually dissolved in water or alcohol before use. A 2% buffered solution of glutaraldehyde is an effective disinfectant. It is less irritating than formaldehyde and is used to disinfect hospital and laboratory equipment. Glutaraldehyde usually disinfects objects within about 10 minutes but may require as long as 12 hours to destroy all spores.

Sterilizing Gases

Many heat-sensitive items such as disposable plastic petri dishes and syringes, heart-lung machine components, sutures, and catheters are now sterilized with ethylene oxide gas (figure 7.7). Ethylene oxide (EtO) is both microbicidal and sporicidal and kills by combining with cell proteins. It is a particularly effective sterilizing agent because it rapidly penetrates packing materials, even plastic wraps.

Sterilization is carried out in a special ethylene oxide sterilizer, very much resembling an autoclave in appearance, that controls the EtO concentration, temperature, and humidity. Because pure EtO is explosive, it is usually supplied in a 10 to 20% concentration mixed with either CO_2 or dichlorodifluoromethane. The ethylene oxide concentration, humidity, and temperature influence the rate of sterilization. A clean object can be sterilized if treated for 5 to 8 hours at 38°C or 3 to 4 hours at 54°C when the relative humidity is maintained at 40 to 50% and the EtO concentration at 700 mg/liter. Extensive aeration of the sterilized materials is necessary to remove residual EtO because it is so toxic.

Betapropiolactone (BPL) is occasionally employed as a sterilizing gas. In the liquid form it has been used to sterilize vaccines and sera. BPL decomposes to an inactive form after several hours and is therefore not as difficult to eliminate as EtO. It also destroys microorganisms more readily than ethylene oxide but does not penetrate materials well and may be carcinogenic. For these reasons, BPL has not been used as extensively as EtO.

Recently vapor-phase hydrogen peroxide has been used to decontaminate biological safety cabinets.

---

1. Why are most antimicrobial chemical agents disinfectants rather than sterilants? What general characteristics should one look for in a disinfectant?

2. Describe each of the following agents in terms of its chemical nature, mechanism of action, mode of application, common uses and effectiveness, and advantages and disadvantages: phenolics, alcohols, halogens (iodine and chlorine), heavy metals, quaternary ammonium compounds, aldehydes, and ethylene oxide.

Evaluation of Antimicrobial Agent Effectiveness

Testing of antimicrobial agents is a complex process regulated by two different federal agencies. The U.S. Environmental Protection Agency regulates disinfectants, whereas agents used on humans and animals are under the control of the Food and Drug Administration. Testing of antimicrobial agents often begins with an initial screening test to see if they are effective and at what concentrations. This may be followed by more realistic in-use testing.

The best-known disinfectant screening test is the **phenol coefficient test** in which the potency of a disinfectant is compared with that of phenol. A series of dilutions of phenol and the experimental disinfectant are inoculated with the test bacteria *Salmonella typhi* and *Staphylococcus aureus,* then placed in a 20 or 37°C water bath. These inoculated disinfectant tubes are next subcultured to regular fresh medium at 5 minute intervals, and the subcultures are incubated for two or more days. The highest dilutions that kill the bacteria after a 10 minute exposure, but not after 5 minutes, are used to calculate the phenol coefficient. The reciprocal of the appropriate test disinfectant dilution is divided by that for phenol to obtain the coefficient. Suppose that the phenol dilution was 1/90 and maximum effective dilution for disinfectant X was 1/450. The phenol coefficient of X would be 5. The higher the phenol coefficient value, the more effective the disinfectant under these test conditions. A value greater than 1 means that the disinfectant is more effective than phenol. A few representative phenol coefficient values are given in table 7.6.

TABLE 7.6   Phenol Coefficients for Some Disinfectants

Disinfectant	Phenol Coefficients[a]	
	Salmonella typhi	*Staphylococcus aureus*
Phenol	1	1
Cetylpyridinium chloride	228	337
O-phenylphenol	5.6 (20°C)	4.0
p-cresol	2.0–2.3	2.3
Hexachlorophene	5–15	15–40
Merthiolate	600	62.5
Mercurochrome	2.7	5.3
Lysol	1.9	3.5
Isopropyl alcohol	0.6	0.5
Ethanol	0.04	0.04
2% I_2 solution in EtOH	4.1–5.2 (20°C)	4.1–5.2 (20°C)

[a]All values were determined at 37°C except where indicated.

The phenol coefficient test is a useful initial screening procedure, but the phenol coefficient can be misleading if taken as a direct indication of disinfectant potency during normal use. This is because the phenol coefficient is determined under carefully controlled conditions with pure bacterial strains, whereas disinfectants are normally used on complex populations in the presence of organic matter and with significant variations in environmental factors like pH, temperature, and presence of salts.

To more realistically estimate disinfectant effectiveness, other tests are often used. The rates at which selected bacteria are destroyed with various chemical agents may be experimentally determined and compared. A **use dilution test** can also be carried out. Stainless steel cylinders are contaminated with specific bacterial species under carefully controlled conditions. The cylinders are dried briefly, immersed in the test disinfectants for 10 minutes, transferred to culture media, and incubated for two days. The disinfectant concentration that kills the organisms in the sample with a 95% level of confidence under these conditions is determined. Disinfectants also can be tested under conditions designed to simulate normal in-use situations. In-use testing techniques allow a more accurate determination of the proper disinfectant concentration for a particular situation.

1. Briefly describe the phenol coefficient test.
2. Why might it be necessary to employ procedures like the use dilution and in-use tests?

Summary

1. Sterilization is the process by which all living cells, viable spores, viruses, and viroids are either destroyed or removed from an object or habitat. Disinfection is the killing, inhibition, or removal of microorganisms (but not necessarily endospores) that can cause disease.

2. Many terms are used to define how microorganisms are controlled: sterilant, disinfectant, sanitization, antisepsis, and antiseptic.

3. Antimicrobial agents that kill organisms often have the suffix -cide, whereas agents that prevent growth and reproduction have the suffix -static.

4. Microbial death is usually exponential or logarithmic.

5. The effectiveness of a disinfectant or sterilizing agent is influenced by population size, population composition, concentration or intensity of the agent, exposure duration, temperature, and nature of the local environment.

6. The efficiency of heat killing is often indicated by the thermal death time or the decimal reduction time.

7. Although treatment with boiling water for 10 minutes kills vegetative forms, an autoclave must be used to destroy endospores by heating at 121°C and 15 pounds of pressure.

8. Moist heat kills by degrading nucleic acids, denaturing enzymes and other proteins, and disrupting cell membranes.

9. Heat-sensitive liquids can be preserved by pasteurization, heating at 63°C for 30 minutes or at 72°C for 15 seconds (flash pasteurization). Heating at 140 to 150°C for 1 to 3 seconds (ultrahigh temperature sterilization) may be used. Tyndallization or fractional steam sterilization is sometimes employed instead of pasteurization.

10. Glassware and other heat-stable items may be sterilized by dry heat at 160 to 170°C for 2 to 3 hours.

11. Microorganisms can be efficiently removed by filtration with either depth filters or membrane filters.

12. Biological safety cabinets with high-efficiency particulate filters sterilize air by filtration.

13. Radiation of short wavelength or high-energy ultraviolet and ionizing radiation can be used to sterilize objects.

14. Chemical agents usually act as disinfectants because they cannot readily destroy bacterial endospores. Disinfectant effectiveness depends on concentration, treatment duration, temperature, and presence of organic material.

15. Phenolics and alcohols are popular disinfectants that act by denaturing proteins and disrupting cell membranes.

16. Halogens (iodine and chlorine) kill by oxidizing cellular constituents; cell proteins may also be iodinated. Iodine is applied as a tincture or iodophor. Chlorine may be added

to water as a gas, hypochlorite, or organic chlorine derivative.

17. Heavy metals tend to be bacteriostatic agents. They are employed in specialized situations such as the use of silver nitrate in the eyes of newborn infants and copper sulfate in lakes and pools.

18. Cationic detergents are often used as disinfectants and antiseptics; they disrupt membranes and denature proteins.

19. Aldehydes such as formaldehyde and glutaraldehyde can sterilize as well as disinfect because they kill spores.

20. Ethylene oxide gas penetrates plastic wrapping material and destroys all life forms

by reacting with proteins. It is used to sterilize packaged, heat-sensitive materials.

21. A variety of procedures can be used to determine the effectiveness of disinfectants, among them the following: phenol coefficient test, measurement of killing rates with germicides, use dilution testing, and in-use testing.

Key Terms

algicide *138*

antimicrobial agent *139*

antisepsis *138*

antiseptics *138*

autoclave *140*

bactericide *138*

bacteriostatic *138*

D value *140*

decimal reduction time (*D*) *140*

depth filters *142*

detergent *148*

disinfectant *137*

disinfection *137*

dry heat sterilization *142*

F value *140*

flash pasteurization *142*

fractional steam sterilization *142*

fungicide *138*

fungistatic *138*

germicide *138*

high-efficiency particulate air (HEPA) filters *142*

iodophor *146*

ionizing radiation *143*

laminar flow biological safety cabinets *142*

membrane filters *142*

pasteurization *142*

phenol coefficient test *148*

sanitization *137*

sterilization *137*

thermal death time (TDT) *140*

tyndallization *142*

ultrahigh temperature (UHT) sterilization *142*

ultraviolet (UV) radiation *143*

use dilution test *149*

viricide *138*

z value *140*

Questions for Thought and Review

1. How can the *D* value be used to estimate the time required for sterilization? Suppose that you wanted to eliminate the risk of *Salmonella* poisoning by heating your food (D_{60} = 0.40 minutes, *z* value = 5.0). Calculate the 12*D* value at 60°C. How long would it take to achieve the same results by heating at 50, 55, and 65°C?

2. How can one alter treatment conditions to increase the effectiveness of a disinfectant?

3. How would the following be best sterilized: glass pipettes and petri plates, tryptic soy broth tubes, Mueller-Hinton agar medium, antibiotic solution, interior of a biological safety cabinet, wrapped package of plastic petri plates?

4. Which disinfectants or antiseptics would be used to treat the following: oral thermometer, laboratory bench top, drinking water, patch of skin before surgery, small medical instruments (probes, forceps, etc.)? List all chemicals suitable for each task.

Additional Reading

General

Barkley, W. E., and Richardson, J. H. 1994. Laboratory safety. In *Methods for general and molecular bacteriology*, P. Gerhardt, et al., editors, 715–34. Washington, D.C.: American Society for Microbiology.

Block, S. S., editor. 1991. *Disinfection, sterilization and preservation*, 4th ed. Philadelphia: Lea and Febiger.

Centers for Disease Control. 1987. Recommendations for prevention of HIV transmission in health-care settings. *Morbid. Mortal. Weekly Rep.* 36(Suppl. 2):3S–18S.

Centers for Disease Control. 1988. Update: Universal precautions for prevention of transmission of human immunodeficiency virus, hepatitis B virus, and other bloodborne pathogens in health-care settings. *Morbid. Mortal. Weekly Rep.* 37(24):377–88.

Centers for Disease Control. 1989. Guidelines for prevention of transmission of human immunodeficiency virus and hepatitis B virus to health-care and public-safety workers. *Morbid. Mortal. Weekly Rep.* 38(Suppl. 6):1–37.

Centers for Disease Control and National Institutes of Health. 1992. *Biosafety in microbiological and biomedical laboratories*, 3d ed. Washington, D.C.: U.S. Government Printing Office.

Collins, C. H., and Lyne, P. M. 1976. *Microbiological methods*, 4th ed. Boston: Butterworths.

Henderson, D. K. 1995. HIV-1 in the health-care setting. In *Principles and practice of infectious diseases*, 4th ed., G. L. Mandell, J. E. Bennett, and R. Dolin editors, 2632–56. New York: Churchill Livingstone.

Marsik, F. J., and Denys, G. A. 1995. Sterilization, decontamination, and disinfection procedures for the microbiology laboratory. In *Manual of clinical microbiology*, 6th ed., P. R. Murray, editor, 86–98. Washington, D.C.: American Society for Microbiology.

Martin, M. A., and Wenzel, R. P. 1995. Sterilization, disinfection, and disposal of infectious waste. In *Principles and practice of infectious diseases*, 4th ed., G. L. Mandell, J. E. Bennett, and R. Dolin editors, 2579–87. New York: Churchill Livingstone.

Perkins, J. J. 1969. *Principles and methods of sterilization in health sciences*, 2d ed. Springfield, Ill.: Charles C Thomas.

Pike, R. M. 1979. Laboratory-associated infections: Incidence, fatalities, causes, and prevention. *Ann. Rev. Microbiol.* 33:41–66.

Russell, A. D.; Hugo, W. B.; and Ayliffe, G. A. J., editors. 1992. *Principles and practice of disinfection, preservation and sterilization*, 2d ed. Oxford: Blackwell Scientific Publications.

Sewell, D. L. 1995. Laboratory-associated infections and biosafety. *Clin. Microbiol. Rev.* 8(3):389–405.

Strain, B A., and Gröschel, D. H. M. 1995. Laboratory safety and infectious waste management. In *Manual of clinical microbiology*, 6th ed., P. R. Murray, editor, 75–85. Washington, D.C.: American Society for Microbiology.

Warren, E. 1981. Laboratory safety. In *Laboratory procedures in clinical microbiology*, J. A. Washington, editor, 729–45. New York: Springer-Verlag.

The Use of Physical Methods in Control

Brock, T. D. 1983. *Membrane filtration: A user's guide and reference manual.* Madison, Wis.: Science Tech Publishers.

The Use of Chemical Agents in Control

Borick, P. M. 1973. *Chemical sterilization.* Stroudsburg, Pa.: Dowden, Hutchinson and Ross.

Russell, A. D. 1990. Bacterial spores and chemical sporicidal agents. *Clin. Microbiol. Rev.* 3(2):99–119.

PART THREE
Microbial Metabolism

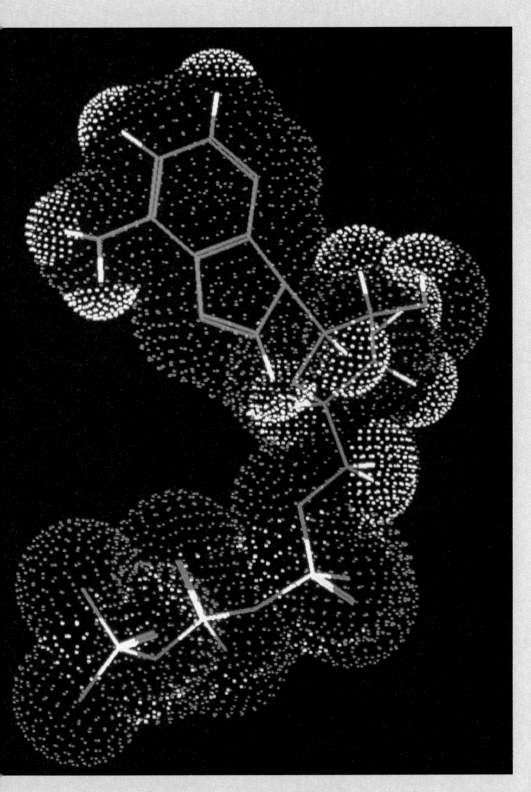

Adenosine 5′-triphosphate (ATP), the cell's
principal energy currency. This computer-generated
graphic shows that ATP is composed of the purine
base adenine, the sugar ribose, and a chain of three
phosphate groups. Different colors indicate the
position of the component atoms: carbon (green),
nitrogen (blue), oxygen (red), phosphorus, and
hydrogen (white).

CHAPTER 8
Metabolism: Energy and Enzymes

The principles and methods of biochemistry now provide the underpinnings of all of the basic biological sciences and are the rational language for discourse in such diverse areas as ecology, clinical medicine, and agriculture.

—E. L. Smith, R. L. Hill,
I. R. Lehman, and
R. J. Lefkowitz

Outline

Concepts

1. Energy is the capacity to do work. Living organisms can perform three major types of work: chemical work, transport work, and mechanical work.

2. Most energy used by living organisms originally comes from sunlight trapped during photosynthesis by photoautotrophs. Chemoheterotrophs then consume autotrophic organic materials and use them as sources of energy and as building blocks.

3. An energy currency is needed to connect energy-yielding exergonic reactions with energy-requiring endergonic reactions. The most commonly used currency is ATP.

4. All living systems obey the laws of thermodynamics.

5. When electrons are transferred from a reductant with a more negative reduction potential to an oxidant with a more positive potential, energy is made available. A reversal of the direction of electron transfer—for example, during photosynthesis—requires energy input.

6. Enzymes are protein catalysts that make life possible by increasing the rate of reactions at low temperatures.

7. Enzymes do not change chemical equilibria or violate the laws of thermodynamics but accelerate reactions by lowering their activation energy.

Chapters 3 and 4 contain many examples of an important principle: that a cell's structure is intimately related to its function. In each instance one can readily relate an organelle's construction to its function (and vice versa). A second unifying principle in biology is that life is sustained by the trapping and use of energy, a process made possible by the action of enzymes. Because this is so crucial to our understanding of microbial function, considerable attention is given to energy and enzymes in this chapter. The metabolism of microorganisms is discussed in subsequent chapters.

This chapter begins with a brief survey of the nature of energy and the laws of thermodynamics. The participation of energy in metabolism and the role of ATP as an energy currency is considered next. The chapter closes with an introduction to the nature and function of enzymes.

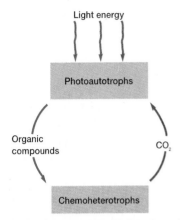

Figure 8.1   The Flow of Carbon and Energy in an Ecosystem. In this example the flow of energy and carbon between photoautotrophs and chemoheterotrophs is shown.

Energy and Work

Energy may be most simply defined as the capacity to do work or to cause particular changes. Thus all physical and chemical processes are the result of the application or movement of energy. Living cells carry out three major types of work, and all are essential to life processes. **Chemical work** involves the synthesis of complex biological molecules required by cells from much simpler precursors; energy is needed to increase the molecular complexity of a cell. Molecules and ions often must be transported across cell membranes against an electrochemical gradient. For example, a molecule sometimes moves into a cell even though its concentration is higher internally. Similarly a solute may be expelled from the cell against a concentration gradient. This process is **transport work** and requires energy input in order to take up nutrients, eliminate wastes, and maintain ion balances. The third type of work is **mechanical work,** perhaps the most familiar of the three. Energy is required to change the physical location of organisms, cells, and structures within cells.

The ultimate source of all biological energy is the visible sunlight impinging on the earth's surface. Light energy is trapped during **photosynthesis,** in which it is absorbed by chlorophyll and other pigments and converted to chemical energy. Chemical energy can then be used by photoautotrophs to transform atmospheric CO_2 into glucose and other biological molecules.

Nutritional types (pp. 100–101).

The complex molecules manufactured by photosynthetic organisms (both plant and microbial producers) serve as a carbon source for chemoheterotrophs, the consumers that require complex organic molecules as a source of material and energy for building their own cellular structures. Chemoheterotrophs

often use O_2 as an electron acceptor when oxidizing glucose and other organic molecules to CO_2. This process, in which O_2 acts as the final electron acceptor and is reduced to water, is called **aerobic respiration.** Much energy is released during this process. Thus, in the ecosystem, light energy is trapped by photoautotrophs; some of this energy subsequently flows to chemoheterotrophs when they use nutrients derived from photoautotrophs (figure 8.1; *see also figure 40.2*). The CO_2 produced during aerobic respiration can be incorporated again into complex organic molecules during photosynthesis. Clearly the flow of carbon and energy in the ecosystem is intimately related.

Cells must efficiently transfer energy from their energy-generating or trapping apparatus to the systems actually carrying out work. That is, cells must have a practical form of energy currency. In living organisms the major currency is **adenosine 5′-triphosphate (ATP;** figure 8.2). When ATP breaks down to **adenosine diphosphate (ADP)** and orthophosphate (P_i), energy is made available for useful work. Later, energy from photosynthesis, aerobic respiration, fermentation, and other processes is used to reform ATP from ADP and P_i. An energy cycle in the cell results (figure 8.3).

Fermentation (pp. 175–76).

The Laws of Thermodynamics

To understand how energy is trapped or generated and how ATP functions as an energy currency, some knowledge of the basic principles of thermodynamics is required. The science of **thermodynamics** analyzes energy changes in a collection of matter called a system. All other matter in the universe is called the surroundings. Thermodynamics focuses on the energy differences between the initial state and the final state of a system. It is not concerned either with how the system gets from start to

(a)

(b)

Figure 8.2   Adenosine Triphosphate and Adenosine Diphosphate. (*a*) Structure of ATP and ADP. The two red bonds are more easily broken or have a high phosphate group transfer potential (see text). The pyrimidine ring atoms have been numbered. (*b*) Space-filling model of ATP. Carbon is in green; hydrogen in light blue; nitrogen in dark blue; oxygen in red; and phosphorus in orange.

Figure 8.3   The Cell's Energy Cycle. ATP is formed from energy released during respiration. Its breakdown to ADP and phosphate (P_i) makes chemical, transport, and mechanical work possible.

finish or with the rate of the process. For instance, if a pan of water is heated to boiling, only the condition of the water at the start and at boiling is important in thermodynamics, not how fast it is heated or on what kind of stove. Two important laws of thermodynamics require elucidation. The **first law of thermodynamics** says that energy can be neither created nor destroyed. The total energy in the universe remains constant although it can be redistributed. For example, many energy exchanges do occur during chemical reactions (e.g., heat is given off by exothermic reactions and absorbed during endothermic reactions), but these heat exchanges do not violate the first law.

Figure 8.4   A Second Law Process. The expansion of gas into an empty cylinder simply redistributes the gas molecules until equilibrium is reached. The total number of molecules remains unchanged.

It is necessary to specify quantitatively the amount of energy used in or evolving from a particular process, and two types of energy units are employed. A **calorie** (cal) is the amount of heat energy needed to raise one gram of water from 14.5 to 15.5° C. The amount of energy also may be expressed in terms of **joules** (J), the units of work capable of being done. One cal of heat is equivalent to 4.1840 J of work. One thousand calories or a kilocalorie (kcal) is enough energy to boil 12 ml or about three thimbles of water. A kilojoule is enough energy to boil about 2.9 ml or 1/2 teaspoon of water, or enable a person weighing 70 kg to climb eight steps. The joule is normally used by chemists and physicists. Because biologists most often speak of energy in terms of calories, this text will employ calories when discussing energy changes.

Although it is true that energy is conserved in the universe, the first law of thermodynamics does not account for many physical and chemical processes. A simple example may help make this clear. Suppose a full gas cylinder is connected to an empty one by a tube with a valve (figure 8.4). If the valve is opened, gas flows from the full to the empty cylinder until the gas pressure is equal on both sides. Energy has not only been redistributed but also conserved. The expansion of gas is explained by the **second law of thermodynamics** and a condition of matter called entropy. **Entropy** may be considered a measure of the randomness or disorder of a system. The greater the disorder of a system, the greater is its entropy. The second law states that physical and chemical processes proceed in such a way that the randomness or disorder of the universe (the system and its surroundings) increases to the maximum possible. Gas will always expand into an empty cylinder.

Free Energy and Reactions

The first and second laws can be combined in a useful equation, relating the changes in energy that can occur in chemical reactions and other processes.

$$\Delta G = \Delta H - T \cdot \Delta S$$

ΔG is the change in free energy, ΔH is the change in enthalpy, T is the temperature in degrees Kelvin ($°C + 273$), and ΔS is the change in entropy occurring during the reaction. The change in **enthalpy** is the change in heat content, which is about the same thing as the change in total energy during the reaction. The **free energy change** is the total energy change in a system available to do useful work. Therefore the change in entropy is a measure of the proportion of the total energy change that the system cannot use in performing work. A reaction will occur spontaneously if the free energy of the system decreases during the reaction or, in other words, if ΔG is negative. It follows from the equation that a reaction with a large positive change in entropy will normally tend to have a negative ΔG value and therefore occur spontaneously. A decrease in entropy will tend to make ΔG more positive and the reaction less favorable.

The change in free energy has a definite, concrete relationship to the direction of chemical reactions. Consider the following simple reaction:

$$A + B \rightleftharpoons C + D$$

If the molecules A and B are mixed, they will combine to form the products C and D. Eventually C and D will become concentrated enough to combine and produce A and B at the same rate as they are formed from A and B. The reaction is now at **equilibrium:** the rates in both directions are equal and no further net change occurs in the concentrations of reactants and products. This situation is described by the **equilibrium constant (K_{eq}),** relating the equilibrium concentrations of products and substrates to one another.

$$K_{eq} = \frac{[C][D]}{[A][B]}$$

If the equilibrium constant is greater than one, the products are in greater concentration than the reactants at equilibrium—that is, the reaction tends to go to completion as written.

The equilibrium constant of a reaction is directly related to its change in free energy. When the free energy change for a process is determined at carefully defined standard conditions of concentration, pressure, pH, and temperature, it is called the **standard free energy change ($\Delta G°$).** If the pH is set at 7.0 (which is close to the pH of living cells), the standard free energy change is indicated by the symbol $\Delta G°'$. The change in standard free energy may be thought of as the maximum amount of energy available from the system for useful work under standard conditions. Using $\Delta G°'$ values allows one to compare reactions without worrying about variations in the ΔG due to differences in environmental conditions. The relationship between $\Delta G°'$ and K_{eq} is given by the following equation:

$$\Delta G°' = -2.303 RT \cdot \log K_{eq}$$

R is the gas constant (1.9872 cal/mole-degree or 8.3145 J/mole-degree), and T is the absolute temperature. Inspection of this equation shows that when $\Delta G°'$ is negative, the equilibrium constant is greater than one and the reaction goes to completion as written. It is said to be an **exergonic reaction** (figure 8.5). In an **endergonic reaction** $\Delta G°'$ is positive and the equilibrium constant is less than one. That is, the reaction is not favorable, and little product will be formed at equilibrium under standard conditions. Keep in mind that the $\Delta G°'$ value shows only where the reaction lies at equilibrium, not how fast the reaction reaches equilibrium.

Figure 8.5 $\Delta G°'$ **and Equilibrium.** The relationship of $\Delta G°'$ to the equilibrium of reactions. Note the differences between exergonic and endergonic reactions.

1. What is energy and what kinds of work are carried out in a cell? Describe the energy cycle and ATP's role in it.
2. What is thermodynamics? Summarize the first and second laws. Define free energy, entropy, and enthalpy.
3. How is the change in standard free energy related to the equilibrium constant for a reaction? What are exergonic and endergonic reactions?

The Role of ATP in Metabolism

Many reactions in the cell are endergonic and will not proceed far toward completion without outside assistance. One of ATP's major roles is to drive such endergonic reactions more to completion. ATP is a **high-energy molecule.** That is, it breaks down or hydrolyzes almost completely to the products ADP and P_i with a $\Delta G°'$ of -7.3 kcal/mole.

$$ATP + H_2O \longrightarrow ADP + P_i$$

In reference to ATP the term high-energy molecule does not mean that there is a great deal of energy stored in a particular bond of ATP. It simply indicates that the removal of the terminal phosphate goes to completion with a large negative standard free energy change, or the reaction is strongly exergonic. In other words, ATP has a high **phosphate group transfer potential;** it readily transfers its phosphate to water. The phosphate group transfer potential is defined as the negative of $\Delta G°'$ for the hydrolytic removal of phosphate. A molecule with a higher group transfer potential will donate phosphate to one with a lower potential.

Thus ATP is ideally suited for its role as an energy currency. It is formed in energy-trapping and -generating processes

Figure 8.6   ATP as a Coupling Agent. The use of ATP to make endergonic reactions more favorable. It is formed by exergonic reactions and then used to drive endergonic reactions.

TABLE 8.1   Selected Biologically Important Redox Couples

Redox Couple	E'_0 (Volts) [a]
$2H^+ + 2e^- \longrightarrow H_2$	−0.42
Ferredoxin$(Fe^{3+}) + e^- \longrightarrow$ ferredoxin(Fe^{2+})	−0.42
$NAD(P)^+ + H^+ + 2e^- \longrightarrow NAD(P)H$	−0.32
$S + 2H^+ + 2e^- \longrightarrow H_2S$	−0.274
Acetaldehyde $+ 2H^+ + 2e^- \longrightarrow$ ethanol	−0.197
Pyruvate$^- + 2H^+ + 2e^- \longrightarrow$ lactate^{2-}	−0.185
$FAD + 2H^+ + 2e^- \longrightarrow FADH_2$	−0.18 [b]
Oxaloacetate$^{2-} + 2H^+ + 2e^- \longrightarrow$ malate^{2-}	−0.166
Fumarate$^{2-} + 2H^+ + 2e^- \longrightarrow$ succinate^{2-}	0.031
Cytochrome b $(Fe^{3+}) + e^- \longrightarrow$ cytochrome b (Fe^{2+})	0.075
Ubiquinone $+ 2H^+ + 2e^- \longrightarrow$ ubiquinone H_2	0.10
Cytochrome c $(Fe^{3+}) + e^- \longrightarrow$ cytochrome c (Fe^{2+})	0.254
$NO_3^- + 2H^+ + 2e^- \longrightarrow NO_2^- + H_2O$	0.421
$NO_2^- + 8H^+ + 6e^- \longrightarrow NH_4^+ + 2H_2O$	0.44
$Fe^{3+} + e^- \longrightarrow Fe^{2+}$	0.771
$O_2 + 4H^+ + 4e^- \longrightarrow 2H_2O$	0.815

[a]E'_0 is the standard reduction potential at pH 7.0.
[b]The value for FAD/FADH$_2$ applies to the free cofactor because it can vary considerably when bound to an apoenzyme.

such as photosynthesis, fermentation, and aerobic respiration. In the cell's economy, exergonic ATP breakdown is coupled with various endergonic reactions to promote their completion (figure 8.6). In other words ATP links energy-generating reactions, which liberate free energy, with energy-using reactions, which require free energy input to proceed toward completion. Facilitation of chemical work is the focus of the preceding example, but the same principles apply when ATP is coupled with endergonic processes involving transport work and mechanical work (figure 8.3).

Oxidation-Reduction Reactions and Electron Carriers

Free energy changes are not only related to the equilibria of "regular" chemical reactions but also to the equilibria of oxidation-reduction reactions. **Oxidation-reduction (redox) reactions** are those in which electrons move from a donor, the **reducing agent** or **reductant,** to an electron acceptor, the **oxidizing agent** or **oxidant.** By convention such a reaction is written with the reductant to the right of the oxidant and the number (n) of electrons (e^-) transferred.

$$\text{Oxidant} + ne^- \rightleftharpoons \text{reductant}$$

The oxidant and reductant pair is referred to as a redox couple (table 8.1). When an oxidant accepts electrons, it becomes the reductant of the couple. The equilibrium constant for the reaction is called the **standard reduction potential** (E_0) and is a measure of the tendency of the reducing agent to lose electrons. The reference standard for reduction potentials is the hydrogen system with an E'_0 (the reduction potential at pH 7.0) of −0.42 volts or −420 millivolts.

$$2H^+ + 2e^- \rightleftharpoons H_2$$

In this reaction each hydrogen atom provides one proton (H^+) and one electron (e^-).

The reduction potential has a concrete meaning. Redox couples with more negative reduction potentials will donate electrons to couples with more positive potentials. Thus electrons will tend to move from reductants at the top of the list in table 8.1 to oxidants at the bottom because they have more positive

potentials. Consider the case of the electron carrier **nicotinamide adenine dinucleotide (NAD⁺).** The NAD$^+$/NADH couple has a very negative E'_0 and can therefore give electrons to many acceptors, including O_2.

$$NAD^+ + 2H^+ + 2e^- \rightleftharpoons NADH + H^+ \qquad E'_0 = -0.32 \text{ volts}$$
$$1/2 \, O_2 + 2H^+ + 2e^- \rightleftharpoons H_2O \qquad E'_0 = +0.82 \text{ volts}$$

Because NAD$^+$/NADH is more negative than $1/2 \, O_2$/H$_2$O, electrons will flow from NADH (the reductant) to O_2 (the oxidant).

$$NADH + H^+ + 1/2 \, O_2 \longrightarrow H_2O + NAD^+$$

When electrons move from a reductant to an acceptor with a more positive redox potential, free energy is released. The $\Delta G^{o\prime}$ of the reaction is directly related to the magnitude of the difference between the reduction potentials of the two couples ($\Delta E'_0$). The larger the $\Delta E'_0$, the greater the amount of free energy made available, as is evident from the equation

$$\Delta G^{o\prime} = -nF \cdot \Delta E'_0$$

in which n is the number of electrons transferred and F is the Faraday constant (23,062 cal/mole-volt or 96,494 J/mole-volt). For every 0.1 volt change in $\Delta E'_0$, there is a corresponding 4.6 kcal change in $\Delta G^{o\prime}$ when a two-electron transfer takes place. This is similar to the relationship of $\Delta G^{o\prime}$ and K_{eq} in other chemical reactions—the larger the equilibrium constant, the greater the $\Delta G^{o\prime}$. The difference in reduction potentials between NAD$^+$/NADH and $1/2 \, O_2$/H$_2$O is 1.14 volts, a large $\Delta E'_0$ value. When electrons move from NADH to O_2 during aerobic

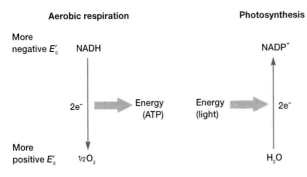

Figure 8.7   Energy Flow in Metabolism. The relationship between electron flow and energy in metabolism. Oxygen and $NADP^+$ serve as electron acceptors for NADH and water, respectively.

respiration, a large amount of free energy is made available to synthesize ATP (figure 8.7). If energy is released when electrons flow from negative to positive reduction potentials, then an input of energy is required to move electrons in the opposite direction, from more positive to more negative potentials. This is precisely what occurs during photosynthesis (figure 8.7). Light energy is trapped and used to move electrons from water to the electron carrier **nicotinamide adenine dinucleotide phosphate ($NADP^+$).**

The cycle of energy flow discussed earlier and illustrated in figure 8.1 can be understood from a different perspective, if the preceding concept is kept in mind. Photosynthetic organisms capture light energy and use it to move electrons from water (and other electron donors) to electron acceptors, such as $NADP^+$, that have more negative reduction potentials. These electrons can then flow back to more positive acceptors and provide energy for ATP production during photosynthesis. Photoautotrophs use ATP and NADPH to synthesize complex molecules from CO_2 (*see chapter 10*). Chemoheterotrophs also make use of energy released during the movement of electrons by oxidizing complex nutrients during respiration to produce NADH. NADH subsequently donates its electrons to O_2, and the energy released during electron transfer is trapped in the form of ATP. The energy from sunlight is made available to all living organisms because of this relationship between electron flow and energy.

Photosynthesis (pp. 181–88).

Respiration and electron transport (pp. 171–74).

Electron movement in cells requires the participation of carriers such as NAD^+ and $NADP^+$, both of which can transport electrons between different locations. The nicotinamide ring of NAD^+ and $NADP^+$ (figure 8.8) accepts two electrons and one proton from a donor, while a second proton is released. There are several other electron carriers of importance in microbial metabolism (table 8.1), and they carry electrons in a variety of ways. **Flavin adenine dinucleotide (FAD)** and **flavin mononucleotide (FMN)** bear two electrons and two protons on the complex ring system shown in figure 8.9. Proteins bearing FAD and FMN are often called flavoproteins. **Coenzyme Q (CoQ)** or **ubiquinone** is a quinone that transports

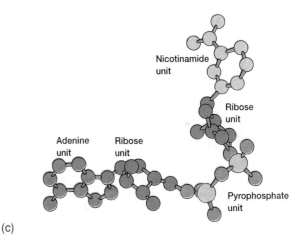

Figure 8.8   The Structure and Function of NAD. (*a*) The structure of NAD and NADP. NADP differs from NAD in having an extra phosphate on one of its ribose sugar units. (*b*) NAD can accept electrons and a hydrogen from a reduced substrate (SH_2). These are carried on the nicotinamide ring. (*c*) Model of NAD^+ when bound to the enzyme lactate dehydrogenase.

electrons and protons in many respiratory electron transport chains (figure 8.10). **Cytochromes** and several other carriers use iron atoms to transport electrons by reversible oxidation and reduction reactions.

$$Fe^{3+} \text{ (ferric iron)} + e^- \rightleftharpoons Fe^{2+} \text{ (ferrous iron)}$$

Figure 8.9 The Structure and Function of FAD. The vitamin riboflavin is composed of the isoalloxazine ring and its attached ribose sugar. FMN is riboflavin phosphate. The portion of the ring directly involved in oxidation-reduction reactions is in color.

Figure 8.10 The Structure and Function of Coenzyme Q or Ubiquinone. The length of the side chain varies among organisms from n = 6 to n = 10.

In the cytochromes these iron atoms are part of a heme group (figure 8.11) or other similar iron-porphyrin rings. Several different cytochromes, each of which consists of a protein and an iron-porphyrin ring, are a prominent part of respiratory electron transport chains. Some iron containing electron-carrying proteins lack a heme group and are called **nonheme iron proteins.** **Ferredoxin** is a nonheme iron protein active in photosynthetic electron transport and several other electron transport processes.

Even though its iron atoms are not bound to a heme group, they still undergo reversible oxidation and reduction reactions.

1. Why is ATP called a high-energy molecule? What is its role in the cell and how does it fulfill this role?
2. Write a generalized equation for a redox reaction. Define reductant, oxidant, and standard reduction potential.
3. How is the direction of electron flow between redox couples related to the standard reduction potential and the release of free energy? Name and briefly describe the major electron carriers found in cells.

Enzymes

Recall that an exergonic reaction is one with a negative $\Delta G^{o\prime}$ and an equilibrium constant greater than one. An exergonic reaction will proceed to completion in the direction written (that is, toward the right of the equation). Nevertheless, one often can combine the reactants for an exergonic reaction with no obvious result, even though products should be formed. It is precisely in these reactions that enzymes play their part.

Structure and Classification of Enzymes

Enzymes may be defined as protein catalysts that have great specificity for the reaction catalyzed and the molecules acted on. A **catalyst** is a substance that increases the rate of a chemical reaction without being permanently altered itself. Thus enzymes speed up cellular reactions. The reacting molecules are called **substrates,** and the substances formed are the **products.**

Protein structure and properties (appendix I).

Many enzymes are indeed pure proteins. However, many enzymes consist of a protein, the **apoenzyme,** and also a nonprotein component, a **cofactor,** required for catalytic activity. The complete enzyme consisting of the apoenzyme and its cofactor is called the **holoenzyme.** If the cofactor is firmly attached to the apoenzyme it is a **prosthetic group.** Often the cofactor is loosely attached to the apoenzyme. It can even

dissociate from the enzyme protein after products have been formed and carry one of these products to another enzyme (figure 8.12). Such a loosely bound cofactor is called a **coenzyme.** For example, NAD^+ is a coenzyme that carries electrons within the cell. Many vitamins that humans require serve as coenzymes or as their precursors. Niacin is incorporated into NAD^+ and riboflavin into FAD. Metal ions may also be bound to apoenzymes and act as cofactors.

Despite the large number and bewildering diversity of enzymes present in cells, they may be placed in one of six general classes (table 8.2). Enzymes usually are named in terms of the substrates they act on and the type of reaction catalyzed. For example, lactate dehydrogenase (LDH) removes hydrogens from lactate.

$$\text{Lactate} + NAD^+ \overset{\text{LDH}}{\rightleftharpoons} \text{pyruvate} + NADH + H^+$$

Lactate dehydrogenase can also be given a more complete and detailed name, L-lactate: NAD oxidoreductase. This name describes the substrates and reaction type with even more precision.

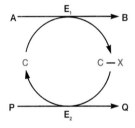

Figure 8.12   Coenzymes as Carriers. The function of a coenzyme in carrying substances around the cell. Coenzyme C participates with enzyme E_1 in the conversion of A to product B. During the reaction, it acquires X from the substrate A. The coenzyme can donate X to substrate P in a second reaction. This will convert it back to its original form, ready to accept another X. The coenzyme is not only participating in both reactions, but is also transporting X to various points in the cell.

Figure 8.11   The Structure of Heme. Heme is composed of a porphyrin ring and an attached iron atom. It is the nonprotein component of many cytochromes. The iron atom alternatively accepts and releases an electron.

TABLE 8.2	Enzyme Classification	
Type of Enzyme	**Reaction Catalyzed by Enzyme**	**Example of Reaction**
Oxidoreductase	Oxidation-reduction reactions	Lactate dehydrogenase: Pyruvate + NADH + H^+ $\rightleftharpoons$ lactate + NAD^+
Transferase	Reactions involving the transfer of groups between molecules	Aspartate carbamoyltransferase: Aspartate + carbamoylphosphate $\rightleftharpoons$ carbamoylaspartate + phosphate
Hydrolase	Hydrolysis of molecules	Glucose-6-phosphatase: Glucose 6-phosphate + H_2O → glucose + P_i
Lyase	Removal of groups to form double bonds or addition of groups to double bonds	Fumarate hydratase: L-malate $\rightleftharpoons$ fumarate + H_2O
Isomerase	Reactions involving isomerizations	Alanine racemase: L-alanine $\rightleftharpoons$ D-alanine
Ligase	Joining of two molecules using ATP energy (or that of other nucleoside triphosphates)	Glutamine synthetase: Glutamate + NH_3 + ATP → glutamine + ADP + P_i

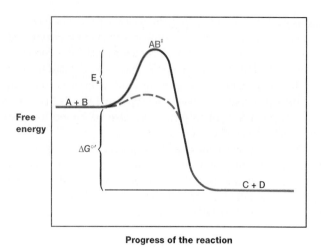

Figure 8.13   Enzymes Lower the Energy of Activation. This figure traces the course of a chemical reaction in which A and B are converted to C and D. The transition-state complex is represented by $AB^{\ddagger}$, and the activation energy required to reach it, by E_a. The colored line represents the course of the reaction in the presence of an enzyme. Note that the activation energy is much lower in the enzyme-catalyzed reaction.

The Mechanism of Enzyme Reactions

It is important to keep in mind that enzymes increase the rates of reactions but do not alter their equilibrium constants. If a reaction is endergonic, the presence of an enzyme will not shift its equilibrium so that more products can be formed. Enzymes simply speed up the rate at which a reaction proceeds toward its final equilibrium.

How do enzymes catalyze reactions? Although a complete answer would be long and complex, some understanding of the mechanism can be gained by considering the course of a normal exergonic chemical reaction.

$$A + B \rightleftharpoons C + D$$

When molecules A and B approach each other to react, they form a **transition-state complex,** which resembles both the substrates and the products (figure 8.13). The **activation energy** is required to bring the reacting molecules together in the correct way to reach the transition state. The transition-state complex can then decompose to yield the products C and D. The difference in free energy level between reactants and products is $\Delta G^{\circ\prime}$. Thus the equilibrium in our example will lie toward the products because $\Delta G^{\circ\prime}$ is negative (i.e., the products are at a lower energy level than the substrates).

Clearly A and B will not be converted to C and D in figure 8.13 if they are not supplied with an amount of energy equivalent to the activation energy. Enzymes accelerate reactions by lowering the activation energy; therefore more substrate molecules will have sufficient energy to come together and form products. Even though the equilibrium constant (or $\Delta G^{\circ\prime}$) is unchanged, equilibrium will be reached more rapidly in the presence of an enzyme because of this decrease in the activation energy.

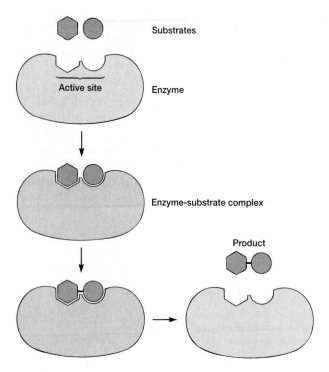

Figure 8.14   Enzyme Function. The formation of the enzyme-substrate complex and its conversion to products is shown.

Researchers have expended much effort in discovering how enzymes lower the activation energy of reactions, and the process is becoming clearer. Enzymes bring substrates together at a special place on their surface called the **active site** or **catalytic site** to form an **enzyme-substrate complex** (figures 8.14, 8.15; *see also AI.19*). The formation of an enzyme-substrate complex can lower the activation energy in many ways. For example, by bringing the substrates together at the active site, the enzyme is, in effect, concentrating them and speeding up the reaction. An enzyme does not simply concentrate its substrates, however. It also binds them so that they are correctly oriented with respect to each other in order to form a transition-state complex. Such an orientation lowers the amount of energy that the substrates require to reach the transition state. These and other catalytic site activities speed up a reaction hundreds of thousands of times, even though the reaction usually takes place at temperatures ranging from 0 to 37°C. These temperatures are not high enough to favor most organic reactions in the absence of enzyme catalysis, yet cells cannot survive at the high temperatures used by an organic chemist in routine organic syntheses. Enzymes make life possible by accelerating specific reactions at low temperatures.

The Effect of Environment on Enzyme Activity

Enzyme activity varies greatly with changes in environmental factors, one of the most important being the substrate concentration. At very low substrate concentrations, an enzyme makes product slowly because it seldom contacts a substrate molecule. If more substrate molecules are present, an enzyme binds

Figure 8.15   An Example of Enzyme-Substrate Complex Formation. (*a*) A space-filling model of yeast hexokinase and its substrate glucose (purple). The active site is in the cleft formed by the enzyme's small lobe (green) and large lobe (gray). (*b*) When glucose binds to form the enzyme-substrate complex, hexokinase changes shape and surrounds the substrate.

substrate more often, and the reaction velocity (usually expressed in terms of the rate of product formation) is greater than at a lower substrate concentration. Thus the rate of an enzyme-catalyzed reaction increases with substrate concentration (figure 8.16). Eventually further increases in substrate concentration do not result in a greater reaction velocity because the available enzyme molecules are binding substrate and converting it to product as rapidly as possible. That is, the enzyme is saturated with substrate and operating at maximal velocity (V_{max}). The

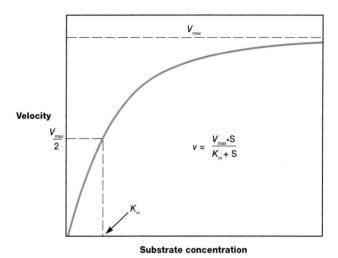

Figure 8.16   Michaelis-Menten Kinetics. The dependence of enzyme activity upon substrate concentration. The maximum velocity (V_{max}) is the rate of product formation when the enzyme is saturated with substrate and making product as fast as possible. The Michaelis constant (K_m) is the substrate concentration required for the enzyme to operate at half its maximal velocity. This substrate curve fits the Michaelis-Menten equation given in the figure, which relates reaction velocity (*v*) to the substrate concentration (S).

resulting substrate concentration curve is a hyperbola (figure 8.16). It is useful to know the substrate concentration an enzyme needs to function adequately. Usually the **Michaelis constant (K_m),** the substrate concentration required for the enzyme to achieve half maximal velocity, is used as a measure of the apparent affinity of an enzyme for its substrate. The lower the K_m value, the lower the substrate concentration at which an enzyme catalyzes its reaction.

Enzymes also change activity with alterations in pH and temperature (figure 8.17). Each enzyme functions most rapidly at a specific pH optimum. When the pH deviates too greatly from an enzyme's optimum, activity slows and the enzyme may be damaged. Enzymes likewise have temperature optima for maximum activity. If the temperature rises too much above the optimum, an enzyme's structure will be disrupted and its activity lost. This phenomenon, known as **denaturation,** may be caused by extremes of pH and temperature or by other factors. The pH and temperature optima of a microorganism's enzymes often reflect the pH and temperature of its habitat. Not surprisingly bacteria growing best at high temperatures often have enzymes with high temperature optima and great heat stability.

Heat stable enzymes in biotechnology (p. 826).

Enzyme Inhibition

Microorganisms can be poisoned by a variety of chemicals, and many of the most potent poisons are enzyme inhibitors. A **competitive inhibitor** directly competes with the substrate at an enzyme's catalytic site and prevents the enzyme from forming product. A classic example of this behavior is seen with the enzyme succinate dehydrogenase, which catalyzes the oxidation

Figure 8.17   pH, Temperature, and Enzyme Activity. The variation of enzyme activity with changes in pH and temperature. The ranges in pH and temperature are only representative. Enzymes differ from one another with respect to the location of their optima and the shape of their pH and temperature curves.

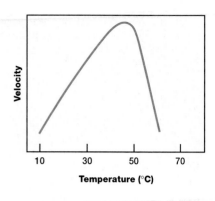

of succinate to fumarate in the tricarboxylic acid cycle (*see chapter 9*). Malonic acid is an effective competitive inhibitor of succinate dehydrogenase because it so closely resembles succinate, the normal substrate (figure 8.18). After malonate binds to the enzyme, it cannot be oxidized and the formation of fumarate is blocked. Competitive inhibitors usually resemble normal substrates, but they cannot be converted to products.

Competitive inhibitors are important in the treatment of many microbial diseases. Sulfa drugs like sulfanilamide resemble p-aminobenzoate, a molecule used in the formation of the coenzyme folic acid. The drugs compete with p-aminobenzoate for the catalytic site of an enzyme involved in folic acid synthesis. This blocks the production of folic acid and inhibits bacterial growth (*see chapter 33*). Humans are not harmed because they cannot synthesize folic acid and must obtain it in their diet.

Destruction of microorganisms by physical and chemical agents (chapter 7).

Inhibitors also can affect enzyme activity by binding to the enzyme at some location other than at the active site. This alters the enzyme's shape, rendering it inactive or less active. These inhibitors are often called **noncompetitive** because they do not directly compete with the substrate. Heavy metal poisons like mercury frequently are noncompetitive inhibitors of enzymes.

Succinic acid   Malonic acid

Figure 8.18   Competitive Inhibition of Succinate Dehydrogenase. A comparison of succinic acid and the competitive inhibitor, malonic acid. The colored atoms indicate the parts of the two molecules that differ.

1. What is an enzyme and how does it speed up reactions? How are enzymes named? Define apoenzyme, holoenzyme, cofactor, coenzyme, prosthetic group, active or catalytic site, and activation energy.

2. How does enzyme activity change with substrate concentration, pH, and temperature? Define Michaelis constant, maximum velocity, and denaturation.

3. What are competitive and noncompetitive inhibitors and how do they inhibit enzymes?

Summary

1. Energy is the capacity to do work. Living cells carry out three major kinds of work: chemical work of biosynthesis, transport work, and mechanical work.

2. The ultimate source of energy is sunlight trapped by photoautotrophs and used to form organic material from CO_2. Photoautotrophs are then consumed by chemoheterotrophs.

3. ATP is the major energy currency and connects energy-generating processes with energy-using processes.

4. The first law of thermodynamics states that energy is neither created nor destroyed.

5. The second law of thermodynamics states that changes occur in such a way that the randomness or disorder of the universe increases to the maximum possible. That is, entropy always increases during spontaneous processes.

6. The first and second laws can be combined to determine the amount of energy made available for useful work.

$$\Delta G = \Delta H - T \cdot \Delta S$$

In this equation the change in free energy (ΔG) is the energy made available for useful work, the change in enthalpy (ΔH) is the change in heat content, and the change in entropy is ΔS.

7. The standard free energy change ($\Delta G^{o\prime}$) for a chemical reaction is directly related to the equilibrium constant.

8. In exergonic reactions $\Delta G^{o\prime}$ is negative and the equilibrium constant is greater than one; the reaction goes to completion as written. Endergonic reactions have a positive $\Delta G^{o\prime}$ and an equilibrium constant less than one.

9. In oxidation-reduction (redox) reactions, electrons move from a donor, the reducing agent or reductant, to an acceptor, the oxidizing agent or oxidant. The standard reduction potential measures the tendency of the reducing agent to give up electrons.

10. Redox couples with more negative reduction potentials donate electrons to those with more positive potentials, and energy is made available during the transfer.

11. Some most important electron carriers in cells are NAD^+, $NADP^+$, FAD, FMN, coenzyme Q, cytochromes, and the nonheme iron proteins.

12. Enzymes are protein catalysts that catalyze specific reactions.

13. Enzymes consist of a protein component, the apoenzyme, and often a nonprotein cofactor that may be a prosthetic group, a coenzyme, or a metal activator.

14. Enzymes speed reactions by binding substrates at their active sites and lowering the activation energy.

15. The rate of an enzyme-catalyzed reaction increases with substrate concentration at low substrate levels and reaches a plateau (the maximum velocity) at saturating substrate concentrations. The Michaelis constant is the substrate concentration that the enzyme requires to achieve half maximal velocity.

16. Enzymes have pH and temperature optima for activity.

17. Enzyme activity can be slowed by competitive and noncompetitive inhibitors.

Key Terms

activation energy   *160*
active site   *160*
adenosine diphosphate (ADP)   *153*
adenosine 5´-triphosphate (ATP)   *153*
aerobic respiration   *153*
apoenzyme   *159*
calorie   *154*
catalyst   *158*
catalytic site   *160*
chemical work   *153*
coenzyme   *159*
coenzyme Q or CoQ (ubiquinone)   *157*
cofactor   *159*
competitive inhibitor   *161*
cytochrome   *157*
denaturation   *161*
endergonic reactions   *155*
energy   *153*

enthalpy   *155*
entropy   *154*
enzyme   *158*
enzyme-substrate complex   *160*
equilibrium   *155*
equilibrium constant (K_{eq})   *155*
exergonic reactions   *155*
ferredoxin   *158*
first law of thermodynamics   *154*
flavin adenine dinucleotide (FAD)   *157*
flavin mononucleotide (FMN)   *157*
free energy change   *155*
high-energy molecule   *155*
holoenzyme   *159*
joule   *154*
mechanical work   *153*
Michaelis constant (K_m)   *161*
nicotinamide adenine dinucleotide (NAD^+)   *156*

nicotinamide adenine dinucleotide phosphate ($NADP^+$)   *157*
noncompetitive inhibitor   *162*
nonheme iron protein   *158*
oxidation-reduction (redox) reaction   *156*
oxidizing agent (oxidant)   *156*
phosphate group transfer potential   *155*
photosynthesis   *153*
product   *159*
prosthetic group   *159*
reducing agent (reductant)   *156*
second law of thermodynamics   *154*
standard free energy change   *155*
standard reduction potential   *156*
substrate   *159*
thermodynamics   *153*
transition-state complex   *160*
transport work   *153*

Questions for Thought and Review

1. Describe in general terms how energy from sunlight is spread throughout the biosphere.

2. Under what conditions would it be possible to create more order in a system without violating the second law of thermodynamics?

3. Suppose that a chemical reaction had a large negative $\Delta G^{\circ\prime}$ value. What would this indicate about its equilibrium constant? If displaced from equilibrium, would it proceed rapidly to completion? Would much or little free energy be made available?

4. Will electrons ordinarily move in an electron transport chain from cytochrome *c* ($E^{\prime}_0 = +210$ mV) to O_2 ($E^{\prime}_0 = +820$ mV) or in the opposite direction?

5. If a person had a niacin deficiency, what metabolic process might well be adversely affected? Why?

6. Draw a diagram showing how enzymes catalyze reactions by altering the activation energy. What is a transition-state complex? Use the diagram to discuss why enzymes do not change the equilibria of the reactions they catalyze.

7. What special properties might an enzyme isolated from a psychrophilic bacterium have? Will enzymes need to lower the activation energy more or less in thermophiles than in psychrophiles?

Additional Reading

General

Becker, W. M. 1977. *Energy and the living cell: An introduction to bioenergetics.* Philadelphia: J. B. Lippincott.
Garrett, R. H., and Grisham, C. H. 1995. *Biochemistry.* New York: Saunders.
Lehninger, A. L.; Nelson, D. L.; and Cox, M. M. 1993. *Principles of biochemistry,* 2d ed. New York: Worth Publishers.
Mathews, C. K., and van Holde, K. E. 1990. *Biochemistry.* Menlo Park, Calif.: Benjamin/Cummings.

Moran, L. A.; Scrimgeour, K. G.; Horton, H. R.; Ochs, R. S.; and Rawn, J. D. 1994. *Biochemistry.* Englewood Cliffs, N.J.: Neil Patterson Publishers/Prentice-Hall.
Stryer, L. 1995. *Biochemistry,* 4th ed. New York: Freeman.
Voet, D., and Voet, J. G. 1995. *Biochemistry,* 2d ed. New York: John Wiley and Sons.
Zubay, G. 1993. *Biochemistry,* 3d ed. Dubuque, Iowa: Wm. C. Brown Communications, Inc.

Enzymes

Boyer, Paul D., editor. 1970–1987. *The enzymes.* San Diego: Academic Press.
Branden, C., and Tooze, J. 1991. *Introduction to protein structure.* New York: Garland Publishing.

Fersht, A. 1984. *Enzyme structure and mechanism,* 2d ed. San Francisco: W. H. Freeman.
International Union of Biochemistry and Molecular Biology. 1992. *Enzyme nomenclature.* San Diego: Academic Press.
Kraut, J. 1988. How do enzymes work? *Science* 242:533–39.
Neidleman, S. L. 1989. Enzymes under stress. *ASM News* 55(2):67–70.
Walsh, C. 1979. *Enzymatic reaction mechanisms.* San Francisco: W. H. Freeman.

CHAPTER 9
Metabolism: The Generation of Energy

It is in the fueling reactions that bacteria display their extraordinary metabolic diversity and versatility. Bacteria have evolved to thrive in almost all natural environments, regardless of the nature of available sources of carbon, energy, and reducing power. . . . The collective metabolic capacities of bacteria allow them to metabolize virtually every organic compound on this planet. . . .

—F. C. Neidhardt,
J. L. Ingraham, and M. Schaechter

Outline

Concepts

1. Metabolism, the sum total of all chemical reactions occurring in the cell, can be divided into catabolism and anabolism. In catabolism, molecules are reduced in complexity and free energy is made available. Anabolism involves the use of free energy to increase the complexity of molecules.

2. During catabolism, nutrients are funneled into a few common pathways for more efficient use of enzymes (a few pathways process a wide variety of nutrients).

3. The tricarboxylic acid cycle is the final pathway for the aerobic oxidation of nutrients to CO_2.

4. The majority of energy released in catabolism is generated by the movement of electrons from electron transport carriers with more negative reduction potentials to ones with more positive reduction potentials. Thus aerobic respiration is much more efficient than anaerobic catabolism.

5. A wide variety of electron acceptors can be used in catabolism: O_2 (aerobic respiration), organic molecules (fermentation), and oxidized inorganic molecules other than O_2 (anaerobic respiration). Furthermore, reduced inorganic molecules as well as organic molecules can serve as electron donors for electron transport and ATP synthesis. Microbial catabolism is unique in the diversity of nutrients and mechanisms employed in energy production.

6. In photosynthesis trapped light energy boosts electrons to more negative reduction potentials or higher energy levels. These energized electrons are then used to make ATP and NAD(P)H during electron transport.

Chapter 8 introduces the basic principles of thermodynamics, the energy cycle and the use of ATP as an energy currency, and the nature and function of enzymes. With this background material in hand, microbial metabolism can be discussed. **Metabolism** is the total of all chemical reactions occurring in the cell. The flow of energy and the participation of enzymes make metabolism possible.

This chapter begins with an overview of metabolism. An introduction to carbohydrate degradation and the aerobic formation of ATP driven by electron transport follows. Next the anaerobic synthesis of ATP through fermentation and anaerobic respiration is described. Then the breakdown of organic substances other than carbohydrates (i.e., lipids, proteins, and amino acids) is briefly surveyed. The chapter concludes with sections on the oxidation of inorganic molecules and the trapping of light energy by photosynthetic light reactions.

An Overview of Metabolism

Metabolism may be divided into two major parts. In **catabolism** [Greek *cata,* down, and *ballein,* to throw] larger and more complex molecules are broken down into smaller, simpler molecules with the release of energy. Some of this energy is trapped and made available for work; the remainder is released as heat. The trapped energy can then be used in anabolism, the second area of metabolism. **Anabolism** [Greek *ana,* up] is the synthesis of complex molecules from simpler ones with the input of energy. An anabolic process uses energy to increase the order of a system (*see chapter 8*). Although the division of metabolism into two major parts is convenient and commonly employed, note that energy-generating processes like chemolithotrophy and photosynthesis are not comfortably encompassed by this definition of catabolism unless it is expanded to include all energy-yielding metabolic processes, whether complex molecules are degraded or not.

Before learning about some of the more important catabolic pathways, it is best to look at the "lay of the land" and get our bearings. Albert Lehninger, a biochemist who worked at Johns Hopkins medical school, helped considerably in this task by pointing out that metabolism may be divided into three stages (figure 9.1). In the first stage of catabolism, larger nutrient molecules (proteins, polysaccharides, and lipids) are hydrolyzed or otherwise broken down into their constituent parts. The chemical reactions occurring during this stage do not release much energy. Amino acids, monosaccharides, fatty acids, glycerol, and other products of the first stage are degraded to a few simpler

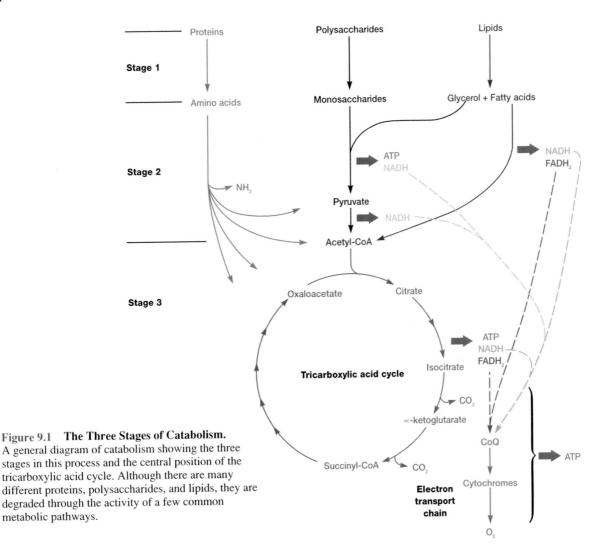

Figure 9.1 The Three Stages of Catabolism. A general diagram of catabolism showing the three stages in this process and the central position of the tricarboxylic acid cycle. Although there are many different proteins, polysaccharides, and lipids, they are degraded through the activity of a few common metabolic pathways.

molecules in the second stage. Usually metabolites like acetyl coenzyme A, pyruvate, and tricarboxylic acid cycle intermediates are formed. The second-stage process can operate either aerobically or anaerobically and often produces some ATP as well as NADH and/or $FADH_2$. Finally, nutrient carbon is fed into the tricarboxylic acid cycle during the third stage of catabolism, and molecules are oxidized completely to CO_2 with the production of ATP, NADH, and $FADH_2$. The cycle operates aerobically and is responsible for the generation of much energy. Much of the ATP derived from the tricarboxylic acid cycle (and stage-two reactions) comes from the oxidation of NADH and $FADH_2$ by the electron transport chain. Oxygen, or sometimes another inorganic molecule, is the final electron acceptor.

Although this picture is somewhat oversimplified, it is useful in discerning the general pattern of catabolism. Notice that the microorganism begins with a wide variety of molecules and reduces their number and diversity at each stage. That is, nutrient molecules are funneled into ever fewer metabolic intermediates until they are finally fed into the tricarboxylic acid cycle. A common pathway often degrades many similar molecules (e.g., several different sugars). These metabolic pathways consist of enzyme-catalyzed reactions arranged so that the product of one reaction serves as a substrate for the next. The existence of a few common catabolic pathways, each degrading many nutrients, greatly increases metabolic efficiency by avoiding the need for a large number of less metabolically flexible pathways. It is in the catabolic phase that microorganisms exhibit their nutritional diversity. Most microbial biosynthetic pathways closely resemble their counterparts in higher organisms. The uniqueness of microbial metabolism lies in the diversity of sources from which it generates ATP and NADH.

Carbohydrates and other nutrients serve two functions in the metabolism of heterotrophic microorganisms: (1) they are oxidized to provide energy, and (2) they supply carbon or building blocks for the synthesis of new cell constituents. Although many anabolic pathways are separate from catabolic routes, there are **amphibolic pathways** [Greek *amphi*, on both sides] that function both catabolically and anabolically. Two of the most important are the glycolytic pathway and the tricarboxylic acid cycle. Most reactions in these two pathways are freely reversible and can be used to synthesize and degrade molecules. The few irreversible catabolic steps are bypassed in biosynthesis with special enzymes that catalyze the reverse reaction (figure 9.2). For example, the enzyme fructose bisphosphatase reverses the phosphofructokinase step when glucose is synthesized from pyruvate (pp. 196–97). The presence of two separate enzymes, one catalyzing the reversal of the other's reaction, permits independent regulation of the catabolic and anabolic functions of these amphibolic pathways.

---

1. Describe metabolism. How are catabolism and anabolism organized? Why are common pathways useful? What is an amphibolic pathway?

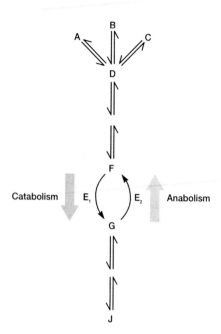

Figure 9.2 **Amphibolic Pathway.** A simplified diagram of an amphibolic pathway such as the glycolytic pathway. Note that the interconversion of intermediates F and G is catalyzed by two separate enzymes: E_1 operating in the catabolic direction and E_2 in the anabolic.

The Breakdown of Glucose to Pyruvate

Microorganisms employ several metabolic pathways to catabolize glucose and other sugars. Because of this metabolic diversity, their metabolism is often confusing. To avoid confusion as much as possible, the ways in which microorganisms degrade sugars to pyruvate and similar intermediates are introduced by focusing on only three routes: (1) glycolysis, (2) the pentose phosphate pathway, and (3) the Entner-Doudoroff pathway. Next the pathways of aerobic and anaerobic pyruvate metabolism are described. For the sake of simplicity, the chemical structures of metabolic intermediates are not used in pathway diagrams.

Diagrams of the glycolytic pathway and other major catabolic pathways with the structures of intermediates and enzyme names (appendix III).

The Glycolytic Pathway

The **Embden-Meyerhof** or **glycolytic pathway** is undoubtedly the most common pathway for glucose degradation to pyruvate in stage two of catabolism. It is found in all major groups of microorganisms and functions in the presence or absence of O_2. **Glycolysis** [Greek *glyco*, sweet, and *lysis*, a loosening] is located in the cytoplasmic matrix when present in eucaryotic microorganisms.

The pathway as a whole may be divided into two parts (figure 9.3 and *appendix III*). In the initial six-carbon stage, glucose is phosphorylated twice and eventually converted to fructose 1,6-bisphosphate. Other sugars are often fed into the pathway

Figure 9.3   **Glycolysis.** The glycolytic pathway for the breakdown of glucose to pyruvate. The two stages of the pathway and their products are indicated.

by conversion to glucose 6-phosphate or fructose 6-phosphate. This preliminary stage does not yield energy; in fact, two ATP molecules are expended for each glucose. These initial steps "prime the pump" by adding phosphates to each end of the sugar. The phosphates will soon be used to make ATP.

The three-carbon stage of glycolysis begins when the enzyme fructose 1,6-bisphosphate aldolase catalyzes the cleavage of fructose 1,6-bisphosphate into two halves, each with a phosphate group. One of the products, glyceraldehyde 3-phosphate, is converted directly to pyruvate in a five-step process. Because the other product, dihydroxyacetone phosphate, can be easily changed to glyceraldehyde 3-phosphate, both halves of fructose 1,6-bisphosphate are used in the three-carbon stage. Glyceraldehyde 3-phosphate is first oxidized with NAD^+ as the electron acceptor, and a phosphate is simultaneously incorporated to give a high-energy molecule called 1,3-bisphosphoglycerate. The high-energy phosphate on carbon one is subsequently

donated to ADP to produce ATP. This synthesis of ATP is called **substrate-level phosphorylation** because ADP phosphorylation is coupled with the exergonic breakdown of a high-energy substrate molecule.

ATP as an energy currency (pp. 155–56).

A somewhat similar process generates a second ATP by substrate-level phosphorylation. The phosphate group on 3-phosphoglycerate shifts to carbon two, and 2-phosphoglycerate is dehydrated to form a second high-energy molecule, phosphoenolpyruvate. This molecule donates its phosphate to ADP forming a second ATP and pyruvate, the final product of the pathway.

The glycolytic pathway degrades one glucose to two pyruvates by the sequence of reactions just outlined. ATP and NADH are also produced. The yields of ATP and NADH may be calculated by considering the two stages separately. In the six-carbon stage two ATPs are used to form fructose 1,6-bisphosphate. For each glyceraldehyde 3-phosphate transformed into pyruvate, one NADH and two ATPs are formed. Because two glyceraldehyde 3-phosphates arise from a single glucose (one by way of dihydroxyacetone phosphate), the three-carbon stage generates four ATPs and two NADHs per glucose. Subtraction of the ATP used in the six-carbon stage from that produced in the three-carbon stage gives a net yield of two ATPs per glucose. Thus the catabolism of glucose to pyruvate in glycolysis can be represented by the following simple equation.

$$\text{Glucose} + 2ADP + 2P_i + 2NAD^+ \longrightarrow$$
$$2 \text{ pyruvate} + 2ATP + 2NADH + 2H^+$$

The Pentose Phosphate Pathway

A second pathway, the **pentose phosphate** or **hexose monophosphate pathway** may be used at the same time as the glycolytic pathway or the Entner-Doudoroff sequence. It can operate either aerobically or anaerobically and is important in biosynthesis as well as in catabolism.

The pentose phosphate pathway begins with the oxidation of glucose 6-phosphate to 6-phosphogluconate followed by the oxidation of 6-phosphogluconate to the pentose ribulose 5-phosphate and CO_2 (figure 9.4 and *appendix III*). NADPH is produced during these oxidations. Ribulose 5-phosphate is then converted to a mixture of three- through seven-carbon sugar phosphates. Two enzymes unique to this pathway play a central role in these transformations: (1) transketolase catalyzes the transfer of two-carbon ketol groups, and (2) transaldolase transfers a three-carbon group from sedoheptulose 7-phosphate to glyceraldehyde 3-phosphate (figure 9.5). The overall result is that three glucose 6-phosphates are converted to two fructose 6-phosphates, glyceraldehyde 3-phosphate, and three CO_2 molecules, as shown in the following equation.

$$3 \text{ glucose 6-phosphate} + 6NADP^+ + 3H_2O \longrightarrow$$
$$2 \text{ fructose 6-phosphate} + \text{glyceraldehyde 3-phosphate} +$$
$$3CO_2 + 6NADPH + 6H^+$$

Figure 9.4   The Pentose Phosphate Pathway.
The conversion of three glucose 6-phosphate
molecules to two fructose 6-phosphates and a
glyceraldehyde 3-phosphate is traced. The
fructose 6-phosphates are changed back to
glucose 6-phosphate. The glyceraldehyde 3-
phosphate can be converted to pyruvate or
combined with a molecule of dihydroxyacetone
phosphate (from the glyceraldehyde 3-phosphate
formed by a second turn of the pathway) to yield
fructose 6-phosphate.

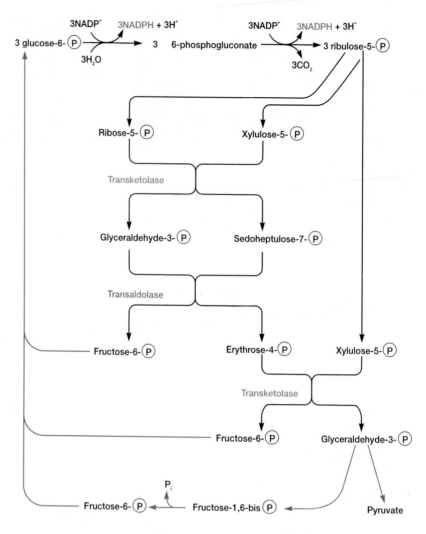

**Figure 9.5   Transketolase and
Transaldolase.** Examples of the transketolase
and transaldolase reactions of the pentose
phosphate pathway. The groups transferred in
these reactions are in color.

The transketolase reaction

CH₂OH	O H	CH₂OH	O H
C=O	C	C=O	C
HO—C—H	H—C—OH	HO—C—H	H—C—OH
H—C—OH	H—C—OH	H—C—OH	CH₂O(P)
CH₂O(P)	H—C—OH	H—C—OH	
Xylulose	CH₂O(P)	H—C—OH	Glyceraldehyde
5-phosphate	Ribose	CH₂O(P)	3-phosphate
	5-phosphate	Sedoheptulose	
		7-phosphate	

The transaldolase reaction

Sedoheptulose 7-phosphate + Glyceraldehyde 3-phosphate → Fructose 6-phosphate + Erythrose 4-phosphate

These intermediates are used in two ways. The fructose 6-phosphate can be changed back to glucose 6-phosphate while glyceraldehyde 3-phosphate is converted to pyruvate by glycolytic enzymes. The glyceraldehyde 3-phosphate also may be returned to the pentose phosphate pathway through glucose 6-phosphate formation. This results in the complete degradation of glucose 6-phosphate to CO_2 and the production of a great deal of NADPH.

$$\text{Glucose 6-phosphate} + 12NADP^+ + 7H_2O \longrightarrow$$
$$6CO_2 + 12NADPH + 12H^+ + P_i$$

The pentose phosphate pathway has several catabolic and anabolic functions that are summarized as follows:

1. NADPH from the pentose phosphate pathway serves as a source of electrons for the reduction of molecules during biosynthesis.

2. The pathway synthesizes four- and five-carbon sugars for a variety of purposes. The four-carbon sugar erythrose 4-phosphate is used to synthesize aromatic amino acids. The pentose ribose 5-phosphate is a major component of nucleic acids, and ribulose 1,5-bisphosphate is the primary CO_2 acceptor in photosynthesis. Note that when a microorganism is growing on a pentose carbon source, the pathway also can supply carbon for hexose production (e.g., glucose is needed for peptidoglycan synthesis).

3. Intermediates in the pentose phosphate pathway may be used to produce ATP. Glyceraldehyde 3-phosphate from the pathway can enter the three-carbon stage of the glycolytic pathway and be converted to ATP and pyruvate. The latter may be oxidized in the tricarboxylic acid cycle to provide more energy. In addition, some NADPH can be converted to NADH, which yields ATP when it is oxidized by the electron transport chain. Because five-carbon sugars are intermediates in the pathway, the pentose phosphate pathway can be used to catabolize pentoses as well as hexoses.

Although the pentose phosphate pathway may be a source of energy in many microorganisms, it is more often of greater importance in biosynthesis. Several functions of the pentose phosphate pathway are mentioned again in chapter 10 when biosynthesis is considered more directly.

The Entner-Doudoroff Pathway

Although the glycolytic pathway is the most common route for the conversion of hexoses to pyruvate, another pathway with a similar role has been discovered. The **Entner-Doudoroff pathway** begins with the same reactions as the pentose phosphate pathway, the formation of glucose 6-phosphate and 6-phosphogluconate (figure 9.6 and *appendix III*). Instead of being further oxidized, 6-phosphogluconate is dehydrated to form 2-keto-3-deoxy-6-phosphogluconate or KDPG, the key intermediate in this pathway. KDPG is then cleaved by KDPG aldolase to pyruvate and glyceraldehyde 3-phosphate. The

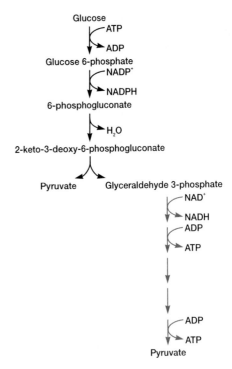

Figure 9.6   The Entner-Doudoroff Pathway. The sequence leading from glyceraldehyde 3-phosphate to pyruvate is catalyzed by enzymes common to the glycolytic pathway.

glyceraldehyde 3-phosphate is converted to pyruvate in the bottom portion of the glycolytic pathway. If the Entner-Doudoroff pathway degrades glucose to pyruvate in this way, it yields one ATP, one NADPH, and one NADH per glucose metabolized.

Most bacteria have the glycolytic and pentose phosphate pathways, but some substitute the Entner-Doudoroff pathway for glycolysis. The Entner-Doudoroff pathway is generally found in *Pseudomonas, Rhizobium, Azotobacter, Agrobacterium,* and a few other gram-negative genera. Very few gram-positive bacteria have this pathway, with *Enterococcus faecalis* being a rare exception.

1. Summarize the major features of the glycolytic pathway, the pentose phosphate pathway, and the Entner-Doudoroff sequence. Include the starting points, the products of the pathways, the critical or unique enzymes, the ATP yields, and the metabolic roles each pathway has.

2. What is substrate-level phosphorylation?

The Tricarboxylic Acid Cycle

Although some energy is obtained from the breakdown of glucose to pyruvate by the pathways previously described, much more is released when pyruvate is degraded aerobically to CO_2 in stage three of catabolism. The multienzyme system called the pyruvate dehydrogenase complex first oxidizes pyruvate to

Figure 9.7 The Tricarboxylic Acid Cycle. The cycle may be divided into three stages based on the size of its intermediates. The three stages are separated from one another by two decarboxylation reactions (reactions in which carboxyl groups are lost as CO_2). The pyruvate dehydrogenase complex forms acetyl-CoA through pyruvate oxidation.

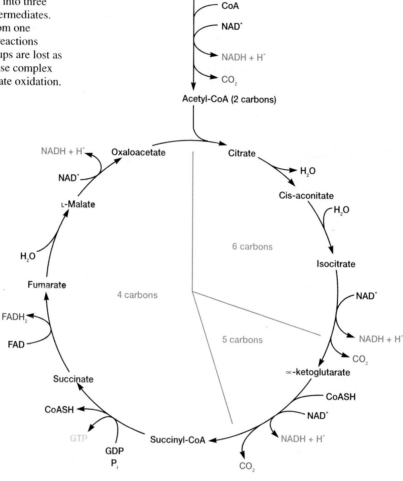

form CO_2 and **acetyl coenzyme A (acetyl-CoA)**, an energy-rich molecule composed of coenzyme A and acetic acid joined by a high-energy thiol ester bond (figure 9.7). Acetyl-CoA arises from the catabolism of many carbohydrates, lipids, and amino acids (figure 9.1). It can be further degraded in the tricarboxylic acid cycle.

The substrate for the **tricarboxylic acid (TCA) cycle, citric acid cycle,** or Krebs cycle is acetyl-CoA (figure 9.7 and *appendix III*). In the first reaction acetyl-CoA is condensed with a four-carbon intermediate, oxaloacetate, to form citrate and to begin the six-carbon stage. Citrate (a tertiary alcohol) is rearranged to give isocitrate, a more readily oxidized secondary alcohol. Isocitrate is subsequently oxidized and decarboxylated twice to yield α-ketoglutarate, then succinyl-CoA. At this point two NADHs are formed and two carbons are lost from the cycle as CO_2. Because two carbons were added as acetyl-CoA at the start, balance is maintained and no net carbon is lost. The cycle now enters the four-carbon stage during which two oxidation steps yield one $FADH_2$ and one NADH per acetyl-CoA. In addition, GTP (a high-energy molecule equivalent to ATP) is produced from succinyl-CoA by substrate-level phosphorylation.

Eventually oxaloacetate is reformed and ready to join with another acetyl-CoA. Inspection of figure 9.7 shows that the TCA cycle generates two CO_2s, three NADHs, one $FADH_2$, and one GTP for each acetyl-CoA molecule oxidized.

TCA cycle enzymes are widely distributed among microorganisms. The cycle appears to be functional in many aerobic bacteria, free-living protozoa, and most algae and fungi. This is not surprising because the cycle is such an important source of energy. However, the facultative anaerobe *E. coli* does not use the complete TCA cycle under anaerobic conditions or when the glucose concentration is high but does at other times. Even those microorganisms that lack the complete TCA cycle usually have most of the cycle enzymes, because one of TCA cycle's major functions is to provide carbon skeletons for use in biosynthesis.

The role of the tricarboxylic acid cycle in biosynthesis (pp. 201–202).

1. Give the substrate and products of the tricarboxylic acid cycle. Describe its organization in general terms. What are its two major functions?

Electron Transport and Oxidative Phosphorylation

Little ATP has been synthesized up to this point. Only the equivalent of four ATP molecules is directly synthesized when one glucose is oxidized to six CO_2 molecules by way of glycolysis and the TCA cycle. Most ATP generated aerobically comes from the oxidation of NADH and $FADH_2$ in the electron transport chain. The electron transport chain itself is examined first, then the mechanism of ATP synthesis is discussed.

The Electron Transport Chain

The **electron transport chain** is composed of a series of electron carriers that operate together to transfer electrons from donors, like NADH and $FADH_2$, to acceptors, such as O_2 (refer to diagram of a mitochondrial electron transport chain in figure 9.8a). The electrons flow from carriers with more negative reduction potentials to those with more positive potentials and eventually combine with O_2 and H^+ to form water. The difference in reduction potentials between O_2 and NADH is large, about 1.14 volts, and makes possible the release of a great deal of energy. Despite the large overall decrease in reduction potential, there are only three places in eucaryotic transport chains where sufficient energy is made available to synthesize ATP: between (1) NADH and coenzyme Q, (2) cytochromes b and c_1, and (3) cytochrome a and O_2. The electron transport chain breaks up the large overall energy release into small steps. Some of the liberated energy is trapped in the form of ATP. As will be seen shortly, electron transport at these points may generate proton and electrical gradients. These gradients can then drive ATP synthesis.

The electron transport chain carriers reside within the inner membrane of the mitochondrion or in the bacterial plasma membrane. The mitochondrial system is arranged into four complexes of carriers, each capable of transporting electrons part of the way to O_2 (figure 9.8b). Coenzyme Q and cytochrome c connect the complexes with each other. The bacterial carriers are also highly organized within the membrane (figure 9.9).

The process by which energy from electron transport is used to make ATP is called **oxidative phosphorylation.** Thus as many as three ATP molecules may be synthesized from ADP and P_i when a pair of electrons pass from NADH to an atom of O_2. This is the same thing as saying that the phosphorus to oxygen (P/O) ratio is equal to 3. Because electrons from $FADH_2$ only pass two oxidative phosphorylation points, the maximum P/O ratio for $FADH_2$ is 2. The actual P/O ratios may be less than 3.0 and 2.0 in eucaryotic mitochondria.

The preceding discussion has focused on the eucaryotic mitochondrial electron transport chain. Although some bacterial chains resemble the mitochondrial chain, they are frequently very different. They vary in their electron carriers (e.g., in their cytochromes) and may be extensively branched.

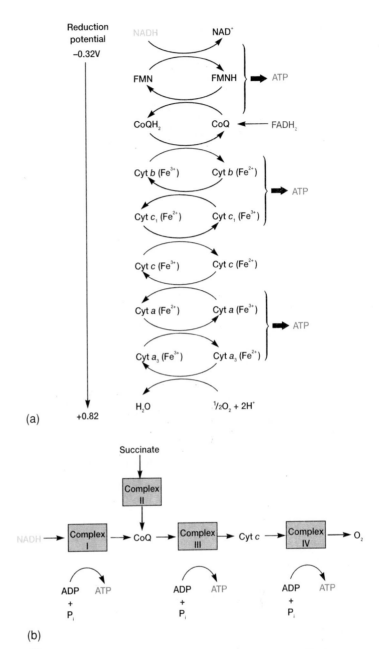

(a)

(b)

Figure 9.8   The Mitochondrial Electron Transport Chain. (a) This diagram is somewhat simplified for the sake of clarity, and several carriers have been excluded. Cyt is an abbreviation for cytochrome and CoQ, for coenzyme Q. (b) The organization of the eucaryotic electron transport chain into four complexes.

Electrons often can enter at several points and leave through several terminal oxidases. The *E. coli* chain even changes composition with alterations in the O_2 level (figure 9.9). Bacterial chains also may be shorter and have lower P/O ratios than mitochondrial transport chains. Thus procaryotic and eucaryotic electron transport chains differ in details of construction although they operate using the same fundamental principles.

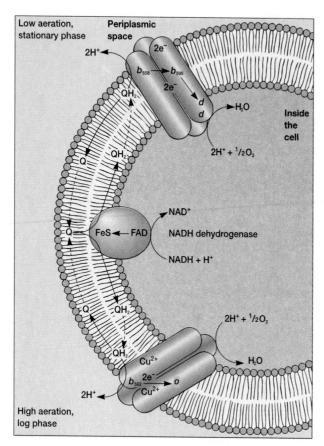

Low aeration, stationary phase

Periplasmic space

Inside the cell

High aeration, log phase

Figure 9.9   The Aerobic Respiratory System of *E. coli.* NADH is the electron source. Ubiquinone-8 (Q) connects the NADH dehydrogenase with two terminal oxidase systems that function at different oxygen levels. At least five cytochromes are involved: b_{558}, b_{595}, b_{562}, *d*, and *o*.

Oxidative Phosphorylation

The mechanism by which oxidative phosphorylation takes place has been studied intensively for years. There are currently several hypotheses about how oxidative phosphorylation occurs; the two most important are the chemiosmotic hypothesis and the conformational change hypothesis.

According to the **chemiosmotic hypothesis,** first formulated in 1961 by the British biochemist Peter Mitchell, the electron transport chain is organized so that, during its operation, protons move outward from the mitochondrial matrix and electrons are transported inward (figures 9.10 and 9.11). Proton movement may result either from carrier loops, as shown in figure 9.11, or from the action of special proton pumps that derive their energy from electron transport. The result is **protonmotive force (PMF),** composed of a gradient of protons and a membrane potential due to the unequal distribution of charges. When protons return to the mitochondrial matrix driven by the protonmotive force, ATP is synthesized in a reversal of the ATP hydrolysis reaction (figure 9.10). A

similar process is believed to take place in bacteria, with electron flow causing the protons to move outward across the plasma membrane (figure 9.9). ATP synthesis occurs when these protons diffuse back into the cell. The protonmotive force also may drive the transport of molecules across membranes (*see chapter 5*) and the rotation of bacterial flagella (*see chapter 3*). The chemiosmotic hypothesis is accepted by most microbiologists. There is considerable evidence for the generation of proton and charge gradients across membranes. However, the evidence for proton gradients as the direct driving force for oxidative phosphorylation is not yet conclusive.

In the **conformational change hypothesis,** energy released by electron transport induces changes in the shape or conformation of the enzyme that synthesizes ATP. These shape changes drive ATP formation, possibly by changing the strength of ATP, ADP, and phosphate binding to the enzyme's catalytic site. The hypothesis is plausible because conformational changes occur in mitochondrial inner membrane proteins during electron transport.

Research in this area is proceeding rapidly, and the precise mechanism of oxidative phosphorylation should soon become clearer. Possibly the final answer will contain features of more than one current hypothesis.

Whatever the precise mechanism, clearly ATP synthesis takes place at the F_1F_0 ATPase or **ATP synthase** (figure 9.12). In mitochondria the F_1 component appears as a spherical structure attached to the surface of the inner membrane by a stalk and the F_0 component, which is embedded in the inner membrane. F_0 is believed to participate in the proton movement across the membrane associated with oxidative phosphorylation. There is evidence that the γ and ε subunits in the stalk change conformation when protons move through F_0. These two subunits could then influence the active site on the β subunit of F_1. In bacteria the F_1F_0 ATPase is on the inner surface of the plasma membrane.

Many chemicals inhibit the aerobic synthesis of ATP. These inhibitors generally fall into two categories. Some directly block the transport of electrons. The antibiotic piericidin competes with coenzyme Q; the antibiotic antimycin A blocks electron transport between cytochromes *b* and *c;* and both cyanide and azide stop the transfer of electrons between cytochrome *a* and O_2 because they are structural analogs of O_2. Another group of inhibitors known as **uncouplers** stops ATP synthesis without inhibiting electron transport itself. Indeed, they may even enhance the rate of electron flow. Normally electron transport is tightly coupled with oxidative phosphorylation so that the rate of ATP synthesis controls the rate of electron transport. The more rapidly ATP is synthesized during oxidative phosphorylation, the faster the electron transport chain operates to supply the required energy. Uncouplers disconnect oxidative phosphorylation from electron transport; therefore the energy released by the chain is given off as heat rather than as ATP. Many uncouplers like

Figure 9.10   Chemiosmosis. The chemiosmotic hypothesis as applied to mitochondrial function. The flow of electrons from NADH to oxygen causes protons to move from the mitochondrial matrix to the intermembrane space. This generates proton and electrical gradients. When protons move back to the matrix through the F_1F_0 complex, F_1 synthesizes ATP.

Figure 9.11   The Chemiosmotic Hypothesis Applied to Eucaryotic Mitochondria. In this scheme the carriers are organized asymmetrically within the inner membrane so that protons are transported across as electrons move along the chain. Proton release into the intermembrane space occurs when electrons are transferred from carriers, such as FMN and coenzyme Q (Q), that carry both electrons and protons to components like nonheme iron proteins (FeS proteins) and cytochromes (Cyt) that transport only electrons. It also is possible that complex IV may pump protons across the membrane as electrons pass from cytochrome a to oxygen. Coenzyme Q transports electrons from complexes I and II to complex III. Cytochrome c moves electrons between complexes III and IV.

Figure 9.12   ATP Synthesis. The spherical structure is sometimes called the F_1 ATPase because it can catalyze the hydrolysis of ATP. This complex structure is the site of ATP synthesis. It is composed of five different subunits (α, β, γ, δ, ε); ATP is synthesized by the β subunit. It rests on a stalk that is connected to the F_0 component embedded in the membrane. F_0 is made of a, b, and c subunits. ATP is synthesized when protons pass through the F_1F_0 complex or ATP synthase under the influence of protonmotive force. ATP, ADP, and P_i are carried across the inner membrane by separate transport proteins.

dinitrophenol and valinomycin may allow hydrogen ions, potassium ions, and other ions to cross the membrane without activating the F_1F_0 ATPase. In this way they destroy the pH and ion gradients. Valinomycin also may bind directly to the F_1F_0 ATPase and inhibit its activity.

The Yield of ATP in Glycolysis and Aerobic Respiration

The maximum ATP yield in eucaryotes from glycolysis, the TCA cycle, and electron transport can be readily calculated. The conversion of glucose to two pyruvate molecules during glycolysis gives a net gain of two ATPs and two NADHs. Because each NADH can yield a maximum of three ATPs during electron transport and oxidative phosphorylation, the total aerobic yield from the glycolytic pathway is eight ATP molecules (table 9.1). Under anaerobic conditions, when the NADH is not oxidized by the electron transport chain, only two ATPs will be generated during the degradation of glucose to pyruvate. The fate of pyruvate under anaerobic conditions is the topic of the next section.

When O_2 is present and the electron transport chain is operating, pyruvate is next oxidized to acetyl-CoA, the substrate for the TCA cycle. This reaction yields 2 NADHs because 2 pyruvates arise from a glucose; therefore 6 more ATPs are formed. Oxidation of each acetyl-CoA in the TCA cycle will yield 1 GTP (or ATP), 3 NADHs, and a single FADH$_2$ for a total of 2 GTPs (ATPs), 6 NADHs, and 2 FADH$_2$s from two

TABLE 9.1	ATP Yield from the Aerobic Oxidation of Glucose by Eucaryotic Cells
Glycolytic Pathway	
Substrate-level phosphorylation (ATP)	2 ATP[a]
Oxidative phosphorylation with 2 NADH	6 ATP
2 Pyruvate to 2 Acetyl-CoA	
Oxidative phosphorylation with 2 NADH	6 ATP
Tricarboxylic Acid Cycle	
Substrate-level phosphorylation (GTP)	2 ATP
Oxidative phosphorylation with 6 NADH	18 ATP
Oxidative phosphorylation with 2 FADH$_2$	4 ATP
Total Aerobic Yield	38 ATP

[a]ATP yields are calculated with an assumed P/O ratio of 3.0 for NADH and 2.0 for FADH$_2$.

acetyl-CoA molecules. As table 9.1 shows, this amounts to 24 ATPs when NADH and FADH$_2$ from the cycle are oxidized in the electron transport chain. Thus the aerobic oxidation of glucose to 6 CO_2 molecules supplies a maximum of 38 ATPs. Only 2 ATP molecules are formed upon the anaerobic conversion of glucose to pyruvate in glycolysis. Because bacterial electron transport systems often have lower P/O ratios than the eucaryotic system being discussed, bacterial aerobic ATP yields are less.

Clearly, aerobic respiration is much more effective than anaerobic processes not involving electron transport and

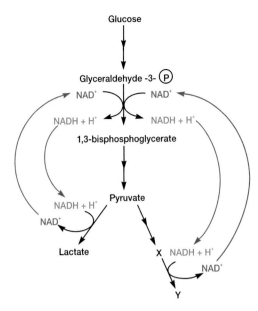

Figure 9.13   Reoxidation of NADH During Fermentation. NADH from glycolysis is reoxidized by being used to reduce pyruvate or a pyruvate derivative (X). Either lactate or reduced product Y result.

oxidative phosphorylation. Many microorganisms, when moved from anaerobic to aerobic conditions, will drastically reduce their rate of sugar catabolism and switch to aerobic respiration, a regulatory phenomenon known as the **Pasteur effect.** This is of obvious advantage to the microorganism as less sugar must be degraded to obtain the same amount of ATP when the more efficient aerobic process can be employed.

---

1. Briefly describe the structure of the electron transport chain and its role in ATP formation.
2. By what two mechanisms might ATP be synthesized during oxidative phosphorylation? What is an uncoupler?
3. Calculate the ATP yield for both glycolysis and the total aerobic oxidation of glucose. Explain your reasoning.

Fermentations

In the absence of O_2, NADH is not usually oxidized by the electron transport chain because no external electron acceptor is available. Yet NADH produced in the glycolytic pathway during the oxidation of glyceraldehyde 3-phosphate to 1,3-bisphosphoglycerate (figure 9.3) must still be oxidized back to NAD^+. If NAD^+ is not regenerated, the oxidation of glyceraldehyde 3-phosphate will cease and glycolysis will stop. Many microorganisms solve this problem by slowing or stopping pyruvate dehydrogenase activity and using pyruvate or one of its derivatives as an electron and hydrogen acceptor in the reoxidation of NADH (figure 9.13). This may lead to the production of more ATP. An energy-yielding process like this, in which organic molecules serve as both electron donors

and acceptors, is called a **fermentation** [Latin *fermentare,* to cause to rise or ferment]. There are many kinds of fermentations, and they often are characteristic of particular microbial groups (figure 9.14). A few more common fermentations are introduced here, and several others are discussed at later points. Two unifying themes should be kept in mind when microbial fermentations are examined: (1) NADH is oxidized to NAD^+, and (2) the electron acceptor is either pyruvate or a pyruvate derivative.

Many fungi and some bacteria, algae, and protozoa ferment sugars to ethanol and CO_2 in a process called **alcoholic fermentation.** Pyruvate is decarboxylated to acetaldehyde, which is then reduced to ethanol by alcohol dehydrogenase with NADH as the electron donor (figure 9.14, number 2). **Lactic acid fermentation,** the reduction of pyruvate to lactate (figure 9.14, number 1), is even more common. It is present in bacteria (lactic acid bacteria, *Bacillus*), algae (*Chlorella*), some water molds, protozoa, and even in animal skeletal muscle. Lactic acid fermenters can be separated into two groups. **Homolactic fermenters** use the glycolytic pathway and directly reduce almost all their pyruvate to lactate with the enzyme lactate dehydrogenase. **Heterolactic fermenters** form substantial amounts of products other than lactate; many produce lactate, ethanol, and CO_2 by way of the phosphoketolase pathway (*see chapter 21*).

The Enterobacteriaceae (pp. 425–27).

Many bacteria, especially members of the family *Enterobacteriaceae,* can metabolize pyruvate to formic acid and other products in a process sometimes called the formic acid fermentation (figure 9.14, number 5). Formic acid may be converted to H_2 and CO_2 by formic hydrogenlyase (a combination of at least two enzymes).

$$HCOOH \longrightarrow CO_2 + H_2$$

There are two types of formic acid fermentation. **Mixed acid fermentation** results in the excretion of ethanol and a complex mixture of acids, particularly acetic, lactic, succinic, and formic acids (table 9.2). If formic hydrogenlyase is present, the formic acid will be degraded to H_2 and CO_2. This pattern is seen in *Escherichia, Salmonella, Proteus,* and other genera. The second type, **butanediol fermentation,** is characteristic of *Enterobacter, Serratia, Erwinia,* and some species of *Bacillus* (figure 9.14, number 4). Pyruvate is converted to acetoin, which is then reduced to 2,3-butanediol with NADH. A large amount of ethanol is also produced, together with smaller amounts of the acids found in mixed acid fermentation.

Formic acid fermentations are very useful in identification of members of the *Enterobacteriaceae.* Butanediol fermenters can be distinguished from mixed acid fermenters in three ways.

1. The Voges-Proskauer test detects the acetoin precursor of butanediol (figure 9.14) and is positive with butanediol fermenters but not with mixed acid fermenters.

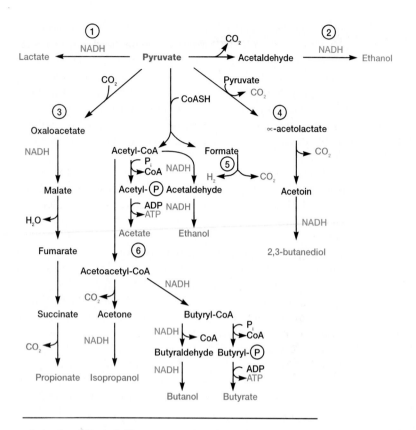

1. Lactic acid bacteria (*Streptococcus, Lactobacillus*), *Bacillus*

2. Yeast, *Zymomonas*

3. Propionic acid bacteria (*Propionibacterium*)

4. *Enterobacter, Serratia, Bacillus*

5. Enteric bacteria (*Escherichia, Enterobacter, Salmonella, Proteus*)

6. *Clostridium*

Figure 9.14   Some Common Microbial Fermentations. Most of these pathways have been simplified by deletion of one or more steps and intermediates. Pyruvate and major end products are shown in color.

TABLE 9.2 Mixed Acid Fermentation Products of *Escherichia coli*		
	Fermentation Balance **(µM Product/100 µM Glucose)**	
	Acid Growth (pH 6.0)	Alkaline Growth (pH 8.0)
Ethanol	50	50
Formic acid	2	86
Acetic acid	36	39
Lactic acid	80	70
Succinic acid	11	15
Carbon dioxide	88	2
Hydrogen gas	75	0.5
Butanediol	0	0

2. Mixed acid fermenters produce four times more acidic products than neutral ones, whereas butanediol fermenters form mainly neutral products. Thus mixed acid fermenters acidify incubation media to a much greater extent. This is the basis of the methyl red test. The test is positive only for mixed acid fermentation because the pH drops below 4.4 and the color of the indicator changes from yellow to red.

3. CO_2 and H_2 arise in equal amounts from formic hydrogenlyase activity during mixed acid fermentation. Butanediol fermenters produce excess CO_2 and the CO_2/H_2 ratio is closer to 5:1.

Formic acid fermenters sometimes generate ATP while reoxidizing NADH. They use acetyl-CoA to synthesize acetyl phosphate, which then donates its phosphate to ADP.

$$\text{Acetyl-CoA} + P_i \longrightarrow \text{CoASH} + \text{acetyl-P}$$
$$\text{Acetyl-P} + \text{ADP} \longrightarrow \text{acetate} + \text{ATP}$$

Microorganisms carry out fermentations other than those already mentioned (Box 9.1). Protozoa and fungi often ferment sugars to lactate, ethanol, glycerol, succinate, formate, acetate, butanediol, and additional products. Substances other than sugars are also fermented. For example, many members of the genus *Clostridium* ferment amino acids, oxidizing one amino acid with a second as the electron acceptor (*see chapter 21*).

---

1. What are fermentations and why are they so useful to many microorganisms? Can ATP be produced during fermentation?

2. Briefly describe alcoholic, lactic acid, and formic acid fermentations. How do mixed acid fermenters and butanediol fermenters differ from each other?

Anaerobic Respiration

Electrons derived from sugars and other organic molecules are usually donated either to organic electron acceptors (fermentation) or to molecular O_2 by way of an electron transport chain (aerobic respiration). However, some bacteria have electron transport chains that can operate with inorganic electron acceptors other than O_2. The anaerobic energy-yielding process in which the electron transport chain acceptor is an oxidized inorganic molecule other than O_2 is called **anaerobic respiration.** The major electron acceptors are nitrate, sulfate, and CO_2, but metals can also be reduced (table 9.3).

Box 9.1

Microbiology and World War I

The unique economic pressures of wartime sometimes provide incentive for scientific discovery. Two examples from the First World War involve the production of organic solvents by the microbial fermentation of readily available carbohydrates, such as starch or molasses.

The German side needed glycerol to make nitroglycerin. At one time the Germans had imported their glycerol, but such imports were prevented by the British naval blockade. The German scientist Carl Neuberg knew that trace levels of glycerol were usually produced during the alcoholic fermentation of sugar by *Saccharomyces cerevisiae*. He sought to develop a modified fermentation in which the yeasts would produce glycerol instead of ethanol. Normally acetaldehyde is reduced to ethanol by NADH and alcohol dehydrogenase (figure 9.14, pathway 2). Neuberg found that this reaction could be prevented by the addition of 3.5% sodium sulfite at pH 7.0. The bisulfite ions reacted with acetaldehyde and made it unavailable for reduction to ethanol. Because the yeast cells still had to regenerate their NAD$^+$ even though acetaldehyde was no longer available, Neuberg suspected that they would simply increase the rate of glycerol synthesis. Glycerol is normally produced by the reduction of dihydroxyacetone phosphate (a glycolytic intermediate) to glycerol phosphate with NADH, followed by the hydrolysis of glycerol phosphate to glycerol. Neuberg's hunch was correct, and German breweries were converted to glycerol manufacture by his procedure, eventually producing 1,000 tons of glycerol per month. Glycerol production by *S. cerevisiae* was not economically competitive under peacetime conditions and was ended. Today glycerol is produced microbially by the halophilic alga *Dunaliella salina,* in which high concentrations of intracellular glycerol accumulate to counterbalance the osmotic pressure

from the high level of extracellular salt. *Dunaliella* grows in habitats such as the Great Salt Lake of Utah and seaside rock pools.

The British side needed the organic solvents acetone and butanol. Butanol was required for the production of artificial rubber, whereas acetone was used as a solvent from nitrocellulose in the manufacture of the smokeless explosive powder cordite. Prior to 1914 acetone was made by the dry heating (pyrolysis) of wood. Between 80 and 100 tons of birch, beech, or maple wood were required to make 1 ton of acetone. When war broke out, the demand for acetone quickly exceeded the existing world supply. However, by 1915 Chaim Weizmann, a young Jewish scientist working in Manchester, England, had developed a fermentation process by which the anaerobic bacterium *Clostridium acetobutylicum* converted 100 tons of molasses or grain into 12 tons of acetone and 24 tons of butanol (most clostridial fermentations stop at butyric acid).

$$2 \text{ pyruvate} \longrightarrow \text{acetoacetate} \longrightarrow \text{acetone} + CO_2$$

$$\text{Acetoacetate} \xrightarrow{\text{NADH}} \text{butyrate} \xrightarrow{\text{NADH}} \text{butanol}$$

This time the British and Canadian breweries were converted until new fermentation facilities could be constructed. Weizmann improved the process by finding a convenient way to select high-solvent producing strains of *C. acetobutylicum*. Because the strains most efficient in these fermentations also made the most heat-resistant spores, Weizmann merely isolated the survivors from repeated 100°C heat shocks. Acetone and butanol were made commercially by this fermentation process until it was replaced by much cheaper petrochemicals in the late 1940s and 1950s. In 1948 Chaim Weizmann became the first president of the State of Israel.

TABLE 9.3	Some Electron Acceptors Used in Respiration		
	Electron Acceptor	**Reduced Products**	**Examples of Microorganisms**
Aerobic	O_2	H_2O	All aerobic bacteria, fungi, protozoa, and algae
Anaerobic	NO_3^-	NO_2^-	Enteric bacteria
	NO_3^-	NO_2^-, N_2O, N_2	*Pseudomonas* and *Bacillus*
	SO_4^{2-}	H_2S	*Desulfovibrio* and *Desulfotomaculum*
	CO_2	CH_4	All methanogens
	S^0	H_2S	*Desulfuromonas* and *Thermoproteus*
	Fe^{3+}	Fe^{2+}	*Pseudomonas* and *Bacillus*

Some bacteria can use nitrate as the electron acceptor at the end of their electron transport chain and still produce ATP. Often this process is called **dissimilatory nitrate reduction.** Nitrate may be reduced to nitrite by nitrate reductase, which replaces cytochrome oxidase.

$$NO_3^- + 2e^- + 2H^+ \longrightarrow NO_2^- + H_2O$$

However, reduction of nitrate to nitrite is not a particularly effective way of making ATP, because a large amount of nitrate is required for growth (a nitrate molecule will accept only two electrons). The nitrite formed is also quite toxic. Therefore nitrate often is further reduced all the way to nitrogen gas, a process known as **denitrification.** Each nitrate will then accept five electrons, and the product will be nontoxic.

$$2NO_3^- + 10e^- + 12H^+ \longrightarrow N_2 + 6H_2O$$

There is considerable evidence that denitrification is a multistep process with four enzymes participating: nitrate reductase, nitrite reductase, nitric oxide reductase, and nitrous oxide reductase.

$$NO_3^- \longrightarrow NO_2^- \longrightarrow NO \longrightarrow N_2O \longrightarrow N_2$$

Interestingly, one of the intermediates is nitric oxide (NO). In mammals this molecule acts as a neurotransmitter, helps regulate blood pressure, and is used by macrophages to destroy bacteria and tumor cells (*see p. 582*). Two types of bacterial nitrite reductases catalyze the formation of NO in bacteria. One contains cytochromes c and d_1 (e.g., *Pseudomonas aeruginosa*), and the other is a copper protein (e.g., *Alcaligenes*). Nitrite reductase seems to be periplasmic in gram-negative bacteria. Nitric oxide reductase catalyzes the formation of nitrous oxide from NO and is a membrane-bound cytochrome bc complex. Nondenitrifying bacteria such as *Serratia marcescens, E. coli,* and *Bacillus cereus* also can produce nitric oxide under certain conditions.

Denitrification is carried out by some members of the genera *Pseudomonas* and *Bacillus*. They use this route as an alternative to normal aerobic respiration and may be considered facultative anaerobes. If O_2 is present, these bacteria use aerobic respiration (the synthesis of nitrate reductase is repressed by O_2). Denitrification in anaerobic soil results in the loss of soil nitrogen and adversely affects soil fertility (*see chapters 40 and 42*).

Two other major groups of bacteria employing anaerobic respiration are obligate anaerobes. Those using CO_2 or carbonate as a terminal electron acceptor are called methanogens because they reduce CO_2 to methane (*see chapter 23*). Sulfate also can act as the final acceptor in bacteria such as *Desulfovibrio*. It is reduced to sulfide (S^{2-} or H_2S), and eight electrons are accepted.

$$SO_4^{2-} + 8e^- + 8H^+ \longrightarrow S^{2-} + 4H_2O$$

Anaerobic respiration is not as efficient in ATP synthesis as aerobic respiration—that is, not as much ATP is produced by oxidative phosphorylation with nitrate, sulfate, or CO_2 as the terminal acceptors. Reduction in ATP yield arises from the fact that these alternate electron acceptors have less positive reduction potentials than O_2 (*see table 8.1*). The reduction potential difference between a donor like NADH and nitrate is smaller than the difference between NADH and O_2. Because energy yield is directly related to the magnitude of the reduction potential difference, less energy is available to make ATP in anaerobic respiration. Nevertheless, anaerobic respiration is useful because it is more efficient than fermentation and allows ATP synthesis by electron transport and oxidative phosphorylation in the absence of O_2.

---

1. Describe the process of anaerobic respiration. Is as much ATP produced in anaerobic respiration as in aerobic respiration? Why or why not? What is denitrification?

Catabolism of Carbohydrates and Intracellular Reserve Polymers

Microorganisms can catabolize many carbohydrates besides glucose. These carbohydrates may come either from outside the cell or from internal sources. Often the initial steps in the degradation of external carbohydrate polymers differ from those employed with internal reserves.

Carbohydrates

Figure 9.15 outlines some catabolic pathways for the monosaccharides (single sugars) glucose, fructose, mannose, and galactose. The first three are phosphorylated using ATP and easily enter the glycolytic pathway. In contrast, galactose must be converted to uridine diphosphate galactose (*see p. 196*) after initial phosphorylation, then changed into glucose 6-phosphate in a three-step process (figure 9.15).

The common disaccharides are cleaved to monosaccharides by at least two mechanisms (figure 9.15). Maltose, sucrose, and lactose can be directly hydrolyzed to their constituent sugars. Many disaccharides (e.g., maltose, cellobiose, and sucrose) are also split by a phosphate attack on the bond joining the two sugars, a process called phosphorolysis.

Polysaccharides, like disaccharides, are cleaved by both hydrolysis and phosphorolysis. Bacteria and fungi degrade external polysaccharides by secreting hydrolytic enzymes that cleave polysaccharides into smaller molecules, which can then be assimilated. Starch and glycogen are hydrolyzed by amylases to glucose, maltose, and other products. Cellulose is more difficult to digest; many fungi and a few bacteria (some gliding bacteria, clostridia, and actinomycetes) produce cellulases that hydrolyze cellulose to cellobiose and glucose. Some members of the genus *Cytophaga,* isolated from marine habitats, excrete an agarase that degrades agar. Many soil bacteria and bacterial plant pathogens degrade pectin, a polymer of galacturonic acid (a galactose derivative) that is an important constituent of plant cell walls and tissues.

Reserve Polymers

Microorganisms often survive for long periods in the absence of exogenous nutrients. Under such circumstances they catabolize intracellular stores of glycogen, starch, poly-β-hydroxybutyrate, and other energy reserves. Glycogen and starch are degraded by phosphorylases. Phosphorylases catalyze a phosphorolysis reaction that shortens the polysaccharide chain by one glucose and yields glucose 1-phosphate.

$$(Glucose)_n + P_i \longrightarrow (glucose)_{n-1} + glucose\text{-}1\text{-}P$$

Glucose 1-phosphate can enter the glycolytic pathway by way of glucose 6-phosphate (figure 9.15).

Monosaccharide interconversions

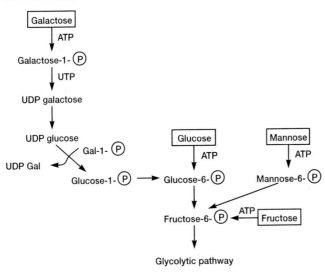

Disaccharide cleavage

1. Maltose + H_2O —maltase→ 2 glucose

   Maltose + P_i —maltose phosphorylase→ β-D-glucose-1-Ⓟ + glucose

2. Cellobiose + P_i —cellobiose phosphorylase→ ∝-D-glucose-1-Ⓟ + glucose

3. Sucrose + H_2O —sucrase→ glucose + fructose

   Sucrose + P_i —sucrose phosphorylase→ ∝-D-glucose-1-Ⓟ + fructose

4. Lactose + H_2O —β-galactosidase→ galactose + glucose

Figure 9.15   Carbohydrate Catabolism. Examples of enzymes and pathways used in disaccharide and monosaccharide catabolism. UDP is an abbreviation for uridine diphosphate.

Poly-β-hydroxybutyrate (PHB) is an important, widespread reserve material. Its catabolism has been studied most thoroughly in *Azotobacter*. This bacterium hydrolyzes PHB to 3-hydroxybutyrate, then oxidizes the hydroxybutyrate to acetoacetate. Acetoacetate is converted to acetyl-CoA, which can be oxidized in the TCA cycle.

Lipid Catabolism

Microorganisms frequently use lipids as energy sources. Triglycerides or triacylglycerols, esters of glycerol and fatty acids (figure 9.16), are common energy sources and will serve as our example. They can be hydrolyzed to glycerol and fatty acids by microbial lipases. The glycerol is then phosphorylated, oxidized to dihydroxyacetone phosphate, and catabolized in the glycolytic pathway (figure 9.3).

Fatty acids from triacylglycerols and other lipids are often oxidized in the **β-oxidation pathway** after conversion to coenzyme A esters (figure 9.17). In this cyclic pathway fatty

Figure 9.16   A Triacylglycerol or Triglyceride. The R groups represent the fatty acid side chains.

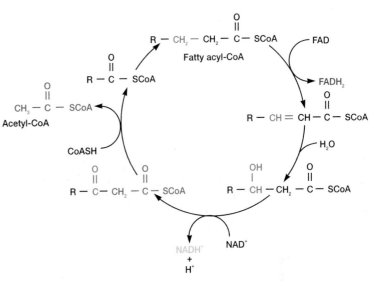

Figure 9.17   Fatty Acid β-Oxidation. The portions of the fatty acid being modified are shown in color.

acids are degraded to acetyl-CoA, which can be fed into the TCA cycle or used in biosynthesis (*see chapter 10*). One turn of the cycle produces acetyl-CoA, NADH, and $FADH_2$; NADH and $FADH_2$ can be oxidized by the electron transport chain to provide more ATP. The fatty acyl-CoA, shortened by two carbons, is ready for another turn of the cycle. Lipid fatty acids are a rich source of energy for microbial growth. In a similar fashion some microorganisms grow well on petroleum hydrocarbons under aerobic conditions.

Protein and Amino Acid Catabolism

Some bacteria and fungi—particularly pathogenic, food spoilage, and soil microorganisms—can use proteins as their source of carbon and energy. They secrete **protease** enzymes that hydrolyze proteins and polypeptides to amino acids, which are transported into the cell and catabolized.

The first step in amino acid use is **deamination,** the removal of the amino group from an amino acid. This is often accomplished by **transamination.** The amino group is transferred from an amino acid to an α-keto acid acceptor (figure 9.18). The organic acid resulting from deamination can be converted to

TABLE 9.4   Representative Chemolithotrophs and Their Energy Sources

Bacteria	Electron Donor	Electron Acceptor	Products
Alcaligenes, Hydrogenophaga, and *Pseudomonas* spp.	H_2	O_2	H_2O
Nitrobacter	NO_2^-	O_2	NO_3^-, H_2O
Nitrosomonas	NH_4^+	O_2	NO_2^-, H_2O
Thiobacillus denitrificans	S^0, H_2S	NO_3^-	SO_4^{2-}, N_2
Thiobacillus ferrooxidans	Fe^{2+}, S^0, H_2S	O_2	Fe^{3+}, H_2O, H_2SO_4

Figure 9.18   Transamination. A common example of this process. The α-amino group (blue) of alanine is transferred to the acceptor α-ketoglutarate forming pyruvate and glutamate. The pyruvate can be catabolized in the tricarboxylic acid cycle or used in biosynthesis.

Alanine   ∝-ketoglutarate   Pyruvate   Glutamate

TABLE 9.5   Energy Yields from Oxidations Used by Chemolithotrophs

Reaction	$\Delta G^{o\prime}$ (kcal/mole)[a]
$H_2 + 1/2\ O_2 \longrightarrow H_2O$	−56.6
$NO_2^- + 1/2\ O_2 \longrightarrow NO_3^-$	−17.4
$NH_4^+ + 1\,1/2\ O_2 \longrightarrow NO_2^- + H_2O + 2H^+$	−65.0
$S^0 + 1\,1/2\ O_2 + H_2O \longrightarrow H_2SO_4$	−118.5
$S_2O_3^{2-} + 2O_2 + H_2O \longrightarrow 2SO_4^{2-} + 2H^+$	−223.7
$2Fe^{2+} + 2H^+ + 1/2\ O_2 \longrightarrow 2Fe^{3+} + H_2O$	−11.2

[a]The $\Delta G^{o\prime}$ for complete oxidation of glucose to CO_2 is −686 kcal/mole. A kcal is equivalent to 4.184kJ.

pyruvate, acetyl-CoA, or a TCA cycle intermediate and eventually oxidized in the TCA cycle to release energy. It also can be used as a source of carbon for the synthesis of cell constituents. Excess nitrogen from deamination may be excreted as ammonium ion, thus making the medium alkaline.

1. Briefly discuss the ways in which microorganisms degrade and use common monosaccharides, disaccharides, and polysaccharides from both external and internal sources.
2. Describe how a microorganism might derive carbon and energy from the lipids and proteins in its diet. What is β-oxidation? Transamination?

Figure 9.19   Patterns of Energy Production. An organic electron donor can donate electrons to an organic acceptor (fermentation), oxygen (aerobic respiration), or an inorganic acceptor other than oxygen (anaerobic respiration). In chemolithotrophy a reduced inorganic molecule donates electrons for energy production.

Oxidation of Inorganic Molecules

As we have seen, microorganisms can oxidize organic molecules such as carbohydrates, lipids, and proteins and synthesize ATP with the energy liberated. The electron acceptor is (1) another more oxidized organic molecule in fermentation, (2) O_2 in aerobic respiration, and (3) an oxidized inorganic molecule other than O_2 in anaerobic respiration (figure 9.19). In both aerobic and anaerobic respiration, ATP is formed as a result of electron transport chain activity. Electrons for the chain can be obtained from inorganic nutrients, and it is possible to derive energy from the oxidation of inorganic molecules rather than from organic nutrients. This ability is confined to a small group of bacteria called **chemolithotrophs** (*see chapter 22*). Each species is rather specific in its preferences for electron

donors and acceptors (table 9.4). The acceptor is usually O_2, but sulfate and nitrate are also used. The most common electron donors are hydrogen, reduced nitrogen compounds, reduced sulfur compounds, and ferrous iron (Fe^{2+}).

Chemolithotrophic bacteria are usually autotrophic and employ the Calvin cycle to fix CO_2 as their carbon source (*see chapter 10*). However, some chemolithotrophs can function as heterotrophs if reduced organic compounds are available. Considerable energy is required to reduce CO_2 to carbohydrate; incorporation of one CO_2 in the Calvin cycle requires three ATPs and two NADPHs. Moreover, much less energy is available from the oxidation of inorganic molecules (table 9.5) than from the complete oxidation of glucose to CO_2, which is accompanied by a standard free energy change of −686 kcal/mole. The P/O ratios for oxidative phosphorylation in chemolithotrophs are probably around 1.0 (in the oxidation of hydrogen it is significantly higher). Because the yield of ATP is so low, chemolithotrophs must oxidize a large quantity of inorganic material to grow and reproduce, and this magnifies their ecological impact.

Several bacterial genera (table 9.4) can oxidize hydrogen gas to produce energy because they possess a hydrogenase enzyme that catalyzes the oxidation of hydrogen.

$$H_2 \longrightarrow 2H^+ + 2e^-$$

The electrons are donated either to an electron transport chain or to NAD^+, depending on the hydrogenase. If NADH is produced, it can be used in ATP synthesis by electron transport and oxidative phosphorylation, with O_2 as the terminal electron acceptor.

The best-studied nitrogen-oxidizing chemolithotrophs are the **nitrifying bacteria** (*see chapter 22*). These are soil and aquatic bacteria of considerable ecological significance. Ammonia oxidation to nitrate depends on the activity of two different genera. For example, *Nitrosomonas* oxidizes ammonia to nitrite.

$$NH_4^+ + 1\tfrac{1}{2} O_2 \longrightarrow NO_2^- + H_2O + 2H^+$$

The nitrite can then be further oxidized by *Nitrobacter* to yield nitrate.

$$NO_2^- + \tfrac{1}{2} O_2 \longrightarrow NO_3^-$$

When these two genera work together, ammonia in the soil is oxidized to nitrate in a process called **nitrification.**

The role of chemolithotrophs in soil and aquatic ecosystems (pp. 814–19).

Energy released upon the oxidation of both ammonia and nitrite is used to make ATP by oxidative phosphorylation. However, microorganisms need a source of electrons (reducing power) as well as a source of ATP in order to reduce CO_2 and other molecules. Since molecules like ammonia and nitrite have more positive reduction potentials than NAD^+, they cannot directly donate their electrons to form the required NADH and NADPH. This is because electrons spontaneously move only from donors with more negative reduction potentials to acceptors with more positive potentials (*see chapter 8*). Sulfur-oxidizing bacteria face the same difficulty. Both types of chemolithotrophs solve this problem by using protonmotive force or ATP to reverse the flow of electrons in their electron transport chains and reduce NAD^+ with electrons from nitrogen and sulfur donors (figure 9.20). Because energy is used to generate NADH as well as ATP, the net yield of ATP is fairly low. Chemolithotrophs can afford this inefficiency as they have no serious competitors for their unique energy sources.

The sulfur-oxidizing bacteria are the third major group of chemolithotrophs. The metabolism of *Thiobacillus* has been best studied. These bacteria oxidize sulfur (S^0), hydrogen sulfide (H_2S), thiosulfate ($S_2O_3^{2-}$), and other reduced sulfur compounds to sulfuric acid; therefore they have a significant ecological impact (Box 9.2). Interestingly they generate ATP by both

Figure 9.20   Reversed Electron Flow. The flow of electrons in the transport chain of *Nitrobacter*. Electrons flowing from nitrite to oxygen (down the reduction potential gradient) will release energy. It requires protonmotive force or ATP energy to force electrons to flow in the reverse direction from nitrite to NAD^+.

oxidative phosphorylation and substrate-level phosphorylation involving **adenosine 5′-phosphosulfate (APS).** APS is a high-energy molecule formed from sulfite and adenosine monophosphate (figure 9.21).

Some of these bacteria are extraordinarily flexible metabolically. For example, *Sulfolobus brierleyi* and a few other species can grow aerobically as sulfur-oxidizing bacteria; in the absence of O_2, they carry out anaerobic respiration with molecular sulfur as an electron acceptor.

Sulfur-oxidizing bacteria, like other chemolithotrophs, can use CO_2 as their carbon source. Many will grow heterotrophically if they are supplied with reduced organic carbon sources like glucose or amino acids.

---

1. How do chemolithotrophs obtain their ATP and NADH? What is their source of carbon?

2. Describe energy production by hydrogen-oxidizing bacteria, nitrifying bacteria, and sulfur-oxidizing bacteria.

Photosynthesis

Microorganisms cannot only derive energy from the oxidation of inorganic and organic compounds, but many can capture light energy and use it to synthesize ATP and NADH or NADPH. This process in which light energy is trapped and converted to chemical energy is called **photosynthesis.** Usually a photosynthetic organism reduces and incorporates CO_2, reactions that are also considered part of photosynthesis. Photosynthesis is one of

Acid Mine Drainage

Each year millions of tons of sulfuric acid flow down the Ohio River from the Appalachian Mountains. This sulfuric acid is of microbial origin and leaches enough metals from the mines to make the river reddish and acidic. The primary culprit is *Thiobacillus ferrooxidans*, a chemolithotrophic bacterium that derives its energy from oxidizing ferrous ion to ferric ion and sulfide ion to sulfate ion. The combination of these two energy sources is important because of the solubility properties of iron. Ferrous ion is somewhat soluble and can be formed at pH values of 3.0 or less in moderately reducing environments. However, when the pH is greater than 4.0 to 5.0, ferrous ion is oxidized to ferric ion by O_2 in the water and precipitates as a hydroxide. If the pH drops below about 2.0 to 3.0 because of sulfuric acid production by spontaneous oxidation of sulfur or sulfur oxidation by thiobacilli and other bacteria, the ferrous ion remains reduced, soluble, and available as an energy source. Remarkably, *T. ferrooxidans* grows well at such acid pHs and actively oxidizes ferrous ion to an insoluble ferric precipitate. The water is rendered toxic for most aquatic life and unfit for human consumption.

The ecological consequences of this metabolic life-style arise from the common presence of pyrite (FeS_2) in coal mines. The bacteria oxidize both elemental components of pyrite for their growth and in the process form sulfuric acid, which leaches the remaining minerals.

Autoxidation or bacterial action
$$2FeS_2 + 7O_2 + 2H_2O \longrightarrow 2Fe^{2+} + 4SO_4^{2-} + 4H^+$$

T. ferrooxidans
$$2Fe^{2+} + \tfrac{1}{2}O_2 + 2H^+ \longrightarrow 2Fe^{3+} + H_2O$$

Pyrite oxidation is further accelerated because the ferric ion generated by bacterial activity readily oxidizes more pyrite to sulfuric acid and ferrous ion. In turn the ferrous ion supports further bacterial growth. It is difficult to prevent *T. ferrooxidans* growth as it requires only pyrite and common inorganic salts. Because *T. ferrooxidans* gets its O_2 and CO_2 from the air, the only feasible method of preventing its damaging growth is to seal the mines to render the habitat anaerobic.

(a) Direct oxidation of sulfite
$$SO_3^{2-} \xrightarrow{\text{sulfite oxidase}} SO_4^{2-} + 2e^-$$

(b) Formation of adenosine 5′-phosphosulfate

$$2SO_3^{2-} + 2AMP \longrightarrow 2APS + 4e^-$$
$$2APS + 2P_i \longrightarrow 2ADP + 2SO_4^{2-}$$
$$2ADP \longrightarrow AMP + ATP$$

$$2SO_3^{2-} + AMP + 2P_i \longrightarrow 2SO_4^{2-} + ATP + 4e^-$$

Adenosine 5′-phosphosulfate

Figure 9.21   Energy Generation by Sulfur Oxidation. (*a*) Sulfite can be directly oxidized to provide electrons for electron transport and oxidative phosphorylation. (*b*) Sulfite can also be oxidized and converted to APS. This route produces electrons for use in electron transport and ATP by substrate-level phosphorylation with APS.

the most significant metabolic processes on earth because almost all our energy is ultimately derived from solar energy (*see* chapter 40). It provides photosynthetic organisms with the ATP and NAD(P)H necessary to synthesize the organic material required for growth. In turn these organisms serve as the base of most food chains in the biosphere. Photosynthesis is also responsible for replenishing our supply of O_2, a remarkable process carried out by a variety of organisms, both eucaryotic and procaryotic (table 9.6). Although most people associate photosynthesis with the more obvious higher plants, over half the photosynthesis on earth is carried out by microorganisms.

Photosynthesis as a whole is divided into two parts. In the **light reactions** light energy is trapped and converted to chemical energy. This energy is then used to reduce or fix CO_2 and synthesize cell constituents in the **dark reactions.** In this section the nature of the light reactions is discussed; the dark reactions are reviewed in the next chapter.

The photosynthetic dark reactions (pp. 193–96).

The Light Reaction in Eucaryotes and Cyanobacteria

All photosynthetic organisms have pigments for the absorption of light. The most important of these pigments are the **chlorophylls.** Chlorophylls are large planar rings composed of four substituted pyrrole rings with a magnesium atom coordinated to the four central nitrogen atoms (figure 9.22). Several chlorophylls are found in eucaryotes, the two most important of which are chlorophyll *a* and chlorophyll *b* (figure 9.22). These two molecules differ slightly in their structure and spectral properties. When dissolved in acetone, chlorophyll *a* has a light absorption peak at 665 nm; the corresponding peak for chlorophyll *b* is at 645 nm. In addition to absorbing red light, chlorophylls also absorb blue light strongly (the second absorption peak for chlorophyll *a* is at 430 nm). Because chlorophylls absorb primarily in the red and blue ranges, green light is transmitted.

TABLE 9.6   Diversity of Photosynthetic Organisms

Eucaryotic Organisms	Procaryotic Organisms
Higher plants	Cyanobacteria (blue-green algae)
Multicellular green, brown, and red algae	Green sulfur bacteria
Unicellular algae (e.g., euglenoids, dinoflagellates, diatoms)	Green nonsulfur bacteria
	Purple sulfur bacteria
	Purple nonsulfur bacteria
	Prochloron

Consequently many photosynthetic organisms are green in color. The long hydrophobic tail attached to the chlorophyll ring aids in its attachment to membranes, the site of the light reactions.

Other photosynthetic pigments also trap light energy. The most widespread of these are the **carotenoids,** long molecules, usually yellowish in color, that possess an extensive conjugated double bond system (figure 9.23). β-carotene is present in *Prochloron* and most divisions of algae; fucoxanthin is found in diatoms, dinoflagellates, and brown algae (*Phaeophyta*). Red algae and cyanobacteria have photosynthetic pigments called **phycobiliproteins,** consisting of a protein with a tetrapyrrole attached (figure 9.23). **Phycoerythrin** is a red pigment with a maximum absorption around 550 nm, and **phycocyanin** is blue (maximum absorption at 620 to 640 nm).

Carotenoids and phycobiliproteins are often called **accessory pigments** because of their role in photosynthesis. Although chlorophylls cannot absorb light energy effectively in the blue-green through yellow range (about 470 to 630 nm), accessory pigments do absorb light in this region and transfer the trapped energy to chlorophyll. In this way they make photosynthesis more efficient over a broader range of wavelengths. Accessory pigments also protect microorganisms

Figure 9.22   Chlorophyll Structure. The structures of chlorophyll *a,* chlorophyll *b,* and bacteriochlorophyll *a.* The complete structure of chlorophyll *a* is given. Only one group is altered to produce chlorophyll *b,* and two modifications in the ring system are required to change chlorophyll *a* to bacteriochlorophyll *a.* The side chain (R) of bacteriochlorophyll *a* may be either phytyl (a 20-carbon chain also found in chlorophylls *a* and *b*) or geranylgeranyl (a 20-carbon side chain similar to phytyl, but with three more double bonds).

Figure 9.23 Representative Accessory Pigments. Beta-carotene is a carotenoid found in algae and higher plants. Note that it has a long chain of alternating double and single bonds called conjugated double bonds. Fucoxanthin is a carotenoid accessory pigment in several divisions of algae. Phycocyanobilin is an example of a linear tetrapyrrole that is attached to a protein to form a phycobiliprotein.

from intense sunlight, which could oxidize and damage the photosynthetic apparatus in their absence.

Chlorophylls and accessory pigments are assembled in highly organized arrays called **antennas,** whose purpose is to create a large surface area to trap as many photons as possible. An antenna has about 300 chlorophyll molecules. Light energy is captured in an antenna and transferred from chlorophyll to chlorophyll until it reaches a special **reaction-center chlorophyll** directly involved in photosynthetic electron transport (figure 9.24). In eucaryotic cells and cyanobacteria, there are two kinds of antennas associated with two different photosystems. **Photosystem I** absorbs longer wavelength light (≥680 nm) and funnels the energy to a special chlorophyll a molecule called P700. The term P700 signifies that this molecule most effectively absorbs light at a wavelength of 700 nm. **Photosystem II** traps light at shorter wavelengths (≤680 nm) and transfers its energy to the special chlorophyll P680.

When the photosystem I antenna transfers light energy to the reaction-center P700 chlorophyll, P700 absorbs the energy and is excited; its reduction potential becomes very negative. It then donates its excited or high-energy electron to a specific acceptor, probably a special chlorophyll a molecule (A) or an iron-sulfur protein (figure 9.25). The electron is eventually transferred to ferredoxin and can then travel in either of two directions. In the cyclic pathway (the dashed lines in figure 9.25), the electron moves in a cyclic route through a series of electron carriers and back to the oxidized P700. The pathway is termed cyclic because the electron from P700 returns to P700 after traveling through the photosynthetic electron transport chain. ATP is formed during cyclic electron transport in the region of cytochrome b_6. This process is called **cyclic photophosphorylation** because electrons travel in a cyclic pathway and ATP is formed. Only photosystem I participates.

Electrons also can travel in a noncyclic pathway involving both photosystems. P700 is excited and donates electrons to ferredoxin as before. In the noncyclic route, however, reduced ferredoxin reduces $NADP^+$ to NADPH (figure 9.25). Because the electrons contributed to $NADP^+$ cannot be used to reduce oxidized P700, photosystem II participation is required. It donates electrons to oxidized P700 and generates ATP in the process. The photosystem II antenna absorbs light energy and excites P680, which then reduces pheophytin a. Pheophytin a is chlorophyll a in which two hydrogen atoms have replaced the central magnesium. Electrons subsequently travel to Q (probably a plastoquinone) and down the electron transport chain to P700. Oxidized P680 then obtains an electron from the oxidation of water to O_2. Thus electrons flow from water all the way to $NADP^+$ with the aid of energy from two photosystems, and ATP is synthesized by **noncyclic photophosphorylation.** It appears that one ATP and one NADPH are formed when two electrons travel through the noncyclic pathway.

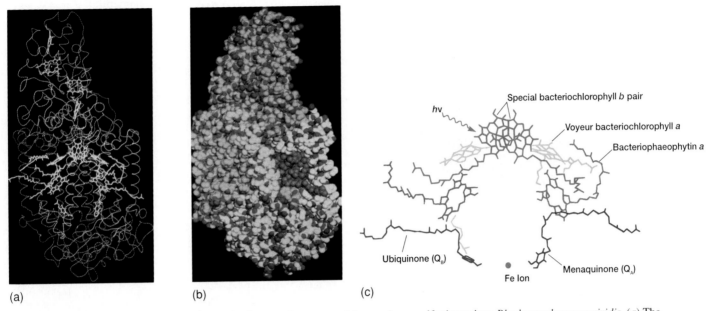

Figure 9.24   A Photosynthetic Reaction Chain. The reaction center of the purple nonsulfur bacterium, *Rhodopseudomonas viridis*. (*a*) The structure of the C_α backbone of the center's polypeptide chains with the bacteriochlorophylls and other prosthetic groups in yellow. (*b*) A space-filling model of the reaction center. The nitrogen atoms are blue; oxygen atoms, red; and sulfur atoms, yellow. The exposed prosthetic group atoms are brown. (*c*) A close-up view of the reaction center prosthetic groups. A photon is first absorbed by the "special pair" of bacteriochlorophyll *a* molecules, thus exciting them. An excited electron then moves to the bacteriopheophytin molecule in the right arm of the system.

Figure 9.25   Green Plant Photosynthesis. Electron flow during photosynthesis in higher plants. Cyanobacteria and eucaryotic algae are similar in having two photosystems, although they may differ in some details. The carriers involved in electron transport are ferredoxin (Fd) and other FeS proteins; cytochromes b_6, b_{563}, and f; plastoquinone (PQ); copper containing plastocyanin (PC); pheophytin *a* (Pheo. *a*); possibly chlorophyll *a* (A); and the unknown quinone Q, which is probably a plastoquinone. Both photosystem I (PS I) and photosystem II (PS II) are involved in noncyclic photophosphorylation; only PS I participates in cyclic photophosphorylation. See the text for further details.

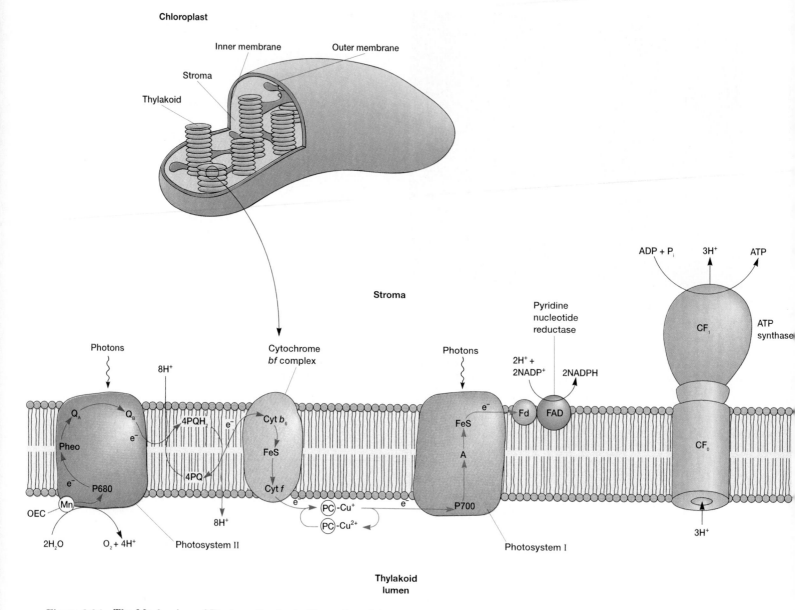

Figure 9.26   The Mechanism of Photosynthesis. An illustration of the chloroplast thylakoid membrane showing photosynthetic electron transport chain function and noncyclic photophosphorylation. The chain is composed of three complexes: PS I, the cytochrome *bf* complex, and PS II. Two diffusible electron carriers connect the three complexes. Plastoquinone (PQ) connects PS I with the cytochrome *bf* complex, and plastocyanin (PC) connects the cytochrome *bf* complex with PS II. The light-driven electron flow pumps protons across the thylakoid membrane and generates an electrochemical gradient, which can then be used to make ATP. Water is the source of electrons and the oxygen-evolving complex (OEC) produces oxygen.

Just as is true of mitochondrial electron transport, photosynthetic electron transport takes place within a membrane. Chloroplast granal membranes contain both photosystems and their antennas. Figure 9.26 shows a thylakoid membrane carrying out noncyclic photophosphorylation by the chemiosmotic mechanism. It is believed that stromal lamellae possess only photosystem I and are involved in cyclic photophosphorylation alone. In cyanobacteria, photosynthetic light reactions are also located in membranes.

The dark reactions require three ATPs and two NADPHs to reduce one CO_2 and use it to synthesize carbohydrate.

$$CO_2 + 3ATP + 2NADPH + 2H^+ + H_2O \longrightarrow$$
$$(CH_2O) + 3ADP + 3P_i + 2NADP^+$$

The noncyclic system generates one NADPH and one ATP per pair of electrons; therefore four electrons passing through the system will produce two NADPHs and two ATPs. A total of 8 quanta of light energy (4 quanta for each photosystem) is needed

TABLE 9.7   Properties of Microbial Photosynthetic Systems

Property	Eucaryotes	Cyanobacteria	Green and Purple Bacteria
Photosynthetic pigment	Chlorophyll *a*	Chlorophyll *a*	Bacteriochlorophyll
Photosystem II	Present	Present	Absent
Photosynthetic electron donors	H_2O	H_2O	H_2, H_2S, S, organic matter
O_2 production pattern	Oxygenic	Oxygenic[a]	Anoxygenic
Primary products of energy conversion	ATP + NADPH	ATP + NADPH	ATP
Carbon source	CO_2	CO_2	Organic and/or CO_2

[a]Some cyanobacteria can function anoxygenically under certain conditions. For example, *Oscillatoria* can use H_2S as an electron donor instead of H_2O.

to propel the four electrons from water to $NADP^+$. Because the ratio of ATP to NADPH required for CO_2 fixation is 3:2, at least one more ATP must be supplied. Cyclic photophosphorylation probably operates independently to generate the extra ATP. This requires absorption of another 2 to 4 quanta. It follows that around 10 to 12 quanta of light energy are needed to reduce and incorporate one molecule of CO_2 during photosynthesis.

The Light Reaction in Green and Purple Bacteria

Green and purple photosynthetic bacteria differ from cyanobacteria and eucaryotic photosynthesizers in several fundamental ways (table 9.7). In particular, green and purple bacteria do not use water as an electron source or produce O_2 photosynthetically—that is, they are **anoxygenic.** In contrast, cyanobacteria and eucaryotic photosynthesizers are almost always **oxygenic.** NADPH is not directly produced in the photosynthetic light reaction of purple bacteria. Green bacteria can reduce NAD^+ directly during the light reaction. To synthesize NAD(P)H, green and purple bacteria must use electron donors like hydrogen, hydrogen sulfide, elemental sulfur, and organic compounds that have more negative reduction potentials than water and are therefore easier to oxidize (better electron donors). Finally, green and purple bacteria possess slightly different photosynthetic pigments, **bacteriochlorophylls** (figure 9.22), many with absorption maxima at longer wavelengths. Bacteriochlorophylls *a* and *b* have maxima in ether at 775 and 790 nm, respectively. In vivo maxima are about 830 to 890 nm (bacteriochlorophyll *a*) and 1,020 to 1,040 nm (Bchl *b*). This shift of absorption maxima into the infrared region better adapts these bacteria to their ecological niches (*see figure 22.3*).

The biology of photosynthetic bacteria (pp. 455–63).

Many differences found in green and purple bacteria are due to their lack of photosystem II; they cannot use water as an electron donor in noncyclic electron transport. Without photosystem II they cannot produce O_2 from H_2O photosynthetically and are restricted to cyclic photophosphorylation. Indeed, almost

Figure 9.27   Purple Nonsulfur Bacterial Photosynthesis. The photosynthetic electron transport system in the purple nonsulfur bacterium, *Rhodobacter sphaeroides*. This scheme is incomplete and tentative. Ubiquinone (Q) is very similar to coenzyme Q. BPh stands for bacteriopheophytin. NAD^+ and the electron source succinate are in color.

all purple and green sulfur bacteria are strict anaerobes. A tentative scheme for the photosynthetic electron transport chain of a purple nonsulfur bacterium is given in figure 9.27. When the special reaction-center chlorophyll P870 is excited, it donates an electron to bacteriopheophytin. Electrons then flow to quinones and through an electron transport chain back to P870 while driving ATP synthesis.

Green and purple bacteria face a further problem because they also require NADH or NADPH for CO_2 incorporation. They may synthesize NADH in at least three ways. If they are growing in the presence of hydrogen gas, which has a reduction potential more negative than that of NAD^+, the hydrogen can be used directly to produce NADH. Like chemolithotrophs, many photosynthetic purple bacteria use ATP or protonmotive force to reverse the flow of electrons in an electron transport chain and move them from inorganic or organic donors to

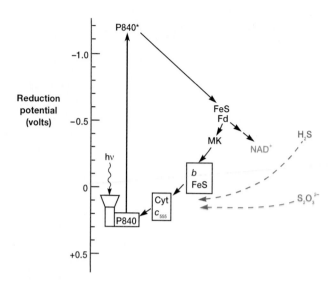

Figure 9.28 NAD Reduction in Green and Purple Bacteria. The use of reversed electron flow to reduce NAD$^+$. The arrow in this diagram represents an electron transport chain that is being driven in reverse by protonmotive force or ATP. That is, electrons are moving from donors with more positive reduction potentials to an acceptor (NAD$^+$) with a more negative potential.

NAD$^+$ (figures 9.27 and 9.28). Green sulfur bacteria appear to carry out a simple form of noncyclic photosynthetic electron flow to reduce NAD$^+$ (figure 9.29).

1. Describe photosynthesis as carried out by eucaryotes and cyanobacteria. How does photosynthesis in green and purple bacteria differ?

2. Define the following terms: light reaction, chlorophyll, carotenoid, phycobiliprotein, accessory pigment, antenna, photosystems I and II, cyclic photophosphorylation, noncyclic photophosphorylation, anoxygenic and oxygenic photosynthesis, and bacteriochlorophyll.

Figure 9.29 Green Sulfur Bacterial Photosynthesis. The photosynthetic electron transport system in the green sulfur bacterium, *Chlorobium limicola*. Light energy is used to make ATP by cyclic photophosphorylation and move electrons from sulfur donors (green and blue) to NAD$^+$ (red). The electron transport chain has a quinone called menaquinone (MK).

Summary

1. Metabolism is the total of all chemical reactions in a cell and may be divided into catabolism and anabolism.

2. Amphibolic pathways like glycolysis and the tricarboxylic acid cycle have both catabolic and anabolic functions.

3. The glycolytic or Embden-Meyerhof pathway occurs by way of fructose 1,6-bisphosphate with the net production of two NADHs and two ATPs, the latter being produced by substrate-level phosphorylation.

4. In the pentose phosphate pathway, glucose 6-phosphate is oxidized twice and converted to pentoses and other sugars. It is a source of NADPH, ATP, pentoses, and tetroses.

5. In the Entner-Doudoroff pathway, glucose is oxidized to 6-phosphogluconate, which is then dehydrated and cleaved to pyruvate and glyceraldehyde 3-phosphate. The latter product can be oxidized by glycolytic enzymes to provide ATP and NADH.

6. The tricarboxylic acid cycle is the final stage of catabolism in most aerobic cells. It oxidizes acetyl-CoA to CO_2 and forms one GTP, three NADHs, and one FADH$_2$ per acetyl-CoA.

7. The NADH and FADH$_2$ produced from the oxidation of carbohydrates, fatty acids, and other nutrients can be oxidized in the electron transport chain. Electrons flow from carriers with more negative reduction potentials to those with more positive potentials, and free energy is released for ATP synthesis by oxidative phosphorylation.

8. Bacterial electron transport chains are often different from eucaryotic chains. In eucaryotes the P/O ratio for NADH is about 3 and that for FADH$_2$ is around 2; P/O ratios are usually much lower in bacteria.

9. Eucaryotic ATP synthesis takes place on small protein spheres, the F_1F_0 ATPases or ATP synthases, located on the inner surface of the inner mitochondrial membrane.

10. The most widely accepted mechanism of oxidative phosphorylation is the chemiosmotic hypothesis in which protonmotive force drives ATP synthesis. However, other mechanisms have been proposed.

11. When the glycolytic pathway operates anaerobically, only 2 ATPs per glucose are formed, whereas aerobic respiration in eucaryotes can yield a maximum of 38 ATPs.

12. In the absence of O_2 a microorganism often uses an oxidized organic molecule as an electron acceptor to reoxidize the NADH formed during glycolysis. A catabolic process that employs an organic molecule as the electron acceptor is called fermentation.

13. Anaerobic respiration is the process of ATP production by electron transport in which the terminal electron acceptor is an oxidized inorganic molecule other than O_2. The most common acceptors are nitrate, sulfate, and CO_2.

14. Microorganisms catabolize many extracellular carbohydrates. Monosaccharides are taken in and phosphorylated; disaccharides may be cleaved to monosaccharides by either hydrolysis or phosphorolysis.

15. External polysaccharides are degraded by hydrolysis and the products are absorbed. Intracellular glycogen and starch are converted to glucose 1-phosphate by phosphorolysis.

16. Fatty acids from lipid catabolism are usually oxidized to acetyl-CoA in the β-oxidation pathway.

17. Proteins are hydrolyzed to amino acids that are then deaminated; their carbon skeletons feed into the TCA cycle.

18. Chemolithotrophs or chemoautotrophs synthesize ATP by oxidizing reduced inorganic compounds—usually hydrogen, reduced nitrogen and sulfur compounds, or ferrous iron—with an electron transport chain and O_2 as the electron acceptor.

19. Chemolithotrophs usually incorporate CO_2 through the Calvin cycle and produce NADH by reversing electron transport.

20. In photosynthesis, eucaryotes and cyanobacteria trap light energy with chlorophyll and accessory pigments and move electrons through photosystems I and II to make ATP and NAD(P)H (the light reactions).

21. Cyclic photophosphorylation involves the activity of photosystem I alone. In noncyclic photophosphorylation photosystems I and II operate together to move electrons from water to $NADP^+$ producing ATP, NADPH, and O_2.

22. Photosynthetic green and purple bacteria differ from eucaryotes and cyanobacteria in possessing bacteriochlorophyll and lacking photosystem II. Thus they cannot use the noncyclic pathway to form NADPH and O_2; they are anoxygenic.

=== **Key Terms** ===

accessory pigments   *183*

acetyl coenzyme A (acetyl-CoA)   *170*

adenosine 5′-phosphosulfate (APS)   *181*

alcoholic fermentation   *175*

amphibolic pathways   *166*

anabolism   *165*

anaerobic respiration   *176*

anoxygenic photosynthesis   *187*

antenna   *184*

ATP synthase   *172*

bacteriochlorophyll   *187*

β-oxidation pathway   *179*

butanediol fermentation   *175*

carotenoids   *183*

catabolism   *165*

chemiosmotic hypothesis   *172*

chemolithotroph   *180*

chlorophylls   *182*

citric acid cycle   *170*

conformational change hypothesis   *172*

cyclic photophosphorylation   *184*

dark reactions   *182*

deamination   *179*

denitrification   *177*

dissimilatory nitrate reduction   *177*

electron transport chain   *171*

Embden-Meyerhof pathway   *166*

Entner-Doudoroff pathway   *169*

fermentation   *175*

glycolysis   *166*

glycolytic pathway   *166*

heterolactic fermenters   *175*

hexose monophosphate pathway   *167*

homolactic fermenters   *175*

lactic acid fermentation   *175*

light reactions   *182*

metabolism   *165*

mixed acid fermentation   *175*

nitrification   *181*

nitrifying bacteria   *181*

noncyclic photophosphorylation   *184*

oxidative phosphorylation   *171*

oxygenic photosynthesis   *187*

Pasteur effect   *175*

pentose phosphate pathway   *167*

photosynthesis   *181*

photosystem I   *184*

photosystem II   *184*

phycobiliproteins   *183*

phycocyanin   *183*

phycoerythrin   *183*

protease   *179*

protonmotive force (PMF)   *172*

reaction-center chlorophyll   *184*

substrate-level phosphorylation   *167*

transamination   *179*

tricarboxylic acid (TCA) cycle   *170*

uncouplers   *172*

=== **Questions for Thought and Review** ===

1. Why might it be desirable for a microorganism with the Embden-Meyerhof pathway and the TCA cycle also to have the pentose phosphate pathway?

2. Of what advantage would it be to a microorganism to possess an electron transport chain and oxidative phosphorylation?

3. Describe two ways in which a microorganism can continue to produce energy when O_2 is absent.

4. Why would it be wasteful for anaerobic microorganisms to operate the complete TCA cycle?

5. How do substrate-level phosphorylation and oxidative phosphorylation differ from one another?

6. Can fermentation products be used in identifying bacteria? Give some examples if the answer is yes.

7. Describe what would happen to microbial metabolism if the enzyme lactate dehydrogenase was inhibited in a homolactic fermenter growing anaerobically on a medium containing glucose as the carbon source. What effects would an inhibitor of the ATP synthase have on an aerobically respiring microorganism? An uncoupler?

8. How would you isolate a thermophilic chemolithotroph that uses sulfur compounds as a source of electrons? What changes in the incubation system would be needed to isolate bacteria using sulfur compounds in anaerobic respiration? How can one tell which process is taking place through an analysis of the sulfur molecules present in the medium?

9. Suppose that you isolated a bacterial strain that carried out oxygenic photosynthesis. What photosystems would it possess and what group of bacteria would it most likely belong to?

Additional Reading

General

Caldwell, D. R. 1995. *Microbial physiology and metabolism.* Dubuque, Iowa: Wm. C. Brown.

Cramer, W. A., and Knaff, D. B. 1991. *Energy transduction in biological membranes. A textbook of bioenergetics.* New York: Springer-Verlag.

Dawes, E. A. 1986. *Microbial energetics.* New York: Chapman.

Dawes, I. W., and Sutherland, I. W. 1992. *Microbial physiology,* 2d ed. London: Blackwell Scientific Publications.

Garrett, R.H., and Grishman, C. M. 1995. *Biochemistry.* New York: Saunders.

Gottschalk, G. 1986. *Bacterial metabolism,* 2d ed. New York: Springer-Verlag.

Jones, C. W. 1982. *Bacterial respiration and photosynthesis.* Washington, D.C.: American Society for Microbiology.

Lehninger, A. L.; Nelson, D. L.; and Cox, M. M. 1993. *Principles of biochemistry,* 2d ed. New York: Worth Publishers.

Mandelstam, J.; McQuillen, K.; and Dawes, I. 1982. *Biochemistry of bacterial growth,* 3d ed. London: Blackwell Scientific Publications.

Mathews, C. K., and van Holde, K. E. 1990. *Biochemistry.* Redwood City, Calif.: Benjamin/Cummings.

Miles, R. J. 1992. Catabolism in mollicutes. *J. Gen. Microbiol.* 138:1773–83.

Moat, A. G., and Foster, J. W. 1995. *Microbial physiology,* 3d ed. New York: John Wiley and Sons.

Moran, L. A.; Scrimgeour, K. G.; Horton, H. R.; Ochs, R. S.; and Rawn, J. D. 1994. *Biochemistry.* Englewood Cliffs, N.J.: Neil Patterson Publishers/Prentice-Hall.

Neidhardt, F. C.; Ingraham, J. L.; and Schaechter, M. 1990. *Physiology of the bacterial cell: A molecular approach.* Sunderland, Mass.: Sinauer Associates.

Voet, D., and Voet, J. G. 1995. *Biochemistry,* 2d ed. New York: John Wiley and Sons.

White, D. 1995. *The physiology and biochemistry of prokaryotes.* New York: Oxford University Press.

Zubay, G. 1993. *Biochemistry,* 3d ed. Dubuque, Iowa: Wm. C. Brown Communications, Inc.

Electron Transport and Oxidative Phosphorylation

Anraku, Y. 1988. Bacterial electron transport chains. *Ann. Rev. Biochem.* 57:101–32.

Capaldi, R. A.; Aggeler, R.; Turina, P.; and Wilkens, S. 1994. Coupling between catalytic sites and the proton channel in F_1F_0-type ATPases. *Trends Biochem. Sci.* 19:284–89.

Grant, W. D. 1987. The enigma of the alkaliphile. *Microbiol. Sci.* 4(8):251–55.

Hatefi, Y. 1985. The mitochondrial electron transport and oxidative phosphorylation system. *Ann. Rev. Biochem.* 54:1015–69.

Pedersen, P. L., and Amzel, L. M. 1993. ATP synthases. *J. Biol. Chem.* 268(14):9937–40.

Trumpower, B. L., and Gennis, R. B. 1994. Energy transduction by cytochrome complexes in mitochondrial and bacterial respiration: The enzymology of coupling electron transfer reactions to transmembrane proton translocation. *Ann. Rev. Biochem.* 63:675–716.

Anaerobic Respiration

Ferguson, S. J. 1987. Denitrification: A question of the control and organization of electron and ion transport. *Trends Biochem. Sci.* 12(9):354–57.

Hochstein, L. I., and Tomlinson, G. A. 1988. The enzymes associated with denitrification. *Ann. Rev. Microbiol.* 42:231–61.

Lovley, D. R. 1993. Dissimilatory metal reduction. *Ann. Rev. Microbiol.* 47:263–90.

Postgate, J. R. 1984. *The sulphate-reducing bacteria,* 2d ed. New York: Cambridge University Press.

Zumft, W. G. 1993. The biological role of nitric oxide in bacteria. *Arch. Microbiol.* 160:253–64.

Oxidation of Inorganic Molecules

Friedrich, B., and Schwartz, E. 1993. Molecular biology of hydrogen utilization in aerobic chemolithotrophs. *Ann. Rev. Microbiol.* 47:351–83.

Kelly, D. P. 1985. Physiology of the thiobacilli: Elucidating the sulphur oxidation pathway. *Microbiol. Sci.* 2(4):105–9.

Smith, A. J., and Hoare, D. S. 1977. Specialist phototrophs, lithotrophs, and methylotrophs: A unity among a diversity of procaryotes? *Bacteriol. Rev.* 41(1):419–48.

Photosynthesis

Deisenhofer, J.; Michel, H.; and Huber, R. 1985. The structural basis of photosynthetic light reactions in bacteria. *Trends Biochem. Sci.* 10(6):243–48.

Govindjee, and Coleman, W. J. 1990. How plants make oxygen. *Sci. Am.* 262(2):50–58.

Grossman, A. R.; Schaefer, M. R.; Chiang, G. G.; and Collier, J. L. 1993. The phycobilisome, a light-harvesting complex responsive to environmental conditions. *Microbiol. Rev.* 57(3):725–49.

Quayle, J. R., and Ferenci, T. 1978. Evolutionary aspects of autotrophy. *Microbiol. Rev.* 42(2):251–73.

Schlegel, H. G., and Bowien, B., editors. 1989. *Autotrophic bacteria.* Madison, Wis.: Science Tech Publishers.

Staehelin, L. A., and Arntzen, C. J., editors. 1986. *Photosynthesis III: Photosynthetic membranes and light harvesting systems. Encyclopedia of plant physiology.* New Series, vol. 19. New York: Springer-Verlag.

Youvan, D. C., and Marrs, B. L. 1987. Molecular mechanisms of photosynthesis. *Sci. Am.* 256(6):42–48.

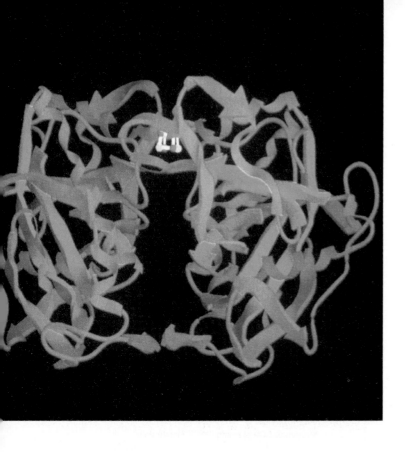

CHAPTER 10

Metabolism:
The Use of Energy
in Biosynthesis

Biological structures are almost always constructed in a hierarchical manner, with subassemblies acting as important intermediates en route from simple starting molecules to the end products of organelles, cells, and organisms.

—W. M. Becker and
D. W. Deamer

Outline

Concepts

1. In anabolism or biosynthesis, cells use free energy to construct more complex molecules and structures from smaller, simpler precursors.
2. Biosynthetic pathways are organized to optimize efficiency by conserving biosynthetic raw materials and energy.
3. Autotrophs use ATP and NADPH from photosynthesis or from oxidation of inorganic molecules to reduce CO_2 and incorporate it into organic material.
4. Catabolic and anabolic pathways may differ in enzymes, regulation, intracellular location, and use of cofactors and nucleoside diphosphate carriers. Although many enzymes of amphibolic pathways participate in both catabolism and anabolism, some pathway enzymes are involved only in one of the two processes.
5. Phosphorus, in the form of phosphate, can be directly assimilated, whereas inorganic sulfur and nitrogen compounds must often be reduced before incorporation into organic material.
6. The tricarboxylic acid (TCA) cycle acts as an amphibolic pathway and requires anaplerotic reactions to maintain adequate levels of cycle intermediates.
7. Most glycolytic enzymes participate in both the synthesis and catabolism of glucose. In contrast, fatty acids are synthesized from acetyl-CoA and malonyl-CoA by a pathway quite different from fatty acid β-oxidation.
8. Peptidoglycan synthesis is a complex, multistep process that is begun in the cytoplasm and completed at the cell wall after the peptidoglycan repeat unit has been transported across the plasma membrane.

As the last chapter makes clear, microorganisms can obtain energy in many ways. Much of this energy is used in biosynthesis or anabolism. During biosynthesis a microorganism begins with simple precursors, such as inorganic molecules and monomers, and constructs ever more complex molecules until new organelles and cells arise (figure 10.1). A microbial cell must manufacture many different kinds of molecules; however, it is possible to discuss the synthesis of only the most important types of cell constituents.

This chapter begins with a general introduction to anabolism, then focuses on the synthesis of carbohydrates, amino acids, purines and pyrimidines, and lipids. It also describes the assimilation of CO_2, phosphorus, sulfur, and nitrogen. The chapter ends with a section on the synthesis of peptidoglycan and bacterial cell walls. Protein and nucleic acid synthesis is so significant and complex that it is described separately in chapter 11.

Because anabolism is the creation of order and a cell is highly ordered and immensely complex, much energy is required for biosynthesis. This is readily apparent from estimates of the biosynthetic capacity of rapidly growing *Escherichia coli* (table 10.1). Although most ATP dedicated to biosynthesis is employed in protein synthesis, ATP is also used to make other cell constituents.

Free energy is required for biosynthesis in mature cells of constant size because cellular molecules are continuously being degraded and resynthesized, a process known as **turnover.** Cells are never the same from instant to instant. Despite the continuous turnover of cell constituents, metabolism is carefully regulated so that the rate of biosynthesis is approximately balanced by that of catabolism. In addition to the energy expended in the turnover of molecules, many nongrowing cells also use energy to synthesize enzymes and other substances for release into their surroundings.

Regulation of metabolism (chapter 12).

Principles Governing Biosynthesis

A microbial cell contains large quantities of proteins, nucleic acids, and polysaccharides, all of which are **macromolecules** or very large molecules that are polymers of smaller units joined together. The construction of large, complex molecules from a few simple structural units or **monomers** saves much genetic storage capacity, biosynthetic raw material, and energy. A consideration of protein synthesis clarifies this. Proteins—whatever size, shape, or function—are made of only 20 common amino acids joined by peptide bonds (*see appendix I*). Different proteins simply have different amino acid sequences but not new

Figure 10.1   The Construction of Cells. The biosynthesis of procaryotic and eucaryotic cell constituents. Biosynthesis is organized in levels of ever greater complexity.

TABLE 10.1 Biosynthesis in *Escherichia coli*			
Cell Constituent	**Number of Molecules per Cell[a]**	**Molecules Synthesized per Second**	**Molecules of ATP Required per Second for Synthesis**
DNA	1[b]	0.00083	60,000
RNA	15,000	12.5	75,000
Polysaccharides	39,000	32.5	65,000
Lipids	15,000,000	12,500.0	87,000
Proteins	1,700,000	1,400.0	2,120,000

From *Bioenergetics* by Albert Lehninger. Copyright © 1971 by the Benjamin/Cummings Publishing Company. Reprinted by permission.

[a]Estimates for a cell with a volume of 2.25 μm³, a total weight of 1×10^{-12}g, a dry weight of 2.5×10^{-13}g, and a 20 minute cell division cycle.

[b]It should be noted that bacteria can contain multiple copies of their genomic DNA.

Figure 10.2   A Hypothetical Biosynthetic Pathway. The routes connecting G with X, Y, and Z are purely anabolic because they are used only for synthesis of the end products. The pathway from A to G is amphibolic—that is, it has both catabolic and anabolic functions. Most reactions are used in both roles; however, the interconversion of C and D is catalyzed by two separate enzymes, E_1 (catabolic) and E_2 (anabolic).

and dissimilar amino acids. Suppose that proteins were composed of 40 different amino acids instead of 20. The cell would then need the enzymes to manufacture twice as many amino acids (or would have to obtain the extra amino acids in its diet). Genes would be required for the extra enzymes, and the cell would have to invest raw materials and energy in the synthesis of these additional genes, enzymes, and amino acids. Clearly the use of a few monomers linked together by a single type of covalent bond makes the synthesis of macromolecules a highly efficient process. Almost all cell structures are built mainly of about 30 small precursors.

The cell often saves additional materials and energy by using many of the same enzymes for both catabolism and anabolism. For example, most glycolytic enzymes are involved in the synthesis and the degradation of glucose.

Although many enzymes in amphibolic pathways (*see chapter 9*) participate in both catabolic and anabolic activities, some steps are catalyzed by two different enzymes. One enzyme catalyzes the reaction in the catabolic direction, the other reverses this conversion (figure 10.2). Thus catabolic and anabolic pathways are never identical although many enzymes are shared. Use of separate enzymes for the two directions of a single step permits independent regulation of catabolism and anabolism. Although this is discussed in more detail in chapter 12, note that the regulation of anabolism is somewhat different from that of catabolism. Both types of pathways can be regulated by

their end products as well as by the concentrations of ATP, ADP, AMP, and NAD^+. Nevertheless, end product regulation generally assumes more importance in anabolic pathways.

To synthesize molecules efficiently, anabolic pathways must operate irreversibly in the direction of biosynthesis. Cells can achieve this by connecting some biosynthetic reactions to the breakdown of ATP and other nucleoside triphosphates. When these two processes are coupled, the free energy made available during nucleoside triphosphate breakdown drives the biosynthetic reaction to completion (*see chapter 8*).

In eucaryotic microorganisms biosynthetic pathways are frequently located in different cellular compartments from their corresponding catabolic pathways (Box 10.1). For example, fatty acid biosynthesis occurs in the cytoplasmic matrix, whereas fatty acid oxidation takes place within the mitochondrion. Compartmentation makes it easier for the pathways to operate simultaneously but independently.

Finally, anabolic and catabolic pathways often use different cofactors. Usually catabolic oxidations produce NADH, a substrate for electron transport. In contrast, when a reductant is needed during biosynthesis, NADPH rather than NADH normally serves as the donor. Fatty acid metabolism provides a second example. Fatty acyl-CoA molecules are oxidized to generate energy, whereas fatty acid synthesis involves acyl carrier protein thioesters (p. 205).

After macromolecules have been constructed from simpler precursors, they are assembled into larger, more complex structures such as supramolecular systems and organelles (figure 10.1). Macromolecules normally contain the necessary information to form spontaneously in a process known as **self-assembly.** For example, ribosomes are large assemblages of many proteins and ribonucleic acid molecules, yet they arise by the self-assembly of their components without the involvement of extra factors.

1. Define biosynthesis or anabolism and turnover.
2. List six principles by which biosynthetic pathways are organized.

The Photosynthetic Fixation of CO_2

Although most microorganisms can incorporate or fix CO_2, at least in anaplerotic reactions (pp. 202–203), only autotrophs use CO_2 as their sole or principal carbon source. The reduction and incorporation of CO_2 requires much energy. Usually autotrophs obtain energy by trapping light during photosynthesis, but some derive energy from the oxidation of reduced inorganic electron donors. Autotrophic CO_2 fixation is crucial to life on earth because it provides the organic matter on which heterotrophs depend.

Photosynthetic light reactions and chemolithotrophy (pp. 180–88).

Almost all microbial autotrophs incorporate CO_2 by a special metabolic pathway called by several names: the

To synthesize fructose 6-phosphate or glucose 6-phosphate from CO_2, the cycle must operate six times to yield the desired hexose and reform the six RuBP molecules.

$$6RuBP + 6CO_2 \longrightarrow 12PGA \longrightarrow$$
$$6RuBP + \text{fructose 6-P}$$

The incorporation of one CO_2 into organic material requires three ATPs and two NADPHs. The formation of glucose from CO_2 may be summarized by the following equation.

$$6CO_2 + 18ATP + 12NADPH + 12H^+ + 12H_2O \longrightarrow$$
$$\text{glucose} + 18ADP + 18P_i + 12NADP^+$$

ATP and NADPH are provided by photosynthetic light reactions or by oxidation of inorganic molecules in chemoautotrophs. Sugars formed in the Calvin cycle can then be used to synthesize other essential molecules.

Synthesis of Sugars and Polysaccharides

Many microorganisms cannot carry out photosynthesis and are heterotrophs that must synthesize sugars from reduced organic molecules rather than from CO_2. The synthesis of glucose from noncarbohydrate precursors is called **gluconeogenesis.** Although the gluconeogenic pathway is not identical with the glycolytic pathway, they do share seven enzymes (figure 10.5). Three glycolytic steps are irreversible in the cell: (1) the conversion of phosphoenolpyruvate to pyruvate, (2) the formation of fructose 1,6-bisphosphate from fructose 6-phosphate, and (3) the phosphorylation of glucose. These must be bypassed when the pathway is operating biosynthetically. For example, the formation of fructose 1,6-bisphosphate by phosphofructokinase is reversed by a separate enzyme, fructose bisphosphatase, which hydrolytically removes a phosphate from fructose bisphosphate. Usually at least two enzymes are involved in the conversion of pyruvate to phosphoenolpyruvate (the reversal of the pyruvate kinase step).

As can be seen in figure 10.5, the pathway synthesizes fructose as well as glucose. Once glucose and fructose have been formed, other common sugars can be manufactured. For example, mannose comes directly from fructose by a simple rearrangement.

$$\text{Fructose 6-phosphate} \rightleftharpoons \text{mannose 6-phosphate}$$

Several sugars are synthesized while attached to a nucleoside diphosphate. The most important nucleoside diphosphate sugar is **uridine diphosphate glucose (UDPG).** Glucose is activated by attachment to the pyrophosphate of uridine diphosphate through a reaction with uridine triphosphate (figure 10.6). The UDP portion of UDPG is recognized by enzymes and carries glucose around the cell for participation in enzyme reactions much like ADP bears phosphate in the form of ATP. UDP-galactose is synthesized from UDPG through a rearrangement of one hydroxyl group. A different enzyme catalyzes the synthesis of UDP-glucuronic acid through the oxidation of UDPG (figure 10.7).

Nucleoside diphosphate sugars also play a central role in the synthesis of polysaccharides such as starch and glycogen. Again, biosynthesis is not simply a direct reversal of catabolism.

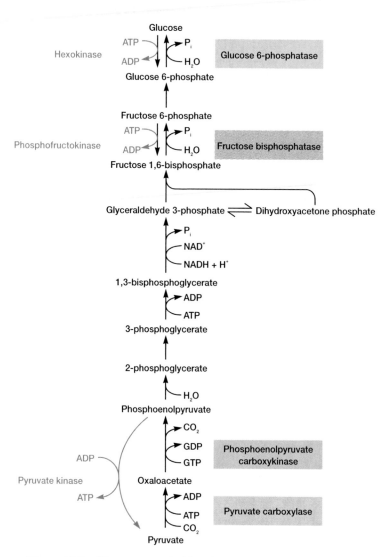

Figure 10.5 **Gluconeogenesis.** The gluconeogenic pathway used in many microorganisms. The names of the four enzymes catalyzing reactions different from those found in glycolysis are in shaded boxes. Glycolytic steps are shown in color for comparison.

Figure 10.6 **Uridine Diphosphate Glucose.** Glucose is in color.

Figure 10.7   Uridine Diphosphate Galactose and Glucuronate Synthesis. The synthesis of UDP-galactose and UDP-glucuronic acid from UDP-glucose. Structural changes are indicated by colored boxes.

Glycogen and starch catabolism (*see chapter 9*) proceeds either by hydrolysis to form free sugars or by the addition of phosphate to these polymers with the production of glucose 1-phosphate. Nucleoside diphosphate sugars are not involved. In contrast, during the synthesis of glycogen and starch in bacteria and algae, adenosine diphosphate glucose is formed from glucose 1-phosphate and then donates glucose to the end of growing glycogen and starch chains.

$$ATP + \text{glucose 1-phosphate} \longrightarrow \text{ADP-glucose} + PP_i$$
$$(\text{Glucose})_n + \text{ADP-glucose} \longrightarrow (\text{glucose})_{n+1} + ADP$$

Nucleoside diphosphate sugars also participate in the synthesis of complex molecules such as bacterial cell walls (pp. 206–9).

---

1. Briefly describe the three stages of the Calvin cycle.
2. What is gluconeogenesis and how does it usually occur? Describe the formation of mannose, galactose, starch, and glycogen. Why are nucleoside diphosphate sugars important?

The Assimilation of Inorganic Phosphorus, Sulfur, and Nitrogen

Besides carbon and oxygen, microorganisms also require large quantities of phosphorus, sulfur, and nitrogen for biosynthesis. Each of these is assimilated, or incorporated into organic molecules, by different routes.

Microbial nutrition (chapter 5).

Microbial participation in biogeochemical cycles (chapter 40).

Phosphorus Assimilation

Phosphorus is found in nucleic acids, proteins, phospholipids, ATP, and coenzymes like NADP. The most common phosphorus sources are inorganic phosphate and organic phosphate esters.

Figure 10.8   Phosphoadenosine 5′-phosphosulfate (PAPS). The sulfate group is in color.

Inorganic phosphate is incorporated through the formation of ATP in one of three ways: by (1) photophosphorylation, (2) oxidative phosphorylation, and (3) substrate-level phosphorylation. Glycolysis provides an example of the latter process. Phosphate is joined with glyceraldehyde 3-phosphate to give 1,3-bisphosphoglycerate, which is next used in ATP synthesis.

$$\text{Glyceraldehyde 3-P} + P_i + NAD^+ \longrightarrow$$
$$\text{1,3-bisphosphoglycerate} + NADH + H^+$$

$$\text{1,3-bisphosphoglycerate} + ADP \longrightarrow$$
$$\text{3-phosphoglycerate} + ATP$$

Microorganisms may obtain organic phosphates from their surroundings in dissolved or particulate form. **Phosphatases** very often hydrolyze organic phosphate esters to release inorganic phosphate. Gram-negative bacteria have phosphatases in the periplasmic space between their cell wall and the plasma membrane, which allows phosphate to be taken up immediately after release. On the other hand, protozoa can directly use organic phosphates after ingestion or hydrolyze them in lysosomes and incorporate the phosphate.

Sulfur Assimilation

Sulfur is needed for the synthesis of amino acids (cysteine and methionine) and several coenzymes (e.g., coenzyme A and biotin) and may be obtained from two sources. Many microorganisms use cysteine and methionine, obtained from either external sources or intracellular amino acid reserves. In addition, sulfate can provide sulfur for biosynthesis. The sulfur atom in sulfate is more oxidized than it is in cysteine and other organic molecules; thus sulfate must be reduced before it can be assimilated. This process is known as **assimilatory sulfate reduction** to distinguish it from the **dissimilatory sulfate reduction** that takes place when sulfate acts as an electron acceptor during anaerobic respiration (*see figure 40.8*).

Anaerobic respiration (pp. 176–78).

Assimilatory sulfate reduction involves sulfate activation through the formation of **phosphoadenosine 5′-phosphosulfate** (figure 10.8), followed by reduction of the sulfate. The process is a complex one (figure 10.9) in which sulfate is first

Figure 10.9   **The Sulfate Reduction Pathway.**

Figure 10.10   **The Ammonia Assimilation Pathway.** Ammonia assimilation by use of glutamate dehydrogenase (GDH) and transaminases. Either NADP- or NAD-dependent glutamate dehydrogenases may be involved.

reduced to sulfite (SO_3^{2-}), then to hydrogen sulfide. Cysteine can be synthesized from hydrogen sulfide in two ways. Fungi appear to combine hydrogen sulfide with serine to form cysteine (process 1), whereas many bacteria join hydrogen sulfide with O-acetylserine instead (process 2).

(1) H_2S + serine $\longrightarrow$ cysteine + H_2O

(2)
$$\text{Serine} \xrightarrow{\text{acetyl-CoA   CoA}} \text{O-acetylserine} \xrightarrow{H_2S \quad \text{acetate}} \text{cysteine}$$

Once formed, cysteine can be used in the synthesis of other sulfur-containing organic compounds.

Nitrogen Assimilation

Because nitrogen is a major component of proteins, nucleic acids, coenzymes, and many other cell constituents, the cell's ability to assimilate inorganic nitrogen is exceptionally important. Although nitrogen gas is abundant in the atmosphere, few microorganisms can reduce the gas and use it as a nitrogen source. Most must incorporate either ammonia or nitrate.

Ammonia Incorporation

Ammonia nitrogen can be incorporated into organic material relatively easily and directly because it is more reduced than other forms of inorganic nitrogen. Some microorganisms form the amino acid alanine in a reductive amination reaction catalyzed by alanine dehydrogenase.

$$\text{Pyruvate} + NH_4^+ + \text{NADH (NADPH)} + H^+$$
$$\rightleftharpoons \text{L-alanine} + NAD^+ (NADP^+) + H_2O$$

The major route for ammonia incorporation often is the formation of glutamate from α-ketoglutarate (a TCA cycle intermediate). Many bacteria and fungi employ **glutamate dehydrogenase,** at least when the ammonia concentration is high.

$$\text{α-ketoglutarate} + NH_4^+ + \text{NADPH (NADH)} + H^+$$
$$\rightleftharpoons \text{glutamate} + NADP^+ (NAD^+) + H_2O$$

Different species vary in their ability to use NADPH and NADH as the reducing agent in glutamate synthesis.

Once either alanine or glutamate has been synthesized, the newly formed α-amino group can be transferred to other carbon skeletons by transamination reactions (*see chapter 9*) to form different amino acids. **Transaminases** possess the coenzyme pyridoxal phosphate, which is responsible for the amino group transfer. Microorganisms have a number of transaminases, each of which catalyzes the formation of several amino acids using the same amino acid as an amino group donor. When glutamate dehydrogenase works in cooperation with transaminases, ammonia can be incorporated into a variety of amino acids (figure 10.10).

A second route of ammonia incorporation involves two enzymes acting in sequence, **glutamine synthetase** and **glutamate synthase** (figure 10.11). Ammonia is used to synthesize glutamine from glutamate, then the amide nitrogen of glutamine is transferred to α-ketoglutarate to generate a new glutamate molecule. Because glutamate acts as an amino donor in transaminase reactions, ammonia may be used to synthesize all common amino acids when suitable transaminases are present (figure 10.12). Both ATP and a source of electrons, such as NADPH or reduced ferredoxin, are required. This route is present in *Escherichia coli, Bacillus megaterium,* and other bacteria. The two enzymes acting in sequence operate very effectively at low ammonia concentrations, unlike the glutamate dehydrogenase pathway.

Assimilatory Nitrate Reduction

The nitrogen in nitrate (NO_3^-) is much more oxidized than that in ammonia. Nitrate must first be reduced to ammonia before the nitrogen can be converted to an organic form. This reduction of nitrate is called **assimilatory nitrate reduction,** which is not the same as that occurring during anaerobic respiration and dissimilatory nitrate reduction (*see chapters 9 and 40*). In assimilatory nitrate reduction, nitrate is incorporated into organic material and does not participate in energy generation. The process is widespread among bacteria, fungi, and algae.

Glutamine synthetase reaction

$$\begin{array}{c} \text{COOH} \\ | \\ \text{CH}_2 \\ | \\ \text{CH}_2 \\ | \\ \text{CH}-\text{NH}_2 \\ | \\ \text{COOH} \end{array} + \text{NH}_3 + \text{ATP} \longrightarrow \begin{array}{c} \text{O} \\ \| \\ \text{C}-\text{NH}_2 \\ | \\ \text{CH}_2 \\ | \\ \text{CH}_2 \\ | \\ \text{CH}-\text{NH}_2 \\ | \\ \text{COOH} \end{array} + \text{ADP} + \text{P}_i$$

Glutamic acid   Glutamine

Glutamate synthase reaction

$$\begin{array}{c} \text{COOH} \\ | \\ \text{C}=\text{O} \\ | \\ \text{CH}_2 \\ | \\ \text{CH}_2 \\ | \\ \text{COOH} \end{array} + \begin{array}{c} \text{COOH} \\ | \\ \text{CH}-\text{NH}_2 \\ | \\ \text{CH}_2 \\ | \\ \text{CH}_2 \\ | \\ \text{C}-\text{NH}_2 \\ \| \\ \text{O} \end{array} + \begin{array}{c} \text{NADPH} + \text{H}^+ \\ \text{or} \\ \text{Fd}_{reduced} \end{array} \longrightarrow \begin{array}{c} \text{COOH} \\ | \\ \text{CH}-\text{NH}_2 \\ | \\ \text{CH}_2 \\ | \\ \text{CH}_2 \\ | \\ \text{COOH} \end{array} + \begin{array}{c} \text{COOH} \\ | \\ \text{CH}-\text{NH}_2 \\ | \\ \text{CH}_2 \\ | \\ \text{CH}_2 \\ | \\ \text{COOH} \end{array} + \begin{array}{c} \text{NADP}^+ \\ \text{or} \\ \text{Fd}_{oxidized} \end{array}$$

α-ketoglutaric acid   Glutamine   Glutamic acid

Figure 10.11 Glutamine Synthetase and Glutamate Synthase. The glutamine synthetase and glutamate synthase reactions involved in ammonia assimilation. Some glutamine synthases use NADPH as an electron source; others use reduced ferredoxin (Fd). The nitrogen being incorporated and transferred is shown in color.

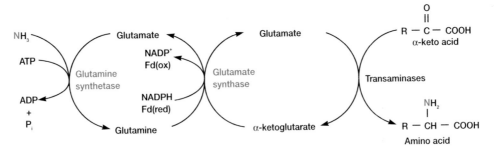

Figure 10.12 Ammonia Incorporation Using Glutamine Synthetase and Glutamate Synthase.

The first step in nitrate assimilation is its reduction to nitrite by **nitrate reductase,** an enzyme that contains both FAD and molybdenum. NADPH is the electron source.

$$\text{NO}_3^- + \text{NADPH} + \text{H}^+ \longrightarrow \text{NO}_2^- + \text{NADP}^+ + \text{H}_2\text{O}$$

Nitrite is next reduced to ammonia with a series of two electron additions catalyzed by **nitrite reductase** and possibly other enzymes. Hydroxylamine may be an intermediate.

$$\text{NO}_2^- \xrightarrow{2e^-} \times \xrightarrow{2e^-} [\text{NH}_2\text{OH}] \xrightarrow[\substack{2\text{H}^+ \\ \text{H}_2\text{O}}]{2e^-} \text{NH}_3$$

The ammonia is then incorporated into amino acids by the routes already described.

Nitrogen Fixation

The reduction of atmospheric gaseous nitrogen to ammonia is called **nitrogen fixation.** Because ammonia and nitrate levels often are low and only a few procaryotes can carry out nitrogen fixation (eucaryotic cells completely lack this ability), the rate of this process limits plant growth in many situations. Nitrogen fixation occurs in (1) free-living bacteria (e.g., *Azotobacter*, *Klebsiella*, *Clostridium*, and *Methanococcus*), (2) bacteria living in symbiotic association with plants such as legumes (*Rhizobium*), and (3) cyanobacteria (*Nostoc* and *Anabaena*). The biological aspects of nitrogen fixation are discussed in chapter 42. The biochemistry of nitrogen fixation is the focus of this section.

The biology of nitrogen-fixing microorganisms (pp. 423, 817, 865–68).

The reduction of nitrogen to ammonia is catalyzed by the enzyme **nitrogenase.** Although the enzyme-bound intermediates in this process are still unknown, it is believed that nitrogen is reduced by two-electron additions in a way similar to that illustrated in figure 10.13. The reduction of molecular nitrogen to ammonia is quite exergonic, but the reaction has a high activation energy because molecular nitrogen is an unreactive gas with a triple bond between the two nitrogen atoms. Therefore

Figure 10.13 **Nitrogen Reduction.** A hypothetical sequence of nitrogen reduction by nitrogenase.

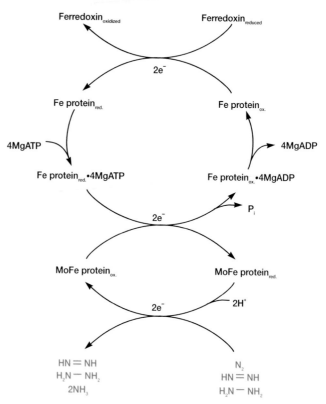

Figure 10.15 **Mechanism of Nitrogenase Action.** The flow of two electrons from ferredoxin to nitrogen is outlined. This process is repeated three times in order to reduce N_2 to two molecules of ammonia, and the substrate-product pairs for each step are shown at the bottom of the figure. See text for a more detailed explanation.

Figure 10.14 **Structure of the Nitrogenase Fe Protein.** The Fe protein's two subunits are arranged like a pair of butterfly wings with the iron sulfur cluster between the wings and at the "head" of the butterfly. The iron sulfur cluster is very exposed, which helps account for nitrogenase's sensitivity to oxygen. The oxygen can readily attack the exposed irons.

nitrogen reduction is expensive and requires a large ATP expenditure. At least 8 electrons and 16 ATP molecules, 4 ATPs per pair of electrons, are required.

$$N_2 + 8H^+ + 8e^- + 16ATP \longrightarrow$$
$$2NH_3 + H_2 + 16ADP + 16P_i$$

The electrons come from ferredoxin that has been reduced in a variety of ways: by photosynthesis in cyanobacteria, respiratory processes in aerobic nitrogen fixers, or fermentations in anaerobic bacteria. For example, *Clostridium pasteurianum* (an anaerobic bacterium) reduces ferredoxin during pyruvate oxidation, whereas the aerobic *Azotobacter* uses electrons from NADPH to reduce ferredoxin.

Nitrogenase is a complex system consisting of two major protein components, a MoFe protein (MW 220,000) joined with one or two Fe proteins (MW 64,000). The MoFe protein contains 2 atoms of molybdenum and 28 to 32 atoms of iron; the Fe protein has 4 iron atoms (figure 10.14). Fe protein is first reduced by ferredoxin, then it binds ATP (figure 10.15). ATP binding lowers the reduction potential of the Fe protein, enabling it to reduce the MoFe protein. ATP is hydrolyzed when this electron transfer occurs. Finally, reduced MoFe protein donates

electrons to atomic nitrogen. Nitrogenase is quite sensitive to O_2 and must be protected from O_2 inactivation within the cell.

The reduction of N_2 to NH_3 occurs in three steps, each of which requires an electron pair (figures 10.13 and 10.15). Six electron transfers take place, and this requires a total 12 ATPs per N_2 reduced. The overall process actually requires at least 8 electrons and 16 ATPs because nitrogenase also reduces H_2O to H_2. The H_2 reacts with diimine (HN $=$ NH) to form N_2 and H_2. This futile cycle operates to some extent even under favorable conditions and makes nitrogen fixation even more expensive. Symbiotic nitrogen-fixing bacteria can consume almost 20% of the ATP produced by the host plant.

Nitrogenase can reduce a variety of molecules containing triple bonds (e.g., acetylene, cyanide, and azide).

$$HC \equiv CH + 2H^+ + 2e^- \longrightarrow H_2C = CH_2$$

The rate of reduction of acetylene to ethylene is even used to measure nitrogenase activity.

Once molecular nitrogen has been reduced to ammonia, the ammonia can be incorporated into organic compounds. In the symbiotic nitrogen fixer *Rhizobium*, it appears that ammonia diffuses out of the bacterial cell and is assimilated in the

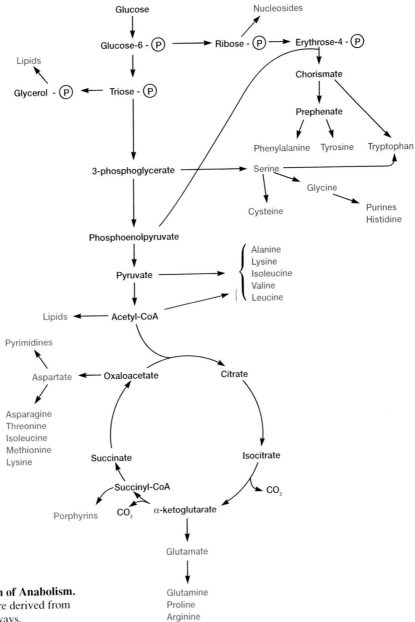

Figure 10.16   The Organization of Anabolism.
Biosynthetic products (in color) are derived from
intermediates of amphibolic pathways.

surrounding legume cell. The primary route of ammonia assimilation seems to be the synthesis of glutamine by the glutamine synthetase–glutamate synthase system (figure 10.11). However, substances such as the purine derivatives allantoin and allantoic acid also are synthesized and used for the transport of nitrogen to other parts of the plant.

The Synthesis of Amino Acids

Microorganisms vary with respect to the type of nitrogen source they employ, but most can assimilate some form of inorganic nitrogen by the routes just described. Amino acid synthesis also requires construction of the proper carbon skeletons, and this is

often a complex process involving many steps. Although individual amino acid biosynthetic pathways are not described in detail, a survey of the general pattern of amino acid biosynthesis is worthwhile. Further details of amino acid biosynthesis may be found in introductory biochemistry textbooks.

The relationship of amino acid biosynthetic pathways to amphibolic routes is shown in figure 10.16. Amino acid skeletons are derived from acetyl-CoA and from intermediates of the TCA cycle, glycolysis, and the pentose phosphate pathway. To maximize efficiency and economy, the precursors for amino acid biosynthesis are provided by a few major amphibolic pathways. Sequences leading to individual amino acids branch off from these central routes. Alanine, aspartate, and glutamate are

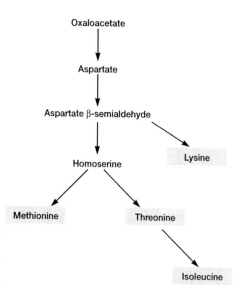

Figure 10.17   A Branching Pathway of Amino Acid Synthesis. The pathways to methionine, threonine, isoleucine, and lysine. Although some arrows represent one step, most interconversions require the participation of several enzymes.

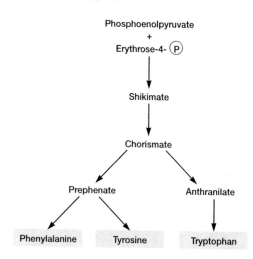

Figure 10.18   Aromatic Amino Acid Synthesis. The synthesis of the aromatic amino acids phenylalanine, tyrosine, and tryptophan. Most arrows represent more than one enzyme reaction.

made by transamination directly from pyruvate, oxaloacetate, and α-ketoglutarate, respectively. Most biosynthetic pathways are more complex, and common intermediates often are used in the synthesis of families of related amino acids for the sake of further economy. For example, the amino acids lysine, threonine, isoleucine, and methionine are synthesized from oxaloacetate by such a branching anabolic route (figure 10.17). The biosynthetic pathways for the aromatic amino acids phenylalanine, tyrosine, and tryptophan also share many intermediates (figure 10.18).

1. How do microorganisms assimilate sulfur and phosphorus?
2. Describe the roles of glutamate dehydrogenase, glutamine synthetase, glutamate synthase, and transaminases in ammonia assimilation. How is nitrate incorporated by assimilatory nitrate reduction?
3. What is nitrogen fixation? Briefly describe the structure and mechanism of action of nitrogenase.
4. Summarize in general terms the organization of amino acid biosynthesis.

Anaplerotic Reactions

Inspection of figure 10.16 will show that TCA cycle intermediates are used in the synthesis of pyrimidines and a wide variety of amino acids. In fact, the biosynthetic functions of this pathway are so essential that most of it must operate anaerobically to supply biosynthetic precursors, even though NADH is not required for electron transport and oxidative phosphorylation in the absence of oxygen. Thus there is a heavy demand

upon the TCA cycle to supply carbon for biosynthesis, and cycle intermediates could be depleted if nothing were done to maintain their levels. However, microorganisms have reactions that replenish cycle intermediates so that the TCA cycle can continue to function when active biosynthesis is taking place. Reactions that replace cycle intermediates are called **anaplerotic reactions** [Greek *anaplerotic,* filling up].

Most microorganisms can replace TCA cycle intermediates by **CO₂ fixation**. It should be emphasized that anaplerotic reactions do not serve the same function as the CO₂ fixation pathway that supplies the carbon required by autotrophs. In autotrophs CO₂ fixation provides most or all of the carbon required for growth. Anaplerotic CO₂ fixation reactions simply replace TCA cycle intermediates and maintain metabolic balance. Usually CO₂ is added to an acceptor molecule, either pyruvate or phosphoenolpyruvate, to form the cycle intermediate oxaloacetate. Some microorganisms (e.g., *Arthrobacter globiformis,* yeasts) use pyruvate carboxylase in this role.

$$\text{Pyruvate} + CO_2 + \text{ATP} + H_2O \xrightarrow{\text{biotin}} \text{oxaloacetate} + \text{ADP} + P_i$$

This enzyme requires the cofactor biotin and uses ATP energy to join CO₂ and pyruvate. Biotin is often the cofactor for enzymes catalyzing carboxylation reactions. Because of its importance, biotin is a vitamin for many species. Other microorganisms, such as the bacteria *Escherichia coli* and *Salmonella typhimurium,* have the enzyme phosphoenolpyruvate carboxylase, which catalyzes the following reaction.

$$\text{Phosphoenolpyruvate} + CO_2 \longrightarrow \text{oxaloacetate} + P_i$$

Some bacteria, algae, fungi, and protozoa can grow with acetate as the sole carbon source by using it to synthesize

Figure 10.19   The Glyoxylate Cycle. The reactions and enzymes unique to the cycle are shown in color. The tricarboxylic acid cycle enzymes that have been bypassed are at the bottom.

TCA cycle intermediates in the **glyoxylate cycle** (figure 10.19). This cycle is made possible by two unique enzymes, isocitrate lyase and malate synthase, that catalyze the following reactions.

$$\text{Isocitrate} \xrightarrow{\text{isocitrate lyase}} \text{succinate + glyoxylate}$$

$$\text{Glyoxylate + acetyl-CoA} \xrightarrow{\text{malate synthase}} \text{malate + CoA}$$

The TCA cycle (pp. 169–70).

The glyoxylate cycle is actually a modified TCA cycle. The two decarboxylations of the latter pathway (the isocitrate dehydrogenase and α-ketoglutarate dehydrogenase steps) are bypassed, making possible the conversion of acetyl-CoA to form oxaloacetate without loss of acetyl-CoA carbon as CO_2. In this fashion acetate and any molecules that give rise to it can contribute carbon to the cycle.

1. Define an anaplerotic reaction and give an example.
2. How does the glyoxylate cycle convert acetyl-CoA to oxaloacetate, and what special enzymes are used?

The Synthesis of Purines, Pyrimidines, and Nucleotides

Purine and pyrimidine biosynthesis is critical for all cells because these molecules are used in the synthesis of ATP, several cofactors, ribonucleic acid (RNA), deoxyribonucleic acid (DNA), and other important cell components. Nearly all microorganisms can sythesize their own purines and pyrimidines as these are so crucial to cell function.

DNA and RNA synthesis (pp. 216–21).

Purines and **pyrimidines** are cyclic nitrogenous bases with several double bonds and pronounced aromatic properties. Purines consist of two joined rings, whereas pyrimidines have only one (figures 10.20 and 10.22). The purines **adenine** and **guanine** and the pyrimidines **uracil, cytosine,** and **thymine** are commonly found in microorganisms. A purine or pyrimidine base joined with a pentose sugar, either ribose or deoxyribose, is a **nucleoside**. A **nucleotide** is a nucleoside with one or more phosphate groups attached to the sugar.

Purine Biosynthesis

The biosynthetic pathway for purines is a complex, 11-step sequence (*see appendix III*) in which seven different molecules contribute parts to the final purine skeleton (figure 10.20). Because the pathway begins with ribose 5-phosphate and the purine skeleton is constructed on this sugar, the first purine product of the pathway is the nucleotide inosinic acid, not a free purine base. The cofactor folic acid is very important in purine biosynthesis. Folic acid derivatives contribute carbons two and eight to the purine skeleton. In fact, the drug sulfonamide inhibits bacterial growth by blocking folic acid synthesis. This interferes with purine biosynthesis and other processes that require folic acid.

Once inosinic acid has been formed, relatively short pathways synthesize adenosine monophosphate and guanosine monophosphate (figure 10.21) and produce nucleoside diphosphates and triphosphates by phosphate transfers from ATP. DNA contains deoxyribonucleotides (the ribose lacks a hydroxyl group on carbon two) instead of the ribonucleotides found in RNA. Deoxyribonucleotides arise from the reduction of nucleoside diphosphates or nucleoside triphosphates by two different routes. Some microorganisms reduce the triphosphates with a system requiring vitamin B_{12} as a cofactor. Others, such as *E. coli*, reduce the ribose in nucleoside diphosphates. Both systems employ a small sulfur-containing protein called thioredoxin as their reducing agent.

Pyrimidine Biosynthesis

Pyrimidine biosynthesis begins with aspartic acid and carbamoyl phosphate, a high-energy molecule synthesized from CO_2 and ammonia (figure 10.22). Aspartate carbamoyltransferase catalyzes the condensation of these two substrates to form carbamoylaspartate, which is then converted to the initial pyrimidine product, orotic acid.

The regulation of aspartate carbamoyltransferase activity (pp. 238–39).

After synthesis of the pyrimidine skeleton, a nucleotide is produced by the ribose 5-phosphate addition using the high-energy intermediate 5-phosphoribosyl 1-pyrophosphate. Thus construction of the pyrimidine ring is completed before ribose is added, in contrast with purine ring synthesis, which begins with ribose 5-phosphate. Decarboxylation of orotidine monophosphate yields uridine monophosphate and eventually uridine triphosphate and cytidine triphosphate.

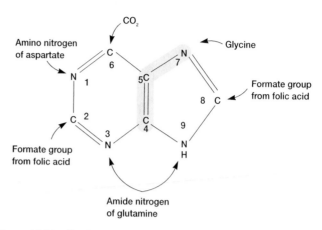

Figure 10.20 Purine Biosynthesis. The sources of purine skeleton nitrogen and carbon are indicated in color.

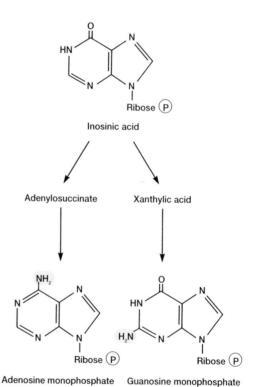

Figure 10.21   Synthesis of Adenosine Monophosphate and Guanosine Monophosphate. The shaded groups are the ones differing from those in inosinic acid.

The third common pyrimidine is thymine, a constituent of DNA. The ribose in pyrimidine nucleotides is reduced in the same way as it is in purine nucleotides. Then deoxyuridine monophosphate is methylated with a folic acid derivative to form deoxythymidine monophosphate (figure 10.23).

1. Define purine, pyrimidine, nucleoside, and nucleotide.
2. Outline the way in which purines and pyrimidines are synthesized. How is the deoxyribose component of deoxyribonucleotides made?

Figure 10.22   Pyrimidine Synthesis. PRPP stands for 5-phosphoribose 1-pyrophosphoric acid, which provides the ribose 5-phosphate chain.

Figure 10.23   Deoxythymidine Monophosphate Synthesis. Deoxythymidine differs from deoxyuridine in having the shaded methyl group.

Lipid Synthesis

A variety of lipids are found in microorganisms, particularly in cell membranes. Most contain **fatty acids** or their derivatives. Fatty acids are monocarboxylic acids with long alkyl chains that usually have an even number of carbons (the average length is 18 carbons). Some may be unsaturated—that is, have one or more double bonds. Most microbial fatty acids are straight chained, but some are branched. Gram-negative

bacteria often have cyclopropane fatty acids (fatty acids with one or more cyclopropane rings in their chains).

Lipid structure and nomenclature (appendix I).

Fatty acid synthesis is catalyzed by the **fatty acid synthetase** complex with acetyl-CoA and malonyl-CoA as the substrates and NADPH as the reductant. Malonyl-CoA arises from the ATP-driven carboxylation of acetyl-CoA (figure 10.24). Synthesis takes place after acetate and malonate have been transferred from coenzyme A to the sulfhydryl group of the **acyl carrier protein (ACP),** a small protein that carries the growing fatty acid chain during synthesis. The synthetase adds two carbons at a time to the carboxyl end of the growing fatty acid chain in a two-stage process (figure 10.24). First, malonyl-ACP reacts with the fatty acyl-ACP to yield CO_2 and a fatty acyl-ACP two carbons longer. The loss of CO_2 drives this reaction to completion. Notice that ATP is used to add CO_2 to acetyl-CoA, forming malonyl-CoA. The same CO_2 is lost when malonyl-ACP donates carbons to the chain. Thus carbon dioxide is essential to fatty acid synthesis but is not permanently incorporated. Indeed, some microorganisms require CO_2 for good growth, but they

Figure 10.24   Fatty Acid Synthesis.
The cycle is repeated until the proper
chain length has been reached. Carbon
dioxide carbon and the remainder of
malonyl-CoA are shown in different
colors. ACP stands for acyl carrier
protein.

can do without it in the presence of a fatty acid like oleic acid
(an 18-carbon unsaturated fatty acid). In the second stage of
synthesis, the β-keto group arising from the initial condensation reaction is removed in a three-step process involving two
reductions and a dehydration. The fatty acid is then ready for
the addition of two more carbon atoms.

Unsaturated fatty acids are synthesized in two ways. Eucaryotes and aerobic bacteria like *Bacillus megaterium* employ
an aerobic pathway using both NADPH and O_2.

$$R - (CH_2)_9 - \overset{\overset{\displaystyle O}{\|}}{C} - SCoA + NADPH + H^+ + O_2 \longrightarrow$$

$$R - CH = CH - (CH_2)_7 - \overset{\overset{\displaystyle O}{\|}}{C} - SCoA + NADP^+ + 2H_2O$$

A double bond is formed between carbons nine and ten, and O_2 is
reduced to water with electrons supplied by both the fatty acid
and NADPH. Anaerobic bacteria and some aerobes create double
bonds during fatty acid synthesis by dehydrating hydroxy fatty
acids. Oxygen is not required for double bond synthesis by this
pathway. The anaerobic pathway is present in a number of common gram-negative bacteria (e.g., *Escherichia coli* and *Salmonella typhimurium*), gram-positive bacteria (e.g., *Lactobacillus
plantarum* and *Clostridium pasteurianum*), and cyanobacteria.

Eucaryotic microorganisms frequently store carbon and energy as **triacylglycerol,** glycerol esterified to three fatty acids.
Glycerol arises from the reduction of the glycolytic intermediate dihydroxyacetone phosphate to glycerol 3-phosphate, which
is then esterified with two fatty acids to give **phosphatidic
acid** (figure 10.25). Phosphate is hydrolyzed from phosphatidic
acid giving a diacylglycerol, and the third fatty acid is attached
to yield a triacylglycerol.

Phospholipids are major components of eucaryotic and most
procaryotic cell membranes. Their synthesis also usually proceeds by way of phosphatidic acid. A special cytidine diphosphate (CDP) carrier plays a role similar to that of uridine and
adenosine diphosphate carriers in carbohydrate biosynthesis. For
example, bacteria synthesize phosphatidylethanolamine, a major
cell membrane component, through the initial formation of CDP-
diacylglycerol (figure 10.25). This CDP derivative then reacts
with serine to form the phospholipid phosphatidylserine, and decarboxylation yields phosphatidylethanolamine. In this way a
complex membrane lipid is constructed from the products of glycolysis, fatty acid biosynthesis, and amino acid biosynthesis.

1. What is a fatty acid? Describe in general terms how the fatty acid
   synthetase manufactures a fatty acid.
2. How are unsaturated fatty acids made?
3. Briefly describe the pathways for triacylglycerol and phospholipid
   synthesis. Of what importance are phosphatidic acid and CDP-
   diacylglycerol?

Peptidoglycan Synthesis

As discussed earlier, most bacterial cell walls contain a large,
complex peptidoglycan molecule consisting of long polysaccharide chains made of alternating N-acetylmuramic acid
(NAM) and N-acetylglucosamine (NAG) residues. Pentapeptide chains are attached to the NAM groups. The polysaccharide
chains are connected through their pentapeptides or by interbridges (*see figures 3.21 and 3.22*).

Peptidoglycan structure and function (pp. 52–4).

Figure 10.25   Triacylglycerol and Phospholipid Synthesis.

Figure 10.26   Bactoprenol Pyrophosphate. Bactoprenol pyrophosphate connected to N-acetylmuramic acid (NAM).

Not surprisingly such an intricate structure requires an equally intricate biosynthetic process, especially because the synthetic reactions occur both inside and outside the cell membrane. Peptidoglycan synthesis is a multistep process that has been best studied in the gram-positive bacterium *Staphylococcus aureus*. Two carriers participate: uridine diphosphate (UDP) and **bactoprenol** (figure 10.26). Bactoprenol is a 55-carbon alcohol that attaches to NAM by a pyrophosphate group and moves peptidoglycan components through the hydrophobic membrane.

The synthesis of peptidoglycan, outlined in figures 10.27 and 10.28, occurs in eight stages.

1. UDP derivatives of N-acetylmuramic acid and N-acetylglucosamine are synthesized in the cytoplasm.

2. Amino acids are sequentially added to UDP-NAM to form the pentapeptide chain (the two terminal D-alanines are added as a dipeptide). ATP energy is used to make the peptide bonds, but tRNA and ribosomes are not involved.

3. The NAM-pentapeptide is transferred from UDP to a bactoprenol phosphate at the membrane surface.

4. UDP-NAG adds NAG to the NAM-pentapeptide to form the peptidoglycan repeat unit. If a pentaglycine interbridge is required, the glycines are added using special glycyl-tRNA molecules, not ribosomes.

5. The completed NAM-NAG peptidoglycan repeat unit is transported across the membrane to its outer surface by the bactoprenol pyrophosphate carrier.

Figure 10.27 Peptidoglycan Synthesis. NAM is N-acetylmuramic acid and NAG is N-acetylglucosamine. The pentapeptide contains L-lysine in *S. aureus* peptidoglycan, and diaminopimelic acid (DAP) in *E. coli.*

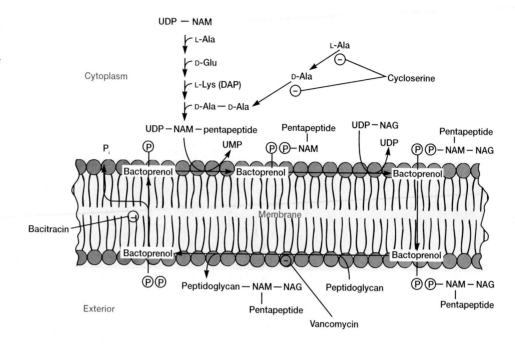

Figure 10.28 Transpeptidation. The transpeptidation reactions in the formation of the peptidoglycans of *Escherichia coli* and *Staphylococcus aureus.*

6. The peptidoglycan unit is attached to the growing end of a peptidoglycan chain to lengthen it by one repeat unit.

7. The bactoprenol carrier returns to the inside of the membrane. A phosphate is released during this process to give bactoprenol phosphate, which can now accept another NAM-pentapeptide.

8. Finally, peptide cross-links between the peptidoglycan chains are formed by **transpeptidation** (figure 10.28). In *E. coli* the free amino group of diaminopimelic acid attacks

the subterminal D-alanine, releasing the terminal D-alanine residue. ATP is used to form the terminal peptide bond inside the membrane. No more ATP energy is required when transpeptidation takes place on the outside. The same process occurs when an interbridge is involved; only the group reacting with the subterminal D-alanine differs.

Peptidoglycan synthesis is particularly vulnerable to disruption by antimicrobial agents. Inhibition of any stage of synthesis weakens the cell wall and can lead to osmotic lysis. Many an-

Figure 10.29   Wall Synthesis Patterns. Patterns of new cell wall synthesis in growing and dividing bacteria. (*a*) Streptococci and some other gram-positive cocci. (*b*) Synthesis in rod-shaped bacteria (*Escherichia coli, Salmonella, Bacillus*). The zones of growth are in blue-green. The actual situation is more complex than indicated because cells can begin to divide again before the first division is completed.

tibiotics interfere with peptidoglycan synthesis. For example, penicillin inhibits the transpeptidation reaction (figure 10.28), and bacitracin blocks the dephosphorylation of bactoprenol pyrophosphate (figure 10.27).

Antibiotic effects on cell wall synthesis (pp. 666–68).

Patterns of Cell Wall Formation

To grow and divide efficiently, a bacterial cell must add new peptidoglycan to its cell wall in a precise and well-regulated way while maintaining wall shape and integrity in the presence of high osmotic pressure. Because the cell wall peptidoglycan is essentially a single enormous network, the growing bacterium must be able to degrade it just enough to provide acceptor ends for the incorporation of new peptidoglycan units. It must also reorganize peptidoglycan structure when necessary. This limited peptidoglycan digestion is accomplished by enzymes known as **autolysins,** some of which attack the polysaccharide chains, while others hydrolyze the peptide cross-links. Autolysin inhibitors keep the activity of these enzymes under tight control.

Control of cell division (pp. 250–51).

Although the location and distribution of cell wall synthetic activity varies with species, there seem to be two general patterns (figure 10.29). Many gram-positive cocci (e.g., *Enterococcus faecalis* and *Streptococcus pyogenes*) have only one to a few zones of growth. The principal growth zone is usually at the site of septum formation, and new cell halves are synthesized back-to-back. The second pattern of synthesis occurs in the rod-shaped bacteria *Escherichia coli, Salmonella,* and *Bacillus.* Active peptidoglycan synthesis occurs at the site of septum formation just as before, but growth sites also are scattered along the cylindrical portion of the rod. Thus growth is distributed more diffusely in rod-shaped bacteria than in the streptococci. Synthesis must lengthen rod-shaped cells as well as divide them. Presumably this accounts for the differences in wall growth pattern.

1. Outline in a diagram the steps involved in the synthesis of peptidoglycan and show their relationship to the plasma membrane. What are the roles of bactoprenol and UDP?

2. What is the function of autolysins in cell wall peptidoglycan synthesis? Describe the patterns of peptidoglycan synthesis seen in gram-positive cocci and in rod-shaped bacteria such as *E. coli.*

Summary

1. In biosynthesis or anabolism, cells use energy to construct complex molecules from smaller, simpler precursors.

2. Many important cell constituents are macromolecules, large polymers constructed of simple monomers.

3. Although many catabolic and anabolic pathways share enzymes for the sake of efficiency, some of their enzymes are separate and independently regulated.

4. Macromolecular components often undergo self-assembly to form the final molecule or complex.

5. Photosynthetic CO_2 fixation is carried out by the Calvin cycle and may be divided into three phases: the carboxylation phase, the reduction

phase, and the regeneration phase. Three ATPs and two NADPHs are used during the incorporation of one CO_2.

6. Gluconeogenesis is the synthesis of glucose and related sugars from nonglucose precursors.

7. Glucose, fructose, and mannose are gluconeogenic intermediates or made directly from them; galactose is synthesized with nucleoside diphosphate derivatives. Bacteria and algae synthesize glycogen and starch from adenosine diphosphate glucose.

8. Phosphorus is obtained from inorganic or organic phosphate.

9. Microorganisms can use cysteine, methionine, and inorganic sulfate as sulfur sources. Sulfate is reduced to sulfide during assimilatory sulfate reduction.

10. Ammonia nitrogen can be directly assimilated by the activity of transaminases and either glutamate dehydrogenase or the glutamine synthetase–glutamate synthase system.

11. Nitrate is incorporated through assimilatory nitrate reduction catalyzed by the enzymes nitrate reductase and nitrite reductase.

12. Nitrogen fixation is catalyzed by the nitrogenase complex. Atmospheric molecular nitrogen is reduced to ammonia, which is then incorporated into amino acids.

13. Amino acid biosynthetic pathways branch off from the central amphibolic pathways.

14. Anaplerotic reactions replace TCA cycle intermediates to keep the cycle in balance while it supplies biosynthetic precursors. Many anaplerotic enzymes catalyze CO_2

fixation reactions. The glyoxylate cycle is also anaplerotic.

15. Purines and pyrimidines are nitrogenous bases found in DNA, RNA, and other molecules. The purine skeleton is synthesized beginning with ribose 5-phosphate and initially produces inosinic acid. Pyrimidine biosynthesis starts with carbamoyl phosphate and aspartate, and ribose is added after the skeleton has been constructed.

16. Fatty acids are synthesized from acetyl-CoA, malonyl-CoA, and NADPH by the fatty acid synthetase system. During synthesis the intermediates are attached to the acyl carrier protein. Double bonds can be added in two different ways.

17. Triacylglycerols are made from fatty acids and glycerol phosphate. Phosphatidic acid is an important intermediate in this pathway.

18. Phospholipids like phosphatidylethanolamine can be synthesized from phosphatidic acid by forming CDP-diacylglycerol, then adding an amino acid.

19. Peptidoglycan synthesis is a complex process involving both UDP derivatives and the lipid carrier bactoprenol, which transports NAM-NAG-pentapeptide units across the cell membrane. Cross-links are formed by transpeptidation.

20. Peptidoglycan synthesis occurs in discrete zones in the cell wall. Existing peptidoglycan is selectively degraded by autolysins so new material can be added.

Key Terms

acyl carrier protein (ACP) *205*
adenine *204*
anaplerotic reactions *202*
assimilatory nitrate reduction *198*
assimilatory sulfate reduction *197*
autolysins *209*
bactoprenol *207*
Calvin cycle *194*
carboxysomes *194*
CO_2 fixation *202*
cytosine *204*
dissimilatory sulfate reduction *197*
fatty acid *205*
fatty acid synthetase *205*

gluconeogenesis *196*
glutamate dehydrogenase *198*
glutamate synthase *198*
glutamine synthetase *198*
glyoxylate cycle *203*
guanine *204*
macromolecule *192*
monomers *192*
nitrate reductase *199*
nitrite reductase *199*
nitrogenase *199*
nitrogen fixation *199*
nucleoside *204*
nucleotide *204*

phosphatase *197*
phosphatidic acid *206*
phosphoadenosine 5′-phosphosulfate *197*
purine *204*
pyrimidine *204*
ribulose-1,5-bisphosphate carboxylase *194*
self-assembly *193*
thymine *204*
transaminases *198*
transpeptidation *208*
triacylglycerol *206*
turnover *192*
uracil *204*
uridine diphosphate glucose (UDPG) *196*

Questions for Thought and Review

1. Discuss the relationship between catabolism and anabolism. How does anabolism depend on catabolism?

2. Suppose that a microorganism was growing on a medium that contained amino acids but no sugars. In general terms how would it synthesize the pentoses and hexoses it required?

3. Activated carriers participate in carbohydrate, lipid, and peptidoglycan synthesis. Briefly describe these carriers and their roles.

4. Which two enzymes discussed in the chapter appear to be specific to the Calvin cycle?

5. Why can phosphorus be directly incorporated into cell constituents whereas sulfur and nitrogen often cannot?

6. What is unusual about the synthesis of peptides that takes place during peptidoglycan construction?

Additional Reading

General

Caldwell, D. R. 1995. *Microbial physiology and metabolism.* Dubuque, Iowa: Wm. C. Brown Communications, Inc.

Dawes, I. W., and Sutherland, I. W. 1992. *Microbial physiology,* 2d ed. Boston, Mass.: Blackwell Scientific Publications.

Garrett, R. H., and Grisham, C. M. 1995. *Biochemistry.* New York: Saunders.

Gottschalk, G. 1986. *Bacterial metabolism,* 2d ed. New York: Springer-Verlag.

Lehninger, A. L.; Nelson, D. L.; and Cox, M. M. 1993. *Principles of biochemistry,* 2d ed. New York: Worth Publishers.

Mandelstam, J.; McQuillen, K.; and Dawes, I. 1982. *Biochemistry of bacterial growth,* 3d ed. London: Blackwell Scientific Publications.

Mathews, C. K., and van Holde, K. E. 1990. *Biochemistry.* Redwood City, Calif.: Benjamin/Cummings.

Moat, A. G., and Foster, J. W. 1995. *Microbial physiology,* 3d ed. New York: John Wiley and Sons.

Neidhardt, F. C.; Ingraham, J. L.; and Schaechter, M. 1990. *Physiology of the bacterial cell: A molecular approach.* Sunderland, Mass.: Sinauer Associates.

Voet, D., and Voet, J. G. 1995. *Biochemistry,* 2d ed. New York: John Wiley and Sons.

White, D. 1995. *The physiology and biochemistry of procaryotes.* New York: Oxford University Press.

Zubay, G. 1993. *Biochemistry,* 3d ed. Dubuque, Iowa: Wm. C. Brown Communications, Inc.

The Photosynthetic Fixation of CO_2

Schlegel, H. G., and Bowien, B., editors. 1989. *Autotrophic bacteria.* Madison, Wis.: Science Tech Publishers.

The Assimilation of Inorganic Phosphorus, Sulfur, and Nitrogen

Brill, W. J. 1977. Biological nitrogen fixation. *Sci. Am.* 236(3):68–81.

Dean, D. R.; Bolin, J. T.; and Zheng, L. 1993. Nitrogenase metalloclusters: Structures, organization, and synthesis. *J. Bacteriol.* 175(21):6737–44.

Dilworth, M., and Glenn, A. R. 1984. How does a legume nodule work? *Trends Biochem. Sci.* 9(12):519–23.

Glenn, A. R., and Dilworth, M. J. 1985. Ammonia movements in rhizobia *Microbiol. Sci.* 2(6):161–67.

Howard, J. B., and Rees, D. C. 1994. Nitrogenase: A nucleotide-dependent molecular switch. *Ann. Rev. Biochem.* 63:235–64.

Luden, P. W. 1991. Energetics of and sources of energy for biological nitrogen fixation. In *Current topics in bioenergetics,* vol. 16, 369–90. San Diego: Academic Press.

Mora, J. 1990. Glutamine metabolism and cycling in *Neurospora crassa. Microbiol. Rev.* 54(3):293–304.

Postgate, J. R. 1982. *The fundamentals of nitrogen fixation.* New York: Cambridge University Press.

Patterns of Cell Wall Formation

Doyle, R. J.; Chaloupka, J.; and Vinter, V. 1988. Turnover of cell walls in microorganisms. *Microbiol. Rev.* 52(4):554–67.

Harold, F. M. 1990. To shape a cell: An inquiry into the causes of morphogenesis of microorganisms. *Microbiol. Rev.* 54(4):381–431.

Koch, A. L. 1990. Growth and form of the bacterial cell wall. *Am. Sci.* 78:327–41.

Photo courtesy of Dr. E. Moudrianakis

The particular field which excites my interest is the division between the living and the non-living, as typified by, say, proteins, viruses, bacteria and the structure of chromosomes. The eventual goal, which is somewhat remote, is the description of these activities in terms of their structure, i.e., the spatial distribution of their constituent atoms, in so far as this may prove possible. This might be called the chemical physics of biology.

—Francis Crick

CHAPTER 11
Metabolism: The Synthesis of Nucleic Acids and Proteins

Concepts

1. The two kinds of nucleic acid, deoxyribonucleic acid (DNA) and ribonucleic acid (RNA), differ from one another in chemical composition and structure. In procaryotic and eucaryotic cells, DNA serves as the repository for genetic information.

2. DNA is associated with basic proteins in the cell. In eucaryotes these are special histone proteins, whereas in procaryotes nonhistone proteins are complexed with DNA.

3. The flow of genetic information usually proceeds from DNA through RNA to protein. A protein's amino acid sequence reflects the nucleotide sequence of its mRNA. This messenger is a complementary copy of a portion of the DNA genome.

4. DNA replication is a very complex process involving a variety of proteins and a number of steps. It is designed to operate rapidly while minimizing errors and correcting those that arise when the DNA sequence is copied.

5. In transcription the RNA polymerase copies the appropriate sequence on the DNA template sense strand to produce a complementary RNA copy of the gene. Transcription differs in a number of ways between procaryotes and eucaryotes, even though the basic mechanism of RNA polymerase action is essentially the same.

6. Translation is the process by which the nucleotide sequence of mRNA is converted into the amino acid sequence of a polypeptide through the action of ribosomes, tRNAs, aminoacyl-tRNA synthetases, ATP and GTP energy, and a variety of protein factors. As in the case of DNA replication, this complex process is designed to minimize errors.

C hapter 10 focuses on the three least complex levels of anabolism, depicted in figure 10.1. The routes for the incorporation of simple precursors, such as CO_2, ammonia, and phosphate, into organic material are introduced, and the pathways responsible for synthesis of the major biological monomers (sugars, amino acids, fatty acids, purines, and pyrimidines) are surveyed. Chapter 10 also discusses polysaccharide, lipid, and peptidoglycan synthesis.

Chapter 11 is concerned with the synthesis of the other two major classes of macromolecules: nucleic acids and proteins. These biosynthetic processes are presented in considerably more detail than polysaccharide biosynthesis, because adequate knowledge of nucleic acids and proteins is essential to understanding molecular biology and microbial genetics. The chapter begins with a general introduction to nucleic acid structure to provide the background necessary for discussion of nucleic acid synthesis and function. This is followed by sections on DNA replication, RNA synthesis, and protein synthesis.

Because deoxyribonucleic acid carries the cell's genetic information and ribonucleic acid directs the synthesis of proteins, the structure and synthesis of nucleic acids are considered before protein synthesis.

Nucleic Acid Structure

The structure and synthesis of purine and pyrimidine nucleotides are introduced in chapter 10. These nucleotides can be combined to form nucleic acids of two kinds (figure 11.1a). **Deoxyribonucleic acid (DNA)** contains the 2′-deoxyribonucleosides (figure 11.1b) of adenine, guanine, cytosine, and thymine. **Ribonucleic acid (RNA)** is composed of the ribonucleosides of adenine, guanine, cytosine, and uracil (instead of thymine). In both DNA and RNA, nucleosides are joined by phosphate groups to form long polynucleotide chains (figure 11.1c). The differences in chemical composition between the chains reside in their sugar and pyrimidine bases: DNA has deoxyribose and thymine; RNA has ribose and uracil in place of thymine.

DNA Structure

Deoxyribonucleic acids are very large molecules, usually composed of two polynucleotide chains coiled together to form a double helix 2.0 nm in diameter (figure 11.2). Each chain contains purine and pyrimidine deoxyribonucleosides joined by phosphodiester bridges (figure 11.1c). That is, two adjacent deoxyribose sugars are connected by a phosphoric acid molecule esterified to a 3′-hydroxyl of one sugar and a 5′-hydroxyl of the other. Purine and pyrimidine bases are attached to the 1′-carbon of the deoxyribose sugars and extend toward the middle of the cylinder formed by the two chains. They are stacked on top of each other in the center, one base pair every 0.34 nm. The

Figure 11.1   The Composition of Nucleic Acids. (a) A diagram showing the relationship of various nucleic acid components. Combination of a purine or pyrimidine base with ribose or deoxyribose gives a nucleoside (a ribonucleoside or deoxyribonucleoside). A nucleotide contains a nucleoside and one or more phosphoric acid molecules. Nucleic acids result when nucleotides are connected together in polynucleotide chains. (b) Examples of nucleosides—the purine nucleoside adenosine and the pyrimidine deoxynucleoside 2′-deoxycytidine. The carbons of nucleoside sugars are indicated by numbers with primes. (c) A segment of a polynucleotide chain showing two nucleosides, deoxyguanosine and thymidine, connected by a phosphodiester linkage between the 3′ and 5′-carbons of adjacent deoxyribose sugars.

Figure 11.2   The Structure of the DNA Double Helix. (*a*) A space-filling model of the B form of DNA with the base pairs, major groove, and minor groove shown. The backbone phosphate groups, shown in color, spiral around the outside of the helix. (*b*) A diagrammatic representation of the double helix. The backbone consists of deoxyribose sugars (S) joined by phosphates (P) in phosphodiester bridges. The arrows at the top and bottom of the chains point in the 5′ to 3′ direction. The ribbons represent the sugar phosphate backbones. (*c*) An end view of the double helix showing the outer backbone and the bases stacked in the center of the cylinder. In the top drawing the ribose ring oxygens are red. The nearest base pair, an AT base pair, is highlighted in white.

purine adenine (A) is always paired with the pyrimidine thymine (T) by two hydrogen bonds. The purine guanine (G) pairs with cytosine (C) by three hydrogen bonds (figure 11.3). This AT and GC base pairing means that the two strands in a DNA double helix are **complementary.** That is, the bases in one strand match up with those of the other according to the base pairing rules. Because the sequences of bases in these strands encode genetic information (*see chapter 13*), considerable effort has been devoted to determining the base sequences of DNA and RNA from many microorganisms.

Nucleic acid sequence comparison and microbial taxonomy (chapter 19).

The two polynucleotide strands fit together much like the pieces in a jigsaw puzzle because of complementary base pairing (Box 11.1). Inspection of figure 11.2*a,b*, depicting the B form of DNA (probably the most common form in cells), shows that the two strands are not positioned directly opposite one another in the helical cylinder. Therefore, when the strands twist about one another, a wide **major groove** and narrower **minor groove** are formed by the backbone. Each base pair rotates 36° around the cylinder with respect to adjacent pairs so that there are 10 base pairs per turn of the helical spiral. Each turn of the helix has a vertical length of 3.4 nm. The helix is right-handed—that is, the chains turn counterclockwise as they approach a viewer looking down the longitudinal axis. The two backbones are antiparallel or run in opposite directions with respect to the orientation of their sugars. One end of each strand has an exposed 5′-hydroxyl group, often with phosphates attached, whereas the other end has a free 3′-hydroxyl group. If the end of a double helix is examined, the 5′ end of one strand and the 3′ end of the other are visible. In a given direction one strand is oriented 5′ to 3′ and the other, 3′ to 5′ (figure 11.2*b*).

RNA Structure

Besides differing chemically from DNA, ribonucleic acid is usually single stranded rather than double stranded like most DNA. An RNA strand can coil back on itself to form a hairpin-shaped structure with complementary base pairing and helical organization. Cells contain three different types of RNA—messenger RNA, ribosomal RNA, and transfer RNA—that differ from one another in function, site of synthesis in eucaryotic cells, and structure.

The Organization of DNA in Cells

Although DNA exists as a double helix in both procaryotic and eucaryotic cells, its organization differs in the two cell types (*see table 4.2*). DNA is organized in the form of a closed circle in almost all procaryotes (the chromosome of *Borrelia* is a linear DNA molecule). This circular double helix is further

Figure 11.3   DNA Base Pairs. DNA complementary base pairing showing the hydrogen bonds (. . .).

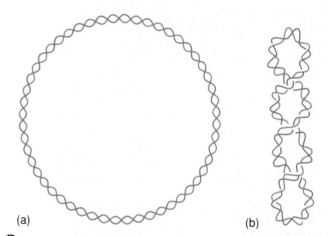

(a)   (b)

Figure 11.4   DNA Forms. (*a*) The DNA double helix of almost all bacteria is in the shape of a closed circle. (*b*) The circular DNA strands, already coiled in a double helix, are twisted a second time to produce supercoils.

twisted into supercoiled DNA (figure 11.4) and is associated with basic proteins but not with the histones found complexed with almost all eucaryotic DNA. These histonelike proteins do appear to help organize bacterial DNA into a coiled chromatin-like structure.

The structure of the bacterial nucleoid (pp. 49–51).

Box 11.1

The Elucidation of DNA Structure

The basic chemical composition of nucleic acids was elucidated in the 1920s through the efforts of P. A. Levene. Despite his major contributions to nucleic acid chemistry, Levene mistakenly believed that DNA was a very small molecule, probably only four nucleotides long, composed of equal amounts of the four different nucleotides arranged in a fixed sequence. Partly because of his influence, biologists believed for many years that nucleic acids were too simple in structure to carry complex genetic information. They concluded that genetic information must be encoded in proteins because proteins were large molecules with complex amino sequences that could vary among different proteins.

As so often happens, further advances in our understanding of DNA structure awaited the development of significant new analytical techniques in chemistry. One development was the invention of paper chromatography by Archer Martin and Richard Synge between 1941 and 1944. By 1948 the chemist Erwin Chargaff had begun using paper chromatography to analyze the base composition of DNA from a number of species. He soon found that the base composition of DNA from genetic material did indeed vary among species just as he expected. Furthermore, the total amount of purines always equaled the total amount of pyrimidines; and the adenine/thymine and guanine/cytosine ratios were always 1. These findings, known as Chargaff's rules, were a key to the understanding of DNA structure.

Another turning point in research on DNA structure was reached in 1951 when Rosalind Franklin arrived at King's College, London, and joined Maurice Wilkins in his efforts to prepare highly oriented DNA fibers and study them by X-ray crystallography. By the winter of 1952–1953, Franklin had obtained an excellent X-ray diffraction photograph of DNA.

The same year that Franklin began work at King's College, the American biologist James Watson went to Cambridge University and met Francis Crick. Although Crick was a physicist, he was very interested in the structure and function of DNA, and the two soon began to work on its structure. Their attempts were unsuccessful until Franklin's data provided them with the necessary clues. Her photograph of fibrous DNA contained a crossing pattern of dark spots, which showed that the molecule was helical. The dark regions at the top and bottom of the photograph showed that the purine and pyrimidine bases were stacked on top of each other and separated by 0.34 nm. Franklin had already concluded that the phosphate groups lay to the outside of the cylinder. Finally, the X-ray data and her determination of the density of DNA indicated that the helix contained two strands, not three or more as some had proposed.

Without actually doing any experiments themselves, Watson and Crick constructed their model by combining Chargaff's rules on base composition with Franklin's X-ray data and their predictions about how genetic material should behave. By building models, they found that a smooth, two-stranded helix of constant diameter could be constructed only when an adenine hydrogen bonded with thymine and when a guanine bonded with cytosine in the center of the helix. They immediately realized that the double helical structure provided a mechanism by which genetic material might be replicated. The two parental strands could unwind and direct the synthesis of complementary strands, thus forming two new identical DNA molecules (figure 11.7). Watson, Crick, and Wilkins received the Nobel Prize in 1962 for their discoveries.

DNA seems much more highly organized in eucaryotic chromatin (*see chapter 4*) and is associated with a variety of proteins, the most prominent of which are **histones.** These are small, basic proteins rich in the amino acids lysine and/or arginine. There are five types of histones in almost all eucaryotic cells studied: H1, H2A, H2B, H3, and H4. Eight histone molecules (two each of H2A, H2B, H3, and H4) form an ellipsoid about 11 nm long and 6.5 to 7 nm in diameter (figure 11.5a). DNA coils around the surface of the ellipsoid approximately $1\frac{3}{4}$ turns or 166 base pairs before proceeding on to the next. This complex of histones plus DNA is called a **nucleosome** (figure 11.5b). Thus DNA gently isolated from chromatin looks like a string of beads. The stretch of DNA between the beads or nucleosomes, the linker region, varies in length from 14 to over 100 base pairs. Histone H1 appears to associate with the linker regions to aid the folding of DNA into more complex chromatin structures (figure 11.5c). When folding reaches a maximum, the chromatin takes the shape of the visible chromosomes seen in eucaryotic cells during mitosis and meiosis (*see figure 4.20*).

---

1. What are nucleic acids? How do DNA and RNA differ in structure?

2. Describe in some detail the structure of the DNA double helix. What does it mean to say that the two strands are complementary and antiparallel?

3. What are histones and nucleosomes? Describe the way in which DNA is organized in the chromosomes of procaryotes and eucaryotes.

Figure 11.5 **Nucleosome Internal Organization and Function.** (*a*) A model of the histone octamer of the nucleosome. It is a tripartite structure composed of two H2A-H2B dimers (blue) and an (H3-H4)$_2$ tetramer (white). (*b*) The histone octamer is shaped somewhat like a flat disk or wedge, and a left-handed helical path circles it. The 146 base pairs of DNA lie in this helical path and the histone-DNA assembly is known as the nucleosome core particle. In this image the DNA is pictured partly as a gray tube coiling about the octamer and partly as a double helix. The histones are blue and white, and the red and orange dots are positive charges on the histones that interact with negative charges on the DNA backbone. (*c*) An illustration of how a string of nucleosomes, each associated with a histone H1, might be organized to form a highly supercoiled chromatin fiber. The nucleosomes are drawn as cylinders. (*a*) *and* (*b*) *courtesy of Dr. E. Moudrianakis.*

DNA Replication

It might be best to review DNA function in a general way before turning to the details of DNA replication. Biologists have long recognized a relationship between DNA, RNA, and protein (figure 11.6), and this recognition has guided a vast amount of research over the past decades. DNA is precisely copied during its synthesis or **replication.** The expression of the information encoded in the base sequence of DNA begins with the synthesis of an RNA copy of the DNA sequence making up a gene (*see chapter 13*). This is called **transcription** because the DNA base sequence is being copied into an RNA base sequence. The RNA that carries information from DNA and directs protein synthesis is **messenger RNA (mRNA).** The last phase of gene expression is **translation** or protein synthesis. The genetic information in the form of an mRNA nucleotide sequence is translated and governs the synthesis of protein. Thus the amino acid sequence

Replication

DNA

Transcription

RNA

Translation

Protein

FIgure 11.6   Relationships between DNA, RNA, and Protein Synthesis.

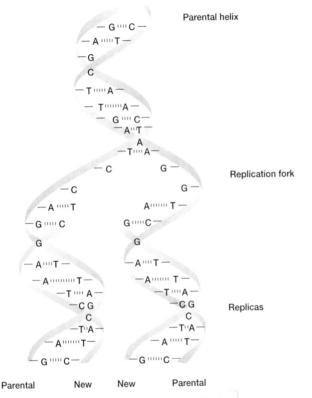

Figure 11.7   Semiconservative DNA Replication. The replication fork of DNA showing the synthesis of two progeny strands. Newly synthesized strands are in maroon. Each copy contains one new and one old strand. This process is called semiconservative replication.

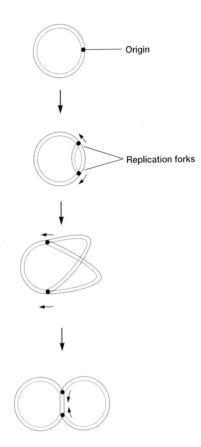

Figure 11.8   Bidirectional Replication. The replication of a circular bacterial genome. Two replication forks move around the DNA forming theta-shaped intermediates. Newly replicated DNA double helix is in red.

of a protein is a direct reflection of the base sequence in mRNA. In turn the mRNA nucleotide sequence is a complementary copy of a portion of the DNA genome.

Pattern of DNA Synthesis

Watson and Crick published their description of DNA structure in April 1953. Almost exactly one month later, a second paper appeared in which they suggested how DNA might be replicated. They hypothesized that the two strands of the double helix unwind from one another and separate (figure 11.7). Free nucleotides now line up along the two parental strands through

complementary base pairing—A with T, G with C (figure 11.3). When these nucleotides are linked together by one or more enzymes, two replicas result, each containing a parental DNA strand and a newly formed strand. The replicating DNA molecule has a Y shape, just as in figure 11.7, and the actual replication process takes place at the **replication fork.** Research in subsequent years has proved Watson and Crick's hypothesis correct.

Replication patterns are somewhat different in procaryotes and eucaryotes. For example, when the circular DNA chromosome of *E. coli* is copied, replication begins at a single point, the origin. Two replication forks move outward from the origin until they have copied the whole **replicon,** that portion of the genome that contains an origin and is replicated as a unit. When the replication forks move around the circle, a structure shaped like the Greek letter theta (θ) is formed (figure 11.8). Finally, since the bacterial chromosome is a single replicon, the forks meet on the other side and two separate chromosomes are released. There is evidence that the replication fork and associated enzymes are attached to the bacterial plasma membrane. If this is the case, the DNA moves through the replication apparatus (which is fixed to the membrane) rather than the replication fork moving around the DNA.

A different pattern of DNA replication occurs during *E. coli* conjugation (*see chapter 14*) and the reproduction of bacteriophages, such as phage lambda (*see chapter 17*). In the **rolling-circle mechanism** (figure 11.9), one strand is nicked and the free 3′-hydroxyl end is extended by replication enzymes. As the 3′ end is lengthened while the growing point rolls around the circular template, the 5′ end of the strand is displaced and forms an ever-lengthening tail. The single-stranded tail may be converted to the double-stranded form by complementary strand synthesis. This mechanism is particularly useful to viruses such as φX174 (p. 360) because it allows the rapid, continuous production of many genome copies from a single initiation event.

Eucaryotic DNA is linear and much longer than procaryotic DNA; *E. coli* DNA is about 1,400 μm in length, whereas the 46

chromosomes in the human nucleus have a total length of 1.8 m (almost 1,400 times longer). Clearly many replication forks must copy eucaryotic DNA simultaneously so that the molecule can be duplicated in a relatively short period, and so many replicons are present that there is an origin about every 10 to 100 μm along the DNA. Replication forks move outward from these sites and eventually meet forks that have been copying the adjacent DNA stretch (figure 11.10). In this fashion a large molecule is copied quickly.

Mechanism of DNA Replication

Because DNA replication is so essential to organisms, a great deal of effort has been devoted to understanding its mechanism. The replication of *E. coli* DNA is probably best understood and is the focus of attention in this section. The process in eucaryotic cells is thought to be similar.

E. coli has three different **DNA polymerase** enzymes, each of which catalyzes the synthesis of DNA in the 5′ to 3′ direction while reading the DNA template in the 3′ to 5′ direction (figures 11.11 and 11.12). The polymerases require deoxyribonucleoside triphosphates (dATP, dGTP, dCTP, and dTTP) as substrates and a DNA template to copy. Nucleotides are added to the 3′ end of the growing chain when the free 3′-hydroxyl group on the deoxyribose attacks the first or alpha phosphate group of the substrate to release pyrophosphate (figure 11.11). DNA polymerase III plays the major role in replication, although it is probably assisted by polymerase I. It is thought that polymerases I and II participate in the repair of damaged DNA (*see chapter 13*).

During replication the DNA double helix must be unwound to generate separate single strands. Unwinding occurs very quickly; the fork may rotate as rapidly as 75 to 100 revolutions per second. **Helicases** are responsible for DNA unwinding. These enzymes use energy from ATP to unwind short stretches of helix just ahead of the replication fork. Once the strands have separated, they are kept single through specific binding with **single-stranded DNA binding proteins (SSBs)** as shown in figure 11.12*a*. Rapid unwinding can lead to tension and formation of supercoils or supertwists in the helix, just as rapid separation of two strands of a rope can lead to knotting or coiling of the rope. The tension generated by unwinding is relieved, and the unwinding process is promoted by enzymes known as

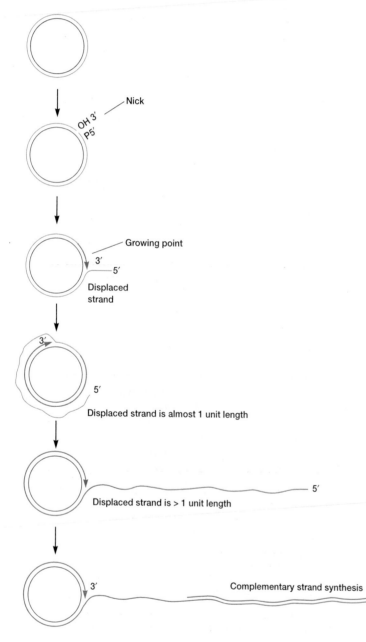

Figure 11.9 The Rolling-Circle Pattern of Replication. A single-stranded tail, often composed of more than one genome copy, is generated and can be converted to the double-stranded form by synthesis of a complementary strand. The "free end" of the rolling-circle strand is probably bound to the primosome.

topoisomerases. These enzymes change the structure of DNA by transiently breaking one or two strands in such a way that it remains unaltered as its shape is changed (e.g., a topoisomerase might tie or untie a knot in a DNA strand). **DNA gyrase** is an *E. coli* topoisomerase that removes the supertwists produced during replication (*see figure 33.6*).

10–100 μm

Replication
forks

Figure 11.10   The Replication of Eucaryotic DNA. Replication is initiated every 10 to 100 μm and the replication forks travel away from the origin. Newly copied DNA is in red.

DNA polymerase reaction

$$n[\text{dATP, dGTP, dCTP, dTTP}] \xrightarrow[\text{DNA template}]{\text{DNA polymerase}} \text{DNA} + n\text{PP}_i$$

The mechanism of chain growth

Figure 11.11   The DNA Polymerase Reaction and Its Mechanism. The mechanism involves a nucleophilic attack by the hydroxyl of the 3′ terminal deoxyribose on the alpha phosphate group of the nucleotide substrate (in this example, cytidine triphosphate).

The details of DNA replication are outlined in a diagram of the replication fork (figure 11.12*b*). The replication process takes place in four stages.

1. Helicases unwind the helix with the aid of topoisomerases like the DNA gyrase. It appears that the DnaB protein is the helicase most actively involved in replication, but the n′ protein also may participate in unwinding. The single strands are kept separate by the DNA binding proteins (SSBs).

2. DNA is probably replicated continuously by DNA polymerase III when the leading strand is copied. Lagging strand replication is discontinuous, and the fragments are synthesized in the 5′ to 3′ direction just as in leading strand synthesis. First, a special RNA polymerase called a **primase** synthesizes a short RNA primer, usually around 10 nucleotides long, complementary to the DNA. This is necessary because DNA polymerases I and III cannot synthesize DNA without a primer. It appears that the primase requires the assistance of several other proteins, and the complex of the primase with its accessory proteins is called the **primosome.** DNA polymerase III holoenzyme then synthesizes complementary DNA beginning at the 3′ end of the RNA primer. Both leading and lagging strand synthesis probably occur on a single multiprotein complex, the replisome. If this is the case, the lagging strand template must be looped around the complex (figure 11.12*b*). The final fragments are around 1,000 to 2,000 nucleotides long in bacteria and approximately 100 nucleotides long in eucaryotic cells. They are called **Okazaki fragments** after their discoverer, Reiji Okazaki.

3. After most of the lagging strand has been duplicated by the formation of Okazaki fragments, DNA polymerase I or RNase H removes the RNA primer. Polymerase I synthesizes complementary DNA to fill the gap resulting from RNA deletion (figure 11.13). The polymerase appears to remove the primer one nucleotide at a time and replace it with the appropriate complementary deoxyribonucleotide. Polymerase III holoenzyme also may be able to fill in the gap.

4. Finally, the fragments are joined by the enzyme **DNA ligase,** which forms a phosphodiester bond between the 3′-hydroxyl of the growing strand and the 5′-phosphate of an Okazaki fragment (figure 11.14). Bacterial ligases use the pyrophosphate bond of NAD$^+$ as an energy source; many other ligases employ ATP instead.

DNA polymerase III holoenzyme, the enzyme complex that synthesizes most of the DNA copy, is a very large entity containing DNA polymerase III and several other proteins. The γδ complex and β subunits bind the holoenzyme to the DNA template and primer. The α subunit can synthesize DNA.

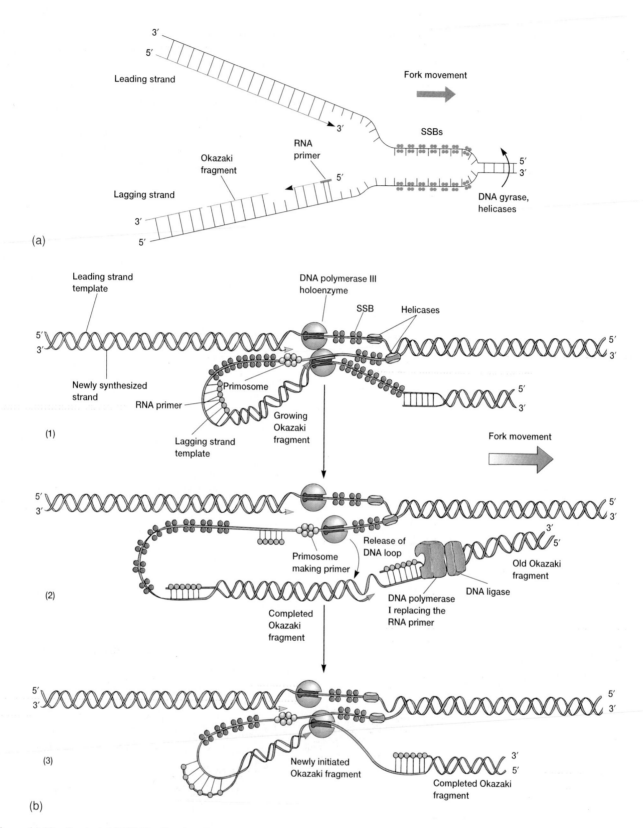

🌀 **Figure 11.12   Bacterial DNA Replication.** The synthesis of DNA in *E. coli* at the replication fork. (*a*) Diagram of the overall process. Bases and base pairs are represented by lines extending outward from the strands. The RNA primer is in red. See text for details. (*b*) A hypothetical model for activity at the replication fork. (*1*) The replisome has two DNA polymerase III holoenzyme complexes. One polymerase continuously copies the leading strand. The lagging strand loops around the other polymerase so that both strands can be replicated simultaneously. (*2*) When DNA polymerase III encounters a completed Okazaki fragment, it releases the lagging strand. The primosome synthesizes another RNA primer on the lagging strand. DNA polymerase I and DNA ligase fill and seal the gaps between completed fragments. (*3*) The holoenzyme binds the lagging strand at the new primer and begins to synthesize another Okazaki fragment.

Figure 11.13   DNA Polymerase I. A model of a large fragment of *E. coli* DNA polymerase I bound to DNA. The protein is in yellow, the DNA template strand in blue, and the primer strand in red. The 3´ end of the primer strand is entering the exonuclease site.

DNA replication is an extraordinarily complex process. At least 30 proteins are required to replicate the *E. coli* chromosome. Presumably much of the complexity is necessary for accuracy in copying DNA. It would be very dangerous for any organism to make many errors during replication because a large number of mutations would certainly be lethal. In fact, *E. coli* makes errors with a frequency of only 10^{-9} or 10^{-10} per base pair replicated (or about 10^{-6} per gene per generation). Part of this precision results from the low error rate of the copying process itself. However, DNA polymerase III (and DNA polymerase I) also can proofread the newly formed DNA. As polymerase III moves along synthesizing a new DNA strand, it recognizes any errors resulting in improper base pairing and hydrolytically removes the wrong nucleotide through a special 3´ to 5´ exonuclease activity (which is found in the ε subunit). The enzyme then backs up and adds the proper nucleotide in its place. Polymerases delete errors by acting much like correcting typewriters.

DNA repair (pp. 272–74).

Despite its complexity and accuracy, replication occurs very rapidly. In bacteria replication rates approach 750 to 1,000 base pairs per second. Eucaryotic replication is much slower, about 50 to 100 base pairs per second. This is not surprising because eucaryotic replication also involves operations like unwinding the DNA from nucleosomes.

1. Define the following terms: replication, transcription, messenger RNA, translation, replicon, replication fork, primosome, and replisome.
2. Be familiar with the nature and functions of the following replication components and intermediates: DNA polymerases I and III, topoisomerase, DNA gyrase, helicase, single-stranded DNA binding protein, Okazaki fragment, DNA ligase, leading strand, and lagging strand.

DNA Transcription or RNA Synthesis

As mentioned earlier, synthesis of RNA under the direction of DNA is called transcription. The RNA product has a sequence complementary to the DNA template directing its synthesis (table 11.1). Thymine is not normally found in mRNA and rRNA. Although adenine directs the incorporation of thymine during DNA replication, it usually codes for uracil during RNA synthesis. Transcription generates three kinds of RNA.

Figure 11.14   The DNA Ligase Reaction. The groups being altered are shaded in color.

TABLE 11.1	RNA Bases Coded for by DNA
DNA Base	**Purine or Pyrimidine Incorporated into RNA**
Adenine	Uracil
Guanine	Cytosine
Cytosine	Guanine
Thymine	Adenine

Figure 11.15   A Polygenic Bacterial Messenger RNA. See text for details.

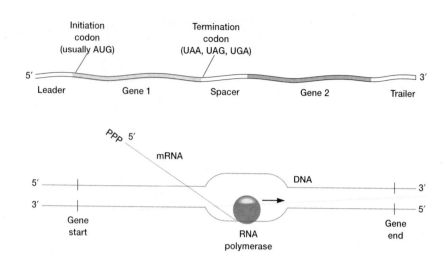

Figure 11.16   mRNA Transcription from DNA. The lower DNA strand is directing mRNA synthesis, and the RNA polymerase is moving from left to right. The polymerase has unwound the DNA double helix to form a transcription bubble and is copying the template or sense strand.

Messenger RNA (mRNA) bears the message for protein synthesis. **Transfer RNA (tRNA)** carries amino acids during protein synthesis, and **ribosomal RNA (rRNA)** molecules are components of ribosomes. Structure and synthesis of procaryotic mRNA is described first.

Transcription in Procaryotes

Procaryotic mRNA is a single-stranded RNA of variable length containing directions for the synthesis of one to many polypeptides. Messenger RNA molecules also have sequences that do not code for polypeptides (figure 11.15). There is a nontranslated **leader sequence** of 25 to 150 bases at the 5′ end preceding the initiation codon. In addition, polygenic mRNAs (those directing the synthesis of more than one polypeptide) have spacer regions separating the segments coding for individual polypeptides. Polygenic messenger polypeptides usually function together in some way (e.g., as part of the same metabolic pathway). At the 3′ end, following the last termination codon, is a nontranslated trailer.

Messenger RNA is synthesized under the direction of DNA by the enzyme **RNA polymerase.** An *E. coli* cell can have as many as 7,000 RNA polymerase molecules; only 2,000 to 5,000 polymerases may be active at any one time. The reaction is quite similar to that catalyzed by DNA polymerase. ATP, GTP, CTP, and UTP are used to produce an RNA copy of the DNA sequence. As mentioned earlier, these nucleotides contain ribose rather than deoxyribose.

$$\text{n[ATP, GTP, CTP, UTP]} \xrightarrow[\text{DNA template}]{\text{RNA polymerase}} \text{RNA} + n\text{PP}_i$$

RNA synthesis, like DNA synthesis, proceeds in a 5′ to 3′ direction with new nucleotides being added to the 3′ end of the growing chain at a rate of about 40 nucleotides per second at 37°C (figure 11.16). It should be noted that pyrophosphate is produced in both DNA and RNA polymerase reactions. Pyrophosphate is then removed by hydrolysis to orthophosphate in a reaction catalyzed by the pyrophosphatase enzyme. Removal of the pyrophosphate product makes DNA and RNA

synthesis irreversible. If the pyrophosphate level were too high, DNA and RNA would be degraded by a reversal of the polymerase reactions.

A **gene** is a DNA segment or sequence that codes for a polypeptide, an rRNA, or a tRNA. Although DNA has two complementary strands, RNA polymerase copies only the **template** or **sense strand** at any particular point on DNA. If both strands of DNA were transcribed, two different mRNAs would result and cause genetic confusion. Thus the sequence corresponding to a gene is located only on one of the two complementary DNA strands, the sense strand. Different genes may be encoded on opposite strands. The RNA polymerase opens or unwinds the double helix to form a transcription bubble and transcribes the sense strand to produce an RNA transcript that is complementary and antiparallel to the DNA template (figure 11.16).

The genetic code and gene structure (pp. 255–61).

The RNA polymerase of *E. coli* is a very large molecule (about 480,000 daltons) containing four types of polypeptide chains: α, β, β′, and σ. The **core enzyme** is composed of four chains (α₂, β, β′) and catalyzes RNA synthesis. The **sigma factor (σ)** has no catalytic activity but helps the core enzyme recognize the start of genes. Once RNA synthesis begins, the sigma factor dissociates from the core enzyme–DNA complex and is available to aid another core enzyme. The precise functions of the α, β, and β′ polypeptides are not yet clear. The α subunit seems to be involved in the assembly of the core enzyme, recognition of promoters (see below), and interaction with some regulatory factors. The binding site for DNA is on β′, and the β subunit binds ribonucleotide substrates. Rifampin, an RNA polymerase inhibitor, binds to the β subunit.

The region to which RNA polymerase binds with the aid of the sigma factor is called the **promoter.** The procaryotic promoter sequence is not transcribed. A 6 base sequence (usually TTGACA), approximately 35 base pairs before the transcription starting point, is present in all bacterial promoters. A TATAAT sequence or **Pribnow box** lies about 10 base pairs before the starting point of transcription or around 16 to 18 base pairs from

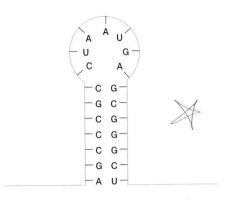

Figure 11.17   Procaryotic Terminators. An example of a hairpin structure formed by an mRNA terminator sequence.

TABLE 11.2	Eucaryotic RNA Polymerases	
Enzyme	**Location**	**Product**
RNA polymerase I	Nucleolus	rRNA (5.8S, 18S, 28S)
RNA polymerase II	Chromatin, nuclear matrix	mRNA
RNA polymerase III	Chromatin, nuclear matrix	tRNA, 5S rRNA

the first hexamer sequence. The RNA polymerase recognizes these sequences, binds to the promoter, and unwinds a short segment of DNA beginning around the Pribnow box. Transcription starts 6 or 7 base pairs away from the 3′ end of the promoter. The RNA polymerase remains at the promoter while it constructs a chain about 9 nucleotides long, then it begins to move down the template strand. The first base used in RNA synthesis is usually a purine, either ATP or GTP. Since these phosphates are not removed during transcription, the 5′ end of procaryotic mRNA has a triphosphate attached to the ribose.

Promoter structure and function (pp. 259–60).

There also must be stop signals to mark the end of a gene or sequence of genes and stop transcription by the RNA polymerase. All procaryotic **terminators** contain a sequence coding for an RNA stretch that can hydrogen bond to form a hairpin-shaped loop and stem structure (figure 11.17). This structure appears to cause the RNA polymerase to pause or stop transcribing DNA. There are two kinds of stop signals or terminators. The first type contains a stretch of about six uridine residues following the mRNA hairpin and causes the polymerase to stop transcription and release the mRNA without the aid of any accessory factors. The second kind lacks a poly-U region after the hairpin and requires the aid of a special protein, the **rho factor** (ρ). It is thought that rho binds to mRNA and moves along the molecule until it reaches the RNA polymerase that has halted at a terminator. The rho factor then causes the polymerase to dissociate from the mRNA, probably by unwinding the mRNA-DNA complex.

Transcription in Eucaryotes

Transcriptional processes in eucaryotic microorganisms (and in other eucaryotic cells) differ in several ways from procaryotic transcription. There are three major RNA polymerases, not one as in procaryotes. RNA polymerase II, associated with chromatin in the nuclear matrix, is responsible for mRNA synthesis. Polymerases I and III synthesize rRNA and tRNA, respectively (table 11.2). The eucaryotic RNA polymerase II is a large aggregate, at least 500,000 daltons in size, with about 10

or more subunits. It is inhibited by the octapeptide α-amanitin. Unlike the bacterial polymerase it requires extra transcription factors to recognize its promoters. The polymerase binds near the start point; the transcription factors bind to the rest of the promoter. Eucaryotic promoters also differ from those in procaryotes. They have combinations of several elements. Three of the most common are the TATA box (located about 30 base pairs before the start point or upstream), the CAAT box (about 75 base pairs upstream), and the GC box (90 base pairs upstream). Recently it has been shown that the TATA-binding protein sharply bends the DNA on attachment. This makes the DNA more accessible to other initiation factors. A variety of general transcription factors, promoter specific factors, and promoter elements have been discovered in different eucaryotic cells. Each eucaryotic gene seems to be regulated differently, and more research will be required to understand the regulation of eucaryotic gene transcription.

Eucaryotic mRNA arises from **posttranscriptional modification** of large RNA precursors, about 5,000 to 50,000 nucleotides long, called **heterogeneous nuclear RNA (hnRNA)** molecules. These are the products of RNA polymerase II activity (figure 11.18). After hnRNA synthesis, the precursor RNA is cleaved by an endonuclease to yield the proper 3′-OH group. The enzyme polyadenylate polymerase then catalyzes the addition of adenylic acid to the 3′ end of hnRNA to produce a poly-A sequence about 200 nucleotides long. The hnRNA finally is cleaved to generate the functional mRNA. Usually eucaryotic mRNA also differs in having a 5′ cap consisting of 7-methylguanosine attached to the 5′-hydroxyl by a triphosphate linkage (figure 11.19). The adjacent nucleotide also may be methylated.

The 3′ poly-A and 5′ capping of eucaryotic mRNAs distinguish them from procaryotic mRNAs. In addition, eucaryotic mRNA normally is monogenic in contrast to procaryotic mRNA, which often is polygenic. The functions of poly-A and capping still are not completely clear. Poly-A protects mRNA from rapid enzymatic degradation. The poly-A tail must be shortened to about 10 nucleotides before mRNA can be degraded. Poly-A also seems to aid in mRNA translation. The 5′ cap on eucaryotic messengers may promote the initial binding of ribosomes to the messenger. The cap also may protect the messenger from enzymatic attack.

Many eucaryotic genes differ from procaryotic genes in being split or interrupted, which leads to another type of

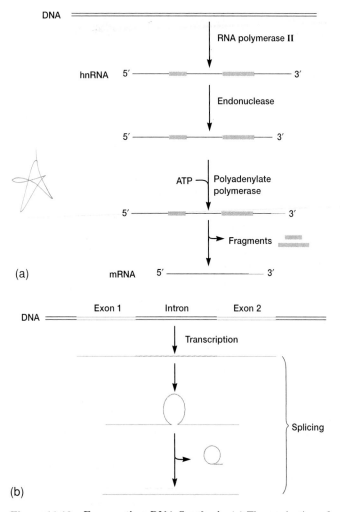

(a)

(b)

Figure 11.18   Eucaryotic mRNA Synthesis. (*a*) The production of eucaryotic messenger RNA. The addition of poly-A to the 3′ end of mRNA is included, but not the capping of the 5′ end. Poly-A sequence and introns are in color. (*b*) The splicing of interrupted genes to produce mRNA. Poly-A sequences and exons are in color. The excised intron is in the shape of a circle or lariat.

posttranscriptional processing. **Split** or **interrupted genes** have **exons** (*ex*pressed sequences), regions coding for RNA that end up in the final RNA product (e.g., mRNA). Exons are separated from one another by **introns** (*in*tervening sequences), sequences coding for RNA that is missing from the final product (figure 11.18*b*). The initial RNA transcript has the intron sequences present in the interrupted gene. Genes coding for rRNA and tRNA may also be interrupted. Except for cyanobacteria and archaeobacteria (*see chapters 22 and 23*), interrupted genes have not been found in procaryotes.

Introns are removed from the initial RNA transcript by a process called **RNA splicing** (figure 11.18*b*). The intron's borders must be clearly marked for accurate removal, and this is the case. Exon-intron junctions have a GU sequence at the intron's 5′ boundary and an AG sequence at its 3′ end. These two sequences define the splice junctions and are recognized by special RNA molecules. The nucleus contains several **small nuclear RNA (snRNA)** molecules, about 60 to 300 nucleotides long. These complex with proteins to form *s*mall *n*uclear *ri*bo*nucleo*protein particles called snRNPs or snurps. Some of the snRNPs recognize splice junctions and ensure splicing accuracy. For example, U1-snRNP recognizes the 5′ splice junction, and U5-snRNP recognizes the 3′ junction. Splicing of pre-mRNA occurs in a large complex called a **spliceosome** that contains the pre-mRNA, at least five kinds of snRNPs, and non-snRNP splicing factors.

As just mentioned, a few rRNA genes also have introns. Some of these pre-rRNA molecules are self-splicing. The RNA actually catalyzes the splicing reaction and now is called a **ribozyme** (Box 11.2). Thomas Cech first discovered that pre-rRNA from the ciliate protozoan *Tetrahymena thermophila* is self-splicing. Sidney Altman then showed that ribonuclease P, which cleaves a fragment from one end of pre-tRNA, contains

7-methylguanosine

Base 1

Base 2

Figure 11.19   The 5′ Cap of Eucaryotic mRNA.

Box 11.2

Catalytic RNA (Ribozymes)

Until recently biologists believed that all cellular reactions were catalyzed by proteins called enzymes (*see chapter 8*). The discovery during 1981–1984 by Cech and Altman that RNA also can sometimes catalyze reactions has transformed our way of thinking about topics as diverse as catalysis and the origin of life. It is now clear that some RNA molecules, called ribozymes, catalyze reactions that alter either their own structure or that of other RNAs.

This discovery has stimulated scientists to hypothesize that the early earth was an "RNA world" in which RNA acted as both the genetic material and a reaction catalyst. Experiments showing that introns from *Tetrahymena thermophila* can catalyze the formation of polycytidylic acid under certain circumstances have further encouraged such speculations. Some have suggested that RNA viruses are "living fossils" of the original RNA world.

The best-studied ribozyme activity is the self-splicing of RNA. This process is widespread and occurs in *Tetrahymena* pre-rRNA; the mitochondrial rRNA and mRNA of yeast and other fungi; chloroplast tRNA, rRNA, and mRNA; and in mRNA from some bacteriophages (e.g., the T4 phage of *E. coli*). The 413-nucleotide rRNA intron of *T. thermophila* provides a good example of the self-splicing reaction. The reaction occurs in three steps and requires the presence of guanosine (Box Figure 11.2). First, the 3′-OH group of guanosine attacks the intron's 5′-phosphate group and cleaves the phosphodiester bond. Second, the new 3′-hydroxyl on the left exon attacks the 5′-phosphate of the right exon. This joins the two exons and releases the intron. Finally, the intron's 3′-hydroxyl attacks the phosphate bond of the nucleotide 15 residues from its end. This releases a terminal fragment and cyclizes the intron. Self-splicing of this rRNA occurs about 10 billion times faster than spontaneous RNA hydrolysis. Just as with enzyme proteins, the RNA's shape is essential to catalytic efficiency. The ribozyme even has Michaelis-Menten kinetics (pp. 160–61).

The discovery of ribozymes has many potentially important practical consequences. Ribozymes act as "molecular scissors" and will enable researchers to manipulate RNA easily in laboratory experiments. It also might be possible to protect hosts by specifically removing RNA from pathogenic viruses, bacteria, and fungi. For example, ribozymes are already being tested against the AIDS, herpes, and tobacco mosaic viruses.

Box Figure 11.2   Ribozyme Action. The mechanism of *Tetrahymena thermophila* pre-rRNA self-splicing. See text for details.

a piece of RNA that catalyzes the reaction. Several other self-splicing rRNA introns have since been discovered. Cech and Altman received the 1989 Nobel Prize in chemistry for these discoveries.

Although the focus has been on mRNA synthesis, it should be noted that both rRNAs and tRNAs also begin as parts of large precursors synthesized by RNA polymerases (table 11.2). The final rRNA and tRNA products result from posttranscriptional processing, as mentioned previously.

1. Define the following terms: leader, trailer, spacer region, polygenic mRNA, RNA polymerase core enzyme, sigma factor, promoter, sense strand, terminator, and rho factor.
2. What is a gene?
3. Define or describe posttranscriptional modification, heterogeneous nuclear RNA, 3′ poly-A sequence, 5′ capping, split or interrupted genes, exon, intron, RNA splicing, snRNA, spliceosome, and ribozyme.

Figure 11.20   Coupled Transcription and Translation in Bacteria. mRNA is synthesized 5′ to 3′ by RNA polymerase while ribosomes are attaching to the newly formed 5′ end of mRNA and translating the message even before it is completed. Polypeptides are synthesized in the N-terminal to C-terminal direction.

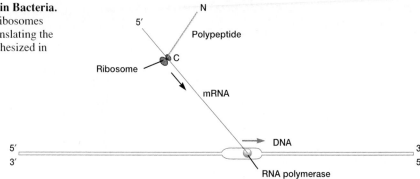

Protein Synthesis

The final step in gene expression is protein synthesis or translation. The mRNA nucleotide sequence is translated into the amino acid sequence of a polypeptide chain in this step. Polypeptides are synthesized by the addition of amino acids to the end of the chain with the free α-carboxyl group (the C-terminal end). That is, the synthesis of polypeptides begins with the amino acid at the end of the chain with a free amino group (the N-terminal) and moves in the C-terminal direction. Protein synthesis is not only quite accurate but also very rapid. In *E. coli* synthesis occurs at a rate of about 300 to 450 residues per minute; eucaryotic translation is slower, about 50 residues per minute.

Polypeptide and protein structure (appendix 1).

Many bacteria grow so quickly that each mRNA must be used with great efficiency to synthesize proteins at a sufficiently rapid rate. Frequently bacterial mRNAs are simultaneously complexed with several ribosomes, each ribosome reading the mRNA message and synthesizing a polypeptide. At maximal rates of mRNA utilization, there may be a ribosome every 80 nucleotides along the messenger or as many as 20 ribosomes simultaneously reading an mRNA that codes for a 50,000 dalton polypeptide. A complex of mRNA with several ribosomes is called a **polyribosome** or polysome. Polysomes are present in both procaryotes and eucaryotes. Bacteria can further increase the efficiency of gene expression through coupled transcription and translation (figure 11.20). While RNA polymerase is synthesizing an mRNA, ribosomes can already be attached to the messenger and involved in polypeptide synthesis. Coupled transcription and translation is possible in procaryotes because a nuclear envelope does not separate the translation machinery from DNA as it does in eucaryotes (*see figure 3.17*).

Transfer RNA and Amino Acid Activation

The first stage of protein synthesis is **amino acid activation,** a process in which amino acids are attached to transfer RNA molecules. These RNA molecules are normally between 73 and 93 nucleotides in length and possess several characteristic structural features. The structure of tRNA becomes clearer when its chain is folded in such a way to maximize the number of normal

base pairs, which results in a cloverleaf conformation of five arms or loops (figure 11.21). The acceptor or amino acid stem holds the activated amino acid on the 3′ end of the tRNA. The 3′ end of all tRNAs has the same —C—C—A sequence; the amino acid is attached to the terminal adenylic acid. At the other end of the cloverleaf is the anticodon arm, which contains the **anticodon triplet** complementary to the mRNA codon triplet. There are two other large arms: the D or DHU arm has the unusual pyrimidine nucleoside dihydrouridine; and the T or TΨC arm has ribothymidine (T) and pseudouridine (Ψ), both of which are unique to tRNA. Finally, the cloverleaf has a variable arm whose length changes with the overall length of the tRNA; the other arms are fairly constant in size.

Transfer RNA molecules are folded into an L-shaped structure (figure 11.22). The amino acid is held on one end of the L, the anticodon is positioned on the opposite end, and the corner of the L is formed by the D and T loops. Because there must be at least one tRNA for each of the 20 amino acids incorporated into proteins, at least 20 different tRNA molecules are needed. Actually more tRNA species than this exist (*see p. 257*).

Amino acids are activated for protein synthesis through a reaction catalyzed by **aminoacyl-tRNA synthetases** (figure 11.23).

$$\text{Amino acid} + \text{tRNA} + \text{ATP} \xrightarrow{\text{Mg}^{2+}}$$
$$\text{aminoacyl-tRNA} + \text{AMP} + \text{PP}_i$$

Just as is true of DNA and RNA synthesis, the reaction is driven to completion when the pyrophosphate product is hydrolyzed to two orthophosphates. The amino acid is attached to the 3′-hydroxyl of the terminal adenylic acid on the tRNA by a high-energy bond (figure 11.24), and is readily transferred to the end of a growing peptide chain. This is why the amino acid is called activated.

There are at least 20 aminoacyl-tRNA synthetases, each specific for a single amino acid and for all the tRNAs (cognate tRNAs) to which each may be properly attached. This specificity is critical because once an incorrect acid is attached to a tRNA, it will be incorporated into a polypeptide in place of the correct amino acid. The protein synthetic machinery recognizes only the anticodon of the aminoacyl-tRNA and cannot tell whether the correct amino acid is attached. Some aminoacyl-tRNA synthetases will even proofread just like DNA polymerases do. If the wrong aminoacyl-tRNA is

Figure 11.21   tRNA Structure. The cloverleaf structure for tRNA in procaryotes and eucaryotes. Bases found in all tRNAs are in diamonds; purine and pyrimidine positions in all tRNAs are labeled Pu and Py respectively.

(a)

(b)

Figure 11.22   Transfer RNA Conformation. (*a*) The three-dimensional structure of tRNA. The various regions are distinguished with different colors. (*b*) A space-filling model of yeast phenylalanine tRNA. The 3′ acceptor end is at the upper right.

Figure 11.23   An Aminoacyl-tRNA Synthetase. A model of *E. coli* glutamyl-tRNA synthetase complexed with its tRNA and ATP. The enzyme is in blue, the tRNA in red and yellow, and ATP in green.

Figure 11.24   Aminoacyl-tRNA. The 3′ end of an aminoacyl-tRNA. The activated amino acid is attached to the 3′- hydroxyl of adenylic acid by a high-energy bond.

Figure 11.25   The 70S Ribosome. The structure of the procaryotic ribosome.

formed, aminoacyl-tRNA synthetases will hydrolyze the amino acid from the tRNA rather than release the incorrect product.

The Ribosome

The actual process of protein synthesis takes place on ribosomes that serve as workbenches, with mRNA acting as the blueprint. Procaryotic ribosomes have a sedimentation value of 70S and a mass of 2.8 million daltons. A rapidly growing *E. coli* cell may have as many as 15,000 to 20,000 ribosomes, about 15% of the cell mass.

Introduction to ribosomal function and the Svedberg unit (p. 48).

The procaryotic ribosome is an extraordinarily complex organelle made of a 30S and a 50S subunit (figure 11.25). Each subunit is constructed from one or two rRNA molecules and many polypeptides. The shape of ribosomal subunits and their association to form the 70S ribosome are depicted in figure 11.26. The region of the ribosome directly responsible for translation is called the translational domain (figure 11.26*d*). Both subunits contribute to this domain, located in the upper half of the small subunit and in the associated areas of the large subunit. For example, the peptidyl transferase (p. 232) is found on the central protuberance of the large subunit. The growing peptide chain emerges from the large subunit at the exit domain. This is located on the side of the subunit opposite the central protuberance in both procaryotes and eucaryotes.

Eucaryotic cytoplasmic ribosomes are 80S, with a mass of 4 million daltons, and are composed of two subunits, 40S and 60S. Many of these ribosomes are found free in the cytoplasmic matrix, whereas others are attached to membranes of the endoplasmic reticulum by their 60S subunit at a site next to the exit domain. The ribosomes of eucaryotic mitochondria and chloroplasts are smaller than cytoplasmic ribosomes and resemble the procaryotic organelle.

Ribosomal RNA is thought to have two roles. It obviously contributes to ribosome structure. The 16S rRNA of the 30S subunit also may aid in the initiation of protein synthesis in procaryotes. There is evidence that the 3′ end of the 16S rRNA complexes with an initiating signal site on the mRNA and helps position the mRNA on the ribosome. It also binds initiation factor 3 (see below) and aminoacyl-tRNA. Because of the discovery of catalytic RNA, some have proposed that ribosomal RNA may have a catalytic role in protein synthesis.

The use of 16S rRNA sequences in the study of bacterial phylogeny (pp. 402–8).

Initiation of Protein Synthesis

Protein synthesis proper may be divided into three stages: initiation, elongation, and termination.

In the initiation stage *E. coli* and most bacteria begin protein synthesis with a specially modified aminoacyl-tRNA, N-formylmethionyl-tRNA^{fMet} (figure 11.27). Since the α-amino is blocked by a formyl group, this aminoacyl-tRNA can be used only for initiation. When methionine is to be added to a growing polypeptide chain, a normal methionyl-tRNA^{Met} is employed. Eucaryotic protein synthesis (except in the mitochondrion and chloroplast) and archaeobacterial protein synthesis begin with a special initiator methionyl-tRNA^{Met}. Although most bacteria start protein synthesis with formylmethionine, the formyl group does not remain but is hydrolytically removed. In fact, one to three amino acids may be removed from the amino terminal end of the polypeptide after synthesis.

Figure 11.28 shows the initiation process in procaryotes. The initiator N-formylmethionyl-tRNA^{fMet} (fMet-tRNA) binds to the free 30S subunit first. Next mRNA attaches to the 30S subunit and is positioned properly through interactions with both the 3′ end of the 16S rRNA and the anticodon of fMet-tRNA. Messengers have a special initiator codon (AUG or

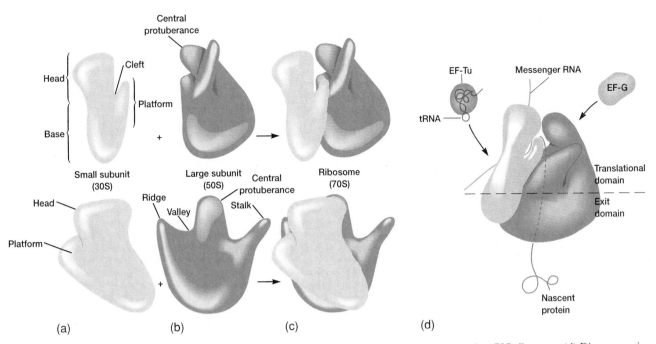

Figure 11.26 **Two Views of the *E. coli* Ribosome.** (*a*) The 30S subunit. (*b*) The 50S subunit. (*c*) The complete 70S ribosome. (*d*) Diagrammatic representation of ribosomal structure with the translational and exit domains shown. The locations of elongation factor and mRNA binding are shown. The growing peptide chain probably remains unfolded and extended until it leaves the large subunit.

$$CH_3 - S - CH_2 - CH_2 - \underset{\underset{\underset{\underset{H}{|}}{C=O}}{\overset{|}{NH}}}{CH} - \overset{\overset{O}{\|}}{C} - tRNA^{fMet}$$

Figure 11.27 **Procaryotic Initiator tRNA.** The initiator aminoacyl-tRNA, N-formylmethionyl-tRNA^fMet, is used by eubacteria. The formyl group is in color. Archaeobacteria use methionyl-tRNA for initiation.

sometimes GUG) that specifically binds with the fMet-tRNA anticodon (*see chapter 13*). Finally, the 50S subunit binds to the 30S subunit-mRNA forming an active ribosome-mRNA complex. The fMet-tRNA is positioned at the P site. There is some uncertainty about the exact initiation sequence, and mRNA may bind before fMet-tRNA in procaryotes. Eucaryotic initiation appears to begin with the binding of a special initiator Met-tRNA to the small subunit, followed by attachment of the mRNA.

In procaryotes three protein **initiation factors** are required (figure 11.28). Initiation factor 3 (IF-3) prevents 30S subunit binding to the 50S subunit and promotes the proper mRNA binding to the 30S subunit. IF-2, the second initiation factor, binds GTP and fMet-tRNA and directs the attachment of fMet-tRNA to the 30S subunit. GTP is hydrolyzed during association of the 50S and 30S subunits. The third initiation factor, IF-1, appears to be needed for release of IF-2 and GDP from the completed 70S

ribosome. IF-1 also may aid in the binding of the 50S subunit to the 30S subunit. Eucaryotes require more initiation factors; otherwise the process is quite similar to that of procaryotes.

The initiation of protein synthesis is very elaborate. Apparently the complexity is necessary to ensure that the ribosome does not start synthesizing a polypeptide chain in the middle of a gene—a disastrous error.

Elongation of the Polypeptide Chain

Every amino acid addition to a growing polypeptide chain is the result of an **elongation cycle** composed of three phases: aminoacyl-tRNA binding, the transpeptidation reaction, and translocation. The process is aided by special protein **elongation factors** (just as with the initiation of protein synthesis). In each turn of the cycle, an amino acid corresponding to the proper mRNA codon is added to the C-terminal end of the polypeptide chain. The procaryotic elongation cycle is described next.

The ribosome has two sites for binding aminoacyl-tRNA and peptidyl-tRNA: (1) the **peptidyl** or **donor site** (the **P site**) and (2) the **aminoacyl** or **acceptor site** (the **A site**). At the beginning of an elongation cycle, the peptidyl site is filled with either N-formylmethionyl-tRNA^fMet or peptidyl-tRNA and the aminoacyl site is empty (figure 11.29). Messenger RNA is bound to the ribosome in such a way that the proper codon interacts with the P site tRNA (e.g., an AUG codon for fMet-tRNA). The next codon (green) is located within the A site and ready to direct the binding of an aminoacyl-tRNA.

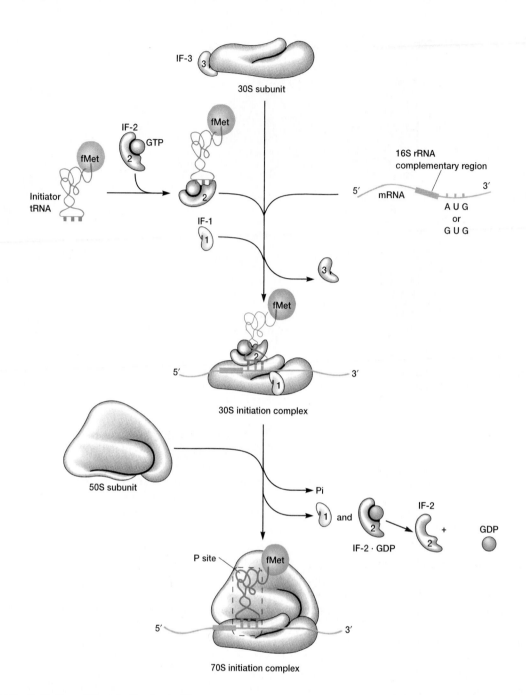

Figure 11.28   Initiation of Protein Synthesis. The initiation of protein synthesis in procaryotes. The following abbreviations are employed: IF-1, IF-2, and IF-3 stand for initiation factors 1, 2, and 3; initiator tRNA is N-formylmethionyl-tRNA^{fMet}. The ribosomal locations of initiation factors are depicted for illustration purposes only. They do not represent the actual initiation factor binding sites. See text for further discussion.

Figure 11.29   Elongation Cycle. The elongation cycle of protein synthesis. The ribosome possesses two sites, a peptidyl or donor site (P site) and an aminoacyl or acceptor site (A site). The arrow below the ribosome in the translocation step shows the direction of mRNA movement. See text for details.

The first phase of the elongation cycle is the aminoacyl-tRNA binding phase. The aminoacyl-tRNA corresponding to the green codon is inserted into the A site. GTP and the elongation factor EF-Tu, which donates the aminoacyl-tRNA to the ribosome, are required for this insertion. When GTP is bound to EF-Tu, the protein is in its active state and delivers aminoacyl-tRNA to the A site. This is followed by GTP hydrolysis, and the EF-Tu·GDP complex leaves the ribosome. EF-Tu·GDP is converted to EF-Tu·GTP with the aid of a second elongation factor, EF-Ts. Subsequently another aminoacyl-tRNA binds to EF-Tu·GTP (figure 11.29).

Aminoacyl-tRNA binding to the A site initiates the second phase of the elongation cycle, the **transpeptidation reaction** (figures 11.29 and 11.30). This is catalyzed by the **peptidyl transferase,** located on the 50S subunit. The α-amino group of the A site amino acid nucleophilically attacks the α-carboxyl group of the C-terminal amino acid on the P site tRNA in this reaction (figure 11.30). The peptide chain grows by one amino acid and is transferred to the A site tRNA. No extra energy source is required for peptide bond formation because the bond linking an amino acid to tRNA is high in energy. Recent evidence suggests that 23S rRNA participates in the peptidyl transferase function. Almost all protein can be removed from the 50S subunit, leaving the 23S rRNA and protein fragments. The remaining complex still has peptidyl transferase activity. Thus the 23S rRNA appears to be a major component of the peptidyl transferase and contributes to both A and P site functions.

The final phase in the elongation cycle is **translocation.** Three things happen simultaneously: (1) the peptidyl-tRNA moves from the A site to the P site; (2) the ribosome moves one codon along mRNA so that a new codon is positioned in the A site; and (3) the empty tRNA leaves the P site. Instead of immediately being ejected from the ribosome, the empty tRNA moves from the P site to the **exit site** (the **E site**) and then leaves the ribosome. The intricate process requires the participation of the EF-G or translocase protein and GTP hydrolysis. The ribosome changes shape as it moves down the mRNA in the 5′ to 3′ direction.

Termination of Protein Synthesis

In the third stage protein synthesis stops when the ribosome reaches one of three special **nonsense codons**—UAA, UAG, and UGA (figure 11.31). Three **release factors** (RF-1, RF-2, and RF-3) aid the ribosome in recognizing these codons. After the ribosome has stopped, peptidyl transferase hydrolyzes the peptide free from its tRNA, and the empty tRNA is released. GTP hydrolysis seems to be required during this sequence of events, although it may not be needed for termination in procaryotes. Next the ribosome dissociates from its mRNA and separates into 30S and 50S subunits. IF-3 binds to the 30S subunit and prevents it from reassociating with the 50S subunit until the proper stage in initiation has been reached. Thus ribosomal subunits come together during protein synthesis and then

Figure 11.30   Transpeptidation. The transpeptidation reaction catalyzed by peptidyl transferase.

separate afterward. The termination of eucaryotic protein synthesis is similar to the process just described except that only one release factor appears to be active.

As a polypeptide leaves the ribosome, it is already folding into its final shape with the aid of molecular chaperones. This folding is possible because protein conformation is a direct function of the amino acid sequence (*see appendix I*). The polypeptide is almost completely folded by the time its synthesis is completed.

Molecular chaperones and their functions (pp. 49–50).

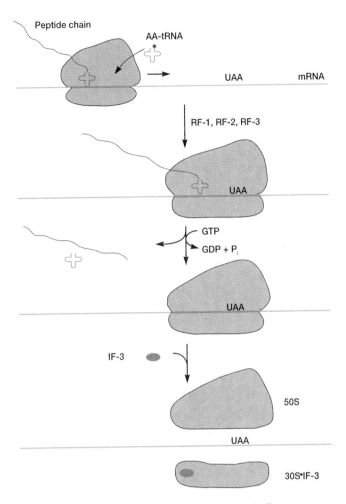

Figure 11.31 Termination of Protein Synthesis in Procaryotes. Although three different nonsense codons can terminate chain elongation, UAA is most often used for this purpose. Three release factors (RF) assist the ribosome in recognizing nonsense codons and terminating translation. GTP hydrolysis is probably involved in termination.

Protein synthesis is a very expensive process. Three GTP molecules probably are used during the elongation cycle, and two ATP high-energy bonds are required for amino acid activation (ATP is converted to AMP rather than to ADP). Therefore five high-energy bonds are required to add one amino acid to a growing polypeptide chain. GTP also is used in initiation and termination of protein synthesis (figures 11.28 and 11.31). Presumably this large energy expenditure is required to ensure the fidelity of protein synthesis. Very few mistakes can be tolerated.

Although the mechanism of protein synthesis is similar in procaryotes and eucaryotes, procaryotic ribosomes differ substantially from those in eucaryotes. This explains the effectiveness of many important chemotherapeutic agents. Either the 30S or the 50S subunit may be affected. For example, streptomycin binding to the 30S ribosomal subunit inhibits protein synthesis and causes mRNA misreading. Erythromycin binds to the 50S subunit and inhibits peptide chain elongation.

The effect of antibiotics on protein synthesis (pp. 663, 668–9).

1. In which direction are polypeptides synthesized? What is a polyribosome and why is it useful?

2. Briefly describe the structure of transfer RNA and relate this to its function. How are amino acids activated for protein synthesis, and why is the specificity of the aminoacyl-tRNA synthetase reaction so important?

3. What are translational and exit domains? Contrast procaryotic and eucaryotic ribosomes in terms of structure. What roles does ribosomal RNA have?

4. Describe the nature and function of the following: fMet-tRNA, initiator codon, IF-3, IF-2, IF-1, elongation cycle, peptidyl and aminoacyl sites, EF-Tu, EF-Ts, transpeptidation reaction, peptidyl transferase, translocation, EF-G or translocase, nonsense codon, and release factors.

Summary

1. DNA differs in composition from RNA in having deoxyribose and thymine rather than ribose and uracil.

2. DNA is double stranded, with complementary AT and GC base pairing between the strands. The strands run antiparallel and are twisted into a right-handed double helix.

3. RNA is normally single stranded, although it can coil upon itself and base pair to form hairpin structures.

4. In almost all procaryotes DNA exists as a closed circle that is twisted into supercoils and associated with histonelike proteins.

5. Eucaryotic DNA is associated with five types of histone proteins. Eight histones associate to form ellipsoidal octamers around which the DNA is coiled to produce the nucleosome.

6. DNA synthesis is called replication. Transcription is the synthesis of an RNA copy of DNA and produces three types of RNA: messenger RNA (mRNA), transfer RNA (tRNA), and ribosomal RNA (rRNA).

7. The synthesis of protein under the direction of mRNA is called translation.

8. Most circular procaryotic DNAs are copied by two replication forks moving around the circle to form a theta-shaped (θ) figure. Sometimes a rolling-circle mechanism is employed instead.

9. Eucaryotic DNA has many replicons and replication origins located every 10 to 100 μm along the DNA.

10. DNA polymerase enzymes catalyze the synthesis of DNA in the 5′ to 3′ direction while reading the DNA template in the 3′ to 5′ direction.

11. The double helix is unwound by helicases with the aid of topoisomerases like the DNA gyrase. DNA binding proteins keep the strands separate.

12. DNA polymerase III holoenzyme synthesizes a complementary DNA copy beginning with a short RNA primer made by a primase enzyme.

13. The leading strand is probably replicated continuously, whereas DNA synthesis on the lagging strand is discontinuous and forms Okazaki fragments.

14. DNA polymerase I excises the RNA primer and fills in the resulting gap. DNA ligase then joins the fragments together.

15. Procaryotic mRNA has nontranslated leader and trailer sequences at its ends. Spacer regions exist between genes when mRNA is polygenic.

16. RNA is synthesized by RNA polymerase that copies the sequence of the DNA template sense strand.

17. The sigma factor helps the procaryotic RNA polymerase bind to the promoter region at the start of a gene.

18. A terminator marks the end of a gene. A rho factor is needed for RNA polymerase release from some terminators.

19. RNA polymerase II synthesizes heterogeneous nuclear RNA, which then undergoes posttranscriptional modification by RNA cleavage and addition of a 3′ poly-A sequence and a 5′ cap to generate eucaryotic mRNA.

20. Many eucaryotic genes are split or interrupted genes that have exons and introns. Exons are joined by RNA splicing. Splicing involves small nuclear RNA molecules, spliceosomes, and sometimes ribozymes.

21. In translation, ribosomes attach to mRNA and synthesize a polypeptide beginning at the N-terminal end. A polysome or polyribosome is a complex of mRNA with several ribosomes.

22. Amino acids are activated for protein synthesis by attachment to the 3′ end of transfer RNAs. Activation requires ATP, and the reaction is catalyzed by aminoacyl-tRNA synthetases.

23. Ribosomes are large, complex organelles composed of rRNAs and many polypeptides. Amino acids are added to a growing peptide chain at the translational domain.

24. Protein synthesis begins with the binding of fMet-tRNA (procaryotes) or an initiator methionyl-tRNA^{Met} (eucaryotes) to an initiator codon on mRNA and to the two ribosomal subunits. This involves the participation of protein initiation factors.

25. In the elongation cycle the proper aminoacyl-tRNA binds to the A site with the aid of EF-Tu and GTP. Then the transpeptidation reaction is catalyzed by peptidyl transferase. Finally, during translocation, the peptidyl-tRNA moves to the P site and the ribosome travels along the mRNA one codon. Translocation requires GTP and EF-G or translocase. The empty tRNA leaves the ribosome by way of the exit site.

26. Protein synthesis stops when a nonsense codon is reached. Procaryotes require three release factors for codon recognition and ribosome dissociation from the mRNA.

Key Terms

amino acid activation   226
aminoacyl or acceptor site (A site)   229
aminoacyl-tRNA synthetases   226
anticodon triplet   226
complementary   214
core enzyme   222
deoxyribonucleic acid (DNA)   212
DNA gyrase   219
DNA ligase   219
DNA polymerase   218
elongation cycle   229
elongation factors   229
exit site (E site)   232
exon   224
gene   222
helicases   218
heterogeneous nuclear RNA (hnRNA)   223
histones   215
initiation factors   229
initiator codon   228
intron   224

leader sequence   222
major groove   214
messenger RNA (mRNA)   216
minor groove   214
nonsense codons   232
nucleosome   215
Okazaki fragment   219
peptidyl or donor site (P site)   229
peptidyl transferase   232
polyribosome   226
posttranscriptional modification   223
Pribnow box   222
primase   219
primosome   219
promoter   222
release factors   232
replication   216
replication fork   217
replicon   217
rho factor   223

ribonucleic acid (RNA)   212
ribosomal RNA (rRNA)   222
ribozyme   224
RNA polymerase   222
RNA splicing   224
rolling-circle mechanism   218
sense strand   222
sigma factor   222
single-stranded DNA binding proteins (SSBs)   218
small nuclear RNA (snRNA)   224
spliceosome   224
split or interrupted genes   224
template   222
terminator   223
topoisomerases   219
transcription   216
transfer RNA (tRNA)   222
translation   216
translocation   232
transpeptidation reaction   232

Questions for Thought and Review

1. How do replication patterns differ between procaryotes and eucaryotes? Describe the operation of replication forks in the generation of theta-shaped intermediates and in the rolling-circle mechanism.

2. Outline the steps involved in DNA synthesis at the replication fork. How do DNA polymerases correct their mistakes?

3. Describe how RNA polymerase transcribes procaryotic DNA. How does the polymerase know where to begin and end transcription?

4. How do eucaryotic RNA polymerases and promoters differ from those in procaryotes? In what ways does eucaryotic mRNA differ from procaryotic mRNA with respect to synthesis and structure? How does eucaryotic synthesis of rRNA and tRNA resemble that of mRNA? How does it differ?

5. Draw diagrams summarizing the sequence of events in the three stages of protein synthesis (initiation, elongation, and termination) and accounting for the energy requirements of translation.

Additional Reading

General

Freifelder, D. 1987. *Molecular biology*, 2d ed. New York: Van Nostrand Reinhold.

Judson, H. F. 1979. *The eighth day of creation: Makers of the revolution in biology*. London: Jonathan Cape.

Kendrew, J., editor. 1994. *The encyclopedia of molecular biology*. Boston: Blackwell Scientific Publications.

Lewin, B. 1994. *Genes*, 5th ed. New York: Oxford University Press.

Stent, G. S., and Calendar, R. 1978. *Molecular genetics*, 2d ed. San Francisco: Freeman.

Watson, J. D.; Hopkins, N. H.; Roberts, J. W.; Steitz, J. A.; and Weiner, A. M. 1987. *Molecular biology of the gene*, 4th ed. Menlo Park, Calif.: Benjamin/Cummings.

Zubay, G. 1993. *Biochemistry*, 3d ed. Dubuque, Iowa: Wm. C. Brown Publishers.

Nucleic Acid Structure

Bauer, W. R.; Crick, F. H. C.; and White, J. H. 1980. Supercoiled DNA. *Sci. Am.* 243(1):118–33.

Darnell, J. E., Jr. 1985. RNA. *Sci. Am.* 253(4):68–78.

Drlica, K., and Rouviere-Yaniv, J. 1987. Histonelike proteins of bacteria. *Microbiol. Rev.* 51(3):301–19.

Felsenfeld, G. 1985. DNA. *Sci. Am.* 253(4):59–67.

Kornberg, R. D., and Klug, A. 1981. The nucleosome. *Sci. Am.* 244(2):52–64.

Rich, A., and Kim, S. H. 1978. The three-dimensional structure of transfer RNA. *Sci. Am.* 238(1):52–62.

DNA Replication

Dickerson, R. E. 1983. The DNA helix and how it is read. *Sci. Am.* 249(6):94–111.

Fox, J. L. 1988. Making and mistaking DNA. *ASM News* 54(3):122–26.

Johnson, K. A. 1993. Conformational coupling in DNA polymerase fidelity. *Ann. Rev. Biochem.* 62:685–713.

Joyce, C. M., and Steitz, T. A. 1994. Function and structure relationships in DNA polymerases. *Ann. Rev. Biochem.* 63:777-822.

Kornberg, A. 1992. *DNA replication*, 2d ed. San Francisco: W. H. Freeman.

Marians, K. J. 1992. Prokaryotic DNA replication. *Ann. Rev. Biochem.* 61:673–719.

Matson, S. W., and Kaiser-Rogers, K. A. 1990. DNA helicases. *Ann. Rev. Biochem.* 59:289–329.

McHenry, C. S. 1988. DNA polymerase III holoenzyme of *Escherichia coli. Ann. Rev. Biochem.* 57:519–50.

Meyer, R. R., and Laine, P. S. 1990. The single-stranded DNA-binding protein of *Escherichia coli. Microbiol. Rev.* 54(4):342–80.

Radman, M., and Walker, R. 1988. The high fidelity of DNA duplication. *Sci. Am.* 259(2):40–46.

Roca, J. 1995. The mechanisms of DNA topoisomerases. *Trends Biochem. Sci.* 20(4):156–60.

Stillman, B. 1994. Smart machines at the DNA replication fork. *Cell* 78:725–28.

Wang, J. C. 1982. DNA topoisomerases. *Sci. Am.* 247(1):94–109.

DNA Transcription or RNA Synthesis

Ahern, H. 1991. Self-splicing introns: Molecular fossils or selfish DNA? *ASM News* 57(5):258–61.

Cech, T. R. 1986. RNA as an enzyme. *Sci. Am.* 255(5):64–75.

Darnell, J. E., Jr. 1983. The processing of RNA. *Sci. Am.* 249(4):90–100.

Das, A. 1993. Control of transcription termination by RNA-binding proteins. *Ann. Rev. Biochem.* 62:893–930.

Guthrie, C. 1991. Messenger RNA splicing in yeast: Clues to why the spliceosome is a ribonucleoprotein. *Science* 253:157–63.

Koleske, A. J., and Young, R. A. 1995. The RNA polymerase II holoenzyme and its implications for gene regulation. *Trends Biochem. Sci.* 20(3):113–16.

McClure, W. R. 1985. Mechanism and control of transcription initiation in prokaryotes. *Ann. Rev. Biochem.* 54:171–204.

Rosbash, M., and Séraphin, B. 1991. Who's on first? The U1 snRNP-5′ splice site interaction and splicing. *Trends Biochem. Sci.* 16(5):187–90.

Sachs, A., and Wahle, E. 1993. Poly(A) tail metabolism and function in eucaryotes. *J. Biol. Chem.* 268(31):22955–58.

Steitz, J. A. 1988. "Snurps." *Sci. Am.* 258(6):56–63.

Tjian, R. 1995. Molecular machines that control genes. *Sci. Am.* 272(2):54–61.

Protein Synthesis

Lake, J. A. 1981. The ribosome. *Sci. Am.* 245(2):84–97.

Lake, J. A. 1985. Evolving ribosome structure: Domains in archaebacteria, eubacteria, eocytes and eukaryotes. *Ann. Rev. Biochem.* 54:507–30.

Merrick, W. C. 1992. Mechanism and regulation of eukaryotic protein synthesis. *Microbiol. Rev.* 56(2):291–315.

Weijland, A., and Parmeggiani, A. 1994. Why do two EF-Tu molecules act in the elongation cycle of protein biosynthesis? *Trends Biochem. Sci.* 19:188–93.

Wek, R. C. 1994. eIF-2 kinases: Regulators of general and gene-specific translation initiation. *Trends Biochem. Sci.* 19:491–96.

CHAPTER 12
Metabolism:
The Regulation of Enzyme Activity and Synthesis

Living cells are self-regulating chemical engines, tuned to operate on the principle of maximum economy.

—A. L. Lehninger

Outline

Concepts

1. Metabolism is regulated in such a way that (a) cell components are maintained at the proper concentrations, even in the face of a changing environment, and (b) energy and material are conserved.

2. The localization of enzymes and metabolites in separate compartments of a cell regulates and coordinates metabolic activity.

3. The activity of regulatory enzymes may be changed through reversible binding of effectors to a regulatory site separate from the catalytic site or through covalent modification of the enzyme. Regulation of enzyme activity operates rapidly and serves as a fine-tuning mechanism to adjust metabolism from moment to moment.

4. A pathway's activity is often controlled by its end products through feedback inhibition of regulatory enzymes located at the start of the sequence and at branch points.

5. The long-term regulation of metabolism in bacteria is achieved through the control of transcription by such mechanisms as sigma factors, repressor proteins during induction and repression, and by the attenuation of many biosynthetic operons.

6. DNA replication and cell division are coordinated in such a way that the distribution of new DNA copies to each daughter cell is ensured.

The organization of microbial metabolism is briefly described in the previous four chapters. Up to this point a metabolic pathway has been treated simply as a sequence of enzymes functioning as a unit, with each enzyme using as its substrate a product of the preceding enzyme-catalyzed reaction. This picture is incomplete because the regulation of pathway operation has been ignored. Both regulation of the activity of individual pathways and coordination of the action of separate sequences are essential to the existence of life. Cells become disorganized and die without adequate control of metabolism. This chapter introduces the principles of metabolic regulation.

The chapter begins with a consideration of the need for metabolic regulation, and subsequently describes metabolic channeling, the regulation of the activity of critical enzymes, and the control of nucleic acid synthesis. Finally, regulation of the bacterial cell cycle is discussed.

The task of the regulatory machinery is exceptionally complex and difficult. Pathways must be regulated and coordinated so effectively that all cell components are present in precisely the correct amounts. Furthermore, a microbial cell must be able to respond effectively to environmental changes by using those nutrients present at the moment and by switching on new catabolic pathways when different nutrients become available. Because all chemical components of a cell usually are not present in the surroundings, microorganisms also must synthesize unavailable components and alter biosynthetic activity in response to changes in nutrient availability. The chemical composition of a cell's surroundings is constantly changing, and these regulatory processes are dynamic and continuously responding to altered conditions.

Regulation is essential for the cell to conserve microbial energy and material and to maintain metabolic balance. If a particular energy source is unavailable, the enzymes required for its use are not needed and their further synthesis is a waste of carbon, nitrogen, and energy. Similarly it would be extremely wasteful for a microorganism to synthesize the enzymes required to manufacture a certain end product if that end product were already present in adequate amounts. Thus both catabolism and anabolism are regulated in such a way as to maximize efficiency of operation.

The drive to maintain balance and conserve energy and material is evident in the regulatory responses of a bacterium like *E. coli*. If the bacterium is grown in a very simple medium containing only glucose as a carbon and energy source, it will synthesize the required cell components in balanced amounts. Addition of a biosynthetic end product (the amino acid tryptophan, for example) to the medium will result in the immediate inhibition of the pathway synthesizing that end product; synthesis of the pathway's enzymes also will slow or cease. If *E. coli* is transferred to medium containing only the sugar lactose, it will synthesize the enzymes required for catabolism of this nutrient.

In contrast, when *E. coli* grows in a medium possessing both glucose and lactose, glucose (the sugar supporting most rapid growth) is catabolized first. The culture will use lactose only after the glucose supply has been exhausted.

The flow of carbon through a pathway may be regulated in three major ways.

1. The localization of metabolites and enzymes in different parts of a cell, a phenomenon called **metabolic channeling,** influences pathway activity.
2. Critical enzymes often are directly stimulated or inhibited to alter pathway activity rapidly.
3. The number of enzyme molecules also may be controlled. The more catalyst molecules present, the greater the pathway's activity. In bacteria regulation is usually exerted at the level of transcription. Control of mRNA synthesis is slower than direct regulation of enzyme activity but does result in the saving of much energy and raw material because enzymes are not synthesized when not required.

Each of these mechanisms is described in detail.

Metabolic Channeling

One of the most common channeling mechanisms is that of **compartmentation,** the differential distribution of enzymes and metabolites among separate cell structures or organelles. Compartmentation is particularly important in eucaryotic microorganisms with their many membrane-bound organelles. For example, fatty acid oxidation is located within the mitochondrion, whereas fatty acid synthesis occurs in the cytoplasmic matrix. Compartmentation makes possible the simultaneous, but separate, operation and regulation of similar pathways. Furthermore, pathway activities can be coordinated through regulation of the transport of metabolites and coenzymes between cell compartments. Suppose two pathways in different cell compartments require NAD for activity. The distribution of NAD between the two compartments will then determine the relative activity of these competing pathways, and the pathway with access to the most NAD will be favored.

Channeling also occurs within compartments such as the cytoplasmic matrix. The matrix is a structured dense material with many subcompartments. In eucaryotes it also is subdivided by the endoplasmic reticulum and cytoskeleton (*see chapter 4*). Metabolites and coenzymes do not diffuse rapidly in such an environment, and metabolite gradients will build up near localized enzymes or enzyme systems. This occurs because enzymes at a specific site convert their substrates to products, resulting in decreases in the concentration of one or more metabolites and increases in others. For example, product concentrations will be high near an enzyme and decrease with increasing distance from it.

Channeling can generate marked variations in metabolite concentrations and therefore directly affect enzyme activity. Substrate levels are generally around 10^{-3} moles/liter (M) to 10^{-6} M or even lower. Thus they may be in the same range as enzyme concentrations and equal to or less than the Michaelis constants (K_m) of many enzymes. Under these conditions the concentration of an enzyme's substrate may control its activity because the substrate concentration is in the rising portion of the hyperbolic substrate saturation curve (figure 12.1). As the substrate level increases, it is converted to product more rapidly; a decline in substrate concentration automatically leads to lower enzyme activity. If two enzymes in different pathways use the same metabolite, they may directly compete for it. The pathway winning this competition—the one with the enzyme having the lowest K_m value for the metabolite—will operate closer to full capacity. Thus channeling within a cell compartment can regulate and coordinate metabolism through variations in metabolite and coenzyme levels.

Enzyme kinetics and the substrate saturation curve (pp. 160–61).

1. Give three ways in which the flow of carbon through a pathway may be regulated.
2. Define the terms metabolic channeling and compartmentation. How are they involved in the regulation of metabolism?

Control of Enzyme Activity

Adjustment of the activity of regulatory enzymes controls the functioning of many metabolic pathways. This section describes these enzymes and discusses their role in regulating pathway activity.

Allosteric Regulation

Usually regulatory enzymes are **allosteric enzymes.** The activity of an allosteric enzyme is altered by a small molecule known as an **effector** or **modulator.** The effector binds reversibly by noncovalent forces to a **regulatory site** separate from the catalytic site and causes a change in the shape or conformation of the enzyme (figure 12.2). The activity of the catalytic site is altered as a result. A positive effector increases enzyme activity, whereas a negative effector decreases activity or inhibits the enzyme. These changes in activity often result from alterations in the apparent affinity of the enzyme for its substrate, but changes in maximum velocity also can occur.

The kinetic characteristics of nonregulatory enzymes (*see chapter 8*) show that the Michaelis constant (K_m) is the substrate concentration required for an enzyme to operate at half its maximal velocity. This constant applies only to hyperbolic substrate saturation curves, not to the sigmoidal curves often seen with allosteric enzymes (figure 12.4). The substrate concentration required for half maximal velocity with allosteric enzymes having sigmoidal substrate curves is called the $[S]_{0.5}$ or $K_{0.5}$ value.

One of the best-studied allosteric regulatory enzymes is the aspartate carbamoyltransferase (ACTase) from *E. coli.* The enzyme catalyzes the condensation of carbamoyl phosphate with aspartate to form carbamoylaspartate (figure 12.3). ACTase catalyzes the

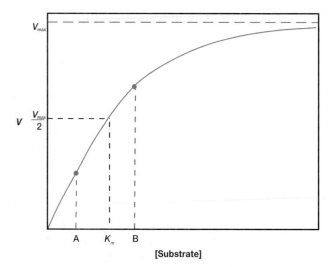

Figure 12.1   Control of Enzyme Activity by Substrate Concentration. An enzyme-substrate saturation curve with the Michaelis constant (K_m) and the velocity equivalent to half the maximum velocity (V_{max}) indicated. The initial velocity of the reaction (v) is plotted against the substrate concentration [Substrate]. The maximum velocity is the greatest velocity attainable with a fixed amount of enzyme under defined conditions. When the substrate concentration is equal to or less than the K_m, the enzyme's activity will vary almost linearly with the substrate concentration. Suppose the substrate increases in concentration from level A to B. Because these concentrations are in the range of the K_m, a significant increase in enzyme activity results. A drop in concentration from B to A will lower the rate of product formation.

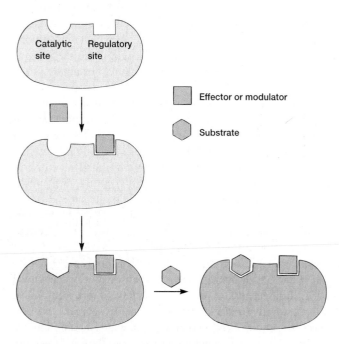

Figure 12.2   Allosteric Regulation. The structure and function of an allosteric enzyme. In this example the effector or modulator first binds to a separate regulatory site and causes a change in enzyme conformation that results in an alteration in the shape of the active site. The active site can now more effectively bind the substrate. This effector is a positive effector because it stimulates substrate binding and catalytic activity.

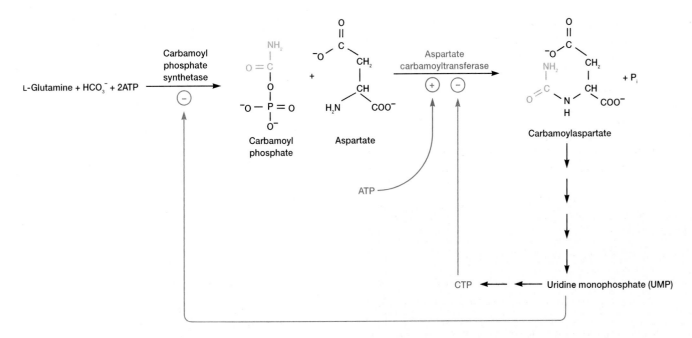

Figure 12.3   ACTase Regulation. The aspartate carbamoyltransferase reaction and its role in the regulation of pyrimidine biosynthesis. The end product CTP inhibits its activity (–) while ATP activates the enzyme (+). Carbamoyl phosphate synthetase is also inhibited by pathway end products such as UMP.

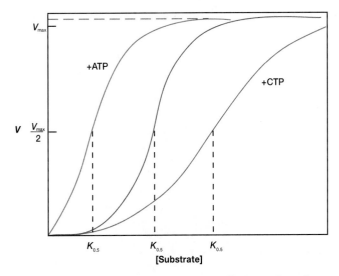

Figure 12.4   The Kinetics of *E. coli* Aspartate Carbamoyltransferase. CTP, a negative effector, increases the $K_{0.5}$ value while ATP, a positive effector, lowers the $K_{0.5}$. The V_{max} remains constant.

rate-determining reaction of the pyrimidine biosynthetic pathway in *E. coli*. The substrate saturation curve is sigmoidal when the concentration of either substrate is varied (figure 12.4). The enzyme has more than one active site, and the binding of a substrate molecule to an active site increases the binding of substrate at the other sites. In addition, cytidine triphosphate (CTP), an end product of pyrimidine biosynthesis, inhibits the enzyme and the purine ATP activates it. Both effectors alter the $K_{0.5}$ value of the enzyme but not its maximum velocity. CTP inhibits by increasing $K_{0.5}$ or by shifting the substrate saturation curve to higher values. This allows the enzyme to operate more slowly at

a particular substrate concentration when CTP is present. ATP activates by moving the curve to lower substrate concentration values so that the enzyme is maximally active over a wider substrate concentration range. Thus when the pathway is so active that the CTP concentration rises too high, ACTase activity decreases and the rate of end product formation slows. In contrast, when the purine end product ATP increases in concentration relative to CTP, it stimulates CTP synthesis through its effects on ACTase.

Pyrimidine and purine biosynthesis (pp. 203–4).

E. coli aspartate carbamoyltransferase provides a clear example of separate regulatory and catalytic sites in allosteric enzymes. The enzyme is a large protein composed of two catalytic subunits and three regulatory subunits (figure 12.5*a*). The catalytic subunits contain only catalytic sites and are unaffected by CTP and ATP. Regulatory subunits do not catalyze the reaction but do possess regulatory sites to which CTP and ATP bind. When these effectors bind to the complete enzyme, they cause conformational changes in the regulatory subunits and subsequently in the catalytic subunits and their catalytic sites. The enzyme can change reversibly between a less active T form and a more active R form (figure 12.5*b,c*). Thus the regulatory site influences a catalytic site about 6.0 nm distant.

Covalent Modification of Enzymes

Regulatory enzymes also can be switched on and off by **reversible covalent modification.** Usually this occurs through the addition and removal of a particular group, one form of the enzyme being more active than the other. For example, glycogen phosphorylase of the bread mold *Neurospora crassa* exists in

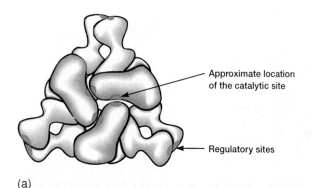

(a)

Figure 12.5   The Structure and Regulation of *E. coli* Aspartate Carbamoyltransferase. (*a*) A schematic diagram of the enzyme showing the six catalytic polypeptide chains (blue), the six regulatory chains (orange), and the catalytic and regulatory sites. The enzyme is viewed from the top. Each catalytic subunit contains three catalytic chains, and each regulatory subunit has two chains. (*b*) The less active T state of ACTase. The lines in the left-hand figure represent the peptide chains. (*c*) The more active R state of ACTase. The regulatory subunits have rotated and pushed the catalytic subunits apart.

(b)

(c)

phosphorylated and dephosphorylated forms called phosphorylase *a* and phosphorylase *b,* respectively (figure 12.6). Phosphorylase *b* is inactive because its required activator AMP is usually not present at sufficiently high levels. Phosphorylase *a,* the phosphorylated form, is active even without AMP. Glycogen phosphorylase is stimulated by phosphorylation of phosphorylase *b* to produce phosphorylase *a.* The attachment of

phosphate changes the enzyme's conformation to an active form. Phosphorylation and dephosphorylation are catalyzed by separate enzymes, which also are regulated.

Phosphorylase and glycogen degradation (p. 178).

Enzymes can be regulated through the attachment of groups other than phosphate. One of the most intensively studied

Figure 12.6   Reversible Covalent Modification of Glycogen Phosphorylase. The active form, phosphorylase *a*, is produced by phosphorylation and is inactivated when the phosphate is removed hydrolytically to produce inactive phosphorylase *b*.

regulatory enzymes is *E. coli* glutamine synthetase, a large, complex enzyme existing in two forms (figure 12.7). When an adenylic acid residue is attached to each of its 12 subunits forming an adenylylated enzyme, glutamine synthetase is not very active. Removal of AMP groups produces the more active deadenylylated glutamine synthetase, and glutamine is formed. The glutamine synthetase system differs from the phosphorylase system in two ways: (1) AMP is used as the modifying agent, and (2) the modified form of glutamine synthetase is less active. Glutamine synthetase also is allosterically regulated.

Glutamine synthetase and its role in nitrogen metabolism (p. 198).

Feedback Inhibition

The rate of many metabolic pathways is adjusted through control of the activity of the regulatory enzymes described in the preceding section. Every pathway has at least one **pacemaker enzyme** that catalyzes the slowest or rate-limiting reaction in the pathway. Because other reactions proceed more rapidly than the pacemaker reaction, changes in the activity of this enzyme directly alter the speed with which a pathway operates. Usually the first step in a pathway is a pacemaker reaction catalyzed by a regulatory enzyme. The end product of the pathway often inhibits this regulatory enzyme, a process known as **feedback inhibition** or **end product inhibition.** Feedback inhibition ensures balanced production of a pathway end product. If the end product becomes too concentrated, it inhibits the regulatory enzyme and slows its own synthesis. As the end product concentration decreases, pathway activity again increases and more product is formed. In this way feedback inhibition automatically matches end product supply with the demand. The previously discussed *E. coli* aspartate carbamoyltransferase is an excellent example of end product or feedback inhibition.

Frequently a biosynthetic pathway branches to form more than one end product. In such a situation the synthesis of pathway end products must be coordinated precisely. It would not do to have one end product present in excess while another is

(a)

(b)

Figure 12.7   The Structure of *E. coli* Glutamine Synthetase. The enzyme contains 12 subunits in the shape of a hexagonal prism. For clarity the subunits are colored alternating green and blue. Each of the six catalytic sites has a pair of Mn^{2+} ions (red). The tyrosine residues to which adenyl groups can be attached are colored red. (*a*) Top view of molecule. (*b*) Side view showing the six nearest subunits.

lacking. Branching biosynthetic pathways usually achieve a balance between end products through the use of regulatory enzymes at branch points (figure 12.8). If an end product is present in excess, it often inhibits the branch-point enzyme on the sequence leading to its formation, in this way regulating its own formation without affecting the synthesis of other products. In figure 12.8 notice that both products also inhibit the initial enzyme in the pathway. An excess of one product slows the flow

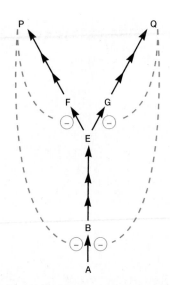

Figure 12.8   Feedback Inhibition. Feedback inhibition in a branching pathway with two end products. The branch-point enzymes, those catalyzing the conversion of intermediate E to F and G, are regulated by feedback inhibition. Products P and Q also inhibit the initial reaction in the pathway. A colored line with a minus sign at one end indicates that an end product, P or Q, is inhibiting the enzyme catalyzing the step next to the minus. See text for further explanation.

of carbon into the whole pathway while inhibiting the appropriate branch-point enzyme. Because less carbon is required when a branch is not functioning, feedback inhibition of the initial pacemaker enzyme helps match the supply with the demand in branching pathways. The regulation of multiply branched pathways is often made even more sophisticated by the presence of **isoenzymes,** different enzymes that catalyze the same reaction. The initial pacemaker step may be catalyzed by several isoenzymes, each under separate and independent control. In such a situation an excess of a single end product reduces pathway activity but does not completely block pathway function because some isoenzymes are still active.

---

1. Define the following: allosteric enzyme, effector or modulator, and $[S]_{0.5}$ or $K_{0.5}$.

2. How can regulatory enzymes be influenced by reversible covalent modification? What groups are used for this purpose with glycogen phosphorylase and glutamine synthetase, and which forms of these enzymes are active?

3. What is a pacemaker enzyme? Feedback inhibition? How does feedback inhibition automatically adjust the concentration of a pathway end product? What are isoenzymes and why are they important in pathway regulation?

Regulation of mRNA Synthesis

The control of metabolism by regulation of enzyme activity is a fine-tuning mechanism: it acts rapidly to adjust metabolic activity from moment to moment. Microorganisms also are able to control the expression of their genome, although over longer intervals. For example, the *E. coli* chromosome can code for about 2,000 to 4,000 peptide chains, yet only around 800 enzymes are present in *E. coli* growing with glucose as its energy source. Regulation of gene expression serves to conserve energy and raw material, to maintain balance between the amounts of various cell proteins, and to adapt to long-term environmental change. Thus control of gene expression complements the regulation of enzyme activity.

Regulation by Sigma Factors

Although the RNA polymerase core enzyme can transcribe any gene to produce a messenger RNA copy, it needs the assistance of a sigma factor to bind the promoter and initiate transcription (*see p. 222*). This provides an excellent means of regulating gene expression. When a complex process requires a radical change in transcription, or the synthesis of several gene products in a precisely timed sequence, it may be regulated by a series of sigma factors. Each sigma factor enables the RNA polymerase core enzyme to recognize a specific set of promoters and transcribe only those genes. Substitution of the sigma factor immediately changes gene expression. Bacteriophages often use sigma factors to control mRNA synthesis during their life cycle (*see chapter 17*). This regulatory mechanism also is common among both gram-negative and gram-positive bacteria. For example, *Escherichia coli* synthesizes several sigma factors. Under normal conditions the sigma factor σ^{70} directs RNA polymerase activity. (The superscript letter or number indicates the function or size of the sigma factor; 70 stands for 70,000 Da.) When flagella and chemotactic proteins are needed, *E. coli* produces σ^{F} (σ^{28}). If the temperature rises too high, σ^{H} (σ^{32}) appears and stimulates the formation of around 17 heat-shock proteins to protect the cell from thermal destruction. As would be expected, the promoters recognized by each sigma factor differ characteristically in sequence at the −10 and −35 positions (*see p. 260*).

One of the best-studied examples of gene regulation by sigma factors is the control of sporulation in the gram-positive *Bacillus subtilis*. When *B. subtilis* is deprived of nutrients, it will form endospores in a complex developmental process lasting about 8 hours.

The bacterial endospore (pp. 67–69).

Normally the *B. subtilis* RNA polymerase uses sigma factor σ^{A} (σ^{43}) to recognize genes. Environmental signals such as nutrient deprivation stimulate a kinase (Kin A or Kin B) to catalyze the phosphorylation of the SpoOF protein. SpoOF transfers the phosphate to SpoOB, which in turn phosphorylates SpoOA. Phosphorylated SpoOA has several effects. It binds to a promoter and represses the expression of the *abrB* gene. *abrB* codes for a protein that inhibits many genes not needed during growth with excess nutrients (e.g., at least three sporulation genes). Phosphorylated SpoOA also activates the production of two sigma factors: an active σ^{F} and an inactive pro-σ^{E}. Sporulation begins when σ^{F} partially replaces σ^{A} in the forespore. The

Figure 12.9   The β-Galactosidase Reaction.

Lactose   Galactose   Glucose

RNA polymerase then transcribes sporulation genes as well as vegetative genes. One of these early sporulation genes codes for another sigma factor, σ^G, that causes RNA polymerase to transcribe late sporulation genes in the forespore. At this point the pro-σ^E protein is activated by cleavage to yield σ^E, which then stimulates the transcription of the gene for pro-σ^K. Pro-σ^K is activated by a protease to yield σ^K and trigger the transcription of late genes in the mother cell. In summary, sporulation is regulated by two cascades of sigma factors, one in the forespore and the other in the mother cell. Each cascade influences the other through a series of signals so that the whole complex developmental process is properly coordinated.

Induction and Repression

The regulation of β-galactosidase synthesis has been intensively studied and serves as a primary example of how gene expression is controlled. This enzyme catalyzes the hydrolysis of the sugar lactose to glucose and galactose (figure 12.9). When *E. coli* grows with lactose as its carbon source, each cell contains about 3,000 β-galactosidase molecules, but has less than three molecules in the absence of lactose. The enzyme β-galactosidase is an **inducible enzyme**—that is, its level rises in the presence of a small molecule called an **inducer** (in this case the lactose derivative allolactose).

The genes for enzymes involved in the biosynthesis of amino acids and other substances often respond differently from genes coding for catabolic enzymes. An amino acid present in the surroundings may inhibit the formation of the enzymes responsible for its biosynthesis. This makes good sense because the microorganism will not need the biosynthetic enzymes for a particular substance if it is already available. Enzymes whose amount is reduced by the presence of an end product are **repressible enzymes,** and metabolites causing a decrease in the concentrations of repressible enzymes are **corepressors.** Generally, repressible enzymes are necessary for synthesis and always are present unless the end product of their pathway is available. Inducible enzymes, in contrast, are required only when their substrate is available; they are missing in the absence of the inducer. It is important to emphasize that although the expression of many biosynthetic genes is regulated by repression, the expression of others is controlled by attenuation.

The Mechanism of Induction and Repression

Although variations in enzyme levels could be due to changes in the rates of enzyme degradation, most enzymes are relatively stable in growing bacteria. Induction and repression result principally from changes in the rate of transcription. When *E. coli* is

growing in the absence of lactose, it often lacks mRNA molecules coding for the synthesis of β-galactosidase. In the presence of lactose, however, each cell has 35 to 50 β-galactosidase mRNA molecules. The synthesis of mRNA is dramatically influenced by the presence of lactose.

DNA transcription mechanism (pp. 222–23).

The rate of mRNA synthesis is controlled by special **repressor proteins** synthesized under the direction of regulator genes. The importance of regulator genes and repressors is demonstrated by mutationally inactivating a regulator gene to form a **constitutive mutant.** The mutant bacterium produces the enzymes in question whether or not they are needed. Thus inactivation of repressor proteins blocks the regulation of transcription. Although the responses to the presence of metabolites are different, both induction and repression are forms of **negative control:** mRNA synthesis proceeds more rapidly in the absence of the controlling factor. The reason for this becomes clear when the mechanism of repressor action is understood.

Gene structure (pp. 257–61).

The repressor protein binds to a specific site on DNA called the **operator,** which it finds in a two-step process. First, the repressor complexes with a DNA molecule, then rapidly slides along the DNA until it reaches its operator and stops. A portion of the repressor fits into the major groove of operator-site DNA, and the N-terminal arm appears to wrap around the DNA. The shape of the repressor protein is ideally suited for specific binding to the DNA double helix.

How does a repressor inhibit transcription? The promoter to which RNA polymerase binds (*see chapter 11*) is located next to the operator; sometimes the operator even overlaps the promoter. Repressor binding at the operator actually stimulates RNA polymerase binding to the promoter in the lactose operon. However, once the polymerase is bound, it is stored at the promoter and does not begin transcription until the repressor leaves the operator. A repressor does not affect the actual rate of transcription once it has begun.

Repressors must exist in both active and inactive forms because transcription would never occur if they were always active. In inducible systems the regulator gene directs the synthesis of an active repressor. The inducer stimulates transcription by reversibly binding to the repressor and causing it to change to an inactive shape (figure 12.10). Just the opposite takes place in a system controlled by repression (figure 12.11). The repressor protein initially is inactive and only assumes an active shape when the corepressor binds to it. The corepressor inhibits transcription by activating the repressor protein.

Figure 12.10 **Gene Induction.** The regulator gene, Reg., synthesizes an active repressor that binds to the operator, *O*, and blocks RNA polymerase binding to the promoter, *P*, unless the inducer inactivates it. In the presence of the inducer, the repressor protein is inactive and transcription occurs.

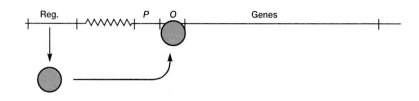

Figure 12.11 **Gene Repression.** The regulator gene, Reg., synthesizes an inactive repressor protein that must be activated by corepressor binding before it can bind to the operator, *O*, and block transcription. In the absence of the corepressor, the repressor is inactive and transcription occurs.

The synthesis of several proteins is often regulated by a single repressor. The **structural genes,** or genes coding for a polypeptide, are simply lined up together on the DNA, and a single mRNA carries all the messages. The sequence of bases coding for one or more polypeptides, together with the operator controlling its expression, is called an **operon.** This arrangement is of great advantage to the bacterium because coordinated control of the synthesis of several metabolically related enzymes

(or other proteins) can be achieved. For example, the lactose or lac operon contains three structural genes. One codes for β-galactosidase; a second gene directs the synthesis of β-galactoside permease, the protein responsible for lactose uptake. The third gene codes for the enzyme β-galactoside transacetylase, whose function still is uncertain. The presence of the first two genes in the same operon ensures that the rates of lactose uptake and breakdown will vary together (Box 12.1, p. 245).

Box 12.1

The Discovery of Gene Regulation

The ability of microorganisms to adapt to their environments by adjusting enzyme levels was first discovered by Emil Duclaux, a colleague of Louis Pasteur. He found that the fungus *Aspergillus niger* would produce the enzyme that hydrolyzes sucrose (invertase) only when grown in the presence of sucrose. In 1900 F. Dienert found that yeast contained the enzymes for galactose metabolism only when grown with lactose or galactose and would lose these enzymes upon transfer to a glucose medium. Such a response made sense because the yeast cells would not need enzymes for galactose metabolism when using glucose as its carbon and energy source. Further examples of adaptation were discovered and by the 1930s H. Karstrom could divide enzymes into two classes: (1) adaptive enzymes that are formed only in the presence of their substrates, and (2) constitutive enzymes that are always present. It was originally thought that enzymes might be formed from inactive precursors and that the presence of the substrate simply shifted the equilibrium between precursor and enzyme toward enzyme formation.

In 1942 Jacques Monod, working at the Pasteur Institute in Paris, began a study of adaptation in the bacterium *E. coli*. It was already known that the enzyme β-galactosidase, which hydrolyzes the sugar lactose to glucose and galactose, was present only when *E. coli* was grown in the presence of lactose. Monod discovered that nonmetabolizable analogues of β-galactosides, such as thiomethylgalactoside, also could induce enzyme production. This discovery made it possible to study induction in cells growing on carbon and energy sources other than lactose so that the growth rate and inducer concentration would not depend on the lactose supply. He next demonstrated that induction involved the synthesis of new enzyme, not just the conversion of already available precursor. Monod accomplished this by making *E. coli* proteins radioactive with ^{35}S, then transferring the labeled bacteria to nonradioactive medium and adding inducer. The newly formed β-galactosidase was nonradioactive and must have been synthesized after addition of inducer.

A study of the genetics of lactose induction in *E. coli* was begun by Joshua Lederberg a few years after Monod had started his work. Lederberg isolated not only mutants lacking β-galactosidase but also a constitutive mutant in which synthesis of the enzyme proceeded in the absence of an inducer (LacI$^-$). During bacterial conjugation (*see chapter 14*), genes from the donor bacterium enter the recipient to temporarily form an organism with two copies of those genes provided by the donor. When Arthur B. Pardee, François Jacob, and Monod transferred the gene for inducibility to a constitutive recipient not sensitive to inducers, the newly acquired gene made the recipient bacterium sensitive to inducer again. This functional gene was not a part of the recipient's chromosome. Thus the special gene directed the synthesis of a cytoplasmic product that inhibited the formation of β-galactosidase in the absence of the inducer. In 1961 Jacob and Monod named this special product the repressor and suggested that it was a protein. They further proposed that the repressor protein exerted its effects by binding to the operator, a special site next to the structural genes. They provided genetic evidence for their hypothesis. The name operon was given to the complex of the operator and the genes it controlled. Several years later in 1967, Walter Gilbert and Benno Müller-Hill managed to isolate the lac repressor and show that it was indeed a protein and did bind to a specific site in the lac operon.

The existence of repression was discovered by Monod and G. Cohen-Bazire in 1953 when they found that the presence of the amino acid tryptophan would repress the synthesis of tryptophan synthetase, the final enzyme in the pathway for tryptophan biosynthesis. Subsequent research in many laboratories showed that induction and repression were operating by quite similar mechanisms, each involving repressor proteins that bound to operators on the genome.

Positive Operon Control and Catabolite Repression

The preceding section shows that operons can be under negative control, resulting in induction and repression. In contrast, some operons (e.g., the lac operon) function only in the presence of a controlling factor—that is, they are under **positive operon control.**

Lac operon function is regulated by the **catabolite activator protein (CAP)** or **cyclic AMP receptor protein (CRP)** and the small cyclic nucleotide **3′, 5′-cyclic adenosine monophosphate (cAMP;** figure 12.12), as well as by the lac repressor protein. The lac promoter contains a CAP site to which CAP must bind before RNA polymerase can attach to the promoter and begin transcription (figure 12.13). The catabolite activator protein is able to bind to the CAP site only when complexed with cAMP (figure 12.14). This positive control system makes lac operon activity dependent on the presence of cAMP as well as on that of lactose.

Figure 12.12   Cyclic Adenosine Monophosphate (cAMP). The phosphate group extends between the 3′ and 5′ hydroxyls of the ribose sugar. The enzyme adenyl cyclase forms cAMP from ATP.

Figure 12.13   Positive Control of the Lac Operon. When cyclic AMP is absent or present at a low level, the CAP protein remains inactive and does not bind to the promoter. In this situation RNA polymerase also does not bind to the promoter and transcribe the operon's genes.

(a)

(b)

Figure 12.14   CAP Structure and DNA Binding. (*a*) A diagram of the CAP dimer, the active form. The C-terminal domains bind to the promoter. The cAMP binding domains are located toward the N-terminal end. (*b*) The CAP dimer binding to DNA at the lac operon promoter. The recognition helices fit into two adjacent major grooves on the double helix. (*c*) A model of the *E. coli* CAP-DNA complex derived from crystal structure studies. The cAMP binding domain is in blue and the DNA binding domain, in purple. The cAMP molecules bound to CAP are in red. Note that the DNA is bent by 90° when complexed with CAP.

(c)

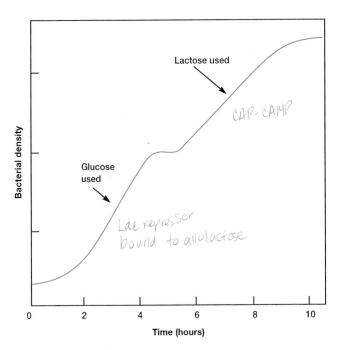

Figure 12.15   Diauxic Growth. The diauxic growth curve of *E. coli* grown with a mixture of glucose and lactose. Glucose is first used, then lactose. A short lag in growth is present while the bacteria synthesize the enzymes needed for lactose use.

The value of this system to bacteria is shown by a simple growth experiment. If *E. coli* grows in a medium containing both glucose and lactose, it uses glucose preferentially until the sugar is exhausted. Then after a short lag, growth resumes with lactose as the carbon source (figure 12.15). This biphasic growth pattern or response is called **diauxic growth.** The cause of diauxic growth or diauxie is complex and not completely understood, but **catabolite repression** or the glucose effect probably plays a part. The enzymes for glucose catabolism are constitutive and unaffected by CAP activity. When the bacterium is given glucose, the cAMP level drops, resulting in deactivation of the catabolite activator protein and inhibition of lac operon expression. The decrease in cAMP may be due to the effect of the phosphoenolpyruvate: phosphotransferase system (PTS) on the activity of adenyl cyclase, the enzyme that synthesizes cAMP. Enzyme III of the PTS donates a phosphate to glucose during its transport; therefore, it enters the cell as glucose 6-phosphate. The phosphorylated form of enzyme III also activates adenyl cyclase. If glucose is being rapidly transported by PTS, the amount of phosphorylated enzyme III is low and the adenyl cyclase is less active, so the cAMP level drops. At least one other mechanism is involved in diauxic growth. When the PTS is actively transporting glucose into the cell, nonphosphorylated enzyme III is more prevalent. Nonphosphorylated enzyme III binds to the lactose permease and allosterically inhibits it, thus blocking lactose uptake.

The phosphoenolpyruvate: phosphotransferase system (p. 105).

Whatever the precise mechanism, such control is of considerable advantage to the bacterium. It will use the most easily catabolized sugar (glucose) first rather than synthesize the enzymes necessary for another carbon and energy source. These control mechanisms are present in a variety of bacteria and metabolic pathways.

Knowledge of microbial regulatory mechanisms is extremely useful in industrial microbiology. A bacterial strain that excretes large amounts of the desired metabolite (e.g., an amino acid or vitamin) often can be developed if mutations are induced that either eliminate feedback inhibition of a critical enzyme in the synthetic pathway or inactivate a regulator gene or operator to yield constitutive enzyme synthesis.

Genetic manipulations and industrial microbiology (pp. 919–21).

Attenuation

Bacteria can regulate transcription in other ways, as may be seen in the tryptophan operon of *E. coli*. The tryptophan operon contains structural genes for five enzymes in this amino acid's biosynthetic pathway. As might be expected, the operon is under the control of a repressor protein coded for by the *trpR* gene (trp stands for tryptophan), and excess tryptophan inhibits transcription of operon genes by acting as a corepressor and activating the repressor protein. Although the operon is regulated mainly by repression, the continuation of transcription also is controlled.

A **leader region** lies between the operator and the first structural gene in the operon, the *trpE* gene, and is responsible for controlling the continuation of transcription after the RNA polymerase has bound to the promoter (figure 12.16a). The leader region contains an **attenuator** and a sequence that codes for the synthesis of a leader peptide. The attenuator is a rho-independent termination site (*see chapter 11*) with a short GC-rich segment followed by a sequence of eight U residues. The four stretches marked off in figure 12.16a have complementary base sequences and can base pair with each other to form hairpin loops. In the absence of a ribosome, segments one and two pair to form a hairpin, while segments three and four generate a second loop next to the poly(U) sequence (figure 12.16b). The hairpin formed by segments three and four plus the poly(U) sequence will terminate transcription. If segment one is prevented from base pairing with segment two, segment two is free to associate with segment three. As a result segment four remains single stranded (figure 12.16c) and cannot serve as a terminator for transcription. It is important to note that the sequence coding for the leader peptide contains two adjacent codons that code for the amino acid tryptophan. Thus the complete peptide can be made only when there is an adequate supply of tryptophan. Since the leader peptide has not been detected, it must be degraded immediately after synthesis.

Ribosome behavior during translation of the mRNA regulates RNA polymerase activity as it transcribes the leader region. This is possible because translation and transcription are tightly coupled. When the active repressor is absent, RNA polymerase binds to the promoter and moves down the leader synthesizing mRNA. If there is no translation of the mRNA after the RNA polymerase has begun copying the leader region,

Figure 12.16 **The Tryptophan Operon Leader.** (*a*) Organization and base pairing of the tryptophan operon leader region. The promoter and operator are to the left of the segment diagrammed, and the first structural gene (*trpE*) begins to the right of the attenuator. (*b* and *c*) The stretches of DNA marked off as *1* through *4* can base pair with each other to form hairpin loops: segment *2* with *1*, and segment *3* with *2* or *4*. See text for details.

segments three and four form a hairpin loop, and transcription terminates before the polymerase reaches the *trpE* gene (figure 12.17*a*). When tryptophan is present, there is sufficient tryptophanyl-tRNA for protein synthesis. Therefore the ribosome will synthesize the leader peptide and continue moving along the mRNA until it reaches a UGA stop codon (*see chapter 13*) lying between segments one and two. The ribosome halts at this codon and projects into segment two far enough to prevent it from pairing properly with segment three (figure 12.17*b*). Segments three and four form a hairpin loop, and the RNA polymerase terminates at the attenuator just as if no translation had taken place. If tryptophan is lacking, the ribosome will stop at the two adjacent tryptophan codons in the leader peptide sequence and prevent segment one from base pairing with segment two, because the tryptophan codons are located within segment one (figures 12.16*a* and 12.17*c*). If this happens while the RNA polymerase is still transcribing the leader region, segments two and three associate before segment four has been synthesized. Therefore segment four will remain single stranded and the terminator hairpin will not form. Consequently, when tryptophan is absent, the RNA polymerase continues on and transcribes tryptophan operon genes. Control of the continuation of transcription by a specific aminoacyl-tRNA is called **attenuation.**

Attenuation's usefulness is apparent. If the bacterium is deficient in an amino acid other than tryptophan, protein synthesis will slow and tryptophanyl-tRNA will accumulate. Transcription of the tryptophan operon will be inhibited by attenuation. When the bacterium begins to synthesize protein rapidly, tryptophan may be scarce and the concentration of tryptophanyl-tRNA may be low. This would reduce attenuation activity and stimulate operon transcription, resulting in larger quantities of the tryptophan biosynthetic enzymes. Acting together, repression and attenuation can coordinate the rate of synthesis of amino acid biosynthetic enzymes with the availability of amino acid end products and with the overall rate of protein synthesis.

When tryptophan is present at high concentrations, any RNA polymerases not blocked by the activated repressor protein probably will not get past the attenuator sequence. Repression decreases transcription about seventyfold and attenuation slows it another eight- to tenfold; when both mechanisms operate together, transcription can be slowed about 600-fold.

Attenuation seems important in the regulation of several amino acid biosynthetic pathways. At least five other operons have leader peptide sequences that resemble the tryptophan system in organization. For example, the leader peptide sequence of the histidine operon codes for seven histidines in a row and is followed by an attenuator that is a terminator sequence.

Gene Regulation by Antisense RNA

Microbiologists have known for many years that gene expression can be controlled by both regulatory proteins (e.g., repressor proteins and CAP) and aminoacyl-tRNA (attenuation). More recently it has been discovered that the activity of some genes is controlled by a special type of small regulatory RNA molecule. The regulatory RNA, called **antisense RNA,** has a base sequence complementary to a segment of another RNA molecule and specifically binds to the target RNA. Antisense RNA binding can block DNA replication, mRNA synthesis, or translation. The genes coding for these RNAs are sometimes called antisense genes.

This mode of regulation appears to be widespread among viruses and bacteria. Examples are the regulation of plasmid replication and Tn10 transposition, osmoregulation of porin protein expression, regulation of λ phage reproduction, and the autoregulation of cAMP-receptor protein synthesis. Antisense RNA regulation has not yet been demonstrated in eucaryotic cells, although there is evidence that it may exist. It is possible that antisense RNAs bind with some eucaryotic mRNAs and stimulate their degradation.

Plasmids and transposons (pp. 280–88).

Figure 12.17   **Attenuation Control.**
The control of tryptophan operon
function by attenuation. See text for
details.

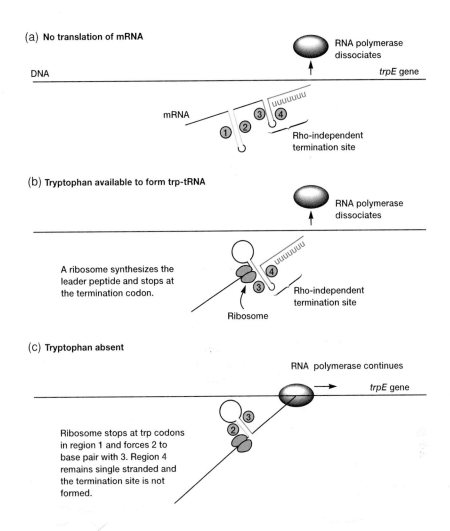

(a) **No translation of mRNA**

DNA

RNA polymerase
dissociates

trpE gene

mRNA

uuuuuuu

Rho-independent
termination site

(b) **Tryptophan available to form trp-tRNA**

RNA polymerase
dissociates

A ribosome synthesizes the
leader peptide and stops at
the termination codon.

uuuuuuu

Rho-independent
termination site

Ribosome

(c) **Tryptophan absent**

RNA polymerase continues

trpE gene

Ribosome stops at trp codons
in region 1 and forces 2 to
base pair with 3. Region 4
remains single stranded and
the termination site is not
formed.

As mentioned previously, antisense RNA can regulate gene expression in several ways. A good example of one mechanism is provided by the ColE1 plasmid of *E. coli* (*see pp. 281, 283*). Plasmid replication is regulated so that about 10 to 30 copies of ColE1 are present in a cell. RNA I, an antisense RNA about 100 nucleotides long, binds with a precursor RNA used to form the RNA primer for plasmid DNA replication. DNA polymerase cannot copy the plasmid without the primer and replication is blocked. Thus RNA I controls DNA replication by adjusting the number of functional RNA primers.

The regulation of *E. coli* outer membrane porin proteins provides a second example of control by antisense RNA. The outer membrane contains channels made of porin proteins (*see p. 56*). The two most important porins in *E. coli* are the OmpF and OmpC proteins. OmpC pores are slightly smaller and are made when the bacterium grows at high osmotic pressures. It is the dominant porin in *E. coli* from the intestinal tract. This makes sense because the smaller pores would exclude many of the toxic molecules present in the intestine. The larger OmpF pores are favored when *E. coli* grows in a dilute environment, and they allow solutes to diffuse into the cell more readily.

The *ompF* and *ompC* genes are partly regulated by a special OmpR protein that represses the *ompF* gene and activates *ompC*. In addition, the *micF* gene produces a 174-nucleotide-long antisense micF RNA that blocks *ompF* action (mic stands for *m*RNA-*i*nterfering *c*omplementary RNA). The micF RNA is complementary to *ompF* at the translation initiation site. It complexes with *ompF* mRNA and represses translation. The *micF* gene is activated by conditions such as high osmotic pressure that favor *ompC* expression. This helps ensure that OmpF protein is not produced at the same time as OmpC protein.

The fact that antisense RNA can bind specifically to mRNA and block its activity has great practical implications. Antisense RNA is already a valuable research tool. Suppose one desires to study the action of a particular gene. An antisense RNA that will bind to the gene's mRNA can be constructed and introduced into the cell, thus blocking gene expression. Changes in the cell are then observed. It also is possible to use the same approach with short strands of antisense DNA that bind to mRNA.

Antisense RNA and DNA may well be effective against a variety of cancers and infectious diseases. Promising preliminary results have been obtained using antisense oligonucleotides directed against *Trypanosoma brucei* (the cause of African sleeping sickness), herpesviruses, the HIV virus, tumor viruses such as the RSV and polyoma viruses, and chronic myelogenous

Figure 12.18 Control of the Cell Cycle in *E. coli*. A 60-minute interval between divisions has been assumed for purposes of simplicity (the actual time between cell divisions may be shorter). *E. coli* requires about 40 minutes to replicate its DNA and 20 minutes after termination of replication to prepare for division. The position of events on the time line is approximate and meant to show the general pattern of occurrences.

leukemia. Although much further research is needed to determine the medical potential of these molecules, they may prove invaluable in the treatment of many diseases.

1. Briefly describe how sigma factors can be used to control gene expression. When might one see this control mechanism used?
2. What are induction and repression and why are they useful? Define inducer, corepressor, repressor protein, regulator gene, constitutive mutant, operator, structural gene, and operon.
3. Define positive control, catabolite activator protein, catabolite repression, and diauxic growth.
4. Define attenuation and describe how it works in terms of a labeled diagram, such as that provided in figure 12.17. What are the functions of the leader region and the attenuator in attenuation?
5. What is antisense RNA? How does it regulate gene expression?

Control of the Cell Cycle

Although much progress has been made in understanding the control of microbial enzyme activity and pathway function, much less is known about the regulation of more complex events such as bacterial sporulation and cell division. This section briefly describes the regulation of bacterial cell division. Attention is focused primarily on *E. coli* because it has been intensively studied.

The complete sequence of events extending from the formation of a new cell through the next division is called the **cell cycle**. A young *E. coli* cell growing at a constant rate will double in length without changing in diameter, then divide into two cells of equal size by transverse fission. Because each daughter cell receives at least one copy of the genetic material, DNA replication and cell division must be tightly coordinated. In fact, if DNA synthesis is inhibited by a drug or a gene mutation, cell division is also blocked and the affected cells continue to elongate, forming long filaments. Termination of DNA replication also seems connected in some way with cell division. Although the growth rate of *E. coli* at 37°C may vary considerably,

division usually takes place about 20 minutes after replication has finished. During this final interval the genetic material must be distributed between the daughter cells. The newly formed DNA copies are attached to adjacent sites on the plasma membrane. Membrane growth first separates them, then a cross wall or septum forms between the two.

Patterns of cell wall formation (p. 209).

Current evidence suggests that two sequences of events, operating in parallel but independently, control division and the cell cycle (figure 12.18). Like eucaryotic cells, bacteria must reach a specific threshold size or initiation mass to trigger DNA replication. *E. coli* also has to reach a threshold length before it can partition its chromosomes and divide into two cells. Thus there seem to be two separate controls for the cell cycle, one sensitive to cell mass and the other responding to cell length. DNA replication and cell division take about 40 minutes to complete.

Some of the *E. coli* cell cycle control mechanisms are becoming clearer, although much remains to be learned. The initiation of DNA replication requires binding of many copies of the DnaA protein to *oriC*, the replication origin site. Active DnaA protein has bound ATP, and the interconversion between DnaA-ATP and DnaA-ADP may help regulate initiation. Other factors also appear to participate in initiating DNA replication. After DNA replication is under way, another round does not immediately begin, partly because the parental DNA strand is methylated right after replication. The methylated replication origin binds to specific areas on the plasma membrane and is inactive.

The initiation of septation is equally complex and tightly regulated. Both termination of DNA replication and the attainment of threshold length are required to trigger septation and cell division. This is at least partly due to the inhibition of septation by the proximity of chromosomes. The presence of DNA damage inhibits septation as well. A cell will complete chromosome replication, repair any DNA damage, and partition the chromosomes into opposite ends before it forms a septum and divides. There probably are one or more regulatory proteins that interact with various division proteins to promote septum formation and division. An adequate supply of peptidoglycan chains or precursors also must be available at the proper time. The FtsZ protein is a division protein essential in initiating septation. This protein is scattered throughout the cell between divisions. At the onset of septation, it forms a ring at the septation site, and the ring then becomes smaller. Because the FtsZ protein hydrolyzes GTP, the ring may be a contractile structure that uses GTP energy. Several other proteins also are required for cell division. For example, the PBP3 or penicillin-binding protein 3 catalyzes peptidoglycan transglycosylation and peptidoglycan transpeptidation to help create the new cell wall (*see pp. 206–9*). Clearly regulation of the bacterial cell cycle is complex and involves several interacting regulatory mechanisms.

The relationship of DNA synthesis to the cell cycle varies with the growth rate. If *E. coli* is growing with a doubling time of about 60 minutes, DNA replication does not take place during the last 20 minutes—that is, replication is a discontinuous process when the doubling time is 60 minutes or longer. When the culture is growing with a doubling time of less than 60 minutes, a second round of replication begins while the first round is still under way (figure 12.19). The daughter cells may actually receive DNA with two or more replication forks, and replication is continuous because the cells are always copying their DNA.

Two decades of research have provided a fairly adequate overall picture of the cell cycle in *E. coli*. Several cell division genes have been identified. Yet it still is not known precisely how the cycle is controlled. Future work should improve our understanding of this important process.

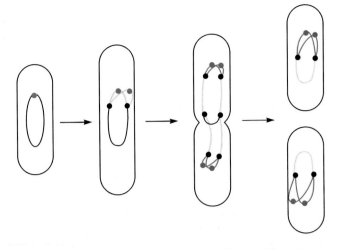

Figure 12.19   DNA Replication in Rapidly Growing Bacteria. A new round of DNA replication is initiated before the original cell divides so that the DNA in daughter cells is already partially replicated. The colored circles at the ends of the DNA loops are replication origins; the black circles along the sides are replication forks. Newly synthesized DNA is in color, with the darker shade representing the most recently synthesized DNA.

1. What is a cell cycle? Briefly describe how the cycle in *E. coli* is regulated and how cycle timing results.

2. How are the two DNA copies separated and apportioned between the two daughter cells?

Summary

1. The regulation of metabolism keeps cell components in proper balance and conserves metabolic energy and material.

2. The localization of metabolites and enzymes in different parts of the cell, metabolic channeling, influences pathway activity. A common channeling mechanism is compartmentation.

3. Regulatory enzymes are usually allosteric enzymes, enzymes in which an effector or modulator binds reversibly to a regulatory site separate from the catalytic site and causes a conformational change in the enzyme to alter its activity.

4. Aspartate carbamoyltransferase is an allosteric enzyme that is inhibited by CTP and activated by ATP.

5. Enzyme activity also can be regulated by reversible covalent modification. Two examples of such regulation are glycogen phosphorylase (phosphate addition) and glutamine synthetase (AMP addition).

6. The first enzyme in a pathway and enzymes at branch points often are subject to feedback inhibition by one or more end products. Excess end product slows its own synthesis.

7. Sigma factors can control the expression of many genes during complex cellular changes.

8. β-galactosidase is an inducible enzyme whose concentration rises in the presence of its inducer.

9. Many biosynthetic enzymes are repressible enzymes whose levels are reduced in the presence of end products called corepressors.

10. Induction and repression result from regulation of the rate of transcription by repressor proteins coded for by regulator genes. A regulator gene mutation can lead to a constitutive mutant, which continuously produces a metabolite.

11. The repressor inhibits transcription by binding to an operator and interfering with the binding of RNA polymerase to its promoter.

12. In inducible systems the newly synthesized repressor protein is active, and inducer binding inactivates it. In contrast, an inactive repressor is synthesized in a repressible system and is activated by the corepressor.

13. Often one repressor regulates the synthesis of several enzymes because they are part of a single operon, a DNA sequence coding for one or more polypeptides and the operator controlling its expression.

14. Positive operon control of the lac operon is due to the catabolite activator protein, which is activated by cAMP. The resulting catabolite repression probably contributes to diauxic growth when *E. coli* is cultured in the presence of both glucose and lactose.

15. In the tryptophan operon a leader region lies between the operator and the first structural gene. It codes for the synthesis of a leader peptide and contains an attenuator, a rho-independent termination site.

16. The synthesis of the leader peptide by a ribosome while RNA polymerase is transcribing the leader region regulates transcription; therefore the tryptophan operon is expressed only when there is insufficient tryptophan available. This mechanism of transcription control is called attenuation.

17. Small antisense RNA molecules regulate the expression of some genes. They can affect DNA replication, RNA transcription, or translation. Antisense RNA and DNA molecules are valuable research tools and may be effective against cancer and infectious disease.

18. The complete sequence of events extending from the formation of a new cell through the next division is called the cell cycle.

19. The end of DNA replication is tightly linked to cell division, so division in *E. coli* usually takes place about 20 minutes after replication is finished. Special division and regulatory proteins likely are involved.

20. The new DNA copies are attached to the plasma membrane and are separated by membrane growth before a cross wall forms.

21. In very rapidly dividing bacterial cells, a new round of DNA replication begins before the cells divide.

Key Terms

allosteric enzyme 238
antisense RNA 248
attenuation 248
attenuator 247
catabolite activator protein (CAP) 245
catabolite repression 247
cell cycle 250
compartmentation 237
constitutive mutant 243
corepressor 243

3′, 5′-cyclic adenosine monophosphate (cAMP) 245
cyclic AMP receptor protein (CRP) 245
diauxic growth 247
effector or modulator 238
end product inhibition 241
feedback inhibition 241
inducer 243
inducible enzyme 243
isoenzymes 242
leader region 247

metabolic channeling 237
negative control 243
operator 243
operon 244
pacemaker enzyme 241
positive operon control 245
regulatory site 238
repressible enzyme 243
repressor proteins 243
reversible covalent modification 239
structural gene 244

Questions for Thought and Review

1. How might a substrate be able to regulate the activity of the enzyme using it?

2. Describe how *E. coli* aspartate carbamoyltransferase is regulated, both in terms of the effects of modulators and the mechanism by which they exert their influence.

3. What is the significance of the fact that regulatory enzymes often are located at pathway branch points?

4. Describe in some detail the organization of the regulatory systems responsible for induction and repression, and the mechanism of their operation.

5. How is *E. coli* able to use glucose exclusively when presented with a mixture of glucose and lactose?

6. Of what practical importance is attenuation in coordinating the synthesis of amino acids and

proteins? Describe how attenuation activity would vary when protein synthesis suddenly rapidly accelerated, then later suddenly decelerated.

7. How does the timing of DNA replication seem to differ between slow-growing and fast-growing cells? Be able to account for the fact that bacterial cells may contain more than a single copy of DNA.

Additional Reading

General

Gottschalk, G. 1986. *Bacterial metabolism,* 2d ed. New York: Springer-Verlag.
Lewin, B. 1994. *Genes,* 5th ed. New York: Oxford University Press.
Lodish, H.; Baltimore, D.; Berk, A.; Zipursky, S.L.; Matsudaira, P.; and Darnell, J. 1995. *Molecular cell biology,* 3d ed. New York: Scientific American Books.
Moat, A. G., and Foster, J. W. 1995. *Microbial physiology,* 3d ed. New York: John Wiley and Sons.
Neidhardt, F. C.; Ingraham, J. L.; and Schaechter, M. 1990. *Physiology of the bacterial cell.* Sunderland, Mass.: Sinauer Associates.
Stent, G. S., and Calendar, R. 1978. *Molecular genetics,* 2d ed. San Francisco: Freeman.
Voet, D., and Voet, J. G. 1995. *Biochemistry,* 2d ed. New York: John Wiley and Sons.
Watson, J. D.; Hopkins, N. H.; Roberts, J. W.; Steitz, J. A.; and Weiner, A. M. 1987. *Molecular biology of the gene,* 4th ed. Menlo Park, Calif.: Benjamin/Cummings.
Zubay, G. 1993. *Biochemistry,* 3d ed. Dubuque, Iowa: Wm. C. Brown Publishers.

Control of Enzyme Activity

Kantrowitz, E. R., and Lipscomb, W. N. 1988. *Escherichia coli* aspartate transcarbamylase: The relation between structure and function. *Science* 241:669–74.
Koshland, D. E., Jr. 1973. Protein shape and biological control. *Sci. Am.* 229(4):52–64.
Saier, M. H., Jr.; Wu, L.-F.; and Reizer, J. 1990. Regulation of bacterial physiological processes by three types of protein phosphorylating systems. *Trends Biochem. Sci.* 15:391–95.

Regulation of mRNA Synthesis

Botsford, J. L., and Harman, J. G. 1992. Cyclic AMP in prokaryotes. *Microbiol. Rev.* 56(1):100–22.
Busby, S., and Buc, H. 1987. Positive regulation of gene expression by cyclic AMP and its receptor protein in *Escherichia coli. Microbiol. Sci.* 4(12):371–75.
Danchin, A. 1987. Membrane integration of carbohydrate transport in bacteria. *Microbiol. Sci.* 4(9):267–69.
Errington, J. 1993. *Bacillus subtilis* sporulation: Regulation of gene expression and control of morphogenesis. *Microbiol. Rev.* 57(1):1–33.
Gold, L. 1988. Posttranscriptional regulatory mechanisms in *Escherichia coli. Ann. Rev. Biochem.* 57:199–233.
Losick, R. 1995. Differentiation and cell fate in a simple organism. *BioScience* 45(6):400–5.
Maniatis, T., and Ptashne, M. 1976. A DNA operator-repressor system. *Sci. Am.* 234(1):64–76.
McClure, W. R. 1985. Mechanism and control of transcription initiation in prokaryotes. *Ann. Rev. Biochem.* 54:171–204.
McKnight, S. L. 1991. Molecular zippers in gene regulation. *Sci. Am.* 264(4):54–64.
Perez-Martin, J.; Rojo, F.; and de Lorenzo, V. 1994. Promoters responsive to DNA bending: A common theme in prokaryotic gene expression. *Microbiol. Rev.* 58(2):268–90.
Ptashne, M. 1992. *A genetic switch,* 2d ed. Cambridge, Mass.: Blackwell Scientific Publications.
Ptashne, M., and Gilbert, W. 1970. Genetic repressors. *Sci. Am.* 222(6):36–44.
Saier, M. H. 1989. Protein phosphorylation and allosteric control of inducer exclusion and catabolite repression by the bacterial phosphoenolpyruvate: sugar phosphotransferase system. *Microbiol. Rev.* 53(1):109–20.

Stock, J. B.; Ninfa, A. J.; and Stock, A. M. 1989. Protein phosphorylation and regulation of adaptive responses in bacteria. *Microbiol. Rev.* 53(4):450–90.
Yanofsky, C. 1981. Attenuation in the control of expression of bacterial operons. *Nature* 289:751–58.

Gene Regulation by Antisense RNA

Green, P. J.; Pines, O.; and Inouye, M. 1986. The role of antisense RNA in gene regulation. *Ann. Rev. Biochem.* 55:569–97.
Nellen, W., and Lichtenstein, C. 1993. What makes an mRNA anti-sense-itive? *Trends Biochem. Sci.* 18:419–23.
Weintraub, H. M. 1990. Antisense RNA and DNA. *Sci. Am.* 262(1):40–46.

Control of the Cell Cycle

D'Ari, R., and Bouloc, P. 1990. Logic of the *Escherichia coli* cell cycle. *Trends Biochem. Sci.* 15:191–94.
de Boer, P. A. J.; Cook, W. R.; and Rothfield, L. I. 1990. Bacterial cell division. *Ann. Rev. Genet.* 24:249–74.
Donachie, W. D. 1993. The cell cycle of *Escherichia coli. Ann. Rev. Microbiol.* 47:199–230.
Marr, A. G. 1991. Growth rate of *Escherichia coli. Microbiol. Rev.* 55(2):316–33.
Murray, A., and Hunt, T. 1993. *The cell cycle: An introduction.* New York: W. H. Freeman.
Sargent, M. G. 1985. How do bacterial nuclei divide? *Microbiol. Sci.* 2(8):235–39.
Whittenbury, R., and Dow, C. S. 1978. Morphogenesis in bacteria. In *Companion to microbiology,* A. T. Bull and P. M. Meadow, editors, 221–63. New York: Longman.

PART FOUR
Microbial Genetics

A bacterial plasmid. This false color transmission electron micrograph shows a bacterial plasmid in which two genes, colored in blue, have been mapped. The DNA-RNA hybrid is red and the rest of the plasmid DNA is yellow.

CHAPTER 13
Microbial Genetics: General Principles

But the most important qualification of bacteria for genetic studies is their extremely rapid rate of growth. . . . a single E. coli *cell will grow overnight into a visible colony containing millions of cells, even under relatively poor growth conditions. Thus, genetic experiments on* E. coli *usually last one day, whereas experiments on corn, for example, take months. It is no wonder that we know so much more about the genetics of* E. coli *than about the genetics of corn, even though we have been studying corn much longer.*

—R. F. Weaver and
P. W. Hedrick

Concepts

1. Genetic information is contained in the nucleotide sequence of DNA (and sometimes RNA). When a structural gene directs the synthesis of a polypeptide, each amino acid is specified by a triplet codon.

2. A gene is a nucleotide sequence that codes for a polypeptide, tRNA, or rRNA. Controlling elements such as promoters and operators often are considered part of the gene.

3. Most bacterial genes have at least four major parts, each with different functions: promoters, leaders, coding regions, and trailers.

4. Mutations are stable, heritable alterations in the gene sequence and usually, but not always, produce phenotypic changes. Nucleic acids are altered in several different ways, and these mutations may be either spontaneous or induced by chemical mutagens or radiation.

5. It is extremely important to keep the nucleotide sequence constant, and microorganisms have several repair mechanisms designed to detect alterations in the genetic material and restore it to its original state. Often more than one repair system can correct a particular type of mutation. Despite these efforts some alterations remain uncorrected and provide material and opportunity for evolutionary change.

The preceding chapters cover background information and concepts helpful in understanding microbial genetics: DNA, RNA, and protein synthesis; and the regulation of metabolism. This chapter reviews some basic concepts of molecular genetics: how genetic information is stored and organized in the DNA molecule, mutagenesis, and DNA repair. In addition, the use of microorganisms to identify potentially dangerous mutagenic agents in the fight against cancer is described. Much of this information will be familiar to those who have taken an introductory genetics course. Because of the importance of bacteria, primary emphasis is placed on their genetics.

Based on the foundation provided by this chapter, chapter 14 contains information on plasmids and the nature of genetic recombination in microorganisms. The section on microbial genetics concludes with an introduction to recombinant DNA technology in chapter 15.

Geneticists, including microbial geneticists, use a specialized vocabulary because of the complexities of their discipline. Some knowledge of basic terminology is necessary at the beginning of this survey of general principles. The experimental material of the microbial geneticist is the **strain or clone.** A clone is a population of cells that are genetically identical. Sometimes a clone is called a pure culture. The term **genome** refers to all the genes present in a cell or virus. Bacteria normally have one set of genes. That is, they are haploid (1N). Eucaryotic microorganisms usually have two sets of genes, or are diploid (2N). The **genotype** of an organism is the specific set of genes it possesses. In contrast, the **phenotype** is the collection of characteristics that are observable by the investigator. All genes are not expressed at the same time, and the environment profoundly influences phenotypic expression. Much genetics research has focused on the relationship between an organism's genotype and phenotype.

Although genetic analysis began with the rediscovery of the work of Gregor Mendel in the early part of this century, the subsequent elegant experimentation involving both bacteria and bacteriophages actually elucidated the nature of genetic information, gene structure, the genetic code, and mutations. A few of these early experiments are reviewed before the genetic code itself is examined.

DNA as the Genetic Material

The early work of Fred Griffith in 1928 on the transfer of virulence in the pathogen *Streptococcus pneumoniae* (figure 13.1) set the stage for the research that first showed that DNA was the genetic material. Griffith found that if he boiled virulent bacteria and injected them into mice, the mice were not affected and no pneumococci could be recovered from the animals. When he injected a combination of killed virulent bacteria and a living nonvirulent strain, the mice died; moreover, he could recover living virulent bacteria from the dead mice. Griffith called this change of nonvirulent bacteria into virulent pathogens **transformation.**

Oswald T. Avery and his colleagues then set out to discover which constituent in the heat-killed virulent pneumococci was responsible for Griffith's transformation. These investigators selectively destroyed cell constituents in extracts of virulent pneumococci, using enzymes that would hydrolyze DNA, RNA, or protein. They then exposed nonvirulent pneumococcal strains to the treated extracts. Transformation of the nonvirulent bacteria was blocked only if the DNA was destroyed, suggesting that DNA was carrying the information required for transformation (figure 13.2). The publication of these studies by O. T. Avery, C. M. MacLeod, and M. J. McCarty in 1944 provided the first evidence that Griffith's transforming principle was DNA and therefore that DNA carried genetic information.

Some years later (1952), Alfred D. Hershey and Martha Chase performed several experiments that proved that DNA was the genetic material in the T2 bacteriophage. Some luck was involved in their discovery, for the genetic material of many viruses is RNA and the researchers happened to select a DNA virus for their studies. Imagine the confusion if T2 had been an RNA virus. The controversy surrounding the nature of genetic information might have lasted considerably longer than it did. Hershey and Chase made the virus DNA radioactive with ^{32}P or labeled the viral protein coat with ^{35}S. When the radioactive viruses were mixed with *E. coli,* the radioactive DNA was injected into the *E. coli* host cell by the virus, while most of the protein remained outside (figure 13.3). Since genetic material was injected and T2 progeny were produced, DNA must have been carrying the genetic information for T2.

The biology of bacteriophages (chapter 17).

Subsequent studies on the genetics of viruses and bacteria were largely responsible for the rapid development of molecular genetics. Furthermore, much of the new recombinant DNA technology (*see chapter 15*) has arisen from recent progress in bacterial and viral genetics. Research in microbial genetics has had a profound impact on biology as a science and on the technology that affects everyday life.

1. Define strain or clone, genome, genotype, and phenotype.
2. Briefly summarize the experiments of Griffith; Avery, MacLeod, and McCarty; and Hershey and Chase. What did each show, and why were these experiments important to the development of microbial genetics?

The Genetic Code

The realization that DNA is the genetic material triggered efforts to understand how genetic instructions are stored and organized in the DNA molecule. Early studies on the nature of the

Figure 13.1 **Griffith's Transformation Experiments.** (*a*) Mice died of pneumonia when injected with pathogenic strains of S pneumococci, which have a capsule and form smooth-looking colonies. (*b*) Mice survived when injected with a nonpathogenic strain of R pneumococci, which lacks a capsule and forms rough colonies. (*c*) Injection with heat-killed strains of S pneumococci had no effect. (*d*) Injection with a live R strain and a heat-killed S strain gave the mice pneumonia, and live S strain pneumococci could be isolated from the dead mice.

Figure 13.2 **Experiments on the Transforming Principle.**
Summary of the experiments of Avery, MacLeod, and McCarty on the transforming principle. DNA alone changed R to S cells, and this effect was lost when the extract was treated with deoxyribonuclease. Thus DNA carried the genetic information required for the R to S conversion or transformation.

R cells + purified S cell polysaccharide ⟶ R colonies

R cells + purified S cell protein ⟶ R colonies

R cells + purified S cell RNA ⟶ R colonies

R cells + purified S cell DNA ⟶ S colonies

S cell extract + protease + R cells ⟶ S colonies

S cell extract + RNase + R cells ⟶ S colonies

Figure 13.3   The Hershey-Chase Experiment.
(*a*) When *E. coli* was infected with a T2 phage containing ^{35}S protein, most of the radioactivity remained outside the host cell. (*b*) When a T2 phage containing ^{32}P DNA was mixed with the host bacterium, the radioactive DNA was injected into the cell and phages were produced. Thus DNA was carrying the virus's genetic information.

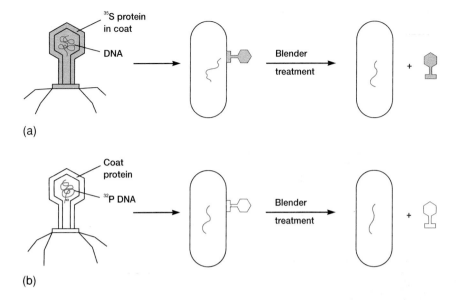

genetic code showed that the DNA base sequence corresponds to the amino acid sequence of the polypeptide specified by the gene. That is, the nucleotide and amino acid sequences are co-linear. It also became evident that many mutations are the result of changes of single amino acids in a polypeptide chain. However, the exact nature of the code was still unclear.

The structure of DNA and RNA (pp. 212–14).

Establishment of the Genetic Code

Since only 20 amino acids normally are present in proteins, there must be at least 20 different code words in a linear, single strand of DNA. The code must be contained in some sequence of the four nucleotides commonly found in the linear DNA sequence. There are only 16 possible combinations (4^2) of the four nucleotides if only nucleotide pairs are considered, not enough to code for all 20 amino acids. Therefore a code word, or **codon,** must involve at least nucleotide triplets even though this would give 64 possible combinations (4^3), many more than the minimum of 20 needed to specify the common amino acids.

The actual codons were discovered in the early 1960s through the experiments carried out by Marshall Nirenberg, Heinrich Matthaei, Philip Leder, and Har Gobind Khorana. In 1968 Nirenberg and Khorana shared the Nobel prize with Robert W. Holley, the first person to sequence a nucleic acid (phenylalanyl tRNA).

Organization of the Code

The genetic code, presented in RNA form, is summarized in table 13.1. Note that there is **code degeneracy.** That is, there are up to six different codons for a given amino acid. Only 61 codons, the **sense codons,** direct amino acid incorporation into protein. The remaining three codons (UGA, UAG, and UAA)

are involved in the termination of translation and are called **stop** or **nonsense codons.** Despite the existence of 61 sense codons, there are not 61 different tRNAs, one for each codon. The 5′ nucleotide in the anticodon can vary, but generally, if the nucleotides in the second and third anticodon positions complement the first two bases of the mRNA codon, an aminoacyl-tRNA with the proper amino acid will bind to the mRNA-ribosome complex. This pattern is evident on inspection of changes in the amino acid specified with variation in the third position (table 13.1). This somewhat sloppy base pairing is known as wobble and relieves cells of the need to synthesize so many tRNAs (figure 13.4). Wobble also decreases the ill effects of DNA mutations.

The mechanism of protein synthesis and tRNA function (pp. 226–33).

---

1. Why must a codon contain at least three nucleotides?
2. Define the following: code degeneracy, sense codon, stop or nonsense codon, and wobble.

Gene Structure

The **gene** has been defined in several ways. Initially geneticists considered it to be the entity responsible for conferring traits on the organism and the entity that could undergo recombination. Recombination involves exchange of DNA from one source with that from another (*see chapter 14*) and is responsible for generating much of the genetic variability found in viruses and living organisms. Genes were typically named for some mutant or altered phenotype. With the discovery and characterization of DNA, the gene was defined more precisely as a linear sequence of nucleotides or codons (this term can be used for RNA as well as DNA) with a fixed start point and end point.

TABLE 13.1   The Genetic Code

		Second Position				
		U	**C**	**A**	**G**	
First Position (5′ End)[a]	**U**	UUU }Phe UUC UUA }Leu UUG	UCU } UCC UCA }Ser UCG	UAU }Tyr UAC UAA }STOP UAG	UGU }Cys UGC UGA STOP UGG Trp	U C A G
	C	CUU } CUC CUA }Leu CUG	CCU } CCC CCA }Pro CCG	CAU }His CAC CAA }Gln CAG	CGU } CGC CGA }Arg CGG	U C A G
	A	AUU } AUC }Ile AUA AUG Met	ACU } ACC ACA }Thr ACG	AAU }Asn AAC AAA }Lys AAG	AGU }Ser AGC AGA }Arg AGG	U C A G
	G	GUU } GUC GUA }Val GUG	GCU } GCC GCA }Ala GCG	GAU }Asp GAC GAA }Glu GAG	GGU } GGC GGA }Gly GGG	U C A G

Third Position (3′ End)

[a]The code is presented in the RNA form. Codons run in the 5′ to 3′ direction. See text for details.

Figure 13.4   Wobble and Coding. The use of wobble in coding for the amino acid glycine. (*a*) Inosine (I) is a wobble nucleoside that can base pair with uracil (U), cytosine (C), or adenine (A). Thus ICC base pairs with GGU, GGC, and GGA in the mRNA. (*b*) Because of the wobble produced by inosine, two tRNA anticodons can recognize the four glycine (Gly) codons. ICC recognizes GGU, GGC, and GGA; CCC recognizes GGG.

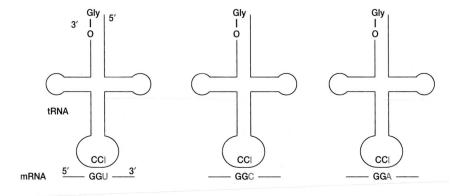

(a) Base pairing of one glycine tRNA with three codons due to wobble

(b) Glycine codons and anticodons (written in the 5′ ⟶ 3′ direction)

Glycine mRNA codons: GGU, GGC, GGA, GGG

Glycine tRNA anticodons: ICC, CCC

At first, it was thought that a gene contained information for the synthesis of one enzyme, the **one gene–one enzyme hypothesis**. This has been modified to the **one gene–one polypeptide hypothesis** because of the existence of enzymes and other proteins composed of two or more different polypeptide chains coded for by separate genes. The segment that codes for a single polypeptide is sometimes also called a **cistron**. More recent results show that even this description is oversimplified. Not all genes are involved in protein synthesis; some code instead for rRNA and tRNA. Thus a gene might be defined as a polynucleotide sequence that codes for a polypeptide, tRNA, or rRNA. Some geneticists think of it as a segment of nucleic acid that is transcribed to give an RNA product. Often, controlling elements such as promoters are considered part of the gene because genes usually have noncoding regulatory sequences associated with them. Most genes consist of discrete sequences of codons that are "read" only one way to produce a single product. That is, the code is not overlapping. Chromosomes therefore usually consist of gene sequences that do not overlap one another (figure 13.5*a*). However, there are exceptions to the rule. Some viruses such as the phage φX174 do have overlapping genes (figure 13.5*b*), and parts of genes overlap in some bacterial genomes.

Procaryotic and viral gene structure differs greatly from that of eucaryotes. In bacterial and viral systems, the coding information within a cistron normally is continuous (some bacterial genes do contain introns); however, in eucaryotic organisms, many if not most genes contain coding information (exons) interrupted periodically by noncoding sequences (introns). An interesting exception to this rule is eucaryotic histone genes, which lack introns. Because procaryotic and viral systems are the best characterized, the more detailed description of gene structure that follows will focus on *E. coli* genes.

Exons and introns in eucaryotic genes (p. 224).

Genes That Code for Proteins

Recall from the discussion of transcription that although DNA is double stranded, only one strand contains coded information and directs RNA synthesis. This strand is called the **sense strand,** and the complementing strand is known as the **antisense strand** (figure 13.6). Since the mRNA is made from the 5′ to the 3′ end, the polarity of the DNA sense strand is therefore 3′ to 5′. Therefore the beginning of the gene is at the 3′ end of the sense strand (also the 5′ end of the antisense strand). An RNA polymerase recognition/binding and regulatory site known as the **promoter** is located at the start of the gene (*see chapter 11*).

The mechanism of transcription (pp. 221–5).

Promoters are sequences of DNA on the sense strand that are usually upstream from the actual coding or transcribed region (figure 13.6)—that is, the promoter is located before or

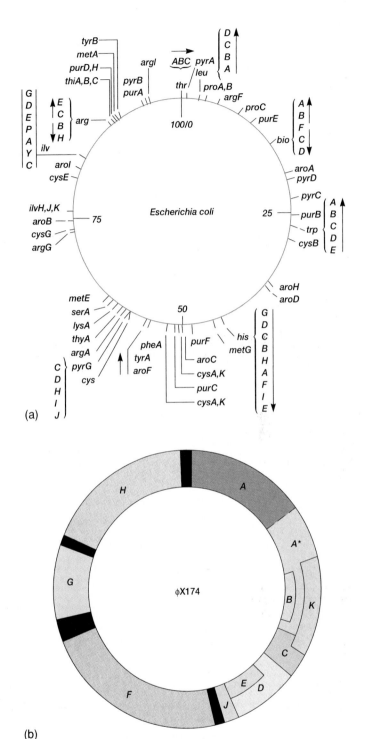

(a)

(b)

Figure 13.5   Chromosomal Organization in Bacteria and Viruses. (*a*) Simplified genetic map of *E. coli.* The *E. coli* map is divided into 100 minutes, the time required to transfer the chromosome from an Hfr cell to F⁻ at 37°C. (*b*) The map of phage øX174 shows the overlap of gene *B* with *A*, *K* with *A* and *C*, and *E* with *D*. The solid regions are spaces lying between genes. Protein A* consists of the last part of protein A and arises from reinitiation of transcription within gene *A*.

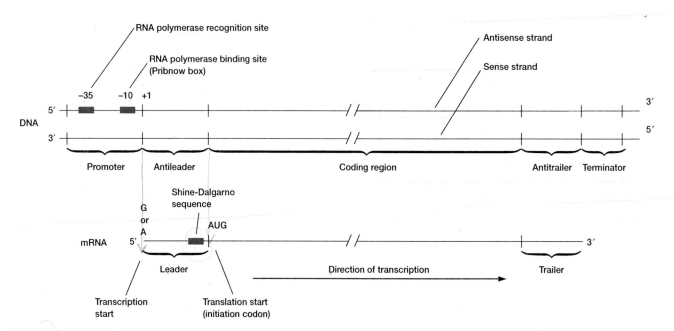

Figure 13.6 A Bacterial Structural Gene. The organization of a typical structural gene in bacteria. Leader and trailer sequences are included even though some genes lack one or both. Transcription begins at the +1 position in DNA, and the first nucleotide incorporated into mRNA is usually GTP or ATP. Translation of the mRNA begins with the AUG initiation codon. Regulatory sites are not shown.

Figure 13.7 A Bacterial Promoter. The lactose operon promoter UV5 and its consensus sequences. The start point for RNA synthesis is labeled +1. The region around –35 is the site at which the RNA polymerase first attaches to the promoter. RNA polymerase binds and begins to unwind the DNA helix at the Pribnow box or RNA polymerase binding site, which is located in the –10 region.

upstream from the coding region in relationship to the direction of transcription (the direction of transcription is referred to as downstream). In *E. coli* the promoter has two important functions, and these relate to two specific segments within the promoter (figures 13.6 and 13.7). Although these two segments do vary slightly in sequence between bacterial strains, they are fairly constant and may be represented by consensus sequences. These are idealized sequences composed of the bases most often found at each position when the sequences from different bacteria are compared. The RNA polymerase recognition site, with a consensus sequence of 5′TTGACA3′ on the antisense strand, is located about 35 base pairs before (the –35 region) the tran-

scriptional start point (labeled as +1) of RNA synthesis. This sequence seems to be the site of the initial association of the RNA polymerase with the DNA. The **RNA polymerase binding site,** also known as the **Pribnow box,** is located at the –10 region and has a consensus sequence 5′TATAAT3′ (a sequence that favors the localized unwinding of DNA). This is where the RNA polymerase begins to unwind the DNA for eventual transcription. The initially transcribed portion of the gene is not necessarily sense or coding material. Rather, a **leader sequence** may be synthesized first. The leader is usually a nontranslated sequence that is important in the initiation of translation and sometimes is involved in regulation of transcription.

The leader (figure 13.6) in procaryotes generally contains a consensus sequence known as the **Shine-Dalgarno sequence,** 5′AGGA3′, the transcript of which complements a sequence on the 16S rRNA in the small subunit of the ribosome. The binding of mRNA leader with 16S rRNA properly orients the mRNA on the ribosome. The leader also sometimes regulates transcription by attenuation (*see chapter 12*). Downstream and next to the leader is the most important part of the structural gene, the coding region.

The **coding region** (figure 13.6) of genes that direct the synthesis of proteins typically begins with the sense DNA sequence 3′TAC5′. This produces the RNA translation initiation codon 5′AUG3′, which codes for N-formylmethionine. This modified form of methionine is the first amino acid incorporated in most procaryotic proteins. The remainder of the coding region of the gene consists of a sequence of codons that specifies the sequence of amino acids for that particular protein. Transcription does not stop at the translation stop codon but rather at a **terminator sequence** (*see chapter 11*). The terminator often lies after a nontranslated **trailer sequence** located downstream from the coding region (figure 13.6). The trailer sequence, like the leader, is needed for the proper expression of the sense portion of the gene.

Besides the basic components described above—the promoter, leader, coding region, trailer, and terminator—many procaryotic genes have a variety of regulatory sites. These are locations where DNA-recognizing regulatory proteins bind to stimulate or prevent gene expression. Regulatory sites often are associated with promoter function, and some consider them to be parts of special promoters. Two such sites, the operator and the CAP binding site, are discussed in chapter 12. Certainly everything is not known about genes and their structure. With the ready availability of purified cloned genes and DNA sequencing technology, major discoveries continue to be made in this area.

The operon and transcription regulation (pp. 242–47).

Genes That Code for tRNA and rRNA

The DNA segments that code for tRNA and rRNA also are considered genes, although they give rise to structurally important RNA rather than protein. In *E. coli* the genes for tRNA are fairly typical, consisting of a promoter and transcribed leader and trailer sequences that are removed during the maturation process (figure 13.8a). The precise function of the leader is not clear; however, the trailer is required for termination. Genes coding for tRNA may code for more than a single tRNA molecule or type of tRNA (figure 13.8a). The segments coding for tRNAs are separated by short spacer sequences that are removed after transcription by special ribonucleases, at least one of which contains catalytic RNA.

RNA splicing and ribozymes (pp. 224–5).

The genes for rRNA also are similar in organization to genes coding for proteins and have promoters, trailers, and terminators (figure 13.8b). Interestingly all the rRNAs are transcribed as a single, large precursor molecule that is cut up by ribonucleases after transcription to yield the final rRNA products. *E. coli* pre-rRNA spacer and trailer regions even contain tRNA genes. Thus the synthesis of tRNA and rRNA involve posttranscriptional modification, a relatively rare process in procaryotes.

1. Define or describe the following: gene, sense and antisense strands, promoter, consensus sequence, RNA polymerase recognition and binding sites, Pribnow box, leader, Shine-Dalgarno sequence, coding region, trailer, and terminator.

2. How do the genes of procaryotes and eucaryotes usually differ from each other?

3. Briefly discuss the general organization of tRNA and rRNA genes. How does their expression differ from that of structural genes with respect to posttranscriptional modification of the gene product?

Mutations and Their Chemical Basis

Considerable information is embedded in the precise order of nucleotides in DNA. For life to exist with stability, it is essential that the nucleotide sequence of genes is not disturbed to any great extent. However, sequence changes do occur and often result in altered phenotypes. These changes are largely detrimental but are important in generating new variability and contribute to the process of evolution. Microbial mutation rates also can be increased, and these genetic changes have been put to many important uses in the laboratory and industry.

Mutations [Latin *mutare,* to change] were initially characterized as altered phenotypes or phenotypic expressions. Long before the existence of direct proof that a mutation is the result of a stable, heritable change in the nucleotide sequence of DNA, geneticists predicted that several basic types of transmitted mutations could exist. They believed that mutations could arise from the alteration of single pairs of nucleotides and from the addition or deletion of one or two nucleotide pairs in the coding regions of a gene. In this section the molecular basis of mutations and mutagenesis is first considered. Then the phenotypic effects of mutations, the detection of mutations, and the use of mutations in carcinogenicity testing are discussed.

Mutations and Mutagenesis

Mutations can alter the phenotype of a microorganism in several different ways. Morphological mutations change the microorganism's colonial or cellular morphology. Lethal mutations, when expressed, result in the death of the microorganism. Since the microorganism must be able to grow in order to be isolated and studied,

Figure 13.8 tRNA and rRNA Genes.
(*a*) A tRNA precursor from *E. coli* that contains
two tRNA molecules. The spacer and extra
nucleotides at both ends are removed during
processing. (*b*) The *E. coli* ribosomal RNA gene
codes for a large transcription product that is
cleaved into three rRNAs and one to three
tRNAs. The 16S, 23S, and 5S rRNA segments
are represented by red lines, and tRNA
sequences are placed in brackets. The seven
copies of this gene vary in the number and kind
of tRNA sequences.

lethal mutations are recovered only if they are recessive in diploid
organisms or conditional (see the following) in haploid organisms.

Conditional mutations are those that are expressed only
under certain environmental conditions. For example, a conditional lethal mutation in *E. coli* might not be expressed under
permissive conditions such as low temperature but would be expressed under restrictive conditions such as high temperature.
Thus the hypothetical mutant would grow normally at the permissive temperature but would die at high temperatures.

Biochemical mutations are those causing a change in the
biochemistry of the cell. Since these mutations often inactivate
a biosynthetic pathway, they frequently make a microorganism
unable to grow on medium lacking an adequate supply of the
pathway's end product (figure 13.15). That is, the mutant cannot grow on minimal medium and requires nutrient supplements. Such mutants are called **auxotrophs,** whereas microbial
strains that can grow on minimal medium are **prototrophs.**

Analysis of auxotrophy has been quite important in microbial
genetics due to the ease of auxotroph selection and the relative
abundance of this mutational type.

Mutant detection and replica plating (pp. 268–70).
Nutrient requirements and nutritional types (pp. 99–102).

A resistant mutant is a particular type of biochemical mutant that acquires resistance to some pathogen, chemical, or antibiotic. Such mutants also are easy to select for and very useful
in microbial genetics.

Mechanisms of drug resistance (pp. 669–71).

Mutations occur in one of two ways. (1) Spontaneous mutations arise occasionally in all cells and develop in the absence of
any added agent. (2) Induced mutations, on the other hand, are
the result of exposure of the organism to some physical or
chemical agent called a **mutagen.**

Figure 13.9   Transition and Transversion Mutations. Errors in replication due to base tautomerization. (*a*) Normally AT and GC pairs are formed when keto groups participate in hydrogen bonds. In contrast, enol tautomers produce AC and GT base pairs. (*b*) Mutation as a consequence of tautomerization during DNA replication. The temporary enolization of guanine leads to the formation of an AT base pair in the mutant, and a GC to AT transition mutation occurs. The process requires two replication cycles. The mutation only occurs if the abnormal first-generation GT base pair is missed by repair mechanisms.

Spontaneous Mutations

Spontaneous mutations arise from several potential sources. This class of mutations may result from errors in DNA replication, damage to DNA from factors such as gamma radiation and heat, or even the action of transposons (*see chapter 14*). A few of the more prevalent mechanisms are described in the following paragraphs.

Generally errors in replication occur when the base of a template nucleotide takes on a rare tautomeric form. Tautomerism is the relationship between two structural isomers that are in chemical equilibrium and readily change into one another. Bases typically exist in the keto form. However, they can at times take on either an imino or enol form (figure

13.9*a*). These tautomeric shifts change the hydrogen-bonding characteristics of the bases, allowing purine for purine or pyrimidine for pyrimidine substitutions that can eventually lead to a stable alteration of the nucleotide sequence (figure 13.9*b*). Such substitutions are known as **transition mutations** and are relatively common, although most of them are repaired by various proofreading functions (*see pp. 221 and 272*). In **transversion mutations,** a purine is substituted for a pyrimidine, or a pyrimidine for a purine. These mutations are rarer due to the steric problems of pairing purines with purines and pyrimidines with pyrimidines.

Spontaneous mutations also arise from **frameshifts,** usually caused by the deletion of DNA segments resulting in an altered codon reading frame. These mutations generally occur where

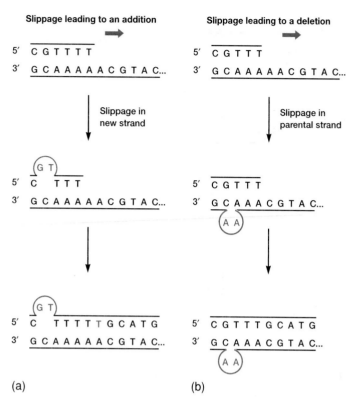

Figure 13.10 Additions and Deletions. A hypothetical mechanism for the generation of additions and deletions during replication. The direction of replication is indicated by the large arrow. In each case there is strand slippage resulting in the formation of a small loop that is stabilized by the hydrogen bonding in the repetitive sequence, the AT stretch in this example. DNA synthesis proceeds to the right in this figure. (*a*) If the new strand slips, an addition of one T results. (*b*) Slippage of the parental strand yields a deletion (in this case, a loss of two Ts).

there is a short repeated nucleotide sequence. In such a location, the pairing of template and new strand can be displaced by the distance of the repeated sequence leading to a deletion of bases in the new strand (figure 13.10).

Spontaneous mutations originate from lesions in DNA as well as from replication errors. For example, it is possible for purine nucleotides to be depurinated—that is, to lose their base. This results in the formation of an **apurinic site,** which will not base pair normally and may cause a transition type mutation after the next round of replication. Cytosine can be deaminated to uracil, which is then removed to form an **apyrimidinic site.**

Induced Mutations

Virtually any agent that directly damages DNA, alters its chemistry, or interferes with repair mechanisms (pp. 272–74) will induce mutations. Mutagens can be conveniently classified according to their mechanism of action. Four common modes of mutagen action are incorporation of base analogs, specific mispairing, intercalation, and bypass of replication.

Base analogs are similar to normal nitrogenous bases and can be incorporated into the growing polynucleotide chain during replication. Once in place, these compounds typically exhibit base pairing properties different from the bases they replace and can eventually cause a stable mutation. A widely used base analog is 5-bromouracil (5-BU), an analog of thymine. It undergoes a tautomeric shift from the normal keto form to an enol much more frequently than does a normal base. The enol forms hydrogen bonds like cytosine and directs the incorporation of guanine rather than adenine (figure 13.11). The mechanism of action of other base analogs is similar to that of 5-bromouracil.

Specific mispairing is caused when a mutagen changes a base's structure and therefore alters its base pairing characteristics. Some mutagens in this category are fairly selective; they preferentially react with some bases and produce a specific kind of DNA damage. An example of this type of mutagen is methyl-nitrosoguanidine, an alkylating agent that adds methyl groups to guanine, causing it to mispair with thymine (figure 13.12). A subsequent round of replication could then result in a GC-AT transition. Other examples of mutagens with this mode of action are the alkylating agents ethyl-methanesulfonate and hydroxylamine. Hydroxylamine hydroxylates the C-4 nitrogen of cytosine, causing it to base pair like thymine. There are many other DNA modifying agents that can cause mispairing.

Intercalating agents distort DNA to induce single nucleotide pair insertions and deletions. These mutagens are planar and insert themselves (intercalate) between the stacked bases of the helix. This results in a mutation, possibly through the formation of a loop in DNA. Intercalating agents include acridines such as proflavin and acridine orange.

Many mutagens, and indeed many carcinogens, directly damage bases so severely that hydrogen bonding between base pairs is impaired or prevented and the damaged DNA can no longer act as a template. For instance, UV radiation generates cyclobutane type dimers, usually thymine dimers, between adjacent pyrimidines (figure 13.13). Other examples are ionizing radiation and carcinogens such as aflatoxin B1 and other benzo(a)pyrene derivatives. Such damage to DNA would generally be lethal but may trigger a repair mechanism that restores much of the damaged genetic material, although with considerable error incorporation (pp. 272–74).

Retention of proper base pairing is essential in the prevention of mutations. Often the damage can be repaired before a mutation is permanently established. If a complete DNA replication cycle takes place before the initial lesion is repaired, the mutation frequently becomes stable and inheritable.

The Expression of Mutations

The expression of a mutation will only be noticed if it produces a detectable, altered phenotype. A mutation from the most prevalent gene form, the **wild type,** to a mutant form is called a

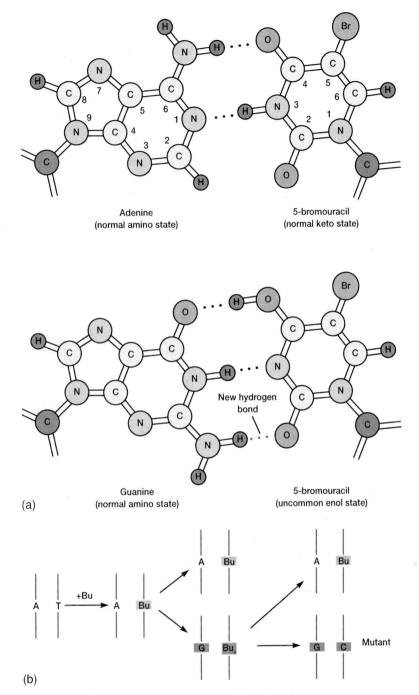

(a)

(b)

Figure 13.11   Mutagenesis by the Base Analog 5-Bromouracil. (*a*) The enol form of 5-BU base pairs with guanine rather than with adenine as might be expected for a thymine analog. (*b*) If the keto form of 5-BU is incorporated in place of thymine, its occasional tautomerization to the enol form (BU$_e$) will produce an AT to GC transition mutation.

forward mutation. Later, a second mutation may make the mutant appear to be a wild-type organism again. Such a mutation is called a **reversion mutation** because the organism seems to have reverted back to its original phenotype. A true **back mutation** converts the mutant nucleotide sequence back to the wild-type arrangement. The wild-type phenotype also can be regained by a second mutation in a different gene, a **suppressor mutation,** which overcomes the effect of the first mutation (table 13.2). If the second mutation is within the same gene, the change may be called a second site reversion or intragenic suppression. Thus, although revertant phenotypes appear to be wild types, the original DNA sequence may not be

Figure 13.12 **Methyl-Nitrosoguanidine Mutagenesis.** Mutagenesis by methyl-nitrosoguanidine due to the methylation of guanine.

Figure 13.13 **Thymine Dimer.** Thymine dimers are formed by ultraviolet radiation. The enzyme photolyase cleaves the two colored bonds during photoreactivation.

restored. In practice, a mutation is visibly expressed when a protein that is in some way responsible for the phenotype is altered sufficiently to produce a new phenotype. However, mutations may occur and not alter the phenotype for a variety of reasons.

Although very large deletion and insertion mutations exist, most mutations affect only one base pair in a given location and therefore are called **point mutations.** There are several types of point mutations (table 13.2).

One kind of point mutation that could not be detected until the advent of nucleic acid sequencing techniques is the **silent mutation.** If a mutation is an alteration of the nucleotide sequence of DNA, mutations can occur and have no visible effect because of code degeneracy. When there is more than one codon for a given amino acid, a single base substitution could result in the formation of a new codon for the same amino acid. For example, if the codon CGU were changed to CGC, it would still code for arginine even though a mutation had occurred. The expression of this mutation would not be detected except at the level of the DNA or mRNA. Since there would be no change in the protein, there would be no change in the phenotype of the organism.

A second type of point mutation is the **missense mutation.** This mutation involves a single base substitution in the DNA that changes a codon for one amino acid into a codon for another. For example, the codon GAG, which specifies glutamic acid, could be changed to GUG, which codes for valine. The expression of missense mutations can vary. Certainly the mutation is expressed at the level of protein structure. However, at the level of protein function, the effect may range from complete loss of activity to no change at all.

Many proteins are still functional after the substitution of a single amino acid, but this depends on the type and location of the amino acid. For instance, replacement of a nonpolar amino acid in the protein's interior with a polar amino acid probably will drastically alter the protein's three-dimensional structure and therefore its function. Similarly the replacement of a critical amino acid at the active site of an enzyme will destroy its activity. However, the replacement of one polar amino acid with another at the protein surface may have little or no effect. Missense mutations may actually play a very important role in providing new variability to drive evolution because they often are not lethal and therefore remain in the gene pool.

Protein structure (appendix I).

A third type of point mutation causes the early termination of translation and therefore results in a shortened polypeptide. Such mutations are called **nonsense mutations** because they involve the conversion of a sense codon to a nonsense or stop codon. Depending on the relative location of the mutation, the phenotypic expression may be more or less severely affected. Most proteins retain some function if they are shortened by only

TABLE 13.2   Summary of Some Molecular Changes from Gene Mutations

Type of Mutation	Result and Example
Forward Mutations	
Single Nucleotide-Pair (Base-Pair) Substitutions	
At DNA Level	
Transition	Purine replaced by a different purine, or pyrimidine replaced by a different pyrimidine (e.g., AT $\longrightarrow$ GC).
Transversion	Purine replaced by a pyrimidine, or pyrimidine replaced by a purine (e.g., AT $\longrightarrow$ CG).
At Protein Level	
Silent mutation	Triplet codes for same amino acid: $\qquad$ AGG $\longrightarrow$ **C**GG $\qquad$ both code for Arg
Neutral mutation	Triplet codes for different but functionally equivalent amino acid: $\qquad$ AAA (Lys) $\longrightarrow$ A**G**A (Arg)
Missense mutation	Triplet codes for a different amino acid.
Nonsense mutation	Triplet codes for chain termination: $\qquad$ CAG (Gln) $\longrightarrow$ UAG (stop)
Single Nucleotide-Pair Addition or Deletion: Frameshift Mutation	Any addition or deletion of base pairs that is not a multiple of three results in a frameshift in reading the DNA segments that code for proteins.
Intragenic Addition or Deletion of Several to Many Nucleotide Pairs	
Reverse Mutations	
True Reversion	AAA (Lys) $\xrightarrow{\text{forward}}$ GAA (Glu) $\xrightarrow{\text{reverse}}$ AAA (Lys) wild type $\qquad\qquad$ mutant $\qquad\qquad$ wild type
Equivalent Reversion	UCC (Ser) $\xrightarrow{\text{forward}}$ UGC (Cys) $\xrightarrow{\text{reverse}}$ AGC (Ser) wild type $\qquad\qquad$ mutant $\qquad\qquad$ wild type CGC (Arg, basic) $\xrightarrow{\text{forward}}$ CCC (Pro, not basic) $\xrightarrow{\text{reverse}}$ CAC (His, basic) wild type $\qquad\qquad\qquad$ mutant $\qquad\qquad\qquad$ pseudo-wild type
Suppressor Mutations	
Intragenic Suppressor Mutations $\quad$ Frameshift of opposite sign at site within gene. Addition of X to the base sequence shifts the reading frame from the CAT codon to XCA followed by TCA codons. The subsequent deletion of a C base shifts the reading frame back to CAT.	CATCATCATCATCATCAT $\quad$ (+) $\quad$ (−) $\qquad\downarrow\qquad\downarrow$ CATXCATATCATCATCAT y $\quad$ x $\quad$ z $\quad$ y $\quad$ y $\quad$ y
Extragenic Suppressor Mutations $\quad$ Nonsense suppressors	Gene (e.g., for tyrosine tRNA) undergoes mutational event in its anticodon region that enables it to recognize and align with a mutant nonsense codon (e.g., UAG) to insert an amino acid (tyrosine) and permit completion of the translation.
$\quad$ Physiological suppressors	A defect in one chemical pathway is circumvented by another mutation—for example, one that opens up another chemical pathway to the same product, or one that permits more efficient uptake of a compound produced in small quantities because of the original mutation.

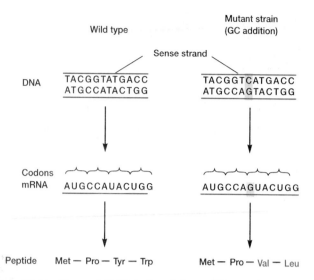

Figure 13.14   Frameshift Mutation. A frameshift mutation resulting from the insertion of a GC base pair. The reading frameshift produces a different peptide after the addition.

one or two amino acids; complete loss of normal function will almost certainly result if the mutation occurs closer to the middle of the gene.

The **frameshift mutation** is a fourth type of point mutation and was briefly mentioned earlier. Frameshift mutations arise from the insertion or deletion of one or two base pairs within the coding region of the gene. Since the code consists of a precise sequence of triplet codons, the addition or deletion of fewer than three base pairs will cause the reading frame to be shifted for all codons downstream. Figure 13.14 shows the effect of a frameshift mutation on a short section of mRNA and the amino acid sequence it codes for.

Frameshift mutations usually are very deleterious and yield mutant phenotypes resulting from the synthesis of nonfunctional proteins. The reading frameshift often eventually produces a nonsense or stop codon so that the peptide product is shorter as well as different in sequence. Of course if the frameshift occurred near the end of the gene, or if there were a second frameshift shortly downstream from the first that restored the reading frame, the phenotypic effect might not be as drastic. A second nearby frameshift that restores the proper reading frame is a good example of an intragenic suppressor mutation.

Mutations also occur in the regulatory sequences responsible for the control of gene expression and in other noncoding portions of structural genes. Constitutive lactose operon mutants in *E. coli* are excellent examples. These mutations map in the operator site and produce altered operator sequences that are not recognized by the repressor protein, and therefore the operon is continuously active in transcription. If a mutation renders the promoter sequence nonfunctional, the coding region of the structural gene will be completely normal, but a mutant phenotype will result due to the absence of

a product. RNA polymerase rarely transcribes a gene correctly without a fully functional promoter.

The lac operon and gene regulation (pp. 245–47).

Mutations also occur in rRNA and tRNA genes and can alter the phenotype through disruption of protein synthesis. In fact, these mutants often are initially identified because of their slow growth. One type of suppressor mutation is a base substitution in the anticodon region of a tRNA that allows the insertion of the correct amino acid at a mutant codon (table 13.2).

1. Define or describe the following: mutation, conditional mutation, auxotroph and prototroph, spontaneous and induced mutations, mutagen, transition and transversion mutations, frameshift, apurinic site, base analog, specific mispairing, intercalating agent, thymine dimer, wild type, forward and reverse mutations, suppressor mutation, point mutation, silent mutation, missense and nonsense mutations, and frameshift mutation.
2. Give three ways in which spontaneous mutations might arise.
3. How do the mutagens 5-bromouracil, methyl-nitrosoguanidine, proflavin, and UV radiation induce mutations?
4. Give examples of intragenic and extragenic suppressor mutations.

Detection and Isolation of Mutants

In order to study microbial mutants, one must be able to detect them readily, even when there are few, and then efficiently isolate them from the parent organism and other mutants. Fortunately this often is easy to do. This section describes some techniques used in mutant detection, selection, and isolation.

Mutant Detection

When collecting mutants of a particular organism, one must know the normal or wild-type characteristics so as to recognize an altered phenotype. A suitable detection system for the mutant phenotype under study also is needed. Since mutations are generally rare, about one per 10^7 to 10^{11} cells, it is important to have a very sensitive detection system so that rare events will not be missed. Geneticists have often induced mutations to increase the probability of obtaining specific changes at high frequency (about one in 10^3 to 10^6); even so, mutations are rare.

Detection systems in bacteria and other haploid organisms are straightforward because any new allele should be seen immediately, even if it is a recessive mutation. Sometimes detection of mutants is direct. If albino mutants of a normally pigmented bacterium are being studied, detection simply requires visual observation of colony color. Other direct detection systems are more complex. For example, the **replica plating** technique is used to detect auxotrophic mutants. It distinguishes between mutants and the wild-type strain based on their ability to grow in the absence of a particular biosynthetic end product (figure 13.15). A lysine auxotroph, for instance,

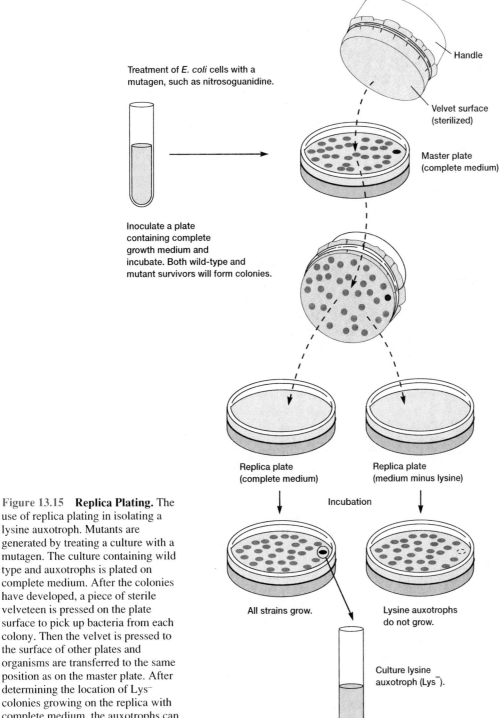

Treatment of *E. coli* cells with a mutagen, such as nitrosoguanidine.

Inoculate a plate containing complete growth medium and incubate. Both wild-type and mutant survivors will form colonies.

Handle

Velvet surface (sterilized)

Master plate (complete medium)

Replica plate (complete medium)

Replica plate (medium minus lysine)

Incubation

All strains grow.

Lysine auxotrophs do not grow.

Culture lysine auxotroph (Lys⁻).

Figure 13.15   Replica Plating. The use of replica plating in isolating a lysine auxotroph. Mutants are generated by treating a culture with a mutagen. The culture containing wild type and auxotrophs is plated on complete medium. After the colonies have developed, a piece of sterile velveteen is pressed on the plate surface to pick up bacteria from each colony. Then the velvet is pressed to the surface of other plates and organisms are transferred to the same position as on the master plate. After determining the location of Lys⁻ colonies growing on the replica with complete medium, the auxotrophs can be isolated and cultured.

will grow on lysine-supplemented media but not on a medium lacking an adequate supply of lysine because it cannot synthesize this amino acid.

Once some sort of detection method is established, mutants are collected. Since a specific mutation is a rare event, it is necessary to look at perhaps thousands to millions of colonies or clones. Using direct detection methods, this could become quite

a task, even with microorganisms. Consider a search for the albino mutants mentioned previously. If the mutation rate were around one in a million, on the average a million or more organisms would have to be tested to find one albino mutant. This probably would require several thousand plates. The task of isolating auxotrophic mutants in this way would be even more taxing with the added labor of replica plating. This difficulty can be partly overcome by using mutagens to increase the mutation rate, thus reducing the number of colonies to be examined. However, it is more efficient to use a selection system employing some environmental factor to separate mutants from wild type microorganisms.

Mutant Selection

An effective selection technique uses incubation conditions under which the mutant will grow, because of properties given it by the mutation, whereas the wild type will not. Selection methods often involve reversion mutations or the development of resistance to an environmental stress. For example, if the intent is to isolate revertants from a lysine auxotroph (Lys⁻), the approach is quite easy. A large population of lysine auxotrophs is plated on minimal medium lacking lysine, incubated, and examined for colony formation. Only cells that have mutated to restore the ability to manufacture lysine will grow on minimal medium (figure 13.16). Thus several million cells can be plated on a single petri dish, and many cells can be tested for mutations by scanning a few petri dishes for growth. This is because the auxotrophs will not grow on minimal medium and confuse the results; only the phenotypic revertants will form colonies. This method has proven very useful in determining the relative mutagenicity of many substances.

Resistance selection methods follow a similar approach. Often wild type cells are not resistant to phage attack or antibiotic treatment, so it is possible to grow the bacterium in the presence of the agent and look for surviving organisms. Consider the example of a phage-sensitive wild type bacterium. When the organism is cultured in medium lacking the virus and then plated out on selective medium containing phages, any colonies that form will be resistant to phage attack and very likely will be mutants in this regard. Resistance selection can be used together with virtually any environmental parameter; resistance to bacteriophages, antibiotics, or temperature are most commonly employed.

Substrate utilization mutations also are employed in bacterial selection. Many bacteria catabolize only a few primary carbon sources. With such bacteria, it is possible to select mutants by a method similar to that used in resistance selection. The culture is plated on medium containing an alternate carbon source. Any colonies that appear can use the substrate and are probably mutants.

Mutant detection and selection methods are used for purposes other than understanding more about the nature of genes or the biochemistry of a particular microorganism. One very important role of mutant selection and detection techniques is in the study of carcinogens. The next section briefly describes one of the first and perhaps best known of the carcinogen testing systems.

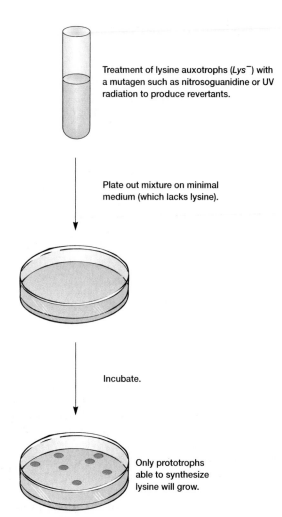

Treatment of lysine auxotrophs (*Lys*⁻) with a mutagen such as nitrosoguanidine or UV radiation to produce revertants.

Plate out mixture on minimal medium (which lacks lysine).

Incubate.

Only prototrophs able to synthesize lysine will grow.

Figure 13.16   Mutant Selection. The production and direct selection of auxotroph revertants. In this example, lysine revertants will be selected after treatment of a lysine auxotroph culture because the agar contains minimal medium that will not support auxotroph growth.

Carcinogenicity Testing

An increased understanding of the mechanisms of mutation and cancer induction has stimulated efforts to identify environmental carcinogens so that they can be avoided. The observation that many carcinogenic agents also are mutagenic is the basis for detecting potential carcinogens by testing for mutagenicity while taking advantage of bacterial selection techniques and short generation times. The **Ames test,** developed by Bruce Ames in the 1970s, has been widely used to test for carcinogens. The Ames test is a mutational reversion assay employing several special strains of *Salmonella typhimurium,* each of which has a different mutation in the histidine biosynthesis operon. The bacteria also have mutational alterations of their cell walls that make them more permeable to test substances. To

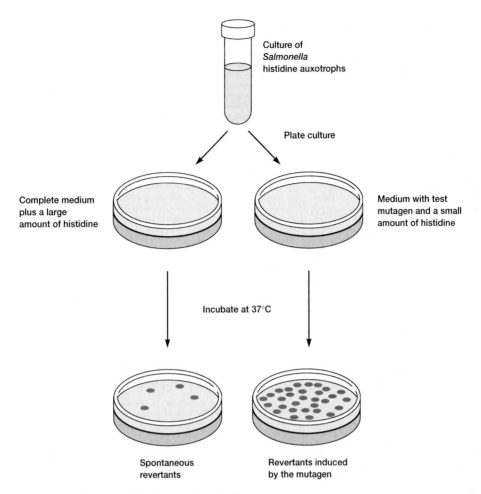

Figure 13.17   The Ames Test for Mutagenicity. See text for details.

further increase assay sensitivity, the strains are defective in the ability to carry out excision repair of DNA (p. 272) and have plasmid genes that enhance error-prone DNA repair.

In the Ames test these special tester strains of *Salmonella* are plated with the substance being tested and the appearance of visible colonies followed (figure 13.17). To ensure that DNA replication can take place in the presence of the potential mutagen, the bacteria and test substance are mixed in dilute molten top agar to which a trace of histidine has been added. This molten mix is then poured on top of minimal agar plates and incubated for 2 to 3 days at 37°C. All of the histidine auxotrophs will grow for the first few hours in the presence of the test compound until the histidine is depleted. Once the histidine supply is exhausted, only revertants that have mutationally regained the ability to synthesize histidine will grow. The visible colonies need only be counted and compared to controls in order to estimate the relative mutagenicity of the compound: the more colonies, the greater the mutagenicity.

A mammalian liver extract is also often added to the molten top agar prior to plating. The extract converts potential carcinogens into electrophilic derivatives that will readily react with DNA. This process occurs naturally when foreign substances are metabolized in the liver. Since bacteria do not have this activation system, liver extract often is added to the test system to promote the transformations that occur in mammals. Many potential carcinogens, such as the aflatoxins (*see pp. 889–90*), are not actually carcinogenic until they are modified in the liver. The addition of extract shows which compounds have intrinsic mutagenicity and which need activation after uptake.

1. Describe how replica plating is used to detect and isolate auxotrophic mutants.

2. Why are mutant selection techniques generally preferable to the direct detection and isolation of mutants?

3. Briefly discuss how reversion mutations, resistance to an environmental factor, and the ability to use a particular nutrient can be employed in mutant selection.

4. What is the Ames test and how is it carried out? What assumption concerning mutagenicity and carcinogenicity is it based upon?

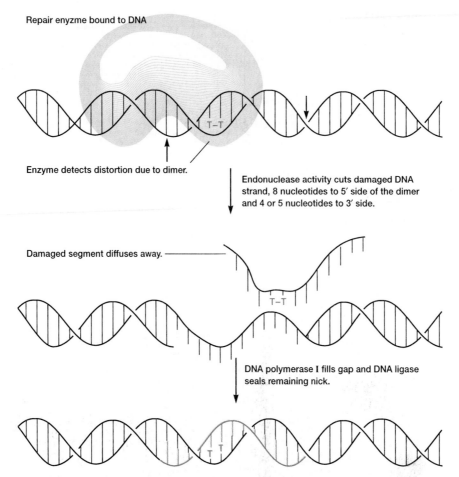

Repair enyzme bound to DNA

Enzyme detects distortion due to dimer.

Endonuclease activity cuts damaged DNA strand, 8 nucleotides to 5′ side of the dimer and 4 or 5 nucleotides to 3′ side.

Damaged segment diffuses away.

DNA polymerase I fills gap and DNA ligase seals remaining nick.

Figure 13.18   Excision Repair. Excision repair of a thymine dimer that has distorted the double helix. The repair endonuclease or uvrABC endonuclease is coded for by the *uvrA, B,* and *C* genes.

DNA Repair

Since replication errors and a variety of mutagens can alter the nucleotide sequence, a microorganism must be able to repair changes in the sequence that might be fatal. DNA is repaired by several different mechanisms besides **proofreading** by replication enzymes (DNA polymerases can remove an incorrect nucleotide immediately after its addition to the growing end of the chain). Repair in *E. coli* is best understood and is briefly described in this section.

DNA replication and proofreading (pp. 216–21).

Excision Repair

Excision repair is a general repair system that corrects damage that causes distortions in the double helix. A repair endonuclease or uvrABC endonuclease removes the damaged bases along with some bases on either side of the lesion (figure 13.18). The resulting single-stranded gap, about 12 nucleotides long, is filled by DNA polymerase I, and DNA ligase joins the fragments (*see chapter 11*). This system can remove thymine dimers (figure 13.13) and repair almost any other injury that produces a detectable distortion in DNA.

Besides this general excision repair system, specialized versions of the system specifically excise sites on the DNA where the sugar phosphate backbone is intact but the bases have been removed to form apurinic or apyrimidinic sites (AP sites). Special endonucleases called AP endonucleases recognize these locations and nick the backbone at the site. Excision repair then commences, beginning with the excision of a short stretch of nucleotides.

Another type of excision repair employs DNA glycosylases. These enzymes remove damaged or unnatural bases yielding AP sites that are then repaired as above. Not all types of damaged bases are repaired in this way, but new glycosylases are being discovered and the process may be of more general importance than first thought.

Removal of Lesions

Thymine dimers and alkylated bases often are directly repaired. **Photoreactivation** is the repair of thymine dimers by splitting them apart into separate thymines with the help of visible light in a photochemical reaction catalyzed by the enzyme photolyase. Since this repair mechanism does not remove and replace nucleotides, it is error free.

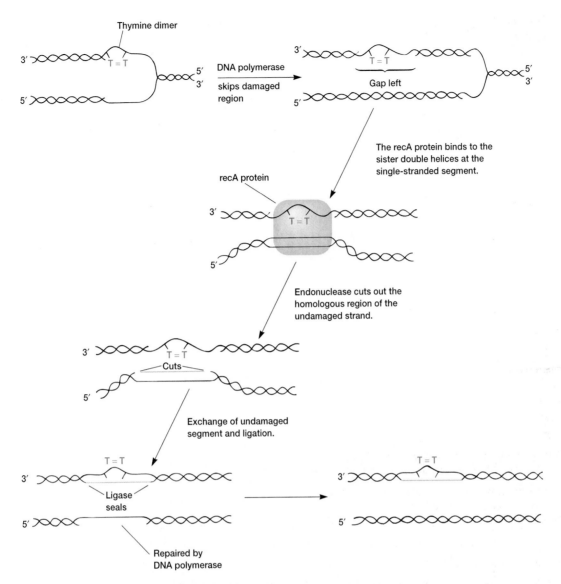

Figure 13.19   Recombination Repair. Recombination repair of the gap caused by a thymine dimer. The recA protein pairs the gap region with the homologous region on a sister helix. An undamaged segment is then cut out and exchanged so that the gap can be filled in.

Sometimes damage caused by alkylation is repaired directly as well. Methyls and some other alkyl groups that have been added to the O–6 position of guanine can be removed with the help of an enzyme known as alkyltransferase or methylguanine methyltransferase. Thus damage to guanine from mutagens such as methyl-nitrosoguanidine (figure 13.12) can be repaired directly.

Postreplication Repair

Despite the accuracy of DNA polymerase action and continual proofreading, errors still are made during DNA replication. Remaining mismatched bases and other errors are usually detected and repaired by the mismatch repair system in *E. coli*. The mismatch correction enzyme scans the newly replicated DNA for mismatched pairs and removes a stretch of newly synthesized DNA around the mismatch. A DNA polymerase then replaces the excised nucleotides, and the resulting nick is sealed with a ligase. Postreplication repair is a type of excision repair (see p. 272).

Recombination Repair

In **recombination repair,** damaged DNA for which there is no remaining template is restored. This situation arises if both bases of a pair are missing or damaged, or if there is a gap opposite a lesion. In this type of repair the **recA protein** cuts a piece of template DNA from a sister molecule (figure 13.19) and puts it into the gap or uses it to replace a damaged strand. Although bacteria are haploid, another copy of the damaged segment often is available because either it has recently been replicated or the cell is growing rapidly and has more than one copy of its chromosome. Once the template is in place, the remaining damage can be corrected by another repair system.

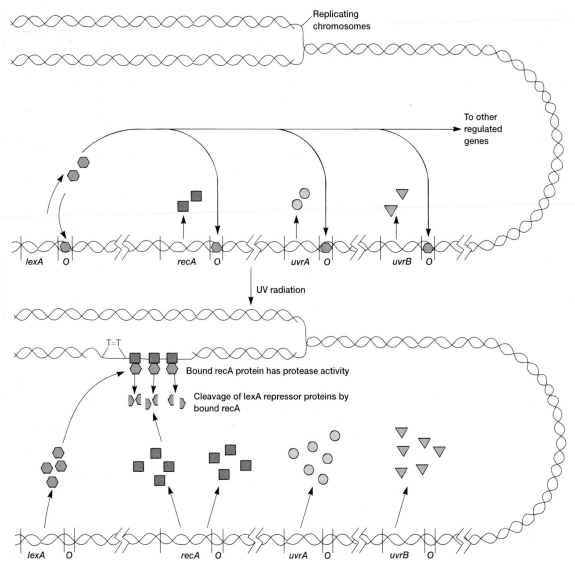

Figure 13.20 **The SOS Repair Process.** In the absence of damage, repair genes are expressed in *E. coli* at low levels due to binding of the lexA repressor protein at their operators (*O*). When the recA protein binds to a damaged region—for example, a thymine dimer created by UV radiation—it destroys lexA and the repair genes are expressed more actively. The *uvr* genes code for the repair endonuclease or uvrABC endonuclease responsible for excision repair.

The recA protein also participates in a type of inducible repair known as **SOS repair.** In this instance the DNA damage is so great that synthesis stops completely, leaving many large gaps. RecA will bind to the gaps and initiate strand exchange. Simultaneously it takes on a proteolytic function that destroys the lexA repressor protein, which regulates the function of many genes involved in DNA repair and synthesis (figure 13.20). As a result many more copies of these enzymes are produced, accelerating the replication and repair processes. The system can quickly repair extensive damage caused by agents such as UV radiation, but it is error prone and does produce mutations. However, it is certainly better to have a few mutations than no DNA replication at all.

1. Define the following: proofreading, excision repair, photoreactivation, methylguanine methyltransferase, mismatch repair, recombination repair, recA protein, SOS repair, and lexA repressor.

2. Describe in general terms the mechanisms of the following repair processes: excision repair, recombination repair, and SOS repair.

Summary

1. The knowledge that DNA is the genetic material came from studies on transformation by Griffith and Avery and from experiments on T2 phage reproduction by Hershey and Chase.

2. Genetic information is carried in the form of 64 nucleotide triplets called codons; sense codons direct amino acid incorporation, and stop or nonsense codons terminate translation.

3. The code is degenerate—that is, there is more than one codon for most amino acids.

4. A gene may be defined as the nucleic acid sequence that codes for a polypeptide, tRNA, or rRNA.

5. The sense strand of DNA carries genetic information and directs the synthesis of the RNA transcript.

6. RNA polymerase binds to the promoter region, which contains RNA polymerase recognition and RNA polymerase binding sites.

7. The gene also contains a coding region and a terminator; it may have a leader and a trailer. Regulatory segments such as operators may be present.

8. The genes for tRNA and rRNA often code for a precursor that is subsequently processed to yield several products.

9. A mutation is a stable, heritable change in the nucleotide sequence of the genetic material, usually DNA.

10. Mutations can be divided into many categories based on their effects on the phenotype; some major types are morphological, lethal, conditional, biochemical, and resistance mutations.

11. Spontaneous mutations can arise from replication errors (transitions, transversions, and frameshifts) and from DNA lesions (apurinic sites).

12. Induced mutations are caused by mutagens. Mutations may result from the incorporation of base analogs, specific mispairing due to alteration of a base, the presence of intercalating agents, and a bypass of replication because of severe damage.

13. The mutant phenotype can be restored to wild type by either a true reverse mutation or a suppressor mutation.

14. There are four important types of point mutations: silent mutations, missense mutations, nonsense mutations, and frameshift mutations.

15. It is essential to have a sensitive and specific detection technique to isolate mutants; an example is replica plating for the detection of auxotrophs (a direct detection system).

16. One of the most effective isolation techniques is to adjust environmental conditions so that the mutant will grow while the wild type organism does not.

17. Because many carcinogens are also mutagenic, one can test for mutagenicity with the Ames test and use the results as an indirect indication of carcinogenicity.

18. Mutations and DNA damage are repaired in several ways; for example: proofreading by replication enzymes, excision repair, removal of lesions (e.g., photoreactivation), postreplication repair (mismatch repair), and recombination repair.

Key Terms

Ames test   *270*

apurinic site   *264*

apyrimidinic site   *264*

auxotrophs   *262*

back mutation   *265*

base analog   *264*

cistron   *259*

code degeneracy   *257*

coding region   *261*

codon   *257*

conditional mutation   *262*

excision repair   *272*

forward mutation   *265*

frameshift   *263*

frameshift mutation   *268*

gene   *257*

genome   *255*

genotype   *255*

intercalating agent   *264*

leader sequence   *260*

missense mutation   *266*

mutagen   *262*

mutation   *261*

nonsense mutation   *266*

phenotype   *255*

photoreactivation   *272*

point mutation   *266*

promoter   *259*

proofreading   *272*

prototrophs   *262*

recA protein   *273*

recombination repair   *273*

replica plating   *268*

reversion mutation   *265*

RNA polymerase binding site or Pribnow box   *260*

sense codons   *257*

sense strand   *259*

Shine-Dalgarno sequence   *261*

silent mutation   *266*

SOS repair   *274*

specific mispairing   *264*

stop or nonsense codons   *257*

strain or clone   *255*

suppressor mutation   *000*

terminator sequence   *261*

trailer sequence   *261*

transformation   *255*

transition mutation   *263*

transversion mutation   *263*

wild type   *264*

wobble   *257*

Questions for Thought and Review

1. Suppose that you have isolated a microorganism from a sample of Martian soil. Describe how you would go about determining the nature of its genetic material.

2. Currently a gene is described in several ways. Which definition do you prefer and why?

3. How could one use small deletion mutations to show that codons are triplet (i.e., that the nucleotide sequence is read three bases at a time rather than two or four)?

4. Sometimes a point mutation does not change the phenotype. List all the reasons you can why this is so.

5. Why might a mutation leading to an amino acid change at a protein's surface not result in a phenotypic change while the substitution of an internal amino acid will?

6. Describe how you would isolate a mutant that required histidine for growth and was resistant to penicillin.

7. How would the following DNA alterations and replication errors be corrected (there may be *more* than one way): base addition errors by DNA polymerase III during replication, thymine dimers, AP sites, methylated guanines, and gaps produced during replication?

===== **Additional Reading** =====

General

Dale, J. W. 1994. *Molecular genetics of bacteria,* 2d ed. New York: John Wiley and Sons.

Gardner, E. J.; Simmons, M. J.; and Snustad, D. P. 1991. *Principles of genetics,* 8th ed. New York: John Wiley and Sons.

Griffiths, A. J. F.; Miller, J. H.; Suzuki, D. T.; Lewontin, R. C.; and Gelbart, W. M. 1993. *An introduction to genetic analysis,* 5th ed. New York: W. H. Freeman.

Holloway, B. W. 1993. Genetics for all bacteria. *Ann. Rev. Microbiol.* 47:659–84.

Joset, F., and Guespin-Michel, J. 1993. *Prokaryotic genetics: Genome organization, transfer and plasticity.* Boston: Blackwell.

Kendrew, J., editor. 1994. *The encyclopedia of molecular biology.* Boston: Blackwell Scientific Publications.

Lewin, B. 1994. *Genes,* 5th ed. New York: Oxford University Press.

Maloy, S. R.; Cronan, J. E., Jr.; and Freifelder, D. 1994. *Microbial genetics,* 2d ed. Boston: Jones and Bartlett.

Russell, P. J. 1992. *Genetics,* 3d ed. New York: Harper Collins.

Scaife, J.; Leach, D.; and Galizzi, A., editors. 1985. *Genetics of bacteria.* New York: Academic Press.

Smith-Keary, P. 1989. *Molecular genetics of Escherichia coli.* New York: Guilford Press.

Stent, G. S., and Calender, R. 1978. *Molecular genetics: An introductory narrative,* 2d ed. San Francisco: W. H. Freeman.

Watson, J. D.; Hopkins, N. H.; Roberts, J. W.; Steitz, J. A.; and Weiner, A. M. 1987. *Molecular biology of the gene,* 4th ed. Menlo Park, Calif.: Benjamin/Cummings.

Weaver, R. F., and Hedrick, P. W. 1992. *Genetics,* 2d ed. Dubuque, Iowa: Wm. C. Brown.

The Genetic Code

Andersson, S. G. E., and Kurland, C. G. 1990. Codon preferences in free-living microorganisms. *Microbiol. Rev.* 54(2):198–210.

Osawa, S.; Jukes, T. H.; Watanabe, K.; and Muto, A. 1992. Recent evidence for evolution of the genetic code. *Microbiol. Rev.* 56(1):229–64.

Gene Structure

Bachmann, B. J. 1990. Linkage map of *Escherichia coli* K–12, Edition 8. *Microbiol. Rev.* 54(2):130–97.

Breathnach, R., and Chambon, P. 1981. Organization and expression of eucaryotic split genes coding for proteins. *Ann. Rev. Biochem.* 50:349–83.

Fournier, M. J., and Ozeki, H. 1985. Structure and organization of the transfer ribonucleic acid genes of *Escherichia coli* K–12. *Microbiol. Rev.* 49(4):379–97.

Girons, I. S.; Old, I. G.; and Davidson, B. E. 1994. Molecular biology of *Borrelia,* bacteria with linear replicons. *Microbiology* 140:1803–16.

Lindahl, L., and Zengel, J. M. 1986. Ribosomal genes in *Escherichia coli. Ann. Rev. Genet.* 20:297–326.

Riley, M. 1993. Functions of the gene products of *Escherichia coli. Microbiol. Rev.* 57(4):862–952.

Weinstock, G. M. 1994. Bacterial genomes: Mapping and stability. *ASM News* 60(2):73–78.

Mutations and Their Chemical Basis

Foster, P. L. 1993. Adaptive mutation: The uses of adversity. *Ann. Rev. Microbiol.* 47:467–504.

Hall, B. G. 1991. Increased rates of advantageous mutations in response to environmental challenges. *ASM News* 57(2):82–86.

Lederberg, J. 1992. Bacterial variation since Pasteur. *ASM News* 58(5):261–65.

Miller, J. H. 1983. Mutational specificity in bacteria. *Ann. Rev. Genet.* 17:215–38.

Singer, B., and Kusmierek, J. T. 1982. Chemical mutagenesis. *Ann. Rev. Biochem.* 52:655–93.

Detection and Isolation of Mutants

Devoret, R. 1979. Bacterial tests for potential carcinogens. *Sci. Am.* 241(2):40–49.

DNA Repair

Bernstein, H. 1983. Recombinational repair may be an important function of sexual reproduction. *BioScience* 33(5):326–31.

Claverys, J. -P., and Lacks, S. A. 1986. Heteroduplex deoxyribonucleic acid base mismatch repair in bacteria. *Microbiol. Rev.* 50(2):133–65.

Friedberg, E. C.; Walker, G. C.; and Siede, W. 1995. *DNA repair and mutagenesis.* Herndon, Va.: ASM Press.

Grossman, L., and Thiagalingam, S. 1993. Nucleotide excision repair, a tracking mechanism in search of damage. *J. Biol. Chem.* 268(23):16871–74.

Howard-Flanders, P. 1981. Inducible repair of DNA. *Sci. Am.* 245(5):72–80.

Sancar, A., and Sancar, G. B. 1988. DNA repair enzymes. *Ann. Rev. Biochem.* 57:29–67.

Sutherland, B. M. 1981. Photoreactivation. *BioScience* 31(6):439–44.

Van Houten, B. 1990. Nucleotide excision repair in *Escherichia coli. Microbiol. Rev.* 54(1):18–51.

Walker, G. C. 1985. Inducible DNA repair systems. *Ann. Rev. Biochem.* 54:425–57.

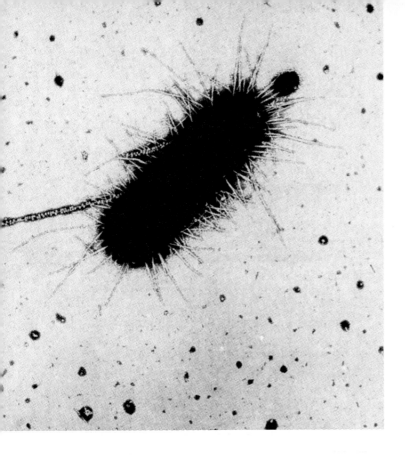

CHAPTER 14
Microbial Genetics: Recombination and Plasmids

Deep in the cavern of the

infant's breast

The father's nature lurks, and

lives anew.

—Horace, *Odes*

Outline

Concepts

1. Recombination is a one-way process in procaryotes: a piece of genetic material (the exogenote) is donated to the chromosome of a recipient cell (the endogenote) and integrated into it.

2. The actual transfer of genetic material between bacteria usually takes place in one of three ways: direct transfer between two bacteria temporarily in physical contact (conjugation), transfer of a naked DNA fragment (transformation), or transport of bacterial DNA by bacteriophages (transduction).

3. Plasmids and transposable elements can move genetic material between bacterial chromosomes and within chromosomes to cause rapid changes in genomes and drastically alter phenotypes.

4. The bacterial chromosome can be mapped with great precision, using Hfr conjugation in combination with transformational and transductional mapping techniques.

5. Recombination of virus genomes occurs when two viruses with homologous chromosomes infect a host cell at the same time.

Chapter 13 introduces the fundamentals of molecular genetics: the way genetic information is organized and stored, the nature of mutations and techniques for their isolation and study, and DNA repair. This chapter focuses on genetic recombination in microorganisms, with primary emphasis placed on recombination in bacteria and viruses.

The chapter begins with a general overview of bacterial recombination and an introduction to both bacterial plasmids and transposable elements. Next, the three types of bacterial recombination—conjugation, transformation, and transduction—are discussed. Because an understanding of the techniques used to locate genes on chromosomes depends on knowledge of recombination mechanisms, mapping the bacterial genome is discussed after the introduction to recombination. The chapter ends with a description of recombination in viruses and a brief discussion of viral chromosome mapping.

Recombination is the process in which a new recombinant chromosome, one with a genotype different from either parent, is formed by combining genetic material from two organisms. It results in a new arrangement of genes or parts of genes and normally is accompanied by a phenotypic change. Most eucaryotes exhibit a complete sexual life cycle, including meiosis, a process of extreme importance in generating new combinations of alleles (alternate forms of a particular gene) through recombination. These chromosome exchanges during meiosis result from **crossing-over** between homologous chromosomes, chromosomes containing identical sequences of genes (figure 14.1). Until about 1945 the primary focus in genetic analysis was on the recombination of genes in plants and animals. The early work on recombination in higher eucaryotes laid the foundations of classical genetics, but it was the development of bacterial and phage genetics between about 1945 and 1965 that really stimulated a rapid advance in our understanding of molecular genetics.

Meiosis (pp. 88–89).

Bacterial Recombination: General Principles

Microorganisms carry out several types of recombination. **General recombination,** the most common form, usually involves a reciprocal exchange between a pair of homologous DNA sequences. It can occur anyplace on the chromosome, and it results from DNA strand breakage and reunion leading to crossing-over (figure 14.2). General recombination is carried out by the products of *rec* genes such as the recA protein so important for DNA repair (*see pp. 273–74*). In bacterial transformation a nonreciprocal form of general recombination

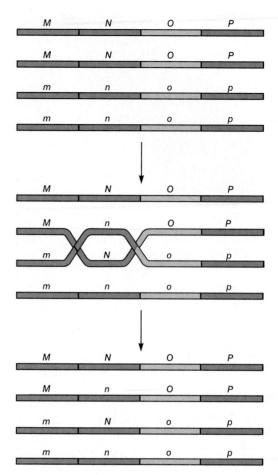

Figure 14.1 Crossing-Over. An example of recombination through crossing-over between homologous eucaryotic chromosomes. The *Nn* gene pair is exchanged. This process usually occurs during meiosis.

takes place (figure 14.3). A piece of genetic material is inserted into the chromosome through the incorporation of a single strand to form a stretch of **heteroduplex DNA.** A second type of recombination, one particularly important in the integration of virus genomes into bacterial chromosomes, is **site-specific recombination.** The genetic material is not homologous with the chromosome it joins, and generally the enzymes responsible for this event are specific for the particular virus and its host. A third type, **replicative recombination,** accompanies the replication of genetic material and does not depend on sequence homology. It is used by genetic elements that move about the chromosome.

DNA replication (pp. 216–21).

Although sexual reproduction with the formation of a zygote and subsequent meiosis is not present in bacteria,

Figure 14.2   The Holliday Model for Reciprocal General Recombination. *Source: From H. Potter and D. Dressler,* Proceedings of the National Academy of Sciences, *73:3000, 1976; after R. Holliday,* Genetics, *78:273, 1974; and previous publication cited therein.*

recombination can take place in several ways. In general, a piece of donor DNA, the **exogenote,** must enter the recipient cell and become a stable part of the recipient cell's genome, the **endogenote.** Two kinds of DNA can move between bacteria. If a DNA fragment is the exchange vehicle, then the exogenote must get into the recipient cell and become incorporated into the endogenote as a replacement piece (or as an "extra" piece) without being destroyed by the host. During replacement of

host genetic material, the recipient cell becomes temporarily diploid for a portion of the genome and is called a **merozygote.** Sometimes the DNA exists in a form that cannot be degraded by the recipient cell's endonucleases. In this case the DNA does not need to be integrated into the host genome but must only enter the recipient to confer its genetic information on the cell. Resistant DNA, such as that in plasmids (see following), usually is circular and has sequences that allow it to maintain

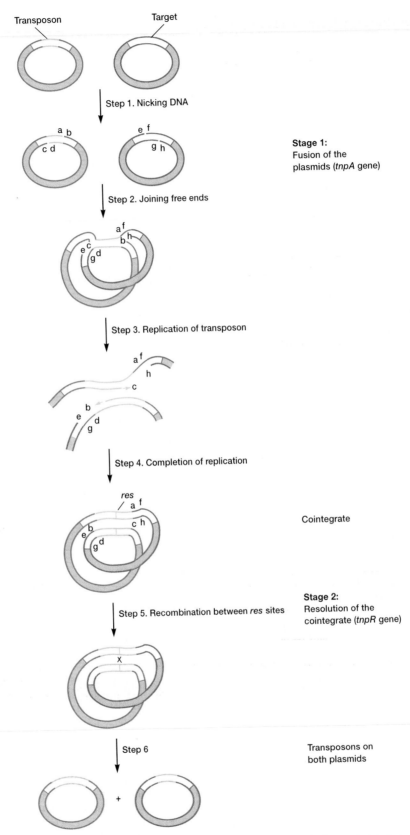

Figure 14.9   Tn3 Transposition Mechanism. Step 1: The two plasmids are nicked to form the free ends labeled a–h. Step 2: Ends a and f are joined, as are g and d. This leaves b, c, e, and h free. Step 3: Two of these remaining free ends (b and c) serve as primers for DNA replication, which is shown in a blowup of the replicating region. Step 4: Replication continues until end b reaches e and end c reaches h. These ends are ligated to complete the cointegrate. Notice that the whole transposon has been replicated. The paired *res* sites are shown for the first time here, even though one *res* site existed in the previous steps. Steps 5 and 6: A crossover occurs between the two *res* sites in the two copies of the transposon, leaving two independent plasmids, each bearing a copy of the transposon.

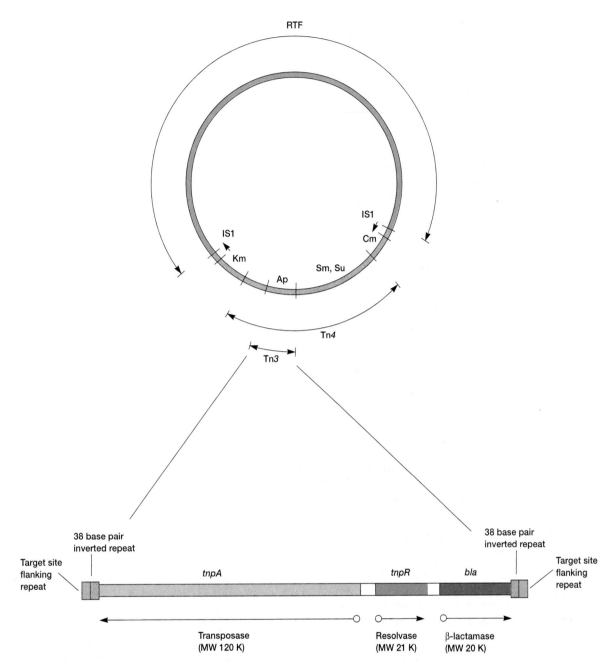

Figure 14.10   The Structure of R Plasmids and Transposons. The R1 plasmid carries resistance genes for five antibiotics: chloramphenicol (Cm), streptomycin (Sm), sulfonamide (Su), ampicillin (Ap), and kanamycin (Km). These are contained in the Tn*3* and Tn*4* transposons. The resistance transfer factor (RTF) codes for the proteins necessary for plasmid replication and transfer. The structure of Tn*3* is shown in more detail. The arrows indicate the direction of gene transcription.

copies of the Tn*3* transposon. In the second stage the cointe-grate is resolved to yield two plasmids, each with a copy of the transposon (figure 14.9, steps 5 and 6). Resolution involves a crossover at the two *res* sites and is catalyzed by a resolvase enzyme coded for by the *tnpR* gene (figure 14.10).

Transposable elements produce a variety of important effects. They can insert within a gene to cause a mutation or stimulate DNA rearrangement, leading to deletions of genetic material. Because some transposons carry stop codons or termination sequences, they may block translation or transcription. Other elements carry promoters and thus activate genes near the point of insertion. Eucaryotic genes as well as procaryotic genes can be turned on and off by transposon movement. Transposons also are located within plasmids and participate in such processes as plasmid fusion and the insertion of F plasmids into the *E. coli* chromosome.

In the previous discussion of plasmids, it was noted that an R plasmid can carry genes for resistance to several drugs. Transposons have antibiotic resistance genes and play a major role in generating these plasmids. Consequently the existence of these elements causes serious problems in the treatment of disease. Since plasmids can contain several different target sites, transposons will move between them; thus plasmids act as both the source and the target for transposons with resistance genes. In fact, it appears that multiple drug resistance plasmids usually are produced by transposon accumulation on a single plasmid (figure 14.10). Because transposons also move between plasmids and primary chromosomes, drug resistance genes can exchange between plasmids and chromosomes, resulting in the further spread of antibiotic resistance.

Transposable elements are widespread in nature. They are present in eucaryotic as well as procaryotic systems. For example, transposable elements have been found in yeast, maize, *Drosophila,* and humans. Clearly, transposable elements play an extremely important role in the generation and transfer of new gene combinations.

---

1. Define the following terms: transposition, transposable element or transposon, insertion sequence, transposase, and composite transposon. Be able to distinguish between an insertion sequence and a composite transposon.

2. How does transposition usually occur in bacteria, and what happens to the target site? What is replicative transposition?

3. Give several important effects transposable elements can have on their possessors.

Bacterial Conjugation

The initial evidence for bacterial **conjugation,** the transfer of genetic information by direct cell to cell contact, came from a simple experiment performed by Joshua Lederberg and Edward L. Tatum in 1946. They mixed two auxotrophic strains, incubated the culture for several hours in nutrient medium, and then

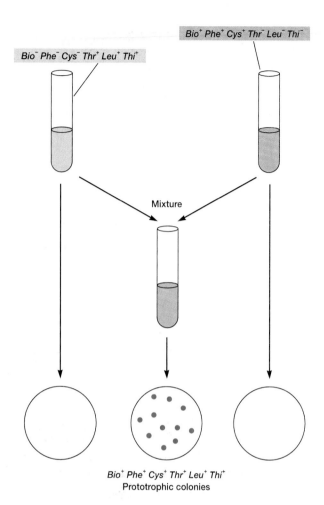

Bio⁻ Phe⁻ Cys⁻ Thr⁺ Leu⁺ Thi⁺

Bio⁺ Phe⁺ Cys⁺ Thr⁻ Leu⁻ Thi⁻

Mixture

Bio⁺ Phe⁺ Cys⁺ Thr⁺ Leu⁺ Thi⁺
Prototrophic colonies

Figure 14.11　Evidence for Bacterial Conjugation. Lederberg and Tatum's demonstration of genetic recombination using triple auxotrophs. See text for details.

plated it on minimal medium. To reduce the chance that their results were due to simple reversion, they used double and triple auxotrophs on the assumption that two or three reversions would not often occur simultaneously. For example, one strain required biotin (Bio⁻), phenylalanine (Phe⁻), and cysteine (Cys⁻) for growth, and another needed threonine (Thr⁻), leucine (Leu⁻), and thiamine (Thi⁻). Recombinant prototrophic colonies appeared on the minimal medium after incubation (figure 14.11). Thus the chromosomes of the two auxotrophs were able to associate and undergo recombination.

Lederberg and Tatum did not directly prove that physical contact of the cells was necessary for gene transfer. This evidence was provided by Bernard Davis (1950), who constructed a U tube consisting of two pieces of curved glass tubing fused at the base to form a U shape with a fritted glass filter between the halves. The filter allows the passage of media but not bacteria. The U tube was filled with nutrient medium and each side inoculated with a different auxotrophic strain of *E. coli* (figure 14.12). During incubation, the medium was pumped back and forth through the filter to ensure medium exchange between the halves. After a

Figure 14.12   The U-Tube Experiment.
The U-tube experiment used to show that genetic recombination by conjugation requires direct physical contact between bacteria. See text for details.

Pressure or suction

Met⁻ Thr⁺ Leu⁺ Thi⁺

Met⁺ Thr⁻ Leu⁻ Thi⁻

Fritted glass filter

4 hour incubation, the bacteria were plated on minimal medium. Davis discovered that when the two auxotrophic strains were separated from each other by the fine filter, gene transfer could not take place. Therefore direct contact was required for the recombination that Lederberg and Tatum had observed.

F⁺ × F⁻ Mating

In 1952 William Hayes demonstrated that the gene transfer observed by Lederberg and Tatum was polar. That is, there were definite donor (F⁺) and recipient (F⁻) strains, and gene transfer was nonreciprocal. He also found that in F⁺ × F⁻ mating the progeny were only rarely changed with regard to auxotrophy (that is, bacterial genes were not often transferred), but F⁻ strains frequently became F⁺.

These results are readily explained in terms of the F factor previously described (figure 14.4). The F⁺ strain contains an extrachromosomal F factor carrying the genes for pilus formation and plasmid transfer. During F⁺ × F⁻ mating or conjugation, the F factor replicates by the rolling-circle mechanism, and a copy of it moves to the recipient (figure 14.13a). The entering strand is copied to produce double-stranded DNA. Because bacterial chromosome genes are rarely transferred with the independent F factor, the recombination frequency is low. It is still not completely clear how the plasmid moves between bacteria. The **sex pilus** or F pilus joins the donor and recipient and may contract to draw them together. The channel for DNA transfer could be either the hollow F pilus or a special conjugation bridge formed upon contact.

The rolling-circle mechanism of DNA replication (p. 218).

Hfr Conjugation

Because certain donor strains transfer bacterial genes with great efficiency and do not usually change recipient bacteria to donors, a second type of conjugation must exist. The F factor is an episome and can integrate into the bacterial chromosome at several different locations by recombination between homologous insertion sequences present on both the plasmid and host chromosomes. When integrated, the F plasmid's *tra* operon is still functional; the plasmid can direct the synthesis of pili, carry out rolling-circle replication, and transfer genetic material to an F⁻ recipient cell. Such a donor is called an **Hfr strain** (for high frequency of recombination) because it exhibits a very high efficiency of chromosomal gene transfer in comparison with F⁺ cells. DNA transfer begins when the integrated F factor is nicked at its site of transfer origin (figure 14.13b). As it is replicated, the chromosome moves through the pilus or conjugation bridge connecting the donor and recipient. Because only part of the F factor is transferred at the start (the initial break is within the F plasmid), the F⁻ recipient does not become F⁺ unless the whole chromosome is transferred. Transfer takes about 100 minutes in *E. coli,* and the connection usually breaks before this process is finished. Thus a complete F factor usually is not transferred, and the recipient remains F⁻.

As mentioned earlier, when an Hfr strain participates in conjugation, bacterial genes are frequently transferred to the recipient. Gene transfer can be in either a clockwise or counterclockwise direction around the circular chromosome, depending on the orientation of the integrated F factor. After the replicated donor chromosome enters the recipient cell, it may be degraded

Figure 14.13 **The Mechanism of Bacterial Conjugation.** (a) F⁺ × F⁻ mating. (b) Hfr × F⁻ mating (the integrated F factor is shown in color).

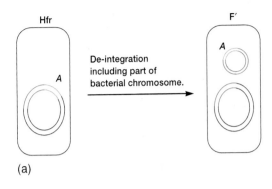

Figure 14.14 **F′ Conjugation.** (*a*) Due to an error in excision, the *A* gene of an Hfr cell is picked up by the F factor. (*b*) The *A* gene is then transferred to a recipient during conjugation.

or incorporated into the F⁻ genome by recombination. Hfr conjugation is perhaps the most efficient natural mechanism of gene transfer between bacteria.

F′ Conjugation

Because the F plasmid is an episome, it can leave the bacterial chromosome. Sometimes during this process the plasmid makes an error in excision and picks up a portion of the chromosomal material to form an **F′ plasmid** (figure 14.14*a*). It is not unusual to observe the inclusion of one or more genes in excised F plasmids. The F′ cell retains all of its genes, although some of them are on the plasmid, and still mates only with an F⁻ recipient. F′ × F⁻ conjugation is virtually identical with F⁺ × F⁻ mating. Once again, the plasmid is transferred, but usually bacterial genes on the chromosome are not (figure 14.14*b*). Bacterial genes on the F′ plasmid are transferred with it and need not be incorporated into the recipient chromosome to be expressed.

The recipient becomes F′ and is a partially diploid merozygote since it has two sets of the genes carried by the plasmid. In this way specific bacterial genes can spread rapidly throughout a bacterial population. Such transfer of bacterial genes is often called sexduction.

F′ conjugation is very important to the microbial geneticist. A partial diploid's behavior shows whether the allele carried by an F′ plasmid is dominant or recessive to the chromosomal gene. The formation of F′ plasmids also is useful in mapping the chromosome since if two genes are picked up by an F factor they must be neighbors.

---

1. What is bacterial conjugation and how was it discovered?
2. Distinguish between F⁺, Hfr, and F⁻ strains of *E. coli* with respect to their physical nature and role in conjugation.
3. Describe in some detail how F⁺ × F⁻ and Hfr conjugation processes proceed, and distinguish between the two in terms of mechanism and the final results.
4. What is F′ conjugation and why is it so useful to the microbial geneticist? How does the F′ plasmid differ from a regular F plasmid? What is sexduction?

DNA Transformation

The second way in which DNA can move between bacteria is through transformation, discovered by Fred Griffith. **Transformation** is the uptake by a cell of a naked DNA molecule or fragment from the medium and the incorporation of this molecule into the recipient chromosome in a heritable form. In natural transformation the DNA comes from a donor bacterium. The process is random, and any portion of a genome may be transferred between bacteria.

The discovery of transformation (pp. 255–56).

When bacteria lyse, they release considerable amounts of DNA into the surrounding environment. These fragments may be relatively large and contain several genes. If a fragment contacts a **competent** cell, one able to take up DNA and be transformed, it can be bound to the cell and taken inside (figure 14.15*a*). The transformation frequency of very competent cells is around 10^{-3} for most genera when an excess of DNA is used. That is, about one cell in every thousand will take up and integrate the gene. Competency is a complex phenomenon and is dependent on several conditions. Bacteria need to be in a certain stage of growth; for example, *S. pneumoniae* becomes competent during the exponential phase when the population reaches about 10^7 to 10^8 cells per ml. When a population becomes competent, bacteria such as pneumococci secrete a small protein called the competence factor that stimulates the production of 8 to 10 new proteins required for transformation. Natural transformation has been discovered so far only in certain gram-positive

Transformation with DNA fragments

DNA fragments

Bacterial chromosome

Uptake of DNA

Integration by nonreciprocal recombination

OR

Degradation

Stable transformation

Unsuccessful transformation

(a)

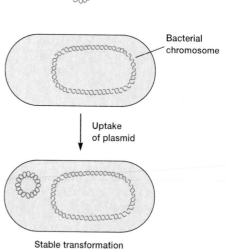

Transformation with a plasmid

DNA plasmid

Bacterial chromosome

Uptake of plasmid

(b) Stable transformation

Figure 14.15 Bacterial Transformation. Transformation with (*a*) DNA fragments and (*b*) plasmids. Transformation with a plasmid often is induced artificially in the laboratory. See text. The transforming DNA is in red.

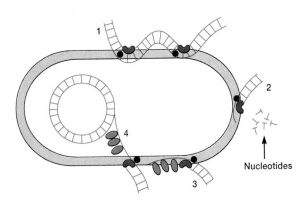

Figure 14.16   The Mechanism of Transformation. (*1*) A long double-stranded DNA molecule binds to the surface with the aid of a DNA-binding protein (●) and is nicked by a nuclease (◠). (*2*) One strand is degraded by the nuclease. (*3*) The undegraded strand associates with a competence-specific protein (◯). (*4*) The single strand enters the cell and is integrated into the host chromosome in place of the homologous region of the host DNA.

and gram-negative genera: *Streptococcus, Bacillus, Thermoactinomyces, Haemophilus, Neisseria, Moraxella, Acinetobacter, Azotobacter,* and *Pseudomonas.* Other genera also may be capable of transformation. Gene transfer by this process occurs in soil and marine environments and may be an important route of genetic exchange in nature.

The mechanism of transformation has been intensively studied in *S. pneumoniae* (figure 14.16). A competent cell binds a double-stranded DNA fragment if the fragment is moderately large (greater than about 5×10^5 daltons); the process is random, and donor fragments compete with each other. DNA uptake requires energy expenditure. One strand is hydrolyzed by an envelope-associated exonuclease; the other strand associates with small proteins and moves through the plasma membrane. The single-stranded fragment can then align with a homologous region of the genome and be integrated, probably by a mechanism similar to that depicted in figure 14.3.

Transformation in *Haemophilus influenzae,* a gram-negative bacterium, differs from that in *S. pneumoniae* in several respects. *Haemophilus* does not produce a competence factor to stimulate the development of competence, and it takes up DNA from only closely related species (*S. pneumoniae* is less particular about the source of its DNA). Double-stranded DNA, complexed with proteins, is taken in by membrane vesicles. The specificity of *Haemophilus* transformation is due to a special 11 base pair sequence (5´AAGTGCGGTCA3´) that is repeated about 600 times in *H. influenzae* DNA. DNA must have this sequence to be bound by a competent cell.

Artificial transformation is carried out in the laboratory by a variety of techniques, including treatment of the cells with calcium chloride, which renders their membranes more permeable to DNA. This approach succeeds even with species that are not naturally competent, such as *E. coli.* Relatively high

concentrations of DNA, higher than would normally be present in nature, are used to increase transformation frequency. When linear DNA fragments are to be used in transformation, *E. coli* usually is rendered deficient in one or more exonuclease activities to protect the transforming fragments. It is even easier to transform bacteria with plasmid DNA since plasmids are not as easily degraded as linear fragments and can replicate within the host (figure 14.15b). This is a common method for introducing recombinant DNA into bacterial cells (*see chapter 15*). DNA from any source can be introduced into bacteria by splicing it into a plasmid before transformation.

1. Define transformation and competence.
2. Describe how transformation occurs in *S. pneumoniae.* How does the process differ in *H. influenzae?*
3. Discuss two ways in which artificial transformation can be used to place functional genes within bacterial cells.

Transduction

Bacterial viruses or bacteriophages participate in the third mode of bacterial gene transfer. These viruses have relatively simple structures in which virus genetic material is enclosed within an outer coat, composed mainly or solely of protein. The coat protects the genome and transmits it between host cells. The morphology and life cycle of bacteriophages is not discussed in detail until chapter 17. Nevertheless, it is necessary to briefly describe the life cycle here as background for a consideration of the bacteriophage's role in gene transfer.

The lytic cycle (pp. 354–61).
Lysogeny (pp. 362–66).

After infecting the host cell, a bacteriophage (phage for short) often takes control and forces the host to make many copies of the virus. Eventually the host bacterium bursts or lyses and releases new phages. This reproductive cycle is called a lytic cycle because it ends in lysis of the host. The cycle has four phases (figure 14.17a). First, the virus particle attaches to a specific receptor site on the bacterial surface. The genetic material, which is often double-stranded DNA, then enters the cell. After adsorption and penetration, the virus chromosome forces the bacterium to make virus nucleic acids and proteins. The third stage begins after the synthesis of virus components. Phages are assembled from these components. The assembly process may be complex, but in all cases phage nucleic acid is packed within the virus's protein coat. Finally, the mature viruses are released by cell lysis.

Bacterial viruses that reproduce using a lytic cycle often are called virulent bacteriophages because they destroy the host cell. Many DNA phages, such as the lambda phage (*p. 363*), are also capable of a different relationship with their host (figure 14.17b).

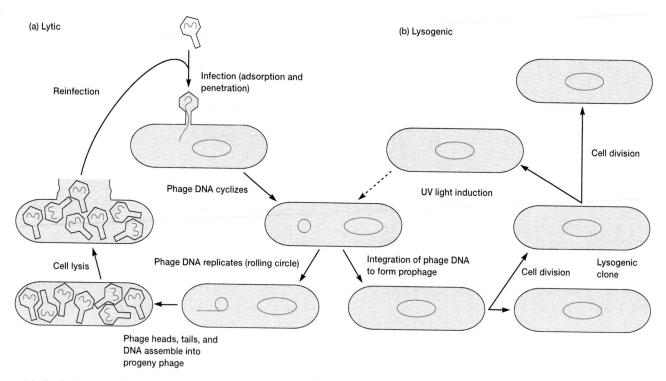

Figure 14.17 Lytic versus Lysogenic Infection by Phage Lambda. (*a*) Lytic infection. (*b*) Lysogenic infection. Viral and prophage DNA are in red.

After adsorption and penetration, the viral genome does not take control of its host and destroy it while producing new phages. Instead the genome remains within the host cell and is reproduced along with the bacterial chromosome. A clone of infected cells arises and may grow for long periods while appearing perfectly normal. Each of these infected bacteria can produce phages and lyse under appropriate environmental conditions. This relationship between the phage and its host is called **lysogeny.** Bacteria that can produce phage particles under some conditions are said to be **lysogens** or **lysogenic,** and phages able to establish this relationship are **temperate phages.** The latent form of the virus genome that remains within the host without destroying it is called the **prophage.** The prophage usually is integrated into the bacterial genome (figure 14.17*b*). Sometimes phage reproduction is triggered in a lysogenized culture by exposure to UV radiation or other factors. The lysogens are then destroyed and new phages released. This phenomenon is called induction.

Transduction is the transfer of bacterial genes by viruses. Bacterial genes are incorporated into a phage capsid because of errors made during the virus life cycle. The virus containing these genes then injects them into another bacterium, completing the transfer. Transduction may be the most common mechanism for gene exchange and recombination in bacteria. There are two very different kinds of transduction: generalized and specialized.

Generalized Transduction

Generalized transduction (figure 14.18) occurs during the lytic cycle of virulent and temperate phages and can transfer any part of the bacterial genome. During the assembly stage,

when the viral chromosomes are packaged into protein capsids, random fragments of the partially degraded bacterial chromosome also may be packaged by mistake. Because the capsid can contain only a limited quantity of DNA, some or all of the viral DNA is left behind. The quantity of bacterial DNA carried depends primarily on the size of the capsid. The P22 phage of *Salmonella typhimurium* usually carries about 1% of the bacterial genome; the P1 phage of *E. coli* carries about 2.0 to 2.5% of the genome. The resulting virus particle often injects the DNA into another bacterial cell but does not initiate a lytic cycle. This phage is known as a **generalized transducing particle** or phage and is simply a carrier of genetic information from the original bacterium to another cell. As in transformation, once the DNA has been injected, it must be incorporated into the recipient cell's chromosome to preserve the transferred genes. The DNA remains double stranded during transfer, and both strands are integrated into the endogenote. About 70 to 90% of the transferred DNA is not integrated but often is able to survive and express itself. **Abortive transductants** are bacteria that contain this nonintegrated, transduced DNA and are partial diploids.

Generalized transduction was discovered in 1951 by Joshua Lederberg and Norton Zinder during an attempt to show that conjugation, discovered several years earlier in *E. coli,* could occur in other bacterial species. Lederberg and Zinder were repeating the earlier experiments with *Salmonella typhimurium.* They found that incubation of a mixture of two multiply auxotrophic strains yielded prototrophs at the level of about one in 10^5. This seemed like good evidence for bacterial recombination, and indeed it was, but their initial conclusion that the transfer resulted from conjugation was not borne out.

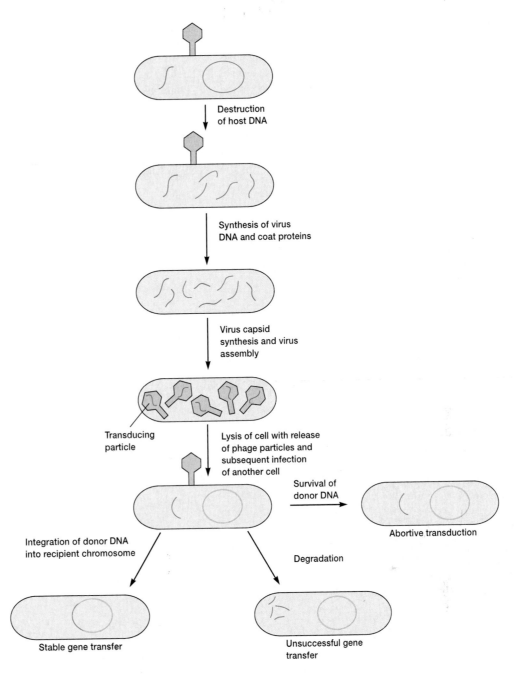

Destruction
of host DNA

Synthesis of virus
DNA and coat proteins

Virus capsid
synthesis and virus
assembly

Transducing
particle

Lysis of cell with release
of phage particles and
subsequent infection
of another cell

Survival of
donor DNA

Abortive transduction

Integration of donor DNA
into recipient chromosome

Degradation

Stable gene transfer

Unsuccessful gene
transfer

Figure 14.18   Generalized Transduction by Bacteriophages. See text for details.

When these investigators performed the U-tube experiment (figure 14.12) with *Salmonella,* they still recovered prototrophs. The filter in the U tube had small enough pores to block the movement of bacteria between the two sides but allowed the phage P22 to pass. Lederberg and Zinder had intended to confirm that conjugation was present in another bacterial species and had instead discovered a completely new mechanism of bacterial gene transfer. The seemingly routine piece of research led to surprising and important results. A scientist must always keep an open mind about results and be prepared for the unexpected.

Specialized Transduction

In **specialized** or **restricted transduction,** the transducing particle carries only specific portions of the bacterial genome. Specialized transduction is made possible by an error in the lysogenic life cycle. When a prophage is induced to leave the host chromosome, excision is sometimes carried out improperly. The resulting phage genome contains portions of the bacterial chromosome (about 5 to 10% of the bacterial DNA) next to the integration site, much like the situation with F′ plasmids (figure 14.19). A transducing phage genome usually is defective and lacks some part of its attachment site. The transducing particle will inject bacterial genes into another bacterium, even though the defective phage cannot reproduce without assistance. The bacterial genes may become stably incorporated under the proper circumstances.

The best-studied example of specialized transduction is the lambda phage. The lambda genome inserts into the host chromosome at specific locations known as attachment or *att* sites (figure 14.20; *see also figures 17.16 and 17.20*). The phage *att* sites and bacterial *att* sites are similar and can complex with each other, although they are not identical. The *att* site for

lambda is next to the *gal* and *bio* genes on the *E. coli* chromosome; consequently, specialized transducing lambda phages most often carry these bacterial genes. The lysate, or product of cell lysis, resulting from the induction of lysogenized *E. coli* contains normal phage and a few defective transducing particles. These particles are called lambda *dgal* because they carry the galactose utilization genes (figure 14.20). Because these lysates contain only a few transducing particles, they often are called **low-frequency transduction lysates (LFT lysates).** Whereas the normal phage has a complete *att* site, defective

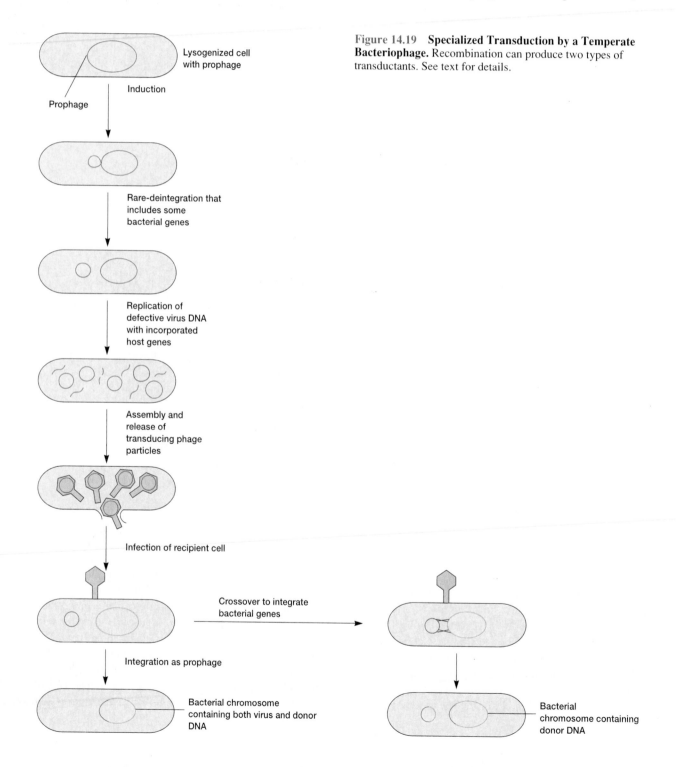

Figure 14.19   Specialized Transduction by a Temperate Bacteriophage. Recombination can produce two types of transductants. See text for details.

Lysogenized cell with prophage

Prophage

Induction

Rare-deintegration that includes some bacterial genes

Replication of defective virus DNA with incorporated host genes

Assembly and release of transducing phage particles

Infection of recipient cell

Crossover to integrate bacterial genes

Integration as prophage

Bacterial chromosome containing both virus and donor DNA

Bacterial chromosome containing donor DNA

transducing particles have a nonfunctional hybrid integration site that is part bacterial and part phage in origin. Integration of the defective phage chromosome does not readily take place. Transducing phages also may have lost some genes essential for reproduction. Stable transductants can arise from recombination between the phage and the bacterial chromosome because of crossovers on both sides of the *gal* site (figure 14.20).

The lysogenic cycle of lambda phage (pp. 363–66).

Defective lambda phages carrying the *gal* gene can integrate if there is a normal lambda phage in the same cell. The normal phage will integrate, yielding two bacterial/phage hybrid *att* sites where the defective lambda *dgal* phage can insert (figure 14.20). It also supplies the genes missing in the defective phage. The normal phage in this instance is termed the **helper phage** because it aids integration and reproduction of the defective phage. These transductants are unstable because the prophages can be induced to excise by agents such as UV radiation. Excision, however, produces a lysate containing a fairly equal mixture of defective lambda *dgal* phage and normal helper phage. Because it is very effective in transduction, the lysate is called a **high-frequency transduction lysate (HFT lysate)**.

Figure 14.20   The Mechanism of Transduction for Phage Lambda and *E. coli*.

Reinfection of bacteria with this mixture will result in the generation of considerably more transductants. LFT lysates and those produced by generalized transduction have one transducing particle in 10^5 or 10^6 phages; HFT lysates contain transducing particles with a frequency of about 0.1 to 0.5.

---

1. Briefly describe the lytic and lysogenic viral reproductive cycles. Define lysogeny, lysogen, temperate phage, prophage, and transduction.

2. Describe generalized transduction, how it occurs, and the way in which it was discovered. What is an abortive transductant?

3. What is specialized or restricted transduction and how does it come about? Be able to distinguish between LFT and HFT lysates and describe how they are formed.

Mapping the Genome

Finding the location of genes in any organism's genome is a very complex task. This section surveys approaches to mapping the bacterial genome, using *E. coli* as an example. All three modes of gene transfer and recombination have been used in mapping.

Gene structure and the nature of mutations (pp. 257–68).

Hfr conjugation is frequently used to map the relative location of bacterial genes. This technique rests on the observation that during conjugation the linear chromosome moves from donor to recipient at a constant rate. In an **interrupted mating experiment** the conjugation bridge is broken and Hfr × F⁻ mating is stopped at various intervals after the start of conjugation by mixing the culture vigorously in a blender (figure 14.21a). The order and timing of gene transfer can be determined because they are a direct reflection of the order of genes on the bacterial chromosome (figure 14.21b). For example, extrapolation of the curves in figure 14.21b back to the x-axis will give the time at which each gene just began to enter the recipient. The result is a circular chromosome map with distances expressed in terms of the minutes elapsed until a gene is transferred. This technique can fairly precisely locate genes 3 minutes or more apart. The heights of the plateaus in figure 14.21b are lower for genes that are more distant from the F factor (the origin of transfer) because there is an ever-greater chance that the conjugation bridge will spontaneously break as transfer continues. Because of the relatively large size of the *E. coli* genome, it is not possible to generate a map from one Hfr strain. Therefore several Hfr strains with the F plasmid integrated at different locations must be used and their maps superimposed on one another. The overall map is adjusted to 100 minutes, although complete transfer may require somewhat more than 100 minutes. Zero time is set at the threonine (*thr*) locus.

Gene linkage also can be determined from transformation by measuring the frequency with which two or more genes simultaneously transform a recipient cell. Consider the case for cotransformation by two genes. In theory, a bacterium could simultaneously receive two genes, each carried on a separate DNA fragment. However, it is much more likely that the genes reside on the same fragment. If two genes are closely linked on the chromosome, then they should be able to cotransform. The

closer the genes are together, the more often they will be carried on the same fragment and the higher will be the frequency of cotransformation. If genes are spaced a great distance apart, they will be carried on separate DNA fragments and the frequency of double transformants will equal the product of the individual transformation frequencies.

Generalized transduction can be used to obtain linkage information in much the same way as transformation. Linkages usually are expressed as cotransduction frequencies, using the argument that the closer two genes are to each other, the more likely they both will reside on the DNA fragment incorporated into a single phage capsid.

Specialized transduction is used to find which phage attachment site is close to a specific gene. The relative locations of specific phage *att* sites are known from conjugational mapping, and the genes linked to each *att* site can be determined by means of specialized transduction. These data allow precise placement of genes on the chromosome.

A simplified genetic map of *E. coli* K12 is given in figure 14.22 and table 14.4. Because conjugation data are not high resolution and cannot be used to position genes that are very close together, the whole map is developed using several mapping techniques. Usually, interrupted mating data are combined with those from cotransduction and cotransformation studies. Data from recombination studies also are used. Normally a new genetic marker in the *E. coli* genome is located within a relatively small region of the genome (10 to 15 minutes long) using a series of Hfr strains with F factor integration sites scattered throughout the genome. Once the genetic marker has been located with respect to several genes in the same region, its position relative to nearby neighbors is more accurately determined using transformation and transduction studies. Recent maps of the *E. coli* chromosome give the locations of more than a thousand genes. Remember that the genetic map only depicts physical reality in a relative sense. A map unit in one region of the genome may not be the same physical distance as a unit in another part.

Genetic maps provide useful information in addition to the order of the genes. For example, there is considerable clustering of genes in *E. coli* K12 (figure 14.22). In the regions around 2, 17, and 27 minutes, there are many genes, whereas relatively few genetic markers are found in the 33 minute region. The areas apparently lacking genes may well have undiscovered genes, but perhaps their function is not primarily that of coding genetic information. One hypothesis is that the 33 minute region is involved in attachment of the *E. coli* chromosome to the plasma membrane during replication and cell division. It is interesting that this region is almost exactly opposite the origin of replication for the chromosome (*oriC*).

Of course a great deal could be learned by comparing a microorganism's genetic map with the actual nucleotide sequence of its genome. Although several virus genomes have been sequenced, in the past it has not been possible to sequence the genomes of bacteria. Recently J. Craig Venter and Hamilton Smith have sequenced the genomes of two free-living bacteria, *Haemophilus influenzae* and *Mycoplasma genitalium.* The genome of *H. influenzae,* the first to be sequenced, contains about 1,743 genes in 1,830,137 base pairs and is much larger than a virus genome. Venter and Smith used an

(a)

(b)

Figure 14.21   The Interrupted Mating Experiment. An interrupted mating experiment on Hfr × F⁻ conjugation. (*a*) The linear transfer of genes is stopped by breaking the conjugation bridge to study the sequence of gene entry into the recipient cell. (*b*) An example of the results obtained by an interrupted mating experiment. The gene order is *lac-tsx-gal-trp.*

approach called **whole-genome shotgun sequencing.** They broke the genomes into fragments using ultrasonic waves and sequenced the fragments. Finally, they ordered the fragments into the proper sequence using sophisticated computer programs. The approach worked so well that it took less than 4 months to sequence the *M. genitalium* genome (about 500,000 base pairs in size). Comparison of the genomes from different bacteria will help deduce which genes are responsible for various bacterial processes. Such information will also aid in the search for human genes because of the similarities between bacterial and human biochemistry.

---

1. Describe how the bacterial genome can be mapped using Hfr conjugation, transformation, generalized transduction, and specialized transduction. Include both a description of each technique and any assumptions underlying its use.

2. Why is it necessary to use several different techniques in genome mapping? How is this done in practice?

Recombination and Genome Mapping in Viruses

Bacteriophage genomes also undergo recombination, although the process is different from that in bacteria. Because phages themselves reproduce within cells and cannot recombine directly, crossing-over must occur inside a host cell. In principle, a virus recombination experiment is easy to carry out. If bacteria are mixed with enough phages, at least two virions will infect each cell on the average and genetic recombination should be observed. Phage progeny in the resulting lysate can be checked for alternate combinations of the initial parental genotypes.

Bacteriophage lytic cycle (pp. 354–61).

Alfred Hershey initially demonstrated recombination in the phage T2, using two strains with differing phenotypes. Two of the parental strains in Hershey's crosses were *h⁺r⁺* and *hr.* The gene *h* influences host range; when gene *h* changes, T2 infects different strains of *E. coli.* Phages with the *r⁺* gene have wild

TABLE 14.4   Selected *E. coli* Genes and Their Map Positions

Symbol	Map Position (Minutes)	Mutation (Affected Substance or Function)	Symbol	Map Position (Minutes)	Mutation (Affected Substance or Function)
aceA,B,E,F	91, 91, 3, 3	Acetate use	*pabA,B*	74, 40	Para-aminobenzoate synthesis
acrA	11	Acridine sensitivity			
ade	See *pur*	Adenine	*panB,C,D*	3, 3, 3	Pantothenate synthesis
ampA	94	Penicillin resistance	*pheA,S*	57, 37	Phenylalanine
araA,B,C,D,E	1, 1, 1, 1, 61	Arabinose use	*pil*	98	Synthesis of fimbriae
argA,B,C,D,E,F,G,	61, 90, 90, 74, 90, 7, 69,	Arginine metabolism	*pit*	77	Inorganic phosphate transport
H,I,P,R,S,	90, 97, 63, 71, 40				
aroA,B,C,D	20, 75, 51, 37	Aromatic compounds	*polA,B,C*	87, 2, 4	DNA polymerases I, II, III
attλ	17	Integration site for λ prophage			
			proA,B,C	6, 6, 9	Proline synthesis
attP2H, II	44, 88	Integration sites for P2	*ptsG,H,I*	25, 52, 52	Phosphotransferase
attφ80	28	Integration site for φ80	*purA,B,C,D,E,*	95, 25, 53, 90, 12,	Purine synthesis
azi	2	Azide resistance	*F,G,H,I*	50, 55, 90, 55	
bioA,B,C,D,F,H	17, 17, 17, 17, 17, 75	Biotin synthesis	*pyrA,B,C,D,E,F*	1, 97, 24, 21, 82, 28	Pyrimidine synthesis
can	63	Canavanine resistance	*recA,B,C,D,E,F*	58, 61, 61, 61, 30, 83	Ultraviolet sensitivity and genetic recombination
cheA,B	42, 41	Chemotactic motility			
cysA,B,C,E,G	52, 28, 59, 81, 74	Cysteine metabolism			
dnaA,B,C,E,G	83, 92, 99, 4, 67	DNA synthesis	*relA,B*	60, 35	RNA synthesis regulation
dsdA,C	51, 51	d-serine			
envA,B	2, 71	Cell envelope formation	*rhaA,B,D,R*	88, 88, 88, 88	Rhamnose
fadA,B,D	87, 87, 40	Fatty acid degradation	*rna, rnb, rnc*	14, 29, 55	Ribonucleases I, II, and III
fuc	60	Fucose metabolism			
galE,K,P,R,T,U	17, 17, 64, 61, 17, 28	Galactose metabolism	*rplA-Y, rpmA-J, rpsA-U*	Various loci	Ribosomal proteins
guaA,B,C	54, 54, 3	Guanine	*rrfA-H, rrlA-H,*	Various loci	rRNA synthesis
hisA,B,C,D,E,F,	44, 44, 44, 44, 44, 44,	Histidine synthesis	*rrnA-H, rrsA-H*		
G,H,I,R	44, 44, 44, 86		*serA,B,C,R,S*	63, 100, 20, 2, 20	Serine metabolism
hsdM,R,S	99, 99, 99	Host specificity	*strC,M*	7, 77	Streptomycin resistance
ilvA,B,C,D,E,F	85, 83, 85, 85, 85, 54	Isoleucine-valine	*supB,C,D,E,F*	16, 27, 43, 15, 27	Suppressors of various mutations
kdgK,R,T	78, 40, 88	Ketodeoxygluconate			
kdpA,B,C,D	16, 16, 16, 16	Potassium dependence	*thiA,B,C*	90, 90, 90	Thiamine (vitamin B$_1$)
lacA,I,Y,Z	8, 8, 8, 8	Lactose metabolism	*thrA,B,C*	0, 0, 0	Threonine
leuA,B,S	2, 2, 15	Leucine synthesis	*tolA,B,C,D*	17, 17, 66, 23	Tolerance to various colicins
lip	15	Lipoate			
lon	10	Filamentous growth	*tonA,B*	4, 28	Resistance to phages, iron uptake
lysA,C	61, 91	Lysine synthesis			
malE,F,G,K,P,Q,T	92, 92, 92, 92,	Maltose uptake and metabolism	*trpA,B,C,D,E,*	28, 28, 28, 28, 28,	Tryptophan synthesis
	75, 75, 75		*R,S,T*	100, 74, 85	
manA	36	Mannose	*tsxB*	9	T6 resistance, nucleoside uptake
metA,B,C,E,F	91, 89, 65, 86, 89	Methionine synthesis			
minB	26	Minicell (no DNA)	*tyrA,R,S*	57, 92, 29, 36	Tyrosine synthesis
mtlA,C,D	81, 81, 81	Mannitol	*ubiA,B,D,E,F,G*	92, 87, 87, 87, 15, 48	Ubiquinone metabolism
mutL,S,T	95, 59, 2	Mutator genes	*uvrA,B,C,D*	92, 18, 42, 86	Repair of ultraviolet damage
nadA,B,C	17, 56, 3	NAD synthesis			
nalA,B	48, 58	Nalidixic acid resistance	*valS*	97	Valine
non	45	Blocks mucoid capsule	*xylA,B*	80	Xylose metabolism
oriC	84	Origin of replication	*zwf*	41	Glucose-6-phosphate dehydrogenase

From Strickberger, Monroe W., *Genetics*, 3rd edition, © 1985, p. 354. Reprinted by permission of Prentice Hall, Upper Saddle River, New Jersey.

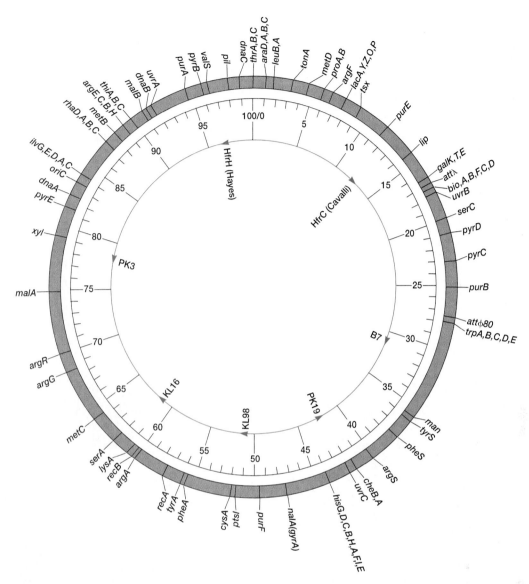

Figure 14.22 *E. coli* **Genetic Map.** A circular genetic map of *E. coli* K12 with the location of selected genes. The inner circle shows the origin and direction of transfer of several Hfr strains. The map is divided into 100 minutes, the time required to transfer the chromosome from an Hfr cell to F⁻ at 37°C.

type plaque morphology, whereas T2 with the *r* genotype has a rapid lysis phenotype and produces larger than normal plaques with sharp edges (figures 14.23*b* and 14.23*c*). In one experiment Hershey infected *E. coli* with large quantities of the *h⁺r⁺* and *hr* T2 strains (figure 14.23*a*). He then plated out the lysates with a mixture of two different host strains and was able to detect significant numbers of *h⁺r* and *hr⁺* recombinants, as well as parental type plaques. As long as there are detectable phenotypes and methods for carrying out the crosses, it is possible to map phage genes in this way.

Plaque formation and morphology (p. 335).

Phage genomes are so small that often it is convenient to map them without determining recombination frequencies. Some techniques actually generate physical maps, which often are most useful in genetic engineering. Several of these meth-

ods require manipulation of the DNA with subsequent examination in the electron microscope. **Denaturation mapping** is an example of this approach. DNA is treated with mildly denaturing conditions so that strand separation occurs in AT-rich sequences because the base pairs are held together with only two hydrogen bonds, but not in GC-rich sequences (three hydrogen bonds per base pair). This causes the formation of bubbles that can be visualized with the electron microscope (figure 14.24). Mutants are compared to one another, and changes in the denaturation bubbles give the physical location of the mutations and their lengths.

A similar approach directly compares wild type and mutant viral chromosomes. In heteroduplex mapping the two types of chromosomes are denatured, mixed, and allowed to rejoin or anneal. When joined, the homologous regions of the different DNA molecules form a regular double helix. In locations where

Figure 14.23 Genetic Recombination in Bacteriophages. (*a*) A summary of a genetic recombination experiment with the *hr* and *h⁺r⁺* strains of the T2 phage. The *hr* chromosome is shown in color. (*b*) The types of plaques produced by this experiment on a lawn of *E. coli.* (*c*) A close-up of the four plaque types.

(a)

(b)

(c)

the bases do not pair due to the presence of a mutation such as a deletion or insertion, bubbles will be visible in the electron microscope. This noncomplementary, single-stranded region is sometimes called a heteroduplex.

Several other direct techniques are used to map viral genomes or parts of them. Restriction endonucleases (*see chapter 15*) are employed together with electrophoresis to analyze DNA fragments and locate deletions and other mutations that affect electrophoretic mobility. Phage genomes also can be directly sequenced to locate particular mutations and analyze the changes that have taken place.

It should be noted that many of these physical mapping techniques also have been employed in the analysis of relatively small portions of bacterial genomes. Furthermore, these methods are useful to the genetic engineer who is concerned with direct manipulation of the DNA.

1. How does recombination in viruses differ from that in bacteria? How did Hershey first demonstrate virus recombination?
2. Describe denaturation mapping and heteroduplex mapping.

Figure 14.24   An Example of Denaturation Mapping. Phage lambda has been partially denatured to produce bubbles in the AT-rich regions. Some of the larger bubbles are labeled with asterisks. At the bottom a denaturation map shows the position of these regions as vertical bars at the proper location on the ruler line. The map begins at the arrow and runs in a counterclockwise direction.

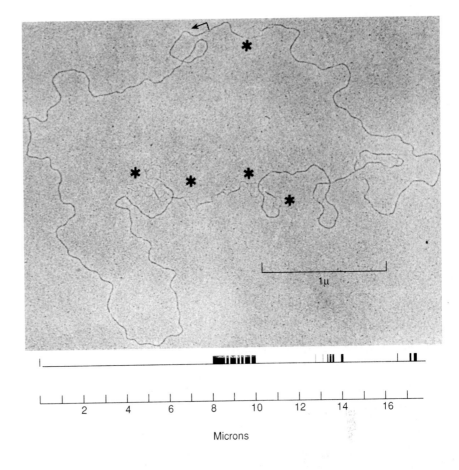

Microns

Summary

1. In recombination, genetic material from two different chromosomes is combined to form a new, hybrid chromosome. There are three types of recombination: general recombination, site-specific recombination, and replicative recombination.

2. Bacterial recombination is a one-way process in which the exogenote is transferred from the donor to a recipient and integrated into the endogenote.

3. Plasmids are small, circular, autonomously replicating DNA molecules that can exist independent of the host chromosome. Their genes are not required for host survival.

4. Episomes are plasmids that can be reversibly integrated with the host chromosome.

5. Many important types of plasmids have been discovered: F factors, R factors, Col plasmids, virulence plasmids, and metabolic plasmids.

6. Transposons or transposable elements are DNA segments that move about the genome in a process known as transposition.

7. There are two types of transposable elements: insertion sequences and composite transposons.

8. Transposable elements cause mutations, block translation and transcription, turn genes on and off, aid F plasmid insertion, and carry antibiotic resistance genes.

9. Conjugation is the transfer of genes between bacteria that depends upon direct cell-cell contact mediated by the F pilus.

10. In F⁺ × F⁻ mating the F factor remains independent of the chromosome and a copy is transferred to the F⁻ recipient; donor genes are not usually transferred.

11. Hfr strains transfer bacterial genes to recipients because the F factor is integrated into the host chromosome. A complete copy of the F factor is not often transferred.

12. When the F factor leaves an Hfr chromosome, it occasionally picks up some bacterial genes to become an F′ plasmid, which readily transfers these genes to other bacteria.

13. Transformation is the uptake of a naked DNA molecule by a competent cell and its incorporation into the genome.

14. Bacterial viruses or bacteriophages can reproduce and destroy the host cell (lytic cycle) or become a latent prophage that remains within the host (lysogenic cycle).

15. Transduction is the transfer of bacterial genes by viruses.

16. In generalized transduction any host DNA fragment can be packaged in a virus capsid and transferred to a recipient.

17. Temperate phages carry out specialized transduction by incorporating bacterial genes during prophage induction and then donating those genes to another bacterium.

18. The bacterial genome can be mapped by following the order of gene transfer during Hfr conjugation; transformational and transductional mapping techniques also may be used. Complete bacterial genomes now are being sequenced and compared with genetic maps.

19. When two viruses simultaneously enter a host cell, their chromosomes can undergo recombination.

20. Virus genomes are mapped by recombination, denaturation, and heteroduplex mapping techniques.

Key Terms

Questions for Thought and Review

1. How does recombination in procaryotes differ from that in most eucaryotes?

2. Distinguish between plasmids, transposons, and temperate phages.

3. How might one demonstrate the presence of a plasmid in a host cell?

4. What effect would you expect the existence of transposable elements and plasmids to have on the rate of microbial evolution? Give your reasoning.

5. How do multiple drug resistant plasmids often arise?

6. Suppose that you carried out a U-tube experiment with two auxotrophs and discovered that recombination was not blocked by the filter but was stopped by treatment with deoxyribonuclease. What gene transfer process is responsible? Why would it be best to use double or triple auxotrophs in this experiment?

7. List the similarities and differences between conjugation, transformation, and transduction.

8. How might one tell whether a recombination process was mediated by generalized or specialized transduction?

9. Why doesn't a cell lyse after successful transduction with a temperate phage?

10. Describe how you would precisely locate the *recA* gene and show that it was between 58 and 58.5 minutes on the *E. coli* chromosome.

Additional Reading

Chapter 13 references also should be consulted, particularly the introductory and advanced textbooks.

General

Berg, D. E., and Howe, M. M., editors. 1989. *Mobile DNA.* Washington, D.C.: American Society for Microbiology.

Brock, T. D. 1990. *The emergence of bacterial genetics.* Cold Spring Harbor, N.Y.: Cold Spring Harbor Laboratory Press.

Levy, S. B., and Marshall, B. M. 1988. Genetic transfer in the natural environment. In *Release of genetically-engineered microorganisms,* M. Sussman, G. H. Collins, F. A. Skinner, and D. E. Stewart-Tall, editors, 61–76. San Diego, Calif.: Academic Press.

Lin, E. C. C.; Goldstein, R.; and Syvanen, M. 1984. *Bacteria, plasmids, and phages: An introduction to molecular biology.* Cambridge, Mass.: Harvard University Press.

Low, K. B., and Porter, R. D. 1978. Modes of gene transfer and recombination in bacteria. *Ann. Rev. Genet.* 12:249–88.

Neidhardt, F. C.; Ingraham, J. L.; and Schaechter, M. 1990. *Physiology of the bacterial cell.* Sunderland, Mass.: Sinauer.

Streips, U. N., and Yasbin, R. E., editors. 1991. *Modern microbial genetics.* New York: Wiley-Liss, Inc.

Bacterial Recombination: General Principles

Dressler, D., and Potter, H. 1982. Molecular mechanisms in genetic recombination. *Ann. Rev. Biochem.* 51:727–61.

Hotchkiss, R. D. 1974. Models of genetic recombination. *Ann. Rev. Microbiol.* 28:445–68.

Kowalczykowski, S. C.; Dixon, D. A.; Eggleston, A. K.; Lauder, S. D.; and Rehrauer, W. M. 1994. Biochemistry of homologous recombination in *Escherichia coli. Microbiol. Rev.* 58(3):401–65.

Kucherlapati, R., and Smith, G. R., editors. 1988. *Genetic recombination.* Washington, D.C.: American Society for Microbiology.

Smith, G. R. 1988. Homologous recombination in procaryotes. *Microbiol. Rev.* 52(1):1–28.

Stahl, F. W. 1987. Genetic recombination. *Sci. Am.* 256(2):91–101.

Bacterial Plasmids

Broda, P. 1979. *Plasmids.* San Francisco: W. H. Freeman.

Elwell, L. P., and Shipley, P. L. 1980. Plasmid-mediated factors associated with virulence of bacteria to animals. *Ann. Rev. Microbiol.* 34:465–96.

Hardy, K. 1986. *Bacterial plasmids,* 2d ed. Washington, D.C.: American Society for Microbiology.

Helinski, D. R. 1973. Plasmid determined resistance to antibiotics: Molecular properties of R factors. *Ann. Rev. Microbiol.* 27:437–70.

Mayer, L. W. 1988. Use of plasmid profiles in epidemiologic surveillance of disease outbreaks and in tracing the transmission of antibiotic resistance. *Clin. Microbiol. Rev.* 1(2):228–43.

Novick, R. P. 1980. Plasmids. *Sci. Am.* 243(6):103–27.

Transposable Elements

Berg, C. M., and Berg, D. E. 1984. Jumping genes: The transposable DNAs of bacteria. *Am. Biol. Teach.* 46(8):431–39.

Cohen, S. N., and Shapiro, J. A. 1980. Transposable genetic elements. *Sci. Am.* 242(2):40–49.

Grindley, N. D. F., and Reed, R. R. 1985. Transpositional recombination in prokaryotes. *Ann. Rev. Biochem.* 54:863–96.

Kleckner, N. 1981. Transposable elements in prokaryotes. *Ann. Rev. Genet.* 15:341–404.

Kleckner, N. 1990. Regulation of transposition in bacteria. *Annu. Rev. Cell Biol.* 6:297–327.

Movable Genetic Elements. 1981. *Cold Spring Harbor symposium on quantitative biology,* vol. 45. Cold Spring Harbor, N.Y.: Cold Spring Harbor Laboratory.

Bacterial Conjugation

Ippen-Ihler, K. A., and Minkley, E. G., Jr. 1986. The conjugation system of F, the fertility factor of *Escherichia coli. Ann. Rev. Genet.* 20:593–624.

DNA Transformation

Lorenz, M. G., and Wackernagel, W. 1994. Bacterial gene transfer by natural genetic transformation in the environment. *Microbiol. Rev.* 58(3):563–602.

McCarty, M. 1985. *The transforming principle: Discovering that genes are made of DNA.* New York: W. W. Norton.

Smith, H. O.; Danner, D. B.; and Deich, R. A. 1981. Genetic transformation. *Ann. Rev. Biochem.* 50:41–68.

Stewart, G. J., and Carlson, C. A. 1986. The biology of natural transformation. *Ann. Rev. Microbiol.* 40:211–35.

Mapping the Genome

Bachmann, B. J. 1990. Linkage map of *Escherichia coli* K–12, Edition 8. *Microbiol. Rev.* 54(2):130–97.

Fleischmann, R. D., et al. 1995. Whole-genome random sequencing and assembly of *Haemophilus influenzae* Rd. *Science* 269(5223):496-512.

Sanderson, K.E.; Hessel, A.; and Rudd, K.E. 1995. Genetic map of *Salmonella typhimurium,* edition VIII. *Microbiol. Rev.* 59(2):241-303.

CHAPTER 15
Recombinant DNA Technology

The recombinant DNA breakthrough has provided us with a new and powerful approach to the questions that have intrigued and plagued man for centuries.

—Paul Berg

The concept and control of the double helix signal a new frontier of biocultural progression. A stereoscopic vision that includes both "creative pessimism" and "creative optimism" is now required.

—Clifford Grobstein

Outline

Concepts

1. Genetic engineering makes use of recombinant DNA technology to fuse genes with vectors and then clone them in host cells. In this way large quantities of isolated genes and their products can be synthesized.
2. The production of recombinant DNA molecules depends on the ability of restriction endonucleases to cleave DNA at specific sites.
3. Plasmids, bacteriophages and other viruses, and cosmids are used as vectors because they can replicate within a host cell while carrying foreign DNA and possess phenotypic traits that allow them to be detected.
4. Genetic engineering is already making substantial contributions to biological research, medicine, industry, and agriculture. Future benefits are probably much greater.
5. Genetic engineering also is accompanied by potential problems in such areas as safety, the ethics of its use with human subjects, environmental impact, and biological warfare.

Chapters 13 and 14 introduce the essentials of microbial genetics. This chapter focuses on the practical applications of microbial genetics and the technology arising from it.

Although human beings have been altering the genetic makeup of organisms for centuries by selective breeding, only recently has the direct manipulation of DNA been possible. The deliberate modification of an organism's genetic information by directly changing its nucleic acid genome is called **genetic engineering** and is accomplished by a collection of methods known as **recombinant DNA technology.** First, the DNA responsible for a particular phenotype is identified and isolated. Once purified, the gene or genes are fused with other pieces of DNA to form recombinant DNA molecules. These are propagated (gene cloning) by insertion into an organism that need not even be in the same kingdom as the original gene donor. Recombinant DNA technology opens up totally new areas of research and applied biology. Thus it is an essential part of **biotechnology,** which is now experiencing a stage of exceptionally rapid growth and development. Although the term has several definitions, in this text biotechnology refers to those processes in which living organisms are manipulated, particularly at the molecular genetic level, to form useful products. The promise for medicine, agriculture, and industry is great; yet the potential risks of this technology are not completely known and may be considerable.

Biotechnology and industrial microbiology (chapter 44).

Recombinant DNA technology is very much the result of several key discoveries in microbial genetics. The first section briefly reviews some landmarks in the development of recombinant technology (table 15.1).

Historical Perspectives

One of the first breakthroughs leading to recombinant DNA (rDNA) technology was the discovery in the late 1960s by Werner Arber and Hamilton Smith of microbial enzymes that make cuts in double-stranded DNA. These enzymes recognize and cleave specific sequences about 4 to 6 base pairs long and are known as **restriction enzymes** or restriction endonucleases (figure 15.1). They normally protect the host cell by destroying phage DNA after its entrance. Restriction enzymes can be used to prepare DNA fragments containing specific genes or portions of genes. For example, the restriction enzyme *Eco*RI, isolated by Herbert Boyer in 1969 from *E. coli,* cleaves the DNA between G and A in the base sequence GAATTC (figure 15.2). Note that in the double-stranded condition, the base sequence GAATTC will base pair with the same sequence running in the opposite direction. *Eco*RI therefore cleaves both DNA strands between the G and the A. When the two DNA fragments separate, they contain single-

Figure 15.1 Restriction Endonuclease Binding to DNA.
(*a*) A schematic diagram of one *Eco*RI endonuclease subunit bound to DNA at the recognition sequence. The enzyme binds to the major groove with the proper sequence. The N-terminal arm of each subunit wraps around the DNA helix. (*b*) A space-filling model of the complete, dimeric *Eco*RI endonuclease bound to DNA. The DNA substrate, with its green and blue strands, faces forward. The endonuclease lies behind the DNA. Its two subunits are colored orange and yellow.

stranded complementary ends, known as sticky ends or cohesive ends. There are hundreds of restriction enzymes that recognize many different specific sequences (table 15.2). Each restriction enzyme name begins with three letters, indicating the bacterium producing it. For example, *Eco*RI is obtained from *E. coli,* whereas *Bam*HI comes from *Bacillus amyloliquefaciens H,* and *Sal*I from *Streptomyces albus.*

In 1970 Howard Temin and David Baltimore independently discovered the enzyme reverse transcriptase that

TABLE **15.1**	Some Milestones in Biotechnology and Recombinant DNA Technology
1970	A complete gene synthesized in vitro Discovery of the first sequence-specific restriction endonuclease and the enzyme reverse transcriptase
1972	First recombinant DNA molecules generated
1973	Use of plasmid vectors for gene cloning
1975	Southern blot technique for detecting specific DNA sequences
1976	First prenatal diagnosis using a gene-specific probe
1977	Methods for rapid DNA sequencing Discovery of "split genes" and somatostatin synthesized using recombinant DNA
1978	Human genomic library constructed
1979	Insulin synthesized using recombinant DNA First human viral antigen (hepatitis B) cloned
1981	Foot-and-mouth disease viral antigen cloned First monoclonal antibody-based diagnostic kit approved for use
1982	Commercial production by *E. coli* of genetically engineered human insulin Isolation, cloning, and characterization of a human cancer gene Transfer of gene for rat growth hormone into fertilized mouse eggs
1983	Engineered Ti plasmids used to transform plants
1985	Tobacco plants made resistant to the herbicide glyphosate through insertion of a cloned gene from *Salmonella* Development of the polymerase chain reaction technique
1987	Insertion of a functional gene into a fertilized mouse egg cures the shiverer mutation disease of mice, a normally fatal genetic disease
1988	The first successful production of a genetically engineered staple crop (soybeans) Development of the gene gun
1989	First field test of a genetically engineered virus (a baculovirus that kills cabbage looper caterpillars)
1990	Production of the first fertile corn transformed with a foreign gene (a gene for resistance to the herbicide bialaphos)
1991	Development of transgenic pigs and goats capable of manufacturing proteins such as human hemoglobin First test of gene therapy on human cancer patients
1994	The Flavr Savr tomato introduced, the first genetically engineered whole food approved for sale Fully human monoclonal antibodies produced in genetically engineered mice

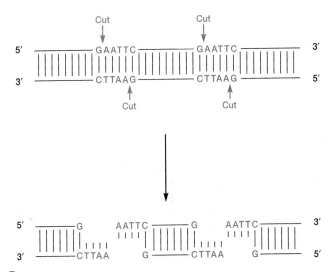

Figure 15.2   Restriction Endonuclease Action. The cleavage catalyzed by the restriction endonuclease *Eco*RI. The enzyme makes staggered cuts on the two DNA strands to form sticky ends.

Figure 15.3   The Synthesis of Double-Stranded cDNA from mRNA. This figure briefly summarizes one procedure for synthesis of cDNA. Reverse transcriptase synthesizes a DNA copy of the mRNA using an oligo-dT primer. The RNA of the resulting RNA-DNA hybrid is degraded to produce single-stranded DNA with a hairpin loop at one end. DNA polymerase I then converts this to a double-stranded form, and S1 nuclease nicks the DNA at the points indicated by the two arrows on the left to generate the final DNA copy.

retroviruses use to produce DNA copies of their RNA genome. This enzyme can be used to construct a DNA copy, called **complementary DNA (cDNA),** of any RNA (figure 15.3). Thus genes or major portions of genes can be synthesized from mRNA.

Reverse transcriptase and retroviruses (p. 376).

The next advance came in 1972, when David Jackson, Robert Symons, and Paul Berg reported that they had successfully generated recombinant DNA molecules. They allowed the sticky ends of fragments to anneal—that is, to base pair with one another—and then covalently joined the fragments with the enzyme DNA ligase.

TABLE 15.2   Some Restriction Endonucleases and Their Recognition Sequences

Enzyme	Microbial Source	Sequence[a]
AluI	Arthrobacter luteus	5´—A—G—↓C—T—3´ 3´—T—C—G↑—A—5´
BamHI	Bacillus amyloliquefaciens H	5´—G—↓G—A—T—C—C—3´ 3´—C—C—T—A—G↑—G—5´
EcoRI	Escherichia coli	5´—G—↓A—A—T—T—C—3´ 3´—C—T—T—A—A↑—G—5´
EcoRII	Escherichia coli	5´—↓C—C—T—G—G—3´ 3´—G—G—A—C—C↑—5´
HaeIII	Haemophilus aegyptius	5´—G—G—↓C—C—3´ 3´—C—C↑—G—G—5´
HindIII	Haemophilus influenzae b	5´—A—↓A—G—C—T—T—3´ 3´—T—T—C—G—A↑—A—5´
PstI	Providencia stuartii	5´—C—T—G—C—A—↓G—3´ 3´—G↑—A—C—G—T—C—5´
SalI	Streptomyces albus	5´—G—↓T—C—G—A—C—3´ 3´—C—A—G—C—T↑—G—5´

[a]The arrows indicate the sites of cleavage on each strand.

Within a year, plasmid **vectors,** or carriers of foreign DNA fragments during gene cloning, had been developed and combined with foreign DNA (figure 15.4). The first such recombinant plasmid capable of being replicated within a bacterial host was the pSC101 plasmid constructed by Stanley Cohen and Herbert Boyer in 1973 (SC in the plasmid name stands for *Stanley Cohen*).

In 1975 Edwin M. Southern published a procedure for detecting specific DNA fragments so that a particular gene could be isolated from a complex DNA mixture. The **Southern blotting technique** depends on the specificity of base complementarity in nucleic acids (figure 15.5). DNA fragments are first separated by size with agarose gel electrophoresis. The fragments are then denatured (rendered single stranded) and transferred to a nitrocellulose filter or nylon membrane so that each fragment is firmly bound to the filter at the same position as on the gel. The filter is bathed with solution containing a radioactive **probe,** a piece of labeled nucleic acid that hybridizes with complementary DNA fragments and is used to locate them. Those fragments complementary to the probe become radioactive and are readily detected by means of **autoradiography.** In this technique a sheet of photographic film is placed over the filter for several hours and then developed. The film is exposed and becomes dark everywhere a radioactive fragment is located because the energy released by the isotope causes the formation of dark-silver grains. Using Southern blotting, one can detect and isolate fragments with any desired sequence from a complex mixture.

More recently, nonradioactive probes to detect specific DNAs have been developed. In one approach the DNA probe is linked to an enzyme such as horseradish peroxidase. After the enzyme-DNA probe has bound to a DNA fragment on the filter, the substrate luminol that will emit light when acted on by the peroxidase is added. The chemiluminescent probe is detected by exposing the filter to photographic film for about 20 minutes. A second technique makes use of the vitamin biotin. A biotin-DNA probe is detected by incubating the filter with either the protein avidin or a similar bacterial protein, streptavidin. The protein specifically attaches to biotin, and is visualized with a

By the late 1970s techniques for easily sequencing DNA, synthesizing oligonucleotides, and expressing eucaryotic genes in procaryotes had also been developed. These techniques could then be used to solve practical problems (table 15.1). The following sections describe how the previously discussed techniques and others are used in genetic engineering.

---

1. Define or describe restriction enzyme, sticky end, cDNA, vector, Southern blotting, probe, and autoradiography.

Synthetic DNA

Oligonucleotides [Greek *oligo*, few or scant] are short pieces of DNA or RNA between about 2 and 20 or 30 nucleotides long. The ability to synthesize DNA oligonucleotides of known sequence is extremely useful. For example, DNA probes can be synthesized and DNA fragments can be prepared for use in a variety of molecular techniques.

DNA structure (pp. 212–14).

DNA oligonucleotides are synthesized by a stepwise process in which single nucleotides are added to the end of the growing chain (figure 15.6). The 3´ end of the chain is attached to a solid support such as a silica gel particle. A DNA synthesizer or "gene machine" carries out the solid-phase synthesis. A specially activated nucleotide derivative is added to the 5´ end of the chain in a series of steps. At the end of an addition cycle, the growing chain is separated from the reaction mixture by filtration or centrifugation. The process is then repeated to attach another nucleotide. It takes about 40 minutes to add a nucleotide to the chain, and chains as large as 50 to 100 nucleotides can be synthesized.

Advances in DNA synthetic techniques have accelerated progress in the study of protein function. One of the most effective ways of studying the relationship of protein structure to function is by altering a specific part of the protein and observing functional changes. In the past this has been accomplished either by chemically modifying individual amino acids or by inducing mutations in the gene coding for the protein under study. There are problems with these two approaches. Chemical modification of a protein is not always specific; several amino acids may be altered, not just the one desired. On the other hand, it is not always possible to produce the proper mutation in the desired gene location. Recently these difficulties have been overcome with a technique called **site-directed mutagenesis.**

In site-directed mutagenesis an oligonucleotide of about 20 residues that contains the desired sequence change is synthesized. The altered oligonucleotide with its artificially mutated sequence is now allowed to bind to a single-stranded copy of the complete gene (figure 15.7). DNA polymerase is added to

Figure 15.4   Recombinant Plasmid Construction. The general procedure used in constructing recombinant plasmid vectors is shown here.

special reagent containing biotin complexed with the enzyme alkaline phosphatase (streptavidin also can be directly attached to the enzyme). The bands with the probe appear blue. These nonradioactive techniques are more rapid and safer than using radioisotopes. On the other hand, they may be less sensitive than radioactively labeled probes.

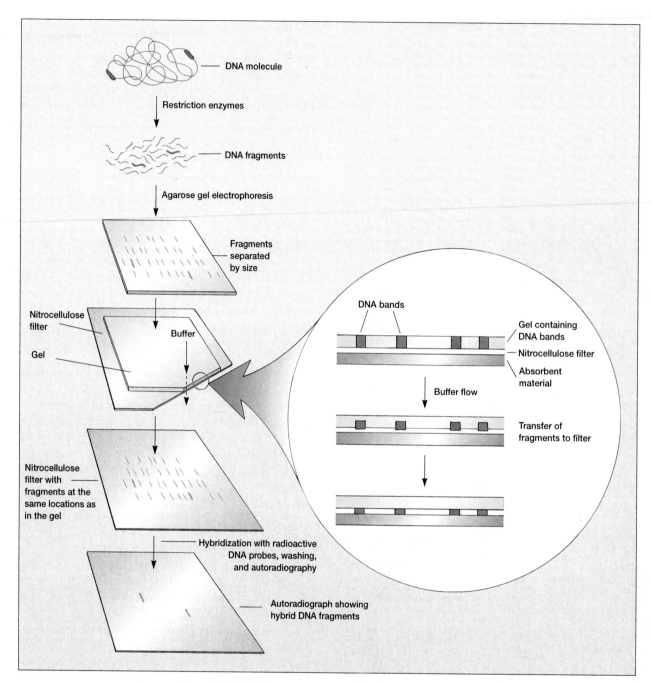

Figure 15.5 **The Southern Blotting Technique.** The insert illustrates how buffer flow transfers DNA bands to the nitrocellulose filter. See text for further details.

the gene-primer complex. The polymerase extends the primer and replicates the remainder of the target gene to produce a new gene copy with the desired mutation. If the gene is attached to a single-stranded DNA bacteriophage (*see pp. 343 and 360*) such as the M13 phage, it can be introduced into a host bacterium and cloned using the techniques to be described shortly. This will yield large quantities of the mutant protein for study of its function.

The Polymerase Chain Reaction

Between 1983 and 1985 Kary Mullis developed a new technique that made it possible to synthesize large quantities of a DNA fragment without cloning it. Although this chapter emphasizes recombinant DNA technology, the **polymerase chain reaction or PCR technique** will be introduced here because of its great practical importance and impact on biotechnology.

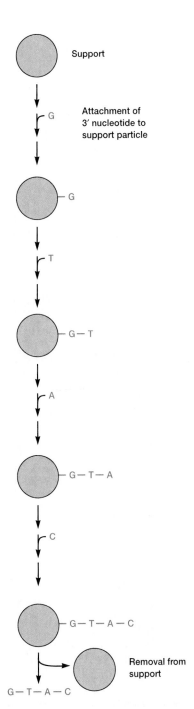

Figure 15.6   The Synthesis of a DNA Oligonucleotide. During each cycle, the DNA synthesizer adds an activated nucleotide (A, T, G, or C) to the growing end of the chain. At the end of the process, the oligonucleotide is removed from its support.

Figure 15.7   Site-Directed Mutagenesis. A synthetic oligonucleotide is used to add a specific mutation to a gene. See text for details.

Figure 15.8 outlines how the PCR technique works. Suppose that one wishes to make large quantities of a particular DNA sequence. The first step is to synthesize fragments with sequences identical to those flanking the targeted sequence. This is readily accomplished with a DNA synthesizer machine as previously described. These synthetic oligonucleotides are usually about 20 nucleotides long and serve as primers for DNA synthesis. The PCR cycle itself takes place in three steps. First, the target DNA containing the sequence to be amplified is heat denatured to separate its complementary strands (step 1). Normally the target DNA is between 100 and 5,000 base pairs in length. Next, primers are added in excess and the temperature lowered so that they can hydrogen bond or anneal to the DNA on both sides of the target sequence (step 2). Because the primers are present in excess, the targeted DNA strands will almost always anneal to the primers rather than to each other. Finally, nucleoside triphosphates and a DNA polymerase are added to the reaction mixture. The DNA polymerase extends the primers and synthesizes copies of the target DNA sequence (step 3). Only polymerases able to function at the high temperatures employed in the PCR technique can be used. Two popular enzymes are the Taq polymerase from the thermophilic bacterium *Thermus aquaticus* and the Vent polymerase from *Thermococcus litoralis*. At the end of one cycle, the targeted sequences on both strands have been copied. When the three-step cycle is repeated (figure 15.8), the four strands from the first cycle are copied to produce eight fragments. The third cycle yields 16 products. Theoretically, 20 cycles will produce about one million copies of the target DNA sequence; 30 cycles yield around one billion copies. Pieces ranging in size from less than 100 base pairs to several thousand base pairs in length can be amplified, and only 10 to 100 picomoles of primer are required. The concentration of target DNA can be as low as 10^{-20} to 10^{-15} M (or 1 to 10^5 DNA copies per 100 µl). The whole reaction mixture is often 100 µl or less in volume.

DNA replication (pp. 216–21).

The polymerase chain reaction technique has now been automated and is carried out by a specially designed machine (figure 15.9). Currently a PCR machine can carry out 25 cycles and amplify DNA 10^5 times in as little as 57 minutes. During a typical cycle the DNA is denatured at 94°C for 15 seconds, then the primers are annealed and extended (steps 2 and 3) at 68°C for 60 seconds.

PCR technology is improving continually. For example, RNA now can be efficiently used in PCR procedures. The r*Tth* DNA polymerase will transcribe RNA to DNA and then amplify the DNA. Cellular RNAs and RNA viruses may be studied even when the RNA is present in very small amounts (as few as 100

Figure 15.8 The Polymerase Chain Reaction. In three cycles the targeted sequence has been amplified to produce eight copies. See text for details.

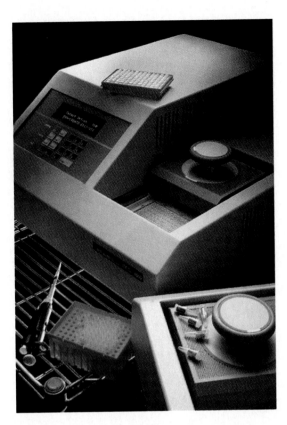

Figure 15.9 A Modern PCR Machine. PCR machines are now fully automated and microprocessor controlled. They can process up to 96 samples at a time.

copies can be transcribed and amplified). PCR also can quantitate DNA products without the use of isotopes. This allows one to find the initial amount of target DNA in less than an hour using automated equipment. Quantitative PCR is quite valuable in virology and gene expression studies. As mentioned earlier, the target DNA to be amplified is normally less than about 5,000 base pairs in length. A "long PCR" technique has been developed that will amplify sequences up to 42 kilobases long.

The PCR technique has already proven exceptionally valuable in many areas of molecular biology, medicine, and biotechnology. It can be used to amplify very small quantities of a specific DNA and provide sufficient material for accurately sequencing the fragment or cloning it by standard techniques. Undoubtedly PCR will be used extensively to aid genetic mapping in the human genome project. PCR-based diagnostic tests for AIDS, Lyme disease, chlamydia, tuberculosis, hepatitis, the human papilloma virus, and other infectious agents and diseases are being developed. The tests are rapid, sensitive, and specific. PCR is particularly valuable in the detection of genetic diseases such as sickle cell anemia, phenylketonuria, and muscular dystrophy. The technique is already having an impact on forensic science where it is being used in criminal cases as a part of DNA fingerprinting technology. It is possible to exclude or incriminate suspects using extremely small samples of biological material discovered at the crime scene.

Preparation of Recombinant DNA

There are three ways to obtain adequate quantities of a DNA fragment. One can extract all the DNA from an organism, cleave the DNA into fragments, isolate the fragment of interest, and finally clone it. Alternatively, all of the fragments can be cloned by means of a suitable vector, and each clone (the population of identical molecules with a single ancestral molecule) can be tested for the desired gene. One also can directly synthesize the desired DNA fragment as described earlier, and then clone it.

Isolating and Cloning Fragments

Long, linear DNA molecules are fragile and easily sheared into fragments by passing the DNA suspension through a syringe needle several times. DNA also can be cut with restriction enzymes. The resulting fragments are either separated by electrophoresis or first inserted into vectors and cloned.

Agarose or polyacrylamide gels usually are used to separate DNA fragments electrophoretically. In **electrophoresis,** charged molecules are placed in an electrical field and allowed to migrate toward the positive and negative poles. The molecules separate because they move at different rates due to their differences in charge and size. In practice, the fragment mixture is usually placed in wells molded within a sheet of gel (figure 15.10). The gel concentration varies with the size of DNA fragments to be separated. Usually 1 to 3% agarose gels or 3 to 20% polyacrylamide gels are used. When an electrical field is generated in the gel, the fragments move through the pores of the gel toward an electrode (negatively charged DNA fragments migrate toward the positive electrode or anode). Each fragment's migration rate is inversely proportional to the log of its molecular weight, and the fragments are separated into size classes (figure 15.10c). A simple DNA molecule might yield only a few bands. If the original DNA was very large and complex, staining of the gel would reveal a smear representing an almost continuous gradient in fragment size. The band or section containing the desired fragment is located with the Southern blotting technique (figure 15.5) and removed. Since a band may represent a mixture of several fragments, it is electrophoresed on a gel with a different concentration to separate similarly sized fragments. The location of the pure fragment is determined by Southern blotting, and it is extracted from the gel.

Once fragments have been separated, they are joined with an appropriate vector, such as a plasmid (*see chapter 14*), to form a recombinant molecule that can reproduce in a host cell. One of the easiest and most popular approaches is to cut the plasmid and donor DNA with the same restriction enzyme so that identical sticky ends are formed (figure 15.11). After a fragment has annealed with the plasmid through complementary base pairing, the breaks are joined by DNA ligase. A second method for creating recombinant molecules can be used with fragments and vectors lacking sticky ends. After cutting the plasmid and donor DNA, one can add poly(dA) to the 3′ ends of the plasmid DNA, using the enzyme terminal transferase (figure 15.12). Similarly, poly(dT) is added to

(a)

(b)   (c)

Figure 15.10   Gel Electrophoresis of DNA. (*a*) A diagram of a vertical gel outfit showing its parts. (*b*) A commercial gel electrophoresis outfit. (*c*) The 1 kilobase ladder is an electrophoretic gel containing a series of DNA fragments of known size. The numbers indicate number of base pairs each fragment contains. The smallest fragments have moved the farthest. The gel on the right shows the fragments that arise when lambda phage DNA is digested with the *Hin*dIII restriction enzyme.

Figure 15.11   Recombinant Plasmid Construction and Cloning. The construction and cloning of a recombinant plasmid vector using an antibiotic resistance gene to select for the presence of the plasmid. The scale of the sticky ends of the fragments and plasmid has been enlarged to illustrate complementary base pairing. (*a*) The electron micrograph shows a plasmid that has been cut by a restriction enzyme and a donor DNA fragment. (*b*) The micrograph shows a recombinant plasmid. See text for details.

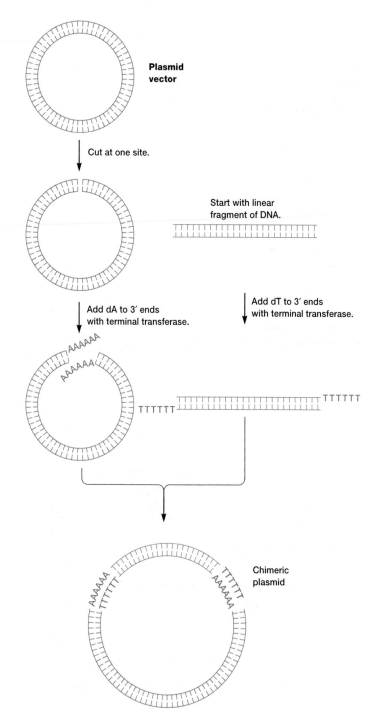

Figure 15.12 Terminal Transferase and the Construction of Recombinant Plasmids. The poly(dA-dT) tailing technique can be used to construct sticky ends on DNA and generate recombinant molecules.

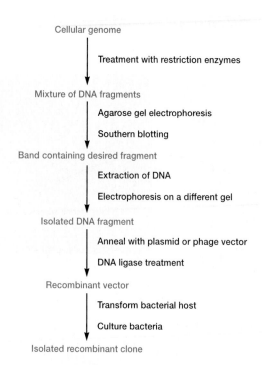

Figure 15.13 Cloning Cellular DNA Fragments. The preparation of a recombinant clone from previously isolated DNA fragments.

The rDNA molecules are cloned by inserting them into bacteria, using transformation or phage injection (*see chapter 14*). Each strain reproduces to yield a population containing a single type of recombinant molecule. The overall process is outlined in figure 15.13. The same cloning techniques can be used with DNA fragments prepared using a DNA synthesizer machine.

Although selected fragments can be isolated and cloned as just described, it often is preferable to fragment the whole genome and clone all the fragments by using a vector. Then the desired clone can be identified. To be sure that the complete genome is represented in this collection of clones, called a genomic library, more than a thousand transformed bacterial strains must be maintained (the larger the genome, the more strains are stored). Libraries of cloned genes also can be generated using phage lambda as a vector and stored as phage lysates.

It is necessary to identify which clone in the library contains the desired gene. If the gene is expressed in the bacterium, it may be possible to assay each clone for a specific protein. However, a nucleic acid probe is normally employed in identification. The bacteria are replica plated on nitrocellulose paper and lysed in place with sodium hydroxide (figure 15.14). This yields a pattern of membrane-bound, denatured DNA corresponding to the colony pattern on the agar plate. The membrane is treated with a radioactive probe as in the original Southern blotting method. The radioactive spots identify colonies on the master plate that contain the desired DNA fragment. This approach also is used to analyze a library of cloned lambda phage (figure 15.15).

the 3′ ends of the fragments. The ends will now base pair with each other and are joined by DNA ligase to form a recombinant plasmid. Some enzymes (e.g., the *Alu*I restriction enzyme from *Arthrobacter luteus*) cut DNA at the same position on both strands to form a blunt end. Fragments and vectors with blunt ends may be joined by T4 DNA ligase (blunt-end ligation).

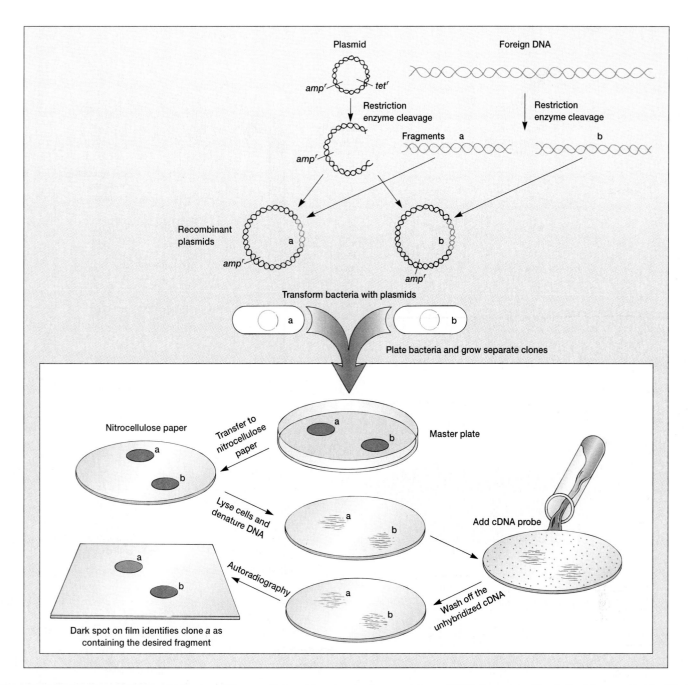

Figure 15.14   Cloning with Plasmid Vectors. The use of plasmid vectors to clone a mixture of DNA fragments. The desired fragment is identified with a cDNA probe. See text for details.

Gene Probes

Success in isolating the desired recombinant clones depends on the availability of a suitable probe. Gene-specific probes are obtained in several ways. Frequently they are constructed with cDNA clones. If the gene of interest is expressed in a specific tissue or cell type, its mRNA is often relatively abundant. For example, reticulocyte mRNA may be enriched in globin mRNA, and pancreatic cells in insulin mRNA. Although mRNA is not available in sufficient quantity to serve as a probe, the desired mRNA species can be converted into cDNA by reverse transcription (figure 15.3). The cDNA copies are purified, spliced into appropriate vectors, and cloned to provide adequate amounts of the required probe.

Probes also can be generated if the gene codes for a protein of known amino acid sequence. Oligonucleotides, about

Figure 15.15   The Use of Lambda Phage as a Vector. (*a*) The preparation of a genomic library. Each colony on the bacterial lawn is a recombinant clone carrying a different DNA fragment. (*b*) Detection and cloning of the desired recombinant phage.

TABLE 15.3   Some Recombinant DNA Cloning Vectors

Type	Vector	Restriction Sequences Present	Features
Plasmid (*E. coli*)	pBR322	*Bam*HI, *Eco*RI, *Hae*III, *Hin*dIII, *Pst*I, *Sal*I, *Xor*II	Carries genes for tetracycline and ampicillin resistance
Plasmid (yeast–*E. coli* hybrid)	pYe(CEN3)41	*Bam*HI, *Bgl*II, *Eco*RI, *Hin*dIII, *Pst*I, *Sal*I	Multiplies in *E. coli* or yeast cells
Cosmid (artificially constructed *E. coli* plasmid carrying lambda cos site)	pJC720	*Hin*dIII	Can be packaged in lambda phage particles for efficient introduction into bacteria; replicates as a plasmid; useful for cloning large DNA inserts
Virus	Charon phage	*Eco*RI, *Hin*dIII, *Bam*HI, *Sst*I	Constructed using restriction enzymes and a ligase, having foreign DNA as central portion, with lambda DNA at each end; carries β-galactosidase gene; packaged into lambda phage particles; useful for cloning large DNA inserts
Virus	Lambda 1059	*Bam*HI	Will carry large DNA fragments (8–21 kb); recombinant can grow on *E. coli* lysogenic for P2 phage, whereas vector cannot
Virus	M13	*Eco*RI	Single-stranded DNA virus; useful in studies employing single-stranded DNA insert and in producing DNA fragments for sequencing
Plasmid	Ti	*Sma*I, *Hpa*I	Maize plasmid

From G. D. Elseth and K. D. Baumgartner, *Genetics*, 1984 Benjamin/Cummings Publishing, Menlo Park, CA. Reprinted by permission of the author.

20 nucleotides or longer, that code for a characteristic amino acid sequence are synthesized. These often are satisfactory probes since they will specifically bind to the gene segment coding for the desired protein.

Sometimes previously cloned genes or portions of genes may be used as probes. This approach is effective when there is a reasonable amount of similarity between the nucleotide sequences of the two genes.

After construction, the probe is labeled to aid detection. Often ^{32}P is added to both DNA strands so that the radioactive strands can be located with autoradiography. Nonradioactively labeled probes may also be used.

Isolating and Purifying Cloned DNA

After the desired clone of recombinant bacteria or phages has been located with a probe, it can be picked from the master plate and propagated. The recombinant plasmid or phage DNA is then extracted and further purified when necessary. The DNA fragment is cut out of the plasmid or phage genome by means of restriction enzymes and separated from the remaining DNA by electrophoresis.

Clearly, DNA fragments can be isolated, purified, and cloned in several ways. Regardless of the exact approach, a key to successful cloning is choosing the right vector. The next section considers types of cloning vectors and their uses.

1. How are oligonucleotides synthesized? What is site-directed mutagenesis?
2. Briefly describe the polymerase chain reaction technique. What is its importance?
3. What is electrophoresis and how does it work?
4. Describe three ways in which a fragment can be covalently attached to vector DNA.
5. Outline in detail two different ways to isolate and clone a specific gene. What is a genomic library?
6. How are gene-specific probes obtained?

Cloning Vectors

There are three major types of vectors: plasmids, bacteriophages and other viruses, and cosmids (table 15.3). Each type has its own advantages. Plasmids are the easiest to work with; rDNA phages and other viruses are more conveniently stored for long periods; larger pieces of DNA can be cloned with cosmids. Besides these three major types, there also are vectors designed for a specific function. For example, shuttle vectors are used in transferring genes between very different organisms and usually contain one replication origin for each host. The pYe type plasmids reproduce in both yeast and *E. coli* and can be used to clone yeast genes in *E. coli*.

Figure 15.16   The pBR322 Plasmid. A map of the *E. coli* plasmid pBR322. The map is marked off in units of 1×10^5 daltons (outer circle) and 0.1 kilobases (inner circle). The locations of some restriction enzyme sites are indicated. The plasmid has resistance genes for ampicillin (*Ap^r*) and tetracycline (*Tet^r*).

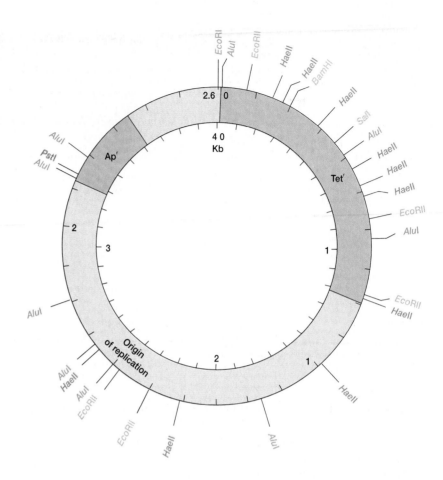

Plasmids

Plasmids were the first cloning vectors. They are easy to isolate and purify, and they can be reintroduced into a bacterium by transformation. Plasmids often bear antibiotic resistance genes, which are used to select their bacterial hosts. A recombinant plasmid containing foreign DNA often is called a **chimera,** after the Greek mythological monster that had the head of a lion, the tail of a dragon, and the body of a goat. One of the most widely used plasmids is pBR322.

The biology of plasmids (pp. 280–83).

Plasmid pBR322 has both resistance genes for ampicillin and tetracycline and many restriction sites (figure 15.16). Several of these restriction sites occur only once on the plasmid

All vectors share several common characteristics. They are typically small, well-characterized molecules of DNA. They contain at least one replication origin and can be replicated within the appropriate host, even when they contain "foreign" DNA. Finally, they code for a phenotypic trait that can be used to detect their presence; frequently it is also possible to distinguish parental from recombinant vectors.

and are located within an antibiotic resistance gene. This arrangement aids detection of recombinant plasmids after transformation (figure 15.17). For example, if foreign DNA is inserted into the ampicillin resistance gene, the plasmid will no longer confer resistance to ampicillin. Thus tetracycline-resistant transformants that lack ampicillin resistance contain a chimeric plasmid.

Phage Vectors

Both single- and double-stranded phage vectors have been employed in recombinant DNA technology. For example, lambda phage derivatives are very useful for cloning and can carry fragments up to about 45 kilobases in length. The genes for lysogeny and integration often are nonfunctional and may be deleted to make room for the foreign DNA. The modified phage genome also contains restriction sequences in areas that will not disrupt replication. After insertion of the foreign DNA into the modified lambda vector chromosome, the recombinant phage genome is packaged into viral capsids and can be used to infect host *E. coli* cells (figure 15.15). These vectors are often used to generate genomic libraries. *E. coli* also can be directly transformed with recombinant lambda

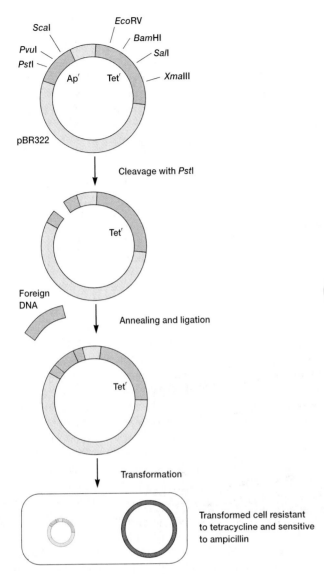

Figure 15.17   Deletion of Recombinant Plasmids. The use of antibiotic resistance genes to detect the presence of recombinant plasmids. Because foreign DNA has been inserted into the ampicillin resistance gene, the recombinant host is only resistant to tetracycline. The restriction enzymes indicated at the top of the figure cleave only one site on the plasmid.

DNA and produce phages. However, this approach is less efficient than the use of complete phage particles. The process is sometimes called **transfection.**

The biology of bacteriophages (chapter 17).

Cosmids

Cosmids are plasmids that contain lambda phage cos sites and can be packaged into phage capsids. The lambda genome contains a cos site at each end. When the genome is to be packaged

in a capsid, it is cleaved at one cos site and the linear DNA is inserted into the capsid until the second cos site has entered. Thus any DNA between the cos sites is packaged. Cosmids typically contain several restriction sites and antibiotic resistance genes. They are packaged in lambda capsids for efficient injection into bacteria, but they also can exist as plasmids within a bacterial host.

---

1. Define shuttle vector, chimera, cosmid, and transfection.
2. Describe how each of the three major types of vectors is used in genetic engineering. Give an advantage of each.
3. How can the presence of two antibiotic resistance genes be used to detect recombinant plasmids?

Inserting Genes into Eucaryotic Cells

Because of its practical importance, much effort has been devoted to the development of techniques for inserting genes into eucaryotic cells. Some of these techniques have also been used successfully to transform bacterial cells. The most direct approach is the use of microinjection. Genetic material directly injected into animal cells such as fertilized eggs is sometimes stably incorporated into the host genome to produce a **transgenic animal,** one that has gained new genetic information from the acquisition of foreign DNA.

Another effective technique that works with mammalian cells and plant cell protoplasts is **electroporation.** If cells are mixed with a DNA preparation and then briefly exposed to pulses of high voltage (from about 250 to 4,000 V/cm for mammalian cells), the cells take up DNA through temporary holes in the plasma membrane. Some of these cells will be transformed.

One of the most effective techniques is to shoot microprojectiles coated with DNA into plant and animal cells. The **gene gun,** first developed at Cornell University, operates somewhat like a shotgun. A .22-caliber blank shell shoots a spray of DNA-coated metallic microprojectiles into the cells (figure 15.18). The device has been used to transform corn and produce fertile corn plants bearing foreign genes. Other guns use either electrical discharges or high-pressure gas to propel the DNA-coated projectiles. These guns are sometimes called biolistic devices, a name derived from biological and ballistic. They have been used to transform microorganisms (yeast, the mold *Aspergillus,* and the alga *Chlamydomonas*), mammalian cells, and a variety of plant cells (corn, cotton, tobacco, onion, and poplar).

Other techniques also are available. Plants can be transformed with *Agrobacterium* vectors as will be described shortly (p. 326). Viruses increasingly are used to insert desired genes into eucaryotic cells. For example, genes may be placed in a retrovirus (*see p. 376*), which then infects the target cell and integrates a DNA copy of its RNA genome into the host chromosome. Recombinant baculoviruses will infect insect cells and promote the production of many proteins.

Figure 15.18   The Gene Gun. DNA-coated microprojectiles are propelled into target cells with a .22-caliber blank shell.

Expression of Foreign Genes in Bacteria

After a suitable cloning vector has been constructed, rDNA enters the host cell, and a population of recombinant microorganisms develops. Most often the host is an *E. coli* strain that lacks restriction enzymes and is *recA⁻* to reduce the chances that the rDNA will undergo recombination with the host chromosome. *Bacillus subtilis* and the yeast *Saccharomyces cerevisiae* also may serve as hosts. Plasmid vectors enter *E. coli* cells by calcium chloride-induced transformation. Electroporation at 3 to 24 kV/cm is effective with both gram-positive and gram-negative bacteria.

A cloned gene is not always expressed in the host cell without further modification of the recombinant vector. To be transcribed, the recombinant gene must have a promoter that is recognized by the host RNA polymerase. Translation of its mRNA depends on the presence of leader sequences and mRNA modifications that allow proper ribosome binding. These are quite different in eucaryotes and procaryotes, and a procaryotic leader must be provided to synthesize eucaryotic proteins in a bacterium. Finally, introns in eucaryotic genes must be removed because the procaryotic host will not excise them after transcription of mRNA; a eucaryotic protein is not functional without intron removal prior to translation.

The problems of expressing recombinant genes in host cells are largely overcome with the help of special cloning vectors called **expression vectors.** These vectors are often derivatives of plasmid pBR322 and contain the necessary transcription and translation start signals. They also have useful restriction sites next to these sequences so that foreign DNA can be inserted with relative ease. Some expression vectors contain portions of the lac operon and can effectively regulate the expression of the cloned genes in the same manner as the operon.

Somatostatin, the 14-residue hypothalamic polypeptide hormone that helps regulate human growth, provides an example of useful cloning and protein production. The gene for somatostatin was initially synthesized by chemical methods. Besides the 42 bases coding for somatostatin, the polynucleotide contained a codon for methionine at the 5′ end (the N-terminal end of the peptide) and two stop codons at the opposite end. To aid insertion into the plasmid vector, the 5′ ends of the synthetic gene were extended to form single-stranded sticky ends complementary to those formed by the *Eco*RI and *Bam*HI restriction enzymes. A modified pBR322 plasmid was cut with both *Eco*RI and *Bam*HI to remove a part of the plasmid DNA. The synthetic gene was then spliced into the vector by taking advantage of its cohesive ends (figure 15.19). Finally, a fragment containing the initial part of the lac operon (including the promoter, operator, ribosome binding site, and much of the β-galactosidase gene) was inserted next to the somatostatin gene. The plasmid now contained the somatostatin gene fused in the proper orientation to the remaining portion of the β-galactosidase gene.

After introduction of this chimeric plasmid into *E. coli*, the somatostatin gene was transcribed with the β-galactosidase gene fragment to generate an mRNA having both messages. Translation formed a protein consisting of the total hormone polypeptide attached to the β-galactosidase fragment by a methionine residue. Cyanogen bromide cleaves peptide bonds at methionine residues. Treatment of the fusion protein with cyanogen bromide broke the peptide chain at the methionine and released the hormone (figure 15.20). Once free, the polypeptide was able to fold properly and become active. Since production of the fusion protein was under the control of the lac operon, it could be easily regulated.

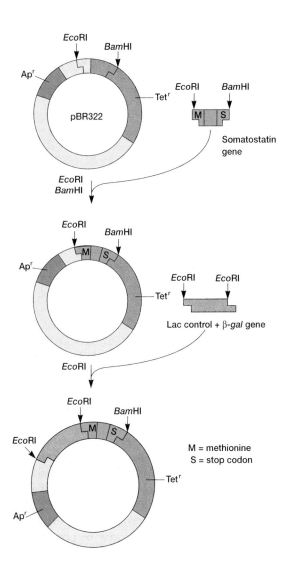

Figure 15.19   Cloning the Somatostatin Gene. An overview of the procedure used to synthesize a recombinant plasmid containing the somatostatin gene. See text for details.

Many proteins have been produced since the synthesis of somatostatin. A similar approach was used to manufacture human insulin. Human growth hormone and some interferons also have been synthesized by rDNA techniques. The human growth hormone gene was too long to synthesize by chemical procedures and was prepared from mRNA as cDNA.

As mentioned earlier, introns in eucaryotic genes are not removed by bacteria and will render the final protein nonfunctional. The easiest solution is to prepare cDNA from processed mRNA that lacks introns and directly reflects the correct amino acid sequence of the protein product. In this instance it is particularly important to fuse the gene with an expression vector since a promoter and other essential sequences will be missing in the cDNA.

Eucaryotic mRNA processing (pp. 223–25).

If the mRNA is scarce, it may not be easy to obtain enough for cDNA synthesis. Often the sequence of the protein coded for by the gene is used to deduce the best DNA sequence for the specific polypeptide segment (a logical procedure called reverse translation). Then the DNA probe is synthesized and used to locate and isolate the desired mRNA after gel electrophoresis. Finally, the isolated mRNA is used to make cDNA.

1. What is a transgenic animal? Describe how electroporation and gene guns are used to insert foreign genes into eucaryotic cells. What other approaches may be used? How are bacteria transformed?
2. How can one prevent rDNA from undergoing recombination in a bacterial host cell?
3. List several reasons why a cloned gene might not be expressed in a host cell. What is an expression vector?
4. Briefly outline the procedure for somatostatin production.
5. Identify a way to eliminate eucaryotic introns during the synthesis of rDNA.

Applications of Genetic Engineering

Genetic engineering and biotechnology will contribute in the future to medicine, industry, and agriculture, as well as to basic research (Box 15.1). In this section some practical applications are briefly discussed.

Medical Applications

Certainly the production of medically useful proteins such as somatostatin, insulin, human growth hormone, and interferon is of great practical importance (table 15.4). This is particularly true of substances that previously only could be obtained from human tissues. For example, in the past, human growth hormone for treatment of pituitary dwarfism was extracted from pituitaries obtained during autopsies and was available only in limited amounts. Interleukin-2 (a protein that helps regulate the immune response) and blood-clotting factor VIII have recently been cloned, and undoubtedly other important peptides and proteins will be produced in the future. A particularly interesting development is the use of transgenic corn and soybean plants to produce monoclonal antibodies for medical uses. Genetically engineered mice now can produce fully human monoclonal antibodies. Synthetic vaccines—for instance, vaccines for malaria and rabies—are also being developed with recombinant techniques. A recombinant hepatitis B vaccine is already commercially available.

cosmetic by introducing genes into human eggs and embryos. In 1983 two Nobel Prize winners, several other prominent scientists, and about 40 religious leaders signed a statement to Congress requesting that such alterations of human eggs or sperm not be attempted. Others argue that the good arising from such interventions more than justifies the risks of misuse and that there is nothing inherently wrong in modifying the human gene pool to reduce the incidence of genetic disease. This discussion extends to other organisms as well. Some argue that we have no right to create new life-forms and further disrupt the genetic diversity that human activities have already severely reduced. United States courts have decided that researchers and companies can patent "nonnaturally occurring" living organisms, whether microbial, plant, or animal.

One of the major current efforts in biotechnology, the human genome project, has the goal of determining the sequences of all human chromosomes. Success in this project and further technical advances in biotechnology will make genetic screening very effective. Physicians will be able to detect critical flaws in DNA long before the resulting genetic disease becomes manifest. In some cases this may lead to immediate treatment. But often nothing can be done about the damage and all sorts of dilemmas arise. Should people be told about the situation? Do they even wish to know about defects that cannot be corrected? What happens if their employers and insurance companies learn of the problem? They may lose their jobs and insurance. The issue of privacy is crucial in a practical sense, especially if genetic data banks are established. There may be increasing pressure on parents to abort fetuses with genetic defects in order not to stress the medical and social welfare systems. Clearly modern biomedical technology holds both threat and promise. Society has not yet faced its implications despite rapid progress in the development of genetic screening technology.

Another area of considerable controversy involves the use of recombinant organisms in agriculture. Many ecologists are worried that the release of unusual, recombinant organisms without careful prior risk assessment may severely disrupt the ecosystem. They point to the many examples of ecosystem disruption from the introduction of foreign organisms. A major concern is the exchange of genes between transgenic plants and viruses or other plants. Viral nucleic acids inserted into plants to make them virus resistant might combine with the genome of an invading virus to make it even more virulent. Weeds might acquire resistance to viruses, pests, and herbicides from transgenic crop plants. Other potential problems trouble critics. For example, insect-killing viruses might destroy many more insect species than expected. Toxin genes inserted into one virus might move to other viruses. Potential risks have been vigorously discussed because several recombinant organisms are either ready for field studies or are already being tested. Thus far, no obvious ecological effects have been observed.

As with any technology, the potential for abuse exists. A case in point is the use of genetic engineering in biological warfare and terrorism. Although there are international agreements that limit the research in this area to defense against other biological weapons, the knowledge obtained in such research can easily be used in offensive biological warfare. Effective vaccines constructed using rDNA technology can protect the attacker's troops and civilian population. Because it is relatively easy and inexpensive to prepare bacteria capable of producing massive quantities of toxins or to develop particularly virulent strains of viral and bacterial pathogens, even small countries and terrorist organizations might acquire biological weapons. Many scientists and nonscientists also are concerned about increases in the military rDNA research carried out by the major powers.

Recombinant DNA technology has greatly enhanced our knowledge of genes and how they function, and it promises to improve our lives in many ways. Yet, as this brief discussion shows, problems and concerns remain to be resolved. Past scientific advances have sometimes led to unanticipated and unfortunate consequences, such as environmental pollution and nuclear weapons. With prudence and forethought we may be able to avoid past mistakes in the use of this new technology.

---

1. Describe four major areas of concern about the application of genetic engineering. In each case give both the arguments for and against the use of genetic engineering.

Summary

1. Genetic engineering became possible after the discovery of restriction enzymes and reverse transcriptase and the development of the Southern blotting technique and other essential methods in nucleic acid chemistry.

2. Oligonucleotides of any desired sequence can be synthesized by a DNA synthesizer machine. This has made possible site-directed mutagenesis.

3. The polymerase chain reaction allows small amounts of DNA to be increased in concentration thousands of times.

4. Agarose gel electrophoresis is used to separate DNA fragments according to size differences.

5. Fragments can be isolated and identified, then joined with plasmids or phage genomes and cloned; one also can first clone all fragments and subsequently locate the desired clone.

6. Three techniques that can be used to join DNA fragments are the creation of similar sticky ends with a single restriction enzyme, the addition of poly(dA) and poly(dT) to create sticky ends, and blunt-end ligation with T4 DNA ligase.

7. Probes for the detection of recombinant clones are made in several ways and usually labeled with ^{32}P. Nonradioactive probes are also used.

8. Several types of vectors may be used, each with different advantages: plasmids, phages and other viruses, cosmids, and shuttle vectors.

9. Genes can be inserted into eucaryotic cells by techniques such as microinjection, electroporation, and the use of a gene gun.

10. The recombinant vector often must be modified by the addition of promoters, leaders, and other elements. Eucaryotic gene introns also must be removed. An expression vector has the necessary features to express any recombinant gene it carries.

11. The hormone somatostatin has been synthesized using recombinant DNA technology.

12. Recombinant DNA technology will provide many benefits in medicine, industry, and agriculture.

13. Despite the great promise of genetic engineering, it also brings with it potential problems in areas of safety, human experimentation, potential ecological disruption, and biological warfare.

Key Terms

autoradiography *308*
biotechnology *306*
chimera *320*
complementary DNA (cDNA) *307*
cosmid *321*
electrophoresis *313*
electroporation *321*

expression vector *322*
gene gun *321*
genetic engineering *306*
oligonucleotides *309*
polymerase chain reaction (PCR technique) *310*
probe *308*
recombinant DNA technology *306*

restriction enzymes *306*
site-directed mutagenesis *309*
Southern blotting technique *308*
Ti plasmid *326*
transfection *321*
transgenic animal *321*
vectors *308*

Questions for Thought and Review

1. Could the Southern blotting technique be applied to RNA and proteins? How might this be done?

2. Why could a band on an electrophoresis gel still contain more than one kind of DNA fragment?

3. What advantage might there be in creating a genomic library first rather than directly isolating the desired DNA fragment?

4. Suppose that one inserted a simple plasmid (one containing an antibiotic resistance gene and a separate restriction site) carrying a human interferon gene into *E. coli*, but none of the transformed bacteria produced interferon. Give as many plausible reasons as possible for this result.

5. In what areas do you think genetic engineering will have the greatest positive impact in the future? Why?

6. What do you consider to be the greatest potential dangers of genetic engineering? Are there ethical problems with any of its potential applications?

Additional Reading

General

Brown, C. M.; Campbell, I.; and Priest, F. G. 1987. *Introduction to biotechnology.* Boston: Blackwell Scientific Publications.

Cohen, S. N. 1975. The manipulation of genes. *Sci. Am.* 233(1):25–33.

Glick, B. R., and Pasternak, J. J. 1994. *Molecular biotechnology: Principles and applications of recombinant DNA.* Washington, D.C.: ASM Press.

Old, R. W., and Primrose, S. B. 1994. *Principles of gene manipulation,* 5th ed. Boston: Blackwell Scientific Publications.

Peters, P. 1993. *Biotechnology: A guide to genetic engineering.* Dubuque, Iowa: Wm. C. Brown.

Watson, J. D.; Gilman, M.; Witkowski, J.; and Zoller, M. 1992. *Recombinant DNA,* 2d ed. San Francisco: W. H. Freeman.

Historical Perspectives

Bickle, T. A., and Krüger, D. H. 1993. Biology of DNA restriction. *Microbiol. Rev.* 57(2):434–50.

Lear, J. 1978. *Recombinant DNA: The untold story.* New York: Crown Publishers.

The Polymerase Chain Reaction

Arnheim, N., and Levenson, C. H. 1990. Polymerase chain reaction. *Chem Eng. News* 68:36–47.

Atlas, R. M. 1991. Environmental applications of the polymerase chain reaction. *ASM News* 57(12):630–32.

Ehrlich, G. D., and Greenberg, S. J. 1994. *PCR-based diagnostics in infectious disease.* Boston: Blackwell Scientific.

Erlich, H. A. 1989. *PCR technology: Principles and applications of DNA amplification.* San Francisco: W. H. Freeman.

Erlich, H. A.; Gelfand, D.; and Sninsky, J. J. 1991. Recent advances in the polymerase chain reaction. *Science* 252:1643–51.

Mullis, K. B. 1990. The unusual origin of the polymerase chain reaction. *Sci. Am.* 262(4):56–65.

Inserting Genes into Eucaryotic Cells

Anderson, W. F., and Diacumakos, E. G. 1981. Genetic engineering in mammalian cells. *Sci. Am.* 245(1):106–21.

Chilton, M.-D. 1983. A vector for introducing new genes into plants. *Sci. Am.* 248(6):51–59.

Karcher, S. J. 1994. Getting DNA into a cell: A survey of transformation methods. *Am. Biol. Teach.* 56(1):14–20.

Applications of Genetic Engineering

Cohen, J. S., and Hogan, M. E. 1994. The new genetic medicines. *Sci. Am.* 271(6):76–82.

Gasser, C. S., and Fraley, R. T. 1992. Transgenic crops. *Sci. Am.* 266(6):62–69.

Gilbert, W., and Villa-Komaroff, L. 1980. Useful proteins from recombinant bacteria. *Sci. Am.* 242(4):74–94.

Hansen, M.; Busch, L.; Burkhardt, J.; Lacy, W. B.; and Lacy, L. R. 1986. Plant breeding and biotechnology. *BioScience* 36(1):29–39.

Hopwood, D. A. 1981. The genetic programming of industrial microorganisms. *Sci. Am.* 245(3):91–102.

Jaenisch, R. 1988. Transgenic animals. *Science* 240:1468–74.

Pestka, S. 1983. The purification and manufacture of human interferons. *Sci. Am.* 249(2):37–43.

Porter, A. G.; Davidson, E. W.; and Liu, J.-W. 1993. Mosquitocidal toxins of bacilli and their genetic manipulation for effective biological control of mosquitoes. *Microbiol. Rev.* 57(4):838–61.

Social Impact of Recombinant DNA Technology

Goodfield, J. 1977. *Playing God: Genetic engineering and the manipulation of life.* New York: Random House.

Grobstein, C. 1977. The recombinant-DNA debate. *Sci. Am.* 237(1):22–33.

Kenney, M. 1986. *Biotechnology: The university-industrial complex.* New Haven, Conn.: Yale University Press.

Krimsky, S. 1982. *Genetic alchemy: The social history of the recombinant DNA controversy.* Cambridge, Mass.: MIT Press.

Moses, P. B. 1987. Strange bedfellows. *BioScience* 37:6–10.

Richards, J., editor. 1978. *Recombinant DNA: Science, ethics, and politics.* New York: Academic Press.

Rifkin, J. 1983. *Algeny.* New York: Viking Press.

Teitelman, R. 1989. *Gene dreams: Wall street, academia, and the rise of biotechnology.* New York: Basic Books, Inc.

Tucker, J. B. Winter 1984–85. Gene wars. *Foreign Policy* 57:58–79.

Wilson, M., and Lindow, S. E. 1993. Release of recombinant microorganisms. *Ann. Rev. Microbiol.* 47:913–44.

Yoxen, E. 1984. *The gene business: Who should control biotechnology?* New York: Harper & Row.

Zilinskas, R. A., and Zimmerman, B. K., editors. 1986. *The gene-splicing wars: Reflections on the recombinant DNA controversy.* New York: Macmillan.

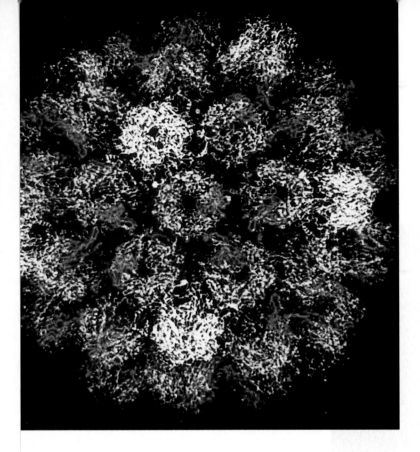

CHAPTER 16
The Viruses: Introduction and General Characteristics

Great fleas have little fleas

upon their backs to bite 'em

And little fleas have lesser

fleas, and so on ad infinitum

—Augustus De Morgan

Concepts

1. Viruses are simple, acellular entities consisting of one or more molecules of either DNA or RNA enclosed in a coat of protein (and sometimes, in addition, substances such as lipids and carbohydrates). They can reproduce only within living cells and are obligately intracellular parasites.

2. Viruses are cultured by inoculating living hosts or cell cultures with a virion preparation. Purification depends mainly on their large size relative to cell components, high protein content, and great stability. Virus concentration may be determined from the virion count or from the number of infectious units.

3. All viruses have a nucleocapsid composed of a nucleic acid surrounded by a protein capsid that may be icosahedral, helical, or complex in structure. Capsids are constructed of protomers that self-assemble through noncovalent bonds. A membranous envelope often lies outside the nucleocapsid.

4. More variety is found in the genomes of viruses than in those of procaryotes and eucaryotes; they may be either single-stranded or double-stranded DNA or RNA. The nucleic acid strands can be linear, closed circle, or able to assume either shape.

5. Viruses are classified on the basis of their nucleic acid's characteristics, capsid symmetry, the presence or absence of an envelope, their host, the diseases caused by animal and plant viruses, and other properties.

The next three chapters focus on the viruses. These are infectious agents with fairly simple, acellular organization. They possess only one type of nucleic acid, either DNA or RNA, and cannot reproduce independently of living cells. Clearly viruses are quite different from procaryotic and eucaryotic microorganisms, and are studied by virologists.

Despite their simplicity in comparison with cellular organisms, viruses are extremely important and deserving of close attention. The study of viruses has contributed significantly to the discipline of molecular biology. Many human viral diseases are already known and more are discovered or arise every year, as demonstrated by the recent appearance of AIDS. The whole field of genetic engineering is based in large part upon discoveries in virology. Thus it is easy to understand why virology (the study of viruses) is such a significant part of microbiology.

This chapter focuses on the broader aspects of virology: its development as a scientific discipline, the general properties and structure of viruses, the ways in which viruses are cultured and studied, and viral taxonomy. Chapter 17 is concerned with the bacteriophages, and chapter 18 is devoted to the viruses of eucaryotes.

Viruses have had enormous impact on humans and other organisms, yet very little was known about their nature until fairly recently. A brief history of their discovery and recognition as uniquely different infectious agents can help clarify their nature.

Early Development of Virology

Although the ancients did not understand the nature of their illnesses, they were acquainted with diseases, such as rabies, that are now known to be viral in origin. In fact, there is some evidence that the great epidemics of A.D. 165 to 180 and A.D. 251 to 266, which severely weakened the Roman Empire and aided its decline, may have been caused by measles and smallpox viruses. Smallpox had an equally profound impact on the New World. Hernán Cortés's conquest of the Aztec Empire in Mexico was made possible by an epidemic that ravaged Mexico City. The virus was probably brought to Mexico in 1520 by the relief expedition sent to join Cortés. Before the smallpox epidemic subsided, it had killed the Aztec King Cuitlahuac (the nephew and son-in-law of the slain emperor, Montezuma II) and possibly 1/3 of the population. Since the Spaniards were not similarly afflicted, it appeared that God's wrath was reserved for the Native Americans, and this disaster was viewed as divine support for the Spanish conquest (Box 16.1).

The first progress in preventing viral diseases came years before the discovery of viruses. Early in the eighteenth century, Lady Wortley Montagu, wife of the English ambassador to Turkey, observed that Turkish women inoculated their children against smallpox. The children came down with a mild case and subsequently were immune. Lady Montagu tried to educate the English public about the procedure but without great success. Later in the century an English country doctor, Edward Jenner, stimulated by a girl's claim that she could not catch smallpox because she had had cowpox, began inoculating humans with

---

Box 16.1

Disease and the Early Colonization of America

Although the case is somewhat speculative, there is considerable evidence that disease, and particularly smallpox, played a major role in reducing Indian resistance to the European colonization of North America. It has been estimated that Indian populations in Mexico declined about 90% within 100 years of initial contact with the Spanish. Smallpox and other diseases were a major factor in this decline, and there is no reason to suppose that North America was any different. As many as 10 to 12 million Indians may have lived north of the Rio Grande before contact with Europeans. In New England alone, there may have been over 72,000 in A.D. 1600; yet only around 8,600 remained in New England by A.D. 1674, and the decline continued in subsequent years.

Such an incredible catastrophe can be accounted for by consideration of the situation at the time of European contact with the Native Americans. The Europeans, having already suffered major epidemics in the preceding centuries, were relatively immune to the diseases they carried. On the other hand, the Native Americans had never been exposed to diseases like smallpox and were decimated by epidemics. In the sixteenth century, before any permanent English colonies had been established, many contacts were made by missionaries and explorers who undoubtedly brought disease with them and infected the natives. Indeed, the English noted at the end of the century that Indian populations had declined greatly but attributed it to armed conflict rather than to disease.

Establishment of colonies simply provided further opportunities for infection and outbreak of epidemics. For example, the Huron Indians decreased from a minimum of 32,000 people to 10,000 in 10 years. Between the time of initial English colonization and 1674, the Narraganset Indians declined from around 5,000 warriors to 1,000, and the Massachusetts Indians, from 3,000 to 300. Similar stories can be seen in other parts of the colonies. Some colonists interpreted these plagues as a sign of God's punishment of Indian resistance: the "Lord put an end to this quarrel by smiting them with smallpox . . . . Thus did the Lord allay their quarrelsome spirit and make room for the following part of his army."

It seems clear that epidemics of European diseases like smallpox laid waste Native American populations and prepared the way for colonization of the North American continent. Many American cities—for example, Boston, Philadelphia, and Plymouth—grew upon sites of previous Indian villages.

material from cowpox lesions. He published the results of 23 successful vaccinations in 1798. Although Jenner did not understand the nature of smallpox, he did manage to successfully protect his patients from the dread disease through exposure to the cowpox virus.

Until well into the nineteenth century, harmful agents were often grouped together and sometimes called viruses [Latin *virus*, poison or venom]. Even Louis Pasteur used the term virus for any living infectious disease agent. The development in 1884 of the porcelain bacterial filter by Charles Chamberland, one of Pasteur's collaborators and inventor of the autoclave, made possible the discovery of what are now called viruses. Tobacco mosaic disease was the first to be studied with Chamberland's filter. In 1892 Dimitri Ivanowski published studies showing that leaf extracts from infected plants would induce tobacco mosaic disease even after filtration to remove bacteria. He attributed this to the presence of a toxin. Martinus W. Beijerinck, working independently of Ivanowski, published the results of extensive studies on tobacco mosaic disease in 1898 and 1900. Because the filtered sap of diseased plants was still infectious, he proposed that the disease was caused by an entity different from bacteria, a filterable virus. He observed that the virus would multiply only in living plant cells, but could survive for long periods in a dried state. At the same time Friedrich Loeffler and Paul Frosch in Germany found that the hoof-and-mouth disease of cattle was also caused by a filterable virus rather than by a toxin. In 1900 Walter Reed began his study of the yellow fever disease whose incidence had been increasing in Cuba. Reed showed that this human disease was due to a filterable virus that was transmitted by mosquitoes. Mosquito control shortly reduced the severity of the yellow fever problem. Thus by the beginning of this century, it had been established that filterable viruses were different from bacteria and could cause diseases in plants, livestock, and humans.

Shortly after the turn of the century, V. Ellermann and O. Bang in Copenhagen reported that leukemia could be transmitted between chickens by cell-free filtrates and was probably caused by a virus. Three years later in 1911, Peyton Rous from the Rockefeller Institute in New York City reported that a virus was responsible for a malignant muscle tumor in chickens. These studies established that at least some malignancies were caused by viruses.

It was soon discovered that bacteria themselves also could be attacked by viruses. The first published observation suggesting that this might be the case was made in 1915 by Frederick W. Twort. Twort isolated bacterial viruses that could attack and destroy micrococci and intestinal bacilli. Although he speculated that his preparations might contain viruses, Twort did not follow up on these observations. It remained for Felix d'Herelle to establish decisively the existence of bacterial viruses. D'Herelle independently isolated bacterial viruses from patients with dysentery, probably caused by *Shigella dysenteriae*. He noted that when a virus suspension was spread on a layer of bacteria growing on agar, clear circular areas containing viruses

and lysed cells developed. A count of these clear zones allowed d'Herelle to estimate the number of viruses present (plaque assay, p. 335). D'Herelle demonstrated that these viruses could reproduce only in live bacteria; therefore he named them bacteriophages because they could eat holes in bacterial "lawns."

The chemical nature of viruses was established when Wendell M. Stanley announced in 1935 that he had crystallized the tobacco mosaic virus (TMV) and found it to be largely or completely protein. A short time later Frederick C. Bawden and Norman W. Pirie managed to separate the TMV virus particles into protein and nucleic acid. Thus by the late 1930s it was becoming clear that viruses were complexes of nucleic acids and proteins able to reproduce only in living cells.

---

1. Describe the major technical advances and discoveries important in the early development of virology.
2. Give the contribution to virology made by each scientist mentioned in this section.

A General Properties of Viruses

Viruses are a unique group of infectious agents whose distinctiveness resides in their simple, acellular organization and pattern of reproduction. A complete virus particle or **virion** consists of one or more molecules of DNA or RNA enclosed in a coat of protein, and sometimes also in other layers. These additional layers may be very complex and contain carbohydrates, lipids, and additional proteins. Viruses can exist in two phases: extracellular and intracellular. Virions, the extracellular phase, possess few if any enzymes and cannot reproduce independently of living cells. In the intracellular phase, viruses exist primarily as replicating nucleic acids that induce host metabolism to synthesize virion components; eventually complete virus particles or virions are released.

In summary, viruses differ from living cells in at least three ways: (1) their simple, acellular organization; (2) the absence of both DNA and RNA in the same virion; and (3) their inability to reproduce independently of cells and carry out cell division as procaryotes and eucaryotes do. Although bacteria such as chlamydia and rickettsia (*see chapter 20*) are obligately intracellular parasites like viruses, they do not meet the first two criteria.

The Cultivation of Viruses

Because they are unable to reproduce independently of living cells, viruses cannot be cultured in the same way as bacteria and eucaryotic microorganisms. For many years researchers have cultivated animal viruses by inoculating suitable host animals or embryonated eggs—fertilized chicken eggs incubated about 6 to 8 days after laying (figure 16.1). To prepare the egg for virus cultivation, the shell surface is first disinfected with iodine and penetrated with a small sterile drill. After inoculation,

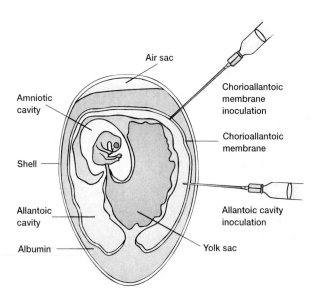

Figure 16.1 Virus Cultivation Using an Embryonated Egg. Two sites that are often used to grow animal viruses are the chorioallantoic membrane and the allantoic cavity. The diagram shows a 9 day chicken embryo.

(a)

(b)

Figure 16.2 Virus Plaques. (*a*) Plaques of viral hepatitis in human liver cells. (*b*) Poliovirus plaques in a monkey kidney cell culture.

the drill hole is sealed with gelatin and the egg incubated. Viruses may be able to reproduce only in certain parts of the embryo; consequently they must be injected into the proper region. For example, the myxoma virus grows well on the chorioallantoic membrane, whereas the mumps virus prefers the allantoic cavity. The infection may produce a local tissue lesion known as a pock, whose appearance often is characteristic of the virus.

More recently animal viruses have been grown in tissue (cell) culture on monolayers of animal cells. This technique is made possible by the development of growth media for animal cells and by the advent of antibiotics that can prevent bacterial and fungal contamination. A layer of animal cells in a specially prepared petri dish is covered with a virus inoculum, and the viruses are allowed time to settle and attach to the cells. The cells are then covered with a thin layer of agar to limit virion spread so that only adjacent cells are infected by newly produced virions. As a result localized areas of cellular destruction and lysis called **plaques** often are formed (figure 16.2) and may be detected if stained with dyes, such as neutral red or trypan blue, that can distinguish living from dead cells. Viral growth does not always result in the lysis of cells to form a plaque. Animal viruses, in particular, can cause microscopic or macroscopic degenerative changes or abnormalities in host cells and in tissues called **cytopathic effects** (figure 16.3). Cytopathic effects may be lethal, but plaque formation from cell lysis does not always occur.

Bacterial viruses or **bacteriophages** (**phages** for short) are cultivated in either broth or agar cultures of young, actively growing bacterial cells. So many host cells are destroyed that turbid bacterial cultures may clear rapidly because of cell lysis. Agar cultures are prepared by mixing the bacteriophage sample with cool, liquid agar and a suitable bacterial culture. The mix-

ture is quickly poured into a petri dish containing a bottom layer of sterile agar. After hardening, bacteria in the layer of top agar grow and reproduce, forming a continuous, opaque layer or "lawn." Wherever a virion comes to rest in the top agar, the virus infects an adjacent cell and reproduces. Eventually, bacterial lysis generates a plaque or clearing in the lawn (figure 16.4). As can be seen in figure 16.4, plaque appearance often is characteristic of the phage being cultivated.

Plant viruses are cultivated in a variety of ways. Plant tissue cultures, cultures of separated cells, or cultures of protoplasts (*see chapter 3*) may be used. Viruses also can be grown in whole plants. Leaves are mechanically inoculated when rubbed with a mixture of viruses and an abrasive such as carborundum. When the cell walls are broken by the abrasive, the viruses directly contact the plasma membrane and infect the exposed host cells. (The role of the abrasive is frequently filled by

Figure 16.3 The Cytopathic Effects of Viruses. (*a*) Normal mammalian cells in tissue culture. (*b*) Appearence of tissue culture cells 18 hours after infection with adenovirus. TEM photomicrographs (×11,000).

(a)

(b)

T1

T2

T3

T4

Figure 16.4 Phage Plaques. Plaques produced on a lawn of *E. coli* by some of the T. coliphages. Note the large differences in plaque appearance. The photographs are about 1/3 full size.

insects that suck or crush plant leaves and thus transmit viruses.) A localized **necrotic lesion** often develops due to the rapid death of cells in the infected area (figure 16.5). Even when lesions do not arise, the infected plant may show symptoms such as changes in pigmentation or leaf shape. Some plant viruses can be transmitted only if a diseased part is grafted onto a healthy plant.

1. What is a virus particle or virion, and how is it different from living organisms?
2. Discuss the ways in which viruses may be cultivated. Define the terms pock, plaque, cytopathic effect, bacteriophage, and necrotic lesion.

Virus Purification and Assays

Virologists must be able to purify viruses and accurately determine their concentrations in order to study virus structure, reproduction, and other aspects of their biology. These methods are so important that the growth of virology as a modern discipline depended on their development.

Virus Purification

Purification makes use of several virus properties. Virions are very large relative to proteins, are often more stable than normal cell components, and have surface proteins. Because of these characteristics, many techniques useful for the isolation of proteins and organelles can be employed in virus isolation. Four of the most widely used approaches are (1) differential and density gradient centrifugation, (2) precipitation of viruses, (3) denaturation of contaminants, and (4) enzymatic digestion of cell constituents.

1. Host cells in later stages of infection that contain mature virions are used as the source of material. Infected cells are first disrupted in a buffer to produce an aqueous suspension or homogenate consisting of cell components and viruses. Viruses can then be isolated by **differential centrifugation,**

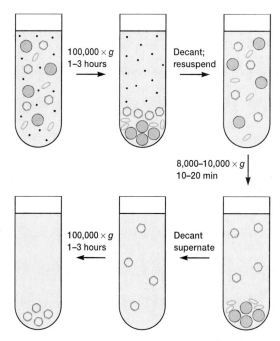

Figure 16.6   The Use of Differential Centrifugation to Purify a Virus. At the beginning the centrifuge tube contains homogenate and icosahedral viruses (in red). First, the viruses and heavier cell organelles are removed from smaller molecules. After resuspension, the mixture is centrifuged just fast enough to sediment cell organelles while leaving the smaller virus particles in suspension; the purified viruses are then collected. This process can be repeated several times to further purify the virions.

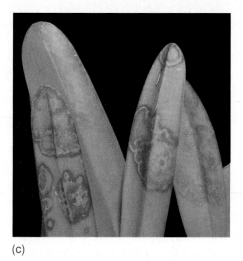

(c)

Figure 16.5   Necrotic Lesions on Plant Leaves. (*a*) Tobacco mosaic virus on *Nicotiana glutinosa.* (*b*) Bean common mosaic disease. Infected leaves show yellow mottling and distortion. (*c*) Tobacco mosaic virus infection of an orchid showing leaf color changes.

the centrifugation of a suspension at various speeds to separate particles of different sizes (figure 16.6). Usually the homogenate is first centrifuged at high speed to sediment viruses and other large cellular particles, and the supernatant, which contains the homogenate's soluble molecules, is discarded. The pellet is next resuspended and centrifuged at a low speed to remove substances heavier than viruses. Higher speed centrifugation then sediments the viruses. This process may be repeated to purify the virus particles further.

Viruses also can be purified based on their size and density by use of **gradient centrifugation** (figure 16.7). A sucrose solution is poured into a centrifuge tube so that its concentration smoothly and linearly increases between the top and the bottom of the tube. The virus preparation, often after purification by differential centrifugation, is layered on top of the gradient and centrifuged. As shown in figure 16.7*a*, the particles settle under centrifugal force until they come to rest at the level where the gradient's density equals theirs (isopycnic gradient centrifugation). Viruses can be separated from other particles only slightly different in density. Gradients also can separate viruses based on differences in their sedimentation rate (rate zonal gradient centrifugation). When this is done, particles are separated

Figure 16.7   Gradient Centrifugation.
(a) A linear sucrose gradient is prepared, *1*, and the particle mixture is layered on top, *2* and *3*. Centrifugation, *4*, separates the particles on the basis of their density and sedimentation coefficient, (the arrows in the centrifuge tubes indicate the direction of centrifugal force). *5*. In isopycnic gradient centrifugation, the bottom of the gradient is denser than any particle, and each particle comes to rest at a point in the gradient equal to its density. Rate zonal centrifugation separates particles based on their sedimentation coefficient, a function of both size and density, because the bottom of the gradient is less dense than the densest particles and centrifugation is carried out for a shorter time so that particles do not come to rest. The largest, most dense particles travel fastest. (b) The densities and sedimentation coefficients of representative viruses (shown in color) and other biological substances.

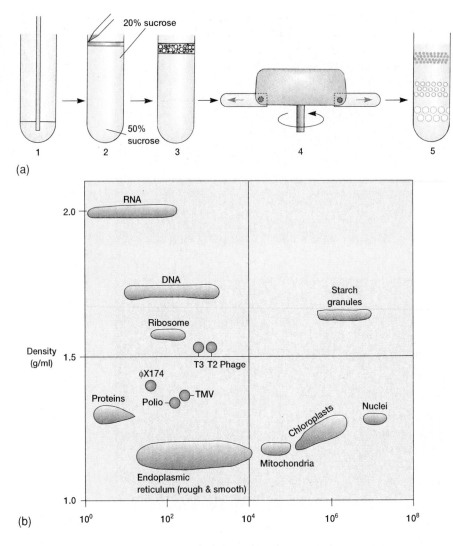

on the basis of both size and density; usually the largest virus will move most rapidly down the gradient. Figure 16.7b shows that viruses differ from one another and cell components with respect to either density (grams per milliliter) or sedimentation coefficient(s). Thus these two types of gradient centrifugation are very effective in virus purification.

2. Viruses, like many proteins, can be purified through precipitation with concentrated ammonium sulfate. Initially, sufficient ammonium sulfate is added to raise its concentration to a level just below that which will precipitate the virus. After any precipitated contaminants are removed, more ammonium sulfate is added and the precipitated viruses are collected by centrifugation. Viruses sensitive to ammonium sulfate often are purified by precipitation with polyethylene glycol.

3. Viruses frequently are less easily denatured than many normal cell constituents. Contaminants may be denatured and precipitated with heat or a change in pH to purify viruses. Because some viruses also tolerate treatment with organic solvents like butanol and chloroform, solvent treatment can be used to both denature protein

contaminants and extract any lipids in the preparation. The solvent is thoroughly mixed with the virus preparation, then allowed to stand and separate into organic and aqueous layers. The unaltered virus remains suspended in the aqueous phase while lipids dissolve in the organic phase. Substances denatured by organic solvents collect at the interface between the aqueous and organic phases.

Denaturation of enzymes (p. 161).

4. Cellular proteins and nucleic acids can be removed from many virus preparations through enzymatic degradation because viruses usually are more resistant to attack by nucleases and proteases than are free nucleic acids and proteins. For example, ribonuclease and trypsin often degrade cellular ribonucleic acids and proteins while leaving virions unaltered.

Virus Assays

The quantity of viruses in a sample can be determined either by a count of particle number or by measurement of the infectious unit concentration. Although most normal virions are probably potentially infective, many will not infect host cells because they do not contact the proper surface site. Thus the total particle count may be from two to one million times the infectious unit number depending on the nature of the virion and the experimental conditions. Despite this, both approaches are of value.

Virus particles can be counted directly with the electron microscope. In one procedure the virus sample is mixed with a known concentration of small latex beads and sprayed on a coated specimen grid. The beads and virions are counted; the virus concentration is calculated from these counts and from the bead concentration (figure 16.8). This technique often works well with concentrated preparations of viruses of known morphology. Viruses can be concentrated by centrifugation before counting if the preparation is too dilute. However, if the beads and viruses are not evenly distributed (as sometimes happens), the final count will be inaccurate.

The most popular indirect method of counting virus particles is the **hemagglutination assay.** Many viruses can bind to the surface of red blood cells (*see figure 32.10*). If the ratio of viruses to cells is large enough, virus particles will join the red blood cells together, forming a network that settles out of suspension or agglutinates. In practice, red blood cells are mixed with a series of virus preparation dilutions and each mixture is examined. The hemagglutination titer is the highest dilution of virus (or the reciprocal of the dilution) that still causes hemagglutination. This assay is an accurate, rapid method for determining the relative quantity of viruses such as the influenza virus. If the actual number of viruses needed to cause hemagglutination is determined by another technique, the assay can be used to ascertain the number of virus particles present in a sample.

A variety of assays analyze virus numbers in terms of infectivity, and many of these are based on the same techniques used for virus cultivation. For example, in the **plaque assay** several dilutions of bacterial or animal viruses are plated out with appropriate host cells. When the number of viruses plated out are much fewer than the number of host cells available for infection and when the viruses are distributed evenly, each plaque in a layer of bacterial or animal cells is assumed to have arisen from the reproduction of a single virus particle. Therefore a count of the plaques produced at a particular dilution will give the number of infectious virions or **plaque-forming units (PFU),** and the concentration of infectious units in the original sample can be easily calculated. Suppose that 0.10 ml of a 10^{-6} dilution of the virus preparation yields 75 plaques. The original concentration of plaque-forming units is

$$PFU/ml = (75\ PFU/0.10\ ml)(10^6) = 7.5 \times 10^8.$$

Viruses producing different plaque morphology types on the same plate may be counted separately. Although the number of PFU does not equal the number of virus particles, their ratios are proportional: a preparation with twice as many viruses will have twice the plaque-forming units.

The same approach employed in the plaque assay may be used with embryos and plants. Chicken embryos can be inoculated with a diluted preparation or plant leaves rubbed with a mixture of diluted virus and abrasive. The number of pocks on

Figure 16.8   Tobacco Mosaic Virus. A tobacco mosaic virus preparation viewed in the transmission electron microscope. Latex beads 264 µm in diameter (white spheres) have been added.

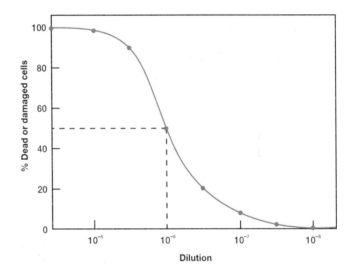

Figure 16.9   A Hypothetical Dose-Response Curve. The LD_{50} is indicated by the dashed line.

embryonic membranes or necrotic lesions on leaves is multiplied by the dilution factor and divided by the inoculum volume to obtain the concentration of infectious units.

When biological effects are not readily quantified in these ways, the amount of virus required to cause disease or death can be determined by the endpoint method. Organisms or cell cultures are inoculated with serial dilutions of a virus suspension. The results are used to find the endpoint dilution at which 50% of the host cells or organisms are damaged or destroyed (figure 16.9). The **lethal dose (LD_{50})** or **infectious dose (ID_{50})** is the dilution that contains a lethal or infectious dose large enough to destroy or damage 50% of the host cells or organisms.

1. Give the four major approaches by which viruses may be purified, and describe how each works. Distinguish between differential and density gradient centrifugation in terms of how they are carried out.

2. How can one find the virus concentration, both directly and indirectly, by particle counts and measurement of infectious unit concentration? Define plaque-forming units, lethal dose, and infectious dose.

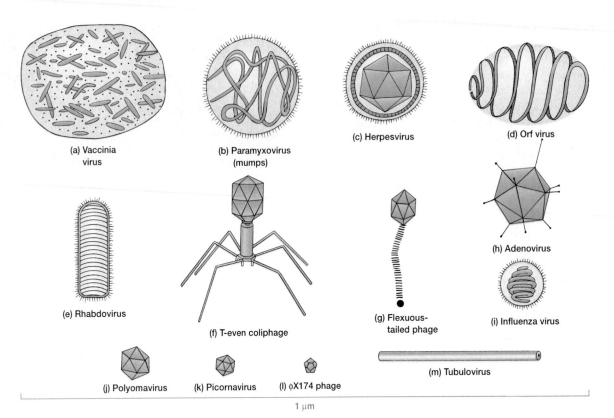

Figure 16.10 **The Size and Morphology of Selected Viruses.** The viruses are drawn to scale. A 1 μm line is provided at the bottom of the figure.

The Structure of Viruses

Virus morphology has been intensely studied over the past decades because of the importance of viruses and the realization that virus structure was simple enough to be understood. Progress has come from the use of several different techniques: electron microscopy, X-ray diffraction, biochemical analysis, and immunology. Although our knowledge is incomplete due to the large number of different viruses, the general nature of virus structure is becoming clear.

Virion Size

Virions range in size from about 10 to 300 or 400 nm in diameter (figure 16.10). The smallest viruses are little larger than ribosomes, whereas the poxviruses, like vaccinia, are about the same size as the smallest bacteria and can be seen in the light microscope. Most viruses, however, are too small to be visible in the light microscope and must be viewed with the scanning and transmission electron microscopes (*see chapter 2*).

General Structural Properties

All virions, even if they possess other constituents, are constructed around a **nucleocapsid** core (indeed, some viruses consist only of a nucleocapsid). The nucleocapsid is composed of a nucleic acid, either DNA or RNA, held within a protein coat called the **capsid,** which protects viral genetic material and aids in its transfer between host cells.

There are four general morphological types of capsids and virion structure.

1. Some capsids are **icosahedral** in shape. An icosahedron is a regular polyhedron with 20 equilateral triangular faces and 12 vertices (figure 16.10*h, j–l*). These capsids appear spherical when viewed at low power in the electron microscope.

2. Other capsids are **helical** and shaped like hollow protein cylinders, which may be either rigid or flexible (figure 16.10*m*).

3. Many viruses have an **envelope,** an outer membranous layer surrounding the nucleocapsid. Enveloped viruses have a roughly spherical but somewhat variable shape even though their nucleocapsid can be either icosahedral or helical (figure 16.10*b, c, i*).

4. **Complex viruses** have capsid symmetry that is neither purely icosahedral nor helical (figure 16.10*a, d, f, g*). They may possess tails and other structures (e.g., many bacteriophages) or have complex, multilayered walls surrounding the nucleic acid (e.g., poxviruses such as vaccinia).

Both helical and icosahedral capsids are large macromolecular structures constructed from many copies of one or a few types of protein subunits or **protomers.** Probably the most important advantage of this design strategy is that the information stored in viral genetic material is used with maximum efficiency. For example, the tobacco mosaic virus (TMV) capsid contains a single type of small subunit possessing 158 amino acids. Only about 474 nucleotides out of 6,000 in the virus RNA are required to code for coat protein amino acids. Unless the same protein is used many times in capsid construction, a large nucleic acid, such as the TMV RNA, cannot be enclosed in a protein coat without using much or all of the available genetic material to code for capsid proteins. If the TMV capsid were composed of six different protomers of the same size as the TMV subunit, about 2,900 of the 6,000 nucleotides would be required for its construction, and much less genetic material would be available for other purposes.

The genetic code and translation (pp. 255–57).

Once formed and exposed to the proper conditions, protomers usually interact specifically with each other and spontaneously associate to form the capsid. Because the capsid is constructed without any outside aid, the process is called self-assembly (p. 61). Some more complex viruses possess genes for special factors that are not incorporated into the virion but are required for its assembly.

Helical Capsids

Helical capsids are shaped much like hollow tubes with protein walls. The tobacco mosaic virus provides a well-studied example of helical capsid structure (figure 16.11). A single type of protomer associates together in a helical or spiral arrangement to produce a long, rigid tube, 15 to 18 nm in diameter by 300 nm long. The RNA genetic material is wound in a spiral and positioned toward the inside of the capsid where it lies within a groove formed by the protein subunits. Not all helical capsids are as rigid as the TMV capsid. Influenza virus RNAs are enclosed in thin, flexible helical capsids folded within an envelope (figures 16.10i and 16.12a, b).

The size of a helical capsid is influenced by both its protomers and the nucleic acid enclosed within the capsid. The diameter of the capsid is a function of the size, shape, and interactions of the protomers. The nucleic acid determines helical capsid length because the capsid does not seem to extend much beyond the end of the DNA or RNA.

Icosahedral Capsids

The icosahedron is one of nature's favorite shapes (the helix is probably most popular). Viruses employ the icosahedral shape because it is the most efficient way to enclose a space. A few

(a)

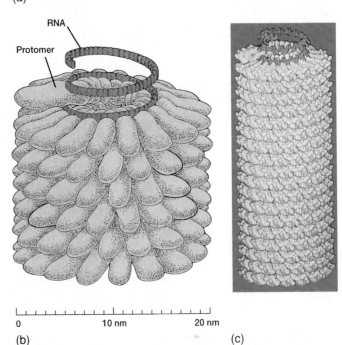

RNA

Protomer

0   10 nm   20 nm

(b)   (c)

Figure 16.11 **Tobacco Mosaic Virus Structure.** (*a*) An electron micrograph of the negatively stained helical capsid (×400,000). (*b*) Illustration of TMV structure. Note that the nucleocapsid is composed of a helical array of promoters with the RNA spiraling on the inside. (*c*) A model of TMV.

genes, sometimes only one, can code for proteins that self-assemble to form the capsid. In this way a small number of linear genes can specify a large three-dimensional structure. Certain requirements must be met to construct an icosahedron. Hexagons pack together in planes and cannot enclose a space, and therefore pentagons must also be used.

When icosahedral viruses are negatively stained and viewed in the transmission electron microscope, a complex icosahedral capsid structure is revealed (figure 16.12). The capsids are constructed from ring- or knob-shaped units called **capsomers,** each usually made of five or six protomers. **Pentamers (pentons)** have five subunits; **hexamers (hexons)** possess six. Pentamers are at the vertices of the icosahedron, whereas hexamers form its edges and triangular

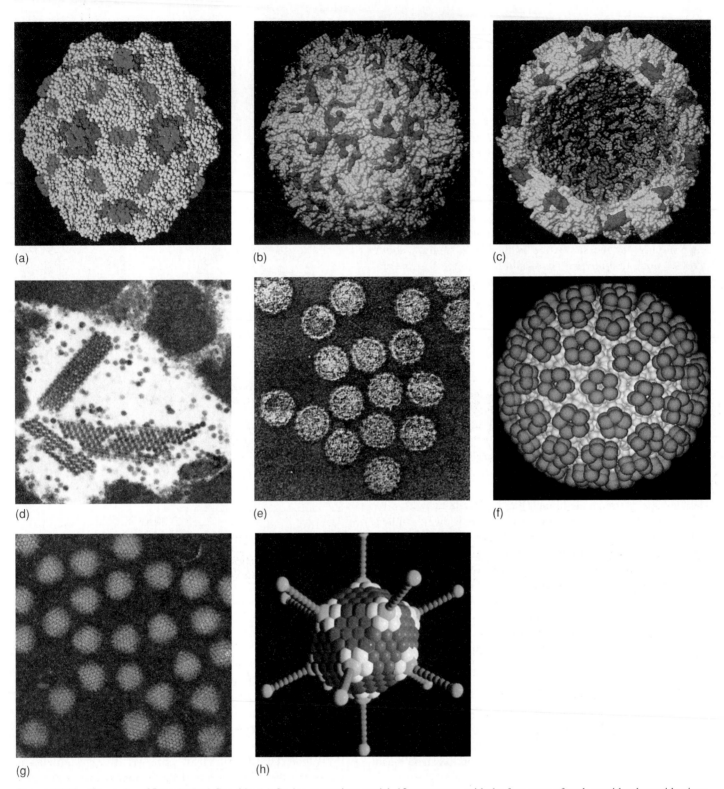

(a)

(b)

(c)

(d)

(e)

(f)

(g)

(h)

Figure 16.12   Examples of Icosahedral Capsids. (*a*) Canine parvovirus model, 12 capsomers, with the four parts of each capsid polypeptide given different colors. (*b*) and (*c*) Poliovirus model, 32 capsomers, with the four capsid proteins in different colors. The capsid surface is depicted in (*b*) and a cross section in (*c*). (*d*) Clusters of the human papilloma virus, 72 capsomers (×80,000). (*e*) Simian virus 40 (SV40), 72 capsomers (×340,000). (*f*) Computer-simulated image of the polyomavirus, 72 capsomers, that causes a rare demyelinating disease of the central nervous system. (*g*) Adenovirus, 252 capsomers (×171,000). (*h*) Computer-simulated model of adenovirus.

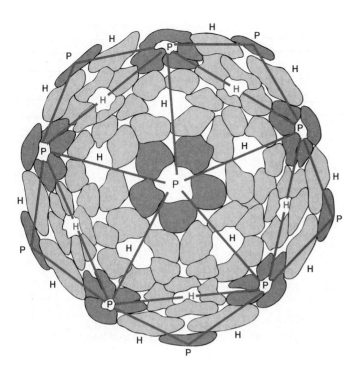

Figure 16.13   The Structure of an Icosahedral Capsid. Pentons are located at the 12 vertices. Hexons form the edges and faces of the icosahedron. This capsid contains 42 capsomers; all protomers are identical.

faces (figure 16.13). The icosahedron in figure 16.13 is constructed of 42 capsomers; larger icosahedra are made if more hexamers are used to form the edges and faces (adenoviruses have a capsid with 252 capsomers as shown in figure 16.12*g*, *h*). In many plant and bacterial RNA viruses, both the pentamers and hexamers of a capsid are constructed with only one type of subunit, whereas adenovirus pentamers are composed of different proteins than are adenovirus hexamers.

Transmission electron microscopy and negative staining (pp. 28–31).

Protomers join to form capsomers through noncovalent bonding. The bonds between proteins within pentamers and hexamers are stronger than those between separate capsomers. Empty capsids can even dissociate into separate capsomers.

Recently it has been discovered that there is more than one way to build an icosahedral capsid. Although most icosahedral capsids appear to contain both pentamers and hexamers, simian virus 40 (SV40), a small double-stranded DNA polyomavirus, has only pentamers (figure 16.14*a*). The virus is constructed of 72 cylindrical pentamers with hollow centers. Five flexible arms extend from the edge of each pentamer (figure 16.14*b*). Twelve pentamers occupy the icosahedron's vertices and associate with five neighbors, just as they do when hexamers also are present. Each of the 60 nonvertex pentamers associates with its six adjacent neighbors in the way

shown in figure 16.14*c* and *d*. An arm extends toward the adjacent vertex pentamer (pentamer 1) and twists around one of its arms. Three more arms interact in the same way with arms of other nonvertex pentamers (pentamers 3 to 5). The fifth arm binds directly to an adjacent nonvertex pentamer (pentamer 6) but does not attach to one of its arms. An arm does not extend from the central pentamer to pentamer 2; other arms hold pentamer 2 in place. Thus an icosahedral capsid is assembled without hexamers by using flexible arms as ropes to tie the pentamers together.

Nucleic Acids

Viruses are exceptionally flexible with respect to the nature of their genetic material and employ all four possible nucleic acid types: single-stranded DNA, double-stranded DNA, single-stranded RNA, and double-stranded RNA. All four types are found in animal viruses. Plant viruses most often have single-stranded RNA genomes. Although phages may have single-stranded DNA or single-stranded RNA, bacterial viruses usually contain double-stranded DNA. Table 16.1 summarizes many variations seen in viral nucleic acids. The size of viral genetic material also varies greatly. The smallest genomes (those of the MS2 and Qβ viruses) are around 1×10^6 daltons, just large enough to code for three to four proteins. At the other extreme, T-even bacteriophages, herpesvirus, and vaccinia virus have genomes of 1.0 to 1.6×10^8 daltons and may be able to direct the synthesis of over 100 proteins. In the following paragraphs the nature of each nucleic acid type is briefly summarized.

Nucleic acid structure (pp. 212–14).

Tiny DNA viruses like φX174 and M13 bacteriophages or the parvoviruses possess single-stranded DNA (ssDNA) genomes (table 16.1). Some of these viruses have linear pieces of DNA, whereas others use a single, closed circle of DNA for their genome (figure 16.15).

Most DNA viruses use double-stranded DNA (dsDNA) as their genetic material. Linear dsDNA, variously modified, is found in many viruses; others have circular dsDNA. The lambda phage has linear dsDNA with cohesive ends—single-stranded complementary segments 12 nucleotides long—that enable it to cyclize when they base pair with each other (figure 16.16).

Besides the normal nucleotides found in DNA, many virus DNAs contain unusual bases. For example, the T-even phages of *E. coli* (*see chapter 17*) have 5-hydroxymethylcytosine (*see figure 17.7*) instead of cytosine. Glucose is usually attached to the hydroxymethyl group.

Most RNA viruses employ single-stranded RNA (ssRNA) as their genetic material. The RNA base sequence may be identical with that of viral mRNA, in which case the RNA strand is called the **plus strand** or **positive strand** (viral mRNA is defined as plus or positive). However, the viral RNA genome may instead

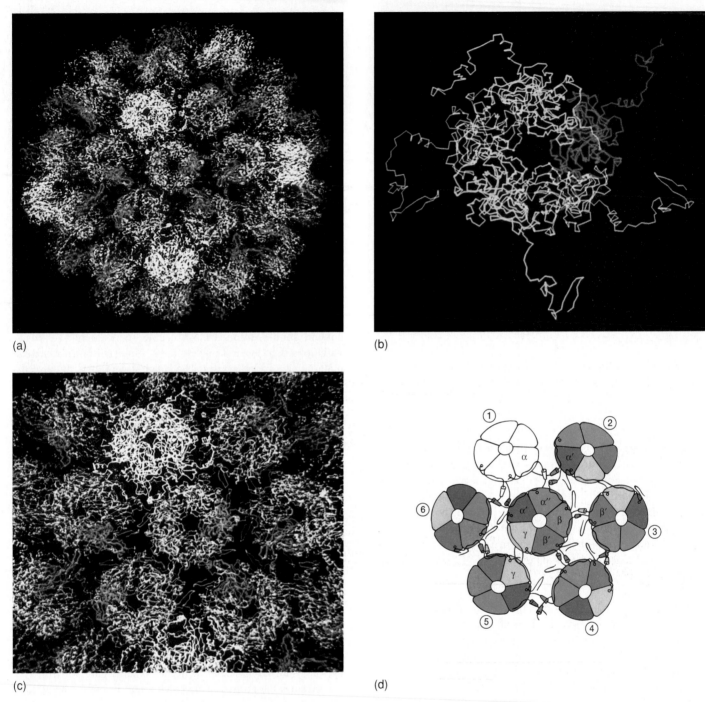

Figure 16.14   An Iscosahedral Capsid Constructed of Pentamers. (*a*) The simian virus 40 capsid. The 12 pentamers at the icosahedron vertices are in white. The nonvertex pentamers are shown with each polypeptide chain in a different color. (*b*) A pentamer with extended arms. (*c*) A close-up view of the virion surface showing a central pentamer associated with six neighbors. (*d*) A schematic diagram of the surface structure depicted in part *c*. The body of each pentamer is represented by a five-petaled flower design. Each arm is shown as a line or a line and cylinder (α-helix) with the same color as the rest of its protomer. The outer protomers are numbered clockwise beginning with the one at the vertex.

TABLE 16.1   Types of Viral Nucleic Acids

Nucleic Acid Type	Nucleic Acid Structure	Virus Examples
DNA		
Single-Stranded	Linear single strand	Parvoviruses
	Circular single strand	φX174, M13, fd phages
Double-Stranded	Linear double strand	Herpesviruses (herpes simplex viruses, cytomegalovirus, Epstein-Barr virus), adenoviruses, T coliphages, lambda phage, and other bacteriophages
	Linear double strand with single chain breaks	T5 coliphage
	Double strand with cross-linked ends	Vaccinia, smallpox
	Closed circular double strand	Papovaviruses (polyomavirus, human papilloma viruses, SV40), PM2 phage, cauliflower mosaic
RNA		
Single-Stranded	Linear, single stranded, positive strand	Picornaviruses (polio, rhinoviruses), togaviruses, RNA bacteriophages, TMV, and most plant viruses
	Linear, single stranded, negative strand	Rhabdoviruses (rabies), paramyxoviruses (mumps, measles)
	Linear, single stranded, segmented, positive strand	Brome mosaic virus (individual segments in separate virions)
	Linear, single stranded, segmented, diploid (two identical single strands), positive strand	Retroviruses (Rous sarcoma virus, human immunodeficiency virus)
	Linear, single stranded, segmented, negative strand	Paramyxoviruses, orthomyxoviruses (influenza)
Double-Stranded	Linear, double stranded, segmented	Reoviruses, wound-tumor virus of plants, cytoplasmic polyhedrosis virus of insects, phage φ6, many mycoviruses

Modified from S. E. Luria, et al., *General Virology*, 3rd edition, 1983. John Wiley & Sons, Inc., New York, NY.

Figure 16.15   Circular Phage DNA. The closed circular DNA of the phage PM2 (×93,000). Note both the relaxed and highly twisted or supercoiled forms.

Figure 16.16   Circularization of Lambda DNA. The linear DNA of the lambda phage can be reversibly circularized. This is made possible by cohesive ends (in color) that have complementary sequences and can base pair with each other.

be complementary to viral mRNA, and then it is called a **minus** or **negative strand.** Polio, tobacco mosaic, brome mosaic, and Rous sarcoma viruses are all positive strand RNA viruses; rabies, mumps, measles, and influenza viruses are examples of negative strand RNA viruses. Many of these RNA genomes are **segmented genomes**—that is, they are divided into separate parts. It is believed that each fragment or segment codes for one protein. Usually all segments are probably enclosed in the same capsid even though some virus genomes may be composed of as many as 10 to 12 segments. However, it is not necessary that all segments be located in the same virion for successful reproduction. The brome mosaic virus genome is composed of four segments distributed among three different virus particles. All three of the largest segments are required for infectivity. Despite this complex and seemingly inefficient arrangement, the different brome mosaic virions manage to successfully infect the same host.

Plus strand viral RNA often resembles mRNA in more than the equivalence of its nucleotide sequence. Just as eucaryotic mRNA usually has a 5′ cap of 7-methylguanosine, many plant and animal viral RNA genomes are capped. In addition, most or all plus strand RNA animal viruses also have a poly-A stretch at the 3′ end of their genome, and thus closely resemble eucaryotic mRNA with respect to the structure of both ends. In fact, plus strand RNAs can direct protein synthesis immediately after entering the cell. Strangely enough, a number of single-stranded plant viral RNAs have 3′ ends that resemble eucaryotic transfer RNA, and the genomes of tobacco mosaic virus will actually accept amino acids. Capping is not seen in the RNA bacteriophages.

Eucaryotic mRNA structure and function (pp. 223–4).

A few viruses have double-stranded RNA (dsRNA) genomes. All appear to be segmented; some, such as the reoviruses, have 10 to 12 segments. These dsRNA viruses are known to infect animals, plants, fungi, and even one bacterial species.

Viral Envelopes and Enzymes

Many animal viruses, some plant viruses, and at least one bacterial virus are bounded by an outer membranous layer called an envelope (figure 16.17). Animal virus envelopes usually arise from host cell nuclear or plasma membranes; their lipids and carbohydrates are normal host constituents. In contrast, envelope proteins are coded for by virus genes and may even project from the envelope surface as **spikes** or **peplomers** (figure 16.17a, b, f). These spikes may be involved in virus attachment to the host cell surface. Since they differ among viruses, they also can be used to identify some viruses. Because the envelope is a flexible, membranous structure, enveloped viruses frequently have a somewhat variable shape and are called pleomorphic. However, the envelopes of viruses like the bullet-shaped rabies virus are firmly attached to the underlying nucleocapsid and endow the virion with a constant, characteristic shape (figure 16.17c). In some viruses the envelope is disrupted by solvents like ether to

such an extent that lipid-mediated activities are blocked or envelope proteins are denatured and rendered inactive. The virus is then said to be "ether sensitive."

Influenza virus (figure 16.17a, b) is a well-studied example of an enveloped virus. Spikes project about 10 nm from the surface at 7 to 8 nm intervals. Some spikes possess the enzyme neuraminidase, which probably aids the virus in penetrating mucous layers of the respiratory epithelium to reach host cells. Other spikes have hemagglutinin proteins, so named because they can bind the virions to red blood cell membranes and cause hemagglutination (*see figure 32.10*). Hemagglutinins probably participate in virion attachment to host cells. Proteins, like the spike proteins that are exposed on the outer envelope surface, are generally glycoproteins—that is, the proteins have carbohydrate attached to them. A nonglycosylated protein, the M or matrix protein, is found on the inner surface of the envelope and helps stabilize it.

Although it was originally thought that virions had only structural capsid proteins and lacked enzymes, some capsid-associated enzymes have been discovered, particularly in enveloped animal viruses. Many of these are involved in nucleic acid replication. For example, the influenza virus uses RNA as its genetic material and carries an RNA-dependent RNA polymerase that acts as an RNA transcriptase and synthesizes mRNA under the direction of its RNA genome. Although viruses lack true metabolism and cannot reproduce independently of living cells, they may carry one or more enzymes essential to the completion of their life cycles.

Nucleic acid replication and transcription (chapter 11).
Animal virus reproduction (pp. 370–77).

Viruses with Capsids of Complex Symmetry

Although most viruses have either icosahedral or helical capsids, many viruses do not fit into either category. The poxviruses and large bacteriophages are two important examples.

The poxviruses are the largest of the animal viruses (about $400 \times 240 \times 200$ nm in size) and can even be seen with a phase-contrast microscope or in stained preparations. They possess an exceptionally complex internal structure with an ovoid- to brick-shaped exterior. The double-stranded DNA is associated with proteins and contained in the nucleoid, a central structure shaped like a biconcave disk and surrounded by a membrane (figure 16.18). Two elliptical or lateral bodies lie between the nucleoid and its outer envelope, a membrane and a thick layer covered by an array of tubules or fibers.

Some large bacteriophages are even more elaborate than the poxviruses. The T2, T4, and T6 phages that infect *E. coli* have been intensely studied. Their head resembles an icosahedron elongated by one or two rows of hexamers in the middle (figure 16.19) and contains the DNA genome. The tail is composed of a collar joining it to the head, a central hollow tube, a sheath surrounding the tube, and a complex baseplate. In T-even phages,

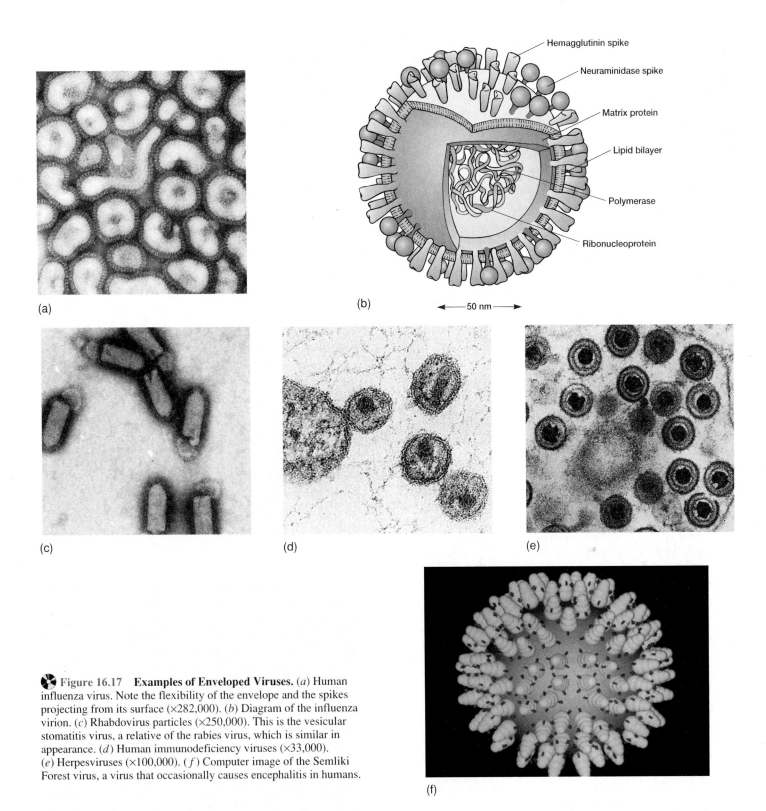

⬥ Figure 16.17 **Examples of Enveloped Viruses.** (*a*) Human influenza virus. Note the flexibility of the envelope and the spikes projecting from its surface (×282,000). (*b*) Diagram of the influenza virion. (*c*) Rhabdovirus particles (×250,000). This is the vesicular stomatitis virus, a relative of the rabies virus, which is similar in appearance. (*d*) Human immunodeficiency viruses (×33,000). (*e*) Herpesviruses (×100,000). (*f*) Computer image of the Semliki Forest virus, a virus that occasionally causes encephalitis in humans.

the baseplate is hexagonal and has a pin and a jointed tail fiber at each corner. The tail fibers are responsible for virus attachment to the proper site on the bacterial surface (*see chapter 17*).

There is considerable variation in structure among the large bacteriophages, even those infecting a single host. In contrast with the T-even phages, many coliphages have true icosahedral heads. T1, T5, and lambda phages have sheathless tails that lack a baseplate and terminate in rudimentary tail fibers. Coliphages T3 and T7 have short, noncontractile tails without tail fibers. Clearly these viruses can complete their reproductive cycles using a variety of tail structures.

Complex bacterial viruses with both heads and tails are said to have **binal symmetry** because they possess a combination of icosahedral (the head) and helical (the tail) symmetry.

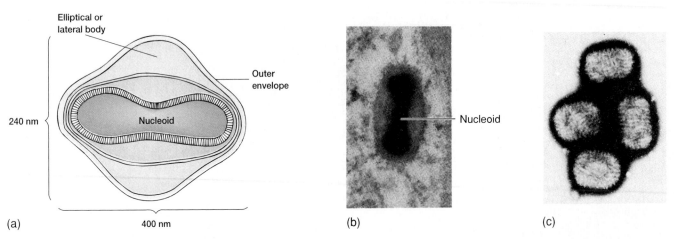

Figure 16.18   Vaccinia Virus Morphology. (*a*) Diagram of vaccinia structure. (*b*) Micrograph of the virion clearly showing the nucleoid (×200,000). (*c*) Vaccinia surface structure. An electron micrograph of four virions showing the thick array of surface fibers (×150,000).

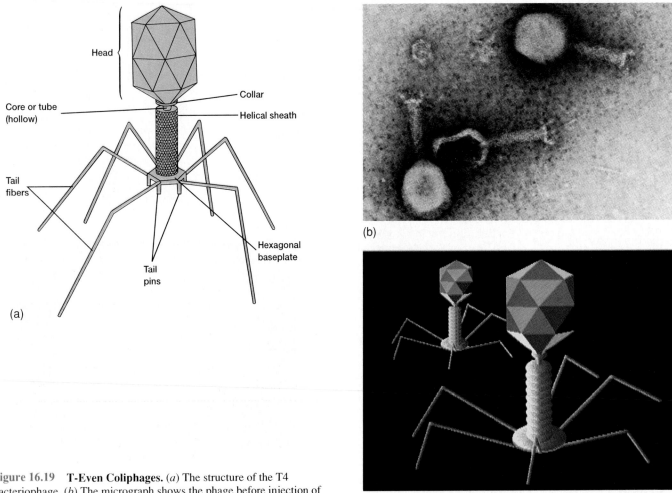

Figure 16.19   T-Even Coliphages. (*a*) The structure of the T4 bacteriophage. (*b*) The micrograph shows the phage before injection of its DNA. (*c*) Color-enhanced computer image of T2 bacteriophages.

Box 16.2

The Origin of Viruses

The origin and subsequent evolution of viruses are shrouded in mystery, in part because of the lack of a fossil record. However, recent advances in the understanding of virus structure and reproduction have made possible more informed speculation on virus origins. At present there are two major hypotheses entertained by virologists. It has been proposed that at least some of the more complex enveloped viruses, such as the poxviruses and herpesviruses, arose from small cells, probably procaryotic, that parasitized larger, more complex cells. These parasitic cells would become ever simpler and more dependent on their hosts, much like multicellular parasites have done, in a process known as retrograde evolution. There are several problems with this hypothesis. Viruses are radically different from procaryotes, and it is difficult to envision the mechanisms by which such a transformation might have occurred or the selective pressures leading to it. In addition, one would expect to find some forms intermediate between procaryotes and at least the more complex enveloped viruses, but such forms have not been detected.

The second hypothesis is that viruses represent cellular nucleic acids that have achieved independence and managed to escape the cell. Possibly a few mutations could convert nucleic acids, which are only synthesized at specific times, into infectious nucleic acids whose replication could not be controlled. This conjecture is supported by the observation that the nucleic acids of retroviruses (*see chapter 18*) and a number of other virions do contain sequences quite similar to those of normal cells, plasmids, and transposons (*see chapter 14*). The small, infectious RNAs called viroids (*see chapter 18*) have base sequences complementary to transposons, the regions around the boundary of mRNA introns (*see chapter 11*), and portions of host DNA. This has led to speculation that they have arisen from introns or transposons.

It is possible that viruses have arisen by way of both mechanisms. Because viruses differ so greatly from one another, it seems likely that they have originated independently many times during the course of evolution. Probably many viruses have evolved from other viruses just as cellular organisms have arisen from specific predecessors. The question of virus origins is complex and quite speculative; future progress in understanding virus structure and reproduction may clarify this question.

---

1. Define the following terms: nucleocapsid, capsid, icosahedral capsid, helical capsid, complex virus, protomer, self-assembly, capsomer, pentamer or penton, and hexamer or hexon. How do pentamers and hexamers associate to form a complete icosahedron; what determines helical capsid length and diameter?

2. All four nucleic acid forms can serve as virus genomes. Describe each, the types of virion possessing it, and any distinctive physical characteristics the nucleic acid can have. What are the following: plus strand, minus strand, and segmented genome?

3. What is an envelope? What are spikes or peplomers? Why are some enveloped viruses pleomorphic? Give two functions spikes might serve in the virus life cycle, and the proteins that the influenza virus uses in these processes.

4. What is a complex virus? Binal symmetry? Briefly describe the structure of poxviruses and T-even bacteriophages.

Principles of Virus Taxonomy

The classification of viruses is in a much less satisfactory state than that of either bacteria or eucaryotic microorganisms. In part, this is due to a lack of knowledge of their origin and evolutionary history (Box 16.2). Usually viruses are separated into several large groups based on their host preferences: animal viruses, plant viruses, bacterial viruses, bacteriophages, and so forth. In the past virologists working with these groups were unable to agree on a uniform system of classification and nomenclature. More recently the International Committee for Taxonomy of Viruses has developed a uniform classification system and divided viruses into 73 families and groups. The committee places greatest weight on three properties to define families: (1) nucleic acid type, (2) nucleic acid strandedness, and (3) presence or absence of an envelope. Virus family names end in *viridae;* subfamily names, in *virinae;* and genus (and species) names, in *virus*. For example, the poxviruses are in the family *Poxviridae;* the subfamily *Chorodopoxvirinae* contains poxviruses of vertebrates. Within the subfamily are several genera that are distinguished on the basis of immunologic characteristics and host specificity. The genus *Orthopoxvirus* contains several species, among them variola major (the cause of smallpox), vaccinia, and cowpox.

Viruses are divided into different taxonomic groups based on a few characteristics, some of the more important of which are:

1. Nature of the host—animal, plant, bacterial, insect, fungal
2. Nucleic acid characteristics—DNA or RNA, single or double stranded, molecular weight, segmentation and number of pieces of nucleic acid (RNA viruses)
3. Capsid symmetry—icosahedral, helical, binal
4. Presence of an envelope and ether sensitivity
5. Diameter of the virion or nucleocapsid
6. Number of capsomers in icosahedral viruses

TABLE 16.2   Some Common Virus Groups and Their Characteristics

Nucleic Acid	Strandedness	Capsid Symmetry[a]	Presence of Envelope	Size of Capsid (nm)[b]	Number of Capsomers	Virus Group	Host Range[c]
RNA	Single	I	−	22–30	32	*Picornaviridae*	A
		I	+	40–75(e)	32	*Togaviridae*	A
		I?	+	100(e)		*Retroviridae*	A
		H	+	9(h), 80–120(e)		*Orthomyxoviridae*	A
		H	+	18(h), 125–250(e)		*Paramyxoviridae*	A
		H	+	14–16(h), 80–160(e)		*Coronaviridae*	A
		H	+	18(h), 70–80 × 130–240 (bullet shaped)		*Rhabdoviridae*	A
		I	−	25		*Bromovirus*	P
		I	−	30		*Cucumovirus*	P
		H	−	18 × 300		*Tobamovirus*	P
		Bacilliform	−	18 × 18–58		Alfalfa mosaic virus	P
		I	−	26–27	32	*Leviviridae* [Qβ]	B
RNA	Double	I	−	70–80	92	*Reoviridae*	A
		I	−	70		Wound tumor virus	P
		I	−	25–50		Mycoviruses	P
		I	+	100(e)		*Cystoviridae*	B
DNA	Single	I	−	20–25	12	*Parvoviridae*	A
		I	−	18 × 30 (paired particles)		*Geminivirus*	P
		I	−	25–35		*Microviridae*	B
		H	−	6 × 900–1,900		*Inoviridae*	B
DNA	Double	I	−	45–55	72	*Papovaviridae*	A
		I	−	60–90	252	*Adenoviridae*	A
		I	−	130–180		*Iridoviridae*	A
		I	+	100, 180–200(e)	162	*Herpesviridae*	A
		C	+	200–260 × 250–290(e)		*Poxviridae*	A
		H	+	40 × 300(e)		*Baculoviridae*	A
		C	+	28 (core), 42(e)	42	*Hepadnaviridae*	A
		I	−	50		*Caulimovirus*	P
		I	−	60		*Corticoviridae*	B
		Bi	−	60, 15[d]		*Pedoviridae*	B
		Bi	−	54–65, 140–170[d]		*Styloviridae*	B
		Bi	−	80 × 110, 110[d]		*Myoviridae*	B

[a]Types of symmetry: I, icosahedral; H, helical; Bacilliform, resembles a rodlike bacterium in shape; C, complex; Bi, binal.

[b]Diameter of helical capsid (h); diameter of enveloped virion (e).

[c]Host range: A, animal; P, plant; B, bacterium.

[d]The first number is the head diameter; the second number, the tail length.

7. Immunologic properties

8. Intracellular location of viral replication

9. The presence or absence of a DNA intermediate (ssRNA viruses)

10. Disease caused and/or special clinical features, method of transmission

Table 16.2 illustrates the use of some of these properties to describe a few common virus groups. Virus classification is further discussed when bacterial, animal, and plant viruses are considered more specifically, and a fairly complete summary of virus taxonomy is presented in appendix V.

1. List some characteristics used in classifying viruses. Which seem to be the most important?

2. What are the endings for virus families, subfamilies, and genera or species?

Summary

1. People were first protected from a viral disease when Edward Jenner developed a smallpox vaccine in 1798.

2. Chamberland's invention of a porcelain filter that could remove bacteria from virus samples enabled microbiologists to show that viruses were different from bacteria.

3. In the late 1930s Stanley, Bawden, and Pirie crystallized the tobacco mosaic virus and demonstrated that it was composed only of protein and nucleic acid.

4. A virion is composed of either DNA or RNA enclosed in a coat of protein (and sometimes other substances as well). It cannot reproduce independently of living cells.

5. Viruses are cultivated using tissue cultures, embryonated eggs, bacterial cultures, and other living hosts.

6. Sites of animal viral infection may be characterized by cytopathic effects such as pocks and plaques. Phages produce plaques in bacterial lawns. Plant viruses can cause localized necrotic lesions in plant tissues.

7. Viruses can be purified by techniques such as differential and gradient centrifugation, precipitation, and denaturation or digestion of contaminants.

8. Virus particles can be counted directly with the transmission electron microscope or indirectly by the hemagglutination assay.

9. Infectivity assays can be used to estimate virus numbers in terms of plaque-forming units, lethal dose (LD_{50}), or infectious dose (ID_{50}).

10. All virions have a nucleocapsid composed of a nucleic acid, either DNA or RNA, held within a protein capsid made of one or more types of protein subunits called protomers.

11. There are four types of viral morphology: naked icosahedral, naked helical, enveloped icosahedral and helical, and complex.

12. Helical capsids resemble long hollow protein tubes and may be either rigid or quite flexible. The nucleic acid is coiled in a spiral on the inside of the cylinder.

13. Icosahedral capsids are usually constructed from two types of capsomers: pentamers (pentons) at the vertices and hexamers (hexons) on the edges and faces of the icosahedron.

14. Viral nucleic acids can be either single stranded or double stranded, DNA or RNA. Most DNA viruses have double-stranded DNA genomes that may be linear or closed circles.

15. RNA viruses usually have ssRNA that may be either plus (positive) or minus (negative) when compared with mRNA (positive). Many RNA genomes are segmented.

16. Viruses can have a membranous envelope surrounding their nucleocapsid. The envelope lipids usually come from the host cell; in contrast, many envelope proteins are viral and may project from the envelope surface as spikes or peplomers.

17. Although viruses lack true metabolism, some contain a few enzymes necessary for their reproduction.

18. Complex viruses (e.g., poxviruses and large phages) have complicated morphology not characterized by icosahedral and helical symmetry. Large phages often have binal symmetry: their heads are icosahedral and their tails, helical.

19. Currently viruses are classified with a taxonomic system placing primary emphasis on the type and strandedness of viral nucleic acids and on the presence or absence of an envelope.

Key Terms

bacteriophage *335*
binal symmetry *347*
capsid *340*
capsomers *341*
complex viruses *340*
cytopathic effects *335*
differential centrifugation *336*
envelope *340*
gradient centrifugation *337*
helical *340*
hemagglutination assay *339*

hexamers (hexons) *341*
icosahedral *340*
infectious dose (ID_{50}) *339*
lethal dose (LD_{50}) *339*
minus or negative strand *346*
necrotic lesion *336*
nucleocapsid *340*
pentamers (pentons) *341*
phage *335*
plaque *335*
plaque assay *339*

plaque-forming units (PFU) *339*
plus strand or positive strand *343*
protomers *341*
segmented genome *346*
spike or peplomer *346*
virion *334*
virologist *333*
virology *333*
virus *334*

Questions for Thought and Review

1. In what ways do viruses resemble living organisms?

2. Why might virology have developed much more slowly without the use of Chamberland's filter?

3. What advantage would an RNA virus gain by having its genome resemble eucaryotic mRNA?

4. A number of characteristics useful in virus taxonomy are listed on page 000. Can you think of any other properties that might be of considerable importance in future studies on virus taxonomy?

Additional Reading

General

Ackermann, H.-W., and Berthiaume, L., editors. 1995. *Atlas of Virus Diagrams*. Boca Raton, Fla.: CRC Press.

Cann, A. J. 1993. *Principles of molecular virology*. San Diego, Calif.: Academic Press.

Dimmock, N. J., and Primrose, S. B. 1994. *Introduction to modern virology*, 4th ed. London: Blackwell Scientific Publications.

Dulbecco, R., and Ginsberg, H. S. 1988. *Virology*, 2d ed. Philadelphia: J. B. Lippincott.

Fields, B. N.; Knipe, D. M.; Chanock, R. M.; Hirsch, M. S.; Melnick, J. L.; Monath, T. P.; and Roizman, B., editors. 1990. *Fields virology*, 2d ed. New York: Raven Press.

Levine, A. J. 1991. *Viruses*. New York: Scientific American Library.

Levy, J. A.; Fraenkel-Conrat, H.; and Owens, R. 1994. *Virology*, 3d ed. Englewood Cliffs, N.J.: Prentice-Hall.

Luria, S. E.; Darnell, J. E., Jr.; Baltimore, D.; and Campbell, A. 1978. *General virology*, 3d ed. New York: John Wiley and Sons.

Matthews, R. E. F. 1991. *Plant virology*, 3d ed. San Diego, Calif.: Academic Press

Scott, A. 1985. *Pirates of the cell: The story of viruses from molecule to microbe*. New York: Basil Blackwell.

Strauss, J. H., and Strauss, E. G. 1988. Evolution of RNA viruses. *Ann. Rev. Microbiol.* 42:657–83.

Voyles, B. A. 1993. *The biology of viruses*. St. Louis: Mosby

Webster, R. G., and Granoff, A., editors 1994. *Encyclopedia of virology*. San Diego, Calif.: Academic Press.

White, D. O., and Fenner, F. J. 1994. *Medical virology*, 4th ed. San Diego, Calif.: Academic Press.

Early Development of Virology

Jennings, F. 1975. *The invasion of America: Indians, colonialism, and the cant of conquest*. Chapel Hill: University of North Carolina Press.

Lechevalier, H. A., and Solotorovsky, M. 1965. *Three centuries of microbiology*. New York: McGraw-Hill.

McNeill, W. H. 1976. *Plagues and peoples*. Garden City, N.Y.: Anchor.

Stearn, E. W., and Stearn, A. E. 1945. *The effect of smallpox on the destiny of the Amerindian*. Boston: Bruce Humphries.

Virus Purification and Assays

Henshaw, N. G. 1988. Identification of viruses by methods other than electron microscopy. *ASM News* 54(9):482–85.

Miller, S. E. 1988. Diagnostic virology by electron microscopy. *ASM News* 54(9):475–81.

The Structure of Viruses

Bradley, D. E. 1971. A comparative study of the structure and biological properties of bacteriophages. In *Comparative virology*, K. Maramorosch and E. Kurstak, editors, 207–53. New York: Academic Press.

Casjens, S. 1985. *Virus structure and assembly*. Boston: Jones and Bartlett.

Harrison, S. C. 1984. Structure of viruses. In *The microbe 1984: Part I, viruses*. 36th Symposium Society for General Microbiology. Cambridge: Cambridge University Press.

Principles of Virus Taxonomy

Eigen, M. 1993. Viral quasispecies. *Sci. Am.* 269(1):42–49.

Lwoff, A., and Tournier, P. 1971. Remarks on the classification of viruses. In *Comparative virology*, K. Maramorosch and E. Kurstak, editors, 1–42. New York: Academic Press.

Matthews, R. E. F. 1985. Viral taxonomy for the nonvirologist. *Ann. Rev. Microbiol.* 39:451–74.

Murphy, F. A.; Fauquet, C. M.; Bishop, D. H. L.; Ghabrial, S. A.; Jarvis, A. W.; and Martelli, G. P. 1995. *The classification and nomenclature of viruses. Sixth report of the international committee on taxonomy of viruses*. New York: Springer-Verlag.

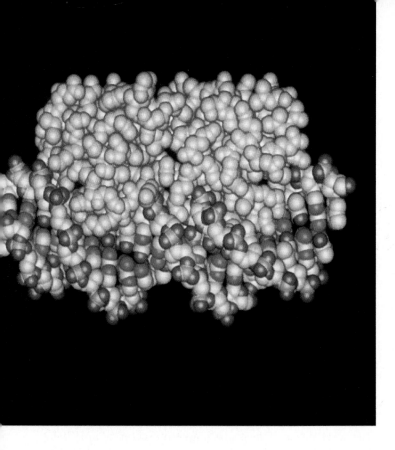

CHAPTER 17
The Viruses: Bacteriophages

You might wonder how such naive outsiders get to know about the existence of bacterial viruses. Quite by accident, I assure you. Let me illustrate by reference to an imaginary theoretical physicist, who knew little about biology in general, and nothing about bacterial viruses in particular. . . . Suppose now that our imaginary physicist, the student of Niels Bohr, is shown an experiment in which a virus particle enters a bacterial cell and 20 minutes later the bacterial cell is lysed and 100 virus particles are liberated. He will say: "How come, one particle has become 100 particles of the same kind in 20 minutes? That is very interesting. Let us find out how it happens! . . . Is this multiplying a trick of organic chemistry which the organic chemists have not yet discovered? Let us find out."

—Max Delbrück

Outline

Concepts

1. Since a bacteriophage cannot independently reproduce itself, the phage takes over its host cell and forces the host to reproduce it.

2. The lytic bacteriophage life cycle is composed of four phases: adsorption of the phage to the host and penetration of virus genetic material, synthesis of virus nucleic acid and capsid proteins, assembly of complete virions, and the release of phage particles from the host.

3. Temperate virus genetic material is able to remain within host cells and reproduce in synchrony with the host for long periods in a relationship known as lysogeny. Usually the virus genome is found integrated into the host genetic material as a prophage. A repressor protein keeps the prophage dormant and prevents virus reproduction.

Chapter 16 introduces many of the facts and concepts underlying the field of virology, including information about the nature of viruses, their structure and taxonomy, and how they are cultivated and studied. Clearly the viruses are a complex, diverse, and fascinating group, the study of which has done much to advance disciplines such as genetics and molecular biology.

Chapters 17 and 18 focus on virus diversity. This chapter is concerned with bacterial viruses or **bacteriophages;** the next surveys animal, plant, and insect viruses. The taxonomy, morphology, and reproduction of each group are covered. Where appropriate, the biological and practical importance of viruses is emphasized (Box 17.1), even though viral diseases are examined in chapter 36.

Since the bacteriophages (or simply phages) have been the most intensely studied viruses and are best understood in a molecular sense, this chapter is devoted to them.

Classification of Bacteriophages

Although properties such as host range and immunologic relationships are used in classifying phages, the most important are phage morphology and nucleic acid properties (figure 17.1). The genetic material may be either DNA or RNA; most bacteriophages have DNA (usually double stranded). Most can be placed in one of a few morphological groups: tailless icosahedral phages, viruses with contractile tails, viruses with noncontractile

tails, and filamentous phages. There are even a few phages with envelopes. The most complex forms are the phages with contractile tails, for example, the T-even phages of *E. coli.*

---

1. Briefly describe in general terms the morphology and nucleic acids of the major phage types.

Reproduction of DNA Phages: The Lytic Cycle

After DNA bacteriophages have reproduced within the host cell, many of them are released when the cell is destroyed by lysis. A phage life cycle that culminates with the host cell bursting and releasing virions is called a **lytic cycle.** The events taking place during the lytic cycle will be reviewed in this section, with the primary focus on the T-even phages of *E. coli.* These are double-stranded DNA bacteriophages with complex contractile tails.

The structure of T-even coliphages (pp. 346–48).

The One-Step Growth Experiment

The development of the one-step growth experiment in 1939 by Max Delbrück and Emory Ellis marks the beginning of modern bacteriophage research. In a **one-step growth experiment,** the reproduction of a large phage population is synchronized so that the molecular events occurring during reproduction can be followed. A culture of susceptible bacteria such as *E. coli* is mixed with bacteriophage particles, and the phages are allowed a short interval to attach to their host cells. The culture is then

---

Box 17.1

An Ocean of Viruses

Microbiologists have previously searched without success for viruses in marine habitats. Thus it has been assumed the oceans probably did not contain many viruses. Recent discoveries have changed this view radically. Several groups have either centrifuged seawater at high speeds or passed it through an ultrafilter and then examined the sediment or suspension in an electron microscope (*see figure 41.8*). They have found that marine viruses are about 10 times more plentiful than marine bacteria. Between 10^6 and 10^9 virus particles per milliliter are present at the ocean's surface. It has been estimated that the top one millimeter of the world's oceans could contain a total of over 3×10^{30} virus particles!

Although little detailed work has been done on marine viruses, it appears that many contain double-stranded DNA. Most are probably bacteriophages and can infect both marine heterotrophs and cyanobacteria. Up to 70% of marine procaryotes may be infected by phages. Viruses that infect diatoms and other major algal components of the marine phytoplankton also have been detected.

Marine viruses may be very important ecologically. Viruses may control marine algal blooms such as red tides (p. 592), and

bacteriophages could account for 1/3 or more of the total aquatic bacterial mortality or turnover. If true, this is of major ecological significance because the reproduction of marine bacteria far exceeds marine protozoan grazing capacity. Virus lysis of procaryotic and algal cells may well contribute greatly to carbon and nitrogen cycling in marine food webs. It could reduce the level of marine primary productivity in some situations.

Bacteriophages also may greatly accelerate the flow of genes between marine bacteria. Virus-induced bacterial lysis could generate most of the free DNA present in seawater. Gene transfer between aquatic bacteria by transformation (*see pp. 292–93*) does occur, and bacterial lysis by phages would increase its probability. Furthermore, such high phage concentrations can stimulate gene exchange by transduction (*see pp. 293–98*). These genetic exchanges could have both positive and negative consequences. Genes that enable marine bacteria to degrade toxic pollutants such as those in oil spills could spread through the native population. On the other hand, antibiotic resistance genes in bacteria from raw sewage released into the ocean also might be dispersed.

greatly diluted so that any virus particles released upon host cell lysis will not immediately infect new cells. This strategy works because phages lack a means of seeking out host cells and must contact them during random movement through the solution. Thus phages are less likely to contact host cells in a dilute mixture. The number of infective phage particles released from bacteria is subsequently determined at various intervals by a plaque count (*see chapter 16*).

A plot of the bacteriophages released from host cells versus time shows several distinct phases (figure 17.2). During the **latent period,** which immediately follows phage addition, there is no release of virions. This is followed by the **rise period** or **burst,** when the host cells rapidly lyse and release infective phages. Finally, a plateau is reached and no more viruses are liberated. The total number of phages released can be used to calculate the **burst size,** the number of viruses produced per infected cell.

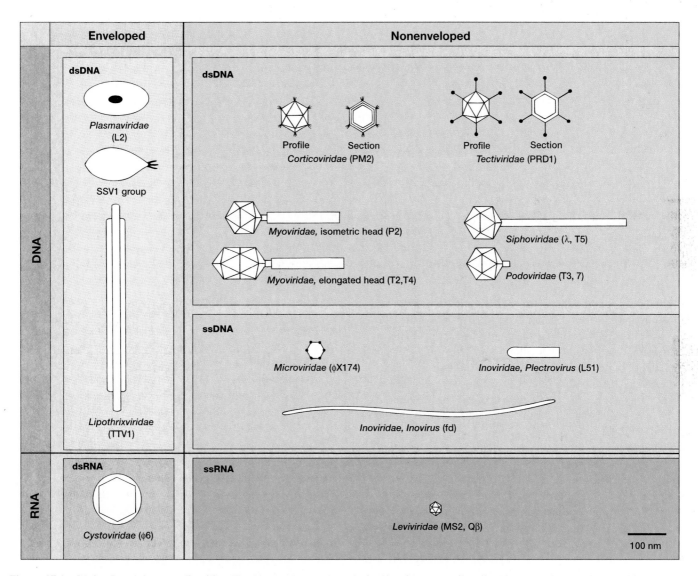

Figure 17.1   Major Bacteriophage Families. The *Myoviridae* are the only family with contractile tails. *Plasmaviridae* are pleomorphic. *Tectiviridae* have distinctive double capsids, whereas the *Corticoviridae* have complex capsids containing lipid.

Figure 17.2   **The One-Step Growth Curve.** In the initial part of the latent period, the eclipse period, the host cells do not contain any complete, infective virions. During the remainder of the latent period, an increasing number of infective virions are present, but none are released. The latent period ends with host cell lysis and rapid release of virions during the rise period or burst. In this figure the blue line represents the total number of complete virions. The red line is the number of free viruses (the unadsorbed virions plus those released from host cells). When *E. coli* is infected with T2 phage at 37°C, the growth plateau is reached in about 30 minutes and the burst size is approximately 100 or more virions per cell. The eclipse period is 11–12 minutes, and the latent period is around 21–22 minutes in length.

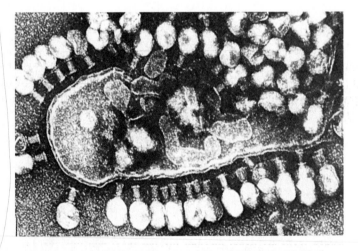

Figure 17.3   **T4 Phage Adsorption and DNA Injection.**

The latent period is the shortest time required for virus reproduction and release. During the first part of this phase, host bacteria do not contain any complete, infective virions. This can be shown by lysing them with chloroform. This initial segment of the latent period is called the **eclipse period** because the virions were detectable before infection but are now concealed or eclipsed. The number of completed, infective phages within the host increases after the end of the eclipse period, and the host cell is prepared for lysis.

The one-step growth experiment with *E. coli* and phage T2 provides a well-studied example of this process. When the experiment is carried out with actively growing cells in rich medium at 37° C, the growth curve plateau is reached in approximately 30 minutes. Bacteriophage reproduction is an exceptionally rapid process, much faster than animal virus reproduction, which may take hours.

Adsorption to the Host Cell and Penetration

Bacteriophages do not randomly attach to the surface of a host cell; rather, they fasten to specific **receptor sites.** The nature of these receptors varies with the phage; cell wall lipopolysaccharides and proteins, teichoic acids, flagella, and pili can serve as receptors. The T-even phages of *E. coli* use cell wall lipopolysaccharides or proteins as receptors. Variation in receptor properties is at least partly responsible for phage host preferences.

The structure of cell walls, flagella, and pili (pp. 51–58, 59–63).

Figure 17.4   **Electron Micrograph of *E. coli* Infected with Phage T4.** Baseplates, contracted sheaths, and tail tubes can be seen (×36,500).

T-even phage adsorption involves several tail structures (*see figure 16.19*). Phage attachment begins when a tail fiber contacts the appropriate receptor site. As more tail fibers make contact, the baseplate settles down on the surface (figures 17.3 and 17.4). Binding is probably due to electrostatic interactions and is influenced by pH and the presence of ions such as Mg^{2+} and Ca^{2+}. After the baseplate is seated firmly on the cell surface, conformational changes occur in the baseplate and sheath, and the tail sheath contracts. Then the central tube or core is

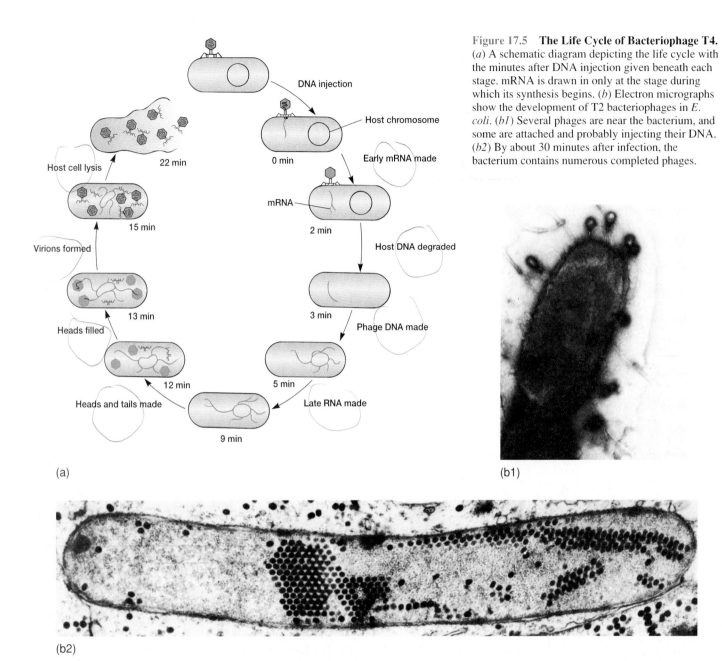

(a)

(b1)

(b2)

Figure 17.5 **The Life Cycle of Bacteriophage T4.**
(*a*) A schematic diagram depicting the life cycle with
the minutes after DNA injection given beneath each
stage. mRNA is drawn in only at the stage during
which its synthesis begins. (*b*) Electron micrographs
show the development of T2 bacteriophages in *E.
coli*. (*b1*) Several phages are near the bacterium, and
some are attached and probably injecting their DNA.
(*b2*) By about 30 minutes after infection, the
bacterium contains numerous completed phages.

pushed through the bacterial wall. The sheath contains ATP,
and this may power the contraction. Finally, the DNA is extruded from the head, through the tail tube, and into the host
cell. The tube may interact with the plasma membrane to form a
pore through which DNA passes. The penetration mechanisms
of other bacteriophages often appear to differ from that of the T-
even phages but have not been studied in much detail.

Synthesis of Phage Nucleic Acids and Proteins

Since the T4 phage of *E. coli* has been intensely studied, its
reproduction will be used as our example (figure 17.5).
Soon after phage DNA injection, the synthesis of host DNA,
RNA, and protein is halted, and the cell is forced to make viral
constituents. *E. coli* RNA polymerase (*see chapter 11*) starts

synthesizing phage mRNA within 2 minutes. This mRNA and
all other **early mRNA** (mRNA transcribed before phage DNA
is made) direct the synthesis of the protein factors and enzymes
required to take over the host cell and manufacture viral nucleic acids. Some early virus-specific enzymes degrade host
DNA to nucleotides, thereby simultaneously halting host gene
expression and providing raw material for virus DNA synthesis. Other early proteins and a few phage head proteins that are
injected along with phage DNA modify the bacterial RNA
polymerase. The polymerase then recognizes promoters on
viral DNA rather than bacterial promoter sequences so that
virus genes are transcribed in preference to host genes. Within
5 minutes, virus DNA synthesis commences.

Promoters and transcription (pp. 222–3).

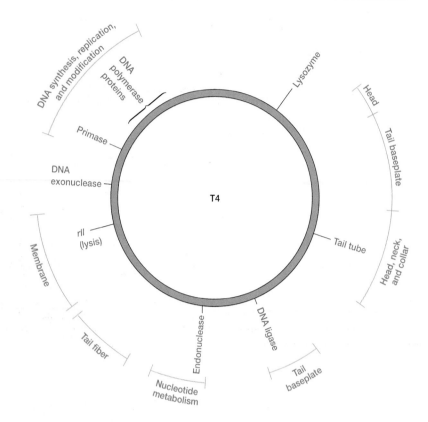

Figure 17.6 A Map of the T4 Genome. Some of its genes and their functions are shown. Genes with related functions tend to be clustered together.

Figure 17.7   5-Hydroxymethylcytosine (HMC). In T4 DNA, the HMC often has glucose attached to its hydroxyl.

It is clear from the sophisticated control of RNA polymerase and the precise order in which events occur in the reproductive cycle that the expression of T4 genes is tightly regulated. Even the organization of the genome appears suited for efficient control of the life cycle. As can be seen in figure 17.6, genes with related functions—such as the genes for phage head or tail fiber construction—are usually clustered together. Early and late genes also are clustered separately on the genome; they are even transcribed in different directions—early genes in the counterclockwise direction and late genes, clockwise. Since transcription always proceeds in the 5′ to 3′ direction, the early and late genes are located on different DNA strands (*see chapters 11 and 13*).

Considerable preparation is required for synthesis of T4 DNA because it contains **hydroxymethylcytosine (HMC)** instead of cytosine (figure 17.7). HMC must be synthesized by two phage-encoded enzymes before DNA replication can begin. After T4 DNA has been synthesized, it is glucosylated by the

addition of glucose to the HMC residues. Glucosylated HMC residues protect T4 DNA from attack by *E. coli* endonucleases called **restriction enzymes,** which would otherwise cleave the viral DNA at specific points and destroy it. This bacterial defense mechanism is called **restriction.** Other groups also can be used to modify phage DNA and protect it against restriction enzymes. For example, methyl groups are added to the amino groups of adenine and cytosine in lambda phage DNA for the same reason. The replication of T4 DNA is an extremely complex process requiring at least seven phage proteins. Its mechanism resembles that described in chapter 11.

Restriction enzymes and genetic engineering (p. 306).

T4 DNA shows what is called terminal redundancy; that is, a base sequence is repeated at both ends of the molecule (figure 17.8). When many DNA copies have been made, about 6 to 10 copies are joined by their terminally redundant ends with the aid of several enzymes (figure 17.8). These very long DNA strands composed of several units linked together are called **concatemers.** During assembly, concatemers are cleaved in such a way that the genome is slightly longer than the T4 gene set. The genetic map is therefore drawn circular (figure 17.6) because T4 DNA is circularly permuted (figure 17.9). The sequence of genes in each T4 virus of a population is the same but starts with a different gene at the 5′ end. If all the linear pieces of DNA were coiled into circles, the DNA circles would have identical gene sequences.

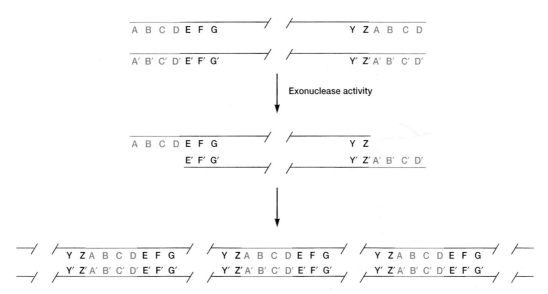

Figure 17.8   An Example of Terminal Redundancy. The gene sequences in color are terminally redundant; they are repeated at each end of the DNA molecule. This makes it possible to join units together by their redundant ends forming a concatemer. For example, if the 3′ ends of each unit were partially digested by an exonuclease, the complementary 5′ ends would be exposed and could base pair to generate a long chain of repeated units. The breaks between terminal sequences indicate that the DNA molecules are longer than shown here.

Figure 17.9   Circularly Permuted Genomes Cut from a Concatemer. The concatemer formed in figure 17.8 can be cut at any point into pieces of equal length that contain a complete complement of genes, even though different genes are found at their ends. If each piece has single-stranded cohesive ends as in figure 17.8, it will coil into a circle with the same gene order as the circles produced by other pieces.

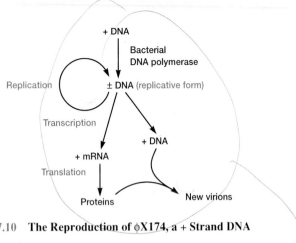

Figure 17.10 The Reproduction of φX174, a + Strand DNA Phage.

Thus far, only double-stranded DNA phage reproduction has been discussed, with the lytic phage T4 as an example. Before the process of phage assembly is described, the reproduction of single-stranded DNA phages will be briefly reviewed. The phage φX174 is a small ssDNA phage using *E. coli* as its host. Its DNA base sequence is the same as that of the viral mRNA (except that thymine is substituted for uracil) and is therefore positive. The phage DNA must be converted to a double-stranded form before either replication or transcription can occur. When φX174 DNA enters the host, it is immediately copied by the bacterial DNA polymerase to form a double-stranded DNA, the **replicative form** or **RF** (figure 17.10). The replicative form then directs the synthesis of more RF copies, mRNA, and copies of the +DNA genome.

The filamentous ssDNA bacteriophages behave quite differently in many respects from φX174 and other ssDNA phages. The fd phage is one of the best studied and is shaped like a long fiber about 6 nm in diameter by 900 to 1,900 nm in length (figure 17.1). The circular ssDNA lies in the center of the filament and is surrounded by a tube made of a small coat protein organized in a helical arrangement. The virus infects male *E. coli* cells by attaching to the tip of the pilus; the DNA enters the host along or possibly through the pilus with the aid of a special adsorption protein. A replicative form is first synthesized and then transcribed. A phage-coded protein then aids in replication of the phage DNA by use of the rolling-circle method (*see chapter 11*).

The Assembly of Phage Particles

The assembly of the T4 phage is an exceptionally complex process. **Late mRNA,** or that produced after DNA replication, directs the synthesis of three kinds of proteins: (1) phage structural proteins, (2) proteins that help with phage assembly without becoming part of the virion structure, and (3) proteins involved in cell lysis and phage release. Late mRNA transcription begins about 9 minutes after T4 DNA injection into *E. coli*. All the proteins required for phage assembly are synthesized simul-

Figure 17.11 The Assembly of T4 Bacteriophage. Note the subassembly lines for the baseplate, tail tube and sheath, tail fibers, and head.

taneously and then used in four fairly independent subassembly lines (figure 17.11). The baseplate is constructed of 15 gene products. After the baseplate is finished, the tail tube is built on it and the sheath is assembled around the tube. The phage head shell is constructed separately of more than 10 proteins and then spontaneously combines with the tail assembly. Tail fibers attach to the baseplate after the head and tail have come together. Although many of these steps occur spontaneously, some require special virus enzymes or host cell factors.

DNA packaging within the T4 head is still a somewhat mysterious process. In some way the DNA is drawn into the completed shell so efficiently that about 500 μm of DNA are packed into a cavity less than 0.1 μm across! It is thought that a long DNA concatemer enters the head shell until it is packed full and contains about 2% more DNA than is needed for the

Figure 17.12   Release of the Pf1 Phage. The Pf1 phage is a filamentous bacteriophage that is released from *Pseudomonas aeruginosa* without lysis. In this illustration the blue cylinders are hydrophobic α-helices that span the plasma membrane, and the red cylinders are amphipathic helices that lie on the membrane surface before virus assembly. In each protomer the two helices are connected by a short, flexible peptide loop (yellow). It is thought that the blue helix binds with circular, single-stranded viral DNA (green) as it is extruded through the membrane. The red helix simultaneously attaches to the growing virus coat that projects from the membrane surface. Eventually the blue helix leaves the membrane and also becomes part of the capsid.

full T4 genome. The concatemer is then cut, and T4 assembly is finished. The first complete T4 particles appear in *E. coli* at 37°C about 15 minutes after infection.

Release of Phage Particles

Many phages lyse their host cells at the end of the intracellular phase. The lysis of *E. coli* takes place after about 22 minutes at 37°C, and approximately 300 T4 particles are released. Several T4 genes are involved in this process. One directs the synthesis of a lysozyme that attacks the cell wall peptidoglycan. Another phage protein both damages the bacterial plasma membrane, allowing lysozyme to reach the cell wall beyond it, and inhibits energy metabolism reactions located in the membrane. The ssDNA virus φX174 is also lytic.

It should be noted that some phages are released without lysis of their host cell. For example, the filamentous fd phages do not kill their host cell but establish a symbiotic relationship in which new virions are continually released by a secretory process. Filamentous phage coat proteins are first inserted into the membrane. The coat then assembles around the viral DNA as

it is secreted through the host plasma membrane (figure 17.12). The host bacteria grow and divide at a slightly reduced rate.

1. How is a one-step growth experiment carried out? Summarize what occurs in each phase of the resulting growth curve. Define latent period, eclipse period, rise period or burst, and burst size.
2. Be able to describe in some detail what is occurring in each phase of the lytic phage life cycle: adsorption and penetration, nucleic acid and protein synthesis, phage assembly, and phage release. Define the following terms: lytic cycle, receptor site, early mRNA, hydroxymethylcytosine, restriction, restriction enzymes, concatemers, replicative form, and late mRNA.
3. How does the reproduction of the ssDNA phages φX174 and fd differ from each other and from the dsDNA T4 phage?

Reproduction of RNA Phages

Many bacteriophages carry their genetic information as single-stranded RNA that can act as a messenger RNA and direct the synthesis of phage proteins. One of the first enzymes synthesized

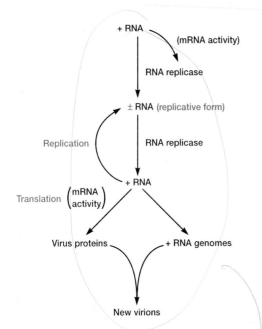

Figure 17.13 The Reproduction of Single-Stranded RNA Bacteriophages.

is **RNA replicase,** an RNA-dependent RNA polymerase (figure 17.13). The replicase then copies the original RNA (a plus strand) to produce a double-stranded intermediate (±RNA), which is called the replicative form and is analogous to the ±DNA seen in the reproduction of ssDNA phages. The replicase next uses this replicative form to synthesize thousands of copies of +RNA. Some of these plus strands are used to make more ±RNA in order to accelerate +RNA synthesis. Other +RNA acts as mRNA and directs the synthesis of phage proteins. Finally, +RNA strands are incorporated into maturing virus particles. The genome of these RNA phages serves as both a template for its own replication and an mRNA.

MS2 and Qβ are small, tailless, icosahedral ssRNA phages of *E. coli,* which have been intensely studied (figure 17.1). They attach to the F-pili of their host and enter by an unknown mechanism. These phages have only three or four genes and are genetically the simplest phages known. In MS2, one protein is involved in phage adsorption to the host cell (and possibly also in virion construction or maturation). The other three genes code for a coat protein, an RNA replicase, and a protein needed for cell lysis.

Only one dsRNA phage has been discovered, the bacteriophage φ6 of *Pseudomonas phaseolicola* (figure 17.1). It is also unusual in possessing a membranous envelope. The icosahedral capsid within its envelope contains an RNA polymerase and three dsRNA segments, each of which directs the synthesis of an mRNA. It is not yet known how the dsRNAs are replicated.

1. How are ssRNA phages reproduced, and what role does RNA replicase play in the process?
2. What is peculiar about the structure of phage φ6?

Temperate Bacteriophages and Lysogeny

Up to this point the primary emphasis has been on **virulent bacteriophages;** these are phages that lyse their host cells during the reproductive cycle. Many DNA phages also can establish a different relationship with their host. After adsorption and penetration, the viral genome does not take control of its host and destroy it while producing new phages. Instead the viral genome remains within the host cell and replicates with the bacterial genome to generate a clone of infected cells that may grow and divide for long periods while appearing perfectly normal. Each of these infected bacteria can produce phages and lyse under appropriate environmental conditions. They cannot, for reasons that will become clear later, be reinfected by the same virus—that is, they have immunity to superinfection. This relationship between the phage and its host is called **lysogeny.** Bacteria having the potential to produce phage particles under some conditions are said to be **lysogens** or **lysogenic,** and phages able to enter into this relationship are **temperate phages.** The latent form of the virus genome that remains within the host but does not destroy it is called the **prophage.** The prophage usually is integrated into the bacterial genome but sometimes exists independently. **Induction** is the process by which phage reproduction is initiated in a lysogenized culture. It leads to the destruction of infected cells and the release of new phages—that is, induction of the lytic cycle. Lysogeny was briefly described earlier in the context of transduction and genetic recombination, but will be discussed in more detail here.

Temperate viruses and transduction (pp. 293–98).

Most bacteriophages that have been studied are temperate, and it appears that there are advantages in being able to lysogenize bacteria. Consider a phage-infected culture that is becoming dormant due to nutrient deprivation. Before bacteria enter dormancy, they degrade their own mRNA and protein. Thus the phage is faced with two problems: it can only reproduce in actively metabolizing bacteria, and phage reproduction is usually permanently interrupted by the mRNA and protein degradation. This predicament can be avoided if the phage becomes dormant (lysogenic) at the same time as its host; in fact, nutrient deprivation does favor lysogeny. Temperate phages also have an advantage in situations where many viruses per cell initiate an infection—that is, where there is a high multiplicity of infection (MOI). When every cell is infected, the last round of replication will destroy all host cells. Thus there is a risk that the phages may be left without a host and directly exposed to environmental hazards for months or years. This prospect is avoided if lysogeny is favored by a high MOI; some bacteria will survive, carry the virus genome, and synthesize new copies as they reproduce. Not surprisingly a high MOI does stimulate lysogeny.

A temperate phage may induce a change in the phenotype of its host cell that is not directly related to completion of its life cycle. Such a change is called a **lysogenic conversion** or a conversion and often involves alterations in bacterial surface characteristics or pathogenic properties. For example, when *Salmonella* is infected by an epsilon phage, the structure of its outer lipopolysaccharide layer (*see pp. 54–6*) may be modified. The phage changes the activities of several enzymes involved in construction of the lipopolysaccharide carbohydrate component and thus alters the antigenic properties of the host. These epsilon-induced changes appear to eliminate surface phage receptors and prevent infection of the lysogen by another epsilon phage. Another example is the temperate phage β of *Corynebacterium diphtheriae,* the cause of diphtheria. Only *C. diphtheriae* that is lysogenized with phage β will produce diphtheria toxin (*see chapters 29 and 37*) because the phage, not the bacterium, carries the toxin gene.

The lambda phage that uses the K12 strain of *E. coli* as its host is the best-understood temperate phage and will serve as our example of lysogeny. Lambda is a double-stranded DNA phage possessing an icosahedral head 55 nm in diameter and a noncontractile tail with a thin tail fiber at its end (figure 17.14). The DNA is a linear molecule with cohesive ends—single-stranded stretches, 12 nucleotides long, that have complementary base sequences and can base pair with each other. Because of these cohesive ends, the linear genome cyclizes immediately

upon infection (figure 17.15). *E. coli* DNA ligase then seals the breaks, forming a closed circle. The lambda genome has been carefully mapped, and over 40 genes have been located (figure 17.16). Most genes are clustered according to their function, with separate groups involved in head synthesis, tail synthesis, lysogeny and its regulation, DNA replication, and cell lysis.

DNA ligase (p. 219).

Lambda phage can reproduce using a normal lytic cycle. Immediately after lambda DNA enters *E. coli*, it is converted to a covalent circle, and transcription by the host RNA polymerase is initiated. As shown in figure 17.16, the polymerase binds to both a rightward promoter (PR) and a leftward promoter (PL) and begins to transcribe in both directions, copying different DNA strands. The first genes that are transcribed code for regulatory proteins that control the lytic cycle: leftward gene *N* and rightward genes *cro* and *cII* (figure 17.16). These and other regulatory genes ensure that virus proteins will be synthesized in an orderly time sequence and will be manufactured only when needed during the life cycle.

Regulation of transcription (pp. 242–48).

Lambda DNA replication and virion assembly are similar to the same processes already described for the T4 phage. One significant difference should be noted. Although initially bidirectional DNA replication is used and theta-shaped intermediates are seen (*see chapter 11*), lambda DNA is primarily synthesized by way of the rolling-circle mechanism to form long concatemers that are finally cleaved to give complete genomes (*see figure 11.9*).

The establishment of lysogeny and the earlier-mentioned immunity of lysogens to superinfection can be accounted for by the presence of the **lambda repressor** coded for by the *cI* gene. The repressor protein chain is 236 amino acids long and folds into a dumbbell shape with globular domains at each end (figure 17.17). One domain is concerned with binding to

Figure 17.14   Bacteriophage Lambda.

Figure 17.15   Lambda Phage DNA. A diagram of lambda phage DNA showing its 12 base, single-stranded cohesive ends (printed in color) and the circularization their complementary base sequences make possible.

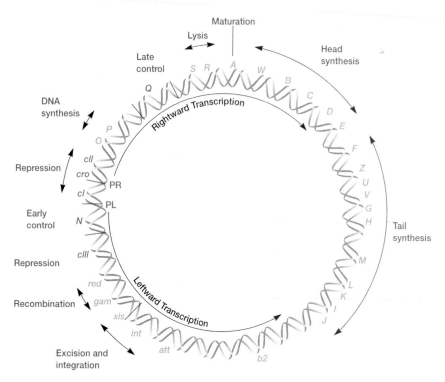

Figure 17.16   The Lambda Phage Genome. The direction of transcription and location of leftward and rightward promoters (PL and PR) are indicated on the inside of the map. The positions of major regulatory sites are shown by lines on the map and regulatory genes are in blue. Lambda DNA is double stranded, and transcription proceeds in opposite directions on opposite strands.

(a)

(b)

Figure 17.17   Lambda Repressor Binding. (*a*) A computer model of lambda repressor binding to the lambda operator. The lambda repressor dimer (brown and green) is bound to DNA (blue and light blue). The arms of the dimer wrap around the major grooves of the double helix. (*b*) A diagram of the lambda repressor-DNA complex. The repressor binds to a 17 bp stretch of the operator. The α3-helices make closest contact with the major grooves of the operator (the helices are labeled in order, beginning at the N terminal of the chain).

DNA, while the other binds with another repressor molecule to generate a dimer (the most active form of the lambda repressor). In a lysogen the repressor is synthesized continuously and binds to the operators O_L and O_R, thereby blocking RNA polymerase activity (figure 17.18c). If another lambda phage tries to infect the cell, its mRNA synthesis also will be inhibited. It should be noted that immunity always involves repressor activity. A potential host cell might remain uninfected due to a mutation that alters its phage receptor site. In such an instance it is said to be resistant, not immune, to the phage.

The sequence of events leading to the initial synthesis of repressor and the establishment of lysogeny is well known (figure

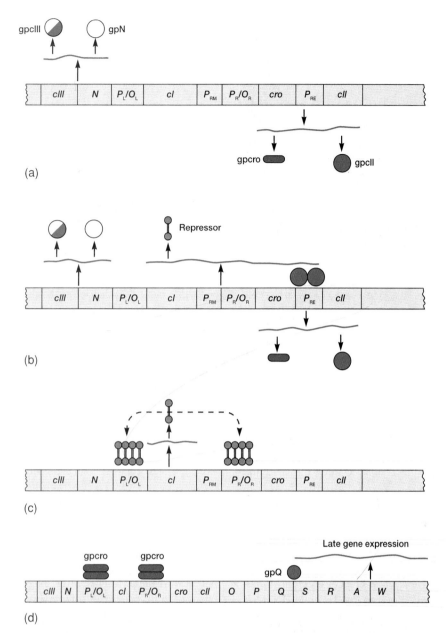

Figure 17.18   Choice Between Lysogeny and Lysis. Events involved in the choice between the establishment of lysogeny and continuation of the lytic cycle. The action gpN is ignored for the sake of simplicity, and the scale in part (*d*) differs from that in parts (*a*) to (*c*). The abbreviation gp stands for gene product (gpcro is the product of the *cro* gene). (*a*) and (*b*) illustrate the initial steps leading to lambda repressor synthesis. (*c*) represents the situation when repressor production overcomes cro synthesis and establishes lysogeny. In part (*d*) the cro protein has accumulated more rapidly than lambda repressor and cro protein dimers (the active form) have bound to O_L and O_R. This blocks both *cI* and *cro* gene function, but not late gene expression, since gpQ has already accumulated and promoted late mRNA synthesis. See text for further details.

17.18). Immediately after lambda DNA has been circularized and transcription has commenced, the cII and cIII proteins accumulate. The cII protein binds next to the promoter for the *cII* gene (P_{RE}, RE stands for repressor establishment), and stimulates RNA polymerase binding. The cIII protein protects cII from degradation by a host enzyme. Lambda repressor (gpcI) is rapidly synthesized and binds to O_R and O_L, thus turning off mRNA synthesis and the production of the cII and cIII proteins. The *cI* gene continues to be transcribed at a low rate because of the activity of a second promoter (P_{RM}, RM stands for repressor

maintenance) that is activated by the repressor itself. This control circuit in which lambda repressor stimulates its own synthesis ensures that lysogeny will normally be stable when once established.

One might expect that lysogeny would be established every time but this is not the case. During this period the **cro protein** (gpcro) has also been accumulating. The cro protein binds to O_R and O_L, turns off the transcription of the repressor gene (as well as inhibiting the expression of other early genes), and represses P_{RM} function (figures 17.18*d* and 17.19). Since

(a)

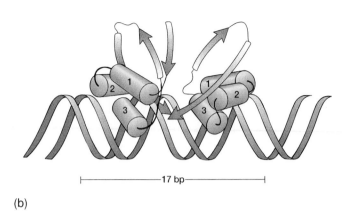

(b)

|———— 17 bp ————|

Figure 17.19   Cro Protein Binding. (*a*) A space-filling model of the cro protein–DNA complex. The cro protein is in light blue. (*b*) A diagram of the cro protein dimer–DNA complex. Like the lambda repressor protein, the cro protein binds to two adjacent DNA major grooves.

the lambda repressor can block *cro* transcription, there is a race between the production of lambda repressor and that of the cro protein. Although cro protein synthesis begins before that of the lambda repressor, gpcro binds to O_R more weakly and must rise to a higher level than the repressor before repressor synthesis is blocked and the lytic cycle started (figure 17.18). The details of this competition are not yet completely clear, but it has been shown that a number of environmental factors influence the outcome of the race and the choice between the lytic and lysogenic pathways.

If the lambda repressor wins the race, the circular lambda DNA is inserted into the *E. coli* genome as first proposed by Alan Campbell. **Integration** or insertion is possible because the cII protein stimulates transcription of the *int* gene at the same time as that of the *cI* gene. The *int* gene codes for the synthesis of an **integrase** enzyme, and this protein becomes plentiful before lambda repressor turns off transcription. Lambda DNA has a phage attachment site (the *att* site) that can base pair with a bacterial attachment site located between the galactose or *gal* operon and the biotin operon on the *E. coli* chromosome. After these two sites match up, the integrase enzyme, with the aid of a special host protein, catalyzes the physical exchange of viral and bacterial DNA strands (figure 17.20). The circular lambda DNA is integrated into the *E. coli* DNA as a linear region next to the *gal* operon and is called a prophage. As can be seen in figure 17.20, the linear order of phage genes has been changed or permuted during integration.

The lambda prophage will leave the *E. coli* genome and begin the production of new phages when the host is unable to survive. The process is known as induction and is triggered by a drop in lambda repressor levels. Occasionally the repressor will spontaneously decline and the lytic cycle commence. However, induction usually is in response to environmental factors such as UV light or chemical mutagens that damage host DNA. This damage causes the recA protein,

which normally plays a role in genetic recombination in *E. coli* (*see chapter 13*), to act as a protease and cleave the repressor chain between the two domains. The separated domains cannot assemble to form the normal active repressor dimer, and the lytic cycle genes become active again. There is some recent evidence that activated recA protein may not directly cleave the repressor. RecA may instead bind to the lambda repressor and stimulate it to proteolytically cleave itself. An early gene located next to the *int* gene, the *xis* gene, codes for the synthesis of an **excisionase** protein that binds to the integrase and enables it to reverse the integration process and free the prophage (figure 17.20). The lytic cycle then proceeds normally.

Most temperate phages exist as integrated prophages in the lysogen. Nevertheless, integration is not an absolute requirement for lysogeny. The *E. coli* phage P1 is similar to lambda in that it circularizes after infection and begins to manufacture repressor. However, it remains as an independent circular DNA molecule in the lysogen and is replicated at the same time as the host chromosome. When *E. coli* divides, P1 DNA is apportioned between the daughter cells so that all lysogens contain one or two copies of the phage genome.

1. Define virulent phage, lysogeny, temperate phage, lysogen, prophage, immunity, and induction.
2. What advantages might a phage gain by being capable of lysogeny?
3. Describe lysogenic conversion and its significance.
4. Precisely how, in molecular terms, is a bacterial cell made lysogenic by a temperate phage like lambda?
5. How is a prophage induced to become active again?
6. Be able to describe the roles of the lambda repressor, cro protein, the recA protein, integrase, and excisionase in lysogeny and induction.
7. How does the temperate phage P1 differ from lambda phage?

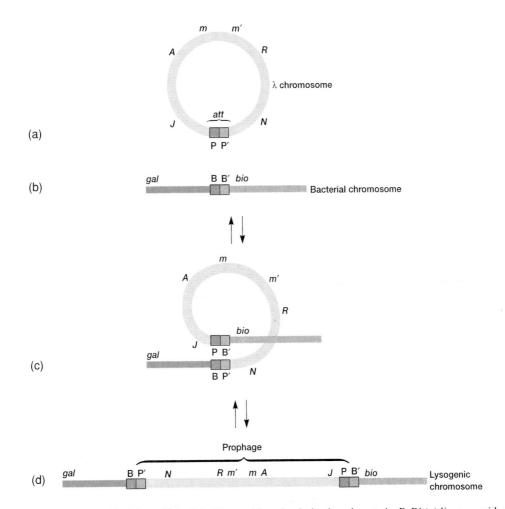

Figure 17.20   Reversible Insertion and Excision of Lambda Phage. After circularization, the *att* site P, P′ (*a*) lines up with a corresponding bacterial sequence B, B′ (*b*) and is integrated between the *gal* and *bio* operons to form the prophage, (*c*) and (*d*). If the process is reversed, the circular lambda chromosome will be restored and can then reproduce.

Summary

1. There are four major morphological groups of phages: tailless icosahedral phages, phages with contractile tails, those with noncontractile tails, and filamentous phages.

2. The lytic cycle of virulent bacteriophages is a life cycle that ends with host cell lysis and virion release.

3. The phage life cycle can be studied with a one-step growth experiment that is divided into an initial eclipse period within the latent period, and a rise period or burst.

4. The life cycle of the T4 phage of *E. coli* is composed of several phases. In the adsorption phase the phage attaches to a specific receptor site on the bacterial surface. This is followed by penetration of the cell wall and insertion of the viral nucleic acid into the cell.

5. Transcription of T4 DNA first produces early mRNA, which directs the synthesis of the protein factors and enzymes required to take control of the host and manufacture phage nucleic acids.

6. T4 DNA contains hydroxymethylcytosine (HMC) in place of cytosine, and glucose is often added to the HMC to protect the phage DNA from attack by host restriction enzymes.

7. T4 DNA replication produces concatemers, long strands of several genome copies linked together.

8. The replication of ssDNA phages proceeds through the formation of a double-stranded replicative form (RF).

9. Late mRNA is produced after DNA replication and directs the synthesis of capsid proteins, proteins involved in phage assembly, and those required for cell lysis and phage release.

10. Complete virions are assembled immediately after the separate components have been constructed.

11. T4, φX174, and many other phages are released upon lysis of the host cell. In contrast, the filamentous ssDNA phages are continually released without host cell lysis.

12. When ssRNA bacteriophage RNA enters a bacterial cell, it acts as a messenger and directs the synthesis of RNA replicase, which then produces double-stranded replicative forms and, subsequently, many +RNA copies.

13. The φ6 phage is the only dsRNA phage known. It is also unusual in having a membranous envelope.

14. Temperate phages, unlike virulent phages, often reproduce in synchrony with the host genome to yield a clone of virus-infected cells. This relationship is lysogeny, and the infected cell is called a lysogen. The latent form of the phage genome within the lysogen is the prophage.

15. Lysogeny is reversible, and the prophage can be induced to become active again and lyse its host.

16. A temperate phage may induce a change in the phenotype of its host cell that is not directly related to the completion of its life cycle. Such a change is called a conversion.

17. Two of the first proteins to appear after infection with lambda are the lambda repressor and the cro protein. The lambda repressor blocks the transcription of both cro protein and those proteins required for the lytic cycle, while the cro protein inhibits transcription of the lambda repressor gene.

18. There is a race between synthesis of lambda repressor and that of the cro protein. If the cro protein level rises high enough in time, lambda repressor synthesis is blocked and the lytic cycle initiated; otherwise, all genes other than the lambda repressor gene are repressed and the cell becomes a lysogen.

19. The final step in prophage formation is the insertion or integration of the lambda genome into the E. coli chromosome catalyzed by a special integrase enzyme.

20. Several environmental factors can lower repressor levels and trigger induction. The prophage becomes active and makes an excisionase protein that causes the integrase to reverse integration, free the prophage, and initiate a lytic cycle.

21. Although most temperate phages exist as integrated prophages, the phage P1 can lysogenize E. coli even though its DNA remains separate from the bacterial chromosome.

Key Terms

bacteriophages 354

burst size 355

concatemer 358

cro protein 365

early mRNA 357

eclipse period 356

excisionase 366

hydroxymethylcytosine (HMC) 358

induction 362

integrase 366

integration 366

lambda repressor 363

late mRNA 360

latent period 355

lysogen 362

lysogenic 362

lysogenic conversion 362

lysogeny 362

lytic cycle 354

one-step growth experiment 354

prophage 362

receptor sites 356

replicative form (RF) 360

restriction 358

restriction enzyme 358

rise period or burst 355

RNA replicase 362

temperate phage 362

virulent bacteriophage 362

Questions for Thought and Review

1. Explain why the T4 phage genome is circularly permuted.

2. Can you think of a way to simplify further the genomes of the ssRNA phages MS2 and Qß? Would it be possible to eliminate one of their genes? If so, which one?

3. No temperate RNA phages have yet been discovered. How might this absence be explained?

4. How might a bacterial cell resist phage infections? Give those mechanisms mentioned in the chapter and speculate on other possible strategies.

Additional Reading

Chapter 16 references also should be consulted, particularly the introductory and advanced texts.

General

Bradley, D. E. 1971. A comparative study of the structure and biological properties of bacteriophages. In *Comparative virology*, K. Maramorosch and E. Kurstak, editors, 207–53. New York: Academic Press.

Freifelder, D. 1987. *Molecular biology: A comprehensive introduction to prokaryotes and eukaryotes*, 2d ed. New York: Van Nostrand Reinhold.

Lewin, B. 1994. *Genes*, 5th ed. New York: Oxford University Press.

Maloy, S. R.; Cronan, Jr., J. E.; and Freifelder, D. 1994. *Microbial Genetics*, 2d ed. Boston: Jones and Bartlett.

Stent, G. S., and Calendar, R. 1978. *Molecular genetics: An introductory narrative*, 2d ed. San Francisco: W. H. Freeman.

Classification of Bacteriophages

Ackermann, H-W. 1987. Bacteriophage taxonomy in 1987. *Microbiol. Sci.* 4(7):214–18.

Reproduction of DNA Phages: The Lytic Cycle

Bazinet, C., and King, J. 1985. The DNA translocating vertex of dsDNA bacteriophage. *Ann. Rev. Microbiol.* 39:109–29.

Black, L. W. 1989. DNA packaging in dsDNA bacteriophages. *Ann. Rev. Microbiol.* 43:267–92.

Campbell, A. M. 1976. How viruses insert their DNA into the DNA of the host cell. *Sci. Am.* 235(6):103–13.

Fiddes, J. C. 1977. The nucleotide sequence of a viral DNA. *Sci. Am.* 237(6):54–67.

Karam, J. D., editor. 1994. *Molecular biology of bacteriophage T4*. Herndon, Va.: ASM Press.

Kellenberger, E. 1980. Control mechanisms governing protein interactions in assemblies. *Endeavour* 4(1):2–14.

Koerner, J. F., and Snustad, D. P. 1979. Shutoff of host macromolecular synthesis after T-even bacteriophage infection. *Microbiol. Rev.* 43(2):199–223.

Kruger, D. H., and Bickel, T. A. 1983. Bacteriophage survival: Multiple mechanisms for avoiding the deoxyribonucleic acid restriction systems of their hosts. *Microbiol. Rev.* 47(3):345–60.

Rabussay, D. 1982. Changes in *Escherichia coli* RNA polymerase after bacteriophage T4 infection. *ASM News* 48(9):398–403.

Russel, M. 1995. Moving through the membrane with filamentous phases. *Trans. Microbiol.* 3(6):223–28.

Wood, W. B., and Edgar, R. S. 1967. Building a bacterial virus. *Sci. Am.* 217(1):60–74.

Temperate Bacteriophages and Lysogeny

Murialdo, H. 1991. Bacteriophage lambda DNA maturation and packaging. *Ann. Rev. Biochem.* 60:125–53

Ptashne, M. 1992. *A genetic switch*, 2d ed. Cambridge, Mass.: Blackwell Scientific Publications.

Ptashne, M.; Johnson, A. D.; and Pabo, C. O. 1982. A genetic switch in a bacterial virus. *Sci. Am.* 247(5):128–40.

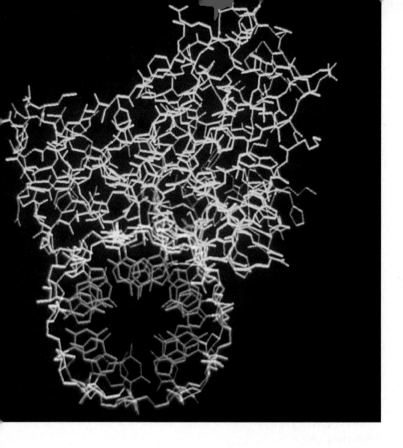

CHAPTER 18

The Viruses: Viruses of Eucaryotes

The Virus

Observe this virus: think

how small

Its arsenal, and yet how

loud its call;

It took my cell, now takes

your cell,

And when it leaves will take

our genes as well.

Genes that are master keys

to growth

That turn it on, or turn it

off, or both;

Should it return to me

or you

It will own the skeleton keys

to do

A number on our tumblers;

stage a coup.

—Michael Newman

Outline

Concepts

1. Although the details differ, animal virus reproduction is similar to that of the bacteriophages in having the same series of phases: adsorption, penetration and uncoating, replication of virus nucleic acids, synthesis and assembly of capsids, and virus release.

2. Viruses may harm their host cells in a variety of ways, ranging from direct inhibition of DNA, RNA, and protein synthesis to the alteration of plasma membranes and formation of inclusion bodies.

3. Not all animal virus infections have a rapid onset and relatively short duration. Some viruses establish long-term chronic infections; others are dormant for a while and then become active again. Slow virus infections may take years to develop.

4. Cancer can be caused by a number of factors, including viruses. Viruses may bring oncogenes into a cell, carry promoters that stimulate a cellular oncogene, or in other ways transform cells into tumor cells.

5. Plant viruses are responsible for many important diseases but have not been intensely studied due to technical difficulties. Most are RNA viruses. Insects are the most important transmission agents, and some plant viruses can even multiply in insect tissues before being inoculated into another plant.

6. Members of at least seven virus families infect insects; the most important belong to the *Baculoviridae, Reoviridae,* or *Iridoviridae.* Many insect infections are accompanied by the formation of characteristic inclusion bodies. A number of these viruses show promise as biological control agents for insect pests.

7. Infectious agents simpler than viruses also exist. Viroids are short strands of infectious RNA responsible for several plant diseases. Prions or virinos are somewhat mysterious proteinaceous particles associated with certain degenerative neurological diseases in humans and livestock.

In chapter 17 the bacteriophages are introduced in some detail because they are very important to the fields of molecular biology and genetics, as well as to virology. The present chapter focuses on viruses that use eucaryotic organisms as hosts. Although plant and insect viruses are discussed, particular emphasis is placed on animal viruses since they are very well studied and the causative agents of so many important human diseases. The chapter closes with a brief summary of what is known about infectious agents that are even simpler in construction than viruses: the viroids and prions.

The chapter begins with a discussion of animal viruses. These viruses are not only of great practical importance but are the best studied of the virus groups to be described in this chapter.

Classification of Animal Viruses

When microbiologists first began to classify animal viruses, they naturally thought in terms of features such as the host preferences of each virus. Unfortunately not all criteria are equally useful. For example, many viruses will infect a variety of animals, and a particular animal can be invaded by several dissimilar viruses. Thus virus host preferences lack the specificity to distinguish precisely among different viruses. Modern classifications are primarily based on virus morphology, the physical and chemical nature of virion constituents, and genetic relatedness.

Morphology is probably the most important characteristic in virus classification. Animal viruses can be studied with the transmission electron microscope while still in the host cell or after release. As mentioned in chapter 16, the nature of virus nucleic acids is also extremely important. Nucleic acid properties such as the general type (DNA or RNA), strandedness, size, and segmentation are all useful. Genetic relatedness can be estimated by techniques such as nucleic acid hybridization, nucleic acid and protein sequencing, and by determining the ability to undergo recombination.

A brief diagrammatic description of DNA and RNA animal virus classification is presented in figures 18.1 and 18.2. Figure 18.3 summarizes pictorially much of the same material.

---

1. List the most important characteristics used in identifying animal viruses (see figures 18.1, 18.2, and 18.3). What other properties can be used to establish the relatedness of different viruses?

Reproduction of Animal Viruses

The reproduction of animal viruses is very similar in many ways to that of phages. Animal virus reproduction may be divided into several stages: adsorption, penetration and uncoating, replication of virus nucleic acids, synthesis and assembly of virus capsids, and release of mature viruses. Each of these stages will be briefly described.

Adsorption of Virions

The first step in the animal virus cycle is adsorption to the host cell surface. This occurs through a random collision of the virion with a plasma membrane receptor site protein, frequently a glycoprotein (a protein with carbohydrate covalently attached). Because the capacity of a virus to infect a cell depends greatly on its ability to bind to the cell, the distribution of these receptor proteins plays a crucial role in the tissue and host specificity of animal viruses. For example, poliovirus receptors are found only in the human nasopharynx, gut, and spinal cord anterior horn cells; in contrast, measles virus receptors are present in most tissues. The dissimilarity in the distribution of receptors for these two viruses helps explain the difference in the nature of polio and measles.

The specific host cell receptor proteins to which viruses attach vary greatly but they are always surface proteins necessary to the cell. As will be discussed shortly, viruses often enter cells by endocytosis. They trick the host cell by attaching to surface molecules that are normally taken up by endocytosis. Thus they are passively carried into the cell. These host cell surface proteins usually are receptors that bind hormones and other important molecules essential to the cell's function and role in the body (table 18.1).

The surface site on the virus can consist simply of a capsid structural protein or an array of such proteins. In some viruses—for example, the poliovirus and rhinoviruses—the binding site is at the bottom of a surface depression or valley. The site can bind to host cell surface projections but cannot be reached by host antibodies. In other cases the virus attaches to the host cell through special projections such as the fibers extending from the corners of adenovirus icosahedrons (see figure 16.12h) or the spikes of enveloped viruses. As mentioned in chapter 16, the influenza virus has two kinds of spikes: hemagglutinin and neuraminidase (see figure 16.17a,b). The hemagglutinin spikes appear to be involved in attachment to the host cell receptor site and recognize sialic acid (N-acetylneuraminic acid). Influenza neuraminidase may aid the virus in penetrating nasal and respiratory tract secretions by degrading mucopolysaccharides; it also may directly aid in adsorption.

Human virus diseases (chapter 36).

Penetration and Uncoating

Viruses penetrate the plasma membrane and enter a host cell shortly after adsorption. Virus uncoating, the removal of the capsid and release of viral nucleic acid, occurs during or shortly after penetration. Both events will be reviewed together. The mechanisms of penetration and uncoating must vary with the type of virus because viruses differ so greatly in structure and mode of reproduction. For example, enveloped viruses may enter cells in a different way than naked or unenveloped virions. Furthermore, some viruses inject only their nucleic acid, whereas others must ensure that a virus-associated RNA or DNA polymerase also enters the

371

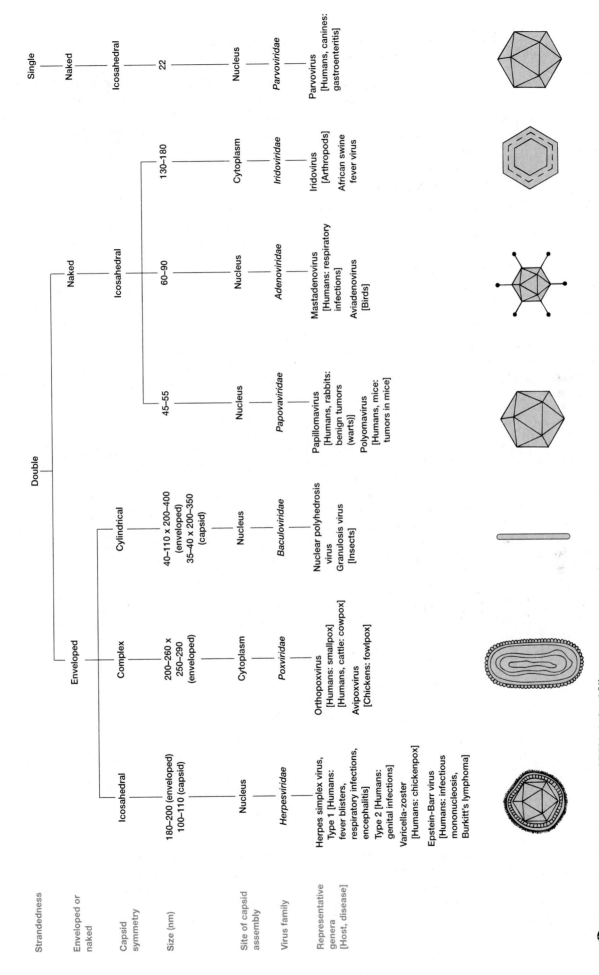

Figure 18.1 The Taxonomy of DNA Animal Viruses.

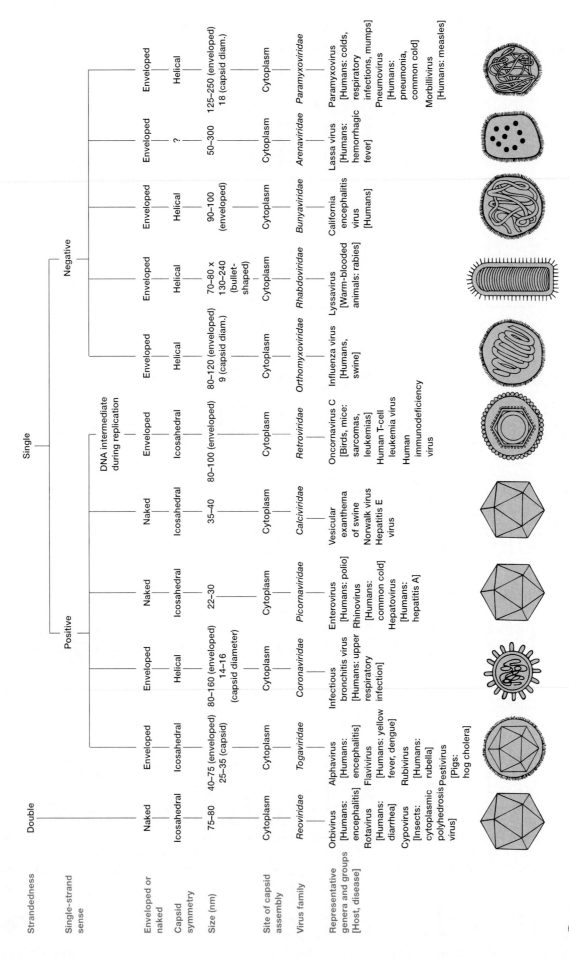

Strandedness	Double		Single								
Single-strand sense			Positive			DNA intermediate during replication	Negative				
Enveloped or naked	Naked		Enveloped	Naked	Naked	Enveloped	Enveloped	Enveloped	Enveloped	Enveloped	Enveloped
Capsid symmetry	Icosahedral		Icosahedral	Icosahedral	Icosahedral	Icosahedral	Helical	Helical	Helical	?	Helical
Size (nm)	75–80		40–75 (enveloped) 25–35 (capsid)	80–160 (enveloped) 14–16 (capsid diameter)	22–30	35–40	80–100 (enveloped)	80–120 (enveloped) 9 (capsid diam.)	70–80 x 130–240 (bullet-shaped)	90–100 (enveloped)	50–300	125–250 (enveloped) 18 (capsid diam.)
Site of capsid assembly	Cytoplasm		Cytoplasm	Cytoplasm	Cytoplasm	Cytoplasm	Cytoplasm	Cytoplasm	Cytoplasm	Cytoplasm	Cytoplasm
Virus family	*Reoviridae*		*Togaviridae*	*Coronaviridae*	*Picornaviridae*	*Calciviridae*	*Retroviridae*	*Orthomyxoviridae*	*Rhabdoviridae*	*Bunyaviridae*	*Arenaviridae* *Paramyxoviridae*
Representative genera and groups [Host, disease]	Orbivirus [Humans: encephalitis] Rotavirus [Humans: diarrhea] Cypovirus [Insects: cytoplasmic polyhedrosis virus]		Alphavirus [Humans: encephalitis] Flavivirus [Humans: yellow fever, dengue] Rubivirus [Humans: rubella] Pestivirus [Pigs: hog cholera]	Infectious bronchitis virus [Humans: upper respiratory infection]	Enterovirus [Humans: polio] Rhinovirus [Humans: common cold] Hepatovirus [Humans: hepatitis A]	Vesicular exanthema of swine Norwalk virus Hepatitis E virus	Oncornavirus C [Birds, mice: sarcomas, leukemias] Human T-cell leukemia virus Human immunodeficiency virus	Influenza virus [Humans, swine]	Lyssavirus [Warm-blooded animals: rabies]	California encephalitis virus [Humans]	Lassa virus [Humans: hemorrhagic fever]	Paramyxovirus [Humans: colds, respiratory infections, mumps] Pneumovirus [Humans: pneumonia, common cold] Morbillivirus [Humans: measles]

⊙ **Figure 18.2   The Taxonomy of RNA Animal Viruses.**

Figure 18.3   A Diagrammatic Description of the Families of Viruses that Infect Vertebrates.

TABLE 18.1	Examples of Host Cell Surface Proteins that Serve as Virus Receptors
Virus	**Cell Surface Protein**
Epstein-Barr virus	Receptor for the C3d complement protein on human B lymphocytes
Hepatitis A virus	Alpha 2-macroglobulin
Herpes simplex virus, type 1	Fibroblast growth factor receptor
Human immunodeficiency virus	CD4 protein on T-helper cells, macrophages, and monocytes
Influenza A virus	Sialic acid–containing glycoprotein
Measles virus	CD46 complement regulator protein
Poliovirus	Immunoglobulin superfamily
Rabies virus	Acetylcholine receptor on neurons
Rhinovirus	Intercellular adhesion molecules (ICAMs) on the surface of respiratory epithelial cells
Reovirus, type 3	β-adrenergic receptor
Rotavirus	Acetylated sialic acid on glycoprotein
Vaccinia virus	Epidermal growth factor receptor

host cell along with the virus genome. The entire process from adsorption to final uncoating may take from minutes to several hours.

The detailed mechanisms of penetration and uncoating are still unclear; it is possible that three different modes of entry may be employed (figure 18.4). At least some naked viruses such as the poliovirus undergo a major change in capsid structure on adsorption to the plasma membrane, so that only their nucleic acids are released in the cytoplasm. The envelope of paramyxoviruses, and possibly some other enveloped viruses, seems to fuse directly with the host cell plasma membrane, and the nucleocapsid is deposited in the cytoplasmic matrix, where uncoating is completed. A virus polymerase, associated with the nucleocapsid, then begins transcribing the virus RNA while it is still within the capsid. Most enveloped viruses may enter cells by a third route: through engulfment by receptor-mediated endocytosis to form **coated vesicles.** The virions attach to coated pits, specialized membrane regions coated on the cytoplasmic side with the protein clathrin. The pits then pinch off to form coated vesicles filled with viruses, and these fuse with lysosomes after the clathrin has been removed. Lysosomal enzymes

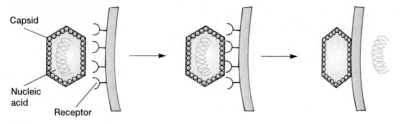

(a) Direct penetration by naked viruses

Capsid
Nucleic acid
Receptor

(b) Enveloped virus fusing with plasma membrane

Capsid protein
Spikes
Nucleic acid
Envelope
Nucleic acid
Capsid

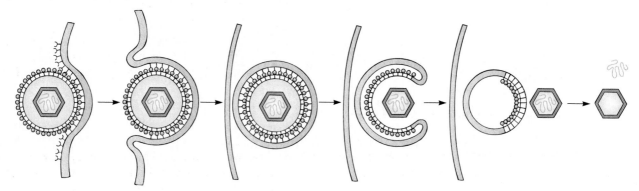

(c) Entry of enveloped virus by endocytosis

Figure 18.4 **Animal Virus Entry.** Mechanisms of animal virus attachment and entry into host cells. See text for description of the three entry modes.

may aid in virus uncoating and low endosomal pHs often trigger the uncoating process. In at least some instances, the virus envelope fuses with the lysosomal membrane, and the nucleocapsid is released into the cytoplasmic matrix (the capsid proteins may have been partially removed by lysosomal enzymes). Once in the cytoplasm, viral nucleic acid may be released from the capsid upon completion of uncoating or may function while still attached to capsid components.

Replication and Transcription in DNA Viruses

The early part of the synthetic phase, governed by the **early genes,** is devoted to taking over the host cell and to the synthesis of viral DNA and RNA. Some virulent animal viruses inhibit host cell DNA, RNA, and protein synthesis, though cellular DNA is not usually degraded. In contrast, nonvirulent viruses may actually stimulate the synthesis of host macromolecules.

DNA replication usually occurs in the host cell nucleus; poxviruses are exceptions since their genomes are replicated in the cytoplasm. Messenger RNA—at least early mRNA—is transcribed from DNA by host enzymes, except for poxvirus early mRNA, which is synthesized by a viral polymerase. Some examples of DNA virus reproduction will help illustrate these generalizations.

Parvoviruses, with a genome composed of one small, single-stranded DNA molecule about 4,800 bases in size, are the simplest of the DNA viruses (*see figure 16.12*a). The genome is so small that it directs the synthesis of only three polypeptides, all capsid components. Even so, the virus must resort to the use of overlapping genes to fit three genes into such a small molecule. That is, the base sequences that code for the three polypeptide chains overlap each other and are read with different reading frames (*see chapter 13*). Since the genome does not code for any enzymes, the virus must use host cell enzymes for all

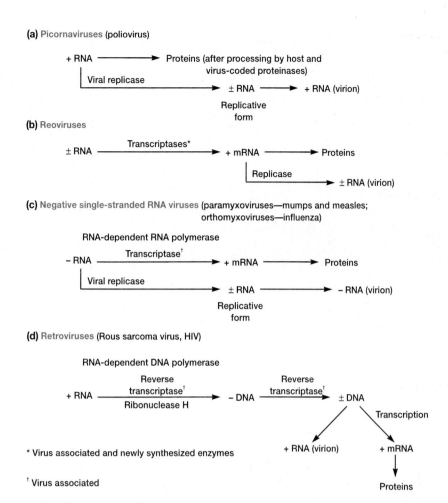

(a) Picornaviruses (poliovirus)

(b) Reoviruses

(c) Negative single-stranded RNA viruses (paramyxoviruses—mumps and measles; orthomyxoviruses—influenza)

(d) Retroviruses (Rous sarcoma virus, HIV)

* Virus associated and newly synthesized enzymes

† Virus associated

Figure 18.5   RNA Animal Virus Reproductive Strategies.

biosynthetic processes. Thus viral DNA can only be replicated in the nucleus during the S period of the cell cycle, when the cell replicates its own DNA.

The genetic code and gene structure (pp. 255–61).

Herpesviruses are a large group of icosahedral, enveloped, double-stranded DNA viruses responsible for many important human and animal diseases (*see figure 16.17*e). The genome is a linear piece of DNA about 160,000 base pairs in size and contains at least 50 to 100 genes. Immediately upon uncoating, the DNA is transcribed by host RNA polymerase to form messengers directing the synthesis of several early proteins, principally regulatory proteins and the enzymes required for virus DNA replication. DNA replication with a virus-specific DNA polymerase begins in the cell nucleus within 4 hours after infection. Host DNA synthesis gradually slows during a lethal virus infection (not all herpes infections result in immediate cell death).

Herpesviruses and disease (pp. 730–33).

Poxviruses such as the vaccinia virus are the largest viruses known and are morphologically complex (*see figure 16.18*). Their double-stranded DNA possesses over 200 genes. The virus enters through receptor-mediated endocytosis in coated

vesicles; the central core escapes from the lysosome and enters the cytoplasmic matrix. The core contains both DNA and a DNA-dependent RNA polymerase that synthesizes early mRNAs, one of which directs the production of an enzyme that completes virus uncoating. DNA polymerase and other enzymes needed for DNA replication are also synthesized early in the reproductive cycle, and replication begins about 1.5 hours after infection. About the time DNA replication commences, late mRNA transcription is initiated. Many late proteins are structural proteins used in capsid construction. The complete reproductive cycle in poxviruses takes about 24 hours.

Poxvirus diseases (p. 720).

Replication and Transcription in RNA Viruses

The RNA viruses are much more diverse in their reproductive strategies than are the DNA viruses. Most RNA viruses can be placed in one of four general groups based on their modes of replication and transcription, and their relationship to the host cell genome. Figure 18.5 summarizes the reproductive cycles characteristic of these groups. Transcription patterns are discussed first, and then mechanisms of RNA replication.

(a)   (b)   (c)

Figure 18.6   Reverse Transcriptase and Ribonuclease H. (*a*) A computer-generated model of reverse transcriptase bound to a nucleic acid complex. Parts *b* and *c* show a model of *E. coli* ribonuclease H complexed with the RNA-DNA hybrid. (*b*) A side view showing the protein (yellow) complexed with the nucleic acid (DNA backbone in lavender, RNA in pink, bases in blue). (*c*) Top view of complex.

Transcription in RNA viruses other than the retroviruses (retroviruses are considered shortly) varies with the nature of the virus genome. The picornaviruses such as poliovirus are the best studied positive strand ssRNA viruses. They use their RNA genome as a giant mRNA, and host ribosomes synthesize an enormous peptide that is then cleaved or processed by both host and viral encoded enzymes to form the proper polypeptides (figure 18.5*a*). In contrast, because their genome is complementary to the mRNA base sequence, negative strand ssRNA viruses (e.g., the orthomyxoviruses and paramyxoviruses) must employ a virus-associated RNA-dependent RNA polymerase or **transcriptase** to synthesize mRNA (figure 18.5*c*). The dsRNA reoviruses carry a virus-associated transcriptase that copies the negative strand of their genome to generate mRNA. Later a new, virus-encoded polymerase continues transcription.

The nature of RNA replication also varies logically with the type of virus genetic material. Viral RNA is replicated in the host cell cytoplasmic matrix. Single-stranded RNA viruses, except retroviruses, use a viral **replicase** that converts the ssRNA into a double-stranded RNA called the **replicative form** (figure 18.5*a,c*). The appropriate strand of this intermediate then directs the synthesis of new viral RNA genomes. Reoviruses differ significantly from this pattern (figure 18.5*b*). The virion contains 10 to 13 different dsRNAs, each coding for an mRNA. Late in the reproductive cycle, a copy of each mRNA associates with the other mRNAs and special proteins to form a large complex. The RNAs in this complex are then copied by the viral replicase to form a double-stranded genome that is incorporated into a new virion.

Retroviruses such as the human immunodeficiency virus (*see figure 16.17*d) possess ssRNA genomes but differ from other RNA viruses in that they synthesize mRNA and replicate their genome by means of DNA intermediates. The virus has an RNA-dependent DNA polymerase or **reverse transcriptase** that copies the +RNA genome to form a –DNA copy (figures 18.5*d* and 18.6*a*). Interestingly, transfer RNA is carried by the virus and serves as the primer required for nucleic acid synthesis (*see chapter 11*). The transformation of RNA into DNA takes place in two steps. First, reverse transcriptase copies the +RNA to form a RNA-DNA hybrid. Then the **ribonuclease H** component of reverse transcriptase degrades the +RNA strand to leave –DNA (figure 18.6*b, c*). After synthesizing –DNA, the reverse transcriptase copies this strand to produce a double-stranded DNA called **proviral DNA,** which can direct the synthesis of mRNA and new +RNA virion genome copies. Notice that during this process genetic information is transferred from RNA to DNA rather than in the normal direction.

The reproduction of retroviruses is remarkable in other ways as well. After proviral DNA has been manufactured, it is converted to a circular form and incorporated or integrated into the host cell chromosome. Virus products are only formed after integration. Sometimes these integrated viruses can change host cells into tumor cells.

Biology of the HIV virus and AIDS (pp. 722–30).

Synthesis and Assembly of Virus Capsids

Some **late genes** direct the synthesis of capsid proteins, and these spontaneously self-assemble to form the capsid just as in bacteriophage morphogenesis. Recently the self-assembly process has been dramatically demonstrated. The addition of

TABLE 18.2	**Intracellular Sites of Animal Virus Reproduction**	

Virus	Nucleic Acid Replication	Capsid Assembly
DNA Viruses		
Parvoviruses	Nucleus	Nucleus
Papovaviruses	Nucleus	Nucleus
Herpesviruses	Nucleus	At nuclear membrane
Adenoviruses	Nucleus	Nucleus
Poxviruses	Cytoplasm	Cytoplasm
RNA Viruses		
Reoviruses	Cytoplasm	Cytoplasm
Togaviruses	Cytoplasm	At membranes
Picornaviruses	Cytoplasm	Cytoplasm
Paramyxoviruses	Cytoplasm	At membranes
Orthomyxoviruses	Nucleus	At membranes
Rhabdoviruses	Cytoplasm	At membranes
Retroviruses	Cytoplasm and nucleus	At plasma membrane

Figure 18.7   **Paracrystalline Clusters.** A crystalline array of adenovirus within the cell nucleus (×35,000).

poliovirus RNA to an extract prepared from uninfected human cells (HeLa cells) results in the formation of new, infectious poliovirus virions. It appears that during icosahedral virus assembly empty **procapsids** are first formed; then the nucleic acid is inserted in some unknown way. The site of morphogenesis varies with the virus (table 18.2). Large paracrystalline clusters of complete virions or procapsids are often seen at the site of virus maturation (figure 18.7). The assembly of enveloped virus capsids is generally similar to that of naked virions, except for poxviruses. These are assembled in the cytoplasm by a lengthy, complex process that begins with the enclosure of a portion of the cytoplasmic matrix through construction of a new membrane. Then newly synthesized DNA condenses, passes through the membrane, and moves to the center of the immature virus. Nucleoid and elliptical body construction takes place within the membrane.

Virion Release

Mechanisms of virion release differ between naked and enveloped viruses. Naked virions appear to be released most often by host cell lysis. In contrast, the formation of envelopes and the release of enveloped viruses are usually concurrent processes, and the host cell may continue virion release for some time. First, virus-encoded proteins are incorporated into the plasma membrane. Then the nucleocapsid is simultaneously released and the envelope formed by membrane budding (figures 18.8 and 18.9). Although most envelopes arise from an altered plasma membrane, in herpesviruses, budding and envelope formation usually involves the nuclear envelope.

1. Describe in detail each stage in animal virus reproduction and contrast it with the same stage in the bacteriophage life cycle (*see chapter 17*).
2. What probably plays the most important role in determining the tissue and host specificity of viruses?
3. Discuss how animal viruses can enter host cells.
4. Describe in general terms the replication and transcription processes that take place during animal virus reproduction.
5. Summarize the ways in which picornaviruses, −ssRNA viruses, and reoviruses carry out transcription and replication.
6. Outline in detail the life cycle of the retroviruses. What is proviral DNA?
7. In what two ways are animal viruses released from their host cells?
8. Define the following terms: uncoating, coated vesicles, overlapping genes, replicative form, reverse transcriptase, and late genes.

Cytocidal Infections and Cell Damage

An infection that results in cell death is a **cytocidal infection.** Animal viruses can harm their host cells in many ways; often this leads to cell death. As mentioned earlier (*see chapter 16*), microscopic or macroscopic degenerative changes or abnormalities in

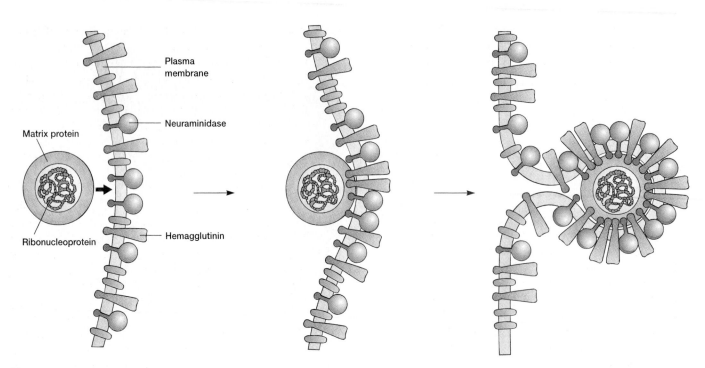

⊗ **Figure 18.8   Release of Influenza Virus by Plasma Membrane Budding.** First, viral envelope proteins (hemagglutinin and neuraminidase) are inserted into the host plasma membrane. Then the nucleocapsid approaches the inner surface of the membrane and binds to it. At the same time viral proteins collect at the site and host membrane proteins are excluded. Finally, the plasma membrane buds to simultaneously form the viral envelope and release the mature virion.

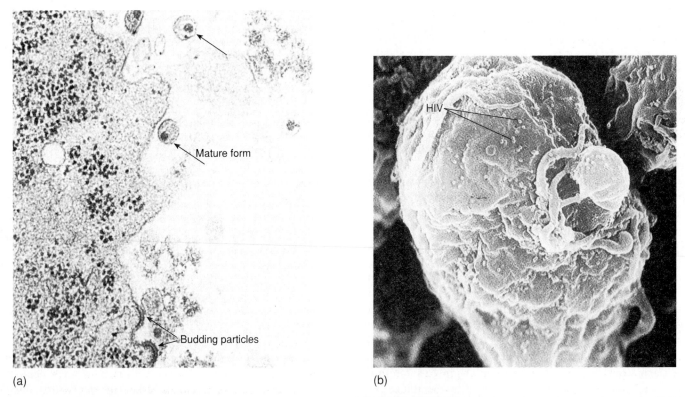

Figure 18.9   Human Immunodeficiency Virus (HIV) Release by Plasma Membrane Budding. (*a*) A transmission electron micrograph of HIV particles beginning to bud, as well as some mature particles. (*b*) A scanning electron micrograph view of HIV particles budding from a lymphocyte.

host cells and tissues are referred to as cytopathic effects (CPEs), and these often result from virus infections. Seven possible mechanisms of host cell damage are briefly described here.

1. Many viruses can inhibit host DNA, RNA, and protein synthesis. Cytocidal viruses (e.g., picornaviruses, herpesviruses, and adenoviruses) are particularly active in this regard. The mechanisms of inhibition are not yet clear.

2. Cell lysosomes may be damaged, resulting in the release of hydrolytic enzymes and cell destruction.

3. Virus infection can drastically alter plasma membranes through the insertion of virus-specific proteins so that the infected cells are attacked by the immune system. When infected by viruses such as herpesviruses and measles virus, as many as 50 to 100 cells may fuse into one abnormal, giant, multinucleate cell called a polykaryocyte. The HIV virus, which causes AIDS (*see chapter 36*), appears to destroy CD4$^+$ T-helper cells at least partly through its effects on the plasma membrane.

4. High concentrations of proteins from several viruses (e.g., mumps virus and influenza virus) can have a direct toxic effect on cells and organisms.

5. Intracellular structures called **inclusion bodies** are formed during many virus infections. These may result from the clustering of subunits or virions within the nucleus or cytoplasm (e.g., the Negri bodies in rabies infections); they also may contain cell components such as ribosomes (arenavirus infections) or chromatin (herpesviruses). Regardless of their composition, these inclusion bodies can directly disrupt cell structure.

6. Chromosomal disruptions result from infections by herpesviruses and others.

7. Finally, the host cell may not be directly destroyed but transformed into a malignant cell. This will be discussed later in the chapter.

It should be emphasized that more than one of these mechanisms may be involved in a cytopathic effect.

---

1. Outline the ways in which viruses can damage host cells during cytocidal infections.

Persistent, Latent, and Slow Virus Infections

Many virus infections (e.g., influenza) are **acute infections**—that is, they have a fairly rapid onset and last for a relatively short time. However, some viruses can establish **persistent** or **chronic infections** lasting many years. The viruses may reproduce at a slow rate without causing disease symptoms; antibodies to the virus will be found in the blood. Examples of this phenomenon are hepatitis B virus (serum hepatitis), the Epstein-Barr virus, and measles virus. In **latent virus infections** the virus stops reproducing and remains dormant for a period

before becoming active again. During latency, no symptoms, antibodies, or viruses are detectable. Herpes simplex type 1 virus often infects children and then becomes dormant within the nervous system ganglia; years later it can be activated to cause cold sores. The varicella-zoster virus causes chickenpox in children and then, after years of inactivity, may produce the skin disease shingles. These and other examples of such infections are discussed in more detail in chapter 36.

The causes of persistence and latency are probably multiple, although the precise mechanisms are still unclear. The virus genome may be integrated into the host genome. Viruses may become less antigenic and thus less susceptible to attack by the immune system. They may mutate to less virulent and slower reproducing forms. Sometimes a deletion mutation (*see chapter 13*) produces a **defective interfering (DI) particle** that cannot reproduce but slows normal virus reproduction, thereby reducing host damage and establishing a chronic infection.

A small group of viruses cause extremely slowly developing infections, often called **slow virus diseases,** in which symptoms may take years to emerge. Measles virus occasionally produces a slow infection. A child may have a normal case of measles, then 5 to 12 years later develop a degenerative brain disease called subacute sclerosing panencephalitis (SSPE). Many slow viruses may not be normal viruses at all. Several neurological diseases of animals and humans develop slowly. The best studied of these diseases are scrapie, a disease of sheep and goats, and the human diseases kuru, fatal familial insomnia, and the Creutzfeldt-Jakob disease. All are thought to be caused by simple, nonviral agents called prions, which are discussed later in the text.

---

1. Define the following: acute infection, persistent or chronic infection, latent virus infection, and slow virus disease.

2. Why might an infection be chronic or latent?

Viruses and Cancer

Cancer [Latin *cancer,* crab] is one of our most serious medical problems and the focus of an immense amount of interest and research. A **tumor** [Latin *tumere,* to swell] is a growth or lump of tissue resulting from **neoplasia** or abnormal new cell growth and reproduction due to a loss of regulation. Tumor cells have aberrant shapes and altered plasma membranes that contain distinctive tumor antigens. They may invade surrounding tissues to form unorganized cell masses. They often lose the specialized metabolic activities characteristic of differentiated tissue cells and rely greatly upon anaerobic metabolism. This reversion to a more primitive or less differentiated state is called **anaplasia.**

There are two major types of tumors with respect to overall form or growth pattern. If the tumor cells remain in place to form a compact mass, the tumor is benign. In contrast, malignant or cancerous tumor cells can actively spread throughout the body in a process known as **metastasis,** often by floating in the blood and establishing secondary tumors. Some cancers are not solid, but cell suspensions. For example, leukemias are composed of malignant white blood cells that circulate throughout the body. Indeed, dozens of kinds of cancers arise from a variety of cell types and afflict all kinds of organisms.

As one might expect from the wide diversity of cancers, there are many causes of cancer, only some of which are directly related to viruses. Possibly as many as 30 to 60% of cancers may be related to diet. Many chemicals in our surroundings are carcinogenic and may cause cancer by inducing gene mutations or interfering with normal cell differentiation.

The Ames test for carcinogens (pp. 270–71).

Carcinogenesis is a complex, multistep process. It can be initiated by a chemical, usually a mutagen, but a cancer does not appear to develop until at least one more triggering event (possibly exposure to another chemical carcinogen or a virus) takes place. Often, as is discussed later, cancer-causing genes, or **oncogenes,** are directly involved and may come from the cell itself or be contributed by a virus. Many of these oncogenes seem to be involved in the regulation of cell growth and differentiation; for example, some code for all or part of growth factors that regulate cell growth. It may be that various cancers arise through different combinations of causes. Not surprisingly the chances of developing cancer rise with age because an older person will have been exposed to carcinogens and other causative factors for a longer time. Immune surveillance and destruction of cancer cells (*see chapter 31*) also may be less effective in older people.

Although viruses are known to cause many animal cancers, it is very difficult to prove that this is the case with human cancers since indirect methods of study must be used. One tries to find virus particles and components within tumor cells, using techniques such as electron microscopy, immunologic tests, and enzyme assays. Attempts are also made to isolate suspected cancer viruses by cultivation in tissue culture or other animals. Sometimes a good correlation between the presence of a virus and cancer can be detected.

At present, viruses have been implicated in the genesis of at least six human cancers. (1) The Epstein-Barr virus (EBV) is one of the best-studied human cancer viruses. EBV is a herpesvirus and the cause of two cancers. Burkitt's lymphoma is a malignant tumor of the jaw and abdomen found in children of central and western Africa. EBV also causes nasopharyngeal carcinoma. Both the virus particles and the EBV genome have been found within tumor cells; Burkitt's lymphoma patients also have high blood levels of antibodies to EBV. Interestingly there is some reason to believe that a person also must have had malaria to develop Burkitt's lymphoma. Environmental factors must play a role, because EBV does not cause much cancer in the United States despite its prevalence. Possibly this is due to a low incidence of malaria in the United States. (2) Hepatitis B virus appears to be associated with one form of liver cancer (hepatocellular carcinoma) and can be integrated into the human genome. (3) Some strains of human papillomaviruses have been linked to cervical cancer. (4) At least two retroviruses, the human T-cell lymphotropic virus I (HTLV-1) and HTLV-2, seem able to cause cancer, adult T-cell leukemia, and hairy-cell leukemia, respectively (*see figure 36.18*). Other retrovirus-associated cancers may well be discovered in the future.

It appears that viruses can cause cancer in several ways. They may bring oncogenes into a cell and insert them into its genome. Rous sarcoma virus (a retrovirus) carries an *src* gene that codes for tyrosine kinase. This enzyme is located mainly in the plasma membrane and phosphorylates the amino acid tyrosine in several cellular proteins. It is thought that this alters cell growth and behavior. Since the activity of many proteins is regulated by phosphorylation and several other oncogenes also code for protein kinases, many cancers may result at least partly from altered cell regulation due to changes in kinase activity. The human T-cell lymphotropic viruses, HTLV-1 and HTLV-2, seem to transform T cells (*see chapter 31*) by producing a regulatory protein that sometimes activates genes involved in cell division as well as stimulating virus reproduction. Some oncogenic viruses carry one or more very effective promoters or enhancers (*see chapter 11*). If these viruses integrate themselves next to a cellular oncogene, the promoter or enhancer will stimulate its transcription, leading to cancer. In this case the oncogene might be necessary for normal cell growth (possibly it codes for a regulatory protein) and only causes cancer when it functions too rapidly or at the wrong time. For example, some chicken retroviruses induce lymphomas when they are integrated next to the *c-myc* cellular oncogene, which codes for a protein that appears to be involved in the induction of either DNA or RNA synthesis. With the possible exception of HTLV-1, it is not yet known how the viruses associated with human cancers actually aid in cancer development.

1. What are the major characteristics of cancer?
2. How might viruses cause cancer? Are there other ways in which a malignancy might develop?
3. Define the following terms: tumor, neoplasia, anaplasia, metastasis, and oncogene.

Plant Viruses

Although it has long been recognized that viruses can infect plants and cause a variety of diseases (*see figure 16.5*), plant viruses generally have not been as well studied as bacteriophages and animal viruses. This is mainly because they are often difficult to cultivate and purify. Some viruses, such as tobacco mosaic virus (TMV), can be grown in isolated protoplasts of plant cells just as phages and some animal viruses are cultivated in cell suspensions. However, many cannot grow in protoplast cultures and must be inoculated into whole plants or tissue preparations. Many plant viruses require insect vectors for transmission; some of these can be grown in monolayers of cell cultures derived from aphids, leafhoppers, or other insects.

Virion Morphology

The essentials of capsid morphology are outlined in chapter 16 and apply to plant viruses since they do not differ significantly in construction from their animal virus and phage relatives. Many have either rigid or flexible helical capsids (tobacco mosaic virus, *see figure 16.11*). Others are icosahedral or have modified the icosahedral pattern with the addition of extra capsomers (turnip yellow mosaic virus, figure 18.10). Most capsids seem composed of one type of protein; no specialized attachment proteins have been detected. Almost all plant viruses are RNA viruses, either single stranded or double stranded (*see tables 16.1 and 16.2*). Caulimoviruses and geminiviruses with their DNA genomes are exceptions to this rule.

Plant Virus Taxonomy

The shape, size, and nucleic acid content of many plant virus groups are summarized in figure 18.11. Like other types of viruses, they are classified according to properties such as nucleic acid type and strandedness, capsid symmetry and size, and envelope presence (*see table 16.2*).

Plant Virus Reproduction

Since tobacco mosaic virus (TMV) has been studied the most extensively, its reproduction will be briefly described. The replication of virus RNA is an essential part of the reproduction

Figure 18.10   Turnip Yellow Mosaic Virus (TYMV). An RNA plant virus with icosahedral symmetry.

Figure 18.11   A Diagrammatic Description of Families and Groups of Viruses that Infect Plants.

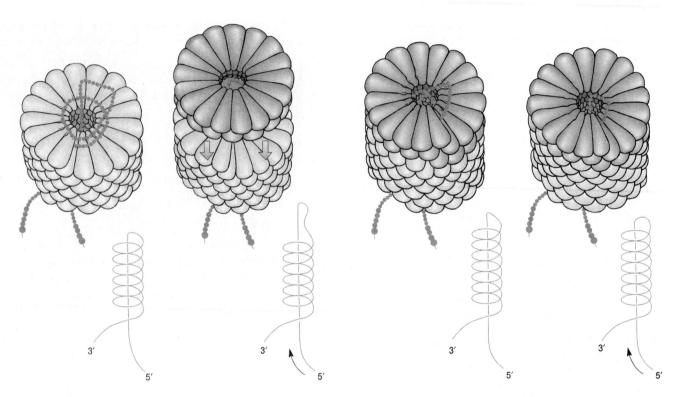

Figure 18.12   TMV Assembly. The elongation phase of tobacco mosaic virus nucleocapsid construction. The lengthening of the helical capsid through the addition of a protein disk to its end is shown in a sequence of four illustrations; line drawings depicting RNA behavior are included. The RNA genome inserts itself through the hole of an approaching disk and then binds to the groove in the disk as it locks into place at the end of the cylinder.

process. Most plants contain RNA-dependent RNA polymerases, and it is possible that these normal constituents replicate the virus RNA. However, some plant virus genomes (e.g., turnip yellow virus and cowpea mosaic virus) appear to be copied by a virus-specific RNA replicase. Possibly TMV RNA is also replicated by a viral RNA polymerase, but the evidence is not clear on this matter. Four TMV-specific proteins, one of them the coat protein, are known to be made. Although TMV RNA is plus single-stranded RNA and could directly serve as mRNA, the production of messenger is complex. The coat protein mRNA has the same sequence as the 3′ end of the TMV genome and arises from it by some sort of intracellular processing.

After the coat protein and RNA genome have been synthesized, they spontaneously assemble into complete TMV virions in a highly organized process (figure 18.12). The protomers (*see chapter 16*) come together to form disks composed of two layers of protomers arranged in a helical spiral. Association of coat protein with TMV RNA begins at a special assembly initiation site close to the 3′ end of the genome. The helical capsid grows by the addition of protomers, probably as disks, to the end of the rod. As the rod lengthens, the RNA passes through a channel in its center and forms a loop at the growing end. In this way the RNA can easily fit as a spiral into the interior of the helical capsid.

Reproduction within the host depends on the virus's ability to spread throughout the plant. Viruses can move long distances

through the plant vasculature; usually they travel in the phloem. The spread of plant viruses in nonvascular tissue is hindered by the presence of tough cell walls. Nevertheless, a virus such as TMV does spread slowly, about 1 mm/day or less. It moves from cell to cell through the plasmodesmata. These are slender cytoplasmic strands extending through holes in adjacent cell walls that join plant cells by narrow bridges. Special viral "movement proteins" are required for movement between cells. The TMV movement protein accumulates in the plasmodesmata, but the way in which it promotes virus movement is not understood.

Several cytological changes can take place in TMV-infected cells. Plant virus infections often produce microscopically visible intracellular inclusions, usually composed of virion aggregates, and hexagonal crystals of almost pure TMV virions sometimes do develop in TMV-infected cells (figure 18.13). The host cell chloroplasts become abnormal and often degenerate, while new chloroplast synthesis is inhibited.

Transmission of Plant Viruses

Since plant cells are protected by cell walls, plant viruses have a considerable obstacle to overcome when trying to establish themselves in a host. TMV and a few other viruses may be carried by the wind or animals and then enter when leaves are mechanically damaged. Some plant viruses are transmitted through contaminated seeds, tubers, or pollen. Soil nematodes

(a)

(b)

Figure 18.13   Intracellular TMV. (*a*) A crystalline mass of tobacco mosaic virions from a 10-day-old lesion in a *Chenopodium amaranticolor* leaf. (*b*) Freeze-fracture view of a crystalline mass of tobacco mosaic virions in an infected leaf cell. In both views the particles can be seen longitudinally and in cross section.

can transmit viruses (e.g., the tobacco ringspot virus) while feeding on roots. Tobacco necrosis virus is transmitted by parasitic fungi. However, the most important agents of transmission are insects that feed on plants, particularly sucking insects such as aphids and leafhoppers.

Insects transmit viruses in several ways. They may simply pick up viruses on their mouth parts while feeding on an infected plant, then directly transfer the viruses to the next plant they visit. Viruses may be stored in an aphid's foregut; the aphid will infect plants when regurgitating while it is feeding. Several plant viruses—for example, the wound tumor virus—

can multiply in leafhopper tissues before reaching the salivary glands and being inoculated into plants (that is, it uses both insects and plants as hosts).

1. Why have plant viruses not been as well studied as animal and bacterial viruses?
2. Describe in molecular terms the way in which TMV is reproduced.
3. How are plant viruses transmitted between hosts?

Viruses of Fungi and Algae

Most of the fungal viruses or mycoviruses isolated from higher fungi such as *Penicillium* and *Aspergillus* (*see chapter 25*) contain double-stranded RNA and have isometric capsids (that is, their capsids are roughly spherical polyhedra), which are approximately 25 to 50 nm in diameter. Many appear to be latent viruses. Some mycoviruses do induce disease symptoms in hosts such as the mushroom *Agaricus bisporus,* but cytopathic effects and toxic virus products have not yet been observed.

Much less is known about the viruses of lower fungi, and only a few have been examined in any detail. Both dsRNA and dsDNA genomes have been found; capsids usually are isometric or hexagonal and vary in size from 40 to over 200 nm. Unlike the situation in higher fungi, virus reproduction is accompanied by host cell destruction and lysis.

Viruses have been detected during ultrastructural studies of eucaryotic algae, but few have been isolated. Those that have been studied have dsDNA genomes and polyhedral capsids. One virus of the green alga *Uronema gigas* resembles many bacteriophages in having a tail.

1. Describe the major characteristics of the viruses that infect higher fungi, lower fungi, and algae. In what ways do they seem to differ from one another?

Insect Viruses

Members of at least seven virus families (*Baculoviridae, Iridoviridae, Poxviridae, Reoviridae, Parvoviridae, Picornaviridae,* and *Rhabdoviridae*) are known to infect insects and reproduce or even use them as the primary host (*see table 16.2*). Of these, probably the three most important are the *Baculoviridae, Reoviridae,* and *Iridoviridae.*

The *Iridoviridae* are icosahedral viruses with lipid in their capsids and a linear double-stranded DNA genome. They are responsible for the iridescent virus diseases of the crane fly and some beetles. The group's name comes from the observation that larvae of infected insects can have an iridescent coloration due to the presence of crystallized virions in their fat bodies.

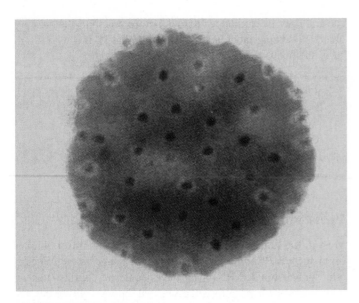

Figure 18.14 Inclusion Bodies. A section of a cytoplasmic polyhedron from a gypsy moth (*Lymantria dispar*). The occluded virus particles with dense cores are clearly visible (×50,000).

Many insect virus infections are accompanied by the formation of inclusion bodies within the infected cells. Granulosis viruses form granular protein inclusions, usually in the cytoplasm. Nuclear polyhedrosis and cytoplasmic polyhedrosis virus infections produce polyhedral inclusion bodies in the nucleus or the cytoplasm of affected cells. Although all three types of viruses generate inclusion bodies, they belong to two distinctly different families. The cytoplasmic polyhedrosis viruses are reoviruses; they are icosahedral with double shells and have double-stranded RNA genomes. Nuclear polyhedrosis viruses and granulosis viruses are baculoviruses—rod-shaped, enveloped viruses of helical symmetry and with double-stranded DNA.

The inclusion bodies, both polyhedral and granular, are protein in nature and enclose one or more virions (figure 18.14). Insect larvae are infected when they feed on leaves contaminated with inclusion bodies. Polyhedral bodies protect the virions against heat, low pH, and many chemicals; the viruses can remain viable in the soil for years. However, when exposed to alkaline insect gut contents, the inclusion bodies dissolve to liberate the virions, which then infect midgut cells. Some viruses remain in the midgut while others spread throughout the insect. Just as with bacterial and vertebrate viruses, insect viruses can persist in a latent state within the host for generations while producing no disease symptoms. A reappearance of the disease may be induced by chemicals, thermal shock, or even a change in the insect's diet.

Much of the current interest in insect viruses arises from their promise as biological control agents for insect pests (*see chapter 44*). Many people hope that some of these viruses may partially replace the use of toxic chemical pesticides. Baculoviruses have received the most attention for at least three reasons. First, they attack only invertebrates and have considerable host specificity; this means that they should be fairly safe for nontarget organisms. Second, because they are encased in protective inclusion bodies, these viruses have a good shelf life and better viability when dispersed in the environment. Finally, they are well suited for commercial production since they often reach extremely high concentrations in larval tissue (as high as 10^{10} viruses per larva). The use of nuclear polyhedrosis viruses for the control of the cotton bollworm, Douglas fir tussock moth, gypsy moth, alfalfa looper, and European pine sawfly has either been approved by the Environmental Protection Agency or is being considered. The granulosis virus of the codling moth also is useful. Usually inclusion bodies are sprayed on foliage consumed by the target insects. More sensitive viruses are administered by releasing infected insects to spread the disease.

1. Summarize the nature of granulosis, nuclear polyhedrosis, and cytoplasmic polyhedrosis viruses and the way in which they are transmitted by inclusion bodies.
2. What are baculoviruses and why are they so promising as biological control agents for insect pests?

Viroids and Prions

Although some viruses are exceedingly small and simple, even simpler infectious agents exist. Over 16 different plant diseases—for example, potato spindle-tuber disease, exocortis disease of citrus trees, and chrysanthemum stunt disease—are caused by a group of infectious agents called **viroids.** These are circular, single-stranded RNAs, about 250 to 370 nucleotides long (figures 18.15 and 18.16), that can be transmitted between plants through mechanical means or by way of pollen and ovules and are replicated in their hosts. The single-stranded RNA normally exists as a closed circle collapsed into a rodlike shape by intrastrand base pairing. Viroids are found principally in the nucleolus of infected cells; between 200 and 10,000 copies may be present. They do not act as mRNAs to direct protein synthesis, and it is not yet known how they cause disease symptoms. A plant may be infected without showing symptoms—that is, it may have a latent infection. The same viroid, when in another species, might well cause a severe disease. Although viroids could be replicated by an RNA-dependent RNA polymerase, they appear to be synthesized from RNA templates by a host RNA polymerase that mistakes them for a piece of DNA. A rolling-circle type mechanism seems to be involved (p. 218).

The potato spindle-tuber disease agent (PSTV) has been most intensely studied. Its RNA is about 130,000 daltons or 359 nucleotides in size, much smaller than any virus genome.

Several PSTV strains have been isolated, ranging in virulence from ones that cause only mild symptoms to lethal varieties. All variations in pathogenicity are due to a few nucleotide changes in two short regions on the viroid. It is believed that these sequence changes alter the shape of the rod and thus its ability to cause disease.

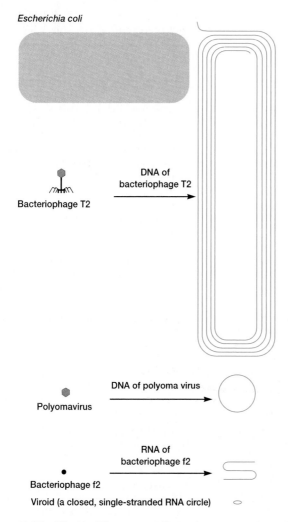

Escherichia coli

DNA of bacteriophage T2

Bacteriophage T2

DNA of polyoma virus

Polyomavirus

RNA of bacteriophage f2

Bacteriophage f2

Viroid (a closed, single-stranded RNA circle)

Figure 18.15   Viroids, Viruses, and Bacteria. A comparison of *Escherichia coli*, several viruses, and the potato spindle-tuber viroid with respect to size and the amount of nucleic acid possessed. (All dimensions are enlarged approximately ×40,000.)

Figure 18.16   Viroid Structure. This schematic diagram shows the general organization of a viroid. The closed single-stranded RNA circle has extensive intrastrand base pairing and interspersed unpaired loops. Viroids have five domains. Most changes in viroid pathogenicity seem to arise from variations in the P and T_L domains.

There is evidence that an infectious agent different from both viruses and viroids can cause disease in livestock and humans. The agent has been called a **prion** (for **pr**oteinaceous **in**fectious particle). The best studied of these prions causes a degenerative disorder of the central nervous system in sheep and goats; this disorder is named scrapie. Afflicted animals lose coordination of their movements, tend to scrape or rub their skin, and eventually cannot walk. The agent has been at least partly purified, and no nucleic acid has yet been detected in it. It seems to be a 33 to 35 kDa hydrophobic membrane protein, often called PrP (for **pr**ion **p**rotein). The *PrP* gene is present in many normal vertebrates and invertebrates. Presumably an altered PrP is at least partly responsible for the disease.

Despite the isolation of PrP, the mechanism of the disease continues to elude researchers. Some believe that the disease is transmitted by the PrP alone. They are convinced that the infective pathogen is an abnormal PrP, one that has been either chemically modified or has changed conformation in some way. When abnormal PrP enters a normal brain, it might bind to normal PrP and induce it to fold into the abnormal conformation. The newly formed abnormal molecule could then attack other normal PrP proteins. Alternatively, PrP could activate enzymes that modify PrP structure. Other researchers feel that this hypothesis is inadequate. They are concerned about the existence of prion strains, presumably genetic, and about the problem of how genetic information can be transmitted between hosts by a protein. They note that proteins have never been known to carry genetic information. Possibly the agent is a **virino,** a tiny scrapie-specific nucleic acid that is coated with PrP. The nucleic acid may not be translated but still interact with host cells in some way to cause disease. This hypothesis is consistent with the finding that many strains of the scrapie agent have been isolated. There is also some evidence that a strain can change or mutate. Finally, some believe that prion diseases are caused by unknown viruses with unusual properties. Presumably further research will clarify the situation.

As mentioned earlier, some slow virus diseases may be due to prions or virinos. This is particularly true of certain neurological diseases of humans and animals. Bovine spongiform encephalopathy ("mad cow disease"), kuru, fatal familial insomnia, the Creutzfeldt-Jakob disease (CJD), and Gerstmann-Sträussler-Scheinker syndrome (GSS) appear to

| Left terminal domain (T_L) | Pathogenicity domain (P) | Central conserved region (CCR) | Variable domain (V) | Right terminal domain (T_R) |

be prion diseases. They result in progressive degeneration of the brain and eventual death. Mad cow disease has reached epidemic proportions in Great Britain and is spread because cattle are fed bone meal made from cattle. CJD and GSS are rare and cosmopolitan in distribution among middle-aged people, while kuru has been found only in the Fore, an eastern New Guinea tribe. This tribe had a custom of consuming dead kinsmen, and women were given the honor of preparing the brain and practicing this ritual cannibalism. They and their children were infected by handling the diseased brain tissue. (Now that cannibalism has been eliminated, the incidence of kuru has decreased and is found only among the older adults.)

1. What are viroids and why are they of great interest?
2. How does a viroid differ from a virus?
3. What is a prion or virino? In what way does a prion appear to differ fundamentally from viruses and viroids?

Summary

1. Animal viruses are classified according to many properties; the most important are their morphology and nucleic acids.

2. The first step in the animal virus life cycle is adsorption of the virus to a target cell receptor site; often special capsid structures are involved in this process.

3. Virus penetration of the host cell plasma membrane may be accompanied by capsid removal from the nucleic acid or uncoating. Most often penetration occurs through engulfment to form coated vesicles, but mechanisms such as direct fusion of the envelope with the plasma membrane also are employed.

4. Early mRNA and proteins are involved in taking over the host cell and the synthesis of viral DNA and RNA. DNA replication usually takes place in the host nucleus and mRNA is initially manufactured by host enzymes.

5. Poxviruses differ from other DNA animal viruses in that DNA replication takes place in host cytoplasm and they carry an RNA polymerase. The parvoviruses are so small that they must conserve genome space by using overlapping genes and other similar mechanisms.

6. The genome of positive ssRNA viruses can act as an mRNA, whereas negative ssRNA virus genomes direct the synthesis of mRNA by a virus-associated transcriptase. Double-stranded RNA reoviruses use both virus-associated and newly synthesized transcriptases to make mRNA.

7. RNA virus genomes are replicated in the host cell cytoplasm. Most ssRNA viruses use a viral replicase to synthesize a dsRNA replicative form that then directs the formation of new genomes.

8. Retroviruses use reverse transcriptase to synthesize a DNA copy of their RNA genome. After the double-stranded proviral DNA has been synthesized, it is integrated into the host genome and directs the formation of virus RNA and protein.

9. Late genes code for proteins needed in (1) capsid construction by a self-assembly process and (2) virus release.

10. Usually, naked virions are released upon cell lysis. In enveloped virus reproduction, virus release and envelope formation normally occur simultaneously after modification of the host plasma membrane, and the cell is not lysed.

11. Viruses can destroy host cells in many ways during cytocidal infections. These include such mechanisms as inhibition of host DNA, RNA, and protein synthesis; lysosomal damage; alteration of host cell membranes; and the formation of inclusion bodies.

12. Although most virus infections are acute or have a rapid onset and last for a relatively short time, some viruses can establish persistent or chronic infections lasting for years. Viruses also can become dormant for a while and then resume activity in what is called a latent infection. Slow viruses may act so slowly that a disease develops over years.

13. Cancer is characterized by the formation of a malignant tumor that metastasizes or invades other tissues and can spread through the body. Carcinogenesis is a complex, multistep process involving many factors.

14. Viruses cause cancer in several ways. For example, they may bring a cancer-causing gene, or oncogene, into a cell, or the virion may insert a promoter next to a cellular oncogene and stimulate it to greater activity.

15. Most plant viruses have an RNA genome and may be either helical or icosahedral. Depending on the virus the RNA genome may be replicated by either a host RNA-dependent RNA polymerase or a virus-specific RNA replicase.

16. The TMV nucleocapsid forms spontaneously by self-assembly when disks of coat protein protomers complex with the RNA.

17. Plant viruses are transmitted in a variety of ways. Some enter through lesions in plant tissues, while others are transmitted by contaminated seeds, tubers, or pollen. Most are probably carried and inoculated by plant-feeding insects.

18. Mycoviruses from higher fungi have isometric capsids and dsRNA, whereas the viruses of lower fungi may have either dsRNA or dsDNA genomes.

19. Members of several virus families—most importantly the *Baculoviridae, Reoviridae,* and *Iridoviridae*—infect insects, and many of these viruses produce inclusion bodies that aid in their transmission.

20. Baculoviruses and other viruses are finding uses as biological control agents for insect pests.

21. Infectious agents simpler than viruses exist. For example, several plant diseases are caused by short strands of infectious RNA called viroids.

22. Prions or virinos are small agents associated with at least six degenerative nervous system disorders: scrapie, bovine spongiform encephalopathy, kuru, fatal familial insomnia, the Gerstmann-Sträussler-Scheinker syndrome, and Creutzfeldt-Jakob disease. The precise nature of prions is not yet clear.

Key Terms

acute infection   *379*

anaplasia   *379*

cancer   *379*

coated vesicles   *373*

cytocidal infection   *377*

defective interfering (DI) particle   *379*

early genes   *374*

inclusion bodies   *379*

late genes   *376*

latent virus infection   *379*

metastasis   *379*

neoplasia   *379*

oncogene   *380*

persistent or chronic infections   *379*

prion   *385*

procapsids   *377*

proviral DNA   *376*

replicase   *376*

replicative form   *376*

retrovirus   *376*

reverse transcriptase   *376*

ribonuclease H   *376*

slow virus diseases   *379*

transcriptase   *376*

tumor   *379*

virino   *385*

viroid   *384*

Questions for Thought and Review

1. Make a list of all the ways you can think of in which viruses differ from their eucaryotic host cells.

2. Compare animal and plant viruses in terms of entry into host cells. Why do they differ so greatly in this regard?

3. Compare lysogeny with the reproductive strategy of retroviruses. What advantage might there be in having one's genome incorporated into that of the host cell?

4. Inclusion bodies are mentioned more than once in the chapter. Where are they found, and what are their functions, if any?

5. Would it be advantageous for a virus to damage host cells? If not, why isn't damage to the host avoided? Is it possible that a virus might become less pathogenic when it has been associated with the host population for a longer time?

6. How does one prove that a virus is causing cancer? Try to think of approaches other than those discussed in the chapter. Give a major reason why it is so difficult to prove that a specific virus causes human cancer.

7. From what you know about cancer, is it likely that a single type of treatment can be used to cure it? What approaches might be effective in preventing cancer?

8. Propose some experiments that might be useful in determining what prions are and how they cause disease.

Additional Reading

Chapter 16 references also should be consulted, particularly the introductory and advanced texts.

General

Davis, B. D.; Dulbecco, R.; Eisen, H. N.; and Ginsberg, H. S. 1990. *Microbiology,* 4th ed. Philadelphia: J.B. Lippincott.

Joklik, W. K.; Willett, H. P.; Amos, D. B.; and Wilfert, C. M. 1992. *Zinsser microbiology,* 20th ed. Norwalk, Conn.: Appleton & Lange.

Lai, M. M. C. 1992. RNA recombination in animal and plant viruses. *Microbiol. Rev.* 56(1):61–79.

McGeoch, D. J. 1989. The genomes of the human herpesviruses: Contents, relationships, and evolution. *Ann. Rev. Microbiol.* 43:235–65.

Strauss, J. H., and Strauss, E. G. 1988. Evolution of RNA viruses. *Ann. Rev. Microbiol.* 42:657–83.

Reproduction of Animal Viruses

Berns, K. I. 1990. Parvovirus replication. *Microbiol. Rev.* 54(3):316–29.

Gallo, R. C. 1986. The first human retrovirus. *Sci. Am.* 255(6):88–98.

Gallo, R. C. 1987. The AIDS virus. *Sci. Am.* 256(1):47–56.

Goodenough, U. W. 1991. Deception by pathogens. *Am. Sci.* 79:344–55.

Greber, U. F.; Singh, I.; and Helenius, A. 1994. Mechanisms of virus uncoating. *Trends Microbiol.* 2(2):52–56.

Hazeltine, W. A. 1991. Molecular biology of the human immunodeficiency virus type 1. *FASEB J.* 5(10):2349–60.

Henle, W.; Henle, G.; and Lennette, E. T. 1979. The Epstein-Barr virus. *Sci. Am.* 241(1):48–59.

Katz, R. A., and Skalka, A. M. 1994. The retroviral enzymes. *Ann. Rev. Biochem.* 63:133–73.

Levy, J. A. 1994. *HIV and the pathogenesis of AIDS.* Washington, D.C.: ASM Press.

Marsh, M., and Helenius, A. 1989. Virus entry into animal cells. *Adv. Virus Res.* 36:107–51.

Norkin, L. C. 1995. Virus receptors: Implications for pathogenesis and the design of antiviral agents. *Microbiol. Rev.* 8(2):293–315.

Simons, K.; Garoff, H.; and Helenius, A. 1982. How an animal virus gets into and out of its host cell. *Sci. Am.* 246(2): 58–66.

Stephens, E. B., and Compans, R. W. 1988. Assembly of animal viruses at cellular membranes. *Ann. Rev. Microbiol.* 42:489–516.

Temin, H. M. 1972. RNA-directed DNA synthesis. *Sci. Am.* 226(1):25–33.

Varmus, H. 1987. Reverse transcription. *Sci. Am.* 257(3):56–64.

Varmus, H. 1988. Retroviruses. *Science* 240:1427–35.

White, J. M., and Littman, D. R. 1989. Viral receptors of the immunoglobulin superfamily. *Cell* 56:725–28.

Cytocidal Infections and Cell Damage

Buller, R. M. L., and Palumbo, G. J. 1991. Poxvirus pathogenesis. *Microbiol. Rev.* 55(1):80–122.

Kaariainen, L., and Ranki, M. 1984. Inhibition of cell functions by RNA-virus infections. *Ann. Rev. Microbiol.* 38:91–109.

Oldstone, M. B. A. 1989. Viral alteration of cell function. *Sci. Am.* 261(2):42–48.

Viruses and Cancer

Bishop, J. M. 1982. Oncogenes. *Sci. Am.* 246(3):81–92.

Hunter, T. 1984. The proteins of oncogenes. *Sci. Am.* 251(2):70–79.

Sherker, A. H., and Marion, P. L. 1991. Hepadnaviruses and hepatocellular carcinoma. *Ann. Rev. Microbiol.* 45:475–508.

Weinberg, R. A. 1983. A molecular basis of cancer. *Sci. Am.* 249(5):126–42.

zur Hausen, H., and de Villiers, E.-M. 1994. Human papillomaviruses. *Ann. Rev. Microbiol.* 48:427–47.

Plant Viruses

Butler, P. J. G., and Klug, A. 1978. The assembly of a virus. *Sci. Am.* 239(5):62–69.

Leisner, S. M., and Howell, S. H. 1993. Long-distance movement of viruses in plants. *Trends Microbiol.* 1(8):314–17.

Matthews, R. E. F. 1991. *Plant virology,* 3d ed. New York: Academic Press.

Insect Viruses

Miller, L. K.; Lingg, A. J.; and Bulla, L. A., Jr. 1983. Bacterial, viral, and fungal insecticides. *Science* 219:715–21.

Wood, H. A., and Granados, R. R. 1991. Genetically engineered baculoviruses as agents for pest control. *Ann. Rev. Microbiol.* 45:69–87.

Viroids and Prions

Aiken, J. M., and Marsh, R. F. 1990. The search for scrapie agent nucleic acid. *Microbiol. Rev.* 54(3):242–46.

Diener, T. O. 1987. *The viroids.* New York: Plenum Press.

Diener, T. O. 1993. The viroid: Big punch in a small package. *Trends Microbiol.* 1(8):289–94.

Eron, C. 1981. *The virus that ate cannibals.* New York: Macmillan.

Medori, R., et al. 1992. Fatal familial insomnia, a prion disease with a mutation at codon 178 of the prion protein gene. *New Eng. J. Med.* 326(7):444–49.

Prusiner, S. B. 1994. Biology and genetics of prion diseases. *Ann. Rev. Microbiol.* 48:655–86.

Prusiner, S. B. 1995. The prion diseases. *Sci. Am.* 272(1):48–57.

Riesner, D., and Gross, H. J. 1985. Viroids. *Ann. Rev. Biochem.* 54:531–64.

PART SIX
The Diversity of the Microbial World

A light micrograph of the cyanobacterium *Anabaena spiroides* (original magnification = ×500). *A. spiroides* is a major component of freshwater plankton and maintains buoyancy using gas vacuoles. This trichome has two terminal heterocysts, specialized cells that are the sites of nitrogen fixation. Heterocysts develop when cyanobacteria are deprived of nitrate and ammonia. Thus these bacteria can flourish in low nitrogen environments and produce massive blooms in lakes. As a result, fish die and the lake's recreational value is lessened.

Source: Data from C.P. Woese, Microbiological Reviews, 51(2):221–227, 1987.

CHAPTER 19
Microbial Taxonomy

What's in a name? that

which we call a rose

By any other name would

smell as sweet. . . .
—W. Shakespeare

It were not best that we

should all think alike; it is

difference of opinion that

makes horse-races.
—Mark Twain

Outline

Concepts

1. In order to make sense of the diversity of organisms, it is necessary to group similar organisms together and organize these groups in a nonoverlapping hierarchical arrangement. Taxonomy is the science of biological classification.

2. The basic taxonomic group is the species, which is defined in terms of either sexual reproduction or general similarity.

3. Morphological, physiological, metabolic, ecological, genetic, and molecular characteristics are all useful in taxonomy because they reflect the organization and activity of the genome. Nucleic acid structure is probably the best indicator of relatedness because nucleic acids are either the genetic material itself or the products of gene transcription.

4. Classifications are based on an analysis of possible evolutionary relationships (phylogenetic or phyletic classification) or on overall similarity (phenetic classification).

5. *Bergey's Manual of Systematic Bacteriology* divides the bacteria between four divisions and seven classes within the kingdom *Procaryotae*.

6. Bacterial taxonomy is rapidly changing due to the acquisition of new data, particularly the use of molecular techniques such as the comparison of ribosomal RNA structure. This has led to new phylogenetic classifications.

7. Organisms can be placed in one of five kingdoms based on cell type, level of organization, and type of nutrition. Alternative systems with two or three domains or empires above the kingdom level have been developed.

O
ne of the most fascinating and attractive aspects of the microbial world is its extraordinary diversity. It seems that almost every possible experiment in shape, size, physiology, and life-style has been tried. The sixth part of the text focuses on this diversity. Chapter 19 introduces the general principles of microbial taxonomy. This is followed by a five-chapter (20–24) survey of the most important bacterial groups. Part six ends with an extensive introduction to the major types of eucaryotic microorganisms: fungi, algae, and protozoa.

Because of the bewildering diversity of living organisms, it is desirable to classify or arrange them into groups based on their mutual similarities. **Taxonomy** [Greek *taxis,* arrangement or order, and *nomos,* law, or *nemein,* to distribute or govern] is defined as the science of biological classification. In a broader sense it consists of three separate but interrelated parts: classification, nomenclature, and identification. **Classification** is the arrangement of organisms into groups or **taxa** (singular, **taxon**) based on mutual similarity or evolutionary relatedness. **Nomenclature** is the branch of taxonomy concerned with the assignment of names to taxonomic groups in agreement with published rules. **Identification** is the practical side of taxonomy, the process of determining that a particular isolate belongs to a recognized taxon. Its practical importance can hardly be overemphasized. For example, it is essential to clinical microbiology (*see chapter 34*); treatment often is exceptionally difficult when the pathogen is unknown.

The term **systematics** often is used for taxonomy. However, many taxonomists define it in more general terms as "the scientific study of organisms with the ultimate object of characterizing and arranging them in an orderly manner." Any study of the nature of organisms, when the knowledge gained is used in taxonomy, is a part of systematics. Thus it encompasses disciplines such as morphology, ecology, epidemiology, biochemistry, molecular biology, and physiology.

Microbial taxonomy is too broad a subject for adequate coverage in a single chapter. Therefore this chapter emphasizes general principles and uses examples primarily from bacterial taxonomy. The taxonomy of each major eucaryotic microbial group is reviewed where the group is introduced in subsequent chapters.

Currently microbial taxonomy is in ferment because of the use of new molecular techniques in classifying microorganisms. Even though these new advances have generated much excitement and are drastically changing microbial taxonomy, more traditional approaches still have value and are extensively used. The chapter begins with a general overview of taxonomy and then briefly introduces classical bacterial taxonomy. Next the results of molecular phylogenetic approaches are summarized. The chapter ends with a discussion of the kingdoms of organisms.

Taxonomic Ranks

In preparing a classification scheme, one places the microorganism within a small, homogeneous group that is itself a member of larger groups in a nonoverlapping hierarchical arrangement. A category in any rank unites groups in the level below it based on shared properties (figure 19.1). The most commonly used levels or ranks (in ascending order) are species, genera, families, orders, classes, phyla or divisions, and kingdoms. Microbial groups at each level or rank have names with endings or suffixes characteristic of that level (table 19.1). Informal names are often

TABLE **19.1**	An Example of Taxonomic Ranks and Names
Rank	**Example**
Kingdom	*Procaryotae*
Division (Phylum)	*Tenericutes*
Class	*Mollicutes*
Order	*Mycoplasmatales*
Family	*Mycoplasmataceae*
Genus	*Mycoplasma*
Species	*M. pneumoniae*

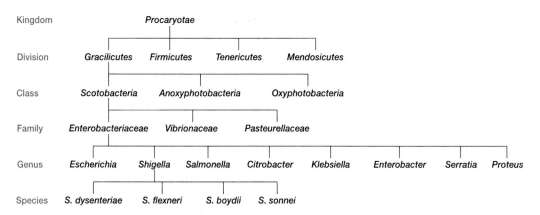

Figure 19.1   Hierarchical Arrangement in Taxonomy. In this example, members of the genus *Shigella* are placed within higher taxonomic ranks according to the system presented in *Bergey's Manual of Systematic Bacteriology.* Not all classification possibilities are given for each rank in order to simplify the diagram.

used in place of formal hierarchical ones. Typical examples of such names are purple bacteria, spirochetes, methane-oxidizing bacteria, sulfate-reducing bacteria, and lactic acid bacteria.

The basic taxonomic group in microbial taxonomy is the **species.** Taxonomists working with higher organisms define the term species differently than microbiologists do. Species of higher organisms are groups of interbreeding or potentially interbreeding natural populations that are reproductively isolated from other groups. This is a satisfactory definition for organisms capable of sexual reproduction, but fails with many microorganisms because they do not reproduce sexually. Bacterial species are characterized by phenotypic and genotypic (*see chapter 13*) differences. A **bacterial species** is a collection of strains that share many stable properties and differ significantly from other groups of strains. A **strain** is a population of organisms that descends from a single organism or pure culture isolate. Strains within a species may differ slightly from one another in many ways. **Biovars** are variant bacterial strains characterized by biochemical or physiological differences, **morphovars** differ morphologically, and **serovars** have distinctive antigenic properties. One strain of a species is designated as the **type strain.** It is usually one of the first strains studied and often is more fully characterized than other strains; however, it does not have to be the most representative member. Only those strains very similar to the type strain are included in a species. Each species is assigned to a genus, the next rank in the taxonomic hierarchy. A **genus** is a well-defined group of one or more species that is clearly separate from other genera. In practice there is considerable subjectivity in assigning species to a genus, and taxonomists may disagree about the composition of genera.

Microbiologists name microorganisms by using the **binomial system** of the Swedish botanist Carl von Linné, or Carolus Linnaeus as he is often called. The Latinized, italicized name consists of two parts. The first part, which is capitalized, is the generic name, and the second is the uncapitalized specific epithet or name (e.g., *Escherichia coli*). The specific name is stable; the oldest epithet for a particular organism takes precedence and must be used. In contrast, a generic name can change if the organism is assigned to another genus because of new information. For example, *Streptococcus pneumoniae* was originally called *Diplococcus pneumoniae*. Often the name will be shortened by abbreviating the genus name with a single capital letter, for example *E. coli*. A list of approved bacterial names was published in 1980 in the *International Journal of Systematic Bacteriology*, and new valid names are published periodically. *Bergey's Manual of Systematic Bacteriology* has been published in four volumes and contains the currently accepted system of bacterial taxonomy. A new edition of *Bergey's Manual of Determinative Bacteriology*, a single volume intended for use in identifying bacteria, has recently been published. (To save space, *Bergey's Manual* often will be used when referring to *Bergey's Manual of Systematic Bacteriology*.)

1. Be able to define the following terms: taxonomy, classification, taxon, nomenclature, identification, systematics, species, strain, type strain, and binomial system.
2. How does the definition of species differ between organisms that reproduce sexually and those not able to do so?

Major Characteristics Used in Taxonomy

Many characteristics are used in classifying and identifying microorganisms. This section briefly reviews some of the most taxonomically important properties. Methods often employed in routine laboratory identification of bacteria are covered in the chapter on clinical microbiology (*see chapter 34*).

Morphological Characteristics

Morphological features are important in microbial taxonomy for many reasons. Morphology is easy to study and analyze, particularly in eucaryotic microorganisms and the more complex procaryotes. In addition, morphological comparisons are valuable because structural features depend on the expression of many genes, are usually genetically stable, and normally (at least in eucaryotes) do not vary greatly with environmental changes. Thus morphological similarity is a good indication of phylogenetic relatedness.

Many different morphological features are employed in the classification and identification of microorganisms (table 19.2). Although the light microscope has always been a very important tool, its resolution limit of about 0.2 µm (*see chapter 2*) reduces its usefulness in viewing smaller microorganisms and structures. The transmission and scanning electron microscopes, with their greater resolution, have immensely aided the study of all microbial groups.

Physiological and Metabolic Characteristics

Physiological and metabolic characteristics are very useful because they are directly related to the nature and activity of microbial enzymes and transport proteins. Since proteins are gene products, analysis of these characteristics provides an indirect comparison of microbial genomes. Table 19.3 lists some of the most important of these properties.

Ecological Characteristics

Many properties are ecological in nature since they affect the relation of microorganisms to their environment. Often these are taxonomically valuable because even very closely related microorganisms can differ considerably with respect to ecological characteristics. Microorganisms living in various parts of the human body markedly differ from one another and from those growing in freshwater, terrestrial, and marine environments.

TABLE 19.2	Some Morphological Features Used in Classification and Identification

Feature	Microbial Groups
Cell shape	All major groups[a]
Cell size	All major groups
Colonial morphology	All major groups
Ultrastructural characteristics	All major groups
Staining behavior	Bacteria, some fungi
Cilia and flagella	All major groups
Mechanism of motility	Gliding bacteria, spirochetes
Endospore shape and location	Endospore-forming bacteria
Spore morphology and location	Bacteria, algae, fungi
Cellular inclusions	All major groups
Color	All major groups

[a]Used in classifying and identifying at least some bacteria, algae, fungi, and protozoa.

TABLE 19.3	Some Physiological and Metabolic Characteristics Used in Classification and Identification

Carbon and nitrogen sources
Cell wall constituents
Energy sources
Fermentation products
General nutritional type
Growth temperature optimum and range
Luminescence
Mechanisms of energy conversion
Motility
Osmotic tolerance
Oxygen relationships
pH optimum and growth range
Photosynthetic pigments
Salt requirements and tolerance
Secondary metabolites formed
Sensitivity to metabolic inhibitors and antibiotics
Storage inclusions

Some examples of taxonomically important ecological properties are life cycle patterns; the nature of symbiotic relationships; the ability to cause disease in a particular host; and habitat preferences such as requirements for temperature, pH, oxygen, and osmotic concentration. Many growth requirements are also considered physiological characteristics (table 19.3).

Genetic Analysis

Because most eucaryotes are able to reproduce sexually, genetic analysis has been of considerable usefulness in the classification of these organisms. As mentioned earlier, the species is defined in terms of sexual reproduction where possible. Although

procaryotes do not reproduce sexually, the study of chromosomal gene exchange through transformation and conjugation is sometimes useful in their classification.

Transformation can occur between different bacterial species but only rarely between genera. The demonstration of transformation between two strains provides evidence of a close relationship since bacterial transformation cannot occur unless the genomes are fairly similar. Transformation studies have been carried out with several genera: *Bacillus, Micrococcus, Haemophilus, Rhizobium,* and others. Despite transformation's usefulness, its results are sometimes hard to interpret because an absence of transformation may result from factors other than major differences in DNA sequence.

Transformation (pp. 255, 292–93).
Conjugation (pp. 288–92).

Conjugation studies also yield taxonomically useful data, particularly with the enteric bacteria (*see chapter 20*). For example, *Escherichia* can undergo conjugation with the genera *Salmonella* and *Shigella,* but not with *Proteus* and *Enterobacter.* These observations fit with other data showing that the first three of these genera are more closely related to one another than to *Proteus* and *Enterobacter.*

Plasmids (*see chapter 14*) are undoubtedly important in taxonomy because they are present in most bacterial genera, and many carry genes coding for phenotypic traits. Because plasmids could have a significant effect on classification if they carried the gene for a trait of major importance in the classification scheme, it is best to base a classification on many characters. When the identification of a group is based on a few characteristics and some of these are coded for by plasmid genes, errors may result. For example, hydrogen sulfide production and lactose fermentation are very important in the taxonomy of the enteric bacteria, yet genes for both traits can be borne on plasmids as well as bacterial chromosomes. One must take care to avoid errors as a result of plasmid-borne traits.

Molecular Characteristics

Some of the most powerful approaches to taxonomy are through the study of proteins and nucleic acids. Because these are either direct gene products or the genes themselves, comparisons of proteins and nucleic acids yield considerable information about true relatedness.

Comparison of Proteins

The amino acid sequences of proteins are direct reflections of mRNA sequences and therefore closely related to the structures of the genes coding for their synthesis. For this reason, comparisons of proteins from different microorganisms are very useful taxonomically. There are several ways to compare proteins. The most direct approach is to determine the amino acid sequence of proteins with the same function. The sequences of proteins with dissimilar functions often change at different rates; some sequences change quite rapidly, whereas others are very stable. Nevertheless, if the sequences of proteins with the same function are similar, the organisms possessing them are probably closely related. The sequences of cytochromes and other electron transport proteins, histones, and a variety of enzymes have been used in taxonomic studies. Because protein sequencing is slow and expensive, more indirect methods of comparing proteins have frequently been employed. The electrophoretic mobility of proteins (*see chapter 15*) is useful in studying relationships at the species and subspecies levels. Antibodies can discriminate between very similar proteins, and immunologic techniques are used to compare proteins from different microorganisms.

Antibody-antigen reactions in vitro (pp. 641–52).

The physical, kinetic, and regulatory properties of enzymes have been employed in taxonomic studies. Because enzyme behavior reflects amino acid sequence, this approach is useful in studying some microbial groups, and group-specific patterns of regulation have been found.

Nucleic Acid Base Composition

Microbial genomes can be directly compared, and taxonomic similarity can be estimated in many ways. The first, and possibly the simplest, technique to be employed is the determination of DNA base composition. DNA contains four purine and pyrimidine bases: adenine (A), guanine (G), cytosine (C), and thymine (T). In double-stranded DNA, A pairs with T, and G pairs with C. Thus the (G + C)/(A + T) ratio or **G + C content,** the percent of G + C in DNA, reflects the base sequence and varies with sequence changes as follows:

$$\text{Mol\% G} + \text{C} = \frac{\text{G} + \text{C}}{\text{G} + \text{C} + \text{A} + \text{T}} \times 100$$

DNA and RNA structure (pp. 212–14).

The base composition of DNA can be determined in several ways. Although the G + C content can be chemically ascertained after hydrolysis of DNA and separation of its bases, physical methods are easier and more often used. The G + C content is often determined from the **melting temperature (T_m)** of DNA. In double-stranded DNA, three hydrogen bonds join GC base pairs, and two bonds connect AT base pairs (*see chapter 11*). As a result DNA with a greater G + C content will have more hydrogen bonds, and its strands will separate only at higher temperatures—that is, it will have a higher melting point. DNA melting can be easily followed spectrophotometrically because the absorbance of 260 nm UV light by DNA increases during strand separation. When a DNA sample is slowly heated, the absorbance increases as hydrogen bonds are broken and reaches a plateau when all the DNA has become single stranded

Figure 19.2   A DNA Melting Curve. The T_m is indicated.

(figure 19.2). The midpoint of the rising curve gives the melting temperature, a direct measure of the G + C content. Since the density of DNA also increases linearly with G + C content, the percent G + C can be obtained by centrifuging DNA in a CsCl density gradient (*see chapter 16*).

The G + C content of many microorganisms has been determined (table 19.4). The G + C content of DNA from animals and higher plants averages around 40% and ranges between 30 and 50%. In contrast, the DNA of both eucaryotic and procaryotic microorganisms varies greatly in G + C content; procaryotic G + C content is the most variable, ranging from around 25 to almost 80%. Despite such a wide range of variation, the G + C content of strains within a particular species is constant. If two organisms differ in their G + C content by more than about 10%, their genomes have quite different base sequences. On the other hand, it is not safe to assume that organisms with very similar G + C contents also have similar DNA base sequences because two very different base sequences can be constructed from the same proportions of AT and GC base pairs. Only if two microorganisms are also alike phenotypically does their similar G + C content suggest close relatedness.

G + C content data are taxonomically valuable in at least two ways. First, they can confirm a taxonomic scheme developed using other data. If organisms in the same taxon are too dissimilar in G + C content, the taxon probably should be divided. Second, G + C content appears to be useful in characterizing bacterial genera since the variation within a genus is usually less than 10% even though the content may vary greatly between genera. For example, *Staphylococcus* has a G + C content of 30 to 38% while *Micrococcus* DNA has 64 to 75% G + C; yet these two genera of gram-positive cocci have many other features in common.

TABLE 19.4 Representative G + C Contents of Microorganisms

Organism	Percent G + C	Organism	Percent G + C	Organism	Percent G + C
Bacteria		*Streptococcus*	33–44	**Slime Molds**	
Actinomyces	59–73	*Streptomyces*	69–73	*Dictyostelium*	22–25
Anabaena	38–44	*Sulfolobus*	31–37	*Lycogala*	42
Bacillus	32–62	*Thermoplasma*	46	*Physarum polycephalum*	38–42
Bacteroides	28–61	*Thiobacillus*	52–68	**Fungi**	
Bdellovibrio	33–52	*Treponema*	25–54	*Agaricus bisporus*	44
Caulobacter	63–67	**Algae**		*Amanita muscaria*	57
Chlamydia	41–44	*Acetabularia mediterranea*	37–53	*Aspergillus niger*	52
Chlorobium	49–58	*Chlamydomonas*	60–68	*Blastocladiella emersonii*	66
Chromatium	48–70	*Chlorella*	43–79	*Candida albicans*	33–35
Clostridium	21–54	*Cyclotella cryptica*	41	*Claviceps purpurea*	53
Cytophaga	33–42	*Euglena gracilis*	46–55	*Coprinus lagopus*	52–53
Deinococcus	62–70	*Nitella*	49	*Fomes fraxineus*	56
Escherichia	48–52	*Nitzschia angularis*	47	*Mucor rouxii*	38
Halobacterium	66–68	*Ochromonas danica*	48	*Neurospora crassa*	52–54
Hyphomicrobium	59–67	*Peridinium triquetrum*	53	*Penicillium notatum*	52
Methanobacterium	32–50	*Scenedesmus*	52–64	*Polyporus palustris*	56
Micrococcus	64–75	*Spirogyra*	39	*Rhizopus nigricans*	47
Mycobacterium	62–70	*Volvox carteri*	50	*Saccharomyces cerevisiae*	36–42
Mycoplasma	23–40	**Protozoa**		*Saprolegnia parasitica*	61
Myxococcus	68–71	*Acanthamoeba castellanii*	56–58		
Neisseria	47–54	*Amoeba proteus*	66		
Nitrobacter	60–62	*Paramecium* spp.	29–39		
Oscillatoria	40–50	*Plasmodium berghei*	41		
Prochloron	41	*Stentor polymorphus*	45		
Proteus	38–41	*Tetrahymena*	19–33		
Pseudomonas	58–70	*Trichomonas*	29–34		
Rhodospirillum	62–66	*Trypanosoma*	45–59		
Rickettsia	29–33				
Salmonella	50–53				
Spirillum	38				
Spirochaeta	51–65				
Staphylococcus	30–38				

Nucleic Acid Hybridization

The similarity between genomes can be compared more directly by use of **nucleic acid hybridization** studies. If a mixture of single-stranded DNA formed by heating dsDNA is cooled and held at a temperature about 25°C below the T_m, strands with complementary base sequences will reassociate to form stable dsDNA, while noncomplementary strands will remain single (figure 19.3). Because strands with similar, but not identical, sequences associate to form less temperature stable dsDNA hybrids, incubation of the mixture at 30 to 50°C below the T_m will allow hybrids of more diverse ssDNAs to form. Incubation at 10 to 15°C below the T_m permits hybrid formation only with almost identical strands.

In one of the more widely used hybridization techniques, nitrocellulose filters with bound nonradioactive DNA strands are incubated at the appropriate temperature with single-stranded DNA fragments made radioactive with ^{32}P, ^3H, or ^{14}C. After radioactive fragments are allowed to hybridize with the membrane-bound ssDNA, the membrane is washed to remove any nonhybridized ssDNA and its radioactivity is measured. The quantity of radioactivity bound to the filter reflects the amount of hybridization and thus the similarity of the DNA sequences. The degree of similarity or homology is expressed as the percent of experimental DNA radioactivity retained on the filter compared with the percent of homologous DNA radioactivity bound under the same conditions (table 19.5 provides examples).

If DNA molecules are very different in sequence, they will not form a stable, detectable hybrid. Therefore DNA-DNA hybridization is only used to study closely related microorganisms. More distantly related organisms are compared by carrying out DNA-RNA hybridization experiments using radioactive ribosomal or transfer RNA. Distant relationships can be detected because rRNA and tRNA genes represent only a small portion of the total DNA genome and have not evolved as rapidly as most other microbial genes. The technique is similar to that employed for DNA-DNA hybridization: membrane-bound DNA is incubated with radioactive rRNA, washed, and

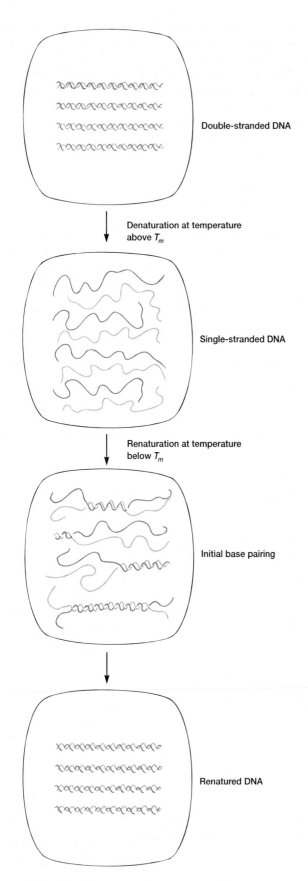

Double-stranded DNA

Denaturation at temperature
above T_m

Single-stranded DNA

Renaturation at temperature
below T_m

Initial base pairing

Renatured DNA

Figure 19.3   Nucleic Acid Melting and Hybridization.
Complementary strands are shown in different colors.

TABLE 19.5	Comparison of *Neisseria* Species by DNA Hybridization Experiments

Membrane-Attached DNA[a]	Percent Homology[b]
Neisseria meningitidis	100
N. gonorrhoeae	78
N. sicca	45
N. flava	35

Source: Data from T. E. Staley and R. R. Colwell, "Applications of Molecular Genetics and Numerical Taxonomy to the Classification of Bacteria" in *Annual Review of Ecology and Systematics*, 8: 282, 1973.

[a]The experimental membrane-attached nonradioactive DNA from each species was incubated with radioactive *N. meningitidis* DNA, and the amount of radioactivity bound to the membrane was measured. The more radioactivity bound, the greater the homology between DNA sequences.

[b] $\dfrac{\textit{N. meningitidis} \text{ DNA bound to experimental DNA}}{\text{Amount bound to membrane-attached } \textit{N. meningitidis} \text{ DNA}} \times 100$

counted. An even more accurate measurement of homology is obtained by finding the temperature required to dissociate and remove half the radioactive rRNA from the membrane; the higher this temperature, the stronger the rRNA-DNA complex and the more similar the sequences.

Ribosomes and ribosomal RNA (p. 228).
Transfer RNA (pp. 226–28).

Nucleic Acid Sequencing

Despite the usefulness of G + C content determination and nucleic acid hybridization studies, genome structures can only be directly compared by sequencing DNA and RNA. Techniques for rapidly sequencing both DNA and RNA are now available; thus far, RNA sequencing has been used more extensively in microbial taxonomy. However, recently complete bacterial genomes have been sequenced (*see pp. 298–9*). Direct comparison of genome sequences undoubtedly will become important in bacterial taxonomy.

Most attention has been given to sequences of the 5S and 16S rRNAs isolated from the 50S and 30S subunits, respectively, of bacterial ribosomes (*see chapters 3 and 11*). The rRNAs are almost ideal for studies of microbial evolution and relatedness since they are essential to a critical organelle found in all microorganisms. Their functional role is the same in all ribosomes. Furthermore, their structure changes very slowly with time, presumably because of their constant and critical role. Ribosomal RNAs can be characterized in terms of partial sequences by the oligonucleotide cataloging method as follows. Purified, radioactive 16S rRNA is treated with the enzyme T_1 ribonuclease, which cleaves it into fragments. The fragments are separated, and all fragments composed of at least six nucleotides are sequenced. The sequences of corresponding 16S rRNA fragments from different bacteria are then aligned and compared using a computer, and association coefficients (S_{ab} values) are calculated. Complete rRNAs can now be sequenced using the enzyme reverse transcriptase and DNA sequencing techniques.

Since rRNA contains variable and stable sequences, both closely related and very distantly related microorganisms can be compared. This is an important advantage because distantly related organisms can only be studied using sequences that change little with time. The results of such studies will be discussed shortly.

---

1. Summarize the advantages of using each major group of characteristics (morphological, physiological/metabolic, ecological, genetic, and molecular) in classification and identification. How is each group related to the nature and expression of the genome? Give examples of each type of characteristic.
2. What two modes of genetic exchange in bacteria have proved taxonomically useful? Why are plasmids of such importance in bacterial taxonomy?
3. Briefly describe some ways in which proteins from different organisms can be compared.
4. What is the G + C content of DNA and how can it be determined through melting temperature studies and density gradient centrifugation?
5. Discuss the use of G + C content in taxonomy. Why is it not safe to assume that two microorganisms with the same G + C content belong to the same species? In what two ways are G + C content data taxonomically valuable?
6. Describe how nucleic acid hybridization studies are carried out using membrane-bound DNA. Why might one wish to vary the incubation temperature during hybridization? What is the advantage of conducting DNA-RNA hybridization studies?
7. How are rRNA sequencing studies conducted and why is rRNA so suitable for determining relatedness?

Classification Systems

Once taxonomically relevant characteristics of microorganisms have been collected, they may be used to construct a classification system. The most desirable classification system, called a **natural classification,** arranges organisms into groups whose members share many characteristics and reflects as much as possible the biological nature of organisms. Linnaeus developed the first natural classification, based largely on anatomical characteristics, in the middle of the eighteenth century. It was a great improvement over previously employed artificial systems because knowledge of an organism's position in the scheme provided information about many of its properties. For example, classification of humans as mammals denotes that they have hair, self-regulating body temperature, and milk-producing mammary glands in the female.

Following the publication in 1859 of Darwin's *On the Origin of Species,* biologists began trying to develop **phylogenetic** or **phyletic classification systems.** These are systems based on evolutionary relationships rather than general resemblance (the term **phylogeny** [Greek *phylon,* tribe or race, and *genesis,* generation or origin] refers to the evolutionary development of a species). This has proven difficult for bacteria and other microorganisms,

primarily because of the lack of a good fossil record. The direct comparison of genetic material and gene products such as RNA and proteins overcomes some of these problems.

Many taxonomists maintain that the most natural classification is the one with the greatest information content or predictive value. A good classification should bring order to biological diversity and may even clarify the function of a morphological structure. For example, if motility and flagella are always associated in particular microorganisms, it is reasonable to suppose that flagella are involved in at least some types of motility. When viewed in this way, the best natural classification system may be a **phenetic system,** one that groups organisms together based on the mutual similarity of their phenotypic characteristics. Although phenetic studies can reveal possible evolutionary relationships, they are not dependent on phylogenetic analysis. They compare many traits without assuming that any features are more phylogenetically important than others—that is, unweighted traits are employed in estimating general similarity. Obviously the best phenetic classification is one constructed by comparing as many attributes as possible. Organisms sharing many characteristics make up a single group or taxon.

---

1. What is a natural classification?
2. What are phylogenetic (phyletic) and phenetic classification systems? How do the two systems differ?

Numerical Taxonomy

The development of computers has made possible the quantitative approach known as **numerical taxonomy.** Peter H. A. Sneath and Robert Sokal have defined numerical taxonomy as "the grouping by numerical methods of taxonomic units into taxa on the basis of their character states." Information about the properties of organisms is converted into a form suitable for numerical analysis and then compared by means of a computer. The resulting classification is based on general similarity as judged by comparison of many characteristics, each given equal weight. This approach was not feasible before the advent of computers because of the large number of calculations involved.

The process begins with a determination of the presence or absence of selected characters in the group of organisms under study. A character usually is defined as an attribute about which a single statement can be made. Many characters, at least 50 and preferably several hundred, should be compared for an accurate and reliable classification. It is best to include many different kinds of data: morphological, biochemical, and physiological.

After character analysis, an association coefficient, a function that measures the agreement between characters possessed by two organisms, is calculated for each pair of organisms in the group. The **simple matching coefficient** (S_{SM}), the most commonly used coefficient in bacteriology, is the proportion of characters that match regardless of whether the attribute is present or absent (table 19.6). Sometimes the **Jaccard coefficient (S_J)** is calculated by ignoring any characters that both organisms lack (table 19.6). Both coefficients increase linearly in value from 0.0 (no matches) to 1.0 (100% matches).

The simple matching coefficients, or other association coefficients, are then arranged to form a **similarity matrix.** This is a matrix in which the rows and columns represent organisms and each value is an association coefficient measuring the similarity of two different organisms so that each organism is compared to every other one in the table (figure 19.4a). Organisms with great similarity are grouped together and separated from dissimilar organisms (figure 19.4b); such groups of organisms are called **phenons** (sometimes called phenoms).

The results of numerical taxonomic analysis are often summarized with a treelike diagram called a **dendrogram** (figures 19.4c and 19.5). The diagram usually is placed on its side with the x-axis or abscissa graduated in units of similarity (figure 19.5). Each branch point is at the similarity value relating the two branches. The organisms in the two branches share so many characteristics that the two groups are seen to be separate only after examination of association coefficients greater than the magnitude of the branch point value. Below the branch point value, the two groups appear to be one. The ordinate in such a dendrogram has no special significance, and the clusters may be arranged in any convenient order.

TABLE 19.6 **The Calculation of Association Coefficients for Two Organisms**

In this example, organisms A and B are compared in terms of the characters they do and do not share. The terms in the association coefficient equations are defined as follows:

		Organism B	
		1	0
Organism A	1	a	b
	0	c	d

a = number of characters coded as present (1) for both organisms
b and c = numbers of characters differing (1,0 or 0,1) between the two organisms
d = number of characters absent (0) in both organisms
Total number of characters compared = $a + b + c + d$

The simple matching coefficient (S_{SM}) = $\dfrac{a+d}{a+b+c+d}$

The Jaccard coefficient (S_J) = $\dfrac{a}{a+b+c}$

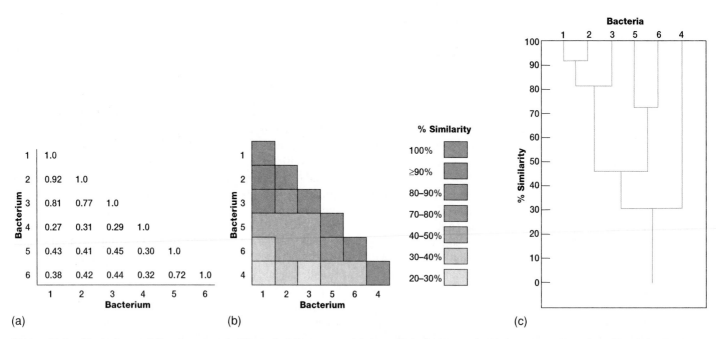

Figure 19.4 Clustering and Dendrograms in Numerical Taxonomy. (a) A small similarity matrix that compares six strains of bacteria. The degree of similarity ranges from none (0.0) to complete similarity (1.0). (b) The bacteria have been rearranged and joined to form clusters of similar strains. For example, strains 1 and 2 are the most similar. The cluster of 1 plus 2 is fairly similar to strain 3, but not at all to strain 4. (c) A dendrogram showing the results of the analysis in part b. Strains 1 and 2 are members of a 90-phenon, and strains 1–3 form an 80-phenon. While strains 1–3 may be members of a single species, it is quite unlikely that strains 4–6 belong to the same species as 1–3.

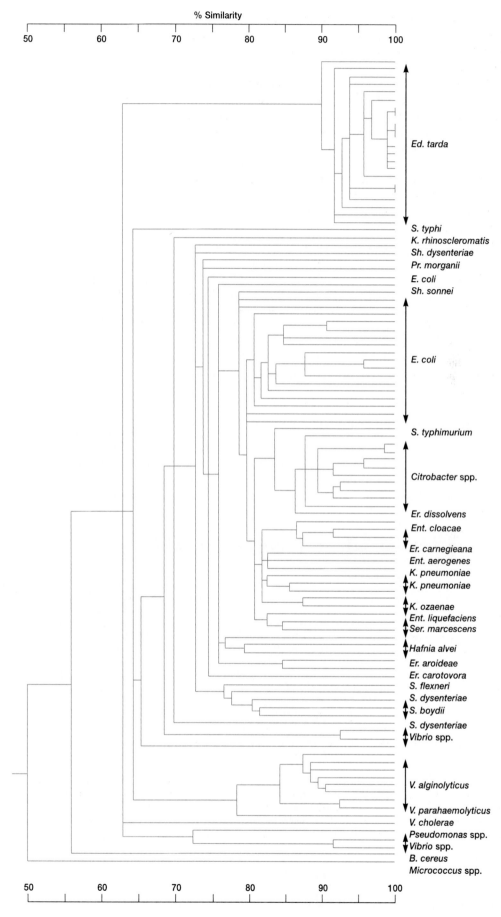

Figure 19.5 The Use of Numerical Taxonomy in Bacteriology. The dendrogram indicates the arrangement of 105 strains of gram-negative fermentative bacteria from the family *Enterobacteriaceae*. The strains were compared using 238 characteristics.

% Similarity

Ed. tarda

S. typhi
K. rhinoscleromatis
Sh. dysenteriae
Pr. morganii
E. coli
Sh. sonnei

E. coli

S. typhimurium

Citrobacter spp.

Er. dissolvens
Ent. cloacae
Er. carnegieana
Ent. aerogenes
K. pneumoniae
K. pneumoniae
K. ozaenae
Ent. liquefaciens
Ser. marcescens
Hafnia alvei
Er. aroideae
Er. carotovora
S. flexneri
S. dysenteriae
S. boydii
S. dysenteriae
Vibrio spp.

V. alginolyticus

V. parahaemolyticus
V. cholerae
Pseudomonas spp.
Vibrio spp.
B. cereus
Micrococcus spp.

The significance of these clusters or phenons in traditional taxonomic terms is not always evident, and the similarity levels at which clusters are labeled species, genera, and so on, are a matter of judgment. Sometimes groups are simply called phenons and preceded by a number showing the similarity level above which they appear (e.g., a 70-phenon is a phenon with 70% or greater similarity among its constituents). Phenons formed at about 80% similarity often are equivalent to bacterial species.

Numerical taxonomy has already proved to be a powerful tool in microbial taxonomy. Although it often has simply reconfirmed already-existing classification schemes, sometimes accepted classifications are found wanting. Its usefulness will continue to increase with the addition of new tests and the accumulation of more data.

1. What is numerical taxonomy and why are computers so important to this approach?
2. Define the following terms: association coefficient, simple matching coefficient, Jaccard coefficient, similarity matrix, phenon, and dendrogram.
3. Which pair of species has more mutual similarity, a pair with an association coefficient of 0.9 or one with a coefficient of 0.6? Why?

Phenetic Classification and *Bergey's Manual*

Because it has not been possible in the past to classify bacteria satisfactorily based on phylogenetic relationships, the system given in *Bergey's Manual of Systematic Bacteriology* is primarily phenetic. Each of the 33 sections in the four volumes contains bacteria that share a few easily determined characteristics and bears a title that either describes these properties or provides the vernacular names of the bacteria included. The characteristics used to define sections are normally features such as general shape and morphology, Gram-staining properties, oxygen relationship, motility, the presence of endospores, the mode of energy production, and so forth. Bacterial groups are divided among the four volumes in the following manner: (1) gram-negative bacteria of general, medical, or industrial importance; (2) gram-positive bacteria other than actinomycetes; (3) gram-negative bacteria with distinctive properties, cyanobacteria, and archaeobacteria; and (4) actinomycetes (gram-positive filamentous bacteria).

Bacterial classification according to Bergey's Manual of Systematic Bacteriology *(appendix IV).*

Although the new edition of *Bergey's Manual* is organized along phenetic and utilitarian lines, R. G. E. Murray, in a theoretical introductory article to the manual, suggests that the kingdom *Procaryotae* be divided into four divisions and seven classes. The four divisions are characterized primarily on (1) the presence or absence of cell walls and (2) the chemical and structural nature of the cell wall. The divisions largely correspond to traditional groups. The **Gracilicutes** (thin skin) have gram-negative walls, and members of the **Firmicutes** (strong or durable skin) have a gram-positive type cell wall. The **Tenericutes** (soft or tender skin) are the mycoplasmas; the archaeobacteria are placed in the **Mendosicutes** (skin with faults). The divisions differ in many other ways. Some of the most important characteristics of the first three divisions are compared in table 19.7; major features of the archaeobacteria are summarized in table 19.8.

TABLE 19.7   Characteristics of *Gracilicutes*, *Firmicutes*, and *Tenericutes*

Property	Gracilicutes	Firmicutes	Tenericutes
Cell wall	Gram-negative type wall with inner 2–10 nm peptidoglycan layer and outer membrane (8–10 nm thick) of lipid, protein, and lipopolysaccharide. (There may be a third outermost layer of protein.)	Gram-positive type wall with a homogeneous, thick cell wall (20–80 nm) composed mainly of peptidoglycan. Other polysaccharides and teichoic acids may be present.	Lack a cell wall and peptidoglycan precursors; enclosed by a plasma membrane
Cell shape	Spheres, ovals, straight or curved rods, helices or filaments; some have sheaths or capsules.	Spheres, rods, or filaments; may show true branching	Pleomorphic in shape; may be filamentous, can form branches
Reproduction	Binary fission, sometimes budding	Binary fission	Budding, fragmentation, and/or binary fission
Metabolism	Phototrophic, chemolithoautotrophic, or chemoorganoheterotrophic	Usually chemoorganoheterotrophic	Chemoorganoheterotrophic; most require cholesterol and long-chain fatty acids for growth.
Motility	Motile or nonmotile. Flagellation can be varied—polar, lophotrichous, peritrichous. Motility may also result from the use of axial filaments (spirochetes) or gliding motility.	Most often nonmotile; have peritrichous flagellation when motile	Usually nonmotile
Appendages	Can produce several types of appendages—pili and fimbriae, prosthecae, stalks	Usually lack appendages (may have spores on hyphae)	Lack appendages
Endospores	Cannot form endospores	Some groups can form endospores.	Cannot form endospores

Comparison of the phylogenetic scheme discussed in the next section and pictured in figures 19.7 to 19.10 with the phenetically based classification of *Bergey's Manual* shows considerable disagreement. Many phenetically defined taxa are not phylogenetically homogeneous and have members distributed among two or more different phylogenetic groups (at least as judged by 16S rRNA studies). Often characteristics given great weight or importance in *Bergey's Manual* do not appear to be phylogenetically significant. For example, photosynthetic bacteria are found in several different phylogenetic groups together with very closely related nonphotosynthetic bacteria. Thus it may not be appropriate to separate all photosynthetic bacteria from the nonphotosynthetic forms as has been done in *Bergey's Manual*. The mycoplasmas are placed in the division *Tenericutes* in *Bergey's Manual*, but rRNA studies show that they are closely related to gram-positive bacteria (division *Firmicutes*) although they lack cell walls. Because rods, cocci, spirals, and other shapes are found scattered among many phylogenetic groups, these morphological variations do not appear to be useful indicators of relatedness.

In the future phenetic classifications will undoubtedly change substantially to reflect more accurately phylogenetic relationships. It also will be necessary to find more phenotypic traits that are true indicators of phylogenetic relatedness and that are easily studied. Some examples of such traits are peptidoglycan type, fatty acid composition, the presence of teichoic acids, glucose fermentation end products, the relationship to oxygen, and the ability to form endospores.

Despite the uncertainties and problems with the classification in *Bergey's Manual*, it is the most widely accepted and used system for the identification of bacteria (Box 19.1). In fact a shorter version, called *Bergey's Manual of Determinative Bacteriology*, is specifically designed to be used in identification. Moreover, *Bergey's Manual* often does provide phylogenetically meaningful information. Because it is the standard in bacterial taxonomy, the following five chapters survey the bacteria in essentially the same order as that followed in *Bergey's Manual of Systematic Bacteriology*.

1. What characteristics are used to place bacteria in different sections of *Bergey's Manual?*
2. Name the four divisions of bacteria given in *Bergey's Manual* and state their distinguishing features.
3. Discuss some characteristics emphasized by *Bergey's Manual* that may not be phylogenetically significant.

Box 19.1

"Official" Nomenclature Lists: A Letter from *Bergey's*[*]

On a number of occasions lately, the impression has been given that the status of a bacterial taxon in *Bergey's Manual of Systematic Bacteriology* or *Bergey's Manual of Determinative Bacteriology* is in some sense official. Similar impressions are frequently given about the status of names in the *Approved List of Bacterial Names* and in the Validation Lists of newly proposed names that appear regularly in the *International Journal of Systematic Bacteriology*. It is therefore important to clarify these matters.

There is no such thing as an official classification. *Bergey's Manual* is not "official"—it is merely the best consensus at the time, and although great care has always been taken to obtain a sound and balanced view, there are also always regions in which data are lacking or confusing, resulting in differing opinions and taxonomic instability. When *Bergey's Manual* disavows that it is an official classification, many bacteriologists may feel that the solid earth is trembling. But many areas are in fact reasonably well established. Yet taxonomy is partly a matter of judgment and opinion, as is all science, and until new information is available, different bacteriologists may legitimately hold different views. They cannot be forced to agree to any "official classification." It must be remembered that, as yet, we know only a small percentage of the bacterial species in nature. Advances in technique also reveal new lights on bacterial relationships. Thus we must expect that existing boundaries of groups will have to be redrawn in the future, and it is expected that molecular biology, in particular, will imply a good deal of change over the next few decades.

The position with the Approved Lists and the Validation Lists is rather similar. When bacteriologists agreed to make a new start in bacteriological nomenclature, they were faced with tens of thousands of names in the literature of the past. The great majority were useless, because, except for about 2,500 names, it was impossible to tell exactly what bacteria they referred to. These 2,500 were therefore retained in the Approved Lists. The names are only approved in the sense that they were approved for retention in the new bacteriological nomenclature. The remainder lost standing in the nomenclature, which means they do not have to be considered when proposing new bacterial names (although names can be individually revived for good cause under special provisions).

The new International Code of Nomenclature of Bacteria requires all new names to be validly published to gain standing in the nomenclature, either by being published in papers in the *International Journal of Systematic Bacteriology* or, if published elsewhere, by being announced in the Validation Lists. The names in the Validation Lists are therefore valid only in the sense of being validly published (and therefore they must be taken account of in bacterial nomenclature). The names do not have to be adopted in all circumstances; if users believe the scientific case for the new taxa and validly published names is not strong enough, they need not adopt the names. For example, *Helicobacter pylori* was immediately accepted as a replacement for *Campylobacter pylori* by the scientific community, whereas *Tatlockia micdadei* had not generally been accepted as a replacement for *Legionella micdadei*. Taxonomy remains a matter of scientific judgment and general agreement.

*From P. H. A. Sneath and D. J. Brenner, "Official" Nomenclature Lists in *ASM News*, 58(4):175, 1992. Copyright © by the American Society for Microbiology. Reprinted by permission.

Escherichia coli   Methanococcus vannielii   Saccharomyces cerevisiae

Figure 19.6 Small Ribosomal Subunit RNA. Representative examples of rRNA secondary structures from the three primary domains or empires: eubacteria (*Escherichia coli*), archaeobacteria (*Methanococcus vannielii*), and eucaryotes (*Saccharomyces cerevisiae*). The red dots mark positions where eubacteria and archaeobacteria normally differ. *Source: Data from C. P. Woese,* Microbiological Reviews, *51(2):221–227, 1987.*

Bacterial Phylogeny and Taxonomy

Bacterial taxonomy is now changing rapidly. This is caused by ever-increasing knowledge of the biology of bacteria and remarkable advances in computers and the use of molecular characteristics to determine true phylogenetic relationships between bacterial groups. In this section phylogenetic and phenetic approaches to bacterial taxonomy are described and compared.

Phylogenetic Studies

Although a variety of molecular techniques are used in estimating the phylogenetic relatedness of bacteria, the comparison of 16S rRNAs isolated from over 2,000 bacteria is of particular importance (figure 19.6). The S_{ab} values obtained from rRNA studies are assumed to be a true measure of relatedness; the higher the S_{ab} values obtained from comparing two organisms, the more closely the organisms are related to each other. If all the T_1 ribonuclease fragments from the 16S rRNAs of two organisms are identical in sequence, the S_{ab} value is 1.0. S_{ab} values are also a measure of evolutionary time. A group of bacteria that branched off from other bacteria a long time ago will exhibit a large range of S_{ab} values because it has had more time to diversify than a group that developed more recently. That is, the narrower the range of S_{ab} values in a group of bacteria, the more modern it is. After S_{ab} values have been determined, a computer calculates the relatedness of the organisms and summarizes their relationships in a dendrogram (figure 19.4c). As mentioned earlier, it is now possible to sequence complete 16S rRNA molecules rather than fragments. This has

given rRNA studies even greater resolving power. Although rRNA comparisons are useful above the species level, DNA similarity studies sometimes are more effective in categorizing individual species.

Ribosomal RNA sequence studies have uncovered a feature of great practical importance. The 16S rRNA of most major phylogenetic groups has one or more characteristic nucleotide sequences called oligonucleotide signatures. **Oligonucleotide signature sequences** are specific oligonucleotide sequences that occur in most or all members of a particular phylogenetic group. They are rarely or never present in other groups, even closely related ones. Thus signature sequences can be used to place microorganisms in the proper group. Signature sequences have been identified for eubacteria, archaeobacteria, eucaryotes, and many major bacterial groups.

Some tentative generalizations arise from the results of the phylogenetic studies summarized in figures 19.7 through 19.10. First, it appears that there are two quite different groups of bacteria, the eubacteria and the archaeobacteria. The **eubacteria** [Greek *eu*, good, and bacteria] or true bacteria comprise the vast majority of bacteria (figure 19.7). Among other properties, eubacteria have cell wall peptidoglycan containing muramic acid (or are related to bacteria with such cell walls) and membrane lipids that resemble eucaryotic membrane lipids in having ester-linked, straight-chained fatty acids (table 19.8). The second group, the **archaeobacteria** [Greek *archaios*, ancient, and *bakterion*, a small rod], differ from eubacteria in many respects and resemble eucaryotes in some ways (table 19.8). It should be noted that the spelling archaeobacteria, which will be

TABLE 19.8 Comparison of Eubacteria, Archaeobacteria, and Eucaryotes

Property	Eubacteria (Bacteria)	Archaeobacteria (Archaea)	Eucaryotes (Eucarya)
Membrane-Enclosed Nucleus	Absent	Absent	Present
Cell Wall	Peptidoglycan containing muramic acid	Variety of types, no muramic acid	No muramic acid
Membrane Lipid	Have ester-linked, straight-chained fatty acids	Have ether-linked, branched aliphatic chains	Have ester-linked, straight-chained fatty acids
Transfer RNA	Thymine present in most tRNAs	No thymine in T or TψC arm of tRNA	Thymine present
	N-formylmethionine carried by initiator tRNA	Methionine carried by initiator tRNA	Methionine carried by initiator tRNA
mRNA Splicing, Capping, and Poly A Tailing	Absent	Absent	Present
Ribosomes			
Size	70S	70S	80S (cytoplasmic ribosomes)
Elongation factor 2	Does not react with diphtheria toxin	Reacts	Reacts
Sensitivity to chloramphenicol and kanamycin	Sensitive	Insensitive	Insensitive
Sensitivity to anisomycin	Insensitive	Sensitive	Sensitive
DNA-Dependent RNA Polymerase			
Number of enzymes	One	Several	Three
Structure	Simple subunit pattern (4 subunits)	Complex subunit pattern similar to eucaryotic enzymes (8–12 subunits)	Complex subunit pattern (12–14 subunits)
Rifampicin sensitivity	Sensitive	Insensitive	Insensitive
Methanogenesis	Absent	Present in some species	Absent
Nitrogen Fixation	Present	Present	Absent

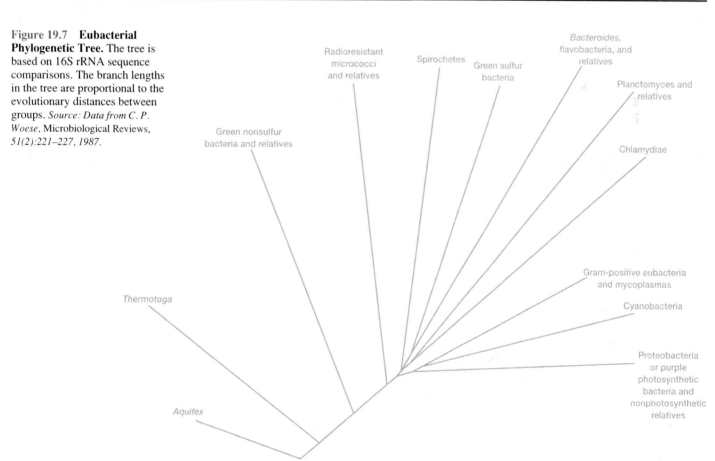

Figure 19.7 Eubacterial Phylogenetic Tree. The tree is based on 16S rRNA sequence comparisons. The branch lengths in the tree are proportional to the evolutionary distances between groups. *Source: Data from C. P. Woese,* Microbiological Reviews, *51(2):221–227, 1987.*

Radioresistant micrococci and relatives

Spirochetes

Green sulfur bacteria

Bacteroides, flavobacteria, and relatives

Planctomyces and relatives

Green nonsulfur bacteria and relatives

Chlamydiae

Thermotoga

Gram-positive eubacteria and mycoplasmas

Cyanobacteria

Aquifex

Proteobacteria or purple photosynthetic bacteria and nonphotosynthetic relatives

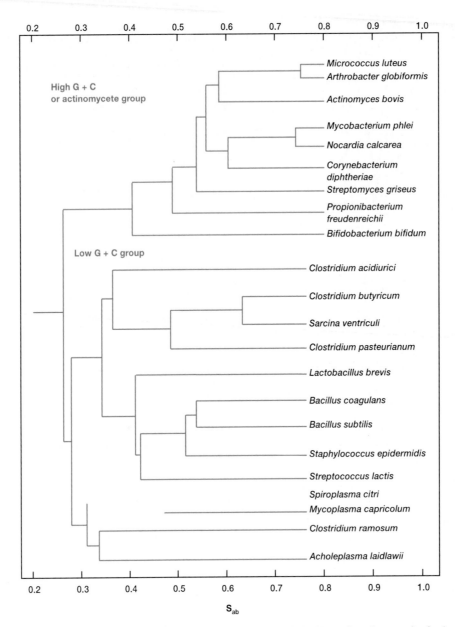

Figure 19.8   The Phylogeny of Gram-Positive Eubacteria. Data are derived from 16S rRNA studies. See text for further details.

used in this text, is equivalent to the more popular spelling, archaebacteria. The letter *o* is inserted because this word is formed by combining two Greek words (a point emphasized in *Bergey's Manual*). Although archaeobacteria are described in more detail at a later point, it should be noted that they differ from eubacteria in lacking muramic acid in their cell walls and in possessing (1) membrane lipids with ether-linked branched aliphatic chains, (2) transfer RNAs without thymidine in the T or TψC arm (*see chapter 11*), (3) distinctive RNA polymerase enzymes, and (4) ribosomes of different composition and shape. Thus although archaeobacteria resemble eubacteria in their procaryotic cell structure, they vary considerably on the molecular level (Box 19.2).

The archaeobacteria (chapter 23).

The 16S rRNA studies suggest that the eubacteria are composed of approximately 11 major groups or divisions, although more may be discovered with further research. Figures 19.7 to 19.9 summarize eubacterial phylogenetic relationships. Because most of these bacteria are described in chapters 20 to 24, the eubacterial groups will be only briefly discussed here and representative members mentioned.

1. Gram-positive eubacteria including the mycoplasmas. These bacteria are divided using DNA base composition into two major subdivisions (figure 19.8). The high G + C or actinomycete group contains genera such as *Actinomyces, Arthrobacter, Bifidobacterium, Corynebacterium, Micrococcus, Mycobacterium, Nocardia, Propionibacterium,* and *Streptomyces.* The low

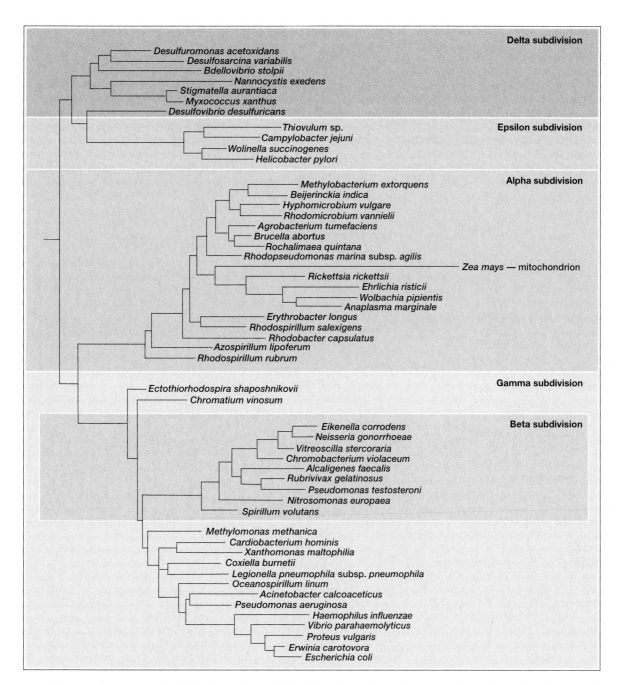

Figure 19.9 **The Phylogeny of the Proteobacteria or Purple Bacteria.** The tree is based on comparisons of 16S rRNA sequences.

G + C group contains *Bacillus, Clostridium, Enterococcus, Lactobacillus, Leuconostoc, Listeria,* mycoplasmas, *Spiroplasma, Staphylococcus,* and *Streptococcus.* The low G + C group is older than the actinomycete group and probably gave rise to it. Mycoplasmas appear to have arisen from the clostridia.

The gram-positive bacteria (chapters 21 and 24).

2. The proteobacteria or purple photosynthetic bacteria and nonphotosynthetic relatives. This division contains most species traditionally thought of as gram-negative

eubacteria (figure 19.9). Sequence data indicate that the ancestral bacterium was a photosynthetic purple bacterium, and photosynthetic genera are scattered throughout the division. Presumably many strains lost photosynthesis when adapting metabolically to new ecological niches. The proteobacteria are exceptionally diverse and are divided into at least five subdivisions as follows:

a. Alpha subdivision: agrobacteria, *Beijerinckia, Caulobacter, Hyphomicrobium, Methylococcus, Nitrobacter,* purple nonsulfur bacteria, rhizobacteria, and rickettsias.

Box 19.2

A New Kingdom of Organisms?

James Lake and his colleagues have found that the ribosomes of sulfur-dependent archaeobacteria differ significantly in shape from those of eubacteria, eucaryotes, and other archaeobacteria. Lake also has analyzed rRNA sequences using a new mathematical procedure and obtained a different phylogentic tree than that shown in figure 19.10 (Box Figure 19.2). He proposes that these sulfur-dependent, thermophilic bacteria constitute a new kingdom of organisms, which he has named eocytes (dawn cells). In this new phylogenetic scheme, there are three major groups of procaryotes. The halobacteria are more closely related to the eubacteria than to the methanogens.

The sulfur-dependent, extremely thermophilic archaeobacteria are renamed eocytes, and the archaeobacteria contain only methanogenic bacteria. In grouping the eocytes with eucaryotes, Lake directly challenges the traditional procaryotic/eucaryotic distinction. He proposes that the early ancestor of the eucaryotes lacked a nucleus, metabolized sulfur, and lived at high temperatures.

It is difficult to forecast the future success of this proposal. Many microbiologists feel that halobacteria and sulfur-dependent, thermophilic bacteria should be kept among the archaeobacteria because differences in ribosome shape and rRNA sequences are not sufficient to justify such a major taxonomic change. This controversy is a good example of the current excitement and rapid progress in the field of bacterial taxonomy.

Box Figure 19.2 **The Eocyte Tree.** The length of the most rapidly evolving branches (eubacteria, eucaryotes, and halobacteria) have been shortened to fit the figure. The tree has two deep branches. One branch, which Lake calls the parkaryote branch, contains the eubacteria, halobacteria, and methanogens (plus *Thermoplasma* and *Archaeoglobus*). The other branch, the karyotes, consists of the eocytes and eucaryotes. *Source: Data from J. A. Lake,* Trends in Biochemical Sciences, *16(2):46–50, 1991, Elsevier Science Publishers, Amsterdam.*

b. Beta subdivision: *Alcaligenes, Bordetella, Leptothrix, Neisseria, Pseudomonas, Rhodocyclus, Spirillum,* and *Thiobacillus.*

c. Gamma subdivision: *Azotobacter, Beggiatoa, Chromatium,* the family *Enterobacteriaceae, Haemophilus, Legionella, Leucothrix, Pasteurella, Pseudomonas,* purple sulfur bacteria, *Thiothrix,* and *Vibrio.*

d. Delta subdivision: *Bdellovibrio, Desulfovibrio* and other sulfur and sulfate reducers, myxobacteria.

e. Epsilon subdivision: *Campylobacter, Wolinella.*

The gram-negative bacteria (chapters 20 and 22).

3. Cyanobacteria. Oxygenic, photosynthetic bacteria. Some representative genera are *Anabaena, Calothrix, Fischerella, Nostoc, Oscillatoria, Scytonema,* and *Synechococcus. Prochloron* is also placed here.

The cyanobacteria (pp. 459–63).

Figure 19.10   Archaeobacterial Phylogenetic Tree. The tree is based on 16S rRNA comparisons and has the same meaning as figure 19.7. *Source: Data from C. P. Woese,* Microbiological Reviews, *51(2):221–271, 1987.*

4. Spirochetes and relatives. Gram-negative bacteria with a spiral shape and distinctive axial filament. *Borrelia, Leptospira, Spirochaeta,* and *Treponema.*

    The spirochetes (pp. 416–18).

5. The Planctomyces group. These walled, budding bacteria resemble other eubacteria, but lack peptidoglycan. *Planctomyces* and *Isosphaera.*

6. *Bacteroides,* flavobacteria, and relatives. Includes *Bacteroides, Cytophaga, Flavobacterium, Flexibacter,* and *Fusobacterium.*

    Bacteroides (pp. 429–30), Cytophaga and Flexibacter (p. 470).

7. Chlamydiae. Intracellular parasites with a unique life cycle and no peptidoglycan. *Chlamydia.*

    The order Chlamydiales (pp. 432–34).

8. Green sulfur bacteria (family *Chlorobiaceae*). Anaerobic, photosynthetic bacteria that can oxidize sulfide and sulfur and have chlorosomes. Includes *Chlorobium* and *Pelodictyon.*

    The green sulfur bacteria (p. 459).

9. Green nonsulfur bacteria and relatives. *Chloroflexus, Herpetosiphon,* and *Thermomicrobium.*

10. Radioresistant micrococci and relatives. Members of the family *Deinococcaceae* are gram-positive cocci with great radiation resistance. The thermophilic, gram-negative rods of the genus *Thermus* appear quite different, although the rRNA sequences indicate that they are related to *Deinococcus.*

    The Deinococcaceae (pp. 442–43).

11. *Thermotoga* and *Thermosipho.* These thermophilic eubacteria from geothermally heated marine sediments differ from all other eubacteria (figure 19.7), but have not been thoroughly studied.

Archaeobacterial phylogeny is depicted in figure 19.10. The archaeobacteria consist of three main groups: the extreme halophiles, the methanogens (bacteria that reduce carbon dioxide to methane), and sulfur-dependent extreme thermophiles. The tree has two main branches: a cluster of extreme thermophiles such as *Sulfolobus* and the methanogen branch. The genus *Thermoplasma* is thermophilic but more closely related to the methanogens than to the other thermophiles. The methanogens are placed in three orders, each of which may contain several genera: *Methanococcales* (*Methanococcus*), *Methanobacteriales* (*Methanobacterium, Methanobrevibacter*), and *Methanomicrobiales* (*Methanosarcina, Methanospirillum*). There are at least six genera of extreme halophiles in the order *Halobacteriales.* Archaeobacterial classification will undoubtedly change with further discoveries.

---

1. Discuss the use of S_{ab} values in determining the relatedness of bacteria and evolutionary age of taxonomic groups. What are oligonucleotide signature sequences?

2. What are eubacteria and archaeobacteria, and in what ways do they differ from each other? Very briefly describe the major eubacterial and archaeobacterial groups.

The Kingdoms of Organisms

Since the beginning of biology, organisms have been classified as either plants or animals. However, discoveries in microbiology over the past century have shown that the two-kingdom system is oversimplified, and many biologists favor dividing organisms among five kingdoms as first suggested by Robert H. Whittaker. A modified version of Whittaker's five-kingdom system is presented in table 19.9 and figure 19.11. Organisms are placed into five kingdoms based on at least three major criteria: (1) cell type—procaryotic or eucaryotic, (2) level of organization—solitary and colonial unicellular organization or multicellular, and (3) nutritional type (Box 19.3 provides a discussion of the origin of the eucaryotic cell). In this system the kingdom *Animalia* contains multicellular animals with wall-less eucaryotic cells and primarily ingestive nutrition, whereas the kingdom *Plantae* is composed of multicellular plants with walled eucaryotic cells and primarily photoautotrophic nutrition. Microbiologists study members of the other three kingdoms. The kingdom *Monera* or *Procaryotae* contains all procaryotic organisms. The kingdom *Protista* is the least homogeneous and hardest to define. **Protists** are eucaryotes with unicellular organization, either in the form of solitary cells or colonies of cells lacking true tissues. They may have ingestive, absorptive, or photoautotrophic nutrition, and they include most of the microorganisms known as algae, protozoa, and many of the simpler fungi. The kingdom *Fungi* contains eucaryotic and predominately multinucleate organisms, with nuclei dispersed in a walled and often septate mycelium (*see chapter 25*); their nutrition is absorptive. The taxonomy of the major protist and fungal phyla is discussed in more detail in chapters 25 to 27.

TABLE 19.9 The Five-Kingdom System

Kingdom *Monera* or *Procaryotae*
Procaryotic cells. Nutrition absorptive, chemosynthetic, photoheterotrophic, or photoautotrophic, with anaerobic, facultative, microaerophilic, or aerobic metabolism. Reproduction asexual; genetic recombination sometimes occurs. Nonmotile or motile by bacterial flagella composed of flagellin proteins or by gliding. Solitary, unicellular, filamentous, colonial, or mycelial.

Kingdom *Protista*
Eucaryotic cells with solitary and colonial unicellular organization. Usually aerobic. Nutrition ingestive, absorptive, or, if photoautotrophic, in photosynthetic plastids. All forms reproduce asexually; many have true sexual reproduction with meiosis and karyogamy. Nonmotile or motile by cilia, flagella, or other means (e.g., pseudopodia). Lack embryos and complex cell junctions.

Kingdom *Fungi*
Primarily multinucleate organisms with eucaryotic nuclei dispersed in a walled and often septate mycelial syncytium (some forms secondarily unicellular). Usually aerobic. Nutrition heterotrophic and absorptive. Flagella lacking, no motility except protoplasmic streaming. Mycelia are haploid or dikaryotic. Meiosis in zygote; haploid spores produced. No pinocytosis or phagocytosis.

Kingdom *Animalia*
Multicellular animals with wall-less eucaryotic cells. Aerobic. Nutrition primarily ingestive with digestion in an internal cavity, but some forms are absorptive and some lack a digestive cavity; phagocytosis and pinocytosis present. Reproduction mainly sexual with meiosis forming gametes; haploid stages other than gametes almost lacking above lowest phyla. Motility based on contractile fibrils. With one exception, zygote develops into a blastula. Extensive cellular and tissue differentiation with complex cell junctions.

Kingdom *Plantae*
Primarily autotrophic plants, multicellular with walled and frequently vacuolate eucaryotic cells and photosynthetic plastids. Aerobic. Simple multicellular to advanced tissue organization; development from solid embryos. Reproduction primarily sexual, with haploid and diploid stages ("alternation of generations"); haploid stage reduced in higher members of the kingdom. Usually nonmotile.

From R. H. Whittaker and L. Margulis, *BioSystems* 10:3–18. Copyright © 1978 Elsevier Scientific Publishers, Ireland. Reprinted by permission.

The five-kingdom system is not accepted by all microbiologists. A major problem is its lack of distinction between archaeobacteria and eubacteria. The kingdom *Protista* also may be too diverse to be taxonomically useful. In addition, the boundaries between the kingdoms *Protista*, *Plantae*, and *Fungi* are ill-defined. For example, the brown algae are probably not closely related to the plants even though the five-kingdom system places them in the *Plantae*. Various alternatives have been

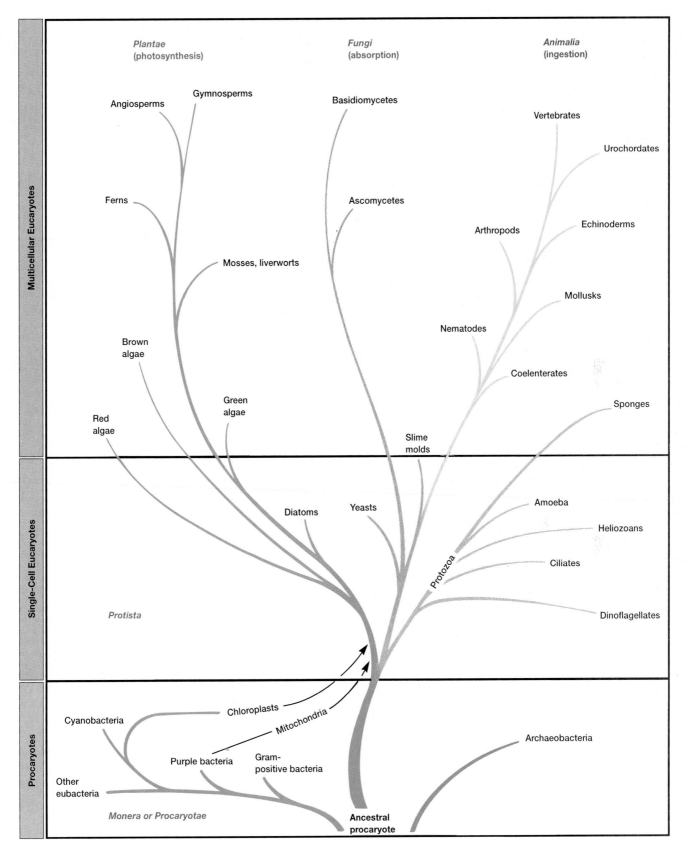

Figure 19.11   The Five-Kingdom System. A simplified diagram showing the five kingdoms, levels of organization, cell types, and some lines of descent. See text for details.

Box 19.3

The Origin of the Eucaryotic Cell

It appears likely that eucaryotic cells arose from procaryotes about 1.4 billion years ago. There has been considerable speculation about how eucaryotes might have developed from procaryotic ancestors, and two hypotheses have been proposed. According to the first, nuclei, mitochondria, and chloroplasts arose by invagination of the plasma membrane to form double-membrane structures containing genetic material and capable of further development and specialization. The similarities between chloroplasts, mitochondria, and modern bacteria are due to conservation of primitive procaryotic features by the slowly changing organelles.

The more popular **endosymbiotic hypothesis** proposes that a free-living fermenting bacterium established a permanent symbiotic relationship with photosynthetic bacteria, which then evolved into chloroplasts. Cyanobacteria have been considered the most likely ancestors of chloroplasts. More recently *Prochloron* has become the favorite candidate. *Prochloron* (pp. 462–63) lives within marine invertebrates and resembles chloroplasts in containing both chlorophyll *a* and *b*, but not phycobilins. The existence of this bacterium suggests that chloroplasts arose from a common ancestor of prochlorophytes and cyanobacteria. Mitochondria arose from an endosymbiotic relationship between the free-living bacterium and bacteria with aerobic respiration (possibly an ancestor of three modern groups: *Agrobacterium, Rhizobium,* and the rickettsias). The symbiotic origin of nuclei has also been proposed, but the evidence is not yet compelling.

The endosymbiotic hypothesis has received support from the discovery of an endosymbiotic cyanobacterium that inhabits the biflagellate protist *Cyanophora paradoxa* and acts as its chloroplast. This endosymbiont, called a cyanelle, resembles the cyanobacteria in its photosynthetic pigment system and fine structure and is surrounded by a peptidoglycan layer. It differs from cyanobacteria in lacking the lipopolysaccharide outer membrane characteristic of gram-negative bacteria. The cyanelle may be a recently established endosymbiont that is evolving into a chloroplast.

At present both hypotheses have supporters. It is possible that new data may help resolve the issue to everyone's satisfaction. However, these hypotheses concern processes that occurred in the distant past and cannot be directly observed. Thus a complete consensus on the matter may never be reached.

---

suggested. The six-kingdom system is the simplest option; it divides the kingdom *Monera* or *Procaryotae* into two kingdoms, the *Eubacteria* and *Archaeobacteria* (figure 19.12*a*). Many attempts have been made to divide the protists into several better-defined kingdoms. The eight-kingdom system of Cavalier-Smith is a good example (figure 19.12*b*). Cavalier-Smith believes that differences in cellular structure and genetic organization are exceptionally important in determining phylogeny; thus he has used ultrastructural characteristics as well as rRNA sequences and other molecular data in developing his classification. He divides all organisms into two empires and eight kingdoms. The empire *Bacteria* contains two kingdoms, the *Eubacteria* and the *Archaeobacteria*. The second empire, the *Eucaryota*, contains six kingdoms of eucaryotic organisms. There are two new kingdoms of eucaryotes. The *Archezoa* are primitive eucaryotic unicellular organisms such as *Giardia* that have 70S ribosomes and lack Golgi bodies, mitochondria, chloroplasts, and peroxisomes. The kingdom *Chromista* contains mainly photosynthetic organisms that have their chloroplasts within the lumen of the rough endoplasmic reticulum rather than in the cytoplasmic matrix (as is the case in the kingdom *Plantae*). Diatoms, brown algae, cryptomonads, and oomycetes are all placed in the *Chromista*. The boundaries of the remaining four kingdoms—*Plantae, Fungi, Animalia,* and *Protozoa*—have been adjusted to better define each kingdom and distinguish it from the others.

Carl Woese and his collaborators primarily employ the comparison of rRNA sequences in developing their classification system and come to quite different conclusions. According to Woese, the kingdom *Procaryotae* is composed of two groups, which differ from one another as much as from the eucaryotes. Figure 19.13 depicts a universal phylogenetic tree that reflects these views. The tree is divided into three major branches representing the three primary groups: eubacteria, archaeobacteria, and eucaryotes. The archaeobacteria and eucaryotes are distant relatives. These three primary groups are called domains or empires and placed above the kingdom level. The traditional kingdoms are distributed among these three domains.

Carl Woese, Otto Kandler, and Mark Wheelis have coined new names for the domains. Eucaryotic organisms with primarily glycerol fatty acyl diester membrane lipids and eucaryotic

(a)

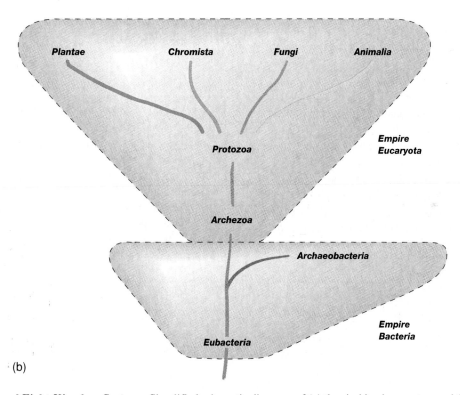

(b)

Figure 19.12   The Six- and Eight-Kingdom Systems. Simplified schematic diagrams of (*a*) the six-kingdom system and (*b*) the eight-kingdom system. See text for details.

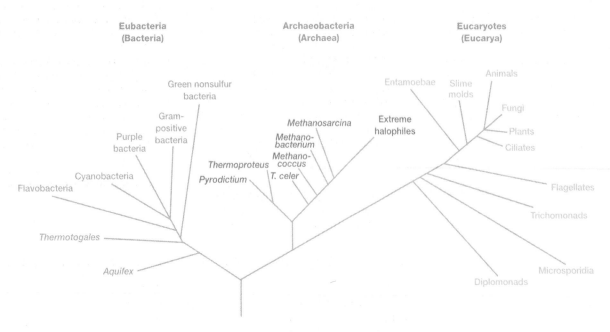

Eubacteria
(Bacteria)

Archaeobacteria
(Archaea)

Eucaryotes
(Eucarya)

Green nonsulfur
bacteria

Gram-
positive
bacteria

Purple
bacteria

Methanosarcina

Methano-
bacterium

Methano-
coccus

Extreme
halophiles

Entamoebae

Slime
molds

Animals

Fungi

Plants

Ciliates

Cyanobacteria

Thermoproteus

Pyrodictium

T. celer

Flavobacteria

Flagellates

Trichomonads

Thermotogales

Aquifex

Microsporidia

Diplomonads

Figure 19.13   Universal Phylogenetic Tree. These relationships were determined from rRNA sequence comparisons. *Source: Data from G. J. Olsen and C. R. Woese, "Ribosomal RNA: A Key to Phylogeny" in* The FASEB Journal, *7:113–123, 1993.*

rRNA belong to the **eucarya.** The domain **bacteria** contains procaryotic cells with membrane lipids that are primarily diacyl glycerol diesters and with eubacterial rRNA. Procaryotes having isoprenoid glycerol diether or diglycerol tetraether lipids (*see pp. 478–79*) in their membranes and archaeobacterial rRNA compose the third domain, **archaea.** These new names are intended to replace the more commonly used terms eucaryotes, eubacteria, and archaeobacteria.

At present there is little agreement about domains and kingdoms. Further research may resolve the issue.

1. With what three major criteria did Whittaker divide organisms into five kingdoms?
2. Give the names and main distinguishing characteristics of the five kingdoms.
3. Briefly describe the six- and eight-kingdom systems. How do they differ from the five-kingdom system?
4. How does Woese divide organisms into domains or empires in his universal phylogenetic tree?

Summary

1. Taxonomy, the science of biological classification, is composed of three parts: classification, nomenclature, and identification.

2. The nature of the species is different for sexually and asexually reproducing organisms. A bacterial species is a collection of strains that have many stable properties in common and differ significantly from other groups of strains.

3. Microorganisms are named according to the binomial system.

4. *Bergey's Manual of Systematic Bacteriology* gives the accepted system of bacterial taxonomy.

5. Morphological, physiological, metabolic, and ecological characteristics are widely used in microbial taxonomy.

6. The study of transformation and conjugation in bacteria is sometimes taxonomically useful. Plasmid-borne traits can cause errors in bacterial taxonomy if care is not taken.

7. Proteins are direct reflections of mRNA sequences and may be used to compare genomes from different organisms.

8. The G + C content of DNA is easily determined and taxonomically valuable because it is an indirect reflection of the base sequence.

9. Nucleic acid hybridization studies are used to compare DNA or RNA sequences and thus determine genetic relatedness.

10. Nucleic acid sequencing is the most powerful and direct method for comparing genomes. The sequences of 16S and 5S rRNA are most often used in phylogenetic studies.

11. The two major types of natural classification are phylogenetic systems and phenetic systems.

12. Classifications are often constructed by means of numerical taxonomy, in which the general similarity of organisms is determined by using a computer to calculate and analyze association coefficients.

13. *Bergey's Manual of Systematic Bacteriology* provides a primarily phenetic classification, and many taxa are not phylogenetically homogeneous. However, it has been proposed that the kingdom *Procaryotae* be divided between four divisions (*Gracilicutes, Firmicutes, Tenericutes,* and *Mendosicutes*) and seven classes.

14. Studies on the sequence of 16S rRNA and other molecular properties suggest that the procaryotes are divided into two very different groups: eubacteria and archaeobacteria.

15. Eubacteria or true bacteria comprise the vast majority of bacteria and can be separated into at least 11 major subgroups based on 16S rRNA similarity.

16. Archaeobacteria differ from eubacteria in cell wall composition, membrane lipids, tRNA structure, ribosomes, and other properties.

17. Organisms may be divided into five kingdoms based on their cell type, level of organization, and nutrition. These are *Animalia, Plantae, Monera* or *Procaryotae, Fungi,* and *Protista.* Other classification systems group organisms into six kingdoms, two empires and eight kingdoms, or three domains and numerous kingdoms.

Key Terms

archaea *412*

archaeobacteria *402*

bacteria *412*

bacterial species *392*

binomial system *392*

biovar *392*

classification *391*

dendrogram *398*

endosymbiotic hypothesis *410*

eubacteria *402*

eucarya *412*

Firmicutes *400*

Fungi *408*

G + C content *394*

genus *392*

Gracilicutes *400*

identification *391*

Jaccard coefficient (S_J) *398*

melting temperature (T_m) *394*

Mendosicutes *400*

Monera *408*

morphovar *392*

natural classification *397*

nomenclature *391*

nucleic acid hybridization *395*

numerical taxonomy *397*

oligonucleotide signature sequences *402*

phenetic system *397*

phenons *398*

phylogenetic or phyletic classification systems *397*

phylogeny *397*

Procaryotae *408*

Protista *408*

protists *408*

serovar *392*

similarity matrix *398*

simple matching coefficient (S_{SM}) *397*

species *392*

strain *392*

systematics *391*

taxon *391*

taxonomy *391*

Tenericutes *400*

type strain *392*

Questions for Thought and Review

1. Why are size and shape often less useful in characterizing bacterial species than eucaryotic microbial species?

2. Why might a phylogenetic approach to microbial classification be preferable to a phenetic approach? Give arguments for the use of a phenetic classification. Which do you think is preferable and why?

3. Would a numerical taxonomist favor the phylogenetic or phenetic approach to obtaining a natural classification? Why might it be best to use unweighted characteristics in classification? Are there reasons to give some properties more weight or importance than others?

4. In what way does the simple matching coefficient (S_{SM}) differ from the Jaccard coefficient (S_J)? Give a reason why the former coefficient might be preferable to the latter.

5. What genetic feature of bacteria makes it advisable to use as many characteristics as convenient in classifying and identifying an organism?

6. What properties of the molecules should be considered when selecting RNAs or proteins to sequence for the purpose of determining the relatedness of microorganisms that are only distantly related?

7. Why is the current bacterial classification system likely to change considerably in the future? How would one select the best features to use in identification of unknown microorganisms and determination of relatedness?

Additional Reading

General

Amann, R. I; Ludwig, W.; and Schleifer, K.-H. 1995. Phylogenetic identification and in situ detection of individual microbial cells without cultivation. *Microbiol. Rev.* 59(1):143–69.

Balows, A.; Truper, H. G.; Dworkin, M.; Harder, W.; and Schleifer, K-H. 1992. *The prokaryotes,* 2d ed. New York: Springer-Verlag.

Goodfellow, M., and O'Donnell, A. G., editors. 1993. *Handbook of new bacterial systematics.* San Diego, Calif.: Academic Press.

Logan, N. A. 1994. *Bacterial systematics.* Boston: Blackwell Scientific.

Margulis, L.; Corliss, J. O.; Melkonian, M.; and Chapman, D. J., editors. 1990. *Handbook of protoctista.* Boston: Jones and Bartlett.

Priest, F., and Austin, B. 1993. *Modern bacterial taxonomy,* 2d ed. New York: Chapman and Hall.

Ragan, M. A., and Chapman, D. J. 1978. *A biochemical phylogeny of the protists.* New York: Academic Press.

Sneath, P. H. A. 1978. Classification of microorganisms. In *Essays in microbiology,* J. R. Norris and M. H. Richmond, editors, 9/1–9/31. New York: John Wiley and Sons.

Major Characteristics Used in Taxonomy

Johnson, J. L. 1984. Nucleic acids in bacterial classification. In *Bergey's manual of systematic bacteriology,* J. G. Holt, vol. 1, N. R. Krieg, editors, 8–11. Baltimore, Md.: Williams & Wilkins.

Jones, D., and Krieg, N. R. 1984. Serology and chemotaxonomy. In *Bergey's manual of systematic bacteriology,* J. G. Holt, vol. 1, N. R. Krieg, editors, 15–18. Baltimore, Md.: Williams & Wilkins.

Classification Systems

Stackebrandt, E. 1992. Unifying phylogeny and phenotypic diversity. In *The prokaryotes,* 2d ed. A. Ballows et al., editors, 19–47. New York: Springer-Verlag.

Numerical Taxonomy

Goodfellow, M.; Jones, D.; and Priest, F. G., editors. 1985. *Computer-assisted bacterial systematics.* New York: Academic Press.

Sneath, P. H. A. 1984. Numerical taxonomy. In *Bergey's manual of systematic bacteriology,* J. G. Holt, vol. 1, N. R. Krieg, editors, 5–7. Baltimore, Md.: Williams & Wilkins.

Sneath, P. H. A., and Sokal, R. R. 1973. *Numerical taxonomy: The principles and practice of numerical classification.* San Francisco: W. H. Freeman.

Staley, T. E., and Colwell, R. R. 1973. Application of molecular genetics and numerical taxonomy to the classification of bacteria. *Ann. Rev. Ecol. Syst.* 8:273–300.

Phenetic Classification and *Bergey's Manual*

Gibbons, N. E., and Murray, R. G. E. 1978. Proposals concerning the higher taxa of bacteria. *Int. J. Syst. Bacteriol.* 28(1):1–6.

Holt, J. G., editor. 1984–1989. *Bergey's manual of systematic bacteriology.* 4 vols. Baltimore, Md.: Williams & Wilkins.

Holt, J. G.; Krieg, N. R.; Sneath, P. H. A.; Staley, J. T.; and Williams, S. T. 1994. *Bergey's manual of determinative bacteriology,* 9th ed. Baltimore, Md.: Williams & Wilkins.

Murray, R. G. E. 1984. The higher taxa, or, a place for everything . . .? In *Bergey's manual of systematic bacteriology,* J. G. Holt, vol. 1, N. R. Krieg, editors, 31–34. Baltimore, Md.: Williams & Wilkins.

Bacterial Phylogeny and Taxonomy

Fox, G. E., et al. 1980. The phylogeny of prokaryotes. *Science* 209:457–63.

Lake, J. A. 1991. Tracing origins with molecular sequences: Metazoan and eukaryotic beginnings. *Trends Biochem. Sci.* 16(2):46–50.

Olsen, G. J., and Woese, C. R. 1993. Ribosomal RNA: A key to phylogeny. *FASEB J.* 7:113–23.

Olsen, G. J.; Woese, C. R.; and Overbeek, R. 1994. The winds of (evolutionary) change: Breathing new life into microbiology. *J. Bacteriol.* 176(1):1–6.

Schleifer, K.-H., and Stackebrandt, E. 1983. Molecular systematics of prokaryotes. *Ann. Rev. Microbiol.* 37:143–87.

Sneath, P. H. A. 1989. Analysis and interpretation of sequence data for bacterial systematics: The view of a numerical taxonomist. *Syst. Appl. Microbiol.* 12:15–31.

Woese, C. R. 1981. Archaebacteria. *Sci. Am.* 244(6):98–122.

Woese, C. R. 1987. Bacterial evolution. *Microbiol. Rev.* 51(2):221–71.

Zavarzin, G. A.; Stackebrandt, E.; and Murray, R.G.E. 1991. A correlation of phylogenetic diversity in the *Proteobacteria* with the influences of ecological forces. *Can. J. Microbiol.* 37:1-6.

Zillig, W.; Palm, P.; Reiter, W.-D.; Gropp, F.; Pühler, G.; and Klenk, H-P. 1988. Comparative evaluation of gene expression in archaebacteria. *Eur. J. Biochem.* 173:473–82.

The Kingdoms of Organisms

Cavalier-Smith, T. 1987. The origin of eukaryote and archaebacterial cells. *Ann. N.Y. Acad. Sci.* 503:17–54.

Cavalier-Smith, T. 1993. Kingdom protozoa and its 18 phyla. *Microbiol. Rev.* 57(4):953–94.

Gray, M. W. 1983. The bacterial ancestry of plastids and mitochondria. *BioScience* 33(11):693–99.

Kabnick, K. S., and Peattie, D. A. 1991. *Giardia:* A missing link between prokaryotes and eukaryotes. *American Scientist* 79:34–43.

Knoll, A. H. 1991. End of the proterozoic eon. *Sci. Am.* 265(1):64–73.

Last, G. A. 1988. Musings on bacterial systematics: How many kingdoms of life? *ASM News* 54(1):22–27.

Margulis, L. 1971. Symbiosis and evolution. *Sci. Am.* 225(2):49–57.

Margulis, L. 1981. How many kingdoms? Current views of biological classification. *Am. Biol. Teach.* 43(9):482–89.

Vidal, G. 1984. The oldest eukaryotic cells. *Sci. Am.* 250(2):48–57.

Whittaker, R. H., and Margulis, L. 1978. Protist classification and the kingdoms of organisms. *BioSystems* 10:3–18.

Woese, C. R.; Kandler, O.; and Wheelis, M. L. 1990. Towards a natural system of organisms: Proposal for the domains archaea, bacteria, and eucarya. *Proc. Natl. Acad. Sci.* 87:4576–79.

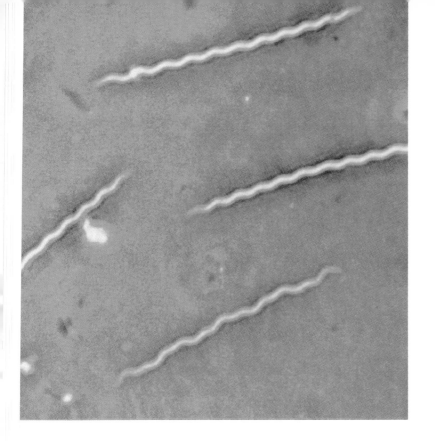

CHAPTER 20
The Bacteria: Gram–Negative Bacteria of General, Medical, or Industrial Importance

Microbes is a vigitable, an'
ivry man is like a conservatory
full iv millyons iv these potted
plants.

—Finley Peter Dunne

Outline

Concepts

1. Volume 1 of *Bergey's Manual* describes 10 groups of gram-negative bacteria having general, medical, or industrial importance. Bacterial endosymbionts are also included in volume 1.

2. The 10 sections are distinguished on the basis of a few major properties: general shape, the presence or absence of motility, mechanism of motility (spirochetes), oxygen relationships, the absence of a cell wall (mycoplasmas), and the requirement for an intracellular existence (rickettsias and chlamydiae).

3. With the exception of the spirochetes (and possibly the mycoplasmas plus a few other groups), the gram-negative bacteria in volume 1 do not vary drastically in general shape or appearance. In contrast, they are very diverse in their metabolism and life-styles, which range from obligately intracellular parasitism to a free-living existence in soil and aquatic habitats.

4. Many bacteria that specialize in predatory or parasitic modes of existence, such as *Bdellovibrio,* the rickettsias, and the chlamydiae, have relinquished some of their metabolic independence through the loss of metabolic pathways. They use the prey's or host's energy supply and/or cell constituents.

5. Many gram-negative bacteria in volume 1 of *Bergey's Manual* are of considerable importance, either as disease agents or in terms of their effects on the habitat. Others, such as *E. coli,* are major experimental organisms studied in many laboratories.

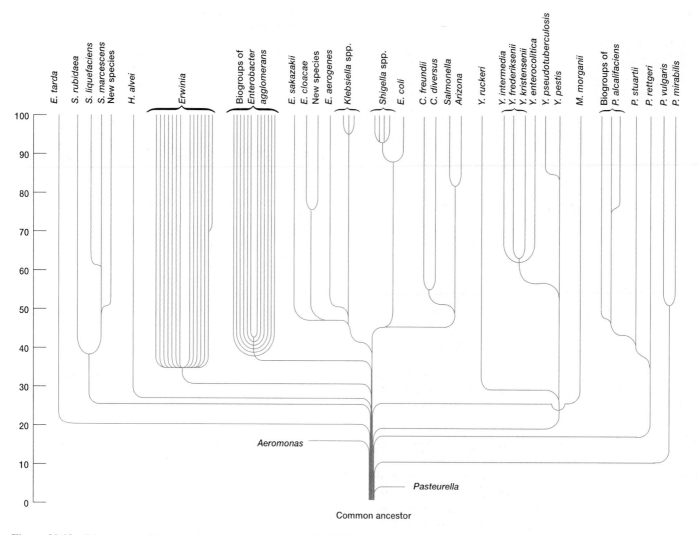

Figure 20.12   Divergence of the Family *Enterobacteriaceae*. The DNA relatedness of genera in the family *Enterobacteriaceae*. The ordinate is the percentage of relatedness, and the horizontal branches depict the degree of relatedness of the branching group to all organisms that have not yet branched off. A common ancestor for the family is assumed.

[Greek *enterikos,* pertaining to the intestine], all degrade sugars by means of the Embden-Meyerhof pathway and cleave pyruvic acid to yield formic acid in formic acid fermentations. Those enteric bacteria that produce large amounts of gas during sugar fermentation, such as *Escherichia* spp., have the formic hydrogenlyase complex that degrades formic acid to H_2 and CO_2. The family can be divided into two groups based on their fermentation products. The majority (e.g., the genera *Escherichia, Proteus, Salmonella,* and *Shigella*) carry out mixed acid fermentation and produce mainly lactate, acetate, succinate, formate (or H_2 and CO_2), and ethanol. In butanediol fermentation the major products are butanediol, ethanol, and carbon dioxide. *Enterobacter, Serratia, Erwinia,* and *Klebsiella* are butanediol fermenters. As mentioned pre-

viously (*see chapter 9*), the two types of formic acid fermentation are distinguished by the methyl red and Voges-Proskauer tests.

Formic acid fermentation and the family Enterobacteriaceae (pp. 175–76).

Because the enteric bacteria are so similar in appearance, biochemical tests are normally used to identify them after a preliminary examination of their morphology, motility, and growth responses (figure 20.13 provides a simple example). Some more commonly used tests are those for the type of formic acid fermentation, lactose and citrate utilization, indole production from tryptophan, urea hydrolysis, and hydrogen sulfide production. For example, lactose fermentation occurs in *Escherichia* and *Enterobacter* but not in *Shigella,*

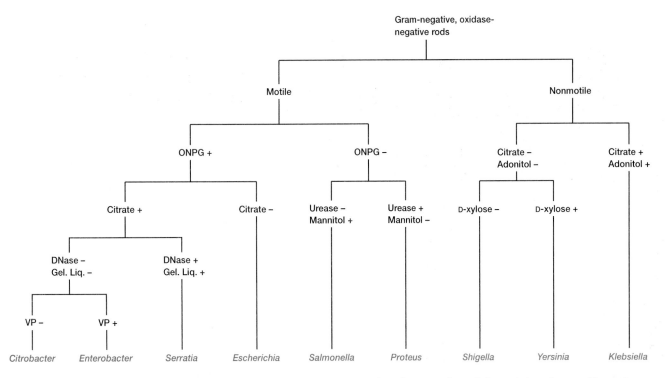

Figure 20.13   Identification of Enterobacterial Genera. A dichotomous key to selected genera of enteric bacteria based on motility and biochemical characteristics. The following abbreviations are used: ONPG, o-nitrophenyl-β-D-galactopyranoside (a test for β-galactosidase); DNase, deoxyribonuclease; Gel. Liq., gelatin liquefaction; and VP, Voges-Proskauer (a test for acetoin).

Salmonella, or *Proteus.* Table 20.5 summarizes a few of the biochemical properties useful in distinguishing between genera of enteric bacteria. The mixed acid fermenters are located on the left in this table and the butanediol fermenters on the right. The usefulness of biochemical tests in identifying enteric bacteria is shown by the popularity of commercial identification systems, such as the Enterotube and API 20-E systems, that are based on these tests.

Commercial rapid identification systems (pp. 690–91).

Members of the *Enterobacteriaceae* are so common, widespread, and important that they are probably more often seen in most laboratories than any other bacteria. *Escherichia coli* is undoubtedly the best-studied bacterium and the experimental organism of choice for many microbiologists. It is a major inhabitant of the colon of humans and other warm-blooded animals, and it is quite useful in the analysis of water for fecal contamination (*see chapter 41*). Some strains cause gastroenteritis or urinary tract infections. Several enteric genera contain very important human pathogens responsible for a variety of diseases: *Salmonella* (figure 20.14), typhoid fever and gastroenteritis; *Shigella,* bacillary dysentery; *Klebsiella,* pneumonia; *Yersinia,* plague. Members of the genus *Erwinia* are major pathogens of crop plants and cause blights, wilts,

and several other plant diseases. These and other members of the family are discussed in more detail at later points in the text (*see chapter 37*).

The *Vibrionaceae*

Members of the family *Vibrionaceae* are gram-negative, straight or curved rods with polar flagella (figure 20.15). Most are oxidase positive, and all use D-glucose as their sole or primary carbon and energy source (see table 20.4). The majority are aquatic microorganisms, widespread in fresh water and the sea. There are five genera in the family: *Vibrio, Photobacterium, Aeromonas, Plesiomonas,* and *Enhydrobacter.* Studies of rRNA homology suggest that *Vibrio* and *Photobacterium* are closely related, whereas *Aeromonas* and *Plesiomonas* differ more widely.

Several vibrios are important pathogens. *V. cholerae* (*see figure 3.1*e) is the causative agent of cholera (*see chapter 37*), and *V. parahaemolyticus* sometimes causes gastroenteritis in humans following consumption of contaminated seafood. *V. anguillarum* and others are responsible for fish diseases.

Several members of the family are unusual in being bioluminescent. *Vibrio fischeri* and at least two species of *Photobacterium* are among the few marine bacteria capable of **bioluminescence**

TABLE 20.5 **Some Characteristics of Selected Genera in the *Enterobacteriaceae***

Characteristics	*Escherichia*	*Shigella*	*Salmonella*	*Citrobacter*	*Proteus*
Methyl red	+	+	+	+	+
Voges-Proskauer	–	–	–	–	d
Indole production	(+)	d	–	d	d
Citrate use	–	–	(+)	+	d
H₂S production	–	–	(+)	d	(+)
Urease	–	–	–	(+)	+
β-galactosidase	(+)	d	d	+	–
Gas from glucose	+	–	(+)	+	+
Acid from lactose	+	–	(–)	d	–
Phenylalanine deaminase	–	–	–	–	+
Lysine decarboxylase	(+)	–	(+)	–	–
Ornithine decarboxylase	(+)	d	(+)	(+)	d
Motility	d	–	(+)	+	+
Gelatin liquifaction (22°C)	–	–	–	–	+
% G + C	48–52	49–53	50–53	50–52	38–41
Other characteristics	1.1–1.5 by 2.0–6.0 µm; peritrichous when motile	No gas from sugars	0.7–1.5 × 2–5 µm; peritrichous flagella	1.0 by 2.0–6.0 µm; peritrichous	0.4–0.8 by 1.0–3.0 µm; peritrichous

ᵃ(+) usually present

ᵇ(–) usually absent

ᶜd, strains or species vary in possession of characteristic

(a)

 Figure 20.15 **The *Vibrionaceae*.** Electron micrograph of *Vibrio alginolyticus* grown on agar, showing a sheathed polar flagellum and unsheathed lateral flagella (×18,000).

(b)

 Figure 20.14 **The *Enterobacteriaceae*.** *Salmonella* treated with fluorescent stains. (*a*) *Salmonella enteritidis* with peritrichous flagella (×500). *S. enteritidis* is associated with gastroenteritis. (*b*) *Salmonella typhi* with acridine orange stain (×2,000). *S. typhi* causes typhoid fever.

Yersinia	Klebsiella	Enterobacter	Erwinia	Serratia
+	(+)[a]	(−)[b]	+	d[c]
− (37°C)	(+)	+	(+)	+
d	d	−	(−)	(−)
(−)	(+)	+	(+)	+
−	−	−	(+)	−
d	(+)	(−)	−	−
+	(+)	+	+	+
(−)	(+)	(+)	(−)	d
(−)	(+)	(+)	d	d
−	−	(−)	(−)	−
(−)	(+)	d	−	d
d	−	(+)	−	d
− (37°C)	−	+	+	+
(−)	−	d	d	(+)
46–50	53–58	52–60	50–58	53–59
0.5–0.8 by 1.0–3.0 μm; peritrichous when motile	0.3–1.0 by 0.6–6.0 μm; capsulated	0.6–1.0 by 1.2–3.0 μm; peritrichous	0.5–1.0 by 1.0–3.0μm; peritrichous; plant pathogens and saprophytes	0.5–0.8 by 0.9–2.0 μm; peritrichous; colonies often pigmented

and emit a blue-green light because of the activity of the enzyme luciferase (Box 20.1). The peak emission of light is usually between 472 and 505 nm, but one strain of *V. fischeri* emits yellow light with a major peak at 545 nm. Although many of these bacteria are free-living, *P. phosphoreum* and *P. leiognathi* live symbiotically in the luminous organs of fish (figure 20.16).

The *Pasteurellaceae*

The *Pasteurellaceae* differ from the other two families in several ways (see table 20.4). Most notably, they are small (0.2 to 0.3 μm in diameter) and nonmotile, normally oxidase positive, have complex nutritional requirements of various kinds, and are parasitic in vertebrates. The family contains three genera: *Pasteurella, Haemophilus,* and *Actinobacillus.*

As might be expected, members of this family are best known for the diseases they cause in humans and many animals. *P. multilocida* and *P. haemolytica* are important animal pathogens. *P. multilocida* is responsible for fowl cholera, which destroys many chickens, turkeys, ducks, and geese each year. *P. haemolytica* is at least partly responsible for pneumonia in cattle, sheep, and goats (e.g., "shipping fever" in cattle). *H. influenzae* type b is a major human pathogen that causes a variety of diseases, including meningitis in children (*see chapter 37*).

1. List the major distinguishing traits of the families *Enterobacteriaceae, Vibrionaceae,* and *Pasteurellaceae.*
2. Into what two groups can the enteric bacteria be placed based on their fermentation patterns?
3. Give two reasons why the enterobacteria are so important.
4. What major human disease is associated with the *Vibrionaceae,* and what species causes it?
5. Briefly describe bioluminescence and the way it is produced.

Anaerobic Gram-Negative Straight, Curved, and Helical Rods

Section 6 of *Bergey's Manual* has one family, the *Bacteroidaceae,* and around 46 genera. This section is exceptionally diverse and contains species that are obligately anaerobic, nonsporing, motile or nonmotile rods of various shapes. They are chemoheterotrophic and usually produce a mixture of organic acids as fermentation end products, but they do not reduce sulfate or other sulfur compounds. The genera are identified according to properties such as general shape, motility and flagellation pattern, and fermentation end products. These bacteria grow in habitats such as the oral cavity and intestinal tract of humans and other animals and the rumen (*see chapter 28*) of ruminants.

Box 20.1

Bacterial Bioluminescence

Several species in the genera *Vibrio* and *Photobacterium* can emit light of a blue-green color. The enzyme luciferase catalyzes the reaction and uses reduced flavin mononucleotide, molecular oxygen, and a long-chain aldehyde as substrates.

$$FMNH_2 + O_2 + RCHO \xrightarrow{luciferase} FMN$$
$$+ H_2O + RCOOH + light$$

The evidence suggests that an enzyme-bound, excited flavin intermediate is the direct source of luminescence. Because the electrons used in light generation are probably diverted from the electron transport chain and ATP synthesis, the bacteria are expending considerable energy on luminescence. Luminescence activity is regulated and can be turned off or on under the proper conditions.

There is much speculation about the role of bacterial luminescence and its value to bacteria, particularly because it is such an energetically expensive process. Luminescent bacteria occupying the special luminous organs of fish do not emit light when they grow as free-living organisms in the seawater. Free-living luminescent bacteria can reproduce and infect young fish. Once settled in a fish's luminous organ, they begin emitting light, which the fish uses for its own purposes. Other luminescent bacteria growing on potential food items such as small crustacea may use light to attract fish to the food source. After ingestion, they could establish a symbiotic relationship in the host's gut.

(a)

(b)

(c)

Figure 20.16 **Bioluminescence.** (*a*) A photograph of the Atlantic flashlight fish *Kryptophanaron alfredi*. The light area under the eye is the fish's luminous organ, which can be covered by a lid of tissue. (*b*) The masses of photobacteria in the SEM view are separated by thin epithelial cells. Line = 6.0 μm. (*c*) Ultrathin section of the luminous organ of a fish, *Equulities novaehollandiae*, with the bioluminescent bacterium *Photobacterium leiognathi*, PL. Bar = 2 μm.

Although the difficulty of culturing these anaerobes has hindered the determination of their significance, they are clearly widespread and important. Often they benefit their host. *Bacteroides succinogenes* and *B. ruminicola* are major components of the rumen flora; the former actively degrades cellulose, and the latter ferments starch, pectin, and other carbohydrates. About 30% of the bacteria isolated from human feces are members of the genus *Bacteroides*, and these organisms may provide extra nutrition by degrading cellulose, pectins, and other complex carbohydrates. The family also is involved in human disease. Members of the genus *Bacteroides* are associated with diseases of major organ systems, ranging from the central nervous system to the skeletal system. *B. fragilis* is a particularly common anaerobic pathogen found in abdominal, pelvic, pulmonary, and blood infections.

1. Give the major properties of the family *Bacteroidaceae*.
2. How do these bacteria benefit and harm their hosts?

Dissimilatory Sulfate- or Sulfur-Reducing Bacteria

Section 7 of *Bergey's Manual* contains a morphologically diverse group of 14 gram-negative genera. These bacteria are united by their strictly anaerobic nature and the ability to use elemental sulfur or sulfate and other oxidized sulfur compounds as electron acceptors during anaerobic respiration (figure 20.17). An electron transport chain generates ATP and reduces sulfur and sulfate to hydrogen sulfide. The best-studied

Figure 20.17   The Dissimilatory Sulfate- or Sulfur-Reducing Bacteria. Representative examples. (*a*) *Desulfotomaculum acetoxidans* with spores, phase contrast (×2,000). (*b*) Phase-contrast micrograph of *Desulfovibrio saprovorans* with PHB inclusions (×2,000). (*c*) *Desulfovibrio gigas;* phase contrast (×2,000). (*d*) *Desulfobacter postgatei;* phase contrast (×2,000).

sulfate-reducing genus is *Desulfovibrio; Desulfuromonas* uses only elemental sulfur as an acceptor. One genus, *Desulfotomaculum* (figure 20.17*a*), is mentioned in section 7 because it reduces sulfate, even though it has a gram-positive cell wall and forms endospores.

Anaerobic respiration (pp. 176–78).

These bacteria are very important in the cycling of sulfur within the ecosystem. Because significant amounts of sulfate are present in almost all aquatic and terrestrial habitats, sulfate-reducing bacteria are widespread and active in locations made anaerobic by microbial digestion of organic materials. *Desulfovibrio* and other sulfate-reducing bacteria thrive in habitats such as muds and sediments of polluted lakes and streams, sewage lagoons and digesters, and waterlogged soils. *Desulfuromonas* is most prevalent in anaerobic marine and estuarine sediments. It also can be isolated from methane digesters and anaerobic hydrogen-sulfide rich muds of freshwater habitats. It uses elemental sulfur, but not sulfate, as its electron acceptor. Often sulfate and sulfur reduction is apparent from the smell of hydrogen sulfide and the blackening of

water and sediment by iron sulfide. Hydrogen sulfide production in waterlogged soils can kill animals, plants, and microorganisms. Sulfate-reducing bacteria negatively impact industry because of their primary role in the anaerobic corrosion of iron in pipelines, heating systems, and other structures (*see chapter 44*).

The sulfur cycle (p. 816).

1. Describe the metabolic specialization that characterizes the bacteria in section 7 of *Bergey's Manual.*
2. Explain the distribution of these bacteria in terms of their metabolic properties. Why are they so important?

Anaerobic Gram-Negative Cocci

The family *Veillonellaceae* contains anaerobic, chemoheterotrophic cocci ranging in diameter from about 0.3 to 2.5 μm. Usually they are diplococci (often with their adjacent sides flattened), but they may exist as single cells, clusters, or chains. All have complex nutritional requirements and ferment substances such as carbohydrates, lactate and other organic acids, and amino acids to produce gas (CO_2 and often H_2) plus a mixture of volatile fatty acids. They are parasites of homeothermic (warm-blooded) animals.

Like many groups of anaerobic bacteria, members of this family have not been thoroughly studied. They are part of the normal flora of the mouth, the gastrointestinal tract, and the urogenital tract of humans and other animals. For example, *Veillonella* is plentiful on the tongue surface and dental plaque of humans, and it can be isolated about 10 to 20% of the time from the vagina. Gram-negative, anaerobic cocci are found in infections of the head, lungs, and the female genital tract, but their precise role in such infections is unclear.

1. What properties distinguish the family *Veillonellaceae?*

The Rickettsias and Chlamydiae

Section 9 of *Bergey's Manual* contains two orders—the *Rickettsiales* and *Chlamydiales*—three families, and ten genera. Despite this diversity, almost all these gram-negative bacteria are obligately intracellular parasites: they grow and reproduce only within host cells. These bacteria are about the same size as the poxviruses (*p. 346*) and resemble viruses in their intracellular existence. They differ from viruses in having both DNA and RNA, a plasma membrane, functioning ribosomes, enzymes participating in metabolic pathways, reproduction by binary fission, and other features.

Order *Rickettsiales*

Almost all members of the order *Rickettsiales* are rod-shaped, coccoid, or pleomorphic bacteria with typical gram-negative walls and no flagella. Although their size varies, these bacteria tend to be very small. For example, *Rickettsia* is 0.3 to 0.5 μm in diameter and 0.8 to 2.0 μm long; *Coxiella* is 0.2 to 0.4 μm by 0.4 to 1.0 μm. All are parasitic or mutualistic. The parasitic forms grow in vertebrate erythrocytes, reticuloendothelial cells (*see figure 29.10*), and vascular endothelial cells. Often they also live in blood-sucking arthropods such as fleas, ticks, mites, or lice, which serve as vectors (*see chapter 35*) or primary hosts. The families *Rickettsiaceae* and *Bartonellaceae* have regular cell walls, and at least some species in each family can be cultured axenically in special media. The *Rickettsiaceae* grow only in nucleated cells, whereas the *Bartonellaceae* also can live in erythrocytes. The third family, the *Anaplasmataceae*, lacks a cell wall, cannot be cultured axenically, and grows only in erythrocytes.

Because the *Rickettsiaceae* are important human pathogens, their reproduction and metabolism have been intensively studied. Rickettsias enter the host cell by inducing phagocytosis. Members of the genus *Rickettsia* immediately

escape the phagosome and reproduce by binary fission in the cytoplasm (figure 20.18). In contrast, *Coxiella* remains within the phagosome after it has fused with a lysosome and actually reproduces within the phagolysosome. Eventually the host cell bursts, releasing new organisms. Besides incurring damage from cell lysis, the host is harmed by the toxic effects of rickettsial cell walls (wall toxicity appears related to the mechanism of penetration into host cells).

Rickettsias are very different from most other bacteria in physiology and metabolism. They lack the glycolytic pathway and do not use glucose as an energy source, but rather oxidize glutamate and tricarboxylic acid cycle intermediates such as succinate. The rickettsial plasma membrane has carrier-mediated transport systems, and host cell nutrients and coenzymes are absorbed and directly used. For example, rickettsias take up both NAD and uridine diphosphate glucose. The membrane also has an adenylate exchange carrier that exchanges ADP for external ATP. Thus host ATP may provide much of the energy needed for growth. This metabolic dependence explains why many of these organisms must be cultivated in the yolk sacs of chick embryos or in tissue culture cells.

This order contains many important pathogens. *Rickettsia prowazekii* and *R. typhi* are associated with typhus fever, and *R. rickettsii*, with Rocky Mountain spotted fever. *Coxiella burnetii* causes Q fever in humans. These diseases are discussed later in some detail (*see chapter 38*). It should be noted that rickettsias are also important pathogens of domestic animals such as dogs, horses, sheep, and cattle.

Order *Chlamydiales*

The order *Chlamydiales* has only one family, *Chlamydiaceae*, with one genus, *Chlamydia*. Chlamydiae are nonmotile, coccoid, gram-negative bacteria, ranging in size from 0.2 to 1.5 μm. They can reproduce only within cytoplasmic vesicles of host cells by a unique developmental cycle involving the formation of elementary bodies and reticulate bodies. Although their envelope resembles that of other gram-negative bacteria, the wall differs in lacking muramic acid and a peptidoglycan layer. The elementary bodies achieve osmotic stability by cross-linking their outer membrane proteins with disulfide bonds. Chlamydiae are extremely limited metabolically and are obligately intracellular parasites of mammals and birds. (However, chlamydia-like bacteria have recently been isolated from spiders, clams, and freshwater invertebrates.) The size of their genome is 4 to 6×10^8 daltons, one of the smallest of all procaryotes, and the G + C content is 41 to 44%.

Chlamydial reproduction begins with the attachment of an **elementary body (EB)** to the cell surface (figure 20.19). Elementary bodies are 0.2 to 0.4 μm in diameter, contain electron-dense nuclear material and a rigid cell wall, and are infectious (figure 20.20). The host cell phagocytoses the EB, which then prevents the fusion of lysosomes with the phagosome and begins to reorganize itself to form a **reticulate body (RB)** or **initial body.** The RB is specialized for reproduction

Figure 20.18   The *Rickettsiales*. Rickettsial morphology and reproduction. (*a*) A human fibroblast filled with *Rickettsia prowazekii* (×1,200). (*b*) A chicken embryo fibroblast late in infection wih free cytoplasmic *R. prowazekii* (×13,600). (*c*) *Coxiella burnetti* growing within fibroblast vacuoles (×9,000). (*d*) *R. prowazekii* leaving a disrupted phagosome (arrow) and entering the cytoplasmic matrix (×46,000).

Figure 20.19   The Chlamydial Elementary Body. The elementary bodies of *Chlamydia trachomatis* infecting tissue culture cells; fluorescent stain (×1,000).

rather than infection. Reticulate bodies are 0.6 to 1.5 μm in diameter and have less dense nuclear material and more ribosomes than EBs; their walls are also more flexible. About 8 to 10 hours after infection, the reticulate body begins to divide and RB reproduction continues until the host cell dies. A chlamydia-filled vacuole or inclusion can become large enough to be seen in a light microscope and even fill the host cytoplasm. After 20 to 25 hours, RBs begin changing back into infectious EBs and continue this process until the host cell lyses and releases the chlamydiae 48 to 72 hours after infection.

Chlamydial metabolism is very different from that of other gram-negative bacteria. Chlamydiae cannot catabolize carbohydrates or other substances and synthesize ATP. *Chlamydia psittaci*, one of the best-studied species, lacks both flavoprotein and cytochrome electron transport chain carriers, but does have a membrane translocase that acquires host ATP in exchange for ADP. Thus chlamydiae seem to be energy parasites that are completely dependent on their hosts for ATP. When supplied with precursors from the host, RBs can synthesize DNA, RNA, glycogen, lipids, and proteins. Presumably the RBs have porins and active membrane transport proteins, but little is known about these. They also can synthesize at least some amino acids and coenzymes. The EBs lack much metabolic activity and cannot take in ATP or synthesize proteins. They seem to be dormant forms concerned exclusively with transmission and infection.

Three chlamydial species are important pathogens of humans and other warm-blooded animals. *C. trachomatis* infects humans and mice. In humans it causes trachoma, nongonococcal urethritis, and other diseases (*see chapter 38*).

Elementary body

Size about 0.3 μm
Rigid cell wall
Relatively resistant to sonication
Resistant to trypsin
Subunit in cell envelope
RNA:DNA content = 1:1
Toxic for mice
Isolated organisms infectious
Adapted for extracellular survival

Reticulate body (initial body)

Size 0.5–1.0 μm
Fragile cell wall
Sensitive to sonication
Lysed by trypsin
No subunit in envelope
RNA:DNA content = 3:1
Nontoxic for mice
Isolated organisms not infectious
Adapted for intracellular growth

(a)

(b)   Hours after infection

Figure 20.20   The Chlamydial Life Cycle. (*a*) An electron micrograph of a microcolony of *Chlamydia trachomatis* in the cytoplasm of a host cell (×160,000). Three developmental stages are visible: the elementary body, EB; reticulate body, RB; and intermediate body, IB, a chlamydial cell intermediate in morphology between the first two forms. (*b*) A schematic representation of the infectious cycle of chlamydiae.

C. psittaci causes psittacosis in humans. However, unlike *C. trachomatis*, it also infects many other animals (e.g., parrots, turkeys, sheep, cattle, and cats) and invades the intestinal, respiratory, and genital tracts; the placenta and fetus; the eye; and the synovial fluid of joints. *Chlamydia pneumoniae* is a common cause of human pneumonia. Recently a fourth species, *C. pecorum*, has been recognized.

1. Compare rickettsias and chlamydiae with viruses.
2. Give the major characteristics of the orders *Rickettsiales* and *Chlamydiales*.
3. Briefly describe the life cycle of rickettsias.
4. In what way does the physiology and metabolism of the rickettsias differ from that of other bacteria?
5. Name some important rickettsial diseases.
6. What are elementary and reticulate bodies? Briefly describe the steps in the chlamydial life cycle.
7. How does chlamydial metabolism differ from that of other bacteria, including the rickettsias?
8. List two or three chlamydial diseases of humans.

The Mycoplasmas

Section 10 of *Bergey's Manual* describes the division *Tenericutes*, which contains only one class, *Mollicutes*, and one order, *Mycoplasmatales*, with three families and six genera.

Mycoplasmas lack cell walls and cannot synthesize peptidoglycan precursors. Thus they are penicillin resistant but susceptible to lysis by osmotic shock and detergent treatment. Because they are bounded only by a plasma membrane, these procaryotes are pleomorphic and vary in shape from spherical or pear-shaped organisms, about 0.3 to 0.8 μm in diameter, to branched or helical filaments (figure 20.21). They are the smallest bacteria capable of self-reproduction. Although most are nonmotile, some can glide along liquid-covered surfaces. Most species differ from the vast majority of bacteria in requiring sterols for growth. They usually are facultative anaerobes, but a few are obligate anaerobes. When growing on agar, most species will form colonies with a "fried-egg" appearance because they grow into the agar surface at the center while spreading outward on the surface at the colony edges (figure 20.22). Their genome is one of the smallest found in procaryotes, about 5 to 10×10^8 daltons; the G + C content ranges from 23 to 41%. Mycoplasmas can be saprophytes, commensals, or parasites, and many are pathogens of plants, animals, or insects.

The taxonomy of the class *Mollicutes* has changed significantly since volume 1 of *Bergey's Manual* was published in 1984. At that time the class contained one order, three families, and six genera. Table 20.6 outlines a recently proposed taxonomy of the class that includes four orders, five families, and eight genera. Families and genera are distinguished by characteristics such as oxygen dependence, the requirement for cholesterol, cell shape, habitat, and the presence of urease.

(a)

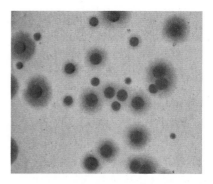

Figure 20.22 *Mycoplasma* **Colonies.** Note the "fried egg" appearance, colonies stained before viewing (×100).

(b)

Figure 20.21   **The Mycoplasmas.** Electron micrographs of *Mycoplasma pneumoniae* showing its pleomorphic nature. (*a*) A transmission electron micrograph of several cells (×47,880). (*b*) A scanning electron micrograph (×26,000).

Comparisons of molecular properties (e.g., 16S rRNA sequences, genome size and composition, and codon use) also are increasingly employed in classification. Further taxonomic changes undoubtedly will occur as more is learned about these important organisms.

It is currently believed that the mollicutes arose from gram-positive bacteria, probably from the clostridia. *Bergey's Manual* discusses the genus *Thermoplasma* in the context of both the mycoplasmas (volume 1) and the archaeobacteria (volume 3). Recent evidence shows that it is a member of the archaeobacteria, and not a mycoplasma.

The metabolism of mycoplasmas is not particularly unusual, although they are deficient in several biosynthetic sequences and often require sterols, fatty acids, vitamins, amino acids, purines, and pyrimidines. Those genera needing sterols incorporate them into the plasma membrane. Mycoplasmas are usually more osmotically stable than eubacterial protoplasts, and their membrane sterols may be a stabilizing factor. Some produce ATP by the Embden-Meyerhof pathway and lactic acid fermentation. Others catabolize arginine or urea to generate ATP. The pentose phosphate pathway seems functional in at least some mycoplasmas; none appear to have the complete tricarboxylic acid cycle.

Mycoplasmas appear to be very important, even though their roles in nature and disease are not yet completely clear. They are remarkably widespread and can be isolated from animals, plants, the soil, and even compost piles. Indeed, about 10% of the cell cultures in use are probably contaminated with mycoplasmas, which seriously interfere with tissue culture experiments and are difficult to detect and eliminate. In animals, mycoplasmas colonize mucous membranes and joints and often are associated with diseases of the respiratory and urogenital tracts. Mycoplasmas cause several major diseases in livestock, for example, contagious bovine pleuropneumonia in cattle (*M. mycoides*), chronic respiratory disease in chickens (*M. gallisepticum*), and pneumonia in swine (*M. hyopneumoniae*). *M. pneumoniae* causes primary atypical pneumonia in humans, and there is increasing evidence that *M. hominis* and *Ureaplasma urealyticum* also are human pathogens. Spiroplasmas have been isolated from insects, ticks, and a variety of plants. They cause disease in citrus plants, cabbage, broccoli, corn, honey bees, and other hosts. Arthropods probably often carry the spiroplasmas between plants. Presumably many more pathogenic mollicutes will be discovered as techniques for their isolation and study improve.

1. What morphological feature distinguishes the mycoplasmas? In what division and class are they found?
2. Give other distinguishing properties of the class *Mollicutes*.
3. What might mycoplasmas use sterols for?
4. Where are mycoplasmas found in animals? List several animal and human diseases caused by them. What kinds of organisms do spiroplasmas usually infect?

TABLE 20.6   Taxonomy and Properties of Members of the Class *Mollicutes*

Classification	No. of Recognized Species	Guanine-Plus-Cytosine Content (mol%)	Genome Size (kpb)	Cholesterol Requirement	Habitat	Other Distinctive Features
Order I, *Mycoplasmatales*						
Family I, *Mycoplasmataceae*						
Genus I, *Mycoplasma*	85	23–40	600–1,350	Yes	Humans, animals	Optimum growth usually at 37°C
Genus II, *Ureaplasma*	5	27–30	760–1,170	Yes	Humans, animals	Urea hydrolysis
Order II, *Entomoplasmatales*						
Family I, *Entomoplasmataceae*						
Genus I, *Entomoplasma*	5	27–29	790–1,140	Yes	Insects, plants	Optimum growth, 30°C
Genus II, *Mesoplasma*	4	27–30	870–1,100	No	Insects, plants	Optimum growth, 30°C; sustained growth in serum-free medium only with 0.04% Tween 80
Family II, *Spiroplasmataceae*						
Genus I, *Spiroplasma*	11	25–30	940–2,200	Yes	Insects, plants	Helical filaments; optimum growth at 30–37°C
Order III, *Acholeplasmatales,*						
Family I, *Acholeplasmataceae*						
Genus I, *Acholeplasma*	9	26–36	1,500–1,650	No	Animals, some plants and insects	Optimum growth at 30–37°C
Order IV, *Anaeroplasmatales*						
Family I, *Anaeroplasmataceae*						
Genus I, *Anaeroplasma*	4	29–34	1,500–1,600	Yes	Bovine or ovine rumen	Oxygen-sensitive anaerobes
Genus II, *Asteroleplasma*	1	40	1,500	No	Bovine or ovine rumen	Oxygen-sensitive anaerobes

From J. G. Tully, et al., "Revised Taxonomy of the Class *Mollicutes*" in *International Journal of Systematic Bacteriology*, 43(2):378-85. Copyright © 1993 American Society for Microbiology, Washington, D.C. Reprinted by permission.

Summary

1. The spirochetes are slender, long, helical, gram-negative bacteria that are motile because of the axial filament underlying an outer sheath or outer membrane.

2. Section 2 of *Bergey's Manual* contains genera that are helical or vibrioid, motile, respiratory, and aerobic or microaerophilic (e.g., *Campylobacter* and *Bdellovibrio*).

3. Section 3 has one family and several other genera of nonmotile (or rarely motile), curved, gram-negative bacteria.

4. Section 4 of *Bergey's Manual* contains a large and diverse collection of gram-negative aerobic rods and cocci.

5. The genus *Pseudomonas* contains straight or slightly curved, gram-negative, aerobic rods that are motile by one or several polar flagella and do not have prosthecae or sheaths.

6. The pseudomonads participate in natural mineralization processes, are major experimental subjects, cause many diseases, and often spoil refrigerated food.

7. *Azotobacter* and *Rhizobium* carry out nitrogen fixation, whereas *Agrobacterium* causes the development of plant tumors.

8. The *Methylococcaceae* are methylotrophs; they use methane and methanol as their sole carbon and energy sources.

9. Section 5 of *Bergey's Manual* contains facultatively anaerobic, gram-negative rods divided into three families and seven other genera.

10. The *Enterobacteriaceae*, often called enterobacteria or enteric bacteria, are gram-negative, peritrichously flagellated or nonmotile, facultatively anaerobic, straight rods with simple nutritional requirements.

11. The enteric bacteria are usually identified by use of a variety of physiological tests and are very important experimental organisms and pathogens of plants and animals.

12. The two other families in section 5 are the oxidase-positive *Vibrionaceae* and the parasitic *Pasteurellaceae.*

13. Obligately anaerobic, gram-negative rods are placed in the family *Bacteroidaceae* when they do not reduce sulfate or other sulfur compounds. Some of these bacteria benefit their hosts (e.g., rumen inhabitants), whereas others are pathogenic.

14. Section 7 of *Bergey's Manual* contains gram-negative bacteria that are anaerobic and can use elemental sulfur and oxidized sulfur compounds as electron acceptors in anaerobic respiration. They are very important in sulfur cycling in the ecosystem.

15. The family *Veillonellaceae* contains anaerobic, gram-negative cocci that are parasites of vertebrates.

16. Section 9 of *Bergey's Manual* describes the orders *Rickettsiales* and *Chlamydiales*, all obligately intracellular parasites responsible for many diseases.

17. Rickettsias have many plasma membrane carriers and make extensive use of host cell nutrients, coenzymes, and ATP.

18. Chlamydiae are nonmotile, coccoid, gram-negative bacteria that reproduce within the cytoplasmic vacuoles of host cells by a life cycle involving elementary bodies (EBs) and reticulate bodies (RBs). They are energy parasites and cannot make ATP.

19. Mycoplasmas are gram-negative bacteria that lack cell walls and cannot synthesize peptidoglycan precursors. Many species require sterols for growth. They are the smallest bacteria capable of self-reproduction and usually grow on agar to give colonies a "fried-egg" appearance.

Key Terms

axial fibrils   *417*
axial filament   *417*
bioluminescence   *427*
elementary body (EB)   *432*
enteric bacteria (enterobacteria)   *425*

enteritis   *420*
initial body   *432*
methylotroph   *424*
mineralization process   *423*

mycoplasmas   *434*
periplasmic flagella   *417*
reticulate body (RB)   *432*
septicemia   *420*

Questions for Thought and Review

1. Suppose that you were a microbiologist working on the Mars probe project. How would you determine whether microorganisms are living on the planet, and why?

2. How does *Bdellovibrio* take advantage of its host's metabolism to grow rapidly and efficiently? In what ways does its reproduction resemble that of bacteriophages?

3. Why might the ability to form dormant cysts be of great advantage to *Agrobacterium* but not as much to *Rhizobium?*

4. Some bacterial groups contain a great diversity of bacteria, whereas other groups have fewer species that resemble each other very closely. What might account for this?

5. What advantage might a species gain by specializing in its nutrient requirements or habitat as methylotrophs and halobacteria do? Are there disadvantages to this strategy?

6. Why are biochemical tests of such importance in identifying enteric bacteria?

7. A pond with black water and a smell like rotten eggs might contain an excess of what kind of bacteria?

8. How might one try to culture a newly discovered rickettsia or chlamydia outside its host?

Additional Reading

General

Balows, A.; Trüper, H. G.; Dworkin, M.; Harder, W.; and Schleifer, K.-H. 1992. *The prokaryotes,* 2d ed. New York: Springer-Verlag.

Holt, J. G., editor-in-chief. 1984. *Bergey's manual of systematic bacteriology,* vol. 1, N. R. Krieg, editor. Baltimore, Md.: Williams & Wilkins.

Holt, J. G., editor-in-chief. 1994. *Bergey's Manual of Determinative Bacteriology,* 9th ed. Baltimore, Md.: Williams & Wilkins.

Lederberg, J., editor. 1992. *Encyclopedia of Microbiology.* San Diego, Calif.: Academic Press.

The Spirochetes

Canale-Parola, E. 1978. Motility and chemotaxis of spirochetes. *Ann. Rev. Microbiol.* 32:69–99.

Harwood, C. S., and Canale-Parola, E. 1984. Ecology of spirochetes. *Ann. Rev. Microbiol.* 38:161–92.

Holt, S. C. 1978. Anatomy and chemistry of spirochetes. *Microbiol. Rev.* 42(1):114–60.

Radolf, J. D. 1994. Role of outer membrane architecture in immune evasion by *Treponema pallidum* and *Borrelia burgdorferi. Trends Microbiol.* 2(9):307–11.

Saint Girons, I.; Old, I. G.; and Davidson, B. E. 1994. Molecular biology of the *Borrelia,* bacteria with linear replicons. *Microbiology* 140:1803–16.

Aerobic/Microaerophilic, Motile, Helical/Vibroid Gram-Negative Bacteria

Cover, T. L., and Blaser, M. J. 1995. *Helicobacter pylori:* A bacterial cause of gastritis, peptic ulcer disease, and gastric cancer. *ASM News* 61(1):21–26.

Diedrich, D. L. 1988. Bdellovibrios: Recycling, remodelling and relocalizing components from their prey. *Microbiol. Sci.* 5(4):100–103.

Walker, R. I., et al. 1986. Pathophysiology of *Campylobacter* enteritis. *Microbiol. Rev.* 50(1):81–94.

Gram-Negative Aerobic Rods and Cocci

Chilton, M-D. 1983. A vector for introducing new genes into plants. *Sci. Am.* 248(6):51–59.

Haber, C. L.; Allen, L. N.; Zhao, S.; and Hanson, R. S. 1983. Methylotrophic bacteria: Biochemical diversity and genetics. *Science* 221:1147–53.

Facultatively Anaerobic Gram-Negative Rods

MacDonell, M. T.; Swartz, D. G.; Ortiz-Conde, B. A.; Last, G. A.; and Colwell, R. R. 1986. Ribosomal RNA phylogenies for the vibrio-enteric group of eubacteria. *Microbiol. Sci.* 3:172–78.

Meighen, E. A. 1991. Molecular biology of bacterial bioluminescence. *Microbiol. Rev.* 55(1):123–42.

Neidhardt, F. C.; Ingraham, J. L.; Low, K. B.; Magasanik, B.; Schaechter, M.; and Umbarger, H. E., editors. 1987. *Escherichia coli and Salmonella typhimurium.* Washington, D.C.: ASM Press

Starr, M. P., and Chatterjee, A. K. 1972. The genus *Erwinia:* Enterobacteria pathogenic to plants and animals. *Ann. Rev. Microbiol.* 26:389–426.

Anaerobic Gram-Negative Straight, Curved, and Helical Rods

Salyers, A. A. 1984. *Bacteroides* of the human lower intestinal tract. *Ann. Rev. Microbiol.* 38:293–313.

The Rickettsias and Chlamydiae

Beatty, W. L.; Morrison, R. P.; and Byrne, G. I. 1994. Persistent chlamydiae: from cell culture to a paradigm for chlamydial pathogenesis. *Microbiol. Rev.* 58(4):686–99.

Hase, T. 1985. Developmental sequence and surface membrane assembly of rickettsiae. *Ann. Rev. Microbiol.* 39:69–88.

McClarty, G. 1994. Chlamydiae and the biochemistry of intracellular parasitism. *Trends Microbiol.* 2(5):157–64.

Moulder, J. W. 1984. Looking at chlamydiae without looking at their hosts. *ASM News* 50(8):353–62.

Moulder, J. W. 1991. Interaction of chlamydiae and host cells in vitro. *Microbiol. Rev.* 55(1):143–90.

Winkler, H. H. 1990. Rickettsia species (as organisms). *Ann. Rev. Microbiol.* 44:131–53.

The Mycoplasmas

Maniloff, J. 1983. Evolution of wall-less procaryotes. *Ann. Rev. Microbiol.* 37:477–99.

Miles, R. J. 1992. Catabolism in mollicutes. *J. Gen. Microbiol.* 138:1773–83.

Sears, B. B., and Kirkpatrick, B. C. 1994. Unveiling the evolutionary relationships of plant-pathogenic mycoplasmalike organisms. *ASM News* 60(6):307–12.

Stanbridge, E. J. 1976. A reevaluation of the role of mycoplasmas in human disease. *Ann. Rev. Microbiol.* 30:169–87.

Tully, J. G.; Bové, J. M.; Laigret, F.; and Whitcomb, R. F. 1993. Revised taxonomy of the class *Mollicutes:* Proposed elevation of a monophyletic cluster of arthropod-associated mollicutes to ordinal rank (*Entomoplasmatales* ord. nov.), with provision for familial rank to separate species with nonhelical morphology (*Entomoplasmataceae* fam. nov.) from helical species (*Spiroplasmataceae*), and emended descriptions of the order *Mycoplasmatales,* family *Mycoplasmataceae. Int. J. Syst. Bacteriol.* 43(2):378–85.

Whitcomb, R. F. 1980. The genus *Spiroplasma. Ann. Rev. Microbiol.* 34:677–709.

CHAPTER 21

The Bacteria: Gram–Positive Bacteria Other Than Actinomycetes

We noted, after having grown the bacterium through a series of such cultures, each fresh culture being inoculated with a droplet from the previous culture, that the last culture of the series was able to multiply and act in the body of animals in such a way that the animals developed anthrax with all the symptoms typical of this affection.

Such is the proof, which we consider flawless, that anthrax is caused by this bacterium.

—Louis Pasteur

Concepts

1. Volume 2 of *Bergey's Manual* contains six sections covering all gram-positive bacteria except the actinomycetes. Bacteria are distributed among these sections on the basis of their shape, the ability to form endospores, acid fastness, oxygen relationships, the ability to temporarily form mycelia, and other properties.

2. Peptidoglycan structure varies among different groups in ways that are often useful in identifying specific groups.

3. The six endospore-forming genera are probably not closely related and are grouped together for convenience.

4. Mycobacteria can form filaments, but unlike the actinomycetes, their filaments are easily fragmented into rods or coccoid elements. Actinomycete hyphae are more permanent.

5. Although most gram-positive bacteria are harmless free-living saprophytes, species from all six sections of volume 2 of *Bergey's Manual* are major pathogens of humans, other animals, and plants. Other gram-positive bacteria are very important in the food and dairy industries.

C hapter 21 surveys the bacteria found in volume 2 of *Bergey's Manual of Systematic Bacteriology*. The chapter describes all major groups of gram-positive bacteria except the actinomycetes, which are covered in chapter 24. As in chapter 20, much of this chapter discusses the most significant biological features of each group and describes representatives selected because of their practical importance or their extensive use in microbiological research.

Volume 2 of *Bergey's Manual* contains six sections that describe all gram-positive bacteria except the actinomycetes. Most of these bacteria are distributed among the first four sections on the basis of their general shape (whether they are rods, cocci, or irregular) and their ability to form endospores. The rod-shaped, acid-fast bacteria known as mycobacteria are placed in section 16 (the sections in *Bergey's Manual* are numbered consecutively, beginning with the first section in volume 1). The last section describes the nocardioforms. These are gram-positive filamentous bacteria that form a branching network called a mycelium, which can break into rod-shaped or coccoid forms. The nocardioforms are discussed in more detail in volume 4, section 26, of *Bergey's Manual*. Therefore they are described in chapter 24, which covers the actinomycetes.

The gram-positive bacteria are placed in the division *Firmicutes* as previously noted (*see chapter 19*). Analysis of the phylogenetic relationships within the gram-positive bacteria by comparison of 16S rRNA sequences (*see figure 19.8*) shows that they are divided into a low G + C group and high G + C or actinomycete group. The distribution of genera within and between the groups shown in figure 19.8 differs markedly from the classification in volume 2 of *Bergey's Manual*. Thus the classification system given in volume 2 of the manual is a practical, convenient one that probably does not always reflect true phylogenetic relatedness. At points, the treatment in this chapter will differ from that given in *Bergey's Manual*.

> *Bacterial evolution and taxonomy (pp. 402–8).*
> *The structure of the gram-positive cell wall (p. 54).*

Peptidoglycan structure varies considerably among different gram-positive groups. Most gram-negative bacteria have a peptidoglycan structure in which *meso*-diaminopimelic acid in position 3 is directly linked through its free amino group with the free carboxyl of the terminal D-alanine of an adjacent peptide chain (figure 21.1*a*; *see also figure 3.21*). This same peptidoglycan structure is present in many gram-positive genera, for example, *Bacillus, Clostridium, Lactobacillus,*

Corynebacterium, Mycobacterium, and *Nocardia*. In other gram-positive bacteria, lysine is substituted for diaminopimelic acid in position 3, and the peptide subunits of the glycan chains are cross-linked by interpeptide bridges containing monocarboxylic L-amino acids or glycine, or both (figure 21.1*b*). Many genera including *Staphylococcus, Streptococcus, Micrococcus, Lactobacillus,* and *Leuconostoc* have this type of peptidoglycan. The genus *Streptomyces* and several other actinomycete genera have replaced *meso*-diaminopimelic acid with L, L-diaminopimelic acid in position 3 and have one glycine residue as the interpeptide bridge. The plant pathogenic corynebacteria provide a final example of peptidoglycan variation. In some of these bacteria, the interpeptide bridge connects positions 2 and 4 of the peptide subunits rather than 3 and 4 (figure 21.1*c*). Because the interpeptide bridge connects the carboxyl groups of glutamic acid and alanine, a diamino acid such as ornithine is used in the bridge. Many other variations in peptidoglycan structure are found, including other interbridge structures and large differences in the frequency of cross-linking between glycan chains. Bacilli and most gram-negative bacteria have fewer cross-links between chains than do gram-positive bacteria such as *Staphylococcus aureus* in which almost every muramic acid is cross-linked to another. These structural variants are often characteristic of particular groups and are therefore taxonomically useful.

---

1. Describe, in a diagram, the chemical composition and structure of the peptidoglycan found in gram-negative bacteria and many gram-positive genera.

2. Briefly discuss the ways in which three other peptidoglycan types differ from the gram-negative peptidoglycan.

3. How do bacilli and most gram-negative bacteria differ from gram-positive bacteria such as *S. aureus* with respect to cross-linking frequency?

Gram-Positive Cocci

Although section 12 of *Bergey's Manual* contains only chemoheterotrophic, mesophilic, nonsporing gram-positive cocci, these bacteria are phylogenetically very diverse. Currently there are at least two families and 34 genera: 6 aerobic, 13 facultatively anaerobic, and 5 anaerobic. Genera are distinguished according to properties such as oxygen relationships, cell arrangement, the presence of catalase and cytochromes, peptidoglycan structure, and G + C content (table 21.1). A few of the most important groups are now described.

Figure 21.1   Representative Examples of Peptidoglycan Structure. (*a*) The peptidoglycan with a direct cross-linkage between positions 3 and 4 of the peptide subunits, which is present in most gram-negative and many gram-positive bacteria. (*b*) Peptidoglycan with lysine in position 3 and an interpeptide bridge. The bracket contains six typical bridges: (*1*) *Staphylococcus aureus*, (*2*) *S. epidermidis*, (*3*) *Micrococcus roseus* and *Streptococcus thermophilus*, (*4*) *Lactobacillus viridescens*, (*5*) *Streptococcus salvarius*, and (*6*) *Leuconostoc cremoris*. The arrows indicate the polarity of peptide bonds running in the C to N direction. (*c*) An example of the cross-bridge extending between positions 2 and 4 from *Corynebacterium poinsettiae*. The interbridge contains an L-diamino acid like ornithine, and L-homoserine (L-Hsr) is in position 3. The abbreviations and structures of amino acids in the figure are found in appendix I. See text for more detail.

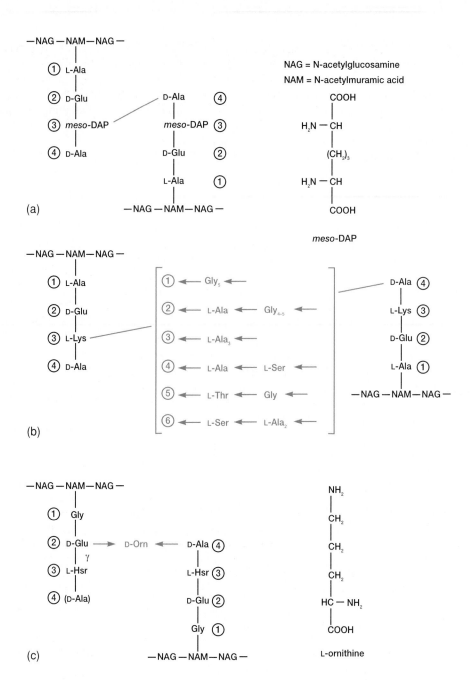

The *Micrococcaceae*

The family *Micrococcaceae* contains gram-positive cocci, 0.5 to 2.5 μm in diameter, that divide in more than one plane to form regular or irregular clusters of cells. All are aerobic or facultatively anaerobic. The peptidoglycan diamino acid is L-lysine. The two most important genera are *Micrococcus* and *Staphylococcus*. These genera differ in such aspects as cell arrangement, oxygen relationships, ability to ferment glucose, the presence of oxidase and teichoic acid, and G + C content (tables 21.1 and 21.2).

The genus *Micrococcus* contains aerobic, catalase-positive cocci that occur mainly in pairs, tetrads, or irregular clusters and are usually nonmotile. Micrococci often are yellow, orange, or red in color. They are widespread in soil, water, and on mammalian skin; the last habitat may be their normal one. Despite their frequent presence on the skin, micrococci do not seem particularly pathogenic. Their G + C content, cell wall composition, and 16S rRNA sequence indicate that they may be more closely related to the genus *Arthrobacter* than to other gram-positive cocci.

TABLE 21.1   Differential Properties of Representative Genera of Gram-Positive Cocci

Characteristics	Aerobic Genera		Facultatively Anaerobic Genera			Anaerobic
	Micrococcus	*Deinococcus*	*Staphylococcus*	*Streptococcus, Enterococcus, Lactococcus*	*Leuconostoc*	*Sarcina*
Predominant arrangement of cells, commonest listed first[a]	Clusters, tetrads	Pairs, tetrads	Clusters, pairs	Chains, pairs	Pairs, chains	Cuboidal packets
Strict aerobes	+[b]	+	–	–	–	–
Facultative anaerobes or microaerophiles	–	–	+	+	+	–
Strict anaerobes	–	–	–	–	–	+
Catalase reaction	+	+	+	–	–	–
Cytochromes present	+	+	+	–	–	ND
Major anaerobic carbohydrate fermentation products	NA	NA	Lactate	Lactate	Lactate, ethanol	Acetate, ethanol, H_2, CO_2, or butyrate
Mol% G + C of DNA	64–75	62–70	30–39	34–46	38–44	28–31

Modified from *Bergey's Manual of Systematic Bacteriology*, Vol. 2, edited by P. H. A. Sneath, et al. Copyright © 1986 Williams and Wilkins, Baltimore, MD. Reprinted by permission.

[a]Only arrangements other than single cells are listed.

[b]Symbols: +, 90% or more of strains positive; –, 10% or less of strains positive; NA, not applicable; ND, not determined.

TABLE 21.2   Properties of the Genera *Micrococcus* and *Staphylococcus*

Characteristics	*Micrococcus*	*Staphylococcus*
Irregular clusters	+[a]	+
Tetrads	+	–
Capsule	–	–
Motility	–	–
Anaerobic glucose fermentation	–	+
Oxidase and benzidine tests	+	–
Resistance to lysostaphin	R	S
Glycine present in peptidoglycan	–	+
Teichoic acid present in cell wall	–	+
Mol% G + C of DNA	64–75	30–39

Modified from *Bergey's Manual of Systematic Bacteriology*, Vol. 2, edited by P. H. A. Sneath, et al. Copyright © 1986 Williams and Wilkins, Baltimore, MD. Reprinted by permission.

[a]Symbols: +, 90% or more of strains positive; –, 10% or less of strains positive; R, resistant; and S, sensitive.

Members of the genus *Staphylococcus* are facultatively anaerobic, nonmotile, gram-positive cocci that usually form irregular clusters (figure 21.2*a,b; see also figure 3.1*a). Like the micrococci, they are catalase positive but differ from the former in several ways: they are oxidase negative, ferment glucose anaerobically, and have teichoic acid in their cell walls and DNA with a much lower G + C content (30 to 39%). Staphylococci are normally associated with the skin, skin glands, and mucous membranes of warm-blooded animals.

Staphylococci are responsible for many human diseases. *S. epidermidis* is a common skin resident that is sometimes responsible for endocarditis and infections of patients with lowered resistance (e.g., wound infections, surgical infections, urinary tract infections). *S. aureus* is the most important human staphylococcal pathogen and causes boils, abscesses, wound infections, pneumonia, toxic shock syndrome, and other diseases.

It also is a major cause of food poisoning as the following case illustrates. In March 1986 an outbreak of acute gastrointestinal illness occurred following a buffet supper served to 855 people at a New Mexico country club. At least 67 people were ill with diarrhea, nausea, or vomiting, and 24 required emergency medical treatment or hospitalization. The problem was *S. aureus* growing in the turkey and dressing served at the

(a)

(b)

(c)

(d)

(e)

(f)

Figure 21.2 **Representative Gram-Positive Cocci.** (*a*) *Staphylococcus aureus*, Gram-stained smear (×1,500). (*b*) Staphylococci arranged like a cluster of grapes; scanning electron micrograph (×34,000). (*c*) *Streptococcus pyogenes* (×900). (*d*) Scanning electron micrograph of *Streptococcus* (×33,000). (*e*) *Micrococcus luteus.* Methylene blue stain (×1,000). (*f*) *Streptococcus pneumoniae* (×900).

buffet. One or more of the food handlers carried *S. aureus* and had contaminated the food. Because the turkey was cooled for 3 hours after cooking, there was sufficient time for the bacterium to grow and produce toxins (*see chapters 37 and 43*). This case is not uncommon. Turkey accounts for 10 to 21% of all bacterial food poisoning cases in which the source of poisoning is known, and it must be cooked and handled carefully.

Unlike other common staphylococci, *S. aureus* produces the enzyme **coagulase,** which causes blood plasma to clot. Growth patterns on blood agar are also useful in identifying these staphylococci (figure 21.4). *S. aureus* usually grows on the nasal membranes and skin; it also is found in the gastrointestinal and urinary tracts of warm-blooded animals.

Staphylococcal diseases (pp. 759–62, 770).

The *Deinococcaceae*

The family *Deinococcaceae* contains only one genus and is quite different from other cocci in 16S rRNA sequence, cell wall structure, and other properties. The deinococci are grampositive cocci associated in pairs or tetrads (figure 21.3) that are

Figure 21.3 **The *Deinococcaceae*.** A *Deinococcus radiodurans* microcolony showing cocci arranged in tetrads (average cell diameter 2.5 μm).

TABLE 21.3   Classification of the Streptococci, Enterococci, and Lactococci

Characteristics	*Streptococcus*	*Enterococcus*	*Lactococcus*
Predominant arrangement (most common first)	Chains, pairs	Pairs, chains	Pairs, short chains
Capsule/slime layer	+	–	–
Habitat	Mouth, respiratory tract	Gastrointestinal tract	Dairy products
Growth at 45°C	Variable	+	–
Growth at 10°C	Variable	Usually +	+
Growth at 6.5% NaCl broth	Variable	+	–
Growth at pH 9.6	Variable	+	–
Hemolysis	Usually β (pyogenic) or α (oral)	α, ß, –	Usually –
Serological group (Lancefield)	Variable (A–O)	Usually D	Usually N
Mol% G + C (normal range)	34–46	34–42	38–40

Representative species	Pyogenic streptococci
	  S. agalactiae
	  S. pyogenes
	  S. equi
	  S. dysgalactiae
	
	Oral streptococci
	  S. gordonii
	  S. salvarius
	  S. sanguis
	  S. oralis
	  S. pneumoniae
	  S. mitis
	  S. mutans
	
	Other streptococci
	  S. bovis
	  S. thermophilus
		E. faecalis
		E. faecium
		E. avium
		E. durans
		E. gallinarum
			L. lactis
			L. raffinolactis
			L. plantarum

aerobic, catalase positive, and usually able to produce acid from few sugars. Unlike most gram-positive bacteria, their plasma membrane contains a large amount of palmitoleic acid and lacks phosphatidylglycerol phospholipids. The cell wall contains L-ornithine and has several distinct layers, including an outer membrane. Almost all strains of *Deinococcus* are unusually resistant to both desiccation and radiation.

Much remains to be discovered about the biology of these bacteria. Deinococci can be isolated from ground meat, feces, air, fresh water, and other sources, but their natural habitat is not yet known. The cause of their great radiation resistance is also not understood; it is possible to obtain radiation-sensitive mutants, and this should prove useful in clarifying the mechanism of resistance.

Streptococcus, Enterococcus, and *Lactococcus*

Bergey's Manual contains nine or more genera of facultatively anaerobic or microaerophilic, catalase-negative, gram-positive cocci. The genus *Streptococcus* is an important member of this group. The streptococci and their close relatives, the enterococci and lactococci, occur in pairs or chains when grown in liquid media (figure 21.2c,d,f; *see also figure 3.1*b), do not

form endospores, and usually are nonmotile. They are all chemoheterotrophs that ferment sugars with lactic acid, but no gas, as the major product—that is, they carry out homolactic fermentation (*see chapter 9*). A few species are anaerobic rather than facultative.

The genus *Streptococcus* is large and complex. Currently *Bergey's Manual* lists 38 species clustered in four groups: pyogenic streptococci, oral streptococci, anaerobic streptococci, and other streptococci. Many bacteria that were placed within the genus have been moved to two new genera, *Enterococcus* (16 species) and *Lactococcus* (5 species). Some major characteristics of these three closely related genera are summarized in table 21.3. Table 21.4 lists a few properties of selected genera.

Many characteristics are used to identify these cocci. One of their most important taxonomic characteristics is the ability to lyse erythrocytes when growing on blood agar, an agar medium containing 5% sheep or horse blood (figure 21.4). In α-**hemolysis** a 1 to 3 mm greenish zone of incomplete hemolysis forms around the streptococcal colony; β-**hemolysis** is characterized by a zone of clearing or complete lysis without a marked color change. In addition, other hemolytic patterns are sometimes seen. Serological studies (*see chapters 32 and 34*) are also

TABLE 21.4   Properties of Selected Streptococci and Relatives

Characteristics	Pyogenic Streptococci	Oral Streptococci			Enterococci	Lactic Acid Streptococci
	S. pyogenes	S. pneumoniae	S. sanguis	S. mutans	E. faecalis	L. lactis
Growth at 10°C	–[a]	–	–	–	+	+
Growth at 45°C	–	–	d	d	+	–
Growth at 6.5% NaCl	–	–	–	–	+	–
Growth at pH 9.6	–	–	–	–	+	–
Growth with 40% bile	–	–	d	d	+	+
α-hemolysis	–	+	+	–	–	d
β-hemolysis	+	–	–	–	+	–
Arginine hydrolysis	+	+	+	–	+	d
Hippurate hydrolysis	–	–	–	–	+	d
Mol% G + C of DNA	35–39	30–39	40–46	36–38	34–38	39

Modified from *Bergey's Manual of Systematic Bacteriology*, Vol. 2, edited by P. H. A. Sneath, et al. Copyright © 1986 Williams and Wilkins, Baltimore, MD. Reprinted by permission.

[a]Symbols: +, 90% or more of strains positive; –, 10% or less of strains positive; d, 11–89% of strains are positive.

Figure 21.4   Streptococcal and Staphylococcal Hemolytic Patterns. (*a*) *Streptococcus pyogenes* on blood agar, illustrating β-hemolysis. (*b*) *Streptococcus pneumoniae* on blood agar, illustrating α-hemolysis. (*c*) *Staphylococcus aureus* on blood agar, illustrating β-hemolysis. (*d*) *Staphylococcus epidermidis* on blood agar with no hemolysis.

(a)   (b)   (c)   (d)

very important in identification because streptococci often have distinctive cell wall antigens. Polysaccharide and teichoic acid antigens found in the wall or between the wall and the plasma membrane are used to identify these cocci, particularly pathogenic β-hemolytic streptococci, by the **Lancefield grouping system** (*see Box 32.3*). Biochemical and physiological tests are essential in identification (e.g., growth temperature preferences, carbohydrate fermentation patterns, acetoin production, reduction of litmus milk, sodium chloride and bile salt tolerance, and the ability to hydrolyze arginine, esculin, hippurate, and starch). Sensitivity to bacitracin, sulfa drugs, and optochin (ethylhydrocuprein) also are used to identify particular species.

Members of the three genera have considerable practical importance. Pyogenic streptococci usually are pathogens and associated with pus formation (pyogenic means pus producing). Most species produce β-hemolysis on blood agar and form chains of cells. The major human pathogen in this group is *S. pyogenes* (streptococcal sore throat, acute glomeru-

Figure 21.5 *Leuconostoc. Leuconostoc mesenteroides;* phase-contrast micrograph. Bar = 10 μm.

lonephritis, rheumatic fever). The normal habitat of oral streptococci is the oral cavity and upper respiratory tract of humans and other animals. In other respects oral streptococci are not necessarily similar. *S. pneumoniae* is α-hemolytic and grows as pairs of cocci (figures 21.2*f* and 21.4*b*). It is associated with lobar pneumonia. *S. mutans* is associated with the formation of dental caries (*see chapter 38*). The enterococci such as *E. faecalis* are normal residents of the intestinal tracts of humans and most other animals. *E. faecalis* is an opportunistic pathogen that can cause urinary tract infections and endocarditis. Unlike the streptococci the enterococci will grow in 6.5% sodium chloride. The lactococci ferment sugars to lactic acid and can grow at 10°C but not at 45°C. *L. lactis* is widely used in the production of buttermilk and cheese (*see chapter 43*) because it can curdle milk and add flavor through the synthesis of diacetyl and other products.

Streptococcal diseases (pp. 746–50).

Leuconostoc

The genus *Leuconostoc* contains facultative gram-positive cocci, which may be elongated or elliptical and arranged in pairs or chains (figure 21.5). Leuconostocs lack catalase and cytochromes and carry out **heterolactic fermentation** (*see chapter 9*) by converting glucose to D-lactate and ethanol or acetic acid by means of the phosphoketolase pathway (figure 21.6). They can be isolated from plants, silage, and milk. The genus is used in wine production, in the fermentation of vegetables such as cabbage (sauerkraut; *see figure 43.20*) and cucumbers (pickles), and in the manufacture of buttermilk, butter, and cheese. *L. mesenteroides* synthesizes dextrans from sucrose and is important in industrial dextran production. *Leuconostoc* species are involved in food spoilage and tolerate high sugar concentrations so well that they grow in syrup and are a major problem in sugar refineries.

Figure 21.6 Heterolactic Fermentation and the Phosphoketolase Pathway. The phosphoketolase pathway converts glucose to lactate, ethanol, and CO_2.

A variety of gram-positive bacteria produce lactic acid as their major or sole fermentation product and are sometimes collectively called **lactic acid bacteria.** *Streptococcus, Enterococcus, Lactococcus,* and *Leuconostoc* are all members of this group. The genus *Lactobacillus* also is included (see p. 449). Lactic acid bacteria are nonsporing and usually nonmotile. They lack cytochromes and obtain energy by substrate-level phosphorylation rather than electron transport and oxidative phosphorylation. They normally depend on sugar fermentation for energy. Nutritionally they are fastidious and many vitamins, amino acids, purines, and pyrimidines must be supplied because of their limited biosynthetic capabilities. Lactic acid bacteria usually are categorized as facultative anaerobes, but some classify them as aerotolerant anaerobes.

Box 21.1

Spores in Space

During the nineteenth-century argument over the question of the evolution of life, the panspermia hypothesis became popular. According to this hypothesis, life did not evolve from inorganic matter on earth but arrived as viable bacterial spores that had escaped from another planet. More recently, the British astronomer Fred Hoyle has revived the hypothesis based on his study of the absorption of radiation by interstellar dust. Hoyle maintains that dust grains were initially viable bacterial cells and that the beginning of life on earth was due to the arrival of bacterial endospores that had survived their trip through space.

Even more recently Peter Weber and J. Mayo Greenberg from the University of Leiden in the Netherlands have studied the effect of very high vacuum, low temperature, and UV radiation on the survival of *Bacillus subtilis* endospores. Their data suggest that endospores within an interstellar molecular cloud might be able to survive between 4.5 to 45 million years. Molecular clouds move through space at speeds sufficient to transport spores between solar systems in this length of time. Although these results do not prove the panspermia hypothesis, they are consistent with the possibility that bacteria might be able to travel between planets capable of supporting life.

Oxygen relationships (pp. 130–31).

1. Describe the major distinguishing characteristics of the following taxa: *Micrococcaceae, Micrococcus, Staphylococcus, Deinococcus, Streptococcus, Enterococcus, Lactococcus,* and *Leuconostoc.*

2. How does the pathogen *S. aureus* differ from the common skin resident *S. epidermidis,* and where is it normally found?

3. What are α-hemolysis, β-hemolysis, and the Lancefield grouping system?

4. Give a representative species and its importance for *Streptococcus, Enterococcus,* and *Lactococcus.* Distinguish between pyogenic and oral streptococci.

5. Of what practical importance is *Leuconostoc?* What are lactic acid bacteria?

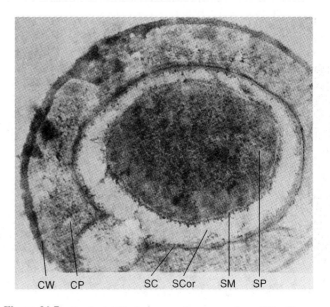

Figure 21.7   Bacterial Endospores. A cross section of *Bacillus megaterium* and its endospore within the vegetative cell wall, CW; cell protoplast, CP; spore coat, SC; spore cortex, SCor; spore membrane, SM; and spore protoplast, SP (×120,000).

Endospore-Forming Gram-Positive Rods and Cocci

Section 13 of *Bergey's Manual* contains almost all the gram-positive, endospore-forming bacteria. (Because of its resemblance to the actinomycetes, the endospore-forming genus *Thermoactinomyces* is placed in volume 4 of the manual.) Bacterial endospores are round or oval intracellular objects that have a complex structure with a spore coat, cortex, and inner spore membrane surrounding the protoplast (figure 21.7). They contain dipicolinic acid, are very heat resistant, and can remain dormant and viable for very long periods (Box 21.1). In one well-documented experiment, endospores remained viable for about 70 years. A recent research article reports that viable endospores have been recovered from Dominican bees that were encased in 25- to 40-million-year-old amber. If this result is confirmed, endospores from an ancestor of *Bacillus sphaericus* have survived for more than 25 million years! Usually endospores are observed in the light microscope after spore staining (*see chapter 2*). They

also can be detected by heating a culture at 70 to 80°C for 10 minutes followed by incubation in the proper growth medium. Because only endospores and some thermophiles would survive such heating, bacterial growth tentatively confirms their presence.

Bacterial endospore structure (pp. 67–9).

Although endospore-forming bacteria are distributed widely, they are primarily soil inhabitants. Soil conditions are often extremely variable, and endospores are an obvious advantage in surviving periods of dryness or nutrient deprivation.

At present there are 10 genera of endospore-forming bacteria, 5 of which are described in table 21.5. The two best-studied and most important genera are *Bacillus* and *Clostridium* (figure 21.8; *see also figures 2.8*d, *2.14*a, *3.1*c, and *3.11*).

TABLE 21.5   Differential Characteristics of Selected Endospore-Forming Genera

Characteristics	*Bacillus*	*Clostridium*	*Sporolactobacillus*	*Desulfotomaculum*	*Sporosarcina*
Rod-shaped	+[a]	+	+	+	–
Filaments	–	D	–	–	–
Rods or filaments curved	–	D	–	D	NA
Cocci in tetrads or packets	–	–	–	–	+
Motile	+	+	+	+	+
Stain gram positive, at least in young cultures	+	+	+	d	+
Strict aerobes	D	–	–	–	+
Facultative anaerobes or microaerophiles	D	–	+	–	–
Strict anaerobes	–	+	–	+	–
Homolactic fermentation	D	–	+	–	–
Dissimilatory sulfate reduction to sulfide	–	–	–	+	–
Catalase	+	–	–	–	+
Oxidase	D	–	ND	ND	+
Marked acidity from glucose	+	D	+	–	–
Nitrate reduced to nitrite	D	D	–	ND	D
Mol% G + C of DNA	32–69	24–54	38–40	37–50	40–42

Modified from *Bergey's Manual of Systematic Bacteriology*, Vol. 2, edited by P. H. A. Sneath, et al. Copyright © 1986 Williams and Wilkins Co., Baltimore, MD. Reprinted by permission.

[a]Symbols: +, 90% or more of strains positive; –, 10% or less of strains positive; d, 11–89% strains positive; D, substantial proportion of species differ; NA, not applicable; ND, not determined.

(a)

(b)

(c)

(d)

Figure 21.8   Endospore-Forming Bacteria. Representative members of the genera *Bacillus* and *Clostridium*. (*a*) *B. anthracis*, spores elliptical and central (×1,600). (*b*) *B. subtilis*, spores elliptical and central. (*c*) *C. botulinum*, spores elliptical and subterminal, cells slightly swollen (×1,100). (*d*) *C. tetani*, spores round and terminal (×1,100).

Both contain gram-positive, endospore-forming, chemo-heterotrophic rods that are either peritrichously flagellated or nonmotile. The genus *Clostridium* is obligately anaerobic and lacks both catalase and a complete electron transport chain. The genus *Bacillus* is aerobic, or sometimes facultative, and catalase positive.

The taxonomy of these bacteria is still somewhat uncertain. The rRNA data indicate that *Bacillus* and *Clostridium* belong to a procaryotic branch that contains nonsporing bacteria such as *Streptococcus, Staphylococcus, Lactobacillus,* and possibly the mycoplasmas (*see figure 19.8*). Many nonsporing bacteria are similar to spore formers except for their lack of endospores. Furthermore, *Bacillus* and *Clostridium* are very diverse phenotypically and genotypically; currently both groups are being subdivided to form new genera. For example, rRNA sequence data has been used to divide the genus *Bacillus* into at least five separate lines. Several species already have been moved to two new genera. The genus *Alicyclobacillus* contains acidophilic, sporing, gram-positive or gram-variable rods having ω-alicyclic fatty acids with 6- or 7-carbon rings in their membranes. Members are aerobic or facultative and have a G + C content of about 51 to 60%. The second new genus is *Paenibacillus* [Latin *paene,* almost, and *bacillus*]. This genus contains gram-positive rods from rRNA group 3 that are facultative, motile by peritrichous flagella, have ellipsoidal endospores and swollen sporangia, produce acid and sometimes gas from glucose and various sugars, and have a G + C content of 40 to 54%. Some examples of organisms that were formerly in the genus *Bacillus* are *Paenibacillus alvei, P. macerans,* and *P. polymyxa.*

These genera are of considerable practical importance. Members of the genus *Bacillus* produce the antibiotics bacitracin, gramicidin, and polymyxin. *B. cereus* causes some forms of food poisoning and can infect humans. *B. anthracis* is the causative agent of the disease anthrax, which can affect both farm animals and humans (*see chapter 37*). Several species are used as insecticides. For example, *B. thuringiensis* and *B. sphaericus* form a solid protein crystal, the **parasporal body,** next to their spores during endospore formation (figure 21.9). The *B. thuringiensis* parasporal body contains protein toxins that will kill over 100 species of moths by dissolving in the alkaline gut contents of caterpillars and destroying their gut epithelium (*see chapter 44*). The solubilized toxin proteins are cleaved by midgut proteases to smaller toxic polypeptides that attack the epithelial cells. The alkaline gut contents escape into the blood, causing paralysis and death. One of these toxins has been isolated and shown to form pores in the plasma membrane. These channels allow monovalent cations such as potassium to pass through. Most *B. thuringiensis* toxin genes are carried on large plasmids. The *B. sphaericus* parasporal body contains proteins toxic for mosquito larvae and may be useful in controlling the mosquitos that carry the malaria parasite *Plasmodium.*

Members of the genus *Clostridium* also have great practical impact. Because they are anaerobic and form heat-resistant endospores, they are responsible for many cases of food spoilage, even in canned foods. *C. botulinum* is the causative

(a)

(b)

Figure 21.9   The Parasporal Body. (*a*) An electron micrograph of a *B. sphaericus* sporulating cell containing a parasporal body just beneath the endospore. Bar = 400 nm. (*b*) The crystalline parasporal body at a higher magnification. The crystal is surrounded by a two-layered envelope (arrows). Bar = 200 nm.

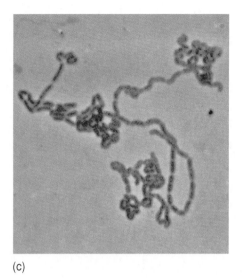

(a)   (b)   (c)

Figure 21.10   Regular, Nonsporing Gram-Positive Rods. Representative lactobacilli. (*a*) *L. acidophilus* (×1,000). (*b*) *L. lactis.* Gram stain (×1,000). (*c*) *L. bulgaricus;* phase contrast (×600).

agent of botulism (*see chapter 37*). Clostridia often can ferment amino acids to produce ATP by oxidizing one amino acid and using another as an electron acceptor in a process called the Stickland reaction. This generates ammonia, hydrogen sulfide, fatty acids, and amines during the anaerobic decomposition of proteins. These products are responsible for many unpleasant odors arising during putrefaction. Several clostridia produce toxins and are major disease agents. *C. tetani* is the causative agent of tetanus, and *C. perfringens,* of gas gangrene and food poisoning. Clostridia also are industrially valuable; for example, *C. acetobutylicum* is used to manufacture butanol.

Microbiology of food (chapter 43).
Microbial insecticides (chapter 44).

1. What is a bacterial endospore? Give its most important properties and two ways to demonstrate its presence.
2. List the major properties of the genera *Bacillus* and *Clostridium.*
3. What practical impacts do the two genera have on society? Define parasporal body and Stickland reaction.

Regular, Nonsporing, Gram-Positive Rods

Section 14 of *Bergey's Manual* contains seven genera of regular, nonsporing, and usually nonmotile, gram-positive rods that are mesophilic and chemoheterotrophic and grow only in complex media. The largest genus, *Lactobacillus,* contains rods and sometimes coccobacilli that lack catalase and cytochromes, are usually facultative or microaerophilic, produce lactic acid as their main or sole fermentation product, and have complex nutritional requirements (figure 21.10). Lactobacilli carry out either a homolactic fermentation using the Embden-Meyerhof

pathway or a heterolactic fermentation with the pentose phosphate pathway (*see chapter 9*). They grow optimally under slightly acid conditions, when the pH is between 4.5 to 6.4. The genus is found on plant surfaces and in dairy products, meat, water, sewage, beer, fruits, and many other materials. Lactobacilli also are part of the normal flora of the human body in the mouth, intestinal tract, and vagina. They usually are not pathogenic.

Lactobacillus is indispensable to the food and dairy industries (*see chapter 43*). Lactobacilli are used in the production of fermented vegetable foods (sauerkraut, pickles, silage), beverages (beer, wine, juices), sour dough, Swiss cheese and other hard cheeses, yogurt, and sausage. Yogurt is probably the most popular fermented milk product in the United States and is produced commercially and by individuals using yogurt-making kits. In commercial production, nonfat or low-fat milk is pasteurized, cooled to 43°C or lower, inoculated with *Streptococcus thermophilus* and *Lactobacillus bulgaricus. S. thermophilus* grows more rapidly at first and renders the milk anaerobic and weakly acidic. *L. bulgaricus* then acidifies the milk even more. Acting together, the two species ferment almost all the lactose to lactic acid and flavor the yogurt with diacetyl (*S. thermophilus*) and acetaldehyde (*L. bulgaricus*). Fruits or fruit flavors to be added are pasteurized separately and then combined with the yogurt.

Lactobacilli also create problems. They sometimes are responsible for spoilage of beer, milk, and meat.

Several other genera in this section are of interest. The genus *Listeria* has short rods that are aerobic or facultative, catalase positive, and motile by peritrichous flagella. They are widely distributed in nature, particularly in decaying matter. *Listeria monocytogenes* is a pathogen of humans and other animals and causes listeriosis, an important food infection (*p. 885–86*). *Caryophanon* is strictly aerobic, catalase positive, and motile by peritrichous flagella. Its normal habitat is cow dung.

Figure 21.11 *Caryophanon* **Morphology.** *Caryophanon latum* in a trichome chain. Note the disk-shaped cells stacked side by side; phase contrast (×3,450).

Figure 21.12 **Irregular, Nonsporing, Gram-Positive Rods.** *Corynebacterium diphtheriae* (×1,000). Note the irregular shapes of individual cells, the angular associations of pairs of cells, and palisade arrangements.

Caryophanon morphology is distinctive. Individual cells are disk-shaped (1.5 to 2.0 µm wide by 0.5 to 1.0 µm long) and joined to form rods about 10 to 20 µm long (figure 21.11).

---

1. List the major properties of the genus *Lactobacillus*. Why is it important in the food and dairy industries?
2. What is distinctive about the morphology of *Caryophanon*?

Irregular, Nonsporing, Gram-Positive Rods

Section 15 of *Bergey's Manual* contains an exceptionally diverse group of 37 genera. Most are irregularly shaped, nonsporing, gram-positive rods with aerobic or facultative metabolism. (However, 11 genera are anaerobic.) The rods may be straight or slightly curved and usually have swellings, club shapes, or other deviations from normal rod morphology. Three representative genera are briefly described: *Corynebacterium*, *Arthrobacter*, and *Actinomyces*.

The genus *Corynebacterium* contains aerobic and facultative, catalase-positive, straight to slightly curved rods, often with tapered ends. Club-shaped forms are also seen. The bacteria often remain partially attached after a distinctive type of binary fission called snapping division, resulting in angular arrangements of the cells, somewhat like Chinese letters, or a palisade arrangement in which rows of cells are lined up side by side (figure 21.12).

(a)   (b)

(c)   (d)

Figure 21.13 **The Rod-Coccus Growth Cycle.** The rod-coccus cycle of *Arthrobacter globiformis* when grown at 25°C. (*a*) Rods are outgrowing from cocci 6 hours after inoculation. (*b*) Rods after 12 hours of incubation. (*c*) Bacteria after 24 hours. (*d*) Cells after reaching stationary phase (3 days' incubation). The cells used for inoculation resembled these stationary-phase cocci. Bars = 10 µm.

(a)

(b)

Figure 21.14   Representatives of the Genus *Actinomyces.*
(*a*) *A. naeslundii;* Gram stain (×1,000). (*b*) *Actinomyces;* scanning
electron micrograph (×18,000). Note filamentous nature of the colony.

Figure 21.15   The Mycobacteria. *Mycobacterium leprae.* Acid-
fast stain (×400). Note the masses of red mycobacteria within blue-
green host cells.

Corynebacteria form metachromatic granules, and their walls
have *meso*-diaminopimelic acid. Although some species are
harmless soil and water saprophytes, many corynebacteria
are plant or animal pathogens. For example, *C. diphtheriae* is
the causative agent of diphtheria in humans (*see chapter 37*).

The genus *Arthrobacter* contains aerobic, catalase-
positive rods with respiratory metabolism and lysine in its
peptidoglycan. Its most distinctive feature is a rod-coccus
growth cycle (figure 21.13). When *Arthrobacter* is growing in
exponential phase, the bacteria are irregular, branched rods
and may reproduce by snapping division. As they enter stationary
phase, the cells change to a coccoid form. Upon transfer to
fresh medium, the coccoid cells produce outgrowths and
again form actively growing rods. Although arthrobacters
often are isolated from fish, sewage, and plant surfaces, their
most important habitat is the soil, where they constitute a significant component of the microbial flora. They are well
adapted to this niche because they are very resistant to desiccation and nutrient deprivation. This genus is unusually flexi-

ble nutritionally and can even degrade some herbicides and
pesticides; it is probably important in the mineralization of
complex organic molecules.

The mechanism of **snapping division** has been studied in
Arthrobacter. These bacteria have a two-layered cell wall, and
only the inner layer grows inward to generate a transverse wall
dividing the new cells. The completed transverse wall or septum
next thickens and puts tension on the outer wall layer, which
still holds the two cells together. Eventually, increasing tension
ruptures the outer layer at its weakest point, and a snapping
movement tears the outer layer apart around most of its circumference. The new cells now rest at an angle to each other and
are held together by the remaining portion of the outer layer that
acts as a hinge.

Members of the genus *Actinomyces* are straight or slightly
curved rods that vary considerably in shape and slender filaments with true branching (figure 21.14). The rods and filaments
may have swollen, clubbed, or clavate ends. They are either facultative anaerobes or anaerobes, and require CO_2 for best
growth. The cell walls contain lysine but not diaminopimelic
acid or glycine. *Actinomyces* species are normal inhabitants of
mucosal surfaces of humans and other warm-blooded animals;
the oral cavity is their preferred habitat. *A. bovis* causes lumpy
jaw in cattle. *Actinomyces* is responsible for actinomycoses, ocular infections, and periodontal disease in humans. The most important human pathogen is *A. israelii.*

1. What properties are common to all genera in section 15 of
   Bergey's Manual? Describe the major features of the genera
   Corynebacterium, Arthrobacter, and *Actinomyces,* including their
   normal habitat and the importance of each.

2. What is snapping division? The rod-coccus growth cycle?

The Mycobacteria

The family *Mycobacteriaceae* contains one genus, *Mycobacterium,* composed of slightly curved or straight rods that sometimes branch or form filaments (figure 21.15). Mycobacterial filaments differ from those of actinomycetes in readily fragmenting into rods and coccoid bodies when disturbed. They are aerobic and catalase positive. Mycobacteria grow very slowly and must be incubated for 2 to 40 days after inoculation of a solidified complex medium to form a visible colony. Their cell walls have a very high lipid content and contain waxes with 60 to 90 carbon **mycolic acids.** These are complex fatty acids with a hydroxyl group on the β-carbon and an aliphatic chain attached to the α-carbon. The presence of mycolic acids and other lipids outside the peptidoglycan layer makes mycobacteria **acid-fast** (basic fuchsin dye cannot be removed from the cell by acid alcohol treatment). Extraction of wall lipid with alkaline ethanol destroys acid fastness.

Actinomycete morphology (pp. 492–94).
Acid-fast staining (pp. 26–27).

Although some mycobacteria are free-living saprophytes, they are best known as animal pathogens. *M. bovis* causes tuberculosis in cattle, other ruminants, and primates. Because this bacterium can produce tuberculosis in humans, dairy cattle are tested for the disease yearly; milk pasteurization kills the pathogen and affords further protection against disease transmission. Thus *M. tuberculosis* is the chief source of tuberculosis in humans. The other major mycobacterial human disease is leprosy, caused by *M. leprae.*

Tuberculosis (pp. 750–52), leprosy (pp. 757–58).

1. Give the distinctive properties of the genus *Mycobacterium,* and indicate how mycobacteria differ from actinomycetes.
2. Define mycolic acid and acid-fast.
3. List two human mycobacterial diseases and their causative agents. Which pathogen causes tuberculosis in cattle?

Summary

1. Bacteria are placed in sections of volume 2 of *Bergey's Manual* according to characteristics such as general shape, the possession of endospores, and their response to acid-fast staining.
2. Based on 16S rRNA analysis, gram-positive bacteria are divided into low G + C and high G + C groups; this system is not always directly related to that in *Bergey's Manual.*
3. Peptidoglycan structure often differs between groups in taxonomically useful ways. Most variations are in amino acid 3 of the peptide subunit or in the interpeptide bridge.
4. Mesophilic, nonsporing, gram-positive cocci are placed in section 12 of *Bergey's Manual.*
5. Members of the genus *Staphylococcus* are facultatively anaerobic, nonmotile, gram-positive cocci that form irregular clusters. They grow on the skin and mucous membranes of warm-blooded animals, and some are important human pathogens.

6. Members of the family *Deinococcaceae* are aerobic, gram-positive cocci that are distinctive in their unusually great resistance to desiccation and radiation.
7. The genera *Streptococcus, Enterococcus,* and *Lactococcus* contain gram-positive cocci arranged in pairs and chains that are usually facultative and carry out homolactic fermentation. Some important species are the pyogenic coccus *S. pyogenes,* the oral streptococci *S. pneumoniae* and *S. mutans,* the enterococcus *E. faecalis,* and the lactococcus *L. lactis.*
8. *Leuconostoc* carries out heterolactic fermentation using the phosphoketolase pathway and is involved in the production of fermented vegetable products, buttermilk, butter, and cheese.
9. Section 13 has endospore-forming gram-positive rods and cocci. *Bacillus* synthesizes antibiotics and insecticides, and causes food poisoning and anthrax. *Clostridium* is responsible for botulism, tetanus, food spoilage, and putrefaction.

10. Section 14 of *Bergey's Manual* contains regular, nonsporing, gram-positive rods such as *Lactobacillus, Listeria,* and *Caryophanon.* *Lactobacillus* carries out lactic acid fermentation and is extensively used in the food and dairy industries.
11. Most bacteria in section 15 (e.g., *Corynebacterium, Arthrobacter,* and *Actinomyces*) are irregularly shaped, nonsporing, gram-positive rods with aerobic or facultative metabolism.
12. Corynebacteria use snapping division, and many cause disease (e.g., diphtheria) in plants and animals. The genus *Arthrobacter* has a rod-coccus life cycle and also shows snapping division. Several species of *Actinomyces* are pathogenic for humans and other animals.
13. Mycobacteria form either rods or filaments that readily fragment. Their cell walls have a high lipid content and mycolic acids; the presence of these lipids makes them acid-fast. Mycobacteria cause tuberculosis and leprosy in humans.

Key Terms

acid-fast   *452*
coagulase   *442*
α-hemolysis   *443*
β-hemolysis   *443*

heterolactic fermentation   *445*
lactic acid bacteria   *445*
Lancefield grouping system   *444*

mycolic acid   *452*
parasporal body   *448*
snapping division   *451*

Questions for Thought and Review

1. On the basis of the treatment of gram-positive bacteria given in volume 2 of *Bergey's Manual,* is it best to consider the volume as an identification guide, a description of phylogenetic relationships, or some of both? Explain your reasoning and conclusion.

2. Draw a diagram illustrating how gram-positive cocci might divide to produce the various clustering patterns observed (chains, tetrads, cubical packets, grapelike clusters).

3. Describe the characteristics most important in distinguishing between members of the following groups of genera: *Micrococcus, Staphylococcus,* and *Streptococcus; Bacillus* and *Clostridium; Corynebacterium, Arthrobacter,* and *Actinomyces; Actinomyces* and *Mycobacterium.*

4. Which bacteria are noted for each of the following: the rod-coccus growth cycle, radiation resistance, formation of dental caries, metachromatic granules and snapping division, coagulase, 60 to 90 carbon mycolic acids and acid fastness, pathogenicity for insects, production of fermented foods and cheeses, use of the phosphoketolase pathway, and the generation of odors during putrefaction?

5. Account for the ease with which anaerobic clostridia can be isolated from soil and other generally aerobic niches.

6. Propose an explanation for the observation that mycobacteria are normally only weakly gram positive.

Additional Reading

General

Balows, A.; Truper, H. G.; Dworkin, M.; Harder, W.; and Schleifer, K.-H. 1992. *The prokaryotes,* 2d ed. New York: Springer-Verlag.

Braun, V., and Hantke, K. 1974. Biochemistry of bacterial cell envelopes. *Ann. Rev. Biochem.* 43:89–121.

Holt, J. G., editor-in-chief. 1986. *Bergey's Manual of Systematic Bacteriology,* vol. 2, P. H. A. Sneath, N. S. Mair, and M. E. Sharpe, editors. Baltimore, Md.: Williams & Wilkins.

Holt, J. G., editor-in-chief. 1994. *Bergey's Manual of Determinative Bacteriology,* 9th ed. Baltimore, Md.: Williams and Wilkins.

Lederberg, J., editor. 1992. *Encyclopedia of microbiology.* San Diego, Calif.: Academic Press.

Schleifer, K.-H., and Kandler, O. 1972. Peptidoglycan types of bacterial cell walls and their taxonomic implications. *Bacteriol. Rev.* 36(4):407–77.

Ward, J. B. 1981. Teichoic and teichuronic acids: Biosynthesis, assembly, and location. *Microbiol. Rev.* 45(2):211–43.

Gram-Positive Cocci

Kawamura, Y.; Hou, X.-G.; Sultana, F.; Miura, H.; and Ezaki, T. 1995. Determination of 165 rRNA sequences of *Streptococcus mitis* and *Streptococcus gordonii* and phylogenetic relationships among members of the genus *Streptococcus. Int. J. Syst. Bacteriol.* 45(2):406–8.

Loesche, W. J. 1986. Role of *Streptococcus mutans* in human dental decay. *Microbiol. Rev.* 50(4):353–80.

Murray, B. E. 1990. The life and times of the enterococcus. *Clin. Microbiol. Rev.* 3(1):46–65.

Ross, P. W. 1985. Streptococcal infections in man. *Microbiol. Sci.* 2(6):174–78.

Schleifer, K.-H., and Kilpper-Balz, R. 1987. Molecular and chemotaxonomic approaches to the classification of streptococci, enterococci, and lactococci: A review. *Syst. Appl. Microbiol.* 10:1–19.

Endospore-Forming Gram-Positive Rods and Cocci

Aronson, A. I.; Beckman, W.; and Dunn, P. 1986. *Bacillus thuringiensis* and related insect pathogens. *Microbiol. Rev.* 50(1):1–24.

Ash, C.; Priest, F. G.; and Collins, M. D. 1993. Molecular identification of rRNA group 3 bacilli (Ash, Farrow, Wallbanks, and Collins) using a PCR probe test: Proposal for the creation of a new genus *Paenibacillus. Antonie van Leeuwenhoek* 64:253–60.

Collins, M. D.; Lawson, P.A.; Willems, A.; Cordoba, J.J.; Fernandez-Garayzabal, J.; Garcia, P.; Cai, J.; Hippe, H.; and Farrow, J.A.E. 1994. The phylogeny of the genus *Clostridium:* Proposal of five new genera and eleven new species combinations. *Int. J. Syst. Bacteriol.* 44(4):812–26.

Drobniewski, F. A. 1993. *Bacillus cereus* and related species. *Clin. Microbiol. Rev.* 6(4):324–38.

Hoyle, F., and Wickramasinghe, C. 1981. Where microbes boldly went. *New Scientist* (13 Aug.): 412–15.

Lambert, B., and Peferoen, M. 1992. Insecticidal promise of *Bacillus thuringiensis:* Facts and mysteries about a successful biopesticide. *BioScience* 42(2):112–22.

Sonenshein, A. L.; Hoch, J. A.; and Losick, R., editors. 1993. *Bacillus subtilis and other gram-positive bacteria.* Washington, D.C.: ASM Press.

Weber, P., and Greenberg, J. M. 1985. Can spores survive in interstellar space? *Nature* 316:403–7.

Whiteley, H. R., and Schnepf, H. E. 1986. The molecular biology of parasporal crystal body formation in *Bacillus thuringiensis. Ann. Rev. Microbiol.* 40:549–76.

The Mycobacteria

Goren, M. B. 1972. Mycobacterial lipids: Selected topics. *Bacteriol. Rev.* 36(1):1–32.

Krulwich, T. A., and Pelliccione, N. J. 1979. Catabolic pathways of coryneforms, nocardias, and mycobacteria. *Ann. Rev. Microbiol.* 33:95–111.

CHAPTER 22

The Bacteria: Remaining Gram-Negative Bacteria and Cyanobacteria

There are wide areas of the bacteriological landscape in which we have so far detected only some of the highest peaks, while the rest of the beautiful mountain range is still hidden in the clouds and the morning fogs of ignorance. The gold is still lying on the ground, but we have to bend down to grasp it.

—Preface to *The Prokaryotes*

Concepts

1. Volume 3 of *Bergey's Manual* contains eight sections, each describing a different group of gram-negative bacteria. Although few of these bacteria are of major medical or industrial significance, they are interesting biologically and often of considerable ecological importance.

2. There are three major groups of photosynthetic procaryotes: the purple bacteria, the green bacteria, and the cyanobacteria. The cyanobacteria are placed in a separate section because they resemble eucaryotic phototrophs in that they possess photosystem II and carry out oxygenic photosynthesis. The purple and green bacteria use electron donors other than water and carry out anoxygenic photosynthesis.

3. Chemolithotrophic bacteria obtain energy and electrons by oxidizing inorganic compounds rather than the organic nutrients employed by most bacteria. They often have substantial ecological impact because of their ability to oxidize many forms of inorganic nitrogen and sulfur.

4. Bacteria do not always have simple, unsophisticated morphology but may produce prosthecae, stalks, buds, sheaths, or complex fruiting bodies.

5. Gliding motility is widely distributed among bacteria and is very useful to organisms that digest insoluble nutrients or move over the surfaces of solid substrata.

6. A group of bacteria is often placed in a particular section of volume 3 on the basis of a few carefully selected properties such as nutritional type even though it may share many characteristics (e.g., morphological traits) with organisms in a different section.

C hapter 22 introduces the bacteria described in volume 3 of *Bergey's Manual of Systematic Bacteriology.* This volume contains bacteria that differ in some distinctive way from the more common gram-negative and gram-positive bacteria of volumes 1 and 2, respectively. The differences include motility (gliding bacteria), metabolism (phototrophic and chemolithotrophic bacteria), and reproduction or morphology (budding, appendaged, and sheathed bacteria). Although volume 3 also covers the archaeobacteria, this group is so distinctive that it will be described separately in chapter 23. This chapter describes the major biological features of each group and a few selected representative forms of particular interest.

Volume 3 of *Bergey's Manual* contains eight sections, each covering a different group of bacteria. This chapter briefly describes the major features of the organisms in the first seven sections.

Anoxygenic Photosynthetic Bacteria

There are three groups of photosynthetic procaryotes: the purple bacteria, the green bacteria, and the cyanobacteria. The cyanobacteria differ most fundamentally from the green and purple photosynthetic bacteria in being able to carry out **oxygenic photosynthesis.** They use water as an electron donor and generate oxygen during photosynthesis. In contrast, purple and green bacteria use **anoxygenic photosynthesis.** Because they are unable to use water as an electron source, they employ reduced molecules such as hydrogen sulfide, sulfur, hydrogen, and organic matter as their electron source for the generation of NADH and NADPH. Consequently, purple and green bacteria do not produce oxygen but many form sulfur granules. Purple sulfur bacteria accumulate granules within their cells, whereas green sulfur bacteria deposit the sulfur granules outside their cells. The purple nonsulfur bacteria use organic molecules as an electron source. There also are differences in photosynthetic pigments, the organization of photosynthetic membranes, nutritional requirements, and oxygen relationships. These are summarized in table 22.1 and will be discussed further in the context of

---

TABLE 22.1   Characteristics of the Major Groups of Photosynthetic Procaryotes

Characteristic	Anoxygenic Photosynthetic Bacteria				Oxygenic Photosynthetic Bacteria
	Green Sulfur	Green Nonsulfur[a]	Purple Sulfur	Purple Nonsulfur	Cyanobacteria
Major photosynthetic pigments	Bacteriochlorophylls *a* plus *c, d,* or *e* (the major pigment)	Bacteriochlorophylls *a* and *c*	Bacteriochlorophyll *a* or *b*	Bacteriochlorophyll *a* or *b*	Chlorophyll *a* plus phycobiliproteins
Morphology of photosynthetic membranes	Photosynthetic system partly in chlorosomes that are independent of the plasma membrane	Chlorosomes present when grown anaerobically	Photosynthetic system contained in spherical or lamellar membrane complexes that are continuous with the plasma membrane	Photosynthetic system contained in spherical or lamellar membrane complexes that are continuous with the plasma membrane	Membranes lined with phycobilisomes
Photosynthetic electron donors	H_2, H_2S, S	Photoheterotrophic donors—a variety of sugars, amino acids, and organic acids; photoautotrophic donors—H_2S, H_2	H_2, H_2S, S	Usually organic molecules; sometimes reduced sulfur compounds or H_2	H_2O
Sulfur deposition	Outside of the cell		Inside the cell[b]	Outside of the cell	
Nature of photosynthesis	Anoxygenic	Anoxygenic	Anoxygenic	Anoxygenic	Oxygenic (sometimes facultatively anoxygenic)
General metabolic type	Obligately anaerobic photolithoautotrophs	Usually photoheterotrophic; sometimes photoautotrophic or chemoheterotrophic (when aerobic and in the dark)	Obligately anaerobic photolithoautotrophs	Usually anaerobic photoorganoheterotrophs; some facultative photolithoautotrophs (in dark, chemoorganoheterotrophs)	Aerobic photolithoautotrophs
Motility	Nonmotile; some have gas vesicles	Gliding	Motile with polar flagella; some are peritrichously flagellated	Motile with polar flagella or nonmotile; some have gas vesicles	Nonmotile or with gliding motility; some have gas vesicles
Percent G + C	48–58	53–55	45–70	61–72	35–71

[a]Characteristics of *Chloroflexus.*

[b]With the exception of *Ectothiorhodospira.*

individual groups. Because cyanobacterial photosynthesis differs so greatly from that of purple and green bacteria (*see chapter 9*), cyanobacteria are placed in section 19 of *Bergey's Manual* and are described later in this chapter. Section 18 of the manual contains the purple and green anoxygenic photosynthetic bacteria.

The mechanism of bacterial photosynthesis (pp. 181–88).

Purple Bacteria

There are three groups of purple bacteria: the families *Chromatiaceae* (the purple sulfur bacteria) and *Ectothiorhodospiraceae*, and the purple nonsulfur bacteria. The family *Ectothiorhodospiraceae* has only one genus, *Ectothiorhodospira*, with red, spiral-shaped, polarly flagellated cells that deposit sulfur globules externally (figure 22.1; *see also figure 3.9*b). The other two

Figure 22.1   Purple Bacteria. *Ectothiorhodospira mobilis;* light micrograph. Bar = 10 μm.

groups are much larger. All purple bacteria use anoxygenic photosynthesis, possess bacteriochlorophylls *a* or *b,* and have their photosynthetic apparatus in membrane systems that are continuous with the plasma membrane. Most are motile by polar flagella.

The **purple sulfur bacteria** are strict anaerobes and usually photolithoautotrophs. They oxidize hydrogen sulfide to sulfur and deposit it internally as sulfur granules (usually within invaginated pockets of the plasma membrane); often they eventually oxidize the sulfur to sulfate. Hydrogen also may serve as an electron donor. *Thiospirillum, Thiocapsa,* and *Chromatium* are typical purple sulfur bacteria (figure 22.2). They are found in anaerobic, sulfide-rich zones of lakes (*see chapter 41*).

Because purple and green photosynthetic bacteria grow best in deeper anaerobic zones of aquatic habitats, they cannot effectively use parts of the visible spectrum normally employed by photosynthetic organisms. There often is a dense surface layer of cyanobacteria and algae in lakes and ponds that absorbs a large amount of blue and red light. The bacteriochlorophyll pigments of purple and green bacteria absorb longer wavelength, far-red light (table 22.2) not used by other photosynthesizers (figure 22.3). In addition, the bacteriochlorophyll absorption peaks at about 350 to 550 nm enable them to grow at greater depths because shorter wavelength light can penetrate water farther. As a result, when the water is sufficiently clear, a layer of green and purple bacteria develops in the anaerobic, hydrogen sulfide-rich zone (figures 22.4 and 22.5).

The microbial ecology of lakes (pp. 840–43).

The **purple nonsulfur bacteria** are exceptionally flexible in their choice of an energy source. Normally they grow anaerobically as photoorganoheterotrophs; they trap light energy and employ organic molecules as both electron and carbon sources. Although they are called nonsulfur bacteria, some

(a)

(b)

(c)

Figure 22.2   Typical Purple Sulfur Bacteria. (*a*) *Chromatium vinosum* with intracellular sulfur granules. Bar = 10 μm. (*b*) Electron micrograph of *C. vinosum.* Note the intracytoplasmic vesicular membrane system. The large white areas are the former sites of sulfur globules. Bar = 0.3 μm. (*c*) *Thiocapsa roseopersicina.* Bar = 10 μm.

TABLE 22.2	Procaryotic Bacteriochlorophyll and Chlorophyll Absorption Maxima	

| | **Long Wavelength Maxima (nm)** | |
Pigment	**In Ether or Acetone**	**Approximate Range of Values in Cells**
Chlorophyll *a*	665	680–685
Bacteriochlorophyll *a*	775	850–910 (purple bacteria)[a]
Bacteriochlorophyll *b*	790	1,020–1,035
Bacteriochlorophyll *c*	660	745–760
Bacteriochlorophyll *d*	650	725–745
Bacteriochlorophyll *e*	647	715–725

[a]The spectrum of bacteriochlorophyll *a* in green bacteria has a different maximum, 805–810 nm.

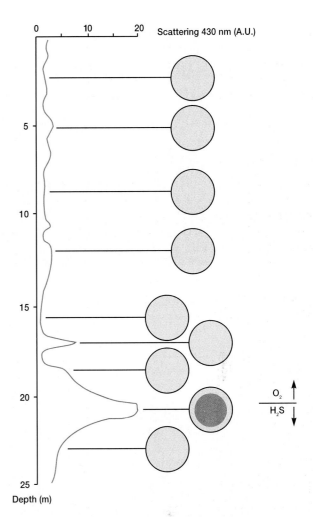

Figure 22.4   The Distribution of Photosynthetic Microorganisms in a Norwegian Fjord. Fifty-milliliter samples taken at various depths have been filtered and the filters mounted on a graph showing the light scattering (an indication of turbidity due to bacterial growth) as a function of depth. The peak at 20.8 m is caused by the purple bacteria *Chromatium* spp. growing at the interface of the aerobic zone and the anaerobic, sulfide-rich layer. The smaller peak at 17.1 m is caused by dinoflagellate growth. A.U. stands for absorbance units.

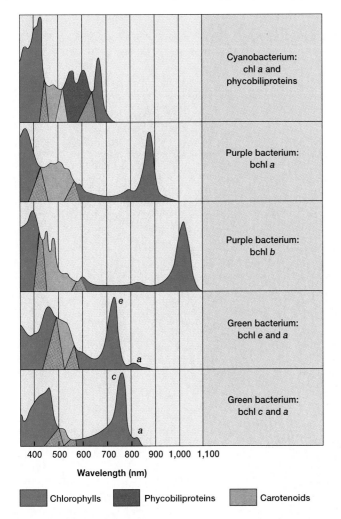

Figure 22.3   Photosynthetic Pigments. Absorption spectra of five photosynthetic bacteria showing the differences in absorption maxima and the contributions of various accessory pigments.

species can oxidize very low, nontoxic levels of sulfide to sulfate, but they do not oxidize elemental sulfur to sulfate. In the absence of light, most purple nonsulfur bacteria can grow aerobically as chemoorganoheterotrophs, but some species carry out fermentations and grow anaerobically. Oxygen inhibits bacteriochlorophyll and carotenoid synthesis so that cultures growing aerobically in the dark are colorless.

Purple nonsulfur bacteria vary considerably in morphology (figure 22.6). They may be spirals (*Rhodospirillum*), rods (*Rhodopseudomonas*), half circles or circles (*Rhodocyclus*), or they may even form prosthecae and buds (*Rhodomicrobium*). Because of their metabolism, they are most prevalent in the mud and water of lakes and ponds with abundant organic matter and low sulfide levels. There also are marine species.

Figure 22.5   Purple Photosynthetic Sulfur Bacteria. (*a*) Purple photosynthetic sulfur bacteria growing in a bog. (*b*) A sewage lagoon with a bloom of purple photosynthetic bacteria.

Figure 22.6   Typical Purple Nonsulfur Bacteria. (*a*) *Rhodospirillum rubrum;* phase contrast (× 410). (*b*) *R. rubrum* grown anaerobically in the light. Note vesicular invaginations of the cytoplasmic membranes; transmission electron micrograph (× 51,000). (*c*) *Rhodopseudomonas acidophila;* phase contrast. (*d*) *Rhodocyclus purpureus;* phase contrast. Bar = 10 μm. (*e*) *Rhodomicrobium vannielii* with vegetative cells and buds; phase contrast.

Figure 22.7   Typical Green Sulfur Bacteria. (*a*) *Chlorobium limicola* with extracellular sulfur granules. (*b*) *Pelodictyon clathratiforme.* (*c*) An electron micrograph of *P. clathratiforme* (× 105,000). Note the chlorosomes (dark gray areas) and gas vesicles (light gray areas with pointed ends).

(a)

(b)

(c)

Green Bacteria

As with the purple bacteria, there are two groups of green bacteria: the **green sulfur bacteria** and the **green nonsulfur bacteria.** The green sulfur bacteria are a small group of obligately anaerobic photolithoautotrophs that use hydrogen sulfide, elemental sulfur, and hydrogen as electron sources. The elemental sulfur produced by sulfide oxidation is deposited outside the cell. Their photosynthetic pigments are located in ellipsoidal vesicles called **chlorosomes** or chlorobium vesicles, which are attached to the plasma membrane but are not continuous with it. The chlorosome membrane is not a normal lipid bilayer or unit membrane. Chlorosomes contain accessory bacteriochlorophyll pigments, but the reaction center bacteriochlorophyll is located in the plasma membrane and must be able to obtain energy from chlorosome pigments. These bacteria flourish in the anaerobic, sulfide-rich zones of lakes. Although they lack flagella and are nonmotile, some species have gas vesicles (figure 22.7c) to adjust their depth for adequate light and hydrogen sulfide. Those forms without vesicles are found in sulfide-rich muds at the bottom of lakes and ponds.

Like the purple bacteria the green sulfur bacteria are very diverse morphologically. They may be rods, cocci, or vibrios; some grow singly, and others form chains and clusters (figure 22.7a,b). They are either grass-green or chocolate-brown in color. Representative genera are *Chlorobium, Prosthecochloris,* and *Pelodictyon.*

Chloroflexus is the major representative of the green nonsulfur bacteria (*Bergey's Manual* calls the group multicellular filamentous green bacteria). It is a filamentous, gliding, thermophilic bacterium that often is isolated from neutral to alkaline hot springs where it grows in the form of orange-reddish mats, usually in association with cyanobacteria. Although it resembles the green bacteria in ultrastructure and photosynthetic pigments, its metabolism is more similar to that of the purple nonsulfur bacteria. *Chloroflexus* can carry out anoxygenic photosynthesis with organic compounds as carbon sources or grow aerobically as a chemoheterotroph. It doesn't appear closely related to any eubacterial group based on 16S rRNA studies, and its taxonomic position is uncertain.

1. How do oxygenic and anoxygenic photosynthesis differ from each other and why?

2. Give the major characteristics of the following groups: purple sulfur bacteria, purple nonsulfur bacteria, and green sulfur bacteria. How do purple and green bacteria differ? Compare *Chloroflexus* with the green sulfur bacteria.

3. Why are the bacteriochlorophylls of purple and green bacteria so useful beneath the surface of lakes?

4. What are chlorosomes or chlorobium vesicles?

Oxygenic Photosynthetic Bacteria

Two groups of oxygenic photosynthesizers are found in section 19 of *Bergey's Manual,* the cyanobacteria and the order *Prochlorales.* The cyanobacteria are discussed first, followed by a brief survey of the prochlorophytes.

The **cyanobacteria** are the largest and most diverse group of photosynthetic bacteria. There is little agreement about the number of cyanobacterial species. Older classifications had as many as 2,000 or more species. In one recent system this has been reduced to 62 species and 24 genera. *Bergey's Manual of Determinative Bacteriology* describes 34 genera in some detail. The G + C content of the group ranges from 35 to 71%. Although cyanobacteria are true procaryotes, their photosynthetic system closely resembles that of the eucaryotes because they have chlorophyll *a* and photosystem II, and carry out oxygenic photosynthesis. Like the red algae, cyanobacteria use phycobiliproteins as accessory pigments. Photosynthetic pigments and electron transport chain components are located in thy-lakoid membranes lined with particles called **phycobilisomes** (figure 22.8). These contain phycobilin pigments, particularly phycocyanin, and transfer energy to photosystem II. Carbon dioxide is assimilated through the Calvin cycle, and the reserve carbohydrate is glycogen. Sometimes they will store extra ni-trogen as polymers of arginine or aspartic acid in cyanophycin granules. Since cyanobacteria lack the enzyme α-ketoglutarate dehydrogenase, they do not have a fully functional citric acid cycle. The pentose phosphate pathway plays a central role in their carbohydrate metabolism. Although many cyanobacteria are obligate photolithoautotrophs, some can grow slowly in the dark as chemoheterotrophs by oxidizing glucose and a few other sugars. Under anaerobic conditions *Oscillatoria limnetica* oxidizes hydrogen sulfide instead of water and carries out anoxygenic photosynthesis much like the green photosynthetic bacteria. As these examples illustrate, cyanobacteria are capa-ble of considerable metabolic flexibility.

Cyanobacteria also vary greatly in shape and appearance. They range in diameter from about 1 to 10 μm and may be unicellular, exist as colonies of many shapes, or form filaments called trichomes (figure 22.9; *see also figures 3.13–3.15*). A

Figure 22.8   Cyanobacterial Thylakoids and Phycobilisomes. *Synechococcus lividus* with an extensive thylakoid system. The phycobilisomes lining these thylakoids are clearly visible as granules at location t (× 60,000).

(a)

(b)

(c)

(d)

Figure 22.9   Oxygenic Photosynthetic Bacteria. Representative cyanobacteria. (*a*) *Chroococcus turgidus,* two colonies of four cells each (× 600). (*b*) *Nostoc* with heterocysts (× 550). (*c*) *Oscillatoria* trichomes seen with Nomarski interference-contrast optics (× 250). (*d*) The cyanobacteria *Anabaena spiroides* and *Microcystis aeruginosa*. The spiral *A. spiroides* is covered with a thick gelatinous sheath (× 1,000).

(a)

(b)

Figure 22.10 **Cyanobacterial Cell Structure.** (*a*) Schematic diagram of a vegetative cell. The insert shows an enlarged view of the envelope with its outer membrane and peptidoglycan. (*b*) Thin section of *Synechocystis* during division. Bar = 1 μm. Many structures are visible.

trichome is a row of bacterial cells that are in close contact with one another over a large area. In contrast, adjacent cells in a simple chain (such as those commonly found in the genus *Bacillus*) associate by only a small area of contact. Although most appear blue-green because of phycocyanin, a few are red or brown in color because of the red pigment phycoerythrin. Despite this variety, cyanobacteria have typical procaryotic cell structures and a normal gram-negative type cell wall (figure 22.10). They often use gas vesicles to move vertically in the water, and many filamentous species have gliding motility (Box 22.1; *see also chapter 3*). Although cyanobacteria lack flagella, several strains of the marine genus *Synechococcus* are able to move at rates of up to 25 μm/second by means of an unknown mechanism.

Cyanobacteria reproduce by binary fission, budding, fragmentation, and multiple fission. In the last process a cell enlarges and then divides several times to produce many smaller progeny, which are released upon the rupture of the parental cell. Fragmentation of filamentous cyanobacteria can generate small, motile filaments called **hormogonia.** Some species develop **akinetes,** specialized, dormant, thick-walled resting cells that are resistant to desiccation. Often these germinate to form new filaments.

Many filamentous cyanobacteria fix atmospheric nitrogen by means of special cells called **heterocysts** (figure 22.11). Around 5 to 10% of the cells can develop into heterocysts

when cyanobacteria are deprived of both nitrate and ammonia, their preferred nitrogen sources. When transforming themselves into heterocysts, cyanobacterial cells synthesize a very thick new wall, reorganize their photosynthetic membranes, discard their phycobiliproteins and photosystem II, and synthesize the nitrogen-fixing enzyme nitrogenase. Photosystem I is still functional and produces ATP, but no oxygen arises from noncyclic photophosphorylation because photosystem II is absent. This inability to generate O_2 is critical because the nitrogenase is extremely oxygen sensitive. The heterocyst wall slows or prevents O_2 diffusion into the cell, and any O_2 present is consumed during respiration. The structure and physiology of the heterocyst ensures that it will remain anaerobic; it is totally dedicated to nitrogen fixation. It obtains nutrients from adjacent vegetative cells and contributes fixed nitrogen in the form of the amino acid glutamine.

The biochemistry of nitrogen fixation (pp. 191–201).

The classification of cyanobacteria is still in an unsettled state, partly due to the lack of pure cultures. At present all taxonomic schemes must be considered tentative. Group I of section 19 in *Bergey's Manual* contains the cyanobacteria and is divided into five orders with 34 genera. Table 22.3 briefly summarizes the major characteristics of the five orders. These are distinguished using colony or trichome morphology and reproductive patterns. Some other properties important in cyanobacterial characterization

(a)

(b)

(c)

Figure 22.11   Examples of Heterocysts and Akinetes.
(*a*) *Cylindrospermum* with terminal heterocysts (H) and subterminal akinetes (A) (×500). (*b*) *Anabaena,* with heterocysts. (*c*) An electron micrograph of an *Anabaena* heterocyst. Bar = 1 μm. Note the cell wall (W), additional outer walls (E), membrane system (M), and a pore channel to the adjacent cell (P).

are cell morphology, ultrastructure, genetic characteristics, physiology and biochemistry, and habitat/ecology (preferred habitat and growth habit). The authors consider genera as tentative and often do not provide species names.

The *Chroococcales* are unicellular rods or cocci that are almost always nonmotile and reproduce by binary fission or budding. The *Pleurocapsales* are also unicellular, though several individual cells may be held together in an aggregate by an outer wall. Members of this order reproduce by multiple fission to form spherical, very small, reproductive cells, often called **baeocytes,** which escape when the outer wall ruptures. Some baeocytes disperse through gliding motility. The other three orders contain filamentous cyanobacteria. Usually the trichome is unbranched and often is surrounded by a sheath or slime layer. The *Oscillatoriales* form unbranched trichomes composed only of vegetative cells, whereas the *Nostocales* and *Stigonematales* produce heterocysts in the absence of an adequate nitrogen source (they also may form akinetes). Heterocystous cyanobacteria are subdivided into those that form linear filaments (*Nostocales*) and cyanobacteria that divide in a second plane to produce branches or aggregates (*Stigonematales*).

Cyanobacteria are very tolerant of environmental extremes and are present in almost all aquatic and terrestrial environments. Thermophilic species may grow at temperatures of up to 75°C in neutral to alkaline hot springs. Some unicellular forms even grow in the fissures of desert rocks. In nutrient-rich warm ponds and lakes, surface cyanobacteria such as *Anacystis* and *Anabaena* can reproduce rapidly to form blooms (figure 22.12). The release of large amounts of organic matter upon the death of the bloom microorganisms stimulates the growth of chemoheterotrophic bacteria that subsequently depletes the available oxygen. This kills fish and other organisms (*see chapter 41*). Some species can produce toxins that kill livestock and other animals that drink the water. Other cyanobacteria, for example, *Oscillatoria,* are so pollution resistant and characteristic of fresh water with high organic matter content that they act as water pollution indicators.

Cyanobacteria are particularly successful in establishing symbiotic relationships with other organisms. They are the photosynthetic partner in most lichen associations. Cyanobacteria are symbionts with protozoa and fungi, and nitrogen-fixing species form associations with a variety of plants (liverworts, mosses, gymnosperms, and angiosperms).

Types of symbiotic relationships (chapter 28).
Lichens (pp. 548–49).

Group II in section 19 contains the order *Prochlorales* with two genera, *Prochloron* and *Prochlorothrix.* These are oxygenic phototrophic procaryotes that have both chlorophyll *a* and *b* but lack phycobilins. Thus although they resemble cyanobacteria with respect to chlorophyll *a,* they differ in also possessing chlorophyll *b,* the only procaryotes to do so. Because prochlorophytes lack phycobilin pigments and phycobilisomes, they are grass-green in color. They resemble chloroplasts in their pigments and thylakoid structure, but their 5S

TABLE 22.3   Characteristics of the Cyanobacterial Orders

Order	General Shape	Reproduction and Growth	Heterocysts	% G + C	Other Properties	Representative Genera
Chroococcales	Unicellular rods or cocci; nonfilamentous aggregates	Binary fission, budding	–	40–49	Almost always nonmotile	*Chamaesiphon* *Gloeobacter* *Gloeothece* *Gleocapsa* *Synechococcus*
Pleurocapsales	Unicellular rods or cocci; may be held together in aggregates	Multiple fission to form baeocytes	–	40–46	Only some baeocytes are motile	*Pleurocapsa* *Dermocarpa* *Chroococcidiopsis*
Oscillatoriales	Filamentous, unbranched trichome with only vegetative cells	Binary fission in a single plane, fragmentation	–	40–67	Usually motile	*Lyngbya* *Oscillatoria* *Spirulina* *Pseudanabaena*
Nostocales	Filamentous, unbranched trichome may contain specialized cells	Binary fission in a single plane, fragmentation to form hormogonia	+	38–47	Often motile, may produce akinetes	*Anabaena* *Cylindrospermum* *Aphanizomenon* *Nostoc* *Scytonema* *Calothrix*
Stigonematales	Filamentous trichomes either with branches or composed of more than one row of cells	Binary fission in more than one plane, hormogonia formed	+	42–46	May produce akinetes, greatest morphological complexity and differentiation in cyanobacteria	*Fischerella* *Stigonema* *Geitleria*

Figure 22.12   Bloom of Cyanobacteria and Algae in a Eutrophic Pond.

and 16S rRNAs show affinities with the cyanobacteria. Possibly a common ancestor gave rise to prochlorophytes, cyanobacteria, and plant chloroplasts.

The two recognized prochlorophyte genera are quite different from one another. *Prochloron* was first discovered as an extracellular symbiont growing either on the surface or within the cloacal cavity of marine colonial ascidian invertebrates (figure 22.13). These bacteria are single-celled, spherical, and from 8 to 30 µm in diameter. Their mol% of G + C is 31 to 41.

Prochlorothrix is free living, has cylindrical cells that form filaments, and has been found in Dutch lakes. Its DNA has a higher G + C content (53 mol%). Unlike *Prochloron* it has been cultured in the laboratory.

Another small prochlorophyte, less than 1 µm in diameter, has recently been discovered flourishing about 100 meters below the ocean surface. It differs from other prochlorophytes in having divinyl chlorophyll *a* and α-carotene instead of chlorophyll *a* and β-carotene. During the summer, it reaches concentrations of 5×10^5 cells per milliliter. It is one of the most numerous of the marine plankton and a significant component of the marine microbial food web.

1. Summarize the major characteristics of the cyanobacteria that distinguish them from other photosynthetic organisms.
2. Define or describe the following: phycobilisomes, hormogonia, akinetes, heterocysts, and baeocytes.
3. What is a trichome and how does it differ from a simple chain of cells?
4. Briefly discuss the ways in which cyanobacteria reproduce.
5. How are heterocysts modified to carry out nitrogen fixation? When do cyanobacteria develop heterocysts?
6. Give the features of the five major cyanobacterial groups.
7. List some important positive and negative impacts cyanobacteria have on humans and the environment.
8. Compare the prochlorophytes with cyanobacteria and chloroplasts. Where does one find them?

(a)

(b)

Figure 22.13   *Prochloron*. (*a*) A scanning electron micrograph of *Prochloron* cells on the surface of a *Didemnum candidum* colony. (*b*) *Prochloron didemni* section; transmission electron micrograph (× 23,500).

Aerobic Chemolithotrophic Bacteria and Associated Organisms

Section 20 of *Bergey's Manual* is devoted to chemolithotrophic bacteria, those bacteria that derive energy and electrons from reduced inorganic compounds. Normally they employ CO_2 as their carbon source or are chemolithoautotrophs, but some can function as chemolithoheterotrophs and use reduced organic carbon sources. Currently these bacteria are divided into three groups based on the inorganic compounds they prefer to oxidize: nitrifiers, colorless sulfur bacteria (sulfur oxidizers), and iron- and manganese-oxidizing bacteria. Magnetotactic bacteria (*see Box 3.3 and chapter 40*) are also included in section 20 of *Bergey's Manual*. The metabolism of chemolithotrophs is described in chapter 9 of this text; the focus here is on the biology of the first two groups. Their ecological impact is discussed further in the context of microbial ecology.

The biochemistry of chemolithotrophy (pp. 180–81).

Nitrifying Bacteria

Nine genera of **nitrifying bacteria** are placed in the family *Nitrobacteraceae*. Although all are aerobic, gram-negative organisms without endospores and able to oxidize either ammonia or nitrite, they differ considerably in other properties (table 22.4). Nitrifiers may be rod-shaped, ellipsoidal, spherical, spirillar or lobate, and they may possess either polar or peritrichous flagella (figure 22.14). Often they have extensive membrane complexes lying in their cytoplasm. Identification is based on properties such as their preference for nitrite or ammonia, their general shape, and the nature of any cytomembranes present.

Nitrifying bacteria are very important ecologically and can be isolated from soil, sewage disposal systems, and freshwater and marine habitats. The genera *Nitrobacter* and *Nitrococcus* oxidize nitrite to nitrate; *Nitrosomonas*, *Nitrosospira*, *Nitrosococcus*, and *Nitrosolobus* oxidize ammonia to nitrite. When two genera such as *Nitrobacter* and *Nitrosomonas* grow together in a niche, ammonia is converted to nitrate, a process called **nitrification** (*see chapter 9*). Nitrification occurs rapidly in soils treated with fertilizers containing ammonium salts. Nitrate nitrogen is readily used by plants, but it is also rapidly lost through leaching of the water soluble nitrate and by denitrification to nitrogen gas. Thus nitrification is a mixed blessing.

The nitrogen cycle and microorganisms (pp. 816–17).

Colorless Sulfur Bacteria

Like the nitrifiers, colorless sulfur bacteria are a highly diverse group. A number of genera capable of oxidizing hydrogen sulfide are placed in section 23 because they contain filamentous gliding bacteria (*Beggiatoa, Thioploca, Thiothrix*). Unicellular rod-shaped or spiral sulfur-oxidizing bacteria that are nonmotile or motile by flagella are located here (table 22.5). Only some of these bacteria have been isolated and studied in pure culture. Most is known about the genera *Thiobacillus* and *Thiomicrospira*. *Thiobacillus* is a gram-negative rod, and *Thiomicrospira* is a long spiral cell (figure 22.15); both are polarly flagellated. They differ from many of the nitrifying bacteria in that they lack extensive internal membrane systems.

The metabolism of *Thiobacillus* has been intensely studied (*see p. 181*). It grows aerobically by oxidizing a variety of

TABLE 22.4	Selected Characteristics of Representative Nitrifying Bacteria					
Species	**Cell Morphology and Size**	**Reproduction**	**Motility**	**Cytomembranes**	**G + C (Mol%)**	**Habitat**
Ammonia-Oxidizing Bacteria						
Nitrosomonas europaea	Rod; 0.8–1.0 × 1.0–2.0 μm	Binary fission	+ or –; 1 or 2 subpolar flagella	Peripheral, lamellar	47.4–51.0	Soil, sewage, fresh water, marine
Nitrosococcus oceanus	Coccoid; 1.8–2.2 μm in diameter	Binary fission	+; 1 or more subpolar flagella	Centrally located parallel bundle, lamellar	50.5–51.0	Obligately marine
Nitrosospira briensis	Spiral; 0.3–0.4 μm in diameter	Binary fission	+ or –; 1 to 6 peritrichous flagella	Lacking	54.1 (1 strain)	Soil
Nitrosolobus multiformis	Lobular; 1.0–1.5 μm in diameter	Binary fission	+; 1 to 20 peritrichous flagella	Internal, compartmentalizing the cell	53.6–55.1	Soil
Nitrite-Oxidizing Bacteria						
Nitrobacter winogradskyi	Rod; 0.6–0.8 × 1.0–2.0 μm	Budding	+ or –; 1 polar flagellum	Polar cap of flattened vesicles in peripheral region of the cell	60.7–61.7	Soil, fresh water, marine
Nitrococcus mobilis	Coccoid; 1.5–1.8 μm in diameter	Binary fission	+; 1 or 2 subpolar flagella	Tubular cytomembranes randomly arranged throughout cytoplasm	61.2 (1 strain)	Marine

From S. W. Watson, et al., "The Family Nitrobacteraceae" in *The Prokaryotes,* Vol. 1, edited by M. P. Starr et al., 1981. Copyright © 1981 Springer-Verlang New York, Inc., New York, NY. Reprinted by permission.

(a)

(b)

(c)

(d)

Figure 22.14   Representative Nitrifying Bacteria. (*a*) *Nitrobacter winogradskyi;* phase contrast (× 2,500). (*b*) *N. winogradskyi.* Note the polar cap of cytomembranes (× 213,000). (*c*) *Nitrosomonas europaea;* phase contrast (× 2,500). (*d*) *N. europaea* with extensive cytoplasmic membranes (× 81,700).

TABLE 22.5   Sulfur-Oxidizing Genera in Section 20 of *Bergey's Manual*

Genus	Cell Shape	Motility, Flagella	% G + C	Location of Sulfur Deposit[a]	Nutritional Type
Thiobacillus	Rods	+; polar	52–68	Extracellular	Obligate or facultative chemolithotroph
Thiomicrospira	Spirals	+; polar	36–44	Extracellular	Obligate chemolithotroph
Thiobacterium	Rods embedded in gelatinous masses	–	N.A.[b]	Intracellular	Probably chemoorganoheterotroph
Thiospira	Spiral rods, usually with pointed ends	+; polar (single or in tufts)	N.A.	Intracellular	Unknown
Macromonas	Rods, cylindrical or bean shaped	+; polar	67	Intracellular	Probably chemoorganoheterotroph

[a]When hydrogen sulfide is oxidized to elemental sulfur.
[b]N.A., data not available.

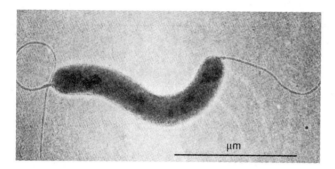

Figure 22.15   Colorless Sulfur Bacteria. *Thiomicrospira pelophila* with polar flagella. Bar = 1 μm.

inorganic sulfur compounds (elemental sulfur, hydrogen sulfide, thiosulfate) to sulfate. ATP is produced with a combination of oxidative phosphorylation and substrate-level phosphorylation by means of adenosine 5´-phosphosulfate (*see figure 9.21*). Although *Thiobacillus* normally uses CO_2 as its major carbon source, *T. novellus* and a few other strains can grow heterotrophically. Some species are very flexible metabolically. For example, *Thiobacillus ferrooxidans* also uses ferrous iron as an electron donor and produces ferric iron as well as sulfuric acid. *T. denitrificans* even grows anaerobically by reducing nitrate to nitrogen gas. It should be noted that sulfur-oxidizing bacteria such as *Thiobacterium* and *Macromonas* probably do not derive energy from sulfur oxidation. They may use the process to detoxify metabolically produced hydrogen peroxide.

Sulfur-oxidizing bacteria have a wide distribution and great practical importance. *Thiobacillus* grows in soil and aquatic habitats, both freshwater and marine. In marine habitats *Thiomicrospira* is more important than *Thiobacillus*. Because of their great acid tolerance (*T. thiooxidans* grows at pH 0.5 and cannot grow above pH 6), these bacteria prosper in habitats they have acidified by sulfuric acid production, even though most other organisms are dying. The production of large quantities of sulfuric acid and ferric iron by *T. ferrooxidans* corrodes concrete and pipe structures. Thiobacilli often cause extensive acid and metal pollution when they release metals from mine wastes. However, sulfur-oxidizing bacteria also are beneficial. They may increase soil fertility when they release unavailable sulfur by oxidizing it to sulfate. Thiobacilli are used in processing low-grade metal ores because of their ability to leach metals from ore.

The sulfur cycle and microorganisms (p. 816).

1. What are chemolithotrophic bacteria? On what basis are they subdivided into three groups?

2. Give the major characteristics of the nitrifying bacteria and discuss their ecological importance. How does the metabolism of *Nitrosomonas* differ from that of *Nitrobacter*?

3. Characterize the colorless sulfur bacteria in section 20 of *Bergey's Manual*.

4. How do colorless sulfur bacteria obtain energy by oxidizing sulfur compounds? What is adenosine 5´-phosphosulfate?

5. List several positive and negative impacts sulfur-oxidizing bacteria have on the environment and human activities.

Budding and/or Appendaged Bacteria

Section 21 of *Bergey's Manual* contains a heterogeneous group of bacteria with at least one of three features: a prostheca, a stalk, or reproduction by budding. A **prostheca** (pl., prosthecae) is an extension of the cell, including the plasma membrane and cell wall, that is narrower than the mature cell. A **stalk** is a nonliving appendage produced by the cell and extending from it. **Budding** is distinctly different from the **binary fission** normally used by bacteria (*p. 209*). The bud first appears as a small protrusion at a single point and enlarges to form a mature cell. Most or all of the bud's cell envelope is newly synthesized. In contrast, portions of the parental cell envelope are shared with the progeny cells during binary fission. Finally, the parental cell retains its identity during budding, and the new cell is often smaller than its parent. In binary fission the parental cell disappears as it forms the progeny.

The majority of genera in this section are placed in one of two subsections: prosthecate bacteria and nonprosthecate bacteria. Each subsection is further divided into budding and nonbudding types. Chemoheterotrophic genera are described in section 21; members of different nutritional categories are usually placed in other sections with bacteria of the same nutritional type. Two examples of prosthecate bacteria are described here: *Hyphomicrobium* and *Caulobacter.*

The genus *Hyphomicrobium* contains chemoheterotrophic, aerobic, budding bacteria that frequently attach to solid objects in freshwater, marine, and terrestrial environments. (They even grow in laboratory water baths.) The vegetative cell measures about 0.5 to 1.0 by 1 to 3 μm (figure 22.16). At the beginning of the reproductive cycle, the mature cell produces a hypha or prostheca, 0.2 to 0.3 μm in diameter, that grows to several μm in length (figure 22.17). The nucleoid divides, and a copy moves into the hypha while a bud forms at its end. As the bud matures, it produces one to three flagella, and a septum divides the bud from the hypha. The bud is finally released as an oval- to pear-shaped swarmer cell, which swims off, then settles down and begins budding. The mother cell may bud several times at the tip of its hypha.

Hyphomicrobium also is distinctive in its physiology and nutrition. Sugars and most amino acids do not support abundant growth; instead *Hyphomicrobium* grows on ethanol and acetate and flourishes with one-carbon compounds such as methanol, formate, and formaldehyde. That is, it is a facultative **methylotroph** and can derive both energy and carbon from reduced one-carbon compounds. It is so efficient at acquiring one-carbon molecules that it grows in a medium without an added carbon source (presumably the medium absorbs sufficient atmospheric carbon compounds). *Hyphomicrobium* may comprise up to 25% of the total bacterial population in oligotrophic or nutrient-poor freshwater habitats.

Bacteria in the genus *Caulobacter* may be polarly flagellated rods or possess a prostheca and **holdfast,** by which they attach to solid substrata (figure 22.18). Caulobacters are usually isolated from freshwater and marine habitats with low nutrient levels, but they also are present in the soil. They often adhere to bacteria, algae, and other microorganisms and may absorb nutrients released by their hosts. The prostheca differs from that of *Hyphomicrobium* in that it lacks cytoplasmic components and is composed almost totally of the plasma membrane and cell wall. It grows longer in nutrient-poor media and can reach more than 10 times the length of the cell body. The prostheca may improve the efficiency of nutrient uptake from dilute habitats by increasing surface area; it also gives the cell extra buoyancy.

Figure 22.16   Prosthecate, Budding Bacteria. *Hyphomicrobium* morphology. *Hyphomicrobium facilis* with hypha and young bud. Bar = 1 μm.

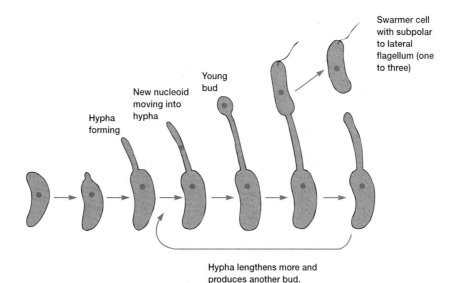

Figure 22.17   The Life Cycle of *Hyphomicrobium*.

(a)

(b)

(c)

(d)

Figure 22.18 *Caulobacter* **Morphology and Reproduction.**
(*a*) Rosettes of cells adhering to each other by their prosthecae; phase contrast (× 600). (*b*) A cell dividing to produce a swarmer (× 6,030). Note prostheca and flagellum. (*c*) A cell with a flagellated swarmer (× 6,030). (*d*) A rosette as seen in the electron microscope. Bar = 1 μm.

The life cycle of *Caulobacter* is unusual (figure 22.19). When ready to reproduce, the cell elongates and a single polar flagellum forms at the end opposite the prostheca. The cell then undergoes asymmetric transverse binary fission to produce a flagellated swarmer cell that swims off. The swarmer eventually comes to rest and forms a new prostheca on the flagellated end while its flagellum disappears. The cycle takes only about two hours for completion.

1. Define the following terms: prostheca, stalk, budding, swarmer cell, methylotroph, and holdfast.
2. Briefly describe the morphology and life cycles of *Hyphomicrobium* and *Caulobacter.*
3. What is unusual about the physiology of *Hyphomicrobium?* How does this influence its ecological distribution?

Sheathed Bacteria

Some bacteria have a **sheath,** a hollow tubelike structure surrounding a chain of cells. Sheaths often are close fitting, but they are never in intimate contact with the cells they enclose

and may contain ferric or manganic oxides. They have at least two functions. Sheaths help bacteria attach to solid surfaces and acquire nutrients from slowly running water as it flows past, even if it is nutrient-poor. Sheaths also protect against predators such as protozoa and *Bdellovibrio* (*see chapter 20*). Section 22 of *Bergey's Manual* contains seven genera of gram-negative, aerobic, heterotrophic, sheathed bacteria.

Two of the best-studied genera are *Sphaerotilus* and *Leptothrix* (figures 22.20 and 22.21). *Sphaerotilus* forms long sheathed chains of rods, 0.7 to 2.4 by 3 to 10 μm, attached to submerged plants, rocks, and other solid objects, often by a holdfast (figure 22.20). The sheaths are not usually encrusted by metal oxides. Single swarmer cells with a bundle of subpolar flagella escape the filament and form a new chain after attaching to a solid object at another site. *Sphaerotilus* prefers slowly running fresh water polluted with sewage or industrial waste. It grows so well in activated sewage sludge that it sometimes forms tangled masses of filaments and interferes with the proper settling of sludge (*see chapter 41*). *Leptothrix* characteristically deposits large amounts of iron and manganese oxides in its sheath. This seems to protect it and allow *Leptothrix* to grow in the presence of high concentrations of soluble iron compounds.

Figure 22.19 *Caulobacter* **Life Cycle.** See text for details.

Figure 22.20   Sheathed Bacteria, *Sphaerotilus natans.* (*a*) Sheathed chains of cells and empty sheaths. (*b*) Chains with holdfasts (*a*) and individual cells containing poly-β-hydroxybutyrate granules. Bars = 10 μm.

Figure 22.21   Sheathed Bacteria, *Leptothrix* Morphology. (*a*) *L. lopholea* trichomes radiating from a collection of holdfasts. (*b*) *L. cholodnii* sheaths encrusted with MnO_2. Bars = 10 μm.

Box 22.1

The Mechanism of Gliding Motility

Gliding motility varies greatly in rate (from about 2 μm per minute to over 600 μm per minute) and in the nature of the motion. Bacteria such as *Myxococcus* and *Flexibacter* glide along in a direction parallel to the longitudinal axis of their cells. Others (*Saprospira*) travel with a screwlike motion or even move in a direction perpendicular to the long axis of the cells in their trichome (*Simonsiella*). *Beggiatoa*, cyanobacteria, and some other bacteria rotate around their longitudinal axis while gliding, but this is not always seen. Many will flex or twitch as well as glide. Such diversity in gliding movement may indicate that more than one mechanism for motility exists. This conclusion is supported by the observation that some gliders (e.g., *Cytophaga* and *Flexibacter*) move attached latex beads over their surface, whereas others such as *Myxococcus* do not. (That is, not all gliding bacteria have moving cell-surface components.) Although slime is required for gliding, it does not appear to propel bacteria directly; rather, it probably attaches them to the substratum and lubricates the surface for more efficient movement.

A variety of mechanisms for gliding motility have been proposed. Cytoplasmic fibrils or filaments are associated with the envelope of many gliding bacteria. In *Oscillatoria* they seem to be contractile and may produce waves in the outer membrane, resulting in movement. Ringlike protein complexes or rotary assemblies resembling flagellar basal bodies are present in some envelopes. These assemblies may spin and move the bacterium along. There is some evidence that differences in surface tension can propel *Myxococcus xanthus*. *Myxococcus* may secrete a surfactant at its posterior end (the end opposite the direction of movement) that lowers the surface tension at the rear end of the rod. The cell would be pulled forward by the greater surface tension exerted on its anterior end.

---

1. What is a sheath and of what advantage is it?
2. How does *Sphaerotilus* maintain its position in running water? How does it reproduce and disperse its progeny?

Nonphotosynthetic, Nonfruiting, Gliding Bacteria

Gliding motility is present in a wide diversity of taxa: fruiting and nonfruiting aerobic chemoheterotrophs, cyanobacteria, green nonsulfur bacteria, and at least two gram-positive genera (*Heliobacterium* and *Desulfonema*). Gliding bacteria lack flagella and are stationary while suspended in liquid medium. When in contact with a surface, they glide along, leaving a slime trail; the gliding mechanism is unknown (Box 22.1). Movement can be very rapid; some cytophagas travel 150 μm in a minute, whereas filamentous gliding bacteria may reach speeds of more than 600 μm/minute. Young organisms are the most motile, and motility often is lost with age. Low nutrient levels usually stimulate gliding.

Gliding motility gives a bacterium many advantages. Many aerobic chemoheterotrophic gliding bacteria actively digest insoluble macromolecular substrates such as cellulose and chitin, and gliding motility is ideal for searching these out. Gliding movement is well adapted to drier habitats and to movement within solid masses such as soil, sediments, and rotting wood that are permeated by small channels. Finally, gliding bacteria, like flagellated bacteria, can position themselves at optimal conditions of light intensity, oxygen, hydrogen sulfide, temperature, and other factors that influence growth.

Seven genera containing rods or individual filaments are placed in the order *Cytophagales* and family *Cytophagaceae*. Bacteria of the genus *Cytophaga* are slender rods, often with pointed ends (figure 22.22*b*). They differ from the gliding, fruiting myxobacteria in lacking fruiting bodies and having a low G + C ratio. *Sporocytophaga* is similar to *Cytophaga* but forms spherical resting cells called microcysts (figure 22.22*c,d*). Another related genus, *Flexibacter*, produces long, flexible threadlike cells when young (figure 22.22*a*) and is unable to use complex polysaccharides.

Members of the genera *Cytophaga* and *Sporocytophaga* are aerobes that actively degrade complex polysaccharides. Soil cytophagas digest cellulose; both soil and marine forms attack chitin, pectin, and keratin. Some marine species even degrade agar. Cytophagas play a major role in the mineralization of organic matter and can cause great damage to exposed fishing gear and wooden structures. They also are a major component of the bacterial population in sewage treatment plants and presumably contribute significantly to this waste treatment process.

Although most cytophagas are free-living, some can be isolated from vertebrate hosts and are pathogenic. *Cytophaga columnaris* and others cause diseases such as columnaris disease, cold water disease, and fin rot in freshwater and marine fish. *Capnocytophaga* flourishes in the gingival crevice (the space between the tooth surface and the gum) of humans and may be involved in the development of periodontal disease. It is unusual among the cytophagas in being anaerobic.

The order *Beggiatoales* contains gliding bacteria that form trichomes (figure 22.23). Two of the best-studied genera in this

(a)   (b)   (c)

(d)

Figure 22.22 Nonphotosynthetic, Nonfruiting, Gliding Bacteria. Representative members of the order *Cytophagales.* (*a*) Long thread cells of *Flexibacter elegans* (× 1,100). (*b*) *Cytophaga* sp. (× 1,150). (*c*) *Sporocytophaga myxococcoides,* vegetative cells on agar (× 1,170). (*d*) *Sporocytophaga myxococcoides,* mature microcysts (× 1,750).

Figure 22.23   *Thiothrix.* A *Thiothrix* colony viewed with phase-contrast microscopy (× 1,000).

order are *Beggiatoa* and *Leucothrix* (figures 22.24 and 22.25). *Beggiatoa* is microaerophilic and grows in sulfide-rich habitats such as sulfur springs, fresh water with decaying plant material, rice paddies, salt marshes, and marine sediments. Its filaments contain short, disklike cells and lack a sheath. *Beggiatoa* is very versatile metabolically. It oxidizes hydrogen sulfide to form large sulfur grains located in pockets formed by invaginations

50μm

Figure 22.24   *Beggiatoa.* *Beggiatoa* sp. colony growing on agar.

(a)

(b)

(c)

(d)

(e)

Figure 22.25 **Morphology and Reproduction of *Leucothrix mucor*.** (*a*) Life cycle of *L. mucor*. (*b*) Separation of gonidia from the tip of a mature filament; phase contrast (× 1,400). (*c*) Gonidia aggregating to form rosettes; phase contrast (× 950). (*d*) Young developing rosettes (× 1,500). (*e*) A knot formed by a *Leucothrix* filament.

Figure 22.26   Gliding, Fruiting Bacteria (Myxobacteria). Myxobacterial cells and myxospores. (*a*) *Stigmatella aurantiaca* (× 1,200). (*b*) *Chondromyces crocatus* (× 950). (*c*) Myxospores of *Myxococcus xanthus* (× 1,100). All photographs taken with a phase-contrast microscope.

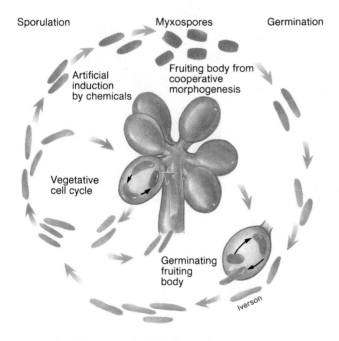

Figure 22.27   Myxobacterial Life Cycle. The outermost sequence depicts the chemical induction of myxospore formation followed by germination. Regular fruiting body production and myxospore germination are also shown.

of the plasma membrane. *Beggiatoa* can subsequently oxidize the sulfur to sulfate. The sulfur electrons are used by the electron transport chain in energy production. Many strains also can grow heterotrophically with acetate as a carbon source, and some may incorporate CO_2 autotrophically.

1. Give three advantages of gliding motility.
2. Briefly describe the following genera: *Cytophaga, Sporocytophaga,* and *Beggiatoa.*
3. Why are the cytophagas ecologically important?

Gliding, Fruiting Bacteria

The **myxobacteria** are gram-negative, aerobic soil bacteria characterized by gliding motility, a complex life cycle with the production of fruiting bodies, and the formation of dormant myxospores. In addition, their G + C content is around 67 to 71%, significantly higher than that of most gliding bacteria. Myxobacterial cells are rods, about 0.6 to 0.9 by 3 to 8 μm long, and may be either slender with tapered ends or stout with rounded, blunt ends (figure 22.26). The approximately 30 species of myxobacteria in the order *Myxobacteriales* are divided into four families based on the shape of vegetative cells, myxospores, and sporangia.

Most myxobacteria are micropredators or scavengers. They secrete an array of digestive enzymes that lyse bacteria and yeasts. Many myxobacteria also secrete antibiotics, which may kill their prey. The digestion products, primarily small peptides, are absorbed. Most myxobacteria use amino acids as their major source of carbon, nitrogen, and energy. All are chemoheterotrophs with respiratory metabolism.

The myxobacterial life cycle is quite distinctive and in many ways resembles that of the cellular slime molds (figure 22.27). In the presence of a food supply, myxobacteria migrate along a solid surface, feeding and leaving slime trails. During

Figure 22.28 **Myxobacterial Fruiting Bodies.**
(*a*) An illustration of typical fruiting body structure.
(*b*) *Myxococcus fulvus.* Fruiting bodies are about
150–400 μm high. (*c*) *Myxococcus stipitatus.* The
stalk is as tall as 200 μm. (*d*) *Chondromyces
crocatus* viewed with the SEM. The stalk may
reach 700 μm or more in height.

(a)

(b)

(c)

(d)

this stage the cells often form a swarm and move in a coordinated fashion. Some species congregate to produce a sheet of cells that moves rhythmically to generate waves or ripples. When their nutrient supply is exhausted, the myxobacteria aggregate and differentiate into a **fruiting body.** Such bodies range in height from 50 to 500 μm and often are attractively colored red, yellow, or brown by carotenoid pigments. They vary in complexity from simple globular objects (*Myxococcus*) to the elaborate, branching, treelike structures formed by *Stigmatella* and *Chondromyces* (figure 22.28). Some cells develop into dormant **myxospores** that frequently are enclosed in walled structures called sporangioles or sporangia. Each species forms a characteristic fruiting body.

Myxospores are not only dormant but desiccation-resistant, and they may survive up to 10 years under adverse conditions. They enable myxobacteria to survive long periods of dryness and nutrient deprivation. The use of fruiting bodies provides further protection for the myxospores and assists in their dispersal. (The myxospores often are suspended above the soil surface.) Because myxospores are kept together within the fruiting body, a colony of myxobacteria automatically develops when the myxospores are released and germinate. This communal organization may be advantageous because myxobacteria obtain nutrients by secreting hydrolytic enzymes and absorbing soluble digestive products. A mass of myxobacteria can produce enzyme concentrations sufficient to digest their prey more easily than can an individual cell. Extracellular enzymes diffuse away from their source, and an individual cell will have more difficulty overcoming diffusional losses than a swarm of cells.

Myxobacteria are found in soils worldwide. They are most commonly isolated from neutral soils or decaying plant material such as leaves and tree bark, and from animal dung. Although they grow in habitats as diverse as tropical rain forests and the arctic tundra, they are most abundant in warm areas.

1. Give the major distinguishing characteristics of the myxobacteria. How do they obtain most of their nutrients?
2. Briefly describe the myxobacterial life cycle. What are fruiting bodies, myxospores, and sporangioles?

Summary

1. Cyanobacteria carry out oxygenic photosynthesis, and purple and green bacteria use anoxygenic photosynthesis.
2. The four most important groups of purple and green photosynthetic bacteria are the purple sulfur bacteria, the purple nonsulfur bacteria, the green sulfur bacteria, and the green nonsulfur bacteria.
3. The bacteriochlorophyll pigments of purple and green bacteria enable them to live in deeper, anaerobic zones of aquatic habitats.
4. Cyanobacteria carry out oxygenic photosynthesis by means of a photosynthetic apparatus similar to that of the eucaryotes. Like the red algae, they have phycobilisomes.
5. Cyanobacteria reproduce by binary fission, budding, multiple fission, and fragmentation

to form hormogonia. Some will produce a dormant akinete.
6. Nitrogen-fixing cyanobacteria usually form heterocysts, specialized cells in which nitrogen fixation occurs.
7. Chemolithotrophic bacteria derive energy and electrons from reduced inorganic compounds. Nitrifying bacteria are aerobic bacteria that oxidize either ammonia or nitrite to nitrate and are responsible for nitrification. Colorless sulfur bacteria oxidize elemental sulfur, hydrogen sulfide, and thiosulfate to sulfate.
8. Section 21 of *Bergey's Manual* contains bacteria with at least one of three features: a prostheca, a stalk, or reproduction by budding.
9. Two examples of the budding and/or appendaged bacteria are *Hyphomicrobium* (budding bacteria that produce swarmer cells),

and *Caulobacter* (bacteria with prosthecae and holdfasts).
10. Section 22 contains bacteria such as *Sphaerotilus* and *Leptothrix* with sheaths, hollow tubelike structures that surround chains of cells without being in intimate contact.
11. Gliding motility is present in a diversity of bacteria.
12. Cytophagas degrade proteins and complex polysaccharides and are active in the mineralization of organic matter.
13. Many bacteria exist as trichomes.
14. Myxobacteria are gram-negative, aerobic soil bacteria with gliding motility and a complex life cycle that leads to the production of dormant myxospores held within fruiting bodies.

Key Terms

akinete *461*

anoxygenic photosynthesis *455*

baeocyte *462*

binary fission *466*

budding *466*

chlorosomes *459*

cyanobacteria *459*

fruiting body *474*

gliding motility *470*

green nonsulfur bacteria *459*

green sulfur bacteria *459*

heterocyst *461*

holdfast *467*

hormogonia *461*

methylotroph *467*

myxobacteria *473*

myxospore *474*

nitrification *464*

nitrifying bacteria *464*

oxygenic photosynthesis *455*

phycobilisome *460*

prostheca *466*

purple nonsulfur bacteria *456*

purple sulfur bacteria *456*

sheath *468*

stalk *466*

trichome *461*

Questions for Thought and Review

1. How do cyanobacteria differ from the purple and green photosynthetic bacteria?
2. Relate the physiology of each major group of photosynthetic bacteria to its preferred habitat.
3. Why is nitrogen fixation an oxygen-sensitive process? How are cyanobacteria able to fix

nitrogen when they also carry out oxygenic photosynthesis?
4. A number of bacterial groups are found in particular habitats to which they are well adapted. Give the habitat for each of the following groups and discuss the reasons for its preference: *Beggiatoa*, purple and green sulfur bacteria, *Hyphomicrobium*, and *Thiobacillus*.

5. Why are gliding and budding and/or appendaged bacteria distributed among so many different sections in *Bergey's Manual*?
6. Discuss the ways in which the distinctive myxobacterial life cycle may be of great advantage to these organisms.

Additional Reading

General

Balows, A.; Trüper, H. G.; Dworkin, M.; Harder, W.; and
Schleifer, K.-H. 1992. *The prokaryotes,* 2d ed. New
York: Springer-Verlag.

Holt, J. G., editor-in-chief. 1989. *Bergey's Manual of
Systematic Bacteriology,* vol. 3, J. T. Staley, M. P.
Bryant, and N. Pfennig, editors. Baltimore, Md.:
Williams & Wilkins.

Holt, J. G., editor-in-chief. 1994. *Bergey's Manual of
Determinative Bacteriology,* 9th ed. Baltimore, Md.:
Williams and Wilkins.

Lederberg, J., editor. 1992. *Encyclopedia of microbiology.*
San Diego, Calif.: Academic Press.

Mayer, F. 1986. *Cytology and morphogenesis of bacteria.*
Berlin: Gebrüder Borntraeger.

Schlegel, H. G., and Bowien, B. 1989. *Autotrophic
bacteria.* Madison, Wis.: Science Tech Publishers.

Shapiro, J. A. 1988. Bacteria as multicellular organisms.
Sci. Am. 258(6):82–89.

Oxygenic Photosynthetic Bacteria

Adams, D. G. 1992. Multicellularity in cyanobacteria. In
Prokaryotic structure and function, S. Mohan, C. Dow,
and J. A. Coles, editors, 341–84. New York:
Cambridge University Press.

Allen, M. M. 1984. Cyanobacterial cell inclusions. *Ann.
Rev. Microbiol.* 38:1–25.

Armstrong, G. A. 1994. Eubacteria show their true colors:
Genetics of carotenoid pigment biosynthesis from
microbes to plants. *J. Bacteriol.* 176(16):4795–802.

Bullerjahn, G. S., and Post, A. F. 1993. The
prochlorophytes: Are they more than just chlorophyll
a/b-containing cyanobacteria? *Crit. Rev. Microbiol.*
19(1):43–59.

Carmichael, W. W. 1994. The toxins of cyanobacteria. *Sci.
Am.* 270(1):78–86.

Glazer, A. N. 1983. Comparative biochemistry of
photosynthetic light-harvesting systems. *Ann. Rev.
Biochem.* 52:125–57.

Grossman, A. R.; Schaefer, M. R.; Chiang, G. G.; and
Collier, J. L. 1993. The phycobilisome, a light-
harvesting complex responsive to environmental
conditions. *Microbiol. Rev.* 57(3):725–49.

Peters, G. A. 1978. Blue-green algae and algal associations.
BioScience 28(9):580–85.

Rogers, L. J., and Gallon, J. R., editors. 1988. *Biochemistry
of the algae and cyanobacteria.* New York: Oxford
University Press.

Stanier, R. Y., and Cohen-Bazire, G. 1977. Phototrophic
prokaryotes: The cyanobacteria. *Ann. Rev. Microbiol.*
31:225–74.

Stewart, W. D. P. 1980. Some aspects of structure and
function in N_2-fixing cyanobacteria. *Ann. Rev.
Microbiol.* 34:497–536.

Aerobic Chemolithotrophic Bacteria and
Associated Organisms

Ghiorse, W. C. 1984. Biology of iron- and manganese-
depositing bacteria. *Ann. Rev. Microbiol.* 38:515–50.

Budding and/or Appendaged Bacteria

Brun, Y. V.; Marczynski, G.; and Shapiro, L. 1994. The
expression of asymmetry during caulobacter cell
differentiation. *Ann. Rev. Biochem.* 63:419–50.

Gober, J. W., and Marques, M. V. 1995. Regulation of
cellular differentiation in *Caulobacter crescentus.*
Microbiol. Rev. 59(1):31–47.

Moore, R. L. 1981. The biology of *Hyphomicrobium* and
other prosthecate, budding bacteria. *Ann. Rev.
Microbiol.* 35:567–94.

Poindexter, J. S. 1981. The Caulobacters: Ubiquitous
unusual bacteria. *Microbiol. Rev.* 45:123–79.

Nonphotosynthetic, Nonfruiting, Gliding
Bacteria

Burchart, R. P. 1981. Gliding motility of prokaryotes:
Ultrastructure, physiology, and genetics. *Ann. Rev.
Microbiol.* 35:497–529.

Larkin, J. M., and Strohl, W. R. 1983. *Beggiatoa,
Thiothrix,* and *Thioploca. Ann. Rev. Microbiol.*
37:341–67.

Reichenbach, H. 1981. Taxonomy of the gliding bacteria.
Ann. Rev. Microbiol. 35:339–64.

Gliding, Fruiting Bacteria

Dworkin, M., and Kaiser, D. 1985. Cell interactions in
myxobacterial growth and development. *Science*
230:18–24.

Shimkets, L. J. 1990. Social and developmental biology of
the myxobacteria. *Microbiol. Rev.* 54(4):473–501.

CHAPTER 23
The Bacteria:
The Archaeobacteria

As is often the case, epoch-making ideas carry with them implicit, unanalyzed assumptions that ultimately impede scientific progress until they are recognized for what they are. So it is with the prokaryote-eukaryote distinction. Our failure to understand its true nature set the stage for the sudden shattering of the concept when a "third form of life" was discovered in the late 1970s, a discovery that actually left many biologists incredulous. Archaebacteria, as this third form has come to be known, have revolutionized our notion of the prokaryote, have altered and refined the way in which we think about the relationship between prokaryotes and eukaryotes . . . and will influence strongly the view we develop of the ancestor that gave rise to all extant life.
—C. R. Woese and R. S. Wolfe

Outline

Concepts

1. Archaeobacteria differ in many ways from both eubacteria and eucaryotes. These include differences in cell wall structure and chemistry, membrane lipid structure, molecular biology, and metabolism.
2. Archaeobacteria grow in a few restricted or specialized habitats: anaerobic, hypersaline, and high temperature.
3. *Bergey's Manual* divides the archaeobacteria into five major groups: methanogenic archaeobacteria, sulfate reducers, extreme halophiles, cell wall-less archaeobacteria, and extremely thermophilic S^0-metabolizers.
4. Methanogenic and sulfate-reducing archaeobacteria have unique cofactors that participate in methanogenesis.
5. Archaeobacteria have special structural, chemical, and metabolic adaptations that enable them to grow in extreme environments.

Chapters 20 and 21 introduce the bacteria described in volumes 1 and 2 of *Bergey's Manual.* Chapter 22 covers the bacteria found in sections 18 to 24 of volume 3. Although the archaeobacteria are described in section 25 of volume 3, they are placed in a separate chapter here because they are so different from other procaryotes.

This chapter begins with a general introduction to the archaeobacteria. Then it briefly discusses the biology of each major archaeobacterial group.

Figure 23.1   Cell Envelopes of Archaeobacteria. Schematic representations and electron micrographs of (*a*) *Methanobacterium formicicum,* a typical gram-positive organism, and (*b*) *Thermoproteus tenax,* a gram-negative archaeobacterium. CW, cell wall; SL, surface layer; CM, cell membrane or plasma membrane; CPL, cytoplasm.

Comparison of the sequences of rRNA from a great variety of organisms shows that organisms may be divided into three major groups: the eubacteria, archaeobacteria, and eucaryotes (*see figure 19.13*). Some of the most important differences are summarized in table 19.8 (*p. 403*). Because archaeobacteria are different from both eubacteria and eucaryotes, their most distinctive properties will first be described in some detail and compared with those of the latter two groups.

rRNA sequences and the archaeobacteria (pp. 402, 407–8).

Introduction to the Archaeobacteria

As a group the **archaeobacteria** [Greek *archaios,* ancient, and *bakterion,* a small rod] are quite diverse, both in morphology and physiology. They can stain either gram positive or gram negative and may be spherical, rod-shaped, spiral, lobed, plate-shaped, irregularly shaped, or pleomorphic. Some are single cells, whereas others form filaments or aggregates. They range in diameter from 0.1 to over 15 μm, and some filaments can grow up to 200 μm in length. Multiplication may be by binary fission, budding, fragmentation, or other mechanisms. Archaeobacteria are just as diverse physiologically. They can be aerobic, facultatively anaerobic, or strictly anaerobic. Nutritionally they range from chemolithoautotrophs to organotrophs. Some are mesophiles; others are hyperthermophiles that can grow above 100°C.

Archaeobacteria usually prefer restricted or extreme aquatic and terrestrial habitats. They are often present in anaerobic, hypersaline, or high-temperature environments. Recently archaeobacteria have been discovered in cold environments. It appears that they constitute up to 34% of the procaryotic biomass in coastal Antarctic surface waters (*see chapter 41*). A few are symbionts in animal digestive systems.

Archaeobacterial Cell Walls

Although archaeobacteria can stain either gram positive or gram negative, their wall structure and chemistry differ from that of the eubacteria. There is considerable variety in archaeobacterial wall structure. Many gram-positive archaeobacteria have a wall with a single thick homogeneous layer like gram-positive eubacteria (figure 23.1*a*). Gram-negative archaeobacteria lack the

outer membrane and complex peptidoglycan network or sacculus of gram-negative eubacteria. Instead they usually have a surface layer of protein or glycoprotein subunits (figure 23.1*b*).

Peptidoglycan structure and chemistry (pp. 52–54).

The chemistry of archaeobacterial cell walls is also quite different from that of the eubacteria. None have the muramic acid and D-amino acids characteristic of eubacterial peptidoglycan. Not surprisingly, all archaeobacteria resist attack by lysozyme and β-lactam antibiotics such as penicillin. Gram-positive archaeobacteria can have a variety of complex polymers in their walls. *Methanobacterium* and some other methanogens have walls containing **pseudomurein,** a peptidoglycanlike polymer that has L-amino acids in its cross-links, N-acetyltalosaminuronic acid instead of N-acetylmuramic acid, and β(1→3) glycosidic bonds instead of β(1→4) glycosidic bonds (figure 23.2). *Methanosarcina* and *Halococcus* contain complex polysaccharides similar to the chondroitin sulfate of animal connective tissue. Other heteropolysaccharides are also found in gram-positive walls.

Gram-negative archaeobacteria have a layer of protein or glycoprotein outside their plasma membrane. The layer may be as thick as 20 to 40 nm. Sometimes there are two layers, a sheath surrounding an electron-dense layer. The chemical content of these walls varies considerably. Some methanogens (*Methanolobus*), *Halobacterium,* and several extreme thermophiles (*Sulfolobus, Thermoproteus,* and *Pyrodictium*) have glycoproteins in their walls. In contrast, other methanogens (*Methanococcus, Methanomicrobium,* and *Methanogenium*) and the extreme thermophile *Desulfurococcus* have protein walls.

Archaeobacterial Lipids and Membranes

As emphasized in table 19.8, a most distinctive feature of the archaeobacteria is the nature of their membrane lipids. They differ from both eubacteria and eucaryotes in having branched chain

Figure 23.2   The Structure of Pseudomurein. The components in parentheses are not always present. Ac represents the acetyl group.

Figure 23.3   Archaeobacterial Membrane Lipids. An illustration of the difference between archaeobacterial lipids and those of other bacteria. Archaeobacterial lipids are derivatives of isopranyl glycerol ethers rather than the usual glycerol fatty acid esters. Three examples of common archaeobacterial glycerolipids are given.

hydrocarbons attached to glycerol by ether links rather than fatty acids connected by ester links (figure 23.3). Sometimes two glycerol groups are linked to form an extremely long tetraether. Usually the diether side chains are 20 carbons in size, and the tetraether chains are 40 carbons. The overall length of the tetraethers can be adjusted by cyclizing the chains to form pentacyclic rings (figure 23.3), and biphytanyl chains may contain from one to four cyclopentyl rings. Polar lipids are also present in archaeobacterial membranes: phospholipids, sulfolipids, and glycolipids. From 7 to 30% of the membrane lipids are nonpolar lipids, which usually are derivatives of squalene (figure 23.4). These lipids can be combined in various ways to yield membranes of different rigidity and thickness. For example, the C_{20} diethers can be used to make a regular bilayer membrane (figure 23.5a). A much more rigid monolayer membrane may be constructed of C_{40} tetraether lipids (figure 23.5b). Of course archaeobacterial membranes may contain a mix of di-

ethers, tetraethers, and other lipids. As might be expected from their need for stability, the membranes of extreme thermophiles such as *Thermoplasma* and *Sulfolobus* are almost completely tetraether monolayers.

Genetics and Molecular Biology

Some features of archaeobacterial genetics are similar to those in eubacteria. Their chromosome is a single closed DNA circle. However, the genomes of some archaeobacteria are significantly smaller than the normal eubacterium. *E. coli* DNA has a size of about 2.5×10^9 daltons, whereas *Thermoplasma acidophilum* DNA is about 0.8×10^9 daltons and *Methanobacterium thermoautotrophicum* DNA is 1.1×10^9 daltons. The variation in G + C content is great, from about 21 to 68 mol%, and is another sign of archaeobacterial diversity. Archaeobacteria have few plasmids.

Squalene

Tetrahydrosqualene

Figure 23.4   Nonpolar Lipids of Archaeobacteria. Two examples of the most predominant nonpolar lipids are the C_{30} isoprenoid squalene and one of its hydroisoprenoid derivatives, tetrahydrosqualene.

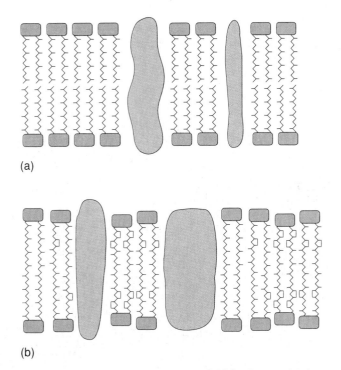

(a)

(b)

Figure 23.5   Examples of Archaeobacterial Membranes. (*a*) A membrane composed of integral proteins and a bilayer of C_{20} diethers. (*b*) A rigid monolayer composed of integral proteins and C_{40} tetraethers.

Archaeobacterial mRNA appears similar to that of eubacteria rather than to eucaryotic mRNA. Polygenic mRNA has been discovered, and there is no evidence for mRNA splicing. Archaeobacterial promoters are similar to those in eubacteria.

Despite these and other similarities, there are also many differences between archaeobacteria and other organisms. Unlike both eubacteria and eucaryotes, the T ψ C arm (*p. 226*) of archaeobacterial tRNA lacks thymine and contains pseudouridine or 1-methylpseudouridine. The archaeobacterial initiator

tRNA carries methionine like the eucaryotic initiator tRNA. Although archaeobacterial ribosomes are 70S like eubacterial ribosomes, electron microscopic studies show that their shape is quite variable and sometimes differs from that of both eubacterial and eucaryotic ribosomes. They do resemble eucaryotic ribosomes in their sensitivity to anisomycin and insensitivity to chloramphenicol and kanamycin. Furthermore, their elongation factor 2 reacts with diphtheria toxin like the eucaryotic EF–2. Finally, archaeobacterial DNA-dependent RNA polymerases resemble the eucaryotic enzymes, not the eubacterial RNA polymerase. They are large, complex enzymes and are insensitive to the drugs rifampin and streptolydigin. These and other differences distinguish the archaeobacteria from both eubacteria and eucaryotes.

Ribosomes and the mechanism of protein synthesis (pp. 228–33).
DNA transcription (pp. 221–25).

Metabolism

Not surprisingly in view of the variety of their life-styles, archaeobacterial metabolism varies greatly between the members of different groups. Some archaeobacteria are organotrophs; others are autotrophic. A few even carry out an unusual form of photosynthesis.

Archaeobacterial carbohydrate metabolism is best understood. The enzyme 6-phosphofructokinase has not been found in archaeobacteria, and they do not appear to degrade glucose by way of the Embden-Meyerhof pathway. Extreme halophiles and thermophiles catabolize glucose using a modified form of the Entner-Doudoroff pathway (*p. 169 and appendix III*) in which the initial intermediates are not phosphorylated. The halophiles have slightly different modifications of the pathway than do the extreme thermophiles but still produce pyruvate and NADH or NADPH. Methanogens are autotrophs and do not catabolize glucose to any significant extent. In contrast with glucose degradation, gluconeogenesis proceeds by a reversal of the Embden-Meyerhof pathway in halophiles and methanogens. All archaeobacteria that have been studied can oxidize pyruvate to acetyl-CoA. They lack the pyruvate dehydrogenase complex present in eucaryotes and respiratory eubacteria and use the enzyme pyruvate oxidoreductase for this purpose. Halophiles and the extreme thermophile *Thermoplasma* do seem to have a functional tricarboxylic acid cycle. No methanogen has yet been found with a complete tricarboxylic acid cycle. Evidence for functional cytochrome chains has been obtained in halophiles and thermophiles.

The Embden-Meyerhof pathway and tricarboxylic acid cycle (pp. 166–67, 169–70, and appendix III).

Very little is known in detail about biosynthetic pathways in the archaeobacteria. Preliminary data suggest that the synthetic pathways for amino acids, purines, and pyrimidines are

Figure 23.6 **Mechanisms of Autotrophic CO_2 Fixation.** (*a*) The reductive tricarboxylic acid cycle. The cycle is reversed with ATP and reducing equivalents [H] to form acetyl-CoA from CO_2. The acetyl-CoA may be carboxylated to yield pyruvate, which can then be converted to glucose and other compounds. This sequence appears to function in *Thermoproteus neutrophilus*. (*b*) The synthesis of acetyl-CoA and pyruvate from CO_2 in *Methanobacterium thermoautotrophicum*. One carbon comes from the reduction of CO_2 to a methyl group, and the second is produced by reducing CO_2 to carbon monoxide through the action of the enzyme CO dehydrogenase (E_1). The two carbons are then combined to form an acetyl group. Corrin-E_2 represents the cobamide-containing enzyme involved in methyl transfers. Special methanogen coenzymes and enzymes are described in figures 23.8 and 23.9.

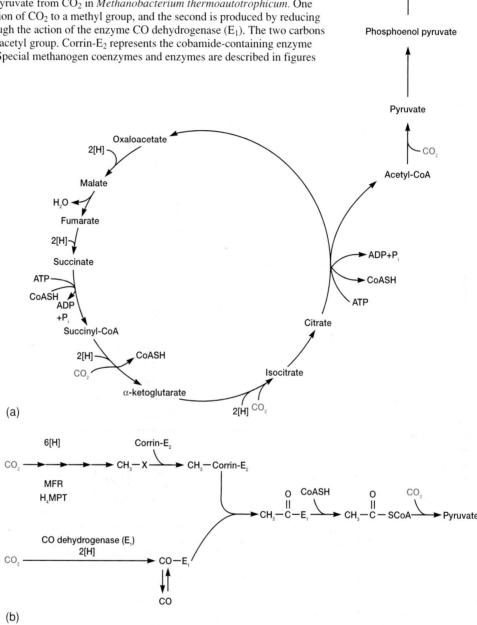

similar to those in other organisms. Some methanogens can fix atmospheric dinitrogen. Not only do many archaeobacteria use a reversal of the Embden-Meyerhof pathway to synthesize glucose, but at least some methanogens and extreme thermophiles employ glycogen as their major reserve material.

Autotrophy is widespread among the methanogens and extreme thermophiles, and CO_2 fixation occurs in more than one

way. *Thermoproteus* and possibly *Sulfolobus* incorporate CO_2 by the reductive tricarboxylic acid cycle (figure 23.6*a*). This pathway is also present in the green sulfur bacteria (*see p. 459*). Methanogenic bacteria and probably most extreme thermophiles incorporate CO_2 by the reductive acetyl-CoA pathway (figure 23.6*b*). A similar pathway also is present in acetogenic bacteria and autotrophic sulfate-reducing bacteria.

TABLE 23.1 Characteristics of the Major Archaeobacterial Groups

Group	General Characteristics	Representative Genera
Methanogenic archaeobacteria	Strict anaerobes. Methane is the major metabolic end product. S^0 may be reduced to H_2S without yielding energy production. Cells possess coenzyme M, factors 420 and 430, and methanopterin.	*Methanobacterium* *Methanococcus* *Methanomicrobium* *Methanosarcina*
Archaeobacterial sulfate reducers	Irregular gram negative coccoid cells. H_2S formed from thiosulfate and sulfate. Autotrophic growth with thiosulfate and H_2. Can grow heterotrophically. Traces of methane also formed. Extremely thermophilic and strictly anaerobic. Possess factor 420 and methanopterin but not coenzyme M or factor 430.	*Archaeoglobus*
Extremely halophilic archaeobacteria	Coccoid or irregularly shaped rods. Gram-negative or gram-positive, primarily aerobic chemoorganotrophs. Require high sodium chloride concentrations for growth (≥1.5 M). Colonies are various shades of red. Neutrophilic or alkalophilic. Mesophilic or slightly thermophilic. Some species contain bacteriorhodopsin and use light for ATP synthesis.	*Halobacterium* *Halococcus* *Natronobacterium*
Cell wall-less archaeobacteria	Pleomorphic cells lacking a cell wall. Thermoacidophilic and chemoorganotrophic. Facultatively anaerobic. Plasma membrane contains a mannose-rich glycoprotein and a lipoglycan.	*Thermoplasma*
Extremely thermophilic S^0-metabolizers	Gram-negative rods, filaments, or cocci. Obligately thermophilic (optimum growth temperature between 70–110°C). Usually strict anaerobes but may be aerobic or facultative. Acidophilic or neutrophilic. Autotrophic or heterotrophic. Most are sulfur metabolizers. S^0 reduced to H_2S anaerobically; H_2S or S^0 oxidized to H_2SO_4 aerobically.	*Desulfurococcus* *Methanopyrus* *Pyrodictium* *Pyrococcus* *Sulfolobus* *Thermococcus* *Thermoproteus*

Archaeobacterial Taxonomy

As shown in figure 19.13, the archaeobacteria are quite distinct from other living organisms. Within the group, however, there is great diversity (*see figure 19.10*). Section 25 of *Bergey's Manual* divides the archaeobacteria into five major groups: methanogenic archaeobacteria, archaeobacterial sulfate reducers, extremely halophilic archaeobacteria, cell wall-less archaeobacteria, and extremely thermophilic S^0-metabolizers. Table 23.1 summarizes some major characteristics of these five groups and gives representatives of each. The last part of the chapter briefly describes the biology of these groups.

1. What are the archaeobacteria? Briefly describe the major ways in which they differ from eubacteria and eucaryotes.
2. How do archaeobacterial cell walls differ from those of the eubacteria? What is pseudomurein?
3. In what ways do archaeobacterial membrane lipids differ from those of eubacteria and eucaryotes? How might these differences lead to stronger membranes?
4. List the differences between archaeobacteria and other organisms with respect to tRNA, ribosome structure and behavior, EF–2 sensitivity, and RNA polymerases.
5. Briefly describe the way in which archaeobacteria degrade and synthesize glucose. In what two unusual ways do they incorporate CO_2?
6. Characterize the five different groups of archaeobacteria and distinguish between them.

Methanogenic Archaeobacteria

Methanogens are strict anaerobes that obtain energy by converting CO_2, H_2, formate, methanol, acetate, and other compounds to either methane or methane and CO_2. They are autotrophic when growing on H_2 and CO_2. This is the largest group of archaeobacteria. There are at least three orders and 18 genera, which differ greatly in overall shape, 16S rRNA sequence, cell wall chemistry and structure, membrane lipids, and other features. For example, methanogens construct three different types of cell walls. Several genera have walls with pseudomurein as mentioned earlier (figure 23.2). Other walls contain either proteins or heteropolysaccharides. The morphology of typical methanogens is shown in figure 23.7, and selected properties of representative genera are presented in table 23.2.

As might be inferred from the methanogens' ability to produce methane anaerobically, their metabolism is unusual. These bacteria contain several unique cofactors: tetrahydromethanopterin (H_4MPT), methanofuran (MFR), coenzyme M (2-mercaptoethanesulfonic acid), coenzyme F_{420}, and coenzyme F_{430} (figure 23.8). The first three cofactors bear the C_1 unit when CO_2 is reduced to CH_4. F_{420} carries electrons and hydrogens, and F_{430} is a nickel tetrapyrrole serving as a cofactor for the enzyme methyl-CoM methylreductase. The pathway for methane synthesis is thought to function as shown in figure 23.9. It appears that ATP synthesis is linked with methanogenesis by electron transport, proton pumping, and a chemiosmotic mechanism (*see pp. 172–74*). Some methanogens

TABLE 23.2   Selected Characteristics of Representative Genera of Methanogens

Genus	Morphology	% G + C	Wall Composition	Gram Reaction	Motility	Methanogenic Substrates Used
Order I. *Methanobacteriales*						
Methanobacterium	Long rods or filaments	32–61	Pseudomurein	+ to variable	–	$H_2 + CO_2$, formate
Methanothermus	Straight to slightly curved rods	33	Pseudomurein with an outer protein S-layer	+	+	$H_2 + CO_2$
Order II. *Methanococcales*						
Methanococcus	Irregular cocci	29–34	Protein	–	–	$H_2 + CO_2$, formate
Order III. *Methanomicrobiales*						
Methanomicrobium	Short curved rods	45–49	Protein	–	+	$H_2 + CO_2$, formate
Methanogenium	Irregular cocci	52–61	Protein or glycoprotein	–	–	$H_2 + CO_2$, formate
Methanospirillum	Curved rods or spirilla	45–50	Protein	–	+	$H_2 + CO_2$, formate
Methanosarcina	Irregular cocci, packets	36–43	Heteropolysaccharide or protein	+ to variable	–	$H_2 + CO_2$, methanol, methylamines, acetate

(a)

(b)

(c)

(d)

(e)

(f)

Figure 23.7   Selected Methanogenic Bacteria. (*a*) *Methanospirillum hungatei;* phase contrast (×2,000). (*b*) *Methanobrevibacter smithii.* (*c*) *Methanosarcina barkeri* from sewage digester; TEM (×6,000). (*d*) *Methanosarcina mazei;* SEM. Bar = 5 μm. (*e*) *Methanobacterium bryantii;* phase contrast (×2,000). (*f*) *Methanogenium marisnigri;* electron micrograph (×45,000).

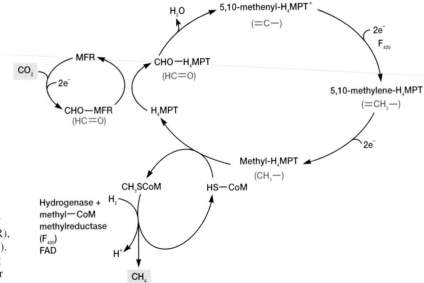

Figure 23.8 Methanogen Coenzymes. The portion of F_{420} (d) that is reversibly oxidized and reduced is shown in color. MFR (a), H_4MPT (b), and coenzyme M (c) carry one-carbon units during methanogenesis (MFR and MPT also participate in the synthesis of acetyl-CoA). The places where the carbon units are attached are in color. H_4MPT carries carbon units on nitrogens 5 and 10 in the same way as the coenzyme tetrahydrofolate. Coenzyme F_{430} (e) is a coenzyme for methyl-CoM methylreductase.

Figure 23.9 Methane Synthesis. Pathway for CH_4 synthesis from CO_2 in *M. thermoautotrophicum.* Cofactor abbreviations: methanopterin (MPT), methanofuran (MFR), and 2-mercaptoethanesulfonic acid or coenzyme M (CoM). The nature of the carbon-containing intermediates leading from CO_2 to CH_4 are indicated in parentheses. See text for further details.

can live autotrophically by forming acetyl-CoA from two molecules of CO_2 and then converting the acetyl-CoA to pyruvate and other products (figure 23.6).

Methanogens thrive in anaerobic environments rich in organic matter: the rumen and intestinal system of animals, freshwater and marine sediments, swamps and marshes, hot springs, anaerobic sludge digesters, and even within anaerobic protozoa. Recently an extremely thermophilic rod-shaped methanogen has been isolated from a marine hydrothermal vent. *Methanopyrus kandleri* has a temperature minimum at 84°C and an optimum of 98°C; it will grow up to 110°C (above the boiling point of water). Methanogens often are of ecological significance. The rate of methane production can be so great that bubbles of methane will sometimes rise to the surface of a lake or pond. Rumen methanogens are so active that a cow can belch 200 to 400 liters of methane a day.

Methanogenic bacteria are potentially of great practical importance since methane is a clean-burning fuel and an excellent energy source. For many years sewage treatment plants have been using the methane they produce as a source of energy for heat and electricity (*see figure 41.16*). An anaerobic digester will degrade particulate wastes like sewage sludge to H_2, CO_2, and acetate. CO_2-reducing methanogens form CH_4 from CO_2 and H_2, while aceticlastic methanogens cleave acetate to CO_2 and CH_4 (about 2/3 of the methane produced by an anaerobic digester comes from acetate). A kilogram of organic matter can yield up to 600 liters of methane. It is quite likely that future research will greatly increase the efficiency of methane production and make methanogenesis an important source of pollution-free energy.

Methanogenesis also can be an ecological problem (*see chapter 40*). Methane absorbs infrared radiation and thus is a greenhouse gas. There is evidence that atmospheric methane concentrations have been rising over the last 200 years. Methane production may significantly promote future global warming (*see Box 42.3*). Recently it has been discovered that methanogens can oxidize Fe^0 and use it to produce methane and energy. This means that methanogens growing around buried or submerged iron pipes and other objects may contribute significantly to iron corrosion.

Archaeobacterial Sulfate Reducers

This group contains gram-negative, irregular coccoid cells with walls consisting of glycoprotein subunits. At present, only the genus *Archaeoglobus* is known. It can extract electrons from a variety of electron donors (e.g., H_2, lactate, glucose) and reduce sulfate, sulfite, or thiosulfate to sulfide. Elemental sulfur is not used as an acceptor. *Archaeoglobus* is extremely thermophilic (the optimum is about 83°C) and can be isolated from marine hydrothermal vents. The organism is not only unusual in being able to reduce sulfate, unlike other archaeobacteria, but it also possesses the methanogen coenzymes F_{420} and methanopterin. It seems intermediate between the methanogens and the extremely thermophilic S^0-metabolizers.

(a)

(b)

Figure 23.10   Examples of Halobacteria. (*a*) *Halobacterium salinarium.* A young culture that has formed long rods; SEM. Bar = 1 μm. (*b*) *Halococcus morrhuae;* SEM. Bar = 1 μm.

---

1. Generally characterize methanogenic archaeobacteria and distinguish them from other groups.
2. Briefly describe how methanogens produce methane and the roles of their unique cofactors in this process.
3. Where does one find methanogens? Discuss their ecological and practical importance.
4. Characterize *Archaeoglobus.* In what way is it similar to the methanogens and how does it differ from other extreme thermophiles?

Extremely Halophilic Archaeobacteria

The **extreme halophiles** are a third major group of archaeobacteria, currently with six genera in one family, the *Halobacteriaceae* (figure 23.10). They are aerobic chemoheterotrophs with respiratory metabolism and require complex nutrients, usually proteins and amino acids, for growth. They are either nonmotile or motile by lophotrichous flagella.

The most obvious distinguishing trait of this family is its absolute dependence on a high concentration of NaCl. These bacteria require at least 1.5 M NaCl (about 8%, wt/vol), and usually have a growth optimum at about 3 to 4 M NaCl (17 to 23%). They will grow at salt concentrations approaching saturation (about 36%). *Halobacterium's* cell wall is so dependent on the presence of NaCl that it disintegrates when the NaCl concentration drops to about 1.5 M. Thus halobacteria only grow in high-salinity habitats such as marine salterns (*see figure 40.15*a) and salt lakes like the Dead Sea between Israel and Jordan, and the Great Salt Lake in Utah. They also can grow in food products such as salted fish and cause spoilage. Halobacteria often have red-to-yellow pigmentation from carotenoids that are probably used as protection against strong sunlight. They can reach such high population levels that salt lakes, salterns, and salted fish actually turn red.

The effect of solutes on halophile growth (pp. 124–25).

Probably the best-studied member of the family is *Halobacterium salinarium* (*H. halobium*). This procaryote is unusual because it can trap light energy photosynthetically without the presence of chlorophyll. When exposed to low oxygen levels, some strains of *Halobacterium* synthesize a modified cell membrane called the **purple membrane,** which contains the protein **bacteriorhodopsin.** ATP is produced by a unique type of photosynthesis without the participation of bacteriochlorophyll or chlorophyll (Box 23.1). *Halobacterium* actually has four rhodopsins, each with a different function. As already mentioned, bacteriorhodopsin drives outward proton transport for purposes of ATP synthesis. Halorhodopsin uses light energy to transport chloride ions into the cell and maintain a 4 to 5 M intracellular KCl concentration. Finally there are two rhodopsins that act as photoreceptors, one for red light and one for blue. They control flagellar activity to position the bacterium optimally in the water column. *Halobacterium* moves to a location of high light intensity, but one in which ultraviolet light is not sufficiently intense to be lethal.

Cell Wall-less Archaeobacteria

Thermoplasma grows in refuse piles of coal mines. These piles contain large amounts of iron pyrite (FeS), which is oxidized to sulfuric acid by chemolithotrophic bacteria. As a result the piles become very hot and acidic. This is an ideal habitat for *Thermoplasma* since it grows best at 55 to 59°C and pH 1 to 2. Although it lacks a cell wall, its plasma membrane is strengthened by large quantities of diglycerol tetraethers, lipopolysaccharides, and glycoproteins. The organism's DNA is stabilized by association with a special histonelike protein that condenses the DNA into particles resembling eucaryotic nucleosomes. At 59°C, *Thermoplasma* takes the form of an irregular filament, whereas at lower temperatures it is spherical (figure 23.11). The cells may be flagellated and motile.

Figure 23.11 *Thermoplasma.* Transmission electron micrograph. Bar = 0.5 μm.

Extremely Thermophilic S⁰-Metabolizers

The last group of archaeobacteria contains extremely thermophilic bacteria, many of which are acidophiles and sulfur dependent. The sulfur may be used either as an electron acceptor in anaerobic respiration or as an electron source by lithotrophs. Almost all are strict anaerobes. They grow in geothermally heated water or soils that contain elemental sulfur. These environments are scattered all over the world. Examples are the sulfur-rich hot springs in Yellowstone National Park, Wyoming, and the waters surrounding areas of submarine volcanic activity (figure 23.12). Such habitats are sometimes called solfatara. These archaeobacteria can be very thermophilic and often are classified as hyperthermophiles (p. 130). The most extreme example is *Pyrodictium,* a bacterium isolated from geothermally heated sea floors. *Pyrodictium* has a temperature minimum of 82°C, a growth optimum at 105°C, and a maximum at 110°C. Both organotrophic and lithotrophic growth occur in this group. Sulfur and H_2 are the most common electron sources for lithotrophs. There are three orders (*Thermococcales, Thermoproteales,* and *Sulfolobales*) and at least nine genera. Two of the better-studied genera are *Thermoproteus* and *Sulfolobus.*

Box 23.1

Photosynthesis in *Halobacterium salinarium*

The halophilic bacterium *Halobacterium salinarium* (formerly *H. halobium*) normally depends on respiration for the production of energy. However, under conditions of low oxygen and high light intensity, the bacteria synthesize a deep-purple pigment called bacteriorhodopsin, which closely resembles the sensory pigment rhodopsin from the rods and cones of vertebrate eyes. Bacteriorhodopsin's chromophore is the carotenoid derivative retinal (the aldehyde of vitamin A) that is covalently attached to the pigment protein by a Schiff base with the amino group of a lysine. The protein extends through the plasma membrane, and its retinal rests in the center of the membrane. Individual bacteriorhodopsins aggregate in the membrane to form crystalline patches called the purple membrane.

Bacteriorhodopsin functions as a light-driven proton pump. When retinal absorbs light, the double bond between carbons 13 and 14 changes from a *trans* to a *cis* configuration and the Schiff base loses a proton. Protons are moved across the plasma membrane to the periplasmic space (*see chapter 3*) during these alterations, and it is thought that the Schiff base changes are directly involved in this movement (Box Figure 23.1). The bacteriorhodopsin protein switches between two conformations during the photocycle. Presumably these conformational changes also are involved in proton transport. The light-driven proton pumping generates a pH gradient that can be used to power the synthesis of ATP by a chemiosmotic mechanism (*see chapter 9*).

This photosynthetic capacity is particularly useful to *Halobacterium* because oxygen is not very soluble in concentrated salt solutions and may decrease to an extremely low level in its habitat. When the surroundings become temporarily anaerobic, the bacterium uses light energy to synthesize sufficient ATP to remain alive until the oxygen level rises again. *Halobacterium* cannot grow anaerobically because it requires oxygen for continued retinal synthesis, but it can survive the stress of temporary oxygen limitation by means of photosynthesis.

Box Figure 23.1 The Photocycle of Bacteriorhodopsin. In this hypothetical mechanism the retinal component of bacteriorhodopsin is buried in the membrane and retinal interacts with two amino acids, A_1 and A_2, that can reversibly accept and donate protons. A_2 is connected to the cell exterior, and A_1 is closer to the cell interior. The two amino acids may be special aspartic acid residues in bacteriorhodopsin. Light absorption by retinal in step 1 triggers an isomerization from 13-*trans*-retinal to 13-*cis*-retinal. The retinal then donates a proton to A_2 in steps 2 and 3, while A_1 is picking up another proton from the interior and A_2 is releasing a proton to the outside. In steps 4 and 5, retinal obtains a proton from A_1 and isomerizes back to the 13-*trans* form. The cycle is then ready to begin again after step 6.

(a)

(b)

Figure 23.12 Habitats for Thermophilic Bacteria. (*a*) The pump geyser in Yellowstone National Park. The orange color is due to the carotenoid pigments of thermophilic bacteria. (*b*) The sulfur cauldron in Yellowstone National Park. The water is at its boiling point and very rich in sulfur. *Sulfolobus* grows well in such habitats.

Members of the genus *Sulfolobus* are gram-negative, aerobic, irregularly lobed spherical bacteria with a temperature optimum around 70 to 80°C and a pH optimum of 2 to 3 (figure 23.13*a, b*). For this reason, they are classed as **thermoacidophiles,** so called because they grow best at acid pH values and high temperatures. Their cell wall contains lipoprotein and carbohydrate but lacks peptidoglycan. They grow lithotrophically on sulfur granules in hot acid springs and soils while oxidizing the sulfur to sulfuric acid (figures 23.12*b* and 23.13*b*). Oxygen is the normal electron acceptor, but ferric iron may be used. Sugars and amino acids such as glutamate also serve as carbon and energy sources.

Thermoproteus is a long thin rod that can be bent or branched (figure 23.13*c*). The cell wall is composed of glycoprotein. *Thermoproteus* is a strict anaerobe and grows at temperatures from 70 to 97°C and pH values between 2.5 and 6.5. It is found in hot springs and other hot aquatic habitats rich in

sulfur. It can grow organotrophically and oxidize glucose, amino acids, alcohols, and organic acids with elemental sulfur as the electron acceptor. That is, *Thermoproteus* can carry out anaerobic respiration. It will also grow chemolithotrophically using H_2 and S^0. Carbon monoxide or CO_2 can serve as the sole carbon source.

1. Where are the extreme halophiles found and what is unusual about their cell walls and growth requirements?
2. Briefly describe how *Halobacterium* carries out photosynthesis. What is the purple membrane and what pigment does it contain?
3. How is *Thermoplasma* able to live in acidic, very hot coal refuse piles when it lacks a cell wall? How is its DNA stabilized?
4. What are thermoacidophiles and where do they grow? In what ways do they use sulfur in their metabolism? Briefly describe *Sulfolobus* and *Thermoproteus*.

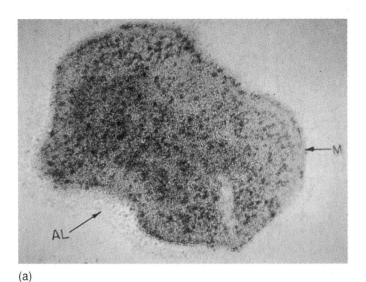

(a)

(b)

Figure 23.13 *Sulfolobus* **and** *Thermoproteus*. (*a*) A thin section of *Sulfolobus brierleyi*. The bacterium, about 1 µm in diameter, is surrounded by an amorphous layer (AL) instead of a well-defined cell wall; the plasma membrane (M) is clearly visible. (*b*) A scanning electron micrograph of a colony of *Sulfolobus* growing on the mineral molybdenite (MoS$_2$) at 60°C. At pH 1.5–3, the bacterium oxidizes the sulfide component of the mineral to sulfate and solubilizes molybdenum. (*c*) Electron micrograph of *Thermoproteus tenax*. Bar = 1 µm.

(c)

Summary

1. Organisms may be divided into three major groups: archaeobacteria, eubacteria, and eucaryotes.

2. Archaeobacteria are highly diverse with respect to morphology, reproduction, physiology, and ecology. They grow in anaerobic, hypersaline, and high-temperature habitats.

3. Archaeobacterial cell walls do not contain peptidoglycan and differ from eubacterial walls in structure. They may be composed of pseudomurein, polysaccharides, glycoproteins, or unaltered proteins.

4. The membrane lipids differ from those of other organisms in having branched chain hydrocarbons connected to glycerol by ether links. Eubacterial and eucaryotic lipids have glycerol connected to fatty acids by ester bonds.

5. Their tRNA, ribosomes, elongation factors, RNA polymerases, and other components distinguish archaeobacteria from eubacteria and eucaryotes.

6. Although much of archaeobacterial metabolism appears similar to that of other organisms, they do differ with respect to glucose catabolism, pathways for CO$_2$ fixation, and the ability of some to synthesize methane.

7. Archaeobacteria may be divided into five groups: methanogenic bacteria, sulfate reducers, extreme halophiles, cell wall-less archaeobacteria, and extreme thermophiles.

8. Methanogenic bacteria are strict anaerobes that can obtain energy through the synthesis of methane. They have several unique cofactors that are involved in methanogenesis.

9. The extreme thermophile *Archaeoglobus* differs from other archaeobacteria in using a variety of electron donors to reduce sulfate. It also contains the methanogen cofactors F$_{420}$ and methanopterin.

10. Extreme halophiles are aerobic chemoheterotrophs that require at least 1.5 M NaCl for growth. They are found in habitats such as salterns, salt lakes, and salted fish.

11. *Halobacterium salinarium* can carry out photosynthesis without chlorophyll or bacteriochlorophyll by using bacteriorhodopsin, which employs retinal to pump protons.

12. The thermophilic archaeobacterium *Thermoplasma* grows in hot, acid coal refuse piles and survives despite the lack of a cell wall.

13. The extremely thermophilic S^0-metabolizers depend on sulfur for growth and are frequently acidophiles. The sulfur may be used as an electron acceptor in anaerobic respiration or as an electron donor by lithotrophs. They are almost always strict anaerobes and grow in geothermally heated soil and water that is rich in sulfur.

=== **Key Terms** ===

archaeobacteria   *478*

bacteriorhodopsin   *486*

extreme halophiles   *485*

methanogens   *482*

pseudomurein   *478*

purple membrane   *486*

thermoacidophiles   *488*

=== **Questions for Thought and Review** ===

1. Discuss how the unusual properties of extreme halophiles and thermophiles reflect the habitats in which they grow.

2. Why do we not classify *Thermoplasma* as a mycoplasma rather than as an archaeobacterium?

3. Discuss what adaptations extremely thermophilic archaeobacteria require to grow at 100°C.

4. Do you think archaeobacteria should be separated from the eubacteria although both groups are procaryotic? Give your reasoning and evidence.

5. Supposing you wished to isolate bacteria from a hot spring in Yellowstone Park, how would you go about it?

=== **Additional Reading** ===

Introduction to the Archaeobacteria

Balows, A.; Truper, H. G.; Dworkin, M.; Harder, W.; and Schleifer, K. -H. 1992. *The prokaryotes,* 2d ed. New York: Springer-Verlag.

Cramer, W. A., and Knaff, D. B. 1991. *Energy transduction in biological membranes.* New York: Springer-Verlag.

Danson, M. J. 1988. Archaebacteria: The comparative enzymology of their central metabolic pathways. In *Advances in microbial physiology,* A. H. Rose and D. W. Tempest, editors, 165–231. New York: Academic Press.

Fewson, C. A. 1986. Archaebacteria. *Biochem. Ed.* 14(3):103–15.

Holt, J. G., editor-in-chief. 1989. *Bergey's manual of systematic bacteriology,* vol. 3, J. T. Staley, M. P. Bryant, and N. Pfennig, editors. Baltimore, Md.: Williams & Wilkins.

Holt, J. G., editor-in-chief. 1994. *Bergey's manual of determinative bacteriology,* 9th ed. Baltimore, Md.: Williams & Wilkins.

Kandler, O., and Zillig, W., editors. 1986. *Archaebacteria '85.* New York: Gustav Fischer Verlag.

Lederberg, J., editor. 1992. *Encyclopedia of microbiology,* San Diego, Calif.: Academic Press.

Schlegel, H. G., and Bowien, B. 1989. *Autotrophic bacteria.* Madison, Wis.: Science Tech Publishers.

Woese, C. R. 1981. Archaebacteria. *Sci. Am.* 244(6):98–122.

Woese, C. R., and Wolfe, R. S., editors. 1985. *Archaebacteria.* Volume VIII of *The bacteria: A treatise on structure and function.* New York: Academic Press.

Wood, H. G.; Radsdale, S. W.; and Pezacka, E. 1986. The acetyl-CoA pathway: A newly discovered pathway of autotrophic growth. *Trends Biochem. Sci.* 11(1):14–17.

Zillig, W.; Palm, P.; Reiter, W.-D.; Gropp, F.; Pühler, G.; and Klenk, H. -P. 1988. Comparative evaluation of gene expression in archaebacteria. *Eur. J. Biochem.* 173:473–82.

Methanogenic Archaeobacteria

Balch, W. E.; Fox, G. E.; Magrum, L. J.; Woese, C. R.; and Wolfe, R. S. 1979. Methanogens: Reevaluation of a unique biological group. *Microbiol. Rev.* 43:260–96.

DiMarco, A. A.; Bobik, T. A.; and Wolfe, R. S. 1990. Unusual coenzymes of methanogenesis. *Ann. Rev. Biochem.* 59:355–94.

Jones, W. J.; Nagle, D. P., Jr.; and Whitman, W. B. 1987. Methanogens and the diversity of archaebacteria. *Microbiol. Rev.* 51:135–77.

Zinder, S. H. 1984. Microbiology of anaerobic conversion of organic wastes to methane: Recent developments. *ASM News* 50(7):294–98.

Extremely Halophilic Archaeobacteria

Birge, R. R. 1990. Nature of the primary photochemical events in rhodopsin and bacteriorhodopsin. *Biochem. Biophys. Acta* 1016:293–327.

Norton, C. F. 1992. Rediscovering the ecology of halobacteria. *ASM News* 58(7):363–67.

Oesterhelt, D. 1985. Light-driven proton pumping in halobacteria. *BioScience* 35(1):18–21.

Oesterhelt, D., and Tittor, J. 1989. Two pumps, one principle: Light-driven ion transport in halobacteria. *Trends Biochem. Sci.* 14:57–61.

Spudich, J. L. 1993. Color sensing in the *Archaea:* a eukaryotic-like receptor coupled to a procaryotic transducer. *J. Bacteriol.* 175(24):7755–61.

Stoeckenius, W. 1976. The purple membrane of salt-loving bacteria. *Sci. Am.* 234(6):38–46.

Extremely Thermophilic S⁰-Metabolizers

Pley, U.; Schipka, J.; Gambacorta, A.; Jannasch, H. W.; Fricke, H.; Rachel, R.; and Stetter, K. O. 1991. *Pyrodictium abyssi* sp. nov. represents a novel heterotrophic marine archaeal hyperthermophile growing at 110°C. *System. Appl. Microbiol.* 14:245–53.

CHAPTER 24

The Bacteria:
The Actinomycetes

*Actinomycetes are very
important from a medical
point of view. . . . They
may be a nuisance, as when
they decompose rubber
products, grow in aviation
fuel, produce odorous
substances that pollute water
supplies, or grow in sewage-
treatment plants where they
form thick clogging
foams. . . . In contrast,
actinomycetes are the producers
of most of the antibiotics.*
　　—H. A. Lechevalier and M. P.
　　　　　Lechevalier

Outline

Concepts

1. Volume 4 of *Bergey's Manual* contains aerobic, gram-positive bacteria—
   the actinomycetes—that form branching hyphae and asexual spores.
2. The morphology and arrangement of spores, cell wall chemistry, and
   the types of sugars present in cell extracts are particularly important
   in actinomycete taxonomy and are used to divide these bacteria into
   different groups.
3. Actinomycetes have considerable practical impact because they play a
   major role in the mineralization of organic matter in the soil and are the
   primary source of most naturally synthesized antibiotics.
4. Actinomycete taxonomy is still developing. The classification of these
   organisms will probably change considerably in the future.

Chapter 24, the last of the survey chapters on bacteria, describes the bacteria placed in volume 4 of *Bergey's Manual of Systematic Bacteriology*. Actinomycetes are gram positive like the bacteria in volume 2, but are distinctive because they have filamentous hyphae that do not normally undergo fragmentation and produce asexual spores. They closely resemble fungi in overall morphology. Presumably this resemblance results partly from adaptation to the same habitats. First, the general characteristics of the actinomycetes are summarized. Then representatives are described, with emphasis on morphology, taxonomy, reproduction, and general importance.

Volume 4 of *Bergey's Manual* contains the **actinomycetes** [s., actinomycete], aerobic, gram-positive bacteria that form branching filaments or hyphae and asexual spores. Although they are a diverse group, the actinomycetes do share many properties.

General Properties of the Actinomycetes

When growing on a solid substratum such as agar, the branching network of hyphae developed by actinomycetes grows both on the surface of the substratum and into it to form a substrate mycelium. Septa usually divide the hyphae into long cells (20 μm and longer) containing several nucleoids.

Sometimes a tissuelike mass results and may be called a **thallus.** Many actinomycetes also have an aerial mycelium that extends above the substratum and forms asexual, thin-walled spores called **conidia** [s., conidium] or **conidiospores** on the ends of filaments (figure 24.1). If the spores are located in a sporangium, they are called **sporangiospores.** The spores can vary greatly in shape (figure 24.2). Actinomycete spores develop by septal formation at filament tips, usually in response to nutrient deprivation. Most are not particularly heat resistant but do withstand desiccation well and thus have considerable adaptive value.

Most actinomycetes are not motile. When motility is present, it is confined to flagellated spores.

Actinomycete cell wall composition varies greatly among different groups and is of considerable taxonomic importance. Four major cell wall types can be distinguished according to three features of peptidoglycan composition and structure: the amino acid in tetrapeptide side chain position 3, the presence of glycine in interpeptide bridges, and peptidoglycan sugar content (table 24.1). Cell extracts of actinomycetes with wall types II, III, and IV also contain characteristic sugars that are useful in identification (table 24.2). Some other taxonomically valuable properties are the morphology and color of mycelia and sporangia, the surface features and arrangement of conidiospores, the percent G + C in DNA, the phospholipid composition of cell membranes, and spore heat resistance. Newer techniques are being applied to actinomycete taxonomy. Comparison of 16S rRNA sequences has proven valuable. Another useful technique is the

Chain of conidiospores

Agar surface

Figure 24.1   An Actinomycete Colony. The cross section of an actinomycete colony with living (colored) and dead (white) hyphae. The substrate mycelium and aerial mycelium with chains of conidiospores are shown.

TABLE 24.1   Actinomycete Cell Wall Types

Cell Wall Type	Diaminopimelic Acid Isomer	Glycine in Interpeptide Bridge	Characteristic Sugars	Representative Genera
I	L, L	+		*Nocardioides, Streptomyces, Streptoverticillium*
II	*Meso*	+		*Micromonospora, Pilimelia, Actinoplanes*
III	*Meso*	–		*Thermoactinomyces, Actinomadura, Frankia*
IV	*Meso*	–	Arabinose, galactose	*Saccharomonospora, Faenia, Nocardia*

Figure 24.2   Examples of Actinomycete Spores as Seen in the Scanning Electron Microscope. (*a*) Sporulating *Faenia* hyphae (×3,000). (*b*) Sporangia of *Pilimelia columellifera* on mouse hair (×520). (*c*) *Micromonospora echinospora*. Bar = 0.5 μm. (*d*) A chain of hairy streptomycete spores. Bar = 1.0 μm. (*e*) *Microbispora rosea*, paired spores on hyphae. Bar = 10 μm. (*f*) Aerial spores of *Kitasatosporia setae*. Bar = 5 μm.

(a)   (b)   (c)

(d)   (e)   (f)

production of large DNA fragments by restriction enzyme digestion and their separation and comparison using pulsed-field electrophoresis.

Gram-positive peptidoglycan structure and chemistry (pp. 439–40).

Actinomycetes have considerable practical significance. They are primarily soil inhabitants and are very widely distributed. They can degrade an enormous number and variety of organic compounds and are extremely important in the mineral-ization of organic matter. Actinomycetes produce most of the medically useful natural antibiotics. Although most actinomycetes are free-living microorganisms, a few are pathogens of humans, other animals, and some plants.

Bergey's Manual divides the actinomycetes into seven sections, primarily based on properties such as cell wall type, conidia arrangement, and the presence or absence of a sporangium (table 24.3). A few genera are not easily placed in sections 26 through 32 and are collected in section 33 under the title "Other Genera."

The sections in volume 4 are not homogeneous, and their composition will undoubtedly change in the future. They do not always fit with 16S rRNA sequence results (figure 24.3). For example, the genus *Actinomadura* is clearly heterogeneous. *Dermatophilus, Promicromonospora,* and *Oerskovia* seem more closely related to bacteria placed in volume 2 than they do to those in volume 4. It has been proposed that the irregular, nonsporing gram-positive rods in volume 2 be clustered with these three genera to form a group called the actinobacteria (figure 24.3).

1. Define actinomycete, thallus, substrate mycelium, aerial mycelium, conidium, conidiospore, and sporangiospore.
2. Describe how cell wall structure and sugar content are used to classify the actinomycetes. Include a brief description of the four major wall types.
3. Why are the actinomycetes of such practical interest?

TABLE 24.2   Actinomycete Whole Cell Sugar Patterns

Sugar Pattern Types[a]	Characteristic Sugars	Representative Genera
A	Arabinose, galactose	*Nocardia, Rhodococcus, Saccharomonospora*
B	Madurose[b]	*Actinomadura, Streptosporangium, Dermatophilus*
C	None	*Thermomonospora, Actinosynnema, Thermoactinomyces, Geodermatophilus*
D	Arabinose, xylose	*Micromonospora, Actinoplanes*

[a]Characteristic sugar patterns are present only in wall types II–IV, those actinomycetes with *meso*-diaminopimelic acid.

[b]Madurose is 3-O-methyl-D-galactose.

TABLE 24.3   Some Characteristics of Major Actinomycete Groups

Group	Wall Type	Sugar Pattern	Mol% G + C	Spore Arrangement	Presence of Sporangia	Selected Genera
Nocardioform actinomycetes[a]	I, IV, VI[b]	A	59–79	Varies	–	*Nocardia, Rhodococcus, Nocardioides, Faenia, Oerskovia, Saccharomonospora*
Actinomycetes with multilocular sporangia	III	B, C, D	57–75	Clusters of spores	+(–)[c]	*Geodermatophilus, Dermatophilus, Frankia*
Actinoplanetes	II	D	71–73	Varies	Usually +	*Actinoplanes, Pilimelia, Dactylosporangium, Micromonospora*
Streptomyces and related genera	I	Not of taxonomic value	69–78	Chains of 5 to more than 50 spores	–	*Streptomyces, Streptoverticillium, Sporichthya*
Maduromycetes	III	B, C	64–74	Varies	+ or –	*Actinomadura, Microbispora, Planomonospora, Streptosporangium*
Thermomonospora and related genera	III	C (sometimes B)	64–73	Varies	–	*Thermomonospora, Actinosynnema, Nocardiopsis*
Thermoactinomycetes	III	C	52–55	Single, heat-resistant endospores	–	*Thermoactinomyces*

[a]Several genera have mycolic acid. Filaments readily fragment into rods and coccoid elements.

[b]Type VI cell wall has lysine instead of diaminopimelic acid.

[c]Members have clusters of spores that may not always be surrounded by a sporangial wall.

Nocardioform Actinomycetes

Most of the genera in section 26 of *Bergey's Manual* are also described in section 17 of volume 2. Because they all resemble members of the genus *Nocardia* (named after Edmond Nocard [1850–1903], French bacteriologist and veterinary pathologist), they are collectively called **nocardioforms.** *Bergey's Manual of Determinative Bacteriology* divides 19 genera into four subgroups of nocardioforms. These bacteria develop a substrate mycelium that readily breaks into rods and coccoid elements (figure 24.4). Several genera also form an aerial mycelium that rises above the substratum and may produce conidia. All genera have a high G + C content like other actinomycetes, and almost all are strict aerobes. Most species have peptidoglycan with *meso*-diaminopimelic acid and no peptide interbridge. The wall usually contains a carbohydrate composed of arabinose and galactose; mycolic acids are present in the genera *Nocardia* and *Rhodococcus.*

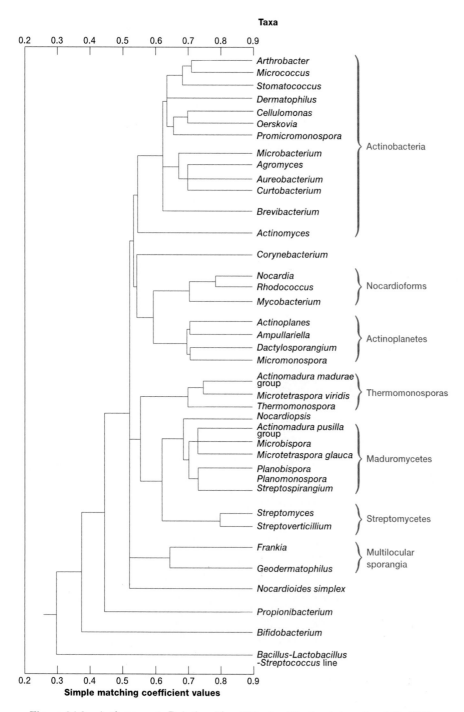

Figure 24.3   Actinomycete Relationships. This classification is based on 16S rRNA sequence comparisons.

Figure 24.4   Nocardioform Actinomycetes. (*a*) *Nocardia asteroides,* substrate mycelium and aerial mycelia with conidia illustration and light micrograph (×1,250). (*b*) A photomicrograph (×650) and illustration of the substrate mycelium of *Faenia rectivirgula* near the growing margin.

The genus *Nocardia* is distributed worldwide in soil and also is found in aquatic habitats. Nocardiae are involved in the degradation of hydrocarbons and waxes and can contribute to the biodeterioration of rubber joints in water and sewage pipes. Although most are free-living saprophytes, some species, particularly *N. asteroides,* are opportunistic pathogens that cause nocardiosis in humans and other animals. People with low resistance due to other health problems are most at risk. The lungs are most often infected, but the central nervous system and other organs may be invaded.

1. What is a nocardioform, and how can the group be distinguished from other actinomycetes?
2. Where is *Nocardia* found, and what problems may it cause? Consider both environmental and public health concerns.

(a)

(b)

Actinomycetes with Multilocular Sporangia

The actinomycetes in section 27 of *Bergey's Manual* form clusters of spores when a hypha divides both transversely and longitudinally. (Multilocular means having many cells or compartments.) All three genera in this section have type III cell walls (table 24.1), although the cell extract sugar patterns differ. The G + C content varies from 57 to 75 mol%. *Geodermatophilus* (type IIIC) has motile spores and is an aerobic soil organism. *Dermatophilus* (type IIIB) also forms packets of motile spores with tufts of flagella, but it is a facultative anaerobe and a parasite of mammals responsible for the skin infection streptothrichosis. *Frankia* (type IIID) forms nonmotile sporangiospores in a sporogenous body (figure 24.5). It grows in symbiotic association with the roots of at least eight families of higher nonleguminous plants (e.g., alder trees) and is a microaerophile able to fix atmospheric nitrogen (*see chapter 42*).

The roots of infected plants develop nodules that fix nitrogen so efficiently that a plant such as an alder can grow in the absence of combined nitrogen when nodulated. Within the nodule cells, *Frankia* forms branching hyphae with globular vesicles at their ends (figure 24.5c). These vesicles may be the sites of nitrogen fixation. The nitrogen-fixation process resembles that of *Rhizobium* in that it is oxygen sensitive and requires molybdenum and cobalt.

1. Give the major distinguishing characteristics of the actinomycetes in section 27 of *Bergey's Manual.*
2. How does *Frankia* differ from *Geodermatophilus* and *Dermatophilus?*

(c)

Figure 24.5   Actinomycetes with Multilocular Sporangia. *Frankia.* (*a*) An interference contrast micrograph showing hyphae, sporangia, and spores. Bar = 5 µm. (*b*) A scanning electron micrograph of a sporangium surrounded by hyphae. (*c*) A nodule of the alder *Alnus rubra* showing cells filled with vesicles of *Frankia.* Scanning electron microscopy. Bar = 5 µm.

Actinoplanetes

Actinoplanetes [Greek *actinos,* a ray or beam, and *planes,* a wanderer], section 28, have an extensive substrate mycelium and are wall type IID. Often the hyphae are highly colored and diffusible pigments may be produced. Normally an aerial mycelium is absent or rudimentary. Conidiospores are usually formed within a sporangium raised above the surface of the substratum at the end of a special hypha called a sporangiophore. The spores can be either motile or nonmotile. These bacteria vary in the arrangement and development of their spores. Some genera (*Actinoplanes, Ampullariella, Pilimelia*) have spherical, cylindrical, or irregular

sporangia with a few to several thousand spores per sporangium (figures 24.2*b* and 24.6). The sporangium develops above the substratum at the tip of a sporangiophore; the spores are arranged in coiled or parallel chains (figure 24.7). *Dactylosporangium* forms club-shaped, fingerlike, or pyriform sporangia with one to six spores (figure 24.6*d,e*). *Micromonospora* bears single spores, which often occur in branched clusters of sporophores (figure 24.2*c*). The section presently contains five genera.

Actinoplanetes [s., actinoplanete] grow in almost all soil habitats, ranging from forest litter to beach sand. They also flourish in fresh water, particularly in streams and rivers (probably because of abundant oxygen and plant debris). Some have been isolated from

(a) *Actinoplanes*

(b)

(c)

Dactylosporangium

(d)

(e)

Figure 24.6   Actinoplanetes. Actinoplanete morphology. (*a*) *Actinoplanes* structure. (*b*) A scanning electron micrograph of mature *Actinoplanes* sporangia. Bar = 5 μm. (*c*) A section of an *Actinoplanes rectilineatus* sporangium. Bar = 3 μm. (*d*) *Dactylosporangium* structure. (*e*) A *Dactylosporangium* colony covered with sporangia (×700).

Figure 24.7 Sporangium Development in an Actinoplanete. The developing sporangium is shown in color.

the ocean. The soil-dwelling species may have an important role in the decomposition of plant and animal material. *Pilimelia* grows in association with keratin. *Micromonospora* actively degrades chitin and cellulose, and it can produce antibiotics such as gentamicin.

---

1. Give the distinguishing properties of the actinoplanetes.
2. Briefly describe the variations in sporangia and sporophore organization of the actinoplanetes.

Streptomyces and Related Genera

Section 29 of *Bergey's Manual* contains genera whose aerial hyphae divide in a single plane to form chains of 5 to 50 or more nonmotile conidiospores with surface texture ranging from smooth to spiny and warty (figures 24.2*d*, 24.8, and 24.9). All have a type I cell wall and a G + C content of around 69 to 78%. The substrate mycelium, when present, does not undergo fragmentation. Members of the section as a whole are often called **streptomycetes** [Greek *streptos,* bent or twisted, and *myces,* fungus]. Currently there are five genera of streptomycetes and their relatives.

Streptomyces is an enormous genus; 378 species are given in the Approved Lists of Bacterial Names. *Bergey's Manual* describes 142 species in detail. Undoubtedly the number of accepted species will decrease with further progress in streptomycete taxonomy. Members of the genus are strict aerobes, are wall type I, and form chains of nonmotile spores within a thin, fibrous sheath (figure 24.9). The three to many conidia in each chain are often pigmented and can be smooth, hairy, or spiny in texture. *Streptomyces* species are determined by means of a mixture of morphological and physiological characteristics, including the following: the color of the aerial and substrate mycelia, spore arrangement, surface features of individual spores, carbohydrate use, antibiotic production, melanin synthesis, nitrate reduction, and the hydrolysis of urea and hippuric acid.

Streptomycetes are very important, both ecologically and medically. The natural habitat of most streptomycetes is the soil, where they may constitute from 1 to 20% of the culturable population. In fact, the odor of moist earth is largely the result of streptomycete production of volatile substances such as **geosmin.** Streptomycetes play a major role in mineralization. They are flexible nutritionally and can aerobically degrade resistant substances such as pectin, lignin, chitin, keratin, latex, and aromatic compounds. Streptomycetes are best known for their synthesis of a vast array of antibiotics, some of which are useful in medicine and biological research. Examples include amphotericin B, chloramphenicol, erythromycin, neomycin, nystatin, streptomycin, tetracycline, and so forth (figure 24.10*a*). Although most streptomycetes are nonpathogenic saprophytes, a few are associated with plant and animal diseases. *Streptomyces scabies* causes scab disease in potatoes and beets (figure 24.10*b*). *S. somaliensis* is the only streptomycete known to be pathogenic for humans. It is associated with **actinomycetoma,** an infection of subcutaneous tissues that produces lesions and leads to swelling, abscesses, and even bone destruction if untreated. *S. albus* and other species have been isolated from patients with various ailments and may be pathogenic.

Figure 24.8 *Streptomyces* **and Related Genera.** Streptomycete conidia arrangement. (*a*) An illustration of typical *Streptomyces* morphology; a light micrograph of *S. carpinesis* spore chains. Bar = 5 μm. (*b*) *Streptoverticillium* morphology; a scanning electron micrograph of *Sv. salmonis* with developing spore chains. Bar = 2 μm.

Figure 24.9 **Streptomycete Spores.** (*a*) Smooth spores of *S. niveus;* scanning electron micrograph. Bar = 0.25 μm. (*b*) Spiney spores of *S. viridochromogenes*. Bar = 0.5 μm. (*c*) Warty spores of *S. pulcher*. Bar = 0.25 μm.

Figure 24.10   Streptomycetes with Practical Importance. (*a*) *Streptomyces griseus.* Colonies of the actinomycete that produces streptomycin. (*b*) *Streptomyces scabies,* an actinomycete growing on a potato.

(a)

(b)

Actinomadura madurae

(a)

(b)   *Streptosporangium*

Figure 24.11   Maduromycetes. (*a*) *Actinomadura madurae* morphology; illustration and electron micrograph of a spore chain (×16,500). (*b*) *Streptosporangium* morphology; illustration and micrograph of *S. album* on oatmeal agar with sporangia and hyphae; SEM. Bar interval = 10 μm.

Antibiotics and their properties (chapter 33).

Other genera in this section are easily distinguished from *Streptomyces. Streptoverticillium* has an aerial mycelium with a whorl of three to six short branches that are produced at fairly regular intervals. These branches have secondary branches bearing chains of spores (figure 24.8*b*). *Sporichthya* is one of the strangest of the actinomycetes. It lacks a substrate mycelium. The hyphae remain attached to the substratum by holdfasts and grow upward to form aerial mycelia that release motile, flagellate conidia in the presence of water.

1. What characteristics do the actinomycetes of section 29 have in common? What is a streptomycete?
2. Describe the major properties of the genus *Streptomyces.*
3. Give three ways in which *Streptomyces* is of ecological and medical importance.
4. How do *Streptoverticillium* and *Sporichthya* differ from *Streptomyces*?

Maduromycetes

A collection of seven genera called the maduromycetes are placed in section 30. All genera have type III cell walls and the sugar derivative **madurose** (3-O-methyl-D-galactose) in whole cell homogenates. Their G + C content is 64 to 74 mol%. Aerial mycelia bear pairs or short chains of spores, and the substrate mycelium is branched. Some genera form sporangia. Typical genera are *Actinomadura, Microbispora, Planomonospora,* and *Streptosporangium* (figures 24.2*e* and 24.11). *Actinomadura* is another actinomycete associated with the disease actinomycetoma.

Thermomonospora and Related Genera

The four genera in section 31 all have type III cell walls and usually a type C sugar pattern (*Thermomonospora* can be IIIB). Their spores are not heat resistant. There is considerable variation in morphology and life-style. *Thermomonospora* produces single spores on the aerial mycelium or on both the aerial and substrate mycelia. It has been isolated from high-temperature habitats such as compost piles and hay and can grow at 40 to 48°C. *Actinosynnema* produces a compact cluster of erect hyphae that resembles a

(a)

(b)

(c)

Figure 24.12 **Thermoactinomycetes.** Thermoactinomycete structure. (*a*) *Thermoactinomyces vulgaris* aerial mycelium with developing endospores at tips of hyphae. Bar = 10 μm. (*b*) Scanning electron micrograph of *T. sacchari* spores. Bar = 1 μm. (*c*) Thin section of a *T. sacchari* endospore. Bar = 0.1 μm. E, exosporium; OC, outer spore coat; IC, inner spore coat; CO, cortex; IM, inner forespore membrane; C, core.

stem; chains of motile conidia are attached to the top of the cluster, much like the branches of a bush or small tree. *Nocardiopsis* has a substrate mycelium that fragments like the nocardioforms, but it lacks mycolic acid and has a different cell wall chemistry. The aerial hyphae assume a zigzag shape and then form long chains of spores.

Thermoactinomycetes

Section 32 of *Bergey's Manual* contains one genus, *Thermoactinomyces*. This genus is thermophilic and wall type IIIC, grows between 45 and 60°C, and forms single spores on both its aerial and substrate mycelia (figure 24.12). Its G + C content is lower than that of other actinomycetes (52 to 55 mol%), and its 16S rRNA sequence suggests a relationship with the genus *Bacillus*. It is commonly found in damp haystacks, compost piles, and other high-temperature habitats. Unlike other actinomycete spores, *Thermoactinomyces* spores (figure 24.12*b,c*) are

true endospores and very heat resistant; they can survive at 90°C for 30 minutes. They are formed within hyphae and appear to have typical endospore structure, including the presence of calcium and dipicolinic acid. *Thermoactinomyces vulgaris* (figure 24.12*a*) from haystacks, grain storage silos, and compost piles is a causative agent of farmer's lung, an allergic disease of the respiratory system in agricultural workers. Recently spores from *Thermoactinomyces vulgaris* were recovered from the mud of a Minnesota lake and found to be viable after about 7,500 years of dormancy.

The bacterial endospore (pp. 67–69).

1. List the major properties of bacteria in sections 30 and 31 of *Bergey's Manual*. What is madurose?
2. Briefly describe the genus *Thermoactinomyces*, with particular emphasis on its unique features.

Summary

1. Actinomycetes are aerobic, gram-positive bacteria that form branching, usually nonfragmenting, hyphae and asexual spores.

2. The asexual spores borne on aerial mycelia are called conidiospores or conidia if they are at the tip of hyphae and sporangiospores when they are within sporangia.

3. Actinomycetes have several distinctively different types of cell walls and often also vary in terms of the sugars present in cell extracts. Properties such as color and morphology are also taxonomically useful.

4. *Bergey's Manual* divides the actinomycetes into eight groups based on such properties as conidia arrangement, the presence of a sporangium, cell wall type, and cell extract sugars.

5. Nocardioform actinomycetes, in section 26 of *Bergey's Manual,* have hyphae that readily fragment into rods and coccoid elements, and often form aerial mycelia with spores.

6. Bacteria in section 27 produce clusters of spores at hyphal tips and have type III cell walls. *Frankia,* one of the three genera in this section, grows in symbiotic association with higher nonleguminous plants and fixes nitrogen.

7. Section 28 contains the actinoplanetes. All are wall type IID and have an extensive substrate mycelium. They are present in the soil, fresh water, and the ocean. The soil forms are probably important in decomposition.

8. *Streptomyces* and related genera are placed in section 29. Their aerial hyphae bear chains of 5 to 50 or more nonmotile conidiospores, and all have a type I cell wall.

9. *Streptomyces,* the largest genus of actinomycetes, has chains of nonmotile conidia within a thin sheath.

10. Streptomycetes are important in the degradation of more resistant organic material in the soil and produce many useful antibiotics. A few cause diseases in plants and animals.

11. Section 30 contains the maduromycetes, actinomycetes with type III cell walls and the sugar derivative madurose.

12. *Thermomonospora* and related genera are normally wall type IIIC (a few are IIIB), and they are placed in section 31.

13. Section 32 contains *Thermoactinomyces,* a thermophilic actinomycete that forms true endospores.

Key Terms

actinomycete   *492*

actinomycetoma   *498*

conidia   *492*

conidiospores   *492*

geosmin   *498*

madurose   *500*

nocardioforms   *495*

sporangiospores   *492*

streptomycetes   *498*

thallus   *492*

Questions for Thought and Review

1. Suppose that you discovered a nodulated plant that could fix atmospheric nitrogen. How might you show that a bacterial symbiont was involved and that *Frankia* rather than *Rhizobium* was responsible?

2. How could one decide whether a newly isolated spore-forming bacterium was producing conidiospores or true endospores?

3. List the properties that are most useful in actinomycete taxonomy and give some indication of their relative importance.

Additional Reading

General Properties of the Actinomycetes

Balows, A.; Trüper, H. G.; Dworkin, M.; Harder, W.; and Schleifer, K.-H. 1992. *The prokaryotes,* 2d ed. New York: Springer-Verlag.

Beaman, B. L.; Saubolle, M. A.; and Wallace, R. J. 1995. *Nocardia, Rhodococcus, Streptomyces, Oerskovia,* and other aerobic actinomycetes of medical importance. In *Manual of Clinical Microbiology,* 6th ed., P. R. Murray, editor-in-chief, 379–99. Washington, D.C.: American Society for Microbiology.

Dietz, A. 1994. The lives and times of industrial actinomycetes. *ASM News* 60(7):366–69.

Embley, T. M., and Stackebrandt, E. 1994. The molecular phylogeny and systematics of the actinomycetes. *Ann. Rev. Microbiol.* 48:257–89.

Ensign, J. C. 1978. Formation, properties, and germination of actinomycete spores. *Ann. Rev. Microbiol.* 32:185–219.

Goodfellow, M.; Modarski, M.; and Williams, S. T. editors. 1984. *The biology of the actinomycetes.* New York: Academic Press.

Holt, J. G., editor-in-chief. 1989. *Bergey's manual of systematic bacteriology,* vol. 4, S. T. Williams and M. E. Sharpe, editors. Baltimore, Md.: Williams & Wilkins.

Holt, J. G., editor-in-chief. 1994. *Bergey's manual of determinative bacteriology,* 9th ed. Baltimore, Md.: Williams & Wilkins.

Kalakoutskii, L. V., and Agre, N. S. 1976. Comparative aspects of development and differentiation in actinomycetes. *Bacteriol. Rev.* 40(2):469–524.

Lechevalier, M. P., and Lechevalier, H. A. 1980. The chemotaxonomy of actinomycetes. In *Actinomycete taxonomy,* A. Dietz and D. W. Thayer, editors, *Actinomycete taxonomy,* special publication 6, 227–91. Arlington, Va.: Society for Industrial Microbiology.

Actinomycetes with Multilocular Sporangia

Benson, D. R., and Silvester, W. B. 1993. Biology of *Frankia* strains, actinomycete symbionts of actinorhizal plants. *Microbiol. Rev.* 57(2):293–319.

Schwintzer, C. R., and Tjepkema, J. D., editors. 1990. *The biology of Frankia and actinorhizal plants.* San Diego, Calif.: Academic Press.

Actinoplanetes

Parenti, F., and Coronelli, C. 1979. Members of the genus *Actinoplanes* and their antibiotics. *Ann. Rev. Microbiol.* 33:389–411.

CHAPTER 25

The Fungi, Slime Molds, and Water Molds

Yeasts, molds, mushrooms, mildews, and the other fungi pervade our world. They work great good and terrible evil. Upon them, indeed, hangs the balance of life; for without their presence in the cycle of decay and regeneration, neither man nor any other living thing could survive.

—Lucy Kavaler

Outline

Concepts

1. Fungi are widely distributed and are found wherever moisture is present. They are of great importance to humans in both beneficial and harmful ways.

2. Fungi exist primarily as filamentous hyphae. A mass of hyphae is called a mycelium.

3. Like some bacteria, fungi digest insoluble organic matter by secreting exoenzymes, then absorbing the soluble nutrients.

4. Two reproductive structures occur in the fungi: (1) sporangia form asexual spores, and (2) gametangia form sexual gametes.

5. The zygomycetes are characterized by resting structures called zygospores—cells in which zygotes are formed.

6. The ascomycetes form zygotes within a characteristic club-shaped structure, the ascus.

7. Yeasts are unicellular fungi—mainly ascomycetes.

8. Basidiomycetes possess dikaryotic hyphae with two nuclei, one of each mating type. The hyphae divide uniquely, forming basidiocarps within which basidia can be found.

9. The deuteromycetes (Fungi Imperfecti) have either lost the capacity for sexual reproduction, or it has never been observed.

10. The slime and water molds resemble the fungi only in appearance and life-style. The three divisions (*Myxomycota, Acrasiomycota,* and *Oomycota*) are now widely regarded as protists.

This chapter is an introduction to the fungi. It describes the members of the kingdom *Fungi,* surveys their diversity, discusses their ecological and commercial impact, and presents some of their typical life cycles. A brief overview is also given of the protists that resemble the fungi—the slime molds and water molds.

According to the five-kingdom system of Whittaker (*see chapter 19*), fungi are placed in the kingdom *Fungi.* Microbiologists use the term **fungus** [pl., fungi; Latin *fungus,* mushroom] to include eucaryotic, spore-bearing organisms with absorptive nutrition, no chlorophyll, and that reproduce sexually and asexually. Scientists who study fungi are known as **mycologists** [Greek *mykes,* mushroom, and *logos,* discourse], and the scientific discipline dealing with fungi is called **mycology.** The study of fungal toxins and their effects on various organisms is called **mycotoxicology,** and the diseases caused in animals are known as **mycoses** [s., mycosis].

Distribution

Fungi are primarily terrestrial organisms, although a few are freshwater or marine. Many are pathogenic and infect plants and animals. Fungi also form beneficial relationships with other organisms. For example, about three-fourths of all vascular plants form associations (called mycorrhizae) between their roots and fungi. Fungi also are found in the upper portions of many plants. These endophytic fungi affect plant reproduction and palatability to herbivores. Lichens are associations of fungi and either algae or cyanobacteria.

Mycorrhizae (pp. 869–70).
Endophytic fungi (pp. 871–72).
Lichens (pp. 548–49).

Importance

Fungi are important to humans in both beneficial and harmful ways. With bacteria and a few other groups of heterotrophic organisms, fungi act as decomposers, a role of enormous significance. They degrade complex organic materials in the environment to simple organic compounds and inorganic molecules. In this way carbon, nitrogen, phosphorus, and other critical constituents of dead organisms are released and made available for living organisms (*see chapter 40*).

Fungi are the major cause of plant diseases (figure 25.1*a*). Over 5,000 species attack economically valuable crops and garden plants and also many wild plants. In like manner many diseases of animals (figure 25.1*b* and table 25.1) and humans (*see chapter 39*) are caused by fungi.

Fungi, especially the yeasts, are essential to many industrial processes involving fermentation (*see chapter 43*). Examples include the making of bread, wine, and beer. Fungi also play a major role in the preparation of some cheeses, soy sauce, and sufu; in the commercial production of many organic acids

(a)

(b)

Figure 25.1   Fungal Diseases. (*a*) Apple scab. Midsummer scab lesions on fruit and leaf, which developed from spring infections. These lesions produce spores capable of causing further spread of the disease. (*b*) Dermatomycosis. Ringworm in a kitten due to *Microsporum canis.*

(citric, gallic) and certain drugs (ergometrine, cortisone); and in the manufacture of many antibiotics (penicillin, griseofulvin) and the immunosuppressive drug cyclosporine.

In addition, fungi are important research tools in the study of fundamental biological processes. Cytologists, geneticists, biochemists, biophysicists, and microbiologists regularly use fungi in their research. Based on this research the yeast *Saccharomyces cerevisiae* is the best understood eucaryotic cell.

Structure

The body or vegetative structure of a fungus is called a **thallus** [pl., thalli]. It varies in complexity and size, ranging from the single-cell microscopic yeasts to multicellular molds, macroscopic puffballs, and mushrooms (figure 25.2). The fungal cell usually is encased in a cell wall of **chitin.** Chitin is a strong but flexible nitrogen-containing polysaccharide consisting of N-acetylglucosamine residues.

TABLE 25.1   Some Mycotoxicoses[a] Produced by Fungal Mycotoxins in Domestic Animals

Disease	Fungus	Mycotoxin	Contaminated Foodstuff	Animals Affected
Aflatoxicosis	*Aspergillus flavus*	Aflatoxins	Rice, corn, sorghum, cereals, peanuts, soybeans	Poultry, swine, cattle, sheep, dogs
Ergotism	*Claviceps purpurea*	Ergot alkaloids	Seedheads of many grasses, grains	Cattle, horses, swine, poultry
Mushroom poisoning	*Amanita verna*	Amanitins	Eaten from pastures	Cattle
Poultry hemorrhagic syndrome	*Aspergillus flavus* and others	Aflatoxins	Toxic grain and meal	Chickens
Slobbers	*Rhizoctonia*	Alkaloid slaframine	Red clover	Sheep, cattle
Tall fescue toxicosis	*Acremonium coenophialum* (an endophytic fungus)	Ergot alkaloids	Endophyte-infected tall fescue plants	Cattle, horses

[a]A mycotoxicosis [pl., mycotoxicoses] is a poisoning caused by a fungal toxin.

(a)

(b)

(c)

(d)

Figure 25.2   Fungal Thalli. (*a*) The microscopic, unicellular yeast, *Saccharomyces cerevisiae,* the yeast that makes bread dough rise; SEM (×21,000). Notice that some of the yeasts are reproducing by budding. (*b*) The multicellular common mold, *Penicillium,* growing on an apple. (*c*) A large group of puffballs, *Lycoperdon,* growing on a log. (*d*) A mushroom is made up of densely packed hyphae that form the mycelium or visible structure (thallus).

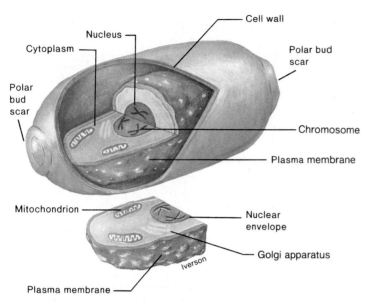

Figure 25.3   A Yeast. Diagrammatic drawing of a yeast cell showing typical morphology. For clarity, the plasma membrane has been drawn separated from the cell wall. In a living cell the plasma membrane adheres tightly to the cell wall.

(a)

(b)

Figure 25.4   Mold Mycelia. (*a*) Scanning electron micrograph of a young mycelial aggregate forming over a leaf stoma (×1,000). (*b*) A very large macroscopic mycelium of a basidiomycete growing on the soil.

A **yeast** is a unicellular fungus that has a single nucleus and reproduces either asexually by budding and transverse division (figure 25.2*a*) or sexually through spore formation. Each bud that separates can grow into a new yeast, and some group together to form colonies. Generally yeast cells are larger than bacteria, vary considerably in size, and are commonly spherical to egg shaped. They have no flagella but do possess most of the other eucaryotic organelles (figure 25.3).

A **mold** consists of long, branched, threadlike filaments of cells called **hyphae** [s., hypha; Greek *hyphe,* web] that form a **mycelium** [pl., mycelia], a tangled mass or tissuelike aggregation (figure 25.4). In some fungi, protoplasm streams through hyphae, uninterrupted by cross walls. These hyphae are called **coenocytic** (figure 25.5*a*). The hyphae of other fungi (figure 25.5*b*) have cross walls called **septa** [s., septum] with either a single pore (figure 25.5*c*) or multiple pores (figure 25.5*d*) that permit protoplasmic streaming. These hyphae are termed **septate.**

Hyphae are composed of an outer cell wall and an inner lumen, which contains the protoplasm and organelles (figure 25.6). A plasma membrane surrounds the protoplasm and lies next to the cell wall.

Many fungi, especially those that cause diseases in humans and animals, are dimorphic (table 25.2)—that is, they have two forms. Dimorphic fungi can change from (1) the yeast (Y) form in the animal to (2) the mold or mycelial form (M) in the external environment in response to changes in various environmental factors (nutrients, CO_2 tension, oxidation-reduction poten-

tials, temperature). This shift is called the **YM shift.** In plant-associated fungi the opposite type of dimorphism exists: the mycelial form occurs in the plant and the yeast form in the external environment.

1. How can a fungus be defined?
2. Where are fungi found?
3. Why are fungi important as decomposers?
4. What are some forms represented by different fungal thalli?
5. What organelles would you expect to find in the cytoplasm of a typical fungus?
6. Describe a typical yeast; a typical mold.

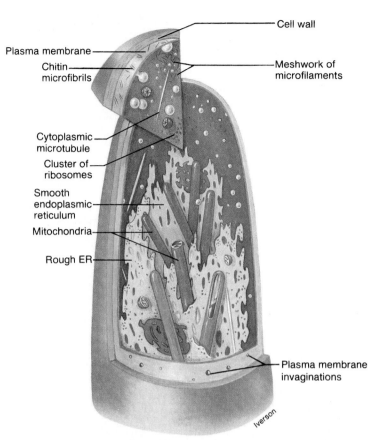

Figure 25.6   **Hyphal Morphology.** Diagrammatic representation of hyphal tip showing typical organelles and other structures.

⬤ Figure 25.5   **Hyphae.** Drawings of (*a*) coenocytic hyphae and (*b*) hyphae divided into cells by septa. (*c*) Electron micrograph (×40,000) of a section of *Drechslera sorokiniana* showing wall differentiation and a single pore. (*d*) Drawing of a multiperforate septal wall structure.

TABLE 25.2	Some Medically Important Dimorphic Fungi
Fungus	**Disease[a]**
Blastomyces dermatitidis	Blastomycosis
Candida albicans	Candidiasis
Coccidioides capsulatum	Coccidioidomycosis
Histoplasma capsulatum	Histoplasmosis
Sporothrix schenckii	Sporotrichosis
Paracoccidioides brasiliensis	Paracoccidioidomycosis

[a]See chapter 39 for a discussion of each of these diseases.

Nutrition and Metabolism

Fungi grow best in dark, moist habitats, but they are found wherever organic material is available. Most fungi are **saprophytes,** securing their nutrients from dead organic material. Like many bacteria, fungi can secrete hydrolytic enzymes that digest external substrates. They then absorb the soluble products. They are chemoorganoheterotrophs and use organic material as a source of carbon, electrons, and energy.

Glycogen is the primary storage polysaccharide in fungi. Most fungi use carbohydrates (preferably glucose or maltose) and nitrogenous compounds to synthesize their own amino acids and proteins.

Fungi usually are aerobic. Some yeasts, however, are facultatively anaerobic and can obtain energy by fermentation, such as in the production of ethyl alcohol from glucose. Obligately anaerobic fungi are found in the rumen of cattle.

Reproduction

Reproduction in fungi can be either asexual or sexual. Asexual reproduction is accomplished in several ways:

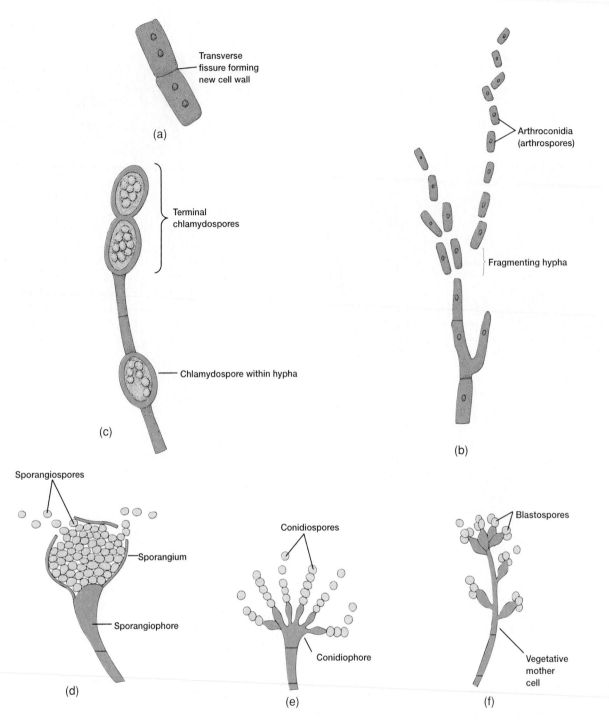

Transverse fissure forming new cell wall

(a)

Terminal chlamydospores

Chlamydospore within hypha

(c)

Arthroconidia (arthrospores)

Fragmenting hypha

(b)

Sporangiospores

Sporangium

Sporangiophore

(d)

Conidiospores

Conidiophore

(e)

Blastospores

Vegetative mother cell

(f)

Figure 25.7   Diagrammatic Representation of Asexual Reproduction in the Fungi and Some Representative Spores. (*a*) Transverse fission. (*b*) Hyphal fragmentation resulting in arthroconidia (arthrospores) and (*c*) chlamydospores. (*d*) Sporangiospores in a sporangium. (*e*) Conidiospores arranged in chains at the end of a conidiophore. (*f*) Blastospores are formed from buds off of the parent cell.

1. A parent cell can divide into two daughter cells by central constriction and formation of a new cell wall (figure 25.7*a*).

2. Somatic vegetative cells (figure 25.2*a*) may bud to produce new organisms. This is very common in the yeasts.

3. The most common method of asexual reproduction is the production of spores. Asexual spore formation occurs in an individual fungus through mitosis and subsequent cell division. There are several types of asexual spores:

a. A hypha can fragment (by the separation of hyphae through splitting of the cell wall or septum) to form cells that behave as spores. These cells are called **arthroconidia** or **arthrospores** (figure 25.7b).

b. If the cells are surrounded by a thick wall before separation, they are called **chlamydospores** (figure 25.7c).

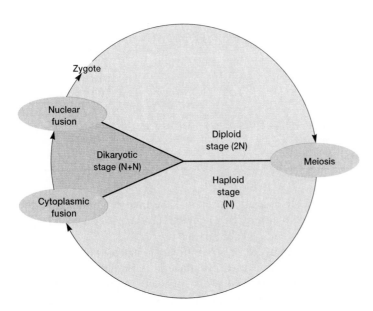

Figure 25.8   Reproduction in Fungi. A drawing of the generalized life cycle for fungi showing the alternation of haploid and diploid stages. Some fungal species do not pass through the dikaryotic stage indicated in this drawing. The asexual (haploid) stage is used to produce spores that aid in the dissemination of the species. The sexual (diploid) stage involves spores that are used for surviving adverse environmental conditions (e.g., cold, dryness, heat).

c. If the spores develop within a sac [**sporangium;** pl., sporangia] at a hyphal tip, they are called **sporangiospores** (figure 25.7*d*).

d. If the spores are not enclosed in a sac but produced at the tips or sides of the hypha, they are termed **conidiospores** (figures 25.7*e* and 25.11).

e. Spores produced from a vegetative mother cell by budding (figure 25.7*f*) are called **blastospores.**

Sexual reproduction in fungi involves the union of compatible nuclei. Some fungal species are self-fertilizing and produce sexually compatible gametes on the same mycelium. Other species require outcrossing between different but sexually compatible mycelia. Depending on the species, sexual fusion may occur between haploid gametes, gamete-producing bodies called **gametangia,** or hyphae. Sometimes both the cytoplasm and haploid nuclei fuse immediately to produce the diploid zygote. Usually, however, there is a delay between cytoplasmic and nuclear fusion. This produces a **dikaryotic stage** in which cells contain two separate haploid nuclei, one from each parent (figure 25.8). Sexual reproduction also can yield spores. For example, the zygote sometimes develops into a **zygospore** (figure 25.9), an **ascospore** (figure 25.12), or a **basidiospore** (figure 25.14).

Figure 25.9   Division *Zygomycota*. Diagrammatic representation of the life cycle of *Rhizopus stolonifer*. Both the sexual and asexual phases are illustrated.

Fungal spores are important for several reasons. The size, shape, color, and number of spores are useful in the identification of fungal species. The spores are often small and light and can remain suspended in air for long periods. Thus they frequently aid in fungal dissemination, a significant factor that explains the wide distribution of many fungi. Fungal spores often spread by adhering to the bodies of insects and other animals. The bright colors and fluffy textures of many molds often are due to their aerial hyphae and spores.

Characteristics of the Fungal Divisions

The taxonomic scheme used in this chapter classifies the fungi into four divisions (table 25.3), based primarily on variations in sexual reproduction. In mycology a division is equivalent to a phylum in animal classification schemes.

Division *Zygomycota*

The division *Zygomycota* contains the fungi called **zygomycetes.** Most live on decaying plant and animal matter in the soil; a few are parasites of plants, insects, animals, and humans. The hyphae of zygomycetes are coenocytic, with many haploid nuclei. Asexual spores, usually wind dispersed, develop in sporangia at the tips of aerial hyphae. Sexual reproduction produces tough, thick-walled zygotes called zygospores that can remain dormant when the environment is too harsh for growth of the fungus.

The bread mold, *Rhizopus stolonifer,* is a very common member of this division. This fungus grows on the surface of moist, carbohydrate-rich foods, such as breads, fruits, and vegetables. On breads, for example, *Rhizopus's* hyphae rapidly cover the surface. Special hyphae called rhizoids extend into the bread, and absorb nutrients (figure 25.9). Other hyphae (stolons) become erect, then arch back into the substratum forming new rhizoids. Still others remain erect and produce at their tips asexual sporangia filled with the black spores, giving the mold its characteristic color. Each spore, when liberated, can start a new mycelium.

Rhizopus usually reproduces asexually, but if food becomes scarce or environmental conditions unfavorable, it begins sexual reproduction. Sexual reproduction requires compatible strains of opposite mating types (figure 25.9). These have traditionally been labeled + and − strains because they are not morphologically distinguishable as male and female. When the two mating strains are close, hormones are produced that cause their hyphae to form projections called **progametangia** [Greek, *pro,* before], and then mature gametangia. After fusion of the gametangia, the nuclei of the two gametes fuse, forming a zygote. The zygote develops a thick, rough, black coat and becomes a dormant zy-

TABLE 25.3   Divisions of Fungi

Division	Common Name	Approximate Number of Species
Zygomycota	Zygomycetes	600
Ascomycota	Sac fungi	35,000
Basidiomycota	Club fungi	30,000
Deuteromycota	Fungi Imperfecti	30,000

gospore. Meiosis often occurs at the time of germination; the zygospore then cracks open and produces an asexual sporangium and the cycle begins anew.

The zygomycetes also contribute to human welfare. For example, *Rhizopus* is used in Indonesia to produce a food called tempeh from boiled, skinless soybeans. Another zygomycete (*Mucor* spp.) is used with soybeans in the Orient to make a cheese called sufu (*see chapter 43*). Others are employed in the commercial preparation of some anesthetics, birth control agents, industrial alcohols, meat tenderizers, and the yellow coloring agent in margarine and butter substitutes.

1. What generally governs the ecological distribution of fungi?
2. How is asexual reproduction accomplished in the fungi? Sexual reproduction?
3. Describe each of the following types of asexual fungal spores: sporangiospore, conidiospore, and blastospore.
4. Describe how a typical zygomycete reproduces.
5. Describe some beneficial uses for zygomycetes.

Division *Ascomycota*

The division *Ascomycota* contains the fungi called **ascomycetes,** commonly known as the sac fungi. Many species are quite familiar and economically important (figure 25.10). For example, most of the red, brown, and blue-green molds that cause food spoilage are ascomycetes. The powdery mildews that attack plant leaves and the fungi that cause chestnut blight and Dutch elm disease are ascomycetes. Many yeasts as well as edible morels and truffles are ascomycetes. The pink bread mold *Neurospora crassa,* also an ascomycete, has been a most important research tool in genetics and biochemistry.

Many ascomycetes are parasites on higher plants. *Claviceps purpurea* parasitizes rye and other grasses, causing the disease **ergot. Ergotism,** the toxic condition in humans and animals who eat grain infected with the fungus, is often accompanied by gangrene, psychotic delusions, nervous spasms, abortion, and

(a)

(b)

(c)

Figure 25.10   Division *Ascomycota*. (*a*) The common morel, *Morchella esculenta,* is one of the choicest edible fungi. It fruits in the spring. (*b*) Scarlet cups, *Sarcoscypha coccinea,* with open ascocarps (apothecia). (*c*) The black truffle, *Tuber brumale,* is highly prized for its flavor by gourmet cooks. Technically truffles are mycorrhizal associations on oak trees.

convulsions. During the Middle Ages ergotism, then known as St. Anthony's fire, killed thousands of people. For example, over 40,000 deaths from ergot poisoning were recorded in France in the year 943. It has been suggested that the widespread accusations of witchcraft in Salem Village (now Danvers) and other Massachusetts communities in the late 1690s may have resulted from outbreaks of ergotism. The pharmacological activities of ergot are due to its active ingredient, lysergic acid diethylamide (LSD). In controlled dosages ergot can be used to induce labor, lower blood pressure, and ease migraine headaches.

The ascomycetes are named for their characteristic reproductive structure, the club- or sac-shaped **ascus** [pl., asci; Greek *askos,* sac]. The mycelium of the ascomycetes is composed of septate hyphae. Asexual reproduction is common in the ascomycetes and takes place by way of conidiospores (figure 25.11).

Sexual reproduction in the ascomycetes always involves the formation of an ascus containing haploid ascospores (figure 25.12*a*). In the more complex ascomycetes, ascus formation is preceded by the development of special **ascogenous hyphae** into which pairs of nuclei migrate (figure 25.12*b*). One nucleus of each pair originates from a "male" mycelium (**antheridium**) or cell and the other from a "female" organ or cell (**ascogonium**) that has fused with it. As the ascogenous hyphae grow, the paired nuclei divide so that there is one pair of nuclei in each cell. After the ascogenous hyphae have matured, nuclear fusion occurs at the hyphal tips in the ascus mother cells. The diploid zygote nucleus then undergoes meiosis, and the resulting four haploid nuclei divide mitotically again to produce a

Figure 25.11   Asexual Reproduction in Ascomycetes. Characteristic conidiospores of *Aspergillus* as viewed with the electron microscope (×1,200).

row of eight nuclei in each developing ascus. These nuclei are walled off from one another. Thousands of asci may be packed together in a cup- or flask-shaped **ascocarp** (figure 25.10*b*). When the ascospores mature, they often are released from the asci with great force. If, perchance, the mature ascocarp is jarred, it may appear to belch puffs of "smoke" consisting of thousands of ascospores. Upon reaching a suitable environment, the ascospores germinate and start the cycle anew.

Although the term yeast is used in a general sense to refer to all unicellular fungi that reproduce asexually by either budding or binary fission (figure 25.13*a,b*), many yeast genera

(a)

(b)

Figure 25.12   The Life Cycle of Ascomycetes. Sexual reproduction involves the formation of asci and ascospores. Within the ascus, karyogamy is followed by meiosis to produce the ascospores. (*a*) Sexual reproduction and ascocarp morphology of a cup fungus. (*b*) The details of sexual reproduction in ascogenous hyphae. The nuclei of the two mating types are represented by unfilled and filled circles. See text for details.

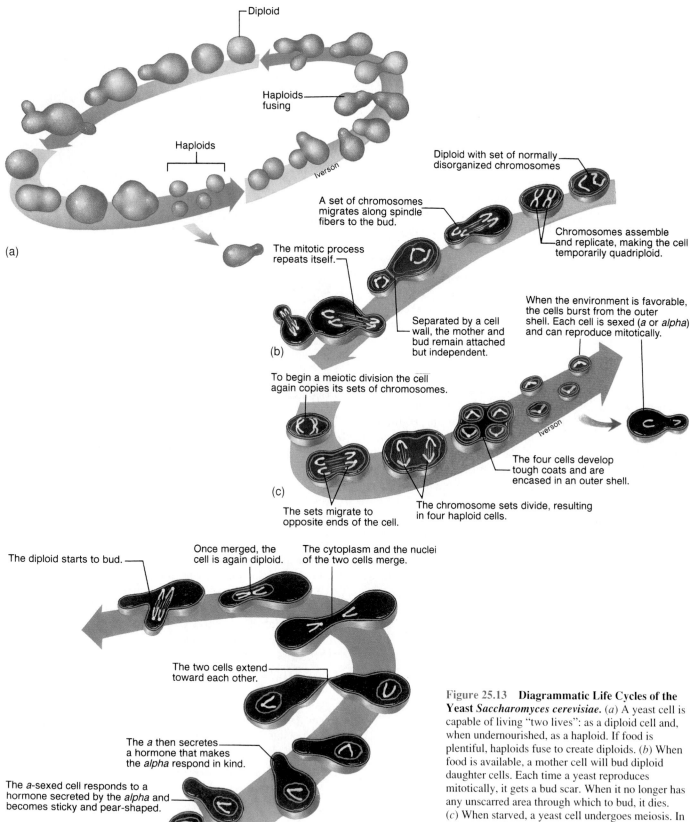

(a)

Diploid

Haploids
fusing

Haploids

Iverson

Diploid with set of normally
disorganized chromosomes

A set of chromosomes
migrates along spindle
fibers to the bud.

Chromosomes assemble
and replicate, making the cell
temporarily quadriploid.

The mitotic process
repeats itself.

Separated by a cell
wall, the mother and
bud remain attached
but independent.

When the environment is favorable,
the cells burst from the outer
shell. Each cell is sexed (a or alpha)
and can reproduce mitotically.

(b)

To begin a meiotic division the cell
again copies its sets of chromosomes.

The four cells develop
tough coats and are
encased in an outer shell.

(c)

The sets migrate to
opposite ends of the cell.

The chromosome sets divide, resulting
in four haploid cells.

Iverson

The diploid starts to bud.

Once merged, the
cell is again diploid.

The cytoplasm and the nuclei
of the two cells merge.

The two cells extend
toward each other.

The a then secretes
a hormone that makes
the alpha respond in kind.

The a-sexed cell responds to a
hormone secreted by the alpha and
becomes sticky and pear-shaped.

a

alpha

(d)

Iverson

**Figure 25.13   Diagrammatic Life Cycles of the
Yeast *Saccharomyces cerevisiae*.** (*a*) A yeast cell is
capable of living "two lives": as a diploid cell and,
when undernourished, as a haploid. If food is
plentiful, haploids fuse to create diploids. (*b*) When
food is available, a mother cell will bud diploid
daughter cells. Each time a yeast reproduces
mitotically, it gets a bud scar. When it no longer has
any unscarred area through which to bud, it dies.
(*c*) When starved, a yeast cell undergoes meiosis. In
this process four cells develop within the mother cell
as a unit and remain in their rugged cell walls until
food is available once more. (*d*) Two haploids in
proximity can fuse and produce a diploid. Once
formed, it buds even before it attains its final shape.
(*See chapter 4 and figures 4.19 and 4.22 for a
complete discussion of mitosis and meiosis.*)

are classified specifically within the ascomycetes because of their sexual reproduction (figure 25.13*c,d*). Yeasts are present in both terrestrial and aquatic habitats in which a suitable carbon source is available.

Division *Basidiomycota*

The division *Basidiomycota* contains the **basidiomycetes,** commonly known as the club fungi. Examples include the smuts, jelly fungi, rusts, shelf fungi, stinkhorns, puffballs, toadstools, mushrooms, and bird's nest fungi.

Basidiomycetes are named for their characteristic structure or cell, the **basidium,** that is involved in sexual reproduction (figure 25.14). A basidium [Greek *basidion,* small base] is produced at the tip of hyphae and normally is club shaped. Basidiospores are produced by the basidium, and basidia may be held within fruiting bodies called **basidiocarps.**

The basidiomycetes affect humans in many ways. Most are saprophytes and decompose plant debris, especially cellulose and lignin. Many mushrooms are used as food throughout the world. The cultivation of *Agaricus campestris* is a multimillion-dollar business (*see figure 43.21*).

Many mushrooms produce specific alkaloids that act as either poisons or hallucinogens. One such example is the "destroying angel" mushroom, *Amanita phalloides.* Two toxins isolated from this species are phalloidin and α-amanitin. Phalloidin primarily attacks liver cells where it binds to cell membranes, causing them to rupture and leak their contents. Alpha-amanitin attacks the cells lining the stomach and small intestine and is responsible for the severe gastrointestinal symptoms associated with mushroom poisoning.

The basidiomycete *Cryptococcus neoformans* is an important human pathogen. It produces the disease called **cryptococcosis,** a systemic infection primarily involving the lungs and central nervous system. Other basidiomycetes, the smuts and rusts, are virulent plant pathogens that cause extensive damage to cereal crops; millions of dollars worth of crops are destroyed annually. In these fungi, large basidiocarps are not formed. Instead the small basidia arise from hyphae at the surface of the host plant. The mycelia grow either intra- or extracellularly in plant tissue.

Fungal diseases (pp. 791–99).

Division *Deuteromycota*

To a large degree fungal taxonomy is based on specific patterns of sexual reproduction. When a fungus lacks the sexual phase (perfect stage), or if this phase has not been observed, it is placed within the division *Deuteromycota,* commonly called the Fungi Imperfecti. Once a perfect stage is observed, the fungus is transferred to its proper division.

Most Fungi Imperfecti are terrestrial, with only a few being reported from freshwater and marine habitats. The majority are either saprophytes or parasites of plants. A few are parasitic on other fungi.

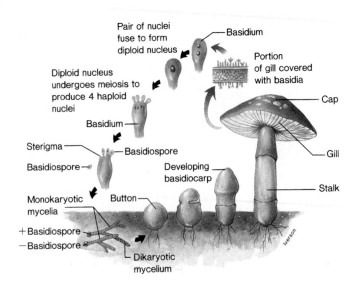

Figure 25.14 Division *Basidiomycota*. The life cycle of a typical soil basidiomycete starts with a basidiospore germinating to produce a monokaryotic mycelium (one with a single nucleus in each septate cell). The mycelium quickly grows and spreads throughout the soil. When this primary mycelium meets another monokaryotic mycelium of a different mating type, the two fuse to initiate a new dikaryotic secondary mycelium. The secondary mycelium is divided by septa into cells, each of which contains two nuclei, one of each mating type. This dikaryotic mycelium is eventually stimulated to produce basidiocarps. A solid mass of hyphae forms a button that pushes through the soil, elongates, and develops a cap. The cap contains many platelike gills, each of which is coated with basidia. The two nuclei in the tip of each basidium fuse to form a diploid zygote nucleus, which immediately undergoes meiosis to form four haploid nuclei. These nuclei push their way into the developing basidiospores, which are then released at maturity.

Many imperfect fungi directly affect human welfare. Several are human pathogens, causing such diseases as athlete's foot, ringworm, and histoplasmosis (*see chapter 39*). The chemical activities of many Fungi Imperfecti are important industrially. For example, some species of *Penicillium* (*see figure 4.1d,e*) synthesize the well-known antibiotics penicillin and griseofulvin. Other species give characteristic aromas to cheeses such as Gorgonzola, Camembert, and Roquefort. Different species of *Aspergillus* are used to ferment soy sauce and to manufacture citric, gluconic, and gallic acids. *Aspergillus flavus* and *A. parasiticus* produce secondary metabolites, called aflatoxins, that are highly toxic and carcinogenic to animals and humans (*see chapter 43*). Another group of fungal toxins, the trichothecenes, are strong inhibitors of protein synthesis in eucaryotic cells.

1. Describe the ascomycete life cycle. How are the ascomycetes important to humans?

2. How do yeasts reproduce sexually? Asexually?

3. Describe the life cycle of a typical basidiomycete.

4. How do some Fungi Imperfecti affect humans?

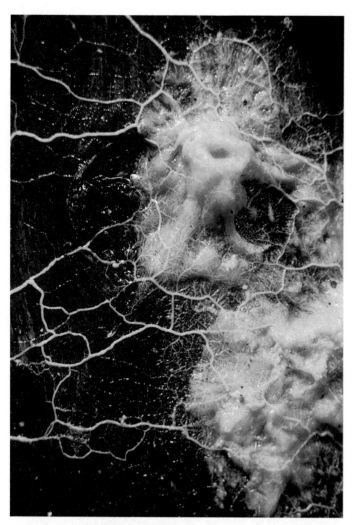

Figure 25.15   Slime Molds. Plasmodium of the slime mold *Physarum;* light micrograph (×175).

Slime Molds and Water Molds

The **slime molds** and **water molds** resemble fungi in appearance and life-style. In their cellular organization, reproduction, and life cycles, they are related most closely to the protists (*see figure 19.11*).

Division *Myxomycota*

Under appropriate conditions **plasmodial slime molds** exist as thin, streaming masses of colorful protoplasm that creep along in an amoeboid fashion over moist, rotting logs, leaves, and other organic matter. Feeding is by phagocytosis. Because this streaming mass lacks cell walls, it is called a **plasmodium** (figures 25.15 and 25.16a). The plasmodium contains many nuclei, and as the organism grows, the diploid nuclei divide repeatedly (*see figure 4.20*).

When the plasmodium matures or when food and/or moisture are scarce, it moves into a lighted area and develops delicate ornate fruiting bodies (figure 25.16b–e). As the fruiting bodies mature, they form spores with cellulose walls that are resistant to environmental extremes. The spores germinate in the presence of adequate moisture to release either nonflagellated amoeboid **myxamoebae** or flagellated **swarm cells.** Initially the myxamoebae or swarm cells feed and are haploid (figure 25.16a); eventually they fuse to form a diploid zygote. The zygote feeds, grows, and multiplies its nuclei through synchronous mitotic division to form the multinucleate plasmodium.

Division *Acrasiomycota*

The vegetative stage of **cellular slime molds** consists of individual irregular amoeboid cells termed myxamoebae (figure 25.17a). The myxamoebae feed phagocytically on bacteria and yeasts. When food is plentiful, they divide repeatedly by mitosis and cytokinesis, producing new daughter myxamoebae. As their food supply is exhausted, the myxamoebae begin to secrete cyclic adenosine monophosphate (cAMP). This attracts other myxamoebae that move toward the cAMP chemotactic source and in turn secrete more cAMP. When the individual myxamoebae aggregate (figure 25.17b), they secrete a slimy sheath around themselves and form a sluglike **pseudoplasmodium** (figure 25.17c). The pseudoplasmodium may move around as a unit for a while, leaving a slime trail, but it eventually becomes sedentary. In the culmination of the asexual phase, pseudoplasmodial cells begin to differentiate into prestalk cells and prespore cells (figure 25.17b). A structure called a **sorocarp** forms (figure 25.17d) and matures into a sporangium that produces spores (figure 25.17e). The spores are eventually released, and when conditions become favorable, they germinate to release haploid amoebae and start the cycle anew.

Division *Oomycota*

Members of the division *Oomycota* are collectively known as **oomycetes** or water molds. Oomycetes resemble fungi in appearance, consisting of finely branched filaments called hyphae. However, oomycetes have cell walls of cellulose, whereas the walls of most fungi are made of chitin.

Oomycota means "egg fungi," a reference to the mode of sexual reproduction in water molds. A relatively large egg cell is fertilized by either a sperm cell or a smaller antheridium. Most oomycetes also produce asexual zoospores that bear two flagella.

Water molds such as *Saprolegnia* and *Achlya* are saprophytes that grow as cottony masses on dead algae and small animals, mainly in freshwater environments. They are important decomposers in aquatic ecosystems. Some water molds are parasitic on the gills of fish. The water mold *Peronospora hyoscyami* is currently responsible for the troublesome "blue mold" of tobacco plants throughout the world. In the United States alone, blue mold produces millions of dollars of damage yearly to tobacco crops. Other oomycetes cause late blight of potatoes (*Phytophthora infestans*) and grape downy mildew (*Plasmopara viticola*).

1. What is a plasmodium?
2. How do the plasmodial slime molds reproduce? Describe the cellular slime mold life cycle.
3. Where would you look for an oomycete?

Figure 25.16   Reproduction in the *Myxomycota*. (*a*) The life cycle of a plasmodial slime mold. (Different parts of the life cycle are drawn at different magnifications.) Plasmodial slime mold fruiting bodies: (*b*) *Hemitrichia* (×100), (*c*) *Stemonitis* (×100), (*d*) *Physarum polycephalum*, and (*e*) *Arcyria denudata*.

Figure 25.17 Division *Acrasiomycota*. *Dictyostelium discoideum,* a cellular slime mold. In the free-living stage a myxamoeba resembles an irregularly shaped amoeba. (*a*) Aggregating myxamoebae become polar and begin to move in an oriented direction due to the influence of cAMP. (*b*) Diagrammatic drawing of cell migration involved in the formation of the sorocarp from (*c*) an initial pseudoplasmodium. (*d*) Light micrograph of a mature fruiting body (sorocarp). (*e*) Electron micrograph of a sporangium showing spores (×1,800).

Summary

1. A fungus is a eucaryotic, spore-bearing organism that has absorptive nutrition and lacks chlorophyll; that reproduces asexually, sexually, or by both methods; and that normally has filamentous hyphae surrounded by cell walls, which usually contain chitin.

2. Fungi are omnipresent in the environment, being found wherever water and suitable organic nutrients occur. They secrete enzymes outside their body structure and absorb the digested food.

3. Fungi are important decomposers that break down organic matter; live as parasites on animals, humans, and plants; play a role in many industrial processes; and are used as research tools in the study of fundamental biological processes.

4. The body or vegetative structure of a fungus is called a thallus.

5. A mold consists of long, branched, threadlike filaments of cells, the hyphae, that form a tangled mass of tissue called a mycelium. Hyphae may be either septate or coenocytic (nonseptate). The mycelium can produce reproductive structures.

6. Yeasts are unicellular fungi that have a single nucleus and reproduce either asexually by budding and transverse division or sexually through spore formation.

7. Many fungi are dimorphic—they alternate between a yeast and a mold form.

8. Asexual reproduction often occurs in the fungi by the production of specific types of spores. Spores are easily dispersed and are resistant to adverse environmental conditions.

9. Sexual reproduction is initiated in the fungi by the fusion of hyphae of different mating strains. In some fungi the nuclei in the fused hyphae immediately combine to form a zygote. In others the two genetically distinct nuclei remain separate, forming pairs that divide synchronously. Eventually some nuclei fuse.

10. The zygomycetes are coenocytic. Most are saprophytic. One example is the common bread mold, *Rhizopus stolonifer*. Sexual reproduction occurs through a form of conjugation involving + and – strains.

11. The ascomycetes are known as the sac fungi because they form a sac-shaped reproductive structure called an ascus. Sexual reproduction involves + and – strains. In asexual reproduction conidiophores produce conidia.

12. The basidiomycetes are the club fungi. They are named after their basidium that produces basidiospores.

13. The deuteromycetes (Fungi Imperfecti) are fungi with no known sexual (perfect) phase.

14. The plasmodial slime molds move about as a plasmodium containing many nuclei that are not separated from one another by membranes. When food or moisture is scarce, these slime molds form sporangia within which spores are produced.

15. The cellular slime molds consist of a vegetative stage called a myxamoeba. The myxamoebae feed until their food supply dwindles, then the cells come together to form a plantlike multicellular structure called a sorocarp. The sorocarp produces haploid spores that germinate when conditions are favorable to form new myxamoebae.

16. The *Oomycota* (water molds), consisting of branched filaments called hyphae, resemble fungi only in appearance.

Key Terms

antheridium *511*

arthroconidia *508*

arthrospore *508*

ascocarp *511*

ascogenous hypha *511*

ascogonium *511*

ascomycetes *510*

ascospore *509*

ascus *511*

basidiocarp *514*

basidiomycetes *514*

basidiospore *509*

basidium *514*

blastospore *509*

cellular slime molds *515*

chitin *504*

chlamydospore *508*

coenocytic *506*

conidiospore *509*

cryptococcosis *514*

dikaryotic stage *509*

ergot *510*

ergotism *510*

fungus *504*

gametangium *509*

hypha *506*

mold *506*

mycelium *506*

mycologist *504*

mycology *504*

mycosis *504*

mycotoxicology *504*

myxamoeba *515*

oomycetes *515*

plasmodial slime mold *515*

plasmodium *515*

progametangium *510*

pseudoplasmodium *515*

saprophyte *507*

septa *506*

septate *506*

slime mold *515*

sorocarp *515*

sporangiospore *509*

sporangium *509*

swarm cell *515*

thallus *504*

water mold *515*

yeast *506*

YM shift *506*

zygomycetes *510*

zygospore *509*

Questions for Thought and Review

1. Some fungi can reproduce both sexually and asexually. What are the advantages and disadvantages of each?

2. Why is nutrition in the fungi a property of membrane function?

3. Why are most fungi confined to a specific ecological niche?

4. Some authorities believe that the slime molds and water molds should be placed in the kingdom *Fungi,* whereas others believe that they should be placed in the kingdom *Protista.* What are the characteristics that these organisms share with both the fungi and protists that have led to this ambiguity?

5. At the present time those fungi that have no sexual reproduction cannot be classified with their sexually reproducing relatives. Why? Do you think that this is likely to change in the future?

6. Because asexual spores are such a rapid way of reproducing for some fungi, what adaptive "use" is there for an additional sexual phase?

7. Some fungi can be viewed as coenocytic organisms that exhibit little differentiation. When it does occur, such as in the formation of reproductive structures, differentiation is preceded by septum formation. Why does this occur?

8. The term mushrooming is a proverbial description for expanding rapidly. Why is this an accurate metaphor?

9. Both bacteria and fungi are major environmental decomposers. Obviously competition exists in a given environment, but fungi usually have an advantage. What is this advantage?

10. Very few antibiotics can control fungi in medicine and agriculture, though there are many for bacterial control. Why is this so?

Additional Reading

General

Alexopoulos, C. J., and Mims, C. W. 1979. *Introductory mycology,* 3d ed. New York: John Wiley and Sons.

Ball, D. M.; Pedersen, J. F.; and Lacefield, G. D. 1993. The tall-fescue endophyte. *Am. Sci.* 81(4):370–79.

Batra, S. W. T., and Batra, R. L. 1967. The fungus gardens of insects. *Sci. Am.* 217:112–20.

Broada, P.; Odds, C.; and Ferridge, D., editors. 1993. *Dimorphic fungi in biology and medicine.* New York: Plenum Publishing Company.

Deacon, J. W. 1980. *Introduction to modern mycology.* New York: John Wiley and Sons.

Klionsky, D. J.; Herman, P. K.; and Emr, S. D. 1990. The fungal vacuole: Composition, function, and biogenesis. *Microbiol. Rev.* 54(3):266–92.

Moore-Landecker, E. 1991. *Fundamentals of fungi,* 3d ed. Englewood Cliffs, N.J.: Prentice-Hall.

Ross, I. K. 1979. *Biology of the fungi: Their development, regulation, and associations.* New York: McGraw-Hill.

Stevens, R. B., editor. 1974. *Mycology guidebook.* Seattle: University of Washington Press.

van der Rest, M. E.; Kamminga, A. H.; Nakano, A.; Anraku, Y.; Poolman, B.; and Konings, W. N. 1995. The plasma membrane of *Saccharomyces cerevisiae:* Structure, function, and biogenesis. *Microbiol. Rev.* 59(2):304–22.

Webster, J. 1980. *Introduction to fungi,* 2d ed. New York: Cambridge University Press.

Characteristics of the Fungal Divisions

Bossche, H., editor. 1993. *Dimorphic fungi in biology and medicine.* New York: Plenum Publishing Company.

Cole, G. T. 1986. Models of cell differentiation in conidial fungi. *Microbiol. Rev.* 50:95–132.

Gold, M. H., and Alic, M. 1993. Molecular biology of the lignin-degrading basidiomycete *Phanerochaete chrysosporium. Microbiol. Rev.* 57(3):605–22.

Herskowitz, I. 1988. Life cycle of the budding yeast *Saccharomyces cerevisiae. Microbiol. Rev.* 52(4):536–53.

Jackson, S. L., and Heath, I. B. 1993. Roles of calcium ions in hyphal tip growth. *Microbiol. Rev.* 57(2):367–82.

Lipke, P., and Kurjan, J. 1992. Sexual agglutination in budding yeast: structure, function, and regulation of adhesion glycoproteins. *Microbiol. Rev.* 56(1):180–94.

Litten, W. 1975. The most poisonous mushrooms. *Sci. Am.* 232:14–22.

Maresca, B., and Kobayashi, G. S. 1989. Dimorphism in *Histoplasma capsulatum:* A model for the study of cell differentiation in pathogenic fungi. *Microbiol. Rev.* 53(2):186–209.

Matossian, M. K. 1982. Ergot and the Salem witchcraft affair. *Am. Scientist* 70:355–71.

Mitchell, A. 1994. Control of meiotic gene expression in *Saccharomyces cerevisiae. Microbiol. Rev.* 58(1):56–70.

Newhouse, J. R. 1990. Chestnut blight. *Sci. Am.* 263:106–11.

Newlon, C. S. 1988. Yeast chromosome replication and segregation. *Microbiol. Rev.* 52(4)568–606.

Orlowski, M. 1991. *Mucor* dimorphism. *Microbiol. Rev.* 55(2):234–58.

Phaff, H. J. 1986. My life with yeasts. *Ann. Rev. Microbiol.* 40:1–28.

Phaff, H. J.; Miller, M. W.; and Mrak, E. M. 1978. *The life of yeasts,* 2d ed. Cambridge, Mass.: Harvard University Press.

Smith, A. H.; Smith, H.; and Weber, N. S. 1979. *Gilled mushrooms.* Dubuque, Iowa: Wm. C. Brown.

Smith, A. H.; Smith, H.; and Weber, N. S. 1981. *Nongilled mushrooms,* 2d ed. Dubuque, Iowa: Wm. C. Brown.

Strobel, G. A., and Lanier, G. N. 1981. Dutch elm disease. *Sci. Am.* 245:56–66.

Werner-Washburne, M.; Braun, E.; Johnston, G.; and Singer, R. 1993. Stationary phase in the yeast *Saccharomyces cerevisiae. Microbiol. Rev.* 57(2):383–401.

Slime Molds and Water Molds

Gross, J. D. 1994. Developmental decisions in *Dictyostelium discoideum. Microbiol. Rev.* 58(3):330–51.

Kessin, R. H. 1988. Genetics of early *Dictyostelium discoideum* development. *Microbiol. Rev.* 52(1):29–49.

Martin, G. W., and Alexopoulos, C. J. 1969. *The myxomycetes.* Iowa City: University of Iowa Press.

CHAPTER 26
The Algae

The term algae means different things to different people, and even the professional botanist and biologist find algae embarrassingly elusive of definition. Thus, laymen have given such names as "pond scums," "frog spittle," "water mosses," and "seaweeds," while some professionals shrink from defining them.

—Harold C. Bold and
Michael J. Wynne

Outline

Concepts

1. Most algae are found in freshwater and marine environments; a few grow in terrestrial habitats.

2. The algae are not a single, closely related taxonomic group but, instead, are a diverse assemblage of unicellular, colonial, and multicellular eucaryotic organisms.

3. Although algae can be autotrophic or heterotrophic, most are photoautotrophs. They store carbon in a variety of forms, including starch, oils, and various sugars.

4. The body of an alga is called the thallus. Algal thalli range from small solitary cells to large, complex multicellular structures.

5. Algae reproduce asexually and sexually.

6. The following divisions of the algae are discussed: *Chlorophyta* (green algae), *Charophyta* (stoneworts/brittleworts), *Euglenophyta* (euglenoids), *Chrysophyta* (golden-brown and yellow-green algae; diatoms), *Phaeophyta* (brown algae), *Rhodophyta* (red algae), and *Pyrrhophyta* (dinoflagellates).

hapter 26 presents some general features of algae. The algae are a very diverse group of organisms, ranging in size from microscopic single cells to kelps over 75 m in length. Algae vary considerably in the structure of their cells, the arrangement of cells in colonial and multicellular forms, and in their pigmentation and color. Algae grow in a wide variety of habitats. Within the oceans and freshwater environments, they are the major producers of O_2 and organic material. A few algae live on land in moist soils and other environments.

Phycology or **algology** is the study of algae. The word phycology is derived from the Greek *phykos,* meaning seaweed. The term **algae** [s., alga] was originally used to define simple "aquatic plants." It no longer has any formal significance in classification schemes. Instead the algae can be described as eucaryotic organisms that lack roots, stems, and leaves but have chlorophyll and other pigments for carrying out oxygen-producing photosynthesis. Many can be placed in the kingdom *Protista.*

Distribution of Algae

Algae most commonly occur in water (fresh, marine, or brackish) in which they may be suspended (**planktonic**) or attached and living on the bottom (**benthic**). A few algae live at the water-atmosphere interface and are termed **neustonic. Plankton** [Greek *plankos,* wandering] consists of free-floating, mostly microscopic aquatic organisms. **Phytoplankton** is made up of algae and small plants, whereas **zooplankton** consists of animals and nonphotosynthetic protists. Some algae grow on moist rocks, wood, trees, and on the surface of moist soil. Algae also live as endosymbionts in various protozoa, mollusks, worms, and corals. Several algae grow as endosymbionts within plants, some are attached to the surface of various structures, and a few lead a parasitic existence. Algae also associate with fungi to form lichens.

Zooxanthella symbiosis (p. 549).

Lichen symbiosis (pp. 548–49).

Classification of Algae

According to the five-kingdom system of Whittaker, the algae belong to seven divisions distributed between two different kingdoms (table 26.1). The primary classification of algae is based on cellular, not organismal, properties. Some more important properties include: (1) cell wall (if present) chemistry and morphology; (2) form in which food or assimilatory products of photosynthesis are stored; (3) chlorophyll molecules and accessory pigments that contribute to photosynthesis; (4) flagella number and the location of their insertion in motile cells; (5) morphology of the cells and/or body (thallus); (6) habitat;

TABLE 26.1   Classification of Algae

Division (Common Name)	Kingdom
Chrysophyta (yellow-green and golden-brown algae; diatoms)	*Protista* (single cell; eucaryotic)
Euglenophyta (photosynthetic euglenoid flagellates)	*Protista*
Pyrrhophyta (dinoflagellates)	*Protista*
Charophyta (stoneworts)	*Protista*
Chlorophyta (green algae)	*Protista*
Phaeophyta (brown algae)	*Plantae* (multicellular; eucaryotic)
Rhodophyta (red algae)	*Plantae*

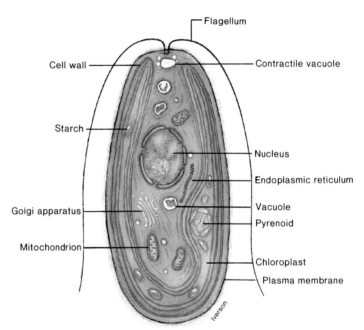

Figure 26.1   Algal Morphology. Schematic drawing of a typical eucaryotic algal cell showing some of its organelles and other structures.

(7) reproductive structures; and (8) life history patterns. Based on these properties the algae are arranged by divisions in table 26.2, which summarizes their more significant characteristics.

Ultrastructure of the Algal Cell

The eucaryotic algal cell (figure 26.1) is surrounded by a thin, rigid cell wall. Some algae have an outer matrix lying outside the cell wall. This usually is flexible and gelatinous, similar to bacterial capsules. When present, the flagella are the locomotor organelles. The nucleus has a typical nuclear envelope with pores; within the nucleus are a nucleolus, chromatin, and karyolymph. The chloroplasts have membrane-bound sacs called thylakoids that carry out the light reactions of photosynthesis. These organelles are embedded in the stroma where the dark

(b)

Figure 26.7 *Phaeophyta* (**Brown Algae**). (*a*) Diagram of the parts of the brown alga, *Nereocystis*. Due to the holdfast organ, the heaviest tidal action and surf seldom dislodge brown algae from their substratum. The stipe is a stalk that varies in length; the bladder is a gas-filled float. (*b*) Giant kelp, *Sargassum*, from which the Sargasso Sea got its name.

Rhodophyta (Red Algae)

The division *Rhodophyta*, the red algae [Greek *rhodon,* rose], includes most of the seaweeds (figure 26.8). A few reds are unicellular but most are filamentous and multicellular. Some red algae are up to 1 m long. The stored food is the carbohydrate called floridean starch (composed of α-1,4 and α-1,6 linked glucose residues).

The red algae contain the red pigment phycoerythrin, one of the two types of phycobilins that they possess. The other accessory pigment is the blue pigment phycocyanin. The presence of these pigments explains how the red algae can live at depths of 100 m or more. The wavelengths of light (green, violet, and blue) that penetrate these depths are not absorbed by chlorophyll *a* but instead by these phycobilins. Not surprisingly the concentrations of these pigments often increases with depth as light intensity decreases. The phycobilins, after absorbing the light energy, pass it on to chlorophyll *a*. The algae appear decidedly red when phycoerythrin predominates over the other pigments. When phycoerythrin undergoes photodestruction in bright light, other pigments predominate and the algae take on shades of blue, brown, and dark green.

The cell walls of most red algae include a rigid inner part composed of microfibrils and a mucilaginous matrix. The matrix is composed of sulfated polymers of galactose called agar, funori, porphysan, and carrageenan. These four polymers give the red algae their flexible, slippery texture. Agar is used extensively in the laboratory as a culture medium component (*see chapter 5*). Many red algae also deposit calcium carbonate in their cell walls and play an important role in building coral reefs.

Pyrrhophyta (Dinoflagellates)

The *Pyrrhophyta* or **dinoflagellates** are unicellular, photosynthetic protistan algae. Most dinoflagellates are marine, but some live in fresh water. Along with the chrysophytes and diatoms,

Figure 26.8 *Rhodophyta* (**Red Algae**). These algae (e.g., *Corallina gracilis*) are much smaller and more delicate than the brown algae. Most red algae have a filamentous, branched morphology as seen here.

the dinoflagellates make up a large part of freshwater and marine plankton and are at the base of many food chains. Species of *Noctiluca, Pyrodinium, Gonyaulax,* and other genera can produce light and are responsible for much of the luminescence (phosphorescence) seen in ocean waters at night. Sometimes dinoflagellate populations reach such high levels that poisonous red tides result (Box 26.2).

The flagella, protective coats (plates), and biochemistry of the dinoflagellates are distinctive. Many dinoflagellates are armored or clad in stiff, patterned, cellulose plates or thecae, which may become encrusted with silica (figure 26.9). Most

Box 26.2

Toxic Algal Blooms

The Bible reports that the first plague Moses visited on the Egyptians was a blood-red tide that killed fish and fouled water. The Red Sea probably is named after these toxic algal blooms. Thousands of years later we still have problems with this plague.

The poisonous and destructive **red tides** that occur frequently in coastal areas often are associated with population explosions, or "blooms," of dinoflagellates. *Gymnodinium* and *Gonyaulax* species are the dinoflagellates most often involved. The pigments in the dinoflagellate cells are responsible for the red color of the water. Under these bloom conditions, the dinoflagellates produce a powerful neurotoxin called saxitoxin. The toxin paralyzes the striated respiratory muscles in many vertebrates by inhibiting sodium transport, which is essential to the function of their nerve cells. The toxin does not harm the shellfish that feed on the dinoflagellates. However, the shellfish do accumulate the toxin and are themselves highly poisonous to organisms, such as humans, who consume the shellfish, resulting in a condition known as **paralytic shellfish poisoning** or neurotoxic shellfish poisoning. Paralytic shellfish poisoning is characterized by numbness of the mouth, lips, face, and extremities. Duration of the illness ranges from a few hours to a few days and usually is not fatal.

Another type of poisoning in humans is called **ciguatera.** It results from eating marine fishes (e.g., grouper, snapper) that have consumed the dinoflagellate *Gambierdiscus toxicus*. The alga's toxin, called ciguatoxin, accumulates in the flesh of fish. This is one of the most powerful toxins known and remains in the flesh even after it has been cooked. Unfortunately it cannot be detected in the fishes and they are not visibly affected. In humans the toxin may cause gastrointestinal disturbances, profuse diarrhea, central nervous system involvement, and respiratory failure.

In 1988 a red tide that has long plagued the Gulf Coast of Florida spread northward to North Carolina. The dinoflagellates released a neurotoxin called brevetoxin, and this prompted state health authorities to shut down all shellfishing for three months. In 1987, in Prince Edward Island, Canada, several people died and hundreds became sick from eating mussels contaminated with domoic acid. The domoic acid was traced to a bloom of diatoms, golden-brown algae once thought to be innocent of all toxicity. The resulting disease, called **amnesic shellfish poisoning,** produces short-term memory loss in its victims. In 1991 pelicans eating anchovies off the California coast were found to be dying from domoic acid poisoning. Shellfish and crab fisheries from Washington to California were closed for several months, resulting in losses reaching hundreds of millions of dollars. In 1993 saxitoxin was found for the first time in crabs from Alaska. Unfortunately there are no treatments for the above types of poisonings. Supportive measures are the only therapy.

Overall, toxic algal blooms are on the rise. No one is certain why this is occurring, but most phycologists believe that the blooms are caused by the continuous pumping of nutrients such as nitrogen and phosphorus into coastal waters. Sewage and agricultural runoff are probably the major sources. Another possibility is world trade: oceangoing ships are unintentionally trafficking in harmful algae, giving the algae a free ride to foreign ports and new habitats in which they can flourish. With people eating more seafood, this toxic menace in the world's oceans will become increasingly more common by the year 2000.

Algae and eutrophication (pp. 840–42).

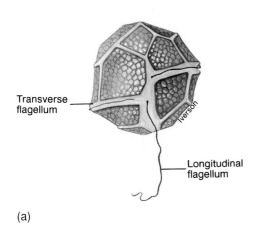

Transverse flagellum

Longitudinal flagellum

(a)

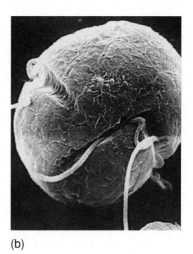

(b)

Figure 26.9   Dinoflagellates. (*a*) *Ceratium.* (*b*) Scanning electron micrograph of *Gymnodinium* (×4,000). Notice the plates of cellulose and the two flagella: one in the transverse groove and the other projecting outward.

have two flagella. In armored dinoflagellates these flagella beat in two grooves that girdle the cell—one transverse (the cingulum), the other longitudinal (the sulcus). The longitudinal flagellum extends posteriorly as a rudder; the flattened, ribbonlike transverse flagellum propels the cell forward while causing it to spin. Due to this spinning, the dinoflagellates received their name from the Greek *dinein,* "to whirl."

Most dinoflagellates have chlorophylls *a* and *c,* in addition to carotenoids and xanthophylls. As a result they usually have a yellowish-green to brown color.

Some dinoflagellates can ingest other cells; others are colorless and heterotrophic. A few even occur as symbionts in many groups of jellyfish, sea anemones, mollusks, and corals. When dinoflagellates form symbiotic relationships, they lose their cellulose plates and flagella, become spherical golden-brown globules in the host cells, and are then termed **zooxanthellae** (*see figure 28.3*a).

1. What is a seaweed?
2. Describe a kelp.
3. Why do red algae appear red?
4. What is unique about the flagellar arrangement in the dinoflagellates?

Summary

1. Phycology (or algology) is the study of algae. Algae are eucaryotic plants or protists that lack roots, stems, and leaves; they have chlorophyll and other pigments for carrying out oxygen-producing photosynthesis (or are closely related to photosynthetic species).

2. Algae are found in almost every environmental niche—in fresh, marine, and brackish water; in certain terrestrial environments; and on moist inanimate objects. They may be endosymbionts, parasites, or components of lichens, and make up a large part of the phytoplankton on the earth.

3. Algal classification is based primarily on cellular properties. Examples include cell wall structure and chemistry, form in which nutrients are stored, types of chlorophyll molecules and accessory pigments, flagella, morphology of the thallus, habitat, reproductive structures, and life history patterns.

4. The vegetative structure of the algae varies from the relative simplicity of a single cell to the more striking complexity of organisms such as the giant kelps. Algae can be unicellular, colonial, filamentous, membranous, or tubular.

5. Both sexual and asexual reproduction occur in the algae. Asexual reproduction includes fragmentation, the production of spores, and binary fission. In sexual reproduction gametes fuse to form a zygote.

6. Green algae (*Chlorophyta*) are a highly diverse group of organisms abundant in the sea, fresh water, and damp terrestrial habitats.

7. Stoneworts/brittleworts (*Charophyta*) have more complex structures than the green algae. They contain whorls of short branches that arise regularly at their nodes. Their gametangia are complex and multicellular. Stoneworts are abundant in fresh to brackish water and are common as fossils.

8. Euglenoids (*Euglenophyta*) have chloroplasts that are biochemically similar to those of the green algae. They have a flexible proteinaceous pellicle inside the plasma membrane.

9. The *Chrysophyta* contain golden-brown algae, yellow-green algae, and diatoms, and vary greatly in pigment composition, cell wall structure, and type of flagellated cells.

10. Brown algae (*Phaeophyta*) are multicellular marine algae, some of which reach lengths of 75 m. The kelps—the largest of the brown algae—contribute greatly to the productivity of the sea as well as to many human needs.

11. Red algae (*Rhodophyta*) have chlorophyll *a* and phycobilins and usually grow at great depths. Some produce agar.

12. Dinoflagellates (*Pyrrhophyta*) are unicellular motile algae responsible for the red tides that are poisonous to many forms of life.

Key Terms

algae *521*

algology *521*

amnetic shellfish poisoning *529*

antheridium *522*

aplanospore *522*

benthic *521*

binary fission *522*

chrysolaminarin *526*

ciguatera *529*

diatom *526*

diatomaceous earth *527*

dinoflagellate *528*

epitheca *526*

euglenoid *524*

fragmentation *522*

frustule *526*

hypotheca *526*

kelp *526*

laminarin *526*

neustonic *521*

oogonium *522*

paralytic shellfish poisoning *529*

pellicle *524*

phycology *521*

phytoplankton *521*

plankton *521*

planktonic *521*

protothecosis *523*

pyrenoid *522*

red tides *530*

scale *526*

spore *522*

stigma (eyespot) *523*

stonewort *523*

thallus *522*

zooplankton *521*

zoospore *522*

zooxanthellae *530*

zygote *522*

Questions for Thought and Review

1. How can algae be distinguished from the photosynthetic bacteria?

2. What characteristics of algae are used as a basis for algal classification?

3. Although multicellularity must have originated many times in the different multicellular groups, how do present-day green algae provide an especially clear example of the probable origin of multicellularity?

4. What characteristics do euglenoids share with higher plants? With animals? Why are they considered protists?

5. Why do marine algae vary so much more in shape and size than those found in fresh water?

6. The phytoplankton of the open oceans (e.g., the Sargasso Sea) are predominantly algae, and these cells form the base of oceanic food chains. Which morphological characteristics of algae are adaptations to a floating (planktonic) habitat?

7. Freshwater algae are distributed worldwide. They rapidly colonize artificial lakes and water impoundments. How do algae accomplish such widespread dispersal?

8. What problem do diatoms have with continued asexual reproduction? How is this problem solved?

9. How are the red algae similar to the cyanobacteria?

10. What are some important characteristics of the green algae? The red algae? The brown algae? The dinoflagellates?

Additional Reading

General

Abbott, I. A., and Dawson, E. Y. 1978. *How to know the seaweeds,* 2d ed. Dubuque, Iowa: Wm. C. Brown Publishers.

Bold, H. C., and Wynne, M. J. 1985. *Introduction to the algae,* 2d ed. Englewood Cliffs, N.J.: Prentice-Hall.

Collins, M. 1978. Algal toxins. *Microbiol. Rev.* 42(3):725–46.

Darley, W. M. 1982. *Algal biology: A physiological approach.* Boston: Blackwell Scientific Publications.

Lee, R. E. 1989. *Phycology,* 2d ed. New York: Cambridge University Press.

Lembi, C. A., and Waaland, J. R., editors. 1988. *Algae and human affairs.* New York: Cambridge University Press.

Noble, R. C. 1990. Death on the half-shell: The health hazards of eating shellfish. *Perspect. Biol. Med.* 33:313–22.

Prescott, G. W. 1978. *How to know the freshwater algae,* 3d ed. Dubuque, Iowa: Wm. C. Brown Publishers.

Round, F. E. 1973. *The biology of algae,* 2d ed. New York: St. Martin's Press.

Scagel, R. F.; Bandoni, R. J.; Maze, J. R.; Rouse, G. E.; Schofield, W. B.; and Stein, J. R. 1982. *Nonvascular plants: An evolutionary survey.* Belmont, Calif.: Wadsworth.

Sze, P. 1993. *A biology of the algae,* 2d ed. Dubuque, Iowa: Wm. C. Brown Publishers.

Trainor, F. R. 1978. *Introductory phycology.* New York: John Wiley and Sons.

Distribution of Algae

Round, F. E. 1984. *The ecology of algae.* New York: Cambridge University Press.

Classification of Algae

Perasso, R. 1989. Origin of the algae. *Nature* 339:142–44.

Ultrastructure of the Algal Cell

Dodge, J. D. 1973. *The fine structure of algal cells.* New York: Academic Press.

Gibor, A. 1966. *Acetabularia:* A useful giant cell. *Sci. Am.* 215:118–24.

Gray, M. W. 1994. One plus one equals one: The making of a crytomonad alga. *ASM News* 60(8):423–27.

Grossman, A.; Schaefer, M.; Chiang, G.; and Collier, J. 1993. The phycobilisome, a light-harvesting complex responsive to environmental conditions. *Microbiol. Rev.* 57(3):725–49.

Jacobs, W. P. 1994. *Caulerpa. Sci. Am.* 271(6):100–05.

Stewart, W. D. P., editor. 1975. *Algal physiology and biochemistry.* Oxford: Blackwell Scientific Publications.

Van Etten, J.; Lane, L.; and Meints, R. 1991. Viruses and viruslike particles of eukaryotic algae. *Microbiol. Rev.* 55(4):586–620.

Characteristics of the Algal Divisions

Anderson, D. 1994. Red tides. *Sci. Am.* 271(2):62–69.

Hallegraeff, G. M. 1993. A review of harmful algal blooms and their apparent global increase. *Phycologia* 32(2):79–99.

Lobban, C. S., and Wynne, M. J., editors. 1981. *The biology of seaweeds.* Oxford: Blackwell Scientific Publications.

McCourt, R. M. 1995. Green algal phylogeny. *Trends Ecol. Evol.* 10(4):159–63.

Pickett-Heaps, J. D. 1975. *Green algae: Structure, reproduction and evolution in selected genera.* Sunderland, Mass.: Sinauer Associates.

Saffo, M. B. 1987. New light on seaweeds. *BioScience* 37:170–80.

Steidinger, K. A., and Haddad, K. 1981. Biologic and hydrographic aspects of red tides. *BioScience* 31:814–19.

Van der Meer, J. P. 1983. The domestication of seaweeds. *BioScience* 33:172–76.

CHAPTER 27
The Protozoa

And a pleasant sight they are indeed. Their shapes range from teardrops to bells, barrels, cups, cornucopias, stars, snowflakes, and radiating suns, to the common amoebas, which have no real shape at all. Some live in baskets that look as if they were fashioned of exquisitely carved ivory filigree. Others use colored bits of silica to make themselves bright mosaic domes. Some even form graceful transparent containers shaped like vases or wine glasses of fine crystal in which they make their homes.
—Helena Curtis

Outline

Concepts

1. Protozoa are protists exhibiting heterotrophic nutrition and various types of locomotion. They occupy a vast array of habitats and niches and have organelles similar to those found in other eucaryotic cells, and also specialized organelles.

2. The protozoa are divided into seven phyla: *Sarcomastigophora, Labyrinthomorpha, Apicomplexa, Microspora, Ascetospora, Myxozoa,* and *Ciliophora.* These phyla represent four major groups: flagellates, amoebae, ciliates, and sporozoa.

3. Protozoa usually reproduce asexually by binary fission. Some have sexual cycles, involving meiosis and the fusion of gametes or gametic nuclei resulting in a diploid zygote. The zygote is often a thick-walled, resistant, and resting cell called a cyst. Some protozoa undergo conjugation in which nuclei are exchanged between cells.

4. All protozoa have one or more nuclei; some have a macro- and micronucleus.

5. Various protozoa feed by holophytic, holozoic, or saprozoic means; some are predatory or parasitic.

hapter 27 presents the major biological features of the protists, known as protozoa. The most important groups are the flagellates, amoebae, sporozoa, and ciliates. These protists demonstrate the great adaptive potential of the basic single eucaryotic cell.

The microorganisms called **protozoa** [s., protozoan; Greek *protos,* first, and *zoon,* animal] are studied in the discipline called **protozoology.** A protozoan can be defined as a usually motile eucaryotic unicellular protist. Protozoa are directly related only on the basis of a single negative characteristic—they are not multicellular. All, however, demonstrate the basic body plan of a single protistan eucaryotic cell.

Distribution

Protozoa grow in a wide variety of moist habitats. Moisture is absolutely necessary for the existence of protozoa because they are susceptible to desiccation. Most protozoa are free living and inhabit freshwater or marine environments. Many terrestrial protozoa can be found in decaying organic matter, in soil, and even in beach sand; some are parasitic in plants or animals.

Importance

Protozoa play a significant role in the economy of nature. For example, they make up a large part of **plankton**—small, free-floating organisms that are an important link in the many aquatic food chains and food webs of aquatic environments. A **food chain** is a series of organisms, each feeding on the preceding one. A **food web** is a complex interlocking series of food chains. Protozoa are also useful in biochemical and molecular

biological studies. Many biochemical pathways used by protozoa are present in all eucaryotic cells. Finally, some of the most important diseases of humans (*see chapter 39*) and animals (table 27.1) are caused by protozoa.

Microbial populations in natural environments (pp. 811–13).

Morphology

Because protozoa are eucaryotic cells, in many respects their morphology and physiology are the same as the cells of multicellular animals (*see figures 4.2 and 4.3*). However, because all of life's various functions must be performed within the individual protozoan, some morphological and physiological features are unique to protozoan cells. In some species the cytoplasm immediately under the plasma membrane is semisolid or gelatinous, giving some rigidity to the cell body. It is termed the **ectoplasm.** The bases of the flagella or cilia and their associated fibrillar structures are embedded in the ectoplasm. The plasma membrane and structures immediately beneath it are called the **pellicle.** Inside the ectoplasm is the area referred to as the **endoplasm,** which is more fluid and granular in composition and contains most of the organelles. Some protozoa have one nucleus, others have two or more identical nuclei. Still other protozoa have two distinct types of nuclei—a macronucleus and one or more micronuclei. The **macronucleus,** when present, is typically larger and associated with trophic activities and regeneration processes. The **micronucleus** is diploid and involved in both genetic recombination during reproduction and the regeneration of the macronucleus.

One or more vacuoles are usually present in the cytoplasm of protozoa. These are differentiated into contractile, secretory, and food vacuoles. **Contractile vacuoles** function as osmoregulatory organelles in those protozoa that live in a hypotonic environment, such as a freshwater lake. Osmotic balance is maintained by

TABLE 27.1   Pathogenic Protozoa That Cause Major Diseases of Domestic Animals

Protozoan Group	Genus	Host	Preferred Site of Infection	Disease
Amoebae	*Entamoeba*	Mammals	Intestine	Amebiasis
	Iodamoeba	Swine	Intestine	Enteritis
Sporozoa	*Babesia*	Cattle	Blood cells	Babesiosis
	Theileria	Cattle, sheep, goats	Blood cells	Theileriasis
	Sarcocystis	Mammals, birds	Muscles	Sarcosporidiosis
	Toxoplasma	Cats	Intestine	Toxoplasmosis
	Isospora	Dogs	Intestine	Coccidiosis
	Eimeria	Cattle, cats, chickens, swine	Intestine	Coccidiosis
	Plasmodium	Many animals	Bloodstream, liver	Malaria
	Leucocytozoon	Birds	Spleen, lungs, blood	Leucocytozoonosis
	Cryptosporidium	Mammals	Intestine	Cryptosporidiosis
Ciliates	*Balantidium*	Swine	Large intestine	Balantidiasis
Flagellates	*Leishmania*	Dogs, cats, horses, sheep, cattle	Spleen, bone marrow, mucous membranes	Leishmaniasis
	Trypanosoma	Most animals	Blood	Trypanosomiasis
	Trichomonas	Horses, cattle	Genital tract	Trichomoniasis (abortion)
	Histomonas	Birds	Intestine	Blackhead disease
	Giardia	Mammals	Intestine	Giardiasis

continuous water expulsion. Most marine protozoa and parasitic species are isotonic to their environment and lack such vacuoles. **Phagocytic vacuoles** are conspicuous in holozoic and parasitic species and are the sites of food digestion (*see figure 4.9*). **Secretory vacuoles** usually contain specific enzymes that perform various functions (such as excystation).

Most anaerobic protozoa (such as *Trichonympha,* which lives in the gut of termites; *see figure 28.1*b) have no mitochondria, no cytochromes, and an incomplete tricarboxylic acid cycle. However, some do have small, membrane-delimited organelles termed **hydrogenosomes.** These structures contain a unique electron transfer pathway in which hydrogenase transfers electrons to protons (which act as the terminal electron acceptors), and molecular hydrogen is formed.

1. Describe a typical protozoan.
2. What roles do protozoa play in the trophic structure of their communities and in the organisms with which they associate?
3. What is unique about the nuclei of some protozoa?
4. Where can protozoa be found?
5. What are the functions of contractile, phagocytic, and secretory vacuoles?

Nutrition

Most protozoa are chemoheterotrophic. Two types of heterotrophic nutrition are found in the protozoa: holozoic and saprozoic. In **holozoic nutrition,** solid nutrients such as bacteria are acquired by phagocytosis and the subsequent formation of phagocytic vacuoles. Some ciliates have a specialized structure for phagocytosis called the **cytostome** (cell mouth). In **saprozoic nutrition,** soluble nutrients such as amino acids and sugars cross the plasma membrane by pinocytosis, diffusion, or carrier-mediated transport (facilitated diffusion or active transport).

Encystment and Excystment

Many protozoa are capable of **encystation.** They develop into a resting stage called a **cyst,** which is a dormant form marked by the presence of a wall and by the reduction of metabolic activity to a very low level. Cyst formation is particularly common among aquatic, free-living protozoa and parasitic forms. Cysts serve three major functions: (1) they protect against adverse changes in the environment, such as nutrient deficiency, desiccation, adverse pH, and low partial pressure of O_2; (2) they are

Figure 27.1 **Protozoan Reproduction.** Binary fission in *Paramecium caudatum* (×100).

sites for nuclear reorganization and cell division (reproductive cysts); and (3) they serve as a means of transfer between hosts in parasitic species.

Although the exact stimulus for **excystation** (escape from the cysts) is unknown, excystation generally is triggered by a return to favorable environmental conditions. For example, cysts of parasitic species excyst after ingestion by the host and form the vegetative form called the **trophozoite.**

Locomotory Organelles

A few protozoa are nonmotile. Most, however, can move by one of three major types of locomotory organelles: pseudopodia, flagella, or cilia. **Pseudopodia** [s., pseudopodium; false feet] are cytoplasmic extensions found in the amoebae that are responsible for the movement and food capture. There are many types of pseudopodia. Flagellates and ciliates move by flagella and cilia. Electron microscopy has shown that protozoan flagella and cilia are structurally the same and identical in function to those of other eucaryotic cells (*see figures 4.23–4.27*).

Reproduction

Most protozoa reproduce asexually, and some also carry out sexual reproduction. The most common method of asexual reproduction is **binary fission.** During this process the nucleus first undergoes mitosis, then the cytoplasm divides by cytokinesis to form two identical individuals (figure 27.1).

Figure 27.2   Conjugation in *Paramecium caudatum*, Schematic Drawing. Follow the arrows. After the conjugants separate, only one of the exconjugants is followed; however, a total of eight new protozoa result from the conjugation.

The most common method of sexual reproduction is **conjugation.** In this process there is an exchange of gametes between paired protozoa of complementary mating types (**conjugants;** *see figure 2.12*b). Conjugation is most prevalent among ciliate protozoa. A well-studied example is *Paramecium caudatum* (figure 27.2). At the beginning of conjugation, two ciliates unite, fusing their pellicles at the contact point. The macronucleus in each is degraded. The individual micronuclei divide twice by meiosis to form four haploid pronuclei, three of which disintegrate. The remaining pronucleus divides again mitotically to form two gametic nuclei, a stationary one and a migratory one. The migratory nuclei pass into the respective conjugates. Then the ciliates separate, the gametic nuclei fuse, and the resulting diploid zygote nucleus undergoes three rounds of mitosis. The eight resulting nuclei have different fates: one nucleus is retained as a micronucleus; three others are destroyed; and the four remaining nuclei develop into macronuclei. Each separated conjugant now undergoes cell division. Eventually progeny with one macronucleus and one micronucleus are formed.

1. What specific nutritional types exist among protozoa?
2. What functions do cysts serve for a typical protozoan? What causes excystation to occur?
3. What is a pseudopodium?
4. How do protozoa reproduce asexually?
5. How do protozoa reproduce sexually? Describe the process of ciliate conjugation.

Classification

Protozoologists now regard the *Protozoa* as a subkingdom, which contains seven of the 14 phyla found within the kingdom *Protista* (table 27.2). In 1993 T. Cavalier-Smith proposed that the protozoa be elevated to kingdom status with 18 phyla (*see chapter 19, p. 410*). The acceptance of this new classification by protozoologists, however, remains to be determined. The phylum *Sarcomastigophora* consists of flagellates and amoebae with a single type of nucleus. The phyla *Labyrinthomorpha, Apicomplexa, Microspora, Ascetospora,* and *Myxozoa* have either saprozoic or parasitic species. The phylum *Ciliophora* has ciliated protozoa with two types of nuclei. The classification of this subkingdom into phyla is based primarily on types of nuclei, mode of reproduction, and mechanism of locomotion.

Representative Types

This section describes some representatives of each group of protozoan protists to present an overview of protozoan diversity and to provide a basis for comparing different groups.

Phylum *Sarcomastigophora*

Protists that have a single type of nucleus and possess flagella (subphylum *Mastigophora*) or pseudopodia (subphylum *Sarcodina*) are placed in the phylum *Sarcomastigophora*. Both sexual and asexual reproduction are seen in this phylum.

The subphylum *Mastigophora* contains both phytoflagellates, chloroplast-bearing flagellates and close relatives, and **zooflagellates.** Zooflagellates do not have chlorophyll and are either holozoic, saprozoic, or symbiotic. Asexual reproduction occurs by longitudinal binary fission along the major body axis. Sexual reproduction is known for a few species, and encystment is common. Zooflagellates are characterized by the presence of one or more flagella. Most members are uninucleate. One major group, the kinetoplastids, has its mitochondrial DNA in a special region called the **kinetoplast** (figure 27.3a; *see also figures 2.12c and 4.12*).

Some zooflagellates are free living. The choanoflagellates are a distinctive example in that they have one flagellum, are solitary or colonial, and are on stalks. Other zooflagellates form symbiotic relationships. For example, *Trichonympha* species (*see figure 28.1b*) are found in the intestine of termites and produce enzymes that the termite needs to digest the wood particles on which it feeds.

The protozoan-termite relationship (pp. 547–48).

Many zooflagellates are important human parasites. *Giardia lamblia* (*see figure 39.18*) can be found in the human intestine where it may cause severe diarrhea. It is transmitted through water that has been contaminated with feces (*see chapters 39 and 41*). Trichomonads, such as *Trichomonas vaginalis,* live in the vagina and urethra of women and in the prostate, seminal vesicles, and urethra of men. They are transmitted primarily by sexual intercourse.

Giardiasis (pp. 801–2).
Trichomoniasis (pp. 806–7).

The zooflagellates called **trypanosomes** are important blood pathogens of humans and animals in certain parts of the world. Because they live in the blood, they are also called hemoflagellates. These parasites (figure 27.3a; *see also figure 4.1c*) have a typical zooflagellate structure. A major human trypanosomal disease is African sleeping sickness caused by *Trypanosoma brucei rhodesiense* or *T. brucei gambiense* (*see chapter 39*).

Trypanosomiasis (pp. 805–6).

The subphylum *Sarcodina* contains the amoeboid protists. They are found throughout the world in both fresh and salt water and are abundant in the soil. Several species are parasites of mammals. Simple amoebae [s., amoeba] move almost continually using their pseudopodia (**amoeboid movement**). Many have no definite shape, and their internal structures (figure 27.3b) occupy no particular position. The single nucleus, contractile and phagocytic vacuoles, and ecto- and endoplasm shift as the amoebae move. Amoebae engulf a variety of materials (small algae, bacteria, other protozoa) through phagocy-

TABLE 27.2 Current Abbreviated Classification of the Subkingdom *Protozoa*

Taxonomic Group	Characteristics	Examples
Phylum: *Sarcomastigophora*	Locomotion by flagella, pseudopodia, or both; when present, sexual reproduction is essentially syngamy (union of gametes external to the parents); single type of nucleus	
Subphylum: *Mastigophora*	One or more flagella; division by longitudinal binary fission; sexual reproduction in some groups	
Class: *Zoomastigophorea*	Chromatophores absent; one to many flagella; amoeboid forms, with or without flagella; sexuality known in some groups; mainly parasitic	*Trypanosoma* *Giardia* *Trichomonas* *Leishmania* *Trichonympha*
Subphylum: *Sarcodina*	Locomotion primarily by pseudopodia; shells (tests) often present; flagella restricted to reproductive stages when present; asexual reproduction by fission; mostly free living	
Superclass: *Rhizopoda*	Locomotion by pseudopodia or by protoplasmic flow with discrete pseudopodia; some contain tests	*Amoeba* *Elphidium* *Coccodiscus*
Phylum: *Labyrinthomorpha*	Spindle-shaped cells capable of producing mucous tracks; trophic stage as ectoplasmic network; nonamoeboid cells; saprozoic and parasitic on algae and seagrass	*Labyrinthula*
Phylum: *Apicomplexa*	All members have a spore-forming stage in their life cycle; contain an apical complex; sexuality by syngamy; all species parasitic; cysts often present; cilia absent; often called the Sporozoa	*Plasmodium* *Toxoplasma* *Eimeria* *Cryptosporidium*
Phylum: *Microspora*	Unicellular spores with spiroplasm containing polar filaments; obligatory intracellular parasites	*Nosema*
Phylum: *Ascetospora*	Spore with one or more spiroplasms; no polar capsules or polar filaments; all parasitic in invertebrates	*Haplosporidium*
Phylum: *Myxozoa*	Spores of multicellular origin; one or more polar capsules; all parasitic, especially in fish	*Myxosoma*
Phylum: *Ciliophora*	Simple cilia or compound ciliary organelles in at least one stage in the life cycle; two types of nuclei; contractile vacuole present; binary fission transverse; sexuality involving conjugation; most species free living, but many commensal, some parasitic	*Didinium* *Stentor* *Vorticella* *Tetrahymena* *Paramecium* *Tokophrya* *Entodinium* *Nyctotherus* *Balantidium* *Ichthyophthirius*

tosis. Some material moves into and out of the plasma membrane by pinocytosis. Reproduction in the amoebae is by simple asexual binary fission. Some amoebae can form cysts.

Many free-living forms are more complex than simple amoebae. *Arcella* manufactures a loose-fitting shell or **test** for protection (figure 27.4*a*). These amoebae extend their pseudopodia from the test aperture to either feed or creep along. The foraminiferans and radiolarians primarily are marine amoebae, with a few occurring in fresh and brackish water. Most foraminiferans live on the seafloor, whereas radiolarians are usually found in the open sea (Box 27.1). Foraminiferan tests and radiolarian skeletons have many unique and beautiful shapes (figure 27.4*b,c*). They range in diameter from about 20 μm to several cm.

Finally, there are many symbiotic amoebae, most of which live in other animals. Two common genera are *Endamoeba* and *Entamoeba*. *Endamoeba blattae* is common in the intestine of cockroaches, and related species are present in termites. *Entamoeba histolytica* (see figure 39.17) is an important parasite of humans, in whom it often produces severe amebic dysentery, which may be fatal. Free-living amoebae from two genera, *Naegleria* and *Acanthamoeba*, can cause disease in humans and other mammals (see chapter 39).

Amebiasis (pp. 799–801).

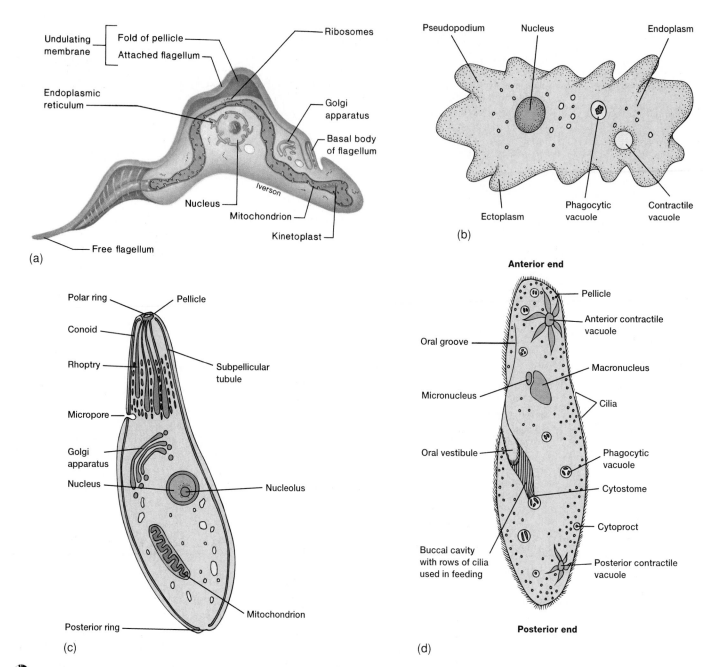

Figure 27.3   Drawings of Some Representative Protozoa. (*a*) Structure of the flagellate, *Trypanosoma brucei rhodesiense.* (*b*) The structure of the amoeboid protist, *Amoeba proteus.* (*c*) Structure of an apicomplexan sporozoite. (*d*) Structure of the ciliate *Paramecium caudatum.*

1. What characteristics would be exhibited by a protozoan that belongs to the phylum *Sarcomastigophora?* What two subphyla does it contain?
2. How would you characterize a zooflagellate? An amoeba?
3. What two human diseases are caused by zooflagellates?
4. Where can two different symbiotic amoebae be found?

Phylum *Labyrinthomorpha*

The very small phylum *Labyrinthomorpha* consists of protists that have spindle-shaped or spherical nonamoeboid vegetative cells. In some genera, amoeboid cells move within a network of mucous tracks using a typical gliding motion. Most members are marine and either saprozoic or parasitic on algae. Several years ago *Labyrinthula* killed most of the "eel grass" on the Atlantic coast, depriving ducks of their food and starving many of them.

(a)

(b)

(c)

Figure 27.4   Some Free-Living Sarcodines. (*a*) An illustration of *Arcella,* showing the test or shell that is made of chitinlike material secreted by the protist. (*b*) The test of the foraminiferan, *Elphidium cristum* (×100). (*c*) A group of siliceous radiolarian shells, light micrograph (×63).

Box 27.1

The Importance of Foraminiferans

Of over 40,000 described species of foraminiferans, about 90% are fossil. During the Tertiary period (about 230 million years ago), the foraminiferans contributed massive shell accumulations to geologic formations. They were so abundant that they formed thick deposits, which became uplifted over time and exposed as great beds of limestone in Europe, Asia, and Africa. The White Cliffs of Dover, the famous landmark of southern England, are made up almost entirely of foraminiferan shells. The Egyptian pyramids of Gizeh, near Cairo, are built of foraminiferan limestone. Currently, foraminiferans are important aids to geologists in identifying and correlating rock layers as they search for oil-bearing strata. The calcareous shells of abundant planktonic foraminiferans are today settling and accumulating over much of the ocean floor as thick deposits called "globigerina ooze"—limestone of the distant future.

Phylum *Apicomplexa*

The **apicomplexans,** often collectively called the sporozoans, have a spore-forming stage in their life cycle and lack special locomotory organelles (except in the male gametes, and the zygote or ookinete). They are either intra- or intercellular parasites of animals and are distinguished by a unique arrangement of fibrils, microtubules, vacuoles, and other organelles, collectively called the apical complex, which is located at one end of the cell.

The **apical complex** contains several components (figure 27.3*c*). One or two polar rings are at the apical end. The **conoid** consists of a cone of spirally arranged fibers lying next to the polar rings. Subpellicular microtubules radiate from the polar rings and probably serve as support elements. Two or more **rhoptries** extend to the plasma membrane and secrete their contents at the cell surface. These secretions aid in the penetration of the host cell. One or more micropores are thought to function in the intake of nutrients.

Apicomplexans have complex life cycles in which certain stages occur in one host (the mammal) and other stages in a different host (often a mosquito). The life cycle has both asexual and sexual phases and is characterized by an alternation of haploid and diploid generations. At some point an asexual reproduction process called schizogony occurs. **Schizogony** is a rapid series of mitotic events producing many small infective organisms through the formation of uninuclear buds. Sexual reproduction involves the fertilization of a large female macrogamate by a small, flagellated male gamete. The resulting zygote becomes a thick-walled cyst called an **oocyst.** Within the oocyst, meiotic divisions produce infective haploid spores.

The three most important sporozoan parasites are *Plasmodium* (the causative agent of malaria), *Toxoplasma* (the causative agent of toxoplasmosis), and *Eimeria* (the causative agent of coccidiosis).

Malaria (pp. 802–4).

Phylum *Microspora*

The small microsporans (3 to 6 μm) are obligatory intracellular parasites. Included in these protozoa are several species of some economic importance because they parasitize beneficial insects. *Nosema bombycis* parasitizes silkworms (figure 27.5) causing the disease **pebrine,** and *Nosema apis* causes serious dysentery (foul brood) in honeybees. There has been an increased interest in these parasites because of their possi-

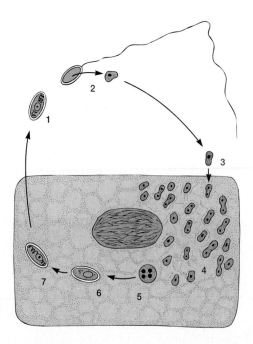

Figure 27.5   The Microsporean *Nosema bombycis*, Which is Fatal to Silkworms. (*1*) A typical spore with one coiled filament. (*2*) When ingested, it extrudes the filament. (*3*) The parasite enters an epithelial cell in the intestine of the silkworm and (*4*) divides many times to form small amoebae that eventually fill the cell and kill it. During this phase, some of the amoebae with four nuclei become spores (*5, 6, 7*). Silkworms are infected by eating leaves contaminated by the feces of infected worms.

ble role as biological control agents for certain insects. For example, *Nosema locustae* has been approved and registered by the United States Environmental Protection Agency for use in residual control of rangeland grasshoppers. Recently five microsporidian genera (*Nosema, Encephalitozoon, Pleistophora, Microsporidium,* and *Enterocytozoon*) have been implicated in human diseases with immunosuppressed and AIDS patients.

Phylum *Ascetospora*

Ascetospora is a relatively small phylum that consists exclusively of parasitic protists characterized by spores lacking polar caps or polar filaments. Ascetosporans such as *Haplosporidium* are parasitic primarily in the cells, tissues, and body cavities of mollusks.

Phylum *Myxozoa*

The myxozoans are all parasitic, most on freshwater and marine fish. They have a resistant spore with one to six coiled polar filaments. The most economically important myxozoan is *Myxosoma cerebralis,* which infects the nervous system and auditory organ of trout and salmon (salmonids). Infected fish lose their sense of balance and tumble erratically—thus the name whirling or tumbling disease. **Proliferative kidney disease,** caused by an unclassified myxozoan, has become one of the most important diseases of cultured salmon throughout the world.

---

1. Describe a typical apicomplexan (sporozoan) protist, including its apical complex.
2. Summarize the sporozoan life cycle. What is schizogony?
3. What are the four most important sporozoan parasites and the diseases they cause?
4. Give one economically important disease caused by a microsporan. What group of animals do myxosporans usually parasitize?

Phylum *Ciliophora*

The phylum *Ciliophora* is the largest of the seven protozoan phyla. There are about 8,000 species of these unicellular, heterotrophic protists that range from about 10 to 3,000 μm long. As their name implies, ciliates employ many cilia as locomotory organelles. The cilia are generally arranged either in longitudinal rows (figure 27.3d; *see also figure 4.26*) or in spirals around the body of the organism. They beat with an oblique stroke; therefore the protist revolves as it swims. Coordination of ciliary beating is so precise that the protist can go either forward or backward.

There is great variation in ciliate shape, and most do not look like the slipper-shaped *Paramecium* (*see figures 2.8e and 4.1a*). In some species (*Vorticella*) the protozoan attaches itself to the substrate by a long stalk. *Stentor* attaches to a substrate and stretches out in a trumpet shape to feed (*see figure 4.1f*). A few species have tentacles for the capture of prey. Some can discharge toxic threadlike darts called toxicysts, which are used in capturing prey.

A most striking feature of ciliates is their ability to capture many particles in a short time by the action of the cilia around the buccal cavity. Food first enters the cytostome and passes into phagocytic vacuoles that fuse with lysosomes after detachment from the cytostome. A vacuole's contents are digested when the vacuole is acidified and lysosomes release digestive enzymes into it. After the digested material has been absorbed into the cytoplasm, the vacuole fuses with a special region of the pellicle called the **cytoproct** and empties its waste material to the outside. Contractile vacuoles are used for osmoregulation and are present chiefly in freshwater species.

Solutes and water activity (pp. 124–25).

Most ciliates have two types of nuclei: a large macronucleus and a smaller micronucleus. The micronucleus is diploid and contains the normal somatic chromosomes. It divides by mitosis and transmits genetic information through meiosis and sexual reproduction. Macronuclei are derived from micronuclei by a complex series of steps. Within the macronucleus are many chromatin bodies, each containing many copies of only one or two genes. Macronuclei are thus polyploid and divide by elongating and then by constricting. They produce mRNA to direct protein synthesis, maintain routine cellular functions, and control normal cell metabolism.

Some ciliates reproduce asexually by transverse binary fission, forming two equal daughter protozoa. Many ciliates also reproduce by conjugation as previously described.

Although most ciliates are free living, symbiotic forms do exist. Some ciliated protozoa live as harmless commensals. For example, *Entodinium* is found in the rumen of cattle, and *Nyctotherus* occurs in the colon of frogs. Other ciliates are strict parasites. For example, *Balantidium coli* lives in the intestine of mammals, including humans, where it can produce dysentery. *Ichthyophthirius* lives in fresh water where it can attack many species of fish, producing a disease known as "ick."

---

1. Describe the morphology of a typical ciliated protozoan.
2. Describe the food-gathering structures found in the ciliated protozoa.
3. In ciliates, what is the function of the macronucleus? The micronucleus?
4. How do ciliates reproduce?
5. Where can the following ciliates be found: *Entodinium, Nyctotherus, Balantidium, Ichthyophthirius?*

Summary

1. Protozoa are protists that can be defined as usually motile eucaryotic unicellular microorganisms.

2. Protozoa are found wherever other organisms exist. They are important components of food chains and food webs. Many are parasitic in humans and animals, and some have become very useful in the study of molecular biology.

3. Because protozoa are eucaryotic cells, in many respects their morphology and physiology resemble those of multicellular animals. However, because all their functions must be performed within the individual protist, many morphological and physiological features are unique to protozoan cells.

4. Some protozoa can secrete a resistant covering and go into a resting stage (encystation) called a cyst. Cysts protect the organism against adverse environments, function as a site for nuclear reorganization, and serve as a means of transmission in parasitic species.

5. Protozoa move by one of three major types of locomotory organelles: pseudopodia, flagella, or cilia. Some have no means of locomotion.

6. Most protozoa reproduce asexually, some use sexual reproduction, and some employ both methods.

7. There are seven protozoan phyla. The phylum *Sarcomastigophora* is characterized by protists that have a single type of nucleus, possess flagella (subphylum *Mastigophora*), pseudopodia (subphylum *Sarcodina*), or both types of locomotory organelles.

8. The subphylum *Sarcodina* consists of the amoeboid protists. They are found throughout the world in both fresh and salt water and in the soil. Some species are parasitic.

9. The phylum *Labyrinthomorpha* contains protists that have spindle-shaped or spherical nonamoeboid vegetative cells. Most members are marine and either saprozoic or parasitic on algae.

10. The phylum *Apicomplexa* consists of sporozoan protists that possess an apical complex, a unique arrangement of fibrils, microtubules, vacuoles, and other organelles at one end of the cell. Representative members include the *Plasmodium* parasites, which cause malaria;

Toxoplasma, which causes toxoplasmosis; and *Eimeria,* the agent of coccidiosis.

11. The phylum *Microspora* consists of very small protists that are intracellular parasites of every major animal group. They are transmitted from one host to the next as a spore, the form from which the group obtains its name.

12. The phylum *Ascetospora* contains protists that produce spores lacking polar capsules. These protists are primarily parasitic in mollusks.

13. The phylum *Myxozoa* consists entirely of parasitic species, usually found in fish. The spore is characterized by one to six polar filaments.

14. The phylum *Ciliophora* comprises a group of protists that have cilia and two types of nuclei. Conjugation in the *Ciliophora* is a form of sexual reproduction that involves exchange of micronuclear material.

Key Terms

amoeboid movement *536*

apical complex *540*

apicomplexan *540*

binary fission *534*

conjugant *535*

conjugation *535*

conoid *540*

contractile vacuole *533*

cyst *534*

cytoproct *541*

cytostome *534*

ectoplasm *533*

encystation *534*

endoplasm *533*

excystation *534*

food chain *533*

food web *533*

holozoic nutrition *534*

hydrogenosomes *534*

kinetoplast *536*

macronucleus *533*

micronucleus *533*

oocyst *540*

pebrine *540*

pellicle *533*

phagocytic vacuoles *534*

plankton *533*

proliferative kidney disease *541*

protozoa *533*

protozoology *533*

pseudopodia *534*

rhoptry *540*

saprozoic nutrition *534*

schizogony *540*

secretory vacuole *534*

test *537*

trophozoite *534*

trypanosome *536*

zooflagellate *536*

Questions for Thought and Review

1. What is the economic impact or human relevance of the protozoa?

2. What criteria are now used in the classification of protozoa?

3. Seven protozoan phyla have been discussed. What are their distinguishing characteristics?

4. What are some typical organelles found in the protozoa?

5. What advantage is there to those protozoa that are capable of forming a cyst?

6. How do protozoa move? Reproduce?

7. If the diversity within the seven protozoan phyla is considered, which one shows the greatest evolutionary advancement? Defend your answer.

8. The protozoa are said to be grouped together for a negative reason. What does this mean?

9. Describe how DNA is distributed to daughter cells when the ciliate *Paramecium* divides. Include a discussion of both conjugation and binary fission.

10. How does a protozoan cyst differ from a bacterial endospore?

Additional Reading

General

Corliss, J. O. 1991. *Microscopic anatomy of the invertebrates,* In *Protozoa,* vol. 2, New York: Wiley-Liss.

Grell, K. G. 1973. *Protozoology.* Heidelberg: Springer-Verlag.

Jahn, T. L.; Bovee, E. C.; and Jahn, F. F. 1979. *How to know the protozoa.* Dubuque, Iowa: Wm. C. Brown.

Krier, J. P. 1978. *Parasitic protozoa.* New York: Academic Press.

Kudo, R. R. 1966. *Protozoology,* 5th ed. Springfield, Ill.: Charles C. Thomas.

Laybourn-Parry, J. 1984. *A functional biology of free-living protozoa.* Berkeley: University of California Press.

Lee, J. J.; Hunter, S. H.; and Bovee, E. C. 1985. *An illustrated guide to the protozoa.* Lawrence, Kan.: Allen Press, Society of Protozoologists.

Levandowsky, M., and Hunter, S. E., editors. 1979. *Biochemistry and physiology of protozoa.* New York: Academic Press.

Margulis, L.; Corliss, J. O.; and Melkonian, M. 1990. *Handbook of Protoctista.* Boston: Jones and Bartlett.

Sleigh, M. 1978. *The biology of protozoa.* Amsterdam and New York: Elsevier Biomedical Press.

Distribution

Fenchel, T. 1987. *Ecology of protozoa.* New York: Springer-Verlag.

Morphology

Dunelson, J. E., and Turner, M. J. 1985. How the trypanosome changes its coat. *Sci. Am.* 252(2):44–51.

McFadeen, G.; Gilson, P.; Hofmann, G.; Adcock, G.; and Maier, U.-G. 1994. Evidence that an amoeba acquired a chloroplast by retaining part of an engulfed eukaryotic alga. *Proc. Natl. Acad. Sci.* 91:3690–94.

Muller, M. 1975. Biochemistry of protozoan microbodies: Peroxisomes, α-glycerophosphate oxidase bodies, hydrogenosomes. *Ann. Rev. Microbiol.* 29:467–83.

Prescott, D. M. 1994. The DNA of ciliated protozoa. *Microbiol. Rev.* 58(2):233–67.

Vanhamme, L., and Pays, E. 1995. Control of gene expression in trypanosomes. *Microbiol. Rev.* 59(2):223–40.

Nutrition

Barker, J., and Brown, M. R. W. 1994. Trojan horses of the microbial world: Protozoa and the survival of bacterial pathogens in the environment. *Microbiology* 140:1253–59.

Rudzinska, M. A. 1973. Do suctoria really feed by suction? *BioScience* 23(2):87–94.

Classification

Cavalier-Smith, T. 1993. Kingdom protozoa and its 18 phyla. *Microbiol. Rev.* 57(4):953–94.

Lee, R. E., and Kugrens, P. 1992. Relationship between the flagellates and the ciliates. *Microbiol. Rev.* 56(4):529–42.

Levine, N. D., et al. 1980. A newly revised classification of the protozoa. *J. Protozool.* 27(1):37–58.

Representative Types

Adam, R. 1991. The biology of *Giardia* spp. *Microbiol. Rev.* 55(4):706–32.

Band, R. D., editor. 1983. Symposium—the biology of small amoebae. *J. Protozool.* 30:192–214.

Bonner, J. T. 1969. Hormones in social amoebae and mammals. *Sci. Am.* 220:78–91.

Clew, H. R.; Saha, A. K.; Siddhartha, D.; and Remaley, A. T. 1988. Biochemistry of *Leishmania* species. *Microbiol. Rev.* 52(4):412–32.

Corliss, J. O. 1979. *The ciliated protozoa: Characterization, classification and guide to the literature,* 2d ed. New York: Pergamon Press.

Wichterman, R. 1986. *The biology of Paramecium,* 2d ed. New York: Plenum Press.

Wolfe, M. 1992. Giardiasis. *Clin. Microbiol. Rev.* 5(1):93–100.

Part Seven
The Nature of Symbiotic Associations

Lichens are one of the best known and most successful symbiotic associations. These colorful and attractive British soldier lichens (*Cladonia cristatella*) are an excellent example of a fruticose lichen. In this photograph, the fungal partner is producing a bright red ascocarp.

CHAPTER 28
Symbiotic Associations: Commensalism, Mutualism, and Normal Microbiota of the Human Body

Symbiosis, particularly

parasitism, is frequently

regarded in distasteful terms

by the hygiene-conscious

citizen. We have a tendency

to think of it as a peculiar and

abnormal association of some

lower organism with a higher

one. There is an element of

snobbishness in such a view,

which must be quickly

abandoned when a discerning

look is taken of the living

world. In nature there is

probably no such thing as a

symbiote-free organism. The

phenomenon of symbiosis is

quite as common as life itself.
—Clark P. Reed

Outline

Concepts

1. Symbiosis is the living together in close association of two or more dissimilar organisms.
2. Commensalism is a relationship in which one organism benefits while the other is neither harmed nor helped.
3. Mutualism is the living together of two or more organisms in an association in which both members benefit.
4. Germfree animals or those that live in association with a few known microorganisms are gnotobiotic.
5. Most microorganisms associated with the human body are bacteria; they normally prefer to colonize specific sites.

S ymbiosis is the living together in close association of two or more dissimilar organisms. This chapter focuses on these associations, emphasizing two of the three types of symbiosis: commensalism and mutualism. Parasitism, the third type of symbiosis, is covered in chapter 29. Within the present chapter examples are given that illustrate each specific symbiotic relationship. However, it also is possible experimentally to have no microorganisms on or in an animal, a condition termed germfree or gnotobiotic life. This special case is discussed. The chapter closes with an overview of the normal microorganisms associated with the human body.

Many microorganisms live much of their lives in a special ecological relationship: an important part of their environment is a member of another species. Any microorganism that spends a portion or all of its life associated with another organism of a different species is called a **symbiont** (or **symbiote**), and the relationship is designated as **symbiosis** [Greek *sym*, together, and *bios*, life].

Types of Symbiosis, Functions, and Examples

There are three types of symbiotic relationships: commensalism, mutualism, and parasitism. Within each category the association may be either ectosymbiotic or endosymbiotic. In **ectosymbiosis** one organism remains outside the other. In **endosymbiosis** one organism is present within the other.

Commensalism

Commensalism [Latin *com*, together, and *mensa*, table] is a relationship in which one organism, the **commensal**, benefits while the other, the host, is neither harmed nor helped. Often both the host and the commensal "eat at the same table." The spatial proximity of the two partners permits the commensal to feed on substances captured or ingested by the host. Typically the commensal also obtains shelter by living either on or in the host. The commensal is not directly dependent on the host metabolically and causes it no particular harm. When the commensal is separated from its host experimentally, it can survive without being provided some factor or factors of host origin. For example, the common, nonpathogenic strain of *Escherichia coli* lives in the human colon and benefits from the nutrients, warmth, and shelter found there but usually causes no disease or discomfort. Later in this chapter the normal bacteria of the human body are considered, and examples of how they may benefit humans in various commensal relationships are presented.

Mutualism

Mutualism [Latin *mutuus*, borrowed or reciprocal] defines the relationship in which some reciprocal benefit accrues to both partners. In this relationship the **mutualist** and the host are metabolically dependent on each other. Several examples of mutualism are presented next.

The Protozoan-Termite Relationship

A classic example of mutualism is the flagellated protozoa that live in the gut of termites and wood roaches. These flagellates exist entirely on a diet of carbohydrates, acquired as cellulose wood chips ingested by their host (figure 28.1). The protozoa engulf wood particles, digest the cellulose, and metabolize it to acetate and other products. Termites oxidize the acetate released by their flagellates. Because the host is incapable of synthesizing cellulases (enzymes that catalyse the hydrolysis of cellulose), it is dependent on the mutualistic protozoa for its existence.

This mutualistic relationship can be readily tested in the laboratory if wood roaches are placed in a bell jar containing wood and a high concentration of O_2. Because O_2 is toxic to the

(a)

(b)

Figure 28.1 **Mutualism.** Light micrographs of (*a*) worker termite of the genus *Reticulitermes* eating wood (×10), and (*b*) *Trichonympha*, a flagellated protozoan from the termite's gut (×135). Notice the many flagella over most of its length. The ability of *Trichonympha* to break down cellulose enables termites to use wood as a food source.

Figure 28.2   Lichens. (*a*) Crustose (encrusting) lichens growing on a granite post. (*b*) Foliose (leafy) lichens growing on a rock wall. (*c*) Fruticose (shrubby) lichens commonly called reindeer moss. (*d*) A variety of crustose and fruticose rock lichens, several brightly colored. (*e*) British soldier lichens (fruticose).

flagellates, they die. The wood roaches are unaffected by the high O_2 concentration and continue to ingest wood, but they soon die of starvation due to a lack of cellulases.

---

1. Define commensalism and mutualism.
2. Give one example of commensalism and mutualism.
3. What is the specific biochemical relationship between termites and their protozoan partners?
4. What would happen to a termite if its gut protozoa were destroyed by a specific antibiotic?

Lichens

Another excellent example of mutualism is provided by lichens. **Lichens** are the association between ascomycetes (the fungus) and certain genera of either green algae or cyanobacteria. In a lichen the fungal partner is termed the **mycobiont** and the algal or cyanobacterial partner, the **phycobiont.** The remarkable aspect of this mutualistic association is that its morphology and metabolic relationships are so constant that lichens are assigned generic and species names. The characteristic morphology of a given lichen is a property of the mutualistic association and is not exhibited by either symbiont individually. There are three morphological types of lichens: **crustose** lichens are compact and appressed to a substratum such as a rock; **foliose** lichens have a leaflike appearance; and **fruticose** lichens have a shrubby shape. Examples of various lichen types are shown in figure 28.2.

Ascomycetes (pp. 510–14).

Because the phycobiont functions as a photosynthetic organism—dependent only on light, air, and certain mineral nutrients—the fungus can get its organic carbon directly from the alga or cyanobacterium. The fungus often obtains nutrients from its partner by haustoria (projections of fungal hyphae)

Box 28.1

Lichens and Air Pollution

ichens are bioindicators of air quality because they are extremely sensitive to two common atmospheric pollutants—ozone (O_3) and sulfur dioxide (SO_2). This sensitivity arises from the lichen's great ability to absorb and concentrate substances dissolved in rain and dew. Because they have no means of excreting absorbed elements, lichens are particularly sensitive to toxic compounds. Lichen distribution thus is inversely related to the amount of SO_2, O_3, and toxic metals in the atmosphere. These pollutants are easily absorbed by lichens and destroy both chlorophylls *a* and *b*.

This decreases photosynthesis and disrupts the metabolic balance between the fungus and the alga or cyanobacterium. Eventually this leads to the destruction of the lichen.

Lichens vary in their response to pollutants. The best bioindicators are the moderately sensitive species, such as *Hypogymnia enteromorpha*. This species also can be used as a means to assess pollution by radioactive substances that may occur following the crash of satellites, uranium mining operations, and nuclear fallout from weapon testing.

(a)

(b)

Figure 28.3   Zooxanthellae. (*a*) Zooxanthellae (green) within the tip of a hydra tentacle (×150). (*b*) The green color of this rose coral (*Manilina*) is due to the abundant zooxanthellae within its tissues.

that penetrate the phycobiont cell wall. It also uses the O_2 produced during phycobiont photosynthesis in carrying out respiration. In turn the fungus protects the phycobiont from excess light intensities, supplies water and minerals to it, and provides a firm substratum within which the phycobiont can grow protected from environmental stress (Box 28.1).

Zooxanthellae

Many marine invertebrates (sponges, jellyfish, sea anemones, corals, ciliates) harbor endosymbiotic, spherical algal cells called **zooxanthellae** within their tissue (figure 28.3*a*). These algae belong to the following genera: *Chrysophyta, Euglenophyta, Pyrrhophyta, Chlorophyta,* and *Rhodophyta (see tables 26.1 and 26.2).* Because the degree of host dependency on the mutualistic alga is somewhat variable, only one well-known example is presented.

The hermatypic (reef-building) corals (figure 28.3*b*) obtain the bulk of their energy requirements from their zooxanthellae

and are unable to use zooplankton from the water. Pigments produced by the coral protect the algae from the harmful effects of ultraviolet radiation. Clearly the zooxanthellae also benefit the coral because the calcification rate is at least 10 times greater in the light than in the dark. Hermatypic corals lacking zooxanthellae have a very low rate of calcification. Based on their stable carbon isotopic composition, it has been determined that most of the organic carbon in the tissues of the hermatypic corals has come from the zooxanthellae. Because of this coral-algal mutualistic relationship—capturing, conserving, and cycling nutrients and energy—coral reefs are among the most productive and successful of known ecosystems.

The Tube Worm–Bacterial Relationship

Hydrothermal vent communities exist several thousand meters below the surface, where the earth's crustal plates are spreading apart (*see Box 6.1*). Vent fluids are anoxic, contain high concentrations of hydrogen sulfide, and can reach a

Figure 28.4   The Tube Worm–Bacterial Relationship. (*a*) A community of tube worms (*Riftia pachyptila*) at the Galapagos Rift hydrothermal vent site (depth 2,550 m). Each worm is more than a meter in length with an additional 20 cm for the gill plume. (*b, c*) Schematic illustration of the anatomical and physiological organization of the tube worm. The animal is anchored inside its protective tube by the vestimentum. At its anterior end is a respiratory gill plume. Inside the trunk of the worm is a trophosome consisting primarily of endosymbiotic bacteria, associated cells, and blood vessels. At the posterior end of the animal is the opisthosome, which anchors the worm in its tube. (*d*) Oxygen, carbon dioxide, and hydrogen sulfide are absorbed through the gill plume and transported to the blood cells of the trophosome. Hydrogen sulfide is bound to the worm's hemoglobin ($HSHbO_2$) and carried to the endosymbiont bacteria. The bacteria oxidize the hydrogen sulfide and use some of the released energy to fix CO_2 in the Calvin cycle. Some fraction of the reduced carbon compounds synthesized by the endosymbionts also is translocated to the animal's tissues.

temperature of 350°C. The seawater surrounding these vents has sulfide concentrations around 250 μM and temperatures 10 to 20°C above the normal seawater temperature of 2.1°C. Many free-living chemolithotrophic bacteria are present, including the genera *Beggiatoa* and *Thiomicrospira*. Several dominant animal species (tube worms, see below) are supported primarily by endosymbiotic chemolithotrophic bacteria that seem most closely related to the genus *Thiomicrospira*. Thus the main energy source of these communities is not sunlight but reduced sulfur produced geologically. Hydrogen sulfide is the "sunlight" of the animals, and sulfide-oxidizing bacteria are the "green plants" of the hydrothermal vent communities. The bacteria oxidize sulfides rather than trap light energy (*see pp. 180–81*).

The discovery of giant (>1 m in length), red, gutless tube worms near these hydrothermal vents has provided an example of a unique form of mutualism and animal nutrition in which chemolithotrophic bacterial endosymbionts are main-

tained within specialized cells of the tube worm host (figure 28.4). Because these bacteria live within the host they are called **endosymbionts.** To date all attempts to culture them have been unsuccessful.

The tube worm takes up hydrogen sulfide from the seawater and binds it to hemoglobin (the reason the worms are bright red). The hydrogen sulfide is then transported in this form to the bacteria, which use the sulfide-reducing power to fix carbon dioxide in the Calvin cycle (*see figure 10.4*). The CO_2 required for this cycle is transported to the bacteria freely dissolved in the blood, bound to hemoglobin, and in the form of organic acids such as malate and succinate. These acids are then decarboxylated to release CO_2 in the trophosome, the tissue containing bacterial symbionts. Using these mechanisms, the bacteria synthesize reduced organic material from inorganic substances. The organic material is then supplied to the tube worm through its circulatory system and serves as the main nutritional source for the tissue cells.

Box 28.2

Modifying the Rumen Microbiota to Benefit Humans

Cattle raisers can often modify the diet of their animals for specific purposes. For example, if beef cattle are given the antibiotic monensin, most of the normal rumen protozoa that produce hydrogen and CO_2 are killed. Thus the methane-producing bacteria (e.g., *Methanobrevibacter ruminantium*) do not have hydrogen and CO_2 from which to produce methane. This lack in turn increases the production of propionate by the starch decomposers (*Succinimonas amylolytica*) and decreases methane production. Because more propionate is produced and not wasted as CO_2 and hydrogen, more nutrients are available to the cattle; therefore they eat less food than the non-monensin-treated animals but gain weight at about the same rate. Increased propionate production stimulates gluconeogenesis, which leads to an increase in animal protein.

The number and variety of animals known to contain hydrogen sulfide-oxidizing chemolithotrophic symbionts has increased tremendously during the past few years. In fact a new symbiosis in which methane is oxidized instead of reduced sulfur compounds has recently been discovered in hydrothermal vent giant clams near natural gas seeps in the Gulf of Mexico (*see figure 40.3*).

Marine and freshwater hydrothermal vents (pp. 811–12).

1. What three morphological forms can a lichen have?
2. Describe the mutualistic relationship between zooxanthellae and corals.
3. Give an example of a chloroplast symbiosis.
4. How do tube worms synthesize organic molecules?
5. Define an endosymbiont.

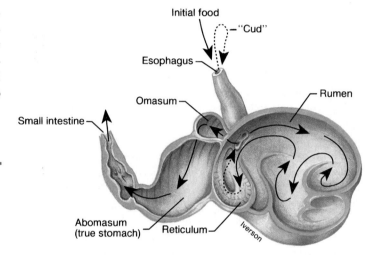

Figure 28.5   Ruminant Stomach. Stomach compartments of a cow. Arrows indicate direction of food movement.

The Rumen Ectosymbiosis

Ruminants are a group of herbivorous animals that have a stomach divided into four compartments and chew a cud consisting of regurgitated, partially digested food. Examples include cattle, deer, camels, sheep, goats, and giraffes. This feeding method has evolved in animals that need to eat large amounts of food quickly, chewing being done later at a more comfortable or safer location. More importantly, by using microorganisms to degrade the thick cellulose walls of grass and other vegetation, ruminants can digest vast amounts of otherwise unavailable forage. Because ruminants cannot synthesize cellulases, they have evolved a symbiotic relationship with microorganisms that produce these enzymes. Cellulases hydrolyze the β $(1 \rightarrow 4)$ linkages between successive D-glucose residues of cellulose and release glucose, which is then fermented to organic acids. These organic acids (acetate, butyrate, propionate) are the true energy source for the ruminant (Box 28.2; *see also figure 9.14*).

The upper portion of a ruminant's stomach expands to form a large pouch called the **rumen** (figure 28.5) and a smaller honeycomblike reticulum. The bottom portion of the stomach consists of an antechamber called the omasum, with the "true" stomach (abomasum) behind it.

The insoluble polysaccharides and cellulose eaten by the ruminant are mixed with saliva and enter the rumen. Within the rumen, food is churned in a constant rotary motion and eventually reduced to a pulpy mass. The rumen contains a large microbial population (10^{12} cells per milliliter) that partially digests and ferments the food it is processing. Later the food moves into the reticulum. Mouthfuls are then regurgitated as a "cud," which is thoroughly chewed for the first time. The food is mixed with saliva, reswallowed, and reenters the rumen while another cud is passed up to the mouth. As this process continues, the partially digested plant material becomes more liquid in nature. The liquid then begins to flow out of the reticulum and into the lower parts of the stomach: first the omasum and then the abomasum (the true stomach). It is in the abomasum that the food encounters the host's normal digestive enzymes and the digestive process continues in the regular mammalian way.

Figure 28.6 Rumen Biochemistry.
(*a*) An overview of the biochemical-physiological processes occurring in various parts of a cow's digestive system. (*b*) More specific biochemical pathways involved in rumen fermentation of the major plant carbohydrates. The top boxes represent substrates and the bottom boxes some of the end products.

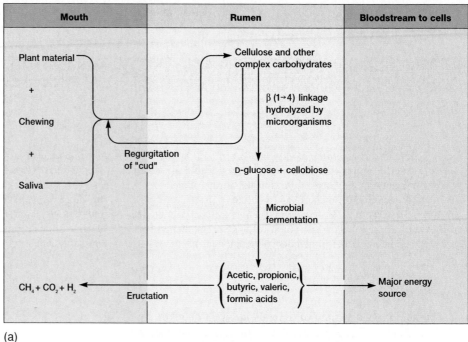

(a)

Food entering the rumen is quickly attacked by the cellulolytic anaerobic bacteria, fungi, and protozoa. Microorganisms break down the plant material, as illustrated in figure 28.6. Because the oxidation-reduction potential in the rumen is −30 mV, all indigenous microorganisms engage in anaerobic metabolism. Specifically the eubacteria ferment carbohydrates to fatty acids, carbon dioxide, and hydrogen. The archaeobacteria (methanogens) produce methane (CH_4) from CO_2 and H_2.

The dietary carbohydrates degraded in the rumen include soluble sugars, starch, pectin, hemicellulose, and cellulose. The largest percentage of each carbohydrate is fermented to volatile fatty acids (acetic, propionic, butyric, formic, and valeric), CO_2, H_2, and methane. The fatty acids produced by the rumen organisms are absorbed into the bloodstream and are oxidized by the animal as its main source of energy. The CO_2 and methane, produced at a rate of 2 liters/minute, are released by **eructation** [Latin *eructare,* to belch], a continuous, scarcely audible reflex process similar to belching. Energy trapped as ATP during fermentation is used to maintain the rumen microorganisms and to support their growth. These microorganisms in turn produce most of the vitamins needed by the ruminant. In the remaining two stomachs, the microorganisms, having performed their symbiotic task, are digested by stomach enzymes to yield amino acids, sugars, and other products. These also are absorbed and used as nutrients by the ruminant.

---

1. What structural features of the rumen make it suitable for a herbivorous type of diet?

2. Why does a cow chew its cud?

3. What biochemical role do the rumen microorganisms play in this type of symbiosis?

(b)

(a)   (b)

Figure 28.7   Raising Germfree Animals. (*a*) Schematic of a germfree isolator. (*b*) Germfree isolators for rearing colonies of small mammals.

Germfree (Gnotobiotic) Animals

The microorganisms normally associated with a particular tissue or structure are called the **microbiota**, the **indigenous microbial population**, the **microflora**, or the **microbial flora.** For consistency, the term microbiota is used in the following discussion.

The human fetus in utero (as is the case in most mammals) is free from bacteria and other microorganisms. Within hours after birth, it begins to acquire a normal microbiota, which stabilizes during the first week or two of life. From then on, enormous numbers of microorganisms are associated with the human body. Under normal conditions the human lives in symbiosis with billions of microorganisms.

Louis Pasteur first suggested that animals could not live in the absence of microorganisms. Attempts between 1899 and 1908 to grow germfree chickens had limited success because the birds died within a month. Thus it was believed that intestinal bacteria were essential for the adequate nutrition and health of the chickens. It was not until 1912 that germfree chickens were shown to be as healthy as normal birds when they were fed an adequate diet.

Since then, germfree animals have become commonplace in research laboratories. Germfree animals or those living in association with one or more known microorganisms are called **gnotobiotic** [Greek *gnotos,* known, and *biota,* the fauna and flora of a region]. A gnotobiotic system is one in which the composition of the microbiota is known (figure 28.7).

Germfree animals and gnotobiotic techniques provide good experimental systems to investigate the interactions of animals and specific microorganisms. The comparison of animals containing normal microbiota with germfree animals helps one better understand the many complex symbiotic associations that exist between the host and its microorganisms.

Establishing a germfree colony of rats, mice, hamsters, rabbits, guinea pigs, or monkeys begins with cesarean sections on pregnant females. The operation is performed under aseptic conditions in a germfree isolator. The newborn animals are then transferred to other germfree isolators in which all entering air, water, and food is sterile. Once the gnotobiotic animals become established under these germfree conditions, normal mating among themselves maintains the germfree colony.

Establishment of a germfree colony is much easier with chickens and other birds than with mammals. Fertile eggs from the birds are first sterilized with a germicide and then placed in sterile, germfree isolators. When the chick hatches, it is sterile and able to feed by itself. To make sure that a colony is germfree, periodic bacteriologic examination of all exhaust air, waste material, and cages must be done. Microorganisms should not be found.

Germfree animals are not anatomically and physiologically normal. For example, they possess poorly developed lymphoid tissue, a thin intestinal wall, an enlarged cecum (where one is present), and a low antibody titer (*see chapter 30*). They require high amounts of vitamin K and the B complexes. Germfree animals also have reduced cardiac output and lower metabolic rates.

Germfree animals are usually more susceptible to pathogens. With the normal protective microbiota absent, foreign and pathogenic microorganisms establish themselves very easily. The number of microorganisms necessary to infect a germfree animal and produce a diseased state also is much smaller. Conversely germfree animals are almost completely resistant to the intestinal protozoan *Entamoeba histolytica* that causes amebic dysentery. This resistance results from the absence of the bacteria that *E. histolytica* uses as a food source. Germfree animals also do not show any dental caries or plaque formation (*see chapter 38*). However, if they are inoculated with cariogenic (caries or cavity causing) streptococci of the *Streptococcus mutans–Streptococcus gordonii* group and fed a high-sucrose diet, they will develop caries. (*S. gordonii* was formerly considered a subpopulation of *S. sanguis.*)

Entamoeba histolytica (pp. 799–801)

1. Define microbiota and gnotobiotic.
2. How would you establish a germfree colony of mice? Of chickens?
3. Compare a germfree mouse to a normal one with respect to overall health. What benefits does an animal gain from its microbiota?

Distribution of the Normal Microbiota of the Human Body

In the balance of this chapter, the distribution and occurrence of the normal microbiota of the human body are considered (figure 28.8). An overview of the microorganisms native to different regions of the body and an introduction to the microorganisms one can expect to find on culture reports is presented (*see figure 34.3*). Because bacteria make up most of the normal microbiota, they are emphasized, whereas the fungi (mainly yeasts) and protozoa are given less attention.

There are four main reasons to acquire knowledge of the normal human microbiota:

1. An understanding of the different microorganisms at specific locations provides greater insight into the possible infections that might result from injury to these body sites.
2. A knowledge of the native organisms in the infected part of the body gives the physician-investigator perspective on the possible source and significance of microorganisms isolated from the infection site.
3. A knowledge of the indigenous microbiota helps the physician-investigator understand the causes and consequences of overgrowth by microorganisms normally absent at a specific body site.
4. An increased awareness of the role that these indigenous microbiota play in stimulating the host immune response can be gained. This awareness is important because the immune system provides protection against potential pathogens.

Skin

Commensal microorganisms living on or in the skin can be either resident (normal) or transient microbiota. Resident organisms normally grow on or in the skin. Their presence becomes fixed in well-defined distribution patterns. Those microorganisms that are temporarily present are transients. Transients usually do not become firmly entrenched; they are unable to multiply and normally die in a few hours.

The anatomy and physiology of the skin vary from one part of the body to another, and resident microbiota reflect these variations. The skin surface or epidermis is not a favorable environment for colonization by microorganisms. Several factors are responsible for this hostile microenvironment. First, the skin is subject to periodic drying. Lack of moisture drives many resident microbiota into a dormant state. However, in certain parts of the body (scalp, ears, axillary areas, genitourinary and anal regions, perineum, palms), moisture is sufficiently high to support a resident microbiota. Second, the skin has a slightly acidic pH due to the organic acids produced by normal staphylococci and secretions from skin oil and sweat glands. The acidic pH (4 to 6) discourages colonization by many microorganisms. Third, sweat contains a high concentration of sodium chloride. This makes the skin surface hyperosmotic and osmotically stresses most microorganisms. Finally, certain inhibitory substances (bactericidal and/or bacteriostatic) on the skin help control colonization, overgrowth, and infection from resident microorganisms. For example, the sweat glands release lysozyme (muramidase), an enzyme that lyses *Staphylococcus epidermidis* and other gram-positive bacteria by hydrolyzing the β $(1 \rightarrow 4)$ glycosidic bond connecting N-acetylmuramic acid and N-acetylglucosamine in the bacterial cell wall peptidoglycan (*see figure 29.8*). The oil glands secrete complex lipids that may be partially degraded by the enzymes from certain gram-positive bacteria (*Propionibacterium acnes*). These bacteria can change the secreted lipids to unsaturated fatty acids such as oleic acid that have strong antimicrobial activity against gram-negative bacteria and some fungi. Some of these fatty acids are volatile and may be associated with a strong odor. Therefore many body deodorants contain antibacterial substances that act selectively against gram-positive bacteria to reduce the production of aromatic unsaturated fatty acids and body odor. However, deodorants can shift the microbiota to predominately gram-negative bacteria and precipitate subsequent infections.

Most skin bacteria are found on the superficial cells, colonizing dead cells, or closely associated with the oil and sweat glands. Excretions from these glands provide the water, amino acids, urea, electrolytes, and specific fatty acids that serve as nutrients primarily for *Staphylococcus epidermidis* and aerobic corynebacteria. Gram-negative bacteria generally are found in the moister regions. The yeasts *Pityrosporum ovale* and *P. orbiculare* normally occur on the scalp. Some dermatophytic fungi may colonize the skin and produce athlete's foot and ringworm.

Athlete's foot disease (p. 793).

The most prevalent bacterium in skin glands is the gram-positive, anaerobic, lipophilic rod *Propionibacterium acnes*. This bacterium usually is harmless; however, it has been associated with the skin disease **acne vulgaris.** Acne commonly occurs during adolescence when the endocrine system is very active. Although the male hormone testosterone appears to be the most potent stimulator of oil glands, adrenal and ovarian hormones also can stimulate sebaceous secretions. Hormonal activity stimulates an overproduction of sebum, a fluid secreted by the oil glands. A large volume of sebum accumulates within the glands and provides an ideal microenvironment for *P. acnes*. In

Normal microbiota of the conjunctiva
1. Coagulase-negative staphylococci
2. *Haemophilus* spp.
3. *Staphylococcus aureus*
4. Streptococci (various species)

Normal microbiota of the nose
1. Coagulase-negative staphylococci
2. Viridans streptococci
3. *Staphylococcus aureus*
4. *Neisseria* spp.
5. *Haemophilus* spp.
6. *Streptococcus pneumoniae*

Normal microbiota of the outer ear
1. Coagulase-negative staphylococci
2. Diphtheroids
3. *Pseudomonas* spp.
4. *Enterobacteriaceae* (occasionally)

Normal microbiota of the mouth and oropharynx
1. Viridians streptococci
2. Coagulase-negative staphylococci
3. *Veillonella* spp.
4. *Fusobacterium* spp.
5. *Treponema* spp
6. *Porphyromonas* spp. and *Prevotella* spp.
7. *Neisseria* spp. and *Branhamella catarrhalis*
8. *Streptococcus pneumoniae*
9. Beta-hemolytic streptococci (not group A)
10. *Candida* spp.
11. *Haemophilus* spp.
12. Diphtheroids
13. *Actinomyces* spp.
14. *Eikenella corrodens*
15. *Staphylococcus aureus*

Normal microbiota of the stomach
1. *Streptococcus*
2. *Staphylococcus*
3. *Lactobacillus*
4. *Peptostreptococcus*

Normal microbiota of the skin
1. Coagulase-negative staphylococci
2. Diphtheroids (including *Propionibacterium acnes*)
3. *Staphylococcus aureus*
4. Streptococci (various species)
5. *Bacillus* spp.
6. *Malassezia furfur*
7. *Candida* spp.
8. *Mycobacterium* spp. (occasionally)

Normal microbiota of the small intestine
1. *Lactobacillus* spp.
2. *Bacteroides* spp.
3. *Clostridium* spp.
4. *Mycobacterium* spp.
5. Enterococci
6. *Enterobacteriaceae*

Normal microbiota of the urethra
1. Coagulase-negative staphylococci
2. Diphtheroids
3. Streptococci (various species)
4. *Mycobacterium* spp.
5. *Bacteroides* spp. and *Fusobacterium* spp.
6. *Peptostreptococcus* spp.

Normal microbiota of the vagina
1. *Lactobacillus* spp.
2. *Peptostreptococcus* spp.
3. Diphtheroids
4. Streptococci (various)
5. *Clostridium* spp.
6. *Bacteroides* spp.
7. *Candida* spp.
8. *Gardnerella vaginalis*

Normal microbiota of the large intestine
1. *Bacteroides* spp.
2. *Fusobacterium* spp.
3. *Clostridium* spp.
4. *Peptostreptococcus* spp.
5. *Escherichia* coli
6. *Klebsiella* spp.
7. *Proteus* spp.
8. *Lactobacillus* spp.
9. Enterococci
10. Streptococci (various species)
11. *Pseudomonas* spp.
12. *Acinetobacter* spp.
13. Coagulase-negative staphylococci
14. *Staphylococcus aureus*
15. *Mycobacterium* spp.
16. *Actinomyces* spp.

Figure 28.8   Normal Microbiota of a Human. A compilation of microorganisms which constitute normal microbiota encountered in various body sites.

some individuals this accumulation triggers an inflammatory response that causes redness and swelling of the gland's duct and produces a **comedo** [pl., comedones], a plug of sebum and keratin in the duct. Inflammatory lesions (papules, pustules, nodules) commonly called "blackheads" or "pimples" can result. *P. acnes* has been identified as the organism that produces lipases which hydrolyse the sebum triglycerides into free fatty acids. Free fatty acids are especially irritating because they can enter the dermis and promote inflammation. Because *P. acnes* is extremely sensitive to tetracycline, this antibiotic may aid acne sufferers. Accutane, a synthetic form of vitamin A, is also used.

Some pathogens found on or in the skin are transient residents that colonize the area around orifices. *Staphylococcus aureus* is the best example. It resides in the nostrils and perianal region but survives poorly elsewhere. In like manner *Clostridium perfringens* usually colonizes only the perineum and thighs, especially in those patients who suffer from diabetes.

---

1. What are four reasons why knowledge of the normal human microbiota is important?
2. Why is the skin not usually a favorable microenvironment for colonization by bacteria?
3. How do microorganisms contribute to body odor?
4. What physiological role does *Propionibacterium acnes* play in the establishment of acne vulgaris?

Nose and Nasopharynx

The normal microbiota of the nose is found just inside the nares. *Staphylococcus aureus* and *S. epidermidis* are the predominant bacteria present and are found in approximately the same numbers as on the skin of the face.

The nasopharynx, that part of the pharynx lying above the level of the soft palate, may contain small numbers of *Streptococcus pneumoniae*, *Neisseria meningitidis*, and *Haemophilus influenzae*. Most of these bacteria lack the capsules present in strains causing clinical infection. Diphtheroids, a large group of gram-positive bacteria (*see chapter 21*), are commonly found in both the nose and nasopharynx.

Oropharynx

The oropharynx is that division of the pharynx lying between the soft palate and the upper edge of the epiglottis. Like the nose, large numbers of *Staphylococcus aureus* and *S. epidermidis* inhabit this region. The most important bacteria found in the oropharynx are the various alpha-hemolytic streptococci (*S. oralis, S. milleri, S. gordonii, S. salivarius*); large numbers of diphtheroids; *Branhamella catarrhalis;* and small gram-negative cocci related to *Neisseria meningitidis*. It should be noted that the palatine and pharyngeal tonsils harbor a similar microbiota, except that within the tonsillar crypts, there is an

increase in *Micrococcus* and the anaerobes *Porphyromonas, Prevotella,* and *Fusobacterium*. (*Porphyromonas* spp. and *Prevotella* spp. were formerly classified as *Bacteroides*.)

Respiratory Tract

The upper and lower respiratory tracts (trachea, bronchi, bronchioles, alveoli) do not have a normal microbiota. This is because microorganisms are removed by (1) the continuous stream of mucus generated by the ciliated epithelial cells and (2) the phagocytotic action of the alveolar macrophages. In addition, a bactericidal effect is exerted by the enzyme lysozyme, present in nasal mucus.

Mucociliary blanket (p. 575).

Oral Cavity (Mouth)

The normal microbiota of the oral cavity contains organisms able to resist mechanical removal by adhering to surfaces like the gums and teeth. Those that cannot attach are removed by the mechanical flushing of the oral cavity contents to the stomach where they are destroyed by hydrochloric acid. The continuous desquamation (shedding) of epithelial cells also removes microorganisms. Those microorganisms able to colonize the mouth find a very comfortable environment due to the availability of water and nutrients, the suitability of pH and temperature, and the presence of many other growth factors.

The oral cavity is colonized by microorganisms from the surrounding environment within hours after a human is born. Initially the microbiota consists mostly of the genera *Streptococcus, Neisseria, Actinomyces, Veillonella,* and *Lactobacillus.* Some yeasts also are present. Most microorganisms that invade the oral cavity initially are aerobes and obligate anaerobes. As the first teeth erupt, the anaerobes (*Porphyromonas, Prevotella,* and *Fusobacterium*) become dominant due to the anaerobic nature of the gingival groove. As the teeth grow, *Streptococcus parasanguis* and *S. mutans* attach to their enamel surfaces; *S. salivarius* attaches to the buccal and gingival epithelial surfaces and colonizes the saliva. These streptococci produce a glycocalyx and various other adherence factors that enable them to attach to oral surfaces. The presence of these bacteria contributes to the eventual formation of dental plaque, caries, gingivitis, and periodontal disease.

Periodontal disease (pp. 784–85).

Eye

At birth and throughout human life, bacterial commensals are found on the conjunctiva of the eye. The predominant bacterium is *Staphylococcus epidermidis* followed by *S. aureus*, aerobic corynebacteria (diphtheroids), and *Streptococcus pneumoniae*. Cultures from the eyelids or conjunctiva also yield *Branhamella catarrhalis, Escherichia, Klebsiella, Proteus, Enterobacter, Neisseria,* and *Bacillus* species. Few anaerobic organisms are present.

External Ear

The basic microbiota of the external ear resemble that of the skin, with coagulase-negative staphylococci and *Corynebacterium* predominating. Less frequently found are *Bacillus, Micrococcus,* and *Neisseria* species. Gram-negative rods such as *Proteus, Escherichia,* and *Pseudomonas* are occasionally seen. Mycological studies show the following fungi to be normal microbiota: *Aspergillus, Alternaria, Penicillium, Candida,* and *Saccharomyces.*

Stomach

As noted earlier, many microorganisms are washed from the oral cavity into the stomach. Owing to the very acidic pH (2 to 3) of the gastric contents, most microorganisms are killed. As a result the stomach usually contains less than 10 viable bacteria per milliliter of gastric fluid. These are mainly *Streptococcus, Staphylococcus, Lactobacillus, Peptostreptococcus,* and yeasts such as *Candida* spp. Microorganisms may survive if they pass rapidly through the stomach or if the organisms ingested with food are particularly resistant to gastric pH (mycobacteria). Normally the number of microorganisms increases after a meal but quickly falls as the acidic pH takes its toll. Changes in the gastric microbiota also occur if there is an increase in gastric pH following intestinal obstruction, which permits a reflux of alkaline duodenal secretions into the stomach. If the gastric pH increases, the microbiota of the stomach are likely to reflect that of the oropharynx and, in addition, contain both gram-negative aerobic and anaerobic bacteria.

Small Intestine

The small intestine is divided into three anatomical areas: the duodenum, jejunum, and ileum. The duodenum (the first 25 cm of the small intestine) contains few microorganisms because of the combined influence of the stomach's acidic juices and the inhibitory action of bile and pancreatic secretions. Of the bacteria present, gram-positive cocci and rods comprise most of the microbiota. *Enterococcus faecalis,* lactobacilli, diphtheroids, and the yeast *Candida albicans* are occasionally found in the jejunum. In the distal portion of the small intestine (ileum), the microbiota begin to take on the characteristics of the colon microbiota. It is within the ileum that the pH becomes more alkaline. As a result anaerobic gram-negative bacteria and members of the family *Enterobacteriaceae* become established.

Large Intestine (Colon)

The colon has the largest microbial population in the body. Microscopic counts of feces approach 10^{12} organisms per gram wet weight. Over 300 different species have been isolated from human feces. The colon can be viewed as a large fermentation vessel, and the microbiota consist primarily of anaerobic, gram-negative, nonsporing bacteria and gram-positive, spore-forming, and nonsporing rods. Not only are the vast majority of microorganisms anaerobic, but many different species are present in large numbers. Several studies have shown that the ratio of anaerobic to facultative anaerobic bacteria is approximately 300 to 1. Even the most abundant of the latter, *Escherichia coli,* is only about 0.1% of the total population.

Besides the many bacteria in the large intestine, the yeast *Candida albicans* and certain protozoa may occur as harmless commensals. *Trichomonas hominis, Entamoeba hartmanni, Endolimax nana,* and *Iodamoeba butschlii* are common inhabitants.

Protozoan diseases (pp. 799–807).

Various physiological processes move the microbiota through the colon so that an adult excretes about 3×10^{13} microorganisms daily. These processes include peristalsis and segmentation, desquamation of the surface epithelial cells to which microorganisms are attached, and continuous flow of mucus that carries adhering microorganisms with it. To maintain homeostasis of the microbiota, the body must continually replace those lost microorganisms. The bacterial population in the human colon usually doubles once or twice a day. Under normal conditions the resident microbial community is self-regulating. Competition and mutualism between different microorganisms and between the microorganisms and their host serve to maintain a status quo. However, if the intestinal environment is disturbed, the normal flora may change greatly. Disruptive factors include stress, altitude changes, starvation, parasitic organisms, diarrhea, and use of antibiotics or probiotics (Box 28.3). Finally, it should be emphasized that the actual proportions of the individual bacterial populations within the indigenous microbiota depend largely on a person's diet.

The initial residents of the colon of breast-fed infants are members of the gram-positive *Bifidobacterium* genus, because human milk contains a disaccharide amino sugar that *Bifidobacterium* species require as a growth factor. In formula-fed infants, *Lactobacillus* species (also gram positive) predominate because formula lacks the required growth factor. With the ingestion of solid food, these initial colonizers of the colon are eventually displaced by a typical gram-negative microbiota. Ultimately the composition of the adult's microbiota is established.

Genitourinary Tract

The upper genitourinary tract (kidneys, ureters, and urinary bladder) is usually free of microorganisms. In both the male and female, a few bacteria (*Staphylococcus epidermidis,*

Box 28.3

Probiotics for Humans and Animals

The large intestine of humans and animals contains a very complex and balanced microbiota. These microorganisms normally prevent infection and have a positive effect on nutrition. Any abrupt change in diet, stress, or antibiotic therapy can upset this microbial balance, making the host susceptible to disease and reducing the efficient use of food. **Probiotics** [Greek *pro,* for, and *bios,* life], the oral administration of either living microorganisms or substances to promote health and growth, has the potential to reestablish the natural balance and return the host to normal health and nutrition.

Probiotic microorganisms are host specific; thus a strain selected as a probiotic in one animal may not be suitable in another species. Furthermore, microorganisms selected for probiotic use should exhibit the following characteristics:

1. Adhere to the intestinal mucosa of the host
2. Be easily cultured
3. Be nontoxic and nonpathogenic to the host
4. Exert a beneficial effect on the host
5. Produce useful enzymes or physiological end products that the host can use
6. Remain viable for a long time
7. Withstand HCl in the host's stomach and bile salts in the small intestine

There are several possible explanations of how probiotic microorganisms displace pathogens and enhance the development and stability of the microbial balance in the large intestine. These include:

1. Competition with pathogens for nutrients and adhesion sites
2. Inactivation of pathogenic bacterial toxins or metabolites
3. Production of inhibitory substances (substances that inhibit pathogen growth)
4. Stimulation of nonspecific immunity

A wide variety of probiotic preparations have been patented for cattle, goats, horses, pigs, poultry, sheep, and domestic animals (Box Table 28.3). Most of these preparations contain lactobacilli and/or streptococci; a few contain bifidobacteria.

Evidence has accumulated that certain probiotic microorganisms also offer considerable health benefits for humans. Potential benefits include:

1. Anticarcinogenic activity
2. Control of intestinal pathogens
3. Improvement of lactose use in individuals who have lactose intolerance
4. Reduction in the serum cholesterol concentration

Although the development of probiotics is still in its early stages, a better understanding of the normal microbiota in the large intestine of both animals and humans will be forthcoming as more microbiologists investigate probiotic activity.

Food microbiology (chapter 43).

BOX TABLE 28.3   Commercial Probiotics for Animals[a]

Product	Microorganism(s)
TOPTECH P	*Lactobacillus delbrueckii* ss. *delbrueckii*
	Lactobacillus spp.
	Enterococcus faecium
	Pediococcus pentosaceus
TOYOCERIN	*Bacillus cereus* "toyoi"
METABION	*Lactobacillus* spp.
CERNIVET-LBC	*Enterococcus faecium*
PROBIOS	*Lactobacillus plantarum*
	Lactobacillus acidophilus
	Lactobacillus casei
	Enterococcus faecium

[a]Commercial preparations are generally food supplements in the form of paste or liquid drench material, which is added to feed or drinking water, or pellets of freeze-dried material, which is swallowed whole by the animal.

Enterococcus faecalis, and *Corynebacterium* spp.) usually are present in the distal portion of the urethra. *Neisseria* and some members of the *Enterobacteriaceae* are occasionally found.

In contrast, the adult female genital tract, because of its large surface area and mucous secretions, has a complex microbiota that constantly changes with the female's menstrual cycle. The major microorganisms are the acid-tolerant lactobacilli, primarily *Lactobacillus acidophilus,* often called **Döderlein's bacilli.** They ferment the glycogen produced by the vaginal epithelium, forming lactic acid. As a result the pH of the vagina and cervical os is maintained between 4.4 and 4.6.

1. What are the most common microorganisms found in the nose? The oropharynx? The nasopharynx? The tonsillar crypts? The lower respiratory tract? The oral cavity? The eye? The external ear? The stomach? The small intestine? The colon? The genitourinary tract?
2. Why is the colon considered a large fermentation vessel?
3. What physiological processes move the microbiota through the gastrointestinal tract?
4. How do the initial residents of breast-fed infants differ from bottle-fed infants?
5. Describe the microbiota of the upper and lower female genitourinary tract.

Summary

1. Symbiosis is a class of relationships in which organisms of different species closely interact with one another for much of their lives.

2. Commensalism occurs when two species live in a relationship in which one benefits while the other is neither harmed nor helped.

3. When two organisms interact so that both benefit, the relationship is termed mutualism. In the protozoan-termite relationship, the termites provide food and shelter for the protozoa. The latter digest cellulose. Lichens are the mutualistic association between ascomycetes (the fungus) and either green algae or cyanobacteria. The fungal partner is the mycobiont and the algal partner, the phycobiont. The fungus provides protection and some nutrients; the phycobiont provides photosynthetic products. Zooxanthellae are specific algal cells found within various marine invertebrates. The coral-algal mutualistic relationship is a good example. The tube worm–bacterial relationship is another example of mutualism where chemolithotrophic bacterial endosymbionts are maintained within specialized cells of the tube worm host.

4. Ruminants are herbivorous animals that possess a special organ, the rumen, within which the digestion of cellulose and other plant polysaccharides occurs through special activity of anaerobic symbiotic bacteria, protozoa, and fungi.

5. Animals that are germfree or those living with one or more known microorganisms are termed gnotobiotic. Methods are available for rearing germfree colonies. Germfree animals and techniques provide good experimental systems with which the interactions of animals and specific species of microorganisms can be investigated.

6. Commensal microorganisms living on or in the skin can be characterized as either transients or residents. The most prominent bacterium found on the skin is *Staphylococcus epidermidis*. *Propionibacterium acnes* is associated with the skin glands, and *Staphylococcus aureus* is usually found around orifices.

7. *Staphylococcus aureus* and *S. epidermidis* are the predominant bacteria of the nose, whereas *Streptococcus pneumoniae, Neisseria meningitidis,* and *Haemophilus influenzae* predominate in the nasopharynx. *Staphylococcus aureus, S. epidermidis,* and various alpha-hemolytic streptococci are most common in the oropharynx.

8. The normal microbiota of the oral cavity is composed of those organisms able to resist mechanical removal.

9. The predominant bacterium found in the conjunctiva of the eye is *S. epidermidis.*

10. The microbiota of the external ear resembles that of the skin.

11. The stomach contains very few microorganisms due to its acidic pH.

12. The distal portion of the small intestine and the entire large intestine have the largest microbial populations in the body. Over 300 species have been identified, the vast majority of them anaerobic.

13. The upper genitourinary tract is usually free of microorganisms. In contrast, the adult female genital tract has a complex microbiota.

14. Probiotic microorganisms are used to promote animal health and nutrition through their effects on the gastrointestinal tract.

Key Terms

acne vulgaris *554*

comedo *556*

commensal *547*

commensalism *547*

crustose *548*

Döderlein's bacillus *558*

ectosymbiosis *547*

endosymbiont *550*

endosymbiosis *547*

eructation *552*

foliose *548*

fruticose *548*

gnotobiotic *553*

hydrothermal vent communities *549*

lichen *548*

microbiota (indigenous microbial population, microflora, or microbial flora) *553*

mutualism *547*

mutualist *547*

mycobiont *548*

phycobiont *548*

probiotics *558*

rumen *551*

ruminant *551*

symbiont or symbiote *547*

symbiosis *547*

zooxanthellae *549*

Questions for Thought and Review

1. Describe and give an example of mutualism and commensalism.

2. From a biochemical point of view, how do termites benefit the protozoa they harbor in their intestines? How do the protozoa benefit the termites?

3. Describe the symbiotic association between algae and fungi. What does the mycobiont contribute to this relationship?

4. Why is the coral-algal mutualistic relationship such a beneficial one?

5. What is unique about hydrothermal vent communities?

6. How can probiotic microorganisms benefit humans and their domestic animals?

7. List and describe some benefits of the rumen symbiosis with respect to both partners.

8. Suggest some experiments that might be useful in a germfree colony kept under gnotobiotic conditions. Give details.

9. What are some specific obstacles that must be overcome before a microorganism can colonize a body site?

10. Describe why the microenvironment of the skin is both favorable to some microorganisms and unfavorable to others. Explain your answer.

Additional Reading

General

Cano, R. J. 1994. *Bacillus* DNA in amber: A window to ancient symbiotic relationships. *ASM News* 60(3):129–34.

Margulis, L. 1981. *Symbiosis in cell evolution: Life and its environment on the early earth.* San Francisco: Freeman.

Margulis, L., and Fester, R., editors. 1991. *Symbiosis as a source of evolutionary innovation: Speciation and morphogenesis.* Cambridge, Mass.: MIT Press.

Symbiosis: 29th symposium of the society for experimental biology. 1975. New York: Cambridge University Press.

Types of Symbiosis, Functions, and Examples

Ahmadjian, V. 1963. The fungi of lichens. *Sci. Am.* 208:122–27.

Ahmadjian, V. 1986. *Symbiosis.* Hanover, N.H.: University Press of New England.

Ahmadjian, V., editor. 1993. *The lichen symbiosis.* New York: John Wiley and Sons.

Baldwin, R. L. 1984. Digestion and metabolism of ruminants. *BioScience* 34(4):244–49.

Breznak, J. A. 1982. Intestinal microbiota of termites and other xylophagus insects. *Ann. Rev. Microbiol.* 36:323-43.

Childress, J. J.; Felbeck, H.; and Somero, G. N. 1987. Symbiosis in the deep sea. *Sci. Am.* 256(2):114–20.

Gaill, F. 1993. Aspects of life development at deep sea hydrothermal vents. *FASEB* 7(6): 558–65.

Gargas, A.; DePriest, P.; Grube, M.; and Tehler, A. 1995. Multiple origins of lichen symbioses in fungi suggested by SSU ribosomal rRNA phylogeny. *Science* 268(5216):1492-95.

Hungate, R. E. 1975. The rumen microbial ecosystem. *Ann. Rev. Microbiol.* 29:39–47.

Lee, J. J., and Corliss, J. O. 1985. Symposium on "symbiosis" in protozoa. *J. Protozool.* 32(3):371–423.

Stetter, K. 1995. Microbial life in hyperthermal environments. *ASM News* 61(6):285-90.

Taylor, D. L. 1973. Algal symbionts of invertebrates. *Ann. Rev. Microbiol.* 27:171.

Williams, A. G. 1986. Rumen holotrich ciliate protozoa. *Microbiol. Rev.* 50(1):25–49.

Wolin, M. J. 1981. Fermentation in the rumen and large intestine. *Science* 213:1463–68.

Germfree (Gnotobiotic) Animals

Gordon, H. A., and Pesti, L. 1971. The gnotobiotic animal as a tool in the study of host microbial relationship. *Bacteriol. Rev.* 35:390–429.

Distribution of the Normal Microbiota of the Human Body

Drasar, B. S., and Barrow, P. A. 1985. *Intestinal microbiology.* Washington, D.C.: American Society for Microbiology.

Mackowiak, P. A. 1982. The normal microbial flora. *N. Engl. J. Med.* 307:83.

Marples, M. J. 1969. Life on the human skin. *Sci. Am.* 220:108–29.

Marsh, P., and Martin, M. 1984. *Oral microbiology,* 2d ed. Washington, D.C.: American Society for Microbiology.

Savage, D. C. 1977. Microbial ecology of the gastrointestinal tract. *Ann. Rev. Microbiol.* 31:107–33.

CHAPTER 29

Symbiotic Associations: Parasitism, Pathogenicity, and Resistance

Pathogenicity is not the rule. Indeed, it occurs so infrequently and involves such a relatively small number of species, considering the huge population of bacteria on earth, that it has a freakish aspect. Disease usually results from inconclusive negotiations for symbiosis, an overstepping of the line by one side or the other, a biological misinterpretation of borders.
—Lewis Thomas

Disease is very old and nothing about it has changed. It is we who change as we learn to recognize what was formerly imperceptible.
—John Martin Charcot

Outline

Concepts

1. If a symbiont either harms or lives at the expense of another organism, it is called a parasitic organism and the relationship is termed parasitism. In this relationship the body of the animal is referred to as the host.

2. Those parasitic organisms capable of causing disease are called pathogens. Disease is any change in the host from a healthy to an unhealthy, abnormal state in which part or all of the host's body is not properly adjusted or capable of carrying on its normal functions.

3. For a parasitic organism to cause disease, it must be transmitted to a suitable host, attach to and/or colonize the host, grow and multiply within or on the host, and interfere with or impair the normal physiological activities of the host. When a parasitic organism is growing and multiplying within or on a host, the host is said to have an infection.

4. The host's ability to resist infection depends on a continuous defense against parasitic invasion.

5. Resistance arises from both innate and acquired body defense mechanisms. The innate, general, or nonspecific immune mechanisms are those with which a host is genetically endowed and include general, physical, chemical, and biological barriers. Acquired resistance mechanisms are those that the host acquires on contact with either the parasitic organism or its products.

Chapter 28 introduces the concept of symbiosis and deals with two of its subordinate categories: commensalism and mutualism. In chapter 29 the third category, parasitism, is presented along with one of its possible consequences—pathogenicity. Fortunately most higher animals possess defense mechanisms that offer resistance to the continuous onslaught of parasitic organisms. Some innate, nonspecific forms of resistance include general, physical, chemical, and biological barriers.

The parasitic way of life is so successful that it has evolved independently in nearly all groups of microorganisms. In recent years concerted efforts to understand microorganisms and their relationships with their hosts have developed within the disciplines of virology, rickettsiology, chlamydiology, bacteriology, mycology, parasitology (protozoology and helminthology), entomology, and zoology. This chapter examines the parasitic way of life in terms of health and disease in the animal body.

Host-Parasite Relationships (Parasitism)

If a symbiont either harms or lives at the expense of another organism (the **host**), it is a **parasitic organism,** and the relationship is called **parasitism.** In this relationship the body of the host can be viewed as a microenvironment that shelters and supports the growth and multiplication of the parasitic organism. The parasitic organism is usually the smaller of the two partners and is metabolically dependent on the host. There are many parasitic agents or organisms among the viruses, procaryotes, fungi, plants, and animals (table 29.1). By convention, when the word **parasite** is used without qualification, it refers specifically to a protozoan or helminthic (nematode, trematode, cestode) organism.

Several types of parasitism are recognized. If an organism lives on the surface of its host, it is an **ectoparasite;** if it lives internally, it is an **endoparasite.** The host on or in which the parasitic organism either attains sexual maturity or reproduces is the **final host.** A host that serves as a temporary but essential environment for development is an **intermediate host.** In contrast, a **transfer host** is not necessary for the completion of the organism's life cycle but is used as a vehicle for reaching a final host. A host infected with a parasitic organism that also can infect humans is called a **reservoir host.**

Because, by definition, parasitic organisms are dependent on their hosts, the symbiotic relationship between the host and parasite is a dynamic one (figure 29.1). When a parasite is growing and multiplying within or on a host, the host is said to have an **infection.** The nature of an infection can vary widely with respect to severity, location, and number of organisms involved (table 29.2). An infection may or may not result in overt disease. An **infectious disease** is any change from a state of health in which part or all of the host body is not properly adjusted or capable of carrying on its normal functions due to the presence of a parasitic organism or its products. Any parasitic organism or agent that produces such a disease is a **pathogen.** Its ability to cause disease is called **pathogenicity.**

At times an infectious organism can enter a latent state in which there is no shedding of the organism and no symptoms present within the host. This latency can be either intermittent or quiescent. Intermittent latency is exemplified by the herpesvirus that causes cold sores (fever blisters). After an initial infection, the symptoms subside. However, the virus remains in

TABLE 29.1	Categorization of Parasitic Organisms and Agents by Size		
Discipline	**Parasitic Group**		**Approximate Size**
Virology	Prions	Agents	350,000 Da
	Viroids	25–400 nm	130,000 Da
	Viruses		
Bacteriology	Chlamydiae		0.2–1.5 µm
	Mycoplasmas		0.3–0.8 µm
	Rickettsias		0.5–2 µm
	Other bacteria	Microorganisms (microbiota)	1–10 µm
Mycology	Fungi		5–10 µm diameter
Protozoology	Protozoa		1–150 µm
Helminthology	Nematodes	Parasites	3 mm–30 cm
	Platyhelminthes (cestodes, trematodes)		1 mm–10 m
Entomology	Ticks and mites		0.1–15 mm
Zoology	Horsehair worms	Ectoparasites	10–20 cm
	Mesozoa		Up to 100 cm
	Leeches		1–5 cm

Figure 29.1 **Symbiosis.** All symbiotic relationships are dynamic, and shifts among them can occur as indicated by the arrows. The most beneficial relationship is mutualism; the most destructive is parasitism. Host susceptibility, virulence of the parasitic organism, and number of parasites are factors that influence these relationships. Disease can result from a shift from either mutualism or commensalism to parasitism. Health may be regained by the reestablishment of mutualism or commensalism.

local nerve tissue and can be activated weeks or months later by factors such as stress or sunlight. In a quiescent latency the organism persists but remains inactive for long periods of time, usually for years. For example, the varicella-zoster virus causes chickenpox in children and remains after the disease has subsided. In adulthood, under certain conditions, the same virus may erupt into a disease called shingles.

Cold sores (p. 730).

Chickenpox (varicella) and shingles (herpes zoster) (p. 716).

The outcome of most host-parasite relationships is dependent on three main factors: (1) the number of organisms present in or on the host, (2) the virulence of the organism, and (3) the host's defenses or degree of resistance. Usually the greater the number of parasitic organisms within a given host, the greater the likelihood of disease. However, a few organisms can cause disease if they are extremely virulent or if the host's resistance is low. A host's resistance can drop so much that its own microbiota may cause disease. Such a disease is sometimes called an endogenous disease because the agent originally comes from within the host's own body. Endogenous diseases can be a serious problem among hospitalized patients with very low resistance (*see chapter 38*).

The term **virulence** [Latin *virulentia,* from *virus,* poison] refers to the degree or intensity of pathogenicity. It is determined by three characteristics of the pathogen: invasiveness, infectivity, and pathogenic potential. **Invasiveness** is the ability of the organism to spread to adjacent or other tissues. **Infectivity** is the ability of the organism to establish a focal point of infection. **Pathogenic potential** refers to the degree that the pathogen causes morbid symptoms. A major aspect of pathogenic potential is toxigenicity. **Toxigenicity** is the pathogen's ability to produce toxins, chemical substances that will damage the host and produce disease. Virulence is often measured experimentally by determining the **lethal dose 50 (LD$_{50}$)** or the **infective dose 50 (ID$_{50}$)**. These values refer to the dose or number of pathogens that will either kill or infect, respectively, 50% of an experimental group of hosts within a specified period (figure 29.2).

---

1. Define parasitic organism, parasitism, infection, infectious disease, pathogenicity, virulence, invasiveness, infectivity, pathogenic potential, and toxigenicity.
2. What types of parasitism are recognized?
3. What factors determine the outcome of most host-parasite relationships?

Determinants of Infectious Disease

To induce an infectious disease, a pathogen must be able to

1. Initially be transported to the host
2. Adhere to, colonize, or invade the host
3. Multiply (grow) or complete its life cycle on or in the host
4. Initially evade host defense mechanisms
5. Possess the mechanical, chemical, or molecular ability to damage the host

The first four factors influence the degree of infectivity and invasiveness. Toxigenicity plays a major role in the fifth. Each determinant is now discussed in more detail.

Transmissibility of the Pathogen

An essential feature in the development of an infectious disease is the initial transport of the pathogen to the host. The most obvious means is direct contact—from host to host (coughing, sneezing, body contact). Pathogens also are transmitted indirectly in a variety of ways. Infected hosts shed pathogens into their surroundings. Once in the environment pathogens can be deposited on various surfaces, from which they can be either resuspended into the air or directly transmitted to a host later. Soil, water, and food are indirect vehicles that harbor and transmit pathogens to hosts. **Vectors** (organisms that transmit pathogens from one host to another) and **fomites** (inanimate objects that harbor and transmit pathogens) also are involved in the spread of many pathogens.

Attachment and Colonization by the Pathogen

After being transmitted to an appropriate host, the pathogen must be able to adhere to and colonize host cells and tissues. Colonization depends on the ability of the pathogen to compete successfully with the host's normal microbiota for essential nutrients. Specialized structures that allow the pathogen to compete for surface attachment sites also are necessary for colonization.

Pathogens and many nonpathogens adhere with a high degree of specificity to particular tissues. Adherence factors called adhesins (table 29.3) are one reason for this specificity. Adhesins are specialized molecules or structures on the pathogen's cell surface that bind to complementary receptor sites on the host cell surface (figure 29.3).

TABLE 29.2   Various Types of Infections Associated with Parasitic Organisms

Type	Definition
Abscess	A localized infection with a collection of pus surrounded by an inflamed area
Acute	Short but severe course
Bacteremia	Presence of viable bacteria in the blood
Chronic	Persists over a long time
Covert	Subclinical, no symptoms
Cross	Transmitted between hosts infected with different organisms
Focal	Exists in circumscribed areas
Fulminating	Infectious agent multiplies with great intensity
Iatrogenic	Caused as a result of health care
Latent	Persists in tissues for long periods, during most of which there are no symptoms
Localized	Restricted to a limited region or to one or more anatomical areas
Mass	Infectious agent occurs in large numbers in systemic circulation
Mixed	More than one organism present simultaneously
Nosocomial	Develops during a stay at a hospital or other clinical care facility
Opportunistic	Due to an agent that does not harm a healthy host but takes advantage of an unhealthy one
Overt	Symptomatic
Phytogenic	Caused by plant pathogens
Primary	First infection that often allows other organisms to appear on the scene
Pyogenic	Results in pus formation
Secondary	Caused by an organism following an initial or primary infection
Sepsis	(1) The condition resulting from the presence of bacteria or their toxins in blood or tissues; the presence of pathogens or their toxins in the blood or other tissues
	(2) Systemic response to infection; this systemic response is manifested by two or more of the following conditions as a result of infection: temperature, >38 or <36°C; heart rate, >90 beats per min; respiratory rate, >20 breaths per min, or pCO_2, <32 mm Hg; leukocyte count, >12,000 cells per ml^3, or >10% immature (band) forms
Septicemia	Blood poisoning associated with persistence of pathogenic organisms or their toxins in the blood
Septic shock	Sepsis with hypotension despite adequate fluid resuscitation, along with the presence of perfusion abnormalities that may include, but are not limited to, lactic acidosis, oliguria, or an acute alteration in mental status
Severe sepsis	Sepsis associated with organ dysfunction, hypoperfusion, or hypotension; hypoperfusion and perfusion abnormalities may include, but are not limited to, lactic acidosis, oliguria, or an acute alteration in mental status
Sporadic	Occurs only occasionally
Subclinical (inapparent or covert)	No detectable symptoms or manifestations
Systemic	Spread throughout the body
Toxemia	Condition arising from toxins in the blood
Zoonosis	Caused by a parasitic organism that is normally found in animals other than humans

Figure 29.2   Determination of the LD_{50} of a Pathogenic Microorganism. Various doses of a specific pathogen are injected into experimental host animals. Deaths are recorded and a graph constructed. In this example, the graph represents the susceptibility of host animals to two different strains of a pathogen—strain A and strain B. For strain A the LD_{50} is 30, and for strain B it is 50. Hence strain A is more virulent than strain B.

TABLE 29.3   Adherence Factors (Adhesins) That Play a Role in Infectious Diseases

Adherence Factor	Description
Filamentous hemagglutinin	Causes adherence to erythrocytes
Fimbriae	Filamentous structures that help attach bacteria to solid surfaces
Glycocalyx or capsule	A layer of exopolysaccharide fibers with a distinct outer margin that surrounds many cells; it inhibits phagocytosis and aids in adherence
Lectin	Any carbohydrate-binding protein or glycoprotein of nonimmune origin
Ligand	A low molecular weight molecule that exhibits specific binding to a complementary site on a high molecular weight molecule such as a protein
Mucous gel	The glycoprotein layer or mucopolysaccharide layer of glycosaminoglycans covering animal cell mucosal surfaces
Pili	Filamentous structures that bind procaryotes together for the transfer of genetic material
Receptors	Complementary binding sites that bind specific ligands or adhesins
S layer	The outermost regularly structured layer of cell envelopes of some archaeobacteria and eubacteria that may promote adherence to surfaces
Slime	A tenacious bacterial film that is less compact than a capsule
Teichoic and lipoteichoic acids	Cell wall components in gram-positive bacteria that aid in adhesion

(a)

(b)

(c)

⬤ **Figure 29.3   Microbial Adherence.** (*a*) Electron micrograph of fimbriated *Escherichia coli* (×16,625). (*b*) Scanning electron micrograph of epithelial cells with adhering vibrios (×1,200). (*c*) *Candida albicans* fimbriae (arrow) are used to attach the fungus to vaginal epithelial cells. Bar = 0.20 μm.

TABLE 29.4   Microbial Products Involved in Pathogen Dissemination Throughout a Mammalian Host

Product	Organism Involved	Mechanism of Action
Coagulase	*Staphylococcus aureus*	Coagulates (clots) the fibrinogen in plasma. The clot protects the pathogen from phagocytosis and isolates it from other host defenses.
Collagenase	*Clostridium* spp.	Breaks down collagen that forms the framework of connective tissues; allows the pathogen to spread.
Deoxyribonuclease (along with calcium and magnesium)	Group A streptococci, staphylococci, *Clostridium perfringens*	Lowers viscosity of exudates, giving the pathogen more mobility.
Elastase and alkaline protease	*Pseudomonas aeruginosa*	Cleaves laminin associated with basement membranes.
Exotoxin B (cysteine protease)	Group A streptococci, *Streptococcus pyogenes*	Degrades proteins.
Hemolysins	Staphylococci, streptococci, *Escherichia coli*, *Clostridium perfringens*	Lyse erythrocytes, causing anemia and weakened host defenses; make iron available for microbial growth.
Hyaluronidase	Groups A, B, C, and G streptococci, staphylococci, clostridia	Hydrolyzes hyaluronic acid, a constituent of the intercellular ground substance that cements cells together and renders the intercellular spaces amenable to passage by the pathogen.
Hydrogen peroxide (H_2O_2) and ammonia (NH_3)	*Mycoplasma* spp., *Ureaplasma* spp.	Are produced as metabolic wastes. These are toxic and damage epithelia in respiratory and urogenital systems.
Immunoglobulin A protease	*Streptococcus pneumoniae*	Cleaves immunoglobulin A into Fab and Fc fragments.
Lecithinase	*Clostridium* spp.	Destroys the lecithin (phosphatidycholine) component of plasma membranes, allowing pathogen to spread.
Leukocidins	Staphylococci, pneumococci, streptococci	Cause degranulation of lysosomes within leukocytes, which decreases host resistance; also kill leukocytes.
Porins	*Salmonella typhimurium*	Inhibit leukocyte phagocytosis by activating the adenylate cyclase system.
Protein A	*Staphylococcus aureus*	Located on cell wall. Immunoglobulin G (IgG) binds to protein A by its Fc end, thereby preventing complement from interacting with bound IgG.
Streptokinase (fibrinolysin, staphylokinase)	Group A, C, and G streptococci, staphylococci	A protein that binds to plasminogen and activates the production of plasmin, thus digesting fibrin clots; this allows the pathogen to move from the clotted area.

Entry of the Pathogen

Entry into host cells and tissues is a specialized strategy used by many pathogens for survival and multiplication. Pathogens often actively penetrate the host's epithelium after attachment to the epithelial surface. This may be accomplished through production of lytic substances that alter the host tissue by (1) attacking the ground substance and basement membranes of integuments and intestinal linings, (2) degrading carbohydrate-protein complexes between cells or on the cell surface (the glycocalyx), or (3) disrupting the cell surface.

At times a pathogen can penetrate the epithelial surface by passive mechanisms not related to the pathogen itself. Examples include (1) small breaks, lesions, or ulcers in a mucous membrane that permit initial entry; (2) wounds, abrasions, or burns on the skin's surface; (3) arthropod vectors that create small wounds while feeding; (4) tissue damage caused by other organisms; and (5) existing eucaryotic internalization pathways (e.g., endocytosis).

Once inside the epithelia the pathogen may penetrate to deeper tissues and continue disseminating throughout the body of the host. One way the pathogen accomplishes this is by producing specific products and/or enzymes that promote spreading (table 29.4). The pathogen also may enter the small terminal lymphatic capillaries that surround epithelial cells. These capillaries merge into large lymphatic vessels that eventually drain into the circulatory system. Once the circulatory system is reached, the pathogen has access to all organs and systems of the host.

Growth and Multiplication of the Pathogen

For a pathogen to be successful in growth and reproduction, it must find an appropriate environment (nutrients, pH, temperature, redox potential) within the host. Those areas of the host's body that provide the most favorable conditions will harbor the pathogen and allow it to grow and multiply to produce an infection. Some pathogens invade specific cells in which they grow and multiply. Many of these intracellular pathogens have evolved such elaborate nutrient-gathering mechanisms that they have become totally dependent on the host's cells. Finally, some pathogens can actively grow and multiply specifically in the blood plasma. Their metabolic waste products often are toxic and produce a condition known as **septicemia** [Greek *septikos*, produced by putrefaction, and *haima*, blood].

---

1. What are some ways in which pathogens are transmitted to their hosts? Define vector and fomite.
2. Describe several specific adherence factors by which pathogens attach to host cells.

3. How do pathogens actively enter host tissues? How do they passively enter?

4. Once inside the epithelial surface, what are some mechanisms that pathogens possess to promote their dissemination throughout the body of a host?

Toxigenicity

Two distinct categories of disease can be recognized based on the pathogen's role in the disease-causing process: infections and intoxications. An infectious disease results partly from the pathogen's growth and reproduction (or invasiveness) that often produce tissue alterations.

Intoxications are diseases that result from the entrance of a specific toxin into the body of a host. Toxins can even induce disease in the absence of the organism that produced them. A **toxin** [Latin *toxicum,* poison] is a specific substance, often a metabolic product of the organism, that damages the host. The term **toxemia** refers to the condition caused by toxins that have entered the blood of the host. Toxins produced by organisms can be divided into two main categories: exotoxins and endotoxins.

Exotoxins

Exotoxins are soluble, heat-labile, proteins that usually are released into the surroundings as the pathogen grows. Often exotoxins may travel from the site of infection to other body tissues or target cells in which they exert their effects. Exotoxins usually are

1. Synthesized by specific pathogens that often have plasmids or prophages bearing the exotoxin genes

2. Heat-labile proteins inactivated at 60 to 80°C

3. Among the most lethal substances known (toxic in very small doses [microgram per kilogram amounts]; e.g., the botulinum toxin)

4. Associated with specific diseases and have specific mechanisms of action

5. Highly immunogenic and stimulate the production of neutralizing antibodies (**antitoxins**)

6. Easily inactivated by formaldehyde, iodine, and other chemicals to form immunogenic **toxoids**

7. Unable to produce a fever in the host directly

8. Often given the name of the disease they produce (e.g., the diphtheria toxin)

9. Usually categorized as neurotoxins, cytotoxins, or enterotoxins according to their mechanism of action

Although toxins occur in many forms, there is a general structural model to which they frequently conform—the **AB model.** In this model each toxin is composed of an enzymatic subunit or fragment (A) that is responsible for the toxic effect once inside the host cell and a binding subunit or fragment (B). Isolated A subunits are enzymatically active but lack binding and cell entry capability, whereas isolated B subunits bind to target cells but are nontoxic and biologically inactive. The B subunit interacts with specific receptors on the target cell or tissue such as the gangliosides GM_1 for cholera toxin, GT_1 and/or GD_1 for tetanus toxin, and GD_1 for botulinum toxin.

Several mechanisms for the entry of A subunits or fragments into target cells have been proposed. In one mechanism the B subunit inserts into the plasma membrane and creates a pore through which the A subunit enters (figure 29.4*a*). In another mechanism entry is by receptor-mediated endocytosis (figure 29.4*b*).

Exotoxins exert their effects in a variety of ways—for instance, by (1) inhibition of protein synthesis, (2) inhibition of nerve synapse function, (3) disruption of membrane transport, or (4) damage to plasma membranes. The mechanism of action can be quite complex, as shown by the example of diphtheria toxin (figure 29.4*b*). The diphtheria toxin is a protein of about 62,000 mol wt. It binds to cell surface receptors by the B fragment portion and is taken into the cell through the formation of a clathrin-coated vesicle (*see p. 373*). The toxin then enters the vesicle membrane and is cleaved into two parts, one of which, the A fragment, escapes into the cytoplasm. The A fragment is an enzyme and catalyzes the addition of an ADP-ribose group to the eucaryotic elongation factor EF2 that aids in translocation during protein synthesis (*see p. 232*). The substrate for this reaction is the coenzyme NAD^+.

$$NAD^+ + EF2 \rightarrow ADP\text{-ribosyl-}EF2 + nicotinamide$$

The modified EF2 protein cannot participate in the elongation cycle of protein synthesis, and the cell dies because it can no longer synthesize proteins.

Exotoxins vary widely in their relative contribution to the disease process with which they are associated. The general properties of some exotoxins are presented in table 29.5.

Exotoxins may be divided into three categories on the basis of the site affected: **neurotoxins** (nerve tissue), **enterotoxins** (intestinal mucosa), and **cytotoxins** (general tissues). Some of the enteric pathogens that produce these exotoxins are presented in table 29.6.

Neurotoxins usually are ingested as preformed toxins that affect the nervous system and indirectly cause enteric (pertaining to the small intestine) symptoms. Examples include staphylococcal enterotoxin B, *Bacillus cereus* emetic toxin [Greek *emetos,* vomiting], and botulinum toxin (for other neurotoxins *see Box 26.2*).

True enterotoxins [Greek *enter,* intestine] have a direct affect on the intestinal mucosa and elicit profuse fluid secretion. The classic enterotoxin, cholera toxin (choleragen), has been studied extensively. It is an AB toxin. The B subunit is made of five parts arranged as a donut-shaped ring. The B subunit ring anchors itself to the epithelial cell's plasma membrane and then inserts the smaller A subunit into the cell. The A subunit activates tissue adenylate cyclase to increase intestinal cyclic AMP (cAMP) concentrations. High concentrations of cAMP provoke the movement of massive quantities of water and electrolytes across the intestinal cells into the lumen of the gut. The genes for this enterotoxigenicity reside on the *Vibrio cholera* chromosome.

Toxins in food (pp. 766–70, 886–87).

(a)

(b)

Figure 29.4   **Diagrammatic Representation of Two Exotoxin Transport Mechanisms.** (*a*) Domain B of the dimeric exotoxin (AB) binds to a specific membrane receptor of a target cell [*1*]. A conformational change [*2*] generates a pore [*3*] through which the A domain crosses the membrane and enters the cytosol, followed by recreation [*4*] of the binding site. (*b*) Receptor-mediated endocytosis of the diphtheria toxin involves the dimeric exotoxin binding to a receptor-ligand complex that is internalized in a clathrin-coated pit that pinches off to become a coated vesicle. The clathrin coat depolymerizes resulting in an uncoated endosome vesicle. The pH in the endosome decreases due to the H$^+$-ATPase activity. The low pH causes A and B components to separate. An endosome in which this separation occurs is sometimes called a CURL (compartment of uncoupling of receptor and ligand). The B domain is then recycled to the cell surface. The A domain moves through the cytosol, catalyzes the ADP-ribosylation of EF-2, and inhibits protein synthesis, leading to cell death.

TABLE 29.5   Properties of Some AB Model Bacterial Exotoxins

Toxin	Organism	Genetic Control	Subunit Structure	Target Cell Receptor	Enzymatic Activity	Biologic Effects
Anthrax toxins	B. anthracis	Plasmid	Three separate proteins (EF, LF, PA)[a]	Unknown, probably glycoprotein	EF is a calmodulin-dependent adenylate cyclase; LF enzyme activity is unknown	EF + PA: increase in target cell cAMP level, localized edema; LF + PA: death of target cells and experimental animals
Bordetella adenylate cyclase toxin	Bordetella spp.	Chromosomal	A-B[b]	Unknown, probably glycolipid	Calmodulin-activated cyclase	Increase in target cell cAMP level; modified cell function or cell death
Botulinum toxin	C. botulinum	Phage	A-B[c]	Possibly ganglioside (GD_{1b})	Zinc-dependent endopeptidase	Decrease in peripheral, presynaptic acetylcholine release; flaccid paralysis
Cholera toxin	V. cholera	Chromosomal	A-5B[d]	Ganglioside (GM_1)	ADP ribosylation of adenylate cyclase regulatory protein, G_S	Activation of adenylate cyclase, increase in cAMP level; secretory diarrhea
Diphtheria toxin	C. diphtheriae	Phage	A-B[e]	Heparin-binding, EGF-like growth factor precursor	ADP ribosylation of elongation factor 2	Inhibition of protein synthesis; cell death
Heat-labile enterotoxins[f]	E. coli	Plasmid	————————— Similar or Identical to Cholera Toxin —————————			
Pertussis toxin	B. pertussis	Chromosomal	A-5B[g]	Unknown, probably glycoprotein	ADP ribosylation of signal-transducing G proteins	Block of signal transduction mediated by target G proteins
Pseudomonas exotoxin A	P. aeruginosa	Chromosomal	A-B	α_2-Macroglobulin/LDL receptor	————— Similar or Identical to Diphtheria Toxin —————	
Shiga toxin	S. dysenteriae	Chromosomal	A-5B[h]	Globotriaosylceramide (Gb_3)	RNA N-glycosidase	Inhibition of protein synthesis, cell death
Shiga-like toxins	Shigella spp., E. coli	Phage	————————— Similar or Identical to Shiga Toxin —————————			
Tetanus toxin	C. tetani	Plasmid	A-B[c]	Ganglioside (GT_1 and/or GD_{1b})	Zinc-dependent endopeptidase	Decrease in neurotransmitter release from inhibitory neurons; spastic paralysis

From G. L. Mandell, et al., *Principles and Practice of Infectious Diseases,* 3rd edition Copyright © 1990 Churchill-Livingstone, Inc., Medical Publishers, New York, NY. Reprinted by permission.

[a]The binding component (known as protective antigen [PA]) catalyzes/facilitates the entry of either edema factor (EF) or lethal factor (LF).

[b]Apparently synthesized as a single polypeptide with binding and catalytic (adenylate cyclase) domains.

[c]Holotoxin is apparently synthesized as a single polypeptide and cleaved proteolytically as diphtheria toxin; subunits are referred to as L: light chain, A equivalent; H: heavy chain, B equivalent.

[d]The A subunit is proteolytically cleaved into A_1 and A_2, with A_1 possessing the ADP-ribosyl transferase activity; the binding component is made up of five identical B units.

[e]Holotoxin is synthesized as a single polypeptide and cleaved proteolytically into A and B components held together by disulfide bonds.

[f]The heat-labile enterotoxins of *E. coli* are now recognized to be a family of related molecules with identical mechanisms of action.

[g]The binding portion is made up of two dissimilar heterodimers labeled S2-S3 and S2-S4 that are held together by a bridging peptide, SS.

[h]Subunit composition and structure similar to cholera toxin.

TABLE 29.6   Examples of Some Bacteria That Produce Exotoxins

Category	Bacterial Examples
Neurotoxin	*Clostridium botulinum* *C. tetani* *Staphylococcus aureus* *Bacillus cereus*
Enterotoxin	*Vibrio cholerae* *Escherichia coli* (enteropathogenic strains) *Salmonella* spp. *Klebsiella* spp. *Clostridium perfringens*
Cytotoxin	*Shigella* spp. *Vibrio parahaemolyticus* *S. aureus* *Clostridium difficile*

1. What is the difference between an infectious disease and an intoxication? Define toxemia.

2. Describe some general characteristics of exotoxins.

3. How do exotoxins get into host cells?

4. Describe the biological effects of several bacterial exotoxins.

5. Discuss the mechanisms by which exotoxins can damage cells.

6. What are the three categories of exotoxins?

Endotoxins

Most gram-negative bacteria have a lipopolysaccharide (LPS) in the outer membrane of their cell wall that, under certain circumstances, is toxic to specific hosts. This LPS (*see figures 3.26 and 3.27*) is called an **endotoxin** because it is bound to the bacterium

Box 29.1

Detection and Removal of Endotoxins

Bacterial endotoxins have plagued the pharmaceutical industry and medical device producers for years. For example, administration of drugs contaminated with endotoxins can result in complications—even death—to patients. Recently endotoxins have become a problem for individuals and firms working with cell cultures and genetic engineering. The result has been the development of sensitive tests and methods to identify and remove these endotoxins. The procedures must be very sensitive to trace amounts of endotoxins. Most firms have set a limit of 0.25 **endotoxin units (E.U.),** 0.025 ng/ml, or less as a release standard for their drugs, media, or products.

One of the most accurate tests for endotoxins is the in vitro *Limulus* amoebocyte lysate (LAL) assay. The assay is based on the observation that when an endotoxin contacts the clot protein from circulating amoebocytes of *Limulus,* a gel-clot forms. The assay kits available today contain calcium, proclotting enzyme, and procoagulogen. The proclotting enzyme is activated by bacterial endotoxin lipopolysaccharide and calcium to form active clotting enzyme (Box Figure 29.1). Active clotting enzyme then catalyzes the cleavage of procoagulogen into polypeptide subunits (coagulogen). The subunits join by disulfide bonds to form a gel-clot. Spectrophotometry is then used to measure the protein precipitated by the lysate. The LAL test is sensitive at the nanogram level but must be standardized against Food and Drug Administration Bureau of Biologics endotoxin reference standards. Results are reported in endotoxin units per milliliter and reference made to the particular reference standards used.

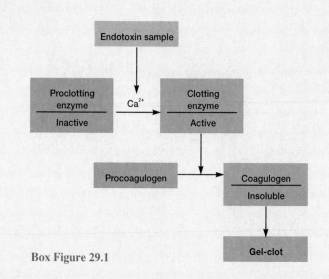

Box Figure 29.1

Removal of endotoxins presents more of a problem than their detection. Those present on glassware or medical devices can be inactivated if the equipment is heated at 250°C for 30 minutes. Soluble endotoxins range in size from molecules with molecular weights of 20,000 to large aggregates with diameters up to 0.1 μm. Thus they cannot be removed by conventional filtration systems. Manufacturers are currently developing special filtration systems and filtration cartridges that retain these endotoxins and help alleviate contamination problems.

---

and is released when the microorganism lyses (Box 29.1). Some is also released during bacterial multiplication. The toxic component of the LPS is the lipid portion, called lipid A. Lipid A is not a single macromolecular structure but appears to be a complex array of lipid residues. The lipid A component exhibits all the properties (see characteristic 5 following) associated with endotoxicity and gram-negative bacteremia.

Gram-negative cell wall (pp. 54–56).

Besides the preceding characteristics, endotoxins are

1. Heat stable
2. Toxic only at high doses (milligram per kilogram amounts)
3. Weakly immunogenic
4. Generally similar, despite source
5. Usually capable of producing general systematic effects: fever (are pyrogenic), shock, blood coagulation, weakness, diarrhea, inflammation, intestinal hemorrhage, and fibrinolysis (enzymatic breakdown of fibrin, the major protein component of blood clots)

Endotoxins probably exert their effects in several ways. Endotoxins can initially activate Hageman Factor (blood clotting factor XII), which in turn activates up to four humoral systems: coagulation, complement, fibrinolytic, and kininogen systems.

Activated Hageman Factor can initiate **coagulation** (the intrinsic blood clotting cascade) through the conversion of plasma thromboplastin antecedent to active thromboplastin antecedent (figure 29.5). The rest of the cascade is classic, leading to cross-linked fibrin formation and a blood clot. If this system goes unchecked, thrombosis (plugs in blood vessels) produces disseminated intravascular coagulation. This in turn causes depletion of blood platelets as well as clotting factors II, V, and VII. In disseminated intravascular coagulation, the use of platelets and coagulation factors exceeds hemostatic production rates, which leads to internal hemorrhaging and organ failure (lungs, kidneys, liver).

Serum contains a group of proteins called complement that plays an important role in the removal of pathogens. Because the complement system normally acts in concert with specific immune responses, a detailed discussion of it is presented in chapter 32. As a general overview the activation of complement (figure 29.5b) causes the release of inflammatory mediators to induce (1) smooth muscle contraction in postcapillary venules, (2) movement of neutrophils to the infection

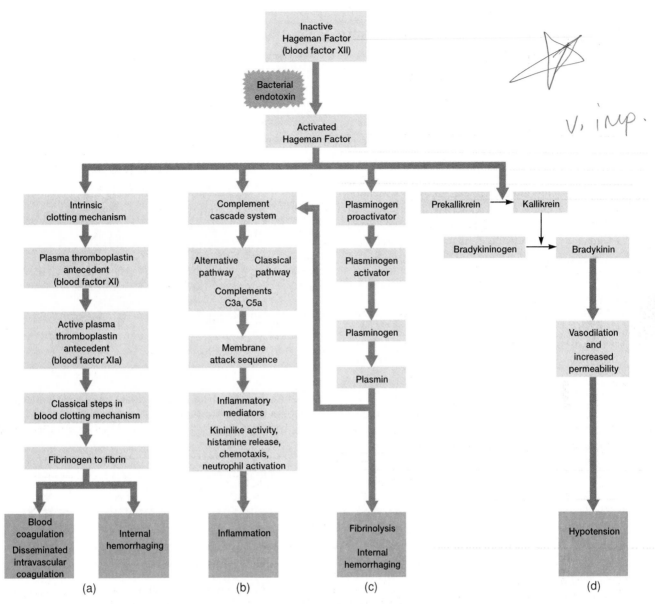

Figure 29.5 **Endotoxins.** The pathophysiological effects of gram-negative bacterial endotoxins (Lipid A) on a mammalian host. Both in vivo and in vitro endotoxins exert four systemic effects. (*a*) They can start the blood clotting cascade that leads to blood coagulation, thrombosis, and acute disseminated intravascular coagulation, which in turn depletes platelets and various clotting factors and results in clinical bleeding; (*b*) they can activate the complement system, which leads to inflammation; (*c*) they can activate fibrinolysis; and (*d*) they can trigger a series of enzymatic reactions that lead to the release of bradykinins and other vasoactive peptides, which causes hypotension.

site, (3) neutrophils to produce leukotrienes, (4) increased vascular permeability, and (5) mast cell degranulation. Inflammation is the result.

The complement system (pp. 633–36).

While Hageman Factor is activating the coagulation and complement pathways, it also activates a countervening fibrinolytic system (figure 29.5*c*). Activated Hageman Factor causes the conversion of plasminogen proactivator to plasminogen activator. The latter stimulates the conversion of plasminogen to plasmin. Plasmin is a very potent fibrinolytic protein that mediates blood clot lysis, which contributes further to internal hemorrhaging. The plasmin formed also can initiate

the complement cascade and exacerbate the inflammatory reaction initially triggered by the gram-negative endotoxin.

Finally, the endotoxin-activated Hageman Factor can activate the kininogen system (figure 29.5*d*), leading to hypotension (lowered blood pressure). In this pathway Hageman Factor causes the conversion of prekallikrein to activated kallikrein, which in turn catalyzes the conversion of bradykininogen to bradykinin. Bradykinin is a potent vasoactive peptide that causes pain, vasodilation, and increased vascular permeability. This vasoactive effect leads to leakage of the fluid portion of the blood into interstitial spaces; therefore blood volume decreases and hypovolemic hypotension results. Circulatory shock and death can follow.

Gram-negative endotoxins also indirectly induce a fever in the host by causing macrophages to release **endogenous pyrogens** that reset the hypothalamic thermostat. One important endogenous pyrogen is the cytokine interleukin-1. Other cytokines released by macrophages, such as the tumor necrosis factor, also produce fever.

Recent evidence indicates that LPS affects macrophages and monocytes by binding to special plasma proteins called **LPS-binding proteins.** The LPS-LPS-binding protein complex then attaches to receptors on monocytes, macrophages, and other cells. This triggers several events, including the production of cytokines IL-1, IL-6, and tumor necrosis factor. As mentioned previously, IL-1 and tumor necrosis factor induce fever. These cytokines also promote other endotoxin effects: complement activation, coagulation, prostaglandin formation, and so forth.

1. Describe the chemical structure of the LPS endotoxin.
2. List some general characteristics of endotoxins.
3. Describe how each of the following operate after being activated by a gram-negative bacterial endotoxin: intrinsic clotting cascade, complement activation, fibrinolytic system, and kininogen system.
4. How do gram-negative endotoxins induce fever in a mammalian host?

Leukocidins and Hemolysins

Some pathogens produce extracellular toxins that may kill phagocytic leukocytes and are termed **leukocidins** [*leuko*cyte and Latin *caedere,* to kill]. Most leukocidins are produced by pneumococci, streptococci, and staphylococci. Once liberated, the toxin attaches to leukocyte membranes and triggers a series of changes leading to degranulation and release of lysosomal enzymes into the leukocyte's cytosol (these events may result from changes in plasma membrane permeability due to pore formation). This destroys the leukocyte and in turn decreases host resistance.

Other toxins called **hemolysins** [*haima, blood,* and Greek *lysis,* dissolution] also can be secreted by pathogenic bacteria. Many hemolysins probably form pores in the plasma membrane of erythrocytes through which hemoglobin and/or ions are released (the erythrocytes lyse or, more specifically, hemolyze). **Streptolysin-O (SLO)** is a hemolysin, produced by *Streptococcus pyogenes,* that is inactivated by O_2 (hence the "O" in its name). SLO causes beta hemolysis of erythrocytes on agar plates incubated anaerobically. A complete zone of clearing around the bacterial colony growing on blood agar is called **beta hemolysis,** and a partial clearing of the blood is called **alpha hemolysis. Streptolysin-S (SLS)** is also produced by *S. pyogenes* but is insoluble and bound to the bacterial cell. It is O_2 stable (hence the "S" in its name) and causes beta hemolysis on aerobically incubated blood-agar plates. SLS also can act as a leukocidin by killing leukocytes that phagocytose the bacterial cell to which it is bound. It should be noted that hemolysins attack the plasma membranes of many cells, not just erythrocytes and leukocytes.

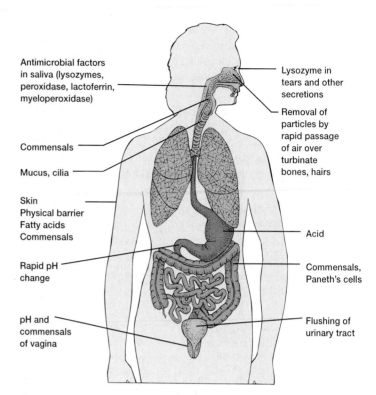

Antimicrobial factors in saliva (lysozymes, peroxidase, lactoferrin, myeloperoxidase)

Commensals

Mucus, cilia

Skin
Physical barrier
Fatty acids
Commensals

Rapid pH change

pH and commensals of vagina

Lysozyme in tears and other secretions

Removal of particles by rapid passage of air over turbinate bones, hairs

Acid

Commensals, Paneth's cells

Flushing of urinary tract

Figure 29.6 **Host Defenses.** Some nonspecific host defense mechanisms that help prevent entry of microorganisms into the host's tissues.

1. What is the mode of action of a leukocidin? Of a hemolysin?
2. Name two specific hemolysins.

General or Nonspecific Host Immune Defense Mechanisms

With few exceptions a potential pathogen invading a human host immediately confronts a vast array of general or nonspecific host immune [Latin *immunis,* safe] defense mechanisms (figure 29.6). Although the effectiveness of some mechanisms is not great, collectively their defense against infection is formidable. These general or nonspecific defense mechanisms can be grouped into four major categories: general, physical, chemical, and biological barriers.

General Barriers

Many direct factors (nutrition, physiology, fever, age, genetics, race) and equally as many indirect factors (personal hygiene, socioeconomic status, living conditions) influence all host-parasite relationships. At times they favor the establishment of the parasitic organism; at other times they provide some measure of defense to the host. Because these factors are so ill-defined, they can be viewed as either general or nonspecific barriers to the establishment of an organism. A few of the better known direct factors are discussed.

Nutrition

In general the more malnourished the host, the greater will be its susceptibility to, and the severity of, infections. This is especially true for very young hosts.

Acute-Phase Reactants

The term **acute-phase reactants** refers to the rapid qualitative and quantitative changes that occur in the host's blood plasma during an acute infection. These changes can decrease the virulence of the pathogen and increase the overall general defense of the host. For example, the virulence of many organisms is enhanced with increased iron availability. Evidence suggests that some hosts are able to redistribute iron in an attempt to withhold it **(hypoferremia)** from the microorganism. Conversely **hyperferremia** can lead to infections with certain normally harmless organisms or provide for their dissemination. Gonococci, for example, spread most often during menstruation, a time in which there is an increased concentration of free iron available to these bacteria.

Gonorrhea (pp. 756–57).

The plasma concentrations of some proteins increase drastically during the acute-phase reaction. For example, the C-reactive protein can increase up to 1,000-fold in concentration. It aids in phagocytosis of bacteria and can activate complement.

Fever

From a physiological point of view, fever results from disturbances in hypothalamic thermoregulatory activity, leading to an increase of the thermal "set point." In adult humans fever is defined as an oral temperature above 98.6°F (37°C) or a rectal temperature above 99.5°F (37.5°C). In almost every instance there is a specific constituent, the exogenous pyrogen [Greek *pyr,* fire, and *gennan,* to produce] of the infecting organism, that directly triggers fever production. Examples include the gram-negative bacterial endotoxins; the N-acetylglucosamine-N-acetylmuramic acid polymer and/or a derived portion of the peptidoglycan cell wall material of both gram-positive and gram-negative bacteria; the soluble enterotoxin secreted by staphylococci; the erythrogenic toxin of group A streptococci; interleukin-1; and many yet undefined soluble carbohydrate and protein moieties produced by viruses, bacteria, fungi, protozoa, and helminths.

The fever induced by a parasitic organism augments the host's defenses by three complementary pathways (figure 29.7): (*a*) it stimulates leukocytes so that they can destroy the organism; (*b*) it enhances microbiostasis (growth inhibition) by decreasing iron available to the organism; and (*c*) it enhances the specific activity of the immune system.

Age

Generally when the host is either very young or very old, susceptibility to infection increases. Babies are at a particular risk after their maternal immunity (*see chapter 30*) has disappeared and before their own immune system has matured. In very old persons there is a decline in the immune system and in the homeostatic functioning of many organs that reduces host defenses.

Genetic Factors

The importance of genetic factors—both the pathogen's and the host's—to susceptibility has been recognized for many years. Often one species is resistant to the infectious diseases that affect another species. For example, humans are not susceptible to canine distemper or hog cholera. Species resistance may depend upon genetically determined factors such as (1) temperature of the host (endothermic versus heterothermic animals); (2) genetically controlled metabolic, physiological, and anatomic differences that affect the ability of an organism to cause infection; and (3) food procurement mechanisms (herbivore, carnivore, omnivore). Even within species genetically associated racial resistance factors affect the host's defense at times. For example, African-Americans, when compared to whites, are more resistant to falciparum malaria due to the sickle cell trait that they often possess.

Malaria (pp. 802–4).

---

1. Describe how each of the following general barriers contribute to the general defense of the host: nutrition, acute-phase reactants, fever, age, and genetics.
2. What are some nonspecific host defense mechanisms that help prevent entry of microorganisms into host tissues?

Physical Barriers

Physical or mechanical barriers, along with the host's secretions (flushing mechanisms), are the first line of defense against parasitic organisms. Protection of the most important body surfaces by this mechanism is discussed next.

Skin and Mucous Membranes

The intact skin forms a very effective mechanical barrier to parasitic invasion. There are several reasons for this:

1. Few organisms have the innate ability to penetrate the skin because its outer layer consists of thick, closely packed keratinized cells (keratins are scleroproteins comprising the main components of hair, nails, and outer skin cells) that organisms cannot enzymatically attack.
2. Continuous shedding of the outer squamous epithelial cells (desquamation) removes those organisms that do manage to adhere.
3. Relative dryness of the skin slows microbial growth.
4. Mild acidity (pH 5 to 6, due to the breakdown of lipids into fatty acids by the normal skin microbiota) inhibits the growth of many organisms.
5. The normal skin microbiota acts antagonistically against many pathogens; it also occupies attachment sites and competes for nutrients.
6. Sebum liberated from the oil (sebaceous) glands forms a protective film over the surface of the skin.
7. Normal washing (by humans) continually removes organisms.

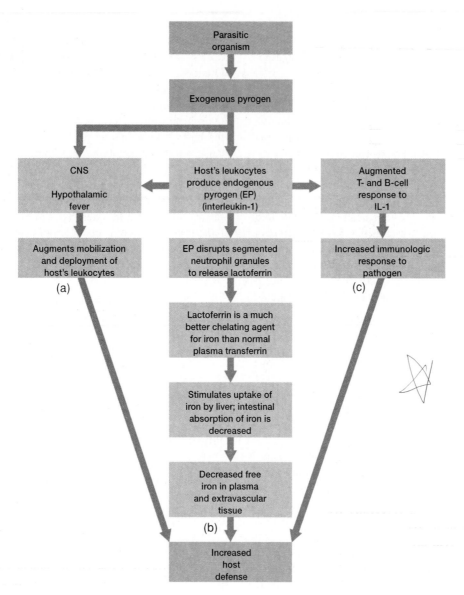

Figure 29.7   Effects of Fever in a Mammalian Host. (*a*) One of the first effects of a fever is to accelerate and augment the mobilization and deployment of the host's leukocytes, which in turn isolate and help destroy the pathogen. (*b*) A fever leads to a reduction of free plasma iron, which limits the growth of those organisms that require a narrow, crucial concentration of iron for their replication and synthesis of toxins. (*c*) One of the immunologic consequences of fever is the production of endogenous pyrogen (IL-1). IL-1 causes the proliferation, maturation, and activation of lymphocytes (T and B cells), which in turn augment the immunologic response of the host to the pathogen.

The mucous membranes of the respiratory, digestive, and urogenital systems withstand parasitic organisms because the intact stratified squamous epithelium and mucous secretions form a protective covering that resists penetration and traps many microorganisms. Furthermore, many mucosal surfaces are bathed in specific antiparasitic secretions. For example, cervical mucus, prostatic fluid, and tears are toxic to many bacteria. One antibacterial substance is lysozyme (muramidase), an enzyme that lyses bacteria by hydrolyzing the $\beta(1 \rightarrow 4)$ bond connecting N-acetylmuramic acid and N-acetylglucosamine in the bacterial cell wall peptidoglycan—especially in gram-positive organisms

(figure 29.8). These mucous secretions also contain specific immune proteins that help prevent the attachment of organisms and significant amounts of iron-binding proteins (lactoferrin) that sequester iron away from these organisms.

Respiratory System

The mammalian respiratory system has formidable defense mechanisms. The average person inhales at least eight microorganisms a minute, or 10,000 each day. Once inhaled, a microorganism must first survive and penetrate the air filtration system of the upper and lower respiratory tracts. Because

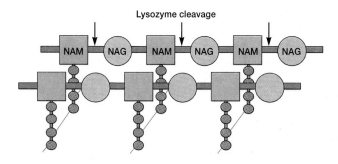

Lysozyme cleavage

Figure 29.8   Action of Lysozyme on the Cell Wall of Gram-Positive Bacteria. In the structure of the cell wall peptidoglycan backbone, the β (1→4) bonds connect alternating N-acetylglucosamine (NAG) and N-acetylmuramic acid (NAM) residues. The chains are cross-linked through the tetrapeptide side chains. Lysozyme splits the molecule at the places indicated by the arrows.

the airflow in these tracts is very turbulent, microorganisms are deposited on the moist, sticky mucosal surfaces. The cilia in the nasal cavity beat toward the pharynx, so that mucus with its trapped microorganisms is moved toward the mouth and expelled. Humidification of the air within the nasal cavity causes many hygroscopic microorganisms to swell and aids the phagocytic process.

The **mucociliary blanket** of the respiratory epithelium traps microorganisms less than 10 μm in diameter that are deposited on the mucosal surface and transports them by ciliary action away from the lungs. Microorganisms larger than 10 μm usually are trapped by hairs and cilia lining the nasal cavity. Coughing and sneezing reflexes clear the respiratory system of microorganisms by expelling air forcefully from the lungs through the mouth and nose, respectively. Salivation also washes microorganisms from the mouth and nasopharyngeal areas into the stomach.

Gastrointestinal Tract

Once parasitic organisms reach the stomach, many are killed by its gastric juice (a mixture of hydrochloric acid, enzymes, and mucus). The very high acidity of gastric juice (pH 2 to 3) is usually sufficient to destroy most organisms and their toxins, although exceptions exist (protozoan cysts, *Clostridium* and *Staphylococcus* toxins). However, many organisms are protected by food particles and reach the small intestine.

Once in the small intestine, pathogens often are damaged by various pancreatic enzymes, bile, enzymes in intestinal secretions, and secretory IgA antibody (*see chapter 30*). **Peristalsis** [Greek *peri*, around, and *stalsis,* contraction] and the normal loss of columnar epithelial cells act in concert to purge intestinal microorganisms. In addition, the normal microbiota of the large intestine (*see figure 28.8*) is extremely important in preventing the establishment of pathogenic organisms. For example, many normal commensals in the intestinal tract produce

metabolic products, such as fatty acids, that prevent "unwanted" microorganisms from becoming established. Other normal microbiota take up attachment sites and compete for nutrients.

Secretory IgA (p. 595).

Genitourinary Tract

Under normal circumstances the kidneys, ureters, and urinary bladder of mammals are sterile. Urine within the urinary bladder is also sterile. However, in both the male and female, a few bacteria are usually present in the distal portion of the urethra (*see figure 28.8*).

The factors responsible for this sterility are complex. For example, urine kills some bacteria due to its low pH and the presence of urea and other metabolic end products (uric acid, hippuric acid, indican, fatty acids, mucin, enzymes). The kidney medulla is so hypertonic that few organisms can survive. The lower urinary tract is flushed with urine and some mucus 4 to 10 times each day, eliminating potential pathogens. In males the anatomical length of the urethra (20 cm) provides a distance barrier that excludes microorganisms from the urinary bladder. Conversely the short urethra (5 cm) in females is more readily traversed by microorganisms; this explains why general urinary tract infections are 14 times more common in females than in males.

The vagina has another unique defense. Under the influence of estrogens, the vaginal epithelium produces increased amounts of glycogen that acid-tolerant *Lactobacillus acidophilus* bacteria called Döderlein's bacilli (*see chapter 28*) degrade to form lactic acid. Normal vaginal secretions contain up to 10^8 Döderlein's bacilli per ml. Thus an acidic environment (pH 3 to 5) unfavorable to most organisms is established. Cervical mucus also has some antibacterial activity.

The Eye

The conjunctiva is a specialized mucus-secreting epithelial membrane that lines the interior surface of each eyelid and the exposed surface of the eyeball. It is kept moist by the continuous flushing action of tears (lacrimal fluid) from the lacrimal glands. Tears contain large amounts of lysozyme and other antimicrobial substances.

---

1. Why is the skin such a good first line of defense against parasitic invasion?

2. How do intact mucous membranes resist parasitic invasion of the host?

3. Describe the different antimicrobial defense mechanisms that operate within the respiratory system of mammals.

4. What factors operate within the gastrointestinal system that help prevent the establishment of pathogenic microorganisms?

5. Except for the anterior portion of the urethra, why is the genitourinary tract a sterile environment?

Chemical Barriers

Mammalian hosts have a chemical arsenal with which to combat the continuous onslaught of parasitic organisms. Some of these chemicals (gastric juices, salivary glycoproteins, lysozyme, oleic acid on the skin, urea) have already been discussed with respect to the specific body site(s) they protect. In addition, tissue extracts, blood, lymph, and other body fluids contain a potpourri of defensive chemicals such as antibodies and complement (*see chapter 32*), fibronectin, hormones, beta-lysin and other polypeptides, interferons, and bacteriocins.

Fibronectin

Fibronectin is a high molecular weight glycoprotein that can interact with certain bacteria. For example, it binds to surface components of *S. aureus* and groups A, C, and G streptococci. This aids in the nonspecific clearance of the bacteria from the body. Fibronectin also covers the receptors of certain epithelial cells to block the attachment of many bacteria.

Hormones

The effects of various mammalian hormones on host defense mechanisms are just beginning to be unraveled. The depressive effects of the corticosteroids on the inflammatory response and immune system are well known. Estrogen's effect on the microbiota of the vagina varies during the menstrual cycle; nonspecific resistance increases with rises in estrogen concentration. The activity of testosterone and adrenal hormones in acne vulgaris caused by *Propionibacterium acnes* has been previously discussed (*see chapter 28*).

Beta-Lysin and Other Polypeptides

Beta-lysin is a cationic polypeptide released from blood platelets; it can kill some gram-positive bacteria by disrupting their plasma membranes. Other cationic polypeptides include leukins, plakins, and phagocytin. A zinc-containing polypeptide, known as the prostatic antibacterial factor, is an important antimicrobial substance secreted by the prostate gland in males.

Interferons

Interferons are a family of related low molecular weight, regulatory glycoproteins (cytokines; *see table 31.1*) produced by many eucaryotic cells in response to numerous inducers: a virus infection, double-stranded RNA, endotoxins, antigenic stimuli, mitogenic (stimulating mitosis) agents, and many parasitic organisms capable of intracellular growth (*Listeria monocytogenes*, chlamydiae, rickettsias, protozoa). Interferons usually are species specific but virus nonspecific. Currently human interferons are classified as IFN-α, IFN-β, and IFN-γ. Some of the ways in which interferon renders cells resistant to virus infections are described in figure 29.9.

Tumor Necrosis Factor Alpha

Tumor necrosis factor alpha (TNF-α) is released from mononuclear phagocytes and some T cells in response to lipopolysaccharides or bacteria such as *Mycobacterium tuberculosis*. In addition to tumoricidal activity, TNF-α has many biological activities and has been recognized as an important inflammatory mediator. It affects a variety of cell types, including polymorphonuclear cells, endothelial cells, fibroblasts, and macrophages. Recently it has been found that TNF-α can activate macrophages, which inhibits pathogen multiplication in vitro.

Bacteriocins

As previously noted, the first line of defense against parasitic organisms is the host's anatomical barrier, consisting of the skin and mucous membranes. These surfaces are colonized by normal microbiota, which by themselves provide a biological barrier against uncontrolled proliferation of foreign microorganisms. Many of these normal bacteria synthesize and release plasmid-encoded substances called **bacteriocins** that are lethal to related species. Bacteriocins may give their producers an adaptive advantage against other bacteria. Sometimes they may increase bacterial virulence by damaging host cells such as mononuclear phagocytes.

Most bacteriocins that have been identified are peptides or proteins and are produced by gram-negative bacteria. (However, recently it has been discovered that some gram-positive bacteria produce bacteriocin-like peptides.) For example, *E. coli* synthesizes bacteriocins called **colicins**, which are coded for by several different plasmids (ColB, ColE1, ColE2, ColI, and ColV). Some colicins bind to specific receptors on the cell envelope of sensitive target bacteria and cause cell lysis, attack specific intracellular sites such as ribosomes, or disrupt energy production.

---

1. How does fibronectin prevent the establishment of a parasitic organism?
2. Give a specific example of how a hormone can affect the establishment of a parasitic organism.
3. How does beta-lysin function against gram-positive bacteria?
4. Describe how interferon confers a protective role to a mammalian cell against viral attack.
5. How do bacteriocins function?

Biological Barriers

Once an organism has breached the first line of defense, it encounters the host's second line of defense: cells that can be mobilized against invading organisms to form a living barrier. In addition to the normal indigenous microbiota, this second line of defense involves cells initially derived from bone marrow cells. Most of these are phagocytes of either the **mononuclear phagocyte system (MPS)** or the **reticuloendothelial system (RES).**

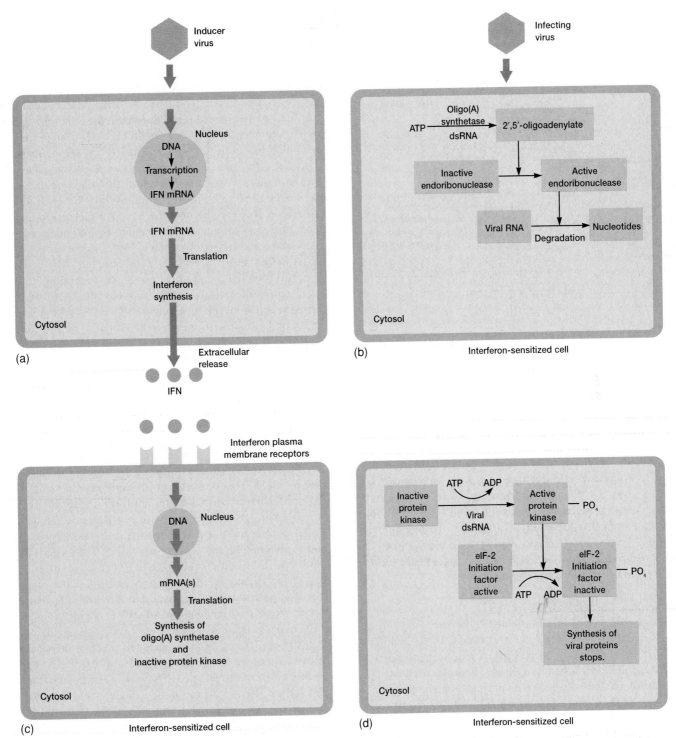

Figure 29.9   The Antiviral Action of Interferon. (*a*) Interferon (IFN) synthesis and release is often induced by a virus infection or double-stranded RNA (dsRNA). (*b*) Interferon binds to a ganglioside receptor on the plasma membrane of a second cell and triggers the production of enzymes that render the cell resistant to virus infection. The two most important such enzymes are oligo(A) synthetase and a special protein kinase. (*c*) When an interferon-stimulated cell is infected, viral protein synthesis is inhibited by an active endoribonuclease that degrades viral RNA. (*d*) An active protein kinase phosphorylates and inactivates the initiation factor eIF-2 required for viral protein synthesis.

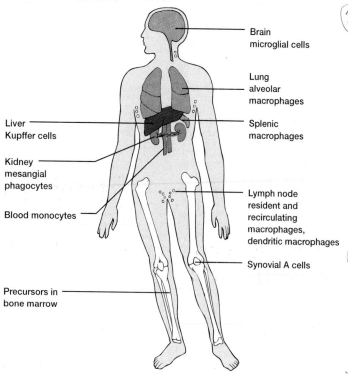

Figure 29.10   The Reticuloendothelial System. This system consists of tissue (such as found within the liver, spleen, and lymph nodes) containing "fixed" or immobile phagocytes that have specific names depending on their location.

The RES is the collection of fixed macrophages, monocytes, and associated cells that are located in the liver, spleen, lymph nodes, and bone marrow (figures 29.10 and 29.11a,b). The mononuclear phagocyte system also includes the blood monocytes and the free macrophages that wander about in body tissues. When pathogens in the blood or lymph pass by these cells, they usually are phagocytosed and destroyed.

Normal Indigenous Microbiota

As already mentioned briefly, another important factor in the general defense of the host against infections is the biological barrier established by the normal microbiota. In many parts of the body, conditions favor the growth of normal residents (*see figure 28.8*) and thus suppress colonization by pathogens. Some ways by which normal microbiota may inhibit pathogen colonization include

1. Producing bacteriocins toxic to other bacteria
2. Competing with potential pathogens for space and nutrients
3. Inhibiting infection by preventing pathogens from attaching to host surfaces
4. Influencing specific clearing mechanisms to rid the body or a particular area of pathogens

Inflammation

Inflammation [Latin, *inflammatio,* to set on fire] is an important nonspecific defense reaction to tissue injury, such as that caused by a pathogen, wound, complement C5a, or immune reactions. The inflammatory response begins when injured tissue cells release chemical signals (inflammatory mediators) that activate the inner lining (endothelium) of nearby capillaries (figure 29.12). Within the capillaries **selectins** (a family of cell adhesion molecules) are displayed on the activated endothelial cells—first P-selectin and then E-selectin. These adhesion molecules randomly attract and attach neutrophils to the endothelial cells, slow the neutrophils down, and cause them to roll along the endothelium. As the neutrophils roll along the endothelium, they encounter the inflammatory mediators that act as activating signals. These signals activate **integrins (adhesion receptors) on the neutrophils.** The neutrophil integrins then tightly attach to endothelial adhesion molecules such as the intracellular adhesion molecule-1 (ICAM-1) and vascular cell adhesion molecule-1 (VCAM-1). This causes the neutrophils to stick to the endothelium and stop rolling. The neutrophils now undergo dramatic shape changes, squeeze through the endothelial wall (extravasation) into the interstitial tissue fluid, migrate to the site of injury, and attack the pathogen or other cause of the tissue damage. Depending on the severity and nature of tissue damage, other types of leukocytes (e.g., lymphocytes, monocytes, and macrophages) may follow the neutrophils.

The inflammatory mediators that are released by the injured tissue cells also raise the acidity in the surrounding extracellular fluid. This decrease in pH activates the extracellular enzyme kallikrein, which splits bradykinin from its long precursor chain. Bradykinin binds to receptors on the capillary wall, opening the junctions between cells and allowing fluid and infection-fighting leukocytes to leave the capillary and enter the infected tissue (figure 29.12). Simultaneously bradykinin (figure 29.13) binds to mast cells in the connective tissue associated with most small blood vessels. This activates the mast cells by causing an influx of calcium ions, which leads to degranulation and release of preformed mediators such as histamine. If nerves in the infected area are damaged, they release substance P, which also binds to mast cells, boosting preformed-mediator release. Histamine in turn makes the intercellular junctions in the capillary wall wider so that more fluid, leukocytes, kallikrein, and bradykinin precursors move out, causing edema. Bradykinin then binds to nearby capillary cells and stimulates the production of prostaglandins (PGE_2 and $PGF_{2\alpha}$) to promote tissue swelling in the infected area. Prostaglandins also bind to free nerve endings, making them fire and start a pain impulse.

The change in mast cell plasma membrane permeability associated with activation allows phospholipase A_2 to release arachidonic acid. Arachidonic acid is then metabolized by the

(a)   Granular leukocytes   Agranular leukocytes

(b)

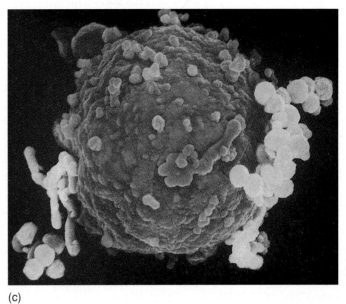

(c)

Figure 29.11   Cells of the Immune System. (*a*) An illustration of the five common types of leukocytes. (*b*) A light micrograph of neutrophils phagocytosing streptococci (×1,000). (*c*) An artificially colored scanning electron micrograph of a B lymphocyte (red) covered with bacteria (yellow) (×8,500).

cyclooxygenase or lipoxygenase pathways, depending on mast cell type. The newly synthesized mediators include prostaglandins E_2 and $F_{2\alpha}$, thromboxane A_2, slow reacting substance (SRS), and leukotrienes (LTC_4 + LTD_4). All play various roles in the inflammatory response.

These physiological changes lead to five classic symptoms of inflammation: redness, heat, pain, swelling, and altered function. The net effect is defensive in nature. In summary, the offending pathogen is neutralized and eliminated by a series of events, the most important of which are the following:

1. The increase in blood flow and capillary dilation bring into the area more antimicrobial factors and leukocytes. Dead cells also release antimicrobial factors.

2. The rise in temperature stimulates the inflammatory response and may inhibit microbial growth.

3. A fibrin clot often forms and may limit the spread of the invaders so that they remain localized.

4. Phagocytes collect in the inflamed area and phagocytose the pathogen. In addition, leukocytosis-promoting factors such as IL-1, complement fragments, and epinephrine stimulate the bone marrow to release neutrophils and increase the rate of granulocyte production.

---

1. What are some ways by which the normal indigenous microbiota of a host prevent colonization by pathogens?

2. What major events occur during an inflammatory reaction, and how do they contribute to pathogen destruction?

3. What causes degranulation of mast cells?

4. How do the cyclooxygenase and lipoxygenase pathways work?

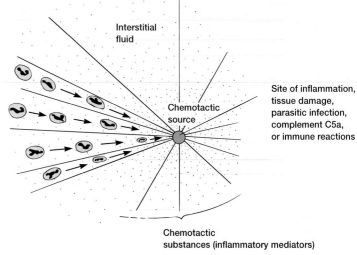

Interstitial fluid

Chemotactic source

Site of inflammation, tissue damage, parasitic infection, complement C5a, or immune reactions

Chemotactic substances (inflammatory mediators)

(d) **Chemotaxis of leukocytes**

🔴 Figure 29.12   **Physiological Events of the Acute Inflammatory Response.** (*a*) The diameter of the arteriole controls the normal blood flow into a capillary bed. Inflammation causes arteriole dilation and increased blood flow (hyperemia) into the affected tissue. The capillary becomes distended and produces the gross red-purple discoloration of tissue due to an increased blood content. (*b*) The junctions between capillary endothelial cells are sufficiently tight to keep large molecules within the lumen. In inflammation, gaps are created that allow leakage of fluid and large molecules into the interstitial space. The fluid leakage produces swelling (edema). (*c*) Normally leukocytes (such as neutrophils) are carried in the lumen of a capillary. During inflammation selectins and integrins cause leukocytes to adhere to the capillary wall, insert pseudopods between the endothelial cells, and squeeze through (extravasation) by a process called diapedesis. (*d*) After leaving the capillary, the leukocytes are attracted (chemotaxis) to the source of the inflammation by various chemotactic or chemokinetic substances (extravascular stimulus). Once at the infection site, leukocytes phagocytose the microorganisms or dead tissue.

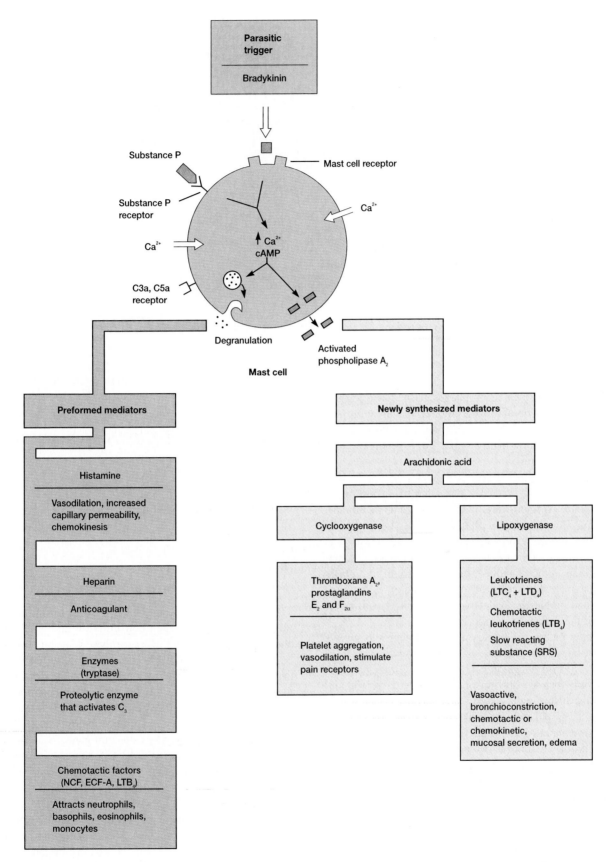

Figure 29.13   Biochemical Events of Inflammation. Mast cell activation and physiological effects of mast cell–derived preformed and newly synthesized mediators that lead to the inflammatory response.

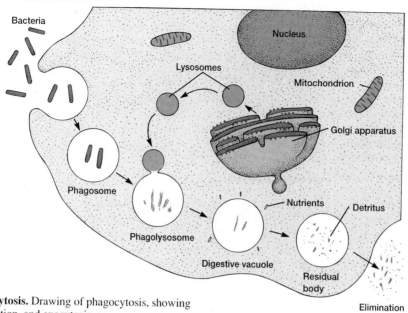

Figure 29.14   Phagocytosis. Drawing of phagocytosis, showing ingestion, intracellular digestion, and exocytosis.

Phagocytosis

Phagocytic cells (monocytes, macrophages, tissue macrophages, and neutrophils) must be able to recognize the invading microorganism (see figure 29.11a,b). They do this with receptors on their surface that allow them to attach nonspecifically to a variety of microorganisms.

Once ingested by phagocytosis, the membrane-enveloped microorganism(s) is delivered to a lysosome by fusion of the phagocytic vacuole (**phagosome**) with the lysosome membrane, forming a new vacuole called a **phagolysosome** (figure 29.14). Because lysosomes seem to disappear from the cytoplasm during this fusion, the process often is called degranulation. Lysosomes contribute to the phagolysosome a variety of hydrolases such as lysozyme, phospholipase A$_2$, ribonuclease, deoxyribonuclease, and proteases. Collectively these participate in the destruction of the entrapped microorganism.

Intracellular pathogens employ diverse strategies to survive within macrophages. For example, *Mycobacterium tuberculosis, M. leprae, Legionella pneumophila,* and *Toxoplasma gondii* inhibit phagosome fusion with lysosomes, thereby preventing exposure to toxic lysosomal contents. In contrast, *Trypanosoma cruzi, Listeria monocytogenes,* and *Shigella flexneri* lyse the phagosomal membrane and escape into the cytoplasm. A third group of microorganisms (*Leishmania* spp., *Mycobacterium leprae, Salmonella typhimurium*) is found within the macrophage phagolysosomal compartment, where they apparently resist inactivation by lysosomal factors.

Lysosomes and endocytosis (pp. 79–80).

Besides the oxygen-independent lysosomal hydrolases, macrophage and neutrophil lysosomes contain oxygen-dependent enzymes that can produce reactive oxygen intermediates (ROIs) such as the superoxide radical (O$_2^-\cdot$), hydrogen peroxide (H$_2$O$_2$), singlet oxygen (^{1}O$_2$), and hydroxyl radical (OH$\cdot$). Neutrophils also contain myeloperoxidase and produce hypochlorous acid. Some reactions forming these toxic products are shown here.

Superoxide formation

$$NADPH + 2O_2 \xrightarrow{\text{NADPH oxidase}} 2O_2^-\cdot + H^+ + NADP^+$$

Hydrogen peroxide formation

$$2O_2^-\cdot + 2H^+ \xrightarrow{\text{superoxide dismutase}} H_2O_2 + O_2$$

Hypochlorous acid formation

$$H_2O_2 + Cl^- \xrightarrow{\text{myeloperoxidase}} HOCl + OH^-$$

Singlet oxygen formation

$$ClO^- + H_2O_2 \rightarrow {}^1O_2 + Cl^- + H_2O$$

Hydroxyl radical formation

$$O_2^-\cdot + H_2O_2 \rightarrow OH\cdot + OH^- + O_2$$

These reactions result from the **respiratory burst** that accompanies the increased oxygen consumption and ATP generation needed for phagocytosis. Because these reactions occur as soon as the phagosome is formed, lysosome fusion is not necessary for the respiratory burst. The toxic oxygen products and hypohalites thus produced are effective in killing invading microorganisms.

Recently macrophages, neutrophils, and mast cells have been shown to form reactive nitrogen intermediates (RNIs) when stimulated by interferons or the tumor necrosis factor. These molecules include nitric oxide (NO) and its oxidized forms, nitrite (NO$_2^-$) and nitrate (NO$_3^-$). The RNIs are very potent cytotoxic agents, and may be either released from cells or generated within cell vacuoles. Nitric oxide is probably the most effective RNI. Macrophages produce it from the amino acid arginine when stimulated by cytokines. Nitric oxide can block cellular respiration by complexing with the iron in electron transport proteins. Macrophage killing of the herpes simplex virus, the protozoa *Toxoplasma gondii* and *Leishmania major,* the opportunistic fungus *Cryptococcus neoformans,* the metazoan pathogen *Schistosoma mansoni,* and tumor cells involves RNIs.

Neutrophils also produce a newly defined family of broad-spectrum antimicrobial peptides called **defensins**. There are four human defensins, called human neutrophil proteins (HNPs): HNP-1, 2, 3, and 4. These defensins are synthesized by myeloid precursor cells during their sojourn in the bone marrow, and are then stored in the cytoplasmic granules of mature cells. This compartmentation strategically locates defensins for extracellular secretion or delivery to phagocytic vacuoles. Susceptible microbial targets include a variety of gram-positive and gram-negative bacteria, yeasts and molds, and some viruses. Defensins act against bacteria and fungi by permeabilizing cell membranes. They form voltage-dependent membrane channels that allow ionic efflux. Antiviral activity involves direct neutralization of enveloped viruses; nonenveloped viruses are not affected by defensins.

---

1. Once a phagolysosome forms, how is the entrapped organism destroyed?
2. What is the purpose of the respiratory burst that occurs within macrophages?
3. What are defensins? How do they function?

Summary

1. Parasitism is a type of symbiosis between two species in which the smaller organism is physiologically dependent on the larger one, termed the host. The parasitic organism usually harms its host in some way.

2. An infection is the colonization of the host by a parasitic organism. An infectious disease is the result of the interaction between the parasitic organism and its host, causing the host to change from a state of health to a diseased state. Any organism that produces such a disease is a pathogen.

3. Pathogenicity refers to the quality or ability of an organism to produce pathological changes or disease. Virulence refers to the degree or intensity of pathogenicity of an organism and is measured experimentally by the LD_{50} or ID_{50}.

4. Pathogens or their products can be transmitted to a host by either direct or indirect means. Transmissibility is the initial requisite in the establishment of an infectious disease.

5. Special adherence factors allow pathogens to bind to specific receptor sites on host cells and colonize the host.

6. Pathogens can enter host cells by both active and passive mechanisms. Once inside, they can produce specific products and/or enzymes that promote dissemination throughout the body of the host.

7. The pathogen generally seeks out the area of the host's body that provides the most favorable conditions for its growth and multiplication.

8. Intoxications are diseases that result from the entrance of a specific toxin into a host. The toxin can induce the disease in the absence of the toxin-producing organism. Toxins produced by pathogens can be divided into two main categories: exotoxins and endotoxins.

9. Exotoxins are soluble, heat-labile, toxic proteins produced by the pathogen as a result of its normal metabolism. They can be categorized as neurotoxins, cytotoxins, or enterotoxins. Most exotoxins conform to the AB model in which the A subunit or fragment is enzymatic and the B subunit or fragment, the binding portion. Several mechanisms exist by which the A component enters target cells.

10. Endotoxins are heat-stable, toxic substances that are part of the cell wall lipopolysaccharide of some gram-negative bacteria. Most endotoxins function by initially activating Hageman Factor, which in turn activates one to four humoral systems. These include the intrinsic blood clotting cascade, complement activation, fibrinolytic system, and kininogen system. Endotoxins also stimulate macrophages to release cytokines such as IL-1, IL-6, and TNF-α.

11. Many direct factors (nutrition, physiology, age, fever, genetics, race) and equally as many indirect factors (personal hygiene, socioeconomic status, environmental conditions) contribute in some degree to all host-parasite relationships. At times they favor the establishment of the parasitic organism; at other times they provide some measure of defense to the host.

12. Physical (mechanical) barriers along with host secretions (flushing mechanisms) are the host's first line of defense against pathogens. Examples include the skin and mucous membranes; and the epithelia of the respiratory, gastrointestinal, and genitourinary systems.

13. Mammalian hosts have specific chemical barriers that help combat the continuous onslaught of parasitic organisms. Examples include fibronectin, hormones, interferons, beta-lysin and other cationic polypeptides.

14. Many normal bacteria found on the skin and mucous membranes produce plasmid-encoded proteins called bacteriocins. These are released into the host's environment and are lethal to related bacterial species.

15. The second line of defense that mammalian hosts possess against parasitic invasion consists predominately of phagocytotic cells of the reticuloendothelial system and are widely dispersed throughout the body of the host.

16. Inflammation is one of the host's nonspecific defensive reactions to a tissue injury that may be caused by a pathogen. Some major events that occur are increased blood supply, increased capillary permeability, leukocyte migration, and phagocytosis.

17. Phagocytosis involves the recognition, ingestion, and destruction of parasitic organisms by lysosomal enzymes, superoxide radicals, hydrogen peroxide, defensins, RNIs, and hypohalite ions.

Key Terms

Questions for Thought and Review

1. Why does a parasitic organism not have to be a parasite?

2. In general, infectious diseases that are commonly fatal are newly evolved relationships between the parasitic organism and the host. Why is this so?

3. What does an organism require to be parasitic?

4. What determinants provide the parasitic organism with the ability to colonize and invade the host?

5. How do the degree and lethality of the pathogenic alterations induced by a parasitic organism influence its establishment?

6. How do most exotoxins enter a host cell?

7. What is the mode of action of several well-known exotoxins?

8. What is the difference between the general properties of endotoxins and exotoxins?

9. How can general nonspecific barriers help to prevent the establishment of a parasitic organism within a host? How could they aid in the establishment of a parasitic organism?

10. What constitutes a mammalian host's first line of defense against invasion by a pathogen?

11. Describe how the physiology of each of the following systems contributes to the defense of the host: respiratory, gastrointestinal, and genitourinary.

12. How do specific chemicals produced by a mammalian host contribute to its general defense against infectious organisms?

13. How can inflammation be beneficial to a host and detrimental to a parasitic organism?

14. What is the relationship between the respiratory burst and phagocytosis?

15. What role do defensins play in host defense?

Additional Reading

General

Ayoub, E. M., editor. 1990. *Microbial determinants of virulence and host response.* Washington, D.C.: ASM Press.

Ewald, P. 1994. *Evolution of infectious diseases.* New York: Oxford University Press.

Groisman, E., and Ochman, H. 1994. How to become a pathogen. *Trends in Microbiol.* 2(8):289–93.

Hoeprich, P. D.; Jordan, M.; and Ronald, A. 1994. *Infectious diseases: A treatise of infectious processes.* Philadelphia: J. B. Lippincott Company.

Ofek, I., and Doyle, R. 1994. *Bacterial adhesion to cells and tissues.* New York: Chapman & Hall.

Salyers, A., and Whitt, D. 1994. *Bacterial pathogenesis: A molecular approach.* Washington, D.C.: ASM Press.

Smith, H. 1977. Microbial surfaces in relation to pathogenicity. *Bacteriol. Rev.* 41:475–500.

Host-Parasite Relationships (Parasitism)

Ewald, P. W. 1993. The evolution of virulence. *Sci. Am.* 268(4):86–93.

Isenberg, H. D. 1988. Pathogenicity and virulence: Another view. *Clin. Microbiol. Rev.* 1(1):40–53.

Toft, C. A. 1991. *Parasite-host associations: Coexistence or conflict?* New York: Oxford University Press.

Determinants of Infectious Disease

Bhakdi, S., and Tranum-Jensen, J. 1991. Alpha-toxin of *Staphylococcus aureus. Microbiol. Rev.* 55(4):733–51.

Bizzini, B. 1979. Tetanus toxin. *Microbiol. Rev.* 43:224–40.

Brubaker, R. R. 1985. Mechanisms of bacterial virulence. *Ann. Rev. Microbiol.* 39:21–50.

Eidels, L. R. L., and Hart, D. A. 1983. Membrane receptors for bacterial toxins. *Microbiol. Rev.* 47:596–614.

Gill, D. M. 1982. Bacterial toxins: A table of lethal amounts. *Microbiol. Rev.* 46:86–88.

Hurley, J. C. 1995. Endotoxemia: Methods of detection and clinical correlates. *Clin. Microbiol. Rev.* 8(2):268–92.

Jack, R.; Tagg, J.; and Ray, B. 1995. Bacteriocins of gram-positive bacteria. *Microbiol. Rev.* 59(2):171–200.

Johnson, J. R. 1991. Virulence factors in *Escherichia coli* urinary tract infection. *Clin. Microbiol. Rev.* 4(1):80–128.

Krueger, K. M., and Barbieri, J. T. 1995. The family of bacterial ADP-ribosylating exotoxins. *Clin. Microbiol. Rev.* 8(1):34-47.

Moss, J., and Vaughan, M. 1990. *ADP-ribosylating toxins and G proteins.* Washington, D.C.: ASM Press.

Petri, W. A. 1991. Invasive amebiasis and the galactose-specific lectin of *Entamoeba histolytica. ASM News* 57(6):299–306.

Rietschel, E. T., and Brade, H. 1992. Bacterial endotoxins. *Sci. Am.* 267(2):54–61.

Schantz, E. J., and Johnson, E. A. 1992. Properties and use of botulinum toxin and other microbial neurotoxins in medicine. *Microbiol. Rev.* 56(1):80–99.

Slifkin, M., and Doyle, R. J. 1990. Lectins and their application to clinical microbiology. *Clin. Microbiol. Rev.* 3(5):197–218.

Spangler, B. D. 1992. Structure and function of cholera toxin and the related *Escherichia coli* heat-labile enterotoxin. *Microbiol. Rev.* 56(4):622–47.

Sugiyama, H. 1980. *Clostridium botulinum* neurotoxin. *Microbiol. Rev.* 44:419–48.

Waters, V. L., and Crosa, J. H. 1991. Colicin V virulence plasmids. *Microbiol. Rev.* 55(3):437–50.

General or Nonspecific Host Immune Defense Mechanisms

Baggiolini, M.; Boulay, F.; Badwey, J.; and Curnutte, J. 1993. Activation of neutrophil leukocytes: Chemoattractant receptors and respiratory burst. *FASEB J.* 7(11):1004–10.

Beaman, L., and Beaman, B. L. 1984. The role of oxygen and its derivatives in microbial pathogenesis and host defense. *Ann. Rev. Microbiol.* 38:27–48.

Collier, H. J. O. 1962. Kinins. *Sci. Am.* 207:111–18.

Costerton, J. W.; Geesey, G. G.; and Cheng, K.-J. 1978. How bacteria stick. *Sci. Am.* 238:86–91.

Edelson, R. L., and Fink, J. M. 1985. The immunologic function of the skin. *Sci. Am.* 256:46–53.

Gordon, J., and Minks, M. A. 1981. The interferon renaissance: Molecular aspects of induction and action. *Microbiol. Rev.* 45:244–51.

Johnson, H. M.; Bazer, F. W.; Szente, B. E.; and Jarpe, M. A. 1994. How interferons fight disease. *Sci. Am.* 270:68–75.

Lancaster, Jr., J. R. 1992. Nitric oxide in cells. *Am. Scientist* 80(3):248–59.

Law, D. 1994. Adhesion and its role in the virulence of enteropathogenic *Escherichia coli. Clin. Microbiol. Rev.* 7(2):152–73.

Lehrer, R. I.; Ganz, T.; and Selsted, M. E. 1990. Defensins: Natural peptide antibiotics from neutrophils. *ASM News* 56(6):315–18.

McNabb, D. C., and Tomasi, T. B. 1981. Host defense mechanisms at mucosal surfaces. *Ann. Rev. Microbiol.* 33:477–96.

Mock, M., and Ullmann, A. 1993. Calmodulin-activated bacterial adenylate cyclases as virulence factors. *Trends in Microbiol.* 1(5):187–91.

Nagai, Y. 1993. Protease-dependent virus tropism and pathogenicity. *Trends in Microbiol.* 1(3):81–87.

Roberts, N. J. 1979. Temperature and host defense. *Microbiol. Rev.* 43:241–48.

Rumyantsev, S. N. 1992. Observations on constitutional resistance to infection. *Immunol. Today* 13(5):184–87.

Taylor, P. W. 1983. Bactericidal and bacteriolytic activity against gram-negative bacteria. *Microbiol. Rev.* 47:46–65.

PART EIGHT

The Fundamentals of Immunology

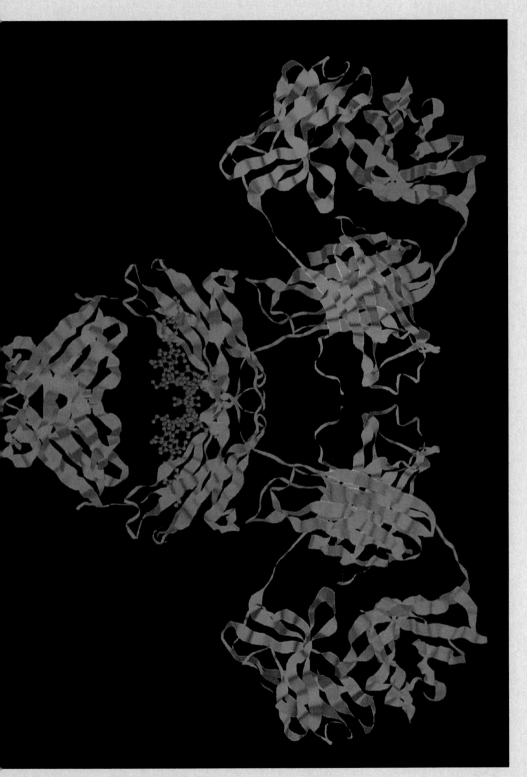

Antibodies, or immunoglobulins, are a group of
glycoproteins present in the blood serum and tissue
fluids of mammals. This is a photograph of the
computer-generated model of human IgGl antibody.
In this model, the heavy chains of the antibody are
colored pink, the light chains are blue, and the
carbohydrate chain is red.

CHAPTER 30
The Immune Response: Antigens and Antibodies

The remarkable capacity of the immune system to respond to many thousands of different substances with exquisite specificity saves us all from certain death by infection.
—Martin C. Raff

The three greatest discoveries of medicine in the last 200 years are antisepsis, antibiotics, and vaccines.
—Stanley A. Plotkin

Outline

Concepts

1. The major function of the immune response in higher animals is to provide specific protection (immunity) against harmful microorganisms, cancer cells, and certain macromolecules that are collectively termed antigens (immunogens).

2. Terminology has been developed to describe the different types of immunity: nonspecific versus specific, natural versus artificial, and active versus passive.

3. Two major classes of lymphocytes, B cells and T cells, specifically recognize and respond to antigens. B cells form immune products called immunoglobulins (antibodies), whereas T cells become activated or sensitized to perform several functions.

4. Immunoglobulins specifically interact with free antigens or cells bearing antigens and mark them for subsequent removal.

5. The basic structure of the immunoglobulin molecule consists of four polypeptide chains: two heavy chains and two light chains.

6. There are five immunoglobulin (Ig) classes based on physicochemical and biological properties. These are IgG, IgM, IgA, IgD, and IgE.

7. Immunoglobulins can be produced naturally in an animal's body in response to immunizations. They, along with catalytic antibodies, also can be produced in vitro through the use of hybridomas.

8. Two distinguishing characteristics of immunoglobulins are their diversity and specificity.

The nonspecific host defenses that are integral components of an animal's resistance to infectious agents or their products were discussed in the previous chapter. In this chapter another aspect of defense called specific immunity is examined. Specific immunity, in contrast to nonspecific defenses or resistance, involves the production of a highly specific defensive response when the host is exposed to an infectious agent or microorganism, its products, some tumor cells, or certain macromolecules. The body recognizes these substances as not belonging to itself and develops a specific immune response, leading to destruction or neutralization of the substances. The necessity of this system to human survival and the problems that it can cause under certain circumstances are presented in this chapter and the next.

As discussed in chapter 29, vertebrates (including humans) are continuously exposed to microorganisms, their metabolic products, or other foreign macromolecules that can cause disease. Fortunately these animals are equipped with an immune system that protects them against this exposure. **Immunity** [Latin *immunis,* free of burden] refers to the general ability of a host to resist a particular disease. The **immune response** that results is a specific and complex series of defensive actions widely distributed throughout the animal's body (figure 30.1). Each particular immune response is a unique local sequence of events, shaped by the nature of the challenge. **Immunology** is the science that deals with these immune responses and the many phenomena responsible for this type of defense.

Nonspecific Resistance

Nonspecific resistance refers to those general mechanisms inherited as part of the innate structure and function of each animal. These act in concert as a first line of defense against infectious microorganisms, their products (toxins), or foreign macromolecules before they cause disease. The various nonspecific defense mechanisms include biological barriers (inflammation, phagocytosis), chemical barriers (enzymatic action, interferons, beta-lysin, fibronectin, complement), general barriers (fever), and physical barriers (skin, mucous membranes). These are discussed in detail in chapter 29.

Specific Immunity

If the nonspecific defenses are breached, specific immunity (the immune response) is called on to protect the host. This system consists of several immunologic mechanisms in which lymphocytes recognize the presence of particular foreign agents or substances termed antigens (immunogens) and act to eliminate them. Elimination can occur by direct lymphocytic destruction

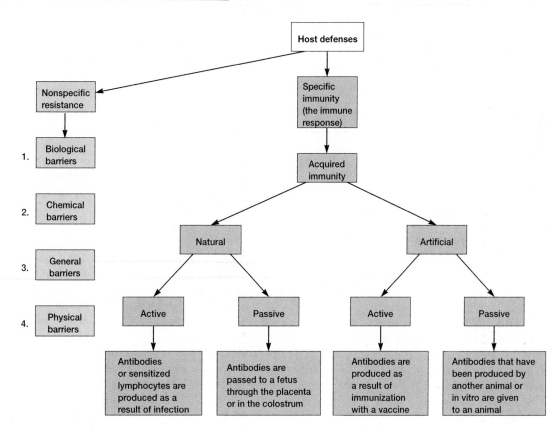

Figure 30.1   Different Host Defenses. This illustration can be used as a guide to the text description of the immune response. The details of the human immune system are astonishingly complex; many events have been omitted so that the overall organization of the immune response can be clearly seen.

TABLE 30.1   Examples of Vaccines Used by the United States to Prevent Viral and Bacterial Diseases in Humans

Disease	Vaccine	Booster	Recommendation
Viral Diseases			
Chickenpox	Attenuated Oka strain (Varivax)	None	Children 12-18 months; older children who have not had chickenpox
Hepatitis A	Inactivated virus (Havrix)	6–12 months	International travelers
Hepatitis B	HB viral antigen (Engerix-B, Recombivax HB)	None	High-risk medical personnel
Influenza	Inactivated virus or viral components	Yearly	Chronically ill individuals and those over 65
Measles, Mumps, Rubella	Attenuated viruses (combination MMR vaccine)	None	Children 15–19 months old
Poliomyelitis	Attenuated (oral poliomyelitis vaccine, OPV) or inactivated vaccine	Adults as needed	Children 2–3 years old
Rabies	Inactivated virus	None	For individuals in contact with wildlife, animal control personnel, veterinarians
Yellow fever	Attenuated virus	10 years	Military personnel and individuals traveling to endemic areas
Bacterial Diseases			
Cholera	Fraction of *Vibrio cholerae*	6 months	Individuals in endemic areas
Diphtheria, Pertussis, Tetanus	Diphtheria toxoid, killed *Bordetella pertussis,* tetanus toxoid (DPT vaccine)	10 years	Children 2–3 months old
Haemophilus influenzae type b	Polysaccharide-protein conjugate (HbCV) or bacterial polysaccharide (HbPV)	None	Children under 5 years of age
Meningococcal infections	Bacterial polysaccharides of serotypes A/C/Y/W-135	None	Military; high-risk individuals
Plague	Fraction of *Yersinia pestis*	Yearly	Individuals in contact with rodents in endemic areas
Pneumococcal pneumonia	Purified *S. pneumoniae* polysaccharide of 23 pneumococcal types	None	Adults over 50 with chronic disease
Tuberculosis	Attenuated *Mycobacterium bovis* (BCG vaccine)	3–4 years	Individuals exposed to TB for prolonged periods of time
Typhoid fever	Killed *Salmonella typhi*	3–4 years	Individuals in endemic areas
Typhus fever	Killed *Rickettsia prowazekii*	Yearly	Scientists and medical personnel in areas where typhus is endemic

of the antigens or by formation of specialized proteins called antibodies that either disrupt antigen function or target the antigen for destruction by other cells.

Acquired Immunity

Acquired immunity refers to the type of specific immunity a host develops after exposure to a suitable antigen, or after transfer of antibodies or lymphocytes from an immune donor. Acquired immunity can be obtained actively or passively by natural or artificial means.

Naturally Acquired Immunity

Naturally acquired active immunity occurs when an individual's immune system contacts an antigenic stimulus such as an infection during normal activities. The immune system responds by producing antibodies and sensitized lymphocytes that inactivate or destroy the antigen. The immunity produced can be either lifelong, as with measles or chickenpox, or last for only a few years, as with tetanus (*see chapter 37*).

Naturally acquired passive immunity involves the transfer of antibodies from one host to another. For example, some of a pregnant woman's antibodies pass across the placenta to her fetus. If the female is immune to diseases such as polio or diphtheria, this placental transfer also gives the newborn immunity to these diseases. Certain other antibodies can pass from the female to her offspring in the first secretions (called **colostrum**) from the mammary glands. Unfortunately naturally acquired passive immunity generally lasts only a short time (weeks or months at the most).

Artificially Acquired Immunity

Artificially acquired active immunity results when an animal is immunized with a vaccine. A **vaccine** consists of a preparation of killed microorganisms; living, weakened (attenuated) microorganisms; or inactivated bacterial toxins (toxoids) that are administered to an animal to induce immunity artificially. Table 30.1 summarizes the principal vaccines used to prevent viral and bacterial diseases in humans.

Artificially acquired passive immunity results when antibodies that have been produced either in an animal or by specific methods in vitro are introduced into a host. Although this type of immunity is immediate, it is short lived (about three weeks). An example would be botulinum antitoxin produced in a horse and given to a human suffering from botulism food poisoning (*see chapter 37*).

1. Give several examples of nonspecific resistance.
2. Contrast active and passive immunity.
3. How does naturally acquired immunity occur?
4. What is a vaccine?
5. Why is artificially acquired passive immunity at times lifesaving?

Origin of Lymphocytes

Lymphocytes are small, nonphagocytic, mononuclear leukocytes that are immunologically competent, or are precursors of such cells (*see figure 29.11*). They lack stainable cytoplasmic granules and are formed in lymphatic tissues. Because lymphocytes (table 30.2) are the pivotal cells of the specific immunologic response, their origin, movement, processing, and general function are summarized here. In chapter 31 their biology is described more thoroughly. Undifferentiated lymphocytes are derived from bone marrow stem cells. They are produced in the bone marrow at a very high rate (10^9 cells per day). Some lymphocytes migrate through the circulatory or lymphatic systems to the secondary lymphoid tissue (thymus, spleen, aggregated lymph nodules in the intestines, and lymph nodes) where they produce lymphocyte colonies.

The undifferentiated lymphocytes that migrate to the thymus undergo special processing and become **T cells** or **T lymphocytes.** (The letter T represents the thymus.) Some T cells are transported away from the thymus and enter the bloodstream where they comprise 70 to 80% of the circulating lymphocytes. Other T cells tend to reside in various organs of the lymphatic system, such as the lymph nodes and spleen. This thymus-dependent differentiation of T cells (or **thymocytes**) occurs during early childhood, and by adolescence the secondary lymphoid organs of the body generally contain a full complement of T cells.

B cells or **B lymphocytes** differentiate in the fetal liver and adult bone marrow (*see figure 29.11c*). (The letter B was originally derived from the **b**ursa of Fabricius, a specialized appendage of the cloaca of chickens where these lymphocytes differentiate.) B cells are distributed by the blood and make up 20 to 30% of the circulating lymphocytes. They also settle in the various lymphoid organs along with the T cells. Figure 30.2 summarizes the development of the T and B cells.

There is a population of lymphoid cells that do not have characteristics of either T or B cells. These are called **null cells** because they lack the specific surface markers of B or T cells, and can be distinguished from them by the presence of cytoplasmic granules. It is currently believed that this population of cells contains most **natural killer (NK) cells** and antibody-dependent cytotoxic T cells (*see chapter 31*). Although these cells are probably of bone marrow origin, their exact lineage is uncertain.

Function of Lymphocytes

Plasma cells are fully differentiated antibody-synthesizing cells that are derived from B lymphocytes. They respond to antigens by secreting antibodies into the blood and lymph. Antibodies are glycoproteins produced by plasma cells after the B cells in

TABLE 30.2   Classes of Lymphocytes

Lymphocyte	Role
T Cells	
T_H (helper) cells; also called CD4[a] cells	Provide "assistance," or potentiate expression of immune function by other lymphocytes
T_S (suppressor) cells; also called CD8 cells	Suppress or impair expression of immune function by other lymphocytes
T_C (cytotoxic) cells; also called CD8 cells	Bring about cytolysis and cell death of "targets"
T_D cells or T_{DTH} (delayed-type hypersensitivity) cells; also called CD4 cells	Recruit and regulate a variety of nonspecific blood cells and macrophages in expression of delayed (Type IV) hypersensitivity reactions
B Cells	
B lymphocytes	Proliferate and mature into antibody-producing cells
Plasma cells	Are mature, active antibody-producing cells
Null Cells	
Natural killer cells	Bring about cytolysis and death of target cells

From Ronald M. Coleman, et al., *Fundamental Immunology*, 2d edition. Copyright © 1992 Wm. C. Brown Communications, Inc. Reprinted by permission of Times Mirror Higher Education Group, Inc., Dubuque, Iowa. All Rights Reserved.

[a]CD stands for *c*luster of *d*ifferentiation, *c*ell-associated *d*ifferentiation antigens, or *c*ell-*d*efined antigens and is the nomenclature recommended by the World Health Organization.

their lineage have been exposed to antigens. Antibodies are specifically directed against the antigen that caused their formation. Because antibodies are soluble in blood and lymph fluids, they provide **humoral** [Latin *humor,* a liquid] **immunity** or **antibody-mediated immunity.** The humoral immune response defends mostly against bacteria, bacterial toxins, and viruses that enter the body's various fluid systems.

T cells do not secrete antibodies. Instead they attack (1) host cells that have been parasitized by viruses or microorganisms, (2) tissue cells that have been transplanted from one host to another, and (3) cancer cells. They also produce **cytokines,** chemical mediators that play specific augmenting and regulatory roles in the immune system (*see pp. 607–11*). Since T cells must physically contact foreign cells or infected cells in order to destroy them, they are said to provide **cell-mediated immunity.**

Null cells (and particularly natural killer cells) destroy tumor cells and virus- and other parasite-infected cells. They also help regulate the immune response. Null cells often exhibit antibody-dependent cellular cytotoxicity. The biology of these three major lymphocyte categories is discussed in more detail in chapter 31 (*pp. 611–20*).

Antigens

The immune system distinguishes between "self" and "nonself" through an elaborate recognition process. Prior to birth the body somehow makes an inventory of the proteins and various other large molecules present (self) and removes most T cells specific for self-determinants. Subsequently self-substances can be distinguished from nonself substances, and lymphocytes can produce specific immunologic reactions against the latter, leading to their removal.

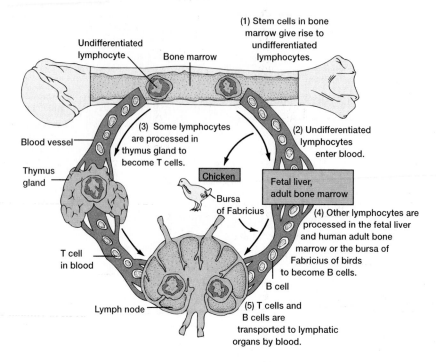

Figure 30.2 **Schematic Representation of Lymphocyte Development.** Bone marrow releases undifferentiated lymphocytes, which, after processing, become T and B cells.

Nonself foreign substances, such as proteins, nucleoproteins, polysaccharides, and some glycolipids, to which lymphocytes respond are called **antigens** (**anti**body **gen**erator). Most antigens are large, complex molecules with a molecular weight generally greater than about 10,000. The ability of a molecule to function as an antigen depends on its size and structural complexity.

In most contexts the term **immunogen** is synonymous with antigen. Sometimes immunogen is used to draw a distinction between a substance capable of reacting only with antibody (antigens or haptens) and a toleragen, a form that induces tolerance. Noting this distinction, for conformity, the term immunogen is used synonymously with antigen in this textbook.

Each antigen can have several **antigenic determinant sites** or **epitopes** (figure 30.3). These areas of the molecule stimulate production of and combine with specific antibodies. Antibodies are formed most readily in response to determinants that project from the foreign molecule or to terminal residues of a specific polymer chain. Chemically determinants include sugars, organic acids and bases, amino acid side chains, hydrocarbons, and aromatic groups.

The number of antigenic determinant sites on the surface of an antigen is its **valence.** If one determinant site is present, the antigen is monovalent. Most antigens, however, have more than one determinant site and are termed multivalent. Multivalent antigens generally elicit a stronger immune response than do monovalent antigens. Sometimes a multivalent antigen, termed a **heterophile** or **heterologous antigen,** can react with antibodies produced in response to a different antigen.

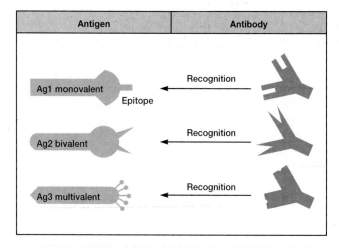

Figure 30.3 **Antigens.** Antigen (Ag) molecules each have a set of antigenic determinants or epitopes. The epitope on monovalent antigen (Ag1) is different from those on the bivalent antigen (Ag2). Some antigens (Ag3) have repeating epitopes; however, each antibody only recognizes one epitope rather than the whole multivalent antigen.

Haptens

Many small organic molecules are not antigenic by themselves but can become antigenic if they bond to a larger carrier molecule such as a protein (figure 30.4). They cannot stimulate antibody formation by themselves but can react with antibodies once formed. Such small molecules are termed **haptens** [Latin *haptein,* to grasp]. When lymphocytes are stimulated by the

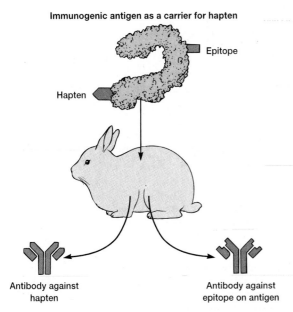

Figure 30.4   Effect of Carrier On Immunogenicity of Hapten.
Haptens, as small molecules, are incapable of inducing an immune response when presented alone. This is markedly different from what occurs when a more complex, immunogenic antigen is presented. In this instance antibody directed against the specific epitope(s) of that antigen can be produced. However, if hapten is linked to a large molecule (which is used as a carrier), the hapten may then be immunogenic. Under these conditions antibody is produced, not only against the epitope of the immunogenic carrier molecule but against the epitope of the hapten as well.

combined molecule, they can react to either the hapten or the larger carrier molecule. One example of a hapten is penicillin. By itself penicillin is not antigenic. However, when it combines with certain serum proteins of sensitive individuals, the resulting molecule does initiate a severe and sometimes fatal allergic immune reaction. In these instances the hapten is acting as an antigenic determinant on the carrier molecule.

Figure 30.5   Electrophoresis of Human Serum. Schematic representation of electrophoretic results illustrating the distribution of serum proteins and the four major classes of immunoglobulins.

1. What is the function of the thymus gland? The bursa of Fabricius? Null cells?
2. Contrast the humoral and cell-mediated immune responses.
3. Distinguish between "self" and "nonself" substances.
4. Define and give several examples of an antigen.
5. What is an antigenic-determinant site or epitope?
6. Describe a hapten.

Antibodies

Antibodies or **immunoglobulins (Ig)** are a group of glycoproteins present in the blood serum and tissue fluids of mammals. Serum glycoproteins can be separated according to their charge in an electric field (*see chapter 15, electrophoresis*) and classified as albumin, alpha-1 globulin, alpha-2 globulin, beta globulin, and gamma globulin (figure 30.5). Notice that the gamma globulin band is wide because it contains a heterogenous class of immunoglobulins; namely, IgG, IgA, IgM, and IgD. These differ from each other in molecular size, structure, charge, amino acid composition, and carbohydrate content. Electrophoretically the immunoglobulins show a broad range of heterogeneity, which extends from the gamma to the alpha-2 fraction of serum. Most antibodies in serum are in the IgG class. A fifth immunoglobulin class, the IgE class, has a similar mobility to IgD but cannot be represented quantitatively because of its low concentration in serum.

Immunoglobulin Structure

Antibodies have more than one antigen-combining site. For example, most human antibodies have two combining sites or are bivalent (figure 30.3). Some bivalent antibody molecules can combine to form multimeric antibodies that have up to 10 combining sites.

All immunoglobulin molecules have a basic structure composed of four polypeptide chains (figure 30.6a) connected to each other by disulfide bonds. Each light chain usually consists of about 220 amino acids and has a molecular weight of approximately 25,000. Each heavy chain consists of about 440 amino acids and has a molecular weight of 50,000 to 77,000. The heavy chains are structurally distinct for each immunoglobulin class or subclass. Both light and heavy chains contain two different regions. **Constant regions (C_L and C_H)** have amino acid sequences that do not vary significantly between antibodies of the same subclass. The **variable regions (V_L and V_H)** from different antibodies do have different sequences (figure 30.6b).

The four chains are arranged in the form of a flexible Y with a hinge region (figure 30.6b). This hinge region allows the antibody molecule to assume a T shape. The stalk of the Y is termed the **crystallizable fragment (Fc)** and contains the site at which the antibody molecule can bind to a cell. The top of the Y consists of two **antigen-binding fragments (Fab)** that bind with compatible antigenic determinant sites. The Fc fragments are composed only of constant regions, whereas the Fab fragments have both constant and variable regions. Intrachain disulfide bonds create loops within each antibody chain (figure 30.6c). A loop along with approximately 25 amino acids on each side is a **domain.** Interchain disulfide bonds link heavy and light chains together.

More specifically the light chain exists in two distinct forms called kappa (κ) and lambda (λ). These can be distinguished by the amino-acid sequence of the carboxyl portion of the chain (figure 30.7a). In human immunoglobulins, the carboxyl-terminal portion of all κ chains is identical; thus, this region is termed the constant (C_L) domain. With respect to the lambda chains, there are four very similar sequences that define the subtypes $\lambda_1, \lambda_2, \lambda_3,$ and λ_4 with their corresponding constant regions $C_\lambda 1, C_\lambda 2, C_\lambda 3,$ and $C_\lambda 4$. Within the light chain variable domain are hypervariable regions or complementarity-determining regions (CDRs) that differ in amino acid sequence more frequently than the rest of the variable domain.

In the heavy chain the NH_2-terminal domain has a pattern of variability similar to that of the V_κ and V_λ domains and is termed the V_H domain. However, notice in figure 30.7b that this domain contains four hypervariable regions. The other domains of the heavy chains are termed constant domains and are numbered $C_H1, C_H2, C_H3,$ and sometimes C_H4, starting with the domain next to the variable domain (figure 30.6c). The constant domains of the heavy chain form the constant (C_H) region. The amino acid sequence of this region determines the classes of heavy chains. In humans there are five classes of heavy chains called gamma (γ), alpha (α), mu (μ), delta (δ), and epsilon (ε).

The properties of these heavy chains determine, respectively, the five immunoglobulin classes—IgG, IgA, IgM, IgD, and IgE. Each immunoglobulin class differs in its general properties, half-life, distribution in the body, and interaction with other components of the host's defensive systems.

Within two of the major immunoglobulin classes, there are variants or subclasses. For example, IgG can be grouped into four subclasses (IgG1, IgG2, IgG3, and IgG4) and IgA into two subclasses (IgA1 and IgA2). Each subclass has differences in the amino acid composition of the heavy chain. These variations can be classified as (1) **isotypes,** referring to the variations in the heavy chain constant regions associated with the different classes and subclasses that are normally present in all individuals (figure 30.8a); (2) **allotypes,** the genetically controlled allelic forms of immunoglobulin molecules that are not present in all individuals (figure 30.8b); and (3) **idiotypes,** referring to individual specific immunoglobulin molecules that differ in the hypervariable region of the Fab portion (figure 30.8c). The many variations of immunoglobulin structure reflect the diversity of antibodies generated by the immune response.

1. Name the two isotypes of antibody light chains.
2. What is the function of the Fc region of an antibody? The Fab region?
3. What is the variable region of an antibody? The hypervariable or complementarity-determining region? The constant region?
4. What determines the class of heavy chain within an antibody?
5. Name the five immunoglobulin classes.
6. What determines the isotype of IgG?

Immunoglobulin Function

All immunoglobulin molecules are bifunctional. The Fab region is concerned with binding to antigen, whereas the Fc region mediates binding to host tissue, various cells of the immune system, some phagocytic cells, or the first component of the complement system. The binding of an antibody with an antigen usually does not cause destruction of the antigen or of the microorganism, cell, or agent to which it is attached. Rather the antibody serves to mark and identify the target for immunologic attack and to activate nonspecific immune responses that can destroy the target. For example, bacteria that are covered with antibodies are better targets for phagocytosis by neutrophils and macrophages. This ability of an antibody to stimulate phagocytosis is termed **opsonization.** Immune destruction also is promoted by antibody-induced activation of the complement system. Both opsonization and the complement system are discussed in detail in chapter 32.

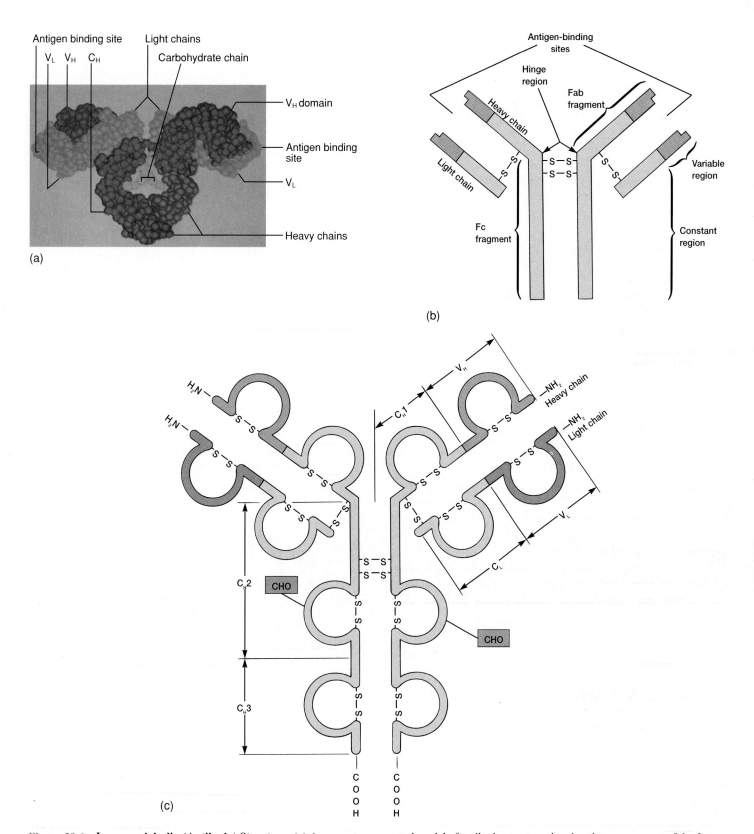

Figure 30.6   Immunoglobulin (Antibody) Structure. (*a*) A computer-generated model of antibody structure showing the arrangement of the four polypeptide chains. (*b*) An immunoglobulin molecule. The molecule consists of two identical light chains and two identical heavy chains held together by disulfide bonds. (*c*) Within the immunoglobulin unit structure, intrachain disulfide bonds create loops that form domains. All light chains contain a single variable domain (V_L) and a single constant domain (C_L). Heavy chains contain a variable domain (V_H) and either three or four constant domains (C_H1, C_H2, C_H3, and C_H4).

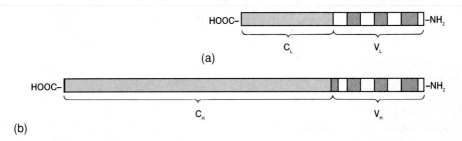

Figure 30.7   Constant and Variable Domains. Location of constant (C) and variable (V) domains within (*a*) light chains and (*b*) heavy chains. The dark yellow bands represent hypervariable regions or complementarity-determining regions within the variable domains.

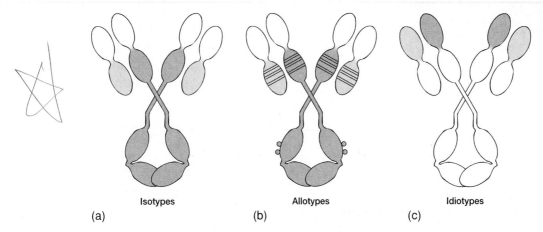

Figure 30.8   Variants or Subclasses of Immunoglobulins. (*a*) Isotypes represent all variants present in serum of a normal individual. (*b*) Allotypes represent alternative forms—genetically controlled so not present in all individuals. (*c*) Idiotypes are individually specific to each immunoglobulin molecule.

Figure 30.9   Immunoglobulin G. (*a*) The basic structure of human IgG. (*b*) The structure of the four human IgG subclasses. Note the arrangement and numbers of disulfide bonds (shown as thin black lines).

Immunoglobulin Classes

IgG is the major immunoglobulin in human serum, accounting for 70 to 75% of the immunoglobulin pool (figure 30.9*a*). IgG is present in blood plasma and tissue fluids. The IgG class acts against bacteria and viruses by opsonizing the invaders and neutralizing toxins. It is also one of the two immunoglobulin classes that activate complement by the classical pathway (*see chapter 32*). IgG is the only immunoglobulin molecule able to cross the placenta and provide naturally acquired passive immunity for the newborn.

There are four IgG subclasses (IgG1, IgG2, IgG3, and IgG4) that vary chemically in their chain composition and the number and arrangement of interchain disulfide bonds (figure 30.9*b*). About 65% of the total serum IgG is IgG1, and 23% is IgG2. Differences in biological function have been noted in these subclasses. For example, IgG2 antibodies are opsonic and develop in response to antitoxins. Anti-Rh antibodies (*see chapter 31*) are of the IgG1 or IgG3 subclass. IgG1 and IgG3 also bind best to monocytes and macrophages and activate

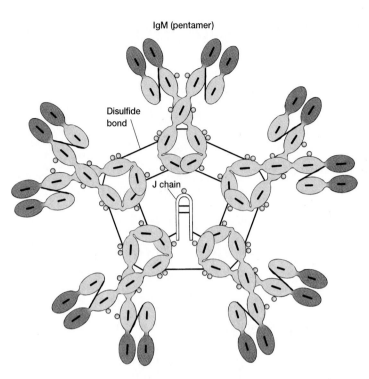

Figure 30.10   Immunoglobulin M. The pentameric structure of human IgM. The disulfide bonds linking peptide chains are shown in black; carbohydrate side chains are in green. Note that 10 antigen-binding sites are present.

Figure 30.11   Immunoglobulin A. The dimeric structure of human secretory IgA. Notice the secretory component (blue) wound around the IgA dimer and attached to the constant domain of each IgA monomer.

complement most effectively. The IgG4 antibodies function as skin-sensitizing immunoglobulins (*see chapter 31*).

IgM accounts for about 10% of the immunoglobulin pool. It is usually a polymer (pentamer) of five monomeric units, each composed of two heavy chains and two light chains (figure 30.10). The monomers are arranged in a pinwheel array with the Fc ends in the center, held together by a special **J** (joining) **chain.** IgM is the first immunoglobulin made during B-cell maturation and the first secreted into serum during a primary antibody response (figure 30.18*b*). Since IgM is so large, it does not leave the bloodstream or cross the placenta. IgM agglutinates bacteria, activates complement by the classical pathway, and enhances the ingestion of pathogens by phagocytic cells. This class also contains special antibodies such as red blood cell agglutinins and heterophile antibodies.

Although most IgM appears to be pentameric, around 5% or less of human serum IgM exists in a hexameric form. This molecule contains six monomeric units but seems to lack a J chain. Hexameric IgM activates complement up to twentyfold more effectively than does the normal pentameric form. It has been suggested that bacterial cell wall antigens such as gram-negative lipopolysaccharides may directly stimulate B cells to form hexameric IgM without a J chain. If this is the case, the immunoglobulins formed during primary immune responses are less homogenous than previously thought.

IgA accounts for about 15% of the immunoglobulin pool. Some IgA is present in the serum as a monomer of two heavy and two light chains. Most IgA, however, occurs in the serum as

a polymerized dimer held together by a J chain (figure 30.11). IgA has special features that are associated with secretory mucosal surfaces (Box 30.1). IgA, when transported from the mucosa-associated lymphoid tissue to mucosal surfaces, acquires a protein termed the secretory component. **Secretory IgA (sIgA),** as the modified molecule is now called, is the primary immunoglobulin of the secretory immune system. This system is found in the gastrointestinal tract, upper and lower respiratory tracts, and genitourinary system. Secretory IgA is also found in saliva, tears, and breast milk. In these fluids and related body areas, sIgA plays a major role in protecting surface tissues against infectious microorganisms by the formation of an immune barrier. For example, in breast milk sIgA helps protect nursing newborns. In the intestine, sIgA attaches to viruses, bacteria, and protozoan parasites such as *Entamoeba histolytica.* This prevents pathogen adherence to mucosal surfaces and invasion of host tissues, a phenomenon known as immune exclusion. In addition, sIgA binds to antigens within the mucosal lamina propria, and the antigen-sIgA complexes are excreted through the adjacent epithelium into the gut lumen. This rids the body of locally formed immune complexes and decreases their access to the circulatory system. Secretory IgA also may neutralize viruses and other intracellular pathogens that reside within epithelial cells. Secretory IgA also plays a role in the alternate complement pathway (*see chapter 32*).

IgD is an immunoglobulin found in trace amounts in the blood serum. It has a monomer structure (figure 30.12) similar to IgG. IgD antibodies do not fix complement and cannot cross the placenta, but they are abundant on the surface of B cells and bind antigens, thus signaling the B cell to start antibody production.

Box 30.1

The GALT System

The major host defense against food-borne pathogens (e.g., *E. coli, S. aureus, Bacillus cereus, Cryptosporidium,* rotavirus, Norwalk virus) and products such as the *C. difficile* cytotoxin is the secretory immune system, whose major immunoglobulin is sIgA. **Gut-associated lymphoid tissue or GALT,** and particularly **Peyer's patches,** contains a full repertoire of lymphoreticular cells, T and B cells, and macrophages. Commitment to antibody synthesis occurs in the Peyer's patches, but antibody production does not (Box Figure 30.1).

In Peyer's patches microfold cells sample foreign antigens from the gut lumen, and macrophages process the antigens for presentation to CD4+ (T4) lymphocytes. The T4 lymphocytes produce the cytokines interleukin-2, interleukin-4, and IFN-γ that stimulate follicle B lymphocytes to become IgA-producing cells. The stimulated B lymphocytes migrate through lymph nodes, the thoracic duct, and the bloodstream before they "home" to the lamina propria of the intestine (a connective tissue layer lying beneath the intestinal epithelium). Here they undergo final differentiation because of the influence of interleukin-6 produced by the predominant T4 lamina propria lymphocytes. When these plasma cells produce IgA dimers, the dimers bind to epithelial cell surface receptors and are transported across the cells in vesicles. They are released into the intestinal lumen as sIgAs (figure 30.11) after being connected to secretory components (SCs) by the epithelial cell. The GALT system is exceptionally complex and is composed of several interactive sites. Some immunologists reserve the term "GALT" for the IgA inductive sites (Peyer's patches, appendix, and solitary lymph nodes) and do not consider the IgA effector sites (lamina propria and intraepithelial lymphocytes) to be a part of GALT.

Enteric pathogens may disrupt the homeostasis of the gut microenvironment and suppress the immune response through toxin production. For example, *E. coli* produces two toxins: (1) a cholera-like, heat-labile toxin (LT) that is an adenylate cyclase activator, and (2) a heat-stable toxin (ST) that activates the intestinal membrane-bound guanylate cyclase. Both LT and ST profoundly affect the immune response by increasing the levels of these cyclic nucleotides, which in turn decreases the ability of Peyer's patch cells to generate GALT sIgA. Thus enterotoxigenic *E. coli* can establish itself and cause disease.

Box Figure 30.1

TABLE 30.3   Physicochemical Properties of Human Immunoglobulin Classes

Property	Immunoglobulin Classes				
	IgG[a]	IgM	IgA[b]	IgD	IgE
Heavy chain	γ_1	μ	α_1	δ	ε
Mean serum concentration (mg/ml)	9	1.5	3.0	0.03	0.00005
Valency	2	5(10)	(2–4)	2	2
Molecular weight of heavy chain (10^3)	51	65	56	70	72
Molecular weight of entire molecule (10^3)	146	970	160[c]	184	188
Placental transfer	+	0	0	0	0
Half-life in blood (days)[d]	21	10	6	3	2
Complement activation					
Classical pathway	++	+++	0	0	0
Alternative pathway	0	0	+	0	0
Major characteristics	Most abundant Ig in body fluids; neutralizes toxins, opsonizes bacteria, activates complement, maternal antibody	First to appear after antigen stimulation; very effective agglutinator	Secretory antibody; protects external surfaces	Present on B-cell surface; B-cell recognition of antigen	Anaphylactic-mediating antibody; resistance to helminths
% carbohydrate	3	7–10	7	12	11

[a]Properties of IgG subclass 1.

[b]Properties of IgA subclass 1.

[c]sIgA = 360 – 400 kDa

[d]Time required for half of the antibodies to disappear.

IgE (figure 30.13) makes up only 0.00005% of the total immunoglobulin pool. The classic skin-sensitizing and anaphylactic antibodies belong to this class. IgE molecules have four constant region domains ($C_\varepsilon 1$, $C_\varepsilon 2$, $C_\varepsilon 3$, and $C_\varepsilon 4$) and two light chains. The Fc portion of the $C_\varepsilon 4$ chain can bind to special Fc_ε receptors on mast cells and basophils. When two IgE molecules on the surface of these cells are cross-linked by binding to the same antigen, the cells degranulate. This degranulation releases histamine and other pharmacological mediators of anaphylaxis. It also stimulates eosinophilia and gut hypermotility (increased rate of movement of the intestinal contents) that aid in the elimination of helminthic parasites. Thus, though IgE is present in small amounts, this class of antibodies has very potent biological capabilities.

Anaphylaxis (pp. 621–22).

Table 30.3 summarizes some of the more important physicochemical properties of the human immunoglobulin classes.

1. What does antigen binding accomplish in a host?
2. What is opsonization?
3. Give the major functions of each immunoglobulin class.
4. Why is the structure of IgG considered the model for all five immunoglobulin classes?
5. Which immunoglobulin can cross the placenta?
6. Which immunoglobulin is most prevalent in the immunoglobulin pool? The least prevalent?

Figure 30.12   Immunoglobulin D. The structure of human IgD. The disulfide bonds linking protein chains are shown in black; carbohydrate side chains are in green.

Figure 30.13   Immunoglobulin E. The structure of human IgE. $C_\varepsilon 1$—$C_\varepsilon 4$ represent the constant domain regions.

Figure 30.14   Gene Shuffling and Antibody Diversity. Antibody diversity is partly the result of the shuffling of gene sequences that code for both heavy and light chains. This drawing shows the shuffling, cutting, and splicing process used to produce an assembled light chain of the antibody molecule.

Diversity of Antibodies

One unique property of antibodies is their remarkable diversity. According to current estimates each human or mouse can synthesize more than 10 million different kinds of antibodies. How is this diversity generated? The answer is threefold: rearrangement of antibody gene segments, somatic mutations, and generation of different codons during antibody gene splicing.

Immunoglobulin genes are split or interrupted genes with many exons (*see pp. 223–24*). Embryonic B cells contain a small number of exons, close together on the same chromosome, that determine the constant (C) region of the light chains (figure 30.14). Separated from them, but on the same chromosome, is a larger cluster of exons that determines the variable (V) region of the light chains. During B-cell differentiation, one exon for the constant region is spliced (cut and joined) onto one exon for the variable region. This splicing produces a complete light-chain antibody gene. A similar splicing mechanism also occurs to join the constant and variable exons of the heavy chains.

Because the light-chain genes actually consist of three parts, and the heavy-chain genes consist of four, the formation of a finished antibody molecule is slightly more complicated than previously outlined (figure 30.15). The germ line DNA for the light-chain gene contains multiple coding sequences called V and J (joining) regions. During the differentiation of a B cell, a deletion (which is variable in length) occurs that joins one V exon with one J exon. This DNA joining process is termed combinatorial joining since it can create many combinations of the V and J regions. When the light-chain gene is transcribed, transcription continues through the DNA region that encodes for the constant portion of the gene. RNA splicing subsequently joins the VJ and C regions creating mRNA.

Combinatorial joining in the formation of a heavy-chain gene occurs by means of DNA splicing of the heavy-chain

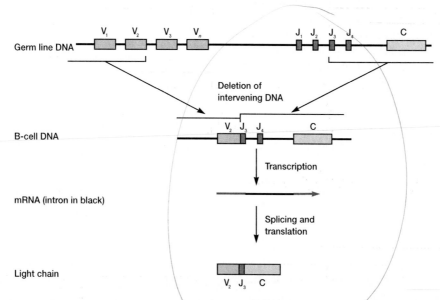

Figure 30.15   Light Chain Production in a Mouse. One V exon is randomly joined with one J-C region by deletion of the intervening DNA. The remaining J exons are eliminated from the RNA transcript during RNA processing. An intron is a segment of DNA occurring between expressed regions of genes.

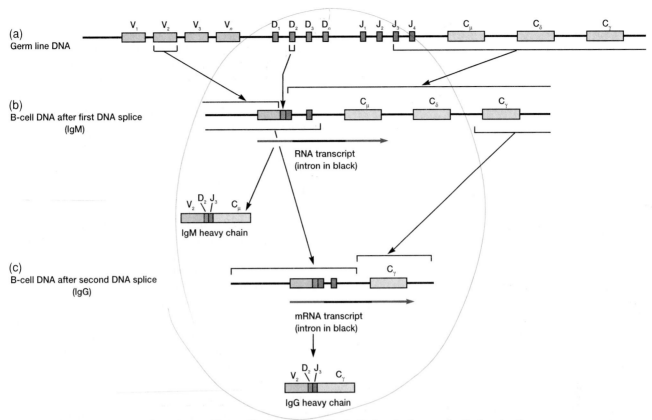

Figure 30.16   The Formation of a Gene for the Heavy Chain of an Antibody Molecule. See text for further details.

TABLE 30.4	Number of Antibodies Possible through the Combinatorial Joining of Mouse Germ Line Genes[a]
λ light chains	V regions = 2 J regions = 3 Combinations = $2 \times 3 = 6$
κ light chains	V_κ regions = 250–350 J_κ regions = 4 Combinations = $250 \times 4 = 1,000$ $= 350 \times 4 = 1,400$
Heavy chains	V_H = 250–1,000 D = 10–30 J_H = 4 Combinations = $250 \times 10 \times 4 = 10,000$ $= 1,000 \times 30 \times 4 = 120,000$
Diversity of antibodies	κ-containing: $1,000 \times 10,000 = 10^7$ $1,400 \times 120,000 = 2 \times 10^8$ λ-containing: $6 \times 10,000 = 6 \times 10^4$ $6 \times 120,000 = 7 \times 10^5$

[a]Approximate values.

counterparts of V and J along with a third set of D (diversity) sequences (figure 30.16a). Initially, all heavy chains have the µ type of constant region. This corresponds to antibody class IgM (figure 30.16b). Another DNA splice joins the VDJ region with a different constant region that can subsequently change the class of antibody produced by the B cell (figure 30.16c).

The amount of antibody diversity in the mouse that can be generated by combinatorial joining is shown in table 30.4. In this animal the κ light chains are formed from combinations of about 250–350 V_κ and 4 J_κ regions giving a maximum of approximately 1,400 different κ chains. The λ chains have their own V_λ and J_λ regions but smaller in number than their κ counterparts (6 different λ chains). The heavy chains have approximately 250–1,000 V_H, 10–30 D, and 4 J_H regions giving a maximum of 120,000 different combinations. Because any light chain can combine with any heavy chain, there will be a maximum of 2×10^8 possible κ chain antibody types.

The value of 2×10^8 different antibodies is actually an underestimate because antibody diversity is further augmented by two processes:

1. The V regions of germ line DNA are susceptible to a high rate of somatic mutation during B-cell development in the bone marrow. These mutations allow B-cell clones to produce different polypeptide sequences.

2. The junction for either VJ or VDJ splicing in combinatorial joining can occur between different nucleotides and thus generate different codons in the spliced gene. For example, one VJ splicing event can join the V sequence CCTCCC with the J sequence TGGTGG in two ways:

$$CCTCCC + TGGTGG = CCGTGG,$$

which codes for the amino acids proline and tryptophan; and

$$CCTCCC + TGGTGG = CCTCGG,$$

which codes for proline and arginine. Thus the same VJ joining could produce polypeptides differing in a single amino acid.

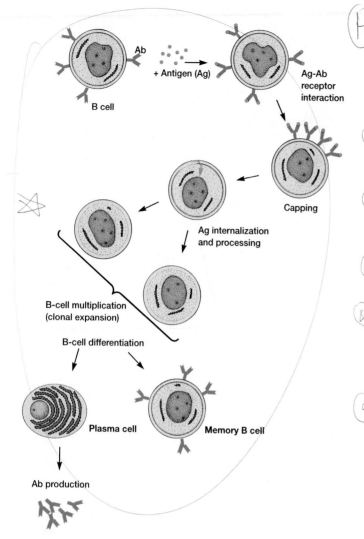

Figure 30.17   Clonal Selection. It is through clonal selection that the immune system can respond specifically to a myriad of possible antigens, whether they are individual molecules or are attached to pathogens and abnormal cells such as cancer cells. B cells or B lymphocytes constantly roam the body, particularly the blood and lymphoid tissues. Each B cell synthesizes only one of the millions of possible antibodies and displays this antibody on its surface. When the antigen meets a B cell having a surface antibody of the proper specificity (top left), it complexes with the antibody and capping occurs. (Capping is the regional aggregation of antibodies on the surface of the cell following Ag-Ab interaction.) The antigen is then internalized; the B cell swells and begins to divide rapidly, producing a B-cell clone. The activated B-cell clone differentiates into plasma cells and memory cells. Plasma cells form the specific antibody that immediately attacks the antigen that provoked its formation. Memory B cells persist in the body and boost the immune system's readiness to eliminate the same antigen if it presents itself in the future.

1. How many chromosomes encode for antibody production in humans?
2. What is the name of each exon set that encodes for the different regions of antibody chains?
3. Describe what is meant by combinatorial joining of VJ and C regions of a chromosome.
4. In addition to combinatorial joining, what other two processes play a role in antibody diversity?

Specificity of Antibodies

As noted previously, combinatorial joinings, somatic mutations, and variations in the splicing process generate the great variety of antibodies produced by B cells. From a large, diverse B-cell pool, specific cells are stimulated by antigens to reproduce and form a B-cell clone that contains the same genetic information. This is known as the **clonal selection theory,** a hypothesis to explain immunologic specificity and memory.

The existence of a small B-cell **clone** (a population of cells derived asexually from a single parent) that can respond to one or a few antigens by producing the correct antibody is the first tenet of this theory. The lymphoid system is thus considered to contain many B-cell clones, each clone able to recognize a specific antigen. The antigen selects the appropriate clone of B cells (hence the phrase "clonal selection"), and the cells from the other clones are unaffected.

According to the second tenet, each B-cell clone is genetically programmed to respond to its own distinctive antigen before the antigen is introduced. The particular antibody for which an individual B cell is genetically competent is integrated into the plasma membrane of that B cell and acts as a specific surface receptor for the corresponding antigen molecule. The reaction of the antibody and antigen initiates the differentiation and multiplication of the B cell to form two different cell populations: plasma cells and **memory B cells** (figure 30.17).

Plasma cells are literally protein factories that produce about 2,000 antibodies per second in their brief five- to seven-day life span. Memory B cells can initiate the antibody-mediated immune response upon detecting the particular antigen molecule for which they are genetically programmed (i.e., they have specificity). These memory cells circulate more actively from blood to lymph and live much longer (years or even decades) than plasma cells. Memory cells are responsible for the immune system's rapid secondary antibody response (figure 30.18b) to the same antigen.

Sources of Antibodies

The need for pure homogeneous antibodies has increased dramatically in recent years. Currently antibodies are produced either naturally by immunization or artificially through hybridoma formation.

Immunization

Specific antibodies can be produced naturally by the immunization of domestic animals or human volunteers (Box 30.2). Whatever the source, purified antigen is injected into the host. The host's immune system recognizes and responds to the antigen, and its B cells proliferate and differentiate to produce specific antibodies. To promote the efficiency of antigen stimulation of antibody production, the antigen is mixed with an **adjuvant** [Latin *adjuvans,* aiding], which enhances the rate and quantity of antibody produced. Following repeated antigen injections at regular intervals, blood is withdrawn

Box 30.2

The First Immunizations

Since the time of the ancient Greeks, it has been recognized that people who have recovered from plague, smallpox, yellow fever, and various other infectious diseases rarely contract the diseases again. The first scientific attempts at artificial immunizations were made in the late eighteenth century by the English physician Edward Jenner (1749–1823). Jenner investigated the basis for the widespread belief of the English peasants that anyone who had vaccinia (cowpox) never contracted smallpox. Smallpox was often fatal—10 to 40% of the victims died—and those who recovered had disfiguring pockmarks. Yet most English milkmaids, who were readily infected with cowpox, had clear skin because cowpox was a relatively mild infection that left no scars.

Jenner's first immunization recipient was a healthy, eight-year-old boy known never to have had either cowpox or smallpox. As Jenner expected, immunization with the cowpox virus caused only mild symptoms in the boy. When he subsequently inoculated the boy with smallpox virus, the boy showed no symptoms of the disease.

Jenner then inoculated large numbers of his patients with cowpox pus, as did other physicians in England and on the European continent (Box Figure 30.2). By 1800 the practice known as vaccination [Latin *vacca*, cow] had begun in America, and by 1805 Napoleon Bonaparte had ordered all French soldiers to be vaccinated.

Further work on immunization was carried out by Louis Pasteur (1822–1895). Pasteur discovered that if cultures of chicken cholera bacteria were allowed to age for two or three months the bacteria produced only a mild attack of cholera when inoculated

Box Figure 30.2   Nineteenth-century physicians performing vaccinations on children.

into chickens. Somehow the old cultures had become less pathogenic (attenuated) for the chickens. He then found that fresh cultures of the bacteria failed to produce cholera in chickens that had been previously inoculated with old, attenuated cultures. To honor Jenner, Pasteur gave the name vaccine to any preparation of a weakened pathogen that was used (as was Jenner's "vaccine virus") to immunize against infectious disease.

---

from the host and allowed to clot. The fluid that remains after the blood clots is the **serum.** Because this serum has been obtained from an immunized host and contains the desired antibodies, it is called **antiserum.**

Although antiserum is a major and convenient source of antibodies, its usefulness is limited in three ways:

1. Antibodies obtained by this method are polyclonal; they are produced by several B-cell clones and have different specificities. This decreases their sensitivity to particular antigens and results in some degree of cross-reaction with closely related antigen molecules.

2. Second or repeated injections of antiserum from one species to another can cause serious allergic or hypersensitivity reactions.

3. Antiserum contains a mixture of antibodies (figure 30.5), not all of which are of interest in a given immunization.

The Primary Antibody Response

During immunization procedures (and also in naturally acquired immunity), there is an initial lag phase of several days following a primary challenge with an antigen. During the lag phase no

antibody can be detected (figure 30.18*a*). The antibody **titer,** which is the reciprocal of the highest dilution of an antiserum that gives a positive reaction in the test being used, rises logarithmically to a plateau during the second or log phase. In the plateau phase the antibody titer stabilizes. This is followed by a decline phase, during which antibodies are naturally metabolized or bound to the antigen and cleared from the circulation. During the primary antibody response, IgM appears first, then IgG (figure 30.18*b*). The affinity of the antibodies for the antigen's determinants is low to moderate during this primary antibody response.

The Secondary Antibody Response

The primary antibody response primes the immune system so that it possesses specific immunologic memory through its clones of memory B cells. Upon secondary antigen challenge (figure 30.18*b*), the B cells mount a heightened or **anamnestic** [Greek *anamnesis,* remembrance] **response** to the same antigen. Compared to the primary antibody response, the secondary antibody response has a shorter lag phase, a more rapid log phase, persists for a longer plateau period, attains a higher IgG titer, and produces antibodies with a higher affinity for the antigen (affinity maturation).

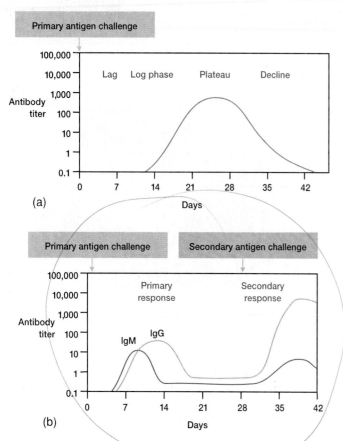

Figure 30.18 **Primary and Secondary Antibody Responses.** (*a*) The four phases of a primary antibody response. (*b*) Comparison of primary and secondary antibody responses. Note that the secondary response to the antigen is both faster and greater than the primary response. T-cell–mediated responses exhibit similar immunologic memory.

Figure 30.19 **Technique for the Production of Monoclonal Antibodies.** Antigen-stimulated spleen cells are fused with special mutant myeloma cells, yielding hybridomas. Each of them secretes a single, "monoclonal" antibody. Once the hybridoma secreting the desired antigen is identified, it is cloned to generate many antibody-secreting cells that yield the huge quantity of a single antibody needed in medicine or science. Some hybridoma cells may be stored frozen and later cloned for antibody production or kept alive in laboratory animals.

Hybridomas

The limitations of antiserum as a source of antibodies have been overcome with the development of hybridoma techniques to manipulate and culture various mammalian cells that synthesize antibodies in vitro. Each cell and its progeny normally produce a **monoclonal antibody (MAb)** of a single specificity.

The methodology of one of these techniques is illustrated in figure 30.19. Animals (usually mice or rats) are immunized with antigens as discussed previously. Once the animals are producing a large quantity of antibodies, their spleens are removed. The spleen cells are separated from each other and fused with **myeloma cells** by the addition of polyethylene glycol, which promotes membrane fusion. Myeloma cells are cancerous plasma cells that can readily be cultivated. Mutant myeloma cells incapable of producing immunoglobulins are used. These fused cells, derived from spleen cells and myeloma cells, are called **hybridomas** (they are hybrids of the two cells).

The fusion mixture is then transferred to a culture medium containing a combination of **h**ypoxanthine, **a**minopterin, and **t**hymidine (HAT). Aminopterin is a poison that blocks a specific metabolic pathway in cells. Myeloma cells lack an enzyme that allows their growth in the presence of aminopterin. How-

ever, the pathway is by-passed in spleen cells provided with the intermediate metabolites hypoxanthine and thymidine. As a result the hybridomas grow in the HAT medium but the myeloma cells die because they have a metabolic defect and cannot employ the bypass pathway.

When the culture is initially established using the HAT medium, it contains spleen cells, myeloma cells, and hybridomas. The unfused spleen cells die naturally in culture within a week or two, and the myeloma cells die in the HAT as just described. In contrast, the fused cells survive because they have the immortality of the myeloma and the metabolic bypass of the spleen cells. Some hybridomas that have the antibody-producing capacity of the original spleen cells are randomly placed in culture wells. The wells are individually tested for production of the desired antibody, and, if positive, the cells within the well are cloned. The clone is immortal and produces monoclonal antibody.

Box 30.3

Immunotoxins

One result of hybridoma research is the production of **immunotoxins** (Box Figure 30.3). Immunotoxins are monoclonal antibodies that have been attached to a specific toxin or toxic agent (antibody + toxin = immunotoxin). Immunotoxins kill target cells and no others, because the antibody binds specifically to plasma membrane surface antigens found only on the target cells. This approach is being used to treat certain types of cancer.

In this procedure cancer cells from a person are injected into mice or rats to stimulate the production of specific antibodies against their plasma membrane antigens. Monoclonal antibodies are produced using hybridomas, purified, and attached to an agent toxic to the cancer cells.

When the immunotoxin is given to a cancer patient, it circulates through the body and binds only to the cancer cells that have the appropriate surface antigens. After binding to the surface, the immunotoxin is taken into cancer cells by receptor-mediated endocytosis, and released inside. The immunotoxin then interferes with the metabolism of the target cells and kills them. Although this procedure is still experimental, it holds great promise in the treatment of certain types of cancer.

Box Figure 30.3

Monoclonal antibodies currently have many applications. For example, they are routinely used in the typing of tissue, in the identification and epidemiological study of infectious microorganisms, in the identification of tumor and other surface antigens, in the classification of leukemias, and in the identification of functional populations of different types of T cells. Anticipated future uses include (1) passive immunizations against infectious agents and toxic drugs, (2) tissue and organ graft protection, (3) stimulation of tumor rejection and elimination, (4) manipulation of the immune response, (5) preparation of more specific and sensitive diagnostic procedures, and (6) delivery of antitumor agents (immunotoxins) to tumor cells (Box 30.3).

Catalytic Antibodies

In the past several years immunologists have applied the principles of enzymology to create a new class of antibody molecules—**catalytic antibodies.** Catalytic antibodies accelerate specific chemical reactions by lowering the free energy of transition states (*see figure 8.13*). They accomplish this by binding reactants in a cleft or crevice on their surfaces and inducing structural changes in the substrate molecules. Because antibodies directed against a huge array of biopolymers, natural products, and synthetic molecules can be formed, catalytic antibodies offer a unique approach for generating tailor-made, enzymelike catalysts.

Catalytic antibodies are made by coupling a transition-state analogue to a carrier protein and injecting the combination into an experimental animal. Antibody-secreting spleen cells are taken from the animal and fused with myeloma cells (figure 30.19). The hybrid antibody-secreting cells divide indefinitely and generate clones of cells, each hybrid clone secreting a monoclonal catalytic antibody with a unique antigen binding pocket. A clone that makes catalytic antibody specific for the analogue is then selected.

Currently available catalytic antibodies transform relatively simple compounds. Much of the potential of catalytic antibodies for biotechnology and molecular biology depends on the development of catalytic antibodies able to act on proteins or nucleic acids. If this can be accomplished, catalytic antibodies could extend the immune system's innate capacity to defend the body. For example, one might stimulate the immune system of a patient with heart disease to produce antibodies that would break up the proteins in blood clots, forestalling heart attacks.

1. What are the two tenets of the clonal selection theory?
2. What two cell populations are produced by a B cell after it initially reacts with an antigen?
3. What is the function of an adjuvant?
4. What is the difference between serum and antiserum?
5. What is a hybridoma? A catalytic antibody? How is each made?

Summary

1. Nonspecific resistance refers to those general mechanisms inherited as part of the innate structure and function of each animal. Examples include biological, chemical, general, and physical barriers. The specific immune response system consists of certain cells (lymphocytes) that can recognize foreign molecules (antigens) and respond to them by forming specialized proteins (antibodies) that target the molecule or agent for interaction with other components of the immune system.

2. Acquired immunity refers to the type of specific immunity that a host develops after exposure to a suitable antigen. It can be obtained actively or passively by natural or artificial means.

3. Undifferentiated lymphocytes are produced in adult bone marrow by stem cells. Some of these lymphocytes migrate to the thymus and undergo special processing to become T cells (T lymphocytes). Other lymphocytes differentiate in other places in the body to form B cells (B lymphocytes). A third population of lymphoid cells that do not have characteristics of either T or B cells are called null cells. This population contains natural killer and cytotoxic T cells.

4. B cells provide defense against antigens by differentiating into plasma cells that secrete antibodies into the blood and lymph, providing humoral or antibody-mediated immunity. T cells attack host cells that have been parasitized by microorganisms, tissue cells transplanted from one host to another, and cancer cells. Because they must physically contact these foreign cells to destroy them, they are said to provide cell-mediated immunity. T cells also play a major role in regulating the immune response. Null cells nonspecifically kill tumor cells, virus- and other parasite-infected cells.

5. An antigen or immunogen is a foreign substance to which lymphocytes respond. Each antigen can have several antigenic determinant sites or epitopes that stimulate production of and combine with specific antibodies. Haptens are small organic molecules that are not antigenic by themselves but can become antigenic if bonded to a larger carrier molecule.

6. Antibodies or immunoglobulins are a group of glycoproteins present in the blood and tissue fluids of all mammals. All immunoglobulins have a basic structure composed of four chains of polypeptides (two light and two heavy)

connected to each other by disulfide bonds. In humans five immunoglobulin classes exist: IgG, IgA, IgM, IgD, and IgE.

7. Antibody diversity results from the rearrangement of the individual exons on the antibody coding chromosomes, somatic mutations, the generation of different codons during splicing, and the independent assortment of light- and heavy-chain genes.

8. The clonal selection theory is a hypothesis that explains immunologic specificity and memory.

9. Specific antibodies can be produced naturally from immunized animals. The primary antibody response in a host occurs following initial exposure to the antigen. This response has lag, log, plateau, and decline phases. Upon secondary antigen challenge, the B cells mount a heightened and accelerated anamnestic response.

10. Hybridomas result from the fusion of spleen cells with myeloma cells. Some of these cells produce a single monoclonal antibody. Monoclonal antibodies have many uses.

11. Catalytic antibodies catalyze specific chemical reactions by lowering the free energy of transition states. These antibodies can be produced by hybridoma techniques.

Key Terms

acquired immunity *588*	domain *592*	lymphocytes *589*
adjuvant *600*	epitope *590*	memory B cells *600*
allotype *592*	GALT *596*	monoclonal antibody (MAb) *602*
anamnestic response *601*	hapten *590*	myeloma cell *602*
antibody *591*	heterologous antigen *590*	natural killer (NK) cell *589*
antibody-mediated immunity *589*	heterophile *590*	naturally acquired active immunity *588*
antigen *590*	humoral immunity *589*	naturally acquired passive immunity *588*
antigen-binding fragment (Fab) *592*	hybridoma *602*	nonspecific resistance *587*
antigenic determinant site *590*	idiotype *592*	null cell *589*
antiserum *601*	IgA *595*	opsonization *592*
artificially acquired active immunity *588*	IgD *595*	Peyer's patches *596*
artificially acquired passive immunity *588*	IgE *597*	plasma cells *589*
B cell *589*	IgG *594*	secretory IgA (sIgA) *595*
B lymphocyte *589*	IgM *595*	serum *601*
catalytic antibodies *603*	immune response *587*	T cell *589*
cell-mediated immunity *589*	immunity *587*	thymocyte *589*
clonal selection theory *600*	immunogen *590*	titer *601*
clone *600*	immunoglobulin (Ig) *591*	T lymphocyte *589*
colostrum *588*	immunology *587*	vaccine *588*
constant region (C_L and C_H) *592*	immunotoxin *603*	valence *590*
crystallizable fragment (Fc) *592*	isotype *592*	variable region (V_L and V_H) *592*
cytokines *589*	J chain *595*	

Questions for Thought and Review

1. What is the difference between a nonspecific host defense mechanism and a specific defense mechanism? Give examples.

2. What are the major differences between natural and artificial immunity and between active and passive immunity?

3. What is the difference between antibody-mediated and cell-mediated immunity?

4. How does the clonal selection theory explain the ability of the immune response to distinguish between self and nonself antigens?

5. All antibodies are proteins. Is there some property of proteins that has favored the evolution of antibodies within this group of biomolecules rather than within others?

6. How did the clonal selection theory inspire the development of monoclonal antibody techniques?

7. What is the difference in the kinetics of antibody formation in response to a first and second exposure to the same antigen?

8. What is the basis of immunologic memory?

9. What is the derivation of B and T cells?

Additional Reading

General

Abbas, A. K.; Lichtman, A. H.; and Pober, J. S. 1994. *Cellular and molecular immunology,* 2d ed. Philadelphia: W. B. Saunders.

Coleman, R. M.; Lombard, M. F.; and Sicard, R. E. 1992. *Fundamental immunology,* 2d ed. Dubuque, Iowa: Wm. C. Brown.

Cruse, J. M., and Lewis, R. E., Jr. 1995. *Illustrated dictionary of immunology.* Boca Raton, Fla.: CRC Press.

Desowitz, R. S. 1987. The thorn in the starfish: How the human immune system works. New York: Norton.

Janeway, C. A., Jr., and Travers, P. 1994. *Immunobiology: The immune system in health and disease.* New York: Garland Publishing.

Kuby, J. 1994. *Immunology,* 2d ed. New York: W. H. Freeman.

Paul, W. E. 1990. *Fundamental immunology,* 2d ed. New York: Raven Press.

Roitt, I. M. 1994. *Essential immunology,* 8th ed. Boston: Blackwell Scientific Publications.

Roitt, I. M.; Brostoff, J.; and Male, D. K. 1993. *Immunology,* 3d ed. St. Louis: C. V. Mosby.

Scientific American Special Issue. 1993. Life, death, and the immune system. *Sci. Am.* 269(3).

Silverstein, A. M. 1989. *A history of immunology.* San Diego: Academic Press.

Tizard, I. R. 1995. *Immunology: An introduction,* 4th ed. Philadelphia: W. B. Saunders.

Acquired Immunity

Johnson, A. G. 1994. Molecular adjuvants and immunomodulators: New approaches to immunization. *Clin. Microbiol. Rev.* 7(3):227–89.

Origin of Lymphocytes

Boyd, R., et al. 1993. The thymic environment. *Immunol. Today* 14(9):445–59.

Cooper, M. D., and Lawton, A. R. 1974. The development of the immune system. *Sci. Am.* 231(5):124–30.

Antibodies

Brewer, J. 1994. IgM hexamers? *Immunol. Today* 15(4):165–68.

Edelman, G. M. 1970. The structure and function of antibodies. *Sci. Am.* 223(1):34–41.

Lascombe, M. B., and Poljak, R. J. 1988. Three-dimensional structure of antibodies. *Ann. Rev. Immunol.* 6:555–80.

Mazanec, M. B.; Nedrud, J. G.; Kaetzel, C. S.; and Lamm, M. E. 1993. A three-tiered view of the role of IgA in mucosal defense. *Immunol. Today* 14(9):430–35.

Tonegawa, S. 1985. The molecules of the immune system. *Sci. Am.* 252(3):41–57.

Diversity of Antibodies

Ada, G. L., and Nossal, G. 1987. The clonal-selection theory. *Sci. Am.* 257(5):62–69.

Kennedy, R. C.; Melnick, J. L.; and Dreesman, G. R. 1986. Anti-idiotypes and immunity. *Sci. Am.* 255(5):48–56.

Tauber, A. 1994. The immune self: Theory or metaphor? *Immunol. Today* 15(3):134–36.

Tonegawa, S. 1983. Somatic generation of antibody diversity. *Nature* 302:575–77.

Sources of Antibodies

Collier, R. J., and Koplan, D. A. 1984. Immunotoxins. *Sci. Am.* 251(2):56–64.

Milstein, C. 1980. Monoclonal antibodies. *Sci. Am.* 243(4):66–74.

Payne, W. J., Jr.; Marshall, D. L.; Shockley, R. K.; and Martin, W. J. 1988. Clinical laboratory applications of monoclonal antibodies. *Clin. Microbiol. Rev.* 1(3):313–29.

Catalytic Antibodies

Lerner, R. A., and Tramontano, A. 1988. Catalytic antibodies. *Sci. Am.* 258(6):58–70.

CHAPTER 31

The Immune Response:
Chemical Mediators,
B- and T-Cell Biology,
and Immune Disorders

Autoimmunity—in which

the immune system recognizes

and attacks the self's own

tissues—is not as simple as it

seemed. Self-recognition

appears to be at the heart of

health as well as of certain

diseases.

—Irun R. Cohen

Concepts

1. The term cytokine is a generic term for a protein released by one cell population that acts as an intracellular (between cells) mediator. Cytokines exert their effects primarily by binding to specific plasma membrane receptors called CDs on target cells. Cytokines function as mediators of natural immunity; activators of effector cells; mediators of mature lymphocyte activation, differentiation, and growth; and mediators of immature leukocyte growth and differentiation. Examples of cytokines include monokines, lymphokines, interleukins, and colony-stimulating factors.

2. B cells have receptor immunoglobulins on their plasma membrane surface that are specific for given antigenic determinants. Contact with the antigenic determinant causes the B cell to divide and differentiate into plasma cells and memory B cells. Plasma cells secrete antibodies that specifically interact with the antigens eliciting their production.

3. Antigen binding on the surface of a T cell causes that cell to proliferate and form sensitized lymphocytes that then interact with the antigen, leading to its removal. T cells also have roles as helper and suppressor cells.

4. Lymphocytes often participate in the immune response by secreting nonspecific glycoproteins called cytokines. In addition, lymphocytes are affected by other cytokines such as interleukins and interferons.

5. Microorganisms often strongly elicit the immune response. Microbial antigens usually stimulate the response through the activity of antigen-presenting cells and T cells. Some stimulate B cells without T-cell cooperation.

6. Some individuals experience harmful overreactions of the immune system known as hypersensitivities or allergies. There are four types: type I or anaphylactic hypersensitivities are characterized by the release of physiological mediators from IgE-bound mast cells and basophils; type II or cytotoxic hypersensitivities result from complement-dependent lysis of cells; type III hypersensitivities involve the formation of immune complexes that are deposited on basement membranes; and type IV hypersensitivities arise from the reaction of T_{DTH} cells, cytokines, and macrophages.

7. At times the body loses tolerance for its own antigens and attacks them. This attack produces autoimmune diseases. At other times the immune system can become defective, leading to immunodeficiencies.

8. Both type I and type IV hypersensitivities are involved in tissue transplantation rejection. Rejection reactions also play an important role in eliminating cancer cells, in blood transfusion accidents, and in Rh incompatibility.

9. Cell-associated differentiation antigens (CDs) are functional cell surface proteins or receptors that can be measured in situ, from peripheral blood, biopsy samples, or other body fluids. Some examples of CDs include interleukin-2 receptor (IL-2r), CD2, CD4, CD8, CD35, and intercellular adhesion molecule-1 (ICAM-1).

Chapter 30 presents a basic discussion of antigens and antibodies. Chapter 31 continues this discussion by presenting other chemical mediators that enter into the immune response, B- and T-cell biology, the processing and presentation of microbial antigens, hypersensitivities (allergies), and several types of immune phenomena that can cause problems in the animal body under certain circumstances. For example, there are genetic, developmental, or induced defects in the immune system that may cause over- or underreaction and lead to various immune disorders. In addition, graft rejection, cancer, and blood transfusion reactions are discussed, because these problems are immune response mediated.

Two major classes of lymphocytes, B cells and T cells, specifically recognize immunogens and form immune proteins called immunoglobulins and sensitized lymphocytes, respectively. These are the immune products that interact with antigens. This defense is further aided by chemical mediators, such as cytokines, that transmit messages between immune system cells. As with all systems in an animal's body, problems can occur within the immune system leading to defects (immunodeficiencies) or overreactions (hypersensitivities). These aspects of immune system function are presented in this chapter in the continuing discussion of the immune response.

Cytokines

Defense against viruses, microorganisms and their products, parasites, and cancer cells is mediated by both natural and specific immunity. Cytokines are required for immunoregulation of both of these immune responses. The term **cytokine** [Greek *cyto,* cell, and *kinesis,* movement] is a generic term for the protein released by one cell population that acts as an intercellular (between cells) mediator. When released from mononuclear phagocytes, these proteins are called **monokines;** when released from T lymphocytes they are called **lymphokines;** when produced by a leukocyte and the action is on another leukocyte, they are **interleukins;** and if their effect is to stimulate the growth and differentiation of immature leukocytes in the bone marrow, they are called **colony-stimulating factors (CSFs).** Figure 31.1 outlines the sources of many of these cytokines.

It should be noted that information on the cytokine network is changing daily. Many new properties of known cytokines are being revealed and new cytokines continually are being discovered. Trials are under way to evaluate the therapeutic potential of cytokines in cancer therapy, AIDS, and septic shock (*see chapter 37, pp. 771–73*).

Cytokines can affect the same cell responsible for their production (an autocrine function), nearby cells (a paracrine function), or can be distributed by the circulatory system to their target cells (an endocrine function). Their production is induced by nonspecific stimuli such as a viral, bacterial, or parasitic infection; cancer; inflammation; or the interaction between a T-cell and antigen. Some cytokines also can induce the production of other cytokines. When needed, they are synthesized by new, transient gene transcription and rapidly secreted. Cytokines exert their effects primarily by binding to specific plasma membrane receptors called CDs (cell-associated differentiation antigens; *see pp. 628–29*) on target cells. Cytokines function in four different ways: (1) as mediators of natural or nonspecific immunity; (2) activators of effector cells; (3) mediators of mature cell (e.g., leukocyte) activation, differentiation, behavior, and growth; and (4) mediators of immature cell (e.g., leukocyte) growth and differentiation. Each of these functions is now briefly discussed along with salient cytokine examples.

Mediators of Natural or Nonspecific Immunity to Microorganisms

Table 31.1 lists the cytokines that are involved with natural or nonspecific immunity. Natural immunity helps protect animals against viral infections and, through the inflammatory reaction (*see figure 29.12*) and other mechanisms, against bacterial, fungal, and parasitic infections.

Two of the best studied cytokines in this group are interferons α and β. **Interferon-α (IFN-α)** is produced by macrophages and monocytes. When IFN-α is secreted by virus-infected cells, it protects neighboring cells that have not yet been infected by inhibiting viral DNA or RNA and protein synthesis (the antiviral state, *see figure 29.9*). IFN-α also increases the expression of class I MHC molecules and Fc receptors on macrophages, and activates the lytic activity of NK cells.

Interferon-β (IFN-β), produced by fibroblasts and other cells, induces the antiviral state; modulates antibody production in concert with other cytokines; induces NK activity; modulates cytotoxic T-cell activity; induces differentiation of fibroblasts; stimulates lymphocytes and monocytes to express MHC molecules; and enhances the secretion of IL-1α and TNF-α by macrophages.

Two other well-studied cytokines of natural immunity are tumor necrosis factor-α and interleukin-1. **Tumor necrosis factor-α (TNF-α)** is the major cytokine along with interferon-γ that mediates the host response to gram-negative bacteria. Mononuclear phagocytes release TNF-α following stimulation by gram-negative LPS (*see figure 3.27*). Once TNF-α is released, it stimulates an increase in the number of B cells, initiation of leukocyte adherence to vascular endothelial cells (*see figure 29.12*), and stimulation of other mononuclear phagocytes to secrete IL-1, IL-6, or IL-8. All of these events are critical for the inflammatory response to occur (*see figure 29.13*). TNF-α also acts as an endogenous pyrogen on the hypothalamus to induce fever (*see figure 29.7*) and is cytotoxic for tumor cells.

Interleukin-1 is a cytokine with two forms (**IL-1α, IL-1β**). These forms appear to mediate both natural and specific acquired immunity. IL-1 is produced by a variety of cells. It is well known that IL-1 is synthesized either following stimulation by CD4+ T-helper cells after contact with antigen (*figure 31.1*) or on direct stimulation by LPS or TNF-α. IL-1 acts with antigen and other cytokines as a second signal to induce the proliferation of CD4+ cells. The second signal is absolutely necessary

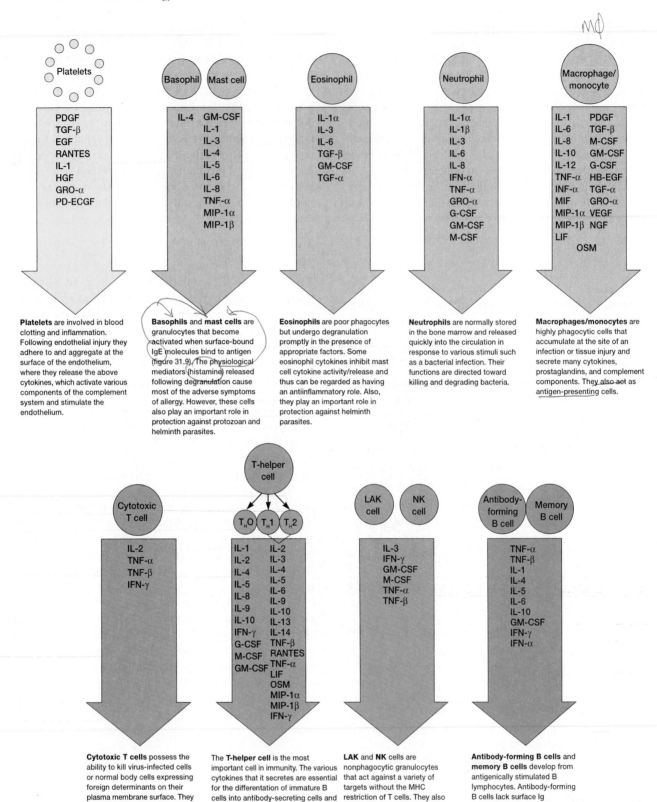

Figure 31.1 **Cytokine Release From Various Leukocytes and Platelets.** It should be noted that the above list of cytokines is not complete and is changing daily. Various other cells such as fibroblasts, endothelial cells, keratinocytes, hepatocytes, epithelial cells, glial cells, and neuronal cells also produce cytokines.

TABLE 31.1   Some of the Cytokines That Mediate Natural Immunity

Cytokine	Cell Source	Functions
IFNs α/β (interferons α/β)	T cells, B cells, monocytes/macrophages, fibroblasts	Antiviral activity, stimulates macrophage activity, modulates MHC class I and II protein expression on various cells, regulates the development of the specific immune response
TNF-α (tumor necrosis factor-α [cachectin])	Neutrophils, activated lymphocytes, NK cells, astrocytes, endothelial cells, smooth muscle cells	A wide variety of effects due to its ability to mediate expression of genes for growth factors and cytokines, transcription factors, receptors, inflammatory mediators, and acute-phase proteins; plays a role in host resistance to infection by serving as an immunostimulant and mediator of the inflammatory response; cytotoxic for tumor cells
TNF-β (tumor necrosis factor-β [lymphotoxin])	Lymphocytes	Same as TNF-α
IL-1α (interleukin-1α)	Monocytes/macrophages, endothelial cells, fibroblasts, neuronal cells, glial cells, keratinocytes, epithelial cells	Produces a wide variety of effects on the differentiation and function of cells involved in inflammatory and immune responses; also affects central nervous and endocrine systems
IL-1β (interleukin-1β)	Same as IL-1α	Same as IL-1α
IL-6 (interleukin-6, cytotoxic T-cell differentiation factor, B-cell differentiation factor)	T cells, monocytes/macrophages, fibroblasts, hepatocytes, endothelial cells, neuronal cells	Activates hematopoietic cells; induces growth of T cells, B cells, hepatocytes, keratinocytes, and nerve cells; stimulates the production of acute-phase proteins
Chemokines (low molecular weight proinflammatory cytokines)	Monocytes/macrophages, endothelial cells, fibroblasts, T cells, platelets	Leukocyte chemotaxis

for the generation of the specific immune response. IL-1 also causes endothelial cells and macrophages to synthesize and release IL-6 and IL-8, which are involved in inflammation. Since IL-1 is identical with the endogenous pyrogen derived from PMNs, it also stimulates the hypothalamus and induces fever.

Low molecular weight proinflammatory cytokines are paracrines called **chemokines** [Greek *chemo,* and *kinesis,* movement]. Chemokines play important roles in acute and chronic inflammation and have immunoregulatory functions. These cytokines participate in the inflammatory process by stimulating leukocytes. It is now believed that chemokines may be one of the principal secondary proinflammatory mediators. They can be divided into two subfamilies: the α subfamily, which includes interleukin-8, melanoma growth stimulation factor (GRO/MGSA), platelet factor-4 (PF-4), and β-thromboglobulin (β-TG); and the β subfamily, which includes monocyte chemoattractant protein-1 (MCP-1/JE), RANTES, and macrophage inflammatory protein-1α (MIP-1α) and MIP-1β. The cytokines of the α subfamily display chemotactic activity mainly for neutrophils, whereas the members of the β subfamily generally attract monocytes.

Activators of Effector Cells

Table 31.2 lists those cytokines that are involved in activating the specific effector cells necessary in enhancing T-helper-cell-mediated immune reactions. For example, **interferon-γ (IFN-γ)**

TABLE 31.2   Some of the Cytokines That Affect Cell Activation

Cytokine	Cell Source	Functions
IFN-γ (interferon-γ)	T cells (T_H1, T_C), NK cells	Activation of T cells, macrophages, neutrophils, and NK cells; increases class I and II MHC molecules
Lymphotoxin	T cells	Activation of neutrophils and endothelial cells
IL-5	T cells (T_H2)	Growth and activation of B cells and eosinophils; activation of eosinophil function; chemotactic for eosinophils
Migration inhibition factor	T cells (T_{DTH})	Conversion from motile to immotile state (macrophages/monocytes)

is produced by IL-2-secreting T cells (T_H1), T_C cells, and NK cells. IFN-γ has many functions: amplifying the activation of T-helper cells; increasing the expression of both class I and II MHC molecules; promoting the differentiation of both T and B cells; regulating the release of immunoglobulins from B cells; promoting the maturation of cytotoxic T cells (CD8+); inducing the antiviral state; and activating macrophages, neutrophils, and NK cells.

Migration inhibition factor (MIF) is produced by macrophages and inhibits the migration of macrophages and

TABLE 31.3   Some of the Cytokines That Mediate Mature Cell Activation, Growth, and Differentiation

Cytokine	Cell Source	Functions
IL-2 (interleukin-2, T-cell growth factor)	T cells (T_H1)	Stimulates T-cell proliferation and differentiation, enhances cytolytic activity of NK cells, promotes proliferation and immunoglobin secretion of activated B cells
IL-4 (interleukin-4, B-cell growth factor-1 [BCGF-1], B-cell stimulatory factor-1 [BCSF-1])	T cells (T_H2), macrophages, mast cells, basophils, B cells	Induces the differentiation of naive CD4+ T cells into T-helper-like cells; induces the proliferation and differentiation of B cells; exhibits diverse effects on T cells, monocytes, granulocytes, fibroblasts, endothelial cells
TGF-β (transforming growth factor-β)	Bone cells, platelets, fibroblasts, monocytes	Inhibits growth of many cell types, inhibits NK cell activity and T-cell proliferation, with IL-4 stimulates IgA secretion

TABLE 31.4   Some of the Cytokines That Mediate Immature Cell Growth and Activation

Cytokine	Cell Source	Functions
IL-3 (interleukin-3)	T cells, keratinocytes, neuronal cells, mast cells	Stimulates the production and differentiation of macrophages, neutrophils, eosinophils, basophils, mast cells
IL-7 (interleukin-7, lymphopoietin-1)	Bone marrow stromal cells, fetal liver cells	Supports the growth of pre-B and pro-B cells; supports the proliferation of T cells; stimulates the cytotoxic activity, proliferation, and generation of cytotoxic T cells
G-CSF (granulocyte colony-stimulating factor)	T cells, macrophages, neutrophils	Enhances the differentiation and activation of neutrophils
M-CSF (macrophage colony-stimulating factor)	T cells, neutrophils, macrophages, fibroblasts, endothelial cells	Stimulates various functions of monocytes and macrophages, promotes the growth and development of macrophage colonies from undifferentiated precursors

monocytes away from the site of infection. MIF secretion is induced by stimuli such as lipopolysaccharides and TNF-α and IFN-γ from T cells. It is easy to see that the activation of only a few macrophages at the site of a delayed hypersensitivity reaction (e.g., a positive tuberculin skin test; *see figure 31.14*) would greatly amplify the reaction by releasing MIF. The released MIF would cause the accumulation of the many nonspecific macrophages characteristic of this type of reaction and also activate them. Because activated macrophages are far more potent than resting macrophages in phagocytosis and intracellular killing, such a recruitment mechanism would greatly amplify the nonspecific defense against microorganisms and other nonself factors. Recently it has been shown that the anterior pituitary gland releases MIF in response to lipopolysaccharides or endotoxins.

Mediators of Mature Cell Activation, Growth, and Differentiation

Table 31.3 lists those cytokines that mediate mature lymphocyte activation, growth, and differentiation. One of these mediators is **interleukin-2 (IL-2).** IL-2 is produced by T-helper cells (T_H1) for a short period of time following antigen activation (figure 31.1). It helps trigger the proliferation of T_H and T_C cells, and its concentration determines the magnitude and duration of the succeeding immune response. IL-2 binds to specific IL-2 receptors on T-helper cells; both the expression of these receptors and that of the IL-2 cytokine are closely linked to antigen activation. IL-2 also can induce the synthesis of IFN-γ, can activate NK cells, and stimulates the growth of some B cells. Recently IL-2 immunotherapy has been used successfully against various tumors such as renal cell carcinoma and malignant melanoma.

Mediators of Immature Cell Growth and Differentiation

Table 31.4 lists some of the better known cytokines that influence hematopoiesis. For example, **interleukin-3 (IL-3)** regulates the proliferation of stem cells and influences the differentiation of bone-marrow–derived mast cells and other leukocytes (macrophages and granulocytes). The colony-stimulating factors (CSFs), such as the granulocyte and macrophage CSFs, stimulate the production of specific cell lines. The capacity of

TABLE 31.5   Selected Characteristics of Interleukins 8 to 14

Cytokine	Cell Source	Functions
IL-8 (interleukin-8)	Monocytes, endothelial cells, fibroblasts, alveolar epithelium, T cells, keratinocytes, neutrophils, hepatocytes	Chemoattractant for PMNs; causes degranulation and expression of receptors; inhibits adhesion of PMNs to endothelium; promotes migration of PMNs through endothelium; induces adhesion of neutrophils to endothelial cells
IL-9 (interleukin-9)	T cells (T_H2 cells)	Causes growth of T_H cells independent of IL-2 or IL-4; enhances response of mast cells to IL-3; stimulates erythropoiesis and production of platelets
IL-10 (interleukin-10)	T cells (T_H2), B cells, macrophages, keratinocytes	Reduces the production of IFN-γ, IL-1, TNF-α, and IL-6 by macrophages; in combination with IL-3 and IL-4, causes mast cell growth; in combination with IL-2, causes growth of cytotoxic T cells and differentiation of $CD8^+$ cells
IL-11 (interleukin-11)	Bone marrow cells	Stimulates development of B cells; in combination with IL-3, induces platelet development; in combination with IL-3, stimulates hematopoiesis
IL-12 (interleukin-12)	B lymphoblastoid cell line, macrophages	Activates T cells and the development of T_H1 cells; stimulates the growth and development of NK cells
IL-13 (interleukin-13)	T cells (T_H2)	Has anti-inflammatory activity; suppresses the cytotoxic functions of monocytes and macrophages, including their ability to kill intracellular parasites
IL-14 (interleukin-14)	T cells	Induces the proliferation of activated B cells, but not resting B cells

CSFs to increase the number of these defensive cells is one mechanism for killing invading microorganisms. In addition to increasing phagocyte numbers, some CSFs also modulate secretory capacity, chemotaxis, and phagocyte killing.

Interleukins 8 Through 14

Recently seven other interleukins have been identified and some of their functions elucidated (table 31.5). All of these play some role in mediating the body's responses to infections, cell growth, and cell development.

1. What is a cytokine? Define monokine, lymphokine, interleukin, and colony-stimulating factor. Describe several characteristics of cytokines.
2. How do cytokines play a role as mediators of natural immunity?
3. How do cytokines activate effector cells?
4. Describe how interleukin-2 (IL-2) and interleukin-3 (IL-3) function.

B-Cell Biology

Stem cells in the bone marrow produce uncommitted B cells that migrate to lymphoid tissue where some mature into cells that circulate freely in the bloodstream and lymph (*see figure 30.2*). These circulating B cells contain surface immunoglobulin M (IgM) and IgD. It is the IgM, however, that is the major receptor for its appropriate antigen. B cells also have receptors for the Fc part of some immunoglobulin classes and may have receptors for complement (C3b). Each B cell may have as many as 10,000 of these specific antigen-receptor sites. The total mature B-cell population of each individual thus carries receptors specific for many antigens; however, each mature B cell possesses receptors specific only for a particular antigenic determinant.

B-Cell Activation

The mature B cell divides and/or secretes antibody when triggered by the appropriate signals. Triggering can be antigen specific, activating a particular clone of cells, or nonspecific and polyclonal through B-cell mitogens.

T-Dependent Antigen Triggering

Most antigens have more than one type of antigenic determinant site (epitope) on each molecule (*see figure 30.3*). B cells specific for a given epitope on the antigen (e.g., epitope X) often cannot develop into plasma cells secreting antibody (anti-X) without the collaboration of helper T cells. A **T-helper (T_H) cell** is a T lymphocyte that helps initiate an effector (B cell or macrophage) function. In other words, binding of the epitope X to the B cell may be necessary, but it is not sufficient for B-cell activation. Antigens that are processed with the aid of T-helper cells are called **T-dependent antigens.** Examples include bacteria, foreign red blood cells, certain proteins, and hapten-carrier combinations (*see figure 30.4*).

The basic mechanism for T-dependent antigen triggering of a B cell is illustrated in figure 31.2 and involves three cells: (1) a macrophage or other antigen-presenting cell to process and present the antigen; (2) a T-helper cell able to recognize the antigen and respond to it; and (3) a B cell specific for the antigen. When all cells and the antigen are present, the following sequence takes place. The macrophage presents part of the antigen and its own class II MHC to the T-helper cell (signal #1 to the T cell) and also makes interleukins 1 and 6 (signal #2 to the T cell). Interleukins 1 and 6 stimulate the T-cell clone to divide and make interleukins 2 and 4–6. Interleukin-1 also stimulates the hypothalamus to raise the body temperature (causing a fever), which enhances the activity of T cells. Interleukins 2 and 4–6 stimulate T-helper cells to multiply. These helper cells directly associate with B cells that display the correct antigen-class II MHC complex and secrete

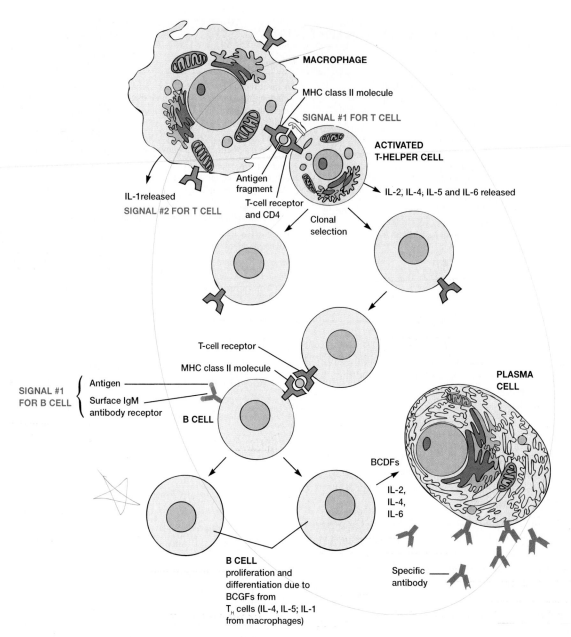

Figure 31.2   T-Dependent Antigen Triggering of a B Cell. Schematic diagram of the events occurring in the interactions of macrophages, T-helper cells, and B cells that produce cell-mediated immunity. Many cytokines (e.g., IL-1, IL-4, IL-5) stimulate B-cell proliferation and may be called B-cell growth factors (BCGFs). Cytokines such as IL-2, IL-4, and IL-6 stimulate B-cell differentiation and are called B-cell differentiation factors (BCDFs).

cytokines called B-cell growth factors (BCGFs), which cause B cells to multiply. As the number of B cells increases, T-helper cells produce other cytokines, the B-cell differentiation factors (BCDFs). These instruct some B cells to stop replicating, to differentiate into plasma cells, and to start producing antibodies. B cell activation involves more than T-helper cell activity. The B cell also recognizes the antigen through its surface IgM receptors for the antigen (signal #1 for the B cell) and is subsequently triggered to proliferate and differentiate into plasma cells secret-

ing antibody. (One B cell can synthesize more than 10 million antibody molecules per hour!) It should be noted here that B cells are more effective antigen presenters than macrophages at low antigen concentrations. The B cells bind antigen, take it up by receptor-mediated endocytosis, and present it to T-helper cells to activate them. In this situation B cells and T$_H$ cells activate each other. All the immune responses that produce IgG, IgA, and IgE involve this T-dependent antigen triggering.

Fever (p. 573).

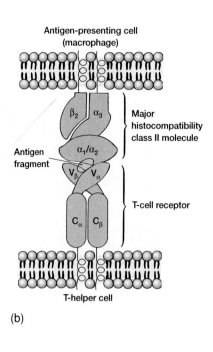

Figure 31.3 **T-cell CD4 Receptor Protein and its Activation.** (*a*) A schematic illustration of the proposed overall structure of the antigen receptor site on a T-cell plasma membrane. (*b*) An antigen-presenting cell begins the activation process by displaying an antigen fragment (e.g., peptide) on its surface as part of a complex with the histocompatibility molecules. A T-helper cell is triggered when the variable region of its CD4 receptor (designated V_α and V_β) reacts with the antigen fragment on the presenting cell surface.

T-Independent Antigen Triggering

Not all antibody responses require T-cell help. There are antigens that trigger B cells into immunoglobulin production without T-cell cooperation. These are called **T-independent antigens.** Examples include certain tumor-promoting agents, anti-immunoglobulin (anti-Ig), antibodies to certain B-cell differentiation antigens, and bacterial lipopolysaccharides. The T-independent antigens are polymeric—that is, composed of repeating polysaccharide or protein subunits. They elicit almost exclusively IgM antibody formation, and there is little switching to IgG; the resulting antibody generally has a low affinity for antigen.

The mechanism for activation by T-independent antigens probably depends on their polymeric structure. These large molecules present a large array of identical epitopes to a B cell specific for that determinant. The multivalent nature of the cell receptor (surface IgM) is such that cell activation occurs and IgM is secreted. Because there is no T-cell help, the B cell cannot switch to high affinity IgG production.

---

1. Name three types of B-cell receptors.
2. What is a mitogen?
3. Name the three cells involved in T-dependent antigen triggering. How does it occur?
4. How does T-independent antigen triggering occur?

T-Cell Biology

T cells are the elements of the cell-mediated immune response. They are immunologically specific, can carry a vast repertoire of immunologic memory and can function in a variety of regulatory and effector ways.

T-Cell Antigen Receptors

T cells have specific receptors for antigen fragments (e.g., peptides with a linear sequence of 8 to 25 amino acids) on their plasma membrane surface. The receptor site is composed of two parts, an alpha polypeptide chain and a beta polypeptide chain (figure 31.3*a*). Each chain is stabilized by disulfide bridges. The receptor is anchored into the plasma membrane by parts of the α and β chains that extend into the cytoplasm. The recognition sites of the T-cell receptor extend out from the membrane and have a terminal variable section complementary to antigen fragments. T cells respond to antigen fragments exposed on the surfaces of **antigen-presenting cells,** most of which are **macrophages, dendritic cells,** and B cells. These cells scavenge foreign materials (antigens or pathogens) from the blood and tissues and digest them. Inside the cells small pieces of the antigens bind to a special class of proteins called major histocompatibility complex (MHC) molecules. The MHC-antigen complex is then transported to the

(b)

(c)

Figure 31.4 Class I Major Histocompatibility Molecule Structure.
(*a*) This side view of the molecule shows the arrangement of its four regions (α_1, α_2, α_3, and β_2). (*b*) Class I MHC molecules can hold only short peptides (shown in blue) because the binding site is closed off. (*c*) Since the binding site is open on both ends, class II MHC molecules can bind to peptides (shown in blue) of different lengths.

surface of the presenting cell and displayed to passing lymphocytes known as T-helper cells (figure 31.3*b*). Some knowledge of these histocompatibility molecules is thus required before T-cell-macrophage interactions and T-cell biology can be understood.

Major Histocompatibility Complex (MHC) Molecules

Almost all human tissue cells contain **major histocompatibility complex (MHC)** molecules on their plasma membrane. This term is derived from the Greek word for tissue [*histo*] and the ability to get along [compatibility]. Human MHC molecules also are known as **human leukocyte antigens (HLAs).** MHC molecules can be divided into two classes: class I molecules are found on almost all types of body cells, whereas class II mole-

cules appear only on those leukocytes involved in the immune response (macrophages, antigen-presenting cells [dendritic cells], and B cells).

Class I molecules (figure 31.4*a*) consist of a complex of two protein chains, one with a mass of 45 kDa (the heavy chain) and the other with a mass of 12 kDa (the light chain). The two chains contain four regions. The outer segment of the heavy chain can be divided into three functional domains, designated α_1, α_2, and α_3. The β_2-microglobulin (β_2m) protein and α_3 segment of the heavy chain are noncovalently associated with one another and are close to the plasma membrane. A small segment of the heavy chain is attached to the membrane by a short amino acid sequence that extends into the cell interior, but the rest of the protein protrudes to the outside. The α_1 and α_2 domains lie to the outside and form the antigen-binding pocket.

Figure 31.5   **Major Histocompatibility Complex.** The MHC region of human chromosome 6 and the gene products associated with each locus. DP, DQ, and DR are not single loci but clusters of several loci encoding α and β chains of class II molecules.

Class II MHC molecules are also transmembrane proteins consisting of α and β chains of mass 34 kDa and 28 kDa, respectively. Both chains are folded to give two domains, the one nearest to the membrane having considerable likeness to the β₂-microglobulin.

Although both types of molecules are structurally distinct, both fold into very similar shapes (figure 31.4b). Each MHC molecule has a deep groove into which a short peptide or other antigen fragment can bind. Because this peptide is not part of the MHC molecule, it can vary from one MHC molecule to the next. On healthy cells all of these peptides come from self-proteins. The presence of foreign peptides (antigen fragments) in the MHC groove alerts the immune system and activates either T cells or macrophages.

Class I and class II molecules present peptides that arise in different places within cells as the result of a process known as antigen processing. Class I molecules bind to peptides that originate in the cytosol (e.g., antigens from replicating viruses). Endogenous antigenic proteins are digested inside the cell as part of the natural process by which a cell continually renews its protein contents. The short peptide fragments that result from this process are pumped by a specific transporter protein from the cytosol into the endoplasmic reticulum. Within the endoplasmic reticulum the class I MHC heavy chain is synthesized and associates with β₂-microglobulin. This dimer appears to bind peptide as soon as it enters the endoplasmic reticulum. The MHC molecule and peptide are then carried to, and incorporated within, the plasma membrane. If the peptide is foreign (e.g., short pieces of viral protein), a passing CD8⁺ T cell (cytotoxic T cell) recognizes it and releases chemicals (cytokines) that directly destroy the entire cell.

Class II MHC molecules bind to fragments that arise from exogenous antigens. This pathway would function with bacteria and viruses that have been taken endocytotically. The antigen-presenting cell takes in the antigen or pathogen by receptor-mediated endocytosis or phagocytosis and produces antigen fragments by digestion in the endosome. Fragments then combine with preformed class II MHC molecules and are delivered to the cell surface. It is here that the peptide is recognized by CD4⁺ T-helper cells (figure 31.2). Unlike CD8⁺ T cells, CD4⁺ T cells do not directly kill target cells. Instead they respond in two distinct ways. One is to enlarge and divide, thereby increasing the number of CD4⁺ cells that react to the antigen. The second response is to secrete cytokines (e.g., interleukin-2) that either directly inhibit the pathogen that produced the antigen or recruit other cells to join in the immune response (figure 31.1).

MHC molecules are coded by a group of genes termed the major histocompatibility complex (MHC) (figure 31.5). The MHC of humans is a cluster of genes located on chromosome 6. These genes are labeled A, B, C, and D and are codominant. Because each individual has two sets of these genes, one from each parent, a person can potentially have any combination of eight different MHC molecules. Only one gene of each pair can code for a MHC molecule in a given individual, and these molecules are different from one individual to another. The closer two individuals are related, the more similar are their MHC molecules.

Class I molecules are made by all cells of the body except red blood cells and comprise MHC types A, B, and C. Class I molecules serve to identify almost all cells of the body as "self." They also stimulate antibody production when introduced into a host with different class I molecules. This is the basis for MHC typing when a patient is being prepared for an organ transplant (Box 31.1).

Class II molecules comprise the D group of the MHC and are produced only by activated macrophages, mature B cells, some T cells, and certain cells of other tissues. Class II molecules do not stimulate antibody production but are required for T-cell communication with macrophages and B cells. As discussed in more detail later, part of the T-cell receptor must recognize a class II molecule on the adjacent cell before the T cell can secrete cytokines necessary for the immune response.

The class III genes encode the second component of complement (C2), factor B, and two forms of the fourth component of complement (C4a and C4b). C2, C4a, and C4b participate in the classical pathway and factor B in the alternative pathway (*see figure 32.2*).

Classes I and II of the MHC are involved in not only the immunologic recognition of microorganisms but also the individual's susceptibility to particular noninfectious diseases. For example, there is evidence of MHC-linked determinants in tuberculoid leprosy, paralytic poliomyelitis, multiple sclerosis, and acute glomerulonephritis (*see chapter 37*). The effect of MHC molecule presence on the frequency of these diseases is just beginning to be understood.

Regulator T Cells

Regulator T cells control the development of effector T cells. Two types exist: T-helper cells (CD4$^+$ cells) and T-suppressor cells. There are three subsets of T-helper cells (see figure 31.1): T_H1, T_H2, and T_HO. **T_H1 cells** produce IL-2, IFN-γ, and other cytokines and are involved in cellular immunity. They are responsible for delayed-type hypersensitivity reactions and macrophage activation. **T_H2 cells** also produce various cytokines (e.g., IL-4, IL-5, IL-10, IL-13) but are involved in humoral immunity. T_H2 cells are helpers for B-cell antibody responses and defense against helminths. Not much is known about **T_HO cells** except that they represent a unique lineage of T-helper cells that produce their own cytokines.

T-helper cells (T_H1) need two signals (costimulation) to be stimulated by an antigen. The first signal is when antigen fragments are effectively presented to a T cell by a macrophage, dendritic cell, or an activated B cell. For example, when a virus infects a host, it is phagocytosed by a macrophage and partially digested (figure 31.6). Some of the viral protein fragments (antigen) are then moved to the surface of the macrophage where they are presented in close association with class II MHC molecules. This combination of viral antigen fragment and MHC class II molecule is needed for recognition by the T cell receptor and CD4 protein on the surface of the T-helper cell and represents activation signal 1. (The abbreviation CD is from cell-associated differentiation antigen; CDs are functional molecules or receptors that serve as surface markers; see pp. 628–29.)

The second signal for activation involves the CD28 protein receptor on T-helper cells (figure 31.7). When CD28 binds to a protein called B7-2 on the surface of a macrophage, it delivers a second signal to the T-helper cell. Both of these signals go into the cytoplasm of the T-helper cell.

Signal 1 activates a tyrosine kinase located in the cytoplasm. Tyrosine kinase adds phosphate groups to the amino acid tyrosine in proteins. When this phosphorylation occurs, the enzyme phospholipase Cγ1 cleaves the phosphatidylinositol bisphosphate located in the T-helper cell plasma membrane. Two cleavage products are formed and two specific pathways now occur.

One of the cleavage products, diacylglycerol, activates protein kinase C. Protein kinase C moves into the nucleus where it catalyzes the formation of a protein complex called AP-1 from separate components. The other cleavage product, inositol trisphosphate, causes a calcium channel to open; calcium ions to rush into the cytoplasm leading to further enzymatic activity and the activation of calmodulin, calcineurin, and nuclear factor of activated T-helper cells (NF-AT). NF-AT then migrates into

Figure 31.6 **Regulator and Effector T Cells.** A virus is phagocytosed by a macrophage and a small antigen fragment (peptide) presented to T-helper cells in association with class II MHC. Once activated, the T_H2 cell secretes IL-2. It is the IL-2 that regulates the proliferation of cytotoxic T cells. Once the T_C cells proliferate, they cause lysis and death of the virus-infected cell.

the nucleus where it binds to the newly formed AP-1 forming a NF-AT/AP-1 complex (a transcription factor). This transcription factor binds tightly to a specific gene sequence in the DNA molecule, causing interleukin-2 mRNA to be transcribed. Interleukin-2 mRNA moves out of the nucleus to the ribosomes where the protein interleukin-2 is produced.

Signal 2 is mediated by the CD28 molecule and B7-2 receptor. This signal activates another tyrosine kinase, causes the formation of the transcription factor CD28RC, and also stabilizes the mRNA transcribed from the interleukin-2 gene. This increases the concentration of interleukin-2.

T-helper cells that are activated by these two signals secrete large amounts of interleukin-2, which activates B cells and cytotoxic T cells (figure 31.6). T-helper cells also secrete gamma-interferon (IFN-γ), which is capable of activating macrophages and increasing enhanced microbicidal activity.

T-suppressor (T_S) cells can suppress B-cell and T-cell responses. Subpopulations of T_S cells specific to a given antigen can be stimulated to proliferate by the interleukin-2 from activated T-helper cells. This proliferation occurs at a slow rate to provide negative feedback control for that part of the immune response known as acquired immune tolerance.

Figure 31.7   Two Signals are Essential for T-helper Cell Activation. The first signal is the presentation of the antigen fragment by a macrophage or other antigen-presenting cell along with the MHC class II molecule to the T-helper cell receptor and CD4 protein. The second signal occurs when the macrophage presents the B7-2 protein to the T-helper cell with its CD28 protein receptor. Both signals send information into the cytoplasm of the T-helper cell. The first signal causes interleukin-2 mRNA to be produced. The second signal boosts the production of interleukin-2 to effective concentrations. It now is known that the gene for interleukin-2 is under very tight regulation. It cannot be transcribed unless the NF-AT/AP-1 complex, AP-1, and at least two other transcription factors (CD28RC and NF-kB) are present. All these factors must be produced anew or activated when the T-helper cell is stimulated through its antigen-specific receptor.

Ⅱ Acquired Immune Tolerance

Acquired immune tolerance (tolerance induction) refers to the body's ability to produce antibodies against nonself antigens such as microbial antigens, while "tolerating" (not producing antibodies against) self-antigens. This tolerance arises early in embryonic life when immunologic competence is being established. Three major mechanisms have been proposed for tolerance induction: clonal deletion, functional inactivation, and clonal anergy. It is likely that all three are partly responsible for immune tolerance.

According to the clonal deletion mechanism, tolerance to self-antigens is achieved by the destruction of lymphocytes that can interact specifically with self-antigens. Recent evidence

TABLE 31.6   Comparison of Lymphocytes Involved in the Immune Response

Property	T Cells	B Cells
Origin	Bone marrow in adults	Bone marrow in adults
Maturation and differentiation	Thymus	Lymphoid tissue or bone marrow; bursa of Fabricius in birds
Longevity	Long (months or years)	Short (days to weeks)
Mobility	Great	Very little
Complement receptors	Absent	Present
Surface immunoglobulins	Absent	Present (IgM, IgD)
Proliferation	Upon antigenic stimulation	Upon antigenic stimulation, differentiate into plasma and memory cells
Immunity type	Cell mediated and humoral	Humoral
Distribution	High in blood, lymph, and lymphoid tissue	High in spleen, lymph nodes, bone marrow, and other lymphoid tissue; low in blood
Secretory product	Lymphokines	Antibodies
Subsets and functions	T-helper (T_H) cell: necessary for B-cell activation by T-dependent antigens and T-effector cells	Memory cell: a long-lived cell responsible for the anamnestic response
	T-suppressor (T_S) cell: blocks induction and/or activation of T_H cells and B cells; helps maintain tolerance	Plasma cell: a cell arising from a B cell that manufactures specific antibodies
	T-regulator (T_R) cell: develops into T_H or T_S cells and controls balance between enhancement and suppression of response to antigen	
	Delayed-type hypersensitivity T (T_{DTH}) cell: provides protection against infectious agents, mediates inflammation, and activates macrophages in delayed-type hypersensitivity	
	Cytotoxic T (T_C) cell: lyses cells recognized as nonself and virus or parasite-infected cells	

shows that deletion of those immature T cells able to recognize self-antigens does take place in the thymus. It is not known precisely how this occurs.

According to the functional inactivation mechanism, some lymphocytes that interact with self-antigens are present throughout life. However, they are normally inhibited from attacking self-antigen by T-suppressor cells.

Recently evidence for a third mechanism—clonal anergy—has accumulated. **Clonal anergy** [immunologists describe an inactive lymphocyte as "anergic" from Greek *an*, negative, and *ergon*, without] turns off self-reactive lymphocytes as a consequence of the way in which certain self-molecules are presented to them.

Effector T Cells

Effector T cells directly attack specific target cells. Several subsets exist: namely, cytotoxic T cells, delayed-type hypersensitivity T cells, and natural killer cells.

Cytotoxic T (T_C) cells ($CD8^+$ cells) are the most-studied and best-understood subset of T cells. They react, using the CD8 receptor protein, with virus-infected cells that display both viral antigens and class I MHC molecules on their surface (figure 31.6). Once recognition has occurred, T_C cells are stimulated to divide and proliferate by the interleukin-2 secreted by T-helper cells. Thus activation of T-helper cells by macrophages

is required for the full immunologic defense provided by the cytotoxic T cells. Once activated, the cytotoxic T cell releases potent cytokines such as gamma-interferon and tumor necrosis factor. These cytokines limit viral replication inside a cell, while also attracting macrophages and other phagocytes that can destroy the virus-infected cell. Recent work indicates that cytotoxic T cells may also produce molecules that elicit a form of cell death called apoptosis—in effect, these molecules signal the infected cell that it should kill itself (see figure 31.6).

Delayed-type hypersensitivity T (T_{DTH}) cells are responsible for initiating type IV delayed-type hypersensitivity reactions through secretion of cytokines. This type of sensitivity and the function of T_{DTH} cells are covered in detail later in the chapter. T_{DTH} cells also activate macrophages and may inhibit their migration away from a site of inflammation.

Table 31.6 summarizes and compares the properties of T and B lymphocytes.

Natural killer (NK) cells are a small population of large nonphagocytic granular lymphocytes with insignificant amounts of surface immunoglobulin and class I MHC. Although they are discussed in the context of effector T cells, NK cells may not be true T cells. Thus they often are referred to as null cells. NK cells are found in animals never exposed to relevant antigens. Asialo GM_1, a glycosphingolipid, is a characteristic antigenic surface marker present in relatively high density on the plasma

(a)

(b)

Figure 31.8   How an Effector T Cell Destroys a Target Cell. (*a*) An effector cell (left) contacts a target cell (right) (×5,700). (*b*) The effector cell secretes a protein (perforin 1) that forms a hole in the target cell's plasma membrane and causes its contents to leak out (cytolysis) (×45,000).

membranes of NK cells. They also have Fc receptors for IgG and have been shown to function in vivo in antibody-dependent cell-mediated cytotoxicity against IgG-coated pathogens or tumor cells.

NK cells help in the host's defense by recognizing and destroying tumor cells, virus-infected cells, fungi, bacteria, protozoa, and helminth parasites. This process is called **immune surveillance.** These cells or organisms are killed by cell-mediated lysis (cytolysis) as follows.

NK cells:

1. Are activated by interferons produced by virus-infected cells and/or interleukin-2 from T cells

2. Recognize a class I MHC on target cells and bind to it (figure 31.8*a*)

3. Then undergo a Ca^{2+}-dependent sequence consisting of

   a.   Microtubule assembly

   b.   Movement of cytoplasmic granules toward the part of the plasma membrane in contact with the target cell

   c.   Reorientation of the Golgi apparatus to the target cell, and

   d.   Movement of the microtubule-organizing center and other cytoskeletal components into the region of the cytoplasm adjacent to the bound target cell

4. Next insert a pore-forming protein, perforin 1, into the target cell's plasma membrane (figure 31.8*b*)

5. Also produce lysosomal secretions.

As a result of these NK cell activities, the attacked cell lyses. Finally, the NK cell recycles its cytoplasmic components in preparation for another attack on a target cell.

1. What is the function of an antigen-presenting cell? What is a T-cell receptor and how is it involved in T-cell activation?

2. What is a histocompatibility molecule? What are MHCs and HLAs? Describe the roles of the three MHC classes.

3. Describe antigen processing. How does this process differ for endogenous and exogenous antigens?

4. Outline the functions of a T-helper cell. Briefly describe how T_H cells are activated by costimulation.

5. What is acquired immune tolerance and in what three ways might it arise?

6. Name three types of effector T cells and describe their roles and the ways in which they operate. What is immune surveillance?

7. Briefly compare and contrast B cells and T cells with respect to their formation, structure, and roles in the immune response (table 31.6).

Superantigens

In recent years there has been an explosion of information concerning microbial superantigens and their roles in human disease processes. **Superantigens** are bacterial proteins that stimulate the immune system much more extensively than do normal antigens. Superantigens nonspecifically stimulate T cells to proliferate by interacting with both class II major histocompatibility complex products on antigen-presenting cells and variable regions on the β chain of the T-cell receptor complex (figure 31.9). Examples of superantigens include pyrogenic toxins (streptococcal scarlet fever toxins of serotypes A, B, and C; staphylococcal toxic shock syndrome toxin-1; staphylococcal

Figure 31.9 **Superantigens.** The most prevalent antigen receptor of T cells is the T-cell receptor. Superantigens stimulate T cells bearing a particular V_β element of the T-cell receptor, regardless of the nature of the V_α receptor element. They bind directly to the outer surface of the MHC class II molecule without being processed by the antigen-presenting cell. Thus a superantigen can activate many more T cells than a normal antigen does.

enterotoxin serotypes A, B, C_n, D, E, and G; streptococcal M protein; staphylococcal serotypes A and B; exfoliative toxin; and the recently identified pyogenic toxins made by groups B, C, F, and G streptococci and *Streptococcus sanguis*. Since superantigens cause symptoms by way of the release of massive quantities of immune cytokines, they should be considered possible chronic associates in such diseases as rheumatic fever, arthritis, Kawasaki syndrome, atopic dermatitis, and one type of psoriasis because of their potent immune system altering capacity.

Hypersensitivities (Allergies)

Hypersensitivity or **allergy** (the terms are used interchangeably) is an exaggerated immune response that results in tissue damage and is manifested in the individual on second or subsequent contact with an antigen. Hypersensitivity reactions can be classified as either immediate or delayed. Obviously immediate reactions appear faster than delayed ones, but the main difference between them is in the nature of the immune response to the antigen. Realizing this fact, Peter Gell and Robert Coombs developed a classification system for reactions responsible for hypersensitivities in 1963. Their system correlates clinical symptoms with information about immunologic events that occur during hypersensitivity reactions. The Gell-Coombs classification system divides hypersensitivity into four types: I, II, III, and IV.

Type I (Anaphylaxis) Hypersensitivity

Type I (anaphylaxis) hypersensitivity is characterized by an allergic reaction occurring immediately following an individual's second contact with the responsible antigen (the **allergen**). Upon initial exposure to an allergen, B cells are stimulated to differentiate into plasma cells and produce specific IgE with the help of T cells (figure 31.10). This IgE is sometimes called a **reagin,** and the individual has a hereditary predisposition for its

production. Once synthesized, IgE binds to the Fc receptors of mast cells and basophils and sensitizes these cells, making the individual allergic to the allergen. When a second exposure to the allergen occurs, the allergen attaches to the surface-bound IgE on the sensitized mast cells causing degranulation. Degranulation releases physiological mediators such as histamine, SRS-A (slow reacting substance of anaphylaxis; also called leukotriene), heparin, prostaglandins, PAF (platelet-activation factor), ECF-A (eosinophil chemotactic factor of anaphylaxis), and proteolytic enzymes. These mediators trigger smooth muscle contractions, vasodilation, increased vascular permeability, and mucous secretion. The inclusive term for these responses is **anaphylaxis** [Greek *ana,* up, back, again; and *phylaxis,* protection]. Anaphylaxis can be divided into systemic and localized reactions.

Systemic anaphylaxis is a generalized response that occurs when an individual sensitized to an allergen receives a subsequent exposure to it. The reaction is immediate due to the large amount of mast cell mediators released over a short period. Usually there is respiratory impairment caused by smooth muscle constriction in the bronchioles. The arterioles dilate, which greatly reduces arterial blood pressure and increases capillary permeability with rapid loss of fluid into the tissue spaces. Because of these reactions the individual can die within a few minutes from reduced venous return, asphyxiation, reduced blood pressure, and circulatory shock. Common examples of allergens that can produce systemic anaphylaxis include drugs (penicillin), passively administered antisera, and insect venom from the stings or bites of wasps, hornets, or bees (figure 31.11).

Localized anaphylaxis is called an atopic ("out of place") allergy. The symptoms that develop depend primarily on the route by which the allergen enters the body. **Hay fever** (allergic rhinitis) is a good example of an atopic allergy involving the upper respiratory tract. Initial exposure involves airborne allergens—such as plant pollen, fungal spores, animal dander, and house

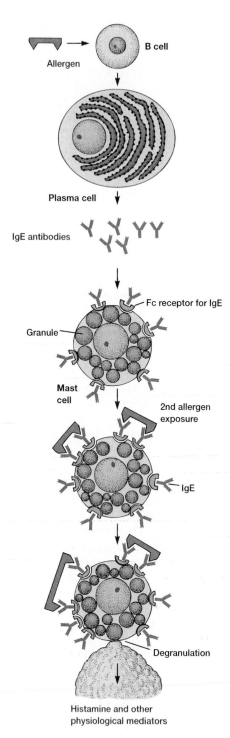

Figure 31.10   Type I (Anaphylaxis) Hypersensitivity. This type of hypersensitivity occurs when IgE antibodies attach to mast cells. The combination of these antibodies with allergens stimulates the mast cell (or basophil) to degranulate and produce the physiological mediators that cause the anaphylactic reaction, asthma, or hay fever.

dust mites—that sensitize mast cells located within the mucous membranes. Reexposure to the allergen causes the typical localized anaphylactic response: itchy and tearing eyes, congested nasal passages, coughing, and sneezing. Antihistamine drugs are used to help alleviate these symptoms.

Figure 31.11   The Anaphylactic Response to Bee Venom. This person has been stung on the arm by a bee, leading to type I (anaphylaxis) hypersensitivity in a generalized area.

Bronchial asthma (asthma means panting) is an example of an atopic allergy involving the lower respiratory tract. Common allergens are the same as for hay fever. In bronchial asthma, however, the air sacs (alveoli) become overdistended and fill with fluid and mucus; the smooth muscle contracts and narrows the walls of the bronchi. Bronchial constriction produces a wheezing or whistling sound during exhalation. Symptomatic relief is obtained from bronchodilators that help relax the bronchial muscles, and from expectorants and liquefacients that dissolve and expel mucous plugs that accumulate.

Allergens that enter the body through the digestive system may cause food allergies. **Hives** (eruptions of the skin) are a good diagnostic sign of a true food allergy. Once established, type I food allergies are usually permanent but can be partially controlled with antihistamines or by avoidance of the allergen.

Skin testing can be used to identify the allergen responsible for allergies. These tests involve inoculating small amounts of suspect allergen(s) into the skin. Sensitivity to the antigen is shown by a rapid inflammatory reaction characterized by redness, swelling, and itching at the site of inoculation (*see figure 32.8*). The affected area in which the allergen-mast cell reaction takes place is called a wheal and flare reaction site.

Once the responsible allergen has been identified, the individual should avoid contact with it. At times this is not possible, and **desensitization** is warranted. This procedure consists of a series of allergen doses injected beneath the skin to stimulate the production of IgG antibodies rather than IgE antibodies. The circulating IgG antibodies can then act as blocking antibodies to intercept and neutralize allergens before they have time to react with mast cell-bound IgE. Recent evidence suggests that suppressor T-cell activity also may cause a decrease in IgE synthesis. Allergy injections are about 65 to 75% effective in individuals whose allergies are caused by inhaled allergens.

Type II (Cytotoxic) Hypersensitivity

Type II (cytotoxic) hypersensitivity is generally called a cytolytic or cytotoxic reaction because it results in the destruction of host cells, either by lysis or toxic mediators. In type II hypersensitivity, IgG or IgM antibodies are directed against cell surface or tissue antigens. They usually stimulate the complement pathway and a variety of effector cells (figure 31.12). The antibodies interact with complement (C1q) and the effector cells through their Fc regions. The damage mechanisms are a reflection of the normal physiological processes involved in interaction of the immune system with pathogens. A classic example of type II hypersensitivity is that resulting when a person receives a transfusion with blood from a donor with a different blood group.

Type III (Immune Complex) Hypersensitivity

Type III (immune complex) hypersensitivity involves the formation of immune complexes (figure 31.13a; see also figure 32.7). Normally these complexes are removed effectively by monocytes of the reticuloendothelial system. In the presence of excess amounts of some antigens, the antigen-antibody complexes may not be efficiently removed. Their accumulation can lead to a hypersensitivity reaction from complement that triggers a variety of inflammatory processes. This inflammation causes damage, especially of blood vessels (vasculitis; figure 31.13b), kidney glomerular basement membranes (glomerulonephritis), joints (arthritis), and skin.

Diseases resulting from type III reactions can be placed into three groups. First, a persistent viral, bacterial, or protozoan infection, together with a weak antibody response, leads to chronic immune complex formation and eventual deposition of the complex in host tissues. Second, the continued production of autoantibody to self-antigen during an autoimmune disease can lead to prolonged immune complex formation. This overloads the reticuloendothelial system, and tissue deposition of the complexes occurs (e.g., in the disease **systemic lupus erythematosus**). Third, immune complexes can form at body surfaces (such as the lungs), following repeated inhalation of allergens from molds, plants, or animals. For example, in Farmer's lung disease, an individual has circulating antibodies to fungi after being exposed repeatedly to moldy hay. These antibodies are primarily IgG. When the allergens (fungal spores) enter the alveoli of the lungs, local immune complexes form, leading to inflammation.

Some group A streptococcal infections can produce an immunologically mediated acute glomerulonephritis (see chapter 37). Although the mechanism is not completely understood, it is believed that complexes of antibody and streptococcal antigen are deposited within the kidney glomeruli and generate a type III hypersensitivity reaction.

Type IV (Cell-Mediated) Hypersensitivity

Type IV (cell-mediated) hypersensitivity involves delayed T-cell-mediated immune reactions. A major factor in the type IV reaction is the time required for a special subset of T cells called delayed-type hypersensitivity T (T_{DTH}) cells to migrate to and accumulate near the antigens. This usually takes a day or more.

Type IV reactions occur when antigens, especially those binding to tissue cells, are phagocytosed by macrophages and then presented to receptors on the T_{DTH} cell surface. Contact between the antigen and T_{DTH} cell causes the cell to proliferate and release cytokines. Cytokines attract lymphocytes, macrophages, and basophils to the affected tissue. Extensive tissue damage may result. Examples of type IV hypersensitivities include tuberculin hypersensitivity (the **TB skin test**), allergic contact dermatitis, some autoimmune diseases, transplantation rejection, and killing of cancer cells.

In tuberculin hypersensitivity a partially purified protein called tuberculin, which is obtained from the bacillus that causes tuberculosis (*Mycobacterium tuberculosis; see chapter 37*), is injected into the skin of the forearm (figure 31.14a). The response in a tuberculin-positive individual begins in about 8 hours, and a reddened area surrounding the injection site becomes indurated (firm and hard) within 12 to 24 hours. The reaction reaches its peak in 48 hours and then subsides. The size of the induration is directly related to the amount of antigen that was introduced and to the degree of hypersensitivity of the tested individual. Other microbial products used in type IV skin testing are histoplasmin for histoplasmosis, coccidioidin for coccidioidomycosis, lepromin for leprosy, and brucellergen for brucellosis.

Allergic contact dermatitis is caused by haptens (see figure 30.4) that combine with proteins in the skin to form the allergen that elicits the immune response. The haptens are the antigenic determinants, and the skin proteins are the carrier molecules for the haptens. Examples of these haptens include cosmetics, plant materials (catechol molecules from poison ivy and poison oak; figure 31.14b), topical chemotherapeutic agents, metals, and jewelry (especially jewelry containing nickel).

Several important chronic diseases involve cell and tissue destruction by type IV hypersensitivity reactions. These diseases are caused by viruses, mycobacteria, protozoa, and fungi that produce chronic infections in which the macrophages and T cells are continually stimulated. Examples are leprosy, tuberculosis, leishmaniasis, candidiasis, and herpes simplex lesions.

---

1. What is a superantigen and how does it function?
2. Discuss the mechanism of type I hypersensitivity reactions and how these can lead to systemic and localized anaphylaxis.
3. What causes a wheal and flare reaction site?
4. Why are type II hypersensitivity reactions called cytolytic or cytotoxic?
5. What characterizes a type III hypersensitivity reaction? Give an example.
6. Characterize a type IV hypersensitivity reaction.
7. What is the TB skin test used for?

Autoimmune Diseases

As discussed earlier, the body is normally able to distinguish its own self-antigens from foreign nonself antigens and does not mount an immunologic attack against the former. This phenomenon is called immune tolerance. At times the body loses tolerance and mounts an abnormal immune attack, either with antibodies or T cells, against a person's own self-tissue antigens.

Figure 31.12   Type II (Cytotoxic) Hypersensitivity. The action of antibody occurs through effector cells or the membrane attack complex, which damages target cell plasma membranes, causing cell destruction.

(a)

Figure 31.13   Type III (Immune Complex) Hypersensitivity. Deposition of immune complexes in blood vessel walls. (*a*) Antibody and antigen combine to form immune complexes. These activate complement, which causes basophils and platelets to degranulate and release histamine and other mediators. These mediators increase vascular permeability. (*b*) The increased permeability allows the immune complexes to be deposited in the blood vessel wall. This induces platelet aggregation to form microthrombi (blood clots) on the vessel wall. PMNs, stimulated by complement, degranulate causing enzymatic damage to the blood vessel wall.

(b)

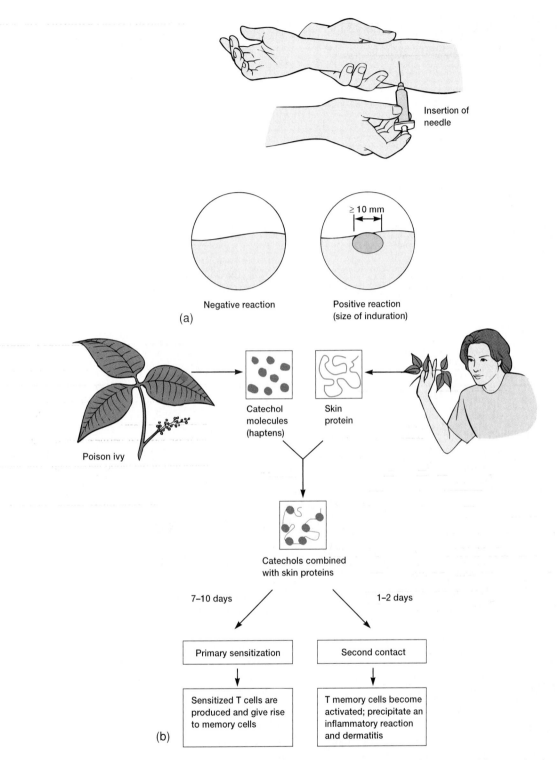

Figure 31.14   Mechanism of Type IV (Cell-mediated) Hypersensitivity. (*a*) The tuberculin skin test. A positive reaction is one in which the diameter of induration is 10 mm or more. (*b*) In this example of contact dermatitis to poison ivy, a person initially becomes exposed to the catechol molecules from the poison ivy plant. The catechols combine with high molecular weight skin proteins and act as haptens. After 7–10 days sensitized T cells are produced and give rise to memory T cells. Upon second contact, the catechols bind to the same skin proteins, and the T memory cells become activated in only 1–2 days, leading to the inflammatory reaction (contact dermatitis).

TABLE 31.7   Some Autoimmune Diseases in Humans

Disorder	Organ or Structure Affected	Evidence or Mechanism
Highly probable[a]		
Autoimmune thrombocytopenic purpura	Platelets	Phagocytosis of antibody-sensitized platelets
Diabetes mellitus (type I)	Pancreas	Cell-mediated and humoral islet cell antibodies
Graves' disease	Thyroid	Thyroid-stimulating hormone receptor antibody
Hashimoto's thyroiditis	Thyroid	Cell-mediated and humoral thyroid cytotoxicity
Insulin resistance	Pancreas	Insulin receptor antibody
Multiple sclerosis	Nerves	Myelin sheath destruction
Myasthenia gravis	Muscles	Acetylcholine receptor antibody
Rheumatoid arthritis	Cartilage	Immune complexes in joints
Systemic lupus erythematosus (SLE)	DNA	Circulating immune complexes
Probable		
Adrenergic drug disorder	Lungs	β-adrenergic receptor antibody
Glomerulonephritis	Kidney	Glomerular basement membrane antibody or immune complexes
Infertility (some cases)	Testes	Antispermatozoal antibodies
Possible		
Chronic active hepatitis	Liver	Smooth muscle antibody
Vasculitis	Blood vessels	Immunoglobulin and complement in vessel walls, low serum complement
Vitiligo	Skin	Melanocyte antibody

[a]Boldfaced headings indicate the likelihood that the disorder is an autoimmune disease.

It is important to distinguish between autoimmunity and autoimmune disease. Autoimmunity often is benign, whereas autoimmune disease often is fatal. **Autoimmunity** is characterized only by the presence of serum autoantibodies; it is a normal consequence of aging; is readily inducible by infectious agents, organisms, or drugs; and is potentially reversible in that it disappears when the offending "agent" is removed or eradicated. **Autoimmune disease** results from the activation of self-reactive T and B cells that, following stimulation by genetic or environmental triggers, cause actual tissue damage (table 31.7).

Four factors influence the development of autoimmune disease. Two major factors are genetic and viral. The third factor is endocrine; the hormone estrogen promotes autoimmune disease, and androgens act as natural immunosuppressive agents. These physiological modulatory effects of sex hormones acting on normal immune responses partially explain the marked female predominance of autoimmune diseases. The fourth factor is psychoneuroimmunological—that is, the influence of stress and neurochemicals on the immune response. Overall, all four of these factors can affect gene expression, which directly or indirectly interferes with important immunoregulatory actions.

Although their causal mechanism is not well known, these diseases are more common in older people and may involve viral or bacterial infections. Some investigators believe that the release of abnormally large quantities of antigens may occur when the infectious agent causes tissue damage. The same agents also may cause body proteins to change into forms that stimulate antibody production or T-cell activation. Simultaneously the activity of T-suppressor cells, which normally limits this type of reaction, seems to be repressed. Many autoimmune diseases have a genetic component. For example, there is a well-known association between an individual's susceptibility to Graves' disease or multiple sclerosis and a specific determinant on the major histocompatibility complex.

Transplantation (Tissue) Rejection

Tissue transplant rejection is the third area (after hypersensitivity and autoimmunity) in which the immune system can act detrimentally. It is occasionally desirable to replace a nonfunctional or damaged body part by transplanting a tissue or organ from one person to another. Transplants between genetically different individuals within a species are termed **allografts** [Greek *allos*, other]. Some transplants do not stimulate an immune response. For example, a transplanted cornea is rarely rejected since lymphocytes do not circulate into the anterior chamber of the eye. This site is considered an immunologically privileged site. Another example of a privileged tissue transplant is the heart valve transplanted from a pig to a human. Such a graft between different species is termed a **xenograft** [Greek *xenos*, strayed].

With allografts, there is the possibility that the recipient's cells will recognize the donor's tissues as foreign. This triggers the recipient's immune mechanisms, which may destroy the donor tissue. Such a response is called a tissue rejection reaction. A tissue rejection reaction can occur by two different mechanisms. First, foreign major histocompatibility complex (MHC) class II molecules on the graft stimulate host T-helper cells to aid cytotoxic T cells in graft destruction (figure 31.15a). Cytotoxic T cells recognize the graft through the foreign MHC

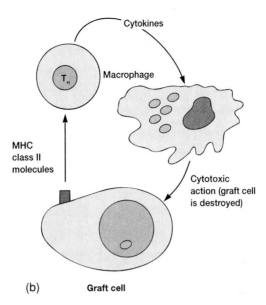

Figure 31.15   Graft Cell Destruction. (*a*) Foreign MHC class II molecules on the graft cell stimulate host T-helper cells to help cytotoxic T cells destroy the target graft cell. Cytotoxic T cells recognize the graft cell by its foreign MHC class I molecules. (*b*) T-helper cells reacting to the graft cell release cytokines that stimulate macrophages to enter the graft and destroy it by cytotoxic action.

TABLE 31.8   Some Congenital Immune Deficiencies in Humans

Condition	Symptoms	Cause
Chronic granulomatous disease	Defective monocytes and neutrophils leading to recurrent bacterial and fungal infections	Failure to produce reactive oxygen intermediates due to defective NADPH oxidase
Common variable hypogammaglobulinemia	Plasma cell or B-cell deficiency and inability to produce adequate specific humoral antibodies	Defective B cell differentiation due to any of several causes
DiGeorge syndrome	T-cell deficiency and very poor cell-mediated immunity	Lack of thymus or a poorly developed thymus
Severe combined immunodeficiency disease (SCID)	Both antibody production and cell-mediated immunity impaired due to a great reduction of B- and T-cell levels	Various mechanisms (e.g., defective B- and T-cell maturation because of X-linked gene mutation; absence of adenosine deaminase in lymphocytes)

class I molecules. A second mechanism involves the T-helper cells reacting to the graft and releasing cytokines (figure 31.15*b*). The cytokines stimulate macrophages to enter the graft and destroy it.

As presented in figure 31.15*a,b*, the major histocompatibility complex molecules play a dominant role in tissue rejection reactions because of their unique association with the recognition system of T cells. Unlike antibodies, T cells cannot recognize or react directly with non-MHC molecules (viruses, allergens). They recognize these molecules only in association with, or complexed to, an MHC molecule. There are two classes of MHC molecules. Class I MHC molecules are present on every cell in the body and consequently are important targets of the rejection reaction. Class II MHC molecules are involved in T-helper cell activation as previously discussed. The greater the antigenic difference between class I molecules of the recipient and donor tissues, the more rapid and severe the rejection reaction is likely to be. However, the reaction can sometimes be minimized if recipient and donor tissues are matched as closely as possible (Box 31.1).

Immunodeficiencies

Defects in one or more components of the immune system can result in its failing to recognize and respond properly to antigens. Such **immunodeficiencies** can make a person more prone to infection than those people capable of a complete and active immune response. Despite the increase in knowledge of functional derangements and cellular abnormalities in the various immunodeficiency disorders, the fundamental biological errors responsible for them remain largely unknown. To date, most genetic errors associated with these immunodeficiencies are located on the X chromosome and produce primary or congenital immunodeficiencies (table 31.8). Other immunodeficiencies can be acquired because of infections by immunosuppressive microorganisms (chronic mucocutaneous candidiasis) or by some viruses (HIV).

Human Blood Types

The surface of human red blood cells (erythrocytes) contains genetically determined sets of antigen molecules. People can be divided into **blood type** groups based on the surface antigens that are present. Red blood cell A and B antigens are primarily

TABLE 31.9   Antigens and Antibodies in Human ABO Blood Groups

Blood Type	Antigen Present on Erythrocyte Membranes	Antibody in Plasma	Can Receive Blood from	Can Give Blood to
A	A	Anti-B	A and O	A and AB
B	B	Anti-A	B and O	B and AB
AB	A and B	Neither anti-A nor anti-B	AB; A; B; O	AB only
O	Neither	Anti-A and anti-B	O	O; A; B; AB

integral plasma membrane glycoproteins that are attached during red blood cell formation by specific A and B transfer enzymes. These surface antigens are genetically determined, and a person's blood type is inherited. Two major classifications of human erythrocytes are ABO grouping and the Rh system.

ABO blood grouping is based on differences in the type of glycoproteins present on erythrocyte surfaces. Type A people have type A glycoproteins on their erythrocyte surfaces; type B people have type B glycoproteins; type AB people have both glycoproteins; type O people have neither of them. The A and B glycoproteins function as antigens and are recognized by specific antibody molecules, usually of the IgM class. Anti-A antibody and anti-B antibody occur in human plasma and combine specifically with A and B antigens, respectively. As a result the erythrocytes clump together. This cell clumping is termed hemagglutination and is discussed in more detail in chapter 32.

Humans normally do not produce antibodies to a particular antigen until they are exposed to it; however, the ABO system is an exception. People routinely produce antibodies to those blood group antigens not present on their own erythrocytes without ever being directly exposed to them. It is thought that these antibodies are formed in response to antigens of the intestinal microbiota that are similar to the blood group substances. The antibodies could then cross-react with antigens on the appropriate erythrocyte. Type A people have the anti-B antibody in their plasma; type B people have anti-A antibody; type AB people have neither of these antibodies; and type O people have both antibodies (table 31.9). It is for this reason that blood is typed before being transfused from one person to another. Donor and recipient must be compatible, or a severe hemagglutination reaction may occur. For example, if type A erythrocytes are given to a type O person, the anti-A antibody in the recipient's blood combines with the A antigen on the surface of the donor erythrocytes and causes them to agglutinate. These agglutinates can block blood vessels, causing serious circulatory problems.

The second major classification of human blood types is the **Rh system.** (Rh stands for rhesus; this factor was first identified in the blood of rhesus monkeys.) Rh classification is based upon the presence or absence of the RhD antigen on

erythrocytes. RhD-positive (RhD+) people have the RhD antigen while RhD-negative (RhD–) people do not. About 85% of all people in the United States are RhD+.

In contrast to the situation with the ABO system, RhD– people do not normally have anti-RhD antibodies in their plasma. They must be exposed to the RhD antigen to develop the antibodies. For example, anti-RhD antibodies develop in an RhD– person over a period of several months following an accidental transfusion of RhD+ blood. Once this has happened, a second transfusion can result in a severe hemagglutination reaction.

The RhD factor is also important in pregnancies. A serious problem may occur when an RhD– mother and an RhD+ father have an RhD+ child. The child produces red blood cells with the RhD antigen, which gains access to the mother's circulation when the placenta ruptures during childbirth, or in some cases during an abortion, miscarriage, or any obstetrical procedure that may cause fetomaternal hemorrhage. Responding to the foreign RhD antigens, the mother produces IgG anti-RhD antibodies that are able to cross the placenta. During any subsequent pregnancy, the RhD– mother may produce enough anti-RhD antibodies to agglutinate and destroy the red blood cells of the RhD+ child, producing **hemolytic disease of the newborn.** Because of the hemolytic process, the liver and spleen enlarge. As these organs enlarge, their function decreases, leading to ascites (fluid accumulation), congestive heart failure, and possibly death.

This disease can be prevented by injecting the RhD– mother with anti-RhD antibodies (RhoGAM) within 72 hours after delivery of the child. These passively administered antibodies prevent the mother from developing antibodies (alloimmunization) against the RhD antigen of the RhD+ red blood cells that have entered her circulation.

Cell-Associated Differentiation Antigens (CDs)

The understanding of how lymphocytes respond to cancer, viruses, and microorganisms provides additional information on the nature, severity, and progression of a particular disease. As noted in many places within this chapter, the involvement of lymphocytes with pathological events can be assessed by the enumeration of specific cells bearing particular **cell-associated differentiation antigens (CDs),** which also are called cluster of

TABLE 31.10   Some Functional Examples of Cell-Associated Differentiation Antigens (CDs)

CD	Function
IL2-R	Interleukin-2 receptor protein mediates the action of IL-2, an immune system growth hormone. Resting T and B cells do not display IL2-R; however, when the cells are stimulated by a challenge to the immune system and begin to divide, molecules of IL2-R are expressed on the cell's plasma membrane and some are released into the fluid surrounding them.
CD2	The CD2 molecule is a glycoprotein present on T cells and NK cells. It functions as an intercellular adhesion molecule and binds to the human leukocyte function-associated antigen-3 or LFA-3 (CD58). CD2 may be involved in the attachment of T_H cells to antigen-presenting cells and cytotoxic T cells to target cells, and in other cellular interactions. It also is a signal-transducing molecule and may help stimulate T cells to proliferate and secrete cytokines.
CD4	The CD4 molecule is a glycoprotein found primarily on T cells associated with helper/inducer regulatory functions and to a lesser degree on monocytes and macrophages. It is a cell adhesion molecule with great affinity for class II MHC. CD4 on T_H cells functions as a signal to activate B cells and induces CD8$^+$ T cells to become cytotoxic/suppressor T cells. CD4 also presents antigen along with class II MHC and functions by transducing signals across the T-cell membrane (figure 31.7).
CD8	The CD8 molecule is a polypeptide found on the surface of cytotoxic/suppressor T cells. It is a cell adhesion molecule and participates in the interaction of class I MHC found on target cells.
CD23	CD23 is a protein found on the surface of IgM-bearing B-cells, eosinophils, macrophages, and some T and NK cells. Soluble CD23 is a B-cell growth factor and upregulates IgE synthesis in conjunction with interleukin-4 released from T cells.
CD35	CD35 is a glycoprotein present on erythrocytes, PMNs, monocytes, all B and some T cells, mast cells, and in soluble form in the plasma. It is the receptor for complement C3b/C4b. The functions of CD35 vary according to the cell type possessing the receptor. For example, CD35 on erythrocytes binds immune complexes bearing C3b or C4b and promotes their clearance from the circulation. On B cells, CD35 augments the maturation of these cells into antibody-secreting cells. It also aids leukocytes in binding and phagocytosing C3b- or C4b-coated objects.
ICAM-1	Intercellular adhesion molecule-1 is a protein that is an important early marker of immune activation and response. It is a ligand for the lymphocyte function-associated antigen-1 (LFA-1) found on lymphocytes, monocytes, and neutrophils. The ICAM-1/LFA-1 interaction is important in T-cell-antigen-specific responses and lymphocyte emigration into inflammatory sites. ICAM-1 also is found on mucosal epithelial cells and erythrocytes. Recent evidence suggests ICAM-1 is the cellular receptor for the human rhinovirus and *Plasmodium falciparum*-infected erythrocytes.

differentiation antigens. CDs are functional cell surface proteins or receptors that can be measured in situ, from peripheral blood, biopsy samples, or other body fluids. They often are used to differentiate between leukocyte subpopulations. Table 31.10 summarizes some of the functions of several CDs.

These CDs have both biological and clinical significance. In normal individuals the concentration of these molecules in serum is very low. When one's immune system is activated in response to disease, the concentration of these soluble molecules usually rises and fluctuates during treatment. Detecting these soluble molecules may help in the management of the disease. For example, circulating levels of ICAM-1 are directly related to the progress of the disease and prognosis of metastatic melanoma. CD23 is a B-cell growth factor, and elevated levels are associated with chronic lymphocytic leukemia. CD35 is a potent agent for the suppression of complement-dependent tissue injury in autoimmune inflammatory disease. Elevated levels of CD8 have been found in childhood lymphoid malignancies and in HIV-infected individuals. It has been established that the CD4 molecule is the cell surface receptor for HIV-1 (*see figure 36.10*), and considerable research is currently being done on the role of both cell-bound and soluble CD4 in AIDS.

1. What is an autoimmune disease and how might it develop?
2. What is an immunologically privileged site and how is it related to transplantation success?
3. How does a tissue rejection reaction occur?
4. Describe an immunodeficiency. How might immunodeficiencies arise?
5. Describe the two major classifications of human red blood cells and explain their importance.
6. What are CDs? Describe their importance.

Summary

1. The term cytokine is a generic term for nonantibody protein released by one cell population that acts as an intercellular mediator. When released from mononuclear phagocytes, these proteins are called monokines; when released from T lymphocytes they are called lymphokines; when produced by a leukocyte and the action is on another leukocyte, they are termed interleukins; and if their effect is to stimulate the growth and differentiation of immature leukocytes in bone marrow, they are called colony-stimulating factors.

2. B cells can be stimulated to divide and/or secrete antibody when triggered by the appropriate signals.

3. T cells are pivotal elements of the cell-mediated immune response. T cells have antigen-specific receptor proteins (as well as major histocompatibility molecules) for antigens on their plasma membrane surface. These MHC molecules are proteins coded by a group of genes termed the major histocompatibility complex.

4. Regulator T cells control the development of effector T cells. Two types exist: T-helper cells (CD4+ cells) and T-suppressor cells. There are three subsets of T-helper cells: T_H1, T_H2, and T_HO. T_H1 cells produce various cytokines and are involved in cellular immunity. T_H2 cells also produce various cytokines but are involved in humoral immunity. Not much is known about T_HO cells except that they represent a unique lineage of T-helper cells that produce their own cytokines.

5. Acquired immune tolerance is the ability of a host to produce antibodies against nonself antigens while tolerating (not producing antibodies against) self-antigens. Three theories have been proposed to account for this tolerance: clonal deletion, functional inactivation, and clonal anergy.

6. Effector T cells directly attack specific target cells. Several subsets exist, namely, cytotoxic T cells, delayed-type hypersensitivity T cells, and natural killer cells.

7. Superantigens are bacterial proteins that stimulate the immune system much more extensively than do normal antigens. They nonspecifically stimulate T cells to proliferate through simultaneous interaction with class II MHC products on antigen-presenting cells and with variable regions on the β chain of the T-cell receptor complex.

8. When the immune response occurs in an exaggerated form and results in tissue damage to the individual, the term hypersensitivity or allergy is applied. There are four types of hypersensitivity reactions, designated as types I through IV.

9. The immune system can act detrimentally and reject tissue transplants. There are different types of transplants. Xenografts involve transplants of privileged tissue between different species, and allografts are transplants between genetically different individuals of the same species.

10. Immunodeficiency diseases are a diverse group of conditions in which an individual's susceptibility to various infections is increased; several severe diseases can arise because of one or more defects in the specific or nonspecific immune response.

11. Two major classifications of human red blood cells (erythrocytes) are ABO grouping and the Rh system. ABO grouping is based on the types of glycoproteins present on erythrocyte surfaces. Rh classification is based on the presence or absence of the RhD antigen on erythrocytes.

12. Cell-associated differentiation antigens (CDs) are functional cell surface molecules or receptors that can be measured in situ, from peripheral blood, biopsy samples, or other body fluids.

Key Terms

ABO blood groups   *628*

acquired immune tolerance (tolerance induction)   *618*

allergen   *621*

allergic contact dermatitis   *623*

allergy   *621*

allograft   *626*

anaphylaxis   *621*

antigen-presenting cells   *613*

autoimmune disease   *626*

autoimmunity   *626*

blood type   *627*

bronchial asthma   *622*

cell-associated differentiation antigens (CDs)   *628*

chemokines   *609*

clonal anergy   *619*

colony-stimulating factors (CSFs)   *607*

cytokine   *607*

cytotoxic T (T_C) cell   *619*

delayed-type hypersensitivity T (T_{DTH}) cell   *619*

dendritic cell   *613*

desensitization   *622*

effector T cell   *619*

hay fever   *621*

hemolytic disease of the newborn   *628*

hives   *622*

human leukocyte antigens (HLAs)   *614*

hypersensitivity   *621*

immune surveillance   *620*

immunodeficiency   *627*

interferon-α (IFN-α)   *607*

interferon-β (IFN-β)   *607*

interferon-γ (IFN-γ)   *609*

interleukin   *607*

interleukin-1 (IL-1α, IL-1β)   *607*

interleukin-2 (IL-2)   *610*

interleukin-3 (IL-3)   *610*

lymphokine   *607*

macrophage   *613*

major histocompatibility complex (MHC)   *614*

migration inhibition factor (MIF)   *609*

monokine   *607*

natural killer (NK) cells   *619*

reagin   *621*

regulator T cell   *616*

Rh system   *628*

superantigen   *620*

systemic lupus erythematosus   *623*

TB skin test   *623*

T-dependent antigen   *611*

T_H1 cell   *616*

T_H2 cell   *616*

T_HO cell   *616*

T-helper (T_H) cell   *611*

T-independent antigen   *613*

T-suppressor (T_S) cell   *617*

tumor necrosis factor-α (TNF-α)   *607*

type I (anaphylaxis) hypersensitivity   *621*

type II (cytotoxic) hypersensitivity   *622*

type III (immune complex) hypersensitivity   *623*

type IV (cell-mediated) hypersensitivity   *623*

xenograft   *626*

Questions for Thought and Review

1. Immunology has contributed a new depth of understanding to our knowledge of plasma membrane function. How has the study of lymphocytes deepened our understanding of plasma membrane biology?
2. A person with AIDS has a low T-helper/T-suppressor cell ratio. What problems does this create?

3. How do B and T cells cooperate in the immune response? What is the role of macrophages?
4. What is the difference between an immunodeficiency and an allergy?
5. Describe how IgE can have both beneficial and detrimental effects in the same host. Why have its detrimental effects persisted? Explain your reasoning.

6. In desensitization procedures, the allergist injects more of the same allergen to which the person is allergic. How can this be beneficial?
7. What is the immunologic basis behind blood transfusions?
8. How does hemolytic disease of the newborn develop?
9. What is agammaglobulinemia? How does it occur?

Additional Reading

General

Frank, M., editor. 1994. *Santer's immunologic diseases.* 5th ed. Boston: Little Brown and Company.
Jerne, N. K. 1973. The immune system. *Sci. Am.* 229(2):52–60.
Scientific American Special Issue. 1993. Life, death, and the immune system. *Sci. Am.* 269(3).

Cytokines

Bernhagen, J.; Calandra, T.; Cerami, A.; and Bucala, R. 1994. Macrophage migration inhibitory factor is a neuroendocrine mediator of endotoxaemia. *Trends in Microbiol.* 2(6):198–201.
Hedges, S. R.; Agace, W.W.; and Svanborg, C. 1995. Epithelial cytokine responses and mucosal cytokine networks. *Trends Microbiol.* 3(7):266–70.
Horuk, R. 1994. The interleukin-8 receptor family: From chemokines to malaria. *Immunol. Today* 15(4):169–73.
Johnson, H. M.; Bazer, F. W.; Szente, B.E.; and Jarpe, M. A. 1994. How interferons fight disease. *Sci. Am.* 270(5):68–75.
Old, L. J. 1988. Tumor necrosis factor. *Sci. Am.* 258(3):59–75.
Smith, R. A. 1990. Interleukin-2. *Sci. Am.* 262(3):50–57.
Verhoef, J., and Mattsson, E. 1995. The role of cytokines in gram-positive bacterial shock. *Trends Microbiol.* 3(4):136–40.
Zurawski, G., and de Vries, J. 1994. Interleukin 13, an interleukin 4-like cytokine that acts on monocytes and B cells, but not on T cells. *Immunol. Today* 15(1):19–26.

B-Cell Biology

Lerner, F. A. 1983. The genetics of antibody diversity. *Sci. Am.* 246(2):102–13.
Metzger, H., editor. 1990. *Fc receptors and the action of antibodies.* Washington, D.C.: American Society for Microbiology.
Milstein, C. 1980. Monoclonal antibodies. *Sci. Am.* 243(4):66–74.

Ramsdell, F., and Fowlkes, B. 1990. Clonal deletion versus clonal anergy. *Science* 248:342–48.

T-Cell Biology

Brown, J.; Jardetzky, T.; Gorga, J.; Stern, L.; Urban, R.; Strominger, J.; and Wiley, D. 1993. Three-dimensional structure of the human class II histocompatibility antigen HLA-DR1. *Nature* 364(6432):33–39.
Engelhard, V. H. 1994. How cells process antigens. *Sci. Am.* 271(2):54–61.
Grey, H.; Alessandro, S.; and Buss, S. 1988. How T cells see antigen. *Sci. Am.* 261(5):56–67.
Marrack, P., and Kappler, J. 1986. The T cell and its receptor. *Sci. Am.* 254(3):36–45.
Raff, M. C. 1976. Cell surface immunology. *Sci. Am.* 234(6):30–38.
Schwartz, R. H. 1993. T cell anergy. *Sci. Am.* 269(2):62–71.
von Boehmer, H. 1988. The developmental biology of T lymphocytes. *Ann. Rev. Immunol.* 6:309–26.
von Boehmer, H., and Kisielow, P. 1991. How the immune system learns about self. *Sci. Am.* 265(4):74–81.
Young, J. D.-E., and Cohn, Z. A. 1988. How killer cells kill. *Sci. Am.* 258(1):38–44.

Superantigens

Held, W. 1994. Superantigens and retroviral infection: Insights from mouse mammary tumor virus. *Immunol. Today* 15(4):184–90.
Johnson, H. M.; Russell, J. K.; and Pontzer, C. H. 1992. Superantigens in human disease. *Sci. Am.* 266(4):92–101.
Murray, D. L.; Ohlendorf, D. H.; Schlievert, P. M. 1995. Staphylococcal and streptococcal superantigens: Their role in human diseases. *ASM News* 61(5):229–35.
Scherer, M. T.; Ignatowicz, L.; Winslow, G. M.; Kappler, J. W.; and Marrack, P. 1993. Superantigens: Bacterial and viral proteins that manipulate the immune system. *Ann. Rev. Cell Biology* 9:101–28.
Schlievert, P. M. 1993. Role of superantigens in human disease. *J. Infect. Dis.* 167(5):997–1002.

Hypersensitivities (Allergies)

Baggiolini, M., and Dahinden, C. 1994. CC chemokines in allergic inflammation. *Immunol. Today* 15(3):127–33.
Bochner, B. S., and Lichtenstein, L. M. 1991. Anaphylaxis. *N. Engl. J. Med.* 324(25):1785–90.
Buisseret, P. D. 1982. Allergy. *Sci. Am.* 247(2):86–95.
Lichtenstein, L. 1993. Allergy and the immune system. *Sci. Am.* 269(3):116–25.

Autoimmune Diseases

Atkinson, M. A., and MacClaren, N. K. 1990. What causes diabetes. *Sci. Am.* 263(1):62–71.
Cohen, I. R. 1988. The self, the world, and autoimmunity. *Sci. Am.* 258(4):52–60.
Koffler, D. 1980. Systemic lupus erythematosus. *Sci. Am.* 243(5):52–61.
Rennie, J. 1990. The body against itself. *Sci. Am.* 263(6):106–15.
Rose, N. R. 1981. Autoimmune diseases. *Sci. Am.* 244(2):80–103.
Steinman, L. 1993. Autoimmune disease. *Sci. Am.* 269(3):106–15.

Transplantation (Tissue) Rejection

Bach, F., and Sachs, D. 1987. Transplantation immunology. *N. Engl. J. Med.* 317(8):402–09.

Immunodeficiencies

Buckley, R. H. 1992. Immunodeficiency diseases. *JAMA* 268(20):2797–2806.

Cell-Associated Differentiation Antigens (CDs)

Fleischer, B. 1994. CD26: A surface protease involved in T-cell activation. *Immunol. Today* 15(4):180–83.
Schlossman, S. F. 1994. CD antigens. *Immunol. Today* 15(3):98–99.

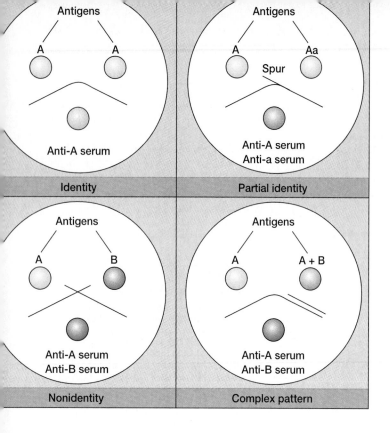

Identity

Partial identity

Nonidentity

Complex pattern

CHAPTER 32
The Immune Response:
Antigen–Antibody Reactions

A foreign cell in the body is

identified by antibody, but the

cell is destroyed by other

agents. Among them is

"complement," an intricate set

of enzymes [proteins].

—Manfred M. Mayer

Outline

Concepts

1. Various types of antigen-antibody reactions occur in higher animals (in vivo) that lead to immune product formation. This union of antigen and antibody initiates the participation of other body elements that determine the ultimate fate of the antigen. For example, the complement system can be activated, leading to cell lysis, phagocytosis, chemotaxis, or stimulation of the inflammatory response. Other defensive reactions include toxin neutralization, antibody-dependent cell-mediated cytotoxicity, opsonization, and immune complex formation.

2. The union of antigen and antibody in vitro produces either a visible reaction or one that can be made visible in a variety of ways. These techniques can be used to identify viruses, microorganisms, and their products; to quantitate and identify antigens and antibodies; to follow the course of a disease; to determine the serotype of a microorganism; and to determine the amount of protection from disease an animal possesses.

3. The older classic tests are named according to what happens to the antigen: agglutination, complement fixation, precipitation, neutralization, and capsular swelling. More recent tests are named according to the technique used: enzyme-linked immunosorbent assay, immunodiffusion, immunoelectrophoresis, immunofluorescence, immunoprecipitation, and radioimmunoassay.

The two previous chapters present a basic discussion of antigens, antibodies, and chemical mediators in both health and disease. This chapter completes the discussion of the immune response by describing the various antigen-antibody reactions that occur in higher animals (in vivo) to protect them against the continuous onslaught of infectious agents, microorganisms and their products, certain macromolecules, and cancer cells. Outside the animal body, tests based on antigen-antibody reactions have become exceptionally useful in the diagnosis of infectious diseases, in the identification of specific viruses and microorganisms, and for monitoring immunologic problems. These in vitro techniques are also discussed.

As presented in chapter 30, the basic Y-shaped antibody molecule is bifunctional. The variable domain (within the Fab regions) binds target antigen while the Fc region interacts with various cells of the immune system, some phagocytic cells, or the first component of the complement system. The principal function of antigen-antibody binding is to identify the target antigen as foreign so that other components of the immune response can react with it and destroy the foreign body.

Antigen-Antibody Binding

An antigen binds to an antibody at the antigen-binding site within the Fab regions of the antibody. More specifically, a pocket is formed by the folding of both the V_H and V_L regions (see figure 30.6c). At this site specific amino acids contact the antigen's epitope determinant or haptenic groups and form multiple noncovalent bonds between the antigen and amino acids of the binding site (figure 32.1).

Because binding is due to weak, noncovalent bonds such as hydrogen bonds and electrostatic attractions, the antigen's shape must exactly match that of the antigen-binding site. If the shapes of the epitope (see figure 30.3) and binding site are not truly complementary, the antibody will not effectively bind the antigen. Although a lock-and-key mechanism normally may operate, in at least one case the antigen-binding site does change shape when it complexes with the antigen (an induced fit mechanism). Regardless of the precise mechanism, antibody specificity results from the nature of antibody-antigen binding.

Antigen-Antibody Reactions in the Animal Body (In Vivo)

Higher animals (vertebrates) possess a highly sophisticated, defensive immunologic response mechanism for developing resistance to specific viruses, microorganisms, macromolecules, foreign agents, and cancer cells. This immunologic response is triggered by specific antigen-antibody reactions. Those reactions occurring in the animal's body are discussed first, and then reactions outside the body.

(a)

(b)

Figure 32.1 Antigen-Antibody Binding. (*a*) An example of antigen-antibody binding is represented in this model of the monoclonal antibody MAb17-IA bound to the surface of the human rhinovirus. The heavy chains are red, light chains are blue, and the antigen capsid protein is yellow. The RNA interior of the virus would be toward the bottom of the diagram. The antibody is bound bivalently across icosahedral twofold axes of the virus. (*b*) Based on X-ray crystallography, the hapten molecule nestles in a pocket formed by the antibody combining site. In the illustration the hapten makes contact with only 10–12 amino acids in the hypervariable regions of the light and heavy chains. The numbers represent contact amino acids.

The Complement System

The **complement system** is composed of a group of serum proteins that play a major role in the animal's defensive immune response. For example, complement proteins can lyse antibody-coated eucaryotic cells and bacteria (cytolysis). Complement can mediate inflammation and attract and activate phagocytic cells. Generally, complement proteins amplify the effects of antibodies (e.g., lysis of cells).

TABLE 32.1 **Proteins of the Complement Cascade in Serum**

Protein	Fragment	Function
Recognition Unit		
C1	q	Binds to the Fc portion of antigen-antibody complexes
	r	Subunit of C1; activates C1s
	s	Cleaves C4 and C2 due to its enzymatic activity
Activation Unit		
C2		Causes viral neutralization
C3	a	Anaphylatoxin, immunoregulatory
	b	Key component of the alternative pathway and major opsonin in serum
	e	Induces leukocytosis
C4	a	Anaphylatoxin
	b	Causes viral neutralization
Membrane Attack Unit		
C5	a	Anaphylatoxin; principal chemotactic factor in serum; induces neutrophil attachment to blood vessel walls
	b	Initiates membrane attack
C6 C7 C8 C9		Participate with C5b in formation of the membrane attack complex that lyses targeted cells
Alternative Pathway		
Factor B		Causes macrophage spreading on surfaces; precursor of C3 convertase
Factor $\overline{D}$		Cleaves Factor B to form active C3Bb in alternative pathway
Properdin		Stabilizes alternative pathway convertase
Regulatory Proteins		
Factor H		Promotes C3b breakdown and regulates alternative pathway
Factor I		Degrades C3b and regulates alternative pathway
C4b binding protein		Inhibits assembly and accelerates decay of C4bC2a
C1 INH complex		Binds to and dissociates C1r and C1s from C1
S protein		Binds fluid-phase C5b67; prevents membrane attachment

The complement cascade is made up of at least 17 complement proteins designated C1 (which has three protein subcomponents) through C9 in addition to Factor B, Factor D, Factor H, Factor I, C4b binding protein, C1 INH complex, S protein, and properdin (table 32.1). The complement system acts in a cascade fashion, the activation of one component resulting in the activation of the next. Collectively the complement proteins make up much of the globulin fraction of serum (*see figure 30.5*). Within plasma and other body fluids, complement proteins are in an inactive state. They usually are activated after the binding of antibodies to antigens and are specifically directed against the target molecules identified by the antibodies. Because complement activation involves binding of the components to antibody-antigen complexes and to each other with their consequent removal from serum, this event is called **complement fixation.**

There are two pathways of complement activation: the **classical** and **alternative pathways** (figure 32.2). Although they employ similar mechanisms, specific proteins are unique to the first part of each pathway.

Activation of the classical pathway requires initiation by the interaction of antibodies with an antigen that is usually cell bound. The order of effectiveness in activating complement is as follows: IgM > IgG3 > IgG1 > IgG2. However, some microbial products (lipid A of endotoxin and staphylococcal protein A) or plasmin (a proteolytic enzyme that dissolves the fibrin of blood clots) may activate C1 directly without antibody participation. Following binding of antigen to antibody, the C1 complement component, which is composed of three proteins (q, r, and s), attaches to the Fc portion of the antibody molecule through its C1q subcomponent. In the presence of calcium ions, a trimolecular complex (C1qrs · Ag · Ab) that has esterase activity is rapidly formed. The activated C1's subcomponent attacks and cleaves its natural substrates in serum (C2 and C4). This leads to binding of a portion of each molecule (C2b and C4b) to the antigen-antibody-complement complex with the release of small C4a and C2a fragments. (The released C2 fragment traditionally has been called C2a. For consistency, this text will label all larger complement fragments that are bound to the target cell as b fragments.) With the binding of C2b to C4b, an enzyme with trypsinlike proteolytic activity is generated. The natural substrate for this enzyme is C3; thus it is termed a C3 convertase. Through the activity of C4b2b (the bar indicates an active enzyme complex), C3 is cleaved into a bound subcomponent C3b and a C3a soluble component. C3b then absorbs to bound C4b2b, forming the complex C4b2b3b which acts as a C5 convertase and cleaves C5 into fragments C5a and C5b. C6 and C7 rapidly bind to C5b, forming a C5b67 complex that possesses an unstable membrane-binding site; once bound to a membrane, this complex is stable. C8 and C9 then bind, forming the **membrane attack complex** (C5b6789) that creates a pore in the plasma membrane of the target cell (figure 32.3). It is believed that the actual pore is a doughnut-shaped polymer of C9. (The perforin pores generated by cytotoxic T cells [*see figure 31.8*] have similar structures since perforin and C9 are homologous but lack an analog of the C5b678 complex.) If the cell is eucaryotic, Na$^+$ and H$_2$O enter through the pore leading to osmotic lysis of the cell. If the cell is a gram-negative bacterium, lysozyme from the blood enters through the pore and digests the peptidoglycan cell wall causing the bacterium to lyse

Figure 32.2   Complement Cascade. Within each pathway the components are arranged in order of their activation and aligned opposite their functional and structural analogs in the opposite pathway.

osmotically. In contrast, gram-positive bacteria are resistant to the cytolytic action of the membrane attack complex because they lack an exposed outer membrane and the thick peptidoglycan prevents an attack on the plasma membrane.

The alternative pathway does not require the binding of antibodies to antigens for its activation. Instead, it plays an important role in the innate, nonspecific immune defense against intravascular invasion by bacteria and some fungi that

occurs before the development of specific antibodies. The alternative pathway begins with cleavage of C3 into fragments C3a and C3b by a blood enzyme. These fragments are produced at a slow rate and do not carry out the next step, the cleavage of C5, because free C3b is rapidly cleaved into inactive fragments. However, when C3b binds to lipopolysaccharide (LPS) of bacterial cell outer membranes, to aggregates of IgA or IgE, or to some endotoxins, it becomes stable (figure 32.2). A protein in

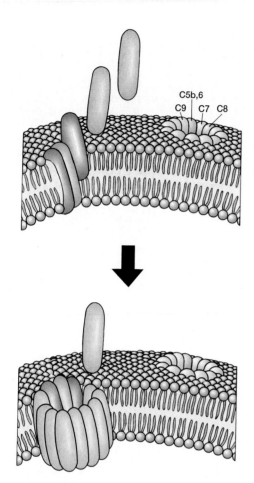

C5b,6
C9 C7 C8

Figure 32.3   The Membrane Attack Complex. The membrane attack complex (MAC) is a tubular structure that forms a transmembrane pore in the target cell's plasma membrane. This representation is the subunit architecture of the membrane attack complex. The transmembrane channel is formed by a C5b678 complex and 10 to 16 polymerized molecules of C9.

blood termed Factor B adsorbs to bound C3b and is cleaved into two fragments by Factor D, leading to the formation of active enzyme C3bBb (this complex is sometimes called the C3 convertase because it cleaves more C3 to C3a and C3b). C3bBb is further stabilized by a second blood protein, properdin, and is changed into the C5 convertase (C3bBb3b). The convertase then cleaves C5 to C5a and C5b. The steps that follow in the alternative pathway are identical with the last steps of the classical pathway.

This overview of complement activation provides a basis for consideration of the function of complement as an integrated system during an animal's defensive effort. Gram-negative bacteria arriving at a local tissue site will interact with components

of the alternative pathway, resulting in the generation of biologically active fragments, opsonization of the bacteria, and initiation of the lytic sequence (however, see Box 32.1). If the bacteria persist or if they invade the animal a second time, antibody responses also will activate the classical pathway. The classical pathway is much more rapid and efficient in mediating opsonization and complement fragment generation.

The generation of complement fragments C3a and C5a leads to several important inflammatory effects. Mast cells release their contents, and the blood supply to the area increases markedly (hyperemia) as blood vessels dilate due to released histamine (*see figure 29.13*). These fragments also cause the release of neutrophils from the bone marrow into the circulation. Neutrophils then make their way to the site of hyperemia where, in the presence of C5a, they attach to the endothelium and leave the blood vessels (*see figure 29.12*). C5a induces a directed, chemotactic migration of neutrophils to the site of complement activation. Macrophages in the area can synthesize even more complement components to interact with the bacteria. All these defensive events promote the ingestion and ultimate destruction of the bacteria by the neutrophils and macrophages.

1. How does antigen-antibody binding occur? What is the basis for antibody specificity?
2. List the major functions of complement.
3. How is the classical complement pathway activated? The alternative pathway?
4. What is the membrane attack complex, and how does its formation lead to cell lysis?
5. What role do complement fragments C3a and C5a play in an animal's defense against gram-negative bacteria?

Toxin Neutralization

Immunity to a disease like diphtheria (*see p. 744*) depends on the production of specific antibodies that inactivate the toxins produced by the bacteria. This process is termed **toxin neutralization** (figure 32.4a,b). Once neutralized, the toxin-antibody complex is either unable to attach to receptor sites on host target cells or unable to enter the cell. For example, diphtheria toxin inhibits protein synthesis after binding to the cell surface by the B fragment and subsequent passage of the active A fragment into the cytosol of the target cell (*see figure 29.4a,b*). Thus the antibody blocks the toxic effect by inhibiting the entry of the A fragment or the binding of the B fragment. An antibody capable of neutralizing a toxin or antiserum containing neutralizing antibody against a toxin is called **antitoxin.**

Exotoxins (pp. 567–69).

Box 32.1

Complement-Resistant Cells

Resistance to complement attack is an attribute of some procaryotic and eucaryotic cells. Complement resistance may be due to factors affecting either the formation of lytic channels or how well the channels lyse target cells. Recent experimental evidence supports both mechanisms. For example, cells may activate complement inefficiently in both the presence and the absence of antibody. Changes in the concentration of membrane antigens may alter the capacity of bound IgG to activate C1. Even after C1 activation, membrane factors may inhibit completion of the activation sequence at the step of C3 and C5 convertase formation or at the steps of C8 and C9 binding. Finally, factors affecting membrane lipid composition and hormonal regulation of lipid synthesis can alter the capacity of the membrane attack complex to insert itself effectively into the plasma membrane and create a lytic channel. Once the latter step has occurred, eucaryotic cells have devised several mechanisms to avoid the lethal effect of the complement channel. The membrane attack complex may be endocytosed by a calcium-dependent process. The complement channel can be actively extruded from neutrophils by exocytosis. Finally, platelets bearing the membrane attack complex demonstrate increased activity of the sodium-potassium pump, a process that repolarizes the membrane during sublethal complement attack.

Recently investigators have shown that specific membrane-bound proteins also help protect normal host cells from the detrimental effects of complement activation. These proteins provide the basis for discrimination between self and nonself. In addition, nucleated eucaryotic cells are highly resistant to complement-mediated cytolysis. Resistance is correlated with a high replacement rate of plasma membrane lipids and the ability to shed the membrane attack complex from the cell surface. Thus the complement cascade participates in the development of the inflammatory response, elimination of pathogens, and removal of immune complexes without destroying host cells at the same time.

Viral Neutralization

IgG, IgM, and IgA antibodies can bind to some viruses during their extracellular phase and inactivate them. This antibody-mediated viral inactivation is called **viral neutralization.** Fixation of classical pathway complement component C4b to the virus aids the neutralization process. Viral neutralization prevents a viral infection due to the inability of the virus to bind to its target cell (figure 32.4c).

Adherence Inhibiting Antibodies

The capacity of bacteria to colonize the mucosal surfaces of mammalian hosts is dependent in part on their ability to adhere to mucosal epithelial cells. Recent studies have demonstrated that secretory IgA (sIgA; see p. 595) antibodies inhibit certain bacterial adherence promoting factors. Thus sIgA has a unique role in protecting the host against infection from certain pathogenic bacteria and perhaps from other microorganisms on mucosal surfaces.

Antibody-Dependent Cell-Mediated Cytotoxicity

When cells are infected with viruses, changes often occur in the plasma membrane proteins that lead to the development of an antibody response against the cells. Cytolysis of the infected cell may be achieved by the membrane attack complex—that is, by the combined action of antibody and complement—or by the binding of natural killer cells to the altered cell after attachment to the Fc region of the antibody (figure 32.5). Antibodies provide the bridging mechanism for an intimate association between the effector and target cell. After attachment the natural killer cell destroys the target cell by releasing cytotoxic mediators or directly causing cytolysis. The process is called **antibody-dependent cell-mediated cytotoxicity.**

IgE and Parasitic Infections

Immune reactions against protozoan and helminthic parasites are only partially understood. Parasites that have a tissue invasive phase in their life cycle often are associated with both eosinophilia (an excessive number of eosinophils in the blood) and elevated IgE levels. Recent evidence shows that, in the presence of elevated IgE, eosinophils can bind to the parasites and discharge their lysosomal granules. Degranulation releases lytic and inflammatory mediators that destroy the parasites (see figure 31.10).

Opsonization

As noted in chapter 29, phagocytes have an intrinsic ability to bind directly to microorganisms by nonspecific cell surface receptors, form phagosomes, and digest the microorganisms (figure 32.6a). This phagocytic process can be greatly enhanced by opsonization. **Opsonization** [Greek opson, prepare victims for] is the process in which microorganisms or other particles are coated by antibody and/or complement, thereby being prepared

Figure 32.4   Neutralization Reactions. (*a*) Effects of a dimeric exotoxin on a susceptible cell. (*b*) Neutralization of the toxin by antitoxin. (*c*) In viral neutralization the specific antibodies neutralize the virus and prevent it from attaching to the susceptible cells.

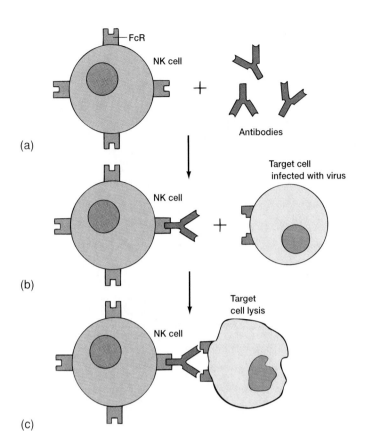

FcR

NK cell

Antibodies

(a)

NK cell

Target cell
infected with virus

(b)

Target
cell lysis

NK cell

(c)

Figure 32.5 Antibody-dependent Cell-mediated Cytotoxicity.
(*a*) In this mechanism, the FcR antibody binds to the Fc receptor (FcR) on the Fc region of natural killer (NK) cells. (*b*) These antibodies then can bind to target cell antigens against which they are specific. (*c*) Subsequently, lysis of the target cells occurs.

for "recognition" and ingestion by phagocytic cells. Opsonizing antibodies belong to the IgG1, IgG3, or IgM isotypes and have a site in the C_H3 domain (*see figure 30.6c*) that binds to an Fc receptor on the surface of macrophages. This binding forms a bridge between the phagocyte and the antigen (figure 32.6*b*).

Phagocytosis (pp. 582–83).

IgG or IgM antibodies also can activate the complement system, most often through the classical pathway but sometimes by the alternative pathway. The net result is to assemble the C3 convertase on the surface of the microorganism together with opsonic fragment C3b. Both neutrophils and macrophages possess surface receptors for C3b. A bridge that aids phagocytosis eventually forms between the microorganism, complement, and phagocytic cell (figure 32.6*c*). If both antibody and C3b opsonize, binding is greatly enhanced (figure 32.6*d*).

Opsonizing antibodies can be directed against surface components of bacteria (e.g., capsular polysaccharides, M protein of group A streptococci, and peptidoglycans of staphylococci).

Figure 32.6 Opsonization. (*a*) A phagocytic cell has some intrinsic ability to bind directly to a microorganism through nonspecific receptors. (*b*) This binding ability is enhanced if the microorganism elicits the formation of antibodies (Ab) that act as a bridge to attach the microorganism to the Fc receptor on the phagocytic cell. (*c*) If the microorganism has activated complement (C3b), the degree of binding is further enhanced by the C3b receptor. (*d*) If both antibody and C3b opsonize, binding is greatly enhanced.

Phagocytic cell	Degree of binding	Opsonin
(a) **Attachment by nonspecific receptors** — Microorganism	±	–
(b) Ab / Fc receptor	+	Antibody
(c) C3b / C3b receptor	+ +	Complement C3b
(d)	+ + + +	Antibody and complement C3b

Components that can trigger deposition of C3b on the surface of microorganisms by the alternative complement pathway include lipopolysaccharides of gram-negative bacteria, lipoteichoic acids of gram-positive cocci, and fungal polysaccharides. Finally, some bacterial components, staphylococcal protein A or lipid A of endotoxin, can activate the classical complement pathway without the participation of antibody.

Inflammation

The inflammatory response can be triggered by nonimmune and immune factors. In the latter case the binding of an antibody to an antigen triggers inflammation by two different routes. The first is mediated by IgE's attachment to the surface of mast cells and basophils, which results in the release of histamine (*see figure 31.10*). Histamine causes most of the characteristic events of the inflammatory response.

Inflammation (pp. 578–81).

The second route involves the complement pathway in which the C3a and C5a fragments (figure 32.2) bind to mast cells and platelets, triggering the release of histamine. C5a also contributes to inflammation by being a very potent chemotactic factor. It attracts macrophages, neutrophils, and basophils to the site of complement fixation.

---

1. How does toxin neutralization occur? Viral neutralization?
2. How does antibody-dependent cell-mediated cytotoxicity occur?
3. Describe the role of IgE in resistance to parasitic infections.
4. What is opsonization and how can it take place?
5. What are the two routes by which antibodies stimulate inflammation?

Immune Complex Formation

Because antibodies have at least two antigen-binding sites and most antigens have at least two antigenic determinants, cross-linking can occur, producing large aggregates termed **immune complexes** (figure 32.7; *see also figure 31.13*). If the immune complex becomes large enough to settle out of solution, a **precipitation** [Latin *praecipitare*, to cast down] or **precipitin reaction** occurs, with the **precipitin** antibody being responsible for the reaction. When the immune complex involves the cross-linking of cells or particles, an **agglutination reaction** occurs and the responsible antibody is an **agglutinin.** Agglutination specifically involving red blood cells is a **hemagglutination** reaction and is caused by a **hemagglutinin.** These immune complexes are more readily phagocytosed in vivo than are free antigens.

The extent of immune complex formation, whether within an animal or in vitro, depends on the relative concentrations of

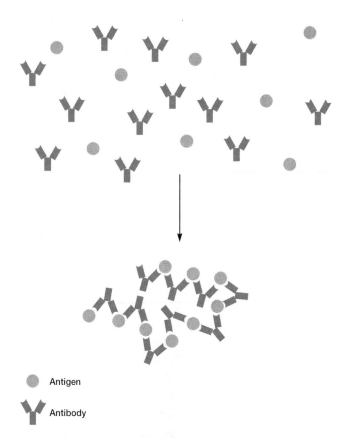

Antigen

Antibody

Figure 32.7   Immune Complex Formation. Antibodies cross-link antigens forming an aggregate of antibody and antigen termed an immune complex.

antibody and antigen. If there is a large excess of antibody, separate antibody molecules usually bind to each antigenic determinant and a less insoluble network or lattice forms (figure 32.17*a*). When antigen is present in excess, two different antigen molecules bind to each antibody and network development also is inhibited. In the equivalence zone the ratio of antibody and antigen concentrations is optimal for the formation of a large network of interconnected antibody and antigen molecules (figures 32.7 and 32.17). All antibody and antigen molecules precipitate or agglutinate as an insoluble complex. Precipitin reactions can occur in both solutions and agar gel media. In either case, antibody-antigen equivalence is required for optimal results (e.g., in the immunodiffusion and immunoprecipitation techniques to be discussed later in the chapter).

In Vivo Testing

Immunologic reactions in vivo can be of diagnostic value and are the basis for some skin tests. Skin testing is the oldest and most valuable tool for the clinical allergist/immunologist. Two basic tests are used: immediate and delayed.

(a)

(b)

Figure 32.8 **In Vivo Skin Testing.** (*a*) Skin prick tests with grass pollen in a person with summer hay fever. Notice the various reactions with increasing dosages (from top to bottom). (*b*) Skin patch test. The surface of the skin (left) is abraded and the suspect allergic extract placed on the skin. After 48 hours (right) it is eczematous and positive for the suspect antigen.

In immediate skin testing the antigens to be tested are introduced into the skin of the patient's forearm or back by pricking or scratching the skin with test antigens. Positive tests appear within 20 minutes as raised erythematous (red) lesions (figure 32.8*a*). The degree of sensitivity is determined from measurement of lesion size. Immediate skin testing is done primarily to determine respiratory allergies.

Delayed or cell-mediated skin tests are used to test for the presence of T-cell reactivity to a specific antigen. They are done by either patch testing or intradermal injection of antigen. Patch testing is used to diagnose contact dermatitis or food allergies. The test allergen is applied to an abraded area of the skin and covered with a patch. The patch is removed 48 hours later, and

the skin site is evaluated (figure 32.8*b*). If positive, a typical contact eczema or inflammation of the skin characterized by redness, itching, and oozing vesicular lesions develops at the site. Some commercially available antigens for evaluating delayed reactions include those for *Candida albicans, Trichophyton,* mumps, diphtheria-tetanus fluid toxoid, streptokinase-streptodornase, PPD (tuberculin), and histoplasmin.

Antigen-Antibody Reactions In Vitro

The first part of this chapter deals with those antigen-antibody reactions that occur within an animal (in vivo). Many of these same reactions can take place outside the animal (in vitro) under controlled laboratory conditions and are extensively used in diagnostic testing. The branch of immunology concerned with reactions in vitro is **serology** [serum and -ology].

During the last two decades, there has been a marked increase in the number, sensitivity, and specificity of serological tests. The increase results from a better understanding of the cell surface of various lymphocytes, the production of monoclonal antibodies (*see figure 30.19*), the development of radioactive and enzyme-linked assays, and the use of fluorescence technology. In this section some more common serological tests employed in the diagnosis of microbial and immunologic diseases are presented.

Agglutination

As noted in figure 32.7, when an immune complex is formed by cross-linking cells or particles with specific antibodies, it is called an agglutination reaction. Agglutination reactions usually form visible aggregates or clumps (**agglutinates**) that can be seen with the unaided eye. Direct agglutination reactions are very useful in the diagnosis of certain diseases. For example, the **Widal test** is a reaction involving the agglutination of typhoid bacilli when they are mixed with serum containing typhoid antibodies from an individual who has typhoid fever.

Recently techniques have been developed that employ microscopic synthetic latex spheres coated with antigens. These coated microspheres are extremely useful in diagnostic agglutination reactions. For example, the modern pregnancy test detects the elevated level of human chorionic gonadotropin (HCG) hormone that occurs in female urine and blood early in pregnancy (figure 32.9*a,b*). Latex agglutination tests are also used to detect antibodies that develop during certain mycotic, helminthic, and bacterial infections, and in drug testing (Box 32.2).

Hemagglutination usually results from antibodies cross-linking red blood cells through attachment to surface antigens and is routinely used in blood typing (*see table 31.9*). In addition, some viruses can accomplish **viral hemagglutination.** For example, if a person has a certain viral disease, such as measles,

Box 32.2

The Rapid Detection of Drugs in Urine

Historically, drug-screening assays were done by classical chemical methods such as thin-layer chromatography (TLC) or liquid chromatography (LC) which, though accurate, are laborious procedures. Assays currently employed in limited drug testing are based on antigen-antibody reactions. Some examples are the radioimmunoassay, enzyme immunoassay, and fluorescence immunoassay methods, all of which depend on sophisticated instrumentation. With the requirement for massive drug testing in sports, federal civilian employment, the military, and other fields, rapid procedures are needed.

One such rapid procedure uses a latex agglutination immunoassay for the detection of cocaine, morphine, barbiturates, THC (marijuana), methadone, pencyclidine, and amphetamines. This method provides the accuracy of the immunoassay approach without the need for expensive equipment, and gives accurate on-site "yes" and "no" results within 3 minutes.

The latex agglutination-inhibition test relies on competition for the antibody between a latex-drug conjugate and any drug that may be present in the urine. A urine sample is placed in the mixing well of a slide containing antibody reagent, buffer, and latex reagent. If the drug is absent, the latex-drug conjugate binds to the antibody and forms large particles that agglutinate. Therefore agglutination is evidence for the absence of drugs in the urine specimen (Box Figure 32.2a,c). If a drug is present in the urine sample, it competes with the latex conjugate for the small amount of available antibody. A sufficient quantity of the drug will prevent the formation of particles and agglutination (Box Figure 32.2b), and a positive urine sample does not change the smooth milky appearance of the test mixture (Box Figure 32.2c).

Box Figure 32.2   Rapid Urine Testing for Drugs. (*a*) An illustration of a reaction with a negative urine sample. (*b*) A positive urine sample. (*c*) One rapid test for cocaine is called Abuscreen and is manufactured by Roche Diagnostic Systems.

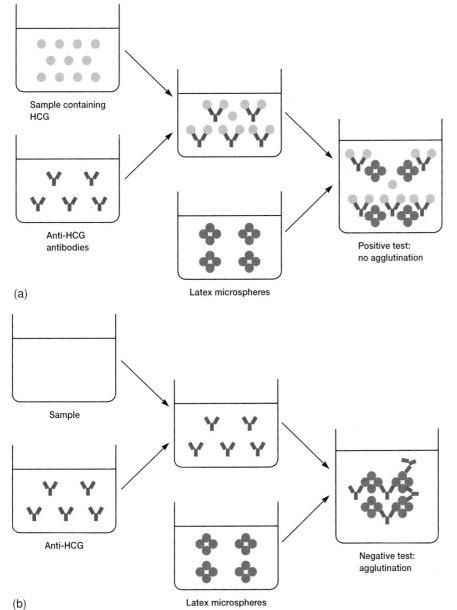

(a)

Sample containing
HCG

Anti-HCG
antibodies

Latex microspheres

Positive test:
no agglutination

(b)

Sample

Anti-HCG

Latex microspheres

Negative test:
agglutination

Figure 32.9   Latex Agglutination Test for Pregnancy. (*a*) In a positive test, urine from the female containing HCG is mixed with a solution of antibody specific for HCG. In the second step latex microspheres coated with HCG are added. If HCG is present, it binds to HCG-specific antibodies, thereby preventing them from agglutinating the microspheres. (*b*) In a negative test the microspheres coated with antigen are agglutinated by HCG-specific antibody.

antibodies will be present in the serum to react with the measles viruses and neutralize them. In a positive test, hemagglutination occurs when measles viruses and red blood cells are mixed but is not seen when the person's serum is added to the mixture; this shows that the serum antibodies have neutralized the measles viruses (figure 32.10*a,b*). This hemagglutination inhibition test is widely used to diagnose influenza, measles, mumps, mononucleosis, and other viral infections (*see chapter 36*).

Agglutination tests are also used to measure antibody titer (*see figure 30.18*). In the tube or well agglutination test, a specific amount of antigen is added to a series of tubes (figure 32.11*a*) or shallow wells in a microtiter plate (figure 32.11*b*). Serial dilutions of serum (1/20, 1/40, 1/80, 1/160, etc.) containing the antibody are then added to each tube or well. The greatest dilution of serum showing an agglutination reaction is determined, and the reciprocal of this dilution is the serum antibody titer.

Agar gel with antibody

Ag1 = 10 mg/dl
Ag2 = 50
Ag3 = 200

(a)

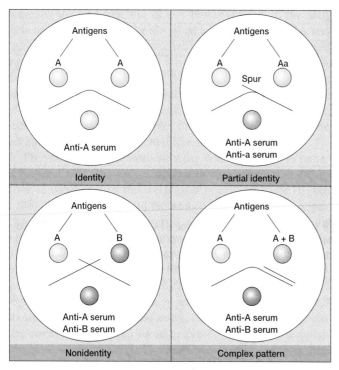

(b)

Figure 32.14   Immunodiffusion. (*a*) Single radial immunodiffusion assay. Three standard solutions of different antigen concentrations (Ag1, Ag2, Ag3) and an unknown (AgX) are placed on agar. After equilibriation the ring diameters are measured. Usually the square of the diameter of the standard rings is plotted on the x-axis and the antigen concentration on the y-axis. From this standard curve the concentration of an unknown can be determined. (*b*) Double diffusion agar assay showing characteristics of identity (top left), reaction of nonidentity (bottom left), partial identity (top right), and a complex pattern (bottom right).

(a)

(b)

(c)

(d)

Figure 32.15   Immunoelectrophoresis. (*a*) Antigens are separated in an agar gel by an electrical charge. (*b*) Antibody (antiserum) is then placed in a trough cut parallel to the direction of the antigen migration. (*c*) The antigens and antibodies diffuse through the agar and form precipitin arcs. (*d*) After staining, better visualization is possible.

The **double diffusion agar assay (Ouchterlony technique)** is based on the principle that diffusion of both antibody and antigen (hence, double diffusion) through agar can form stable and easily observable immune complexes. Test solutions of antigen and antibody are added to the separate wells punched in agar. The solutions diffuse outward, and when antigen and the appropriate antibody meet, they combine and precipitate at the equivalence zone, producing an indicator line (or lines) (figure 32.14*b*). The visible line of precipitation permits a comparison of antigens for identity (same antigenic determinants), partial identity (cross-reactivity), or nonidentity against a given selected antibody. For example, if a V-shaped line of precipitation forms, this demonstrates that the antibodies bind to the same antigenic determinants in each antigen sample and are identical. If one well is filled with a different antigen that shares some but not all determinants with the first antigen, a Y-shaped line of precipitation forms, demonstrating partial identity. In this reaction the stem of the Y, called a spur, is formed if those antigen or antigenic determinants absent in the first well but present in the second one (antigen A in figure 32.14*b*) react with the diffusing antibodies. If two completely unrelated

antigens are added to the wells, either a single straight line of precipitation forms between the two wells, or two separate lines of precipitation form, creating an X-shaped pattern, a reaction of nonidentity.

Immunoelectrophoresis

Some antigen mixtures are too complex to be resolved by simple diffusion and precipitation. Greater resolution is obtained by the technique of **immunoelectrophoresis** in which antigens are first separated based on their electrical charge, then visualized by the precipitation reaction. In this procedure antigens are separated by electrophoresis in an agar gel. Positively charged proteins move to the negative electrode, and negatively charged proteins move to the positive electrode (figure 32.15*a*). A trough is then cut next to the wells (figure 32.15*b*) and filled with antibody. If the plate is incubated, the antibodies and antigens will diffuse and eventually form precipitation bands (figure 32.15*c*) that can be better visualized by staining (figure 32.15*d*). This assay is used to separate the major blood proteins in serum for certain diagnostic tests.

Electrophoresis (p. 313).

Immunofluorescence

Immunofluorescence is a process in which dyes called fluorochromes are exposed to UV, violet, or blue light to make them fluoresce or emit visible light. Dyes such as rhodamine B or fluorescein isothiocyanate can be coupled to antibody molecules without changing the antibody's capacity to bind to a specific antigen. Fluorochromes also can be attached to antigens. There are two main kinds of fluorescent antibody assays: direct and indirect.

Direct immunofluorescence involves fixing the specimen (cell or microorganism) containing the antigen of interest onto a slide (figure 32.16*a*). Fluorescein-labeled antibodies are then added to the slide and incubated. The slide is washed to remove any unbound antibody and examined with the fluorescence microscope (*see figure 2.11*) for a yellow-green fluorescence. The pattern of fluorescence reveals the antigen's location. Direct immunofluorescence is used to identify antigens such as those found on the surface of group A streptococci and to diagnose enteropathogenic *Escherichia coli, Neisseria meningitidis, Salmonella typhi* (figure 32.16*c*), *Shigella sonnei, Listeria monocytogenes, Haemophilus influenzae* type b, and the rabies virus.

The fluorescence microscope (pp. 23–25).

Indirect immunofluorescence (figure 32.16*b*) is used to detect the presence of antibodies in serum following an individual's exposure to microorganisms. In this technique a known antigen is fixed onto a slide. The test antiserum is then added, and if the specific antibody is present, it reacts with antigen to

(a)

(b)

(c)

🔬 **Figure 32.16   Direct and Indirect Immunofluorescence.** (*a*) In the direct fluorescent-antibody (FA) technique, the specimen containing antigen is fixed to a slide. Fluorescenated antibodies that recognize the antigen are then added, and the specimen is examined under a UV microscope for yellow-green fluorescence. (*b*) Indirect fluorescent-antibody technique (IFA). The antigen on a slide reacts with an antibody directed against it. The antigen-antibody complex is located with a fluorescent antibody that recognizes immunoglobulins. (*c*) *Salmonella typhi* stained with acridine orange (×2,000).

form a complex. When fluorescein-labeled anti-immunoglobulin is added, it reacts with the fixed antibody. After incubation and washing, the slide is examined with the fluorescence microscope. The occurrence of fluorescence shows that antibody specific to the test antigen is present in the serum. Indirect immunofluorescence is used to identify the presence of *Treponema pallidum* antibodies in the diagnosis of syphilis (treponemal antibody absorption, FTA-ABS; *see figure 37.19*) as well as antibodies produced in response to other microorganisms.

Immunoprecipitation

The **immunoprecipitation** technique detects soluble antigens that react with antibodies called precipitins. The precipitin reaction occurs when bivalent or multivalent antibodies and antigens are mixed in the proper proportions. The antibodies link the antigen to form a large antibody-antigen network or lattice that settles out of solution when it becomes sufficiently large (figure 32.17*a*). Immunoprecipitation reactions occur only at the equivalence zone when there is an optimal ratio of antigen to antibody so that a lattice forms. If the precipitin reaction takes place in a test tube (figure 32.17*b*), a precipitation ring forms in the area in which the optimal ratio or equivalence zone develops.

---

1. What does a negative complement fixation test show? A positive test?
2. What are the two types of ELISA methods and how do they work? What is a chromogen?
3. Name two types of immunodiffusion tests and describe how they operate.
4. Describe the immunoelectrophoresis technique.
5. What is a fluorochrome?
6. Name and describe the two kinds of fluorescent antibody assays.
7. Specifically, when do immunoprecipitation reactions occur?

Neutralization

Neutralization tests are antigen-antibody reactions that determine whether the activity of a toxin or virus has been neutralized by antibody. Laboratory animals or tissue culture cells are used as "indicator systems" in these tests. The toxin or virus to be assayed has known effects on the indicator system. The effect in the animal might be death, paralysis, or skin lesions. For example, when the exotoxin of *Clostridium botulinum* is suspected of causing food poisoning in a person (*see chapter 37*), a sample of either the suspect food or the serum, stools, or vomitus of the ill patient is collected. Two groups of indicator mice are used. The control group receives the botulinum antitoxin and the experimental group does not. Filtrates from the source samples are injected into both groups of mice. If toxin is present, all mice except those receiving the antitoxin will die, and the test is positive for food botulism.

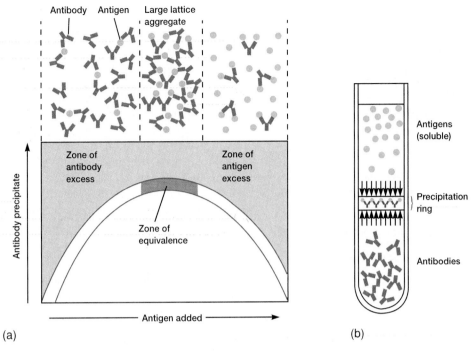

Figure 32.17   Immunoprecipitation. (*a*) Graph showing that a precipitation curve is based on the ratio of antigen to antibody. The zone of equivalence represents the optimal ratio for precipitation. (*b*) A precipitation ring test. Antibodies and antigens diffuse toward each other in a test tube. A precipitation ring is formed at the zone of equivalence.

Box 32.3

History and Importance of Serotyping

In the early 1930s Rebecca Lancefield (1895–1981) recognized the importance of serological tests. She developed a classification system for the streptococci based on the antigenic nature of cell wall carbohydrates. Her system is now known as the **Lancefield system** in which each different serotype is a Lancefield group and identified by a letter (A through O). This scheme is based on specific antibody agglutination reactions with cell wall carbohydrate antigens (C polysaccharides) extracted from the streptococci. Lancefield also showed that further subdividing of the group A streptococci into specific serological types was possible, based on the presence of type-specific M (protein) antigens.

More recently *Escherichia coli, Salmonella* and other bacteria have been serotyped with specific antigen-antibody reactions involving flagella (H) antigens, capsular (K) antigens, and somatic (cell wall or O) antigens. Within *E. coli* there are over 167 different O antigens.

The current value of serotyping may be seen in the fact that *E. coli* O55, O111, and O127 serotypes are the ones most frequently associated with infantile diarrhea. Thus the serotype of *E. coli* from stool samples of infants with infantile diarrhea is of diagnostic value and aids in identifying the source of the infection.

Viral neutralization assays are frequently used to detect viral infections. Suspected blood serum containing viral antibodies can be introduced into tissue culture cells or embryonated eggs (*see figure 16.1 and table 34.1*). If antibodies are present against the virus, viral neutralization will occur and prevent the virus from infecting the culture cells. No cytopathic effects will be seen.

Radioimmunoassay

The **radioimmunoassay (RIA)** technique has become an extremely important tool in biomedical research and clinical practice (e.g., in cardiology, blood banking, diagnosis of allergies, and endocrinology). Indeed, Rosalyn Yalow won the 1977 Nobel Prize in Physiology or Medicine for its development. RIA uses a purified antigen that is radioisotope-labeled and competes for antibody with unlabeled standard or antigen in experimental samples. The radioactivity associated with the antibody is then detected by means of radioisotope analyzers and autoradiography (photographic emulsions that show areas of radioactivity). If there is much antigen in an experimental sample, it will compete with the radioisotope-labeled antigen for antigen-binding sites on the antibody, and little radioactivity will be bound. A large amount of bound radioactivity indicates that there is little antigen present in the experimental sample.

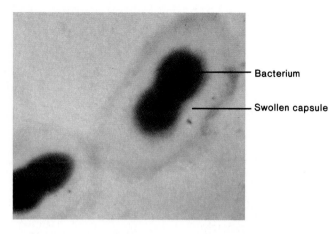

Bacterium

Swollen capsule

Figure 32.18   Serotyping. *Streptococcus pneumoniae* have reacted with a specific pneumococcal antiserum leading to swelling of their capsules (the Quellung reaction). The capsules seen around the pairs indicate potential virulence.

Serotyping

Serotyping refers to serological procedures used to differentiate strains (serovars or serotypes) of microorganisms that differ in the antigenic composition of a structure or product (Box 32.3). The serological identification of a strain of a pathogen has diagnostic value. Often the symptoms of infections depend on the nature of cell products released by a pathogen. Genes for virulence often occur in the same clone with genes for antigenic cell wall material. Therefore it is possible to identify a pathogen serologically by testing for cell wall antigens. For example, there are 84 strains of *Streptococcus pneumoniae*, each differing in the nature of its capsular material. These differences can be detected by capsular swelling (termed the **Quellung reaction**) if antisera specific for the capsular types are used (figure 32.18).

1. What is a neutralization test?
2. Describe the RIA technique. What is serotyping?

Summary

1. The major function of an antigen-antibody reaction is to mark and target antigens for destruction by other components of the immune system.

2. Antigen-antibody complexes activate a system of proteins called the complement system. This results in the formation of a membrane attack complex that can lyse antibody-coated cells. Other components of this pathway mediate inflammation, act as attractants for and activators of phagocytic cells, and can act as opsonins.

3. In toxin neutralization, antibodies react with toxins, preventing either the B fragment of the toxin from binding to the target cell or the A fragment from entering the cytosol of the cell.

4. IgG, IgM, and IgA antibodies can bind to some extracellular viruses and inactivate them by what is termed viral neutralization.

5. Secretory IgA inhibits certain bacterial adherence promoting factors to prevent bacteria from colonizing mucosal surfaces and initiating the disease process.

6. Antibody-dependent cell-mediated cytotoxicity involves an antibody response to certain virus-infected cells. Cytolysis of the infected cells can then be achieved by the action of complement and formation of the membrane attack complex.

7. Elevated IgE levels cause eosinophils to bind to tissue parasites causing degranulation and release of lytic and inflammatory mediators that destroy the parasites.

8. Opsonization is the process by which microorganisms or other particles are coated with antibody and/or complement and thus prepared for ingestion by phagocytic cells.

9. An antigen can trigger inflammation by binding to IgE on the surface of either mast cells or basophils and stimulating histamine release. Antibodies also can trigger inflammation through the production of complement fragments C3a and C5a that bind to mast cells and platelets and trigger the release of histamine.

10. The cross-linking of large numbers of antigens and antibodies produces immune complexes. The responsible antibodies are precipitins, agglutinins, or hemagglutinins, depending on the cell or molecule involved.

11. Two types of immunologic reactions are used for in vitro skin testing: immediate and delayed. Immediate tests detect antibodies and are used to test for respiratory allergies, whereas delayed tests detect cell-mediated immunity and are used to detect contact dermatitis and food allergies.

12. Agglutination reactions in vitro usually form aggregates or clumps (agglutinates) that are visible with the naked eye. Tests have been developed, such as the Widal test, latex microsphere agglutination reaction, hemagglutination, and viral hemagglutination, to detect antigen as well as to determine antibody titer.

13. The complement fixation test can be used to detect a specific antibody for a suspect microorganism in an individual's serum.

14. The enzyme-linked immunosorbent assay (ELISA) involves the linking of various enzymes to either antigens or antibodies. Two basic methods are involved: the double antibody sandwich method and the indirect immunosorbent assay. The first method detects antigens and the latter, antibodies.

15. Immunodiffusion refers to a precipitation reaction that occurs between antibody and antigen in an agar gel medium. Two techniques are routinely used: double diffusion in agar and single radial diffusion.

16. In immunoelectrophoresis antigens are separated based on their electrical charge, then visualized by precipitation and staining.

17. Immunofluorescence is a process in which certain dyes called fluorochromes are irradiated with UV, violet, or blue light to make them fluoresce. These dyes can be coupled to antibody. There are two main kinds of fluorescent antibody assays: direct and indirect.

18. Immunoprecipitation reactions occur only when there is an optimal ratio of antigen and antibody to produce a lattice at the zone of equivalence, which is evidenced by a visible precipitate.

19. Neutralization tests are antigen-antibody reactions that determine whether the activity of a toxin or virus has been neutralized by antibody. Laboratory animals or tissue culture cells are used as indicator systems in these tests.

20. Radioimmunoassays use a purified antigen that is radioisotope-labeled and used to compete with unlabeled standard or antigen in experimental samples for a specific antibody.

21. Serotyping refers to serological procedures used to differentiate strains (serovars or serotypes) of microorganisms that have differences in the antigenic composition of a structure or product.

Key Terms

agglutinates   *641*

agglutination reaction   *640*

agglutinin   *640*

alternative pathway   *634*

antibody-dependent cell-mediated cytotoxicity   *637*

antitoxin   *636*

chromogen   *646*

classical pathway   *634*

complement fixation   *634*

complement system   *633*

double diffusion agar assay (Ouchterlony technique)   *649*

enzyme-linked immunosorbent assay (ELISA)   *644*

hemagglutination   *640*

hemagglutinin   *640*

immune complex   *640*

immunodiffusion   *646*

immunoelectrophoresis   *649*

immunofluorescence   *649*

immunoprecipitation   *650*

Lancefield system   *651*

membrane attack complex   *634*

opsonization   *637*

precipitation or precipitin reaction   *640*

precipitin   *640*

Quellung reaction   *652*

radioimmunoassay (RIA)   *651*

serology   *641*

serotyping   *652*

single radial immunodiffusion (RID) assay   *646*

toxin neutralization   *636*

viral hemagglutination   *641*

viral neutralization   *637*

Widal test   *641*

Questions for Thought and Review

1. If excess complement is added in a complement fixation test, how will this affect the results?

2. Compare and contrast the characteristics of four types of in vivo immunologic reactions. Give an example of each type.

3. Specific chemical markers are conjugated to antibodies or antigens in some immunologic tests. What are the advantages of this technique over such tests as agglutination or complement fixation?

4. Even though the complement system is not an integral part of the immune system, it is usually described in association with it. Why is this so?

5. What is the value in determining an antibody titer?

6. Of what value are neutralization tests performed in animals or in tissue culture cells?

7. What is the importance of serotyping?

8. What are some features of an in vitro antigen-antibody reaction that makes the reaction so useful for identification and monitoring tests?

9. How are chromogens used as biological markers?

10. What is the benefit of opsonization?

11. Even though *Streptococcus pneumoniae* has over 80 different serotypes, serological reactions are sensitive enough to differentiate one serotype from another. How is this possible?

12. How can electrophoresis be used in immunologic testing procedures?

Additional Reading

General

Kuby, J. 1994. *Immunology,* 2d ed. New York: Freeman.

Murray, P. R., editor-in-chief. 1995. *Manual of clinical microbiology,* 6th ed. Washington, D.C.: ASM Press.

Rose, N. R.; Macario, E.; Fahey, J.; Friedman, H.; and Penn, G., editors. 1992. *Manual of clinical laboratory immunology,* 4th ed. Washington, D.C.: American Society for Microbiology.

Ross, G. D., editor. 1986. *Immunobiology of the complement system.* New York: Academic Press.

Sci. Am. 1993. 269(3). Life, death, and the immune system. (Thematic Issue).

Stites, D. P.; Terr, A. I.; and Parslow, T. G. 1994. *Basic and clinical immunology,* 8th ed. Norwalk, Conn.: Appleton and Lange.

Antigen-Antibody Binding

Engelhard, V. H. 1994. How cells present antigens. *Sci. Am.* 271(2):54–61.

Antigen-Antibody Reactions in the Animal Body (In Vivo)

Figueroa, J. E., and Densen, P. 1991. Infectious diseases associated with complement deficiencies. *Clin. Microbiol. Rev.* 4(3):359–95.

Joiner, K. A. 1988. Complement evasion by bacteria and parasites. *Ann. Rev. Microbiol.* 42:201–30.

Mayer, M. 1973. The complement system. *Sci. Am.* 229(5):54–66.

Müller-Eberhard, H. J. 1988. Molecular organization and function of the complement system. *Ann. Rev. Biochem.* 57:321–47.

Antigen-Antibody Reactions In Vitro

Herrmann, J. E. 1986. Enzyme-linked immunoassays for the detection of microbial antigens and their antibodies. *Adv. Appl. Microbiol.* 31:271–89.

Herzenberg, L. A.; Sweet, R. G.; and Herzenberg, L. A. 1976. Fluorescence activated cell sorting. *Sci. Am.* 234(3):108–17.

Jackson, J. B., and Balfour, H. H., Jr. 1988. Practical diagnostic testing for human immunodeficiency virus. *Clin. Microbiol. Rev.* 1(1):124–38.

Lopez, M.; Fleisher, T.; and deShazo, R. D. 1992. Use and interpretation of diagnostic immunologic laboratory tests. *JAMA* 268(20):2970–90.

PART NINE
Microbial Diseases and Their Control

A colored transmission electron micrograph
(×44,000) of a dividing virulent *Streptococcus
pyogenes* group A bacterium. This is the resistant
strain that causes necrotizing fascitis, which destroys
the sheath covering skeletal muscles, and myositis,
the inflammation and destruction of skeletal muscle
and fat tissue. (A virulent strep A infection killed
Sesame Street Muppeteer Jim Henson in 1990, and
in 1994 the press wrote headlines with articles on
"the flesh-eating invasive disease.")

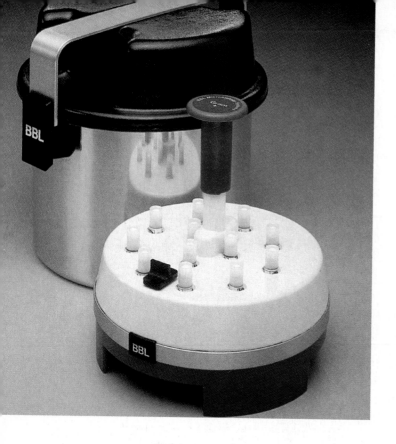

CHAPTER 33
Antimicrobial Chemotherapy

The Lord created medicines out of the earth, and the sensible will not despise them.
—Ecclesiasticus 38:4

It was the knowledge of the great abundance and wide distribution of actinomycetes, which dated back nearly three decades, and the recognition of the marked activity of this group of organisms against other organisms that led me in 1939 to undertake a systematic study of their ability to produce antibiotics.
—Selman A. Waksman

Concepts

1. Many diseases are treated with chemotherapeutic agents, such as antibiotics, that inhibit or kill the pathogen while harming the host as little as possible.

2. Ideally, antimicrobial agents disrupt microbial processes or structures that differ from those of the host. They may damage pathogens by hampering bacterial cell wall synthesis, inhibiting microbial protein and nucleic acid synthesis, disrupting microbial membrane structure and function, or blocking metabolic pathways through inhibition of key enzymes.

3. The effectiveness of chemotherapeutic agents depends on many factors: the route of administration and location of the infection, the presence of interfering substances, the concentration of the drug in the body, the nature of the pathogen, the presence of drug allergies, and the resistance of microorganisms to the drug.

4. The increasing number and variety of drug-resistant pathogens is a serious public health problem.

5. Although antibacterial chemotherapy is more advanced, drugs for the treatment of fungal and viral infections are also becoming increasingly available.

The control of microorganisms is critical for the prevention and treatment of disease. Chapter 7 is concerned principally with the chemical and physical agents used to treat inanimate objects in order to destroy microorganisms or inhibit their growth. In addition, the use of antiseptics is briefly reviewed. Microorganisms also grow on and within other organisms, and microbial colonization can lead to disease, disability, and death. Thus the control or destruction of microorganisms residing within the bodies of humans and other animals is of great importance.

When disinfecting or sterilizing an inanimate object, one naturally must use procedures that do not damage the object itself. The same is true for the treatment of living hosts. The most successful drugs interfere with processes that differ between the pathogen and host, and seriously damage the target microorganism while harming its host as little as possible. This chapter introduces the principles of chemotherapy and briefly reviews the characteristics of selected antibacterial, antifungal, and antiviral drugs.

Modern medicine is dependent on **chemotherapeutic agents,** chemical agents that are used to treat disease. Chemotherapeutic agents destroy pathogenic microorganisms or inhibit their growth at concentrations low enough to avoid undesirable damage to the host. Most of these agents are **antibiotics** [Greek *anti,* against, and *bios,* life], microbial products or their derivatives that can kill susceptible microorganisms or inhibit their growth. Drugs such as the sulfonamides are sometimes called antibiotics although they are synthetic chemotherapeutic agents, not microbially synthesized.

The Development of Chemotherapy

The modern era of chemotherapy began with the work of the German physician Paul Ehrlich (1854–1915). Ehrlich reasoned that a chemical with selective toxicity that would kill pathogens and not human cells might be effective in treating disease. He hoped to find a toxic dye molecule, a "magic bullet," that would specifically bind to pathogens and destroy them; therefore he began experimenting with dyes. By 1904 Ehrlich found that the dye Trypan Red was active against the trypanosome that causes African sleeping sickness (*see figure 27.3*a) and could be used therapeutically. Subsequently Ehrlich and a young Japanese scientist named Sahachiro Hata tested a variety of arsenicals on syphilis-infected rabbits and found that compound number 606, arsphenamine, was active against the syphilis spirochete (*see figure 20.1*b). Arsphenamine was made available in 1910 under the trade name Salvarsan. Ehrlich's successes in the chemotherapy of sleeping sickness and syphilis established his concept of selective toxicity and led to the testing of hundreds of compounds for their therapeutic potential.

In 1927 the German chemical industry giant, I. G. Farbenindustrie, began a long-term search for chemotherapeutic agents under the direction of Gerhard Domagk. The company provided vast numbers of dyes and other chemicals that

Domagk tested for activity against pathogenic bacteria and for toxicity in animals. During this screening program Domagk discovered that Prontosil Red, a new dye for staining leather, was nontoxic for animals and completely protected mice against pathogenic streptococci and staphylococci. These results were published in 1935, and in the same year the French scientists Jacques and Therese Trefouel showed that Prontosil Red was converted in the body to sulfanilamide, the true active factor. Domagk had actually discovered sulfonamides or sulfa drugs and for this discovery he received the Nobel Prize in 1939.

The story of the discovery and development of penicillin, the first antibiotic to be used therapeutically, is complex and fascinating. Although penicillin was actually discovered in 1896 by a 21-year-old French medical student named Ernest Duchesne, his work was forgotten, and penicillin was rediscovered and brought to the attention of scientists by the Scottish physician Alexander Fleming. Fleming had been interested in finding something that would kill pathogens ever since working on wound infections during the First World War. One day in September 1928, a *Penicillium notatum* spore accidentally landed on the surface of an exposed petri dish before it had been inoculated with staphylococci, and a new medical era was born. Although the precise events are still unclear, Ronald Hare has suggested that Fleming left the contaminated plate on a laboratory bench while he was on vacation. Because the first few days of the vacation were cool, the fungus grew more rapidly than the bacteria and produced penicillin. When the weather then turned warm, the bacteria began to grow and were lysed. On his return, Fleming noticed that a *Penicillium* colony was growing at one edge and that the staphylococci surrounding it had been destroyed. Rather than discarding the contaminated plate, he correctly deduced that his mold contaminant was producing a diffusible substance lethal to staphylococci. He began efforts to characterize what he called penicillin. He found that broth from a *Penicillium* culture contained penicillin and that the antibiotic could destroy several pathogenic bacteria. Unfortunately Fleming's next experiments convinced him that penicillin would not remain active in the body long enough after injection to destroy pathogens. After giving a talk on penicillin and publishing several papers on the subject between 1929 and 1931, he dropped the research.

In 1939 Howard Florey, a professor of pathology at Oxford University, was in the midst of testing the bactericidal activity of many substances, including lysozyme and the sulfonamides. After reading Fleming's paper on penicillin, one of Florey's coworkers, Ernst Chain, obtained the *Penicillium* culture from Fleming and set about culturing it and purifying penicillin. Florey and Chain were greatly aided in this by the biochemist Norman Heatley. Heatley devised the original assay, culture, and purification techniques needed to produce crude penicillin for further experimentation. When the purified penicillin was injected into mice infected with streptococci or staphylococci, practically all the mice survived. Florey and Chain's success was reported in 1940, and subsequent human trials were equally successful. Fleming, Florey, and Chain received the Nobel prize in 1945 for the discovery and production of penicillin.

Box 33.1

The Use of Antibiotics in Microbiological Research

Although the use of antibiotics in the treatment of disease is emphasized in this chapter, it should be noted that antibiotics are extremely important research tools. For example, they aid the cultivation of viruses by preventing bacterial contamination. When eggs are inoculated with a virus sample, antibiotics often are included in the inoculum to maintain sterility. Usually a mixture of antibiotics (e.g., penicillin, amphotericin, and streptomycin) also is added to tissue cultures used for virus cultivation and other purposes.

Researchers often use antibiotics as instruments to dissect metabolic processes by inhibiting or blocking specific steps and observing the consequences. Although selective toxicity is critical when antibiotics are employed therapeutically, specific toxicity is more important in this context: the antibiotic must act by a specific and precisely understood mechanism. A clinically useful antimicrobial agent such as penicillin sometimes may be employed in research, but often an agent with specific toxicity and excellent research potential is too toxic for therapeutic use. The actinomycins, discovered in 1940 by Selman Waksman, are a case in point. They are so toxic to higher organisms that it was suggested they be used as rat poison. Today actinomycin D is a standard research tool specifically used to

block RNA synthesis. Other examples of antibiotics useful in research, with the process inhibited, are the following: chloramphenicol (bacterial protein synthesis), cycloserine (peptidoglycan synthesis), nalidixic acid and novobiocin (bacterial DNA synthesis), rifampin (bacterial RNA synthesis), cycloheximide (eucaryotic protein synthesis), daunomycin (fungal RNA synthesis), mitomycin C (DNA synthesis), polyoxin D (fungal cell wall chitin synthesis), and cerulenin (fatty acid synthesis).

In practice, the antibiotic is administered and changes in cell function are monitored. If one desired to study the dependence of bacterial flagella synthesis on RNA transcription, the flagella could be removed by high-speed mixing in a blender, followed by actinomycin D addition to the incubation mixture. The bacterial culture would then be observed for flagella regeneration in the absence of RNA synthesis. The results of such experiments must be interpreted with caution. Flagella synthesis may have been blocked because actinomycin D inhibited some other process, thus affecting flagella regeneration indirectly rather than simply inhibiting transcription of a gene required for flagella synthesis. Furthermore, not all microorganisms respond in the same way to a particular drug.

The discovery of penicillin stimulated the search for other antibiotics. Selman Waksman announced in 1944 that he had found a new antibiotic, streptomycin, produced by the actinomycete *Streptomyces griseus*. This discovery arose from the patient screening of about 10,000 strains of soil bacteria and fungi. Waksman received the Nobel Prize in 1952, and his success led to a worldwide search for other antibiotic-producing soil microorganisms. Microorganisms producing chloramphenicol, neomycin, terramycin, and tetracycline were isolated by 1953.

The discovery of chemotherapeutic agents and the development of newer, more powerful drugs has transformed modern medicine and greatly alleviated human suffering. Furthermore, antibiotics have proven exceptionally useful in microbiological research (Box 33.1).

---

1. What are chemotherapeutic agents? Antibiotics?
2. What contributions to chemotherapy were made by Ehrlich, Domagk, Fleming, Florey and Chain, and Waksman?

General Characteristics of Antimicrobial Drugs

As Ehrlich so clearly saw, to be successful a chemotherapeutic agent must have **selective toxicity:** it must kill or inhibit the microbial pathogen while damaging the host as little as possible.

The degree of selective toxicity may be expressed in terms of (1) the therapeutic dose, the drug level required for clinical treatment of a particular infection, and (2) the toxic dose, the drug level at which the agent becomes too toxic for the host. The **therapeutic index** is the ratio of the toxic dose to the therapeutic dose. The larger the therapeutic index, the better the chemotherapeutic agent (all other things being equal).

A drug that disrupts a microbial function not found in eucaryotic animal cells often has a greater selective toxicity and a higher therapeutic index. For example, penicillin inhibits bacterial cell wall peptidoglycan synthesis but has little effect on host cells because they lack cell walls; therefore penicillin's therapeutic index is high. A drug may have a low therapeutic index because it inhibits the same process in host cells or damages the host in other ways. These undesirable effects on the host, called **side effects,** are of many kinds and may involve almost any organ system (table 33.1). Because side effects can be severe, chemotherapeutic agents should be administered with great care.

Drugs vary considerably in their range of effectiveness. Many are **narrow-spectrum drugs**—that is, they are effective only against a limited variety of pathogens (table 33.1). Others are **broad-spectrum drugs** and attack many different kinds of pathogens. Drugs may also be classified based on the general microbial group they act against: antibacterial, antifungal, antiprotozoan, and antiviral. Some agents can be used

TABLE 33.1	Properties of Some Common Antibacterial Drugs		
Drug	**Primary Effect**	**Spectrum**	**Side Effects**[a]
Ampicillin	Cidal	Broad (gram +, some −)	Allergic responses (diarrhea, anemia)
Bacitracin	Cidal	Narrow (gram +)	Renal injury if injected
Carbenicillin	Cidal	Broad (gram +, many −)	Allergic responses (nausea, anemia)
Cephalosporins	Cidal	Broad (gram +, some −)	(Allergic responses, thrombophlebitis, renal injury)
Chloramphenicol	Static	Broad (gram +, −; rickettsia and chlamydia)	Depressed bone marrow function, allergic reactions
Ciprofloxacin	Cidal	Broad (gram +, −)	Gastrointestinal upset, allergic responses
Clindamycin	Static	Narrow (gram +, anaerobes)	Diarrhea
Dapsone	Static	Narrow (mycobacteria)	(Anemia, allergic responses)
Erythromycin	Static	Narrow (gram +, mycoplasma)	(Gastrointestinal upset, hepatic injury)
Gentamicin	Cidal	Narrow (gram −)	(Allergic responses, nausea, loss of hearing, renal damage)
Isoniazid	Static or cidal	Narrow (mycobacteria)	(Allergic reactions, gastrointestinal upset, hepatic injury)
Methicillin	Cidal	Narrow (gram +)	Allergic responses (renal toxicity, anemia)
Penicillin	Cidal	Narrow (gram +)	Allergic responses (nausea, anemia)
Polymyxin B	Cidal	Narrow (gram −)	(Renal damage, neurotoxic reactions)
Rifampin	Static	Broad (gram +, mycobacteria)	(Hepatic injury, nausea, allergic responses)
Streptomycin	Cidal	Broad (gram +, −; mycobacteria)	(Allergic responses, nausea, loss of hearing, renal damage)
Sulfonamides	Static	Broad (gram +, −)	Allergic responses (renal and hepatic injury, anemia)
Tetracyclines	Static	Broad (gram +, −; rickettsia and chlamydia)	Gastrointestinal upset, teeth discoloration (renal and hepatic injury)
Trimethoprim	Cidal	Broad (gram +, −)	(Allergic responses, rash, nausea, leukopenia)
Vancomycin	Cidal	Narrow (gram +)	Hypotension, neutropenia, kidney damage, allergic reactions

[a]Occasional side effects are in parentheses. Other side effects not listed may also arise.

against more than one group; for example, sulfonamides are active against bacteria and some protozoa.

Chemotherapeutic agents can be synthesized by microorganisms or manufactured by chemical procedures independent of microbial activity. A number of the most commonly employed antibiotics are natural—that is, totally synthesized by one of a few bacteria or fungi (table 33.2). In contrast, several important chemotherapeutic agents are completely synthetic. The synthetic antibacterial drugs in table 33.1 are the sulfonamides, trimethoprim, chloramphenicol, ciprofloxacin, isoniazid, and dapsone. Many antiviral and antiprotozoan drugs are synthetic. An increasing number of antibiotics are semisynthetic. Semisynthetic antibiotics are natural antibiotics that have been chemically modified by the addition of extra chemical groups to make them less susceptible to inactivation by pathogens. Ampicillin, carbenicillin, and methicillin (table 33.1) are good examples.

Chemotherapeutic agents, like disinfectants, can be either **cidal** or **static.** As mentioned earlier (*see chapter 7*), static agents reversibly inhibit growth; if the agent is removed, the microorganisms will recover and grow again.

Although a cidal agent kills the target pathogen, its activity is concentration dependent and the agent may be only static at

TABLE 33.2	Microbial Sources of Some Antibiotics
Microorganism	**Antibiotic**
Bacteria	
Streptomyces spp.	Amphotericin B
	Chloramphenicol (also synthetic)
	Erythromycin
	Kanamycin
	Neomycin
	Nystatin
	Rifampin
	Streptomycin
	Tetracyclines
Micromonospora spp.	Gentamicin
Bacillus spp.	Bacitracin
	Polymyxins
Fungi	
Penicillium spp.	Griseofulvin
	Penicillin
Cephalosporium spp.	Cephalosporins

Figure 33.2 Interpretation of Kirby-Bauer Test Results. The relationship between the minimal inhibitory concentrations with a hypothetical drug and the size of the zone around a disk in which microbial growth is inhibited. As the sensitivity of microorganisms to the drug increases, the MIC value decreases and the inhibition zone grows larger. Suppose that this drug varies from 7–28 µg/ml in the body during treatment. Dashed line A shows that any pathogen with a zone of inhibition less than 12 mm in diameter will have an MIC value greater than 28 µg/ml and will be resistant to drug treatment. A pathogen with a zone diameter greater than 17 mm will have an MIC less than 7 µg/ml and will be sensitive to the drug (see line B). Zone diameters between 12 and 17 mm indicate intermediate sensitivity and usually signify resistance.

TABLE 33.4 Mechanisms of Antibacterial Drug Action

Drug	Mechanism of Action
Cell Wall Synthesis Inhibition	
Penicillin Ampicillin Carbenicillin Methicillin Cephalosporins	Inhibit transpeptidation enzymes involved in the cross-linking of the polysaccharide chains of the bacterial cell wall peptidoglycan. Activate cell wall lytic enzymes.
Vancomycin	Binds directly to the D-Ala-D-Ala terminus and inhibits transpeptidation.
Bacitracin	Inhibits cell wall synthesis by interfering with action of the lipid carrier that transports wall precursors across the plasma membrane.
Protein Synthesis Inhibition	
Streptomycin Gentamicin	Binds with the 30S subunit of the bacterial ribosome to inhibit protein synthesis and causes misreading of mRNA.
Chloramphenicol	Binds to the 50S ribosomal subunit and blocks peptide bond formation through inhibition of peptidyl transferase.
Tetracyclines	Bind to the 30S ribosomal subunit and interfere with aminoacyl-tRNA binding.
Erythromycin and clindamycin	Bind to the 50S ribosomal subunit and inhibit peptide chain elongation.
Fusidic acid	Binds to EF-G and blocks translocation.
Nucleic Acid Synthesis Inhibition	
Ciprofloxacin and other quinolones	Inhibit bacterial DNA gyrase and thus interfere with DNA replication, transcription, and other activities involving DNA.
Rifampin	Blocks RNA synthesis by binding to and inhibiting the DNA-dependent RNA polymerase.
Cell Membrane Disruption	
Polymyxin B	Binds to the plasma membrane and disrupts its structure and permeability properties.
Metabolic Antagonism	
Sulfonamides	Inhibit folic acid synthesis by competition with p-aminobenzoic acid.
Trimethoprim	Blocks tetrahydrofolate synthesis through inhibition of the enzyme dihydrofolate reductase.
Dapsone	Interferes with folic acid synthesis.
Isoniazid	May disrupt pyridoxal or NAD metabolism and functioning. Inhibits the synthesis of the mycolic acid "cord factor."

Streptomycin, gentamicin, spectinomycin, clindamycin, chloramphenicol, tetracyclines, erythromycin, and many other antibiotics inhibit protein synthesis by binding with the procaryotic ribosome. Because these drugs discriminate between procaryotic and eucaryotic ribosomes, their therapeutic index is fairly high, but not as favorable as that of cell wall synthesis inhibitors. Some drugs bind to the 30S (small) subunit, while others attach to the 50S (large) ribosomal subunit. Several different steps in the protein synthesis mechanism can be affected: aminoacyl-tRNA binding, peptide bond formation, mRNA reading, and translocation. For example, fusidic acid binds to EF-G and blocks translocation, whereas mucopirocin inhibits isoleucyl-tRNA synthetase (*see chapter 11*).

The antibacterial drugs that inhibit nucleic acid synthesis or damage cell membranes often are not as selectively toxic as other antibiotics. This is because procaryotes and eucaryotes do not differ as greatly with respect to nucleic acid synthetic mechanisms or cell membrane structure. Good examples of drugs that affect nucleic acid synthesis or membrane structure are quinolones and polymyxins. Quinolones inhibit the DNA gyrase and thus interfere with DNA replication, repair, and transcription. Polymyxins act as detergents or surfactants and disrupt the bacterial plasma membrane.

Several valuable drugs act as **antimetabolites:** they block the functioning of metabolic pathways by competitively inhibiting the use of metabolites by key enzymes. Sulfonamides and several other drugs inhibit folic acid metabolism. Sulfonamides (e.g., sulfanilamide, sulfamethoxazole, and sulfacetamide) have a high therapeutic index because humans cannot synthesize folic acid and must obtain it in their diet. Most bacterial pathogens synthesize their own folic acid and are therefore susceptible to inhibitors of folate metabolism. Antimetabolite drugs also can inhibit other pathways. For example, isoniazid interferes with either pyridoxal or NAD metabolism.

---

1. Give five ways in which chemotherapeutic agents kill or damage bacterial pathogens.
2. What are antimetabolites?

Factors Influencing the Effectiveness of Antimicrobial Drugs

It is crucial to recognize that drug therapy is not a simple matter. Drugs may be administered in several different ways, and they do not always spread rapidly throughout the body or immediately kill all invading pathogens. A complex array of factors influence the effectiveness of drugs.

First, the drug must actually be able to reach the site of infection. The mode of administration plays an important role. A drug such as penicillin G is not suitable for oral administration because it is relatively unstable in stomach acid. Some antibiotics—for example, gentamicin and other aminoglycosides—are not well absorbed from the intestinal tract and must be injected intramuscularly or given intravenously. Other antibiotics (neomycin, bacitracin) are applied topically to skin lesions. Nonoral routes of administration often are called **parenteral routes.** Even when an agent is administered properly, it may be excluded from the site of infection. For example, blood clots or necrotic tissue can protect bacteria from a drug, either because body fluids containing the agent may not easily reach the pathogens or because the agent is absorbed by materials surrounding it.

Second, the pathogen must be susceptible to the drug. Bacteria in abscesses may be dormant and therefore resistant to chemotherapy, because penicillins and many other agents affect pathogens only if they are actively growing and dividing. A pathogen, even though growing, may simply not be susceptible to a particular agent. For example, penicillins and cephalosporins, which inhibit cell wall synthesis (table 33.4), do not harm mycoplasmas, which lack cell walls.

Third, the chemotherapeutic agent must exceed the pathogen's MIC value if it is going to be effective. The concentration reached will depend on the amount of drug administered, the route of administration and speed of uptake, and the rate at which the drug is cleared or eliminated from the body. It makes sense that a drug will remain at high concentrations longer if it is absorbed over a long period and excreted slowly.

Finally, chemotherapy has been rendered a less effective and much more complex matter by the spread of drug resistance plasmids. The nature of drug resistance is dealt with later in this chapter.

---

1. Briefly discuss the factors that influence the effectiveness of antimicrobial drugs.
2. What is parenteral administration of a drug?

Antibacterial Drugs

With the background provided by the preceding introduction to chemotherapy, a few of the more important chemotherapeutic agents can be briefly described. Table 33.5 lists some drugs used to treat a few of the more common bacterial pathogens. Tables 33.1 and 33.4 summarize the properties and mechanism of action of a variety of drugs.

Sulfonamides or Sulfa Drugs

A good way to inhibit or kill pathogens is by use of compounds that are **structural analogues,** molecules structurally similar to metabolic intermediates. These analogues compete with metabolites in metabolic processes because of their similarity, but are just different enough so that they cannot function normally in cellular metabolism. The first antimetabolites to be used successfully as chemotherapeutic agents were the sulfonamides, discovered by G. Domagk. **Sulfonamides** or sulfa

TABLE 33.5 Chemotherapy of Some Representative Bacterial Pathogens

Pathogen	Representative Diseases	Drugs of Choice[a]
Gram-Positive Bacteria		
Corynebacterium diphtheriae	Diphtheria	Erythromycin, penicillin G
Staphylococcus, penicillinase-positive	Boils, pneumonia, wound infections	A cephalosporin, cloxacillin, dicloxacillin, vancomycin
Staphylococcus, penicillinase-negative	Boils, pneumonia, wound infections	Penicillin G or V, a cephalosporin, vancomycin
Streptococcus, hemolytic	Strep throat, skin infections, sepsis, rheumatic fever	Penicillin[b] G or V, erythromycin, or a cephalosporin
Streptococcus pneumoniae	Pneumonia	Penicillin[b] G or V, erythromycin, or a cephalosporin
Gram-Negative Bacteria		
Bordetella pertussis	Whooping cough or pertussis	Erythromycin, ampicillin
Escherichia coli	Urinary tract infections	Sulfonamide, cephalosporin, ampicillin
Haemophilus influenzae type b	Meningitis, pneumonia	Cefotaxime, ceftriaxone, ampicillin
Klebsiella pneumoniae	Pneumonia, urinary tract infections	Newer cephalosporins[c], gentamicin
Legionella pneumophila	Pneumonia	Erythromycin
Neisseria gonorrhoeae	Gonorrhea	Amoxicillin, ceftriaxone, spectinomycin
Pseudomonas aeruginosa	Urinary tract and burn infections, pneumonia	Carbenicillin or ticarcillin
Salmonella typhi	Typhoid fever, septicemia, gastroenteritis	Ceftriaxone, chloramphenicol, ampicillin
Shigella dysenteriae	Dysentery	Trimethoprim-sulfamethoxazole, quinolones, ciprofloxacin
Vibrio cholerae	Cholera	Tetracycline, trimethoprim-sulfamethoxazole
Acid-Fast Bacteria		
Mycobacterium-avium complex	Pneumonia	Rifampin plus ethambutol
Mycobacterium tuberculosis	Tuberculosis	Isoniazid plus rifampin ± pyrazinamide
Other Bacteria		
Chlamydia trachomatis	Nongonococcal urethritis, trachoma	Doxycycline, azithromycin, erythromycin
Mycoplasma pneumoniae	Pneumonia	Tetracycline, erythromycin
Rickettsia spp.	Rocky Mountain spotted fever, typhus fever	Tetracycline, chloramphenicol
Treponema pallidum subsp. *pallidum*	Syphilis	Penicillin G, tetracycline, ceftriaxone

[a]The drugs are listed in approximate order of preference. A number of others are also used.

[b]Penicillinase-resistant penicillins may have to be employed (e.g., methicillin, nafcillin, oxacillin).

[c]Many newer cephalosporins are available: ceftriaxone, cefoperazone, cefuroxime, among others.

Figure 33.3 **Sulfanilamide.** Sulfanilamide and its relationship to the structure of folic acid.

drugs are structurally related to sulfanilamide, an analogue of p-aminobenzoic acid (figures 33.3 and 33.4). The latter substance is used in the synthesis of the cofactor folic acid.

When sulfanilamide or another sulfonamide enters a bacterial cell, it competes with p-aminobenzoic acid for the active site of an enzyme involved in folic acid synthesis, and the folate concentration decreases. The decline in folic acid is detrimental to the bacterium because folic acid is essential to the synthesis of purines and pyrimidines, the bases used in the construction of DNA, RNA, and other important cell constituents. The resulting inhibition of purine and pyrimidine synthesis leads to cessation of bacterial growth or death of the pathogen. Sulfonamides are selectively toxic for many pathogens because these bacteria manufacture their own folate and cannot effectively take up the cofactor. In contrast, humans cannot synthesize folate and must obtain it in the diet; therefore sulfonamides will not affect the host.

Sulfonamides and the competitive inhibition of enzymes (pp. 161–62).

Purine and pyrimidine structure and synthesis (pp. 203–4).

Sulfamethoxazole

Sulfisoxazole

Figure 33.4   Two Modern Sulfonamide Drugs. The shaded areas
are side chains substituted for a hydrogen in sulfanilamide (figure 33.3).

Nalidixic acid

Norfloxacin

Ciprofloxacin

Figure 33.5   Quinolone Antimicrobial Agents. Ciprofloxacin
and norfloxacin are newer generation fluoroquinolones. The 4-
quinolone ring in nalidixic acid has been numbered.

The effectiveness of sulfonamides is limited by the increasing sulfonamide resistance of many bacteria. Furthermore, as
many as 5% of the patients receiving sulfa drugs experience adverse side effects, chiefly in the form of allergic responses
(fever, hives, and rashes).

Quinolones

A second group of synthetic antimicrobial agents are increasingly used to treat a wide variety of infections. The **quinolones**
are synthetic drugs that contain the 4-quinolone ring. The first
quinolone, nalidixic acid (figure 33.5), was synthesized in 1962.
More recently a family of fluoroquinolones has been produced.
Three of these—ciprofloxacin, norfloxacin, and ofloxacin—are
currently used in the United States, and more fluoroquinolones
are being synthesized and tested. Quinolones are effective when
administered orally. They sometimes cause adverse side effects,
particularly gastrointestinal upset.

Quinolones act by inhibiting the bacterial DNA gyrase or
topoisomerase II, probably by binding to the DNA gyrase complex. This enzyme introduces negative twists in DNA and helps
separate its strands (figure 33.6). DNA gyrase inhibition disrupts
DNA replication and repair, transcription, bacterial chromosome
separation during division, and other cell processes involving
DNA. It is not surprising that quinolones are bactericidal.

The mechanism of DNA replication (pp. 218–21).

The quinolones are broad-spectrum drugs. They are highly
effective against enteric bacteria such as *E. coli* and *Klebsiella
pneumoniae.* They can be used with *Haemophilus, Neisseria,
Pseudomonas aeruginosa,* and other gram-negative pathogens.
The quinolones also are active against gram-positive bacteria

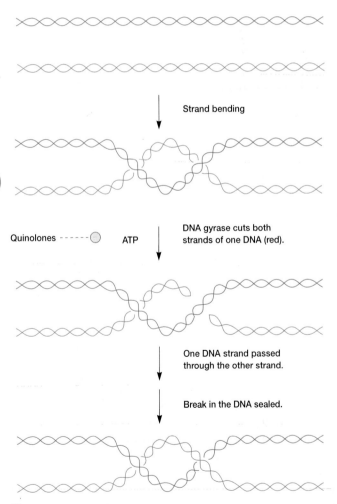

Strand bending

Quinolones ----- ◯   ATP

DNA gyrase cuts both
strands of one DNA (red).

One DNA strand passed
through the other strand.

Break in the DNA sealed.

Figure 33.6   DNA Gyrase Action and Quinolone Inhibition.

Penicillin G

High activity against most gram-positive bacteria, low against gram-negative; destroyed by acid and penicillinase

Penicillin V

More acid resistant than penicillin G

Ampicillin

Active against gram-positive and gram-negative bacteria; acid stable

Carbenicillin

Active against gram-negative bacteria like *Pseudomonas* and *Proteus*; acid stable; not well absorbed by small intestine

Methicillin

Penicillinase-resistant, but less active than penicillin G; acid-labile

Ticarcillin

Similar to carbenicillin, but more active against *Pseudomonas*

Figure 33.7   Penicillins. The structures and characteristics of representative penicillins. All are derivatives of 6-aminopenicillanic acid; in each case the shaded portion of penicillin G is replaced by the side chain indicated. The β-lactam ring is also shaded, and an arrow points to the bond that is hydrolyzed by penicillinase.

such as *Staphylococcus aureus, Streptococcus pyogenes,* and *Mycobacterium tuberculosis.* Currently they are used in treating urinary tract infections, sexually transmitted diseases caused by *Neisseria* and *Chlamydia,* gastrointestinal infections, respiratory tract infections, skin infections, and osteomyelitis.

Penicillins

Penicillin G or benzylpenicillin, the first antibiotic to be widely used in medicine, has the structural properties characteristic of the penicillin family (figure 33.7). Most **penicillins** are deriva-

tives of 6-aminopenicillanic acid and differ from one another only with respect to the side chain attached to its amino group. The most crucial feature of the molecule is the β-lactam ring, which appears to be essential for activity. **Penicillinase,** the enzyme synthesized by many penicillin-resistant bacteria, destroys penicillin activity by hydrolyzing a bond in this ring (figure 33.7).

The mechanism of action of penicillins is still not completely known. Their structures do resemble that of the terminal D-alanyl-D-alanine found on the peptide side chain of the peptidoglycan subunit. It has been proposed that penicillins inhibit

First-generation cephalosporin

Cephalothin

7-aminocephalosporanic acid

Second-generation cephalosporin

Cefoxitin

Third-generation cephalosporins

Cefoperazone

Ceftriaxone

Figure 33.8   Cephalosporin Antibiotics. These drugs are derivatives of 7-aminocephalosporanic acid and contain a β-lactam ring.

the enzyme catalyzing the transpeptidation reaction because of their structural similarity, which would block the synthesis of a complete, fully cross-linked peptidoglycan and lead to osmotic lysis. The mechanism is consistent with the observation that penicillins act only on growing bacteria that are synthesizing new peptidoglycan. However, more recently it has been discovered that penicillins bind to several penicillin-binding proteins and may destroy bacteria by activating their own autolytic enzymes (autolysins). Apparently the mechanism of penicillin action is more complex than previously imagined.

Bacterial peptidoglycan structure (pp. 52–54).
Peptidoglycan synthesis (pp. 206–9).

Penicillins differ from each other in several ways. Penicillin G is effective against gonococci, meningococci, and several gram-positive pathogens such as streptococci and staphylococci (table 33.5), but must be administered parenterally because it is destroyed by stomach acid. Penicillin V is similar to penicillin G, but it is more acid resistant and can be given orally. Ampicillin can be administered orally and has a broader spectrum of activity as it is effective against gram-negative bacteria such as *Haemophilus, Salmonella,* and *Shigella.* Carbenicillin and ticarcillin also are broad spectrum and particularly potent against *Pseudomonas* and *Proteus.*

An increasing number of bacteria are penicillin resistant. Penicillinase-resistant penicillins such as methicillin (figure 33.7), nafcillin, and oxacillin are frequently employed against these bacterial pathogens.

Although penicillins are the least toxic of the antibiotics, about 1 to 5% of the adults in the United States are allergic to them. Occasionally a person will die of a violent allergic response; therefore patients should be questioned about penicillin allergies before treatment is begun.

Cephalosporins

Cephalosporins are a family of antibiotics originally isolated in 1948 from the fungus *Cephalosporium,* and their β-lactam structure is very similar to that of the penicillins (figure 33.8). As might be expected from their structural similarities, cephalosporins resemble penicillins in inhibiting the transpeptidation reaction during peptidoglycan synthesis. They are broad-spectrum drugs frequently given to patients with penicillin allergies.

Many cephalosporins are in use. There are three groups or generations of these drugs that differ in their spectrum of activity (figure 33.8). First-generation cephalosporins are more effective against gram-positive than gram-negative pathogens. Second-generation drugs act against many gram-negative as well as gram-positive pathogens. Third-generation drugs are particularly effective against gram-negative pathogens, and often also reach the central nervous system.

Most cephalosporins (including cephalothin, cefoxitin, ceftriaxone, and cefoperazone) are administered parenterally. Cefoperazone is resistant to destruction by β-lactamases and effective against many gram-negative bacteria, including *Pseudomonas aeruginosa*. Cephalexine and cefixime are given orally rather than by injection.

The Tetracyclines

The **tetracyclines** are a family of antibiotics with a common four-ring structure to which a variety of side chains are attached (figure 33.9). Oxytetracycline and chlortetracycline are naturally produced by some species of the actinomycete genus *Streptomyces;* others are semisynthetic drugs. These antibiotics inhibit protein synthesis by combining with the small (30S) subunit of the ribosome and inhibiting the binding of aminoacyl-tRNA molecules to the ribosomal A site. Because their action is only bacteriostatic, the effectiveness of treatment depends on active host resistance to the pathogen.

The mechanism of protein synthesis (pp. 226–33).

Tetracyclines are broad-spectrum antibiotics active against gram-negative bacteria, gram-positive bacteria, rickettsias, chlamydiae, and mycoplasmas (table 33.5). High doses may result in nausea, diarrhea, yellowing of teeth in children, and damage to the liver and kidneys.

Aminoglycoside Antibiotics

There are several important **aminoglycoside antibiotics. Streptomycin,** kanamycin, neomycin, and tobramycin are synthesized by *Streptomyces,* whereas gentamicin comes from a related bacterium, *Micromonospora purpurea.* Although there is considerable variation in structure among the different aminoglycosides, all contain a cyclohexane ring and amino sugars (figure 33.10). Aminoglycosides bind to the small ribosomal subunit and interfere with protein synthesis in at least two ways. They directly inhibit protein synthesis and also cause misreading of the genetic message carried by mRNA.

The aminoglycosides are bactericidal and tend to be most active against gram-negative pathogens. Streptomycin's usefulness has decreased greatly due to widespread drug resistance, but it is still effective against tuberculosis and plague. Gentamicin is used to treat *Proteus, Escherichia, Klebsiella,* and *Serratia* infections. Aminoglycosides are quite toxic and can cause deafness, renal damage, loss of balance, nausea, and allergic responses.

Erythromycin and Other Macrolides

Erythromycin, the most frequently used **macrolide antibiotic,** is synthesized by *Streptomyces erythraeus.* The macrolides contain a 12- to 22-carbon lactone ring linked to one or more sugars

Figure 33.9 **Tetracyclines.** Three members of the tetracycline family. Tetracycline lacks both of the groups that are shaded. Chlortetracycline (aureomycin) differs from tetracycline in having a chlorine atom (blue); doxycycline consists of tetracycline with an extra hydroxyl (purple).

Figure 33.10 **Representative Aminoglycoside Antibiotics.**

(figure 33.11). Erythromycin is usually bacteriostatic and binds with the 23S rRNA of the 50S ribosomal subunit to inhibit peptide chain elongation during protein synthesis.

Erythromycin is a relatively broad-spectrum antibiotic effective against gram-positive bacteria, mycoplasmas, and a few gram-negative bacteria. It is used with patients allergic to penicillins and in the treatment of whooping cough, diphtheria, diarrhea caused by *Campylobacter,* and pneumonia from *Legionella* or *Mycoplasma* infections.

Newer macrolides are now in use. Clindamycin is effective against a variety of bacteria including staphylococci and anaerobes such as *Bacteroides.* Azithromycin is particularly effective against *Chlamydia trachomatis.*

Chloramphenicol

Although **chloramphenicol** (figure 33.12) was first produced from cultures of *Streptomyces venezuelae,* it is now made through chemical synthesis. Like erythromycin, chloramphenicol binds to 23S rRNA on the 50S ribosomal subunit. It inhibits the peptidyl transferase and is bacteriostatic.

Figure 33.11   Erythromycin, a Macrolide Antibiotic. The 14-member lactone ring is connected to two sugars.

Figure 33.12   Chloramphenicol.

This antibiotic has a very broad spectrum of activity but unfortunately is quite toxic. One may see allergic responses or neurotoxic reactions. The most common side effect is a temporary or permanent depression of bone marrow function, leading to aplastic anemia and a decreased number of blood leukocytes. Chloramphenicol is used only in life-threatening situations when no other drug is adequate.

---

1. For each class of antibacterial drugs presented, give the following information: general chemical composition or structure, mechanism of action, behavioral properties (cidal or static, broad spectrum or narrow spectrum), route of administration and therapeutic uses, significant problems with side effects, and drug resistance.

2. Define structural analogue, 6-aminopenicillanic acid, penicillinase, aminoglycoside, and macrolide.

Drug Resistance

The spread of drug-resistant pathogens is one of the most serious threats to the successful treatment of microbial disease (Box 33.2). This section describes the ways in which bacteria acquire drug resistance and how resistance spreads within a bacterial population.

Mechanisms of Drug Resistance

Bacteria become drug resistant in several different ways. It should be noted at the beginning that a particular type of resistance mechanism is not confined to a single class of drugs. Two bacteria may use different resistance mechanisms to withstand the same chemotherapeutic agent. Furthermore, resistant mutants arise spontaneously and are then selected. Mutants are not created directly by exposure to a drug.

Pathogens often become resistant simply by preventing entrance of the drug. Many gram-negative bacteria are unaffected by penicillin G because it cannot penetrate the envelope's outer membrane. Changes in penicillin binding proteins also render a cell resistant. A decrease in permeability can lead to sulfonamide resistance. Mycobacteria resist many drugs because of the high content of mycolic acids (*p. 452*) in a complex lipid layer outside their wall peptidoglycan. This layer is impermeable to most drugs.

A second resistance strategy is to pump the drug out of the cell after it has entered. Some pathogens have plasma membrane translocases that expel drugs. Because they are relatively nonspecific and can pump many different drugs, these transport proteins often are called multidrug-resistance pumps. Many are drug/proton antiporters—that is, protons enter the cell as the drug leaves. Such systems are present in *E. coli, Pseudomonas aeruginosa,* and *Staphylococcus aureus.*

Many bacterial pathogens resist attack by inactivating drugs through chemical modification. The best-known example is the hydrolysis of the β-lactam ring of many penicillins by the enzyme penicillinase. Drugs also are inactivated by the addition of groups. Resistant organisms may phosphorylate or acetylate aminoglycosides and acetylate chloramphenicol.

Because each chemotherapeutic agent acts on a specific target, resistance arises when the target enzyme or organelle is modified so that it is no longer susceptible to the drug. For example, the affinity of ribosomes for erythromycin and chloramphenicol can be decreased by a change in the 23S rRNA to which they bind. The effects of antimetabolites may be resisted through alteration of susceptible enzymes. In sulfonamide-resistant bacteria the enzyme that uses p-aminobenzoic acid during folic acid synthesis (the tetrahydropteroic acid synthetase) often has a much lower affinity for sulfonamides. *Mycobacterium tuberculosis* becomes resistant to the drug rifampin due to mutations that alter the β subunit of its RNA polymerase. Rifampin cannot bind to the mutant RNA polymerase and block the initiation of transcription.

Resistant bacteria may either use an alternate pathway to bypass the sequence inhibited by the agent or increase the production of the target metabolite. For example, some bacteria are resistant to sulfonamides simply because they use preformed folic acid from their surroundings rather than synthesize it themselves. Other strains increase their rate of folic acid production and thus counteract sulfonamide inhibition.

The Origin and Transmission of Drug Resistance

The genes for drug resistance are present on both the bacterial chromosome and **plasmids,** small circular DNA molecules that can exist separate from the chromosome or be integrated in it.

Plasmids (pp. 280–83).

Box 33.2

Antibiotic Misuse and Drug Resistance

The sale of antimicrobial drugs is big business. In the United States millions of pounds of antibiotics valued at many millions of dollars are produced annually. Approximately 40 to 50% of these antibiotics are added to livestock feed.

Because of the massive quantities of antibiotics being prepared and used, an increasing number of diseases are resisting treatment due to the spread of drug resistance. A good example is *Neisseria gonorrhoeae,* the causative agent of gonorrhea. Gonorrhea was first treated successfully with sulfonamides in 1936, but by 1942 most strains were resistant and physicians turned to penicillin. Within 16 years a penicillin-resistant strain had emerged in the Far East. A penicillinase-producing gonococcus reached the United States in 1976 and is still spreading in this country.

In late 1968 an epidemic of dysentery caused by *Shigella* broke out in Guatemala and affected at least 112,000 persons; 12,500 deaths resulted. The strains responsible for this devastation carried an R plasmid giving them resistance to chloramphenicol, tetracycline, streptomycin, and sulfonamide. In 1972 a typhoid epidemic swept through Mexico producing 100,000 infections and 14,000 deaths. It was due to a *Salmonella typhi* strain with the same multiple drug-resistance pattern seen in the previous *Shigella* outbreak.

Haemophilus influenzae type b is responsible for many cases of childhood pneumonia and middle ear infections, as well as respiratory infections and meningitis. It is now becoming increasingly resistant to tetracyclines, ampicillin, and chloramphenicol. The same situation is occurring with *Streptococcus pneumoniae.*

In 1946 almost all strains of *Staphylococcus* were penicillin sensitive. Today most hospital strains are resistant to penicillin G, and some are now also resistant to methicillin and/or gentamicin and only can be treated with vancomycin. Some strains of *Enterococcus* have become resistant to most antibiotics, including vancomycin. If the vancomycin resistance gene spreads to *Staphylococcus aureus,* this virulent pathogen well may become untreatable.

It is clear from these and other examples (e.g., multiresistant *Mycobacterium tuberculosis*) that drug resistance is an extremely serious public health problem. Much of the difficulty arises from drug misuse. Drugs frequently have been overused in the past. It has been estimated that over 50% of the antibiotic prescriptions in hospitals are given without clear evidence of infection or adequate medical indication. Many physicians have administered antibacterial drugs to patients with colds, influenza, viral pneumonia, and other viral diseases. Frequently antibiotics are prescribed without culturing and identifying the pathogen or without determining bacterial sensitivity to the drug. Toxic, broad-spectrum antibiotics are sometimes given in place of narrow-spectrum drugs as a substitute for culture and sensitivity testing, with the consequent risk of dangerous side effects, superinfections, and the selection of drug-resistant mutants. The situation is made worse by patients not completing their course of medication. When antibiotic treatment is ended too early, drug-resistant mutants may survive. Drugs are available to the public in many countries; people may practice self-administration of antibiotics and further increase the prevalence of drug-resistant strains.

The use of antibiotics in animal feeds is undoubtedly another contributing factor to increasing drug resistance. The addition of low levels of antibiotics to livestock feeds does raise the efficiency and rate of weight gain in cattle, pigs, and chickens (partially because of infection control in overcrowded animal populations). However, this also increases the number of drug-resistant bacteria in animal intestinal tracts. There is evidence for the spread of bacteria such as *Salmonella* from animals to human populations. In 1983, 18 people in four midwestern states were infected with a multiple-drug-resistant strain of *Salmonella newport.* Eleven were hospitalized for salmonellosis and one died. All 18 patients had recently been infected by eating hamburger from beef cattle fed subtherapeutic doses of chlortetracycline for growth promotion. Elimination of antibiotic food supplements might well aid in slowing the spread of drug resistance.

Spontaneous mutations in the bacterial chromosome, although they do not occur very often, will make bacteria drug resistant. Usually such mutations result in a change in the drug receptor; therefore the antibiotic cannot bind and inhibit (e.g., the streptomycin receptor protein on bacterial ribosomes). Many mutants are probably destroyed by natural host resistance mechanisms. However, when a patient is being treated extensively with antibiotics, some resistant mutants may survive and flourish because of their competitive advantage over nonresistant strains.

Frequently a bacterial pathogen is drug resistant because it has a plasmid bearing one or more resistance genes; such plasmids are called **R plasmids** (resistance plasmids). Plasmid resistance genes often code for enzymes that destroy or modify drugs; for example, the hydrolysis of penicillin or the acetylation of chloramphenicol and aminoglycoside drugs. Plasmid-associated genes have been implicated in resistance to the aminoglycosides, choramphenicol, penicillins and

cephalosporins, erythromycin, tetracyclines, sulfonamides, and others. Once a bacterial cell possesses an R plasmid, the plasmid may be transferred to other cells quite rapidly through normal gene exchange processes such as conjugation, transduction, and transformation. Because a single plasmid may carry genes for resistance to several drugs, a pathogen population can become resistant to several antibiotics simultaneously, even though the infected patient is being treated with only one drug.

Conjugation (pp. 288–92).
Transduction (pp. 293–98).
Transformation (pp. 292–93).

Extensive drug treatment favors the development and spread of antibiotic-resistant strains because the antibiotic destroys normal, susceptible bacteria that would usually compete with drug-resistant strains. The result may be the emergence of drug-resistant pathogens leading to a **superinfection.** Superinfections are a significant problem because of the existence of multiple-drug-resistant bacteria that often produce drug-resistant respiratory and urinary tract infections. A classic example of a superinfection resulting from antibiotic administration is the disease pseudomembranous enterocolitis caused by *Clostridium difficile*. When a patient is treated with clindamycin, ampicillin, or cephalosporin, many intestinal bacteria are killed, but *C. difficile* is not. This intestinal inhabitant, which is normally a minor constituent of the population, flourishes in the absence of competition and produces a toxin that stimulates the secretion of a pseudomembrane by intestinal cells. If the superinfection is not treated early with vancomycin, the pseudomembrane must be surgically removed or the patient will die. Fungi, such as the yeast *Candida albicans,* also produce superinfections when bacterial competition is eliminated by antibiotics.

Several strategies can be employed to discourage the emergence of drug resistance. The drug can be given in a high enough concentration to destroy susceptible bacteria and most spontaneous mutants that might arise during treatment. Sometimes two different drugs can be administered simultaneously with the hope that each drug will prevent the emergence of resistance to the other. Finally, chemotherapeutic drugs, particularly broad-spectrum drugs, should be used only when definitely necessary. If possible, the pathogen should be identified, drug sensitivity tests run, and the proper narrow-spectrum drug employed.

Despite efforts to control the emergence and spread of drug resistance, the situation continues to worsen. Of course, as mentioned previously, antibiotics should be used in ways that reduce the development of resistance. Another approach is to search for new antibiotics that microorganisms have never encountered (*see Box 41.2*). Pharmaceutical companies collect and analyze samples from around the world in a search for completely new antimicrobial agents. Structure-based or rational drug design is a third option. If the three-dimensional structure of a susceptible target molecule such as an enzyme essential to microbial function is known, computer programs can be used to design drugs that precisely fit the target molecule. These drugs might be able to bind to the target and disrupt its function sufficiently to destroy the pathogen. Pharmaceutical companies are using this approach to attempt to develop drugs for the treatment of AIDS, cancer, and the common cold. At least one company is developing "enhancers." These are cationic peptides that disrupt bacterial membranes by displacing their magnesium ions. Antibiotics then penetrate and rapidly exert their effects.

---

1. Briefly describe the five major ways in which bacteria become resistant to drugs and give an example of each.
2. Define plasmid, R plasmid, and superinfection. How are R plasmids involved in the spread of drug resistance?

Antifungal Drugs

Treatment of fungal infections generally has been less successful than that of bacterial infections largely because eucaryotic fungal cells are much more similar to human cells than are bacteria. Many drugs that inhibit or kill fungi are therefore quite toxic for humans. In addition, most fungi have a detoxification system that modifies many antibiotics, probably by hydroxylation. As a result the added antibiotics are fungistatic only as long as repeated application maintains high levels of unmodified antibiotic. Despite their relatively low therapeutic index, a few drugs are useful in treating many major fungal diseases. Effective antifungal agents frequently either extract membrane sterols or prevent their synthesis. Similarly, because animal cells do not have cell walls, the enzyme chitin synthase is the target for fungal-active antibiotics such as polyoxin D and nikkomycin.

Fungal infections are often subdivided into infections of superficial tissues or superficial mycoses and systemic mycoses. Treatment for these two types of disease is very different. Several drugs are used to treat superficial mycoses. Three drugs containing imidazole—miconazole, ketoconazole (figure 33.13), and clotrimazole—are broad-spectrum agents available as creams and solutions for the treatment of dermatophyte infections such as athlete's foot, and oral and vaginal candidiasis (*see chapter 39*). They are thought to disrupt fungal membrane permeability and inhibit sterol synthesis. Tolnaftate is used topically for the treatment of cutaneous infections, but is not as effective against infections of the skin and hair. **Nystatin** (figure 33.13), a polyene antibiotic from *Streptomyces,* is used to control *Candida* infections of the skin, vagina, or alimentary tract. **Griseofulvin** (figure 33.13), an antibiotic formed by *Penicillium,* is given orally to treat chronic dermatophyte infections. It is thought to disrupt the mitotic spindle and inhibit cell division; it also may inhibit protein and nucleic acid synthesis. Side effects of griseofulvin include headaches, gastrointestinal upset, and allergic reactions.

Superficial and systemic mycoses (pp. 791, 795–97).

Systemic infections are very difficult to control and can be fatal. Three drugs commonly used against systemic mycoses are **amphotericin B,** 5-flucytosine, and fluconazole (figure 33.13). Amphotericin B from *Streptomyces* sp. binds to the sterols in fungal membranes, disrupting membrane permeability and causing leakage of cell constituents. It is quite toxic and used only for serious, life-threatening infections. The synthetic oral antimycotic agent 5-flucytosine (5-fluorocytosine) is effective against most systemic fungi, although drug resistance often develops rapidly. The drug is converted to 5-fluorouracil by the fungi, incorporated into RNA in place of uracil, and disrupts RNA function. Its side effects include skin rashes, diarrhea, nausea, aplastic anemia, and liver damage. Fluconazole is used in the treatment of candidiasis, cryptococcal meningitis, and coccidioidal meningitis. Adverse effects are relatively uncommon.

---

1. Summarize the mechanism of action and the therapeutic use of the following antifungal drugs: miconazole, nystatin, griseofulvin, amphotericin B, and 5-flucytosine.

Amphotericin B

Nystatin

Miconazole

5-flucytosine

Griseofulvin

Ketoconazole

Figure 33.13 **Antifungal Drugs.** Six commonly used drugs are shown here.

Antiviral Drugs

For many years the possibility of treating viral infections with drugs appeared remote because viruses enter host cells and make use of host cell enzymes and constituents to a large extent. A drug that would block virus reproduction also was thought to be toxic for the host. Inhibitors of virus-specific enzymes and life cycle processes have now been discovered, and several drugs are used therapeutically (table 33.6). Four are shown in figure 33.14.

Animal virus life cycles (pp. 370–77).

Most antiviral drugs disrupt either critical stages in the virus life cycle or the synthesis of virus-specific nucleic acids. **Amantadine** and rimantadine can be used to prevent influenza A infections. When given in time, it will reduce the incidence of influenza by 50 to 70% in an exposed population. Amantadine blocks the penetration and uncoating of influenza virus particles (*see chapter 18*). **Adenine arabinoside** or **vidarabine** disrupts the activity of DNA polymerase and several other enzymes involved in DNA and RNA synthesis and function. It is given intravenously or applied as an ointment to treat herpes infections. A

TABLE 33.6 **Chemotherapy of Some Viral Pathogens**

Pathogen	Drugs of Choice
Cytomegalovirus	Ganciclovir, foscarnet
Hepatitis B or C virus	α_{2a}- or α_{2b}-interferons
Herpes simplex virus	Acyclovir, vidarabine, foscarnet, trifluridine, idoxuridine
Human immunodeficiency virus	Azidothymidine (zidovudine, AZT), didanosine
Influenza A virus	Amantadine, rimantadine
Respiratory syncytial virus	Ribavirin
Varicella-zoster virus	Acyclovir

third drug, **acyclovir,** is also used in the treatment of herpes infections. Upon phosphorylation, acyclovir resembles deoxy-GTP and inhibits the virus DNA polymerase. Unfortunately acyclovir-resistant strains of herpes are already developing. Although these three drugs act only against herpes or influenza viruses, drugs active against other viruses are being developed. A notable example is **azidothymidine (AZT)** or **zidovudine,** which interferes with

Amantadine

Azidothymidine
(AZT) or zidovudine

Adenine arabinoside
(Ara-A, vidarabine)

Acyclovir

🔵 Figure 33.14   **Representative Antiviral Drugs.**

the reverse transcriptase enzyme, and therefore with the reproduction of retroviruses such as the human immunodeficiency virus. AZT is one of the few effective AIDS treatments (*see p. 729*). Several newer drugs are now commercially available: didanosine, foscarnet, idoxuridine, and trifluridine.

Probably the most publicized antiviral agents are **interferons.** These small proteins, produced by the host, inhibit virus replication and may be clinically useful in the treatment of influenza, hepatitis, herpes, and colds. The mechanism of interferon action is discussed in more detail in chapter 29.

1. Describe two different ways in which antiviral drugs disrupt virus reproduction and give an example of each.

Summary

1. Chemotherapeutic agents are compounds that destroy pathogenic microorganisms or inhibit their growth and are used in the treatment of disease. Most are antibiotics: microbial products or their derivatives that can kill susceptible microorganisms or inhibit their growth.

2. The modern era of chemotherapy began with Paul Ehrlich's work on drugs against African sleeping sickness and syphilis. Other early pioneers were Gerhard Domagk, Alexander Fleming, Howard Florey, Ernst Chain, Norman Heatley, and Selman Waksman.

3. An effective chemotherapeutic agent must have selective toxicity. A drug with great selective toxicity has a high therapeutic index and usually disrupts a structure or process unique to the pathogen. It has fewer side effects.

4. Antibiotics can be classified in terms of the range of target microorganisms (narrow spectrum versus broad spectrum); their source (natural, semisynthetic, or synthetic); and their general effect (static versus cidal).

5. Antibiotic effectiveness can be estimated through the determination of the minimal inhibitory concentration and the minimal lethal concentration with dilution susceptibility tests. Disk diffusion tests like the Kirby-Bauer test are often used to estimate a pathogen's susceptibility to drugs quickly.

6. Chemotherapeutic agents can damage bacterial pathogens in several ways: inhibition of cell wall synthesis, protein synthesis, or nucleic acid synthesis; disruption of membrane structure; and inhibition of key enzymes.

7. A variety of factors can greatly influence the effectiveness of antimicrobial drugs during actual use.

8. Sulfonamides or sulfa drugs resemble p-aminobenzoic acid and competitively inhibit folic acid synthesis.

9. Quinolones are a family of bactericidal synthetic drugs that inhibit DNA gyrase and thus inhibit such processes as DNA replication.

10. Members of the penicillin family contain a β-lactam ring and disrupt bacterial cell wall synthesis, resulting in cell lysis. Some, like penicillin G, are usually administered by injection and are most effective against gram-positive bacteria. Others can be given orally (penicillin V), are broad spectrum (ampicillin, carbenicillin), or are penicillinase resistant (methicillin).

11. Cephalosporins are similar to penicillins, and are given to patients with penicillin allergies.

12. Tetracyclines are broad-spectrum antibiotics having a four-ring nucleus with attached groups. They bind to the small ribosomal subunit and inhibit protein synthesis.

13. Aminoglycoside antibiotics like streptomycin and gentamicin bind to the small ribosomal subunit, inhibit protein synthesis, and are bactericidal.

14. Erythromycin is a bacteriostatic macrolide antibiotic that binds to the large ribosomal subunit and inhibits protein synthesis.

15. Chloramphenicol is a broad-spectrum, bacteriostatic antibiotic that inhibits protein synthesis. It is quite toxic and used only for very serious infections.

16. Bacteria can become resistant to a drug by excluding it from the cell, pumping the drug out of the cell, enzymatically altering it, modifying the target enzyme or organelle to make it less drug sensitive, and so forth. The genes for drug resistance may be found either on the bacterial chromosome or on a plasmid called an R plasmid.

17. Chemotherapeutic agent misuse fosters the increase and spread of drug resistance, and may lead to superinfections.

18. Because fungi are more similar to human cells than bacteria, antifungal drugs generally have a lower therapeutic index than antibacterial agents and produce more side effects.

19. Superficial mycoses can be treated with miconazole, ketoconazole, clotrimazole, tolnaftate, nystatin, and griseofulvin. Amphotericin B, 5-flucytosine, and fluconazole are used for systemic mycoses.

20. Antiviral drugs interfere with critical stages in the virus life cycle (amantadine and rimantadine) or inhibit the synthesis of virus-specific nucleic acids (zidovudine, adenine arabinoside, acyclovir). Interferon proteins inhibit virus replication and may be therapeutically useful in the future.

Key Terms

acyclovir 672
adenine arabinoside or vidarabine 672
amantadine 672
aminoglycoside antibiotic 668
amphotericin B 671
antibiotic 657
antimetabolites 663
azidothymidine (AZT) 672
broad-spectrum drugs 658
cephalosporin 667
chemotherapeutic agent 657
chloramphenicol 668
cidal 659

dilution susceptibility tests 660
erythromycin 668
griseofulvin 671
interferon 673
Kirby-Bauer method 660
macrolide antibiotic 668
minimal inhibitory concentration (MIC) 660
minimal lethal concentration (MLC) 660
narrow-spectrum drugs 658
nystatin 671
parenteral route 663
penicillinase 666
penicillins 666

plasmids 669
quinolones 665
R plasmids 670
selective toxicity 658
side effects 658
static 659
streptomycin 668
structural analogues 663
sulfonamide 663
superinfection 670
tetracycline 668
therapeutic index 658
zidovudine 672

Questions for Thought and Review

1. What advantage might soil bacteria and fungi gain from the synthesis of antibiotics?

2. Why do penicillins and cephalosporins have a much higher therapeutic index than most other antibiotics? What are antimetabolites?

3. Would there be any advantage to administering a bacteriostatic agent along with penicillin? Any disadvantage?

4. Why do antifungal drugs have a much lower therapeutic index than do most antibacterial drugs? Why does one often have to apply antifungal medications frequently?

5. What advantages would there be from administering two chemotherapeutic agents simultaneously?

6. Why might it be desirable to prepare a variety of semisynthetic antibiotics?

7. Give several ways in which the development of antibiotic-resistant pathogens can be slowed or prevented.

8. Why is it so difficult to find or synthesize effective antiviral drugs?

===== **Additional Reading** =====

General

Abramowicz, M., editor. 1990. The choice of antimicrobial drugs. *Medical Letter on Drugs and Therapeutics* 32(817):41–48.

Aharonowitz, Y., and Cohen, G. 1981. The microbiological production of pharmaceuticals. *Sci. Am.* 245(3):141–52.

Brooks, G. F.; Jawetz, E.; Melnick, J. L.; and Adelberg, E. A. 1991. *Medical microbiology,* 19th ed. Norwalk, Conn.: Appleton & Lange.

Bugg, C. E.; Carson, W. M.; and Montgomery, J A. 1993. Drugs by design. *Sci. Am.* 269(6):92–98.

Eliopoulos, G. M., and Eliopoulos, C. T. 1988. Antibiotic combinations: Should they be tested? *Clin. Microbiol. Rev.* 1(2):139–56.

Franklin, T. J., and Snow, G. A. 1981. *Biochemistry of antimicrobial action,* 3d ed. New York: John Wiley and Sons.

Gootz, T. D. 1990. Discovery and development of new antimicrobial agents. *Clin. Microbiol. Rev.* 3(1):13–31.

Mandell, G. L.; Bennett, J. E.; and Dolin, R. 1995. *Principles and practice of infectious diseases,* 4th ed. New York: Churchill Livingstone.

Murray, P. R., editor-in-chief. 1995. *Manual of clinical microbiology,* 6th ed. Washington, D.C.: ASM Press.

Physicians's desk reference. Oradell, N.J.: Medical Economics Books (published annually).

The Development of Chemotherapy

Bottcher, H. M. 1964. *Wonder drugs: A history of antibiotics.* Philadelphia: J. B. Lippincott.

Hare, R. 1970. *The birth of penicillin.* Atlantic Highlands, NJ: Allen and Unwin.

Determining the Level of Antimicrobial Activity

Courvalin, P. 1992. Interpretive reading of antimicrobial susceptibility tests. *ASM News* 58(7):368–75.

Sanders, C. C. 1991. A problem with antimicrobial susceptibility tests. *ASM News* 57(4):187–90.

Antibacterial Drugs

Abraham, E. P. 1981. The beta-lactam antibiotics. *Sci. Am.* 244(6):76–86.

Hooper, D. C., and Wolfson, J. S. 1991. Fluoroquinolone antimicrobial agents. *N. Engl. J. Med.* 324(6):384–92.

Nayler, J. H. C. 1991. Semi-synthetic approaches to novel penicillins. *Trends Biochem. Sci.* 16:234–37.

Piddock, L. J. V. 1993. Newer fluoroquinolones and gram-positive bacteria. *ASM News* 59(12):603–8.

Sepkowitz, K. A.; Raffalli, J.; Riley, L.; Kiehn, T. E.; and Armstrong, D. 1995. Tuberculosis in the AIDS era. *Clin. Microbiol. Rev.* 8(2):180–99.

Tomasz, A. 1979. The mechanism of the irreversible antimicrobial effects of penicillins: How the beta-lactam antibiotics kill and lyse bacteria. *Ann. Rev. Microbiol.* 33:113–37.

Wolfson, J. S., and Hooper, D. C. 1989. Fluoroquinolone antimicrobial agents. *Clin. Microbiol. Rev.* 2(4):378–424.

Drug Resistance

Clowes, R. C. 1973. The molecule of infectious drug resistance. *Sci. Am.* 228(4):19–27.

Davies, J. 1994. Inactivation of antibiotics and the dissemination of resistance genes. *Science* 264:375–82.

Field, A. K., and Biron, K. K. 1994. "The end of innocence" revisited: Resistance of herpes viruses to antiviral drugs. *Clin. Microbiol. Rev.* 7(1):1–13.

Hardy, K. 1986. *Bacterial plasmids,* 2d ed. Washington, D.C.: American Society for Microbiology.

Jacoby, G. A., and Archer, G. L. 1991. New mechanisms of bacterial resistance to antimicrobial agents. *N. Engl. J. Med.* 324(9):601–9.

Klugman, K. P. 1990. Pneumococcal resistance to antibiotics. *Clin. Microbiol. Rev.* 3(2):171–96.

Lappe, M. 1982. *Germs that won't die: The medical consequences of the misuse of antibiotics.* Garden City, N.Y.: Doubleday.

Laughlin, C. A.; Black, R. J.; Feinberg, J.; Freeman, D. J.; Ramsey, J.; Ussery, M. A.; and Whitley, R. J. 1991. Resistance to antiviral drugs. *ASM News* 57(10):514–17.

Levy, S. B. 1992. *The Antibiotic Paradox.* New York: Plenum.

Lewis, K. 1994. Multidrug resistance pumps in bacteria: Variations on a theme. *Trends Biochem. Sci.* 19:119–23.

Ma, D.; Cook, D. N.; Hearst, J. E.; and Nikaido, H. 1994. Efflux pumps and drug resistance in gram-negative bacteria. *Trends Microbiol.* 2(12):489–93.

Nikaido, H. 1994. Prevention of drug access to bacterial targets: Permeability barriers and active efflux. *Science* 264:382–88.

Spratt, B. G. 1994. Resistance to antibiotics mediated by target alterations. *Science* 264:388–93.

Trends Microbiol. 1994. 2(10). Drug resistance: The new apocalypse. (Thematic Issue).

Antifungal Drugs

Fromtling, R. A. 1988. Overview of medically important antifungal azole derivatives. *Clin. Microbiol. Rev.* 1(2):187–217.

Georgopapadakou, N. H., and Tkacz, J. S. 1995. The fungal cell wall as a drug target. *Trends Microbiol.* 3(3):98–104.

Walsh, T. J. 1988. Recent advances in the treatment of systemic fungal infections: A brief review. *ASM News* 54(5):240–43.

Antiviral Drugs

Bean, B. 1992. Antiviral therapy: Current concepts and practices. *Clin. Microbiol. Rev.* 5(2):146–82.

De Clercq, E. 1995. Antiviral therapy for human immunodeficiency virus infections. *Clin. Microbiol. Rev.* 8(2):200–39

Dolin, R. 1985. Antiviral chemotherapy and chemoprophylaxis. *Science* 227:1296–1303.

Hirsch, M. S., and Kaplan, J. C. 1987. Antiviral therapy. *Sci. Am.* 256(4):76–85.

Parkman, P. D., and Hopps, H. E. 1988. Viral vaccines and antivirals: Current use and future prospects. *Ann. Rev. Public Health* 9:203–21.

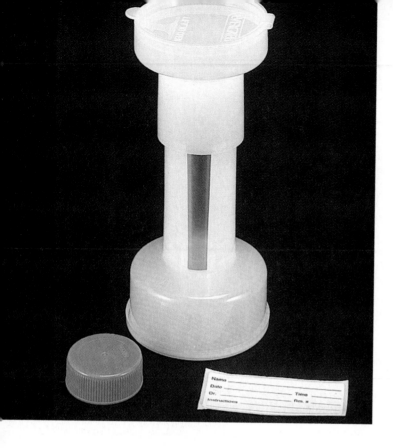

CHAPTER 34
Clinical Microbiology

The specimen is the beginning. All diagnostic information from the laboratory depends upon the knowledge by which specimens are chosen and the care with which they are collected and transported.

—Cynthia A. Needham

I'm not afraid when we have to bring a risky, hazardous virus into the lab. I just hope and pray, and I don't think about it.

—Akinyele Fabiyi

Outline

Concepts

1. Clinical microbiologists and clinical microbiology laboratories perform many services, all related to the identification and control of microorganisms.

2. Success in clinical microbiology depends on (1) using the proper aseptic technique; (2) correctly obtaining the clinical specimen from the infected patient by swabs, needle aspiration, intubation, or catheters; (3) correctly handling the specimen; and (4) quickly transporting the specimen to the laboratory.

3. Once the clinical specimen reaches the laboratory, it is cultured and identified. Identification measures include microscopy; growth on enrichment, selective, differential, or characteristic media; specific biochemical tests; rapid test methods; immunologic techniques; bacteriophage typing; and molecular methods such as nucleic acid-based detection methods, gas-liquid chromatography, and plasmid fingerprinting.

4. After the microorganism has been isolated, cultured, and/or identified, samples are used in susceptibility tests to find which method of control will be most effective. The results are provided to the physician as quickly as possible.

5. Computer systems in clinical microbiology are designed to speed identification of the pathogen and communication of results back to the physician.

P athogens, particularly bacteria, coexist with harmless microorganisms on or in the host. These pathogens must be properly identified as the actual cause of infectious diseases. This is the purpose of clinical microbiology. The clinical microbiologist identifies agents and organisms (hereafter referred to as microorganisms) based on morphological, biochemical, immunologic, and molecular procedures. Time is a significant factor in the identification process, especially in life-threatening situations. Computers and commercially available methods of rapid identification have greatly aided the clinical microbiologist. Molecular methods allow identification of microorganisms based on highly specific biochemical properties. Once isolated and identified, the microorganism can then be subjected to antimicrobial sensitivity tests. In the final analysis the patient's well-being and health benefit most from clinical microbiology—the subject of this chapter.

The major concern of the **clinical microbiologist** is to isolate and identify microorganisms from clinical specimens rapidly. The purpose of the clinical microbiology laboratory is to provide the physician with information concerning the presence or absence of microorganisms that may be involved in the infectious disease process (figure 34.1). These individuals and facilities also determine the susceptibility of microorganisms to antimicrobial agents. Clinical microbiology makes use of information obtained from research on such diverse topics as microbial biochemistry and physiology, immunology, molecular biology, and the host-parasite relationships involved in the infectious disease process. Computers and rapid tests have assumed an ever-increasing role in clinical microbiology.

Specimens

In clinical microbiology a clinical specimen (hereafter, specimen) represents a portion or quantity of human material that is tested, examined, or studied to determine the presence or absence of particular microorganisms. Several guidelines on the safety precautions to be used in selection, collection, and handling of the specimen need emphasis (Box 34.1):

1. The specimen selected should adequately represent the infectious disease.
2. A quantity of specimen adequate for complete examination should be obtained.
3. Attention must be given to specimen collection in order to avoid contamination from the many varieties of microorganisms indigenous to the skin and mucous membranes (*see figure 28.8*).
4. The specimen should be forwarded promptly to the clinical laboratory.
5. If possible, the specimen should be obtained before antimicrobial agents have been administered to the patient.

Collection

Specimens may be collected by several methods using aseptic technique. Aseptic technique refers to specific procedures used to prevent unwanted microorganisms from contaminating the clinical specimen. Each method is designed to ensure that only the proper material will be sent to the clinical laboratory.

The most common method used to collect specimens (e.g., pus) from the skin and mucous membranes (eye, ear, nose, throat, open wounds) is the sterile **swab.** A sterile swab is a rayon- or dacron-tipped polystyrene applicator. Manufacturers of swabs have their own unique container design and instructions for proper use. For example, many commercially manufactured swabs contain a transport medium designed to preserve a variety of microorganisms and to prevent multiplication of rapidly growing members of the population (figure 34.2a).

Specimens from blood and cerebrospinal fluid may be collected by **needle aspiration.** After the proper aseptic technique has been followed, a sample of fluid is drawn into a sterile tube that has been treated with an anticoagulant such as heparin or potassium oxalate (figure 34.2b). The anticoagulant prevents the microorganisms from being entrapped in a clot, which would make isolation difficult.

Intubation [Latin *in,* into and *tuba,* tube] is the inserting of a tube into a body canal or hollow organ. For example, intubation can be used to collect specimens from the stomach. In this procedure a long sterile tube is attached to a syringe (figure 34.2c), and the tube is either swallowed by the patient or passed through a nostril into the patient's stomach. Specimens are then withdrawn periodically into the sterile syringe. The most common intubation tube is the Levin tube.

A **catheter** is a tubular instrument used for withdrawing or introducing fluids from or into a body cavity. For example, urine specimens may be collected with catheters. Three types are commonly used. The hard catheter is used when the urethra is very narrow or has strictures. The French catheter is a soft tube used to obtain a single specimen sample. If multiple samples are required over a prolonged period, a Foley catheter is used (figure 34.2d).

Urine also may be obtained by the clean-catch method. After the patient has cleansed the urethral meatus (opening), a small container is used to collect the urine. In the clean-catch midstream method, the first urine voided is not collected because it will be contaminated with those microorganisms normally occurring in the lower portion of the urethra. Only the midstream portion is collected since it most likely will contain those microorganisms found in the urinary bladder.

Most lower respiratory specimens are obtained from sputum. Specifically, **sputum** is the mucous secretion expectorated from the lungs, bronchi, and trachea through the mouth, in contrast to saliva, which is the secretion of the salivary glands. Sputum is collected in specially designed sputum cups (figures 34.1a and 34.2e).

(a) The identification of the microorganism begins at the patient's bedside. The nurse is giving instructions to the patient on how to obtain a sputum specimen.

(b) The specimen is sent to the laboratory to be processed. Notice that the specimen and worksheet are in different Ziplock bags.

(c) Specimens such as sputum are plated on various types of media under a laminar airflow hood. This is to prevent specimen aerosols from coming in contact with the microbiologist.

(d) Sputum and other specimens are usually Gram stained to determine whether or not bacteria are present and to obtain preliminary results on the nature of any bacteria found.

(e) After incubation, the plates are examined for significant isolates. The Gram stain may be reexamined for correlation.

(f) Suspect colonies are picked for biochemical, immunologic, or molecular testing.

(g) Colonies are prepared for identification by rapid test systems.

(h) In a short period of time, sometimes 4 hours, computer-generated information is obtained that will consist of biochemical identification and antibiotic susceptibility results.

(i) All information about the specimen is now entered into a computer and the data are transmitted directly to the hospital ward.

Figure 34.1   Isolation and Identification of Microorganisms in a Clinical Laboratory.

Box 34.1

Universal Precautions for Health-Care Professionals

Since medical history and examination cannot reliably identify all patients infected with HIV or other blood-borne pathogens, blood and body-fluid precautions should be consistently used for *all* patients. Workers in microbiological research laboratories also are exposed to health hazards and need to employ universal precautions (*see Box 7.2, p. 145*).

1. All health-care workers should routinely use appropriate barrier precautions to prevent skin and mucous-membrane exposure when contact with blood or other body fluids of any patient is anticipated. Gloves should be worn for touching blood and body fluids, mucous membranes, or non-intact skin of all patients, for handling items or surfaces soiled with blood or body fluids, and for performing venipuncture and other vascular access procedures. Gloves should be changed after contact with each patient. Masks and protective eyewear or face shields should be worn during procedures that are likely to generate droplets of blood or other body fluids to prevent exposure of mucous membranes of the mouth, nose, and eyes. Gowns or aprons should be worn during procedures that are likely to generate splashes of blood or other body fluids.

2. Hands and other skin surfaces should be washed immediately and thoroughly if contaminated with blood or other body fluids. Hands should be washed immediately after gloves are removed.

3. All health-care workers should take precautions to prevent injuries caused by needles, scalpels, and other sharp instruments or devices during procedures; when cleaning used instruments; during disposal of used needles; and when handling sharp instruments after procedures. To prevent needlestick injuries, needles should not be recapped, purposely bent or broken by hand, removed from disposable syringes, or otherwise manipulated by hand. After they are used, disposable syringes and needles, scalpel blades, and other sharp items should be placed in puncture-resistant containers for disposal.

4. Although saliva has not been implicated in HIV transmission, to minimize the need for emergency mouth-to-mouth resuscitation, mouthpieces, resuscitation bags, or other ventilation devices should be available for use in areas in which the need for resuscitation is predictable.

5. Health-care workers who have exudative lesions or weeping dermatitis should refrain from all direct patient care and from handling patient-care equipment.

6. The following procedure should be used to clean up spills of blood or blood-containing fluids. (1) Put on gloves and any other necessary barriers. (2) Wipe up excess material with disposable towels and place the towels in a container for sterilization. (3) Disinfect the area with either a commercial EPA-approved germicide or household bleach (sodium hypochlorite). The latter should be diluted from 1/100 (smooth surfaces) to 1/10 (porous or dirty surfaces); the dilution should be no more than 24 hours old. When dealing with large spills or those containing sharp objects such as broken glass, first cover the spill with disposable toweling. Then saturate the toweling with commercial germicide or a 1/10 bleach solution and allow it to stand for at least 10 minutes. Finally clean as described above.

Source: Adapted from *Morbidity and Mortality Weekly Report,* 36 (Suppl. 2S) 5S–10S. Centers for Disease Control and Prevention Guidelines, Atlanta, GA.

Handling

Immediately after collection the specimen must be properly labeled and handled. The person collecting the specimen is responsible for ensuring that the name, registration number, and location of the patient are correctly and legibly written or imprinted on the culture request form (figure 34.3*a*). This information must correspond to that written or imprinted on a label affixed to the specimen container. The type or source of the sample and the choice of tests to be performed also must be specified on the request form (figure 34.3*b*).

Transport

Speed in transporting the specimen to the clinical laboratory after it has been obtained from the patient is of prime importance. Some laboratories refuse to accept specimens if they have been in transit too long.

Microbiological specimens may be transported to the laboratory by various means (figure 34.1*b*). For example, certain specimens should be transported in a medium that preserves the microorganisms and helps maintain the ratio of one organism to another. This is especially important for specimens in which normal microorganisms may be mixed with microorganisms foreign to the body location.

Special treatment is required for specimens when the microorganism is thought to be anaerobic. The material is aspirated with a needle and syringe. Most of the time it is practical to remove the needle, cap the syringe with its original seal, and bring the specimen directly to the clinical laboratory. Transport of these specimens should take no more than 10 minutes; otherwise, the specimen must be injected immediately into an anaerobic transport vial (figure 34.4). Vials should contain a transport medium with an indicator, such as resazurin, to show that the interior of the vial is anaerobic at the time the specimen is introduced. Swabs for anaerobic

Figure 34.2 Collection of Clinical Specimens. (*a*) A drawing of a sterile swab with a specific transport medium. (*b*) A sterile collection tube for cerebrospinal fluid. (*c*) Nasotracheal intubation. (*d*) A drawing of a Foley catheter. Notice that three separate lumens are incorporated within the round shaft of the catheter for drainage of urine, inflation, and introducing irrigating solutions into the urinary bladder. After the Foley catheter has been introduced into the urinary bladder, the tip is inflated to prevent it from being expelled. (*e*) This specially designed sputum cup allows the patient to expectorate a clinical specimen directly into the cup. In the laboratory, the cup can be opened from the bottom to reduce the chance of contamination from extraneous pathogens.

(a)

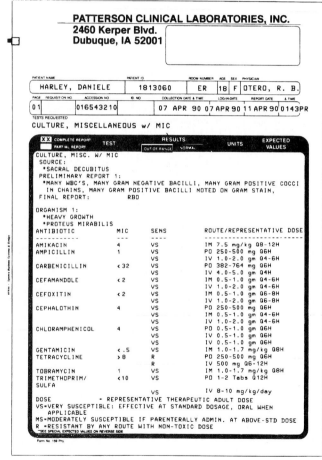

(b)

Figure 34.3   Culture Request Forms. (*a*) A typical handwritten culture request form that accompanies a clinical specimen to the laboratory. (*b*) A typical computerized culture request form from a clinical laboratory showing various information such as tests ordered (culture [white blood cell count], MIC [minimal inhibitory concentration]), source of clinical specimen (sacral decubitus ulcer), preliminary microorganisms, final microorganisms, and antibiotic choice with route and dose.

Figure 34.4   Some Anaerobic Transport Systems. A vial and syringe. These systems contain a nonnutritive transport medium that retards diffusion of oxygen after specimen addition and helps maintain microorganism viability up to 72 hours. A built-in color indicator system is clear and turns lavender in the presence of oxygen.

culture usually are less satisfactory than aspirates or tissues, even if they are transported in an anaerobic vial.

Many clinical laboratories insist that stool specimens (the fecal discharge from the bowels) for culture be transported in various buffered preservatives. Preparation of these transport media is described in various manuals (see Additional Reading).

Transport of urine specimens to the clinical laboratory must be done as soon as possible. No more than 1 hour should elapse between the time the specimen is obtained and the time it is examined. If this time schedule cannot be followed, the urine sample must be refrigerated immediately.

1. What is the function of the clinical microbiologist? The clinical microbiology laboratory?

2. What general guidelines should be followed in collecting and handling clinical specimens?

3. Define the following terms: specimen, swab, catheter, and sputum.

4. What are some transport problems associated with stool specimens? Anaerobic cultures? Urine specimens?

5. What must be done to a clinical specimen before it goes to the clinical laboratory?

Identification of Microorganisms from Specimens

The clinical microbiology laboratory can provide preliminary or definitive identification of microorganisms based on (1) microscopic examination of specimens, (2) study of the growth and biochemical characteristics of isolated microorganisms, (3) immunologic tests that detect antibodies or microbial antigens, (4) bacteriophage typing, and (5) molecular methods.

Microscopy

Wet-mount, heat-fixed, or chemical-fixed specimens can be examined with an ordinary bright-field microscope. These preparations can be enhanced with either phase-contrast or dark-field microscopy. The latter is the procedure of choice for the detection of spirochetes in skin lesions associated with early syphilis or in the cerebrospinal fluid and urine of people with early leptospirosis. The fluorescence microscope can be used to identify certain microorganisms (*Mycobacterium tuberculosis*) after they are stained with fluorochromes (*see* chapter 2). (Some morphological features used in classification and identification of microorganisms are presented in chapter 19 and in table 19.3.)

The light microscope (pp. 18–25).

Many stains that can be used to examine specimens for specific microorganisms have been described. Two of the more widely used are the Gram stain and acid-fast stain. Because these stains are based on the chemical composition of cell walls, they are not useful in identifying bacteria without walls. Refer to standard references, such as the *Manual of Clinical Microbiology* published by the American Society for Microbiology, for details about other reagents and staining procedures.

Growth or Biochemical Characteristics

Typically microorganisms have been identified by their particular growth or biochemical characteristics. These characteristics vary depending on whether the clinical microbiologist is dealing with viruses, rickettsias, chlamydiae, mycoplasmas, gram-positive or gram-negative bacteria, fungi, or parasites.

Viruses

Viruses are identified by isolation in living cells (table 34.1), by serological tests, or by nucleic acid technology (techniques such as restriction endonuclease analysis, nucleic acid hybridization,

TABLE 34.1   Some Different Cell Types Used to Isolate and Identify Human Viruses

Virus Family	Virus Type	Suckling Mouse	Embryonated Egg	Primary Monkey Kidney	Primary Human Embryonic Kidney	Semi-continuous Human Diploid	Continuous Human Heteroploid
Picornaviridae	Rhinovirus	0[a]	0	+	±	++	0 to ++
	Polioviruses 1–3	0	0	++	++	++	++
	Coxsackievirus A1–24	++	0	0 to ++[b]	0 to ++[b]	0 to ++[b]	0 to ++[b]
	Coxsackievirus B1–6	++	0	++	++	±	++
	Echovirus 1–32	0 to ++	0	++	+	++	0
Orthomyxoviridae	Influenza A, B, C	0	++	0 to ++	±	±	0
Paramyxoviridae	Parainfluenza 1, 2, 3, 4	0	0	++	+	+	±
	Respiratory syncytial	0	0	+	+	++	++
	Measles	0	0	+	++	0	±
	Mumps	0	++	++	++	+	0
Adenoviridae	Adenovirus types 1–34	0	0	± to +	++	+	+
Herpesviridae	Herpes simplex	0	++	0 to ++	++	++	0 to ++
	Varicella-zoster	0	0	0	+	++	0
	Cytomegalovirus	0	0	0	0	++	0
Togaviridae	Rubella virus	0	0	+	+	0	0
	Eastern equine encephalitis	++	0	±		0	±
	Western equine encephalitis	++	0	+		0	+
	St. Louis encephalitis	++	0	0		0	+
Bunyaviridae	California encephalitis virus	++	0	0		0	+
Reoviridae	Rotavirus	0	0	0	0	0	0
	Colorado tick fever	+	0	0	0	0	0
Poxviridae	Vaccinia	+	++	++	++	++	++
Arenaviridae	Lymphocytic choriomeningitis	++	++	++	++	++	++

From G. L. Mandell, et al., *Principles and Practice of Infectious Diseases*, 3d edition. Copyright © 1990 Churchill Livingstone, Inc., Medical Publishers, New York, NY. Reprinted by permission.

[a]0: not suitable for isolation; ±: some strains may be recovered; +: many strains may be recovered; ++: most strains will be recovered.

[b]Coxsackievirus A–7, A–9, and A–16 strains grow well in cell culture.

and the polymerase chain reaction [*see chapter 15*]). Several types of living cells are available: cell cultures, embryonated hen's eggs, and experimental animals.

Cell cultures are divided into three general classes:

1. Primary cultures consist of cells derived directly from tissues like the kidney and lung.

2. Semicontinuous cell cultures are obtained from subcultures of a primary culture and usually consist of diploid fibroblasts that undergo a finite number of divisions.

3. Continuous cell cultures are derived from transformed cells that are generally epithelial in origin. These cultures grow rapidly, are heteroploid (having a chromosome number that is not a simple multiple of the haploid number), and can be subcultured indefinitely.

Each type of cell culture favors the growth of a different array of viruses, just as bacterial culture media have differing selective and restrictive properties for growth of bacteria (table 34.2).

Viral replication in cell cultures is detected in three ways: (1) by observing cytopathic effects, (2) by hemadsorption, and (3) by using interference.

A **cytopathic effect** is an observable change that occurs in cells because of viral replication (*see chapter 16*). Examples include ballooning, binding together, clustering, or even death of the culture cells (*see figure 16.3*). During the incubation period of a cell culture, red blood cells can be added. Several viruses alter the plasma membrane of infected culture cells so that red blood cells adhere firmly to them. This phenomenon is called **hemadsorption** (*see figure 32.10*). Some viruses produce no cytopathic effects and no hemadsorption.

TABLE 34.2   Isolation of Pure Bacterial Cultures from Specimens

Selective Media

A selective medium is prepared by the addition of specific substances to a culture medium that will permit growth of one group of bacteria while inhibiting growth of some other groups. The following are examples:

Salmonella-Shigella agar (SS) is used to isolate *Salmonella* and *Shigella* species. Its bile salt mixture inhibits many groups of coliforms. Both *Salmonella* and *Shigella* species produce colorless colonies because they are unable to ferment lactose. Lactose-fermenting bacteria will produce pink colonies.

Mannitol salt agar (MS) is used for the isolation of staphylococci. The selectivity is obtained by the high (7.5%) salt concentration that inhibits growth of many groups of bacteria. The mannitol in this medium helps in differentiating the pathogenic from the nonpathogenic staphylococci, in that the former ferment mannitol to form acid while the latter do not.

Bismuth sulfite agar (BS) is used for the isolation of *Salmonella typhi*, especially from stool and food specimens. *S. typhi* reduces the sulfite to sulfide, resulting in black colonies with a metallic sheen.

Differential Media

The incorporation of certain chemicals into a medium may result in diagnostically useful growth or visible change in the medium after incubation. The following are examples:

Eosin methylene blue agar (EMB) differentiates between lactose fermenters and nonlactose fermenters. EMB contains lactose, salts, and two dyes—eosin and methylene blue. *E. coli*, which is a lactose fermenter, will produce a dark colony or one that has a metallic sheen. *S. typhi*, a nonlactose fermenter, will appear colorless.

MacConkey agar is used for the selection and recovery of *Enterobacteriaceae* and related gram-negative rods. The bile salts and crystal violet in this medium inhibit the growth of gram-positive bacteria and some fastidious gram-negative bacteria. Because lactose is the sole carbohydrate, lactose-fermenting bacteria produce colonies that are various shades of red, whereas nonlactose fermenters produce colorless colonies.

Hektoen enteric agar is used to increase the yield of *Salmonella* and *Shigella* species relative to other microbiota. The high bile salt concentration inhibits the growth of gram-positive bacteria and retards the growth of many coliform strains.

Enrichment Media

The addition of blood, serum, or extracts to tryptic soy agar or broth will support the growth of many fastidious bacteria. These media are used primarily to isolate bacteria from cerebrospinal fluid, pleural fluid, sputum, and wound abscesses. The following are examples:

Blood agar (can also be a differential medium): addition of citrated blood to tryptic soy agar makes possible variable hemolysis, which permits differentiation of some species of bacteria. Three hemolytic patterns can be observed on blood agar.

1. α-hemolysis—greenish to brownish halo around the colony (e.g., *Streptococcus gordonii*, *Streptococcus pneumoniae*).

2. β-hemolysis—complete lysis of blood cells resulting in a clearing effect around growth of the colony (e.g., *Staphylococcus aureus* and *Streptococcus pyogenes*).

3. γ-hemolysis—no change in medium (e.g., *Staphylococcus epidermidis* and *Staphylococcus saprophyticus*).

Chocolate agar is made from heated blood, which provides necessary growth factors to support bacteria such as *Haemophilus influenzae* and *Neisseria gonorrhoeae*.

Characteristic Media

Characteristic media are used to test bacteria for particular metabolic activities, products, or requirements. The following are examples:

Urea broth is used to detect the enzyme urease. Some enteric bacteria are able to break down urea, in the presence of urease, into ammonia and CO_2.

Triple sugar iron (TSI) agar contains lactose, sucrose, and glucose plus ferrous ammonium sulfate and sodium thiosulfate. TSI is used for the identification of enteric organisms by their ability to attack glucose, lactose, or sucrose and to liberate sulfides from ammonium sulfate or sodium thiosulfate.

Citrate agar contains sodium citrate, which serves as the sole source of carbon, and ammonium phosphate, the sole source of nitrogen. Citrate agar is used to differentiate enteric bacteria on the basis of citrate utilization.

Lysine iron agar (LIA) is used to differentiate bacteria that can either deaminate or decarboxylate the amino acid lysine. LIA contains lysine, which permits enzyme detection, and ferric ammonium citrate for the detection of H_2S production.

Sulfide, indole, motility (SIM) medium can perform three different tests. One can observe the production of sulfides, formation of indole (a metabolic product from tryptophan utilization), and motility. This medium is generally used for the differentiation of enteric organisms.

(a)

(b)

Figure 34.5   Viral Identification Test Using Immunofluorescence in Tissue Culture. (*a*) Two infected nuclei in a cytomegalovirus (CMV) positive tissue culture. (*b*) Several infected cells in a herpes simplex virus positive tissue culture.

Cells infected with these viruses, although they appear normal, are resistant to infection with certain other viruses, a phenomenon called **interference.**

Embryonated hen's eggs can be used for virus isolation. There are three main routes of egg inoculation for virus isolation: (1) the allantoic cavity, (2) the amniotic cavity, and (3) the chorioallantoic membrane (*see figure 16.1*). Virus replication is recognized by the development of pocks on the chorioallantoic membrane, by the development of hemagglutinins (*see figure 32.10*) in the allantoic and amniotic fluid, and by death of the embryo.

Laboratory animals, especially suckling mice, can be used for virus isolation. Inoculated animals are observed for specific signs of disease or death.

Several new serological tests for viral identification make use of monoclonal antibody-based immunofluorescence (*see figure 32.16*). These tests (figure 34.5) detect viruses such as the cytomegalovirus and herpes simplex virus in tissue-vial cultures.

---

1. Name two specimens for which microscopy would be used in the initial diagnosis of an infectious disease.
2. Name three general classes of cell cultures.
3. Give the three ways by which the presence of viral replication is detected in cell culture.
4. What are the three main routes of egg inoculation for virus isolation?

Fungi

In the clinical laboratory, yeasts can be quickly identified to species using rapid methods such as the API 20C yeast identification system (Analytab Products, Plainview, N.Y.) or the Microring YT system (Medical Wire & Equipment Co., Victory Gardens, N.J.), or BIOLOG (figure 34.10).

The identification of molds often can be made if a portion of the specimen is mixed with a drop of 10% potassium hydroxide on a glass slide. A coverslip is applied, the slide is flamed gently, and the material is examined microscopically. Culture procedures also can be used, but these require many days for growth and collection of results. Antibody testing on blood plasma is also routinely done, as is the direct examination of specimens by immunofluorescence when specific antisera are available.

Parasites

Wet mounts of stool specimens or urine can be examined microscopically for the presence of eggs, cysts, larvae, or vegetative cells of parasites. Blood smears for malaria and trypanosome parasites are stained with Giemsa stain. Some serological tests are also available.

Rickettsias

Although rickettsias, chlamydiae, and mycoplasmas are bacteria, they differ from other bacterial pathogens in a variety of ways. Therefore the identification of these three groups is discussed separately. Rickettsias can be diagnosed serologically or by isolation of the microorganism. Because isolation is both hazardous to the clinical microbiologist and expensive, serological methods are preferred. Isolation of rickettsias is generally confined to reference and specialized research laboratories.

Chlamydiae

Chlamydiae can be demonstrated in tissues and cell scrapings with Giemsa staining, which detects the characteristic intracellular inclusion bodies (*see figure 20.20*). Immunofluorescent staining of tissues and shed cells with monoclonal antibody reagents is a more sensitive and specific means of diagnosis (Box 34.2; *see also figure 30.19*). The most sensitive method for demonstrating chlamydiae in clinical specimens is growth in cell culture (McCoy cells).

Immunofluorescence (pp. 649–50).

Mycoplasmas

The most routinely used techniques for identification of the mycoplasmas are serological (hemagglutinin) or complement-fixing antigen-antibody reactions. These microorganisms are slow growing; therefore positive results from isolation procedures are rarely available before 30 days—a long delay with an approach that offers little advantage over standard serological techniques. Recently nucleic acid hybridization has been applied to the detection of *Mycoplasma pneumoniae* in clinical specimens.

Agglutination (pp. 641–43).

Box 34.2

Monoclonal Antibodies in Clinical Microbiology

An important application of monoclonal antibody technology (*see hybridomas, chapter 30*) is the identification of microorganisms. Monoclonal antibodies have been prepared for a wide range of viruses, bacteria, fungi, and parasites in many research laboratories. If specific monoclonal antibodies are selected, immunologic assays can be created for different types of analyses. For example, monoclonal antibodies of cross-species or cross-genus reactivity have applications in the taxonomy of microorganisms. Those monoclonal antibodies that define species-specific antigens are extremely valuable in diagnostic reagents. Monoclonal antibodies that exhibit more restrictive specificity can be used to identify strains or biotypes within a species, to aid in studies of antigenic drift, and in epidemiological studies involving the matching of microbial strains. In addition, individual antigenic determinants on protein molecules can be mapped.

In the clinical microbiology laboratory, monoclonal antibodies to viral or bacterial antigens are replacing polyclonal antibodies for use in culture confirmation when accurate, rapid identification is required. With the use of sensitive techniques such as fluorescent antibody assays (*see chapter 32*), it is possible to perform culture identifications with improved accuracy, speed, and fewer microorganisms. The formulation of direct assays with monoclonal antibody reagents, which contain no contaminating antibodies and produce a minimum of artifacts, is now reality. The highly defined and reproducible properties of monoclonal antibodies invite their incorporation into immunoassays being developed for the next generation of instruments that will detect microbial antigens and serum antibodies for the clinical microbiologist.

✻ Serological Tests ✻

TABLE 34.3 Some Biochemical Tests Used to Identify an Unknown Bacterium Taken from the Patient's Specimen

Biochemical Test	Description
Casein hydrolysis	Detects the presence of caseinase, an enzyme able to hydrolyze milk protein casein.
Catalase	Detects the presence of catalase, which converts hydrogen peroxide to water and O_2.
Coagulase	Detects the presence of coagulase. Coagulase causes plasma to clot. This is an important test for *Staphylococcus aureus*.
Lipid hydrolysis	Detects the presence of lipase. Lipase breaks down lipids into simple fatty acids and glycerol.
Gelatin liquefaction	Detects whether or not a bacterium can produce an enzyme that hydrolyzes the gelatin protein.
Hydrogen sulfide production	Detects the formation of hydrogen sulfide from the amino acid cysteine by the presence of cysteine desulfurase; important in *Salmonella* identification.
IMViC (indole; methyl red; Voges-Proskauer; citrate)	The indole test detects the production of indole from the amino acid tryptophan. Methyl red is a pH indicator to determine whether the bacterium has produced acid. Vi (Voges-Proskauer) detects the production of acetoin. The citrate test determines whether or not the bacterium can use sodium citrate as a sole source of carbon.
Litmus milk	Determines five bacterial characteristics: (1) a pink color indicates an acid reaction and the fermentation of lactose; (2) a blue color arises from an alkaline reaction and no proteolysis (hydrolysis of protein); (3) a white color indicates litmus has acted as an electron acceptor; (4) curd formation indicates acid produced from lactose or rennin production; and (5) peptonization shows curd (coagulated protein) digestion by casein hydrolysis.
Nitrate reduction	Detects whether a bacterium can use nitrate as an electron acceptor.
Oxidase	Detects the presence of cytochrome *c* oxidase that is able to reduce O_2; especially important in detecting *Neisseria* spp. and pseudomonads.
Starch hydrolysis	Detects the presence of the enzyme amylase, which is able to hydrolyze starch.

Bacteria

The presence of bacterial growth usually can be recognized by the development of colonies on solid media or turbidity in liquid media. The time for visible growth to occur is an important variable in the clinical laboratory. For example, most pathogenic bacteria require only a few hours to produce visible growth, whereas it may take weeks for colonies of mycobacteria to become evident. The clinical microbiologist as well as the clinician should be aware of reasonable reporting times for various cultures.

The initial identity of a bacterial organism may be suggested by (1) the source of the culture specimen; (2) its microscopic appearance; (3) its pattern of growth on selective, differential, enrichment, or characteristic media (table 34.2; *see also figure 5.10*); and (4) its hemolytic, metabolic, and fermentative properties on the various media (table 34.2; *see also table 19.3*). After the microscopic and growth characteristics of a bacterium are examined, specific biochemical tests can be performed. These tests represent the most common method for identifying bacteria (table 34.3 and figure 34.6).

Microbial nutrition and types of media (chapter 5).

(a) **Methyl Red Test.** This is a qualitative test of the acidity produced by bacteria grown in MR-VP broth. After the addition of several drops of methyl red solution, a bright red color (left tube) is a positive test; a yellow or orange color (right tube) is a negative test.

(b) **Voges-Proskauer Reaction.** VP-positive bacteria produce acetylmethylcarbinol or acetoin, which reacts with the reagents to produce a red color (left tube); a VP-negative tube is shown on the right.

(c) **Starch Hydrolysis.** After incubation on starch agar, plates are flooded with iodine solution. A positive test is indicated by the colorless area around the growth (left); a negative test is shown on the right.

(d) **Tube Catalase Test.** After incubation of slant cultures, 1 ml of 3% hydrogen peroxide is trickled down the slants. Catalase converts hydrogen peroxide to water and oxygen bubbles (left tube); a negative catalase test is shown in the right tube.

(e) **Slide Catalase Test.** A wooden applicator stick (or nichrome wire loop) is used to pick up a colony from a culture plate and place it in a drop of hydrogen peroxide on a glass slide. A positive catalase reaction (left slide) shows gas bubbles; a negative catalase reaction reveals an absence of gas bubbles (right slide).

(f) **Nitrate Reduction.** After 24–48 hours of incubation, nitrate reagents are added to culture tubes. The tube on the left illustrates gas formation (a positive reaction for nitrate reduction); the tube in the middle is a positive reaction for nitrate reduction to nitrite as indicated by the red color; the tube on the right is a negative broth control.

Figure 34.6 Some Common Diagnostic Tests Used in Microbiology.

(g) **Test for Amino Acid Decarboxylase.** The tube on the left is an uninoculated control; the second tube from the left is lysine decarboxylase negative; the third tube is lysine decarboxylase positive; and the tube on the right is lysine deaminase positive.

(h) **Test for Indole.** Tryptophan can be broken down to indole by some bacteria. The presence of indole is detected by adding Kovacs' reagent. A red color on the surface is a positive test for indole (left tube) and an orange-yellow color is a negative test for indole (right tube).

(i) **Gelatin Hydrolysis.** If gelatin is hydrolyzed by the enzyme gelatinase, it does not gel when cooled but remains a liquid. Thus it flows when the culture is tilted backward (right tube). A negative control is on the left. Note that the solid gelatin does not flow when the tube is tilted.

Figure 34.6   *continued*

(j) **Phenylalanine Deamination Test.** When 10% ferric chloride is added to a phenylalanine deaminase agar slant culture, a dark green color (tube on the right) is a positive test for the enzyme. The tube on the left is an uninoculated control, and the tube in the middle is a negative test.

(k) **Bile Solubility for Pneumococcus.** The tube on the left and the one in the middle contain cultures of pneumococcus (*Streptococcus pneumoniae*). The tube on the right contains α-streptococcus. One-tenth ml of deoxycholate was added to the middle and right tubes, and 0.1 ml of distilled water was added to the left tube. The pneumococci in the center are bile soluble, as indicated by the clear suspension. The bacteria in the right tube are not bile soluble, as indicated by the turbidity.

(l) **Stormy Fermentation of Litmus Milk.** The tube on the left shows fermentation; the tube on the right is negative for stormy fermentation.

(m) **Fermentation Reactions.** The tube on the left shows acid (yellow color) and gas in the Durham tube. The center tube shows no carbohydrate utilization to produce acid or gas. The tube on the right has less acid formation than that on the left.

(n) **Oxidase Test.** Filter paper is moistened with a few drops of 1% tetramethyl-p-phenylenediamine dihydrochloride. With a wooden applicator, growth from an agar medium is smeared on the paper. A positive test is the development of a purple color within 10 seconds.

(o) **Bacitracin Sensitivity Test.** The bacteria on the left are presumptively identified as group A streptococci because of inhibition by the antibiotic bacitracin. The bacteria on the right are bacitracin resistant.

Figure 34.6 *continued*

(p) **Identification of *Neisseria gonorrhoeae*.** Oxidase-positive test on filter paper is shown on the left. Identicult reaction for *N. gonorrhoeae* (center). The plate on the right shows characteristic growth on modified Thayer-Martin medium.

(q) **Optochin Sensitivity.** An optochin disk on blood agar specifically inhibits pneumococci. The optochin test for *Streptococcus pneumoniae* is based on the zone of inhibition.

(r) **Urease Production.** Christensen urea agar slants. Urease-positive bacteria hydrolyze urea to ammonia, which turns the phenol red indicator red-violet. From left to right: uninoculated control, delayed positive (>24 hrs), rapidly positive (<4 hrs), negative reaction.

(s) **Triple Sugar Iron Agar Reactions.** Left to right: left tube is an uninoculated control, second tube is K/K (nonfermenter), third tube is A/A with gas indicating lactose or sucrose fermentation, and the right tube is K/A plus H$_2$S production. A stands for acid production, and K indicates that the medium becomes alkaline.

(t) **Lowenstein-Jensen Medium.** Growth of *Mycobacterium tuberculosis* showing nodular and nonpigmented growth.

Figure 34.6   *continued*

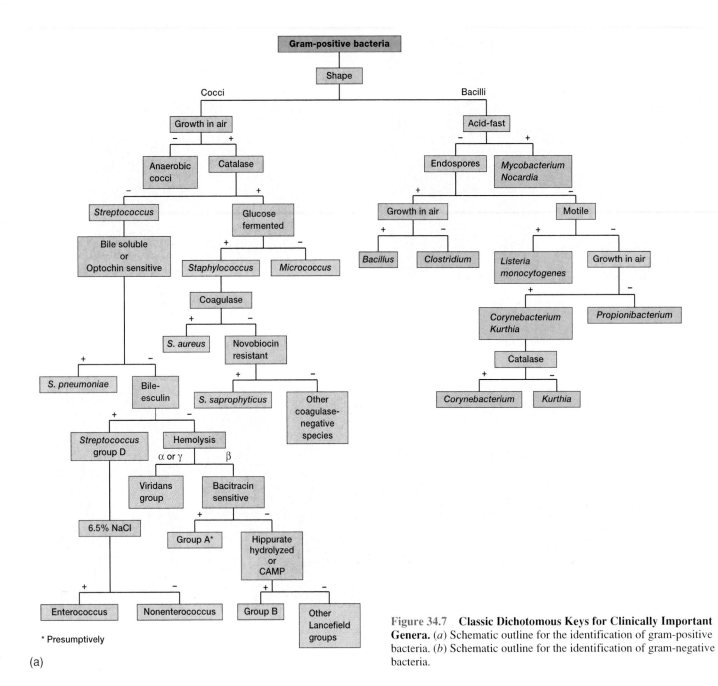

Figure 34.7 **Classic Dichotomous Keys for Clinically Important Genera.** (*a*) Schematic outline for the identification of gram-positive bacteria. (*b*) Schematic outline for the identification of gram-negative bacteria.

* Presumptively

(a)

Classic dichotomous keys are coupled with the biochemical tests for the identification of bacteria from specimens. Generally, fewer than 20 tests are required to identify clinical bacterial isolates to the species level (figure 34.7).

1. How can fungi and parasites be detected in a clinical specimen? Rickettsias? Chlamydiae? Mycoplasmas?

2. Why must the clinical microbiologist know what are reasonable reporting times for various microbial specimens?

3. How can a clinical microbiologist determine the initial identity of a bacterium?

4. Describe a dichotomous key that would be used to identify a bacterium.

Rapid Methods of Identification

Clinical microbiology has benefited greatly from technological advances in equipment, computer programs and data bases, molecular biology, and immunochemistry. With respect to the detection of microorganisms in specimens, there has been a shift from the multistep methods previously discussed to unitary procedures and systems that incorporate standardization, speed, reproducibility, miniaturization, mechanization, and automation. These rapid identification methods can be divided into three categories: (1) manual biochemical systems, (2) mechanized/automated systems, and (3) immunologic systems.

One of the most popular manual biochemical systems for the rapid identification of members of the family *Enterobacte-*

(b)

Figure 34.7 *continued*

(a)

(b)

Figure 34.8   Rapid Tests. The API 20E manual biochemical system for microbial identification. (*a*) Positive and (*b*) negative results.

riaceae and other gram-negative bacteria is the API 20E system. It consists of a plastic strip (figure 34.8) with 20 microtubes containing dehydrated substrates that can detect certain biochemical characteristics. The test substrates in the 20 microtubes are inoculated with bacteria in sterile physiological saline. After 5 hours or overnight incubation, the 20 test results are converted to a seven- or nine-digit profile (figure 34.9). This profile number can be used with a computer or a book called the *API Profile Index* to find the name of the bacterium.

A popular mechanized/automated system is BIOLOG (figure 34.10). This system identifies microorganisms on their ability to catabolize 95 different carbon sources. The system can identify over 1,100 gram-negative and gram-positive bacteria as well as yeasts.

Immunologic Techniques

The culturing of certain viruses, bacteria, fungi, and parasites from clinical specimens may not be possible because the methodology remains undeveloped (*Treponema pallidum;* hepatitis A, B, C; and Epstein-Barr virus), is unsafe (rickettsias), or is impractical for all but a few clinical microbiology laboratories (mycobacteria). Cultures also may be negative because of prior antimicrobial therapy or because of the chronic state of the infectious disease in the patient. Under these circumstances, detection of antibodies or antigens may be of considerable diagnostic use.

Immunologic systems for the detection and identification of pathogens from clinical specimens are easy to use, give relatively

(a) Normal 7-digit code 5 144 572 = *E. coli*.

(b) 9-digit code 2 212 004 63 = *Pseudomonas aeruginosa*

Construction of a 9-digit profile

To the 7-digit profile illustrated in part *a*, 2 digits are added corresponding to the following characteristics:

NO₂:	Reduction of nitrate to nitrite only
N₂ GAS:	Complete reduction of nitrate to N₂ gas or amines
MOT:	Observation of motility
MAC:	Growth on MacConkey medium
OF/O:	Oxidative utilization of glucose (OF-open)
OF/F:	Fermentative utilization of glucose (OF-closed)

Figure 34.9   The API 20E Profile Number. The conversion of API 20E test results to the codes used in identification of unknown bacteria. The test results read top to bottom (and right to left in part *b*) correspond to the 7- and 9-digit codes when read in the right-to-left order. The tests required for obtaining a 7-digit code take an 18–24 hour incubation and will identify most members of the *Enterobacteriaceae*. The longer procedure that yields a 9-digit code is required to identify many gram-negative nonfermenting bacteria. The following tests are common to both procedures: ONPG (β-galactosidase); ADH (arginine dihydrolase); LDC (lysine decarboxylase); ODC (ornithine decarboxylase); CIT (citrate utilization); H₂S (hydrogen sulfide production); URE (urease); TDA (tryptophane deaminase); IND (indole production); VP (Voges-Proskauer test for acetoin); GEL (gelatin liquefaction); the fermentation of glucose (GLU), mannitol (MAN), inositol (INO), sorbitol (SOR), rhamnose (RHA), sucrose (SAC), melibiose (MEL), amygdalin (AMY), and arabinose (ARA); and OXI (oxidase test).

Figure 34.10   The BIOLOG System for the Rapid Screening of Specimens. BIOLOG's data base is the world's largest. With just three test panels covering gram-negative bacteria, gram-positive bacteria, and yeasts, this system can now identify over 1,100 species. The library includes virtually all pathogenic aerobic species and a wide range of environmental aerobic species. The identification of yeasts and lactic acid bacteria are of particular importance in both human health and food microbiology. To use this system, a cell suspension is prepared, a panel is inoculated, incubated for 4–72 hours, and read. User-friendly computer software interprets the test results.

[handwritten: 1) Western | go to genetics text for detail]

TABLE 34.4   Some Common Rapid Immunologic Test Kits for the Detection of Bacteria and Viruses in Clinical Specimens

Bactigen (Wampole Laboratories, Cranburg, N.J.)
The Bactigen kit is used for the detection of *Streptococcus pneumoniae, Haemophilus influenzae* type b, and *Neisseria meningitidis* groups A, B, C, and Y from cerebrospinal fluid, serum, and urine.

Culturette Group A Strep ID Kit (Marion Scientific, Kansas City, Mo.)
The Culturette kit is used for the detection of group A streptococci from throat swabs.

Directigen (Hynson, Wescott, and Dunning, Baltimore, Md.)
The Directigen Meningitis Test kit is used to detect *H. influenzae* type b, *S. pneumoniae,* and *N. meningitidis* groups A and C.

The Directigen Group A Strep Test kit is used for the direct detection of group A streptococci from throat swabs.

Fluortec-F and -M (General Diagnostics, Morris Plains, N.J.)
These two kits are used for the rapid detection of the *Bacteroides fragilis* and *Bacteroides melaninogenicus* groups in clinical specimens.

Gono Gen (Micro-Media Systems, San Jose, Calif.)
The Gono Gen kit detects *Neisseria gonorrhoeae.*

QuickVue *H. pylori* Test (Quidel, San Diego, Calif.)
A seven minute test for detection of IgG antibodies against *Helicobacter pylori* in human serum or plasma.

Staphaurex (Wellcome Diagnostics, Research Triangle Park, N.C.)
Staphaurex screens and confirms *Staphylococcus aureus* in 30 seconds.

Directigen RSV (Becton Dickinson Microbiology Systems, Cockeysville, Md.)
By using a nasopharyngeal swab, the respiratory syncytial virus can be detected in 15 minutes.

SureCell Herpes (HSV) Test (Kodak, Rochester, N.Y.)
Detects the herpes (HSV) 1 and 2 viruses in minutes.

SUDS HIV-1 Test (Murex Corporation, Norcross, Ga.)
Detects antibodies to HIV-1 antigens in about 10 minutes.

---

rapid reaction endpoints, and are sensitive and specific. Some of the more popular immunologic rapid test kits for viruses and bacteria are presented in table 34.4.

Each individual's immunologic response to a microorganism is quite variable. As a result the interpretation of serological tests is sometimes difficult. For example, a single elevated antibody titer usually does not distinguish between active and past infections. Furthermore, the lack of a measurable antibody titer may reflect either a microorganism's lack of immunogenicity or an insufficient time for an antibody response to develop following the onset of the infectious disease. For these reasons, test selection and timing of specimen collection are essential to the proper interpretation of immunologic tests.

Antibody titer (pp. 601 and 643).

The most widely used immunologic techniques available to detect microorganisms in clinical specimens are covered in detail in chapter 32. Examples include direct immunofluorescence, complement fixation, neutralization, agglutination, immunoelectrophoresis, and immunoassay; monoclonal antibodies are especially useful in some of these techniques (Box 34.2). No single technique is universally applicable for measuring an individual's immunologic response to all microorganisms. Techniques are therefore chosen based on their selectivity, specificity, ease, speed of performance, and cost-effectiveness.

One of the newest immunologic techniques being used in the clinical microbiology laboratory is **immunoblotting.** Immunoblotting involves polyacrylamide gel electrophoresis of a protein specimen followed by transfer of the separated proteins to nitrocellulose sheets (*see figure 15.5*). Protein bands are then visualized by treating the nitrocellulose sheets with solutions of dye-tagged antibodies. This procedure demonstrates the presence of common and specific proteins among different strains of microorganisms (figure 34.11). Immunoblotting also can be used to show strain-specific immune responses to microorganisms, to serve as an important diagnostic indicator of a recent infection with a particular strain of microorganism, and to allow for prognostic implications with severe infectious diseases.

Another new immunologic technique uses liposomes. A **liposome** [Greek *lipos,* fat, and *soma,* body] is an artificially created microscopic spherical vesicle formed by a lipid bilayer enclosing an aqueous compartment. The aqueous compartment contains a colored dye. The liposome is then sensitized by coupling a specific antibody (or antigen) to the liposome surface. Antibodies specific for the desired pathogen are attached to a membrane in a particular shape, such as a triangle. After a patient's sample is added to the test well (figure 34.12*a*), the analyte (the substance undergoing analysis) instantly binds to the capture antibodies (or antigen). When the dye-filled sensitized liposomes are added (figure 34.12*b*) and link to the immobilized analyte (figure 34.12*c*), a triangle (or other shape) indicates a positive reaction (figure 34.12*d*). If the sample is negative and contains no analyte, no binding occurs in the test area when the liposomes are added and either a blank or some other shape appears. Liposome tests are currently available for group A streptococci and the respiratory syncytial virus.

---

1. Describe in general how biochemical tests are used in the API 20E system to identify bacteria.
2. Name the two basic immunologic procedures used in rapid test kits to identify microorganisms.
3. Why might cultures for some microorganisms be unavailable?
4. Why are test selection and timing of specimen collection essential to the proper interpretation of immunologic tests?
5. What are some uses of immunoblotting in the clinical microbiology laboratory? The use of liposomes?

Bacteriophage Typing

Bacteriophages (phages) are viruses that attack members of a particular bacterial species, or strains within a species (*see chapter 17*). **Bacteriophage (phage) typing** is based on the specificity of phage surface receptors for cell surface receptors. Only those bacteriophages that can attach to these surface receptors can infect bacteria and cause lysis. On a petri dish culture,

Figure 34.11   Immunoblotting (Western Blot).
Immunoblot of the standard strains of *Clostridium difficile*. Arrows indicate strain-specific bands (specific molecular weights) of the various proteins in different lanes (A–E). The molecular weight of the protein is indicated on the left.

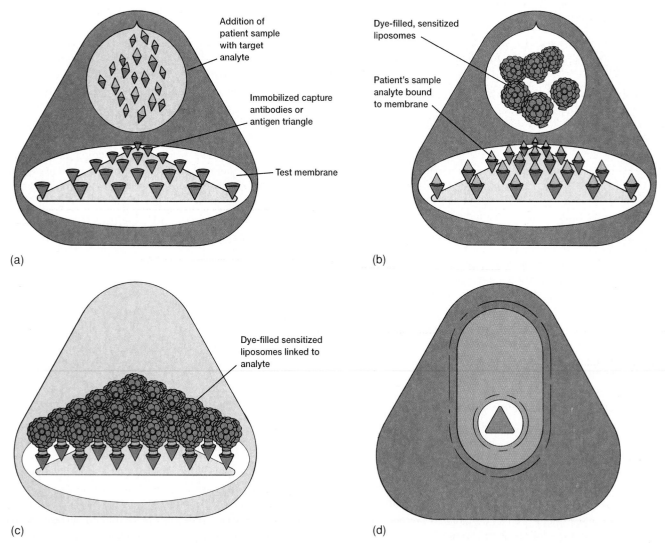

Figure 34.12   The Use of Liposomes in Diagnostic Testing. (*a*) The addition of a patient's sample containing the target analyte. Immobilized capture antibodies or antigen on a test membrane form a geometric design such as a triangle. (*b*) When dye-filled liposomes are added, they bind to the patient's bound analyte (*c*) giving a triangle. (*d*) The triangle indicates a positive reaction.

lytic bacteriophages cause plaques on lawns of sensitive bacteria. These plaques represent infection by the virus (*see figures 16.2 and 16.4*).

In bacteriophage typing the clinical microbiologist inoculates the bacterium to be tested onto a petri plate. The plate is heavily and uniformly inoculated with a cotton swab so that the bacteria will grow to form a solid sheet or lawn of cells. No uninoculated areas should be left. The plate is then marked off into squares (15 to 20 mm per side), and each square is inoculated with a drop of suspension from the different phages available for typing. After the plate is incubated for 24 hours, it is observed for plaques. The phage type is reported as a specific genus and species followed by the types that can infect the bacterium. For example, the series 10/16/24 indicates that this bacterium is sensitive to phages 10, 16, and 24, and belongs to a collection of strains, called a **phagovar,** that have this particular phage sensitivity. Bacteriophage typing remains a tool of the research and reference laboratory.

Molecular Methods

With the application of new molecular technology, it is now possible to analyze the molecular characteristics of microorganisms in the clinical laboratory. Some of the most accurate approaches to microbial identification are through the analysis of proteins and nucleic acids. Examples previously discussed (*see chapter 19*) include comparison of proteins; physical, kinetic, and regulatory properties of microbial enzymes; nucleic acid-base composition (*see table 19.4*); nucleic acid hybridization; and nucleic acid sequencing. Three other molecular methods being widely used are nucleic acid probes, gas-liquid chromatography, and plasmid fingerprinting.

Nucleic Acid-Based Detection Methods

A recent development in clinical microbiology is the use of nucleic acids (DNA and RNA) as a diagnostic tool for the detection and identification of microorganisms. For example, DNA probe technology (*see chapter 15*) identifies a microorganism by probing its genetic composition. The use of cloned DNA as a probe is based upon the capacity of single-stranded DNA to bind (hybridize) with a complementary nucleic acid sequence present in test specimens to form a double-stranded DNA hybrid (figure 34.13). Thus a single-stranded sequence derived from one microorganism (the probe) is used to search for others containing the same sequence. This hybridization reaction may be applied to purified DNA preparations, to bacterial colonies, or to clinical specimens such as tissue, serum, sputum, and pus. Recently DNA probes have been developed that bind to complementary strands of ribosomal RNA. These DNA:rRNA hybrids are more sensitive than conventional DNA probes, give results in 2 hours or less, and require the presence of fewer microorganisms. DNA probe sensitivity can be increased by over

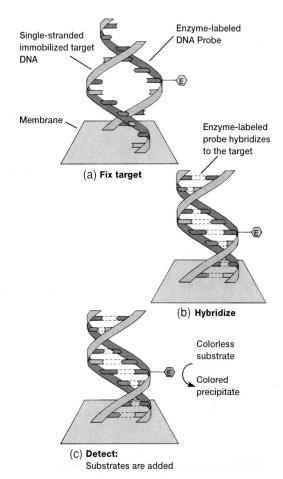

(a) **Fix target**

(b) **Hybridize**

(c) **Detect:** Substrates are added

Figure 34.13   Basic Steps in A DNA Probe Hybridization Assay. (*a*) Single-stranded target nucleic acid is bound to a membrane. A DNA probe with attached enzyme (E) also is employed. (*b*) The probe is added to the membrane. If the probe hybridizes to the target DNA, a double-stranded DNA hybrid is formed. (*c*) A colorless substrate is added. The enzyme attached to the probe converts the substrate to a colored precipitate. This detection system is semiquantitative, in that color intensity is proportional to the quantity of hybridized target nucleic acid present.

one million if the target DNA is first amplified using the polymerase chain reaction (*see pp. 310–13*). DNA:rRNA probes are available or are currently being developed for most clinically important microorganisms.

Ribosomal RNA from *E. coli* can be used to type bacterial strains by probing chromosomal DNA in Southern blots (*see figure 15.5*). This method of strain typing, called **ribotyping,** makes use of the fact that rRNA genes are scattered throughout the chromosome of most bacteria. When the chromosomes of several strains are cleaved using restriction endonucleases and the digests are analyzed with the Southern blotting procedure, rRNA probes will produce different patterns with different strains. Cloned rRNA genes also can be used as probes instead of *E. coli* rRNA, and similar banding patterns result.

Gas-Liquid Chromatography

During chromatography a chemical mixture carried by a liquid or gas is separated into its individual components because of processes such as adsorption, ion-exchange, and partitioning between different solvent phases. In gas-liquid chromatography (GLC), specific microbial metabolites, cellular fatty acids, and products from the pyrolysis (a chemical change caused by heat) of whole bacterial cells are analyzed and identified. These compounds are easily removed from growth media by extraction with an organic solvent such as ether. The ether extract is then injected into the GLC system. Both volatile and nonvolatile acids can be identified. Based on the pattern of fatty acid production, common anaerobes isolated from clinical specimens can be identified.

The reliability, precision, and accuracy of GLC have been improved significantly with continued advances in instrumentation; the introduction of instruments for high-performance liquid chromatography; and the use of mass spectrometry, nuclear magnetic resonance spectroscopy, and associated analytical techniques for the identification of components separated by the chromatographic process. These combined techniques have recently been used to discover specific chemical markers of various infectious disease agents by direct analysis of body fluids.

Plasmid Fingerprinting

As presented in chapter 14, a plasmid is an autonomously replicating extrachromosomal molecule of DNA in bacteria. **Plasmid fingerprinting** identifies microbial isolates of the same or similar strains; related strains often contain the same number of plasmids with the same molecular weights and similar phenotypes. In contrast, microbial isolates that are phenotypically distinct have different plasmid fingerprints. Plasmid fingerprinting of many *E. coli, Salmonella, Campylobacter,* and *Pseudomonas* strains and species has demonstrated that this method often is more accurate than other phenotyping methods such as biotyping, antibiotic resistance patterns, phage typing, and serotyping.

The technique of plasmid fingerprinting involves five steps:

1. The bacterial strains are grown in broth or on agar plates.
2. The cells are harvested and lysed with a detergent.
3. The plasmid DNA is separated from the chromosomal DNA.
4. The plasmid DNA is applied to agarose gels and electrophoretically separated.
5. The gel is stained with ethidium bromide, which binds to DNA, causing it to fluoresce under UV light. The plasmid DNA bands are then located.

Figure 34.14 Plasmid Fingerprinting. Agarose gel electrophoresis of plasmid DNA.

Because the migration rate of plasmid DNA in agarose is inversely proportional to the molecular weight, plasmids of a different size appear as distinct bands in the stained gel. The molecular weight of each plasmid species can then be determined from a plot of the log of the distance that each species has migrated versus the log of the molecular weights of plasmids of known size that have been electrophoresed simultaneously in the same gel (figure 34.14).

Susceptibility Testing

Many clinical microbiologists believe that determining the susceptibility of a microorganism to specific chemotherapeutic agents (drugs) is one of the most important tests performed in the clinical microbiology laboratory. Results (figure 34.6*o*) can show the agents to which a microorganism is most susceptible and the proper therapeutic dose needed to treat the infectious

A#problems w/ Resistance ?

disease. (Dilution susceptibility tests, disk diffusion tests [Kirby-Bauer method], and drug concentration measurements in the blood are discussed in detail in chapter 33.)

---

1. What is the basis for bacteriophage typing?
2. How can nucleic acid-based detection methods be used by the clinical microbiologist? Gas-liquid chromatography?
3. How can a suspect bacterium be plasmid fingerprinted? Why is susceptibility testing so important in clinical microbiology?

Computers in Clinical Microbiology

Computer systems in the clinical microbiology laboratory are designed primarily to replace the handwritten mode of information acquisition and transmission. Computers improve the efficiency of the laboratory operation and increase the speed and clarity with which results can be reported to physicians. From a work-flow standpoint, the major functions involving the computer are test ordering, result entry, analysis of results, and report preparation (figure 34.1*h,i*).

Test orders may be entered into the computer from the hospital unit or laboratory. Ordering routines should include specific requests (e.g., rule out *Nocardia* and diphtheria), all pertinent patient data, and an accession number. Once the test order has been placed, the system should allow the usual work flow to proceed with the labeled clinical specimen, date, test order, and computer accession number.

After clinical results are obtained in the laboratory, they are entered into a written log and then into the computer. To meet the many needs of microbiological entry, the computer system must be rapid and flexible in its entry modes.

Printed reports of the patient's laboratory findings (figure 34.3*b*) are the product of the computer system. Print programs also should allow flexible formatting of the reports so that additional data can be generated. For example, the computer system should be able to generate cumulative reports that summarize days to weeks of an inpatient stay.

Besides reporting routine laboratory tests, computers can manage specimen logs, reports of overdue tests, quality control statistics, antimicrobial susceptibility probabilities, hospital epidemiological data, and many other items. The computer can be interfaced with various automated instruments for rapid and accurate calculation and transfer of clinical data.

---

1. What are some different ways in which computers can be used in the clinical microbiology laboratory?
2. From the standpoint of work flow, how can computers be specifically used in a clinical microbiology laboratory?

Summary

1. The major concern of the clinical microbiologist is to isolate and identify microorganisms from clinical specimens rapidly. A clinical specimen represents a portion or quantity of biological material that is tested, examined, or studied to determine the presence or absence of specific microorganisms.

2. Specimens may be collected by various methods that include swabs, needle aspiration, intubation, catheters, and clean-catch techniques. Each method is designed to ensure that only the proper material will be sent to the clinical laboratory.

3. Immediately after collection the specimen must be properly handled and labeled. Speed in transporting the specimen to the clinical laboratory after it has been collected is of prime importance.

4. The clinical microbiology laboratory can provide preliminary or definitive identification of microorganisms based on (1) microscopic examination of specimens; (2) growth or biochemical characteristics of microorganisms

isolated from cultures; and (3) immunologic techniques that detect antibodies or microbial antigens.

5. Viruses are identified by isolation in living cells or serological tests. Several types of living cells are available: cell culture, embryonated hen's eggs, and experimental animals. Rickettsial disease can be diagnosed serologically or by isolation of the organism. Chlamydiae can be demonstrated in tissue and cell scrapings with Giemsa stain, which detects the characteristic intracellular inclusion bodies. The most routinely used techniques for identification of the mycoplasmas are serological. Identification of fungi often can be made if a portion of the specimen is mixed with a drop of 10% potassium hydroxide. Wet mounts of stool specimens or urine can be examined microscopically for the presence of parasites.

6. The initial identity of a bacterial organism may be suggested by (1) the source of the culture specimen; (2) its microscopic appearance; (3) its pattern of growth on

selective, differential, enrichment, or characteristic media; and (4) its hemolytic, metabolic, and fermentative properties.

7. Rapid methods for microbial identification can be divided into three categories: (1) manual biochemical systems, (2) mechanized/automated systems, and (3) immunologic systems.

8. Bacteriophage typing for bacterial identification is based on the fact that phage surface receptors bind to specific cell surface receptors. On a petri plate culture, bacteriophages cause plaques on lawns of bacteria with the proper receptors.

9. Various molecular methods also can be used to identify microorganisms. Examples include nucleic acid-based detection, gas-liquid chromatography, and plasmid fingerprinting.

10. Computer systems in clinical microbiology are designed to replace handwritten information exchange and to speed data evaluation and report preparation.

Key Terms

bacteriophage (phage) typing *693*

catheter *677*

clinical microbiologist *677*

cytopathic effect *683*

hemadsorption *683*

immunoblotting *693*

interference *684*

intubation *677*

liposome *693*

needle aspiration *677*

phagovar *695*

plasmid fingerprinting *696*

ribotyping *695*

sputum *677*

swab *677*

Questions for Thought and Review

1. How can clinical specimens be taken from a patient with various infectious diseases? Give specific examples of procedures used.

2. As more new ways for identifying the characteristics of microorganisms emerge, the number of distinguishable microbial strains also seems to increase. Why do you think this occurs?

3. What precaution must be observed when a culture is obtained from the respiratory system?

4. Why are miniaturized identification systems used in clinical microbiology? Describe one such system and its advantage over classic dichotomous keys.

5. Why is gas-liquid chromatography a useful approach to the identification of anaerobes?

6. How does a clinical microbiologist convert an API 20E test result to a numerical code for bacterial identification?

7. How is a dichotomous key used in bacterial identification?

8. What are some different ways in which biochemical reactions can be used to identify microorganisms?

9. What are some advantages of automation in the clinical microbiology laboratory?

10. When should laboratory animals be used in the identification of microorganisms?

11. Why is plasmid fingerprinting such an accurate method for the identification of microorganisms?

Additional Reading

General

Baron, E. J.; Peterson, L. R.; and Finegold, S. M. 1994. *Bailey and Scott's diagnostic microbiology,* 9th ed. St. Louis: C. V. Mosby.

Coonrod, J. D.; Kunz, L. J.; and Ferraro, M. J. 1983. *The direct detection of microorganisms in clinical specimens.* San Diego, Calif.: Academic Press.

Finegold, S. M., and Smith, W. J. 1982. *Diagnostic microbiology,* 6th ed. St. Louis: C. V. Mosby.

Gerhardt, P.; Murray, R. G. E.; Wood, W. A.; and Krieg, N. R., editors. 1994. *Methods for general and molecular bacteriology.* Washington, D.C.: American Society for Microbiology.

Isenberg, H. D., editor. 1992. *Clinical microbiology procedures handbook.* Washington, D.C.: American Society for Microbiology.

Koneman, E. W.; Allen, S. D.; Dowell, V. R., Jr.; Janda, W. M.; Sommers, H. M.; and Winn, W. C., Jr. 1988. *Color atlas and textbook of diagnostic microbiology,* 3d ed. Philadelphia: J. B. Lippincott.

Murray, P. R., editor-in-chief. 1995. *Manual of clinical microbiology,* 6th. ed. Washington, D.C.: ASM Press.

Persing, D. H., editor. 1993. *Diagnostic molecular microbiology.* Washington, D.C.: American Society for Microbiology.

Pike, R. M. 1979. Laboratory-associated infections: Incidence, fatalities, causes, and prevention. *Ann. Rev. Microbiol.* 33:41–66.

Turgeon, M. L. 1990. *Immunology and serology in laboratory medicine.* St. Louis: C. V. Mosby Co.

Washington, J. A., II. 1981. *Laboratory procedures in clinical microbiology.* New York: Springer-Verlag.

Specimens

Bartlett, R.; Mazens-Sullivan, M.; Tetreault, J.; Lobel, S.; and Nivard, J. 1994. Evolving approaches to management of quality in clinical microbiology. *Clin. Microbiol. Rev.* 7(1):55–88.

Chonmaitree, T.; Baldwin, C.; and Lucia, H. 1989. Role of the virology laboratory in diagnosis and management of patients with central nervous system disease. *Clin. Microbiol. Rev.* 2(1):1–14.

Emori, T., and Gaynes, R. 1993. An overview of nosocomial infections, including the role of the microbiology laboratory. *Clin. Microbiol. Rev.* 6(4):428–42.

Forney, J. E., editor. 1987. *Collection, handling, and shipment of microbiological specimens* (PHS Publication No. 976). Washington, D.C.: U.S. Public Health Service.

Johnson, F. B. 1990. Transport of viral specimens. *Clin. Microbiol. Rev.* 3(2):120–31.

Mayer, L. W. 1988. Use of plasmid profiles in epidemiologic surveillance of disease outbreaks and in tracing the transmission of antibiotic resistance. *Clin. Microbiol. Rev.* 1(2):228–43.

Identification of Microorganisms

Amann, R. I.; Ludwig, W.; and Schleifer, K.-H. 1995. Phylogenetic identification and in situ detection of individual microbial cells without cultivation. *Microbiol. Rev.* 59(1):143–69.

Belkum, A. 1994. DNA fingerprinting of medically important microorganisms by use of PCR. *Clin. Microbiol. Rev.* 7(2):174–84.

Gray, L. D., and Fedorko, D. P. 1992. Laboratory diagnosis of bacterial meningitis. *Clin. Microbiol. Rev.* 5(2):130–45.

Manafi, M.; Kneifel, W.; and Bascomb, S. 1991. Fluorogenic and chromogenic substrates used in bacterial diagnostics. *Microbiol. Rev.* 55(3):335–48.

Pezzlo, M. 1988. Detection of urinary tract infections by rapid methods. *Clin. Microbiol. Rev.* 1(3):268–80.

Reiss, E., and Morrison, C. 1993. Nonculture methods for diagnosis of disseminated candidiasis. *Clin. Microbiol. Rev.* 6(4):311–23.

Rotbart, H. A. 1991. Nucleic acid detection systems for enteroviruses. *Clin. Microbiol. Rev.* 4(2):156–68.

Stager, C. E., and Davis, J. R. 1993. Automated systems for identification of microorganisms. *Clin. Microbiol. Rev.* 5(3):302–27.

Tenover, F. C. 1988. Diagnostic deoxyribonucleic acid probes for infectious diseases. *Clin. Microbiol. Rev.* 1(1):82–101.

Weiss, J. B. 1995. DNA probes and PCR for diagnosis of parasitic infections. *Clin. Microbiol. Rev.* 8(1):113–30.

Wolcott, M. J. 1992. Advances in nucleic acid-based detection methods. *Clin. Microbiol. Rev.* 5(4):370–86.

Computers in Clinical Microbiology

Ryon, K. J., and Peebles, J. E. 1982. On-line computer entry of routine and AutoMicrobic System bacteriology results. In *Rapid methods and automation in microbiology,* R. C. Tilton, editor, 23–27. Washington, D.C.: American Society for Microbiology.

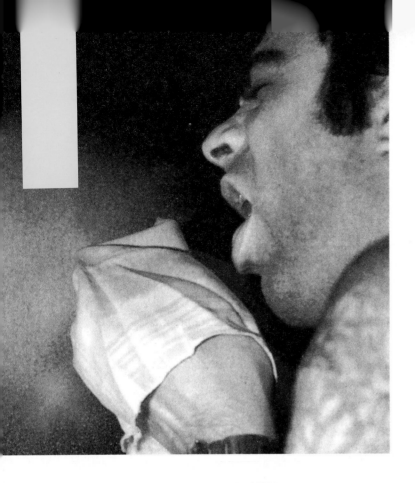

CHAPTER 35
The Epidemiology of Infectious Disease

Epidemics of infectious disease are often compared with forest fires. Once fire has spread through an area, it does not return until new trees have grown up. Epidemics in humans develop when a large population of susceptible individuals is present. If most individuals are immune, then an epidemic will not occur.
— Andrew Cliff and Peter Haggett

Infectious diseases constitute the major cause of death worldwide and will not be conquered during our lifetimes.
— Emerging Infections. Microbial Threats to Health in the United States. IOM. 1992.

Concepts

1. The science of epidemiology deals with the occurrence and distribution of disease within a given population. Infectious disease epidemiology is concerned with organisms or agents responsible for the spread of infectious diseases in human and other animal populations.

2. Because numbers and time are major epidemiological parameters, statistics is an important working tool in this discipline. Statistics is used to determine morbidity, frequency, and mortality rates.

3. To trace the origin and manner of spread of an infectious disease outbreak, it is necessary to learn what pathogen is responsible.

4. Epidemiologists investigate five links in the infectious disease cycle: (1) characteristics of the pathogen, (2) source and/or reservoir of the pathogen, (3) mode of transmission, (4) susceptibility of the host, and (5) exit mechanisms.

5. The control of nosocomial (hospital) infections has received increasing attention in recent years because of the number of individuals involved, increasing costs, and the length of hospital stays.

This chapter describes the epidemiological parameters that are studied in the infectious disease cycle. The practical goal of epidemiology is to establish effective control, prevention, and eradication measures within a given population. Because nosocomial (hospital) acquired infections have increased in recent years, a brief synopsis of their epidemiology also is presented.

The science of epidemiology originated and evolved in response to the great epidemic diseases such as cholera, typhoid fever, smallpox, and yellow fever (Box 35.1). Today its scope embraces all diseases: infectious diseases and those resulting from anatomical deformities, genetic abnormalities, metabolic dysfunction, malnutrition, neoplasms, psychiatric disorders, and aging. This chapter emphasizes only infectious disease epidemiology.

By definition, **epidemiology** [Greek *epi,* upon, and *demos,* people or population, and *logy,* study] is the science that evaluates the occurrence, determinants, distribution, and control of health and disease in a defined human population. **Health** is the condition in which the organism (and all of its parts) performs its vital functions normally or properly. It is a state of physical and mental well-being and not merely the absence of disease. A **disease** [French *des,* from, and *aise,* ease] is an impairment of the normal state of an organism or any of its components that hinders the performance of vital functions. It is a response to environmental factors (e.g., malnutrition, industrial hazards, climate), specific infective agents (e.g., viruses, bacteria, fungi, protozoa, helminths), inherent defects of the body (e.g., various genetic or immunologic anomalies), or combinations of these.

Any individual who practices epidemiology is an **epidemiologist.** Epidemiologists are, in effect, disease detectives. Their major concerns are the discovery of the factors essential to disease occurrence and the development of methods for disease prevention.

Epidemiological Terminology

When a disease occurs occasionally, and at irregular intervals in a human population, it is a **sporadic disease** (e.g., typhoid fever). When it maintains a steady, low-level frequency at a moderately regular interval, it is an **endemic** [Greek *endemos,* dwelling in the same people] **disease** (e.g., the common cold). **Hyperendemic diseases** gradually increase in occurrence frequency beyond the endemic level but not to the epidemic level (e.g., the common cold during winter months). An **epidemic** [Greek *epidemios,* upon the people] is a sudden increase in the occurrence of a disease above the expected level (figure 35.1). The AIDS epidemic is an excellent example. The first case in an epidemic is called the **index case.** An **outbreak,** on the other hand, is the sudden, unexpected occurrence of a disease, usually focally or in a limited segment of a population (e.g., Legionnaires' disease). Although the epidemiology of an outbreak may be no different from that of an epidemic, the community regards the outbreak as less serious. A **pandemic** [Greek *pan,* all] is an increase in disease occurrence within a large population over a

---

Box 35.1

John Snow: The First Epidemiologist

Much of what we know today about the epidemiology of cholera is based on the classic studies conducted by the British physician John Snow between 1849 and 1854. During this period a series of cholera outbreaks occurred in London, England, and Snow set out to find the source of the disease. Some years earlier when he was still a medical apprentice, Snow had been sent to help during an outbreak of cholera among coal miners. His observations convinced him that the disease was usually spread by unwashed hands and shared food, not by "bad" air or casual direct contact.

Thus when the outbreak of 1849 occurred, Snow believed that cholera was spread among the poor in the same way as among the coal miners. He suspected that water, and not unwashed hands and shared food, was the source of the cholera infection among the wealthier residents. Snow examined official death records and discovered that most of the victims in the Broad Street area had lived close to the Broad Street pump or had been in the habit of drinking from it. He concluded that cholera was spread by drinking water from the Broad Street pump, which was contaminated with raw sewage containing the disease agent. When the pump handle was removed, the number of cholera cases dropped dramatically.

In 1854 another cholera outbreak struck London. Part of the city's water supply came from two different suppliers: the Southwark and Vauxhall Company and the Lambeth Company. Snow interviewed cholera patients and found that most of them purchased their drinking water from the Southwark and Vauxhall Company. He also discovered that this company obtained its water from the Thames River below locations where Londoners had discharged their sewage. In contrast, the Lambeth Company took its water from the Thames before the river reached the city. The death rate from cholera was over eightfold lower in households supplied with Lambeth Company water. Water contaminated by sewage was transmitting the disease. Finally, Snow concluded that the cause of the disease must be able to multiply in water. Thus he nearly recognized that cholera was caused by a microorganism, though Robert Koch didn't discover the causative bacterium (*Vibrio cholerae*) until 1883.

To commemorate these achievements, the John Snow Pub now stands at the site of the old Broad Street pump. Those who complete the Epidemiologic Intelligence Program at the Centers for Disease Control and Prevention receive an emblem bearing a replica of a barrel of Whatney's Ale—the brew dispensed at the John Snow Pub.

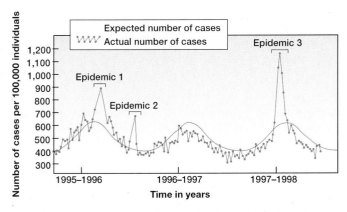

Figure 35.1　A Graph Illustrating Three Epidemics. The solid line indicates the expected number of endemic cases. The connected dots indicate the actual number of cases. Epidemics (marked by brackets) are sharp increases in the number of cases of a disease above that which is normally expected (solid line).

very wide region (usually the world). Usually, pandemic diseases spread among continents. The influenza outbreak of the 1960s is a good example.

The discipline that deals with the factors that influence the frequency of a disease in an animal population is known as **epizootiology.** Moderate prevalence of a disease in animals is termed **enzootic,** a sudden outbreak is **epizootic,** and wide dissemination is **panzootic.** Animal diseases that can be transmitted to humans are termed **zoonoses** [Greek *zoon,* animal, and *nosos,* disease].

Measuring Frequency: The Tools of Epidemiologists

Besides epidemiology's historical ties to epidemic diseases, it also has been closely allied with statistics. **Statistics is the** branch of mathematics dealing with the collection, organization, and interpretation of numerical data. As a science particularly concerned with rates and the comparison of rates, epidemiology was the first medical field in which statistical methods were extensively used.

Measures of frequency usually are expressed as fractions. The numerator is the number of individuals experiencing the event—infection or other problem—and the denominator is the number of individuals in whom the event could have occurred, that is, the population at risk. The fraction is a proportion or ratio but is commonly called a rate because a time period is always specified. (A rate also can be expressed as a percentage if the fraction is multiplied by 100.) In population statistics, rates usually are stated per 1,000 individuals, although other powers of 10 may be used for particular diseases (e.g., per 100 for very common diseases and per 10,000 or 100,000 for uncommon diseases).

A **morbidity rate** measures the number of individuals that become ill because of a specific disease within a susceptible population during a specific period. It is an incidence rate and reflects the number of new cases in a period. The rate is commonly determined when the number of new cases of illness in the general population is known from clinical reports. It is calculated as follows:

$$\text{Morbidity rate} = \frac{\text{number of new cases of a disease during a specific period}}{\text{number of individuals in the population}}$$

For example, if there were 700 new cases of influenza per 100,000 individuals, then the morbidity rate would be expressed as 700 per 100,000 or 0.7%.

The **prevalence rate** refers to the total number of individuals infected in a population at any one time no matter when the disease began. The prevalence rate depends on both the incidence rate and the duration of the illness.

The **mortality rate** is the relationship of the number of deaths from a given disease to the total number of cases of the disease. The mortality rate is a simple statement of the proportion of all deaths that are assigned to a single cause. It is calculated as follows:

$$\text{Mortality rate} = \frac{\text{number of deaths due to a given disease}}{\text{size of the total population with the same disease}}$$

For example, if there were 15,000 deaths due to AIDS in a year, and the total number of people infected was 30,000, the mortality rate would be 15,000 per 30,000 or 1 per 2 or 50%.

The determination of morbidity, prevalence, and mortality rates aids public health personnel in directing health-care efforts to control the spread of infectious diseases. For example, a sudden increase in the morbidity rate of a particular disease may forecast a need for action to prevent a future rise in mortality.

---

1. What is epidemiology?
2. What terms are used to describe the occurrence of a disease in a human population? In an animal population?
3. Define morbidity rate, prevalence rate, and mortality rate.
4. How would you define a disease? Health?

Infectious Disease Epidemiology

An infectious disease is due to agents such as viruses, bacteria, fungi, protozoa, and helminths that can be transmitted from one host to another. It can vary from mild to severe to deadly for the host. An epidemiologist studying an infectious disease is concerned with the causative agent, the source and/or reservoir of the disease agent, how it was transmitted, what host and environmental factors could have aided development of the disease within a defined population, and how best to control or eliminate the disease. These factors describe the natural history or cycle of an infectious disease.

Box 35.2

"Typhoid Mary"

I n the early 1900s there were thousands of typhoid fever cases, and many died of the disease. Most of these cases arose when people drank water contaminated with sewage or ate food handled by or prepared by individuals who were shedding the typhoid fever bacterium (*Salmonella typhi*). The most famous carrier of the typhoid bacterium was Mary Mallon.

Between 1896 and 1906 Mary Mallon worked as a cook in seven homes in New York City. Twenty-eight cases of typhoid fever occurred in these homes while she worked in them. As a result the New York City Health Department had Mary arrested and admitted to a city hospital for patients with infectious diseases. Examination of

Mary's stools showed that she was shedding large numbers of typhoid bacteria though she exhibited no external symptoms of the disease. An article published in 1908 in the *Journal of the American Medical Association* referred to her as "Typhoid Mary," an epithet by which she is still known today. After being released when she pledged not to cook for others or serve food to them, Mary changed her name and began to work as a cook again. For five years she managed to avoid capture while continuing to spread typhoid fever. Eventually the authorities tracked her down. She was held in custody for 23 years until she died in 1938. As a lifetime carrier, Mary Mallon was positively linked with 10 outbreaks of typhoid fever, 53 cases, and 3 deaths.

Recognition of an Infectious Disease in a Population

Epidemiologists can recognize an infectious disease in a population by using various surveillance methods. Surveillance is a dynamic activity that includes gathering information on the development and occurrence of a disease, collating and analyzing the data, summarizing the findings, and using the information to select control methods. Some combination of the following surveillance methods is used most often:

1. Generation of morbidity data from case reports
2. Collection of mortality data from death certificates
3. Investigation of actual cases
4. Collection of data from reported epidemics
5. Field investigation of epidemics
6. Review of laboratory results: surveys of a population for antibodies against the agent and specific microbial serotypes, skin tests, cultures, stool analyses, etc.
7. Population surveys using valid statistical sampling to determine who has the disease
8. Use of animal and vector disease data
9. Collection of information on the usage of specific biologics—antibiotics, antitoxins, vaccines, and other prophylactic measures
10. Use of demographic data on population characteristics such as human movements during a specific time of the year

As just noted, surveillance may not always require the direct examination of cases. However, to accurately interpret surveillance data and study the course of a disease in individuals, epidemiologists and other medical professionals must be aware of the pattern of infectious diseases. Often infectious diseases have characteristic signs and symptoms. **Signs** are objective changes in the body, such as a fever or rash, that can be directly observed. **Symptoms** are subjective changes, such as pain and loss of appetite, that are personally experienced by the patient.

Sometimes the term symptom is used more broadly to include the signs. A **disease syndrome** is a set of signs and symptoms that are characteristic of the disease. Frequently extra laboratory tests are required for an accurate diagnosis because symptoms and readily observable signs are not sufficient for diagnosis.

The course of an infectious disease has a characteristic pattern and can be divided into several phases. Knowledge of the pattern is essential in accurately diagnosing the disease.

1. The **incubation period** is the period after pathogen entry and before signs and symptoms appear. The pathogen is spreading but has not yet reached a sufficiently high level to cause difficulties. This period's length varies with disease.
2. The **prodromal stage** is the period in which there is an onset of signs and symptoms, but they are not yet clear enough to make a diagnosis. The patient often is contagious. Innate resistance mechanisms are active.
3. The period of illness is the phase in which the disease is most severe and has characteristic signs and symptoms. The immune response has been triggered; B and T cells are becoming active.
4. In the period of decline, the signs and symptoms begin to disappear. The recovery stage often is referred to as convalescence.

Correlation with a Single Causative Agent

After an infectious disease has been recognized in a population, epidemiologists correlate the disease outbreak with a specific organism—its exact cause must be discovered (Box 35.2). At this point the clinical or diagnostic microbiology laboratory enters the investigation. Its purpose is to isolate and identify the organism that caused the disease and to determine the pathogen's susceptibility to control agents or methods that may assist in its eradication.

Role of clinical microbiology in disease diagnosis (chapter 34).

Figure 35.2   Epidemic Curves. (*a*) In a common-source epidemic, there is a rapid increase up to a peak in the number of individuals infected and then a rapid but more gradual decline. Cases usually are reported for a period that equals approximately one incubation period of the disease. (*b*) In a propagated epidemic the curve has a gradual rise and then a gradual decline. Cases usually are reported over a time interval equivalent to several incubation periods of the disease.

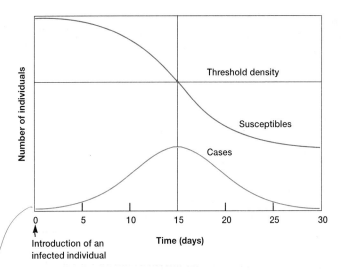

Figure 35.3   Diagrammatic Representation of the Spread of An Imaginary Propagated Epidemic. The lower curve represents the number of cases and the upper curve the number of susceptible individuals. Notice the coincidence of the peak of the epidemic wave with the threshold density of susceptibles.

Recognition of an Epidemic

As previously noted, an infectious disease epidemic is usually a short-term increase in the occurrence of the disease in a particular population (figure 35.1). Two major types of epidemic are recognized: common source and propagated.

A **common-source epidemic** is characterized by a sharp rise to a peak and then a rapid, but not as pronounced, decline in the number of individuals infected (figure 35.2*a*). This type of epidemic usually results from a single common contaminated source such as food (food poisoning) or water (Legionnaires' disease).

A **propagated epidemic** is characterized by a relatively slow and prolonged rise and then a gradual decline in the number of individuals infected (figure 35.2*b*). This type of epidemic usually results from the introduction of a single infected individual into a susceptible population. The initial infection is then propagated to others in a gradual fashion until many individuals within the population are infected. An example is the increase in mumps or chickenpox cases that coincides with new populations of sensitive children who arrive in classrooms each fall. Only one infected child is necessary to propagate the epidemic.

To understand how epidemics are propagated, consider figure 35.3. At time 0, all individuals in this population are susceptible to a hypothetical pathogen. The introduction of an infected individual initiates the epidemic outbreak (lower curve), which spreads and reaches a peak (day 15). As individuals recover from the disease, they become immune and no longer transmit the pathogen (upper curve). The number of susceptible individuals therefore decreases. The decline in the number of susceptibles to the threshold density (the minimum number of individuals necessary to continue propagating the disease) coincides with the peak of the epidemic wave, and the incidence of new cases declines because the pathogen cannot propagate itself.

Herd immunity is the resistance of a population to infection and pathogen spread because of the immunity of a large percentage of the population (figure 35.4). The larger the proportion of those immune, the smaller the probability of effective contact between infective and susceptible individuals—that is, many contacts will be with immunes, and thus the population will exhibit a group resistance. A susceptible member of such an immune population enjoys an immunity that is not of his or her own making (not self-made) but instead arises because of membership in the group.

At times public health officials immunize large portions of the susceptible population in an attempt to maintain a high level of herd immunity. Any increase in the number of susceptibles may result in an endemic disease becoming epidemic. The proportion of immune to susceptible individuals must be constantly monitored because new susceptible individuals continually enter a population through migration and birth. In addition, pathogens can change so much through processes such as antigenic shift (see next paragraph) that immune individuals become susceptible again.

Many pathogens do not ordinarily change in nature. They cause endemic diseases because infected humans perpetually transfer them to others (e.g., sexually transmitted diseases) or because they continually reenter the human population from animal reservoirs (e.g., rabies). Other pathogens continue to evolve and may produce epidemics (e.g., AIDS, influenza virus [A strain], and *Legionella* bacteria). One way in which a pathogen changes is by **antigenic shift,** a major genetically determined change in the antigenic character of a pathogen. An antigenic shift can be so extensive that the pathogen is no longer recognized by the host's immune system. For example, influenza viruses frequently change by mutation from one antigenic type to another. Antigenic shift

Figure 35.4   Herd Immunity. The kinetics of the spread of an infectious disease and the effect of increasing the number of immune individuals in the population in limiting the disease. On day 1, a single infected individual enters the population. The incubation period is 1 day, and recovery occurs in 2 days. The number of susceptibles is the total population on day 1. The number of infected and recovered are illustrated in the two graphs.

Figure 35.5   Infectious Disease Cycle. See text for further details.

also occurs through the hybridization of different influenza virus serovars; two serovars of a virus intermingle to form a new antigenic type. Hybridization may occur between an animal strain and a human strain of the virus. Even though resistance in the human population becomes so high that the virus can no longer spread (herd immunity), it can be transmitted to animals, where the hybridization takes place. Smaller antigenic changes also can take place in pathogen strains and help the pathogen avoid host immune responses. These smaller changes are called **antigenic drift.**

Whenever antigenic shift or drift occurs, the population of susceptibles increases because the immune system has not been exposed to the new mutant strain. If the percentage of susceptibles is above the threshold density (figure 35.3), the level of protection provided by herd immunity will decrease and the morbidity rate will increase. For example, the morbidity rates of diphtheria and measles among school children may reach epidemic levels if the number of susceptibles rises above 30% for the whole population. As a result the goal of public health agencies is to make sure that at least 70% of the population is immunized against these diseases to provide the herd immunity necessary for protection of those who are not immunized.

---

1. How can epidemiologists recognize an infectious disease in a population? Define sign, symptom, and disease syndrome. What are the four phases seen during the course of an infection?
2. Differentiate between common-source and propagated epidemics.
3. Explain herd immunity.
4. What is the significance of antigenic shift and drift in epidemiology?

The Infectious Disease Cycle: Story of a Disease

To continue to exist, a pathogen must reproduce and be disseminated among its hosts. Thus an important aspect of infectious disease epidemiology is a consideration of how reproduction and dissemination occur. The **infectious disease cycle** or **chain** represents these events in the form of an intriguing epidemiological mystery story (figure 35.5).

What Pathogen Caused the Disease?

The first link in the infectious disease cycle is the pathogen. After an infectious disease has been recognized in a population, epidemiologists must correlate the disease outbreak with a specific pathogen. The disease's exact cause must be discovered. This is where Koch's postulates (*see chapter 1*), and modifications of them, are used to determine the etiology or cause of an infectious disease. At this point the clinical or diagnostic microbiology laboratory enters the investigation (*see chapter 34*). Its purpose is to isolate and identify the pathogen that caused the disease and to determine the pathogen's susceptibility to control agents or methods that may assist in its eradication.

Many pathogens can cause infectious diseases in humans and will be discussed in detail in chapters 36 to 39. Often these pathogens are transmissible from one individual to another. A **communicable disease** is one caused by a pathogen that can be transmitted from one host to another. Pathogens have the potential to produce disease (pathogenicity); however, this potential is a function of such factors as the number of pathogens, their virulence, and the nature and magnitude of host defenses (*pp. 562–63*).

What Was the Source and/or Reservoir of the Pathogen?

The source and/or reservoir of a pathogen is the second link in the infectious disease cycle. Identifying the source and/or reservoir is an important aspect of epidemiology. If the source or reservoir of the infection can be eliminated or controlled, the infectious disease cycle itself will be interrupted and transmission of the pathogen will be prevented (Boxes 35.1 and 35.2).

A **source** is the location from which the pathogen is immediately transmitted to the host, either directly through the environment or indirectly through an intermediate agent. The source can be either animate (e.g., humans or animals) or inanimate (e.g., water, soil, or food). The **period of infectivity** is the time during which the source is infectious or is disseminating the pathogen.

The **reservoir** is the site or natural environmental location in which the pathogen is normally found living and from which infection of the host can occur. Reservoirs also can be animate or inanimate.

Much of the time, human hosts are the most important animate sources of the pathogen and are called carriers. A **carrier** is an infected individual who is a potential source of infection for others. Carriers play an important role in the epidemiology of disease. Four types of carriers are recognized:

1. An **active carrier** is an individual who has an overt clinical case of the disease.
2. A **convalescent carrier** is an individual who has recovered from the infectious disease but continues to harbor large numbers of the pathogen.

TABLE 35.1 Infectious Organisms in Nonhuman Reservoirs That May Be Transmitted to Humans

Disease	Etiologic Agent	Usual or Suspected Nonhuman Host	Usual Method of Human Infection
Anthrax	*Bacillus anthracis*	Cattle, horses, sheep, swine, goats, dogs, cats, wild animals, birds	Inhalation or ingestion of spores; direct contact
Babesiosis	*Babesia bovis, B. divergens, B. microti, B. equi*	*Ixodes* ticks of various species	Bite of infected tick
Brucellosis (undulant fever)	*Brucella melitensis, B. abortus, B. suis*	Cattle, goats, swine, sheep, horses, mules, dogs, cats, fowl, deer, rabbits	Milk; direct or indirect contact
Campylobacteriosis	*Campylobacter fetus, C. jejuni*	Cattle, sheep, poultry, swine, pets, other animals	Contaminated water and food
Cat-scratch disease	*Bartonella henselae*	Cats, dogs	Cat or dog scratch
Colorado tick fever	*Coltivirus*	Squirrels, chipmunks, mice, deer	Tick bite
Cowpox	Cowpox virus	Cattle, horses	Skin abrasions
Cryptosporidiosis	*Cryptosporidium* spp.	Calves	Contact with infected calves
Encephalitis (California)	Arbovirus	Rats, squirrels, horses, deer, hares, cows	Mosquito
Encephalitis (St. Louis)	Arbovirus	Birds	Mosquito
Encephalomyelitis (Eastern equine)	Arbovirus	Birds, ducks, fowl, horses	Mosquito
Encephalomyelitis (Venezuelan equine)	Arbovirus	Rodents, horses	Mosquito
Encephalomyelitis (Western equine)	Arbovirus	Birds, snakes, squirrels, horses	Mosquito
Giardiasis	*Giardia lamblia*	Rodents, deer, cattle, dogs, cats	Contaminated water
Glanders	*Pseudomonas mallei*	Horses	Skin contact; inhalation
Hantavirus pulmonary syndrome	Pulmonary syndrome hantavirus	Deer mice	Contact with the saliva, urine, or feces of deer mice; aerosolized viruses
Herpes B viral encephalitis	*Herpesvirus simiae*	Monkeys	Monkey bite; contact with material from monkeys
Leptospirosis	*Leptospira interrogans*	Dogs, rodents, wild animals	Direct contact with urine, infected tissue, and contaminated water
Listeriosis	*Listeria monocytogenes*	Sheep, cattle, goats, guinea pigs, chickens, horses, rodents, birds, crustaceans	Food-borne
Lyme disease	*Borrelia burgdorferi*	Ticks (*Ixodes scapularis*) or related ticks	Bite of infected tick
Lymphocytic choriomeningitis	Arbovirus	Mice, rats, dogs, monkeys, guinea pigs	Inhalation of contaminated dust; ingestion of contaminated food
Mediterranean fever (boutonneuse fever, African tick typhus)	*Rickettsia conorii*	Dogs	Tick bite

From Guy Youmans, et al., *The Biologic and Clinical Basis of Infectious Diseases.* Copyright © 1985 W. B. Saunders, Philadelphia, PA. Reprinted by permission.

3. A **healthy carrier** is an individual who harbors the pathogen but is not ill.

4. An **incubatory carrier** is an individual who is incubating the pathogen in large numbers but is not yet ill.

Convalescent, healthy, and incubatory carriers may harbor the pathogen for only a brief period (hours, days, or weeks) and then are called **casual, acute,** or **transient carriers.** If they harbor the pathogen for long periods (months, years, or life), they are called **chronic carriers.**

As noted earlier, infectious diseases called zoonoses occur in animals and are occasionally transmitted to humans; thus these animals also can serve as reservoirs. Humans contract the pathogen by several mechanisms: coming into direct contact with diseased animal flesh (tularemia); drinking contaminated cow's milk (tuberculosis and brucellosis); inhaling dust particles contaminated by animal excreta or products (Q fever, anthrax); or eating insufficiently cooked infected flesh (anthrax, trichinosis). In addition, being bitten by insect **vectors** (organisms that spread disease from one host to another) such as mosquitoes, ticks, fleas, mites, or biting flies (equine encephalomyelitis and malaria, Lyme disease, Rocky Mountain spotted fever, plague, scrub typhus, and tularemia); or being bitten by a diseased animal (rabies) can lead to infection.

Table 35.1 lists some more common zoonoses found in the Western Hemisphere. This table is noninclusive in scope; it merely abbreviates the enormous spectrum of zoonotic diseases that are relevant to human epidemiology. Domestic animals are

TABLE 35.1   Continued

Disease	Etiologic Agent	Usual or Suspected Nonhuman Host	Usual Method of Human Infection
Melioidosis	*Pseudomonas pseudomallei*	Rats, mice, rabbits, dogs, cats	Arthropod vectors, water, food
Orf (contagious ecthyma)	Virus	Sheep, goats	Through skin abrasions
Pasteurellosis	*Pasteurella multocida*	Fowl, cattle, sheep, swine, goats, mice, rats, rabbits	Animal bite
Plague (bubonic)	*Yersinia pestis*	Domestic rats, many wild rodents	Flea bite
Psittacosis	*Chlamydia psittaci*	Birds	Direct contact, respiratory aerosols
Q fever	*Coxiella burnetii*	Cattle, sheep, goats	Inhalation of infected soil and dust
Rabies	Rabies virus (Rhabdovirus group)	Dogs, bats, opposums, skunks, foxes, cats, cattle	Bite of rabid animal
Rat bite fever	*Spirillum minus*	Rats, mice, cats	Rat bite
	Streptobacillus moniliformis	Rats, mice, squirrels, weasels, turkeys, contaminated food	Rat bite
Relapsing fever (borreliosis)	*Borrelia* spp.	Rodents, porcupines, opposums, armadillos, ticks, lice	Tick or louse bite
Rickettsialpox	*Rickettsia akari*	Mice	Mite bite
Rocky Mountain spotted fever	*Rickettsia rickettsii*	Rabbits, squirrels, rats, mice, groundhogs	Tick bite
Salmonellosis	*Salmonella* spp. (except *S. typhosa*)	Fowl, swine, sheep, cattle, horses, dogs, cats, rodents, reptiles, birds, turtles	Direct contact; food
Scrub typhus	*Rickettsia tsutsugamushi*	Wild rodents, rats	Mite bite
Tuberculosis	*Mycobacterium bovis*	Cattle, horses, cats, dogs	Milk; direct contact
Tularemia	*Francisella tularensis*	Wild rabbits, most other wild and and domestic animals	Direct contact with infected carcass, usually rabbit; tick bite, biting flies
Typhus fever (endemic)	*Rickettsia mooseri*	Rats	Flea bite
Vesicular stomatitis	Virus (Rhabdovirus group)	Cattle, swine, horses	Direct contact
Weil's disease (leptospirosis)	*Leptospira interrogans*	Rats, mice, skunks, opposums, wildcats, foxes, raccoons, shrews, bandicoots, dogs, cattle, swine	Through skin, drinking water, eating food
Yellow fever (jungle)	Yellow fever virus	Monkeys, marmosets, lemurs, mosquitoes	Mosquito

the most common source of zoonoses because they live in greater proximity to humans than do wild animals. Diseases of wild animals that are transmitted to humans tend to occur sporadically because close contact is infrequent. Other major reservoirs of pathogens are water, soil, and food. These reservoirs are discussed in detail in chapters 41 to 43.

How Was the Pathogen Transmitted?

To maintain an infectious disease in a human population, the pathogen must be transmitted from one host or source to another. Transmission is the third link in the infectious disease cycle and occurs by four main routes: airborne, contact, vehicle, and vector-borne.

Airborne Transmission

Because air is not a suitable medium for the growth of a pathogen, any pathogen that is airborne must have originated from a source such as humans, other animals, plants, soil, food, or water. In **airborne transmission** the pathogen is truly suspended in the air and travels over a meter or more from the source to the host. The pathogen can be contained within droplet nuclei or dust. **Droplet nuclei** are small particles, 1 to 4 μm in diameter, that represent what is left from the evaporation of larger particles (10 μm or more in diameter) called droplets. Droplet nuclei can remain airborne for hours or days and travel long distances.

When animals or humans are the source of the airborne pathogen, it usually is propelled from the respiratory tract into the air by an individual's coughing, sneezing, or vocalization. For example, enormous numbers of moisture droplets are aerosoled during a typical sneeze (figure 35.6). Each droplet is about 10 μm in diameter and initially moves about 100 m/second or more than 200 mi/hour!

Dust also is an important route of airborne transmission. At times a pathogen adheres to dust particles and contributes to the number of airborne pathogens when the dust is resuspended by some disturbance. A pathogen that can survive for relatively long periods in or on dust creates an epidemiological problem, particularly in hospitals, where dust can be the source of hospital-acquired infections. Table 35.2 summarizes some human airborne pathogens and the diseases they cause.

Box 35.3

The First Indications of Person–to–Person Spread of an Infectious Disease

In 1773 Charles White, an English surgeon and obstetrician, published his "Treatise on the Management of Pregnant and Lying-In Women." In it, he appealed for surgical cleanliness to combat childbed or puerperal fever. (**Puerperal fever** is an acute febrile condition that can follow childbirth and is caused by streptococcal infection of the uterus and/or adjacent regions.) In 1795 Alexander Gordon, a Scottish obstetrician, published his "Treatise on the Epidemic Puerperal Fever of Aberdeen," which demonstrated for the first time the contagiousness of the disease. In 1843 Oliver Wendell Holmes, a noted physician and anatomist in the United States, published a paper entitled "On the Contagiousness of Puerperal Fever" and also appealed for surgical cleanliness to combat this disease.

However, the first person to realize that a pathogen could be transmitted from one person to another was the Hungarian physician Ignaz Phillip Semmelweis. Between 1847 and 1849 Semmelweis observed that women who had their babies at the hospital with the help of medical students and physicians were four times as likely to contract puerperal fever as those who gave birth with the help of midwives. He concluded that the physicians and students were infecting women with material remaining on their hands after autopsies and other activities. Semmelweis thus began washing his hands with a calcium chloride solution before examining patients or delivering babies. This simple procedure led to a dramatic decrease in the number of cases of puerperal fever and saved the lives of many women. As a result Semmelweis is credited with being the pioneer of antisepsis in obstetrics. Unfortunately, in his own time, most of the medical establishment refused to acknowledge his contribution and adopt his procedures. After years of rejection Semmelweis had a nervous breakdown in 1865. He died a short time later of a wound infection. It is very probable that it was a streptococcal infection, arising from the same pathogen he had struggled against his whole professional life.

TABLE 35.2   Some Airborne Pathogens and the Diseases They Cause in Humans

Microorganism	Disease	Microorganism	Disease
Viruses		**Bacteria**	
Varicella	Chickenpox	*Actinomyces* spp.	Lung infections
Influenza	Flu	*Bordetella pertussis*	Whooping cough
Rubeola	Measles	*Chlamydia* spp.	Psittacosis
Rubella	German measles	*Cornyebacterium diphtheriae*	Diphtheria
Mumps	Mumps	*Mycoplasma pneumoniae*	Pneumonia
Poliomyelitis	Polio	*Mycobacterium tuberculosis*	Tuberculosis
Acute respiratory viruses	Viral pneumonia	*Neisseria* spp.	Meningitis
Pulmonary syndrome hantavirus	Hantavirus pulmonary syndrome	*Streptococcus* spp.	Pneumonia, sore throat
		Fungi	
		Blastomyces spp.	Lung infections
		Candida spp.	Disseminated infections
		Coccidioides spp.	Coccidioidomycosis
		Histoplasma capsulatum	Histoplasmosis

Figure 35.6   A Sneeze. High-speed photograph of an aerosol generated by an unstifled sneeze. The particles seen are comprised of saliva and mucus laden with microorganisms. These airborne particles may be infectious when inhaled by a susceptible host. Even a surgical mask will not prevent the spread of all particles.

Contact Transmission

Contact transmission implies the coming together or touching of the source or reservoir of the pathogen and the host (Box 35.3). Contact can be direct, indirect, or by droplet spread. Direct contact implies an actual physical interaction with the infectious source. This route is frequently called person-to-person contact. Person-to-person transmission occurs primarily by touching, kissing, or sexual contact (sexually transmitted diseases); by contact with oral secretions or body lesions (herpes and boils); by nursing mothers (staphylococcal infections); and through the placenta (AIDS, syphilis). Some infectious pathogens also can be transmitted by direct contact with animals or animal products (*Salmonella* and *Campylobacter*).

Indirect contact refers to the transmission of the pathogen from the source to the host through an intermediary—most often an inanimate object. The intermediary is usually contaminated by an animate source. Common examples of intermediary inanimate objects include thermometers, eating utensils, drinking cups, and bedding. *Pseudomonas* bacteria are easily transmitted by this route. This mode of transmission is often also considered a form of vehicle transmission (see next section).

In droplet spread the pathogen is carried on particles larger than 5 μm. The route is through the air but only for a very short distance—usually less than a meter. Because these particles are large, they quickly settle out of the air. As a result droplet transmission of a pathogen depends on the proximity of the source and the host. Measles is an example of a droplet-spread disease.

Vehicle Transmission

Inanimate materials or objects involved in pathogen transmission are called **vehicles.** In **common vehicle transmission** a single inanimate vehicle or source serves to spread the pathogen to multiple hosts but does not support its reproduction. Examples include surgical instruments, bedding, and eating utensils. In epidemiology these common vehicles are called **fomites** [s., fomes or fomite]. A single source containing pathogens (blood, drugs, IV fluids) can contaminate a common vehicle that causes multiple infections. Food and water are important common vehicles for many human diseases (*see tables 37.3 and 43.5*).

Vector-Borne Transmission

As noted earlier, living transmitters of a pathogen are called vectors. Most vectors are arthropods (insects, ticks, mites, fleas) or vertebrates (dogs, cats, skunks, bats). **Vector-borne transmission** can be either external or internal. In external (mechanical) transmission the pathogen is carried on the body surface of a vector. Carriage is passive, with no growth of the pathogen during transmission. An example would be flies carrying *Shigella* organisms on their feet from a fecal source to a plate of food that a person is eating.

In internal transmission the pathogen is carried within the vector. Here it can go into either a harborage or biologic transmission phase. In **harborage transmission** the pathogen does not undergo morphological or physiological changes within the vector. An example would be the transmission of *Yersinia pestis* (the etiologic agent of plague) by

the rat flea from rat to human. **Biologic transmission** implies that the pathogen does go through a morphological or physiological change within the vector. An example would be the developmental sequence of the malarial parasite inside its mosquito vector.

Malaria (pp. 802–4).

IV Why Was the Host Susceptible to the Pathogen?

The fourth link in the infectious disease cycle is the host. The susceptibility of the host to a pathogen depends on both the pathogenicity of the organism and the nonspecific and specific defense mechanisms of the host. These susceptibility factors are the basis for chapters 29 through 32 and are outside the realm of epidemiology.

V How Did the Pathogen Leave the Host?

The fifth and last link in the infectious disease cycle is release or exit of the pathogen from the host. It is equally important that the pathogen escapes from its host as it is that the pathogen originally contacts and enters the host. Unless a successful escape occurs, the disease cycle will be interrupted and the pathogenic species will not be perpetuated. Escape can be active or passive, although often a combination of the two occurs. Active escape takes place when a pathogen actively moves to a portal of exit and leaves the host. Examples include the many parasitic helminths that migrate through the body of their host, eventually reaching the surface and exiting. Passive escape occurs when a pathogen or its progeny leaves the host in feces, urine, droplets, saliva, or desquamated cells. Microorganisms usually employ passive escape mechanisms.

---

1. What are some epidemiologically important characteristics of a pathogen? What is a communicable disease?
2. Define source, reservoir, period of infectivity, and carrier.
3. What types of infectious disease carriers does epidemiology recognize?
4. Describe the four main types of infectious disease transmission and give examples of each.
5. Define the terms droplet nuclei, vehicle, fomite, and vector.

Virulence and the Mode of Transmission

There is evidence that a pathogen's virulence may be strongly influenced by its mode of transmission and ability to live outside its host. When the pathogen uses a mode of transmission such as direct contact, it cannot afford to make the host so ill that it will not be transmitted effectively. This is the case with the common cold, which is caused by rhinoviruses and several other respiratory viruses. If the virus reproduced too rapidly and damaged its host extensively, the person would be bedridden and not contact others. The efficiency of transmission would drop because rhinoviruses shed from the cold sufferer could not contact new hosts and would be inactivated by exposure. Cold sufferers must be able to move about and directly contact others. Thus virulence is low and people are not incapacitated by the common cold.

On the other hand, if a pathogen uses a mode of transmission not dependent on host health and mobility, then the person's health will not be a critical matter. The pathogen might be quite successful—that is, transmitted to many new hosts even though it kills its host relatively quickly. Host death will mean the end of any resident pathogens, but the species as a whole can spread and flourish as long as the increased transmission rate outbalances the loss due to host death. This situation may arise in several ways.

When a pathogen is transmitted by a vector, it will be benefitted by extensive reproduction and spread within the host. If pathogen levels are very high in the host, a vector such as a biting insect has a better chance of picking up the pathogen and transferring it to a new host. Indeed, pathogens transmitted by biting arthropods such as mosquitoes often are very virulent (e.g., malaria, typhus, sleeping sickness). It is important that such pathogens take good care of their vectors, and the vector generally remains healthy, at least long enough for pathogen transmission.

Virulence also is often directly correlated with a pathogen's ability to survive in the external environment. If a pathogen cannot survive well outside its host and does not use a vector, it depends on host survival and will tend to be less virulent. When a pathogen can survive for long periods outside its host, it can afford to leave the host and simply wait for a new one to come along. This seems to promote increased virulence. Host health is not critical, but extensive multiplication within the host will increase the efficiency of transmission. Good examples are tuberculosis and diphtheria. *Mycobacterium* and *Corynebacterium* survive for a long time, at least weeks to months, outside human hosts.

Human cultural patterns and behavior almost certainly also affect pathogen virulence. Waterborne pathogens such as *Vibrio cholerae* (which causes diarrhea) are transmitted through drinking-water systems. They can be virulent because immobile hosts still shed pathogens, which frequently reach the water. By this argument, establishment of uncontaminated drinking water supplies should reduce cholera virulence, and this seems to be the case. The same appears to be true of *Shigella* and bacillary dysentery. Often one of the best ways to reduce virulence may be to reduce the frequency of transmission.

The Emergence of New Diseases

By the late 1960s and early 1970s, many thought that the threat of disease had almost been eliminated because of the development of antibiotics and public health measures. However, more recently many new diseases have arisen: AIDS, hepatitis C and E, hantavirus pulmonary syndrome, Lyme disease, Legionnaires' disease, toxic-shock syndrome, and others. It is now clear that society will be continually faced with both new infectious diseases and the resurgence of older diseases once thought

to be conquered (e.g., tuberculosis). This realization has stimulated the establishment of a field called systematic epidemiology, which focuses on the ecological and social factors that influence the development and emergence of disease.

Several factors characteristic of the modern world undoubtedly favor the development and spread of new diseases. Potential pathogens must find a susceptible host population in order to reproduce. The spread of a new disease often used to be limited by the travel time needed to reach a new host population. If the travel time was sufficiently long, as when a ship crossed the ocean, the infected travelers would either recover or die before reaching a new population. Now the world is linked by a very rapid transportation system, and an infected person or animal can be transported almost anywhere within hours to days. Large populations can even move rapidly across continents in response to economic distress and political oppression. When they do, they carry along their diseases. International commerce can rapidly transport vectors and infected plants and animals anywhere in the world.

New diseases more easily arise and take root when ecosystems are disrupted or changing. When a habitat has low biological diversity or has been disturbed by natural events or human activity, there are fewer predators and competitors for the new pathogens and vectors. If the host population is compromised and put at risk by nutritional factors or personal habits such as sexual practices and smoking, a new disease may take hold more readily. For example, the spread of Lyme disease in New England probably was due partly to ecological disruption that eliminated predators of deer. An increase in deer and the deer tick provided a favorable situation for pathogen spread to people. The clearing of tropical forests for economic development may trigger the spread of new diseases as pathogens move from old hosts to the human population. Both narcotic use and sexual promiscuity greatly stimulate the spread of disease. The development of drug-resistant pathogens is another example of human activity affecting the evolution of virulence.

It is probably best to view disease frequency changes as the outcome of many different factors, ranging from mutational events in individual pathogens to changes in economic and behavioral patterns in populations worldwide. Broad-scale ecological and climatic alterations also profoundly influence the development and spread of new diseases. Because the world is now so interrelated, we cannot wall ourselves off from other countries and continents. Changes in the disease status of one part of the world may well affect the health of the remainder.

Control of Epidemics

The development of an infectious disease is a complex process involving many factors, as is the design of specific epidemiological control measures. Epidemiologists must consider available resources and time constraints, adverse effects of potential control measures, and human activities that might influence the spread of the infection. Many times control activities reflect compromises among alternatives. To proceed intelligently, one must identify components of the infectious disease cycle that are primarily responsible for a particular epidemic. Control measures should be directed toward that part of the cycle that is most susceptible to control—the weakest link in the chain.

There are three kinds of control measures. The first kind is directed toward reducing or eliminating the source or reservoir of infection:

1. Quarantine and isolation of cases and carriers
2. Destruction of an animal reservoir of infection
3. Treatment of sewage to reduce water contamination
4. Therapy that reduces or eliminates infectivity of the individual

The second kind of control measure is designed to break the connection between the source of the infection and susceptible individuals. Examples include general sanitation measures:

1. Chlorination of water supplies
2. Pasteurization of milk
3. Supervision and inspection of food and people who handle food
4. Destruction of vectors by spraying with insecticides

The third type of control measure reduces the number of susceptible individuals and raises the general level of herd immunity by immunization. Examples include the following:

1. Passive immunization to give a temporary immunity following exposure to a pathogen or when a disease threatens to take an epidemic form
2. Active immunization to protect the individual from the pathogen and the host population from the epidemic

The Role of the Public Health System: Epidemiological Guardian

The control of an infectious disease relies heavily on a well-defined network of nurses, physicians, epidemiologists, and infection control personnel who supply epidemiological information to a network of local, state, national, and international organizations. These individuals and organizations comprise the public health system. For example, each state has a public health laboratory that is involved in infection surveillance and control. The communicable disease section of a state laboratory includes specialized laboratory services for the examination of specimens or cultures submitted by physicians, the local health department, hospital laboratories, sanitarians, epidemiologists, and others. These groups share their findings with other health agencies in the state, the Centers for Disease Control and Prevention (CDC), and the World Health Organization (WHO).

The Centers for Disease Control and Prevention (formerly the Centers for Disease Control) of the United States Public Health Service (USPHS), located in Atlanta, Georgia, has a central role in the U.S. public health system. The CDC (1) supports all United States laboratories, the World Health Organization,

and other countries in their efforts against infectious diseases; (2) publishes the accumulated state reports in the *Morbidity and Mortality Weekly Reports (MMWR)*; (3) establishes criteria for the safe and accurate performance of microbiological procedures; (4) trains public health personnel; and (5) conducts seminars on all aspects of epidemiology.

Nosocomial Infections

Nosocomial infections [Greek *nosos,* disease, and *komeion,* to take care of] are produced by infectious pathogens that develop within a hospital or other type of clinical care facility and are acquired by patients while they are in the facility. Besides harming patients, nosocomial infections can affect nurses, physicians, aides, visitors, salespeople, delivery personnel, custodians, and anyone who has contact with the hospital. Most nosocomial infections become clinically apparent while patients are still hospitalized; however, disease onset can occur after patients have been discharged. Infections that are incubating when patients are admitted to a hospital are not nosocomial; they are community acquired. However, because such infections can serve as a ready source or reservoir of pathogens for other patients or personnel, they are also considered in the total epidemiology of nosocomial infections.

Source

Pathogens that cause nosocomial diseases come from either endogenous or exogenous sources. Endogenous sources are the patient's own microbiota; exogenous sources are microbiota other than the patient's. Endogenous pathogens are either brought into the hospital by the patient or are acquired when the patient becomes colonized after admission. In either case the pathogen colonizing the patient may subsequently cause a nosocomial disease (e.g., when the pathogen is transported to another part of the body or when the host's resistance drops). If it cannot be determined that the specific pathogen responsible for a nosocomial disease is exogenous or endogenous, then the term autogenous is used. An **autogenous infection** is one that is caused by an agent derived from the microbiota of the patient, despite whether it became part of the patient's microbiota following his or her admission to the hospital.

There are many potential exogenous sources in a hospital. Animate sources are the hospital staff, other patients, and visitors. Some examples of inanimate exogenous sources are food, urinary catheters, intravenous and respiratory therapy equipment, and water systems (softeners, dialysis units, and hydrotherapy equipment).

Control, Prevention, and Surveillance

In the United States nosocomial infections occur in approximately 5% of the patients admitted to acute-care hospitals, prolong hospital stays by 4 to 13 days, result in over 4.5 billion dollars a year in direct hospital charges, and lead to over

20,000 direct and 60,000 indirect deaths annually. The enormity of this problem has led most hospitals to allocate substantial resources to the development of methods and programs for the surveillance, prevention, and control of nosocomial infections.

All personnel involved in the care of patients should be familiar with basic infection control measures such as isolation policies of the hospital; aseptic techniques; proper handling of equipment, supplies, food, and excreta; and surgical wound care and dressings. To adequately protect their patients, hospital personnel must practice proper aseptic technique and hand-washing procedures, and must wear gloves when contacting mucous membranes, secretions, and "moist body substances." Patients also should be monitored with respect to the frequency, distribution, symptomatology, and other characteristics common to nosocomial infections. This constant control and surveillance can be invaluable in preventing many nosocomial infections, patient discomfort, extra stays, and further expense. Nosocomial infections are discussed further in chapter 38 (*pp. 785–86*).

The Hospital Epidemiologist

Because of nosocomial infections, all hospitals desiring accreditation by the Joint Commission on Accreditation of Healthcare Organizations (JCAHO) must have a designated individual directly responsible for developing and implementing policies governing control of infections and communicable diseases. This individual may be a registered nurse and is known as a hospital epidemiologist, nurse epidemiologist, infection control nurse, or infection control practitioner. In larger hospitals, a physician is the hospital epidemiologist and should be trained in infectious diseases. He or she oversees a staff that includes nurse epidemiologists, quality assurance specialists, fellows in infectious disease/hospital epidemiology, and epidemiology technicians. The hospital epidemiologist must meet with an infection control committee composed of various professionals who have expertise in the different aspects of infection control and hospital operation. The infection control committee periodically evaluates laboratory reports, patients' charts, and surveys done by the hospital epidemiologist to determine whether there has been any increase in the frequency of particular infectious diseases or potential pathogens.

Overall, the services provided by the hospital epidemiologist should include at least the following:

1. Research in infection control
2. Evaluation of disinfectants, rapid test systems, and other products
3. Efforts to encourage appropriate legislation related to infection control, particularly at the state level
4. Efforts to contain hospital operating costs, especially those related to fixed expenses such as the DRGs (diagnosis-related groups)
5. Surveillance and comparison of endemic and epidemic infection frequencies
6. Direct participation in a variety of hospital activities relating to infection control and maintenance of employee health
7. Education of hospital personnel in communicable disease control and disinfection and sterilization procedures
8. Establishment and maintenance of a system for identifying, reporting, investigating, and controlling infections and communicable diseases of patients and hospital personnel
9. Maintenance of a log of incidents related to infections and communicable diseases
10. Monitoring trends in the antimicrobial drug resistance of infectious agents

Recently computer software packages have been developed to aid the infection control practitioner. Such packages generate routine reports, cause-and-effect tabulations, and graphics for the daily epidemiological monitoring that must be done.

---

1. Describe how virulence and the mode of transmission may be related. What might cause the development of new human diseases?
2. In what three general ways can epidemics be controlled? Give one or two specific examples of each type of control measure.
3. What role does the public health system play in epidemiology?
4. Describe a nosocomial infection.
5. What two general sources are responsible for nosocomial infections? Give some specific examples of each general source.
6. Why are nosocomial infections important?
7. What does a hospital epidemiologist do to control nosocomial infections?

Summary

1. Epidemiology is the science that evaluates the determinants, occurrence, distribution, and control of health and disease in a defined population.

2. The epidemiology of an infectious disease involves the determination of what pathogen caused a specific disease, the source and/or reservoir of the pathogen, how the disease was transmitted, and what host and environmental factors could have caused the disease to develop within a defined population. Disease control and elimination are also important functions of the epidemiologist.

3. Epidemiological data can be obtained from such factors as morbidity, prevalence, and mortality rates.

4. Surveillance is necessary for recognizing a specific infectious disease within a given population. This consists of gathering data on the occurrence of the disease, collating and analyzing the data, summarizing the findings, and applying the information to control measures.

5. One purpose of the clinical microbiology laboratory is to isolate and identify the pathogen responsible for an infectious disease outbreak.

6. A common-source epidemic is characterized by a sharp rise to a peak and then a rapid, but not as pronounced, decline in the number of individuals infected. A propagated epidemic is characterized by a relatively slow and prolonged rise and then a gradual decline in the number of individuals infected.

7. Herd immunity is the resistance of a population to infection and pathogen spread because of the immunity of a large percentage of the individuals within the population.

8. The infectious disease cycle or chain involves the characteristics of the pathogen, the source and/or reservoir of the pathogen, the transmission of the pathogen, the susceptibility of the host, the exit mechanism of the pathogen from the body of the host, and its spread to a new reservoir or host.

9. There are four major modes of transmission: airborne, contact, vehicle, and vector-borne.

10. The degree of virulence may be influenced by the pathogen's preferred mode of transmission. New human diseases may arise and spread because of ecosystem disruption, rapid transportation, human behavior, and other factors.

11. The public health system consists of individuals and organizations that function in the control of infectious diseases and epidemics.

12. Epidemiological control measures can be directed toward reducing or eliminating infection sources, breaking the connection between sources and susceptible individuals, or isolating the susceptible individuals and raising the general level of herd immunity by immunization.

13. Nosocomial infections are infections that develop within a hospital and are produced by a pathogen acquired during a patient's stay. These infections come from either endogenous or exogenous sources.

14. Hospitals must designate an individual to be responsible for identifying and controlling nosocomial infections. This person is known as a hospital epidemiologist, nurse epidemiologist, infection control nurse, or infection control practitioner.

Key Terms

active carrier
acute carrier
airborne tran
antigenic d
antigenic
autogen
biolog
carrie
cas
ch
c

enzootic *701*
epidemic *700*
epidemiologist *700*
epidemiology *700*
-izootic *701*
701

709

03
ease *700*
d *702*
ier *706*
700
sease cycle (chain) *705*
rate *701*
rate *701*
nial infection *711*

outbreak *700*
pandemic *700*
panzootic *701*
period of infectivity *705*
prevalence rate *701*
prodromal stage *702*
propagated epidemic *703*
puerperal fever *708*
reservoir *705*
signs *702*
source *705*
sporadic disease *700*
statistics *701*
symptoms *702*
transient carrier *706*
vector *706*
vector-borne transmission *709*
vehicle *709*
zoonoses *701*

Questions for Thought and Review

1. Why is international cooperation a necessity in the field of epidemiology?

2. Are any risks involved in attempting to eliminate a pathogen from the world?

3. Why is a knowledge of statistics important to an epidemiologist?

4. What is the role of the Centers for Disease Control and Prevention in epidemiology?

5. What common sources of infectious disease are found in your community? How can the etiologic agents of such infectious diseases spread from their source or reservoir to members of your community?

6. How could you experimentally prove the cause of an infectious disease?

7. How do epidemiologists recognize an infectious disease within a given population?

8. How could you prove that an epidemic of a given infectious disease was occurring?

9. Why is the infectious disease cycle also referred to as a chain?

10. How can epidemics be controlled?

11. Why are nosocomial infections so important to humans?

12. Why do some epidemiologists believe it is impossible to prevent all nosocomial infections?

13. What is the difference between a fomite and a vector? A reservoir and a source?

14. How can changes in herd immunity contribute to an outbreak of a disease on an island?

15. What is an index case?

16. Where in the world would you expect new human diseases to arise? Why?

Additional Reading

General

Axnick, K. J., and Yarbrough, M. 1984. *Infection control: An integrated approach.* St. Louis: C. V. Mosby.

Ayliffe, G. A., and Taylor, L. J. 1982. *Hospital acquired infections: Principles and prevention.* Littleton, Mass.: John Wright-PSG.

Bennett, J. V., and Brachman, P. S., editors. 1986. *Hospital infections,* 2d ed. Boston: Little, Brown.

Centers for Disease Control and Prevention. *Morbidity and mortality reports.* (A weekly report that discusses infectious diseases.) Atlanta: Centers for Disease Control and Prevention.

Centers for Disease Control and Prevention. *Surveillance.* (Annual summaries of specific infectious diseases.) Atlanta: Centers for Disease Control and Prevention.

Epidemiology issue. 1986. *Science* 234 (November 24): 921.

Ewald, P. W. 1993. The evolution of virulence. *Sci. Am.* 268(4):86–93.

Garrett, L. 1994. *The coming plague: Newly emerging diseases in a world out of balance.* New York: Farrar, Straus and Giroux.

Jaret, P. 1991. The disease detectives. *Nat. Geographic* 179(1):114–40.

Kahn, H. A. 1983. *An introduction to epidemiologic methods.* New York: Oxford University Press.

Kiple, K. F., editor. 1993. *The Cambridge world history of human disease.* New York: Cambridge University Press.

Levins, R.; Awerbuch, T.; Brinkmann, U.; Eckardt, I.; Epstein, P.; Makhoul, N.; de Possas, C. A.; Puccia, C.; Spielman, A.; and Wilson, M. E. 1994. The emergence of new diseases. *American Scientist* 82:52–60.

Lilienfeld, D. E., and Stolley, P. 1994. *Foundations of epidemiology,* 3d ed. New York: Oxford University Press.

Mack, A., editor. 1991. *In time of plague: The history and social consequences of lethal epidemic disease.* New York: New York University Press.

Mandell, G. L.; Bennett, J. E.; and Dolin, R. 1995. *Principles and practice of infectious diseases,* 4th ed. New York: Churchill Livingstone.

Mausner, J. S., and Kramer, S. 1985. *Epidemiology: An introductory text,* 2d ed. Philadelphia: W. B. Saunders.

Preston, R. 1994. *The hot zone.* New York: Random House.

Roueche, B. 1984. *The medical detectives.* 2 vols. New York: Truman Talley Books.

Thomas, G., and Morgan-Witts, M. 1982. *Anatomy of an epidemic.* New York: Doubleday.

The Infectious Disease Cycle: Story of a Disease

Anderson, R. M., and May, R. M. 1992. Understanding the AIDS pandemic. *Sci. Am.* 266(5):58–67.

Chesney, P. J.; Bergdoll, M. S.; and Davis, J. P. 1984. The disease spectrum, epidemiology, and etiology of toxic-shock syndrome. *Ann. Rev. Microbiol.* 38:315–38.

Cliff, A., and Haggett, P. 1984. Island epidemics. *Sci. Am.* 250(5):138–47.

Fraser, D. W., and McDade, J. E. 1979. Legionellosis. *Sci. Am.* 241(4):82–99.

Heyward, W. L., and Curran, J. W. 1988. The epidemiology of AIDS in the U.S. *Sci. Am.* 259(4):72–81.

Kaplan, M. M., and Webster, R. G. 1977. The epidemiology of influenza. *Sci. Am.* 237(6):88–92.

Mann, J. M.; Chin, J.; Piot, P.; and Quinn, T. 1988. The international epidemiology of AIDS. *Sci. Am.* 259(4):82–89.

McEvedy, C. 1988. The bubonic plague. *Sci. Am.* 258(2):118–23.

Moore, P., and Broome, C. 1994. Cerebrospinal meningitis epidemics. *Sci. Am.* 271(5):38–47.

Riley, L. W. 1987. The epidemiologic, clinical, and microbiological features of hemorrhagic colitis. *Ann. Rev. Microbiol.* 41:383–407.

Control of Epidemics

Haley, R. W.; Quade, D.; Freeman, H. E.; and Bennett, J. V. 1980. Conceptual model of an infection surveillance and control program. *Am. J. Epidemiol.* 111:608–12.

Salk, D. 1980. Eradication of poliomyelitis in the United States. *Rev. Infect. Dis.* 2:228–30.

Schultz, M. G. 1983. Emerging zoonoses. *N. Engl. J. Med.* 308:1285.

Nosocomial Infections

Beyt, B. E., Jr.; Troxler, S. H.; and Guidry, J. L. 1984. Computer assisted hospital surveillance and control of nosocomial infections. *Clinical Research* 32:291A.

Harris, A. A.; Levin, S.; and Rrenholme, G. 1984. Selected aspects of nosocomial infections in the 1980s. *Am. J. Med.* 77(1B):3–11.

CHAPTER 36
Human Diseases Caused by Viruses

It is a modern plague: the first great pandemic of the second half of the 20th century. The flat, clinical-sounding name given to the disease by epidemiologists—acquired immune deficiency syndrome—has been shortened to the chilling acronym AIDS.

—Robert C. Gallo

Outline

Concepts

1. Some viruses can be transmitted through the air and directly or indirectly involve the respiratory system. Most of these viruses are highly communicable and cause diseases such as chickenpox, influenza, measles, mumps, respiratory syndromes and viral pneumonia, rubella, hantavirus pulmonary syndrome, and the now extinct smallpox.

2. The arthropod-borne diseases are transmitted by arthropod vectors from human to human or animal to human. Examples include the various encephalitides, Colorado tick fever, and historically important yellow fever.

3. Some viruses are so sensitive to environmental influences that they are unable to survive for significant periods of time outside their hosts. These viruses are transmitted from host to host by direct contact and cause diseases such as AIDS, cold sores, the common cold, cytomegalovirus inclusion disease, genital herpes, human herpesvirus 6 infections, human parvovirus B19 infections, certain leukemias, infectious mononucleosis, rabies, and viral hepatitides.

4. Viruses that can be transmitted by food and water and usually either grow in or pass through the intestinal system leave the body in the feces and are acquired through the oral route. Examples of such diseases include viral gastroenteritis, hepatitis A and E, and poliomyelitis.

5. The slow virus diseases represent progressive pathological processes caused by viruses or prions that remain clinically silent during a prolonged period of months or years, after which progressive clinical disease becomes apparent, usually ending months later in profound disability or death. Examples include Creutzfeldt-Jakob disease, kuru, progressive multifocal leukoencephalopathy, Gerstmann-Sträussler-Scheinker syndrome, and subacute sclerosing panencephalitis.

6. One other disease associated with viruses but that does not fit into any of the foregoing categories is warts.

Chapters 16, 17, and 18 provide a review of the general biology of viruses and an introduction to basic virology. Chapter 36 continues this coverage by discussing viruses that are pathogenic to humans. Viruses are grouped according to their mode of acquisition and transmission, and viral diseases that occur in the United States are emphasized.

More than 400 different viruses can infect humans. Human diseases caused by viruses are unusually interesting, considering the small amount of genetic information introduced into a host cell. This apparent simplicity belies the severe pathological features, clinical consequences, and death that result from many viral diseases. With few exceptions, only prophylactic or supportive treatment is available. Collectively these diseases are some of the most common and yet most puzzling of all infectious diseases. The resulting frustration is compounded when year after year familiar diseases of unknown etiology become linked to virus infections.

Airborne Diseases

Because air does not support virus growth, any virus that is airborne must have originated from a source such as another human. When humans are the source of the airborne virus, it usually is propelled from the respiratory tract by an individual's coughing, sneezing, or vocalizing.

Chickenpox (Varicella) and Shingles (Zoster)

Chickenpox (varicella) is a highly contagious skin disease primarily of children 2 to 7 years of age. It is estimated that about 4 million cases occur annually in the United States. The causative agent is the varicella-zoster virus, a member of the family *Herpesviridae,* which is acquired by droplet inhalation into the respiratory system. Following an incubation period of from 10 to 23 days, small vesicles erupt on the face or upper trunk, fill with pus, rupture, and become covered by scabs (figure 36.1). Healing of the vesicles occurs in about 10 days. During this time intense itching often occurs; treatment is with acyclovir. Chickenpox can be prevented or the infection shortened with an attenuated varicella vaccine (Varivax).

Individuals who recover from chickenpox are subsequently immune to this disease; however, they are not free of the virus, as viral DNA resides in a dormant (latent) state within the nuclei of sensory neurons in the dorsal root ganglia. This viral DNA is maintained in infected cells but virions cannot be detected (figure 36.2a). When the infected person becomes immunocompromised by such factors as age, AIDS, or psychological or physiological stress, the viruses may become activated (figure 36.2b). They migrate down sensory nerves, initiate viral replication, and produce painful vesicles (figure 36.2c) because of sensory nerve damage. This reactivated form of chickenpox is called **shingles (zoster).** Most cases occur in people over 50 years of age. Shingles does not require specific therapy; however, in immunocompromised individuals, acyclovir (Zovirax; *see figure 33.14*) or famciclovir (Famvir) are recommended.

(a)

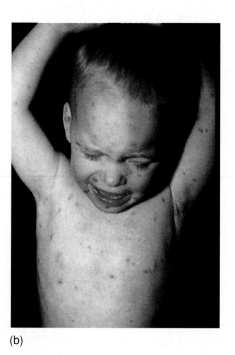

(b)

Figure 36.1   **Chickenpox (Varicella).** (*a*) Pathogenesis. (*b*) Typical vesicular skin rash.

(a) **Primary infection**

(b) **Recurrence**

(c)

Figure 36.2   Pathogenesis of the Varicella-Zoster Virus.
(*a*) After an initial infection with varicella (chickenpox), the viruses migrate up sensory peripheral nerves to their ganglia, producing a latent infection. (*b*) When a person becomes immunocompromised or is under psychological or physiological stress, the viruses may be activated. (*c*) They migrate down sensory nerve axons, initiate viral replication, and produce painful vesicles.

Influenza (Flu)

Influenza [Italian, to influence], or the **flu,** is a respiratory system disease caused by orthomyxoviruses. Influenza viruses are classified into A, B, and C groups based on the antigens (H and N) of their protein coats. One unique feature of the influenza viruses is the frequency with which changes in antigenicity occur. These changes are called antigenic variation. If the variation is small, it is called antigenic drift; if it is large, it is called antigenic shift. Antigenic variation occurs almost yearly with the influenza A virus, less frequently with the B virus, and has not been demonstrated with the C virus. Alteration of the virus's antigenic structure can lead to infection in humans by variants of the virus to which little or no resistance is present in the population at risk. This explains why influenza continues to be a major epidemic disease and frequently produces worldwide pandemics.

The virus (*see figure 16.17*a) is acquired by inhalation or ingestion of virus-contaminated respiratory secretions. During an incubation period of 1 to 2 days, the virus adheres to the epithelium of the respiratory system and hydrolyzes the mucus that covers it using the neuraminidase present in the viral en-

velope spikes. The virus then attaches to the epithelial cell by its hemagglutinin spike protein, causing part of the cell's plasma membrane to bulge inward, seal off, and form a vesicle (receptor-mediated endocytosis). This encloses the virus in an endosome. The hemagglutinin molecule in the virus envelope undergoes a dramatic conformational change when the endosomal pH decreases. The hydrophobic ends of the hemagglutinin spring outward and extend toward the endosomal membrane. After they contact the membrane, fusion occurs and the RNA nucleocapsid is released into the cytoplasmic matrix.

Influenza is characterized by chills, fever, headache, malaise, and general muscular aches and pains. These symptoms arise from the death of respiratory epithelial cells, probably due to attacks by activated T cells. Recovery usually occurs in 3 to 7 days, during which coldlike symptoms appear as the fever subsides. Influenza alone usually is not fatal. However, death may result from pneumonia caused by secondary bacterial invaders such as *Staphylococcus aureus, Streptococcus pneumoniae,* and *Haemophilus influenzae.* Influenza epidemics vary greatly in severity and mortality rate. For example, the pandemics of 1918 and 1957 were severe, whereas the 1977 pandemic was mild. The most disastrous recorded pandemic

Box 36.1

Reye's and Guillain-Barré Syndromes

An occasional complication of influenza, chickenpox, and a few other viral diseases in children under 14 years of age is **Reye's syndrome.** After the initial infection has disappeared, the child suddenly begins to vomit persistently and experiences convulsions followed by delirium and a coma. Pathologically, the brain swells with injury to the neuronal mitochondria, fatty infiltration into the liver occurs, blood ammonia is elevated, and both serum glutamic oxaloacetic transaminase (SGOT) and serum glutamic pyruvic transaminase (SGPT) are elevated in the blood. Diagnosis is made by the measurement of the levels of these enzymes and ammonia.

The relationship between the initial viral infection and the brain and liver damage is unknown. Treatment is nonspecific and directed toward reducing intracranial pressure and correcting metabolic and electrolyte abnormalities. Some children who recover have residual neurological deficits—impaired mental capacity, seizures, and hemi-

plegia (paralysis on one side of the body). Mortality ranges from 10 to 40%. It is suspected that the use of aspirin or salicylate-containing products to lower the initial viral fever increases one's chances of acquiring Reye's syndrome.

Another condition that involves the central nervous system and is associated with influenza infections is **Guillain-Barré syndrome** (sometimes called **French polio**). In this disorder the individual suffers a delayed reaction (usually within 8 weeks) either to the actual virus infection or to vaccines against influenza. The virus or viral antigen in the vaccine damages the Schwann cells that myelinate the peripheral nerves and thus causes demyelination. As a result a prominent feature of this syndrome is a symmetric weakness of the extremities and sensory loss. Fortunately recovery usually is complete because Schwann cells eventually proliferate and wrap around the demyelinated nerves.

occurred in 1918–1919. It has been estimated that 21 million died worldwide, and about 550,000 deaths were recorded in the United States alone. A commercially available identification technique is Directigen FLU-A (an enzyme immunoassay [EIA] rapid test). This test can detect influenza A virus in clinical specimens in less than 15 minutes.

As with many other viral diseases, only the symptoms of influenza usually are treated. However, the antiviral drugs amantadine (*see figure 33.14*) and rimantidine have been shown to reduce the duration and symptoms of type A influenza if administered promptly. Aspirin (salicylic acid) should be avoided in children younger than 14 years to reduce the risk of Reye's syndrome (Box 36.1). The mainstay for prevention of influenza since the late 1940s has been inactivated virus vaccines, especially for the chronically ill, individuals over age 65, residents of nursing homes, and health-care workers in close contact with people at risk.

Antiviral drugs (pp. 672–73).

Currently two influenza A virus subtypes are epidemic in humans: those with H1N1 and H3N2 hemagglutinin and neuraminidase surface glycoproteins. Influenza A infections usually peak in the winter and involve 10% or more of the population, with rates of 50 to 75% in school-age children. Influenza B usually accounts for only 3% of all flu cases in the United States.

Measles (Rubeola)

Measles [rubeola: Latin *rubeus,* red] is a highly contagious skin disease that is endemic throughout the world. The measles virus, a member of the genus *Morbillivirus* and the family *Paramyxoviridae,* enters the body through the respiratory tract or the conjunctiva of the eyes. The receptor for the measles

virus has recently been shown to be the complement regulator CD46, also known as membrane cofactor protein. The incubation period is usually 10 to 21 days, and the first symptoms begin about the tenth day with a nasal discharge, cough, fever, headache, and conjunctivitis. Within 3 to 5 days skin eruptions occur as faintly pink macules (figure 36.3) and normally last about 5 to 10 days. (A macule or macula is a discolored area on the skin that is not raised above the skin surface.) Lesions of the oral cavity include the diagnostically useful bright-red **Koplik's spots** with a bluish-white speck in the center of each. Very infrequently there occurs a progressive degeneration of the central nervous system called subacute sclerosing panencephalitis (table 36.6). Permanent immunity to measles follows the disease, and no specific treatment is available. The use of attenuated measles vaccine (Attenuvax) or in combination (MMR vaccine; measles, mumps, rubella) is recommended for all children. Since public health immunization programs began in 1963, there has been over a 98% decrease in measles cases. Currently there are about 2,000 to 3,000 cases a year in the United States, and 90% of these are among unvaccinated individuals. In less well-developed countries, however, the morbidity and mortality in young children from measles infection remain high. It has been estimated that measles infects 44 million people and kills about 1.5 million a year worldwide. Serious outbreaks of measles are still reported in North America and Europe, especially among college students.

Mumps

Mumps is an acute generalized disease that occurs primarily in school-age children. It is caused by a paramyxovirus that is transmitted in saliva and respiratory droplets to nonimmune contacts. The portal of entry is the respiratory tract. The most

(a)

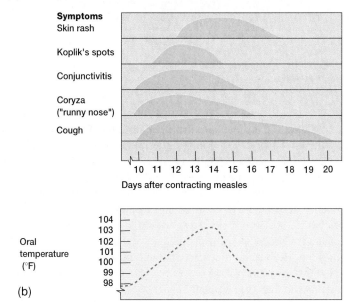

Symptoms

Skin rash

Koplik's spots

Conjunctivitis

Coryza
("runny nose")

Cough

10 11 12 13 14 15 16 17 18 19 20

Days after contracting measles

Oral temperature (°F)

104
103
102
101
100
99
98

(b)

Figure 36.3   Measles (Rubeola). (*a*) The rash of small, raised spots is typical of measles. (*b*) Signs and symptoms of a measles infection.

Figure 36.4   Mumps. Child with diffuse swelling of the salivary (parotid) glands due to the mumps virus.

Respiratory Syndromes and Viral Pneumonia

Acute viral infections of the respiratory system are among the most common causes of human disease. The infectious agents are called the acute respiratory viruses and collectively produce a variety of clinical manifestations, including rhinitis (inflammation of the mucous membrane of the nose), tonsillitis, laryngitis, and bronchitis. The adenoviruses, coxsackievirus A, coxsackievirus B, echovirus, influenza viruses, parainfluenza viruses, poliovirus, respiratory syncytial virus, and reovirus are thought to be responsible. It should be emphasized that for most of these viruses there is a lack of specific correlation between the agent and the clinical manifestation—hence the term syndrome. Immunity is not complete, and reinfection is common. The best treatment is care and rest.

Disease syndromes, signs, and symptoms (p. 702).
Classification of animal viruses (pp. 370–72; Appendix V).

In those cases of pneumonia for which no cause can be identified, viral pneumonia may be assumed if mycoplasmal pneumonia has been ruled out. The clinical picture is nonspecific. Symptoms may be mild, or there may be severe illness and death.

Mycoplasmal disease, primary atypical pneumonia (p. 779–80).

Respiratory syncytial virus (RSV) often is described as the most dangerous cause of respiratory infection in young children. In the United States over 90,000 infants are hospitalized each year and over 4,000 die. The RSV is a member of the RNA virus family, *Paramyxoviridae.* It is a negative single-stranded RNA virus that is enveloped with two virally specific glycoproteins as part of the structure. One of them, the large glycoprotein or G, is responsible for the binding of the virus to the host cell; the other, the fusion protein or F, permits fusion of the viral envelope with the

prominent manifestations of mumps are swelling and tenderness of the salivary (parotid) glands 16 to 18 days after infection of the host by the virus (figure 36.4). The swelling usually lasts for 1 to 2 weeks and is accompanied by a low grade fever. Meningitis and inflammation of the epididymis and testes can be important complications associated with this disease—especially in the postpubescent male. Therapy of mumps is limited to symptomatic and supportive measures. A live, attenuated mumps virus vaccine is available. It usually is given as part of the triple MMR vaccine. There are currently about 3,000 cases of mumps in the United States each year. Recently a few outbreaks have occurred among unvaccinated children in states without a comprehensive school immunization law. Prevention and control involves keeping children with mumps from school or other associated activities for about 2 weeks.

Figure 36.5 German Measles (Rubella). This disease is characterized by a rash of red spots. Notice that the spots are not raised above the surrounding skin as in measles (rubeola; see figure 36.3).

Figure 36.6 Smallpox. Back of hand showing single crop of smallpox vesicles.

host cell plasma membrane, leading to entry of the virus. The F protein also induces the fusion of the plasma membranes of infected cells. RSV thus gets its name from the resulting formation of a syncytium or multinucleated mass of fused cells. The source of the RSV is respiratory secretions of humans. Clinical manifestations consist of an acute onset of fever, cough, rhinitis, and nasal congestion. In infants and young children, this often progresses to bronchitis and viral pneumonia. Diagnosis is by either Directigen RSV or Test-Pack RSV rapid test kits. The virus is found worldwide and causes seasonal (November to May) outbreaks lasting several months. Treatment is with aerosolized ribavirin. A recently developed series of antibody (RSV-immune globulin) injections has been shown to reduce the severity of this disease in infants by 75%. Prevention and control consists of isolation for RSV-infected individuals, use of gowns when contact with secretions is likely, and strict attention to good handwashing practices.

Rubella (German Measles)

Rubella [Latin *rubellus,* reddish] was first described in Germany in the 1800s and was subsequently called **German measles.** It is a moderately contagious skin disease that occurs primarily in children 5 to 9 years of age. It is caused by the rubella virus, a single-stranded RNA virus that is a member of the family *Togaviridae.* Rubella is worldwide in distribution and occurs more frequently during the winter and spring months. This virus is spread in droplets that are shed from the respiratory secretions of infected individuals. Once the virus is inside the body, the incubation period ranges from 12 to 23 days. A rash of small red spots (figure 36.5), usually lasting no more than 3 days, and a light fever are the normal symptoms. The rash appears as immunity develops and the virus disappears from the blood, suggesting that the rash is immunologically mediated and not caused by the virus infecting skin cells.

Rubella can be a disastrous disease (**congenital rubella syndrome**) in the first trimester of pregnancy and can lead to fetal death, premature delivery, or a wide array of congenital defects that affect the heart, eyes, and ears.

Because rubella is usually such a mild infection, no treatment is indicated. All children and women of childbearing age who have not been previously exposed to rubella should be vaccinated. The live attenuated rubella vaccine (RA 27/3) is recommended. Because routine vaccination began in the United States in 1969, fewer than 1,000 cases of rubella and 10 cases of congenital rubella occur annually.

Smallpox (Variola)

Smallpox (variola) was once one of the most prevalent of all diseases (figure 36.6). Since the advent of immunization with the vaccinia virus (*see figure 16.18*), and because of concerted efforts by the World Health Organization, smallpox has been eradicated throughout the world. (The last case from a natural infection occurred in Somalia in 1977.) This was possible because a disease such as smallpox has obvious clinical features, virtually no asymptomatic carriers, only a human reservoir, and a short period of infectivity (3 to 4 weeks). Because humans are the only hosts, the spread of smallpox virus was prevented until no new cases developed, effectively eradicating the disease. The virus, however, is still kept in two locations: the CDC in Atlanta, Georgia, and the Research Institute for Viral Preparations in Moscow. The World Health Organization is now trying to decide whether or not to destroy the remaining smallpox virus stocks.

Disease (smallpox) and the early colonization of America (p. 333).

1. Why are chickenpox and shingles discussed together? What is their relationship?

2. Briefly describe the course of an influenza infection and how the virus causes the symptoms associated with the flu. Why has it been difficult to develop a single flu vaccine?

3. What are some common symptoms of measles?

4. What are Koplik's spots?

5. What is one side effect that mumps can cause in a young postpubescent male?

6. Describe some syndromes caused by the acute respiratory viruses. Respiratory syncytial virus.

7. Is viral pneumonia a specific disease? Explain.

8. When is a German measles infection most dangerous and why?

Arthropod-Borne Diseases

The arthropod-borne viruses (arboviruses) are transmitted by bloodsucking arthropods from one vertebrate host to another. They multiply in the tissues of the arthropod without producing disease, and the vector acquires a lifelong infection. Diseases produced by the arboviruses can be divided into three clinical syndromes: (1) fevers of an undifferentiated type with or without a rash; (2) encephalitis (inflammation of the brain), often with a high fatality rate; and (3) hemorrhagic fevers, also frequently severe and fatal (Box 36.2). Table 36.1 summarizes the six major human arbovirus diseases that occur in the United States. For all these diseases, immunity is believed to be permanent after a single infection. No vaccines are available for humans, although supportive treatment is beneficial.

Box 36.2

Viral Hemorrhagic Fevers: A Microbial History Lesson

Scientists know of several viruses lurking in the tropics that—with a little help from nature—could wreak far more loss of life than will likely result from the AIDS pandemic. Collectively these viruses produce what is known as the **hemorrhagic fevers.** The viruses are passed among wild vertebrates, which serve as reservoir hosts. Arthropods transmit the viruses among vertebrates, and humans are infected when they invade the environment of the natural host. These diseases, distributed throughout the world, are known by over 70 names, usually denoting the geographic area where they were first described.

Viral hemorrhagic fevers most often are fatal. Patients suffer headache, muscle pain, flushing of the skin, massive hemorrhaging either locally or throughout the body, circulatory shock, and death.

Despite the lack of public awareness, recent outbreaks may foreshadow much broader outbreaks in the future. For example, in the late 1960s dozens of scientists in West Germany fell seriously ill, and several died, from a mysterious new disease. Victims suffered from a breakdown of liver function and a bizarre combination of bleeding and blood clots. The World Health Organization traced the outbreak to a batch of fresh monkey cells the scientists had used to grow polioviruses. The cells from the imported Ugandan monkeys were infected with the lethal tropical Marburg virus and the scientists suffered from **Marburg viral hemorrhagic fever.**

In 1977 the *Plebovirus* causing Rift Valley Fever in sheep and cattle moved from these animals into the South African population. The virus, which causes severe weakness, incapacitating headaches, damage to the retina, and hemorrhaging, then made its way to Egypt, where millions of humans became infected and thousands died.

Among the most frightening hemorrhagic outbreak was that of the **Ebola virus hemorrhagic fever** in Zaire and Sudan in 1976. This disease infected more than 1,000 people and left over 500 dead. It became concentrated in hospitals, where it killed many of the Belgian physicians and nurses treating infected patients. A similar outbreak occurred in this same area in 1995 but was quickly contained.

In the United States in 1989, epidemiologists provided new evidence that rats infected with a potentially deadly hemorrhagic virus are prevalent in Baltimore slums. The virus appears to be taking a previously unrecognized toll on the urban poor by causing **Korean hemorrhagic fever.**

In the summer of 1993, reports appeared in the news media about a mysterious illness that had caused over 30 deaths among the Navajo nation in the four-corners area of the southwestern United States. The CDC finally determined the causative agent to be a hantavirus, a negative single-stranded RNA virus that is a member of the family *Bunyaviridae.* Hantaviruses are endemic in rodents, such as deer mice, in many areas of the world. Deer mice shed the virus in their saliva, feces, and urine. Humans contract the disease when they inhale aerosolized particles containing the excreted virus. Throughout Asia and central Eruope, hantaviruses cause hemorrhagic fever with renal syndrome in humans. But the type of virus found in the Southwest had not been previously recognized, and no hantavirus anywhere in the world has been associated with the clinical syndrome initially seen among the Navajo; namely, the **hantavirus pulmonary syndrome** in which the virus destroys the lungs. In 1993 the CDC named this virus **pulmonary syndrome hantavirus** (sometimes called the Sin Nombre or no-name virus) and to date, isolated cases have been reported from almost every state. Prevention involves wearing gloves when handling mice and spraying the feces and urine of all mice with a disinfectant.

Although to date, these epidemics have not become global, they do provide a humbling vision of humankind's viral vulnerability. History shows that the life-threatening viral hemorrhagic outbreaks often have arisen when humans moved into unexplored terrain or when living conditions deteriorated in ways that generated new viral hosts. In each case medical and scientific resources have been reactive, not proactive.

TABLE 36.1   Summary of the Six Major Human Arbovirus Diseases That Occur in the United States

Disease	Distribution	Vectors	Mortality Rate
California encephalitis (La Crosse)	North Central, Atlantic, South	Mosquitoes (*Aedes* spp.)	Fatalities rare
Colorado tick fever	Pacific Coast (mountains)	Ticks (*Dermacentor andersoni*)	Fatalities rare
Eastern equine encephalitis (EEE)	Atlantic, Southern Coast	Mosquitoes (*Aedes* spp.)	50–70%
St. Louis encephalitis (SLE)	Widespread	Mosquitoes (*Culex* spp.)	10–30%
Venezuelan equine encephalitis (VEE)	Southern United States	Mosquitoes (*Aedes* spp. and *Culex* spp.)	20–30% (children) <10% (adults)
Western equine encephalitis (WEE)	Mountains of the West	Mosquitoes (*Culex* spp.)	3–7%

Colorado Tick Fever

Colorado tick fever is caused by an RNA virus of the genus *Coltivirus*. The tick, *Dermacentor andersoni,* is the main vector that transmits the disease to humans. Ground squirrels, rabbits, and deer are the principal reservoirs. The disease occurs in mountainous regions in western states. Approximately 300 cases are reported annually in the United States, but the actual incidence is probably much higher. After a 3 to 6 day incubation period, there is an abrupt onset of fever, chills, severe headache, photophobia, rash, and muscle pain. These symptoms persist for 5 to 8 days and then disappear; complications are rare. Serology is used to confirm the diagnosis. No specific treatments exist for Colorado tick fever; therapy is limited to supportive care. Prevention involves common tick precautions (*see pp. 752–53*).

Yellow Fever

Yellow fever is not endemic to the United States, but because of its historical importance, it is briefly discussed here. Yellow fever was the first human disease found to be caused by a virus. (Walter Reed discovered this in 1901.) It also provided the first confirmation (by Carlos Juan Finley) that an insect could transmit a virus. Yellow fever is caused by a flavivirus that is endemic in many tropical areas, such as Mexico, South America, and Africa.

The disease received its first name, yellow jack, because jaundice is a prominent sign in severe cases. The jaundice is due to the deposition of bile pigments in the skin and mucous membranes because of damage to the liver. The disease is spread through a population in two epidemiological patterns. In the urban cycle human-to-human transmission is by *Aedes aegypti* mosquitoes. In the sylvatic cycle the mosquitoes transmit the virus between monkeys and from monkeys to humans (sylvatic means in the woods or affecting wild animals).

Once inside a person the virus spreads to local lymph nodes and multiplies; from this site it moves to the liver, spleen, kidneys, and heart, where it can persist for days. In the early stages of the disease, the infected person experiences fever, chills, headache, and backache, followed by nausea and vomiting. In severe cases the virus produces lesions in the infected organs and hemorrhaging occurs.

There is no specific treatment for yellow fever. Diagnosis is by serology. An active immunity to yellow fever results from an initial infection or from vaccines containing the attenuated yellow fever 17D strain or the Dakar strain virus. Prevention and control of this disease involves vaccination and control of the insect vector.

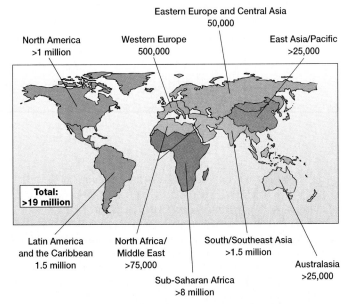

Figure 36.7   Distribution of AIDS in Adults by Continent or Region, as of Mid-1993. The World Health Organization estimates that as of mid-1993, over 19 million individuals have become infected with HIV since the start of the pandemic. *Source: Data from Michael H. Merson, "Slowing the Spread of HIV: Agenda for the 1990's" in Science, Vol. 260, figure 1, page 1266, May 28, 1993.*

Direct Contact Diseases

Acquired Immune Deficiency Syndrome (AIDS)

It is now recognized that **AIDS (acquired immune deficiency syndrome)** is the first great pandemic of the second half of the twentieth century. First described in 1981 AIDS is the result of an infection by the **human immunodeficiency virus (HIV),** a lentivirus within the family *Retroviridae.* The disease appears to have begun in central Africa as early as the 1950s. Recently a virus (the simian immunodeficiency virus or SIV) related to HIV-1 and HIV-2, the strains primarily responsible for AIDS, has been isolated from African green monkeys and is believed to be the ancestor of the AIDS virus. Somehow, this virus entered humans and mutated to the two current African human viruses SBL and LAV-2, which are intermediate viruses; these intermediate viruses may have evolved into the highly virulent HIV-1. Once established, HIV-1 spread to the Caribbean and then to the United States and Europe.

Epidemiologically AIDS occurs worldwide (figure 36.7). The World Health Organization estimates that over 19 million

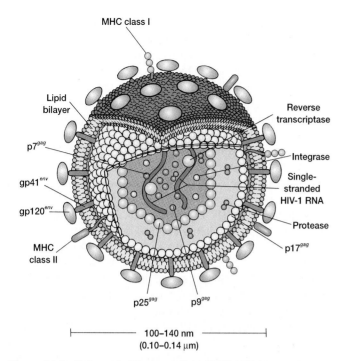

MHC class I

Lipid bilayer

p7^{gag}

gp41^{env}

gp120^{env}

MHC class II

Reverse transcriptase

Integrase

Single-stranded HIV-1 RNA

Protease

p17^{gag}

p25^{gag} p9^{gag}

100–140 nm
(0.10–0.14 μm)

Figure 36.8   Schematic Diagram of the HIV-1 Virion. The HIV-1 virion is an icosahedral structure containing 72 external spikes. These spikes are formed by the two major viral-envelope proteins, gp120 and gp41. (Gp stands for glycoprotein—the proteins are linked to sugars— and the number refers to the mass of the protein, in thousands of daltons.) The HIV-1 lipid bilayer is also studded with various host proteins, including class I and class II major histocompatibility complex molecules, acquired during virion budding. The core of HIV-1 contains four nucleocapsid proteins (p25, p17, p9, p7) each of which is proteolytically cleaved from a 53 kDa *gag* precursor by the HIV-1 protease. The phosphorylated p25 polypeptide forms the chief component of the inner shelf of the nucleocapsid, whereas the p17 protein is associated with the inner surface of the lipid bilayer and stabilizes the exterior and interior components of the virion. The p7 protein binds directly to the genomic RNA through a zinc-finger structural motif and together with p9 forms the nucleoid core. The retroviral core contains two copies of the single-stranded HIV-1 genomic RNA that is associated with the various preformed viral enzymes, including the reverse transcriptase, integrase, ribonuclease, and protease.

people are infected worldwide and that over 40 million people will have been infected by the year 2000. Through June 1995, about 476,899 cases and 295,473 deaths have been reported in the United States. However, according to the Centers for Disease Control and Prevention, the number of U.S. cases will level off during the next 5 years, with about 60,000 to 70,000 new cases each year. The groups most at risk in acquiring AIDS are (in descending order of risk) homosexual/bisexual men; intravenous (IV) drug users; heterosexuals who have intercourse with drug users, prostitutes, and bisexuals; transfusion patients or hemophiliacs who must receive clotting factor preparations made from donated blood; and children born of infected mothers. The mortality rate from AIDS is extremely high. The case-fatality rate averages about 92% for those adults diagnosed with AIDS before 1987. By the year 2000, AIDS will become the major killer of children with an estimated 10 million infections worldwide.

AIDS is caused primarily by the HIV-1 virus (some cases result from an HIV-2 infection). This virus is a retrovirus and closely related to HTLV-1, the cause of adult T-cell leukemia, and HTLV-2, which has been isolated from individuals with hairy-cell leukemia (*see leukemia, p. 733*). HIV-1 is an enveloped lentivirus (*see chapter 18*) with a cylindrical core inside its capsid (figure 36.8). The core contains two copies of its plus single-stranded RNA genome and several enzymes. Thus far 10 virus-specific proteins have been discovered. One of them, the gp120 envelope protein, participates in HIV-1 attachment to CD4+ cells (T-helper cells; see below).

One feature that distinguishes the lentiviruses from other retroviruses is the remarkable complexity of their viral genomes. Most retroviruses that are capable of replication contain only three genes—namely, *gag, pol,* and *env.* The *gag* and *env* genes encode the core nucleocapsid polypeptides and surface-coat proteins of the virus, respectively, whereas the *pol* gene codes for the viral reverse transcriptase, protease, integrase, and ribonuclease enzymes. HIV-1 also contains in its 9 kb RNA genome at least six additional genes (*vif, vpu, vpr, tat, rev,* and *nef*) (figure 36.9). It is the distinct and concerted actions of these six genes that probably underlie

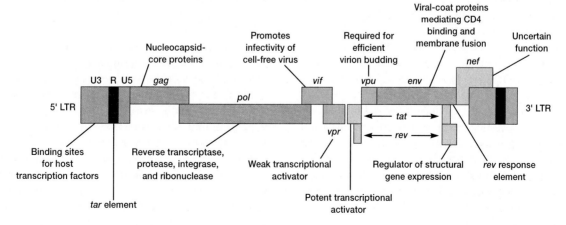

Nucleocapsid-core proteins

Promotes infectivity of cell-free virus

Required for efficient virion budding

Viral-coat proteins mediating CD4 binding and membrane fusion

Uncertain function

U3 R U5   *gag*   *vif*   *vpu*   *env*   *nef*

5' LTR   *pol*   *tat*   3' LTR

vpr   *rev*

Binding sites for host transcription factors

Reverse transcriptase, protease, integrase, and ribonuclease

Weak transcriptional activator

Regulator of structural gene expression

rev response element

tar element

Potent transcriptional activator

Figure 36.9   Genomic Structure of HIV-1. All nine known HIV-1 genes are shown in this schematic drawing and recognized primary functions given. The 5′ and 3′ long terminal repeats (LTRs) containing regulatory sequences recognized by various host transcription factors are also depicted. Notice the positions of the *tat* and *rev* genes, the *tar* (transactivation response) element, and the *rev* response element.

TABLE 36.2   HIV-1 Modes of Transmission

Route

Sexual: homosexual, heterosexual

Parenteral: intravenous drug use, transfusion, needle-stick (rare), invasive medical/dental procedure (very rare)

Maternal to child: perinatal, breast-feeding

Cutaneous (very rare): mucous membrane, skin

Predisposing Factors

Life-style: large number of sexual partners, sexually promiscuous partner (e.g., prostitute contact), drug abuse with needle sharing

Traumatic sexual practices: "fisting," penile/vaginal abrasion

Sexually transmitted diseases: ulcerative genital lesions, exudative genital infection

Source: Data from W. A. Blattner, "HIV Epidemiology: Past, Present, and Future" in *FASEB Journal*, 5(10):2340-48, 1991.

the profound pathogenicity of HIV-1. From a therapeutic standpoint, this same genome may also be the "Achilles' heel" of the virus. This possibility has spurred a broad-based search for antagonists specific for these HIV-1 gene products.

The AIDS virus is acquired by direct exposure of a person's bloodstream to body fluids (blood, semen, vaginal secretions) containing the virus through sexual contact or perinatally from an infected mother to her fetus (table 36.2). It also is possible that a newborn can be infected through breast-feeding. Once inside the body, the virus gp120 envelope protein (figure 36.8) binds to the CD4 glycoprotein plasma membrane receptor on CD4$^+$ T cells, macrophages, dendritic cells, and monocytes (figure 36.10). (Dendritic cells are present throughout the body's mucosal surfaces and bear the CD4 protein. Thus it is possible that these are the first cells infected by the HIV in sexual transmission.) After the envelope has fused with the plasma membrane, the virus releases its core protein and two RNA strands into the cytoplasm.

Inside the infected cell, the core protein remains associated with the RNA as it is copied into a single strand of DNA by the RNA/DNA-dependent DNA polymerase activity of the reverse transcriptase enzyme. The RNA is next degraded by another reverse transcriptase component, ribonuclease H, and the DNA strand is duplicated to form a double-stranded DNA copy of the original RNA genome. A complex of the double-stranded DNA (the provirus) and the integrase enzyme moves into the nucleus. Then the proviral DNA is integrated into the cell's DNA through a complex sequence of reactions catalyzed by the integrase (figure 36.10). The integrated provirus can remain latent, giving no sign of its presence. Alternatively the provirus can force the cell to synthesize viral mRNA. Some of the RNA is translated to produce viral proteins by the cell's own ribosomes. Viral proteins and the complete HIV-1 RNA genome are then assembled into new virions that bud from the infected host cell (*see figure 18.9*). Eventually the host cell lyses.

The precise mechanism of AIDS pathogenesis still is not known, and many hypotheses exist. Many believe that AIDS is

caused primarily by depletion of T cells and disruption of their function, although the exact mechanisms are unclear. The cytopathic effect may be due to the disruption of plasma membrane permeability and function by excessive virus budding. Free gp120 proteins may bind to CD4 proteins on uninfected cells, making them targets for attack by immune system cells. Infected cells do fuse with other cells to form large, multinucleate syncytia that eventually die, and this may contribute greatly to cell destruction. Possibly, insertion of the provirus DNA into the cell's genome and the transposition of the integrated provirus disrupt cell function and destroy the host T cell. Recent research shows that the virus replicates at a furious pace, about a billion virus particles a day. The immune system often manages to hold it off for years but eventually succumbs in almost all cases.

Once a human's CD4$^+$ cells are infected with HIV-1, four types of pathological changes may ensue. First, a mild form of AIDS may develop with symptoms that include fever, malaise, headache, macular rash, weight loss, lymph node enlargement (lymphadenopathy), oral candidiasis (figure 36.11), and the presence of antibodies to HIV-1 (figure 36.12). These symptoms may occur in the first few months after infection, last for 1 to 3 weeks, and recur. This is known as **AIDS-related complex (ARC)**. ARC develops into a full-blown case of AIDS in an undetermined proportion of cases.

Second, a true case of AIDS can develop directly upon infection. The mean interval between HIV infection and the onset of AIDS appears to be about 8 to 10 years, although it varies considerably with each individual. At first, a person's immune system responds to the HIV-1 infection by manufacturing HIV-1 antibodies but not in sufficient quantities to stop the viral attack. The virus becomes established within primarily CD4$^+$ T-helper cells, and HIV accumulates in lymphoid organs in large quantities even before symptoms appear. Initially, CD4$^+$ T-helper cells proliferate abnormally in the lymph nodes. Thereafter the lymph nodes' internal structure collapses due to viral replication. This leads to a decline in the number of lymphocytes within the lymph nodes and results in a selective depletion of the CD4$^+$ T-cell subset that is critical to the propagation of the entire T-cell pool. When this CD4$^+$ population declines, interleukin-2 (IL-2) production also decreases. Because IL-2 stimulates the production of T cells in general (*see figure 31.1*), the whole T-cell population may decline. This leaves the infected person open to opportunistic infections: invasion by pathogens that proliferate widely only because the immune system is defective.

It should be noted that factors other than direct T-cell destruction also may be involved in AIDS pathogenesis. HIV may reduce the immune response by destroying or disabling dendritic cells, which present foreign antigens to T cells. HIV also mutates exceptionally rapidly and thus could evade and eventually overwhelm the immune system. HIV may disrupt the balance between different types of T-helper cells and consequently decrease the killer T-cell population. It is possible that several different mechanisms contribute.

New findings suggest still another potential mechanism for the depletion of CD4$^+$ cells (figure 36.13). In HIV-infected

Figure 36.10   Life Cycle of HIV-1. (*1*) After interaction of gp120 with the CD4+ cell plasma membrane receptor, gp41-mediated membrane fusion occurs. (*2*) This leads to the entry of HIV-1 into the cell. The lck denotes a lymphoid-specific tyrosine kinase that binds to CD4. (*3*) After internalization and uncoating, reverse transcription of viral RNA begins. (*4*) This leads to the production of the double-stranded DNA form of the virus in the presence of appropriate host factors. (*5*) The HIV-1 integrase promotes the insertion of this viral DNA duplex into the CD4 cell's genome after the DNA has entered the nucleus. (*6*) This gives rise to the HIV-1 provirus. (*7*) The expression of the HIV-1 gene is stimulated initially by the action of specific inducible and constitutive host transcription factors with binding sites in the long terminal repeat. Their binding leads to the sequential production of various viral mRNAs. (*8*) The first mRNAs produced correspond to the multiply spliced species of approximately 2.0 kilobases encoding *tat, rev,* and *nef* regulatory proteins. (*9*) Subsequently the viral structural proteins are produced, allowing the (*10*) assembly and morphogenesis of the virions. (*11*) The new HIV-1 virions that are produced by viral budding from the host CD4+ cell can then reinitiate the retroviral life cycle by infecting other CD4+ target cells.

Figure 36.11   Some Pathogens and Diseases Associated with AIDS. (*a*) *Pneumocystis carinii* protozoa in lung tissue; light micrograph (×500). (*b*) *Toxoplasma gondii* protozoa located in a brain cyst; light micrograph (×1,000). (*c*) *Cryptosporidium* attached to the surface of the intestinal epithelium; light micrograph (×1,000). (*d*) *Cryptococcus neoformans* in the central nervous system; light micrograph (×150). (*e*) *Histoplasma capsulatum*, hyphae and sporangia; light micrograph (×125). (*f*) Candidiasis of the oral cavity and tongue (thrush) caused by *Candida albicans*. (*g*) Kaposi's sarcoma on an AIDS patient.

Figure 36.12   The Typical Serological Pattern in an HIV-1 Infection. HIV-1 antigen (HIV Ag) is detectable as early as 2 weeks after infection and typically declines at seroconversion. Seroconversion occurs when HIV-1 antibodies have risen to detectable levels. This usually takes place several weeks to months after the HIV-1 infection. The period between HIV-1 infection and seroconversion often is associated with an acute illness. Whether or not the individual has flulike symptoms, the appearance of circulating HIV-1 antigens typically occurs before IgG antibodies against gp41 and gp25 develop. HIV-1 antigen then usually disappears following seroconversion but reappears in the latter stages of the disease. The reappearance of antigen usually indicates impending clinical deterioration. An asymptomatic HIV-1 antigen-positive individual is six times more likely to develop AIDS within 3 years than a similar individual who is HIV-1 antigen negative. Thus testing for the presence of the HIV-1 antigen assists clinicians in monitoring the progression of the disease.

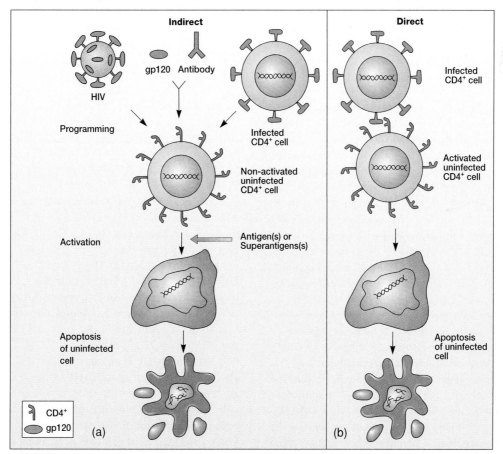

Figure 36.13   Apoptosis and AIDS. Apoptosis is a homeostatic physiological suicide mechanism in which cell death occurs naturally during normal tissue turnover. Usually apoptosis occurs after activation of a calcium-dependent endogenous endonuclease. Cells undergoing apoptosis display profound structural changes such as a decrease in cell volume, blebbing of the plasma membrane, and nuclear fragmentation. The nuclear DNA is cleaved into short oligonucleosomal length DNA fragments. The dying cell sheds small membrane-bound apoptotic bodies, which are phagocytosed and digested. (*a*) There may be several ways in which an HIV infection can indirectly trigger apoptosis. In all cases an initial event would program or prime the target cell so that apoptosis would be triggered by the binding of antigens or superantigens to the cell's T-cell receptors. Possibly the external gp120 envelope glycoprotein of the HIV virion binds to the CD4 protein on lymphocytes and programs the lymphocyte. A combination of free gp120 and antibodies to gp120 also could stimulate programmed cell death. First, the gp120 would bind to CD4 receptors. Then antibodies would attach to the gp120 and cause clustering of the receptors, thus priming the uninfected CD4+ cell. It also is possible that binding of the infected cell's surface gp120 proteins to the CD4 receptors on an uninfected cell will program the uninfected cell for apoptosis in response to antigens. (*b*) Apoptosis may be directly triggered in an uninfected cell. The gp120 envelope proteins on the surface of an infected cell may combine with the CD4 proteins of an uninfected cell and directly stimulate programmed cell death without activation by antigens.

individuals the loss of CD4⁺ cells is associated with lymphocyte activation; however, this activation does not result in cell proliferation, as it does normally, but rather in cell death by a mechanism known as programmed cell death [**apoptosis;** Greek "a falling off" from *apo,* off, and *ptosis,* fall].

In 1993 the CDC revised its definition of AIDS to include all HIV-infected persons who have fewer than 200 CD4⁺ cells per μl or a CD4⁺ cell percentage of total lymphocytes of less than 14. The reason for this current definition is that the development of a particular opportunistic infection is related to the concentration of CD4⁺ cells in the blood. Healthy individuals have about 1,000 such cells in every μl of blood. In HIV-1 infected individuals, the number declines by an average of about 40 to 80 cells per μl a year. When the CD4⁺ cell count falls to between 400 and 200 per μl, the first opportunistic infections and disease processes usually appear (table 36.3). Examples of such opportunistic infections and diseases include *Pneumocystis carinii* pneumonia (figure 36.11*a*); *Mycobacterium avium-intracellulare* pneumonia (*see p. 746*); toxoplasmosis (figure 36.11*b*); herpes zoster infection (figure 36.2*b*); chronic diarrhea caused by *Cyclospora;* cryptococcal meningitis (figure 36.11*d*); *Histoplasma capsulatum* infection (figure 36.11*e*); and tuberculosis (*see figure 37.7*).

The third main type of disease caused by HIV-1 involves the central nervous system since virus-infected macrophages can cross the blood-brain barrier. The classical symptoms of central nervous system disease in AIDS patients are headaches, fevers, subtle cognitive changes, abnormal reflexes, and ataxia (irregularity of muscular action). Dementia and severe sensory and motor changes characterize more advanced stages of the disease. Autoimmune neuropathies, cerebrovascular disease, and brain tumors also are common. Histological changes include inflammation of neurons, nodule formation, and demyelination. All evidence indicates that these neurological changes are correlated with higher levels of HIV-1 antigen (figure 36.12) and/or the HIV-1 genome in central nervous system tissue. In AIDS dementia, macrophages and glial cells (supporting cells of the nervous system) are primarily infected and bud new viruses. However, it is unlikely that direct infection of these cells by HIV-1 is responsible for the symptoms. More probably the symptoms arise through either the secretion of viral proteins or viral induction of cytokines that bind to glial cells and neurons. HIV-1 induction of interleukin-1 and tumor necrosis factor-α (TNF-α) may stimulate further viral reproduction and the induction of other cytokines (e.g., interleukin-6, granulocyte-macrophage colony-stimulating factor [GMCSF]). IL-1 and TNF-α in combination with IL-6 and GMCSF could account for the many clinical and histopathological findings in the central nervous system of AIDS individuals.

The fourth result of an HIV infection is cancer. Individuals infected with HIV-1 have an increased risk of three types of tumors: (1) Kaposi's sarcoma (figure 36.11*g*), (2) carcinomas of the mouth and rectum, and (3) B-cell lymphomas or lymphoproliferative disorders. It seems likely that the depression of the initial immune response enables secondary tumor-causing agents to initiate the cancers.

TABLE 36.3   Disease Processes Associated with AIDS

Candidiasis of bronchi, trachea, or lungs

Candidiasis, esophageal

Cervical cancer, invasive

Coccidioidomycosis, disseminated or extrapulmonary

Cryptosporidiosis, chronic intestinal (>1 month's duration)

Cyclospora, diarrheal disease

Cytomegalovirus disease (other than liver, spleen, or lymph nodes)

Cytomegalovirus retinitis (with loss of vision)

Encephalopathy, HIV-related

Herpes simplex: chronic ulcer(s) (>1 month's duration); or bronchitis, pneumonitis, or esophagitis

Histoplasmosis, disseminated or extrapulmonary

Isosporiasis, chronic intestinal (>1 month's duration)

Kaposi's sarcoma

Lymphoma, Burkitt's (or equivalent term)

Lymphoma, immunoblastic (or equivalent term)

Lymphoma, primary, of brain

Mycobacterium avium complex or *M. kansasii*

Mycobacterium tuberculosis, any site

Mycobacterium, other species or unidentified species

Pneumocystis carinii pneumonia

Pneumonia, recurrent

Progressive multifocal leukoencephalopathy

Salmonella septicemia, recurrent

Toxoplasmosis of brain

Wasting syndrome due to AIDS

Source: Data from *MMWR* 41 (No. RR17). 1993 Revised Classification System for HIV Infection and Expanded Surveillance Case Definition for AIDS Among Adolescents and Adults.

The laboratory diagnosis of AIDS can be by viral isolation and culture or by using assays for viral reverse transcriptase activity or viral antigens (figure 36.12). However, diagnosis is best accomplished through the detection of specific anti-HIV antibodies in the blood. These antibodies may be detected by GENIE or HIVAG-1 (Abbott) rapid tests, the SUDS (Murex) 10 minute assay, enzyme-linked immunoabsorbent assay (*see figure 32.13*), indirect immunofluorescence (*see figure 32.16*), immunoblot (Western Blot; Box 36.3), and radioimmunoprecipitation (*see figure 32.17*) methods.

It has recently been shown that if a person infected with HIV-1 is also infected with HHV-6, the development of AIDS is strikingly enhanced. Both viruses can infect a single CD4⁺ cell; however, the presence of HHV-6 greatly accelerates the process. This is the first clear evidence of a cofactor being involved in AIDS. (A cofactor is an independent agent or substance that acts synergistically to foster disease.) The scenario is as follows. Ordinarily the immune system controls the herpesvirus. However, if the immune system is suppressed by HIV-1, the herpesvirus will start to replicate, unleashing a vicious cycle of destruction. First, it kills T-helper cells; then it infects AIDS virus-infected cells and awakens the virus from its latent state. Thus immune suppression caused by the HIV activates the HHV-6 virus, the HHV-6 virus in turn activates the HIV-1

Box 36.3

AIDS Testing

Although a cure for AIDS is not yet in sight, effective diagnostic methods and graphic documentation techniques have been developed to identify those who have been infected with the HIV virus. To screen for AIDS, a blood sample is centrifuged to separate the serum from the blood cells. The serum is then subjected to an initial ELISA screening procedure to detect antibodies to the HIV virus. Any blood sample that is reactive on repeat testing is presumed positive for antibodies to the virus, indicating the individual has been infected with the virus. Blood samples testing positive two out of three times are then subjected to confirmation testing by either the Western Blot procedure or an indirect immunofluorescence assay (IFA). Each of these tests is designed to detect HIV-1 antibodies.

The Western Blot procedure involves the isolation and testing of individual viral proteins to determine if the antibodies initially discovered in ELISA tests will recognize HIV-specific protein antigen and bind to proteins considered to be definitive signs of infection. In this procedure component proteins of the HIV virus are separated on polyacrylamide gel by electrophoresis and transblotted onto nitrocellulose strips, which are then incubated with the individual's serum. Any HIV antibodies present bind to viral proteins contained on the test strip. The bound antibodies are then visualized using a conjugate enzyme and chromagen. The intensity of protein band staining depends on the degree of antibody binding to the viral proteins on the strip.

Findings are reported as negative if no virus-specific bands are detected and if the results concur with the initial ELISA screening. Results are reported as positive if the characteristic HIV protein bands are present and the ELISA screening test also was positive. When reactions are observed with other viral bands or when the results of the Western Blot do not correlate with the ELISA test, the findings are reported as indeterminate, and the tests are repeated in 6 weeks.

After the Western Blot is completed, a permanent visual documentation can be made of each test strip. For example, Box Figure 36.3 shows three control strips that are negative, strongly positive, and weakly positive. The remaining 40 test strips show results ranging from negative (laboratory number 13) to strongly positive (number 28).

Box Figure 36.3   Western Blot Strips.

A polymerase chain reaction (PCR) test also can be used. It detects the actual virus rather than detecting antibody. PCR provides a direct measure of the presence of viral DNA in the sample. This test is very fast and accurate but expensive. It may be useful in situations in which it is important to find out definitely and immediately whether HIV exposure has taken place.

virus, and both kill CD4+ T-helper cells. HIV-1 also may enter through the genital skin blisters caused by the herpes infection, and this would aid transmission. Genital herpes is probably a cofactor for HIV-2 as well as HIV-1.

The immunologic defect in AIDS is apparently irreversible and has not yet responded to treatment. Primary treatment is directed at the symptoms, opportunistic infections, and malignancies. Although it is difficult to prove, primary treatment of HIV has likely resulted in prolonging the life of HIV-infected individuals and improving their quality of life. Currently four antiretrovirals are approved for use in HIV disease: 3'-azidothymidine (AZT, zidovudine, Retrovir), 2',3'-dideoxyinosine (ddI, didanosine, Videx), 2',3'-dideoxycytidine (ddC, zalcitabine, HIVID), and Zerit (stavudine; formerly known as d4T). All four of these drugs inhibit the reverse transcriptase as it synthesizes DNA. Since they closely resemble thymidine but lack the correct attachment point for the next nucleotide in the chain (due to their lack of a hydroxy group on carbon 3'), they serve as a DNA chain terminator. Thus replication ceases when any of these drugs are incorporated into the growing DNA chain. These drugs appear to slow the progress of the disease, but they do not cure it. They probably do not improve the chances of survival, nor do they have any effect on virus transmission by infected individuals. Unfortunately it is now known that AZT-resistant strains of the virus exist, and most of the drug's modest benefits typically fade within a year. A recent clinical study suggests that treatment with interleukin-2 (*see chapter 31*) very early in an HIV infection may greatly aid individuals in resisting the virus. It substantially increases the CD4+ cell count.

Figure 36.14   Cold Sores. Herpes simplex fever blisters on the lip.

Another avenue of current research is the development of a vaccine that can (1) stimulate the production of neutralizing antibodies which can bind to the virus envelope and prevent it from entering host cells, and (2) promote the destruction of those cells already infected with the virus. The production of an effective vaccine, if possible, is not yet in sight. One difficulty is that the envelope proteins of the virus (figure 36.8) continually change their antigenic properties.

Prevention and control of AIDS involves screening of blood and heat treatment of blood products to destroy the virus. Education and protected sexual behavior and practices, including the use of condoms, are keys to comprehensive community-based prevention programs. Education of intravenous drug users concerning the need to avoid sharing needles and syringes is also very important in prevention programs. It must be emphasized that protected sexual behavior and the use of condoms is safer but not totally safe. The only absolute protection against AIDS and other sexually transmitted diseases is abstinence and completely monogamous relationships.

Cold Sores

Cold sores or **fever blisters (herpes labialis)** are caused by the herpes simplex virus type 1 (HSV-1). Like all herpesviruses, it is a double-stranded DNA virus with an enveloped, icosahedral capsid. The term herpes is derived from the Greek word meaning "to creep," and clinical descriptions of herpes labialis (lips) go back to the time of Hippocrates (circa 400 B.C.). Transmission is through direct contact of epithelial tissue surfaces with the virus. A blister(s) develops at the inoculation site (figure 36.14) because of host- and viral-mediated tissue destruction. Most blisters involve the epidermis and surface mucous membranes of the lips, mouth, and gums. The blisters generally heal within a week. However, after a primary infection, the virus travels to the trigeminal nerve ganglion, where it remains in a latent state for the lifetime of the infected person. Stressful stimuli such as excessive sunlight, fever, trauma, chilling, emotional stress, and

hormonal changes can reactivate the virus. Once reactivated, the virus moves from the trigeminal ganglion down a peripheral nerve to the border of the lip or other parts of the face to produce another fever blister. Primary and recurring infections also may occur in the eyes, causing **herpetic keratitis** (inflammation of the cornea)—currently a major cause of blindness in the United States. The drugs vidarabine and acyclovir are effective against cold sores. By adulthood, 70 to 90% of all people in the United States have been infected and have type 1 herpes antibodies. The "gold standard" for the detection of herpes is cell culture. However, cell culture requires up to 14 days for results. The rapid detection of herpes simplex virus (HSV) antigen is by several enzyme immunoassay kits (HERPCHEK, VIDAS, IDEIA, SureCell). This rapid test is especially useful in cases of pregnant women with genital infections and individuals who are particularly susceptible to severe infections.

Common Cold

The **common cold [coryza:** Greek *koryza,* discharge from the nostrils] is one of the most frequent infections experienced by humans of all ages. About 50% of the cases are caused by rhinoviruses [Greek *rhinos,* nose], single-stranded RNA viruses in the family *Picornaviridae.* There are over 115 distinct serotypes, and each of these antigenic types has a varying capacity to infect the nasal mucosa and to cause a cold. In addition, immunity to many of them is transitory. Several other respiratory viruses are also associated with colds (e.g., coronaviruses and parainfluenza viruses). Thus colds are common because of the diversity of rhinoviruses, the involvement of other respiratory viruses, and the lack of a durable immunity.

Rhinoviruses provide an excellent example of the medical relevance of research on virus morphology. The complete rhinovirus capsid structure has recently been elucidated with the use of X-ray diffraction techniques. The results help explain rhinovirus resistance to human immune defenses. Human antibodies against the virus combine with four types of surface proteins. Unfortunately there are at least 89 variants of these four antigens, making the production of a cold vaccine from them unlikely. The structural studies have shown that the site that recognizes and binds to cell surface molecules during infection lies at the bottom of a surface cleft about 250 nm deep and 1.2 to 3 nm wide. Thus the binding site is well protected from the immune system while it carries out its functions, and the preparation of a vaccine directed against the binding site is not feasible. Possibly drugs that could fit in the cleft and interfere with virus attachment can be designed.

Viral invasion of the upper respiratory tract is the basic mechanism in the pathogenesis of a cold. The virus enters the body's cells by binding to the adhesion molecule ICAM-1 (figure 36.15). The clinical manifestations include the familiar nasal stuffiness and/or partial obstruction, sneezing, scratchy throat, and a watery discharge from the nose. The discharge becomes thicker and assumes a yellowish appearance over several

Figure 36.15   Role of an Integrin (Intercellular Adhesion Molecule-1) in Rhinovirus Uptake Into Cells of the Upper Respiratory Tract. Rhinoviruses are a large family of viruses that are the single major cause of acute respiratory infections in humans. ICAM-1 (intercellular adhesion molecule-1) has been identified as the cell surface receptor for the majority of these rhinoviruses. (*a*) In this computer-generated model, the virus-binding portion of ICAM-1 is shown in orange-red attaching to a rhinovirus (gray icosahedral protein capsid in the center). (*b*) This diagram indicates how the rhinovirus surface proteins (shown in blue-green) bind to the ICAM-1 molecule (red-yellow).

(a)   (b)

days. General malaise is commonly present. The disease usually runs its course in about a week. Diagnosis of the common cold is made from observations of clinical symptoms. There are no procedures for direct examination of clinical specimens or for serological diagnosis.

The source of the cold viruses may be infected individuals excreting viruses in nasal secretions, airborne transmission over short distances by way of moisture droplets, or transmission on contaminated hands or fomites. Epidemiological studies of rhinovirus colds have shown that the familiar explosive, noncontained sneeze (*see figure 35.6*) may not play an important role in virus spread. Rather, hand-to-hand contact between a rhinovirus "donor" and a susceptible "recipient" is more likely. The common cold occurs worldwide with two main seasonal peaks, spring and early autumn. Infection is most common early in life and generally decreases with an increase in age. Nothing is available for treating the common cold except additional rest, extra fluids, and the use of aspirin for alleviating local and systemic discomfort.

Cytomegalovirus Inclusion Disease

Cytomegalovirus inclusion disease is caused by the human cytomegalovirus (HCMV), a member of the herpesvirus family. Most people become infected with this virus at some time during their life; in the United States, as many as 80% of individuals older than 35 years have been exposed to this virus and carry a lifelong infection. Although most HCMV infections are asymptomatic, certain patient groups are at risk to develop serious illness and long-term effects from an HCMV infection. For example, this virus remains the leading cause of congenital virus infection in the United States, a significant cause of transfusion-acquired infections, and a frequent contributor to

morbidity and mortality among organ transplant recipients and immunocompromised individuals (especially AIDS patients). Because the virus persists in the body, it is shed for several years in saliva, urine, semen, and cervical secretions.

The HCMV can infect any cell of the body, where it multiplies slowly and causes the host cell to swell in size—hence the prefix cytomegalo, which means "an enlarged cell." Infected cells contain the unique **intranuclear inclusion bodies** and cytoplasmic inclusions (figure 36.16). In fatal cases, cell damage is seen in the gastrointestinal tract, lungs, liver, spleen, and kidneys. Normally cytomegalovirus inclusion disease symptoms resemble those of infectious mononucleosis.

Laboratory diagnosis is by viral isolation from urine, blood, lung, semen, or tissue. Serological tests (e.g., immunofluorescence, complement fixation, ELISA, and immunohistologic staining) and a rapid test kit (CMV-vue kit) also are available.

Epidemiologically the virus has a worldwide distribution, especially in developing countries where infection is universal by childhood. The prevalence of this disease increases with a lowering of socioeconomic status and hygienic practices. The only drugs available, ganciclovir and foscarnet, are used only for high-risk patients. Infection can be prevented by avoiding close personal contact (including sexual) with an actively infected individual. Transmission by blood transfusion or organ transplantation can be avoided by using blood or organs from seronegative donors.

Genital Herpes

Genital herpes is caused by the herpes simplex virus type 2 (HSV-2) (figure 36.17*a*). It is most frequently transmitted by sexual contact. The primary infection occurs after an incubation

Figure 36.16 Cytomegalovirus Inclusion Disease. Electron micrograph of a single animal cell infected with the cytomegalovirus (×66,000). The intranuclear inclusion body has a typical "owl-eyed" appearance.

(a)

(b)

Figure 36.17 Genital Herpes. (*a*) Herpes simplex virus type 2 (yellow and green) inside an infected cell. (*b*) Herpes vesicles on the penis. The vesicles contain fluid that is infectious.

period of about a week. Fever, a burning sensation, and genital soreness are frequently present. Blisters that appear in the infected area (figure 36.17*b*) are the result of cell lysis and the development of a local inflammatory response; they contain fluid and infectious viruses. Although blisters generally heal spontaneously in a few weeks, the viruses retreat to nerve cells in the sacral spinal plexus, where they remain in a latent form. Periodically the viruses multiply and migrate down nerve fibers to the epithelia, where they produce new blisters. Activation may be due to sunlight, sexual activity, illness accompanied by fever, hormones, or stress.

Besides being transmitted by sexual contact, herpes can be spread to an infant during vaginal delivery, leading to **congenital (neonatal) herpes.** Congenital herpes is one of the most life-threatening of all infections in newborns, affecting approximately 1,500 to 2,200 babies per year in the United States. It can result in neurological involvement as well as blindness. As a result any female who has had genital herpes should have a caesarean section instead of delivering vaginally. For unknown reasons, genital herpes is also associated with a higher-than-normal rate of cervical cancer and miscarriages.

Although there is no cure for genital herpes, oral use of the antiviral drug acyclovir has proven to be effective in ameliorating the recurring blister outbreaks. Topical acyclovir is also effective in reducing virus shedding, the time until the crusting of blisters occurs, and new lesion formation. Idoxuridine and trifluridine are used to treat herpes infections of the eye.

In the United States the incidence of genital herpes has increased so much during the past decade that is now a very common sexually transmitted disease. It is estimated that over 25 million Americans are infected with the herpes simplex virus type 2.

Human Herpesvirus 6 Infections

Human herpesvirus 6 (HHV-6, type A and type B) is the etiologic agent of **exanthem subitum** [Greek *exanthema,* rash] in infants. It was originally termed **roseola infantum** and then given the ordinal designation **sixth disease** to differentiate it from other exanthems and roseolas. Exanthem subitum is a short-lived disease characterized by a high fever of 3 to 4 days' duration, after which the temperature suddenly drops to normal and a macular rash appears on the trunk and then spreads to other areas of the body. HHV-6 infects over 90% of the United States infant population, and most children are seropositive for HHV-6 by 3 years of age. CD4[+] T cells are the main site of viral replication, whereas monocytes are in an infected, latent state. The tropism of HHV-6 appears to be wide, including CD8[+] T cells, natural killer cells, and probably epithelial cells. In adults HHV-6 is commonly found in peripheral-blood mononuclear cells and saliva, suggesting that the infection is lifelong. Since the salivary glands are the major site of latent infection, transmission is probably by way of saliva.

HHV-6 also produces latent and chronic infections and is occasionally reactivated in immunocompromised hosts leading to pneumonitis. Furthermore, HHV-6 has been implicated in several other diseases (lymphadenitis and infectious mononucleosis-like syndrome or chronic fatigue syndrome) in immunocompetent adults. Diagnosis is either by immunofluorescence or enzyme immunoassay. To date, there is no antiviral therapy or prevention.

Human Parvovirus B19 Infections

Since its discovery in 1975, **human parvovirus B19** (family *Parvoviridae,* genus *Parvovirus*) has emerged as a significant human pathogen. B19 virions are uniform, icosahedral, unenveloped particles approximately 23 nm in diameter. The genome of B19 is a single-stranded DNA. There is a spectrum of disease caused by parvovirus B19 infection, ranging from mild symptoms (fever, headache, chills, malaise) in normal persons, **erythema infectiosum** in children (fifth disease), and a joint disease syndrome in adults; to more serious diseases such as aplastic crisis in persons with sickle cell disease and autoimmune hemolytic anemia, and pure red cell aplasia due to persistent B19 virus infection in immunocompromised individuals. The B19 parvovirus can also infect the fetus, resulting in anemia and fetal hydrops (the accumulation of fluid in tissues). It is assumed that the natural mode of infection is by the respiratory route. A variety of techniques are available for the detection of the B19 virus. Antiviral antibodies appear to represent the principal means of defense against B19 parvovirus infection and disease. The treatment of individuals suffering from acute and persistent B19 infections with commercial immunoglobins containing anti-B19 and human monoclonal antibodies to B19 have shown to be effective therapy. Most of the time infection is followed by lifelong immunity.

Leukemia

Certain **leukemias** in humans are caused by two retroviruses: human T-cell lymphotropic virus 1 (HTLV-1) and HTLV-2. The viruses are transmitted by transfusions of contaminated blood, among drug addicts sharing needles, by sexual contact, across the placenta, from the mother's milk, or by mosquitoes.

Retroviruses and their replication (p. 376).

Viruses and cancer (pp. 379–80).

Figure 36.18   Hairy-Cell Leukemia. False-color transmission electron micrograph (×3,100) of abnormal B lymphocytes. Notice that the lymphocytes are covered with characteristic hair-like membrane-derived protrusions.

HTLV-1 causes **adult T-cell leukemia.** Once within the body the HTLV-1 virus enters white blood cells and integrates into the cellular genome, where it activates the growth-promoting genes by acting at a distance from them (trans-acting). The transformed cell proliferates extensively, and death generally results from the explosive proliferation of the leukemia cells or from opportunistic infections. To date, no effective treatment exists.

In 1982 the second human retrovirus (HTLV-2) was shown to be the agent responsible for hairy-cell leukemia. This virus shares the same trans-acting mechanism as HTLV-1. Hairy-cell leukemia gets its name from the many membrane-derived protrusions that give white blood cells the appearance of being "hairy" (figure 36.18). This leukemia is a chronic, progressive lymphoproliferative disease. The malignancy is believed to originate in a stage of B-cell development. The bone marrow, spleen, and liver become infiltrated with malignant cells. This lowers the person's immunity. The primary cause of mortality is bacterial and other opportunistic infections. IFN-α has shown some promise for treatment in certain cases.

Mononucleosis (Infectious)

The Epstein-Barr virus (EBV), a member of the family *Herpesviridae,* is the etiologic agent of **infectious mononucleosis (mono),** a disease whose symptoms closely resemble those of cytomegalovirus-induced mononucleosis. Because the Epstein-Barr virus occurs in oropharyngeal secretions, it can be spread by mouth-to-mouth contact (hence the terminology infectious and kissing disease) or shared drinking bottles and glasses. Once a person has contracted mono, the virus enters lymphatic tissue, multiplies, and infects B cells. Infected B cells rapidly proliferate and take on an atypical appearance (Downey cells) that is useful in diagnosis. The disease is manifested by enlargement of the lymph nodes and spleen, sore throat, headache, nausea, general weakness and tiredness, and a mild fever that usually peaks in the early evening. The disease lasts for 1 to 6 weeks and is self-limited. Like other herpesviruses, EBV does become latent, probably in the B-cell population.

Treatment of mononucleosis is largely supportive and includes plenty of rest. Specific chemotherapy with phosphonoacetic acid, adenine arabinoside, and acyclovir is nearing clinical application. Diagnosis of mononucleosis is made with a serological test for nonspecific (heterophile) antibodies. Several rapid tests are on the market.

The peak incidence of mononucleosis occurs in people 15 to 25 years of age. Collegiate populations, particularly those in the upper-socioeconomic class, have a high incidence of the disease. About 50% of college students have no immunity, and approximately 15% of these can be expected to contract mononucleosis. People in lower-socioeconomic classes tend to acquire immunity to the disease because of early childhood infection. It has been estimated that over 80% of the United States population are eventually infected. The Epstein-Barr virus may well be the most common virus in humans. EBV infections are associated with the cancers Burkitt's lymphoma in tropical Africa and nasopharyngeal carcinoma in Southeast Asia, East and North Africa, and in Eskimos (*see p. 380*).

Rabies

Rabies [Latin *rabere,* rage or madness] has been the object of human fascination, torment, and fear since the disease was first recognized in antiquity. Improvements in prevention during the past 50 years have led to the almost complete elimination of indigenously acquired rabies in the United States. Worldwide, virtually all cases of human rabies are attributable to dog bites in developing countries where canine rabies is still endemic.

Rabies is caused by a number of different strains of highly neurotropic viruses. Most belong to a single serotype in the genus *Lyssavirus,* family *Rhabdoviridae.* The bullet-shaped virion (figure 36.19*a; see also figure 16.17c*) contains a single-stranded, negative-sense RNA genome.

As previously noted, in the United States rabies is primarily a disease of animals. Most wild animals can become infected with rabies, but susceptibility varies according to species. Foxes, coyotes, and wolves are the most susceptible; intermediate are skunks, raccoons, insectivorous bats, and bobcats; and opossums are quite resistant.

The virus multiplies in the salivary glands of an infected host. It is transmitted to humans or other animals by the bite of an infected animal whose saliva contains the virus; by aerosols of the virus that can be spread in caves where bats roost; or by contamination of scratches, abrasions, open wounds, and mucous membranes with saliva from an infected animal.

After inoculation, a region of the virions' glycoprotein envelope spike attaches to the plasma membrane of nearby skeletal muscle cells, the virus enters the cells, and multiplication of the virus occurs. When the concentration of the muscle virus is sufficient, the virus enters the nervous system through unmyelinated sensory and motor terminals; the reported binding site is the nicotinic acetylcholine receptor. Since the virus is sequestered from the immune system, infection can no longer be halted by immunization. The virus spreads by retrograde axonal flow at 8 to 20 mm per day until it reaches the spinal cord, when the first specific symptoms of the disease—pain or paresthesia at the wound site—may occur. A rapidly progressive encephalitis develops as the virus quickly disseminates through the central nervous system. The virus then spreads throughout the body along the peripheral nerves, including those in the salivary glands, where it is shed in the saliva.

Within brain neurons the virus produces characteristic **Negri bodies,** masses of viruses or unassembled viral subunits that are visible in the light microscope. In the past the diagnosis of rabies consisted solely of examining nervous tissue for the presence of these bodies. Today diagnosis is based on direct immunofluorescent-antibody (dIFA) of brain tissue, virus isolation, detection of Negri bodies, and a rapid rabies enzyme-mediated immunodiagnosis test.

Symptoms of rabies usually begin 2 to 16 weeks after viral exposure and include anxiety, irritability, depression, fatigue, loss of appetite, fever, and a sensitivity to light and sound. The disease quickly progresses to a stage of paralysis. In about 50% of all cases, intense and painful spasms of the throat and chest muscles occur when the victim swallows liquids. The mere sight, thought, or smell of water can set off spasms. Consequently rabies has been called hydrophobia (fear of water). Death results from destruction of the regions of the brain that regulate breathing.

Safe and effective vaccines (human diploid-cell rabies vaccine HDCV [Imovax Rabies] or rabies vaccine adsorbed [RVA]) against rabies are available; however, to be effective they must be given soon after the person has been infected. Veterinarians and laboratory personnel, who have a high risk of exposure to rabies, usually are immunized every 2 years and tested for the presence of suitable antibody titer. About 30,000 people annually receive this treatment. In the United States fewer than 10 cases of rabies occur yearly in humans; about 6,000 cases of animal rabies are reported each year from various sources (figure 36.19*b*). Prevention and control involves preexposure vaccination of dogs and cats, postexposure vaccination of humans, and preexposure vaccination of humans at special risk (persons spending a month or more in countries where rabies is common in dogs).

Viral Hepatitides

Any infection that results in inflammation of the liver is called **hepatitis** [pl., hepatitides] [Greek *hepaticus,* liver]. Currently nine viruses are recognized as causing hepatitis. Two are herpesviruses (cytomegalovirus [CMV] and Epstein-Barr virus [EBV]) and seven are hepatotropic viruses.

EBV and CMV cause mild, self-resolving forms of hepatitis with no permanent hepatic damage. Both viruses cause the typical infectious mononucleosis syndrome of fatigue, nausea, and malaise.

(a)

Figure 36.19   Rabies. (*a*) Electron micrograph of the rabies virus (yellow) (×36,700). Note the bullet shape. The external surface of the virus contains spikelike glycoprotein projections that bind specifically to cellular receptors. (*b*) In the United States rabies is found in terrestrial animals in 10 distinct geographic areas. In each area a particular species is the reservoir and one of five antigenic variants of the virus predominates as illustrated by the five different colors. Although not shown, an additional eight viral variants are found in insectivorous bats and cause sporadic cases of rabies in terrestrial animals throughout the country. Absence of a strain does not imply absence of rabies.

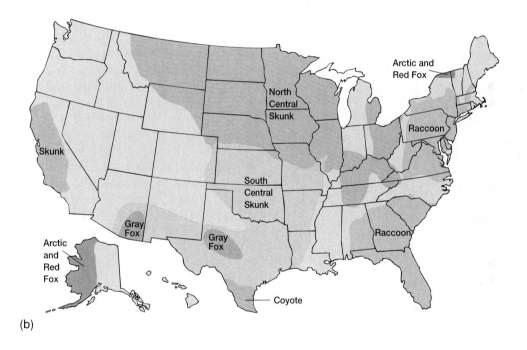

(b)

Of the seven human hepatotropic viruses, only five are well characterized; hepatitis F and G are newly discovered viruses (table 36.4). Hepatitis A (sometimes called infectious hepatitis), and hepatitis E (formerly enteric-transmitted NANB hepatitis), are transmitted by fecal-oral contamination and discussed in the section on food- and waterborne diseases (p. 738). The other major types include hepatitis B (sometimes called serum hepatitis), hepatitis C (formerly non-A, non-B hepatitis), and hepatitis D (formerly delta hepatitis).

Hepatitis B (serum hepatitis) is caused by the hepatitis B virus (HBV), a double-stranded circular DNA virus of complex structure. HBV is classified as an *Orthohepadnavirus* within the family *Hepadnaviridae*. Serum from individuals infected with

hepatitis B contains three distinct antigenic particles: a spherical 22 nm particle, a 42 nm spherical particle (containing DNA and DNA polymerase) called the **Dane particle,** and tubular or filamentous particles that vary in length (figure 36.20). The small spherical and tubular particles are the unassembled components of the Dane particle—the infective form of the virus. The unassembled particles contain **h**epatitis **B** **s**urface **a**ntigen (HBsAg) whose presence in the blood is (1) an indicator of hepatitis B infection, (2) the basis for the large-scale screening of blood for the hepatitis B virus, and (3) the basis for the first vaccine for human use developed by recombinant DNA technology.

The hepatitis B virus is normally transmitted through blood transfusions, contaminated equipment, drug users' unsterile

TABLE 36.4   Characteristics of Hepatitides Caused by Hepatotropic Viruses[a]

Disease (Virus)	Genome	Classification	Transmission	Outcome	Prevention
Hepatitis A (Hepatitis A)	RNA	*Picornaviridae, Hepatovirus*	Fecal-oral	Subclinical, acute infection	Killed HAV (Havrix vaccine)
Hepatitis B (Hepatitis B)	DNA	*Hepadnaviridae, Orthohepadnavirus*	Blood, needles, body secretions, placenta	Subclinical, acute chronic infection; cirrhosis; primary hepatocarcinoma	Recombinant HBV vaccines
Hepatitis C (Hepatitis C)	RNA	*Flaviviridae, Pestivirus, or Flavivirus* (?)	Blood	Subclinical, acute chronic infection; primary hepatocarcinoma	Routine screening of blood
Hepatitis D (Hepatitis D)	RNA	Unclassified	Blood	Superinfection or coinfection with HBV	HBV vaccine
Hepatitis E (Hepatitis E)	RNA	*Caliciviridae* (?)	Fecal-oral	Subclinical, acute infection (but high mortality in pregnant women)	Improve sanitary conditions

[a]Hepatitis F and G have been discovered but not well characterized. Thus they are not included in this table.

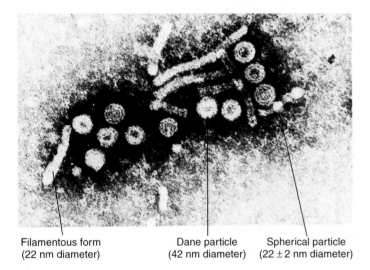

Filamentous form
(22 nm diameter)

Dane particle
(42 nm diameter)

Spherical particle
(22 ± 2 nm diameter)

Figure 36.20   Hepatitis B Virus in Serum. Electron micrograph (×210,000) showing the three distinct types of hepatitis B antigenic particles. The spherical particles and filamentous forms are small spheres or long filaments without an internal structure, and only two of the three characteristic viral envelope proteins appear on their surface. Dane particles are the complete, infectious virion.

dice, the accumulation of bilirubin (a breakdown product of hemoglobin) in the skin and other tissues with a resulting yellow appearance. Chronic hepatitis B infection also causes the development of primary liver cancer, hepatocellular carcinoma (hepatitis B is second in importance only to tobacco as a known human carcinogen).

General measures for prevention and control involve (1) excluding contact with HBV-infected blood and secretions, and minimizing needle-sticks by scrupulous technique; (2) passive prophylaxis with intramuscular injection of hepatitis B immune globulin within 7 days of exposure; and (3) active prophylaxis with two recombinant vaccines: Engerix-B, and Recombivax HB. These vaccines are widely used by health professionals who are at increased risk of contacting the hepatitis B virus. Other individuals who should be vaccinated include:

1. Contacts of HBV carriers (e.g., household members, sex partners, institutional inmates)
2. International travelers
3. Sexually active homosexual males
4. Hemodialysis patients
5. Recipients of blood and related products that are possibly HBV contaminated

needles, or any body secretion (saliva, sweat, semen, breast milk, urine, feces). The virus also can pass from the blood of an infected mother through the placenta to infect the fetus. Each year an estimated 300,000 people in the United States are infected with HBV. About 4,000 persons die yearly from hepatitis-related cirrhosis and about 1,000 die from HBV-related liver cancer. Worldwide, HBV infects over 200 million people.

The clinical signs of hepatitis B vary widely. Most cases are asymptomatic. However, sometimes fever, loss of appetite, abdominal discomfort, nausea, fatigue, and other symptoms gradually appear following an incubation period of 1 to 3 months. The virus infects liver hepatic cells and causes liver tissue degeneration and the release of liver-associated enzymes (transaminases) into the bloodstream. This is followed by jaun-

Another form of hepatitis is called **hepatitis C.** The virion has an 80 nm diameter, has a lipid coat, contains a single strand of linear ssRNA, and has been designated the hepatitis C virus (HCV) within the family *Flaviviridae*. This virus is transmitted by intimate contact with virus-contaminated blood, by the fecal-oral route, by in utero transmission from mother to fetus, or through organ transplantation. Hepatitis C disease is milder than hepatitis B and similar to hepatitis A, but often without the jaundice. Diagnosis is made by a first-generation enzyme-linked immunosorbent assay (ELISA), which detects serum antibody to a recombinant antigen of HCV. Hepatitis C is found worldwide. In recent years hepatitis C has accounted for more than 90% of hepatitis cases developing after a blood transfusion. Treatment is with recombinant IFN-α 2b (Intron A) three times weekly for 6 months.

In 1977 a cytopathic hepatitis agent termed the **Delta agent** was discovered. Later it was called the hepatitis D virus (HDV) and the disease **hepatitis D** designated. HDV is a unique agent in that it is dependent on the hepatitis B virus to provide the envelope protein (HBsAg) for its RNA genome. Thus HDV only replicates in liver cells in which the HBV also is actively replicating. Furthermore, the RNA of the HDV is smaller than the RNA of the smallest picornaviruses and its circular conformation differs from the linear structure typical of animal RNA viruses. (The nearest relatives of the HDV are in the plant kingdom—the viroids, virusoids, and satellite RNAs [*see chapter 18*].) HDV is spread only to persons who are already infected with HBV (superinfection) or to individuals who get both viruses at once (coinfection). The primary laboratory tools for the diagnosis of an HDV infection are serological tests for anti-delta antibodies. Treatment of patients with chronic HDV remains difficult. Early studies indicate that some positive results can be obtained with alpha interferon for 3 months to 1 year. Worldwide, there are approximately 300 million HBV carriers, and available data indicate that no fewer than 5% of these are infected with HDV. Thus because of the propensity of HDV to cause fulminant as well as chronic liver disease, continued incursion of HDV into areas of the world where persistent hepatitis B infection is endemic will have serious implications. Prevention and control involves the widespread use of the hepatitis B vaccine.

Recently the presence of two other forms of hepatitis have been identified: **hepatitis F** (causing fulminant, post-transfusion hepatitis) and a syncytial giant-cell hepatitis (**hepatitis G**) with viruslike particles resembling the measles virus (*Paramyxovirus*). Further virologic, epidemiological, and molecular efforts to clone and characterize these new agents and their diseases are currently being undertaken.

---

1. Describe the AIDS virus and how it cripples the immune system. How is the virus transmitted? What types of pathological changes can result?
2. Why do people periodically get cold sores? Describe the causative agent.
3. Why do people get the common cold so frequently? How are cold viruses spread?
4. Give two major ways in which herpes simplex virus type 2 is spread. Why do herpes infections become active periodically?
5. What two types of leukemias are caused by viruses?
6. Describe the causative agent and some symptoms of mononucleosis and exanthem subitum.
7. How does the rabies virus cause death in humans?
8. What are the different causative viruses of hepatitis and how do they differ from one another? How can one avoid hepatitis? Do you know anyone who is a good candidate for infection with these viruses?

Food-Borne and Waterborne Diseases

Food and water have been recognized as potential carriers of disease since the beginning of recorded history. Collectively

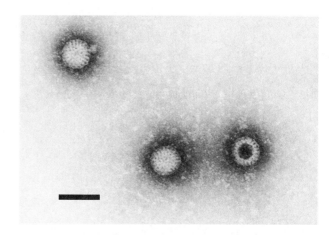

Figure 36.21   Viral Gastroenteritis. Electron micrograph of rotaviruses (reoviruses) in a human gastroenteritis stool filtrate. Note the spokelike appearance of the icosahedral capsids that surround double-stranded RNA within each virion. Bar = 100 nm.

more infectious diseases occur by these two routes than any other. A few human viral diseases that are food- and waterborne are now discussed.

> *Water-based diseases (chapter 41).*
> *Diseases and food (chapter 43).*

Gastroenteritis (Viral)

Acute viral gastroenteritis (inflammation of the stomach or intestines) is caused by four major categories of viruses: rotaviruses (figure 36.21), Norwalk and Norwalk-like viruses, other caliciviruses, and astroviruses. The medical importance of these viruses is summarized in table 36.5.

The viruses responsible for gastroenteritis are probably transmitted by the fecal-oral route. Infection is most common during the cooler months in contrast to bacteria-caused diarrheal diseases, which usually occur in the warmer months of the year. Diarrheal diseases are the leading cause of childhood deaths (5 to 10 million deaths per year) in developing countries where malnutrition is common. Current estimates are that viral gastroenteritis produces 30 to 40% of the cases of infectious diarrhea in the United States, far outnumbering documented cases of bacterial and parasitic diarrhea (the cause of approximately 40% of presumed cases of diarrhea remains unknown). In the United States rotaviruses account for about 3.5 million cases of illness, resulting in 35% of the hospitalizations for gastroenteritis, and 75 to 150 deaths each year.

Viral gastroenteritis is seen most frequently in infants 1 to 11 months of age, where the virus attacks the upper intestinal epithelial cells of the villus, causing malabsorption, impairment of sodium transport, and diarrhea. The clinical manifestations range from asymptomatic to a relatively mild diarrhea with headache and fever to a severe and occasionally fatal dehydrating disease. Vomiting is almost always present.

TABLE 36.5   Medically Important Gastroenteritis Viruses

Virus	Epidemiological Characteristics	Clinical Characteristics
Rotaviruses		
Group A	Endemic diarrhea in infants worldwide	Dehydrating diarrhea for 5–7 days; fever, abdominal cramps, nausea, and vomiting common
Group B	Large outbreaks in adults and children in China	Severe watery diarrhea for 3–5 days
Group C	Sporadic cases in children in Japan	Similar to group A
Norwalk and Norwalk-like viruses	Epidemics of vomiting and diarrhea in older children and adults; occurs in families, communities, and nursing homes; often associated with shellfish, other food, or water and infected food handlers	Acute vomiting, fever, myaglia, and headache lasting 1–2 days
Caliciviruses other than the Norwalk group	Pediatric diarrhea; associated with shellfish and other foods in adults	Rotavirus-like illness in children; Norwalk-like in adults
Astroviruses	Pediatric diarrhea; reported in nursing homes	Watery diarrhea for 1–3 days

Viral gastroenteritis is usually self-limited. Treatment is designed to provide relief through the use of oral fluid replacement with isotonic liquids, analgesics, and antiperistaltic agents.

Hepatitis A

Hepatitis A (infectious hepatitis) usually is transmitted by fecal-oral contamination of food, drink, or shellfish that live in contaminated water and contain the virus in their digestive system. The disease is caused by the hepatitis A virus (HAV). HAV has recently been reclassified as the type species of a new genus (*Hepatovirus*) in the family *Picornaviridae*. The hepatitis A virus is an icosahedral, single-stranded linear, positive-sense RNA virus that lacks an envelope. Once in the digestive system, the viruses multiply within the intestinal epithelium. Usually only mild intestinal symptoms result. Occasionally viremia (the presence of viruses in the blood) occurs and the viruses may spread to the liver. The viruses reproduce in the liver, enter the bile, and are released into the small intestine. This explains why feces are so infectious. Symptoms last from 2 to 20 days and include anorexia, general malaise, nausea, diarrhea, fever, and chills. If the liver becomes infected, jaundice ensues. Laboratory diagnosis is by detection of the hepatitis A antibody. About 30,000 cases are reported annually in the United States. Fortunately the mortality rate is low (less than 1%), and infections in children are usually asymptomatic. Most cases resolve in 4 to 6 weeks and yield a strong immunity. Approximately 40 to 80% of the United States population have serum antibodies though few have been aware of the disease. Control of infection is by simple hygienic measures, the sanitary disposal of excreta, and the killed HAV vaccine (Havrix).

Hepatitis E

Hepatitis E is implicated in many epidemics in certain developing countries in Asia, Africa, and Central and South America. The virus has been cloned and sequenced. The genomic organization suggests it is the prototype human pathogen for a new class of single-stranded linear RNA viruses or perhaps a separate genus within the family *Caliciviridae*. Infection usually is associated with feces-contaminated drinking water. Presumably HEV enters the blood from the gastrointestinal tract, replicates in the liver, is released from hepatocytes into the bile, and is subsequently excreted in the feces. Like hepatitis A, an HEV infection usually runs a benign course and is self-limiting. One exception is that at least 10% of women infected in their last 3 months of pregnancy die of fulminant hepatic failure. Antibody assay systems are under development using HEV cDNA products. There are no specific measures for preventing HEV infections other than those aimed at improving the level of health and sanitation in affected areas. There are only a few documented cases of hepatitis E in the United States and all of them imported.

Poliomyelitis

Poliomyelitis [Greek *polios*, gray, and *myelos*, marrow or spinal cord], **polio**, or **infantile paralysis** is caused by the poliovirus, a member of the family *Picornaviridae* (Box 36.4). The virus is very stable and can remain infectious for relatively long periods in food and water—its main routes of transmission. Once ingested, the virus multiplies in the mucosa of the throat and/or small intestine. From these sites the virus invades the tonsils and lymph nodes of the neck and terminal portion of the small intestine. Generally, there are either no symptoms or a brief illness characterized by fever, headache, sore throat, vomiting, and loss of appetite. The virus sometimes enters the bloodstream and causes a viremia. In most cases (more than 99%), the viremia is transient and clinical disease does not result. In the minority of cases (less than 1%), the viremia persists and the virus enters the central nervous system and causes paralytic polio. The virus has a high affinity for anterior horn motor nerve cells of the spinal cord. Once inside these cells, it multiplies and destroys the cells; this results in motor and muscle paralysis. Since the arrival of the formalin-inactivated Salk vaccine (1954) and the attenuated virus Sabin vaccine (1962), the incidence of polio has decreased markedly. There are fewer than 10 cases per year, and no endogenous reservoir of polioviruses exists in the United States. However, there is a continuing need for vaccination programs in all population groups to limit the spread of poliovirus when it is introduced from other countries. In developing countries, 4 of every 1,000 children born annually have paralytic disease caused by the poliovirus. Prevention and control is by vaccination; global eradication of polio is possible by the year 2000.

Box 36.4

A Brief History of Polio

Like many other infectious diseases, polio is probably of ancient origin. Various Egyptian hieroglyphics dated approximately 2000 B.C. depict individuals with wasting, withered legs and arms (Box Figure 36.4). In 1840 the German orthopedist Jacob von Heine described the clinical features of poliomyelitis and identified the spinal cord as the problem area. Little further progress was made until 1890, when Oskar Medin, a Swedish pediatrician, portrayed the natural history of the disease as epidemic in form. He also recognized that a systemic phase, characterized by minor symptoms and fever, occurred early and was complicated by paralysis only occasionally. Major progress occurred in 1908, when Karl Landsteiner and William Popper successfully transmitted the disease to monkeys. In the 1930s much public interest in polio occurred because of the polio experienced by Franklin D. Roosevelt. This led to the founding of the March of Dimes campaign in 1938; the sole purpose of the March of Dimes was to collect money for research on polio. In 1949 John Enders, Thomas Weller, and Frederick Robbins discovered that the poliovirus could be propagated in vitro in cultures of human embryonic tissues of nonneural origin. This was the keystone that later led to the development of vaccines.

In 1952 David Bodian recognized that there were three distinct serotypes of the poliovirus. Jonas Salk successfully immunized humans with formalin-inactivated poliovirus in 1953, and this vaccine (IPV) was licensed in 1955. In 1962 Albert Sabin and others developed the live attenuated poliovirus vaccines (oral polio vaccines, OPV). Both the Salk and Sabin vaccines led to a dramatic decline of paralytic poliomyelitis in most developed countries and, as such, have been rightfully hailed as two of the great accomplishments of medical science.

Box Figure 36.4   Ancient Egyptian with Polio. Note the withered leg.

---

1. What two virus groups are associated with acute viral gastroenteritis? How do they cause the disease's symptoms?
2. Describe some symptoms of hepatitis A.
3. Why was hepatitis A called infectious hepatitis?
4. At what specific sites within the body can the poliomyelitis virus multiply? What is the usual outcome of an infection?

Slow Virus Diseases

An introduction to slow diseases caused by viruses and prions appears on pp. 379 and 384–86. A **slow virus disease** can be defined as a progressive pathological process caused by a transmissible agent—virus or prion—that remains clinically silent during a prolonged incubation period of months to years, after which progressive clinical disease becomes apparent. This usually ends months later in disability or death. It is probably inappropriate to say that a slow virus disease or slow disease is caused by a slow virus for two reasons. First, the causative agent of some of these diseases does not fit the conventional definition of a virus. Second, even in those diseases caused by

viruses, it is the disease process and not the virus that is slow. Six of these diseases are summarized in table 36.6.

Other Diseases

Several other human diseases (e.g., diabetes mellitus, viral arthritis) have been associated with viruses but do not fit into any of the previous categories. A good example is warts.

Warts

Warts or verrucae [Latin *verruca,* wart] are horny projections on the skin caused by the papillomaviruses (*see figure 16.12*d). At least eight distinct genotypes produce benign epithelial tumors that vary in respect to their location, clinical appearance, and histopathologic features. Warts occur principally in children and young adults and are limited to the skin and mucous membranes. The viruses are spread between people by direct contact; autoinoculation occurs through scratching. Four major kinds of warts are **plantar warts, verrucae vulgaris, flat** or **plane warts,** and **anogenital condylomata (venereal warts)** (figure 36.22). Treatment includes physical

TABLE 36.6   Slow Virus Diseases of Humans

Disease	Agent	Incubation Period	Nature of Disease
Creutzfeldt-Jakob disease	Prion	Months to years	Spongiform encephalopathy (degenerative changes in the central nervous system)
Kuru	Prion	Months to years	Spongiform encephalopathy
Gerstmann-Sträussler-Scheinker Syndrome (GSS)	Prion	Months to years	Genetic neurodegenerative disease
Fatal familial insomnia	Prion	Months to years	Genetic neurodegenerative disease with progressive, untreatable insomnia
Progressive multifocal leukoencephalopathy	Papovavirus	Years	Central nervous system demyelination
Subacute sclerosing panencephalitis (SSPE)	Measle virus variant	2–20 years	Chronic sclerosing (hard tissue) panencephalitis (involving both white and gray matter of the brain)

(a)   (b)   (c)   (d)

Figure 36.22   Warts. (*a*) Common warts on fingers. (*b*) Flat warts on the face. (*c*) Plantar warts on the feet. (*d*) Perianal condyloma acuminata.

destruction of the wart(s) by electrosurgery, cryosurgery with liquid nitrogen or solid CO_2, laser fulguration (drying), direct application of the drug podophyllum to the wart(s), or injection of IFN-α (Intron A).

Anogenital condylomata (venereal warts) are sexually transmitted and caused by types 6, 11, and 42 **h**uman **p**apilloma**v**irus (HPV). Once the virus enters the body, the incubation period is 1 to 6 months. The warts (figure 36.22*d*) are soft, pink, cauliflowerlike growths that occur on the external genitalia, in the vagina, on the cervix, or in the rectum. They often are multiple and vary in size. Evidence exists that cervical cancer and perianal cancer occur more frequently in people who have a history of genital warts.

---

1. Name six slow virus diseases of humans.
2. What kind of viruses cause the formation of warts? Describe venereal warts or anogenital condylomata.

Summary

1. More than 400 different viruses can infect humans. These viruses can be grouped and discussed according to their mode of acquisition/transmission.

2. Most airborne viral diseases involve either directly or indirectly the respiratory system. Examples include chickenpox (varicella), shingles (zoster), rubella (German measles), influenza (flu), measles (rubeola), mumps, the acute respiratory viruses such as the respiratory syncytial virus, the extinct smallpox (variola), and viral pneumonia.

3. The arthropod-borne viral diseases are transmitted by arthropod vectors from human to human or animal to human. Examples include California encephalitis; Colorado tick fever; St. Louis encephalitis; eastern, western, and Venezuelan equine encephalitis; and yellow fever. All these diseases are characterized by fever, headache, nausea, vomiting, and the characteristic encephalitis.

4. Person-to-person contact is another way of acquiring or transmitting a viral disease. Examples of such diseases include AIDS, cold sores, the common cold, cytomegalovirus inclusion disease, genital herpes, human herpesvirus 6 infections, human parvovirus B19 infection, certain leukemias, infectious mononucleosis, rabies, and the five types of hepatitis: hepatitis A (infectious hepatitis); hepatitis B (serum hepatitis); hepatitis C (post-transfusion NANB hepatitis); hepatitis D (delta hepatitis); and hepatitis E (enteric-transmitted NANB hepatitis).

5. The viruses that are transmitted in food and water usually grow in the intestinal system and leave the body in the feces. Acquisition is generally by the oral route. Examples of diseases caused by these viruses include acute viral gastroenteritis (rotavirus and others), infectious hepatitis A, and poliomyelitis.

6. A slow virus disease is a pathological process caused by a transmissible agent (a prion or virus) that remains clinically silent for a prolonged period, after which the clinical disease becomes apparent. Examples include Creutzfeldt-Jakob disease, kuru, subacute sclerosing panencephalitis, Gerstmann-Sträussler-Scheinker syndrome, fatal familial insomnia, and progressive multifocal leukoencephalopathy. These diseases are chronic infections of the central nervous system that result in progressive degenerative changes and eventual death.

7. Common skin warts are caused by viruses and can be spread by autoinoculation through scratching or by direct or indirect contact. Anogenital condylomata (venereal warts) are sexually transmitted.

Key Terms

acute viral gastroenteritis *737*

adult T-cell leukemia *733*

AIDS (acquired immune deficiency syndrome) *722*

AIDS-related complex (ARC) *724*

anogenital condylomata (venereal warts) *739*

apoptosis *728*

chickenpox (varicella) *716*

cold sore *730*

Colorado tick fever *722*

common cold *730*

congenital (neonatal) herpes *732*

congenital rubella syndrome *720*

coryza *730*

cytomegalovirus inclusion disease *731*

Dane particle *735*

Delta agent *737*

Ebola virus hemorrhagic fever *721*

erythema infectiosum *733*

exanthem subitum *732*

fever blister *730*

flat or plane warts *739*

genital herpes *731*

Guillain-Barré syndrome (French Polio) *718*

hantavirus pulmonary syndrome *721*

hemorrhagic fevers *721*

hepatitis *734*

hepatitis A *738*

hepatitis B *735*

hepatitis C *736*

hepatitis D *737*

hepatitis E *738*

hepatitis F *737*

hepatitis G *737*

herpes labialis *730*

herpetic keratitis *730*

human herpesvirus 6 *732*

human immunodeficiency virus (HIV) *722*

human parvovirus B19 *733*

infantile paralysis *738*

infectious mononucleosis (mono) *733*

influenza or flu *717*

intranuclear inclusion body *731*

Koplik's spots *718*

Korean hemorrhagic fever *721*

leukemia *733*

Marburg viral hemorrhagic fever *721*

measles (rubeola) *718*

mumps *718*

Negri bodies *734*

plantar warts *739*

polio *738*

poliomyelitis *738*

pulmonary syndrome hantavirus *721*

rabies *734*

respiratory syncytial virus (RSV) *719*

Reye's syndrome *718*

roseola infantum *732*

rubella (German measles) *720*

shingles (zoster) *716*

sixth disease *732*

slow virus disease *739*

smallpox (variola) *720*

verrucae vulgaris *739*

wart *739*

yellow fever *722*

Questions for Thought and Review

1. Briefly describe each of the major or most common viral diseases in terms of its causative agent, signs and symptoms, the course of infection, mechanism of pathogenesis, epidemiology, and prevention and/or treatment.

2. Why are slow virus diseases not caused by slow viruses?

3. Which virus is responsible for each of the following: shingles, German measles, influenza, measles, mumps, and smallpox?

4. What are respiratory syndromes?

5. From an epidemiological perspective, why are most arthropod-borne viral diseases hard to control?

6. In terms of molecular genetics, why is the common cold such a prevalent viral infection in humans?

7. What are the differences between the five types of hepatitis?

8. Which viral diseases can be transmitted by sexual contact?

9. Why are the herpesviruses called persistent viruses?

10. What viruses specifically attack the nervous system?

11. Why is rabies such a feared disease?

12. Which viruses can cause encephalitis? Hepatitis?

13. What are two types of mononucleosis?

14. Will it be possible to eradicate many viral diseases in the same way as smallpox? Why or why not?

Additional Reading

General

Bean, B. 1992. Antiviral therapy: current concepts and practices. *Clin. Microbiol. Rev.* 5(2):146–82.

Dimmock, N. J., and Primrose, S. B. 1994. *Introduction to modern virology,* 4th ed. Cambridge, Mass.: Blackwell Science, Inc.

Fields, B. N.; Knipe, D. M.; Chanock, R. M.; Hirsch, M. S.; Melnick, J. L.; Monath, T. P.; and Roizman, B., editors. 1990. *Fields virology,* 2d ed. New York: Raven Press.

Galasso, G. J. 1990. *Antiviral agents and viral diseases of man,* 3d ed. New York: Raven Press.

Hirsch, M. S., and Kaplan, J. C. 1987. Antiviral therapy. *Sci. Am.* 256(4):76–85.

Huang, A. S. 1977. Viral pathogenesis and molecular biology. *Microbiol. Rev.* 41:811–19.

Joklik, W. K.; Willett, H. P.; Amos, D. B.; and Wilfert, C. M. 1992. *Zinsser microbiology,* 20th ed. E. Norwalk, Conn.: Appleton & Lange.

Le Guenno, B. 1995. Emerging viruses. *Sci. Am.* 273(4):56–64.

Levy, J. A. 1994. *HIV and the pathogenesis of AIDS.* Washington, D.C.: ASM Press.

Levy, J.; Fraenkel-Conrat, H.; and Owens, R. 1994. *Virology,* 3d ed. Englewood Cliffs, N.J.: Prentice Hall.

Mandell, G. L.; Douglas, R. G., Jr.; and Bennett, J. E. 1994. *Principles and practice of infectious disease,* 4th ed. New York: John Wiley and Sons.

Morse, S. S. 1993. *Emerging viruses.* New York: Oxford University Press.

Shulman, S. T.; Phair, J. P.; and Sommers, H. M. 1992. *The biologic and clinical basis of infectious disease,* 4th ed. Philadelphia: W.B. Saunders.

Waner, J. L. 1994. Mixed viral infections: Detection and management. *Clin. Microbiol. Rev.* 7(2):143–51.

White, D. O., and Fenner, F. J. 1994. *Medical virology,* 4th ed. San Diego: Academic Press.

Airborne Diseases

Agut, H. 1993. Puzzles concerning the pathogenicity of human herpesvirus 6. *N. Engl. J. Med.* 329(3):203–04.

Buller, R. M., and Palumbo, G. 1991. Poxvirus pathogenesis. *Microbiol. Rev.* 55(1):80–122.

Field, A., and Biron, K. 1994. "The end of innocence" revisited: resistance of herpesviruses to antiviral drugs. *Clin. Microbiol. Rev.* 7(1):1–13.

Henderson, D. A. 1976. The eradication of smallpox. *Sci. Am.* 235(1):25–33.

Henle, W.; Henle, G.; and Lennette, E. T. 1979. The Epstein-Barr virus. *Sci. Am.* 241(3):48–59.

Kaplan, M. M., and Webster, R. G. 1977. The epidemiology of influenza. *Sci. Am.* 237(4):88–106.

Morgan, E. M., and Rapp, F. 1977. Measles virus and its associated diseases. *Microbiol. Rev.* 41:636–41.

Shaw, M. W.; Arden, H. H.; and Maassab, H. F. 1992. New aspects of influenza virus. *Clin. Microbiol. Rev.* 5(1):74–92.

Stevens, J. G. 1989. Human herpesviruses: A consideration of the latent state. *Microbiol. Rev.* 53(3):318–32.

Sweet, C., and Smith, H. 1980. Pathogenicity of influenza viruses. *Microbiol. Rev.* 44:303–9.

Vainionpää, R., and Hyypiä, T. 1994. Biology of parainfluenza viruses. *Clin. Microbiol. Rev.* 7(2):265–75.

Welliver, R. C. 1989. Detection, pathogenesis, and therapy of respiratory syncytial virus infections. *Clin. Microbiol. Rev.* 1(2):27–39.

Wood, D. L., and Brunell, P.A. 1995. Measles control in the United States: Problems of the past and challenges for the future. *Clin. Microbiol. Rev.* 8(2):260-67.

Arthropod-Borne Diseases

Calisher, C. H. 1994. Medically important arboviruses of the United States and Canada. *Clin. Microbiol. Rev.* 7(1):89–116.

Strauss, J. H., and Strauss, E. G. 1994. The alphaviruses: Gene expression, replication, and evolution. *Microbiol. Rev.* 58(3):491–562.

Direct Contact Diseases

Alcamo, I. E. 1993. *AIDS: The biological basis.* Dubuque, Iowa: Wm. C. Brown Communications, Inc.

Aral, S. O., and Holmes, K. 1991. Sexually transmitted diseases in the AIDS era. *Sci. Am.* 264(2):62–69.

Atwood, J.; Berger, J. R.; Kaderman, R.; Tornatore, C. S.; and Major, E. O. 1993. Human immunodeficiency virus type 1 infection of the brain. *Clin. Microbiol. Rev.* 6(4):339–66.

DeClercq, E. 1995. Antiviral therapy for human immunodeficiency virus infections. *Clin. Microbiol. Rev.* 8(2):200-39.

Feitelson, M. 1993. Hepatitis B virus infection and primary hepatocellular carcinoma. *Clin. Microbiol. Rev.* 5(3):275–301.

Fishbein, D. B., and Robinson, L. E. 1993. Rabies. *N. Engl. J. Med.* 25(329):1632–38.

Forbes, B. A. 1989. Acquisition of cytomegalovirus infection: An update. *Clin. Microbiol. Rev.* 3(2):204–16.

Friedman-Kien, A. E. 1989. *Color atlas of AIDS.* Philadelphia: W.B. Saunders.

Gallo, R. C., and Montagnier, L. 1988. AIDS in 1988. *Sci. Am.* 259(4):40–48.

Greene, W. C. 1991. The molecular biology of human immunodeficiency virus type 1 infection. *N. Engl. J. Med.* 324(5):308–15.

Hazeltine, W. A., and Wong-Staal, F. 1988. The molecular biology of the AIDS virus. *Sci. Am.* 259(4):52–62.

Hollinger, F. B. 1990. *Viral hepatitis,* 2d ed. New York: Raven Press.

Kaplan, M. M., and Koprowski, H. 1980. Rabies. *Sci. Am.* 242(2):120–34.

Kovacs, J. A., et al. 1995. Increases in CD4 T lymphocytes with intermittent courses of interleukin-2 in patients with human immunodeficiency virus infection. *N. Engl. J. Med.* 332(9):567–75.

Levy, J. 1993. Pathogenesis of human immunodeficiency virus infection. *Microbiol. Rev.* 57(1):183–289.

Masucci, M. G., and Ernberg, I. 1994. Epstein-Barr virus: Adaptation to a life within the immune system. *Trends in Microbiol.* 2(4)125-30.

McDougall, J. K., editor. 1990. *Cytomegaloviruses.* New York: Springer-Verlag.

McLean, A. R. 1993. The balance of power between HIV and the immune system, *Trends in Microbiol.* 1(1):9–13.

Nowak, M. A., and McMichael, A. J. 1995. How HIV defeats the immune system. *Sci Am.* 273(2)58-65.

Okano, M.; Thiele, G. M.; Davis, J. R.; Grierson, H. L.; and Purtilo, D. T. 1988. Epstein-Barr virus and human diseases: Recent advances in diagnosis. *Clin. Microbiol. Rev.* 1(3):300–12.

Polish, L. B.; Gallagher, M.; Fields, H. A.; Hadler, S. C. 1993. Delta hepatitis: Molecular biology and clinical epidemiological features. *Clin. Microbiol. Rev.* 6(3):211–29.

Redfield, R. R., and Burk, D. S. 1988. HIV infection: The clinical picture. *Sci. Am.* 259(4):90–98.

Science. 1993. 260. AIDS: The unanswered questions (Thematic Issue).

Staczek, J. 1990. Animal cytomegaloviruses. *Microbiol. Rev.* 54(3):247–65.

Steffy, K., and Wong-Staal, F. 1991. Genetic regulation of human immunodeficiency virus. *Microbiol. Rev.* 55(2):193–205.

Tiollais, P., and Buendia, M. A. 1991. Hepatitis B virus. *Sci. Am.* 264(4):116–24.

Winkler, W. G., and Vogel, K. 1992. Control of rabies in wildlife. *Sci. Am.* 266(6):86–93.

Food-Borne and Waterborne Diseases

Blacklow, N. R., and Greenberg, H. B. 1991. Viral gastroenteritis. *N. Engl. J. Med.* 325(4):252–64.

Christensen, M. L. 1989. Human viral gastroenteritis. *Clin. Microbiol. Rev.* 2(1):51–89.

Cuthbert, J. A. 1994. Hepatitis C: Progress and problems. *Clin. Microbiol. Rev.* 7(4):505–32.

Hedberg, C. W., and Osterholm, M. T. 1993. Outbreaks of food-borne and waterborne viral gastroenteritis. *Clin. Microbiol. Rev.* 6(3):199–210.

Hogle, J. M.; Chow, M.; and Filman, O. 1987. The structure of the poliovirus. *Sci. Am.* 256(3):42–54.

Spector, D. H., and Baltimore, D. 1975. The molecular biology of poliovirus. *Sci. Am.* 232(1):24–31.

Slow Virus Diseases

Aiken, J. M., and Marsh, R. F. 1990. The search for scrapie agent nucleic acid. *Microbiol. Rev.* 54(3):242–46.

Holland, J. J. 1974. Slow, inapparent, and recurrent viruses. *Sci. Am.* 230(5):32–40.

Prusiner, S. B. 1984. Prions. *Sci. Am.* 251:50–59.

Prusiner, S. B. 1995. The prion diseases. *Sci. Am.* 272(1):48–57.

Other Diseases

Norkin, L. C. 1982. Papoviral persistent infections. *Microbiol. Rev.* 46:384–91.

Roman, A., and Fife, K. 1989. Human papillomaviruses: Are we ready to type? *Clin. Microbiol. Rev.* 2:166–90.

CHAPTER 37
Human Diseases Caused Primarily by Gram-Positive and Gram-Negative Bacteria

Soldiers have rarely won wars. They more often mop up after the barrage of epidemics. And typhus, with its brothers and sisters— plague, cholera, typhoid, dysentery—has decided more campaigns than Caesar, Hannibal, Napoleon, and all the . . . generals of history. The epidemics get the blame for the defeat, the generals the credit for victory. It ought to be the other way around . . . .

—Hans Zinsser

Concepts

1. Bacterial diseases of humans can be discussed according to their mode of acquisition/transmission.

2. Most of the airborne diseases caused by bacteria involve the respiratory system. Examples include diphtheria, Legionnaires' disease and Pontiac fever, pertussis, streptococcal diseases, and tuberculosis. Other airborne bacteria can cause skin diseases, including cellulitis, erysipelas, and scarlet fever, or systemic diseases such as meningitis, glomerulonephritis, and rheumatic fever.

3. Although arthropod-borne bacterial diseases are generally rare, they are of interest either historically (plague) or because they have been newly introduced into humans (Lyme disease).

4. Most of the direct contact bacterial diseases involve the skin, mucous membranes, or underlying tissues. Examples include anthrax, bacterial vaginosis, cat-scratch disease, chancroid, gas gangrene, leprosy, staphylococcal diseases, and syphilis. Others can become disseminated throughout specific regions of the body—for example, gonorrhea, staphylococcal diseases, syphilis, tetanus, and tularemia.

5. The food-borne and waterborne bacterial diseases are contracted when contaminated food or water is ingested. These diseases are essentially of two types: infections and intoxications. An infection occurs when a pathogen enters the gastrointestinal tract and multiplies. Examples include *Campylobacter,* shigellosis, and typhoid fever. An intoxication occurs because of the ingestion of a toxin produced outside the body. Examples include botulism, cholera, and staphylococcal food poisoning.

6. Some microbial diseases and their effects cannot be categorized under a specific mode of transmission. Two important examples are sepsis and septic shock. Gram-positive bacteria, fungi, and endotoxin-containing gram-negative bacteria can initiate the pathogenic cascade of sepsis leading to septic shock.

The first three parts of this textbook cover the general biology of bacteria. Chapters 19 through 24 specifically review bacterial morphology and taxonomy. Chapter 37 continues the coverage of bacteria by discussing some of the more important gram-positive and gram-negative bacteria that are pathogenic to humans.

Of all the known bacterial species, only a few are pathogenic to humans. In the following sections the more important gram-positive and gram-negative disease-causing bacteria are discussed according to their mode of acquisition/transmission.

Airborne Diseases

Most airborne diseases caused by bacteria involve the respiratory system. Other airborne bacteria can cause skin diseases. Some of the better known of these diseases are now discussed.

Diphtheria

Diphtheria [Greek *diphthera,* membrane, and *ia,* condition] is an acute contagious disease caused by the gram-positive *Corynebacterium diphtheriae* (*see figure 21.12*). *C. diphtheriae* is well adapted to airborne transmission by way of nasopharyngeal secretions and is very resistant to drying. Diphtheria mainly affects poor people living in crowded conditions. Once within the respiratory system, bacteria that carry the prophage β containing the *tox* gene (*see chapter 17*) produce diphtheria toxin, an exotoxin that causes an inflammatory response and the formation of a grayish pseudomembrane on the respiratory mucosa (figure 37.1). The exotoxin is also absorbed into the circulatory system and distributed throughout the body, where it may cause destruction of cardiac, kidney, and nervous tissues by inhibiting protein synthesis (*see table 29.5*).

Diphtheria exotoxin (p. 567).

Typical symptoms of diphtheria include a thick mucopurulent (containing both mucus and pus) nasal discharge, fever, and cough. Diagnosis is made by observation of the pseudomembrane in the throat and by bacterial culture. Diphtheria antitoxin is given to neutralize any unabsorbed exotoxin in the patient's tissues; penicillin and erythromycin are used to treat the infection. Prevention is by active immunization with the **DPT** (diphtheria-pertussis-tetanus) **vaccine.**

C. diphtheriae can also infect the skin, usually at a wound or skin lesion, causing a slow-healing ulceration termed **cutaneous diphtheria.** Most cases involve people over 30 years of age who have a weakened immunity to the diphtheria toxin and live in tropical areas.

Fewer than 100 diphtheria cases are reported annually in the United States, and most occur in nonimmunized individuals. Because humans are the only reservoirs for *C. diphtheriae,* total eradication theoretically is possible if worldwide immunization becomes a reality, as it did for smallpox.

Legionnaires' Disease and Pontiac Fever

In 1976 the term **Legionnaires' disease,** or **legionellosis,** was coined to describe an outbreak of pneumonia that occurred at the Pennsylvania State American Legion Convention in Philadelphia. The bacterium responsible for the outbreak was described as *Legionella pneumophila,* a nutritionally fastidious aerobic gram-negative rod (figure 37.2). It is now known that

Figure 37.1   Diphtheria Pathogenesis. Diphtheria is a well-known exotoxin-mediated infectious disease caused by *Corynebacterium diphtheriae.* The disease is an acute, contagious, febrile illness characterized by local oropharyngeal inflammation and pseudomembrane formation. If the exotoxin gets into the blood, it is disseminated and can damage the peripheral nerves, heart, and kidneys.

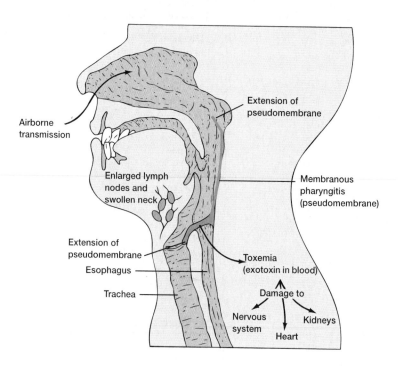

this bacterium is part of the natural microbial community of soil and aquatic ecosystems, and it has been found in large numbers in air-conditioning systems and shower stalls.

Infection with *L. pneumophila* results from the airborne spread of bacteria from an environmental reservoir to the human respiratory system. Males over 50 years of age most commonly contract the disease, especially if they are compromised by heavy smoking, alcoholism, or chronic illness. The bacteria reside within the phagosomes of alveolar macrophages, where they multiply and produce localized tissue destruction through export of a cytotoxic exoprotease. Symptoms include a high fever, nonproductive cough (respiratory secretions are not brought up during coughing), headache, neurological manifestations, and severe bronchopneumonia. Diagnosis depends on isolation of the bacterium, documentation of a rise in antibody titer over time, or a rapid test kit using urine to detect antigens. Treatment begins with supportive measures and the administration of erythromycin or rifampin.

Prevention of Legionnaires' disease depends on the identification and elimination of the environmental source of *L. pneumophila* contamination. Chlorination, the heating of water, and the cleaning of water-containing devices can help control the multiplication and spread of *Legionella*. These control measures are effective because the pathogen does not appear to be spread from person to person.

Since the initial outbreak of this disease in 1976, many outbreaks during summer months have been recognized in all parts of the United States. About 1,000 to 1,400 cases are diagnosed each year, and about 30,000 or more additional mild or subclinical cases are thought to occur. It is estimated that 3 to 6% of all nosocomial pneumonias are due to *L. pneumophila,* especially among immunocompromised patients.

L. pneumophila also causes an illness called **Pontiac fever.** This disease, which resembles an allergic disease more than an infection, is characterized by an abrupt onset of fever, headache, dizziness, and muscle pains. It is indistinguishable clinically from the various respiratory syndromes caused by viruses. Pneumonia does not occur. The disease resolves spontaneously within 2 to 5 days. No deaths from Pontiac fever have been reported.

Pontiac fever was first described from an outbreak in a county health department in Pontiac, Michigan. Ninety-five percent of the employees became ill and eventually showed elevated serum titers against *L. pneumophila*. These bacteria were later isolated from the lungs of guinea pigs exposed to the air of the building. The likely source was water from a defective air conditioner.

Meningitis

Meningitis [Greek *meninx,* membrane, and *-itis,* inflammation] is an inflammation of the brain or spinal cord meninges (membranes). Based on the specific cause, it can be divided into **bacterial (septic) meningitis** and the **aseptic meningitis syndrome** (table 37.1). As shown by the table, there are many causes of the aseptic meningitis syndrome, only some of which can be treated with antimicrobials. Thus accurate identification of the causative agent is essential to proper treatment of the disease. The immediate sources of the bacteria responsible for meningitis are respiratory secretions from carriers or active cases. The bacteria initially colonize the nasopharynx after which they cross the mucosal barrier and enter the bloodstream and cerebrospinal fluid, where they produce inflammation of the meninges.

The usual symptoms of meningitis include an initial respiratory illness or sore throat interrupted by one of the meningeal syndromes: vomiting, headache, lethargy, confusion, and stiffness

Figure 37.2 **Legionnaires' Disease.** *Legionella pneumophila,* the causative agent of Legionnaires' disease, with many lateral flagella; SEM (×10,000).

TABLE 37.1   Causative Agents of Meningitis by Diagnostic Category

Type of Meningitis	Causative Agent
Bacterial Meningitis	*Streptococcus pneumoniae*
	Neisseria meningitidis
	Haemophilus influenzae type b
	Gram-negative bacilli
	Group B streptococci
	Listeria monocytogenes
	Mycobacterium tuberculosis
	Nocardia asteroides
	Staphylococcus aureus
	Staphylococcus epidermidis
Aseptic Meningitis Syndrome	
Agents Requiring Antimicrobials	Fungi
	Amoebae
	Syphilis
	Mycoplasmas
Agents Requiring Other Treatments	Viruses
	Leptospires
	Cancers
	Parasitic cysts
	Chemicals

in the neck and back. Bacterial meningitis can be diagnosed by a Gram stain and culture of the bacteria from cerebrospinal fluid or rapid tests (*see table 34.4*). Once the causative bacterium is identified by culture, specific antibiotics (penicillin, chloramphenicol, cefotaxime, ceftriaxone, ofloxacin) are used for treatment. Vaccines against *S. pneumoniae* and *N. meningitidis* are available for high-risk individuals. Approximately 48% of all bacterial meningitis cases among children in the United States are due to *Haemophilus influenzae* type b (about 15,000 annually), 27% to *Neisseria meningitidis* (about 2,500), and 11% to *Streptococcus pneumoniae* (about 1,000).

A person may have meningitis symptoms, but show no gram-negative bacteria in gram-stained specimens, and have negative cultures. In such a case the diagnosis often is aseptic meningitis syndrome. Aseptic meningitis is more difficult to treat, and the prognosis is usually poor.

Mycobacterium avium–M. intracellulare Pneumonia

During the past decade it has been discovered that there is an extremely large group of mycobacteria that are normal inhabitants of soil, water, and house dust. Two of these have become noteworthy pathogens in the United States. The two, *Mycobacterium avium* and *Mycobacterium intracellulare* are so closely related that they are referred to as the *M. avium* complex (MAC).

These mycobacteria are found worldwide and infect a variety of insects, birds, and animals. Both the respiratory and the gastrointestinal tract have been proposed as the portal of entry for the *M. avium* complex. The gastrointestinal tract is thought to be the most common site of colonization and dissemination in AIDS patients. MAC causes a pulmonary infection in humans similar to *M. tuberculosis*. Pulmonary MAC is more common in non-AIDS patients, particularly in elderly persons with preexisting pulmonary disease.

Shortly after the recognition of AIDS and the associated opportunistic infections (*see table 36.3*), it became apparent that one of the more common infections was caused by MAC. Disseminated infection with MAC occurs in 15 to 40% of persons with AIDS in the United States with CD4$^+$ cell counts of less than 100 per cubic millimeter. Disseminated infection with MAC produces disabling symptoms, including fever, malaise, weight loss, night sweats, and diarrhea. Carefully controlled epidemiological studies have shown that MAC shortens survival by 5 to 7 months among persons with AIDS. With more effective antiviral therapy for AIDS and with prolonged survival, the number of cases of disseminated MAC is likely to increase substantially, and its contribution to AIDS mortality will increase concomitantly.

MAC can be isolated from sputum, blood, and aspirates of bone marrow. Acid-fast stains are of value in making a diagnosis. The most sensitive method for detection is the commercially available lysis-centrifugation blood culture system (Wampole Laboratories). Although no drugs are currently approved by the FDA for the therapy of MAC, every regimen should contain either azithromycin or clarithromycin or ethambutol as a second drug, and one or more of the following: closfazimine, rifabutin, rifampin, ciprofloxacin, and amikacin.

Pertussis

Pertussis [Latin *per*, intensive, and *tussis*, cough], sometimes called "whooping cough," is caused by the gram-negative bacterium *Bordetella pertussis*. (*B. parapertussis* is a closely related species that causes a milder form of the disease.) Pertussis is a highly contagious disease that primarily affects children. It has been estimated that over 95% of the world's population has experienced either mild or severe symptoms of the disease. Around 500,000 die from the disease each year. However, there are less than 5,000 cases and less than 10 deaths annually in the United States.

Transmission occurs by inhalation of the bacterium in droplets released from an infectious person. The incubation period is 7 to 14 days. Once inside the upper respiratory tract, the bacteria attach to the ciliated epithelial cells by producing a specific adherence factor called filamentous hemagglutinin (*see table 29.3*), which recognizes a complementary molecule on the cells. After attachment, the bacteria synthesize several toxins (*see table 29.5*) that are responsible for the symptoms. The most important toxin is pertussis toxin, which causes increased tissue susceptibility to histamine and serotonin, and an increased lymphocyte response. *B. pertussis* also produces an extracytoplasmic invasive adenylate cyclase, and tracheal cytotoxin and dermonecrotic toxin, which destroy epithelial tissue. In addition, the secretion of a thick mucus impedes ciliary action, and often, ciliated epithelial cells die.

Pertussis is divided into three stages. (1) The catarrhal stage, so named because of the mucous membrane inflammation, which is insidious and resembles the common cold. (2) Prolonged coughing sieges characterize the paroxysmal stage. During this stage the infected person tries to cough up the mucous secretions by making 5 to 15 rapidly consecutive coughs followed by the characteristic whoop—a hurried deep inspiration. The catarrhal and paroxysmal stages last about 6 weeks. (3) Final recovery may take several months (the convalescent stage).

Laboratory diagnosis of pertussis is by culture of the bacterium, fluorescent antibody staining of smears from nasopharyngeal swabs, and serological tests. The development of a strong, lasting immunity takes place after an initial infection. Treatment is with erythromycin, tetracycline, or chloramphenicol. Prevention is with the DPT vaccine (*p. 588*); vaccination of children is recommended when they are 2 to 3 months old (*see table 30.1*).

Streptococcal Diseases

Streptococci, commonly called strep, are a heterogeneous group of gram-positive bacteria. In this group *Streptococcus pyogenes* (group A β-hemolytic streptococci; *see pp. 443–45*) is one of the most important bacterial pathogens. The different serotypes produce (1) extracellular enzymes that break down host molecules; (2) streptokinases, proteins that activate a host-blood factor that dissolves blood clots; (3) the cytolysins streptolysin O and streptolysin S, which kill host leukocytes; and (4) capsules and M protein that help retard phagocytosis.

S. pyogenes is widely distributed among humans, but usually people are asymptomatic carriers. Individuals with acute infections may spread the pathogen, and transmission can occur through respiratory droplets, direct, or indirect contact. When highly virulent strains appear in schools, they can cause sharp outbreaks of sore throats and scarlet fever. Due to the cumulative buildup of antibodies to many different *S. pyogenes* serotypes over the years, outbreaks among adults are less frequent.

Diagnosis of a streptococcal infection is based on both clinical and laboratory findings. Several rapid tests are available (*see table 34.4*). Treatment is with penicillin or erythromycin. Vaccines are not available for streptococcal diseases other than streptococcal pneumonia because of the large number of serotypes.

The best control measure is prevention of bacterial transmission. Individuals with a known infection should be isolated and treated. Personnel working with infected patients should follow standard aseptic procedures.

In the following sections some more important human streptococcal diseases are discussed (figure 37.3).

Cellulitis and Erysipelas

Cellulitis is a diffuse, spreading infection of subcutaneous skin tissue. The resulting inflammation is characterized by a defined area of redness (erythema) and the accumulation of fluid (edema).

The most frequently diagnosed skin infection caused by *S. pyogenes* is **impetigo** (impetigo also can be caused by *S. aureus* [*see figure 36.17e*]). Impetigo is a superficial cutaneous infection, most commonly seen in children, usually located on the face, and characterized by crusty lesions and vesicles surrounded by a red border. Impetigo is most common in late summer and early fall. The drugs of choice for impetigo are penicillin or erythromycin in those individuals who are allergic to penicillin.

Erysipelas [Greek *erythros*, red, and *pella*, skin] is an acute infection and inflammation of the dermal layer of the skin. It occurs primarily in infants and people over 30 years of age with a history of streptococcal sore throat. The skin often develops painful reddish patches that enlarge and thicken with a sharply defined edge (figure 37.4). Recovery usually takes a week or longer if no treatment is given. The drugs of choice for the treatment of erysipelas are erythromycin and penicillin. Erysipelas may recur periodically at the same body site for years.

Invasive Streptococcus A Infections

In the nineteenth century invasive *Streptococcus pyogenes* infections were a major cause of morbidity and mortality. However, during this century the incidence of severe group A streptococcal infections has declined, especially since the arrival of antibiotic therapy. In the mid-1980s there was a worldwide increase in group A streptococcal sepsis; clusters of rheumatic fever were reported from locations within the United States, and a streptococcal toxic shock-like syndrome emerged. (A virulent strep A infection killed *Sesame Street* Muppeteer Jim Henson in 1990, and in 1994 the press made headlines with articles on "the flesh-eating invasive disease.")

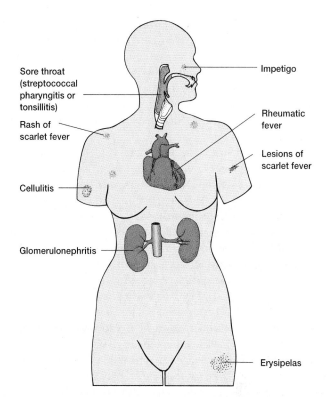

Figure 37.3   Streptococcal Diseases. Some of the more prominent diseases associated with group A streptococcal infections, and the body sites affected.

Figure 37.4   Erysipelas. Notice the bright, raised, rubbery lesion at the site of initial entry (white arrow) and the spread of the inflammation to the foot. The reddening is caused by toxins produced by the streptococci as they invade new tissue.

The development of invasive strep A disease appears to depend on the presence of specific virulent strains (M-1 and M-3 serotypes) and predisposing host factors (surgical or nonsurgical wounds, diabetes, and other underlying medical problems). A life-threatening infection begins when invasive strep A strains penetrate a mucous membrane or take up residence in a skin lesion such as a bruise. This infection can quickly lead either to **necrotizing fasciitis** [Greek *nekrosis*, deadness, Latin *fascis*, band or bandage, and *itis*, inflammation] that destroys the sheath covering skeletal muscles or to **myositis** [Greek *myos*, muscle, and *itis*], the inflammation and destruction of skeletal muscle and fat tissue. Because necrotizing fasciitis and myositis arise and spread so quickly, they have been colloquially called "galloping gangrene."

One reason invasive strep A strains are so deadly is that about 85% of them carry the genes for the production of *s*trepto-coccal *p*yrogenic *e*xotoxins A and B (Spe exotoxins). Exotoxin A acts as a superantigen. This superantigen quickly stimulates T cells to begin producing abnormally large quantities of cytokines. These cytokines damage the endothelial cells that line blood vessels, causing fluid loss and rapid tissue death from a lack of oxygen. Another pathogenic mechanism involves the secretion of exotoxin B (cysteine protease). Cysteine protease is an enzyme that rapidly destroys tissue by breaking down proteins.

Since 1986 it has been recognized that invasive strep A infections can also trigger a *toxic shock-like syndrome* (TSLS), characterized by a precipitous drop in blood pressure, failure of multiple organs, and a very high fever (figure 37.22). TSLS is caused by an invasive strep A that produces one or more of the streptococcal pyrogenic exotoxins. TSLS has a mortality rate of over 30%.

Rapid treatment is necessary to reduce the risk of death, and penicillin G remains the treatment of choice. In addition, surgical removal of dead and dying tissue usually is needed in more advanced cases of necrotizing fasciitis. It is estimated that 10,000 to 15,000 cases of invasive strep A infections occur annually in the United States, and between 5 and 10% of them are associated with necrotizing conditions.

Because group A streptococci are less contagious than cold or flu viruses, infected individuals do not pose a major threat to people around them. The best preventive measures are simple ones such as covering food, washing hands, and cleansing and medicating wounds.

Poststreptococcal Diseases

The poststreptococcal diseases are glomerulonephritis and rheumatic fever. They occur 1 to 4 weeks after an acute streptococcal infection (hence the term post). Today these two diseases are the most serious problems associated with streptococcal infections in the United States.

Glomerulonephritis or **Bright's disease** is an inflammatory disease of the renal glomeruli, membranous structures within the kidney where blood is filtered. Damage probably results from the deposition of antigen-antibody complexes, possibly involving the streptococcal M protein, in the glomeruli. Thus the disease arises from a type III immune complex-mediated hypersensitivity reaction (*see figure 31.13*). The complexes cause destruction of the glomerular membrane, allowing proteins and blood to leak into the urine. Clinically the affected person exhibits edema, fever, hypertension, and hematuria (blood in the urine). The disease occurs primarily among school-age children. Diagnosis is based on the clinical history, physical findings, and confirmatory evidence of prior

▶ **Figure 37.5 Scarlet Fever.** The strawberry-colored tongue of this streptococcal disease.

streptococcal infection. The incidence of glomerulonephritis in the United States is less than 0.5% of streptococcal infections. Penicillin G or erythromycin can be given for any residual streptococci. However, there is no specific therapy once kidney damage has occurred. About 80 to 90% of all cases undergo slow spontaneous healing of the damaged glomeruli, whereas the others develop a chronic form of the disease. The latter may require a kidney transplant or lifelong renal dialysis.

Rheumatic fever is an autoimmune disease characterized by inflammatory lesions involving the heart valves, joints, subcutaneous tissues, and central nervous system. It usually results from a prior streptococcal sore throat infection. The exact mechanism of rheumatic fever development remains unknown. The disease occurs most frequently among children 6 to 15 years of age and manifests itself through a variety of signs and symptoms, making diagnosis difficult. In the United States rheumatic fever has become very rare (less than 0.05% of streptococcal infections), but it occurs 100 times more frequently in tropical countries. Therapy is directed at decreasing the inflammation and fever and controlling cardiac failure. Salicylates and corticosteroids are the mainstays of treatment. Though rheumatic fever is rare, it is still the most common cause of permanent heart valve damage in children.

Scarlet Fever

Scarlet fever (scarlatina) results from a throat infection with a strain of *S. pyogenes* that carries a lysogenic bacteriophage. This codes for the production of an erythrogenic or rash-inducing toxin that causes shedding of the skin. Scarlet fever is a communicable disease spread by inhalation of infective respiratory droplets. After a 2-day incubation period, a scarlatinal rash appears on the upper chest and then spreads to the remainder of the body. This rash represents the skin's generalized reaction to the circulating toxin. Along with the rash, the infected individual experiences a sore throat, chills, fever, headache, and a strawberry-colored tongue (figure 37.5). Treatment is with penicillin.

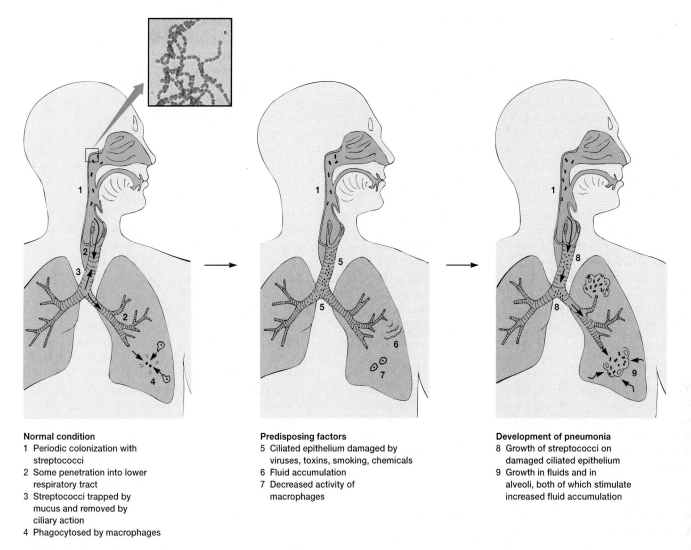

Normal condition
1 Periodic colonization with streptococci
2 Some penetration into lower respiratory tract
3 Streptococci trapped by mucus and removed by ciliary action
4 Phagocytosed by macrophages

Predisposing factors
5 Ciliated epithelium damaged by viruses, toxins, smoking, chemicals
6 Fluid accumulation
7 Decreased activity of macrophages

Development of pneumonia
8 Growth of streptococci on damaged ciliated epithelium
9 Growth in fluids and in alveoli, both of which stimulate increased fluid accumulation

Figure 37.6   Predisposition to and the Development of Streptococcal Pneumonia. The insert shows the morphology of *Streptococcus pneumoniae* (× 1,000).

Streptococcal Sore Throat

Streptococcal sore throat is one of the most common bacterial infections of humans and is commonly called strep throat. The β-hemolytic group A streptococci are spread by droplets of saliva or nasal secretions. The incubation period in humans is 2 to 4 days. The incidence of sore throat is greater during the winter and spring months.

The action of the strep bacteria in the throat (**pharyngitis**) or on the tonsils (**tonsillitis**) stimulates an inflammatory response and the lysis of leukocytes and erythrocytes. An inflammatory exudate consisting of cells and fluid is released from the blood vessels and deposited in the surrounding tissue. This is accompanied by a general feeling of discomfort or malaise, fever (usually above 101°F), and headache. Prominent physical manifestations include redness, edema, and lymph node enlargement in the throat. Several common rapid test kits are available for diagnosing strep throat. In the absence of complications, the disease is self-limited and disappears within a week. However, treatment with penicillin G benzathine (or erythromycin for

penicillin-allergic people) can shorten the infection and clinical syndromes, and is especially important in children for the prevention of complications such as rheumatic fever and glomerulonephritis. Infections in older children and adults tend to be milder and less frequent due in part to the immunity they have developed against the many serotypes encountered in early childhood. Prevention and control measures include proper disposal or cleansing of objects (e.g., Kleenex, handkerchiefs) contaminated by discharges from the infected individual.

Streptococcal Pneumonia

Streptococcal pneumonia is now considered an **endogenous infection**—that is, it is contracted from one's own normal microbiota (*see figure 28.8*). It is caused by the gram-positive *Streptococcus pneumoniae,* found in the upper respiratory tract (figure 37.6). However, disease usually occurs only in those individuals with predisposing factors such as viral infections of the respiratory tract, physical injury to the tract, alcoholism, or diabetes. About 60 to 80% of all respiratory diseases known as pneumonia are caused by *S. pneumoniae.* An estimated 150,000 to 300,000 people in the United States contract this form of pneumonia annually, and between 13,000 to 66,000 deaths result.

The primary virulence factor of *S. pneumoniae* is its capsular polysaccharide. The capsule prevents binding of antibody to the cell wall and thus inhibits phagocytosis. The pathogenesis is due to the rapid multiplication of the bacteria in alveolar spaces. The alveoli fill with blood cells and fluid and become inflamed. The sputum is often rust colored because of blood coughed up from the lungs. The onset of clinical symptoms is usually abrupt, with chills, hard labored breathing, and chest pain. Diagnosis is by chest X ray, biochemical tests, and culture. Penicillin G, cefotaxime, ofloxacin, and ceftriaxone have contributed to a greatly reduced mortality rate. For individuals who are sensitive to penicillin, erythromycin, or tetracycline can be used. Recently a penicillin-resistant strain of *S. pneumoniae* has appeared in the United States. A pneumococcal vaccine (Pneumovax) is available for people who are debilitated (e.g., people in chronic-care facilities). Preventive and control measures include immunization and adequate treatment of infected persons.

Tuberculosis

Over a century ago Robert Koch (*see figure 1.5*) identified *Mycobacterium tuberculosis* as the causative agent of **tuberculosis (TB).** At the time, TB was rampant, causing 1/7 of all deaths in Europe and 1/3 of deaths among productive young adults. Today TB remains a global health problem of enormous dimension. It is estimated that there are 1 billion (20% of the world's human population) infected worldwide, with 8 million new cases and over 3 million deaths per year.

In the United States this disease occurs most commonly among the homeless, elderly, malnourished, or alcoholic poor males, minorities, immigrants, and Native Americans. More than 26,000 new cases of tuberculosis and over 12,000 deaths are reported annually. Most cases in the United States are caused by the acid-fast *Mycobacterium tuberculosis,* acquired from other humans through droplet nuclei and the respiratory route (figure 37.7). It appears that about 1/4 to 1/3 of active TB cases in the United States may be due to recent transmission. The majority of active cases result from the reactivation of old dormant infections. Worldwide, *M. bovis* and *M. africanum* also cause TB. Transmission to humans from susceptible animal species and their products (e.g., milk) is also possible. Recently there has been a steady yearly increase in the number of TB cases as a result of the AIDS epidemic. Available statistics indicate that a close association exists between AIDS and TB. Therefore further spread of HIV infection among the population with a high prevalence of TB infection is resulting in dramatic increases in TB.

AIDS (pp. 722–30).

Once in the lungs the bacteria are phagocytosed by macrophages and a hypersensitivity response forms small, hard nodules called **tubercles,** which are characteristic of tuberculosis and give the disease its name. The disease process usually stops at this stage, but the bacteria often remain alive within the

macrophages. Resistance to oxidative killing, inhibition of phagosome-lysosome fusion, and inhibition of diffusion of lysosomal enzymes are some of the mechanisms that may explain the survival of *M. tuberculosis* inside macrophages. In time the tubercle may change to a cheeselike consistency and is then called a **caseous lesion.** If such lesions calcify, they are termed **Ghon complexes,** which show up prominently in a chest X ray. (Often the primary lesion is called the Ghon's tubercle or Ghon's focus). Sometimes the tubercle lesions liquefy and form air-filled **tuberculous cavities.** From these cavities the bacteria can spread to new foci of infections throughout the body. This spreading is often called **miliary tuberculosis** due to the many tubercles the size of millet seeds that are formed in the infected tissue. It also may be called **reactivation tuberculosis** because the bacteria have been reactivated in the initial site of infection.

Persons infected with *M. tuberculosis* develop a cell-mediated immunity due to the bacteria being phagocytosed by macrophages. This immunity involves sensitized T cells (figure 37.7b) and is the basis for the tuberculin skin test (*see figure 31.14*). In this test a purified protein derivative (PPD) of *M. tuberculosis* is injected intracutaneously (the Mantoux test). If the person has had tuberculosis, sensitized T cells react with these proteins, and a delayed hypersensitivity reaction occurs within 48 hours. This positive skin reaction appears as an induration (hardening) and reddening of the area around the injection site. Multiple puncture tests such as the Tine test are more convenient but not as accurate.

In a young person a positive skin test possibly indicates active tuberculosis. In older persons it may result from previous disease, vaccination, or a false-positive test. In both cases X rays and bacterial isolation should be completed.

The symptoms of tuberculosis are fever, fatigue, and weight loss. A cough, which is characteristic of pulmonary involvement, may result in expectoration of bloody sputum.

Laboratory diagnosis of tuberculosis is by isolation of the acid-fast bacterium, chest X ray, commercially available DNA probes, HPLC, the BACTEC NAP test, and the Mantoux or tuberculin skin test. Both chemotherapy and chemoprophylaxis are carried out by administering isoniazid (INH), plus rifampin, ethambutol, and pyrazinamide. These drugs are administered simultaneously for 12 to 24 months as a way of decreasing the possibility that the patient develops drug resistance.

Recently, new *multi-drug-resistant strains of tuberculosis* **(MDR-TB)** have developed and are spreading. A multi-drug-resistant strain is defined as *M. tuberculosis* resistant to isoniazid and rifampin, with or without resistance to other drugs. Within the United States drug resistance has increased from 2 to 9% in the past three decades, and similar increases have occurred in many other countries. This has resulted in many cases of marginally treatable, often fatal, disease. Inadequate therapy is the most common means by which resistant bacteria are acquired, and patients who have previously undergone therapy should be presumed to harbor MDR-TB until proved otherwise.

The way in which MDR-TB arises is now known. Tubercle bacilli have spontaneous, predictable rates of chromosomally born mutations that confer resistance to drugs. These mutations

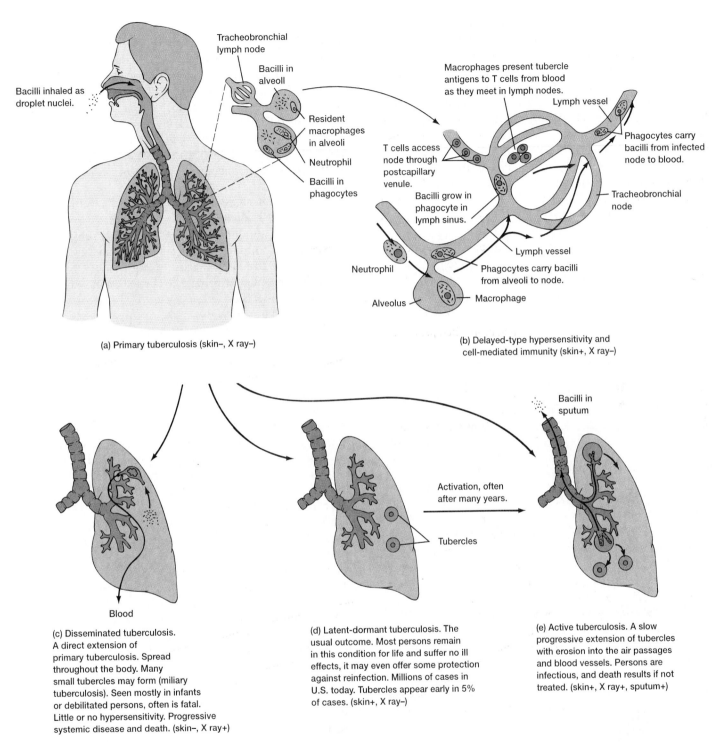

(a) Primary tuberculosis (skin–, X ray–)

(b) Delayed-type hypersensitivity and cell-mediated immunity (skin+, X ray–)

(c) Disseminated tuberculosis. A direct extension of primary tuberculosis. Spread throughout the body. Many small tubercles may form (miliary tuberculosis). Seen mostly in infants or debilitated persons, often is fatal. Little or no hypersensitivity. Progressive systemic disease and death. (skin–, X ray+)

(d) Latent-dormant tuberculosis. The usual outcome. Most persons remain in this condition for life and suffer no ill effects, it may even offer some protection against reinfection. Millions of cases in U.S. today. Tubercles appear early in 5% of cases. (skin+, X ray–)

(e) Active tuberculosis. A slow progressive extension of tubercles with erosion into the air passages and blood vessels. Persons are infectious, and death results if not treated. (skin+, X ray+, sputum+)

Figure 37.7   Tuberculosis. The history of untreated tuberculosis in the human body.

are unlinked; hence resistance to one drug is not associated to resistance to an unrelated drug. The emergence of drug resistance represents the survival of random preexisting mutations, not a change caused by exposure to the drug—that the mutations are not linked is the cardinal principle underlying TB chemotherapy. For example, mutations causing resistance to isoniazid or rifampin occur roughly 1 in 10^8 replications of *M. tuberculosis*. The likelihood of spontaneous mutations caus-

ing resistance to both isoniazid and rifampin is the sum of these probabilities, or 1 in 10^{16}. However, these biological mechanisms of resistance break down when chemotherapy is inadequate. In the circumstances of monotherapy, erratic drug ingestion, omission of one or more drugs, suboptimal dosage, poor drug absorption, or an insufficient number of active drugs in a regimen, a susceptible strain on *M. tuberculosis* may become resistant to multiple drugs within a matter of months.

Prevention and control of tuberculosis requires rapid specific therapy to interrupt infectious spread. Retreatment of patients who have multi-drug-resistant tuberculosis should be carried out in programs with comprehensive microbiological, pharmacokinetic, psychosocial, and nutritional support systems. In many countries individuals, especially infants and children, are vaccinated with **b**acille Calmette-Guérin (BCG) vaccine to prevent complications such as meningitis. The BCG vaccine appears to protect about half of those inoculated. Tuberculosis rates also can be lowered by better public health measures and social conditions, for example, a reduction in homelessness and drug abuse.

1. What causes the typical symptoms of diphtheria and how are individuals protected against this disease? *DT exotoxin / DPT vaccine*
2. What is the environmental source of the bacterium that causes Legionnaires' disease? Pontiac fever? *Soil*
3. What are the two major types of meningitis? Why is it so *Aseptic Septic* important to determine which type a person has?
4. Name the three stages of pertussis. *CATARRHAL, PAROXYSMAL, FINAL recovery*
5. Name the most important human diseases caused by *Streptococcus pyogenes*. How do they differ from one another?
6. How is tuberculosis diagnosed? Describe the various types of lesions and how they are formed. How do multi-drug-resistant strains of tuberculosis develop? *TB Skin test* *Inadequate therapy*

Arthropod-Borne Diseases

Although arthropod-borne bacterial diseases are generally rare, they are of interest either historically (plague) or because they have been newly introduced into humans (Lyme disease). These two diseases are now discussed.

Lyme Disease

Lyme disease (LD, Lyme borreliosis) was first observed and described in 1975 among people of Old Lyme, Connecticut. It has become the most common tick-borne zoonosis in the United States, with more than 10,000 cases being reported annually. In fact, Lyme disease has reached epidemic proportions, and if it were not for AIDS, Lyme disease would be the most important "new" infectious disease of humans in the United States. The disease is also present in Europe and Asia.

The Lyme spirochetes responsible for this disease comprise at least three species, currently designated *Borrelia burgdorferi* (figure 37.8*a*), *B. garinii*, and *B. afzelii*. Deer and field mice are the natural hosts. In the Northeast of the United States, *B. burgdorferi* is transmitted to humans by the bite of the infected deer tick (*Ixodes scapularis*, formerly *I. dammini*; figure 37.8*b*). On the Pacific Coast, especially in California, the reservoir is a dusky-footed woodrat, and the tick, *I. pacificus*.

Clinically Lyme disease is a complex illness with three major stages. The initial, localized stage occurs a week to 10 days after an infectious tick bite. The illness usually begins with an expanding, ring-shaped, skin lesion with a red outer border and partial central clearing (figure 37.8*c*). This often is accompanied by flulike symptoms (malaise and fatigue, headache, fever, and chills). Often the tick bite is unnoticed, and the skin lesion may be missed due to skin coloration or its obscure location such as on the scalp. Thus treatment, which is usually effective at this stage, may not be given because the illness is passed off as "just a touch of the flu."

The second, disseminated stage may appear weeks or months after the initial infection. It consists of several symptoms such as neurological abnormalities, heart inflammation, and bouts of arthritis (usually in the major joints such as the elbows or knees). The inflammation that produces organ damage is initiated and possibly perpetuated by the immune response to one or more spirochetal proteins.

Finally, like syphilis, years later the late stage may appear. Infected individuals may develop demyelination of neurons with symptoms resembling Alzheimer's disease and multiple sclerosis. Behavioral changes also can occur.

Laboratory diagnosis of Lyme's disease is based on (1) the recovery of the spirochete from the patient, (2) use of the polymerase chain reaction (*see chapter 15*) for detection of *B. burgdorferi* DNA in urine, or (3) serological testing (Lyme ELISA or Western Blot) for IgM and IgG antibodies to the pathogen. Treatment with amoxicillin or tetracycline early in the illness results in prompt recovery and prevents arthritis and other complications. If nervous system involvement is suspected, ceftriaxone is used since it can cross the blood-brain barrier.

Prevention and control of Lyme disease involves environmental modification (clearing and burning tick habitat) and the application of acaricidal compounds (agents that destroy mites and ticks). An individual's risk of acquiring Lyme disease may be greatly reduced by education and personal protection. The following points should be kept in mind whenever a person is active in an area where Lyme disease or other tick-borne zoonoses occur:

1. It takes a minimum of 24 hours of attachment and feeding for transmission to occur; thus prompt removal of attached ticks will greatly reduce the risk of infection. To remove an embedded tick, one should use tweezers to grasp the tick as close as possible to the skin and then pull with slow, steady pressure in a direction perpendicular to the skin.
2. Because each deer tick life cycle stage is most abundant at a certain time, there are periods when an individual should be most aware of the risk of infection. The most dangerous times are May through July, when the majority of nymphal deer ticks are present and the risk of transmission is greatest.

(a)

(b)

(c)

Figure 37.8   **Lyme Disease.** (*a*) The etiological agent is the spirochete *Borrelia burgdorferi;* SEM. (*b*) The vector in the Northeast is the tick *Ixodes scapularis.* The youngest (nymphal stage, top) is about the size of a poppy seed. An unengorged adult (bottom), and an engorged adult (center) can reach the size of a jelly bean. (*c*) The typical rash (erythema migrans) showing concentric rings around the initial site of the tick bite.

3. If you must be in the woods, dress accordingly. Wear light-colored pants and good shoes. Tuck the cuffs of your pants into long socks to deny ticks easy entry under your clothes. After coming out of the woods, check all clothes for ticks.

4. Repellents containing high concentrations of DEET (diethyltoluamide) or permanone are available over the counter and are very noxious to ticks. Premethrin kills ticks on contact but is approved only for use on clothing.

5. Immediately after being in a high-risk area, examine your body for bites or itches. Taking a shower and using lots of soap aids in this examination. Areas such as the scalp, armpits, and groin are difficult to examine effectively but are preferred sites for tick attachment. Special attention should be given to these parts of the body.

Plague

In the southwestern part of the United States, **plague** [Latin *plaga,* pest] occurs primarily in wild rodents (ground squirrels and prairie dogs). However, massive human epidemics occurred during the Middle Ages, and the disease was known as the Black Death because one of its characteristics is blackish areas on the skin caused by subcutaneous hemorrhages. Infections now occur in humans only sporadically or in limited outbreaks. In the United States approximately 25 cases are reported annually, and the mortality rate is about 15%.

The disease is caused by the gram-negative bacterium *Yersinia pestis.* It is transmitted from rodent to human by the bite of an infected flea, direct contact with infected animals or their products, or inhalation of contaminated airborne droplets

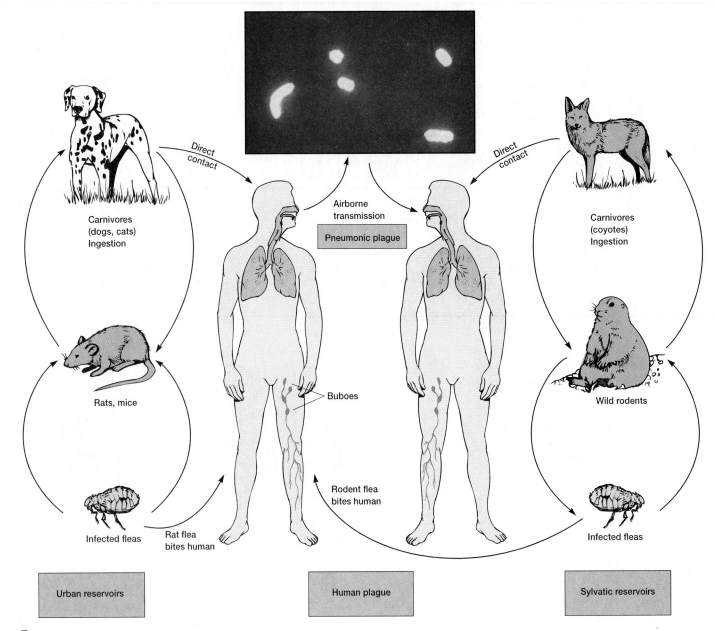

🌀 **Figure 37.9** **Plague.** Plague is spread to humans through (1) the urban cycle and rat fleas, (2) the sylvatic cycle centered on wild rodents and their fleas, or (3) by airborne transmission from an infected person leading to pneumonic plague. Dogs, cats, and coyotes can also acquire the bacterium by ingestion of infected animals. The insert shows *Yersinia pestis* stained with fluorescent antibodies.

(figure 37.9). Once in the human body, the bacteria multiply in the blood and lymph. An important factor in the virulence of *Y. pestis* is its ability to survive and proliferate inside phagocytic cells rather than be killed by them. One of the ways this is accomplished is by the YOPS (**y**ersinal plasmid-encoded **o**uter membrane **p**roteins) that counteract natural defense mechanisms and help the bacteria multiply and disseminate in the host.

Symptoms—besides the subcutaneous hemorrhages—include fever and the appearance of enlarged lymph nodes called **buboes** (hence the old name, **bubonic plague**). In 50 to 70% of the untreated cases, death follows in 3 to 5 days from toxic conditions caused by the large number of bacilli in the blood.

Laboratory diagnosis of plague is by direct microscopic examination, culture of the bacterium, serological tests, the PCR for detection of bacteria in infected fleas, and phage testing. Treatment is with streptomycin or tetracycline, and recovery from the disease gives a good immunity.

If the patient does not receive antibiotics, the bacteria may invade the lungs and produce **pneumonic plague,** a type of pneumonia. The mortality rate for this kind of plague is almost 100% if it is not recognized within 12 to 24 hours. Obviously great care must be taken to prevent the spread of airborne infections to personnel taking care of pneumonic plague patients.

Prevention and control involves ectoparasite and rodent control, isolation of human patients, prophylaxis or abortive therapy of exposed persons, and vaccination (USP Plague vaccine) of persons at high risk.

---

1. What is the causative agent of Lyme disease and how is it transmitted to humans? How does the illness begin? Describe the three stages of Lyme disease. *B. burgdorferi, deer tick bite,*

2. Why is plague sometimes called the Black Death? How is it transmitted? Distinguish between bubonic and pneumonic plague. *black hemmorrhages .*

Direct Contact Diseases

Most of the direct contact bacterial diseases involve the skin or underlying tissues. Others can become disseminated through specific regions of the body. Several examples of each are now discussed.

Anthrax

Anthrax [Greek *anthrax*, coal] is a highly infectious animal disease that can be transmitted to humans by direct contact with infected animals (cattle, goats, sheep) or their products. The causative bacterium is the gram-positive *Bacillus anthracis.* Its endospores can remain viable in soil and animal products for decades. Human infection is usually through a cut or abrasion of the skin, resulting in **cutaneous anthrax;** however, inhaling endospores may result in **pulmonary anthrax,** also known as woolsorter's disease. If endospores reach the intestine, **gastrointestinal anthrax** may result.

In humans the incubation period for cutaneous anthrax is 1 to 15 days. It begins with a skin papule that ulcerates and is called an **eschar** (figure 37.10). Headache, fever, and nausea are the major symptoms. Pulmonary anthrax resembles influenza. If the bacteria invade the bloodstream, the disease can be fatal. The signs and symptoms of anthrax are due to anthrax toxins, a complex exotoxin system composed of three proteins (*see table 29.5*). The genes encoding these factors are found on plasmids.

Diagnosis is by direct microscopic examination, culture of the bacterium, and serology; therapy is with penicillin G or penicillin G plus streptomycin. Vaccination of animals, primarily cattle, is an important control measure. However, people with a high occupational risk, such as those who handle infected animals or their products, including hides and wool, should be immunized with the cell-free vaccine obtainable from the Centers for Disease Control and Prevention. Fewer than 10 cases of anthrax occur annually in the United States.

Bacterial Vaginosis

Bacterial vaginosis is considered a sexually transmitted disease. It has a polymicrobic etiology that includes *Gardnerella vaginalis* (a gram-negative to gram-variable, pleomorphic

Figure 37.10   Anthrax. The malignant (destructive) eschar pustules of anthrax on the arm of an infected person.

nonmotile rod), *Mobiluncus* spp., *Mycoplasma hominis,* and various anaerobic bacteria. The recent finding that these microorganisms inhabit the rectum of 20 to 40% of healthy women indicates a potential source of autoinfection in addition to sexual transmission. Although a mild disease it is a risk factor for obstetric infections, various adverse outcomes of pregnancy, and pelvic inflammatory disease. The vaginosis is characterized by a copious, frothy, fishy-smelling discharge without pain or itching. Diagnosis is based on this fishy odor and the microscopic observation of clue cells in the discharge. **Clue cells** are sloughed-off vaginal epithelial cells covered with bacteria, mostly *G. vaginalis.* Treatment for bacterial vaginosis is with metronidazole (Flagyl), a drug that kills the anaerobes that are needed for the continuation of the disease.

Cat-Scratch Disease

Cat-scratch disease (CSD) is a loosely defined syndrome, the cause of which eluded microbiologists for decades. Currently the etiology centers on two recently described bacteria: *Bartonella* (formerly *Rochalimaea*) *henselae* and *Afipia felis.* Although *A. felis* has been isolated from lymph node biopsies in most patients who have CSD, serological evidence suggests that *B. henselae* (a gram-negative bacillus) is more likely the causative agent of CSD. However, a weak serological response does not exclude *A. felis* as a pathogen because there is evidence to suggest that it may be an intracellular pathogen, in which case humoral immunity may play a minor role in clearing this bacterium from the human body. Also, *Bartonella* species cause human diseases other than CSD (bacillary angiomatosis [a diseased state of blood vessels], parenchymal bacillary peliosis [blood accumulation in the liver], and septicemia).

The diagnosis of CSD is based on the combination of the clinical history of either a cat scratch or bite at the skin surface (primary lesion) and subsequent swelling of the lymph node(s) that drain the inoculation site. This is accompanied by malaise and fever. The typical case of CSD is self-limiting, with abatement of symptoms over a period of days to weeks, and the resolution of the lymphadenopathy over several months.

Chancroid

Chancroid [French *chancre,* a destructive sore, and Greek *eidos,* to form], also known as **genital ulcer disease,** is a sexually transmitted disease caused by the gram-negative bacillus *Haemophilus ducreyi.* The bacterium enters the skin through a break in the epithelium. After an incubation period of 4 to 7 days, a papular lesion develops within the epithelium, causing swelling and white blood cell infiltration. Within several days a pustule forms and ruptures, producing a painful circumscribed ulcer with a ragged edge; hence the term genital ulcer disease. Most of the ulcers in males are on the penis and in females at the entrance of the vagina. Genital ulcer disease occurs commonly in the tropics; however, in the past decade there have been major outbreaks in the United States. Worldwide, genital ulcer disease is an important cofactor in the transmission of the AIDS virus; thus it could be an important cofactor in the United States as well. Diagnosis is by isolating *H. ducreyi* from the ulcers; treatment is with erythromycin or ceftriaxone. Control is by the use of condoms or abstinence.

Gas Gangrene or Clostridial Myonecrosis

Clostridium perfringens, C. novyi, and *C. septicum* are gram-positive spore-forming rods termed the histotoxic clostridia. They can produce a necrotizing infection of skeletal muscle called **gas gangrene** [Greek *gangraina,* an eating sore] or **clostridial myonecrosis** [*myo,* muscles, and *necrosis,* death].

Histotoxic clostridia occur in the soil worldwide and also are part of the normal endogenous microflora of the human large intestine (*see figure 28.8*). Contamination of injured tissue with spores from soil containing histotoxic clostridia or bowel flora is the usual means of transmission. Infections are commonly associated with wounds resulting from abortions, automobile accidents, military combat, or frostbite.

If the spores germinate in anaerobic tissue, the bacteria grow and secrete α-toxin. Growth often results in the accumulation of gas (mainly hydrogen as a result of carbohydrate fermentation), and of the toxic breakdown products of skeletal muscle tissue.

Clinical manifestations include severe pain, edema, drainage, and muscle necrosis. The pathology arises from progressive skeletal muscle necrosis due to the effects of α-toxin. Other enzymes produced by the bacteria degrade collagen and tissue, facilitating spread of the disease.

Gas gangrene is a medical emergency. Laboratory diagnosis is through recovery of the appropriate species of clostridia accompanied by the characteristic disease symptoms. Treatment is extensive surgical debridement (removal of all dead tissue),

the administration of polyvalent antitoxin, and antimicrobial therapy with penicillin and tetracycline. Hyperbaric oxygen therapy (the use of high concentrations of oxygen at elevated pressures) also is considered effective. The oxygen saturates the infected tissue and thereby prevents the growth of the obligately anaerobic clostridia.

Prevention and control includes debridement of contaminated traumatic wounds plus antimicrobial prophylaxis and prompt treatment of all wound infections. Amputation of limbs often is necessary to prevent further spread of the disease.

Gonorrhea

Gonorrhea [Greek *gono,* seed, and *rhein,* to flow] is an acute, infectious, sexually transmitted disease of the mucous membranes of the genitourinary tract, eye, rectum, and throat. It is caused by the gram-negative, oxidase-positive, diplococcus, *Neisseria gonorrhoeae.* These bacteria are also referred to as **gonococci** [pl. of gonococcus; Greek *gono,* seed, and *coccus,* berry] and have a worldwide distribution.

Once inside the body the gonococci attach to the microvilli of mucosal cells by means of pili and protein II, which function as adhesions. This attachment prevents the bacteria from being washed away by normal vaginal discharges or by the strong flow of urine. They are then phagocytosed by the mucosal cells and may even be transported through the cells to the intercellular spaces and subepithelial tissue. Phagocytes, such as neutrophils, also may contain gonococci (figure 37.11) inside vesicles. Because the gonococci are intracellular at this time, the host's defenses have little effect on the bacteria. Following penetration of the bacteria, the host tissue responds locally by the infiltration of mast cells, more PMNs, and plasma cells. These cells are later replaced by fibrous tissue that may lead to urethral closing, or stricture, in males.

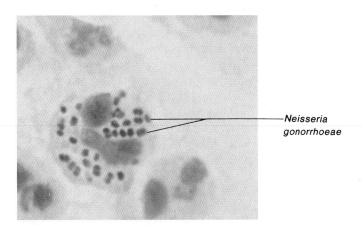

Figure 37.11 **Gonorrhea.** Gram stain of male urethral exudate showing *Neisseria gonorrhoeae* (diplococci) inside a PMN; light micrograph (× 500). Although the presence of gram-negative diplococci in exudates is a probable indication of gonorrhea, the bacterium should be isolated and identified.

In males the incubation period is 2 to 8 days. The onset consists of a urethral discharge of yellow, creamy pus and frequent, painful urination that is accompanied by a burning sensation. In females the disease is more insidious in that few individuals are aware of any symptoms. However, some symptoms may begin 7 to 21 days after infection. These are generally mild; some vaginal discharge may occur. The gonococci also can infect the uterine tubes and surrounding tissues, leading to **pelvic inflammatory disease (PID)**. This occurs in 10 to 20% of infected females. Gonococcal PID is a major cause of sterility and ectopic pregnancies because of scar formation in the uterine tubes. Gonococci disseminate most often during menstruation, a time in which there is an increased concentration of free iron available to the bacteria. In both sexes disseminated gonococcal infection with bacteremia may occur. This can lead to involvement of the joints (gonorrheal arthritis), heart (gonorrheal endocarditis), or pharynx (gonorrheal pharyngitis). Gonorrheal eye infections occur most often in newborns as they pass through an infected birth canal. The resulting disease is called **ophthalmia neonatorum** or **conjunctivitis of the newborn,** which was once a leading cause of blindness in many parts of the world. To prevent this, tetracycline, erythromycin, povidone-iodine, or silver nitrate in dilute solution is placed in the eyes of newborns. This type of treatment is required by law in the United States and many other nations.

Laboratory diagnosis of gonorrhea relies on the successful growth of *N. gonorrhoeae* in culture to determine oxidase reaction, Gram stain reaction, and colony and cell morphology. The performance of confirmation tests also is necessary. Because the gonococci are very sensitive to adverse environmental conditions and survive poorly outside the body, special transport media (e.g., PACE) are necessary. A DNA probe (Gen-Probe Pace) for *N. gonorrhoeae* has been developed and is used to supplement other diagnostic techniques.

The Centers for Disease Control and Prevention consider four treatment regimens to be coequal after sensitivity testing has been done: (1) penicillin G plus probenecid, (2) ampicillin plus probenecid, (3) ceftriaxone or ofloxacin plus doxycycline for 7 days, or (4) spectinomycin.

Penicillin-resistant strains of gonococci have now developed and occur worldwide. Most of these strains carry a plasmid that directs the formation of penicillinase, a β-lactamase enzyme able to inactivate penicillin G and ampicillin. Since 1980 strains of *N. gonorrhoeae* with chromosomally mediated penicillin resistance have developed. Instead of producing a penicillinase, these strains have altered penicillin-binding proteins. Since 1986 tetracycline-resistant *N. gonorrhoeae* also have developed.

The most effective method for control of this sexually transmitted disease is public education, diagnosing and treating the asymptomatic patient, condom use, and treating infected individuals quickly to prevent further spread of the disease. Slightly under a million cases of gonorrhea are reported in the United States each year, but the actual number of cases is thought to be three to four times greater than that. More than 60% of all cases occur in the 15- to 24-year-old age group. Repeated gonococcal infections are common. Protective immunity to reinfection does not arise because of the antigenic variety of gonococci.

Leprosy

Leprosy [Greek *Lepros,* scaly, scabby, rough] or **Hansen's disease** is a severely disfiguring skin disease caused by *Mycobacterium leprae* (*see figure 21.15*). The only reservoirs of proved significance are humans. The disease most often occurs in tropical countries, where there are more than 10 million cases. An estimated 4,000 cases exist in the United States, with approximately 200 to 300 new cases reported annually.

Transmission of leprosy is most likely to occur when children are exposed for prolonged periods to infected individuals who shed large numbers of *M. leprae*. Nasal secretions probably are the infectious material for family contacts.

The incubation period is about 3 to 5 years but may be much longer, and the disease progresses slowly. The bacterium invades peripheral nerve and skin cells and becomes an obligately intracellular parasite. It is most frequently found in the Schwann cells that surround peripheral nerve axons and in mononuclear phagocytes. The earliest symptom of leprosy is usually a slightly pigmented skin eruption several centimeters in diameter. Approximately 75% of all individuals with this early solitary lesion heal spontaneously because of the cell-mediated immune response to *M. leprae*. However, in some individuals this immune response may be so weak that one of two distinct forms of the disease occurs: tuberculoid or lepromatous leprosy (figure 37.12).

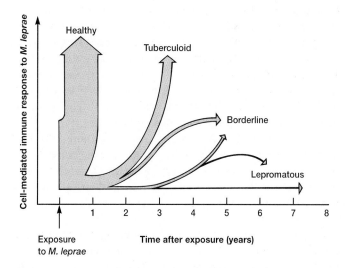

Figure 37.12   Development of Leprosy. A schematic representation of the hypothesis of how the development of subclinical infection and various types of leprosy is related to the time of onset of cell-mediated immune response to *M. leprae* antigens after the initial exposure. The thickness of the lines indicates the proportion of individuals from the exposed population that is likely to fall into each category.

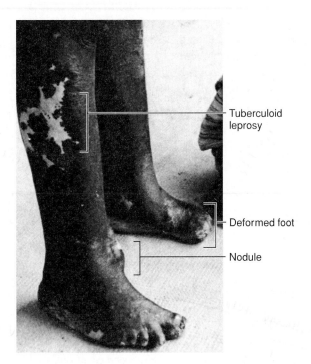

Figure 37.13   Leprosy. In tuberculoid leprosy the skin within the nodule is completely without sensation. The deformed foot is associated with lepromatous leprosy. Note the disfiguring nodule on the ankle.

Figure 37.14   Peptic Ulcer Disease. Scanning electron micrograph (× 3,441) of *Helicobacter pylori* adhering to gastric cells.

Tuberculoid (neural) leprosy is a mild, nonprogressive form of leprosy associated with a delayed-type hypersensitivity reaction (*see chapter 31*) to antigens on the surface of *M. leprae.* It is characterized by damaged nerves and regions of the skin that have lost sensation and are surrounded by a border of nodules (figure 37.13). Afflicted individuals who do not develop hypersensitivity have a relentlessly progressive form of the disease, called **lepromatous (progressive) leprosy,** in which large numbers of *M. leprae* develop in the skin cells. The bacteria kill skin tissue, leading to a progressive loss of facial features, fingers, toes, and other structures. Moreover, disfiguring nodules form all over the body. Nerves are also infected, but usually are less damaged than in tuberculoid leprosy.

Because the leprosy bacillus cannot be cultured in vitro, laboratory diagnosis is supported by the demonstration of the bacterium in biopsy specimens and by acid-fast staining. Serodiagnostic methods, such as the fluorescent leprosy antibody absorption test, DNA amplification, and ELISA have recently been developed.

Treatment is long-term with the sulfone drug dapson and rifampin, with or without clofazimine. Alternative drugs are ethionamide or protionamide. Use of *Mycobacterium w* vaccine in conjunction with the drugs shortens the duration of drug therapy and speeds recovery from the disease.

There is good evidence that the nine-banded armadillo is an animal reservoir for the leprosy bacillus in the United States but plays no role in transmission of leprosy to humans. Identification and treatment of patients with leprosy is the key to control.

Children of presumably contagious parents should be given chemoprophylactic drugs until treatment of the parents has made them noninfectious.

Peptic Ulcer Disease and Gastritis

A gram-negative, microaerophilic spiral bacillus found in gastric biopsy specimens from patients with histologic **gastritis** [Greek *gaster,* stomach, and *itis,* inflammation] was successfully cultured in Perth, Australia, in 1982 and named *Campylobacter pylori.* In 1993 its name was changed to *Helicobacter pylori.* It now appears that this bacterium is responsible for most cases of chronic gastritis not associated with another known primary cause (e.g., autoimmune gastritis or eosinophilic gastritis), and it is a leading factor in the pathogenesis of **peptic ulcer disease.** In addition, there are strong positive correlations between gastric cancer rates and *H. pylori* infection rates in certain populations.

The evidence for *H. pylori* as a gastrointestinal pathogen is now very strong, if not overwhelming. For example, *H. pylori* has been isolated from the gastric mucosa (figure 37.14) of 95% of patients with gastric ulcer disease and virtually 100% of those patients with chronic gastritis, but not in healthy tissue.

H. pylori colonizes only gastric mucus-secreting cells, beneath the gastric mucous layers, and surface pili are believed to be the adhesions associated with this process. *H. pylori* binds to Lewis antigens (which are part of the blood group antigens that determine blood group O) and to the monosaccharide sialic acid, also found in the glycoproteins on the surface of gastric epithelial cells.

H. pylori is also a strong producer of urease. Urease activity may create an alkaline environment by urea hydrolysis to produce ammonia that protects the bacterium from gastric acid until it colonizes under the layer of mucus in the stomach. The potential virulence factors responsible for epithelial cell damage and inflammation probably include proteases, phospholipases, cytokines, and cytotoxins.

H. pylori is most likely transmitted from person to person, although infection from a common exogenous source cannot be completely ruled out. Support for the person-to-person transmission comes from evidence of clustering within families and from reports of higher than expected prevalences in residents of custodial institutions and nursing homes.

Laboratory identification of *H. pylori* is by culture of gastric biopsy specimens, examination of stained biopsies for the presence of bacteria, detection of serum IgG (Pyloriset EIA-G, Malakit *Helicobacter pylori*), the urea breath test, or detection of urease activity in the biopsies. Triple treatment is with bismuth subsalicylate (Pepto-Bismol) combined with metronidazole and either tetracycline or amoxicillin for 2 weeks.

Staphylococcal Diseases

The genus *Staphylococcus* consists of gram-positive cocci, 0.5 to 1.5 μm in diameter, occurring singly, in pairs, and in tetrads, and characteristically dividing in more than one plane to form irregular clusters. The cell wall contains peptidoglycan and teichoic acid. Staphylococci are facultative anaerobes and usually catalase positive.

Staphylococci are among the most important bacteria that cause disease in humans. They are normal inhabitants of the upper respiratory tract, skin, intestine, and vagina (*see figure 28.8*). Staphylococci, with pneumococci and streptococci, are members of a group of invasive gram-positive bacteria known as the pyogenic (or pus-producing) cocci. These bacteria cause various suppurative, or pus-forming diseases (e.g., boils, carbuncles, folliculitis, impetigo contagiosa, scalded-skin syndrome) in humans.

Staphylococci can be divided into pathogenic and relatively nonpathogenic strains based on the synthesis of the enzyme coagulase. Coagulase-positive strains, such as *S. aureus* (*see figure 21.2*a), often produce a yellow carotenoid pigment—which has led to their being commonly called golden staph (*see figure 5.8*)—and cause severe chronic infections. Coagulase-negative staphylococci (CoNS) such as *S. epidermidis* do not produce coagulase, are nonpigmented, and are generally less invasive but have increasingly been associated (as opportunistic pathogens) with serious nosocomial infections.

Staphylococci are further classified into slime producers (SP) and non-slime producers (NSP). The ability to produce slime has been proposed as a marker for pathogenic strains of staphylococci (figure 37.15*a*).

Slime is a viscous extracellular glycoconjugate that allows these bacteria to adhere to smooth surfaces such as prosthetic

(a)

(b)

Figure 37.15   Slime and Biofilms. *S. aureus* and certain coagulase-negative staphylococci produce a viscous extracellular glycoconjugate called slime. (*a*) Bacterial cells of *S. aureus,* one of which produces a slime layer (arrowhead; transmission electron microscopy, × 10,000). (*b*) A biofilm on a venous catheter consisting of *S. epidermidis* and slime. The slime encases and adheres the bacterial colonies to the catheter (scanning electron micrograph, × 6,000).

medical devices and catheters (*also see p. 58*). Scanning electron microscopy has clearly demonstrated that biofilms (figure 37.15*b*) consisting of staphylococci encased in a slimy matrix are formed in association with biomaterial-associated infections (Box 37.1). Slime also appears to inhibit neutrophil chemotaxis, phagocytosis, and the antimicrobial agents vancomycin and teicoplanin.

Staphylococci, harbored by either an asymptomatic carrier or a person with the disease, can be spread by the hands, expelled from the respiratory tract, or transported in or on animate

Box 37.1

Biofilms

Biofilms consist of microorganisms immobilized at a substratum surface and typically embedded in an organic polymer matrix of microbial origin (*see chapter 41*). They develop on virtually all surfaces immersed in natural aqueous environments, including both biological (aquatic plants and animals) and abiological (concrete, metal, plastics, stones). Biofilms form particularly rapidly in flowing aqueous systems where a regular nutrient supply is provided to the microorganisms. Extensive microbial growth, accompanied by excretion of copious amounts of extracellular organic polymers, thus leads to the formation of visible slimy layers (biofilms) on solid surfaces.

Most of the human gastrointestinal tract is colonized by specific groups of microorganisms (the normal indigenous microbiota; *see chapters 28 and 29*) that give rise to natural biofilms. These natural biofilms provide protection from pathogenic species.

Insertion of a prosthetic device into the human body often leads to the formation of biofilms on the surface of the device. The microorganisms primarily involved are *Staphylococcus epidermidis* (figure 37.15*b*), other coagulase-negative staphylococci, and gram-negative bacteria. These normal skin inhabitants possess the ability to tenaciously adhere to the surfaces of inanimate prosthetic devices. Within the biofilms they are protected from the body's normal defense mechanisms and also from antibiotics; thus the biofilm also provides a source of infection for other parts of the body as bacteria detach during biofilm sloughing.

Some examples of biofilms of medical importance include:

1. The deaths following massive infections of patients receiving the Jarvik 7 artificial hearts

2. Cystic fibrosis patients harboring great numbers of *Pseudomonas aeruginosa* that produce large amounts of alginate polymers, which inhibit the diffusion of antibiotics

3. Teeth, where biofilm forms plaque that leads to tooth decay (*see figure 38.4*)

4. Contact lenses, where bacteria may produce severe eye irritation, inflammation, and infection

5. Air-conditioning and other water retention systems where potentially pathogenic bacteria, such as *Legionella* species, may be protected from the effects of chlorination by biofilms

---

and inanimate objects. Staphylococci can produce disease in almost every organ and tissue of the body (figure 37.16). However, it should be emphasized that staphylococcal disease, for the most part, occurs in people whose defensive mechanisms have been compromised, such as those in hospitals.

Staphylococci produce disease through their ability to multiply and spread widely in tissues and through their production of many extracellular substances (table 37.2). Some of these substances are exotoxins, and others are enzymes thought to be involved in staphylococcal invasiveness. Many toxin genes are carried on plasmids; in some cases genes responsible for pathogenicity reside on both a plasmid and the host chromosome.

The pathogenic capacity of a particular *S. aureus* strain is due to the combined effect of extracellular factors and toxins, together with the invasive properties of the strain. At one end of the disease spectrum is staphylococcal food poisoning, caused solely by the ingestion of preformed enterotoxin (table 37.2). At the other end of the spectrum are staphylococcal bacteremia and disseminated abscesses in most organs of the body.

The classic example of the staphylococcal lesion is the localized abscess (figure 37.17*a–d*).When *S. aureus* becomes established in a hair follicle, tissue necrosis results. Coagulase is produced and forms a fibrin wall around the lesion that limits the spread. Within the center of the lesion, liquefaction of necrotic tissue occurs, and the abscess spreads in the direction of least resistance. The abscess may be either a furuncle (boil) or a carbuncle. The central necrotic tissue drains, and healing eventually occurs. However, the bacteria may spread from any focus by the lymphatics and bloodstream to other parts of the body.

Newborn infants and children can develop a superficial skin infection characterized by the presence of encrusted pustules (figure 37.17*e*). This disease, called impetigo contagiosa, is caused by *S. aureus* and group A streptococci. It is contagious and can spread rapidly through a nursery or school. It usually occurs in areas where sanitation and personal hygiene are poor.

Toxic shock syndrome (TSS) is a staphylococcal disease with potentially serious consequences. Most cases of this syndrome have occurred in females who use superabsorbent tampons during menstruation. However, the toxin associated with this syndrome is also produced in men and in nonmenstruating women by *S. aureus* present at sites other than the genital area (e.g., in surgical wound infections). Toxic shock syndrome is characterized by low blood pressure, fever, diarrhea, an extensive skin rash, and shedding of the skin. These symptoms are caused by the **t**oxic **s**hock **s**yndrome **t**oxin-1 (TSST-1) released by the *S. aureus* (table 37.2), but several other enterotoxins (SEB and SEC$_1$) also may be involved. Several hundred cases of toxic shock syndrome are reported annually in the United States.

Staphylococcal scalded skin syndrome (SSSS) is a third example of a common staphylococcal disease (figure 37.17*f*).

| 1 | Tissue where *S. aureus* is often found but does not normally cause disease |
| 6 | Osteomyelitis |

Diseases that may be caused by *S. aureus* are:

2 Pimples and impetigo

3 Boils and carbuncles on any surface area

4 Wound infections and abscesses

5 Spread to lymph nodes and to blood (septicemia), resulting in widespread seeding

6 Osteomyelitis

7 Endocarditis

8 Meningitis

9 Enteritis and enterotoxin poisoning (food poisoning)

10 Nephritis

11 Respiratory infections:
Pharyngitis
Laryngitis
Bronchitis
Pneumonia

Figure 37.16   Staphylococcal Diseases. The sites of the major staphylococcal infections of humans are indicated by the above numbers.

TABLE 37.2   Various Toxins and Enzymes Produced by Staphylococci

Product	Physiological Action
β-lactamase	Breaks down penicillin
Catalase	Converts hydrogen peroxide into water and oxygen and reduces killing by phagocytosis
Coagulase	Reacts with prothrombin to form a complex that can cleave fibrinogen and cause the formation of a fibrin clot; fibrin may also be deposited on the surface of staphylococci, which may protect them from destruction by phagocytic cells; coagulase production is synonymous with invasive pathogenic potential
DNase	Destroys DNA
Enterotoxins	Are divided into heat-stable toxins of six known types (A, B, C1, C2, D, E); responsible for the gastrointestinal upset typical of food poisoning
Exfoliative toxins A and B	Causes loss of the surface layers of the skin in scalded-skin syndrome
Hemolysins	Alpha hemolysin destroys erythrocytes and causes skin destruction
	Beta hemolysin destroys erythrocytes and sphingomyelin around nerves
	Gamma hemolysin destroys erythrocytes
	Delta hemolysin destroys erythrocytes
Hyaluronidase	Also known as spreading factor; breaks down hyaluronic acid located between cells, allowing for penetration and spread of bacteria
Panton-Valentine leukocidin	Inhibits phagocytosis by granulocytes and can destroy these cells by forming pores in their phagosomal membranes
Lipases	Break down lipids
Nuclease	Breaks down nucleic acids
Protein A	Is antiphagocytic by competing with neutrophils for the Fc portion of specific opsonins
Proteases	Break down proteins
Toxic shock syndrome toxin-1	Is associated with the fever, shock, and multisystem involvement of toxic shock syndrome

SSSS is caused by strains of *S. aureus* that carry a plasmid-borne gene for the **exfoliative toxin** or **exfoliatin** (sometimes the toxin gene is on the bacterial chromosome instead). In this disease the epidermis peels off to reveal a red area underneath—thus the name of the disease. SSSS is seen most commonly in infants and children, and neonatal nurseries occasionally suffer large outbreaks of the disease.

The definitive diagnosis of staphylococcal disease can be made only by isolation and identification of the staphylococcus involved. This requires culture, catalase, and coagulase tests;

(a)

(b)

(c)

(d)

(e)

(f)

Figure 37.17   Staphylococcal Skin Infections. (*a*) Superficial folliculitis in which raised, domed pustules form around hair follicles. (*b*) In deep folliculitis the microorganism invades the deep portion of the follicle and dermis. (*c*) A furuncle arises when a large abscess forms around a hair follicle. (*d*) A carbuncle consists of a multilocular abscess around several hair follicles. (*e*) Impetigo on the neck of 2-year-old male. (*f*) Scalded skin syndrome in a 1-week-old premature male infant. Reddened areas of skin peel off, leaving "scalded"-looking moist areas.

serology; DNA fingerprinting; and phage typing. Commercial rapid test kits also are available. There is no specific prevention for staphylococcal disease. The mainstay of treatment is the administration of specific antibiotics: penicillin, cloxacillin, methicillin, vancomycin, oxacillin, cefotaxime, ceftriaxone, a cephalosporin, or rifampin and others. Because of the prevalence of drug-resistant strains, all staphylococcal isolates should be tested for antimicrobial susceptibility (Box 37.3). Cleanliness, hygiene, and aseptic management of lesions are the best means of control.

Syphilis

Venereal syphilis [Greek *syn*, together, and *philein*, to love] is a contagious sexually transmitted disease caused by the spirochete *Treponema pallidum* subsp. *pallidum* (*T. pallidum, see figure 20.1*b). **Congenital syphilis** is the disease acquired in utero from the mother.

T. pallidum enters the body through mucous membranes or minor breaks or abrasions of the skin. It migrates to the regional lymph nodes and rapidly spreads throughout the body. The disease is not highly contagious, and there is only about a 1 in 10 chance of acquiring it from a single exposure to an infected sex partner.

Three recognizable stages of syphilis occur in untreated adults. In the primary stage, after an incubation period of about 10 days to 3 weeks or more, the initial symptom is a small, painless, reddened ulcer, or **chancre** [French *canker*, a destructive sore] with a hard ridge that appears at the infection site (figure 37.18*a*) and contains spirochetes. Contact with the chancre during sexual intercourse may result in disease transmission. In about 1/3 of the cases, the disease does not progress further and the chancre disappears. Serological tests are positive in about 80% of the individuals during this stage (figure 37.19). In the remaining cases the spirochetes enter the bloodstream and are distributed throughout the body.

Box 37.2

Resistant Staphylococci

During the late 1950s and early 1960s, *Staphylococcus aureus* caused considerable morbidity and mortality as a nosocomial, or hospital-acquired, pathogen. Since then, penicillinase-resistant, semisynthetic penicillins have proved to be successful antimicrobial agents in the treatment of staphylococcal infections. Unfortunately **m**ethicillin-**r**esistant *S. aureus* (MRSA) strains have recently emerged as a major nosocomial problem. One way in which staphylococci become resistant is through acquisition of a chromosomal gene (*mecA*) that encodes an alternate target protein which is not inactivated by methicillin. The majority of the strains are resistant to several of the most commonly used antimicrobial agents, including macrolides, aminoglycosides, and the beta-lactam antibiotics, including the latest generation of cephalosporins. Serious infections by methicillin-resistant strains have been most often successfully treated with an older, potentially toxic antibiotic, vancomycin. However, strains of *Enterococcus* recently have become resistant to vancomycin, and it is possible that this resistance will soon spread to *Staphylococcus*.

Recently methicillin-resistant *S. epidermidis* strains also have emerged as a nosocomial problem, especially in individuals with prosthetic heart valves or in people who have undergone other forms of cardiac surgery. Resistance to methicillin also may extend to the cephalosporin antibiotics. Difficulties in performing in vitro tests that adequately recognize cephalosporin resistance of these strains continue to exist. Serious infections due to methicillin-resistant *S. epidermidis* have been successfully treated with combination therapy, including vancomycin plus rifampin or an aminoglycoside.

(a)

(b)

(c)

 Figure 37.18   Syphilis. (*a*) Primary syphilitic chancre of the penis. (*b*) Palmar lesions of secondary syphilis. (*c*) Ruptured gumma and ulcer of upper hard palate of the mouth.

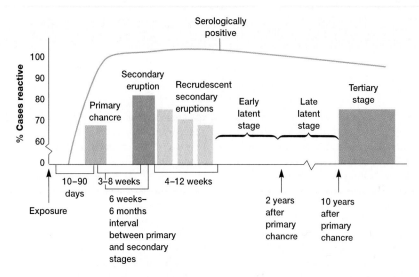

Figure 37.19   The Course of Untreated Syphilis. See text for further details.

Within 2 to 10 weeks after the primary lesion, the disease may enter the secondary stage, which is characterized by a skin rash (figure 37.18*b*). By this time 100% of the individuals are serologically positive. Other symptoms during this stage include the loss of patches of hair, malaise, and fever. Both the chancre and the rash lesions are infectious.

After several weeks the disease becomes latent. During the latent period the disease is not normally infectious, except for possible transmission from mother to fetus (congenital syphilis). After many years a tertiary stage develops in about 40% of untreated individuals with secondary syphilis. During this stage degenerative lesions called **gummas** (figure 37.18*c*) form in the skin, bone, and nervous system as the result of hypersensitivity reactions. This stage also is characterized by a great reduction in the number of spirochetes in the body. Involvement of the central nervous system may result in tissue loss that can lead to mental retardation, blindness, a "shuffle" walk (tabes), or insanity. Many of these symptoms have been associated with such well-known people as Al Capone, Francisco Goya, Henry VIII, Adolf Hitler, Scott Joplin, Friedrich Nietzsche, Franz Schubert, Oscar Wilde, and Kaiser Wilhelm (Box 37.4).

Diagnosis of syphilis is through a clinical history, a thorough physical examination, and dark-field and immunofluorescence examination of fluids from the lesions (except oral lesions) for typical motile or fluorescent spirochetes. Because humans respond to *T. pallidum* with the formation of antitreponemal antibody and a complement-fixing reagin, serological tests are very informative. Examples include tests for nontreponemal antigens (VDRL, **V**enereal **D**isease **R**esearch **L**aboratories test; RPR, **R**apid **P**lasma **R**eagin test; complement fixation or the Wassermann test) and treponemal antibodies (FTA-ABS, **f**luorescent **t**reponemal **a**ntibody-**abs**orption test; TPI, *T. pallidum* **i**mmobilization; *T. pallidum* complement fixation; TPHA, *T. pallidum* **h**em**a**gglutination).

Treatment in the early stages of the disease is easily accomplished with long-acting benzathine penicillin G or aqueous procaine penicillin. Later stages of syphilis are more difficult to treat with drugs and require much larger doses over a longer period. For example, in neurosyphilis cases, treponemes occasionally survive such drug treatment. Immunity to syphilis is not complete, and subsequent infections can occur once the first infection has spontaneously disappeared or has been eliminated with antibiotics.

Prevention and control of syphilis depends on (1) public education (2) prompt and adequate treatment of all new cases, (3) follow-up on sources of infection and contact so they can be treated, (4) sexual hygiene, and (5) prophylaxis (condoms) to prevent exposure. At present, the incidence of syphilis, as well as other sexually transmitted diseases, is rising in most parts of the world. In the United States around 50,000 cases of primary and secondary syphilis in the civilian population and about 1,000 cases of congenital syphilis are reported annually. The highest incidence is among those 20 to 39 years of age.

Tetanus

Tetanus [Greek *tetanos,* to stretch] is caused by *Clostridium tetani,* an anaerobic gram-positive spore former (*see figure 21.8*d). The endospores of *C. tetani* are commonly found in hospital environments, in soil and dust, and in the feces of many farm animals and humans.

Box 37.4

A Brief History of Syphilis

Syphilis was first recognized in Europe near the end of the fifteenth century. During this time the disease reached epidemic proportions in the Mediterranean areas. According to one hypothesis, syphilis is of New World origin and Christopher Columbus (1451–1506) and his crew acquired it in the West Indies and introduced it into Spain after returning from their historic voyage. Another hypothesis is that syphilis had been endemic for centuries in Africa and may have been transported to Europe at the same time that vast migrations of the civilian population were occurring (1500).

Syphilis was initially variously called the Italian disease, the French disease, and the great pox as distinguished from smallpox. In 1530 the Italian physician and poet Girolamo Fracastoro wrote *Syphilis sive Morbus Gallicus* (Syphilis or the French Disease). In this poem a Spanish shepherd named Syphilis is punished for being disrespectful to the gods by being cursed with the disease. Several years later Fracastoro published a series of papers in which he described the possible mode of transmission of the "seeds" of syphilis through sexual contact.

Its venereal transmission was not definitely shown until the eighteenth century. The term venereal is derived from the name Venus, the Roman goddess of love. Recognition of the different stages of syphilis was demonstrated in 1838 by Philippe Ricord, who reported his observations on more than 2,500 human inoculations. In 1905 Fritz Schaudinn and Erich Hoffmann discovered the causative bacterium, and in 1906 August von Wassermann introduced the diagnostic test that bears his name. In 1909 Paul Ehrlich introduced an arsenic derivative, arsphenamine or salvarsan, as therapy. During this period, an anonymous limerick aptly described the course of this disease:

There was a young man from Black Bay
Who thought syphilis just went away
He believed that a chancre
Was only a canker
That healed in a week and a day.

But now he has "acne vulgaris"—
(Or whatever they call it in Paris);
On his skin it has spread
From his feet to his head,
And his friends want to know where his hair is.

There's more to his terrible plight:
His pupils won't close in the light
His heart is cavorting,
His wife is aborting,
And he squints through his gun-barrel sight.

Arthralgia cuts into his slumber;
His aorta is in need of a plumber;
But now he has tabes,
And saber-shinned babies,
While of gummas he has quite a number.

He's been treated in every known way,
But his spirochetes grow day by day;
He's developed paresis,
Has long talks with Jesus,
And thinks he's the Queen of the May.

Transmission to humans is associated with skin wounds. Any break in the skin can allow *C. tetani* endospores to enter, and if the oxygen tension is low enough, the endospores germinate. When the bacteria die and lyse, the neurotoxin tetanospasmin is released. **Tetanospasmin** is an endopeptidase that selectively cleaves the synaptic vesicle membrane protein synaptobrevin. This prevents exocytosis and release of inhibitory neurotransmitters (gammaaminobutyric acid and glycine) at synapses within the spinal cord motor nerves. The result is uncontrolled stimulation of skeletal muscles.

Early in the course of the disease, tetanospasmin causes tension or cramping and twisting in skeletal muscles surrounding the wound and tightness of the jaw muscles. With more advanced disease, there is trismus ("lockjaw"), an inability to open the mouth because of the spasm of the masseter muscles. Facial muscles may go into spasms producing the characteristic expression known as risus sardonicus.

Spasms or contractions of the trunk and extremity muscles may be so severe that there is boardlike rigidity, painful tonic convulsions, and opisthotonos (backward bowing of the back so that the heels and back approach each other [figure 37.20]). Death usually results from spasms of the diaphragm and intercostal respiratory muscles. A second toxin **tetanolysin,** is a hemolysin that aids in tissue destruction.

Testing for tetanus is suggested whenever an individual has a history of wound infection and muscle stiffness. Prevention of tetanus involves the use of the tetanus toxoid. The toxoid, which incorporates an adjuvant (aluminum salts) to increase its immunizing potency, is given routinely with diphtheria toxoid and pertussis vaccine. An initial dose is normally administered a few months after birth, a second dose 4 to 6 months later, and finally a reinforcing dose 6 to 12 months after the second injection. A final booster is given between the ages of 4 to 6 years. For many years booster doses of tetanus toxoid were administered every 3 to 5 years. However, that

Figure 37.20   Tetanus. Painting of a soldier dying of tetanus. Note the extreme contraction of the trunk and extremity muscles producing opisthotonos—backward bowing of the back.

antibiotics. Prevention and control involves public education, protective clothing, and vector control. An attenuated live vaccine is available from the United States Army for high-risk laboratory workers. Fewer than 300 cases of tularemia are reported annually in the United States.

1. How can humans acquire anthrax? Gas gangrene? Gonorrhea? Chancroid? Leprosy? Peptic ulcer disease? Describe the major symptoms of each.
2. Define the following terms: cutaneous anthrax, bacterial vaginosis, pelvic inflammatory disease, ophthalmia neonatorum, tuberculoid and lepromatous leprosy, gummas, and chancre.
3. Describe several diseases caused by the staphylococci.
4. Name and describe the three stages of syphilis.
5. How is the disease tetanus acquired? What are its symptoms and how do they arise?
6. From what animal can tularemia be contracted?

practice has been discontinued since it has been shown that a single booster dose can provide protection for 10 to 20 years. Serious hypersensitivity reactions have occurred when too many doses of toxoid were administered over a period of years. Booster doses today are generally given only when an individual has sustained a wound infection.

Control measures for tetanus are not possible because of the wide dissemination of the bacterium in the soil and the long survival of its endospores. The case fatality rate in generalized tetanus ranges from 30 to 90% because tetanus treatment is not very effective. Therefore prevention is all important and depends on (1) active immunization with toxoid, (2) proper care of wounds contaminated with soil, (3) prophylactic use of antitoxin, and (4) administration of penicillin. Around 100 cases of tetanus are reported annually in the United States—the majority of which are intravenous drug users.

Tularemia

The gram-negative bacterium *Francisella tularensis* subsp. *tularensis* (Jellison type A) is widely found in animal reservoirs in the United States and causes the disease **tularemia** (from Tulare, a county in California where the disease was first described). It may be transmitted to humans by biting arthropods (ticks, deer flies, or mosquitoes), direct contact with infected tissue (rabbits), inhalation of aerosolized bacteria, or ingestion of contaminated food or water. After an incubation period of 2 to 10 days, a primary ulcerative lesion appears at the infection site, lymph nodes enlarge, and a high fever develops. Diagnosis is by PCR or culture of the bacterium and fluorescent antibody and agglutination tests; treatment is with streptomycin, tetracycline, or aminoglycoside

Food-Borne and Waterborne Diseases

Many bacteria contaminating food and water can cause acute gastroenteritis or inflammation of the stomach and intestinal lining. When food is the source of the pathogen, the condition is often called **food poisoning.** Gastroenteritis can arise in two ways. The bacteria may actually produce a **food-borne infection.** That is, they may first colonize the gastrointestinal tract and grow within it, then either invade host tissues or secrete exotoxins. Alternatively the pathogen may secrete an exotoxin that contaminates the food and is then ingested by the host. This is sometimes referred to as a **food intoxication** because the toxin is ingested and the presence of living bacteria is not required. Because these toxins disrupt the functioning of the intestinal mucosa they are called **enterotoxins** (*see table 29.6*). Common symptoms of enterotoxin poisoning are nausea, vomiting, and diarrhea.

Worldwide, diarrheal diseases are second only to respiratory diseases as a cause of adult death; they are the leading cause of childhood death, and in some parts of the world they are responsible for more years of potential life lost than all other causes combined. For example, each year around 5 million children (more than 13,600 a day) die from diarrheal diseases in Asia, Africa, and South America. In the United States estimates exceed 10,000 deaths per year from diarrhea, and an average of 500 childhood deaths are reported.

This section describes several of the more common bacteria associated with gastrointestinal infections, food intoxications, and waterborne diseases. Table 37.3 summarizes many of the bacterial pathogens responsible for food poisoning.

Diseases transmitted by foods (pp. 885–88).

Food spoilage (pp. 889–92).

Cholera

Throughout recorded history **cholera** [Greek *chole,* bile] has caused seven pandemics in various areas of the world, especially in Asia, the Middle East, and Africa. The disease has been rare in the United States since the 1800s, but an endemic focus is believed to exist on the gulf coast of Louisiana and Texas.

Cholera is caused by the gram-negative *Vibrio cholerae* bacterium of the family *Vibrionaceae* (figure 37.21). Although there are many serogroups, only 01 and 0139 have exhibited the ability to cause epidemics. *V. cholerae* 01 is divided into two serotypes, Inaba and Ogawa, and two biotypes, classic and El Tor.

Cholera is acquired by ingesting food or water contaminated by fecal material from patients or carriers. (Shellfish and copepods are natural reservoirs.) In 1961 the El Tor biotype emerged as an important cause of cholera pandemics, and in 1993 the newly identified strain *V. cholerae* 0139 emerged in India. This novel toxigenic strain does not agglutinate with 01 antiserum but possesses epidemic and pandemic potential. In Calcutta, India, serogroup 0139 of *Vibrio cholerae* has displaced El Tor *V. cholerae* serogroup 01 (causative agent of the seventh pandemic), an event that has never happened in the recorded history of cholera.

Once the bacteria enter the body, the incubation period is from 24 to 72 hours. The bacteria adhere to the intestinal mucosa of the small intestine, where they are not invasive but secrete **choleragen,** a cholera toxin. Choleragen is a protein composed of two functional units, an enzymatic A subunit and an intestinal receptor-binding B subunit. The A subunit enters the intestinal epithelial cells and activates the enzyme adenylate cyclase by the addition of an ADP-ribosyl group in a way similar to that employed by diphtheria toxin (*see figure 29.4*). As a result choleragen stimulates hypersecretion of water and chloride ions while inhibiting absorption of sodium ions. The patient experiences an outpouring of fluid and electrolytes, with associated abdominal muscle cramps, vomiting, fever, and watery diarrhea. The diarrhea can be so profuse that a person can lose 10 to 15 liters of fluid during the infection. Death may result from the elevated concentration of blood proteins, caused by reduced fluid levels, which leads to circulatory shock and collapse.

Laboratory diagnosis is by culture of the bacterium from feces and subsequent identification by agglutination reactions with specific antisera. Treatment is by oral rehydration therapy with NaCl plus sucrose to stimulate water uptake by the intestine; the antibiotics of choice are a tetracycline, trimethoprim-sulfamethoxazole, or ciprofloxacin. The most reliable control methods are based on proper sanitation, especially of water supplies. The mortality rate without treatment is often over 50%; with treatment and supportive care, it is less than 1%. Fewer than 20 cases of cholera are reported each year in the United States.

Botulism

Food-borne **botulism** [Latin *botulus,* sausage] is a form of food poisoning caused by *Clostridium botulinum,* which is an obligately anaerobic endospore-forming, gram-positive rod that is found in soil and aquatic sediments. The most common source of infection is home-canned food that has not been heated sufficiently to kill contaminating *C. botulinum* endospores. The endospores can germinate, and a toxin is produced during vegetative growth. If the food is then eaten without adequate cooking, the toxin remains active and the disease results.

The botulinum toxin is a neurotoxin that binds to the synapses of motor neurons. It selectively cleaves the synaptic vesicle membrane protein synaptobrevin, thus preventing exocytosis and release of the neurotransmitter acetylcholine. As a consequence, muscles do not contract in response to motor neuron activity, and flaccid paralysis results. Symptoms of botulism occur within 18 to 24 hours of toxin ingestion and include blurred vision, difficulty in swallowing and speaking, muscle weakness, nausea, and vomiting. Without adequate treatment, 1/3 of the patients may die within a few days of either respiratory or cardiac failure.

Laboratory diagnosis is by a hemagglutination test or inoculation of mice with the patient's serum, stools, or vomitus to prove toxigenicity. Treatment relies on supportive care and polyvalent antitoxin. Fewer than 100 cases of botulism occur in the United States annually.

Infant botulism is the most common form of botulism in the United States and is confined to infants under a year of age. Approximately 100 cases are reported each year. It appears that ingested endospores, which may be present in honey or other baby foods, germinate in the infant's intestine. *C. botulinum* then multiplies and produces the toxin. The infant becomes constipated, listless, generally weak, and eats poorly. Death may result from respiratory failure.

Prevention and control of botulism involves (1) strict adherence to safe food-processing practices by the food industry, (2) educating the public on safe home-preserving (canning) methods for foods, and (3) not feeding honey to infants younger than 1 year of age.

Campylobacter jejuni Gastroenteritis

Campylobacter jejuni is a gram-negative curved rod found in the intestinal tract of animals. Studies with chickens, turkeys, and cattle have shown that as much as 50 to 100% of a flock or herd of these birds or animals secrete *C. jejuni*. These bacteria also can be isolated in high numbers from surface waters. They are transmitted to humans by contaminated food and water, contact with infected animals, or anal-oral sexual activity. *C. jejuni* causes an estimated 2 million cases of *Campylobacter*

TABLE 37.3 Bacteria That Cause Acute Bacterial Diarrheas and Food Poisonings

Organism	Incubation Period (Hours)	Vomiting	Diarrhea	Fever	Epidemiology	Pathogenesis
Staphylococcus	1–8 (rarely, up to 18)	+++	+	–	Staphylococci grow in meats, dairy and bakery products and produce enterotoxins.	Enterotoxins act on receptors in gut that transmit impulse to medullary centers.
Bacillus cereus	2–16	+++	++	–	Reheated fried rice causes vomiting or diarrhea.	Enterotoxins formed in food or in gut from growth of *B. cereus*.
Clostridium perfringens	8–16	±	+++	–	Clostridia grow in rewarmed meat dishes. Huge numbers ingested.	Enterotoxin produced during sporulation in gut, causes hypersecretion.
Clostridium botulinum	18–24	±	Rare	–	Clostridia grow in anaerobic foods and produce toxin.	Toxin absorbed from gut and blocks acetylcholine release at neuromuscular junction.
Escherichia coli (some strains)	24–72	±	++	–	Organisms grow in gut and produce toxin. May also invade superficial epithelium.	Toxin causes hypersecretion in small intestine ("traveler's diarrhea").
Vibrio parahaemolyticus	6–96	+	++	±	Organisms grow in seafood and in gut and produce toxin, or invade.	Toxin causes hypersecretion; vibrios invade epithelium; stools may be bloody.
Vibrio cholerae (mild cases)	24–72	+	+++	–	Organisms grow in gut and produce toxin.	Toxin causes hypersecretion in small intestine. Infective dose >10^7 vibrios.
Shigella spp. (mild cases)	24–72	±	++	+	Organisms grow in superficial gut epithelium. *S. dysenteriae* produces toxin.	Organisms invade epithelial cells; blood, mucus, and neutrophils in stools. Infective dose <10^3 organisms.
Salmonella spp. (gastroenteritis)	8–48	±	++	+	Organisms grow in gut. Do not produce toxin.	Superficial infection of gut, little invasion. Infective dose >10^5 organisms.
Salmonella typhi (typhoid fever)	10–14 days	±	±	++	Bacteria invade the gut epithelium and reach the lymph nodes, liver, spleen, and gallbladder.	Symptoms probably due to endotoxins and tissue inflammation. Infective dose ≥10^7 organisms.
Clostridium difficile	?	–	+++	+	Antibiotic-associated colitis.	Toxin causes epithelial necrosis in colon; pseudomembranous colitis.
Campylobacter jejuni	2–10 days	–	+++	++	Infection by oral route from foods, pets. Organism grows in small intestine.	Invasion of mucous membrane. Toxin production uncertain.
Yersinia enterocolitica	?	±	++	+	Fecal-oral transmission. Food-borne. Animals infected.	Gastroenteritis or mesenteric adenitis. Occasional bacteremia. Toxin produced occasionally.

From Geo. F. Brooks, et al., *Medical Microbiology*, 19th edition. Copyright 1991 Appleton & Lange, Norwalk, CT. Reprinted by permission.

gastroenteritis—inflammation of the intestine—and subsequent diarrhea in the United States each year.

The incubation period is 2 to 10 days. *C. jejuni* invades the epithelium of the small intestine, causing inflammation, and also secretes an exotoxin that is antigenically similar to the cholera toxin. Symptoms include diarrhea, high fever, severe inflammation of the intestine along with ulceration, and bloody stools.

Laboratory diagnosis is by culture in an atmosphere with reduced O_2 and added CO_2. The disease is self-limited, and treatment is supportive; fluids, electrolyte replacement, and erythromycin is used in severe cases. Recovery usually takes from 5 to 8 days. Prevention and control involves good personal hygiene and food handling precautions, including pasteurization of milk and thorough cooking of poultry.

Clinical Features

Abrupt onset, intense vomiting for up to 24 hours, recovery in 24–48 hours. Occurs in persons eating the same food. No treatment usually necessary except to restore fluids and electrolytes.

With incubation period of 2–8 hours, mainly vomiting. With incubation period of 8–16 hours, mainly diarrhea.

Abrupt onset of profuse diarrhea; vomiting occasionally. Recovery usual without treatment in 1–4 days. Many clostridia in cultures of food and feces of patients.

Diplopia, dysphagia, dysphonia, respiratory embarrassment. Treatment requires clear airway, ventilation, and intravenous polyvalent antitoxin. Exotoxin present in food and serum. Mortality rate high.

Usually abrupt onset of diarrhea; vomiting rare. A serious infection in newborns. In adults, "traveler's diarrhea" is usually self-limited in 1–3 days. Use diphenoxylate (Lomotil) but no antimicrobials.

Abrupt onset of diarrhea in groups consuming the same food, especially crabs and other seafood. Recovery is usually complete in 1–3 days. Food and stool cultures are positive.

Abrupt onset of liquid diarrhea in endemic area. Needs prompt replacement of fluids and electrolytes IV or orally. Tetracyclines shorten excretion of vibrios. Stool cultures positive.

Abrupt onset of diarrhea, often with blood and pus in stools, cramps, tenesmus, and lethargy. Stool cultures are positive. Give trimethoprim sulfamethoxazole or ampicillin or chloramphenicol in severe cases. Do not give opiates. Often mild and self-limited. Restore fluids.

Gradual or abrupt onset of diarrhea and low-grade fever. Nausea, headache, and muscle aches common. No antimicrobials unless systemic dissemination is suspected. Stool cultures are positive. Prolonged carriage is frequent.

Initially fever, headache, malaise, anorexia, and muscle pains. Fever may reach 40°C by the end of the first week of illness and lasts for 2 or more weeks. Diarrhea often occurs, and abdominal pain, cough, and sore throat may be prominent. Antibiotic therapy shortens duration of the illness.

Especially after abdominal surgery, abrupt bloody diarrhea and fever. Toxins in stool. Oral vancomycin useful in therapy.

Fever, diarrhea; PMNs and fresh blood in stool, especially in children. Usually self-limited. Special media needed for culture at 43°C. Erythromycin in severe cases with invasion. Usual recovery in 5–8 days.

Severe abdominal pain, diarrhea, fever; PMNs and blood in stool; polyarthritis, erythema nodosum, especially in children. If severe, treat with gentamicin. Keep stool specimen at 4°C before culture.

Figure 37.21   Cholera. *Vibrio cholerae* adhering to intestinal epithelium; scanning electron micrograph (× 12,000). Notice that the bacteria is slightly curved with a single polar flagellum.

The initial source of the bacterium is the intestinal tracts of animals. Humans acquire the bacteria from contaminated foods such as beef products, poultry, eggs, egg products, or water. Around 45,000 cases a year are reported in the United States, but there actually may be as many as 2 to 3 million cases annually.

Once the bacteria are in the body, the incubation time is only about 8 to 48 hours. The disease results from a true foodborne infection because the bacteria multiply and invade the intestinal mucosa. Abdominal pain, cramps, diarrhea, nausea, vomiting, and fever are the most prominent symptoms, which usually persist for 2 to 5 days but can last for several weeks. During the acute phase of the disease, as many as 1 billion salmonella can be found per gram of feces. Most adult patients recover, but the loss of fluids can cause problems for children and elderly people.

Laboratory diagnosis is by isolation of the bacterium from food or patients' stools. Treatment is with fluid and electrolyte replacement. Prevention depends on good food-processing practices, proper refrigeration, and adequate cooking.

Shigellosis

Shigellosis or bacillary dysentery is a diarrheal illness resulting from an acute inflammatory reaction of the intestinal tract caused by the four species of the genus *Shigella* (gram-negative, nonmotile, facultative rods). About 25,000 cases a year are reported in the United States, and around 500,000 deaths a year worldwide are due to bacillary dysentery.

Shigella is restricted to human hosts. *S. sonnei* is the usual pathogen in the United States and Britain, but *S. flexneri* is also fairly common. The organism is transmitted by the fecal-oral

Salmonellosis

Salmonellosis (*Salmonella* gastroenteritis) is caused by over 2,000 *Salmonella* serovars (strains; a subspecies category). The most frequently reported one from humans is *S.* serovar *typhimurium*. This bacterium is a gram-negative, motile, nonspore-forming rod.

route—primarily by food, fingers, feces, and flies (the four "F's")—and is most prevalent among children, especially 1- to 4-year-olds. The infectious dose is only around 10 to 100 bacteria. In the United States shigellosis is a particular problem in day care centers and custodial institutions where there is crowding.

The shigellae are facultative intracellular parasites that multiply within the cells of the colon epithelium. The bacteria induce the mucosal cell to phagocytose them and then disrupt the phagosome membrane. After reproducing in the cytoplasmic matrix, the shigellae invade adjacent mucosal cells. They may produce both endotoxins and exotoxins but do not usually spread beyond the colon epithelium. The watery stools often contain blood, mucus, and pus. In severe cases the colon can become ulcerated.

The incubation period usually ranges from 1 to 3 days and the organisms are shed over a period of 1 to 2 weeks. Identification of isolates is based on biochemical characteristics and serology. The disease normally is self-limiting in adults and lasts an average of 4 to 7 days; in infants and young children it may be fatal. Usually fluid and electrolyte replacement are sufficient, and antibiotics may not be required in mild cases although they can shorten the duration of symptoms. Sometimes, particularly in malnourished infants and children, neurological complications and kidney failure result. When necessary, treatment is with trimethoprim-sulfamethoxazole or fluoroquinolones. Antibiotic-resistant strains are becoming a problem. Prevention is a matter of good personal hygiene and the maintenance of a clean water supply.

Staphylococcal Food Poisoning

Staphylococcal food poisoning is the major type of food intoxication in the United States. It is caused by ingestion of improperly stored or cooked food (particularly foods such as ham, processed meats, chicken salad, pastries, ice cream, and hollandaise sauce) in which *Staphylococcus aureus* has grown.

S. aureus (a gram-positive coccus) is very resistant to heat, drying, and radiation; it is found in the nasal passages and on the skin of humans and other mammals worldwide. From these sources it can readily enter food. If the bacteria are allowed to incubate in certain foods, they produce heat-stable enterotoxins that render the food dangerous even though it appears normal. Six different enterotoxins have been identified and are designated A, B, C_1, C_2, D, and E.

Typical symptoms include severe abdominal pain, cramps, diarrhea, vomiting, and nausea. The onset of symptoms is rapid (usually 1 to 8 hours) and of short duration (usually less than 24 hours). The mortality rate of staphylococcal food poisoning is negligible among healthy individuals.

Diagnosis is based on the symptoms or laboratory diagnosis of the bacteria from foods. Enterotoxins may be detected in foods by animal toxicity tests. Treatment is with fluid and electrolyte replacement. Prevention and control involve avoidance of food contamination, and control of personnel responsible for food preparation and distribution.

Traveler's Diarrhea and *Escherichia coli* Infections

Millions of people travel yearly from country to country. Unfortunately a large percentage of these travelers acquire a rapidly acting, dehydrating condition called **traveler's diarrhea.** This diarrhea results from an encounter with certain viruses, bacteria, or protozoa usually absent from the traveler's environment. One of the major causative agents is *E. coli*. This bacterium circulates in the resident population, typically without causing symptoms due to the immunity afforded by previous exposure. Because many of these bacteria are needed to initiate infection, contaminated food and water are the major means by which the bacteria are spread. This is the basis for the popular warnings to international travelers: "Don't drink the local water" and "Boil it, peel it, cook it, or forget it."

E. coli may cause diarrheal disease by several mechanisms, and six categories or strains of diarrheagenic *E. coli* are now recognized: enterotoxigenic *E. coli* (ETEC), enteroinvasive *E. coli* (EIEC), enteroahemorrhagic *E. coli* (EHEC), enteropathogenic *E. coli* (EPEC), enteroaggregative *E. coli* (EAggEC), and diffusely adhering *E. coli* (DAEC).

The **enterotoxigenic *E. coli* (ETEC)** strains produce one or both of two distinct enterotoxins (*see Box 14.1*), which are responsible for the diarrhea and distinguished by their heat stability: heat-stable enterotoxin (ST) and heat-labile enterotoxin (LT). The genes for ST and LT production and for colonization factors are usually plasmid-borne. ST binds to a glycoprotein receptor that is coupled to guanylate cyclase on the surface of intestinal epithelial cells. Activation of guanylate cyclase stimulates the production of cyclic guanosine monophosphate (cGMP), which leads to the secretion of electrolytes and water into the lumen of the small intestine, manifested as the watery diarrhea characteristic of an ETEC infection. LT binds to specific gangliosides on the epithelial cells and activates membrane-bound adenylate cyclase, which leads to increased production of cyclic adenosine monophosphate (cAMP) through the same mechanism employed by cholera toxin. Again, the result is hypersecretion of electrolytes and water into the intestinal lumen.

The **enteroinvasive *E. coli* (EIEC)** strains cause diarrhea by penetrating and multiplying within the intestinal epithelial cells. The ability to invade the epithelial cells is associated with the presence of a large plasmid; EIEC may also produce a cytotoxin and an enterotoxin.

The **enteropathogenic *E. coli* (EPEC)** strains attach to the brush border of intestinal epithelial cells and cause a specific type of cell damage called effacing lesions. **Effacing lesions** or **attaching-effacing (AE) lesions** represent destruction of brush border microvilli adjacent to adhering bacteria. This cell destruction leads to the subsequent diarrhea. As a result of this pathology, the term AE *E. coli* is used to describe true EPEC strains. It is now known that AE *E. coli* is an important cause of diarrhea in children residing in developing countries.

The **enterohemorrhagic *E. coli* (EHEC)** strains carry the genetic determinants for attaching-effacing lesions and Shiga-like toxin production. The attaching-effacing lesion causes hemorrhagic colitis with severe abdominal pain and cramps followed by bloody diarrhea. The Shiga-like toxins I and II (also called verotoxins 1 and 2) have also been implicated in two extraintestinal diseases; hemolytic uremic syndrome and thrombotic thrombocytopenic purpura. It is believed these toxins kill vascular endothelial cells. A major form of EHEC is the *E. coli* 0157:H7 that has caused many outbreaks of hemorrhagic colitis in the United States since it was first recognized in 1982. Currently there are a minimum of 20,000 *E. coli* 0157:H7 cases and 250 deaths in the United States each year.

The **enteroaggregative *E. coli* (EAggEC)** strains adhere to epithelial cells in localized regions, forming clumps of bacteria with a "stacked brick" appearance. Conventional extracellular toxins have not been detected in EAggEC, but unique lesions are seen in epithelial cells, suggesting the involvement of toxins.

The **diffusely adhering *E. coli* (DAEC)** strains adhere over the entire surface of epithelial cells and usually cause disease in immunologically naive or malnourished children. It has been suggested that DAEC may have an as yet undefined virulence factor.

Diagnosis of traveler's diarrhea caused by *E. coli* is based on past travel history and symptoms. Laboratory diagnosis is by isolation of the specific type of *E. coli* from feces and identification using DNA probes, the determination of virulence factors, and the polymerase chain reaction. Treatment is with fluid and electrolytes plus doxycycline and trimethroprim-sulfamethoxazole. Recovery is usually without complications. Prevention and control involve avoiding contaminated food and water.

Typhoid Fever

Typhoid [Greek *typhodes,* smoke] **fever** is caused by several virulent serovars of *Salmonella typhi* and is acquired by ingestion of food or water contaminated by feces of infected humans or animals. In earlier centuries the disease occurred in great epidemics.

Once in the small intestine the incubation period is about 10 to 14 days. The bacteria colonize the small intestine, penetrate the epithelium, and spread to the lymphoid tissue, blood, liver, and gallbladder. Symptoms include fever, headache, abdominal pain, anorexia, and malaise, which last several weeks.

After approximately 3 months, most individuals stop shedding bacteria in their feces. However, a few individuals continue to shed *S. typhi* for extended periods but show no symptoms. In these carriers, the bacteria continue to grow in the gallbladder and reach the intestine through the bile duct (*see Box 35.2 on "Typhoid Mary," one of the most famous typhoid carriers*).

Laboratory diagnosis of typhoid fever is by demonstration of typhoid bacilli in the blood, urine, or stools and serology (the Widal test). Treatment with ceftriaxone, trimethoprim-sulfamethoxazole, or ampicillin has reduced the mortality rate to less than 1%. Recovery from typhoid confers a permanent immunity. Purification of drinking water, milk pasteurization, prevention of food handling by carriers, and complete isolation of patients are the most successful prophylactic measures. There is a vaccine for high-risk individuals (*see table 30.1*). About 400 to 500 cases of typhoid fever occur annually in the United States.

Sepsis and Septic Shock

Some microbial diseases and their effects cannot be categorized under a specific mode of transmission. Two important examples are sepsis and septic shock. Septic shock is the most common cause of death in intensive care units and the thirteenth most common cause of death in the United States. Unfortunately the incidence of these two disorders continues to rise: 400,000 cases of sepsis and 200,000 episodes of septic shock are estimated to occur annually in the United States, resulting in more than 100,000 deaths.

Sepsis recently has been redefined by physicians as the systemic response to a microbial infection (*see also table 29.2*). This response is manifested by two or more of the following conditions: temperature above 38 or below 36°C; heart rate above 90 beats per minute; respiratory rate above 20 breaths per minute or a pCO_2 below 32 mmHg; leukocyte count above 12,000 cells per ml^3 or below 4,000 cells per ml^3. **Septic shock** is sepsis associated with severe hypotension (low blood pressure due to shock) despite adequate fluid replacement. Gram-positive bacteria, fungi, and endotoxin-containing gram-negative bacteria can initiate the pathogenic cascade of sepsis leading to septic shock. Gram-negative sepsis is most commonly caused by *Escherichia coli,* followed by *Klebsiella* spp., *Enterobacter* spp., and *Pseudomonas aeruginosa.* Endotoxin, or lipopolysaccharide (LPS; *see figure 3.27*), an integral component of the outer membrane of gram-negative bacteria, has been implicated as a primary initiator of the pathogenesis of gram-negative septic shock.

The pathogenesis of sepsis and septic shock begins with the proliferation of the microorganism at the location of infection (figure 37.22). The microorganism may invade the bloodstream directly or may proliferate locally and release various products into the bloodstream. These products include both structural

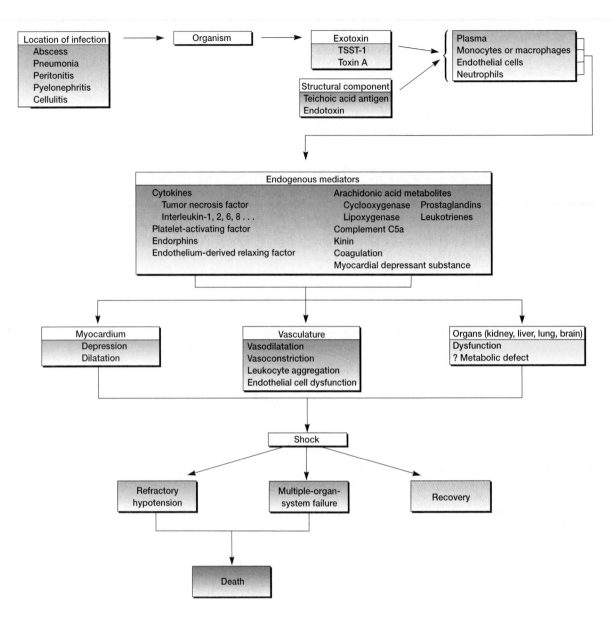

Figure 37.22   The Septic Shock Cascade. Microbial exotoxins (toxic shock syndrome toxin-1 [TSST-1], *Pseudomonas aeruginosa* toxin A [toxin A]) and structural components of the microorganism (teichoic acid antigen, endotoxin) trigger the biochemical events that lead to such serious complications as shock, adult respiratory distress syndrome, and disseminated intravascular coagulation.

components of the microorganisms (endotoxin, teichoic acid antigen) and exotoxins synthesized by the microorganism. All of these products can stimulate the release of the endogenous mediators of sepsis from endothelial cells, plasma cells (monocytes, macrophages, neutrophils), and plasma cell precursors.

The endogenous mediators have profound physiological effects on the heart, vasculature, and other body organs. The consequences are either recovery or septic shock leading to death. Death usually ensues if one or more organ systems fail completely.

Diagnosis of sepsis and septic shock is based primarily on the symptoms (see definition of sepsis) of the patient. A bacteremia (presence of viable bacteria in the blood) is not essential for diagnosis.

The early administration of antimicrobial therapy is an important aspect of effective management of sepsis and septic shock. Because the results of blood culture and susceptibility testing cannot usually be provided in less than 48 to 72 hours, and because more than 50% of the deaths caused by gram-negative sepsis occur during the first 2 days of the illness, broad-spectrum antibacterial therapy is needed. Initial therapy consists of a combination of an aminoglycoside plus either an expanded-spectrum cephalosporin or an antipseudomonal penicillin. Treatment can then be tailored to specific pathogens when culture and susceptibility results are available. Recently, monoclonal antibodies have been developed that bind to endotoxin and significantly attenuate its adverse effects in septic shock patients.

1. Distinguish between food intoxication, food poisoning, and food-borne infection. What is an enterotoxin?
2. Why is cholera the most severe form of gastroenteritis?
3. How does one acquire botulism? Describe how botulinum toxin causes flaccid paralysis.
4. What is the most common form of gastroenteritis in the United States and how are the symptoms caused?
5. What is the usual source of the bacterium responsible for salmonellosis? Shigellosis? Where and how does *Shigella* infect people?
6. Describe the most common type of food intoxication in the United States and how it arises.
7. Describe a typhoid carrier. How does one become a carrier?
8. What are some specific causes of traveler's diarrhea? Briefly describe the six major types of pathogenic *E coli*.
9. What causes sepsis? Septic shock? Why is septic shock so life threatening?

Summary

1. Although only a small percentage of all bacteria are responsible for human illness, the suffering and death they cause are significant. Each year, millions of people are infected by pathogenic bacteria using the four major modes of transmission: airborne, arthropod-borne, direct contact, and food-borne and waterborne.

2. As the fields of microbiology, immunology, pathology, pharmacology, and epidemiology have expanded current understanding of the disease process, the incidence of many human illnesses has decreased. Many bacterial infections, once the leading cause of death, have been successfully brought under control in most developed countries. Alternatively, several are increasing in incidence throughout the world.

3. The bacteria emphasized in this chapter and the diseases they cause are as follows:

   a. Airborne diseases

      diphtheria (*Corynebacterium diphtheriae*)

      Legionnaires' disease and Pontiac fever (*Legionella pneumophila*)

      meningitis (*Haemophilus influenzae* type b, *Neisseria meningitidis,* and *Streptococcus pneumoniae*)

      M. avium-M. intracellulare pneumonia

      Streptococcal diseases (*Streptococcus pyogenes*)

      tuberculosis (*Mycobacterium tuberculosis*)

      pertussis (*Bordetella pertussis*)

   b. Arthropod-borne diseases

      Lyme disease (*Borrelia burgdorferi*)

      plague (*Yersinia pestis*)

   c. Direct contact diseases

      anthrax (*Bacillus anthracis*)

      bacterial vaginosis (*Gardnerella vaginalis*)

      cat-scratch disease (*Bartonella henselae*)

      chancroid (*Haemophilus ducreyi*)

      gas gangrene or clostridial myonecrosis (*Clostridium perfringens*)

      gonorrhea (*Neisseria gonorrhoeae*)

      leprosy (*Mycobacterium leprae*)

      peptic ulcer disease (*Helicobacter pylori*)

      staphylococcal diseases (*Staphylococcus aureus*)

      syphilis (*Treponema pallidum*)

      tetanus (*Clostridium tetani*)

      tularemia (*Francisella tularensis*)

   d. Food-borne and waterborne diseases

      cholera (*Vibrio cholerae*)

      botulism (*Clostridium botulinum*)

      gastroenteritis (*Campylobacter jejuni* and other bacteria)

      salmonellosis (*Salmonella* serovar *typhimurium*)

      shigellosis (*Shigella* spp.)

      staphylococcal food poisoning (*Staphylococcus aureus*)

      typhoid fever (*Salmonella typhi*)

      traveler's diarrhea (*Escherichia coli*)

4. Gram-positive bacteria, fungi, and endotoxin-containing gram-negative bacteria can initiate the pathogenic cascade of sepsis leading to septic shock. Gram-negative sepsis is most commonly caused by *E. coli*, followed by *Klebsiella* spp., *Enterobacter* spp., and *Pseudomonas aeruginosa*.

Key Terms

anthrax *755*

aseptic meningitis syndrome *745*

bacterial (septic) meningitis *745*

bacterial vaginosis *755*

biofilms *760*

botulism *767*

Bright's disease *748*

buboes *754*

bubonic plague *754*

caseous lesion *750*

cat-scratch disease (CSD) *755*

cellulitis *747*

chancre *762*

chancroid *756*

cholera *767*

choleragen *767*

clostridial myonecrosis *756*

clue cells *755*

congenital syphilis *762*

cutaneous anthrax *755*

cutaneous diphtheria *744*

diffusely adhering *E. coli* (DAEC) *771*

diphtheria *744*

DPT vaccine *744*

effacing lesions *770*

endogenous infection *749*

enteroaggregative *E. coli* (EAggEC) *771*

enterohemorrhagic *E. coli* (EHEC) *771*

enteroinvasive *E. coli* (EIEC) *770*

enteropathogenic *E. coli* (EPEC) *770*

enterotoxigenic *E. coli* (ETEC) *770*

enterotoxin *766*

erysipelas *747*

eschar *755*

exfoliative toxin (exfoliatin) *761*

food-borne infection *766*

food intoxication *766*

food poisoning *766*

gas gangrene *756*

gastritis *758*

gastroenteritis *768*

gastrointestinal anthrax *755*

genital ulcer disease *756*

Ghon complex *750*

glomerulonephritis *748*

gonococci *756*

gonorrhea *756*

gummas *764*

impetigo *747*

Legionnaires' disease (legionellosis) *744*

lepromatous (progressive) leprosy *758*

leprosy (Hansen's disease) *757*

Lyme disease (LD, Lyme borreliosis) *752*

meningitis *745*

miliary tuberculosis *750*

multi-drug-resistant strains of tuberculosis (MDR-TB) *750*

myositis *748*

necrotizing fasciitis *748*

ophthalmia neonatorum (conjunctivitis of the newborn) *757*

pelvic inflammatory disease (PID) *757*

peptic ulcer disease *758*

pertussis *746*

pharyngitis *749*

plague *753*

pneumonic plague *754*

Pontiac fever *745*

pulmonary anthrax *755*

reactivation tuberculosis *750*

rheumatic fever *748*

salmonellosis *769*

scarlet fever (scarlatina) *748*

sepsis *771*

septic shock *771*

shigellosis *769*

slime *759*

staphylococcal food poisoning *770*

staphylococcal scalded skin syndrome (SSSS) *760*

streptococcal pneumonia *749*

streptococcal sore throat *749*

tetanolysin *765*

tetanospasmin *765*

tetanus *764*

tonsillitis *749*

toxic shock-like syndrome (TSLS) *748*

toxic shock syndrome (TSS) *760*

traveler's diarrhea *770*

tubercles *750*

tuberculoid (neural) leprosy *758*

tuberculosis (TB) *750*

tuberculous cavities *750*

tularemia *766*

typhoid fever *771*

venereal syphilis *762*

Questions for Thought and Review

1. Briefly describe each of the major or most common bacterial diseases in terms of its causative agent, signs and symptoms, the course of infection, mechanism of pathogenesis, epidemiology, and prevention and/or treatment.

2. Because there are many etiologies of gastroenteritis, how is a definitive diagnosis usually made?

3. Differentiate between the following factors of bacterial intoxication and infection: etiologic agents, onset, duration, symptoms, and treatment.

4. How would you differentiate between salmonellosis and botulism?

5. Why do urinary tract infections frequently occur after catheterization procedures?

6. Why is botulism the most serious form of bacterial food poisoning?

7. Why is tuberculosis still a problem in underdeveloped countries? In the United States?

8. What is the standard treatment for most bacterial diarrheas?

9. Why are most cases of gastroenteritis not treated with antibiotics?

10. How are tetanus, gas gangrene, and botulism related?

11. Give several reasons why treatment of tuberculosis is so difficult.

12. Describe the various bacteria that cause food poisoning and give each one's most probable source.

13. Explain why Lyme disease is receiving increased attention in the United States.

14. You have been assigned the task of eradicating gonorrhea in your community. Explain how you would accomplish this.

15. You are a park employee. How would you prevent people from getting arthropod-borne diseases?

16. Why is septic shock caused by gram-negative sepsis increasing in prevalence in the United States?

Additional Reading

General

Benenson, A. S., editor. 1990. *Control of communicable diseases in man,* 15th ed. Washington, D.C.: American Public Health Association.

Jett, B. D.; Huycke, M. M.; and Gilmore, M. S. 1994. Virulence of enterococci. *Clin. Microbiol. Rev.* 7(4):462–78.

Joklik, W. K.; Willett, H. P.; Amos, D. B.; and Wilfert, C. M. 1992. *Zinsser microbiology,* 20th ed. E. Norwalk, Conn.: Appleton & Lange.

Kotb, M. 1995. Bacterial pyrogenic exotoxins as superantigens. *Clin. Microbiol. Rev.* 8(3)411-26.

Mandell, G. L.; Bennett, J. E.; and Dolan, R. 1995. *Principles and practices of infectious diseases.* 4th ed. New York: Churchill Livingston.

Murray, D. L.; Ohlendorf, D. H.; and Schlievert, P. M. 1995. Staphylococcal and streptococcal superantigens: Their role in human diseases. ASM News 61(5):229-35.

Rosebury, T. 1973. *Microbes and morals: The strange story of venereal disease.* New York: Ballantine.

Salyers, A. A., and Whitt, D. D. 1994. *Bacterial Pathogenesis: A molecular approach.* Washington, D.C.: ASM Press.

Shulman, S. T.; Phair, J. P.; and Sommers, H. M. 1992. *The biologic and clinical basis of infectious diseases,* 4th ed. Philadelphia, PA.: W. B. Saunders.

Airborne Diseases

Barbaree, J. editor. 1993. *Legionella: Current status and emerging perspectives.* Washington, D.C.: American Society for Microbiology.

Barksdale, L. 1970. *Corynebacterium diphtheriae* and its relatives. *Bacteriol. Rev.* 34:378–422.

Bermudez, L. E., and Kaplan, G. 1995. Recombinant cytokines for controlling mycobacterial infections. *Trends Microbiol.* 3(1):22–26.

Bisno, A. L. 1991. Group A streptococcal infections and acute rheumatic fever. *N. Engl. J. Med.* 325(11):783–93.

Bloom, B. 1994. *Tuberculosis: Pathogenesis, protection, and control.* Washington, D.C.: American Society for Microbiology.

Dowling, J. N.; Saha, A. K.; and Glew, R. H. 1992. Virulence factors of the family *Legionellaceae. Microbiol. Rev.* 56(1):32–60.

Fischetti, V. A. 1991. Streptococcal M protein. *Sci. Am.* 264(6):58–65.

Fraser, D. W., and McDade, J. E. 1979. Legionellosis. *Sci. Am.* 241(5):82–89.

Friedman, R. L. 1988. Pertussis: The disease and new diagnostic methods. *Clin. Microbiol. Rev.* 1(4):365–76.

Iseman, M. 1993. Treatment of multidrug-resistant tuberculosis. *N. Engl. J. Med.* 329(11):784–91.

Kloos, W., and Bannerman, T. 1994. Update on clinical significance of coagulase-negative staphylococci. *Clin. Microbiol. Rev.* 7(1):117–40.

Quagliarello, V., and Scheld, W. 1992. Bacterial meningitis: Pathogenesis, pathophysiology, and progress. *N. Engl. J. Med.* 327(12):864–71.

Sepkowitz, K.; Raffalli, J.; Riley, L.; Kiehn, T.; and Armstrong, D. 1995. Tuberculosis in the AIDS era. *Clin. Microbiol. Rev.* 8(2):180-99.

Todd, J. K. 1988. Toxic shock syndrome. *Clin. Microbiol. Rev.* 1(4):432–46.

Tunkel, A. R., and Scheld, W. M. 1993. Pathogenesis and pathophysiology of bacterial meningitis. *Clin. Microbiol. Rev.* 6(2):118–36.

Unny, S. K., and Middlebrook, B. L. 1983. Streptococcal rheumatic carditis. *Microbiol. Rev.* 47:97–109.

Wayne, L. G., and Sramek, H. A. 1992. Agents of newly recognized or infrequently encountered mycobacterial diseases. *Clin. Microbiol. Rev.* 5(1):1–25.

Winn, W. C., 1988. Legionnaires' disease: Historical perspective. *Clin. Microbiol. Rev.* 1(1):60–81.

Arthropod-Borne Diseases

Barbour, A. G. 1988. Laboratory aspects of Lyme borreliosis. *Clin. Microbiol. Rev.* 1(4):399–414.

Habicht, G. S.; Beck, G.; and Benach, J. L. 1987. Lyme disease. *Sci. Am.* 257(3):78–83.

Kantor, F. S. 1994. Disarming lyme disease. *Sci. Am.* 271(3):34–39.

McEvedy, C. 1988. The bubonic plague. *Sci. Am.* 258(2):118–23.

Straley, S. C.; Skrzypek, E.; Plano, G.; and Bliska, J. 1993. Yops of *Yersinia* spp. pathogenic for humans. *Infect. Immun.* 61(8):3105–110.

Szczepanski, A., and Benach, J. 1991. Lyme borreliosis: Host response to *Borrelia burgdorferi. Microbiol. Rev.* 55(1):21–34.

Direct Contact Diseases

Andrews, D. M., and Ruoff, K. L. 1993. *Afipia, Rochalimaea,* and cat-scratch disease. *Clin. Microbiol. News.* 15(1):1–3.

Aral, S. O., and Holmes, K. K. 1991. Sexually transmitted diseases in the AIDS era. *Sci. Am.* 264(2):62–69.

Catlin, B. W. 1993. *Gardnerella vaginalis:* Characteristics, clinical considerations, and controversies. *Clin. Microbiol. Rev.* 5(2):213–37.

Cover, T. L., and Blaser, M. J. 1995. *Helicobacter pylori:* A bacterial cause of gastritis, peptic ulcer disease, and gastric cancer. ASM News 61(1):21–26.

Dennie, C. C. 1962. *A history of syphilis.* Springfield, Ill.: Charles C Thomas.

Hastings, R. C.; Gillis, T. P.; Krahenbuhl, J. L.; and Franzblau, S. G. 1988. Leprosy. *Clin. Microbiol. Rev.* 1(3):330–48.

Hook, E. W., and Marra, C. M. 1992. Acquired syphilis in adults. *N. Engl. J. Med.* 326(16):1060–69.

Isenberg, S.; Apt, L.; and Wood, M. 1995. A controlled trial of povidone-iodine as prophylaxis against ophthalmia neonatorum. *N. Engl. J. Med.* 332(9):562–66.

Larsen, S. A.; Steiner, B. M.; and Rudolph, A. H. 1995. Laboratory diagnosis and interpretation of tests for syphilis. *Clin. Microbiol. Rev.* 8(1):1–21.

Lee, A.; Fox, J.; and Hazell, S. 1993. Minireview. Pathogenicity of *Helicobacter pylori:* A perspective. *Infect. Immun.* 61(5):1601–10.

Peterson, W. 1991. *Helicobacter pylori* and peptic ulcer disease. *N. Engl. J. Med.* 324(15):1043–48.

Stine, G. J. 1992. *The biology of sexually transmitted diseases.* Dubuque, Iowa: Wm. C. Brown Communications, Inc.

Tramont, E. 1989. Gonococcal vaccines. *Clin. Microbiol. Rev.* 2(Suppl.):S74–S77.

Trees, D. L., and Morse, S. A. 1995. Chancroid and *Haemophilus ducreyi:* An update. *Clin. Microbiol. Rev.* 8(3):357-75.

Food-Borne and Waterborne Diseases

Archer, D. L., and Young, F. E. 1988. Contemporary issues: Diseases with a food vector. *Clin. Microbiol. Rev.* 1(4):377–98.

Aronon, S. S. 1980. Infant botulism. *Ann. Rev. Med.* 31:541–50.

Blake, P. A.; Weaver, R. E.; and Hollis, D. G. 1980. Diseases of humans caused by vibrios. *Ann. Rev. Microbiol.* 34:341–67.

Dowell, V. R. J. 1984. Botulism and tetanus: Selected epidemiologic and microbiologic aspects. *Rev. Infect. Dis.* 6:202–7.

Gorbach, S. L. 1975. Traveler's diarrhea and toxigenic *E. coli. N. Engl. J. Med.* 292:933–36.

Guerrant, R., and Bobak, D. 1991. Bacterial and protozoal gastroenteritis. *N. Engl. J. Med.* 325(5):327–40.

Hatheway, C. L. 1990. Toxigenic clostridia. *Clin. Microbiol. Rev.* 3(1):66–98.

Janda, J. M.; Powers, C.; Bryant, R. G.; and Abbott, S. L. 1988. Current perspectives on the epidemiology and pathology of clinically significant *Vibrio* spp. *Clin. Microbiol. Rev.* 1(3)245–67.

Kaper, J. B.; Morris, J. G., Jr.; and Levine, M. M. 1995. Cholera. *Clin. Microbiol. Rev.* 8(1):48–86.

Knoop, F. C.; Owens, M.; and Crocker, I. C. 1993. *Clostridium difficile:* Clinical disease and diagnosis. *Clin. Microbiol. Rev.* 6(3):251–65.

Lyerly, D. M.; Krivan, H. C.; and Wilkins, T. D. 1988. *Clostridium difficile:* Its diseases and toxins. *Clin. Microbiol. Rev.* 1(1):1–18.

Maurelli, A. T. 1992. *Shigella* inside and out: Lifestyles of the invasive and dysenteric. ASM News 58(11):603–08.

Nachamkin, I., editor. 1992. *Campylobacter jejuni: Current status and future trends.* Washington, D.C.: American Society for Microbiology.

Rippey, S. R. 1994. Infectious diseases associated with molluscan shellfish consumption. *Clin. Microbiol. Rev.* 7(4):419–25.

Wachsmuth, K.; Blake, P.; and Olsvik, O. 1994. *Vibrio cholerae and cholera: Molecular to global perspectives.* Washington, D.C.: American Society for Microbiology.

Sepsis and Septic Shock

Bone, R. 1993. Gram-negative sepsis: A dilemma of modern medicine. *Clin. Microbiol. Rev.* 6(1):57–68.

Hoge, C. W.; Schwartz, B.; Talkington, D.; Breiman, R.; MacNeil, E.; and Englender, S. 1993. The changing epidemiology of invasive group A streptococcal infections and the emergence of streptococcal toxic-shock like syndrome. *JAMA* 269:384–89.

Wagner, S.; Friedman, L.; and Dodd, R. 1994. Transfusion-associated bacterial sepsis. *Clin. Microbiol. Rev.* 7(3):290–302.

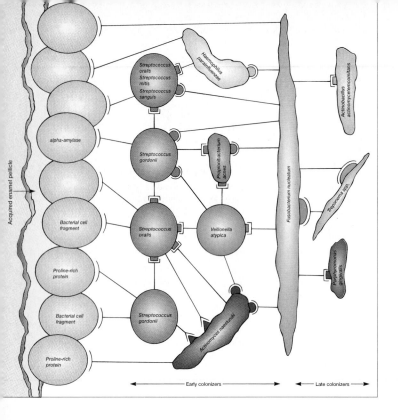

CHAPTER 38

Human Diseases Caused by Other Bacteria (Chlamydiae, Mycoplasmas, Rickettsias); Dental and Nosocomial Infections

The chlamydiae are a genetically diverse group of bacteria with similarities in morphology, intracellular development, and antigenic properties. Chlamydiae are common causes of sexually transmitted disease (STD) and respiratory infection. Millions of cases of C. trachomatis infection occur in the United States each year . . .

—Robert C. Barnes

Outline

Concepts

1. Three chlamydial species cause human disease: *Chlamydia pneumoniae* causes chlamydial pneumonia; *Chlamydia trachomatis* causes inclusion conjunctivitis, lymphogranuloma venereum, nongonococcal urethritis, and trachoma; and *C. psittaci* causes psittacosis.

2. Three species of mycoplasmas are human pathogens: *Mycoplasma hominis* and *Ureaplasma urealyticum* cause genitourinary tract disease, whereas *M. pneumoniae* is a major cause of acute respiratory disease and pneumonia.

3. The rickettsias found in the United States can be divided into the typhus group (epidemic typhus caused by *R. prowazekii* and murine typhus caused by *R. typhi*) and the spotted fever group (Rocky Mountain spotted fever caused by *R. rickettsii* and ehrlichiosis caused by *Ehrlichia chaffeensis*), with Q fever (caused by *Coxiella burnetii*) being an exception because it forms endosporelike structures and does not have to use an insect vector as the other rickettsias do.

4. Several bacterial odontopathogens are responsible for the most common bacterial diseases in humans—tooth decay and periodontal disease. Both are the result of plaque formation and the production of lactic and acetic acids by the odontopathogens.

5. Bacteria are the leading cause of hospital-acquired or nosocomial diseases such as bacteremias, burn wound infections, respiratory tract infections, surgical wound infections, and urinary tract infections. These diseases represent a significant proportion of all infectious diseases acquired by humans.

Chapter 20 reviews the general biology of the chlamydiae, mycoplasmas, and rickettsias; chapter 35 presents the epidemiology of nosocomial diseases. Chapter 38 continues this coverage by discussing the important human pathogens noted in these chapters. The microorganisms involved in dental infections are also described.

Several groups of bacterial pathogens are unusual in many respects and do not fit neatly into traditional classification and discussion schemes. Nevertheless, they are important and worthy of discussion. This chapter contains an assemblage of such bacteria and the diseases they cause.

Chlamydial Diseases

The chlamydiae are obligate intracellular parasites of eucaryotic cells (*see figures 20.19 and 20.20*). Three species cause human disease: *Chlamydia pneumoniae, C. trachomatis,* and *C. psittaci.*

The chlamydiae (pp. 432–34).

The clinical spectrum and organ specificity of chlamydial infections are determined by the method of transmission and antigenic properties of the infecting strain. The 15 serotypes of *C. trachomatis* are divided among three groups: mouse, lymphogranuloma venereum (serotypes L_1–L_3), and trachoma (serotypes A–K). Serotypes A–K infect epithelial cells and are associated with infections of the conjunctiva (membrane covering the eyeball), pharynx, respiratory tract, urethra, cervix of the uterus, and uterine tubes. Serotypes L_1–L_3 cause the systemic disease lymphogranuloma venereum. These human chlamydial diseases are summarized in table 38.1, and several are now discussed in more detail.

Chlamydial Pneumonia

Chlamydial pneumonia is caused by *Chlamydia pneumoniae.* Clinically, infections are generally mild; pharyngitis, bronchitis, and sinusitis commonly accompany some lower respiratory tract involvement. Symptoms include fever, a productive cough, sore throat, hoarseness, and pain on swallowing. Infections with *C. pneumoniae* are common but sporadic; about 50% of adults have antibody to the chlamydiae. Evidence suggests that *C. pneumoniae* is primarily a human pathogen directly transmitted from human to human. Identification of chlamydial pneumonia is based on symptoms and a microimmunofluorescence test. Tetracycline and erythromycin are used for treatment.

Inclusion Conjunctivitis

Inclusion conjunctivitis is an acute infectious disease caused by *C. trachomatis* serotypes D–K, and it occurs throughout the world. It is characterized by a copious mucous discharge from the eye, an inflamed and swollen conjunctiva, and the

TABLE 38.1	Human Diseases Caused by Chlamydiae	
Microorganism and Disease(s)	**Method of Transmission**	**Serotypes**
Chlamydia trachomatis		
Endemic trachoma	Flies, fingers, fomites	A, B, Ba, C
Inclusion conjunctivitis	During birth, sexual contact	D–K
Newborn pneumonia	During birth	D–K
Nongonococcal urethritis	Sexual contact	D–K
Postgonococcal urethritis	Sexual contact	D–K
Epididymitis	Sexual contact	D–K
Proctitis	Sexual contact	D–K
Mucopurulent cervicitis	Sexual contact	D–K
Endometritis	Sexual contact	D–K
Salpingitis	Sexual contact	D–K
Lymphogranuloma venereum	Sexual contact	L_1, L_2, L_3
Chlamydia psittaci		
Psittacosis (ornithosis)	Zoonosis from bird droppings	Many
Chlamydia pneumoniae		
Pneumonia	Person to person	None

presence of large inclusion bodies. In inclusion conjunctivitis of the newborn, the chlamydiae are acquired during passage through an infected birth canal. The disease appears 7 to 12 days after birth. If the chlamydiae colonize an infant's nasopharynx and tracheobronchial tree, pneumonia may result. Adult inclusion conjunctivitis is acquired by contact with infective genital tract discharges.

Without treatment, recovery usually occurs spontaneously over several weeks or months. Therapy involves treatment with tetracycline, erythromycin, or a sulfonamide. The specific diagnosis of *C. trachomatis* can be made by direct immunofluorescence, Giemsa stain, nucleic acid probes, and culture. Genital chlamydial infections and inclusion conjunctivitis are sexually transmitted diseases that are spread by indiscriminate contact with multiple sex partners. Prevention depends upon diagnosis and treatment of all infected individuals.

Lymphogranuloma Venereum

Lymphogranuloma venereum (*LGV*) is a sexually transmitted disease caused by *C. trachomatis* serotypes L_1–L_3. It has a worldwide distribution but is more common in tropical climates.

LGV proceeds through three phases. (1) In the primary phase a small ulcer appears several days to several weeks after a person is exposed to the chlamydiae. The ulcer may appear on the penis in males or on the labia or vagina in females. The ulcer heals quickly and leaves no scar. (2) The secondary phase begins 2 to 6 weeks after exposure, when the chlamydiae infect lymphoid cells, causing the regional lymph nodes to become enlarged and tender; such nodes are called buboes (figure 38.1). Systemic symptoms such as fever, chills, and anorexia

Bubo

Figure 38.1   Lymphogranuloma Venereum. The bubo in the left inguinal area is draining.

are common. (3) If the disease is not treated, a late phase ensues. This results from fibrotic changes and abnormal lymphatic drainage that produces fistulas (abnormal passages leading from an abscess or a hollow organ to the body surface or from one hollow organ to another) and urethral or rectal strictures (a decrease in size). An untreatable fluid accumulation in the penis, scrotum, or vaginal area may result.

The disease is detected by staining infected cells with iodine to observe inclusions (chlamydia-filled vacuoles), culture of the chlamydiae from a bubo, nucleic acid probes, or by the detection of a high micro-IF antibody titer to an LGV serotype. Treatment in the early phases consists of aspiration of the buboes and administration of drugs: tetracycline, doxycycline, erythromycin, or sulfamethoxazole. The late phase may require surgery. The methods used for the control of LGV are the same as for other sexually transmitted diseases: reduction in promiscuity, use of condoms, and early diagnosis and treatment of infected individuals. About 300 cases of LGV occur annually in the United States.

Nongonococcal Urethritis

Nongonococcal urethritis (NGU) is any inflammation of the urethra not due to the bacterium *Neisseria gonorrhoeae*. This condition is caused both by nonmicrobial factors such as catheters and drugs and by infectious microorganisms. The most important causative agents are *C. trachomatis, Ureaplasma urealyticum, Mycoplasma hominis, Trichomonas vaginalis, Candida albicans,* and herpes simplex viruses. Most infections are acquired sexually, and of these, approximately 50% are

Chlamydia infections. NGU from chlamydia is probably the most common sexually transmitted disease in the United States, with over 10 million Americans infected. It is endemic elsewhere throughout the world.

Symptoms of NGU vary widely. Males may have few or no manifestations of disease; however, complications can exist. These include a urethral discharge, itching, and inflammation of the male reproductive structures. Females may be asymptomatic or have a severe infection called **pelvic inflammatory disease (PID)** that often leads to sterility. Chlamydia may account for as many as 200,000 to 400,000 cases of PID annually in the United States. In the pregnant female, a chlamydial infection is especially serious because it is directly related to miscarriage, stillbirth, inclusion conjunctivitis, and infant pneumonia.

Diagnosis of NGU requires the demonstration of a leukocyte exudate and exclusion of urethral gonorrhea by Gram stain and culture. Several rapid tests for detecting *Chlamydia* in urine specimens are also available. Treatment is with tetracycline, doxycycline, erythromycin, or sulfisoxazole.

Psittacosis (Ornithosis)

Psittacosis (ornithosis) is a worldwide infectious disease of birds that is transmissible to humans. It was first described in association with parrots and parakeets, both of which are psittacine birds. The disease is now recognized in many other birds—among them, pigeons, chickens, ducks, and turkeys—and the general term ornithosis [Latin *ornis,* bird] is used.

Ornithosis is caused by *Chlamydia psittaci.* Humans contract this disease either by handling infected birds or by inhaling dried bird excreta that contains viable *C. psittaci.* Ornithosis is recognized as an occupational hazard within the poultry industry, particularly to workers in turkey processing plants.

After entering the respiratory tract, the chlamydiae are transported to the cells of the liver and spleen. They multiply within these cells and then invade the lungs, where they cause inflammation, hemorrhaging, and pneumonia.

Laboratory diagnosis is either by isolation of *C. psittaci* from blood or sputum, or by serological studies. Treatment is with tetracycline. Because of antibiotic therapy, the mortality rate has dropped from 20 to 2%. Between 100 and 200 cases of ornithosis are reported annually in the United States. Prevention and control is by chemoprophylaxis (tetracycline) for pet birds and poultry.

Trachoma

Trachoma [Greek *trachoma,* roughness] is a contagious disease created by *C. trachomatis* serotypes A–C. It is one of the oldest known infectious diseases of humans and is the greatest single cause of blindness throughout the world. Probably over 500 million people are infected and 20 million blinded each year by this chlamydia. In endemic areas most children are chronically infected within a few years of birth. Active disease in adults over age 20 is three times as frequent in females as in males because of mother-child contact. Although uncommon in the United States, except among American Indians in the Southwest, trachoma is widespread in Asia, Africa, and South America.

Trachoma is transmitted by contact with inanimate objects such as soap and towels, by hand-to-hand contact that carries *C. trachomatis* from an infected eye to an uninfected eye, or by flies. The disease begins abruptly with an inflamed conjunctiva. This leads to an inflammatory cell exudate and necrotic eyelash follicles beneath the conjunctival surface (figure 38.2). The disease usually heals spontaneously. However, with reinfection, vascularization of the cornea, or **pannus** formation, occurs, leading to scarring of the conjunctiva. If scar tissue accumulates over the cornea, blindness results.

Diagnosis and treatment of trachoma are the same as for inclusion conjunctivitis (previously discussed). However, prevention and control of trachoma lies more in health education and personal hygiene—such as access to clean water for washing—than in treatment.

Figure 38.2   Trachoma. An active infection showing marked follicular hypertrophy of both eyelids. The inflammatory nodules cover the thickened conjunctiva of the eye.

1. How can one distinguish *Chlamydia trachomatis* from *C. psittaci?* How do chlamydiae differ from most other bacteria?

2. How does an infant acquire inclusion conjunctivitis?

3. Describe the three phases of lymphogranuloma venereum.

4. What is nongonococcal urethritis and what agents can cause it? Describe complications that may develop in the absence of treatment.

5. How do humans contract ornithosis? Chlamydial pneumonia?

6. How does *C. trachomatis,* serotypes A–C, cause trachoma? Describe how it is transmitted, and the way in which blindness may result.

Mycoplasmal Diseases

The mycoplasmas are pleomorphic, nonmotile microorganisms without cell walls and are among the smallest independently living organisms. They are ubiquitous worldwide as saprophytes and parasites of plants, animals, and humans. Three species are pathogenic to humans: *Ureaplasma urealyticum, Mycoplasma hominis,* and *M. pneumoniae.*

The mycoplasmas (pp. 434–36).

Genitourinary Diseases

Ureaplasma urealyticum and *M. hominis* are common parasitic microorganisms of the genital tract, and their transmission is related to sexual activity. Both mycoplasmas can opportunistically cause inflammation of the reproductive organs of males and females. Because mycoplasmas are not usually cultured by clinicians, management and treatment of these infections

depend on a recognition of clinical syndromes and provision for adequate therapy. Tetracyclines are active against most strains; resistant organisms can be treated with erythromycin.

Primary Atypical Pneumonia

Typical pneumonia has a bacterial origin. If a bacterium cannot be isolated, the pneumonia is termed atypical and a virus is usually suspected. If viruses can't be detected, then **mycoplasmal pneumonia** can be considered. This pneumonia is caused by *Mycoplasma pneumoniae (see figure 20.21),* a mycoplasma with worldwide distribution. Spread involves close contact and sometimes airborne droplets. The disease is fairly common and mild in infants and small children; serious disease is seen principally in older children and young adults.

Respiratory syndromes and viral pneumonia (p. 719–20).

M. pneumoniae usually infects the upper respiratory tract, and subsequently moves to the lower respiratory tract, where it attaches to respiratory mucosal cells. It then produces peroxide, which may be a toxic factor, but the exact mechanism of pathogenesis is unknown. An alternation in mucosal cell nucleic acid metabolism has been observed. The manifestations of this disease vary in severity from asymptomatic to a serious pneumonia. The latter is accompanied by death of the surface mucosal cells, lung infiltration, and congestion. Initial symptoms include headache, weakness, a low fever, and a predominant cough. The disease and its symptoms usually persist for weeks. The mortality rate is less than 1%.

Several rapid tests using latex agglutination of *M. pneumoniae* antibodies are available for diagnosis of mycoplasmal pneumonia. When isolated from respiratory secretions, the mycoplasmas form distinct colonies with a "fried egg" appearance (*see figure 20.22*). During the acute stage of the disease, diagnosis must be made by clinical observations. Tetracyclines or erythromycin are effective in treatment. There are no preventive measures.

---

1. What three species of mycoplasmas cause disease in humans?
2. What two mycoplasmas may cause genitourinary disease?
3. What is meant by primary atypical pneumonia? How is this disease spread?

Rickettsial Diseases (Rickettsioses)

The rickettsias, like the chlamydiae, are fastidious bacteria that are obligate intracellular parasites. They are small, pleomorphic coccobacilli maintained in the environment through a cycle involving mammalian reservoirs and insect vectors. Except for epidemic typhus, humans are only incidental hosts and are not useful in propagating the rickettsias (Box 38.1). Rickettsias can be divided into four groups based on characteristics such as their intracellular location, growth temperature optimum, antigenic properties, DNA composition (mol% G + C), and the diseases they cause. These four are (1) the typhus group—epidemic typhus (*Rickettsia prowazekii*) and murine typhus (*R. typhi*); (2) the spotted fever group—Rocky Mountain spotted fever (*R. rickettsii*); (3) the scrub typhus group (*R. tsutsugamushi*); and (4) a miscellaneous group that includes Q fever (*Coxiella burnetii*). Because members of the genus *Ehrlichia* are closely related to the genus *Rickettsia* and placed in the family *Rickettsiaceae*, ehrlichiosis also will be described here. In the following sections, only diseases that occur in the United States are discussed.

The rickettsias (p. 432).

Ehrlichiosis

In 1986 the first case of **ehrlichiosis** was diagnosed in the United States and shown to be caused by a new species of *Rickettsiaceae, Ehrlichia chaffeensis*. Since the initial discovery, more than 250 cases have been reported in the south-central and southern Atlantic states. *E. chaffeensis* is transmitted from unknown animal vectors to humans by *Dermacentor andersoni* and *Amblyomma americanum* ticks. Once inside the human body, *E. chaffeensis* infects circulating leukocytes causing a nonspecific febrile illness that resembles Rocky Mountain spotted fever. Diagnosis involves serological tests and tetracycline is the drug of choice.

In 1994 a new form of ehrlichiosis was discovered. Human granulocytic ehrlichiosis (HGE) is transmitted by deer ticks and possibly dog ticks, and has been found in 30 states, particularly in the southeastern and south central United States. The causative agent is an *Ehrlichia* species different from *E. chaffeensis*. The disease is characterized by the rapid onset of fever, chills, headaches, and muscle aches. Treatment is with doxycycline.

Epidemic (Louse-Borne) Typhus

Epidemic (louse-borne) typhus is caused by *Rickettsia prowazekii*, which is transmitted from person to person by the body louse. In the United States a reservoir of *R. prowazekii* also exists in the southern flying squirrel. When a louse feeds on an infected rickettsemic person, the rickettsias infect the insect's gut and multiply, and large numbers of organisms appear in the feces in about a week. When a louse takes a blood meal, it defecates. The irritation causes the affected individual to scratch the site and contaminate the bite wound with rickettsias. The rickettsias then spread by way of the bloodstream and infect the endothelial cells of the blood vessels, causing a **vasculitis** (inflammation of the blood vessels). This produces an abrupt headache, fever, and muscle aches. A rash begins on the upper trunk, and spreads. Without treatment, recovery takes about 2 weeks, though mortality rates are very high (around 50%), especially in the elderly. Recovery from the disease gives a solid immunity and also protects the person from murine typhus.

Diagnosis is by the characteristic rash, symptoms, and the Weil-Felix reaction (Box 38.2). Chloramphenicol and tetracycline are effective against typhus. Control of the human body louse (*Pediculus humanus corporis*) and the conditions that foster its proliferation are mainstays in the prevention of epidemic typhus, although a typhus vaccine is available for high-risk individuals. The importance of louse control and good public hygiene is shown by the prevalence of typhus epidemics during times of war and famine when there is crowding and little attention to the maintenance of proper sanitation. For example, around 30 million cases of typhus fever and 3 million deaths occurred in the Soviet Union and Eastern Europe between 1918 and 1922. The bacteriologist Hans Zinsser believes that Napoleon's retreat from Russia in 1812 may have been partially provoked by typhus and dysentery epidemics that ravaged the French army. Fewer than 25 cases of epidemic typhus are reported in the United States each year.

Endemic (Murine) Typhus

The etiologic agent of **endemic (murine) typhus** is *Rickettsia typhi*. It occurs in isolated areas around the world, including southeastern and Gulf Coast states, especially Texas. The disease occurs sporadically in individuals who come into contact with rats and their fleas (*Xenopsylla cheopi*). The disease is nonfatal in the rat and is transmitted from rat to rat by fleas.

Box 38.1

The Hazards of Microbiological Research

The investigation of human pathogens often is a very dangerous matter, and several microbiologists have been killed by the microorganisms they were studying. The study of typhus fever provides a classic example. In 1906 Howard T. Ricketts (1871–1910), an associate professor of pathology at the University of Chicago, became interested in Rocky Mountain spotted fever, a disease that had decimated the Nez Percé and Flathead Indians of Montana. By infecting guinea pigs, he established that a small bacterium was the disease agent and was transmitted by ticks. In late 1909 Ricketts traveled to Mexico to study Mexican typhus.

He discovered that a microorganism similar to the Rocky Mountain spotted fever bacillus could cause the disease in monkeys and be transmitted by lice. Despite his careful technique, he was bitten while transferring lice in his laboratory and died of typhus fever on May 3, 1910. The causative agent of typhus fever was fully described in 1916 by the Brazilian scientist H. da Roche-Lima and named *Rickettsia prowazekii* in honor of Ricketts and Stanislaus von Prowazek, a Czechoslovakian microbiologist who died in 1915 while studying typhus.

Today modern equipment to control microorganisms, such as laminar airflow hoods (*see figures 7.6 and 34.1*c), have greatly reduced the risks of research on microbial pathogens.

Box 38.2

The Weil-Felix Reaction

During World War I Austrian bacteriologists Edmond Weil (1880–1922) and Arthur Felix (1887–1956) were in charge of diagnosing typhus in the Austrian Army. In 1916 they developed the agglutination test for typhus that now bears their names.

The **Weil-Felix reaction** is based on the detection of heterophilic antibodies—that is, cross-reactive antibodies. In this test, antibodies produced in response to a particular rickettsial infection agglutinate bacterial strains of *Proteus,* designated OX–19, OX–2,

and OX–K. These cross-reactions occur because rickettsial strains possess cell wall antigens that are similar to polysaccharide O antigens of the *Proteus* strains. The results are considered to give presumptive diagnosis of rickettsial diseases. However, it should be kept in mind that *Proteus* species are indigenous to humans and may cause urinary tract infections. Therefore the presence of antibodies to *Proteus* OX strains is not of itself an absolute criterion for the presence of rickettsial disease.

When an infected flea takes a human blood meal, it defecates. Its feces are heavily laden with rickettsias, which infect humans by contaminating the bite wound.

The clinical manifestations of murine [Latin *mus, muris,* mouse or rat] typhus are similar to those of epidemic typhus except that they are milder in degree and the mortality rate is much lower: less than 5%. Diagnosis and treatment also are the same. Rat control and avoidance of rats are preventive measures for the disease. Fewer than 100 cases of endemic typhus are reported in the United States each year.

Q Fever

Q fever (Q for query because the cause of the fever was not known for some time) is an acute zoonotic disease caused by *Coxiella burnetii. C. burnetii* is different from the other rickettsias in its ability to survive outside host cells by forming a resistant

endosporelike body. This rickettsia infects both wild animals and livestock. In animals, ticks (many species) transmit *C. burnetii,* whereas in humans transmission is primarily by inhalation of dust contaminated with rickettsias from dried animal feces, urine, or milk. The disease is apt to occur in epidemic form among slaughterhouse workers and sporadically among farmers and veterinarians. Each year, fewer than 100 cases of Q fever are reported in the United States.

In humans, after inhalation of the rickettsias, local proliferation occurs in the lungs. This may result in mild respiratory symptoms similar to those of atypical pneumonia or influenza. Q fever itself is an acute illness characterized by the sudden onset of severe headache, myalgia (muscle pain), and fever, which may remain very high for more than a month if not treated. Unlike the other rickettsial diseases, Q fever is not accompanied by a rash. It is rarely fatal, but endocarditis—inflammation of the heart muscle—occurs in about 10% of the cases. Five to ten years may elapse between the initial infection and the appearance of the endocarditis. During this interval the rickettsias live in the liver and often cause hepatitis. Diagnosis is most commonly made serologically. Treatment is with chloramphenicol and tetracycline. Prevention and control consists of vaccinating researchers and others at high occupational risk and in areas of endemic Q fever; cow and sheep milk should be pasteurized before consumption.

Rocky Mountain Spotted Fever

Rocky Mountain spotted fever is caused by *Rickettsia rickettsii*. Although originally detected in the Rocky Mountain area, most cases of this disease now occur east of the Mississippi River. The disease is transmitted by ticks and usually occurs in people who are or have been in tick-infested areas. There are two principal vectors: *Dermacentor andersoni,* the wood tick, is distributed in the Rocky Mountain states and is active during the spring and early summer. *D. variabilis,* the dog tick, has assumed greater importance and is almost exclusively confined to the eastern half of the United States. Unlike the other rickettsias discussed, *R. rickettsii* can pass from generation to generation of ticks through their eggs in a process known as **transovarian passage.** No humans or mammals are needed as reservoirs for the continued propagation of this rickettsia in the environment.

When humans contact infected ticks, the rickettsias are either deposited on the skin (if the tick defecates after feeding) and then subsequently rubbed or scratched into the skin, or the rickettsias are deposited into the skin as the tick feeds. Once inside the skin, the rickettsias enter the endothelial cells of small blood vessels, where they multiply and produce a characteristic vasculitis (inflammation of blood vessels).

The disease is characterized by the sudden onset of a headache, high fever, chills, and a skin rash (figure 38.3) that initially appears on the ankles and wrists and then spreads to the trunk of the body. If the disease is not treated, the rickettsias can destroy the blood vessels in the heart, lungs, or kidneys and cause death. Usually, however, severe pathological changes are avoided by antibiotic therapy (chloramphenicol, chlortetracycline), the development of immune resistance, and supportive therapy. Diagnosis is made through observation of symptoms and signs such as the characteristic rash and by serological tests. The best means of prevention remains the avoidance of tick-infested habitats and animals (*see preventive methods for Lyme disease, pp.752–3*). There are approximately 1,000 reported cases of Rocky Mountain spotted fever annually in the United States.

1. What two antibiotics are used against most rickettsial infections?
2. How is epidemic typhus spread? Ehrlichiosis? Murine typhus? What are their symptoms?
3. What is unique about *Coxiella burnetii* compared to the other rickettsias?
4. Describe the symptoms of Rocky Mountain spotted fever.
5. How does transovarian passage occur?

Figure 38.3 Rocky Mountain Spotted Fever. Typical rash occurring on the arms and chest consists of generally distributed, sharply defined macules.

Dental Infections

Some microorganisms found in the oral cavity are discussed in chapter 28 and presented in figure 28.8. Of this large number, only a few bacteria can be considered true **odontopathogens,** dental pathogens. These few odontopathogens are responsible for the most common bacterial diseases in humans: tooth decay and periodontal disease.

Dental Plaque

The human tooth has a naturally occurring defense mechanism against bacterial colonization that complements the protective role of saliva. The hard enamel surface selectively absorbs acidic glycoproteins (mucins) from saliva, forming a membranous layer called the **acquired enamel pellicle.** This pellicle, or organic covering, contains many sulfate (SO_4^{2-}) and carboxylate (—COO^-) groups that confer a net negative charge to the tooth surface. Because most bacteria also have a net negative charge, there is a natural repulsion between the tooth surface and bacteria in the oral cavity. Unfortunately this natural defense mechanism breaks down when dental plaque formation occurs.

Dental plaque formation begins with the initial colonization of the pellicle by *Streptococcus gordonii, S. oralis,* and *S. mitis.* These bacteria selectively adhere to the pellicle by specific ionic, hydrophobic, and lectinlike interactions. Once the tooth surface is colonized, subsequent attachment of other bacteria results from a variety of specific coaggregation reactions (figure 38.4*a,b*). **Coaggregation** is the result of cell-to-cell recognition between genetically distinct bacteria. Many of these interactions are mediated by a lectin on one bacterium that interacts with a complementary carbohydrate receptor on the other bacterium.

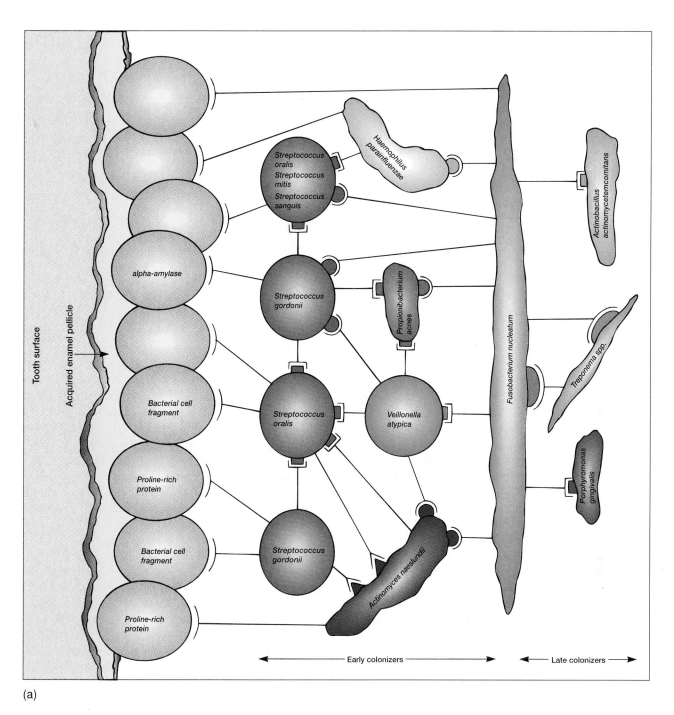

(a)

Figure 38.4   The Formation of Dental Plaque on a Freshly Cleaned Tooth Surface. (*a*) Diagrammatic representation of the proposed temporal relationship of bacterial accumulation and multigeneric coaggregation during the formation of dental plaque on the acquired enamel pellicle. Early colonizers of the tooth surface coaggregate with each other, and late colonizers of the tooth surface coaggregate with each other, but with few exceptions, early colonizers do not recognize late colonizers. After the tooth surface is covered with the earliest colonizers, each newly added bacterium becomes a new surface for recognition by unattached bacteria. (*b*) Dental plaque consisting of bacteria plus polysaccharides such as glucan is shown attached to the enamel surface of a tooth; transmission electron micrograph (×13,600). (*c*) The enzyme glucosyltransferase or dextransucrase (produced by the oral bacteria) cause the assembly of glucose units from sucrose into glucans, and fructose is released. Sucrose, glucose, and fructose molecules also are metabolized by the oral bacteria to produce lactate and other acids. Lactate is responsible for dental caries.

(b)

(c)

The most important species at this stage are *Actinomyces viscosus, A. naeslundii,* and *S. gordonii.* After these species colonize the pellicle, a microenvironment is created that allows *Streptococcus mutans* and *S. sobrinus* to become established on the tooth surface by attaching to these initial colonizers.

These streptococci produce extracellular enzymes (glucosyltransferases) that polymerize the glucose moiety of sucrose into a heterogeneous group of extracellular water-soluble and water-insoluble glucan polymers and other polysaccharides. The fructose by-product can be used in fermentation. **Glucans** are branched-chain polysaccharides composed of glucose units held together by $\alpha(1{\to}6)$ and $\alpha(1{\to}3)$ linkages (figure 38.4c). They act like a cement to bind bacterial cells together, forming a plaque ecosystem. (Dental plaque is one of the most dense collections of bacteria in the body—and perhaps the source of the first human microorganisms to be seen under a microscope by Anton van Leeuwenhoek in the seventeenth century.) Once plaque becomes established, a low oxidation reduction potential is created on the surface of the tooth. This leads to the growth of strict anaerobic bacteria (*Bacteroides melaninogenicus, B. oralis,* and *Veillonella alcalescens*), especially between opposing teeth and the dental-gingival crevices.

After the microbial plaque ecosystem develops, bacteria produce lactic and possibly acetic acids from sucrose and other sugars. Because plaque is not permeable to saliva, the acids are not diluted or neutralized, and they demineralize the enamel to produce a lesion on the tooth. It is this chemical lesion that initiates dental decay.

Dental Decay (Caries)

As described previously, a histologically undetectable chemical lesion caused by the diffusion into the tooth's enamel of undissociated fermentation acids initiates the decay process. Once these acids move below the enamel surface, they dissociate and react with the hydroxyapatite of the enamel to form soluble calcium and phosphate ions. As the ions diffuse outward, some reprecipitate as calcium phosphate salts in the tooth's surface layer to create a histologically sound outer layer overlying a porous subsurface area. Between meals and snacks, the pH returns to neutrality and some calcium phosphate reenters the lesion and crystallizes. The result is a demineralization-remineralization cycle.

When fermentable foods high in sucrose are eaten for prolonged periods, acid production overwhelms the repair process and demineralization is greater than remineralization. This leads to cavitation, otherwise known as dental decay or **caries** [Latin, rottenness]. Once the hard enamel has been breached, bacteria can invade the dentin and pulp of the tooth and cause its death.

No drugs are available to prevent dental caries. The main strategies for prevention include minimal ingestion of sucrose; daily brushing, flossing, and mouthwashes; and professional cleaning at least twice a year to remove plaque. The use of fluorides in toothpaste, drinking water, mouthwashes, or professionally applied to the teeth protects against lactic and acetic acids and reduces tooth decay.

Periodontal Disease

Periodontal disease refers to a diverse group of diseases that affect the periodontium. The **periodontium** is the supporting structure of a tooth and includes the cementum, the periodontal membrane, the bones of the jaw, and the gingivae (gums). The gingiva is the tissue, covered by a mucous membrane, that

Figure 38.5   Periodontal Disease. Notice the plaque on the teeth (arrow), especially at the gum margins, and the inflamed gums.

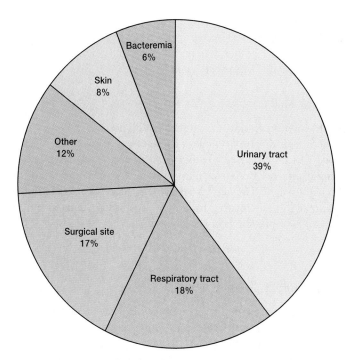

Figure 38.6   Nosocomial Infections. Relative frequency by body site. These data are from the National Nosocomial Infections Surveillance, which is conducted by the Centers for Disease Control and Prevention (CDC).

surrounds the necks of the teeth and covers the jaws. Disease is initiated by the formation of **subgingival plaque,** the plaque that forms at the dentogingival margin and extends down into the gingival tissue. Recent evidence shows that *Porphyromonas* (formerly *Bacteroides*) *gingivalis* is the main species in this plaque and is responsible for the breakdown of tissue through the production of several proteases. The result is an initial inflammatory reaction known as **periodontitis,** which is caused by the host's immune response to both the plaque bacteria and the tissue destruction. This leads to swelling of the tissue and the formation of periodontal pockets. Bacteria colonize these pockets and cause more inflammation, which leads to the formation of a periodontal abscess, bone destruction or **periodontosis,** inflammation of the gingiva or **gingivitis,** and general tissue necrosis (figure 38.5). If the condition is not treated, the tooth may fall out of its socket.

Periodontal disease can be controlled by frequent plaque removal; by brushing, flossing, and mouthwashes; and at times, by oral surgery of the gums and antibiotics.

1. Name some common odontopathogens that are responsible for dental caries, dental plaque, and periodontal disease. Be specific.
2. What is the function of the acquired enamel pellicle?
3. How does plaque formation occur? Dental decay?
4. Describe some pathological manifestations of periodontal disease.
5. How can caries and periodontal diseases be prevented?

Nosocomial Infections

The epidemiology of **nosocomial infections** is presented in chapter 35 (*pp. 711–12*). What follows is a continuation of that

discussion by means of a brief examination of the microorganisms and agents responsible for these infections. The Centers for Disease Control and Prevention estimate that 5 to 10% of all hospital patients acquire some type of nosocomial infection. Because approximately 40 million people are admitted to hospitals annually, about 2 to 4 million people may develop an infection they did not have upon entering the hospital. Thus nosocomial infections represent a significant proportion of all infectious diseases acquired by humans.

Nosocomial diseases are usually caused by bacteria, most of which are noninvasive and part of the normal microbiota; viruses, protozoa, and fungi are rarely involved. Figure 38.6 summarizes the most common types of nosocomial infections. The specific bacteria associated with the various body sites and disease processes follow.

Bacteremia

Bacteremia, the transient presence of bacteria in the blood, accounts for about 6% of all nosocomial infections. Two types of bacteremias of **bloodstream infections (BSIs)** exist: primary and secondary. A primary BSI results from the accidental introduction of bacteria directly into the body by way of intravenous infusion, transducers, respiratory devices, prostheses, catheters, endoscopes, hemodialysis units, or parenteral nutrition. A secondary BSI results from an infection at another body site such as the urinary tract, respiratory tract, or surgical wounds. The microorganisms most responsible for BSIs are **coagulase-negative staphylococci (CoNS)**, *Staphylococcus aureus,* and *Escherichia coli.* BSIs occur primarily in neonates and the elderly, whose natural defense mechanisms are compromised.

Burn Wounds

Burn wounds account for about 8% of all nosocomial infections. A burn wound is tissue damage caused by heat, chemicals, radiation, or electricity. Because the skin is no longer an effective barrier to microorganisms, many now are able to colonize the burned tissue. *Pseudomonas aeruginosa, Staphylococcus aureus,* and many gram-negative bacilli are most often acquired by the burn victim from the hospital environment.

Physical barriers, the skin (p. 573).

Respiratory Tract Infections

The lower respiratory tract ranks second in incidence (about 18%) of nosocomial infections. More patients die of nosocomial pneumonias than of any other nosocomial disease. Most of these pneumonias are related to respiratory devices that aid breathing or administer medications. The microorganisms most responsible are *S. aureus, Pseudomonas aeruginosa,* and *Enterobacter* spp.

Surgical Site Infections

Surgical site infections (SSIs) rank third in incidence (about 17%) of nosocomial infections. It is estimated that 5 to 12% of all surgical patients develop postoperative infections. In operations involving the gastrointestinal, respiratory, and genitourinary tracts, the infection rate approaches 30%. The main reason is that these sites harbor large numbers of microorganisms, and once displaced, they can colonize the surgical wound. The bacteria most frequently isolated from surgical wounds are *S. aureus,* coagulase-negative staphylococci, and *Enterococcus* spp.

Urinary Tract Infections

Urinary tract infections (UTIs) are the most common nosocomial infections. They usually account for about 39% of all cases and are typically related to urinary catheterization. The bacteria most frequently associated with these infections are *E. coli, Enterococcus* spp., and *P. aeruginosa.*

Miscellaneous Infections

In adults, cutaneous infections are among the least common of all nosocomial infections. However, newborns are particularly susceptible to cutaneous and eye infections. As a result most nurseries have an infection rate three times greater than any other hospital area. Cutaneous infections result primarily from *S. aureus,* coagulase-negative staphylococci, *P. aeruginosa,* and *E. coli.*

Another group of individuals particularly susceptible to nosocomial infections are those who are immunosuppressed: cancer and AIDS patients, and organ transplant patients receiving immunosuppressive drug therapy.

1. Of what significance are nosocomial infections compared to infectious diseases as a whole?
2. What are two types of bacteremias? Give an example of how each occurs.
3. What are the most frequent pathogens of burn wounds?
4. What causes most nosocomial pneumonias?
5. Why are surgical wound infections so common?
6. Why are urinary tract infections so common in hospitals?

Summary

1. The chlamydiae are obligate intracellular parasites of eucaryotic cells. Three species cause human disease: *Chlamydia trachomatis, C. pneumoniae,* and *C. psittaci. C. trachomatis* causes inclusion conjunctivitis, lymphogranuloma venereum (LGV), nongonococcal urethritis (NGU), and trachoma. All of these diseases are spread by person-to-person contact. Psittacosis (ornithosis) is an infectious disease of birds that is transmitted to humans. It is caused by *C. psittaci. C. pneumoniae* causes chlamydial pneumonia.

2. The mycoplasmas are procaryotes without cell walls and are the smallest independently living organisms. Two mycoplasmas, *Ureaplasma urealyticum* and *Mycoplasma hominis,* are common pathogens of the genitourinary tract and spread by sexual contact. They cause limited pathology in humans. Mycoplasmal pneumonia, or primary atypical pneumonia, is caused by *Mycoplasma pneumoniae.* Its spread involves airborne droplets and close contact. The lower respiratory tract is usually colonized where mucosal damage is apparent, and this leads to pneumonia.

3. The rickettsias are fastidious bacteria that are obligate intracellular parasites. They are maintained in the environment through a cycle involving mammalian reservoirs and insect vectors. Humans are only occasional, incidental hosts. Ehrlichiosis is caused by *Ehrlichia chaffeensis*. It is spread from animals to humans by ticks. Epidemic (louse-borne) typhus is caused by *R. prowazekii*. It is spread from person to person by the body louse. Murine (endemic) typhus is caused by *R. typhi*. It is spread from rats to humans by rat fleas. Q fever is an acute zoonotic disease caused by *Coxiella burnetii*. Humans inhale the rickettsias from contaminated animal products. This leads to an atypical pneumonia and other problems. Rocky Mountain spotted fever is caused by *R. rickettsii*. It is transmitted to humans by ticks and causes a characteristic skin rash.

4. Dental plaque formation begins on a tooth with the initial colonization of the acquired enamel pellicle by *Streptococcus gordonii*, *S. oralis*, and *S. mitis*. Other bacteria then become attached and form a plaque ecosystem. The bacteria produce acids that cause a chemical lesion on the tooth and initiate dental decay or caries. Periodontal disease is a group of diverse clinical entities that affect the periodontium. Disease is initiated by the formation of subgingival plaque, which leads to tissue inflammation known as periodontitis and to periodontal pockets. Bacteria that colonize these pockets can cause an abscess, periodontosis, gingivitis, and general tissue necrosis.

5. Nosocomial infections are a significant proportion of all infectious diseases acquired by humans. Examples include primary and secondary bacteremias, burn wound infections, respiratory tract infections, surgical wound infections, and urinary tract infections. Newborns are particularly prone to nosocomial cutaneous and eye infections. Individuals who are immunosuppressed, such as cancer and AIDS patients, and organ transplant patients receiving immunosuppressive drug therapy also are particularly susceptible to nosocomial infections.

Key Terms

acquired enamel pellicle *782*

bacteremia *785*

caries *784*

chlamydial pneumonia *777*

coaggregation *782*

dental plaque *782*

ehrlichiosis *780*

endemic (murine) typhus *780*

epidemic (louse-borne) typhus *780*

gingivitis *785*

glucans *784*

inclusion conjunctivitis *777*

lymphogranuloma venereum (LGV) *777*

mycoplasmal pneumonia *779*

nongonococcal urethritis (NGU) *778*

nosocomial infections *785*

odontopathogens *782*

pannus *779*

pelvic inflammatory disease (PID) *778*

periodontal disease *784*

periodontitis *785*

periodontium *784*

periodontosis *785*

psittacosis (ornithosis) *778*

Q fever *781*

Rocky Mountain spotted fever *782*

subgingival plaque *785*

trachoma *778*

transovarian passage *782*

vasculitis *780*

Weil-Felix reaction *781*

Questions for Thought and Review

1. Briefly describe each of the major or most common diseases covered in this chapter in terms of its causative agent, signs and symptoms, the course of infection, mechanism of pathogenesis, epidemiology, and prevention and/or treatment.

2. What specific bacteria are responsible for most nosocomial infections?

3. Why does the use of specific chemotherapeutic agents often give rise to nosocomial infections?

4. What is the major difference between trachoma and inclusion conjunctivitis?

5. Why is Rocky Mountain spotted fever the most important rickettsial disease in the United States?

6. As a group of diseases, what are some of the common characteristic symptoms and pathological results of rickettsial disease?

7. Why is it difficult to definitely isolate the microorganism responsible for nongonococcal urethritis?

8. Why are dental diseases relatively new infectious diseases of humans?

9. What makes Q fever a unique rickettsial disease?

10. What specific chlamydial and mycoplasmal diseases are transmitted sexually?

Additional Reading

General

Mandell, G. L.; Bennett, J. E.; and Dolan, R. 1995. *Principles and practice of infectious diseases,* 4th ed. New York: Churchill Livingstone.

Zinsser, H., 1935. *Rats, lice, and history.* Boston: Little, Brown.

Chlamydial Diseases

Barnes, R. C. 1989. Laboratory diagnosis of human chlamydial infections. *Clin. Microbiol. Rev.* 2(2):119–36.

Beatty, W. L.; Morrison, R. P.; and Byrne, G. I. 1994. Persistent chlamydiae: From cell culture to a paradigm for chlamydial pathogenesis. *Microbiol. Rev.* 58(4):686–99.

Moulder, J. W. 1991. Interaction of chlamydiae and host cells in vitro. *Microbiol. Rev.* 55(1):143–90.

Schachter, J., and Caldwell, H. 1980. Chlamydiae. *Ann. Rev. Microbiol.* 34:285–90.

Mycoplasmal Diseases

Cassell, C. H., and Cole, B. C. 1981. Mycoplasmas as agents of human disease. *N. Engl. J. Med.* 304:80–85.

Inderlied, C. B.; Kemper, C. A.; and Bermudez, L. E. M. 1993. The *Mycobacterium avium* complex. *Clin. Microbiol. Rev.* 6(3):266–310.

Maniloff, J., editor. 1992. *Mycoplasmas: Molecular biology and pathogenesis.* Washington, D.C.: American Society for Microbiology.

Wayne, L. G., and Sramek, H. A. 1992. Agents of newly recognized or infrequently encountered mycobacterial diseases. *Clin. Microbiol. Rev.* 5(1):1–25.

Rickettsial Diseases (Rickettsioses)

Baca, O. G., and Pretsky, D. 1983. Q fever and *Coxiella burnetti:* A model for host-parasite interactions. *Microbiol. Rev.* 47:127–44.

Burgdorfer, W., and Anacker, R. L., editors. 1981. *Rickettsiae and rickettsial disease.* New York: Academic Press.

Reimer, L. G. 1993. Q fever. *Clin. Microbiol. Rev.* 6(3):193–98.

Spach, D. *et al.* 1993. Tick-borne diseases in the United States. *N. Engl. J. Med.* 329(13):936–47.

Walker, D. H. 1989. Rocky Mountain spotted fever: A disease in need of microbial concern. *Clin. Microbiol. Rev.* 2(3):227–40.

Weiss, E. 1982. The biology of the rickettsiae. *Ann. Rev. Microbiol.* 32:345–70.

Dental Infections

Cutler, W. C.; Kalmar, J. R.; and Genco, C. A. 1995. Pathogenic strategies of the oral anaerobe, *Porphyromonas gingivalis. Trends Microbiol.* 3(2):45–51.

Gibbons, R. J., and van Houte, J. 1975. Bacterial adherence in oral microbial ecology. *Ann. Rev. Microbiol.* 29:19–44.

Hamada, S., and Slade, H. D. 1980. Biology, immunology, and cariogenicity of *Streptococcus mutans. Microbiol. Rev.* 44:331–46.

Jenkinson, H. F. 1994. Adherence and accumulation of oral streptcocci. *Trends Microbiol.* 2(6):209–12.

Kolenbrander, P.; Ganeshkumar, N.; Cassels, F.; and Hughes, C. 1993. Coaggregation: Specific adherence among human oral plaque bacteria. *The FASEB Journal* 7(5):406–13.

Loesche, W. J. 1986. Role of *Streptococcus mutans* in human dental decay. *Microbiol. Rev.* 50:353–71.

Marsh, P. D., and Martin, M. V. 1984. *Oral microbiology,* 2d ed. Washington, D.C.: American Society for Microbiology.

Mergenhan, S. E., and Rosan, B. 1985. *Molecular basis of oral microbial adhesion.* Washington, D.C.: American Society for Microbiology.

Nosocomial Infections

Ayliffe, G. A., and Taylor, L. J. 1982. *Hospital-acquired infections. Principles and prevention.* Littleton, Mass.: John Wright-PSG.

Benenson, A. S., editor. 1990. *Control of communicable diseases in man,* 15th ed. Washington, D.C.: American Public Health Association.

Bennett, J., and Brachman, P. S. 1986. *Hospital infections.* Boston: Little, Brown.

Bingen, E.; Denamur, E.; and Elion, J. 1994. Use of ribotyping in epidemiological surveillance of nosocomial outbreaks. *Clin. Microbiol. Rev.* 7(3):311–27.

Gurevich, I. 1989. *Infectious diseases in critical care nursing.* Rockville, Md.: Aspen.

Smith, P. W. 1985. *Infection control in long-term care facilities.* New York: John Wiley and Sons.

Swartz, M. N. 1994. Hospital-acquired infections: Diseases with increasingly limited therapies. *Proc. Natl. Acad. Sci.* 91:2420–27.

Walsh, T. J., and Pizzo, P. A. 1988. Nosocomial fungal infections: A classification for hospital-acquired fungal infections and mycoses arising from endogenous flora or reactivation. *Ann. Rev. Microbiol.* 42:517–45.

Wenzel, R. P. 1988. *Prevention and control of nosocomial infections.* Baltimore: Williams & Wilkins.

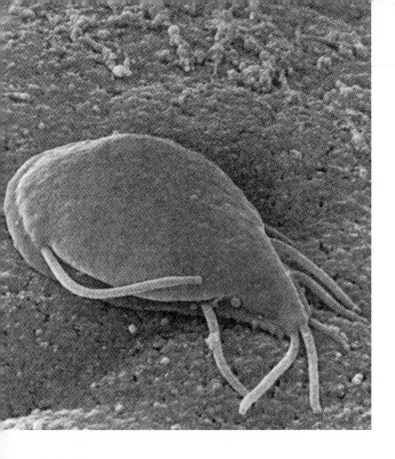

CHAPTER 39
Human Diseases Caused by Fungi and Protozoa

The ability of fungi to cause disease in humans appears to be an accidental phenomenon.
—John W. Rippon

Historians believe that malaria has probably had a greater impact on world history than any other infectious disease, influencing the outcome of wars, various population movements, and the development and decline of various civilizations.
—Lynne S. Garcia

Outline

Concepts

1. Fungal diseases (mycoses) are usually divided into five groups according to the level of infected tissue and mode of entry into the host: (1) superficial, (2) cutaneous, (3) subcutaneous, (4) systemic, and (5) opportunistic infections.

2. The superficial mycoses occur mainly in the tropics and include black piedra, white piedra, and tinea versicolor.

3. The cutaneous mycoses—that is, those of the outer layer of the skin—are generally called ringworms, tineas, or dermatophytoses. These diseases occur worldwide and represent the most common fungal diseases in humans.

4. The dermatophytes that cause the subcutaneous—below the skin—mycoses are normal saprophytic inhabitants of the soil. They must be introduced into the body beneath the cutaneous layer. Examples of these diseases include chromomycosis, maduromycosis, and sporotrichosis.

5. The systemic mycoses are the most serious of the fungal infections in the normal host because the responsible fungi can disseminate throughout the body. Examples include blastomycosis, coccidioidomycosis, cryptococcosis, and histoplasmosis.

6. The opportunistic mycoses can create life-threatening situations in the compromised host. Examples of these diseases include aspergillosis, candidiasis, and *Pneumocystis carinii* pneumonia.

7. About 20 different protozoa cause human diseases that afflict hundreds of millions of people throughout the world. Examples include amebiasis, giardiasis, malaria, the hemoflagellate diseases, toxoplasmosis, and trichomoniasis.

8. Certain fungal and protozoan diseases are increasing in incidence because of organ transplants, immunosuppressive drugs, and AIDS.

The purpose of this chapter is to describe some of the fungal and protozoan microorganisms that are pathogenic to humans and to discuss the clinical manifestations, diagnosis, epidemiology, pathogenesis, and treatment of the diseases caused by them.

Besides the viruses and bacteria, two other major groups of microorganisms cause infectious diseases in humans: fungi and protozoa. The biology of these organisms is covered in chapters 25 and 27, respectively. This chapter emphasizes the diseases they cause.

Fungal Diseases

Although hundreds of thousands of fungal species are found in the environment, only about 50 can produce disease in humans. **Medical mycology** is the discipline that deals with the fungi that causes human disease. These fungal diseases, known as **mycoses** [s., mycosis; Greek *mykes,* fungus], are divided into

five groups according to the type of infected tissue in the host: superficial, cutaneous, subcutaneous, systemic, and opportunistic mycoses (table 39.1).

Superficial Mycoses

The superficial mycoses are extremely rare in the United States, and most occur in the tropics. The fungi responsible are limited to the outer surface of hair and skin and hence are called superficial. Infections of the hair shaft are collectively called **piedras** (Spanish for stone because they are associated with the hard nodules formed by mycelia [*see figure 25.4*] on the hair shaft). For example, **black piedra** is caused by *Piedraia hortae* and forms hard black nodules on the hairs of the scalp (figure 39.1). **White piedra** is caused by the yeast *Trichosporon beigelii* and forms light-colored nodules on the beard and mustache. Some superficial mycoses are called **tineas** [Latin for grub, larva, worm], the specific type being designated by a modifying term. Tineas are superficial fungal infections involving the outer layers of the skin, nails, and hair. **Tinea versicolor** is caused by the yeast *Malassezia furfur* and forms brownish-red scales on the skin of the trunk, neck, face,

TABLE 39.1   Examples of Medically Important Fungi

Group	Pathogen	Location	Disease
Superficial mycoses	*Piedraia hortae*	Scalp	Black piedra
	Trichosporon beigelii	Beard, mustache	White piedra
	Malassezia furfur	Trunk, neck, face, arms	Tinea versicolor
Cutaneous mycoses	*Trichophyton mentagrophytes, T. verrucosum*	Beard hair	Tinea barbae
	Trichophyton, Microsporum	Scalp hair	Tinea capitis
	Trichophyton rubrum, T. mentagrophytes, Microsporum canis	Smooth or bare parts of the skin	Tinea corporis
	Epidermophyton floccosum, T. mentagrophytes, T. rubrum	Groin, buttocks	Tinea cruris (jock itch)
	T. rubrum, T. mentagrophytes, E. floccosum	Feet	Tinea pedis (athlete's foot)
	T. rubrum, T. mentagrophytes	Nails	Tinea unguium (onychomycosis)
Subcutaneous mycoses	*Phialophora verrucosa, Fonsecaea pedrosoi*	Legs, feet	Chromoblastomycosis
	Madurella mycetomatis	Feet, other areas of body	Maduromycosis
	Sporothrix schenckii	Puncture wounds	Sporotrichosis
Systemic mycoses	*Blastomyces dermatitidis*	Lungs, skin	Blastomycosis
	Coccidioides immitis	Lungs, other parts of body	Coccidioidomycosis
	Cryptococcus neoformans	Lungs, skin, bones, viscera, central nervous system	Cryptococcosis
	Histoplasma capsulatum	Within phagocytes	Histoplasmosis
Opportunistic mycoses	*Aspergillus fumigatus, A. flavus*	Respiratory system	Aspergillosis
	Candida albicans	Skin or mucous membranes	Candidiasis
	Pneumocystis carinii	Lungs, sometimes brain	*Pneumocystis* pneumonia

Figure 39.1   Superficial Mycosis: Black Piedra. Hair shaft infected with *Piedraia hortae;* light micrograph (×200).

and arms. Treatment involves removal of the skin scales with a cleansing agent and removal of the infected hairs. Good personal hygiene prevents these infections.

Cutaneous Mycoses

Cutaneous mycoses—also called **dermatophytoses** [Greek *derma,* skin, and *phyto,* plant], **ringworms,** or tineas—occur worldwide and represent the most common fungal diseases in humans. Three genera of cutaneous fungi, or **dermatophytes,** are involved in these mycoses: *Epidermophyton, Microsporum,* and *Trichophyton.* Diagnosis is by microscopic examination of biopsied areas of the skin cleared with 10% potassium hydroxide and by culture on Sabouraud dextrose agar. Treatment is with topical ointments such as miconazole (Monistatderm), tolnaftate (Tinactin), or clotrimazole (Lotrimin) for 2 to 4 weeks. Griseofulvin and itraconazole (Sporanox) are the only oral fungal agents currently approved by the FDA for treating dermatophytoses.

The mode of action of ketoconazole, miconazole, tolnaftate, and griseofulvin (p. 671).

Tinea barbae [Latin *barba,* the beard] is an infection of the beard hair (figure 39.2) caused by *Trichophyton mentagrophytes* or *T. verrucosum.* It is predominantly a disease of men who live in rural areas and acquire the fungus from infected animals.

Tinea capitis [Latin *capita,* the head] is an infection of the scalp hair (figure 39.3*a*). It is characterized by loss of hair, inflammation, and scaling. Tinea capitis is primarily a childhood disease caused by *Trichophyton* or *Microsporum* species. Person-to-person transmission of the fungus occurs frequently when poor hygiene and overcrowded conditions exist. The fungus also occurs in domestic animals, from whom it can be transmitted to humans. A Wood's light (a UV light) can help with the diagnosis of tinea capitis because fungus-infected hair fluoresces when illuminated by UV radiation (figure 39.3*b*).

Figure 39.2   Cutaneous Mycosis: Tinea Barbae. Ringworm of the beard caused by *Trichophyton mentagrophytes.*

(a)

(b)

Figure 39.3   Cutaneous Mycosis: Tinea Capitis. (*a*) Ringworm of the head caused by *Microsporum audouinii.* (*b*) Close-up utilizing a Wood's light.

Figure 39.4   Cutaneous Mycosis: Tinea Corporis. Ringworm of the body—in this case, the forearm—caused by *Trichophyton mentagrophytes*. Notice the circular patches (arrows).

Figure 39.5   Cutaneous Mycosis: Tinea Cruris. Ringworm of the groin caused by *Epidermophyton floccosum*.

Figure 39.6   Cutaneous Mycosis: Tinea Pedis. Ringworm of the foot caused by *Trichophyton rubrum, T. mentagrophytes,* or *Epidermophyton floccosum.*

Tinea corporis [Latin *corpus,* the body] is a dermatophytic infection of the smooth or bare parts of the skin (figure 39.4). The disease is characterized by circular, red, well-demarcated, scaly, vesiculopustular lesions accompanied by itching. Tinea corporis is caused by *Trichophyton rubrum, T. mentagrophytes,* or *Microsporum canis.* Transmission of the disease is by direct contact with infected animals (*see figure 25.1*b) or humans or by indirect contact through fomites.

Tinea cruris [Latin *crura,* the leg] is a dermatophytic infection of the groin (figure 39.5). The pathogenesis and clinical manifestations are similar to those of tinea corporis. The responsible fungi are *Epidermophyton floccosum, T. mentagrophytes,* or *T. rubrum.* Factors predisposing one to recurrent disease are moisture, occlusion, and skin trauma. Wet bathing suits, athletic supporters (**jock itch**), tight-fitting slacks, panty hose, and obesity are frequently contributing factors.

Tinea pedis [Latin *pes,* the foot], also known as **athlete's foot,** and **tinea manuum** [Latin *mannus,* the hand] are dermatophytic infections of the feet (figure 39.6) and hands, respectively. Clinical symptoms vary from a fine scale to a vesiculopustular eruption. Itching is frequently present. Warmth, humidity, trauma, and occlusion increase susceptibility to infection. Most infections are caused by *T. rubrum, T. mentagrophytes,* or *E. floccosum.* Tinea pedis and tinea manuum occur throughout the world, are most commonly found in adults, and increase in frequency with age.

Tinea unguium [Latin *unguis,* nail] is a dermatophytic infection of the nail bed (figure 39.7). In this disease the nail becomes discolored and then thickens. The nail plate rises and separates from the nail bed. *Trichophyton rubrum* or *T. mentagrophytes* are the causative fungi.

Subcutaneous Mycoses

The dermatophytes that cause subcutaneous mycoses are normal saprophytic inhabitants of soil and decaying vegetation. Because they are unable to penetrate the skin, they must be introduced into the subcutaneous tissue by a puncture wound that has been contaminated with soil containing the fungi. Most infections involve barefooted agricultural workers. Once in the subcutaneous tissue, the disease develops slowly—often over a period of years. During this time the fungi produce a nodule that eventually ulcerates and the organisms spread along lymphatic

Figure 39.7 **Cutaneous Mycosis: Tinea Unguium.** Ringworm of the nails caused by *Trichophyton rubrum.*

Figure 39.8 **Subcutaneous Mycosis.** Chromoblastomycosis of the foot caused by *Fonsecaea pedrosoi.*

channels producing more subcutaneous nodules. At times such nodules drain to the skin surface. The administration of oral 5-fluorocytosine, iodides, amphotericin B, and surgical excision are the usual treatments. Diagnosis is accomplished by culture of the infected tissue.

The mode of action of amphotericin B (p. 671).

One type of subcutaneous mycosis is **chromoblastomycosis.** The nodules are pigmented a dark brown. This disease is caused by the black molds *Phialophora verrucosa* or *Fonsecaea pedrosoi.* These fungi exist worldwide, especially in tropical and subtropical regions. Most infections involve the legs and feet (figure 39.8).

Another subcutaneous mycosis is **maduromycosis,** caused by *Madurella mycetomatis,* which is distributed worldwide and is especially prevalent in the tropics. Because the fungus destroys subcutaneous tissue and produces serious deformities, the resulting infection is often called a **eumycotic mycetoma** or fungal tumor (figure 39.9). One form of mycetoma, known as Madura foot, occurs through skin abrasions acquired while walking barefoot on contaminated soil.

Sporotrichosis is the subcutaneous mycosis caused by the dimorphic (*see table 25.2*) fungus *Sporothrix schenckii.* The disease occurs throughout the world and is the most common subcutaneous mycotic disease in the United States. The fungus can be found in the soil, on living plants, such as barberry shrubs and roses, or in plant debris, such as sphagnum moss and pine-bark mulch. Infection occurs by a puncture wound from a thorn or splinter contaminated with the fungus. The disease is an occupational hazard to florists, gardeners, and forestry workers. After an incubation period of 1 to 12 weeks, a small red papule arises and begins to ulcerate (figure 39.10). New lesions appear along lymph channels and can remain localized or spread throughout the body, producing **extracutaneous sporotrichosis.**

Figure 39.9 **Subcutaneous Mycosis.** Eumycotic mycetoma of the foot caused by *Madurella mycetomatis.*

Figure 39.10 **Subcutaneous Mycosis.** Sporotrichosis of the arm caused by *Sporothrix schenckii.*

Systemic Mycoses

Except for *Cryptococcus neoformans,* which has only a yeast form, the fungi that cause the systemic or deep mycoses are dimorphic—that is, they exhibit a parasitic yeastlike phase (Y) and a saprophytic mold or mycelial phase (M). *(See the YM shift, chapter 25.)* Most systemic mycoses are acquired by the inhalation of spores from soil in which free-living fungi reside. If a person inhales enough spores, an infection begins as a lung lesion, becomes chronic, and spreads through the bloodstream to other organs (the target organ varies with the species).

Blastomycosis is the systemic mycosis caused by *Blastomyces dermatitidis,* a fungus that grows as a budding yeast in humans but as a mold on culture media and in the environment. It is found predominately in the soil of the Mississippi and Ohio River basins. The disease occurs in three clinical forms: cutaneous, pulmonary, and disseminated. The initial infection begins when blastospores *(see figure 25.7f)* are inhaled into the lungs. The fungus can then spread rapidly, especially to the skin, where cutaneous ulcers and abscess formation occur (figure 39.11). *B. dermatitidis* can be isolated from pus and biopsy sections. Diagnosis requires the demonstration of thick-walled, yeastlike cells 8 to 15 μm in diameter. Complement-fixation, immunodiffusion, and skin (blastomycin) tests are also useful. Amphotericin B, itraconazole, 2-hydroxystilbamidine, or imidazole derivatives are the drugs of choice for treatment. Surgery may be necessary for the drainage of large abscesses. Approximately 30 to 60 deaths are reported each year in the United States. There are no preventive or control measures.

Coccidioidomycosis, also known as valley fever, San Joaquin fever, or desert rheumatism because of the geographical distribution of the fungus, is caused by *Coccidioides immitis. C. immitis* exists in the dry, highly alkaline soils of North, Central, and South America. It has been estimated that in the United States about 100,000 people are infected annually with 50 to 100 deaths. Endemic areas have been defined by massive skin testing with the antigen coccidioidin *(see chapter 32).* In the soil and on culture media, this fungus grows as a mold that forms arthroconidia *(see figure 25.7b)* at the tips of hyphae. Arthroconidia are so abundant in these endemic areas that by simply moving through such an area, one can acquire the disease by inhaling them. In humans the fungus grows as a yeast-forming, thick-walled spherule filled with endospores (figure 39.12). Most cases of coccidioidomycosis are asymptomatic or indistinguishable from ordinary upper respiratory infections. Almost all cases resolve themselves in a few weeks, and a lasting immunity results. A few infections result in a progressive chronic pulmonary disease. The fungus also can spread throughout the body, involving almost any organ or site. Diagnosis is accomplished by aspiration and identification of the large spherules (approximately 80 μm in diameter) in pus, sputum, and aspirates. Culturing clinical samples in the presence of penicillin and streptomycin on Sabouraud agar also is diagnostic. Newer methods of rapid confirmation include the testing of supernatants of liquid media cultures for antigens, serology, and skin testing.

Figure 39.11   Systemic Mycosis. Blastomycosis of the forearm caused by *Blastomyces dermatitidis.*

Spherules

Figure 39.12   Systemic Mycosis: Coccidioidomycosis. *Coccidioides immitis* mature spherules filled with endospores within a tissue section; light micrograph (×400).

Miconazole, itraconazole, ketoconazole, and amphotericin B are the drugs of choice for treatment. Prevention involves reducing exposure to dust (soil) in endemic areas.

Cryptococcosis is a systemic mycosis caused by *Cryptococcus neoformans.* This fungus always grows as a large budding yeast. In the environment *C. neoformans* is a saprophyte with a worldwide distribution. Aged, dried pigeon droppings are an apparent source of infection. Cryptococcosis is found in approximately 15% of AIDS patients. The fungus enters the body by the respiratory tract, causing a minor pulmonary infection that is usually transitory. Some pulmonary infections spread to the skin, bones, viscera, and central nervous system. Once the nervous system is involved, meningitis usually results *(see table*

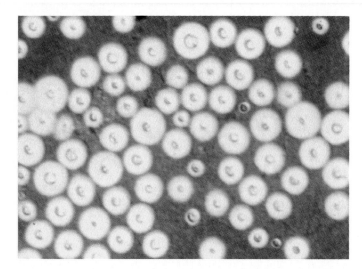

Figure 39.13 Systemic Mycosis: Cryptococcosis. India ink preparation showing *Cryptococcus neoformans.* Although these microorganisms are not budding, they can be differentiated from artifacts by their doubly refractile cell walls, distinctly outlined capsules, and refractile inclusions in the cytoplasm; light micrograph (×150).

(a)

(b)

Figure 39.14   Morphology of *Histoplasma capsulatum* var. *capsulatum.* (*a*) Mycelia, microconidia, and chlamydospores as found in the soil. These are the infectious particles; light micrograph (×125). (*b*) Yeastlike cells in a macrophage. Budding *H. capsulatum* within a vacuole. Tubular structures, *ts*, are observed beneath the cell wall, cw; electron micrograph (×23,000).

*37.1 and figure 36.11*d). Diagnosis is accomplished by detection of the thick-walled spherical yeast cells (figure 39.13) in pus, sputum, or exudate smears using India ink to define the organism. The fungus can be easily cultured on Sabouraud dextrose agar. Identification of the fungus in body fluids is made by immunologic procedures. Treatment includes either a combination of amphotericin B and fluorocytosine, or itraconazole (Sporanox). There are no preventive or control measures.

Histoplasmosis is caused by *Histoplasma capsulatum* var. *capsulatum,* a facultative fungus parasite that grows intracellularly. It appears as a small budding yeast in humans and on culture media at 37°C. At 25°C it grows as a mold, producing small microconidia (1 to 5 μm in diameter) that are borne singly at the tips of short conidiophores. Large macroconidia or chlamydospores (8 to 16 μm in diameter) are also formed on conidiophores (figure 39.14*a; see also figure 36.11*e). In humans the yeastlike form grows within phagocytic cells (figure 39.14*b*). *H. capsulatum* var. *capsulatum* is found as the mycelial form in soils throughout the world and is localized in areas that have been contaminated with bird or bat excrement. The chlamydospores, particularly the microconidia, are easily spread by air currents. Within the United States, histoplasmosis is endemic within the Mississippi, Kentucky, Tennessee, Ohio, and Rio Grande River basins. More than 75% of the people who reside in parts of these areas have antibodies against the fungus. It has been estimated that in endemic areas of the United States, about 500,000 individuals are infected annually: 50,000 to 200,000 become ill; 3,000 require hospitalization; and about 50 die. The total number of infected individuals may be over 40 million in the United States alone. Histoplasmosis is an occupational disease among spelunkers (people who explore caves) and bat guano miners.

Humans acquire histoplasmosis from airborne microconidia that are produced under favorable environmental conditions. Microconidia are most prevalent where bird droppings, or guano—especially from starlings, crows, blackbirds, cowbirds, sea gulls, turkeys, and chickens—have accumulated. It is noteworthy that the birds themselves are not infected because of their high body temperature; their droppings simply provide the nutrients for this fungus. Only bats and humans demonstrate the disease and harbor the fungus.

Histoplasmosis is a disease of the reticuloendothelial system; thus many organs of the body can be infected (*see figure 29.10*). More than 95% of "histo" cases have either no symptoms or mild symptoms such as coughing, fever, and joint pain. Lesions may appear in the lungs and show calcification; most infections resolve on their own. Only rarely does the disease disseminate.

Laboratory diagnosis is accomplished by complement-fixation tests and isolation of the fungus from tissue specimens. Most individuals with this disease exhibit a hypersensitive state that can be demonstrated by the histoplasmin skin test. Currently the most effective treatment is with amphotericin B, ketoconazole, or itraconazole. Prevention and control involves wearing protective clothing and masks before entering or working in infested habitats. Soil decontamination with 3 to 5% formalin is effective where economically and physically feasible.

Opportunistic Mycoses

An **opportunistic organism** is generally harmless in its normal environment but becomes pathogenic in a compromised host. A **compromised host** is seriously debilitated and has a lowered resistance to infection. There are many causes of this condition, among them the following: malnutrition, alcoholism, cancer, diabetes, leukemia, or another infectious disease; trauma from surgery or injury; an altered microbiota from the prolonged use of antibiotics (e.g., in vaginal candidiasis); and immunosuppression by drugs, viruses (HIV), hormones, or genetic deficiencies. The most important opportunistic mycoses include systemic aspergillosis, candidiasis, and *Pneumocystis carinii* pneumonia.

Of all the fungi that cause disease in compromised hosts, none are as widely distributed as the *Aspergillus* species. *Aspergillus* is omnipresent in nature, being found wherever organic debris occurs. *Aspergillus fumigatus* is the usual cause of **aspergillosis.** *A. flavus* is the second most important species, particularly in invasive disease of immunosuppressed patients.

The major portal of entry for *Aspergillus* is the respiratory tract. Inhalation of conidiospores (*see figure 25.7e*) can lead to several types of pulmonary aspergillosis. One type is allergic aspergillosis. Infected individuals may develop an immediate allergic response and suffer typical asthmatic attacks when exposed to fungal antigens on the conidiospores. In bronchopulmonary aspergillosis the major clinical manifestation of the allergic response is a bronchitis resulting from both type I and type III hypersensitivities (*see figures 31.10 and 31.13*). Although tissue invasion seldom occurs in bronchopulmonary aspergillosis, *Aspergillus* often can be cultured from the sputum. A most common manifestation of pulmonary involvement is the occurrence of colonizing aspergillosis, in which *Aspergillus* forms colonies within the lungs that develop into "fungus balls" called aspergillomas. These consist of a tangled mass of mycelia growing in a circumscribed area. From the pulmonary focus, the fungus may spread, producing disseminated aspergillosis in a variety of tissues and organs (figure 39.15). In patients whose resistance has been severely compromised, invasive aspergillosis may occur and fill the lung with fungal mycelia.

Laboratory diagnosis of aspergillosis depends on identification, either by direct examination of pathological specimens or by isolation and characterization of the fungus. Successful therapy depends on treatment of the underlying disease so that host resistance increases. Treatment is with itraconazole.

Figure 39.15 An Opportunistic Mycosis. Aspergillosis of the eye caused by *Aspergillus fumigatus*.

Candidiasis is the mycosis caused by the dimorphic fungus *Candida albicans* (figure 39.16*a*). In contrast to the other pathogenic fungi, *C. albicans* is a member of the normal microbiota within the gastrointestinal tract, respiratory tract, vaginal area, and mouth (*see figures 28.8 and 36.11*f). In healthy individuals *C. albicans* does not produce disease. Growth is suppressed by other microbiota. However, if anything upsets the normal microbiota, *Candida* may multiply rapidly and produce candidiasis. Recently *Candida* species have become important nosocomial pathogens. In some hospitals they may represent almost 10% of nosocomial bloodstream infections.

No other mycotic pathogen produces as diverse a spectrum of disease in humans as does *C. albicans* (Box 39.1). Most infections involve the skin or mucous membranes. This occurs because *C. albicans* is a strict aerobe and finds such surfaces very suitable for growth. Cutaneous involvement usually occurs when the skin becomes overtly moist or damaged.

Oral candidiasis, or **thrush** (figure 39.16*b*), is a fairly common disease in newborns. It is seen as many small white flecks that cover the tongue and mouth. At birth, newborns do not have a normal microbiota in the oropharyngeal area. If the mother's vaginal area is heavily infected with *C. albicans,* the upper respiratory tract of the newborn becomes colonized during passage through the birth canal. Thrush occurs because growth of *C. albicans* cannot be inhibited by the other microbiota. Once the newborn has developed its own normal oropharyngeal microbiota, thrush becomes uncommon. **Paronychia** and **onychomycosis** are associated with *Candida* infections of the subcutaneous tissues of the digits and nails, respectively (figure 39.16*c*). These infections usually result from continued immersion of the appendages in water.

Intertriginous candidiasis involves those areas of the body, usually opposed skin surfaces, that are warm and moist: axillae, groin, skin folds. **Napkin (diaper) candidiasis** is typically

Box 39.1

The Emergence of Candidiasis

Written descriptions of oral lesions that were probably thrush date back to the early 1800s. In 1839 Bernard Langenbeck in Germany described the organism he found in oral lesions of a patient as "Typhus-Leichen" (typhus bodies). By 1841 Emil Berg established the fungal etiology of thrush by infecting healthy babies with what he called "aphthous membrane material." In 1843 Charles Robin gave the organism its first name: *Oidium albicans*. Since then, more than 100 synonyms have been used for this fungus; of them all, *Candida albicans,* the name Roth Berkhout proposed in 1923, has persisted.

In 1861 Albert Zenker described the first well-documented case of systemic candidiasis. Historically, the most interesting period for candidiasis research coincided with the introduction of antibiotics. Since then there have been documented cases of this fungus involving all tissues and organs of the body, as well as an increase in the overall incidence of candidiasis. Some *Candida*-associated infections and diseases include arthritis, endophthalmitis, meningitis, myocarditis, myositis, and peritonitis. Besides the widespread use of antibiotics, other therapeutic modalities and surgical procedures such as organ transplants and prostheses have been important in the expanding worldwide incidence of candidiasis.

(a)

(b)

(c)

Figure 39.16 Opportunistic Mycoses Caused by *Candida albicans.* (*a*) Scanning electron micrograph of the yeast form (×10,000). Notice that some of the cells are reproducing by budding. (*b*) Thrush, or oral candidiasis, is characterized by the formation of white patches on the mucous membranes of the tongue and elsewhere in the oropharyngeal area. These patches form a pseudomembrane composed of spherical yeast cells, leukocytes, and cellular debris. (*c*) Paronychia and onychomycosis of the hands.

found in infants whose diapers are not changed frequently and therefore are not kept dry. **Candidal vaginitis** can result from diabetes, antibiotic therapy, oral contraceptives, pregnancy, or any other factor that compromises the female host. Normally the omnipresent lactobacilli (Döderlein's bacilli) can control *Candida* in this area by the low pH they create. However, if their numbers are decreased by any of the aforementioned factors, *Candida* may proliferate, causing a curdlike, yellow-white discharge from the vaginal area. *Candida* can be transmitted to males during intercourse and lead to **balanitis;** thus it also can be considered a sexually transmitted disease. Balanitis is a *Candida* infection of the male glans penis and occurs primarily in uncircumcised males. The disease begins as vesicles on the penis that develop into patches and are accompanied by severe itching and burning.

Lactobacilli or Döderlein's bacilli (p. 558).

Diagnosis of candidiasis is difficult because (1) this fungus is a frequent secondary invader in diseased hosts, (2) a mixed microbiota is most often found in the diseased tissue, and (3) no completely specific immunologic procedures for the identification of *Candida* currently exist. There is no satisfactory treatment for candidiasis. Cutaneous lesions can be treated with topical agents such as sodium caprylate, sodium propionate, gentian violet, nystatin, miconazole, and trichomycin. Ketoconazole, amphotericin B, fluconazole, itraconazole, and flucytosine also can be used for systemic candidiasis.

Nystatin (p. 671).

Pneumocystis carinii organisms are eucaryotic protists found in the lungs of a wide variety of mammals. The natural history of *P. carinii* is poorly understood, in large part because of a lack of a continuous in vitro culture system for its propagation. Although it was once considered a protozoan parasite, recent comparisons of rRNA and DNA sequences from several genes have shown that *P. carinii* is more closely related to fungi than protozoa.

The disease that this protist causes, **pneumocystis pneumonia** or ***Pneumocystis carinii* pneumonia (PCP),** occurs almost exclusively in immunocompromised hosts. Extensive use of immunosuppressive drugs and irradiation for the treatment of cancers and following organ transplants accounts for the formidable prevalence rates noted recently. This pneumonia also occurs in more than 80% of AIDS patients. Both the organism and the disease remain localized in the lungs—even in fatal cases. Within the lungs *Pneumocystis* causes the alveoli to fill with a frothy exudate.

Laboratory diagnosis of pneumocystis pneumonia can be made definitively only by microscopically demonstrating the presence of the microorganisms in infected lung material (*see figure 36.11a*) or by a PCR analysis. Treatment is by means of oxygen therapy and either a combination of trimethoprim and sulfamethoxazole, atovaquone (Mepron), or inhalable pentamidine. Prevention and control is through prophylaxis with drugs in susceptible persons.

---

1. How are human fungal diseases categorized?
2. What are two piedras that infect humans?
3. Briefly describe the major tineas that occur in humans.
4. Describe the three types of subcutaneous mycoses that affect humans.
5. Why is *Histoplasma capsulatum* found in bird feces but not within the birds themselves?
6. Why are some mycotic diseases of humans called opportunistic mycoses?
7. What parts of the human body can be affected by *Candida* infections?
8. When is *Pneumocystis carinii* pneumonia likely to occur in humans?

Protozoan Diseases

Protozoa have become adapted to practically every type of habitat on the face of the earth, including the human body. Though fewer than 20 genera of protozoa cause disease in humans (table 39.2), their impact is formidable. For example, there are over 150 million cases of malaria in the world each year. In tropical Africa alone, malaria is responsible each year for the deaths of more than a million children under the age of 14. It is estimated that there are at least 8 million cases of trypanosomiasis, 12 million cases of leishmaniasis, and over 500 million cases of amebiasis yearly. The remainder of this chapter discusses some of these protozoan diseases of humans.

Amebiasis

It is now accepted that two species of *Entamoeba* infect humans: the nonpathogenic *E. dispar* and the pathogenic *E. histolytica*. *E. histolytica* is responsible for **amebiasis (amebic dysentery).** This very common parasite is endemic in warm climates where adequate sanitation and effective personal hygiene is lacking. Within the United States about 3,000 to 5,000 cases are reported annually. However, it is a major cause of parasitic death worldwide; about 500 million people are infected and as many as 100,000 die of amebiasis each year.

Infection occurs by ingestion of mature cysts. After excystation in the lower region of the small intestine, the metacyst divides rapidly to produce eight small trophozoites (figure 39.17). These trophozoites move to the large intestine where they can invade the host tissue, live as commensals in the lumen of the intestine, or undergo encystation.

TABLE 39.2 **Examples of Medically Important Protozoa**

Phylum	Group	Pathogen	Disease
Sarcomastigophora	Amoebae	*Entamoeba histolytica*	Amebiasis, amebic dysentery
		Acanthamoeba spp., *Naegleria fowleri*	Amebic meningoencephalitis
Apicomplexa	Coccidia	*Cryptosporidium* spp.	Cryptosporidiosis
Ciliophora	Ciliates	*Balantidium coli*	Balantidiasis
Sarcomastigophora	Blood and tissue flagellates	*Leishmania tropica*	Cutaneous leishmaniasis
		L. braziliensis	Mucocutaneous leishmaniasis
		L. donovani	Kala-azar (visceral leishmaniasis)
		Trypanosoma cruzi	American trypanosomiasis
		T. brucei gambiense, T. brucei rhodesiense	African sleeping sickness
Sarcomastigophora	Digestive and genital organ flagellates	*Giardia lamblia*	Giardiasis
		Trichomonas vaginalis	Trichomoniasis
Apicomplexa	Sporozoa	*Plasmodium falciparum, P. malariae, P. ovale, P. vivax*	Malaria
		Toxoplasma gondii	Toxoplasmosis

🔬 Figure 39.17   **Amebiasis Caused by *Entamoeba histolytica.*** (*a*) Light micrographs of a trophozoite (×1,000) and (*b*) a cyst (×1,000). (*c*) Life cycle. Infection occurs by the ingestion of a mature cyst of the parasite. Excystation occurs in the lower region of the small intestine and the metacyst rapidly divides to give rise to eight small trophozoites (only four are shown). These enter the large intestine, undergo binary fission, and may (1) invade the host tissues, (2) live in the lumen of the large intestine without invasion, or (3) undergo encystation and pass out of the host in the feces.

If the infective trophozoites invade the intestinal tissues, they multiply rapidly and spread laterally, while feeding on erythrocytes, bacteria, and yeasts. The invading trophozoites destroy the epithelial lining of the large intestine by producing proteolytic enzymes. Lesions (ulcers) are characterized by minute points of entry into the mucosa, colonization of the mucosal layer, and extensive enlargement of the lesion after penetration into the submucosa. *E. histolytica* also may invade and produce lesions in extraintestinal foci, especially the liver, to cause hepatic amebiasis. However, all extraintestinal amebic lesions are secondary to the ones established in the large intestine.

The symptoms of amebiasis are highly variable, ranging from an asymptomatic infection to fulminating dysentery, exhaustive diarrhea accompanied by blood and mucus, appendicitis, and abscesses in the liver, lungs, or brain.

Laboratory diagnosis of amebiasis is based on finding trophozoites in fresh warm stools and cysts in ordinary stools. Serological testing also should be done. The therapy for amebiasis is complex and depends on the location of the infection within the host and the host's condition. Asymptomatic cyst passers should always be treated with iodoquinol or paromomycin because they represent the most important reservoir of the parasite in the population. Amebaquin (diiodohydroxyquin) is the drug of choice for carriers. In symptomatic intestinal amebiasis, metronidazole (Flagyl) plus iodoquinol are the drugs of choice. Prevention and control of amebiasis is achieved by avoiding water or food that might be contaminated with human feces in endemic areas. Viable cysts in water can be destroyed by hyperchlorination or iodination.

Freshwater Amoeba Diseases

Free-living amoebae of the genera *Naegleria* and *Acanthamoeba* are facultative parasites responsible for causing **primary amebic meningoencephalitis** (*see table 37.1*) in humans. They are among the most common protozoa found in fresh water and moist soil. In addition, several *Acanthamoeba* spp. are known to infect the eye, causing a chronically progressive ulcerative *Acanthamoeba* **keratitis,** inflammation of the cornea, which may result in blindness. Wearers of soft contact lenses may be predisposed to this infection and should take care to prevent contamination of their lens cleaning and soaking solutions. Diagnosis of these infections is by demonstration of the amoebae in clinical specimens. Most freshwater amoebae are resistant to commonly used antibiotic agents. These amoebae are involved in fewer than 100 human disease cases annually in the United States.

Giardiasis

Giardia lamblia [syn., *G. duodenalis, G. intestinalis*] is a flagellated protozoan (figure 39.18*a*) that causes the very common intestinal disease **giardiasis.** (It was discovered by van Leeuwenhoek when he examined his own stools.) *G. lamblia* is worldwide in distribution, and it affects children more seriously than it does adults. In the United States this protozoan is the most common cause of epidemic waterborne diarrheal disease (about 30,000 cases yearly). Approximately 7% of the population are healthy carriers and shed cysts in their feces. *G. lamblia* is endemic in child day-care centers in the United States, with estimates of 5 to 15% of diapered children being infected.

(a)

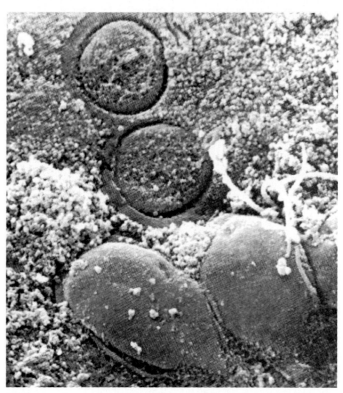

(b)

Figure 39.18   Giardiasis. (*a*) *Giardia lamblia* adhering to the epithelium by its sucking disk; scanning electron micrograph. (*b*) Upon detachment from the epithelium, the protozoa often leave clear impressions on the microvillus surface (upper circles); scanning electron micrograph.

Transmission is most frequent with cyst-contaminated water supplies. Epidemic outbreaks have been recorded in wilderness areas, suggesting that humans may be infected from "clean water" with *Giardia* harbored by rodents, deer, cattle, or household pets. This implies that human infections also can be a zoonosis (*see table 35.1*). As many as 200 million humans may be infected worldwide.

Following ingestion, the cysts undergo excystation in the duodenum, forming trophozoites. The trophozoites inhabit the upper portions of the small intestine, where they attach to the intestinal mucosa by means of their sucking disks (figure 39.18*a*). The ability of the trophozoites to adhere to the intestinal epithelium accounts for the fact that they are rarely found in stools. It is thought that the trophozoites feed on mucous secretions and reproduce to form such a large population that they interfere with nutrient absorption by the intestinal epithelium.

Giardiasis varies in severity, and asymptomatic carriers are common. The disease can be acute or chronic. Acute giardiasis is characterized by severe diarrhea, epigastric pain, cramps, voluminous flatulence ("passing gas"), and anorexia. Chronic giardiasis is characterized by intermittent diarrhea, with periodic appearance and remission of symptoms. A protein like the cholera toxin has been isolated from *G. lamblia* and is responsible for the diarrhea.

Laboratory diagnosis is based on the identification of trophozoites—only in the severest of diarrhea—or cysts in stools. A commercial ELISA test is also available for the detection of *G. lamblia* antigen in stool specimens. Quinacrine hydrochloride (Atabrine) and metronidazole (Flagyl) are the drugs of choice for adults, and furazolidone is used for children because it is available in a pleasant-tasting liquid suspension. Prevention and control involves proper treatment of community water supplies, especially the use of slow sand filtration (*see chapter 41*).

Malaria

The most important human parasite among the sporozoa is *Plasmodium,* the causative agent of **malaria** (Box 39.2). It has been estimated that more than 100 million people are infected, and about one million die annually of malaria in Africa alone. About 1,000 cases are reported each year in the United States.

Box 39.2

A Brief History of Malaria

No other single infectious disease has had the impact on humans that malaria has. The first references to its periodic fever and chills can be found in early Chaldean, Chinese, and Hindu writings. In the late fifth century B.C., Hippocrates described certain aspects of malaria. In the fourth century B.C., the Greeks noted an association between individuals exposed to swamp environments and the subsequent development of periodic fever and enlargement of the spleen (splenomegaly). In the seventeenth century the Italians named the disease *mal' aria* (bad air) because of its association with the ill-smelling vapors from the swamps near Rome. At about the same time, the bark of the quina-quina (cinchona) tree of South America was used to treat the intermittent fevers, although it was not until the mid-nineteenth century that quinine was identified as the active alkaloid. The major epidemiological breakthrough came in 1880, when French army surgeon Charles Louis Alphonse Laveran observed exflagellated gametocytes in fresh blood. Five years later the Italian histologist Camillo Golgi observed the multiplication of the asexual blood forms. In the late 1890s Patrick Manson postulated that malaria was transmitted by mosquitoes. Sir Ronald Ross, a British army surgeon in the Indian Medical Service, subsequently observed developing plasmodia in the intestine of mosquitoes, supporting Manson's theory. Using birds as experimental models, Ross definitively established the major features of the life cycle of *Plasmodium* and received the Nobel Prize in 1902.

Human malaria is known to have contributed to the fall of the ancient Greek and Roman empires. Troops in both the United States Civil War and the Spanish-American War were severely incapacitated by the disease. More than 25% of all hospital admissions during these wars were malaria patients. During World War II malaria epidemics severely threatened both the Japanese and Allied forces in the Pacific. The same can be said for the military conflicts in Korea and Vietnam.

In the twentieth century efforts have been directed toward understanding the biochemistry and physiology of malaria, controlling the mosquito vector, and developing antimalarial drugs. In the 1960s it was demonstrated that resistance to *P. falciparum* among West Africans was associated with the presence of hemoglobin-S in their erythrocytes. Hb-S differs from normal hemoglobin-A by a single amino acid, valine, in each half of the Hb molecule. Consequently these erythrocytes—responsible for sickle cell disease—have a low binding capacity for oxygen. Because the malarial parasite has a very active aerobic metabolism, it cannot grow and reproduce within these erythrocytes.

In 1955 the World Health Organization began a worldwide malarial eradication program that finally collapsed by 1976. Among the major reasons for failure were the development of resistance to DDT by the mosquito vectors and the development of resistance to chloroquine by strains of *Plasmodium.* Currently scientists are directing their efforts at new approaches, such as the development of vaccines and more potent drugs. For example, in 1984 the gene encoding the sporozoite antigen was cloned, permitting the antigen to be mass-produced by genetic engineering techniques. Overall, no greater achievement for molecular biology could be imagined than the control of malaria—a disease that has caused untold misery throughout the world since antiquity and remains one of the world's most serious infectious diseases.

Human malaria is caused by four species of *Plasmodium: P. falciparum, P. malariae, P. vivax,* and *P. ovale.* The life cycle of *P. vivax* is shown in figure 39.19. The parasite first enters the bloodstream through the bite of an infected female *Anopheles* mosquito. As she feeds, the mosquito injects a small amount of saliva containing an anticoagulant along with small haploid sporozoites (*see figure 27.3*c). The sporozoites in the bloodstream immediately enter hepatic cells of the liver. In the liver they undergo multiple asexual fission (schizogony) and produce merozoites. After being released from the liver cells, the merozoites either infect other liver cells, thus continuing the preerythrocytic stage, or attach to erythrocytes and penetrate these cells.

Once inside the erythrocyte, the *Plasmodium* begins to enlarge as a uninucleate cell termed a trophozoite. The trophozoite's nucleus then divides asexually to produce a schizont that has 6 to 24 nuclei. The schizont divides and produces mononucleated merozoites. Eventually the erythrocyte lyses, releasing the merozoites into the bloodstream to infect other erythrocytes. This erythrocytic stage is cyclic and repeats itself approximately every 48 to 72 hours or longer, depending on the species of *Plasmodium* involved. This sudden release of merozoites, toxins, and erythrocyte debris triggers an attack of the chills and

fever, so characteristic of malaria. Occasionally, merozoites differentiate into macrogametocytes and microgametocytes, which do not rupture the erythrocyte. When these are ingested by a mosquito, they develop into female and male gametes, respectively. In the mosquito's gut the infected erythrocytes lyse and the gametes fuse to form a diploid zygote called the ookinete. The ookinete migrates to the mosquito's gut wall, penetrates, and forms an oocyst on its outer surface. In a process called sporogony, the oocyst undergoes meiosis and forms sporozoites that migrate to the salivary glands of the mosquito. The cycle is now complete, but when the mosquito bites another human host, the cycle begins anew.

The pathological changes caused by malaria involve not only the erythrocytes but also the spleen and other visceral organs. Classic symptoms first develop with the synchronized release of merozoites and erythrocyte debris into the bloodstream, resulting in the malarial paroxysms—shaking chills, then burning fever followed by sweating. It may be that the fever and chills are caused partly by a malarial toxin that induces macrophages to release TNF-α and interleukin-1. Several of these paroxysms constitute an attack. After one attack there is a remission that lasts from a few weeks to several months. Then there is a relapse. Between paroxysms, the patient feels normal.

 Figure 39.19   Malaria. Life cycle of *Plasmodium vivax.*

Anemia can result from the loss of erythrocytes, and the spleen and liver often hypertrophy. Children and nonimmune individuals can die of cerebral malaria.

Diagnosis of malaria is made by demonstrating the presence of parasites within Wright- or Giemsa-stained erythrocytes (figure 39.20). When blood smears are negative, serological testing can establish a diagnosis of malaria in individuals. Treatment includes administration of chloroquine, amodiaquine, or mefloquine. These suppressant drugs are effective in eradicating erythrocytic asexual stages. Primaquine has proved satisfactory in eradicating the exoerythrocytic stages. However, because resistance to these drugs is occurring rapidly, more expensive drug combinations are being used. One example is Fansidar, a combination of pyrimethamine and sulfadoxine. It is worth noting that individuals who are traveling to areas where malaria is endemic (figure 39.21) should receive chemoprophylactic treatment with chloroquine. Efforts to develop a vaccine are under way.

Hemoflagellate Diseases

The flagellated protozoa that are transmitted by the bites of infected arthropods and infect the blood and tissues of humans are called **hemoflagellates.** There are two major groups of pathogens: the leishmanias and the trypanosomes.

Figure 39.20   Malaria: Erythrocytic Cycle. Trophozoites of *P. falciparum* in circulating erythrocytes; light micrograph (×1,100). The young trophozoites resemble small rings resting in the erythrocyte cytoplasm.

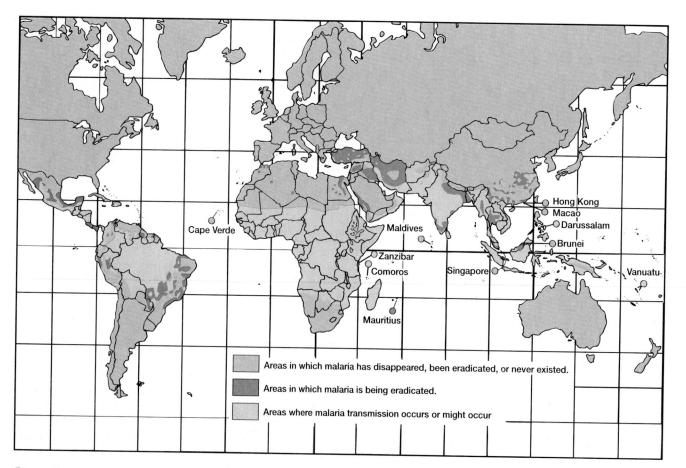

Areas in which malaria has disappeared, been eradicated, or never existed.

Areas in which malaria is being eradicated.

Areas where malaria transmission occurs or might occur

Figure 39.21   Geographic Distribution of Malaria. Notice that malaria is endemic around the equator. *Source: Data from the* World Health Statistics Quarterly, *41:69, 1988, World Health Organization, Switzerland.*

Figure 39.22   Leishmaniasis. (*a*) A person with mucocutaneous leishmaniasis, which has destroyed the nasal septum and deformed the nose and lips. (*b*) A person with diffuse cutaneous leishmaniasis.

(a)

(b)

Leishmaniasis

Leishmanias are flagellated protozoa that cause a group of human diseases collectively called **leishmaniasis.** The primary reservoirs of these parasites are canines and rodents. All species of *Leishmania* use sand flies of the genus *Phlebotomus* as intermediate hosts. The leishmanias are transmitted from animals to humans or between humans by these sand flies. When an infected sand fly takes a human blood meal, it introduces flagellated promastigotes into the skin of the definitive host. Within the skin the promastigotes are engulfed by macrophages, multiply by binary fission and form small nonmotile cells called amastigotes. These destroy the host cell, and are engulfed by other macrophages in which they continue to develop and multiply.

Leishmania braziliensis, which has an extensive distribution in the forest regions of tropical America, causes mucocutaneous leishmaniasis (espundia) (figure 39.22*a*). The disease produces lesions involving the mouth, nose, throat, and skin and results in extensive scarring and disfigurement.

Leishmania donovani is endemic in large areas within northern China, eastern India, the Mediterranean countries, the Sudan, and Latin America. It produces visceral leishmaniasis (kala-azar). The disease involves the reticuloendothelial system and often results in intermittent fever and enlargement of the spleen and liver. Individuals who recover develop a permanent immunity.

Leishmania tropica and *L. mexicana* occur in the more arid regions of the Eastern Hemisphere and cause cutaneous leishmaniasis. In this disease a relatively small red papule forms at the site of each insect bite—the inoculation site. These papules are frequently found on the face and ears. They eventually develop into crustated ulcers (figure 39.22*b*). Healing occurs with scarring and a permanent immunity.

Laboratory diagnosis of leishmaniasis is based on finding the parasite within infected macrophages in stained smears from lesions or infected organs. Treatment includes the pentavalent antimonial compounds (Pentostam, Glucantime). Vector and reservoir control and aggressive epidemiological surveillance are the best options for prevention and containment of this disease.

Trypanosomiasis

Another group of flagellated protozoa called **trypanosomes** (*see figure 27.3a*) cause the aggregate of diseases termed **trypanosomiasis.** *Trypanosoma brucei gambiense,* found in the rain forests of west and central Africa, and *T. brucei rhodesiense,* found in the upland savannas of east Africa, cause African trypanosomiasis. Reservoirs for these trypanosomes are domestic cattle and wild animals, within which the parasites cause severe malnutrition. Both species use tsetse flies as intermediate hosts. The parasites are transmitted through the bite of the fly to humans. Once the protozoa enter the bloodstream, they begin to multiply. They cause interstitial inflammation and necrosis within the lymph nodes and small blood vessels of the brain and heart. In *T. brucei rhodesiense* infection the disease develops so rapidly that infected individuals often die within a year. In *T. brucei gambiense* infection the parasites invade the central nervous system, where necrotic damage causes a variety of nervous disorders, including the characteristic sleeping sickness. The name sleeping sickness is derived from the lethargy of the host—characteristically lying prostrate, drooling from the mouth, and insensitive to pain. Usually the victim dies in 2 to 3 years. Trypanosomiasis is such a problem in parts of Africa that millions of square miles are not fit for human habitation.

T. cruzi causes **American trypanosomiasis (Chagas' disease),** which occurs in the tropics and subtropics of continental America. The parasite uses the triatomid (cone-nosed) bug as a vector. As the triatomid bug takes a blood meal, the parasites are discharged in the insect's feces. Some trypanosomes enter the bloodstream through the wound and invade the liver, spleen, lymph nodes, and central nervous system, where they multiply and destroy the parasitized cells.

Trypanosomiasis is diagnosed by finding motile parasites in fresh blood and by serological testing. Treatment for African trypanosomiasis uses suramin and pentamidine for non-nervous-system involvement and melarsoprol when the nervous system is involved. Currently there is no drug suitable for Chagas' disease, although nifurtimox (Lampit) and benznidazole have shown some value. Vaccines are not useful because the parasite is able to change its protein coat and evade the immunologic response (Box 39.3).

Toxoplasmosis

Toxoplasmosis is a disease caused by the protozoan *Toxoplasma gondii*. This apicomplexan protozoan has been found in nearly all animals and most birds; cats are the definitive host and required for completion of the sexual cycle. Animals shed cysts in the feces; the cysts enter another host by way of the nose or mouth; and the trophozoites colonize the intestine. Toxoplasmosis also can be transmitted by the ingestion of raw or undercooked meat, congenital transfer, blood transfusion, or a tissue transplant. Originally toxoplasmosis gained public notice when it was discovered that in pregnant women the protozoan might also infect the fetus, causing serious congenital defects or death.

Most cases of toxoplasmosis are asymptomatic. Adults usually complain of an "infectious mononucleosislike" syndrome. In immunoincompetent or immunosuppressed individuals, it frequently results in fatal disseminated disease with a heavy cerebral involvement.

Acute toxoplasmosis is usually accompanied by lymph node swelling (lymphadenopathy) with reticular cell enlargement or hyperplasia. Pulmonary necrosis, myocarditis, and hepatitis caused by tissue necrosis are common. Retinitis (inflammation of the retina of the eye) is associated with the necrosis due to the proliferation of the parasite within retinal cells. Currently toxoplasmosis has become a major cause of death in AIDS patients from a unique encephalitis with necrotizing lesions accompanied by inflammatory infiltrates (*see figure 36.11*b). It continues to cause more than 3,000 congenital infections per year in the United States.

Laboratory diagnosis of toxoplasmosis is by serological tests. Epidemiologically, toxoplasmosis is ubiquitous in all higher animals. Treatment of toxoplasmosis is with a combination of pyrimethamine (Daraprim) and sulfadiazine. Prevention and control requires minimizing exposure by the following: avoiding eating raw meat and eggs, washing hands after working in the soil, cleaning cat litterboxes daily, keeping personal cats indoors if possible, and feeding them commercial food.

Trichomoniasis

Trichomoniasis is a sexually transmitted disease caused by the protozoan flagellate *Trichomonas vaginalis* (figure 39.23). It is one of the most common sexually transmitted diseases, with an estimated 7 million cases annually in the United States and 180 million cases annually worldwide. In response to the parasite, the body accumulates leukocytes at the site of the infection. In females this usually results in a profuse purulent vaginal discharge that is yellowish to light cream in color and characterized by a disagreeable odor. The discharge is accompanied by itching. Males are generally asymptomatic because of the trichomonacidal action of prostatic secretions; however, at times a

---

Box 39.3

The Surface Antigens of Parasitic Trypanosomes

Because of the severe debilitation to humans and animals caused by the trypanosomes, a serious effort is now under way by the World Health Organization to produce a vaccine for trypanosome-caused diseases. A major obstacle to such a vaccine is the fact that trypanosomes can change their surface glycoproteins. In the tsetse fly the trypanosomes are noninfective. Before transfer to a mammalian host, they secrete a thick coat of variant surface glycoprotein (VSG) that protects them from the host's antibodies. Furthermore, they can genetically change this coat into a series of variant antigenic types (VATs) more rapidly than the immune system of the mammalian host can respond.

Each trypanosome possesses genes for 2,000 or more different VSGs in its nucleus. Gene expression occurs in a somewhat predictable order. For example, the expression of one VSG usually occurs after the expression of another specific VSG. As a result the VSGs in a population of trypanosomes are heterogeneous in any given infection, but there is a single VSG that predominates and against which the host mounts its defensive antibodies. When a new host is infected, the trypanosomes change their repertoire and new VSG variants appear. The mechanism that controls switching from one VAT, in which one VSG is being expressed, to another VAT, with the expression of a different VSG, is being intensely studied. Although the process is not yet completely understood, activation involves gene rearrangement by at least two mechanisms. It is also clear that the switching is independent of the host's immune system. When the tsetse fly takes the trypanosomes up again, the trypanosomes shed their glycoprotein coats and the genetic cycle starts anew.

burning sensation occurs during urination. Diagnosis is made in females by microscopic examination of the discharge and identification of the parasite. Infected males will demonstrate the parasite in semen or urine. Treatment is by administration of metronidazole (Flagyl).

1. How do infections caused by *Entamoeba histolytica* occur?

2. What is the most common cause of epidemic waterborne diarrheal disease?

3. Describe in detail the life cycle of the malarial parasite.

4. What protozoa are represented within the hemoflagellates? What diseases do they cause?

5. In what two ways does *Toxoplasma* affect human health?

6. How would you diagnose trichomoniasis in a female? In a male?

Figure 39.23   Trichomoniasis. *Trichomonas vaginalis*, showing the characteristic undulating membranes and flagella; scanning electron micrograph (×12,000).

Summary

1. Human fungal diseases, or mycoses, can be divided into five groups according to the level and mode of entry into the host. These are the superficial, cutaneous, subcutaneous, systemic, and opportunistic mycoses.

2. Superficial mycoses of the hair shaft are collectively called piedras. Two major types are black piedra and white piedra. Tinea versicolor is a third common superficial mycosis.

3. The cutaneous fungi that parasitize the hair, nails, and outer layer of the skin are called dermatophytes, and their infections are termed dermatophytoses, ringworms, or tineas. At least seven types can occur in humans: tinea barbae (ringworm of the beard), tinea capitis (ringworm of the scalp), tinea corporis (ringworm of the body), tinea cruris (ringworm of the groin), tinea pedis (ringworm of the feet), tinea manuum (ringworm of the hands), and tinea unguium (ringworm of the nails).

4. The dermatophytes that cause the subcutaneous mycoses are normal saprophytic inhabitants of soil and decaying vegetation.

Three types of subcutaneous mycoses can occur in humans: chromoblastomycosis, maduromycosis, and sporotrichosis.

5. Most systemic mycoses that occur in humans are acquired by inhaling the spores from the soil where the free-living fungi are found. Four types can occur in humans: blastomycosis, coccidioidomycosis, cryptococcosis, and histoplasmosis.

6. An opportunistic organism is one that is generally harmless in its normal environment but that can become pathogenic in a compromised host. The most important opportunistic mycoses affecting humans include systemic aspergillosis, candidiasis, and *Pneumocystis carinii* pneumonia.

7. Protozoa are responsible for some of the most serious human diseases that affect hundreds of millions of people worldwide.

8. *Entamoeba histolytica* is the amoeboid protozoan responsible for amebiasis. This is a very common disease in warm climates throughout the world. It is acquired when one ingests the cysts with contaminated food or water.

9. *Giardia lamblia* is a flagellated protozoan that causes the common intestinal disease giardiasis. This disease is distributed throughout the world, and in the United States it is the most common cause of waterborne diarrheal disease.

10. The most important human parasite among the sporozoa is *Plasmodium*, the causative agent of malaria. Human malaria is caused by four species of *Plasmodium*: *P. falciparum*, *P. vivax*, *P. malariae*, and *P. ovale*.

11. The flagellated protozoa that are transmitted by arthropods and infect the blood and tissues of humans are called hemoflagellates. Two major groups occur: the leishmanias, which cause the diseases collectively termed leishmaniasis, and the trypanosomes, which cause trypanosomiasis.

12. Toxoplasmosis is a disease caused by the protozoan *Toxoplasma gondii*. It is one of the major causes of death in AIDS patients.

13. Trichomoniasis is a sexually transmitted disease caused by the protozoan flagellate *Trichomonas vaginalis*.

Key Terms

amebiasis (amebic dysentery) *799*

American trypanosomiasis (Chagas' disease) *805*

aspergillosis *797*

athlete's foot *793*

balanitis *799*

black piedra *791*

blastomycosis *795*

candidal vaginitis *799*

candidiasis *797*

chromoblastomycosis *794*

coccidioidomycosis *795*

compromised host *797*

cryptococcosis *795*

dermatophyte *792*

dermatophytosis *792*

eumycotic mycetoma *794*

extracutaneous sporotrichosis *794*

giardiasis *801*

hemoflagellate *804*

histoplasmosis *796*

intertriginous candidiasis *797*

jock itch *793*

keratitis *801*

leishmania *805*

leishmaniasis *805*

maduromycosis *794*

malaria *802*

medical mycology *791*

mycosis *791*

napkin (diaper) candidiasis *797*

onychomycosis *797*

opportunistic organism *797*

oral candidiasis *797*

paronychia *797*

piedra *791*

Pneumocystis carinii pneumonia (PCP) *799*

pneumocystis pneumonia *799*

primary amebic meningoencephalitis *801*

ringworm *792*

sporotrichosis *794*

thrush *797*

tinea *791*

tinea barbae *792*

tinea capitis *792*

tinea corporis *793*

tinea cruris *793*

tinea manuum *793*

tinea pedis *793*

tinea unguium *793*

tinea versicolor *791*

toxoplasmosis *806*

trichomoniasis *806*

trypanosome *805*

trypanosomiasis *805*

white piedra *791*

Questions for Thought and Review

1. Briefly describe each of the major or most common fungal and protozoan diseases in terms of its causative agent, signs and symptoms, the course of infection, mechanism of pathogenesis, epidemiology, and prevention and/or treatment.

2. There are no antibiotics—nothing equivalent to penicillin or streptomycin—to control fungi in the human body. Recalling the mechanisms by which antibiotics affect bacteria, why do you think there has been no similar success in the control of fungal pathogens?

3. Give several reasonable explanations why most fungal diseases in humans are not contagious.

4. What factors determine one's susceptibility to fungal diseases?

5. What is the relationship between AIDS and mycotic and protozoan diseases?

6. Why do most protozoan diseases occur in the tropics?

7. Give several reasons why malaria is still one of the most serious of all infectious diseases that affect humans.

8. What morphological property is the most important in the identification of fungi?

9. Compared to fungal parasites, how are protozoan parasites transmitted?

10. What are dermatomycoses?

Additional Reading

General

Cheng, T. C. 1986. *General parasitology,* 2d ed. New York: Academic Press.

Emmons, C. W.; Chapman, H. B.; Utz, J. P.; and Kwon-Chung, K. J. 1977. *Medical mycology,* 3d ed. Philadelphia: Lea and Febiger.

Joklik, W. K.; Willett, H. P.; Amos, D. B.; and Wilfert, C. M. 1992. *Zinsser microbiology,* 20th ed. Norwalk, Conn.: Appleton & Lange.

Mandell, G. L.; Bennett, J. E.; and Dolan, R. 1995. *Principles and practice of infectious diseases,* 4th ed. New York: Churchill Livingstone.

Markell, E. K. 1986. *Medical parasitology,* 6th ed. Philadelphia: W. B. Saunders.

Rippon, J. W. 1988. *Medical mycology,* 3d ed. Philadelphia: W. B. Saunders.

Schmidt, G. D., and Roberts, L. R. 1995. *Foundations of parasitology,* 6th ed. Dubuque, Iowa: Wm. C. Brown Publishers.

Fungal Diseases

Ahearn, D. G. 1978. Medically important yeasts. *Ann. Rev. Microbiol.* 32:59–75.

Bartlett, M., and Smith, J. 1991. *Pneumocystis carinii,* an opportunist in immunocompromised patients. *Clin. Microbiol. Rev.* 4(2):137–49.

Calderone, R. 1993. Recognition between *Candida albicans* and host cells. *Trends Microbiol.* 1(2):55–58.

Jones, J. M. 1990. Laboratory diagnosis of invasive candidiasis. *Clin. Microbiol. Rev.* 3(1):32–45.

Larone, D. H. 1993. *Medically important fungi.* Washington, D.C.: American Society for Microbiology.

McGough, D. 1993. Clinical and laboratory aspects of the "black yeasts." *Clin. Microbiol. Newsletter* 15(19):145–51.

Mills, J., and Masur, H. 1990. AIDS-related infections. *Sci. Am.* 263(2):50–59.

Musial, C. E.; Cockerill, F. R.; and Roberts, G. D. 1988. Fungal infections of the immunocompromised host: Clinical and laboratory aspects. *Clin. Microbiol. Rev.* 1(4):349–64.

Richardson, M., and Warnock, D. 1993. *Fungal infection.* Oxford, England: Blackwell Scientific Publications.

Travassos, L. R., and Lloyd, K. O. 1980. *Sporothrix schenckii* and related species of *Ceratocytis. Microbiol. Rev.* 44(4):683–721.

Weitzman, I., and Summerbell, R. C. 1995. The dermatophytes. *Clin. Microbiol. Rev.* 8(2):240–59.

Wheat, J. 1995. Endemic mycoses in AIDS: A clinical review. *Clin. Microbiol. Rev.* 8(1):146–59.

Zimmer, B. L. 1990. Serology of coccidioidomycosis. *Clin. Microbiol. Rev.* 3(2):247–68.

Protozoan Diseases

Adam, R. D. 1991. The biology of *Giardia* spp. *Microbiol. Rev.* 55(4):706–32.

Borst, P., and Rudenko, G. 1994. Antigenic variation in African trypanosomes. *Science* 264:1872–73.

Bruckner, D. A. 1992. Amebiasis. *Clin. Microbiol. Rev.* 5(4):356–69.

Cabral-Marciano, F. 1988. Biology of *Naegleria* spp. *Microbiol. Rev.* 52(1):114–33.

Cogswell, F. 1992. The hypnozoite and relapse in primate malaria. *Clin. Microbiol. Rev.* 5(1):26–35.

Garcia, L. S., and Bruckner, D. A., editors. 1993. *Diagnostic medical parasitology.* Washington, D.C.: American Society for Microbiology.

Glew, R. H.; Saha, A. K.; Das, S.; and Remaley, A. T. 1988. Biochemistry of the *Leishmania* species. *Microbiol. Rev.* 52(4):412–32.

Grimaldi, G., and Tesh, R. B. 1993. Leishmaniases of the new world: Current concepts and implications for future research. *Clin. Microbiol. Rev.* 6(3):230–50.

John, D. T. 1982. Primary amoebic meningoencephalitis and the biology of *Naegleria fowleri. Ann. Rev. Microbiol.* 36:101–23.

Joiner, K., and Dubrementz, J. F. 1993. *Toxoplasma gondii:* A protozoan for the nineties. *Infect. Immun.* 61(4):1169–72.

Jones, T., and Hoffman, S. 1994. Malaria vaccine development. *Clin. Microbiol. Rev.* 7(3):303–10.

Kretschmer, R. R. 1990. *Amebiasis.* Boca Raton, Fla.: CRC Press.

McCoy, J. 1994. Adherence and cytotoxicity of *Entamoeba histolytica* or how lectins let parasites stick around. *Infect. and Immun.* 62(8):3045–50.

Miller, L. H.; Good, M. F.; and Milon, G. 1994. Malaria pathogenesis. *Science* 264:1878–83.

Mirelman, D. 1987. Ameba-bacterium relationship in amebiasis. *Microbiol. Rev.* 51(2):272–84.

Petri, W. 1991. Invasive amebiasis and the galactose-specific lectin of *Entamoeba histolytica. ASM News* 57(6):299–306.

Tanowitz, H. B.; Kirchhoff, D. S.; Morris, S. A.; Weiss, L. M.; and Wittner, M. 1992. Chagas' disease. *Clin. Microbiol. Rev.* 5(4):400–19.

Weber, R.; Bryan, R. T.; Schwartz, D. A.; and Owen, R. L. 1994. Human microsporidial infections. *Clin. Microbiol. Rev.* 7(4):426–61.

Weiss, J. B. 1995. DNA probes and PCR for diagnosis of parasitic infections. *Clin. Microbiol. Rev.* 8(1):113–30.

Wolfe, M. 1992. Giardiasis. *Clin. Microbiol. Rev.* 5(1):93–100.

PART TEN
Microorganisms and the Environment

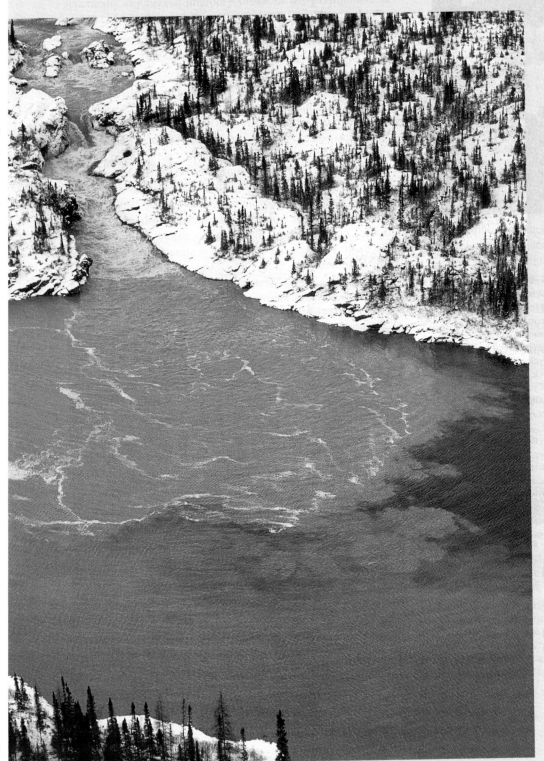

We often think of terrestrial and marine aquatic environments as being distinctly different and separated from one another. In reality these "different" environments interact in interesting ways, often involving microorganisms and their activities. This picture of two rivers in Quebec, Canada, shows water from the smaller river, polluted by an iron ore mining site, flowing into the larger Moisie River. The polluted water impacts both the Moisie River itself and its adjacent forest and soil systems. The plants and soils next to these rivers also release nutrients that influence aquatic microbial processes. As this water moves to the ocean, it will affect microbial processes all along the river, and ultimately it will influence the ocean itself. This illustrates the fact that the environment truly is a single unit. To comprehend the whole and its interacting parts, all components and the linkages that exist between them must be understood.

Figure 40.8   The Sulfur Cycle. Photosynthetic and chemosynthetic microorganisms contribute to the environmental sulfur cycle. Anaerobic sulfate and sulfite reductions carried out by *Desulfovibrio* and other microorganism are dissimilatory processes and are noted with purple arrows. Sulfate reduction also can occur in assimilatory reactions, resulting in organic sulfur forms. Elemental sulfur reduction to sulfide is carried out by *Desulfuromonas*, thermophilic archaeobacteria, or cyanobacteria in hypersaline sediments. Sulfur oxidation can be carried out by a wide range of aerobic chemotrophs and by anaerobic phototrophs.

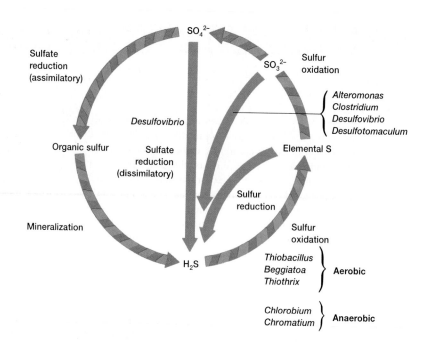

Carbon fixation occurs through the activities of cyanobacteria and green algae, photosynthetic bacteria (e.g., *Chromatium* and *Chlorobium*), and aerobic chemolithoautotrophs.

Sulfur Cycle

Microorganisms contribute greatly to the sulfur cycle, a simplified version of which is shown in figure 40.8. Photosynthetic microorganisms transform sulfur by using sulfide as an electron source. In the absence of light, sulfide can cross into oxidized environments, allowing *Thiobacillus* and similar chemolithoautotrophic genera to function (*pp. 181 and 464–66*). In contrast, when sulfate diffuses into reduced habitats, it provides an opportunity for different groups of microorganisms to carry out **sulfate reduction.** For example, when a usable organic reductant is present, *Desulfovibrio* can derive energy by using sulfate as an oxidant (*pp. 178 and 430–31*). This use of sulfate as an external electron acceptor to form sulfide, which accumulates in the environment, is an example of a **dissimilatory reduction** process and anaerobic respiration. In comparison, the reduction of sulfate for use in amino acid and protein biosynthesis is described as an **assimilatory reduction** process (*see chapter 10*). Other organisms have been found to carry out dissimilatory elemental sulfur reduction. These include *Desulfuromonas* (*p. 431*), thermophilic archaeobacteria (*see chapter 23*), and also cyanobacteria in hypersaline sediments. Sulfite is another critical intermediate that can be reduced to sulfide by a wide variety of microorganisms, including *Alteromonas* and *Clostridium*, as well as *Desulfovibrio* and *Desulfotomaculum*.

When pH and oxidation-reduction conditions are favorable, several key transformations in the sulfur cycle also occur as the result of chemical reactions in the absence of microorganisms. An important example of such an abiotic process is the oxidation of sulfide to elemental sulfur. This takes place rapidly at a neutral pH, with a half-life of approximately 10 minutes for sulfide at room temperature.

Nitrogen Cycle

The nitrogen cycle is "simpler" than the carbon or sulfur cycles because there are no distinct and different uses by photosynthetic microorganisms (figure 40.9). Despite this simplification several important aspects of the nitrogen cycle should be emphasized: the processes of nitrification, denitrification, and nitrogen fixation.

Nitrification is the aerobic process of ammonium ion (NH_4^+) oxidation to nitrite (NO_2^-) and subsequent nitrite oxidation to nitrate (NO_3^-). Bacteria of the genera *Nitrosomonas* and *Nitrosococcus,* for example, play important roles in the first step, and *Nitrobacter* and related chemolithoautotrophic bacteria carry out the second step. In addition, **heterotrophic nitrification** by bacteria and fungi contributes significantly to these processes in more acidic environments.

Nitrification and nitrifiers (p. 181).

The process of **denitrification** requires a different set of environmental conditions. This dissimilatory process, in which nitrate is used as an oxidant in anaerobic respiration, usually involves heterotrophs such as *Pseudomonas denitrificans.* The major products of denitrification include nitrogen gas (N_2) and nitrous oxide (N_2O), although nitrite (NO_2^-) also can accumulate. Nitrite is of environmental concern because it can contribute to the formation of carcinogenic nitrosamines. Finally,

Figure 40.9   The Environmental Nitrogen Cycle.
Flows that occur predominantly under aerobic conditions are noted with open arrows. Anaerobic processes are noted with solid bold arrows. Processes occurring under both aerobic and anaerobic conditions are marked with cross-barred arrows. Important genera contributing to the nitrogen cycle are given as examples.

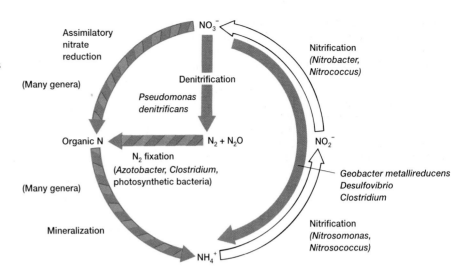

nitrate can be transformed to ammonia in dissimilatory reduction by a variety of bacteria, including *Geobacter metallireducens, Desulfovibrio* spp., and *Clostridium.*

Denitrification and anaerobic respiration (p. 176–78).

Nitrogen assimilation occurs when inorganic nitrogen is used as a nutrient and incorporated into new microbial biomass. Ammonium ion, because it is already reduced, can be directly incorporated without major energy costs. However, when nitrate is assimilated, it must be reduced with a significant energy expenditure. In this process nitrite may accumulate as a transient intermediate.

The biochemistry of nitrogen assimilation (pp. 198–99).

Nitrogen fixation can be carried out by aerobic or anaerobic bacteria. Under aerobic conditions a wide range of free-living microbial genera (*Azotobacter, Azospirillum*) contribute to this process. Under anaerobic conditions the most important free-living nitrogen fixers are members of the genus *Clostridium.* Nitrogen fixation by cyanobacteria such as *Anabaena* and *Oscillatoria* can lead to the enrichment of fresh and marine water with nitrogen. These nutrient-enrichment processes are discussed in chapter 41. In addition, nitrogen fixation can occur through the activities of bacteria that develop symbiotic associations with plants. These associations include *Rhizobium* and *Bradyrhizobium* with legumes, *Frankia* in association with many woody shrubs, and *Anabaena* with *Azolla,* a water fern important in rice cultivation.

The establishment of the Rhizobium-*legume association (pp. 865–68).*

The nitrogen-fixation process involves a sequence of reduction steps that require major energy expenditures. Ammonia, the product of nitrogen reduction, is immediately incorporated into organic matter as an amine. Reductive processes are extremely sensitive to O_2 and must occur under anaerobic conditions even

in aerobic microorganisms. Protection of the nitrogen-fixing enzymes is achieved by means of a variety of mechanisms, including physical barriers, as occurs with heterocysts in some cyanobacteria (*see chapter 22*), O_2 scavenging molecules, and high rates of metabolic activity. Nitrogen fixation is catalyzed by the enzyme **nitrogenase,** using reduced ferredoxin as its immediate source of reducing power. The biochemistry of this process is described in chapter 10.

The biochemistry of nitrogen fixation (pp. 199–201).

Other Cycling Processes

The iron cycle is especially important in terms of microbial functioning (figure 40.10). The major genera that carry out iron oxidations, transforming ferrous ion (Fe^{2+}) to ferric ion (Fe^{3+}), are *Thiobacillus ferrooxidans* under acidic conditions, *Gallionella* under neutral pH conditions, and *Sulfolobus* under acidic, thermophilic conditions. Much of the earlier literature suggested that additional genera could oxidize iron, including *Sphaerotilus* and *Leptothrix.* These two genera are still termed "iron bacteria" by many nonmicrobiologists. Confusion about the role of these genera resulted from the occurrence of the chemical oxidation of ferrous ion to ferric ion (forming insoluble iron precipitates) at neutral pH values, where microorganisms also grow on organic substrates. These microorganisms are now classified as chemoheterotrophs.

Iron reduction occurs under anaerobic conditions resulting in the accumulation of ferrous ion. Although many microorganisms can reduce small amounts of iron during their metabolism, most iron reduction is carried out by specialized iron-respiring microorganisms such as *Geobacter metallireducens* and *Shewanella putrefaciens,* which can obtain energy for growth on organic matter using ferric iron as an oxidant.

Figure 40.10 The Iron Cycle. A simplified iron cycle with examples of microorganisms contributing to these oxidation and reduction processes. In addition to ferrous ion (2+) oxidation and ferric ion (3+) reduction, magnetite (Fe_3O_4), a mixed valence iron compound formed by magnetotactic bacteria is important in the iron cycle. Different microbial groups carry out the oxidation of ferrous ion, depending on environmental conditions.

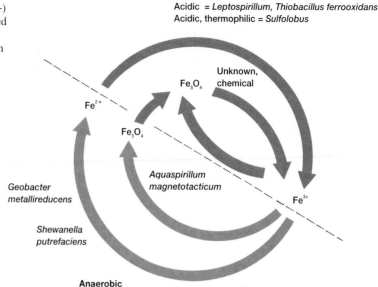

In addition to these relatively simple reductions to ferrous ion, some magnetotactic bacteria such as *Aquaspirillum magnetotacticum* (*see Box 3.3*) transform extracellular iron to the mixed valence iron oxide mineral magnetite (Fe_3O_4) and construct intracellular magnetic compasses. Furthermore, dissimilatory iron-reducing bacteria accumulate magnetite as an extracellular product.

Magnetite has been detected in sediments, where it is present in particles similar to those found in bacteria, indicating a longer-term contribution of bacteria to iron cycling processes. It has been suggested that Fe^{3+} reduction may have been the first globally significant mechanism for organic matter oxidation to CO_2. Magnetite also is important in environmental bioremediation and particularly in medicine. The particles can be coated with immunologic reagents, and with their magnetic properties, they can be recovered from the bloodstream by a simple magnet! In addition, genes for magnetite synthesis have been cloned into other organisms, creating new magnetically sensitive microorganisms.

The importance of microorganisms in manganese cycling is becoming much better appreciated. The manganese cycle (figure 40.11) involves the transformation of manganous ion (Mn^{2+}) to MnO_2 (equivalent to manganic ion [Mn^{4+}]), which occurs in hydrothermal vents, bogs, and as an important part of rock varnishes. *Leptothrix, Arthrobacter,* and *Metallogenium* are important in Mn^{2+} oxidation. *Shewanella, Geobacter,* and other chemoorganotrophs can carry out the complementary manganese reduction process. The manganese cycle is very closely linked to the iron cycle.

Microorganisms also can use a wide variety of metals as electron acceptors. Metals such as europium, tellurium, selenium, and rhodium can be reduced. Important microorganisms that reduce these metals include the photoorganotrophs *Rhodobacter, Rhodospirillum* and *Rhodopseudomonas*. For

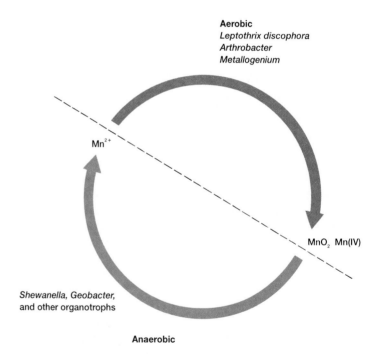

Figure 40.11 The Manganese Cycle. Microorganisms make important contributions to the manganese cycle. Manganous ion (2+) is oxidized to manganic oxide (valence equivalent to 4+). Manganous oxide reduction is noted with a maroon arrow. Examples of organisms carrying out these processes are given.

selenium, *Pseudomonas stutzeri, Thauera selenatis,* and *Wolinella succinogenes* are active. Such reductions can decrease the toxicity of a metal.

The microbial transformation of phosphorus involves primarily the transformation of phosphorus (+5 valence) from simple orthophosphate to various more complex forms, including

TABLE 40.2   Examples of Microorganism-Metal Interactions and Relations to Effects on Microorganisms and Warm-Blooded Animals

Metal Group	Metal		Interactions and Transformations	
			Microorganisms	Warm-blooded Animals
Noble metals	Ag Au Pt	Silver Gold Platinum	Microorganisms can reduce ionic forms to the elemental state, which can become associated with cell walls. Low levels of ionized metals released to the environment have antimicrobial activity.	Many of these metals can be reduced to elemental forms and do not tend to cross the blood-brain barrier. Silver reduction can result in argyria and inert deposits in the skin.
Metals that form stable carbon metal bonds	As Hg Se	Arsenic Mercury Selenium	Microorganisms can transform inorganic and organic forms to methylated forms, some of which tend to bioaccumulate in higher tropic levels.	Methylated forms of some metals can cross the blood-brain barrier, resulting in neurological effects or death.
Other metals	Cu Zn Co	Copper Zinc Cobalt	In the ionized form, at higher concentrations, these metals can directly inhibit microorganisms. They are often required at lower concentrations as trace elements.	At higher levels, clearance from higher organisms occurs by reaction with plasma proteins and other mechanisms. Many of these metals serve as trace elements at lower concentrations.

polyphosphates found in metachromatic granules (*see p. 47*). A unique (and assumed microbial) product is phosphine (PH_3) with a -3 valence, which is liberated from swamps, and which ignites when exposed to air. This can then ignite methane produced in the same environment! The production of methane by anaerobic microorganisms will be discussed in the next chapter.

1. What major oxidized and reduced forms of carbon, nitrogen, and sulfur are important in biogeochemical cycling?
2. By what means can microorganisms contribute to biogeochemical cycling in the environment?
3. Define or describe the following: dissimilatory reduction, assimilatory reduction, nitrification, denitrification, nitrogen fixation, and nitrogenase.

Metals, Microorganisms, and Humans

Many microorganisms can change the physical characteristics of metals and metalloids. Metalloids, such as silicon, selenium, and arsenic, are intermediate in properties between metals and nonmetals. Although most metals are commonly considered toxic, microorganisms and warm-blooded animals do not respond in similar ways. The general groupings of metals and their different effects on warm-blooded animals and microorganisms are summarized in table 40.2.

The "metals" can be considered in broad categories. The "noble metals" tend not to cross the vertebrate blood-brain barrier but can have distinct effects on microorganisms. Microorganisms also can reduce ionic forms of noble metals to their elemental forms.

The second group includes metals or metalloids that microorganisms can methylate to form more mobile products called organometals. Some organometals can cross the blood-brain barrier and affect the central nervous system of vertebrates. Organometals contain carbon-metal bonds. These bonds are their unique identifying characteristics.

The mercury cycle is of particular interest and illustrates many characteristics of those metals that can be methylated. Mercury compounds were widely used in industrial processes over the centuries. One has only to think of Lewis Carroll's allusion to this problem when he wrote of the Mad Hatter in *Alice in Wonderland*. At that time mercury was used in the shaping of felt hats. Microorganisms methylated some of the mercury, thus rendering it more toxic to the hatmakers.

A devastating situation developed in southwestern Japan when large-scale mercury poisoning occurred in the Minamata Bay region because of industrial mercury releases into the marine environment. Inorganic mercury that accumulated in bottom muds of the bay was methylated by anaerobic bacteria of the genus *Desulfovibrio* (figure 40.12). Such methylated mercury forms are volatile and lipid soluble, and the mercury concentrations increased in the food chain (a process known as **biomagnification**). The mercury was ultimately ingested by the human population, the "top consumers," through their primary food source—fish—leading to severe neurological disorders.

There is a similar situation in many of the freshwater lakes in the north-central United States and in Canada, where mercury compounds were used to control microbial growth in pulp mills. Even decades later the fish in lakes downstream from these pulp mills cannot yet be used for food, and fishing is only for recreation.

The third group of metals occurs in ionic forms directly toxic to microorganisms. The metals in this group also can affect more complex organisms. However, plasma proteins react with the ionic forms of these metals and aid in their excretion unless excessive long-term contact and ingestion occur. Relatively high doses of these metals are required to cause lethal effects. At lower concentrations many of these metals serve as required trace elements.

The differing sensitivity of more complex organisms and microorganisms to metals forms the basis of many antiseptic procedures developed over the last 150 years (*see chapter 7*). The noble metals, although microorganisms tend to develop

Figure 40.12   The Mercury Cycle.
Interactions between different
regions— the atmosphere, aerobic
water, and anaerobic sediment—are
critical. Microorganisms in anaerobic
sediments, primarily *Desulfovibrio,*
can transform mercury to methylated
forms that can be transported to water
and the atmosphere. These methylated
forms also undergo biomagnification.
The production of volatile elemental
mercury (Hg^0) releases this metal to
waters and the atmosphere. Sulfide, if
present in the anaerobic sediment, can
react with ionic mercury to produce
less soluble HgS.

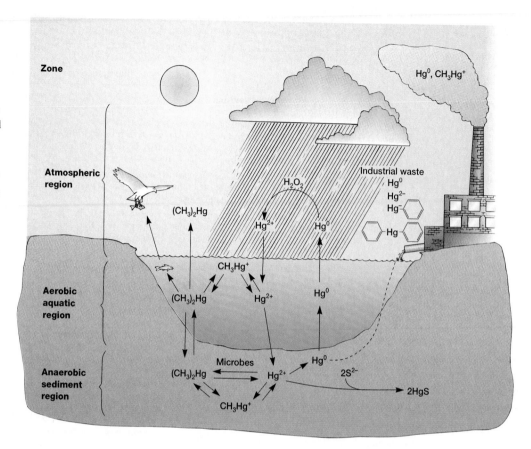

resistance to them, continue to be used in preference to antibiotics in some medical applications. A good example is the treatment of burns with silver-containing antimicrobial compounds.

1. What are different ways in which microorganisms can interact
   with metals? How can microbial activity render metals more toxic
   for humans?

2. Why do metals such as mercury have such major effects on higher
   organisms?

Interactions in Resource Utilization

Microbial succession can occur when organic matter is used as
a source of nutrients and energy. Under aerobic conditions, oxidized products such as nitrate, sulfate, and carbon dioxide (figure 40.13) will result from microbial degradation of complex
organic matter. In comparison, under anaerobic conditions reduced end products tend to accumulate. Use of waste products
of one group of microorganisms by other organisms is a commensalistic relationship (*see chapter 28*) that often is seen in
natural environments. An excellent example of such relationships is the Winogradsky column (*p. 831*).

Microbial interactions and succession also occur when
mixtures of electron acceptors—oxidants—are available, as
noted in table 40.3. If O_2, nitrate, manganese ion, ferric ion,
sulfate, CO_2, or any combination thereof is available in a
particular environment, a predictable sequence of oxidant use

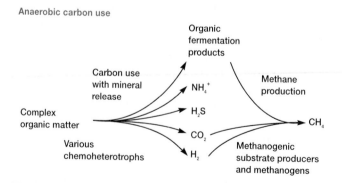

Figure 40.13   The Influence of Oxygen on Decomposition.
Microorganisms form different products when breaking down complex
organic matter aerobically than they do under anaerobic conditions.
Under aerobic conditions oxidized products accumulate, while reduced
products accumulate anaerobically. These reactions also illustrate
commensalistic transformations of a substrate, where the waste
products of one group of microorganisms can be used by a second type
of microorganism.

takes place when an oxidizable substrate is available. Oxygen is employed as an electron acceptor first because it inhibits nitrate use by microorganisms capable of respiration with either O_2 or nitrate. While O_2 is available, sulfate reducers and methanogens are inhibited because these groups are obligate anaerobes.

Electron acceptors and growth energetics (pp. 174–75; 176–78).

Once the O_2 and nitrate are exhausted and fermentation products, including hydrogen, have accumulated, competition for use of other oxidants begins. Oxidized forms of manganese and iron will be used first, followed by competition between sulfate reducers and methanogens. This is influenced by the greater energy yield obtained with sulfate as an electron acceptor. Differences in enzymatic affinity for hydrogen, an important substrate used by both groups, also play a critical role. The sulfate reducer *Desulfovibrio* grows rapidly and uses the available hydrogen at a faster rate than *Methanobacterium*. When the sulfate is exhausted, *Desulfovibrio* no longer oxidizes hydrogen,

and the hydrogen concentration rises. The methanogens finally dominate the habitat and reduce CO_2 to methane.

This order of oxidant use is repeated whenever O_2, nitrate, Mn^{4+}, Fe^{3+}, and/or sulfate again become available. On the basis of these interactions, microbial use of available oxidants can be understood, predicted, and managed.

1. How does commensalism relate to processes of microbial succession?
2. Why is an orderly sequence of oxidant use observed in many natural ecosystems?

Organic Substrate Use by Microorganisms

In the earlier discussion of nutrient cycling, no distinction is made between different types of organic substrates. This is a marked oversimplification because organic substrates differ in their susceptibility to degradation.

Factors influencing degradation include the following:

Elemental composition
Structure of basic repeating units
Linkages between repeating units
Nutrients present in the environment
Abiotic conditions (pH, oxidation-reduction potential, O_2, osmotic conditions)
Microbial community present

The major complex substrates used by microorganisms are summarized in table 40.4. Of these, only previously grown microbial biomass contains all of the nutrients required for microbial growth. Chitin, protein, microbial biomass, and nucleic acids contain nitrogen in large amounts. The remaining complex substrates contain primarily or only carbon, hydrogen, and

TABLE 40.3   Order of Oxidant Use by a Mixed Microbial Community under Successive Aerobic and Anaerobic Conditions (pH 7.0)

Oxidant	Oxidant Couple Reduction Potential (V)	Order of Use	Oxygen Present
Oxygen	+0.814	1	+
Nitrate	+0.741	2	–
Mn^{4+}	+0.401	3	–
Fe^{3+}	–0.185	4	–
SO_4^{2-}	–0.214	5	–
CO_2	–0.244	6	–

Source: Data from K. A. Smith and J. R. M. Arah, "Anaerobic Microenvironments in Soil and the Occurrence of Anaerobic Bacteria" in *FEMS Symposium 33*, p. 258, 1986. Barking, Essex: Elsevier.

TABLE 40.4   Complex Organic Substrate Characteristics That Influence Decomposition and Degradability

Substrate	Basic Subunit	Linkages (if Critical)	C	H	O	N	P	With O_2	Without O_2
			Elements Present in Large Quantity					**Degradation**	
Starch	Glucose	$\alpha(1 \rightarrow 4)$ $\alpha(1 \rightarrow 6)$	+	+	+	–	–	+	+
Cellulose	Glucose	$\beta(1 \rightarrow 4)$	+	+	+	–	–	+	+
Hemicellulose	C6 and C5 sugars	$\beta(1 \rightarrow 4), \beta(1 \rightarrow 3),$ $\beta(1 \rightarrow 6)$	+	+	+	–	–	+	+
Lignin	Phenylpropane	C–C, C–O bonds	+	+	+	–	–	+	–
Chitin	N-acetylglucosamine	$\beta(1 \rightarrow 4)$	+	+	+	+	–	+	+
Protein	Amino acids	Peptide bonds	+	+	+	+	–	+	+
Hydrocarbon	Aliphatic, cyclic, aromatic		+	+	–	–	–	+	+–
Lipids	Glycerol, fatty acids; some contain phosphate and nitrogen	Esters	+	+	+	+	+	+	+
Microbial biomass		Complex	+	+	+	+	+	+	+
Nucleic acids	Purine and pyrimidine bases, sugars, phosphate	Complex	+	+	+	+	+	+	+

oxygen. If microorganisms are to grow by using these substrates, they must acquire the remaining nutrients they need for growth from the environment.

The O_2 relationships for the use of these substrates are also of interest because most of the substrates can be readily degraded with or without oxygen's presence. The major exception is lignin.

Hydrocarbons are unique in that microbial degradation, especially of straight-chained and branched forms, involves the initial addition of molecular O_2. Recently, anaerobic degradation of hydrocarbons with sulfate or nitrate as oxidants has been observed. With sulfate present, organisms of the genus *Desulfovibrio* are active. This occurs only slowly and with microbial communities that have been exposed to these compounds for extended periods. Such degradation may have resulted in the sulfides that are present in "sour gases" associated with petroleum.

Lignin, an important structural component in mature plant materials—composing 1/3 of wood, for example—is a special case in which biodegradability is dependent on O_2 availability. There often is no significant degradation because most filamentous fungi that degrade native lignin in situ can function only with O_2 present. Lignin's lack of biodegradability under anaerobic conditions results in accumulation of lignified materials, including the formation of peat bogs and muck soils. This absence of lignin degradation under anaerobic conditions also is important in construction. Large masonry structures often are built on swampy sites by driving in wood pilings below the water table and placing the building footings on the pilings. As long as the foundations remain water-saturated and anaerobic, the structure is stable. If the water table drops, however, the pilings will begin to rot and the structure will be threatened. Similarly, the cleanup of harbors can lead to decomposition of costly docks built with wooden pilings due to increased aerobic degradation of wood by filamentous fungi. Rumen function provides a final example of the relationship between lignin degradation and oxygen. The rumen (*see chapter 28*), being anaerobic, does not allow significant degradation of lignin present in animal feeds. The use of sugars and carbohydrates in the rumen leaves an inactive residue that can improve soils more effectively than the original feeds.

Patterns of microbial biodegradation are important in many habitats. They contribute to the accumulation of petroleum products, the formation of bogs, and the preservation of valuable historical objects.

---

1. Which common microbial substrate is not degraded to any great extent under anaerobic conditions?

2. Why is decomposition important for the continued functioning of the environment?

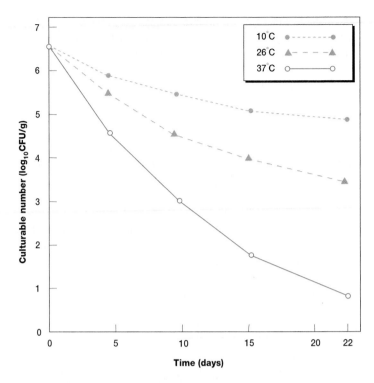

Figure 40.14   Temperature and Microbial Survival. Microorganisms survive better in the environment at lower temperatures. This is illustrated by the survival of *Escherichia coli* in soil at various temperatures.

"Foreign" Derived Microorganisms— Survival and Fate

A characteristic of natural habitats, whether soil, water, or other natural systems, is that most bacteria used as indicators of fecal pollution are not able to survive. The natural habitat for many of these organisms is the nutrient-rich intestinal tract of warm-blooded animals, and these microorganisms appear to have lost the ability to survive in the harsh outside world.

These **allochthonous,** or introduced, microorganisms are constantly being added to natural environments as a part of human and animal wastes and in sewage materials. Generally these vegetative bacteria survive better at lower temperatures (figure 40.14). At extremely low temperatures, including freezing conditions, the survival time of these "foreign" microorganisms is greatly extended.

The detection of low levels of "foreign" bacteria is a major focus in environmental microbiology. If they are pathogens, they may be present at levels below the limit of test sensitivity. *Campylobacter, Legionella,* and *Vibrio* have been of concern, and such microorganisms confront the environmental microbiologist with new challenges (Box 40.2).

Box 40.2

Viable but Uncultured Vegetative Bacteria

For most of microbiology's history, a viable microorganism has been defined as one that is able to grow actively, resulting in the formation of a colony or visible turbidity in a liquid medium. John R. Postgate of the University of Sussex in England was one of the first to note that microorganisms stressed by survival in natural habitats—or in many selective laboratory media—were particularly sensitive to secondary stresses. Such stress can result in the production of viable microorganisms that have lost the ability to grow on media normally used for their cultivation. To determine the growth potential of such microorganisms, Postgate developed what is now called the Postgate Microviability Assay, which allows microorganisms to be cultured in a thin agar film under a coverslip. The potential of a cell to change its morphology, even if it does not grow, indicates that the microorganism does show "life signs."

Since that time many workers, including R. R. Colwell and K. Kogure, have developed additional sensitive microscopic and isotopic procedures to evaluate the presence and significance of these viable but "noncultured" bacteria. As examples, levels of fluorescent antibody and acridine orange-stained cells often are compared with population counts obtained by the most probable number (MPN) method and plate counts using selective and nonselective media. The release of radioactive-labeled cell materials also is used to monitor stress effects on microorganisms. Despite these advances the estimation of substrate-responsive viable cells by the microscopic direct viable count (DVC) method, first proposed by Postgate, is still important. These studies emphasize that even when bacteria such as *Escherichia coli, Vibrio cholerae, Klebsiella pneumoniae, Enterobacter aerogenes,* and *Enterococcus faecalis* have lost their ability to grow on conventional laboratory media using standard cultural techniques, they still might be able to play a role in infectious disease.

The situation in natural environments with mixed populations is much more complex. Here, where often only 1 to 10% of observable cells are able to form colonies, the microbiologist is attempting to culture microorganisms that perhaps never have been cultured or characterized. Their lack of growth leads to the use of another term: that of being "uncultured" or "as yet uncultivated." In the future it is possible that media or proper environmental conditions for their growth will be developed. Examples of "not yet cultured" microorganisms include *Epulopiscium,* many hyperthermophilic archeobacteria in marine environments (*see p. 129–30; 825–30*), and microbes found in hydrothermal vent tube worms (*see pp. 549–51*).

Epulopiscium (See Box 3.1).

Many studies have been directed toward learning why "foreign" microorganisms gradually die after being released to the environment. Among the possibilities are predation by protozoa, parasitism by *Bdellovibrio* (*see chapter 20*) and other organisms, lack of space, lack of nutrients, and the presence of toxic substances. After many years of study, it appears that the major reason "foreign" microorganisms die out is that they cannot compete effectively with indigenous microorganisms for the low amounts of nutrients present in the environment.

Even microorganisms recovered from a particular environment, after growth in the laboratory on rich media, may lose their ability to survive when placed back in their original environment. The cause may be physical or physiological. From a physical standpoint the microorganisms may find themselves outside their protected physical niche, where they can be consumed by protozoa and other predators as noted previously. On the other hand, after growth in rich laboratory media, they may have lost the ability to compete physiologically with the native populations. In bioremediation, added microorganisms usually will only be effective in special situations (*see chapter 44*), and most bioremediation is carried out by stimulating microorganisms that are normally present in the environment.

1. Name several genera of disease-causing microorganisms for which soil or water may be an important route of disease transmission.
2. What is the effect of lower temperatures on the die-out rate of "foreign" microorganisms that are added to the natural environment?
3. Why might microorganisms, after isolation from an environment and growth in the laboratory, lose their ability to survive in the environment they were taken from?
4. Why do "foreign" microorganisms often rapidly die out in a habitat?

Genetically Engineered Microorganisms— Fate and Effects

The procedures by which genetically engineered microorganisms are produced are summarized in chapter 15. These techniques allow the creation of desired microbial variants with specifically altered gene sequences in plasmids or in the genomes of bacteria and fungi. Microorganisms with changed characteristics are now being added to a wide variety of natural environments in an attempt to accelerate processes or to make possible processes that would not normally take place. The assumption is that these microorganisms will be able to function effectively without having longer-term effects on the natural indigenous population.

A major concern has been that such genetically engineered microorganisms (GEMs) will possibly transfer their genetic information to the natural populations, leading to unacceptable changes in microbial community structure or function. Before such modified microorganisms can affect the environment, a series of steps must first occur. These steps include survival and multiplication of the added microorganisms and transfer of the new genetic sequence to other organisms (*see chapter 14*). Each step will occur at a low-probability level, and the possibility of an effect's eventually becoming observable can be estimated.

Assessing the potential effects of genetically engineered microorganisms on the environment and the microbial community is a difficult process, and the uncertainties are many. Several factors limit the rates of genetic exchange in natural habitats, including the general lack of nutrients and the corresponding low growth rates of most microorganisms. The physical separation that occurs when microorganisms attach to surfaces and particles and the tendency of free DNA to adsorb onto clays and other surfaces will limit interactions. Furthermore, DNA, if released from a genetically engineered microorganism, may be degraded by other microorganisms. There is continuing interest in this problem, which is based on the possible occurrence of conjugation, transduction, or transformation (*see chapter 14*). The occurrence of microorganisms that are naturally competent to incorporate free DNA and the stabilization of free DNA on solid surfaces leaves open the possibility of this occurring under natural conditions.

Finally, it should be noted that microorganisms modified by modern genetic engineering techniques may be less fit to compete in low-nutrient-flux natural environments. Additional safeguards can be developed before the microorganisms are released into the environment. Microorganisms can be constructed with special DNA sequences that contain copy blocks, making it less likely that they will replicate the special DNA or exchange it with other microorganisms. Such a modified microorganism also may be less fit to compete and survive because of the additional energetic burden required to maintain the extra DNA. Many genetically modified microorganisms are planned for use in the natural environment. General principles of population genetics, bacterial physiology, epidemiology, and pathogenesis will play important roles in any attempts to assess possible effects of the release of these intentionally altered microorganisms. By the use of molecular techniques, it has been possible to detect transformation and genetic exchange in soils and waters, which increases the importance of studying genetic exchange in the environment.

1. What steps might be required to occur before genetically engineered microorganisms could cause observable effects in the environment?
2. What factors can limit the exchange of genetic information in the environment, especially from genetically engineered microorganisms?

Extreme Environments

Microorganisms vary greatly in their tolerance for pH, oxidized versus reduced environments, temperature, pressure, salinity, water availability, and ionizing radiation (*see chapter 6*). Thus, environmental factors greatly influence the survival of any particular microorganism. An extreme environment is one in which physical or chemical conditions become even more restrictive, resulting in a decrease in the diversity of microbial types that can maintain themselves. Under particularly restrictive conditions (figure 40.15; *see also figure 23.12*) this process can continue until only a single type of microorganism, essentially a monoculture, exists.

From the viewpoint of the successful microorganism, however, there may be less justification for describing such an environment as extreme. The habitat might be optimum for it and actually required for the microorganism's continued existence. Important extreme environments and major microbial groups that have been observed in them are summarized in table 40.5.

Many microbial genera have specific requirements for survival and functioning in such so-called extreme environments. For example, a high sodium ion concentration is required to maintain membrane integrity in many halophilic bacteria, including members of the genus *Halobacterium*. Halobacteria require a sodium ion concentration of at least 1.5 M, and about 3 to 4 M for optimum growth.

Halophilic bacteria (pp. 125; 485–86).

The bacteria found in deep-sea environments have different pressure requirements, depending on the depth from which they are recovered. These bacteria can be described as **barotolerant bacteria** (growth from approximately 1 to 500 atm), **moderately barophilic bacteria** (growth optimum 5,000 meters, and still able to grow at 1 atm), and **extreme barophilic bacteria,** which require approximately 400 atm or higher for growth (*see chapter 41*).

Intriguing changes in basic physiological processes occur in microorganisms functioning under extreme acidic or alkaline conditions. These acidophilic and alkalophilic microorganisms have markedly different problems in maintaining a more neutral internal pH and chemiosmotic processes (*see chapter 9*). Obligately acidophilic microorganisms can grow at a pH of 3.0, and a difference of 4.0 pH units can exist between the interior and exterior of the cell. These acidophiles include members of the genera *Thiobacillus, Bacillus, Sulfolobus,* and

(a)

(b)

(c)

Figure 40.15   Microorganisms Growing in Extreme Environments. Many microorganisms are especially suited to survive in extreme environments. (*a*) Salterns turned red by halophilic algae and halobacteria. (*b*) A hot spring colored green and blue by cyanobacterial growth. (*c*) A source of acid drainage from a mine into a stream. The soil and water have turned red due to the presence of precipitated iron oxides caused by the activity of bacteria such as *Thiobacillus*.

TABLE 40.5	Characteristics of Extreme Environments in which Microorganisms Grow	
Stress	**Environmental Conditions**	**Microorganisms Observed**
Temperature	110–115°C, deep marine trenches	*Methanopyrus kandleri* *Pyrodictium abyssi*
	67–102°C, marine basins	*Pyrococcus abyssi*
	85°C, hot springs	*Thermus* *Sulfolobus*
	75°C, sulfur hot springs	*Thermothrix thiopara*
Osmotic stress	13–15% NaCl	*Chlamydomonas*
	25% NaCl	*Halobacterium* *Halococcus*
Acidic pH	pH 3.0 or lower	*Saccharomyces* *Thiobacillus*
Basic pH	pH 10.0 or above	*Bacillus*
Low water availability	$a_w = 0.6–0.65$	*Torulopsis* *Candida*
Temperature and low pH	85°C, pH 1.0	*Cyanidium* *Sulfolobus acidocaldarum*
Pressure	500–1,035 atm	*Colwellia hadaliensis*

Thermoplasma. The higher relative internal pH is maintained by a net outward translocation of protons. This may occur as the result of unique membrane lipids, hydrogen ion removal during reduction of oxygen to water, or the pH-dependent characteristics of membrane-bound enzymes.

The extreme alkalophilic microorganisms grow at pH values of 10.0 and higher and must maintain a net inward translocation of protons. These obligate alkalophiles cannot grow below a pH of 8.5 and are often members of the genus *Bacillus*; *Micrococcus* and *Exiguobacterium* representatives have also been reported. Some photosynthetic cyanobacteria also have similar characteristics. Increased internal proton concentrations may be maintained by means of coordinated hydrogen and sodium ion fluxes.

Observations of microbial growth at temperatures approaching 115 to 120°C in thermal vent areas (Box 40.3; *see also Box 6.1*) reveal that the limits of temperature on microbial growth have not yet been determined. This area will continue to be a fertile field for investigation. For some successful microorganisms, an extreme environment may not be "extreme" but required and even, perhaps, ideal.

Thermophilic microorganisms (pp. 129–30; 486–88).

Summary

1. Microorganisms exist in populations and communities, and their interactions with the abiotic environment contribute to the development of different ecosystems. The normal physiological state of microorganisms in natural environments is starvation unless resources allowing growth become available.

2. Biogeochemical cycling involves oxidation and reduction processes, and changes in the concentrations of gaseous cycle components, such as carbon, nitrogen, and sulfur can result from microbial activity.

3. Microorganisms serve as primary producers that accumulate organic matter. In addition, many chemoheterotrophs decompose the organic matter that primary producers accumulate and carry out mineralization, the release of inorganic nutrients from organic matter.

4. Major polymers used by microorganisms differ in structure, linkage, elemental composition, and susceptibility to degradation under aerobic and anaerobic conditions. Lignin is degraded only under aerobic conditions, a fact that has important implications in terms of carbon retention in the biosphere.

5. In terms of effects on humans, metals can be considered in three broad groups: the noble metals, metals for which toxic organometallic compounds can be formed, and certain other metals. The second of these groups is of particular concern.

6. Most disease-causing and indicator bacteria from the intestinal tract of humans and other higher organisms do not survive in the environment. At lower temperatures, however, increased survival can occur. Some disease-causing microorganisms may have environmental reservoirs.

7. Genetically engineered microorganisms will be of continuing concern, particularly if foreign DNA is maintained and transferred in the environment.

8. Decreased species diversity usually occurs in extreme environments, and many microorganisms that can function in such habitats have specialized growth requirements. For them, extreme environments can be required environments.

9. Many approaches can be used to study microorganisms in the environment. These include nutrient cycling, biomass, numbers, activity, and community structure analyses. It is now possible to study the genetic characteristics of microorganisms that cannot be grown in the laboratory.

10. Hyperthermophiles, or organisms with optimum growth temperatures above 80°C, can provide a unique source of thermostable enzymes. Such enzymes may have unique biotechnological applications.

Key Terms

allochthonous 822

assimilatory reduction 816

bacteriostasis 814

barotolerant bacteria 824

biochemical oxygen demand (BOD) 826

biofilms 813

biogeochemical cycling 814

biomagnification 819

chemical oxygen demand (COD) 826

community 813

consumer 811

decomposer 811

denitrification 816

dissimilatory reduction 816

ecosystem 811

extreme barophilic bacteria 824

food web 813

fungistasis 814

gradient 813

heterotrophic nitrification 816

Liebig's law of the minimum 813

microbial ecology 813

microbiostasis 814

microenvironment 813

mineralization 813

moderately barophilic bacteria 824

natural environment 811

nitrification 816

nitrogenase 817

nitrogen fixation 817

population 813

primary producer 811

primary production 811

Shelford's law of tolerance 813

sulfate reduction 816

total organic carbon (TOC) 826

Questions for Thought and Review

1. How might you attempt to grow a microorganism in the laboratory to increase its chances of being a strong competitor when placed back in a natural habitat? Do you think that this is even possible?

2. Considering the possibility of microorganisms functioning at temperatures approaching 120°C, what do you think the limiting factor for microbial growth at higher temperatures will be and why?

3. Where in their bodies do most people have noble metals? Why have these been used with more success than nonmetal materials, which have been tested over many decades?

4. Compare the degradation of lignin and cellulose by microorganisms. What different environmental factors are required for these important polymers to be degraded?

5. You have isolated DNA from an environmental sample that contains the genetic information of microorganisms that you cannot grow in the laboratory. What will you do with this DNA and why?

Additional Reading

General

Atlas, R. M., and Bartha, R. 1992. *Microbial ecology: Fundamentals and applications*, 3d ed. Redwood City, Calif.: Benjamin/Cummings.

Fletcher, M., and Gray. T. R. G., editors. 1987. "Ecology of microbial communities." In *41st Symposium of the Society for General Microbiology*. New York: Cambridge University Press.

Margulis, L.; Chase, D.; and Guerrero, R. 1986. Microbial communities. *BioScience* 36(3):160–70.

Margulis, L., and Fester, R. 1991. *Symbiosis as a source of evolutionary innovation: Speciation and morphogenesis*. Cambridge, Mass.: MIT Press.

Mitchell, R., editor. 1992. *Environmental microbiology*. New York: John Wiley.

The Physiological State of Microorganisms in the Environment

Barker, J., and Brown, M. R. W. 1994. Trojan horses of the microbial world: protozoa and the survival of bacterial pathogens in the environment. *Microbiology* 140:1253–59.

Cavanaugh, C. M.; Levering, P. R.; Maki, J. S.; Mitchell, R.; and Lidstrom, M. E. 1987. Symbiosis of methylotropic bacteria and deep-sea mussels. *Nature* 325:346–48.

Deming, J. W.; Somers, L. K.; Straube, W. L.; Swartz, D. G.; and Macdonell, M. T. 1988. Isolation of an obligately barophilic bacterium and description of *Colwellia*, new genus. *Syst. Appl. Microbiol.* 10:152–60.

Fletcher, M. 1991. The physiological activity of bacteria attached to solid surfaces. *Adv. Microb. Physiol.* 32:53–85.

Morita, R. Y. 1993. Bioavailability of energy in the starvation state. In *Starvation in bacteria*, S. Kjelleberg, editor, 1–25. New York: Plenum Press.

Nutrient Cycling Processes

Lovley, D. K. 1991. Dissimilatory Fe (III) and Mn (IV) reduction. *Microbiol. Rev.* 55:259–87.

Lovley, D. K. 1993. Anaerobes into heavy metal: Dissimilatory metal reduction in anoxic environments. *Trends Ecol. & Evol.* 8:213–17.

Moore, M. D., and Kaplan, S. 1994. Members of the family *Rhodospirillaceae* reduce heavy-metal oxyanions to maintain redox poise during photosynthetic growth. *ASM News* 60:17–23.

Stacey, G.; Burris, R. H.; and Evans, H. J. 1991. *Biological nitrogen fixation*. New York: Chapman & Hall.

Whitman, W. B., and Rogers, J. E. 1991. *Microbial production of greenhouse gases: Methane, nitrogen oxides, and halomethanes*. Washington, D.C.: American Society for Microbiology.

Genetically Engineered Microorganisms—Fate and Effects

Lorenz, M. G., and Wackernagel, W. 1994. Bacterial gene transfer by natural genetic transformation in the environment. *Microbiol. Rev.* 58:563–93.

Prosser, J. I. 1994. Molecular marker systems for detection of genetically engineered micro-organisms in the environment. *Microbiology* 140:5–17.

Stewart-Tull, D. E. S., and Sussman, M. 1992. *The release of genetically modified microorganisms—REGEM 2*. New York: Plenum Press.

Stotzky, G.; Broder, M. W.; Doyle, J. D.; and Jones, R. A. 1993. Selected methods for the detection and assessment of ecological effects resulting from the release of genetically engineered microorganisms to the terrestrial environment. *Adv. Appl. Microbiol.* 38:1–98.

Extreme Environments

Adams, M. W. W. 1993. Enzymes and proteins from organisms that grow near and above 100°C. *Ann. Rev. Microbiol.* 47:627–58.

Adams, M. W. W.; Perler, F. B.; and Kelly, R. M. 1995. Extremozymes: Expanding the limits of biocatalysis. *Bio\Technology* 13:662–68.

Gottschal, J. C., and Prins, R. A. 1991. Thermophiles: A life at elevated temperatures. *Trends Ecol. & Evol.* 6:157–62.

Hively, W. 1993. Life beyond boiling. *Discover* May:87–91.

Karl, D. M. 1995. *Microbiology of deep-sea hydrothermal vents*. Boca Raton, Fla.: CRC Press.

Krulwich, T. A., and Ivey, D. M. 1990. Bioenergetics in extreme environments. In *The Bacteria*, vol. XII. I. C. Gunsalus, J. R. Sokatch, and L. N. Ornston, editors, 417–35. New York: Academic Press.

Tunnicliffe, V. 1992. Hydrothermal-vent communities of the deep sea. *American Scientist* 80:336–49.

Methods Used in Environmental Studies

Amann, R. I.; Ludwig, W.; and Schleifer, K.-H. 1995. Phylogenetic identification and in situ detection of individual microbial cells without cultivation. *Microbiol. Rev.* 59(1):143–69.

Atlas, R. M.; Sayler, G. S.; Burlage, R. S.; and Bej, A. K. 1992. Molecular approaches for environmental monitoring of microorganisms. *BioTechniques* 12(5):706–17.

Gerhardt, P.; Murray, R. G. E.; Wood, W. A.; and Krieg, N. R. 1993. *Methods for general and molecular bacteriology*. Washington, D.C.: ASM Press.

Kelleer, G. H., and Manak, M. M., 1994. *DNA probes*, 2d ed. New York: Stockton Press.

Kemp, P. F.; Sherr, B. F.; Sherr, E. B.; and Cole, J. J. 1993. *Handbook of methods in aquatic microbial ecology*. Boca Raton: Lewis Publishers.

Kepner, R. L., and Pratt, J. R. 1994. Use of fluorochromes for direct enumeration of total bacteria in environmental samples: Past and present. *Microbiol. Rev.* 58:603–15.

Towner, K. J., and Cockayne, A. 1993. *Molecular methods for microbial identification and typing*. New York: Chapman & Hall.

Weaver, R. W.; Angle, J. S.; Bottomley, P. J.; Bezdicek, D. F.; Smith, J. S.; Tabatabai, M. A.; and Wollum, A. G. 1994. *Methods of soil analysis: Part 2. Microbiological and biochemical properties*. Madison, Wisc.: Soil Science Soc. Am.

CHAPTER 41
Marine and Freshwater Environments

*W*ater is a very good servant,

but it is a cruel master.

      —John Bullein

*T*he world turns softly

Not to spill its lakes and

rivers

The water is held in its arms

And the sky is held in the

water

What is water, That pours

silver

And can hold the sky?

      —Hilda Conkling

Concepts

1. Marine environments represent the largest volume of water on the earth. Most of this water is at low temperature (2 to 3°C) and under high pressure.

2. Phytoplankton (photosynthetic microorganisms) form the basis of primary production in most marine and freshwater environments.

3. Deep hydrothermal vents and hydrocarbon seeps provide unique environments for microorganisms to function without light.

4. Surfaces and biofilms are important in aquatic and marine environments. Biofilms can vary in complexity, ranging from single cells on surfaces to complex systems with cell aggregates, conduits, and interstitial spaces.

5. Microbial mats are found in many marine and freshwater environments. These complex microbial communities also occur as fossils and represent one of the earliest forms of microbial communities, with some being 3.5 billion years old.

6. Dissolved O_2 is present at low concentrations in most waters. If sufficient organic matter is present, microorganisms use O_2 faster than it can be replenished. This can result in cyclic changes in dissolved O_2 or in complete depletion of O_2.

7. The biological utilization of organic wastes follows regular and predictable sequences. Once these sequences are understood, more efficient sewage treatment systems can be created.

8. Sewage treatment can be carried out using large vessels where mixing and aeration can be controlled (conventional treatment). Constructed wetlands, where aquatic plants and their associated microorganisms are used, now are finding widespread applications in the treatment of liquid wastes.

9. Indicator organisms, which usually die off at slower rates than many disease-causing microorganisms, can be used to evaluate the microbiological quality of water.

10. Disease-causing viruses and protozoa may be present in water that meets drinking water standards set in terms of indicator organisms.

11. Many important disease-causing microorganisms can be present in water not contaminated by human wastes.

12. Groundwater is an important source of water, especially in rural areas. This resource is of increasing concern to microbiologists.

Of all the water found on earth, 97% is marine. Most of this water is at a temperature of 2 to 3°C and devoid of light; 62% is under high pressure (>100 atm). Microscopic phytoplankton and associated bacteria create a complex food web that can extend over long distances and extreme depths. The marine environment seems so vast that it cannot be affected by pollution; however, in coastal areas human activities are increasingly disrupting microbial processes and damaging water quality.

Fresh waters, although a small part of the waters on earth, are extremely important as a source of drinking water. In many locations the contamination of surface and subsurface waters by domestic and industrial wastes causes environmental problems.

Marine and freshwater environments create unique niches for many specialized microorganisms not found in habitats without a continuous water phase.

The Nature of Marine and Freshwater Environments

Nutrient Levels

Nutrient levels in marine and freshwater environments can vary from extremely low, in the range of micrograms of organic matter per liter, to levels approaching those in laboratory culture media. High nutrient levels are found in polluted environments and sewage treatment plants, for example. With changes in nutrient levels, shifts between low-nutrient responsive

(oligotrophic) and high-nutrient responsive (copiotrophic) microorganisms can occur. Nutrient turnover rates also vary. In marine environments the turnover time for nutrient processing may range from hundreds to thousands of years. In contrast, marsh and estuarine areas may have rapid rates of nutrient turnover, and a complex, diverse microbial community of rapidly responding microorganisms is present.

Gradients

A major factor in aquatic environments is the movement of materials, whether they be gaseous, solid, or in solution. These changes in concentration are part of the world of aquatic microorganisms; these microorganisms are able to respond rapidly to select their most suitable environment.

The mixing and movement of nutrients, O_2, and waste products that occur in freshwater and marine environments are the dominant factors controlling the microbial community. For example, in deep lakes or oceans, organic matter from the surface can sink to great depths, creating nutrient-rich zones where decomposition takes place. Gases and soluble wastes produced by microorganisms in these deep marine zones can move into upper waters and stimulate the activity of other microbial groups. Similar processes take place on a lesser scale in nutrient-rich lakes, and even in microbial mats (*see p. 833*) where gradients are established on a scale of millimeters.

The **Winogradsky column** illustrates many of these principles (figure 41.1). In this microcosm a layer of reduced mud is mixed with sodium sulfate, sodium carbonate, and shredded

Figure 41.1   The Winogradsky Column. This forms a microcosm where microorganisms and nutrients interact over a vertical gradient. Fermentation products and sulfide migrate up from the reduced lower zone, and oxygen penetrates from the surface. This creates conditions similar to those in a lake with nutrient-rich sediments. Light is provided to stimulate the penetration of sunlight into the anaerobic lower region, which allows photosynthetic microorganisms to develop.

newspaper—a cellulose source—and additional mud and water are placed in the column, which is then incubated in the light. A series of reactions occurs as the column begins to mature, with particular members of the microbial community developing in specific microenvironments in response to chemical gradients.

In the bottom of the column, cellulose is degraded to fermentation products by the genus *Clostridium* (*see chapter 21*). With these fermentation products available as reductants and using sulfate as an oxidant, *Desulfovibrio* produces hydrogen sulfide (*see chapter 20*). The hydrogen sulfide diffuses upward toward the oxygenated zone, creating a stable hydrogen sulfide gradient. In this gradient the photoautotrophs *Chlorobium* and *Chromatium* develop as visible olive green and purple zones. These microorganisms use hydrogen sulfide as an electron source, and CO_2, from sodium carbonate, as a carbon source. Above this region the purple nonsulfur bacteria of the genera *Rhodospirillum* and *Rhodopseudomonas* can grow. These photoheterotrophs use organic matter as an electron donor under anaerobic conditions and function in a zone where the sulfide level is lower. Both O_2 and hydrogen sulfide may be present higher in the column, allowing specially adapted microorganisms to function. These include *Beggiatoa* and *Thiothrix,* which use reduced sulfur compounds as a reductant and O_2 as an oxidant. In the upper portion of the column, diatoms and cyanobacteria may be visible.

Green and purple photosynthetic bacteria (pp. 456–59).
Cyanobacteria (pp. 459–63).

These commensalistic microorganisms, which develop sequentially, are dependent on the reductant originally provided as the cellulose or plant materials. When this reductant is exhausted, the column gradually becomes oxidized and the sulfide-dependent photosynthetic microorganisms and other anaerobes can no longer maintain themselves in the microcosm.

Oxygen and carbon dioxide, two important gases in water, also develop concentration gradients. Oxygen is only sparingly soluble in water. Its solubility is influenced by the concentration in the gas phase, the water temperature, the gas pressure, and dissolved salts. Temperature and pressure, especially, control the amount of O_2 available for use by microorganisms (table 41.1). At lower temperatures the O_2 concentration can be markedly higher. Rapid depletion of dissolved O_2 can occur when nutrients contaminate the water, often leading to fish kills and the transformation of lakes and rivers into open sewers.

The second major gas in water, CO_2, plays many important roles in chemical and biological processes. The carbon dioxide-bicarbonate-carbonate equilibrium can control the pH in weakly buffered waters, or it can be controlled by the pH of strongly buffered waters. As shown in figure 41.2, the pH of distilled water, which is not buffered, is determined by the dissolved

TABLE 41.1 The Effects of Water Temperature and Elevation on Dissolved O_2 Levels (mg/liter) in Water

Temperature (°C)	Elevation above Sea Level (Meters)			
	0	1,000	2,000	3,000
0	14.6	12.9	11.4	10.2
5	12.8	11.2	9.9	8.9
10	11.3	9.9	8.8	7.9
15	10.0	8.9	7.9	7.1
25	9.1	8.1	7.1	6.4
30	8.2	7.3	6.4	5.8
35	7.5	6.6	5.9	5.3
40	6.9	6.1	5.4	4.9

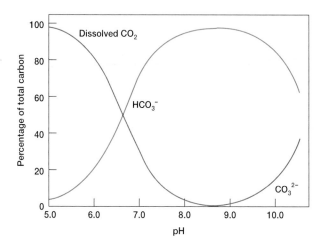

Figure 41.2 The Relationship of pH to Dissolved CO_2.
Atmospheric gases can affect the physical characteristics of water. The pH of water is influenced by the amount of dissolved carbon dioxide in the water, and also by the equilibrium of the dissolved carbon dioxide with bicarbonate and carbonate ions.

CO_2 in equilibrium with the air, and is approximately 5.0 to 5.5. In comparison, water strongly buffered at pH 8.0 contains CO_2 absorbed from the air, which is present primarily as bicarbonate. When autotrophic microorganisms such as algae use CO_2, the pH of many waters will be increased.

1. Why do nutrient levels vary in aquatic environments? What terms are used to describe low- and high-nutrient responsive bacteria?

2. What are the reasons for adding cellulose, sodium sulfate, and sodium carbonate to the Winogradsky column? Discuss this in relation to microbial groups that respond to these materials or their products.

3. What are the two most important gases in aquatic environments? What factors influence their concentrations in water?

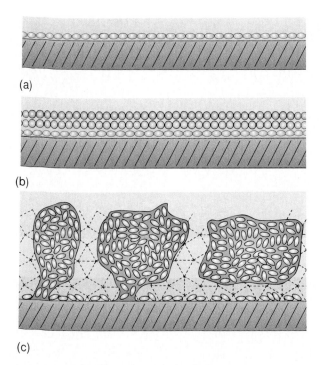

(a)

(b)

(c)

Figure 41.3   The Growth of Biofilms. Biofilms can develop and become extremely complex, depending on the energy sources that are available. (*a*) Initial colonization by a single type of bacterium. (*b*) Development of a more complex biofilm with layered microorganisms of different types. (*c*) A mature biofilm with cell aggregates, interstitial pores, and conduits.

Surfaces and Biofilms

Differences in nutrient levels are accentuated as one moves from the bulk phase of water to surfaces where both nutrients and microorganisms tend to accumulate. As a general rule, the "balance point" at which surface attachment gives microorganisms an advantage is approximately 25 mg of dissolved organic matter per liter. As nutrient levels decrease, the advantage of being associated with surfaces increases. The significance of surfaces for microbial colonization was established by many earlier microbiologists, including N. L. Söhngen (early 1900s) and C. E. Zobell (1940s–1950s). Zobell established the role of macromolecular adhesion for nutrient concentration and also found that bacterial adhesion is a two-step process, involving reversible and irreversible adhesion.

With lower nutrient levels, one can begin to observe the development of simple **biofilms** as an important factor in microbial growth (figure 41.3*a*). With increased time and nutrient accumulation on the surface, microorganisms also can attach and form a monolayer of microorganisms.

Depending on the energetic situation (light, nutrients present) and diffusion rates, these biofilms can become more complex with layers of organisms of different types (figure 41.3*b*). A typical example would involve photosynthetic organisms on the surface, facultative chemoorganotrophs in the middle, and possibly sulfate-reducing microorganisms on the bottom.

More complex biofilms can develop to form a three-dimensional structure with cell aggregates, interstitial pores, and conduit channels (figure 41.3*c*). This structure allows nutrients to reach the biomass, and the channels are shaped by protozoa that graze the bacteria.

Our understanding of the nature of such complex biofilms has been assisted by the use of a new and powerful microscopic technique—that of **confocal scanning laser microscopy (CSLM;** figure 41.4)—as described in Box 41.1. With CSLM it is possible to generate optical "thin sections" of 1 µm thickness or less across a complex biofilm. With computer simulations, models can be constructed from these "thin sections" to better understand how microorganisms function in this complex environment.

The occurrence of such complex biofilms has a wide impact in environmental and public health microbiology. For example, such biofilms may be a major source of *Legionella* in heated waters and also in potable water systems (*see pp. 744–45*).

Biofilms in medicine (p. 760).

Microbial Mats

Depending on environmental conditions, biofilms can become so large that they are visible and have macroscopic dimensions. Bands of microorganisms of different colors are evident as shown in figure 41.5. These thick biofilms, called **microbial mats,** are found in many freshwater and marine environments. These mats are complex layered microbial communities that can form at the surface of rocks or sediments in hypersaline and freshwater lakes, lagoons, hot springs, and beach areas. They consist of microbial filaments, including cyanobacteria. A major characteristic of mats is the extreme gradients that are present. Light only penetrates approximately 1 mm into these communities, and below this photosynthetic zone, anaerobic conditions occur. As is the case at the bottom of a Winogradsky column, sulfate-reducing bacteria play a major role in mats. The sulfide that these organisms produce diffuses to the anaerobic lighted region, allowing sulfur photosynthetic microorganisms to grow. Some believe that microbial mats could have allowed the formation of terrestrial ecosystems prior to the development of vascular plants, and fossil microbial mats, called **stromatolites,** have been dated at over 3.5 billion years old. Molecular techniques and stable isotope measurements (*see table 40.6*) are being used to better understand the development of these unique microbial communities.

---

1. Waters allow microorganisms to exist that are not common in soils. What are some of these microorganisms and what are their unique characteristics?

2. Biofilms can vary in complexity. What are different types of biofilms that can be formed?

3. What are microbial mats? In what types of environments do they occur?

4. What is the estimated age of the oldest stromatolites?

runoff sediment from riparian areas (the edge of a river), or leaves and other organic matter falling directly into the water (figure 41.11*b*). Chemoorganotrophic microorganisms metabolize the available organic material and provide an energy base for the ecosystem. Under most conditions the amounts of organic matter added to streams and rivers will not exceed the system's oxidative capacity, and productive, aesthetically pleasing streams and rivers will be maintained.

The capacity of streams and rivers to process such added organic matter is, however, limited. If too much organic matter is added, the water may become anaerobic. This is especially the case with urban and agricultural areas located adjacent to streams and rivers. The release of inadequately treated municipal wastes and other materials from a specific location along a river or stream represents a **point source of pollution.** Such point source additions of organic matter can produce distinct and predictable changes in the microbial community and available oxygen, creating an oxygen sag curve (figure 41.12). Runoff from fields and feedlots, which causes algal blooms in eutrophic water bodies, is an example of a **nonpoint source of pollution.**

When the amount of organic matter added is not excessive, the algae will grow using the minerals released from the organic matter. This leads to the production of O_2 during the daylight hours, and respiration will occur at night farther down the river, resulting in **diurnal oxygen shifts.** Eventually the O_2 level approaches saturation, completing the self-purification process. This demand for oxygen is usually expressed in terms of the biochemical oxygen demand (BOD) or the chemical oxygen demand (COD).

The biochemical oxygen demand is an indirect measure of organic matter in aquatic environments. It is the amount of dissolved O_2 needed for microbial oxidation of biodegradable organic matter. When O_2 consumption is measured, the O_2 itself must be present in excess and not limit oxidation of the nutrients (table 41.3). To achieve this, the waste sample is diluted to assure that at least 2 mg/liter of O_2 are used while at least 1 mg/liter of O_2 remains in the test bottle. Ammonia released during organic matter oxidation can also exert an O_2 demand in the BOD test, so nitrification or the **nitrogen oxygen demand (NOD)** is often inhibited by nitrapyrin. In the normal BOD test,

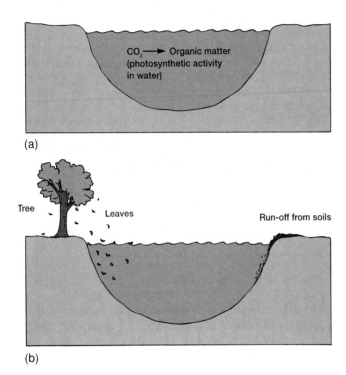

Figure 41.11   Organic Matter Sources for Lakes and Rivers. Organic matter used by microorganisms in lakes and rivers can be synthesized in the water, or can be added to the water from outside. (*a*) Within-stream sources of organic matter (autochthonous sources), primarily photosynthesis, and (*b*) sources of organic matter which enter streams and rivers from outside the water (allochthonous sources). Stream cross sections are shown.

Figure 41.12   The Dissolved Oxygen Sag Curve. Microorganisms and their activities can create gradients over distance and time when nutrients are added to rivers. An excellent example is the dissolved oxygen sag curve, caused when organic wastes are added to a clean river system. During the later stages of self-purification, the phototrophic community will again become dominant, resulting in diurnal changes in river oxygen levels.

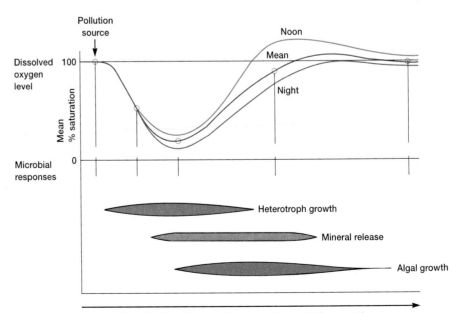

which is run for 5 days at 20°C on untreated samples, nitrification is not a major concern. However, when treated effluents are analyzed, NOD can be a problem.

The chemical oxygen demand test involves measuring the amount of an oxidizing agent such as permanganate that is consumed when the organic matter in a water sample is oxidized completely to CO_2. Organic matter in a sample also can be reacted directly with O_2 at a high temperature to produce CO_2. The CO_2 is then quickly measured by sensitive infrared techniques or potentiometric procedures to give a measure of **total organic carbon** or **TOC.**

Limits on BOD and COD levels in sewage and other waste water have been established. Properly treated low BOD and COD effluents that meet the standards have minimum effects on lakes, rivers, and marine outfall areas.

Microbial biomass measurements, including estimates of the photosynthetic and heterotrophic components in a body of water, are very important in the study of aquatic microorganisms, as noted in table 40.6. The photosynthetic microorganisms in water can be measured by extracting chlorophyll and other photosynthetic pigments. By measuring absorption at a series of wavelengths under neutral and acidic conditions, one can determine the quantity of pheophytin, a chlorophyll molecule with the Mg replaced by two hydrogen atoms. This product is an important index of the physiological stress being placed on the algae.

Microscopic and chemical techniques can also be used to estimate microbial numbers and biomass in water. However, it is often necessary to concentrate the microorganisms by filtration or centrifugation before these measurements can be completed. Bacteria can also be separated from the algae, protozoa, aquatic insects, and animals that may be present by means of centrifugation or filtration.

If too much organic matter (BOD, COD) is added to a river or lake, however, these changes may not be easily reversed and an anaerobic foul-smelling body of water that will not support fish and other oxygen-requiring organisms can be created. At this point the lake or river is almost "dead" biologically, although important groups of microorganisms are still present and functioning. Once a body of water reaches this point, only major remediation efforts will restore it to its original condition and allow fish and other aquatic animals to survive.

TABLE 41.3 **The Biochemical Oxygen Demand (BOD) Test: A System with Excess and Limiting Components**

Components in Excess at the End of the Incubation Period

Nitrogen
Phosphorus
Iron
Trace elements
Microorganisms
Oxygen

Component Limiting at the End of the Incubation Period

Organic matter

---

1. Define the following terms: eutrophication, epilimnion, hypolimnion, and thermocline.
2. Define COD and BOD. What are the major methods used to measure microbial biomass in water?
3. Why is filtration often required to concentrate bacteria from water? What often happens when this technique is used with soils?
4. What components should limit the reactions in a BOD test, and what components should not limit reaction rates? Why?

Nutrient Removal from Water

The aerobic self-purification sequence that occurs when organic matter is added to lakes and rivers can be carried out under controlled conditions where natural processes are intensified. This often involves the use of large basins (conventional sewage treatment) where mixing and gas exchange are carefully controlled. Purification also can be carried out using constructed wetlands where natural reed and aquatic plant communities (and their associated microorganisms) process dissolved nutrients. Such procedures can be effective in minimizing environmental degradation of valuable water resources and in destroying potential human pathogens.

Conventional **sewage treatment** as shown in figure 41.13, normally involves **primary, secondary,** and **tertiary treatment,** as summarized in table 41.4. Physical, or primary, treatment can remove 20 to 30% of the BOD that is present in particulate form. In this treatment, particulate material is removed by screening, precipitation of small particulates, and settling in basins or tanks. The resulting solid material is usually called **sludge.** Biological, or secondary, treatment is then used for the removal of dissolved organic matter. About 90 to 95% of the BOD and many bacterial pathogens are removed by this process.

Figure 41.13 A Modern Conventional Sewage Treatment Plant. Sewage treatment plants allow natural processes of self-purification that occur in rivers and lakes to be carried out under more intense, managed conditions in large concrete vessels. The figure shows a plant in New Jersey. The trickling filters and activated sludge basins are particularly prominent.

TABLE 41.4 **Major Steps in Primary, Secondary, and Tertiary Treatment of Wastes**

Treatment Step	Processes
Primary	Removal of insoluble particulate materials by screening, addition of alum and other coagulation agents, and other physical procedures

Secondary	Biological removal of dissolved organic matter
	  Trickling filters
	  Activated sludge
	  Lagoons
	  Extended aeration systems
	  Anaerobic digesters

Tertiary	Biological removal of inorganic nutrients Chemical removal of inorganic nutrients Virus removal/inactivation Trace chemical removal

Several approaches can be used in secondary treatment to remove dissolved organic matter (figure 41.14). An aerobic activated sludge system (figure 41.14*a*) involves a horizontal flow of materials with a recycle of sludge—the active biomass that is formed when organic matter is oxidized and degraded by microorganisms. Activated sludge systems can be designed with variations in mixing. In addition, the ratio of organic matter added to the active microbial biomass can be varied. A low-rate system (low nutrient input per unit of microbial biomass) will produce a cleaner effluent. A high-rate system (high nutrient input per unit of microbial biomass) will remove more dissolved organic carbon per unit time, but produce a poorer quality effluent.

Aerobic secondary treatment also can be carried out with a trickling filter (figure 41.14*b*). The waste effluent is passed over rocks or other solid materials upon which microbial films have developed, and the microbial community in these films degrades the organic waste. A sewage treatment plant can be operated to pro-

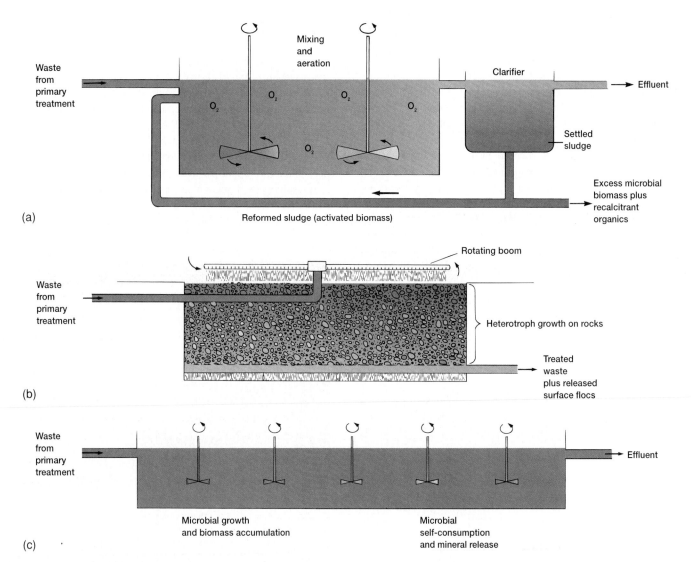

Figure 41.14 Aerobic Secondary Sewage Treatment. Different types of reactors can be used to carry out the microbial use of dissolved organic matter in secondary aerobic treatment of sewage. (*a*) Activated sludge with microbial biomass recycling. (*b*) Trickling filter. (*c*) Extended aeration without biomass recycling.

(a)   (b)   (c)

Figure 41.15   Filamentous Microorganisms in Bulking Sludge. If these form major growths during aerobic sludge treatment, they can cause bulking (nonsettling) of sludges. (*a*) *Sphaerotilus.* (*b*) *Thiothrix* without sulfide present. (*c*) *Thiothrix* with sulfide in the environment showing elemental sulfur granules.

duce less sludge by employing the **extended aeration** process (figure 41.14*c*). Microorganisms grow on the dissolved organic matter, and the newly formed microbial biomass is eventually consumed by **endogenous respiration.** This requires extremely large aeration basins and extended aeration times. In addition, with the biological self-utilization of the biomass, minerals originally present in the microorganisms are again released to the water.

When the activated sludge process is run under less than optimum conditions, particularly with lower O_2 levels or with a microbial community that is too young or too old, unsatisfactory floc formation and settling can occur. The result is a **bulking sludge,** caused by the massive development of filamentous bacteria such as *Sphaerotilus* and *Thiothrix,* together with many poorly characterized filamentous organisms (figure 41.15).

All of the aerobic processes produce excess microbial biomass, or sewage sludge, which contains many recalcitrant organics. Often the sludges from aerobic sewage treatment, together with the materials settled out in primary treatment, are further treated by **anaerobic digestion** (figure 41.16). Anaerobic digesters are large fermentation tanks designed to operate anaerobically with continuous input of untreated sludge and removal of the final, stabilized sludge product. The resulting methane is removed by a vent at the top of the tank. This digestion process involves three steps: (1) the fermentation of the sludge components to form organic acids, including acetate; (2) production of the methanogenic substrates: acetate, CO_2, and hydrogen; and finally, (3) methanogenesis by the methane producers. These methanogenic processes, summarized in table 41.5, involve critical balances between oxidants and reductants. To function most efficiently, the hydrogen concentration must be maintained at a low level. If hydrogen and organic acids accumulate, methane production can be inhibited, resulting in a stuck digester.

Methanogenic bacteria (pp. 482–85).

Anaerobic digestion has many advantages. Most of the microbial biomass produced in aerobic growth is used for methane production. The resulting sludge occupies less volume and can

Figure 41.16   High-Efficiency Anaerobic Bioreactors Used for Sludge Digestion and Methane Production. These "egg-shaped" units are so well insulated that they maintain their temperature (39°C) without outside heating, and the methane can be burned to provide electricity. These reactors are located at Kiel, Germany.

be dried easily. However, heavy metals and other environmental contaminants are often concentrated in the sludge. There may be longer-term environmental and public health effects from disposal of this material on land or in water (Box 41.3).

Tertiary treatment purifies sewage more than is possible with primary and secondary treatments. The goal is to remove such pollutants as nonbiodegradable organic material (e.g., polychlorinated biphenyls), heavy metals, and minerals. It is particularly important to eliminate the salts of nitrogen and phosphorus compounds that can promote eutrophication. Organic pollutants can be removed with activated carbon filters. Phosphate usually is precipitated as calcium or iron phosphate (for example, by the addition of lime). Excess nitrogen may be removed by "stripping," volatilization as NH_3 at high pHs. Ammonia itself can be chlorinated to form dichloramine, which is then converted to molecular nitrogen. In some cases, biological processes also are used to

Water and Disease Transmission

Since the beginning of recorded history water has been recognized as a potential carrier of disease. In order to protect his health, Alexander the Great (356–323 B.C.) had his personal drinking water carried in silver urns. Clearly the association between noble heavy metals such as silver and the prevention of waterborne diseases was early established through chance observation. The connection between a freshwater supply and the health of an urban population was recognized by the time of the Roman Empire (27 B.C.). However, much of the technology for the protection of the water supply was subsequently lost until the middle of the nineteenth century.

Water Purification

Water purification is a critical link in promoting public health and safety. As shown in figure 41.19, water purification can involve a variety of steps, whose nature depends on the type of impurities in the raw water source. For example, if the raw water contains large amounts of iron and manganese, which will often precipitate when water is exposed to air, it may be necessary to aerate the water and employ other methods to remove these ions early in the purification sequence. Usually municipal water supplies are purified by a process that consists of at least three or four steps. If the raw water contains a great deal of suspended material, it often is first routed to a **sedimentation basin** and held so that sand and other very large particles can settle out. The partially clarified water is then mixed with chemicals such as alum

and lime and moved to a **settling basin** where more material precipitates out. This procedure is called **coagulation** or flocculation and removes microorganisms, organic matter, toxic contaminants, and suspended fine particles. After these steps the water is further purified by passing it through a filtration unit (figure 41.20). **Rapid sand filters,** which depend on the physical trapping of fine particles and flocs, are usually used for this purpose. This filtration removes up to 99% of the remaining bacteria. After filtration the water is treated with a disinfectant. This step usually involves chlorination, but ozonation is becoming increasingly popular. When chlorination is employed, the chlorine dose must be large enough to leave residual free chlorine at a concentration of 0.2 to 2.0 mg/liter. A concern is the creation of **halomethanes,** a group of compounds that may be carcinogens, formed when chlorine reacts with organic matter.

The preceding purification process effectively removes or inactivates disease-causing bacteria and indicator organisms (coliforms). Unfortunately, however, the use of coagulants, rapid filtration, and chemical disinfection often does not consistently and reliably remove *Giardia lamblia* cysts, *Cryptosporidium* oocysts, *Cyclospora,* and viruses. *Giardia* is now recognized as the most common identified waterborne pathogen in the United States. The protozoan, first observed by Leeuwenhoek in 1681, has trophozoite and cyst forms. The disease often is called traveler's disease and is transmitted primarily through untreated stream water or undependable municipal water supplies. More consistent removal of *Giardia* cysts can be achieved with **slow sand filters.** This treatment involves the slow passage of water through a bed of sand in which a microbial layer covers the

Water purification steps

Untreated water

Color and precipitate removal

Softening (Ca, Mg removal)

Turbidity removal

Taste and odor removal

Disinfection

Aeration
Chemical oxidation
Ion exchange
Sedimentation

Chemical precipitation
• dosing
• mixing
• flocculation
• settling
Ion exchange

Chemical coagulation
• dosing
• mixing
• flocculation
• settling
Filtration

Aeration
Chemical oxidation
Adsorption

Irradiation
Ozonation
Chlorination

Drinking water

Water purification processes

Figure 41.19 Water Purification. Several alternatives can be used for drinking water treatment depending on the initial quality of the water. A major concern is disinfection: chlorination can lead to the production of potentially carcinogenic trihalomethanes (THMs).

surface of each sand grain. It is biological rather than physical. Waterborne microorganisms are removed by adhesion to the surface microbial layer (Box 41.4).

Giardia *(pp. 536, 801–2).*

In the last few years, *Cryptosporidium* has become of even greater concern than *Giardia*. This protozoan parasite is smaller than *Giardia* and is even more difficult to remove from water. The occurrence of a major *Cryptosporidium* outbreak in Milwaukee, Wisconsin, in the spring of 1993 showed the vulnerability of water supplies to this organism. This protozoan caused few serious problems in the past. Only with the increased recent occurrence of immunocompromised individuals has this organism been able to have major effects on these susceptible populations. *Cryptosporidium* (hidden spore cysts) is found in about 90% of sewage samples, in 75% of river waters, and in 28% of drinking waters. Infection can occur with as few as 10 oocysts, derived

from a wide range of companion and domestic animals, and in a healthy person the infections are self-limiting. However, in immunocompromised individuals, *Cryptosporidium* infections can lead to death. The small size of this protozoan makes it even more dangerous because it cannot be removed easily by sand filters. The oocysts are only 4 to 6 µm in diameter (*Giardia* is approximately 7 to 10 by 8 to 12 µm in size). *Cryptosporidium* also is extremely disinfection resistant. The control of *Cryptosporidium* in drinking waters is a subject of intense research.

Another newly emerging protozoan human pathogen is *Cyclospora*. This coccidian protozoan is larger than *Cryptosporidium* and has 2 sporocysts, each with 2 sporozoites (*Cryptosporidium* has 4 sporozoites in its oocysts). *Cyclospora* causes cyclosporiasis, a self-limiting diarrhea that lasts from 19 to 43 days and can be accompanied by nausea, vomiting, cramps, and fever.

Figure 41.20   Water Filtration. Physical filtration is an important step in drinking water treatment. This is a cross section of a typical sand filter showing layers of sand and graded gravel.

Box 41.4

Waterborne Diseases, Water Supplies, and Slow Sand Filtration: The Return of a Time-Tested Concept in Drinking Water Treatment

Slow sand filtration, in which drinking water is passed through a sand filter that develops a layer of microorganisms on its surface, has had a long and interesting history. After London's severe cholera epidemic of 1849, Parliament, in an act of 1852, required that the entire water supply of London be passed through slow sand filters before use.

The value of this process was shown in 1892, when a major cholera epidemic occurred in Hamburg, Germany, and 10,000 lives were lost. The neighboring town, Altona, which used slow sand

filtration, did not have a cholera epidemic. Slow sand filters were installed in many cities in the early 1900s, but the process fell into disfavor with the advent of rapid sand filters, chlorination, and the use of coagulants such as alum. Slow sand filtration, a time-tested process, is regaining favor because of its filtration effectiveness and lower maintenance costs. Slow sand filtration is particularly effective for the removal of *Giardia* from waters. For this reason slow sand filtration is used in many mountain communities where *Giardia* is a problem.

Viruses in drinking water also must be destroyed or removed. Coagulation and filtration do reduce virus levels about 90 to 99%. Further inactivation of viruses by chemical oxidants, high pH, and photooxidation may yield a reduction as great as 99.9%. None of these processes, however, is considered to provide sufficient protection. New standards for virus inactivation are being developed. Bacteriophages, which can be easily grown and assayed, are being used as disinfection surrogates. If sufficient log reductions of bacteriophage infectivity occur with a given disinfection process, it is assumed that viruses capable of infecting humans will be reduced to satisfactory levels.

Microbiological Analysis of Water Purity

Monitoring and detection of indicator and disease-causing microorganisms are a major part of sanitary microbiology. By chlorinating drinking water supplies, control of most major disease-causing bacteria can be achieved. The inability to consistently remove viruses and protozoa and the lack of water quality standards for these microorganisms are of continuing concern. Additional problems are caused by recreational bodies of water and areas where shellfish are harvested. In these locations water treatment barriers against contamination may not exist, and transmission of microbial diseases can occur.

Bacteria from the intestinal tract generally do not survive in the aquatic environment, are under physiological stress, and gradually lose their ability to form colonies on differential and selective media. Their die-out rate depends on the water temperature, the effects of sunlight, the populations of other bacteria present, and the chemical composition of the water. Procedures have been developed to "resuscitate" these stressed coliforms before they are identified using selective and differential media.

A wide range of viral, bacterial, and protozoan diseases result from the contamination of water with human fecal wastes (*see chapters 36, 37, and 39*). Although many of these pathogens can be detected directly, environmental microbiologists have generally used **indicator organisms** as an index of possible water contamination by human pathogens. Researchers are still searching for the "ideal" indicator organism to use in sanitary microbiology. The following are among the suggested criteria for such an indicator:

1. The indicator bacterium should be suitable for the analysis of all types of water: tap, river, ground, impounded, recreational, estuary, sea, and waste.

2. The indicator bacterium should be present whenever enteric pathogens are present.

3. The indicator bacterium should survive longer than the hardiest enteric pathogen.

4. The indicator bacterium should not reproduce in the contaminated water and produce an inflated value.

5. The assay procedure for the indicator should have great specificity; in other words, other bacteria should not give positive results. In addition, the procedure should have high sensitivity and detect low levels of the indicator.

6. The testing method should be easy to perform.

7. The indicator should be harmless to humans.

8. The level of the indicator bacterium in contaminated water should have some direct relationship to the degree of fecal pollution.

Coliforms, including *Escherichia coli,* are members of the family *Enterobacteriaceae.* These bacteria make up approximately 10% of the intestinal microorganisms of humans and other animals (*see figure 28.8*) and have found widespread use as indicator organisms. They lose viability in fresh water at slower rates than most of the major intestinal bacterial pathogens. When such "foreign" enteric indicator bacteria are not detectable in a specific volume (100 ml) of water, the water is considered **potable** [Latin *potabilis,* fit to drink], or suitable for human consumption.

The Enterobacteriaceae (pp. 425–27).

The coliform group includes *E. coli, Enterobacter aerogenes,* and *Klebsiella pneumoniae.* Coliforms are defined as facultatively anaerobic, gram-negative, nonsporing, rod-shaped bacteria that ferment lactose with gas formation within 48 hours at 35°C. The original test for coliforms that was used to meet this definition involved the presumptive, confirmed, and completed tests, as shown in figure 41.21. The presumptive step is carried out by means of tubes inoculated with three different sample volumes to give an estimate of the **most probable number (MPN)** of coliforms in the water. The complete process, including the confirmed and completed tests, requires at least 4 days of incubations and transfers.

Unfortunately the coliforms include a wide range of bacteria whose primary source may not be the intestinal tract. To deal with this difficulty, tests have been developed that allow waters to be tested for the presence of **fecal coliforms.** These are coliforms derived from the intestine of warm-blooded animals, which can grow at the more restrictive temperature of 44.5°C.

To test for coliforms and fecal coliforms, and more effectively recover stressed coliforms, a variety of simpler and more specific tests have been developed. These include the membrane filtration technique, the **presence-absence (P-A) test** for coliforms and the related Colilert **defined substrate test** for detecting both coliforms and *E. coli.*

The **membrane filtration technique** (*see figure 6.7*), has become a common and often preferred method of evaluating the microbiological characteristics of water. The water sample is passed through a membrane filter. The filter with its trapped bacteria is transferred to the surface of a solid medium or to an absorptive pad containing the desired liquid medium. Use of the proper medium allows the rapid detection of total coliforms, fecal coliforms, or fecal streptococci by the presence of their characteristic colonies (figure 41.22; *see also figure 6.8*). Samples can be placed on a less selective resuscitative medium, or incubated at a less stressful temperature, prior to growth under the final set of selective conditions. An example of a resuscitation step is the use of a 2 hour incubation on a pad soaked with lauryl sulfate broth, as is carried out in the LES Endo procedure. A resuscitation step often is needed with chlorinated samples,

Water sample

Inoculate 15 tubes: 5 with 10 ml of sample, 5 with 1.0 ml of sample, and 5 with 0.1 ml of sample.

Double-strength broth

Single-strength broth

10 10 10 10 10
(ml)

1.0 1.0 1.0 1.0 1.0
(ml)

0.1 0.1 0.1 0.1 0.1
(ml)

Presumptive

Lactose or lauryl tryptose broth

Negative presumptive. The absence of gas in broth tubes indicates coliforms are absent. Incubate an additional 24 hours to be sure.

24 ± 2 hours
35°C

After 24 hours of incubation, the tubes of lactose broth are examined for gas production.

Negative   Positive

Confirmed

No gas produced, coliform group absent.

Positive test: gas production — use positive confirmed tubes to determine MPN.

All positive presumptive cultures used to inoculate tubes of brilliant green lactose bile broth. Incubation for 48 ± 3 hours at 35°C.

Negative   Positive

Completed

Brilliant green lactose bile broth or lauryl tryptose broth

Nutrient agar slant

Use coliform colonies to inoculate nutrient agar slant and a broth tube.

Plates of Levine's EMB or LES Endo agar are streaked from positive tubes and incubated at 35°C for 24 ± 2 hours.

After 24 hours of incubation make a Gram-stained slide from the slant. If the bacteria are gram-negative, nonsporing rods and produce gas from lactose, the completed test is positive.

Figure 41.21   The Multiple-Tube Fermentation Test. The multiple-tube fermentation technique has been used for many years for the sanitary analysis of water. Lactose broth tubes are inoculated with different water volumes in the presumptive test. Tubes that are positive for gas production are inoculated into brilliant green lactose bile broth in the confirmed test, and positive tubes are used to calculate the most probable number (MPN) value. The completed test is used to establish that coliform bacteria are present.

(a)   (b)   (c)

Figure 41.22   Coliform and Enterococcal Colonies. Membrane filters made it possible to more rapidly test waters for the presence of coliforms, fecal coliforms, and fecal enterococci by the use of differential media. (*a*) Coliform reactions on an Endo medium. (*b*) Fecal coliform growth on a bile salt medium (m-FC agar) containing aniline blue dye. (*c*) Fecal enterococci growing on an azide-containing medium (KF agar) with TTC, an artificial electron acceptor, added to allow better detection of colonies.

TABLE 41.6	Advantages and Disadvantages of the Membrane Filter Technique for Evaluation of the Microbial Quality of Water

Advantages

Good reproducibility
Single-step results often possible
Filters can be transferred between different media
Large volumes can be processed to increase assay sensitivity
Time savings are considerable
Ability to complete filtrations on site
Lower total cost in comparison with MPN procedure

Disadvantages

High-turbidity waters limit volumes sampled
High populations of background bacteria cause overgrowth
Metals and phenols can adsorb to filters and inhibit growth

Source: Data from A. E. Greenberg, et al., *Standard Methods for the Examination of Water and Wastewater*, 16th edition, page 886, 1985. American Public Health Association, Washington, D.C.

(a)   (b)   (c)

Figure 41.23   The Defined Substrate Test. This much simpler test is now being used to detect coliforms and fecal coliforms in single 100 ml water samples. The medium uses ONPG and MUG (see text) as defined substrates. (*a*) Uninoculated control. (*b*) Yellow color due to the presence of coliforms. (*c*) Fluorescent reaction due to the presence of fecal coliforms.

where the microorganisms are especially stressed. The advantages and disadvantages of the membrane filter technique are summarized in table 41.6. Membrane filters have been widely used with water that does not contain high levels of background organisms, sediment, or heavy metals.

More simplified tests for detecting coliforms and fecal coliforms are now available. The presence-absence test (P-A test) can be used for coliforms. This is a modification of the MPN procedure, in which a larger water sample (100 ml) is incubated in a single culture bottle with a triple-strength broth containing lactose broth, lauryl tryptose broth, and bromcresol purple indicator. The P-A test is based on the assumption that no coliforms should be present in 100 ml of drinking water. A positive test results in the production of acid (a yellow color) and constitutes a positive presumptive test requiring confirmation.

To test for both coliforms and *E. coli,* the related Colilert defined substrate test can be used. A water sample of 100 ml is added to a specialized medium containing O-nitrophenyl-β-D-

galactopyranoside (**ONPG**) and 4-methylumbelliferyl-β-D-glucuronide (**MUG**) as the only nutrients. If coliforms are present, the medium will turn yellow within 24 hours at 35°C due to the cleavage of the ONPG, as shown in figure 41.23. To check for *E. coli,* the medium is observed under long-wavelength UV light for fluorescence. When *E. coli* is present, the MUG is modified to yield a fluorescent product. If the test is negative for the presence of coliforms, the water is considered acceptable for human consumption. The main change from previous standards is the requirement to have waters free of coliforms and fecal coliforms. If coliforms are present, fecal coliforms or *E. coli* must be tested for.

In the United States a set of general guidelines for microbiological quality of drinking waters has been developed, including standards for coliforms, viruses, and *Giardia* (table 41.7). If unfiltered surface waters are being used, one coliform test must be run each day when the waters have higher turbidities.

Other indicator microorganisms include **fecal enterococci.** The fecal enterococci are increasingly being used as an indicator of fecal contamination in brackish and marine water. In salt water these bacteria die back at a slower rate than the fecal coliforms, providing a more reliable indicator of possible recent pollution.

The genus Enterococcus *(pp. 443–45).*

TABLE 41.7	**Current Microbiological Standards for Water in the United States, 1991**
Agent	**Allowable Limits/100 ml**
Coliforms (MCLᵃ) per 100 ml	1 positive sample found in <40 samples/month <5% positive samples found if >40 samples/month
Fecal coliforms	None
Giardia	None
Enterococci	None
Viruses	None

Source: Data from L. S. Clesceri, et al., *Standard Methods for the Examination of Water and Wastewater.* 17th edition, 1989. American Public Health Association, Washington, D.C.

ᵃMaximum concentration limit. Above this limit, one is in violation of the guidelines. The number of analyses to be completed per month depends on the city population.

Water-Based Diseases

Many important human pathogens are maintained in association with living organisms other than humans, including many wild animals and birds (*see table 35.1*). Some of these bacterial and protozoan pathogens can survive in water and infect humans. As examples, *Vibrio vulnificus, V. parahaemolyticus,* and *Legionella* are of continuing concern. When waters are used for recreation or are a source of seafood that is consumed uncooked, the possibility for disease transmission certainly exists. A food-borne infection may have serious consequences for immunologically compromised individuals.

Some major waterborne diseases are summarized in table 41.8. Waterborne bacterial and viral pathogens are extensively discussed in chapters 36 and 37. The protozoan pathogen *Giardia* is described in chapter 39, and both *Giardia* and *Cryptosporidium* are discussed earlier in this chapter.

Another waterborne protozoan disease of increasing concern worldwide is primary amebic meningoencephalitis (PAM), an infection of the central nervous system caused by *Naegleria fowleri.* The disease usually occurs in children or young adults who have been swimming in lakes or pools or have been waterskiing. After nasal infection the protozoan reaches the brain and initiates an inflammatory response. The result usually is fatal. Warm water and heated industrial effluents promote the growth of this protozoan.

Protozoan diseases (pp. 799–807).
Food-borne and waterborne viral diseases (pp. 737–39).
Food-borne and waterborne bacterial diseases (pp. 766–71).

TABLE 41.8	**Water-Based Microbial Pathogens That Can Be Maintained in the Environment Independent of Humans**	
Organism	**Reservoir**	**Comments**
Bacteria		
Aeromonas hydrophila	Free-living	Sometimes associated with gastroenteritis, cellulitis, and other diseases
Chromobacterium violaceum	Soil runoff	Common saprophyte in water—rare in terms of human infection
Mycobacterium	Infected animals and free-living	Complex recovery procedure required
Leptospira	Infected animals	Hemorrhagic effects, jaundice
Legionella pneumophila	Free-living and associated with protozoa	Found in cooling towers, evaporators, condensers, showers, and other water sources
Vibrio cholera	Free-living	Found in many waters and estuaries
Vibrio parahaemolyticus	Free-living in coastal waters	Causes diarrhea in shellfish consumers
Salmonella enteriditis	Animal intestinal tracts	Common in many waters
Campylobacter	Bird and animal reservoirs	Major cause of diarrhea; common in processed poultry; a microaerophile
Yersinia enterocolitica	Frequent in animals and in the environment	Waterborne gastroenteritis
Pseudomonas aeruginosa	Free-living	Swimmer's ear and related infections
Protozoa		
Giardia lamblia	Beavers, sheep, dogs, cats	Major cause of early spring diarrhea; important in cold mountain water
Cryptosporidium	Many species of domestic and wild animals	Causes acute enterocolitis; important with immunologically compromised individuals; cysts resistant to chemical disinfection
Acanthamoeba	Sewage sludge disposal areas	Can cause granulomatous amebic encephalitis (GAE); keratitis, corneal ulcers
Naegleria fowleri	Warm water (hot tubs), swimming pools, lakes	Inhalation in nasal passages; central nervous system infection; causes primary amebic meningoencephalitis (PAM)

1. Describe the way in which municipal drinking water is usually purified. Why might a slow sand filter be useful in this process?

2. Why might *Cryptosporidium* and *Cyclospora* be of greater concern than *Giardia?*

3. How is a coliform defined? How does this definition relate to presumptive, confirmed, and completed tests? What is an indicator organism, and what properties should it have?

4. How does one differentiate between coliforms and fecal coliforms in the laboratory?

5. In what type of environment is it better to use fecal enterococci rather than fecal coliforms as an indicator organism?

6. What are the advantages and disadvantages of membrane filters for microbiological examinations of water?

7. Why has the defined substrate test with ONPG and MUG been accepted as a test of drinking water quality?

8. Give a few examples of waterborne diseases caused by protozoa and bacteria that are maintained in the environment independent of humans.

9. An infection by *Naegleria fowleri* leads to fatal nervous system effects. What is the portal of entry for this organism?

Groundwater Quality and Home Treatment Systems

Groundwater, or water in gravel beds and fractured rocks below the surface soil, is a widely used but often unappreciated water resource. In the United States groundwater supplies at least 100 million people with drinking water, and in rural and suburban areas beyond municipal water distribution systems, 90 to 95% of all drinking water comes from this source.

The great dependence on this resource has not resulted in a corresponding understanding of microorganisms and microbiological processes that occur in the groundwater environment. Increasing attention is now being given to predicting the fate and effects of groundwater contamination (figure 41.24) on the chemical and microbiological quality of this resource. Pathogenic microorganisms and dissolved organic matter are removed from water during subsurface passage through adsorption and trapping by fine sandy materials, clays, and organic matter. Microorganisms associated with these materials—including predators such as protozoa—can use the trapped pathogens as food. This results in purified water with a lower microbial population.

This combination of adsorption-biological predation is used in home treatment systems (figure 41.25). Conventional **septic tank** systems include an anaerobic liquefaction and digestion step that occurs in the septic tank itself (the tank functions as a simple anaerobic digester). This is followed by organic matter adsorption and entrapment of microorganisms in the aerobic leach-field environment where biological oxidation occurs. A septic tank may not operate correctly for several reasons. It will not function properly if the retention time of the waste in the septic tank is too short. Retention time decreases when the flow is too rapid or when excessive sludge has accumulated in the septic tank. As a result undigested solids move into the leach field, gradually plugging the system. If the leach field floods and becomes anaerobic, biological oxidation does not occur, and effective treatment ceases.

When a suitable soil is not present and the septic tank outflow drains too rapidly to the deeper subsurface, problems can occur. Fractured rocks and coarse gravel materials provide little effective adsorption or filtration. This may result in the contamination of well water with pathogens and the transmission of disease. In addition, phosphorus from the waste will not be retained effectively and may pollute the groundwater. This often leads to nutrient enrichment of ponds, lakes, and rivers as the subsurface water enters these environmentally sensitive water bodies.

Subsurface zones also can become contaminated with pollutants from other sources. Land disposal of sewage sludges, illegal dumping of septic tank pumpage, improper toxic waste disposal, and runoff from agricultural operations all contribute to groundwater contamination with chemicals and microorganisms. Deep-well injection of industrial wastes has raised questions about the longer-term fate and effects of these materials.

The subsurface environment was once thought to be essentially sterile. More recently, viable microorganisms have been discovered down to a depth of 1,000 to 1,200 meters. The subsurface has several distinct limitations for microbial survival and functioning. Nutrients such as nitrogen and phosphorus may be present at very low levels and limit microbial growth. Although most pristine subsurface waters are aerobic, after contamination with organic matter, they quickly become anaerobic. With slow diffusion of O_2 into some subsurface waters, even low levels of organic substrates will render an environment anaerobic. This change cannot be easily corrected. Oxygen, peroxide, nutrients, and microorganisms can be added to enhance degradation in **bioremediation** programs.

Many pollutants that reach the subsurface will persist and may affect the quality of groundwater for extended periods. Much research is being conducted to find ways to treat groundwater, either in place—**in situ treatment**—or above ground. Microorganisms and microbial processes are used in many of these remediation efforts.

1. In rural areas, approximately what percentage of the water used for human consumption is groundwater?

2. What factors can limit microbial activity in subsurface environments? Consider the energetic and nutritional requirements of microorganisms in your answer.

3. How, in principle, are a septic tank system and a leach-field system supposed to work? What factors can reduce the effectiveness of this system?

Figure 41.24   Groundwater Contamination. A wide variety of activities can lead to groundwater contamination, which must be managed and controlled. Contaminants are released by agriculture, landfills, chemical wastes, urban areas, rural septic tanks, deforestation, and mining, as examples. Recharge from rivers also can lead to groundwater contamination. Microorganisms play roles in the release and modification of substances that contaminate groundwaters, particularly nitrates. Microorganisms also are being used to control and mitigate contamination of this vital water resource.

Figure 41.25   The Septic Tank Home Treatment System. This system combines an anaerobic waste liquefaction unit (the septic tank) with an aerobic leach field. Biological oxidation of the liquefied waste takes place in the leach field, unless the soil becomes flooded.

Summary

1. Marine and freshwater environments are dominated by the water phase, and photosynthetic microorganisms provide much organic matter for use by other organisms.

2. Biofilms are critical parts of the aquatic microbial system. These can range from single layers of an organism, to layered systems with different types of microorganisms. Complex systems with channels and interstitial spaces also occur.

3. Microbial mats are complex layered communities found in marine and freshwater environments, and also in the fossil record. Some of these fossil mats are more than 3.5 billion years old. Sessile, prosthecate, and gliding microorganisms are also prominent in marine and freshwater environments.

4. Oxygen is present at low levels in water, and if sufficient organic matter is present, heterotrophic microorganisms can use this dissolved O_2 more rapidly than it can be replenished. Respiration or decomposition of algal and cyanobacterial blooms also can cause O_2 depletion.

5. In marine systems most water is at a low temperature, without light, and under high pressure. Hydrothermal and methane seeps provide unique environments for specialized microorganisms, many of which have developed special symbiotic relations with higher organisms such as clams and tube worms.

6. The biochemical oxygen demand (BOD) test is a crude measure of organic matter that can be oxidized by the aerobic microbial community. In this assay, oxygen should never limit the rate of reaction.

7. With the addition of inorganic nutrients, cyanobacteria and algae play major roles in the accumulation of organic matter in water.

8. Eutrophication can be caused by nutrient releases from rural and urban areas. Sources of nitrogen and phosphorus are particularly important in eutrophication.

9. The addition of organic matter can produce dissolved O_2 sag curves and diurnal O_2 changes in the latter stages of self-purification.

10. Conventional sewage treatment is a controlled intensification of natural self-purification processes, and it can involve primary, secondary, and tertiary treatment.

11. Anaerobic digestion involves the stepwise degradation of organic matter to produce methane.

12. Constructed wetlands involve the use of aquatic plants (floating, emergent, submerged) and their associated microorganisms for the treatment of liquid wastes. These systems are being used in a wide range of environments. Domestic wastes, industrial effluents, and acid mine drainage (AMD) can be treated.

13. Municipal drinking water is purified through processes such as sedimentation, coagulation with chemicals, sand filtration, and chlorination.

14. Indicator organisms are used to indicate the presence of pathogenic microorganisms. Most probable number (MPN) and membrane filtration procedures are employed to estimate the number of indicator organisms present.

15. Enterococci are better indicator organisms for use in salt water than are coliforms because of their slower relative die-back rates.

16. Presence-absence (P-A) tests for coliforms and defined substrate tests for coliforms and *E. coli* allow 100 ml water volumes to be tested with minimum time and materials.

17. Waterborne bacterial and protozoan diseases are very important in recreational water and water containing shellfish that are eaten raw.

18. *Cryptosporidium* is of increasing concern because of its small size, which makes it difficult to remove from water by filtration, and also because of its disinfection resistance. The ability of *Cryptosporidium* to infect immunocompromised individuals makes this protozoan a major threat to water supplies. *Cyclospora* is another newly emerging protozoan pathogen.

19. Home treatment systems operate on general self-purification principles. The septic tank provides anaerobic liquefaction and digestion, whereas the aerobic leach field allows oxidation of the soluble effluent.

20. Groundwater is an important resource that can be affected by pollutants from septic tanks and other sources. This vital water source must be protected and improved.

Key Terms

anaerobic digestion   *845*

barophile   *836*

biofilms   *833*

bioremediation   *854*

bulking sludge   *845*

coagulation   *848*

coliform   *850*

confocal scanning laser microscopy (CSLM)   *833*

constructed wetlands   *845*

defined substrate test   *850*

diurnal oxygen shifts   *842*

endogenous respiration   *845*

epilimnion   *840*

eutrophic   *840*

eutrophication   *840*

extended aeration   *845*

fecal coliforms   *850*

fecal enterococci   *853*

groundwater   *854*

halomethanes   *848*

hypolimnion   *840*

indicator organism   *850*

in situ treatment   *854*

membrane filtration technique   *850*

microbial mat   *833*

most probable number (MPN)   *850*

MUG   *852*

nitrogen oxygen demand (NOD)   *842*

nonpoint source of pollution   *842*

oligotrophic   *840*

ONPG   *852*

phytoplankton   *838*

point source of pollution   *842*

potable   *850*

presence-absence (P-A) test   *850*

primary treatment   *843*

rapid sand filter   *848*

secondary treatment   *843*

sedimentation basin   *848*

septic tank   *854*

settling basin   *848*

sewage treatment   *843*

slow sand filter   *848*

sludge   *843*

stromatolite   *833*

tertiary treatment   *843*

thermocline   *840*

total organic carbon (TOC)   *843*

Winogradsky column   *831*

Questions for Thought and Review

1. You wish to study the characteristics of microorganisms from a 10,000 meter depth in the ocean. What effects might decompression have on these organisms, and what type of equipment would you design to work with them under their natural conditions?

2. Which microorganisms might limit the acceptability of clear high mountain water as drinking water? How could one best treat this water before taking a drink?

3. Discuss the advantages and disadvantages for microorganisms that become attached to surfaces.

4. How might it be possible to rejuvenate an aging eutrophic lake? Consider physical, chemical, and biological approaches.

5. Bulking sludge is a major problem in sewage treatment, especially when activated sludge plants are overloaded. As low O_2 levels are often a cause of this problem, how could existing processes be modified in terms of improving O_2 availability?

6. You wish to develop a constructed wetland for removal of metals from a stream at one site, and at another site, you wish to treat acid mine drainage. How might you approach each of these problems?

7. Why are indicator organisms still being used despite the fact that methods are available for the direct isolation of most pathogens found in water?

8. What alternatives, if any, can one use for protection against microbiological infection when swimming in polluted recreational water? Assume that you are part of a water rescue team.

9. Discuss the threats to the purity of our groundwater and suggest ways in which groundwater quality can be recovered and preserved.

Additional Reading

The references provided at the end of chapter 40 also may be consulted for further information.

The Nature of Marine and Freshwater Environments

Cohen, Y., and Rosenberg, E. 1989. *Microbial mats. Physiological ecology of benthic microbial communities.* Washington, D.C.: American Society for Microbiology.

DesManas, D. J. 1990. Microbial mats and the early evolution of life. *Trends Ecol. & Evol.* 5(5):140–44.

Dixon, A. E.; Samaskinos, S.; and Atkinson, M. R. 1991. A scanning confocal microscope for transmission and reflection imaging. *Nature* 351:551–53.

Fletcher, M. 1991. The physiological activity of bacteria attached to solid surfaces. *Adv. Microbial Phys.* 32:53–85.

Lichtman, J. W. 1994. Confocal microscopy. *Sci. Am.* 271(2):40–45.

Marrao, G.; Verussimo, A.; Bowker, R. G.; and da Costa, M. S. 1993. Biofilms as major sources of *Legionella* spp. in hydrothermal areas and their dispersion into stream water. *FEMS Microbiol. Ecol.* 12:25–33.

Marshall, K. 1992. Biofilms; An overview of bacterial adhesion, activity, and control at surfaces. *ASM News* 58(4):202–7.

Massol-Deya, A. A.; Whallon, J.; Hickey, R. F.; and Tiedje, J. M. 1994. Biofilm architecture: A fortuitous engineering feature. *ASM News* 60(8):406.

Matsumoto, B., and Kramer, T. 1994. Theory and applications of confocal microscopy. *Cell Vision* 1(3):190–98.

Rogers, J. E.; Dowsett, A. B.; Dennis, P. J.; Lee, J. V.; and Keevil, C. W. 1994. Influence of plumbing materials on biofilm formation and growth of *Legionella pneumophila* in potable water systems. *Appl. Environ. Microbiol.* 60(6):1842–51.

Stoodley, P.; deBeer, D.; and Lewandowski, Z. 1994. Liquid flow in biofilm systems. *Appl. Environ. Microbiol.* 60(8):2711–16.

Tijhuis, L.; van Benthum, W. A. J.; van Loosdrecht, M. C. M.; and Heijnen, J. J. 1994. Solids retention time in spherical biofilms in a biofilm airlift suspension reactor. *Biotech. Bioeng.* 44(8):867–79.

Tuchman, N. C., and Peterson, C. G. 1995. Substratum-associated microbiota. *Water Environ. Res.* 67:702–13.

van Gemerden, H. 1993. Microbial mats: A joint venture. *Marine Geology* 113:3–25.

The Microbial Community in Marine and Freshwater Environments

Appenzeller, T. 1991. Fire and ice under the deep-sea floor. *Science* 252:1790–92.

Austin, B. 1988. *Marine microbiology.* New York: Cambridge University Press.

Franklin, C. 1994. "Black smokers" multiply on ocean floor. *New Scientist.* Oct. 22:20.

Myres, F. S., and Anderson, A. 1992. Microbes from 20,000 feet under the sea. *Science* 255:28–29.

Olsen, G. J. 1994. Archaea, archaea, everywhere. *Nature* 371:657.

Perry, K. A.; Kostka, J. E.; Luther III, G. W.; and Nealson, K. H. 1993. Mediation of sulfur speciation by a Black Sea facultative anaerobe. *Science* 259:801–03.

Rheinheimer, G. 1991. *Aquatic microbiology* 4th ed. New York: John Wiley and Sons.

Sherr, E. B., and Sherr, B. F. 1991. Planktonic microbes: Tiny cells at the base of the ocean's food webs. *Trends Ecol. & Evol.* 6(2):50–54.

Thingstad, T. F.; Heldal, M.; Bratbak, G.; and Dundas, I. 1993. Are viruses important partners in pelagic food webs? *Trends Ecol. & Evol.* 8(6):209–13.

The Marine Environment

Anderson, D. M. 1994. Red tides. *Sci Am.* 271(2):62–68.

Culotta, E. 1992. Red menace in the world's oceans. *Science* 257:1476–77.

Fenical, W. 1993. Chemical studies of marine bacteria: Developing a new resource. *Chem. Rev.* 93:1673–83.

Gage, J. D. 1991. Biological rates in the deep sea: A perspective from studies on processes in the benthic boundary layer. *Rev. Aq. Sci.* 5(1):49–100.

Gage, J. D., and Tayler, P. A. 1991. *Deep-sea biology: A natural history of organisms at the deep-sea floor.* Cambridge: Cambridge University Press.

Huyghe, P. 1993. Algae. *Discover* April. 72–75.

Jannasch, H. W., and Taylor, C. D. 1984. Deep-sea microbiology. *Ann. Rev. Microbiol.* 38:487–514.

Karl, D. M. 1994. Accurate estimation of microbial loop processes and rates. *Microb. Ecol.* 28:147–50.

Sayles, F. L. 1992. Biogeochemical processes on the sea floor. *Oceanus* 35:68–75.

Shimizu, Y. 1993. Microalgal metabolites. *Chem. Rev.* 93:1685–89.

Freshwater Microorganisms

Ford, T. E. 1993. *Aquatic microbiology,* Cambridge, Mass.: Blackwell Scientific Publications, Inc.

Giere, O. 1993. *Meiobenthology* New York: Springer Verlag.

Gorrie, P. 1992. Lakes as laboratories. *Can. Geographic* 112(2):69–78.

Shallenberg, M., and Kalff, J. 1992. The ecology of sediment bacteria in lakes and comparison with other aquatic ecosystems. *Experientia* 74:919–34.

Spring, S.; Amann, R.; Ludwig, W.; Schleifer, K-H.; van Gemerden, H.; and Peterson, N. 1993. Dominating role of an unusual magnetotactic bacterium in the microaerobic zone of a freshwater sediment. *Appl. Environ. Microbiol.* 59(8):2397–2403.

Nutrient Removal from Water

Berk, S. G., and Gunderson, J. H. 1995. *Wastewater organisms,* Boca Raton: Lewis Publishers.

Bitton, G. 1994. *Wastewater microbiology,* Gainsville, N.Y.: Wiley-Liss.

Cowan, R. M.; Love, N. G.; Sock, S. M.; and White, K. 1995. Activated sludge and other aerobic suspended culture processes. *Water Environ. Res.* 67:433-50.

Green, B. 1994. Constructed wetlands are big in small communities. *Water, Env. and Technology* 6:51–55.

Mann, R. A., and Bavor, H. J. 1993. Phosphorus removal in constructed wetlands using gravel and industrial waste substrata. *Wat. Sci. Tech.* 27(1):107–13.

Montgomery, J. M. 1985. *Water treatment: Principles and design.* New York: John Wiley and Sons.

Vile, M. A., and Wieder, R. K. 1993. Alkalinity generation by Fe (III) reduction versus sulfate reduction in wetlands constructed for acid mine drainage treatment. *Water, Air, and Soil Poll.* 69:425–41.

Whitmore, T. N., and Robertson, L. J. 1995. The effect of sewage sludge treatment processes on oocysts of *Cryptosporidium parvum. J. Appl. Microbiol.* 78:34–38.

Water-Based Diseases

Atlas, R. M., and Bej, A. K. 1990. Detecting bacterial pathogens in environmental water samples by using PCR and gene probes. In *PCR protocols: A guide to*

TABLE 42.2   Nutrient Resources and Estimated Annual Turnover Times for Major Components in a Grassland Ecosystem

| Component | Nutrient Resources[a] | | C/N Ratio | Approx. Turnover Time (Years) |
	Carbon	Nitrogen		
Above ground green	43.0	1.4	31	1–2
Above ground dead	64.0	2.5	26	2
Plant crowns	160.0	4.6	35	2
Live roots	88.0	2.3	38	3–4
Senescent roots	282.0	14.0	20	3–4
Detrital roots	149.0	9.0	17	4
Soil organic matter	1,327.0	127.0	10	100–1,000 and longer
Bacteria	30.4	7.6	4	0.83
Saprophytic fungi	6.3	0.63	10	0.83
Vesicular arbuscular fungi	0.7	0.07	10	0.50
Protozoa	0.2	0.028	7	0.17
Nematodes	0.217	0.022	10	0.50
Insects	0.068	0.004	8	0.40

Source: Data from R. G. Woodmansee, et al., "Nitrogen Budget of a Shortgrass Prairie" in *Oecologia*, 34:363–82, 1978; and H. W. Hunt, et al., "The Detrital Food Web in a Shortgrass Prairie" in *Biol. Fert. Soils*, 3:57–68, 1987.

[a]In grams per meter square, to a depth of 10 cm.

TABLE 42.3   Easily Cultured Gram-Positive Irregular Branching and Filamentous Bacteria Common in Soils

Bacterial Group	Representative Genera	Comments and Characteristics
Coryneform bacteria	*Arthrobacter*	Rod-coccus cycle
	Cellulomonas	Important in degradation of cellulose
	Corynebacterium	Club-shaped cells
Mycobacteria	*Mycobacterium*	Acid-fast
Nocardioforms	*Nocardia*	Rudimentary branching
Actinomycetes	*Streptomyces*	Aerobic filamentous bacteria
	Thermoactinomyces	Higher temperature growth

materials, and soil humus. In addition, some members of these groups actively degrade pesticides. The filamentous actinomycetes, primarily of the genus *Streptomyces,* produce an odor-causing compound called **geosmin,** which gives soils their characteristic earthy odor.

Actinomycetes (chapter 24).

Soil microorganisms also can be categorized on the basis of their preference for either easily available or more resistant substrates. The microorganisms that respond rapidly to the addition of easily utilizable substrates such as sugars and amino acids—for example, members of the genus *Pseudomonas*—are called **zymogenous** microorganisms. Those indigenous forms that tend to use native organic matter to a greater extent are called **autochthonous** microorganisms. Bacteria of the genus *Arthrobacter* and many soil actino-

mycetes can be considered members of this group. A less understood part of the microbial community in soils and in other natural environments is the **oligotrophs,** microorganisms that can be maintained on media containing less than 15 mg/liter of organic matter (*see Box 40.1*). When soils are inoculated onto such low-nutrient media, oligotrophic microorganisms often are present at higher populations than the easily cultured microorganisms recovered on most normal laboratory media. There is continuing interest in oligotrophs, and the study of these slow-growing organisms is an important area of modern environmental microbiology.

Arthrobacter (pp. 450–51).

Soil insects and other animals such as earthworms also contribute to organic matter transformations in soils. These organisms can serve as **decomposer-reducers.** The term decomposer-reducer indicates that organisms not only carry out decomposition or mineralization but also physically "reduce" the size of organic substrate particles such as plant litter. This increases the surface area and makes organic materials more available for use by bacteria and fungi. The decomposer-reducers mix substrates with their internal gut microflora and enzymes, and this contributes substantially to decomposition.

Protozoa also can influence nutrient cycling by feeding on "palatable" microorganisms. This process of **microbivory,** or use of microorganisms as a food source, results in higher rates of nitrogen and phosphorus mineralization, thus increasing the availability of nutrients for plant growth.

Protozoa (chapter 27).

A significant part of the biological activity of soils arises from enzymes released by plants, insects, and other animals, and from lysed microorganisms. These free enzymes contribute to many hydrolytic degradation reactions, such as proteolysis;

catalase and peroxidase activities also have been detected. Apparently these free enzymes associate with clays and humic materials, which helps to protect them from denaturation and microbial degradation.

---

1. Describe the contributions of bacteria and fungi to soil nutrient levels.
2. What is the role of protozoa, soil insects, and other animals in the functioning of the microbial community in soils?
3. Where is most of the nitrogen located in a temperate grassland soil? Is it readily available?
4. Describe zymogenous, autochthonous, and oligotrophic microorganisms.

Microorganisms and the Formation of Different Soils

In surveying lands across different climatic zones, one can observe significant changes in plant communities and soils. Often, primary production and decomposition are not in equilibrium. The result is excessive accumulation of organic matter or, if decomposition occurs too quickly, a soil that can become poor in organic matter and of low fertility. The production and decomposition of organic matter are influenced by soil temperature, as shown in figure 42.4. Above a mean annual temperature of approximately 25°C in aerobic soils, the rate of decomposition can be higher than that of organic matter accumulation.

These differences can be observed in soils from tropical, grassland, forest, and bog regions (figure 42.5). In moist tropical soils, with their higher mean temperatures, organic matter is decomposed very quickly, and the mobile inorganic nutrients can be leached out of the surface soil environment, causing a rapid loss of fertility. To limit nutrient loss, many tropical plants have root systems that penetrate the rapidly decomposing litter layer. Thus it is possible to "recycle" nutrients before they are lost with water movement through the soil (figure 42.5a). With deforestation, the nutrients are not recycled, leading to their loss from the soil and decreased soil fertility.

In many temperate region soils, in contrast, the decomposition rates are less than that of primary production, leading to litter accumulation. Deep root penetration in temperate grasslands results in the formation of fertile soils, which provide a valuable resource for the growth of crops in intensive agriculture (figure 42.5b).

The soils in many cooler coniferous forest environments suffer from an excessive accumulation of organic matter as plant litter (figure 42.5c). In winter, when moisture is available, the soils are cool, and this limits decomposition. In summer, when the soils are warm, water is not as available for decomposition. Organic acids are produced in the cool, moist litter layer,

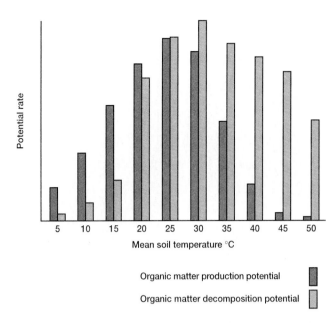

Figure 42.4 Soil Temperature and the Potential for Organic Matter Production and Decomposition. At approximately 25°C the processes of production and decomposition of organic matter are near equilibrium. Above 25°C, in tropical areas, organic matter can be decomposed faster than it can be synthesized. At higher temperatures decomposition often will be limited by drying of soils.

and they leach into the underlying soil. These acids solubilize soil components such as aluminum and iron, and a bleached zone may form. Litter continues to accumulate, and fire becomes the major means by which nutrient cycling is maintained. Controlled burns are becoming a more important part of environmental management in this type of plant-soil system.

Bog soils provide a unique set of conditions for microbial growth (figure 42.5d). In these soils the decomposition rate is slowed by the waterlogged, predominantly anoxic conditions, which lead to peat accumulation. When such areas are drained, they become more aerobic and the soil organic matter is degraded, resulting in soil subsidence. Under aerobic conditions the lignin-cellulose complexes of the accumulated organic matter are more susceptible to decomposition by the filamentous fungi.

This analysis of different soil types shows that the microbial role in soil formation and decomposition is varied but understandable in terms of general environmental factors that influence biological decomposition processes.

---

1. Describe each major soil type discussed in this section in terms of the balance between primary production and organic matter decomposition.
2. If decomposition is slower than primary production, what are the consequences for plant community functioning?

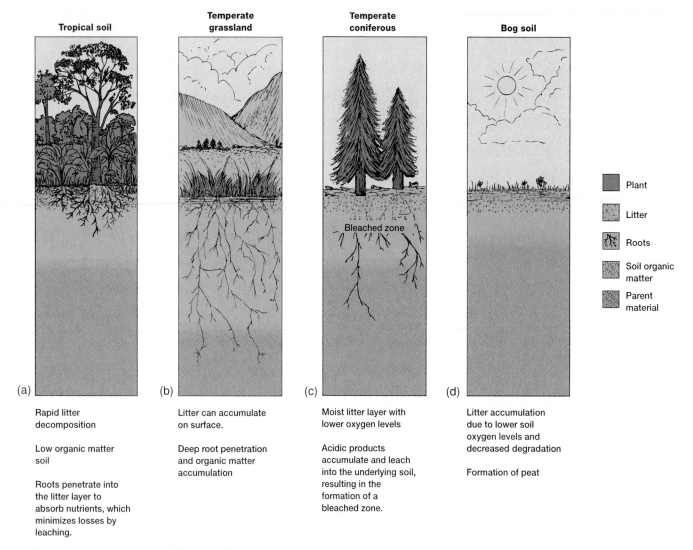

Tropical soil

Temperate grassland

Temperate coniferous

Bog soil

Bleached zone

Plant

Litter

Roots

Soil organic matter

Parent material

(a)

(b)

(c)

(d)

Rapid litter decomposition

Low organic matter soil

Roots penetrate into the litter layer to absorb nutrients, which minimizes losses by leaching.

Litter can accumulate on surface.

Deep root penetration and organic matter accumulation

Moist litter layer with lower oxygen levels

Acidic products accumulate and leach into the underlying soil, resulting in the formation of a bleached zone.

Litter accumulation due to lower soil oxygen levels and decreased degradation

Formation of peat

Figure 42.5   Examples of Plant-Soil Systems. Climate, parent material, plants, and microorganisms interact over time to form different plant-soil systems. In these figures the characteristics of tropical, temperate grassland, temperate forest, and bog soils are illustrated.

Soil Microorganism Associations with Plants

Plants—the major source of organic matter on which soil microorganisms are dependent—are literally covered with microorganisms (figure 42.6). Microorganisms of various kinds are associated with the leaves, stems, flowers, seeds, and roots. The microbial community influences plants in many direct and indirect ways. The presence of microorganisms increases the rate of organic matter release from the roots by **exudation.** With grasses, for example, microorganisms can increase exudation rates by 80 to 100%. Materials released by roots include a wide range of potential microbial substrates, inhibitors, and stimulants (table 42.4). Rates of organic matter release by plants are also increased by stresses, including herbage removal and light changes.

The roles of microorganisms in influencing plant growth through the release of morphology-influencing compounds such as gibberellins and cytokinins are only now beginning to be fully appreciated in terms of their biotechnological potential.

The Rhizosphere

Soil microbial populations respond to the release of organic materials near the plant root, increasing their numbers and changing the characteristics of the microbial community (table 42.5). This region, called the **rhizosphere,** was first described by Lorenz Hiltner in 1904. In more fertile soils, which have higher normal microbial populations, this effect will not be as pronounced as it is with lower fertility soils. The rhizosphere response is especially marked in desert soils

Figure 42.6   Root Surface Microorganisms. Plant roots release nutrients that allow intensive development of bacteria and fungi on and near the plant root surface. A scanning electron micrograph shows bacteria and fungi growing on a root surface.

TABLE 42.4	Compounds Excreted by Axenic Wheat Roots	
Volatile Compounds	**Low-Molecular-Weight Compounds**	**High-Molecular-Weight Compounds**
CO_2	Sugars	Polysaccharides
Ethanol	Amino acids	Enzymes
Isobutanol	Vitamins	
Isoamyl alcohol	Organic acids	
Acetoin	Nucleotides	
Isobutyric acid		

From J. W. Woldendorp, "The Rhizosphere as Part of the Plant-Soil System" in *Structure and Functioning of Plant Populations* (Amsterdam, Holland: Proceedings, Royal Dutch Academy of Sciences, Natural Sciences Section: 2d Series, 1978) 70:243.

that have lower normal populations of microorganisms. A much greater relative stimulation will occur in the nutrient-rich rhizosphere of these soils. Rhizosphere microorganisms serve as a labile source of nutrients and also play a critical role in organic matter synthesis and degradation.

The genus *Rhizobium* (*see p. 423*) is a prominent member of the rhizosphere community. This bacterium can establish a symbiotic association with legumes and fix nitrogen for use by the plant. *Rhizobium* species infect and nodulate specific hosts (figure 42.7).

This complex infection process, presented here in a simplified form, appears to involve a series of molecules produced by the host plant that lead to the exchange of recognition signals. Flavonoid inducers produced by the plant play a major role in this process by stimulating the *Rhizobium* to synthesize specific **Nod factors** that promote the binding of the bacterium to sites on the root hairs (figure 42.7*b–e*). After bacterial attachment, the root hairs curl and the bacteria induce the plant to form an **infection thread** that grows down the root hair.

The *Rhizobium* then travels within the infection thread into the underlying root cells as noted in figure 42.7*f*. At no time does the *Rhizobium* actually enter the plant cell when it is in the infection thread! As the bacteria pass from the infection thread into the host cell (figure 42.7*h*), the *Rhizobium* is enclosed by a plant-derived membrane, called the peribacteroid membrane, to form a **bacteroid** (figure 42.7*i*). Further growth and differentiation lead to the development of a nitrogen-fixing form, a structure called a **symbiosome** (figure 42.7*j*). At this point, specific nodule components such as leghemoglobin, which protect the nitrogen fixation enzymes from oxygen, are produced to complete the nodulation process (figure 42.7*k*). The final **root nodules** are shown in figure 42.7*l*.

TABLE 42.5	Microbial Populations and Their Responses in the Rhizosphere		
Organisms	**Rhizosphere Soil (R)[a]**	**Control Soil (S)**	**Approximate R:S Ratio[b]**
Taxonomic Groups			
Unicellular bacteria	$1,200 \times 10^6$	53×10^6	23:1
Actinomycetes	46×10^6	7×10^6	7:1
Fungi	12×10^5	1×10^5	12:1
Protozoa	24×10^2	10×10^2	2:1
Algae	5×10^3	27×10^3	0.2:1
Bacterial Physiological Groups			
Ammonifiers	500×10^6	4×10^6	125:1
Gas-producing anaerobes	39×10^4	3×10^4	13:1
Anaerobes	12×10^6	6×10^6	2:1
Denitrifiers	126×10^6	1×10^5	1,260:1
Aerobic cellulose decomposers	7×10^5	1×10^5	7:1
Anaerobic cellulose decomposers	9×10^3	3×10^3	1:1
Endospore formers	930×10^3	575×10^3	1:1

Modified from T. R. G. Gray and S. T. Williams, *Soil Micro-Organisms.* Copyright © 1971 Longman Group Limited, Essex, United Kingdom. Reprinted by permission.

[a]Values are expressed in numbers per gram.

[b]R:S ratio equals ratio of organisms in the rhizosphere (R) to those in the control soil (S) not influenced by plant exudates.

Figure 42.15   Fungal Endophytes. Fungi have been found in the upper parts of some plants. A fungal endophyte growing inside the leaf sheath of a grass, tall fescue, is shown.

Figure 42.17   *Agrobacterium*. This bacterium can be used as a vector to incorporate foreign DNA into susceptible plants. *Agrobacterium*-caused tumor on a *Kalanchoe* sp. plant.

Figure 42.16   Parasitic Castration of Plants by Endophytic Fungi. Stroma of the fungus *Atkinsonella hypoxylon* infecting *Danthonia compressa* and causing abortion of the terminal spikelets.

positive or deleterious effect on plant growth or development. The use of these bacteria as microbial delivery systems in agriculture is a current topic in agricultural biotechnology.

Agrobacterium, Plant Tumors, and Molecular Biology

Another exciting plant-soil microorganism interaction is the *Agrobacterium* infection that produces tumorlike growths on plants (figure 42.17). Gall formation involves *Agrobacterium* strains that contain the **Ti** (tumor-inducing) **plasmid.** This plasmid may be modified to allow the transfer of genetic characteristics such as herbicide resistance and bioluminescence to plants. With the ability to modify plant DNA using *Agrobacterium*-borne plasmids, rapid advances are being made in plant molecular biology.

Agrobacterium (Box 15.2 and p. 424).

1. List five characteristics that can be measured in studying the soil microbial community.
2. Define the following terms: rhizosphere, root nodule, bacteroid, associative nitrogen fixation, mycorrhizae, ectomycorrhizae, endomycorrhizae, and actinorhizae.
3. How is a legume root nodule formed by *Rhizobium* activity?
4. What are the major contributions of mycorrhizae and actinorhizae to plant functioning?
5. What are tripartite and tetrapartite associations?
6. What are endophytic microorganisms and why are they important to plants?
7. Discuss the nature and importance of the Ti plasmid.

endophytes may be less susceptible to attack by various chewing insects due to the production of alkaloids, a form of "chemical defense." Not all such relationships are mutualistic (beneficial to both partners). Some are parasitic. Parasitic fungal endophytes can actually reduce the genetic variability of the plant by sterilizing their host (figure 42.16). This decreases the plant's ability to resist the fungal endophyte. The "parasitic castration of plants" by systemic fungi is of major importance in the co-evolution of plants and fungi.

Endophytic bacteria have been discovered in cotton, pears, and potatoes. Some are plant pathogens that can survive for extended periods in a quiescent state. The majority have no known

Soil Organic Matter and Soil Fertility

Soil organic matter helps retain nutrients, maintain soil structure, and hold water for plant use. This important resource is subject to gains and losses, depending on changes in environmental conditions and agricultural management practices. Plowing and other similar disturbances expose the soil organic matter to more oxygen, leading to extensive microbiological degradation of organic matter. Irrigation causes periodic wetting and drying, which can also lead to increased degradation of soil organic matter, especially at higher temperatures.

Several approaches have been developed to overcome management practices that stimulate soil microorganisms and lead to increased degradation of soil organic matter. These approaches include "no till" or "minimum till" agriculture, in which surface soil is minimally disturbed and chemicals are used to control weed growth.

Composting is another, much older but still effective approach to maintaining and augmenting the organic matter content of a soil. In this process plant materials are allowed to decompose under moist, aerobic conditions, in which the readily utilizable plant fractions are rapidly decomposed. If the compost is too moist or too dry, the desired decomposition process will not occur. The compost pile reaches higher temperatures, allowing thermophilic microorganisms to participate actively in these processes. When the composting is completed, the residual lignin and other more resistant plant materials will have been partially transformed to humus. Such biologically stabilized compost, when added to soil, increases the soil organic matter content and does not stimulate the soil microorganisms.

Thermophile characteristics (pp. 129–30).

1. What soil management processes can lead to increases in degradation of soil organic matter?
2. Why might composted materials have different effects on soil microorganisms and organic matter than additions of fresh plant materials?

Pesticides and Microorganisms— Fallibility and Recalcitrance

The use of pesticides to control undesired plants, insects, and other organisms has become widespread in the last 40 years. This fact has provoked important questions concerning the nature of pesticide degradation and the degradability of both natural and synthetic chemicals.

Originally it was assumed, given time and the almost infinite variety of microorganisms, that all organic compounds, including those synthesized in the laboratory, would eventually degrade. This idea of microbial infallibility was suggested in 1952. Observations of natural and synthetic organic compound accumulation in natural environments, however, began to raise questions about the ability of microorganisms to degrade these varied substances and the role of the environment (clays, anaerobic conditions) in protecting some chemicals. With the development of synthetic pesticides, it became distressingly evident that not all organic compounds are immediately biodegradable. This chemical **recalcitrance** (resisting authority or control) resulted from the apparent fallibility of microorganisms, or their inability to degrade some industrially synthesized chemical compounds.

Degradation is subject to several definitions. As shown in figure 42.18, it can be defined as a minor change in a molecule, such as dehalogenation. Degradation can also be thought of as fragmentation, where the original structure of the molecule can still be recognized in the fragments. Finally, biodegradation can mean complete mineralization of a compound to inorganic forms.

Degradation often is promoted by the presence of easily usable energy sources, allowing the modification of an otherwise recalcitrant compound. The recalcitrant molecule is only modified or degraded by the microorganism in the presence of a growth-promoting substrate that is metabolized at the same time. **Cometabolism** is extremely important in managing the

Figure 42.18 **Biodegradability.** Several definitions of biodegradability are used in environmental microbiology. (*a*) A minor change in a molecule (dehalogenation). (*b*) Fragmentation. (*c*) Mineralization.

degradation of biologically resistant chemicals such as trichloroethylene (TCE), a widespread contaminant in subsurface environments. By using methane as an energy source, subsurface microorganisms can modify the normally recalcitrant TCE molecule. If the cometabolic process results in the oxidation of the recalcitrant compound, it is called cooxidation. An example would be the microbial oxidation of short-chained alkanes or alkenes to corresponding alcohols, ketones, acids, or epoxides in the presence of a usable energy source.

Degradation of a complex compound takes place in several stages. In the case of halogenated compounds, dehalogenation often occurs early in the overall process. Dehalogenation of many compounds containing chlorine, bromine, or fluorine occurs faster under anaerobic than under aerobic conditions. Once the dehalogenation steps are completed, degradation of the main structure of many pesticides often proceeds more rapidly in the presence of O_2. It should be noted that the rate of degradation varies with the location of halogens on the molecule. For example, the presence of substituents in the *meta* position on aromatic compounds can cause a marked inhibition of degradation, as is observed with the herbicide 2,4,5-T, in comparison with 2,4-D (figure 42.19).

The soil microbial community also can change its characteristics on exposure to complex organic molecules. After the community's repeated exposure to a given chemical, faster rates of degradation may occur (figure 42.20). A microbial community can become so efficient at rapid herbicide decomposition that herbicide effectiveness is diminished. To counteract this process, herbicides can be changed to throw the microbial community off balance, thus preserving the effectiveness of the chemicals. The degradation of many pesticides may also result in the accumulation of organic fragments that bind with organic matter in the soil. The longer-term fate and possible effects of "bound" pesticide residues on the soil system, plants, and higher organisms are largely unknown.

Soil Microorganism Interactions with the Atmosphere

Soil microorganisms have interesting interactions with the atmosphere. Ice-nucleating bacteria from decomposing litter can increase precipitation. Specific strains of *Pseudomonas syringae* synthesize proteins that can serve as nucleation centers for snow formation in cloud chambers. Such ice-nucleating soil microorganisms may significantly influence weather on a global scale. "Ice-minus" strains of *P. syringae,* developed by recombinant technology, can be sprayed on frost-sensitive crops such as strawberries. In the presence of the "ice-minus" bacteria, water will not freeze on the leaves as readily, making it possible to save frost-sensitive crops from destruction. The release of such genetically modified microorganisms (GEMs) to the environment is of continuing concern (*see pp. 326–28*).

Soil microorganisms also influence the atmosphere by degrading airborne pollutants such as hydrogen, CO, benzene, trichloroethylene (TCE), and formaldehyde. They can substantially improve the air in closed buildings (Box 42.2). Although soil microorganisms cannot completely eliminate these pollutants, they can decrease pollutant concentrations to equilibrium levels of approximately 1 to 2 ppm.

(a)

(b)

Figure 42.19   The *Meta* Effect and Biodegradation. Minor structural differences can have major effects on the biodegradability of chemicals. The *meta* effect is an important example. (*a*) Readily biodegradable 2,4-D, with an exposed *meta* position. (*b*) Recalcitrant 2,4,5-T, with the *meta* position blocked by a chlorine group.

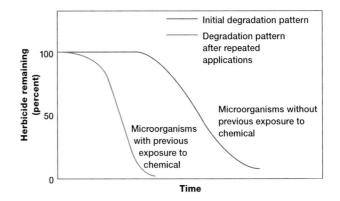

Figure 42.20   Repeated Exposure and Degradation Rate. Addition of an herbicide to a soil can result in changes in the degradative ability of the microbial community. Relative degradation rates for an herbicide after initial addition to a soil, and after repeated exposure to the same chemical.

Soil microorganisms, through their activities in biogeochemical cycling (*see chapter 40*), can have major effects on global fluxes of a variety of gases. These gases can be considered as those that are "relatively stable" and those that are "reactive gases." Relatively stable gases that are influenced by microbial activities include carbon dioxide, nitrous oxide, nitric oxide, and methane. Microorganisms also contribute to the flow of reactive gases such as ammonia, hydrogen sulfide, and dimethylsulfide, as major examples. These reactive gases tend to be produced in more waterlogged environments.

Atmospheric gases such as carbon dioxide, nitrous oxide, nitric oxide, and methane that decrease the loss of radiation from the earth and may produce atmospheric warming are called **greenhouse gases.** The production and consumption of these

Box 42.2

Soil Microorganisms and "Sick Buildings"

A major problem in the development of more energy-efficient homes and office buildings is the potential effect of such closed environments on human health. With many people spending much of their lives in such enclosed environments, the "sick building syndrome" is an increasing concern. While saving energy, these "sick buildings" have higher levels of many volatile compounds, including benzene, trichloroethylene (TCE), formaldehyde, phenolics, and solvents. These are released from rugs, furniture, plastic flooring, paints, and office machines such as photocopiers and printers. An important but still largely unappreciated means of improving the air in such "sick buildings" is through plants and their associated soil microorganisms.

Plants not only produce oxygen, but the soil microorganisms degrade many airborne pollutants. It is recommended that one plant be used per 100 square feet of living area. As noted by B. C. Wolverton, "The ultimate solution to the indoor air pollution problem must involve plants, the plant soils and their associated microorganisms." Soil microorganisms, especially in association with plants, can help keep air in closed environments fresher and more healthful (plants are also nice to look at).

Other microorganisms in buildings can cause problems. These include *Legionella* (*see pp. 744–45*) which can grow in cooling systems, and allergy-causing molds such as *Aspergillus* (*p. 797*), which can grow in damp areas.

Box 42.3

Soils, Termites, and Intestinal Microorganisms: Understanding Atmospheric Methane Changes

Termites are important components of tropical ecosystems, where their use of cellulosic plant materials allows rapid—sometimes too rapid—recycling of plant materials. Termites harbor significant populations of archaeobacteria that use products of cellulose digestion, including CO_2 and hydrogen, to produce methane.

Termites occur on 2/3 of the earth's land surface, and, based on laboratory studies, 0.77% of the carbon ingested by termites can be released as methane. In tropical wet savannas and cultivated areas,

termite populations are increasing rapidly. This increase is being accelerated by the destruction of tropical forests, which results in the accumulation of dead plant materials on the soil surface. This provides an ideal environment for the growth of termites. Termites are estimated to be contributing annually at least 1.5×10^{14} grams of methane, together with hydrogen and CO_2, to the atmosphere. This is believed to be contributing to measurable increases in the atmospheric methane level. Thus unseen termites and their associated gut microorganisms may be affecting global warming.

greenhouse gases can be influenced by a range of human activity-related factors. These include fertilization and automobile use, conversion of soils to agricultural use, and the use of landfills, all of which can cause increases in mineral nitrogen levels.

Methane is a greenhouse gas of increasing concern and can be derived from a variety of sources. These include ruminants, rice paddies, and landfills. Landfills, especially, can release methane to the atmosphere over longer terms (decades, centuries). Another interesting source of atmospheric methane is microorganisms that inhabit the gut of wood-eating termites. As noted in Box 42.3, with increased deforestation and the accumulation of plant residues, populations of termites (and termite gut-inhabiting methanogenic microorganisms) are increasing.

This is not the end of the story. Methane, once released by microorganisms inhabiting various environments, can be consumed by aerobic bacteria called methanotrophs (*see p. 424*), which are present in soils and waters. The question is, can methanotrophs keep up with the increased methane production rates? A critical factor influencing methane consumption in soils is the soil ammonium ion concentration. As ammonium

ion levels increase due to agriculture and atmospheric pollution, the ability of soils to use methane decreases. Recent studies have found that the methane-using microorganisms form toxic nitrite from the ammonium ion, and the inhibitory effect is increased with higher atmospheric methane levels! Thus the equilibrium between methane formation and use by microorganisms may be influenced by an additional factor: soil nitrogen levels.

1. How are recalcitrance, fallibility, and infallibility related?

2. What is the process of cometabolism? Why is it important in microbial modification of many chemicals?

3. What definitions of degradation are used in environmental microbiology?

4. In which ways can soil microorganisms interact with the atmosphere in closed buildings?

5. How do soil microorganisms influence the production and consumption of greenhouse gases? What are critical points that can be managed by humans?

Part Eleven
Food and Industrial Microbiology

Chapter 43

Microbiology of Food

Chapter 44

Industrial Microbiology and Biotechnology

We often see banquet tables filled with a tempting variety of foods for us to sample. Foods such as cheese, bread, and wine are produced by specific microorganisms grown under carefully controlled conditions. In addition, products of industrial microbiology often are used as food flavor enhancers and emulsifiers. If the wrong microorganisms are present during food production, we will be left with unpleasant smells and tastes instead of delightful surprises. Even when food is properly prepared, microorganisms can affect it and our health. If foods are handled or stored improperly high levels of spoilage organisms can develop, especially in foods such as the salad shown here. Pathogens in contaminated food can release toxins and produce a food intoxication or may invade the person eating the food and cause a food-borne infection. Clearly foods and microorganisms are closely linked. This section of the book will familiarize you with foods and their associated microorganisms.

CHAPTER 43
Microbiology of Food

Tell me what you eat, and I will tell you what you are.
—Brillat-Savarin

There is no greater love than love of food. Air and water give us life, but food gives us a way of life.
—George Bernard Shaw

Concepts

1. Foods often provide an ideal environment for microbial survival and growth.

2. Microbial growth in foods involves successional changes, with intrinsic, or food-related, and extrinsic, or environmental, factors interacting with the microbial community over time.

3. Foods can be preserved by physical, chemical, and biological processes. Refrigeration does not significantly reduce microbial populations but only retards spoilage. Pasteurization results in a pathogen-free product with a longer shelf life.

4. Sterilization can be used to control microorganisms in foods. Heat-resistant spore-forming bacteria are used to test sterilization efficiency.

5. Chemicals can be added to foods to control microbial growth. Such chemicals include sugar, salt (decreasing water availability), and many organic chemicals that affect specific groups of microorganisms.

6. Foods can transmit a wide range of diseases to humans. In food infection the food serves as a vehicle for the transfer of the pathogen to the consumer, in whom the pathogen grows and causes disease. With food intoxication the microorganisms grow in the food and produce toxins that can then affect the consumer.

7. Modern molecular techniques are being used to detect disease-causing microorganisms in foods. Rapid and sensitive detection is possible with these procedures.

8. Dairy products, grains, meats, fruits, and vegetables can be fermented. It has been suggested that some fermented dairy products have antimicrobial and anticancer characteristics, especially when particular lactic acid bacteria are used.

9. Fungi, if they grow in moist corn and grain products, can produce a variety of disease-causing chemicals. These can include carcinogenic mycotoxins such as aflatoxins and fumonisins, as well as behavior-altering ergot alkaloids.

10. Wines are produced by the direct fermentation of fruit juices or musts. For fermentation of cereals and grains, starches and proteins contained in these substrates must first be hydrolyzed to provide substrates for the alcoholic fermentation.

11. Microbial cells, grown on a variety of substrates, can be used directly as a nutrient by animals and humans. Popular single-cell protein sources include the cyanobacterium *Spirulina* and yeasts.

Foods and microorganisms have long and interesting associations that developed long before the beginning of recorded history. Foods are not only of nutritional value to those who consume them but often are ideal culture media for microbial growth. Foods may be preserved by fermentation, or they may spoil, depending on the microorganisms present.

Microorganisms can be used to transform raw foods into gastronomic delights, including cheeses, pickles, sausages, and soy sauce. Wines, beers, and other alcoholic products also are produced through microbial activity. On the other hand, foods also can serve as a vehicle for disease transmission, and the detection and control of pathogens and food spoilage microorganisms are important parts of food microbiology. During the entire sequence of food handling, from the producer to the final consumer, microorganisms can affect food quality and human health.

Having wholesome, nutrient-rich foods is important for all people. Microbial growth in foods can result in either preservation or spoilage, depending on the microorganisms involved and the food storage conditions.

Undesirable microorganisms can cause illness in two ways when contaminated foods are eaten. A person may be infected by the food-borne pathogens, which then grow in the consumer. In other cases toxic components may be formed when microorganisms grow in food before it is ingested, resulting in the rapid onset of disease symptoms after consumption. Contamination by disease-causing microorganisms can occur at any point in the food-handling sequence.

Food Spoilage and Preservation Processes

With the beginning of agriculture and a decreasing dependence on hunting and gathering, the need to preserve excess foods became essential to survival. The use of salt as a meat preservative and the production of cheeses and curdled milks was introduced in Near Eastern civilization as early as 3000 B.C. The production of wines and the preservation of fish and meat by smoking also

were common by this time. Despite a long tradition of efforts to preserve food from spoilage, it was not until the nineteenth century that the microbial spoilage of food was studied rigorously. Louis Pasteur established the modern era of food microbiology in 1857, when he showed that microorganisms cause milk spoilage. Pasteur's work in the 1860s proved that heat could be used to control spoilage organisms in wines and beers (*see chapter 1*).

Control of microorganism growth (chapter 7).

A variety of intrinsic and extrinsic factors determine whether microbial growth will preserve or spoil foods, as shown in figure 43.1. The intrinsic or food-related factors include pH, moisture content, water activity or availability, oxidation-reduction potential, physical structure of the food, available nutrients, and the possible presence of natural antimicrobial agents. Extrinsic or environmental factors include temperature, relative humidity, gases (CO_2, O_2) present, and the types and amounts of microorganisms present in the food.

To best manage these potential sources of microbial contamination and growth, from the point of production to the point of use, the hazard analysis critical control point (HACCP) approach is widely used. This approach emphasizes monitoring the quality of food ingredients at critical process handling steps. If these individual steps are properly controlled, a safe product at the point of consumption will result. As a part of this program, microbiological limits have been established for a wide range of raw materials and also for final food products. It is expected that HACCP will become even more important in the future to assure that safe foods are available for the public.

Intrinsic Factors

The pH of a food is critical because a low pH favors the growth of yeasts and molds (*see chapter 6*). In neutral or alkaline pH foods, such as meats, bacteria are more dominant in the spoilage process. This leads to proteolysis and the anaerobic breakdown of proteins or **putrefaction,** which releases foul-smelling amine compounds. Depending on the major substrate present in a food, different types of spoilage may occur, as summarized in table 43.1.

Figure 43.1   Intrinsic and Extrinsic Factors. A variety of intrinsic and extrinsic factors can influence microbial growth in foods. Time-related successional changes occur in the microbial community and the food.

TABLE 43.1   Differences in Spoilage Processes in Relation to Food Characteristics

Substrate	Chemical Reactions or Processes[a]	Typical Products and Effects
Pectin	Pectinolysis	Methanol, uronic acids (loss of fruit structure, soft rots)
Proteins	Proteolysis, deamination	Amino acids, peptides, amines, H_2S, ammonia, indole (bitterness, souring, bad odor, sliminess)
Carbohydrates	Hydrolysis, fermentations	Organic acids, CO_2, mixed alcohols (souring, acidification)
Lipids	Hydrolysis, fatty acid degradation	Glycerol and mixed fatty acids (rancidity, bitterness)

[a]Other reactions also occur during the spoilage of these substrates.

TABLE 43.2   Approximate Minimum Water Activity Relationships of Microbial Groups of Importance in Food Spoilage

Organism Group	a_w
Most spoilage bacteria	0.90
Most spoilage yeasts	0.88
Most spoilage molds	0.80
Halophilic bacteria	0.75
Xerophilic molds	0.61
Osmophilic yeasts	0.60

From James M. Jay, *Modern Food Microbiology*, 4th edition. Reprinted by permission of Chapman & Hall, New York, NY.

The presence and availability of water also affect the ability of microorganisms to colonize foods. Simply by drying a food, one can control or eliminate spoilage processes. Water, even if present, can be made less available by adding solutes such as sugar and salt. Water availability is measured in terms of water activity (a_w). This represents the ratio of relative humidity of the air over a test solution compared with that of distilled water. When large quantities of salt or sugar are added to food, most microorganisms are dehydrated by the hypertonic conditions and cannot grow (table 43.2; *see also table 6.3*). Even under these adverse conditions, osmophilic and xerophilic microorganisms may spoil food. **Osmophilic** [Greek *osmus,* impulse, and *philein,* to love] **microorganisms** grow best in or on media with a high osmotic concentration, whereas **xerophilic** [Greek *xerosis,* dry, and *philein,* to love] **microorganisms** prefer a low a_w environment and may not grow under high a_w conditions.

Water activity and microbial growth (pp. 124–25).

The oxidation-reduction potential of a food also influences spoilage. When meat products, especially broths, are cooked, they often have lower oxidation-reduction potentials. These products with their readily available amino acids, peptides, and growth factors are ideal media for the growth of anaerobes, including *Clostridium* (*see table 37.3*).

The physical structure of a food also can affect the course and extent of spoilage. The grinding and mixing of foods such as sausage and hamburger not only increase the food surface area and alter cellular structure, but also distribute contaminating microorganisms throughout the food. This can result in rapid spoilage if such foods are stored improperly. Vegetables and fruits have outer skins (peels and rinds) that protect them from spoilage. Often spoilage microorganisms have specialized enzymes that help them weaken and penetrate protective peels and rinds, especially after having been bruised.

Many foods contain natural antimicrobial substances, including complex chemical inhibitors and enzymes. Coumarins found in fruits and vegetables exhibit antimicrobial activity. Cow's milk and eggs also contain antimicrobial substances. Eggs are rich in the enzyme lysozyme that can lyse the cell walls of contaminating gram-positive bacteria (*see figure 29.8*). Another interesting food with antimicrobial activities is the hot sauces used on raw oysters and other seafoods. Tabasco and other hot red pepper sauces apparently have particularly desirable antimicrobial characteristics.

Herbs and spices often possess significant antimicrobial substances; generally fungi are more sensitive than most bacteria. Sage and rosemary are two of the most antimicrobial spices. Aldehydic and phenolic compounds are found in cinnamon, mustard, and oregano. These compounds inhibit microbial growth. Other important inhibitors are garlic, which contains allicin, and cloves, which have eugenol. However, spices also can contain pathogenic and spoilage organisms. Coliforms, *B. cereus, C. perfringens,* and *Salmonella* species have been detected in most spices. Microorganisms can be eliminated or reduced by ethylene oxide sterilization. This treatment can result in *Salmonella*-free spices and herbs and a 90% reduction in the levels of general spoilage organisms.

Unfermented green and black teas also have well-documented antimicrobial properties because of their polyphenol contents, which apparently are diminished when the teas are fermented. Such teas are active against bacteria, viruses, and fungi.

Extrinsic Factors

Temperature and relative humidity are important extrinsic factors in determining whether a food will spoil. At higher relative

TABLE 43.3   Basic Approaches to Food Preservation

General Technique	Examples of Process
Asepsis, removal of microorganisms	Avoidance of microbial contamination, filtration, centrifugation
Low temperature	Refrigeration, freezing
High temperature	Partial or complete heat inactivation of microorganisms (pasteurization and canning)
Water removal	Lyophilization, use of a spray dryer or heated drum dryer
Water availability decrease	Addition of solutes such as salt or sugar to decrease a_w values
Chemical preservation	Addition of specific inhibitory substances (e.g., organic acids, nitrates, sulfur dioxide)
Radiation	Use of ionizing (UV) and nonionizing (gamma rays) radiation

humidities microbial growth is initiated more rapidly, even at lower temperatures (especially when refrigerators are not maintained in a defrosted state). When drier foods are placed in moist environments, moisture absorption can occur on the food surface, eventually allowing microbial growth.

The atmosphere in which the food is stored also is important. This is especially true with shrink-packed foods because many plastic films allow oxygen diffusion, which results in increased growth of surface-associated microorganisms. Excess CO_2 can decrease the solution pH, inhibiting microbial growth. Storing meat in a high CO_2 atmosphere inhibits gram-negative bacteria, resulting in a population dominated by the lactobacilli.

1. What are some intrinsic factors that influence food spoilage and how do they exert their effects?
2. Why might sausage and other ground meat products provide a better environment for the growth of food spoilage organisms than raw cuts of meats?
3. List some antimicrobial substances found in foods. What is the mechanism of action of lysozyme?
4. What primary extrinsic factors can determine whether food spoilage will occur?
5. Why might CO_2 concentrations in the food environment modify responses of food spoilage organisms?

Food Preservation Alternatives

Foods can be preserved by a variety of methods (table 43.3). It is vital to eliminate or reduce the populations of spoilage and disease-causing microorganisms and to maintain the microbiological quality of a food with proper storage and packaging.

Contamination often occurs after a package or can is opened and just before the food is served. This can provide an ideal opportunity for growth and transmission of pathogens, if care is not taken. Washing utensils with chemical sanitizers such as chlorine or quaternary ammonium compounds and limiting human contact with foods can make a major difference in the microbiological quality of foods.

Sanitizing agent characteristics (pp. 144–48).

Physical Removal of Microorganisms

Microorganisms can be removed from water, wine, beer, juices, soft drinks, and other liquids by filtration. This can keep bacterial populations low or eliminate them entirely. Prefilters and centrifugation often are used to maximize filter life and effectiveness. Several major brands of beer are filtered rather than pasteurized to better preserve the flavor and aroma of the original product.

Temperature Effects

Refrigeration at 5°C retards microbial growth, although with extended storage, psychrophiles and psychrotrophs will eventually grow and produce spoilage. Slow microbial growth at temperatures below −10°C has been described, particularly with fruit juice concentrates, ice cream, and some fruits. Some microorganisms are very sensitive to cold and their numbers will be reduced, but cold does not lead to significant decreases in overall microbial populations.

Temperature effects on microbial growth (pp. 127–30).

Controlling microbial populations in foods by means of high temperatures can significantly limit disease transmission and spoilage. **Pasteurization** uses high temperatures to eliminate disease-causing organisms and reduce microbial populations (*see chapter 7*). Sterilization relies on high temperatures to eliminate all living organisms. Heating processes, first used by Nicholas Appert in 1809 (Box 43.1), provide a safe means of preserving foods, particularly when carried out in commercial **canning** operations (figure 43.2). Canned food is heated in special containers called retorts at about 115°C for intervals ranging from 25 to over 100 minutes. The precise time and temperature depend on the nature of the food. Sometimes canning does not kill all the microorganisms, but only those that will spoil the food (remaining bacteria are unable to grow due to acidity of the food, as an example). After heat treatment the cans are cooled as rapidly as possible, usually with cold water. Quality control and processing effectiveness are sometimes compromised, however, in home processing of foods, especially with less acid (pH values greater than 4.6) products such as green beans or meats.

Destruction of microorganisms by heat (pp. 139–42).

Box 43.1

An Army Travels on Its Stomach

The movement and maintenance of large numbers of military personnel have always been limited by food supplies. The need to maintain large numbers of troops under hostile and inclement conditions led the French government in 1795 to offer a prize of 12,000 francs to the individual who could preserve foods for use under field conditions. Eventually the prize was awarded to Nicholas Appert, a candy maker, for his development of a heating process in which meats and other products could be preserved under sealed conditions.

Appert's work was based on the assumptions that heating and boiling control "ferments" and that sealing the food in bottles before heating it avoids the effects of air on spoilage. Despite Leeuwenhoek's earlier work, Appert did not have the concept of microorganisms to assist him in explaining the effectiveness of his process. His containers were large glass bottles, sealed with laminated corks and fish glue. With extreme care and attention to detail, he was able to heat these bottles in boiling water to provide food that could be stored for several years. Appert's work was an important foundation for the later studies of Louis Pasteur.

Figure 43.2   A Canning Operation. Microbial control is important in the processing and preservation of many foods. Worker pouring peas into a large, clean vat during the preparation of vegetable soup. After preparation the soup is transferred to cans. Each can is heated for a short period, sealed, processed at temperatures around 110–121°C in a canning retort to destroy spoilage microorganisms, and finally cooled.

Pasteurization involves heating food to a temperature that kills disease-causing microorganisms such as *Mycobacterium tuberculosis* (*see chapters 21 and 37*) and substantially reduces the levels of spoilage organisms. In the processing of milk, beers, and fruit juices by conventional low-temperature holding (LTH) pasteurization, the liquid is maintained at 62.8°C for 30 minutes. Products can also be held at 71°C for 15 seconds, a high-temperature, short-time (HTST) process; milk can be treated at 141°C for 2 seconds for ultra-high-temperature (UHT) processing. Shorter-term processing results in improved flavor and extended product shelf life.

Such heat treatment is based on a statistical probability that the number of remaining viable microorganisms will be below a certain level after a particular heating time at a specific temperature. This process is discussed in detail on pp. 139–42.

Dehydration, used to produce freeze-dried foods, is now a common means of eliminating microbial growth. This modern process is simply an update of older procedures in which grains, meats, fish, and fruits were dried. The combination of free-water loss with an increase in solute concentration in the remaining water makes this type of preservation possible.

Preservation by Chemicals and Radiation

Various chemical agents can be used to preserve foods, and these substances are closely regulated by the U.S. Food and Drug Administration and are listed as being "generally recognized as safe" or GRAS (table 43.4). They include simple organic acids, sulfite, ethylene oxide as a gas sterilant, sodium nitrite, and ethyl formate. The effectiveness of many of these chemical preservatives depends on the food pH. As an example, sodium propionate is most effective at lower pH values, where it is primarily undissociated and able to be taken up by lipids of microorganisms. Breads, with their low pH values, often contain sodium propionate as a preservative. Chemical preservatives are used with grain, dairy, vegetable, and fruit products. Sodium nitrite is an important chemical used to help preserve ham, sausage, bacon, and other cured meats by inhibiting the growth of *Clostridium botulinum* and the germination of its spores. This protects against botulism and reduces the rate of spoilage. Besides increasing meat safety, nitrite decomposes to nitric acid, which reacts with heme pigments to keep the meat red in color. Current concern about nitrite arises from the observation that it can react with amines to form carcinogenic nitrosamines. Nitrite is added in very small amounts, and eventually it may be possible to eliminate its use entirely.

TABLE 43.4   Major Groups of Chemicals Used in Food Preservation

Preservatives	Approximate Maximum Use Range	Organisms Affected	Foods
Propionic acid/propionates	0.32%	Molds	Bread, cakes, some cheeses, inhibitor of ropy bread dough
Sorbic acid/sorbates	0.2%	Molds	Hard cheeses, figs, syrups, salad dressings, jellies, cakes
Benzoic acid/benzoates	0.1%	Yeasts and molds	Margarine, pickle relishes, apple cider, soft drinks, tomato ketchup, salad dressings
Parabens[a]	0.1%	Yeasts and molds	Bakery products, soft drinks, pickles, salad dressings
SO_2/sulfites	200–300 ppm	Insects and microorganisms	Molasses, dried fruits, wine, lemon juice (not to be used in meats or other foods recognized as sources of thiamine)
Ethylene/propylene oxides	700 ppm	Yeasts, molds, vermin	Fumigant for spices, nuts
Sodium diacetate	0.32%	Molds	Bread
Dehydroacetic acid	65 ppm	Insects	Pesticide on strawberries, squash
Nisin	1%	Lactic acid bacteria, clostridia	Certain pasteurized cheese spreads
Sodium nitrite	120 ppm	Clostridia	Meat-curing preparations
Caprylic acid	—	Molds	Cheese wraps
Ethyl formate	15–200 ppm	Yeasts and molds	Dried fruits, nuts

From James M. Jay, *Modern Food Microbiology*, 4th edition. Reprinted by permission of Chapman & Hall, New York, NY.

[a]Methyl-propyl-, and heptyl-esters of *p*-hydroxybenzoic acid.

Nisin is a polypeptide agent that is the most widely used food preparation antibiotic. It is nontoxic and naturally produced by strains of *Lactococcus lactis* and affects mainly gram-positive bacteria, especially *Enterococcus faecalis*. This can be used particularly in low-acid foods to improve inactivation of *Clostridium botulinum* during the canning process or to inhibit germination of any surviving spores. The mode of action is probably inhibition of peptidoglycan synthesis (*see peptidoglycan synthesis, pp. 206–9*), and thus there is no effect on humans.

Radiation, both ionizing and nonionizing, has an interesting history in relation to food preservation. Ultraviolet radiation is used to control populations of microorganisms on the surfaces of laboratory and food-handling equipment, but it does not penetrate food. The major method used for radiation sterilization of food is gamma irradiation from a cobalt-60 source (*see pp. 143–44*). Such electromagnetic radiation has excellent penetrating power and must be used with moist foods because the radiation produces peroxides from water in the microbial cells, resulting in oxidation of sensitive cellular constituents. This process of **radappertization,** named after Nicholas Appert (Box 43.1), can extend the shelf life of seafoods, fruits, and vegetables. To sterilize meat products, commonly 4.5 to 5.6 megarads are used.

Among the more interesting radiation-resistant bacteria that have been studied is *Deinococcus radiodurans* (*see chapter 21*). This bacterium has a complex cell wall structure and tetrad-forming growth patterns (*see figure 21.3*). It also has an extraordinary capacity to withstand high doses of radiation, although the mechanism for its resistance is not understood.

1. Describe the major approaches used in food preservation.
2. What types of chemicals can be used to preserve foods?
3. Nitrite is often used to improve the storage characteristics of prepared meats. What toxicological problems may result from the use of this chemical?
4. Under what conditions can ultraviolet light and gamma radiation be used to control microbial populations in foods and in food preparation? What is radappertization?

Diseases and Foods

Diseases Transmitted by Foods

Many diseases transmitted by foods, or food poisonings, are discussed in chapters 36 and 37, and only a few of the more important food-borne bacterial pathogens are mentioned here. There are two primary types of food-related diseases: food-borne infections and food intoxications.

Food-borne and waterborne diseases (pp. 737–39, 766–71).

TABLE 43.5   Major Food-Borne Infectious Diseases

Disease	Organism	Incubation Period and Characteristics	Major Foods Involved
Salmonellosis	S. typhimurium, S. enteritidis	8–48 hr Enterotoxin and cytotoxins	Meats, poultry, fish, eggs, dairy products
Campylobacteriosis	Campylobacter jejuni	Usually 2–10 days Most toxins are heat-labile	Milk, pork, poultry products, water
Listeriosis	L. monocytogenes	Varying periods Related to meningitis and abortion; newborns and the elderly especially susceptible	Meat products, especially pork and milk
Escherichia coli diarrhea and colitis	E. coli	24–72 hr Enterotoxigenic positive and negative strains; hemorrhagic colitis	Undercooked ground beef, raw milk
Shigellosis	Shigella sonnei, S. flexneri	24–72 hr	Egg products, puddings
Yersiniosis	Yersinia enterocolitica	16–48 hr Some heat-stable toxins	Milk, meat products, tofu
Plesiomonas diarrhea	Plesiomonas shigelloides	1–2 hr	Uncooked mollusks and foreign travel
Vibrio parahaemolyticus gastroenteritis	V. parahaemolyticus	16–48 hr	Seafood, shellfish

---

Box 43.2

Typhoid Fever and Canned Meat

Minor errors in canning have led to major typhoid outbreaks. In 1964 canned corned beef produced in South America was cooled, after sterilization, with nonchlorinated water; the vacuum created when the cans were cooled drew *S. typhi* into some of the cans, which were not completely sealed. This contaminated product was later sliced in an Aberdeen, Scotland, food store, and the meat slicer became a continuing contamination source; the result was a major epidemic that involved 400 people. The *S. typhi* was a South American strain, and eventually the contamination was traced to the contaminated water used to cool the cans.

This case emphasizes the importance of careful food processing and handling to control the spread of disease during food production and preparation.

A **food-borne infection** involves the ingestion of the pathogen, followed by growth accompanied by tissue invasion and/or the release of toxins. The major diseases of this type are summarized in table 43.5 (*see also table 37.3*). Salmonellosis results from ingestion of the organism, and all species and strains of *Salmonella* are pathogenic (*see also chapter 37*). Gastroenteritis is the disease of most concern in relation to foods, occurring after an incubation time as short as 8 hours. Meats, poultry, and dairy products are the primary sources of the pathogen. *Salmonella* infection can arise from contamination by workers in food-processing plants and restaurants (Box 43.2; *see also Box 35.2*).

Campylobacter jejuni is considered a leading cause of acute bacterial gastroenteritis in humans and can affect persons of all ages. This important pathogen is often transmitted by uncooked or poorly cooked poultry products. For example, transmission often occurs when kitchen utensils and containers are used for chicken preparation and then for salads. Contamination with as few as 10 viable *Campylobacter jejuni* can lead to the onset of diarrhea. *Campylobacter jejuni* also is transmitted by raw milk, and the organism has been found on various red meats. Thorough cooking of food prevents this disease transmission problem.

Listeriosis, caused by *Listeria monocytogenes* (*see chapter 21*), is of continuing interest, as shown by the outbreak that occurred in Southern California in 1985. This outbreak was caused by improper pasteurization of milk used in the commercial production of Mexican-style cheeses. At least 86 cases of infection occurred, including 58 cases involving mother-infant pairs. Forty-seven people died. The outbreak was traced to pinhole leaks in the heat exchangers of a pasteurizing unit. The

Box 43.3

Recently there has been an interesting controversy concerning whether wooden or plastic cutting boards are better to minimize survival and possible cross contamination from microorganisms associated with uncooked foods. Suprisingly wooden cutting boards have been found to be antimicrobial! In comparison with plastic cutting boards, where antimicrobial activity does not occur, microorganisms on the surface of wooden cutting boards will die back at rapid rates. It is felt that this effect is due to water-soluble phenolics from the lignin, which makes up approximately 1/3 of wood. It is important, in spite of these antimicrobial characteristics of wood, to avoid accumulation of foods that can protect microorganisms and allow them to grow on cutting boards or any other equipment used in food preparation. Enzymes also are sensitive to these water-soluble wood constituents. Wooden toothpicks often are used to transfer microbial isolates for use in the polymerase chain reaction (*see PCR, pp. 310–13*). Apparently soluble phenolics leached from wooden toothpicks can inhibit these sensitive enzymes. If plastic toothpicks are used, no inhibition of enzymatic activity is observed. These observations suggest that wooden materials, with their antimicrobial and antienzymatic characteristics, may be good or bad, depending on what a person is trying to do.

---

leaks allowed incoming raw milk to contaminate the pasteurized milk before production of the cheese. *Listeria* is difficult to work with because an extended incubation of samples is required for growth and detection.

Escherichia coli is now recognized as an important food-borne disease organism. Enteropathogenic, enteroinvasive, and enterotoxigenic types can cause diarrhea. The enterohemorrhagic *E. coli* 0157:H7, which causes hemorrhagic colitis, is of particular concern.

All these food-borne diseases are associated with poor hygienic practices. Whether by water or food transmission, the fecal-oral route is maintained, with the food providing the vital link between hosts. Fomites, such as sink faucets, drinking cups, and cutting boards also play a role in the maintenance of the fecal-oral route of contamination (Box 43.3).

Microbial growth in food products also can result in a **food intoxication,** as summarized in table 37.3 (*pp. 768–69*). Intoxication produces symptoms shortly after the food is consumed because growth of the disease-causing microorganism is not required. Toxins produced in the food can be associated with microbial cells or can be released from the cells.

Most *Staphylococcus aureus* strains cause a staphylococcal enteritis related to the synthesis of extracellular toxins (*see chapter 37*). These are heat-resistant proteins, and heating will not usually render the food safe. The effects of the toxins are quickly felt, with disease symptoms occurring within 2 to 6 hours. The main reservoir of *S. aureus* is the human nasal cavity. Frequently *S. aureus* is transmitted to a person's hands and then is introduced into food during preparation. Growth and enterotoxin production usually occur when contaminated foods are held at room temperature for several hours.

Three gram-positive rods are known to cause food intoxications: *Clostridium botulinum, C. perfringens,* and *Bacillus cereus* (*see table 37.3*). *C. botulinum* poisoning is discussed in chapter 37, and *C. perfringens* intoxication is described here.

Clostridium perfringens food poisoning is one of the more widespread food intoxications. These microorganisms, which produce exotoxins, must grow to levels of approximately 10^6 bacteria per gram or higher in a food to cause disease. At least 10^8 bacteria must be ingested. They are common inhabitants of soil, water, food, spices, and the intestinal tract. Upon ingestion the cells sporulate in the intestine. The enterotoxin is a spore-specific protein and is produced during the sporulation process. Enterotoxin can be detected in the feces of affected individuals. *Clostridium perfringens* food poisoning is common and occurs after meat products are heated, which results in O_2 depletion. If the foods are cooled slowly, growth of the microorganism can occur. At 45°C, enterotoxin can be detected 3 hours after growth is initiated. Onset of the symptoms—watery diarrhea, nausea, and abdominal cramps—usually occurs in about 8 to 16 hours.

Bacillus cereus can cause two distinct types of illnesses depending on the type of toxin produced: an emetic illness characterized by nausea and vomiting with an incubation time of 1 to 6 hours, and a diarrheal type, with an incubation of 4 to 16 hours. The emetic type is often associated with boiled or fried rice, while the diarrheal type is associated with a wider range of foods.

Disease Microorganism Detection

A major problem in maintaining food safety is the need to rapidly detect microorganisms in order to curb outbreaks that can affect large populations. This is especially important because of widescale distribution of perishable foods. Standard cultural techniques (*see chapter 5*) may require days to weeks for positive identification of pathogens. Identification is often complicated by the low numbers of pathogens compared with the background microflora. Furthermore, the varied chemical

Figure 43.3   Molecular Probes and Food Microbiology. Molecular techniques are finding increasing use in microbial analysis of foods. Autoradiogram of a radioactive-labeled *Listeria monocytogenes* probe against 100 *Listeria* cultures. Only the *Listeria monocytogenes* cultures show sequence homology and binding with the DNA probe, darkening the autoradiogram film. The other *Listeria* spp. do not react with the probe.

Figure 43.4   The Polymerase Chain Reaction and Food Microbiology. The polymerase chain reaction (PCR) can be used to improve detection of microorganisms and their toxins in foods. PCR of *Escherichia coli* strains isolated from soft cheese. Approximately 100,000 cells of each strain were lysed and subjected to 40 cycles of PCR to detect DNA coding for a heat-stable toxin (STI). Lane 1: 595 base pair (bp) *mal*B fragment and 175 bp STI fragment; lane 2: *Hin*fI restriction analysis of the 175 bp fragment in lane 1, indicating that this fragment is the STI DNA; lanes 3 and 4: positive controls with LT-and ST-producing organisms, respectively; lane 5: molecular weight markers.

and physical composition of foods can make isolation difficult. Fluorescent antibody, enzyme-linked immunoassays (ELISAs), and radioimmunoassay techniques have proven of value (*see chapter 32*). These can be used to detect small amounts of pathogen-specific antigens.

New applications of immunologic techniques to the rapid detection of food-borne pathogens continue to appear. One of the newest techniques is the use of immunomagnetic Dynabeads, which allow desired microorganisms to react with immunologic reagents that are bound to magnetic particles. After the immunologic reaction is completed, the attached microorganisms can be removed from the food by use of a magnetic collector for further study. This makes possible much more rapid direct enrichment of desired microorganisms as an alternative to time-consuming growth in selective media, which has been used for enrichment of food-borne pathogens in the past.

Molecular biology techniques also are increasingly used in identification. The basic methods of analysis and manipulation of DNA and RNA are discussed in chapter 15. These methods are valuable for three purposes: (1) to detect the presence of a single, specific pathogen; (2) to detect viruses that cannot be grown conveniently; and (3) to identify slow-growing or non-culturable pathogens.

Pathogens now can be identified by detecting specific DNA or RNA base sequences with **probes.** These may range from 10 to 10,000 bases in length, but usually probes of 14 to 40 bases are employed. They may be created by generating fragments with restriction endonucleases or through direct chemical synthesis. Probes are labeled by linking them to a variety of enzymatic, isotopic, chromogenic, or luminescent/fluorescent markers. A major advantage of their use is the speed with which specific microorganisms can be detected in a set of cultures, as shown in figure 43.3. In this example, a hydrophobic grid-membrane system has been used. The *Listeria monocytogenes* cultures are radioactive, indicating that they have bound the probe, while other *Listeria* species do not show probe binding.

These techniques also have been extended to allow the detection of a few target cells in large populations of background microorganisms. For example, by using the polymerase chain reaction (*see pp. 310–13*), as few as 10 toxin-producing *E. coli* cells can be detected in a population of 100,000 cells isolated from soft cheese samples. A rapid probe of the sample showed that none of the 58 cultures recovered from this cheese produced a heat-labile toxin (LT), whereas one produced a 175 base pair fragment characteristic of the heat-stable toxin (ST). Based on the susceptibility of the *E. coli* DNA to the *Hin*fI restriction enzyme (lane 2 in figure 43.4), this was identified as DNA related to the ST (lane 4).

These brief examples show the exciting potential of applying rapid DNA- and RNA-based techniques to the detection and control of food pathogens.

(a)

(b)

Figure 43.5   Food Spoilage. When foods are not stored properly, microorganisms can cause spoilage. Typical examples are fungal spoilage of (*a*) bread and (*b*) corn. Such spoilage of corn is called ear rot. This can result in major economic losses.

1. Differentiate between food intoxication and food-borne infection, in terms of disease occurrence.

2. What common food is often related to *Campylobacter*-caused gastroenteritis? What means can be used to control the occurrence of this disease from this source?

3. Why is listeriosis of concern from a medical standpoint?

4. Describe *S. aureus* food intoxication and how it occurs. Will thorough heating prevent the disease?

5. Under what conditions does *C. perfringens* poisoning arise and when is its enterotoxin produced?

6. How can molecular techniques improve the detection of disease-causing microorganisms in foods?

7. In what way can immunology and magnetic beads be used in food microbiology?

Food Spoilage

The role of microorganisms in the spoilage and deterioration of specific groups of foods has led to interesting turns in history, occasionally being associated with "miracles" and "witchcraft." *Serratia marcescens*, felt to be the cause of the "Miracle of Bolsena," which occurred in 1263, produces a red pigment when it grows on moist food materials. Some historians feel it may have been responsible for reports of "blood" on communion wafers and other bread. The riddle was eventually solved by Bartolomeo Bizio in 1879. He also named the bacterium *Serratia marcescens*.

Molds can rapidly grow on grains and corn when these products are held under moist conditions (figure 43.5). Infec-

tion of grains by the ascomycete *Claviceps purpura* causes **ergotism,** a toxic condition. Hallucinogenic alkaloids produced by this fungus can lead to altered behavior, abortion, and death if infected grains are eaten. Ergotism is further discussed in chapter 25 (*pp. 510–11*).

Fungus-derived carcinogens include the aflatoxins and fumonisins. **Aflatoxins** are produced in moist grains and nut products. Aflatoxins were discovered in 1960, when 100,000 turkey poults died from eating fungus-infested peanut meal. *Aspergillus flavus* was found in the infected peanut meal, together with alcohol-extractable toxins termed aflatoxins. These flat-ringed planar compounds intercalate with the cells' nucleic acids and act as frameshift mutagens and carcinogens. This occurs primarily in the liver, where they are converted to unstable derivatives. The major aflatoxin types are shown in figure 43.6. These compounds and their derivatives can be separated by chromatographic procedures and can be recognized under UV light by their characteristic fluorescence (figure 43.7). Besides their importance in grains, they have also been observed in milk, beer, cocoa, raisins, and soybean meal. The U.S. Environmental Protection Agency has established allowable action levels, and limits of 20 parts per billion (ppb) are used for feeds and nut products. Milk, in comparison, has been assigned an action level of 0.5 ppb.

More recently discovered fungal contaminants of corn are the fumonisins, first isolated in 1988. These are produced by *Fusarium moniliforme* and cause leukoencephalomalacia in horses, pulmonary edema in pigs, and esophageal cancer in humans. There are six different fumonisins with the basic structure shown in figure 43.8. Corn and corn-based feeds and foods are of concern, including cornmeal and corn grits. As this is such a

Figure 43.6   Aflatoxins. When *Aspergillus flavus* and related fungi grow on foods, carcinogenic aflatoxins can be formed. These have four basic structures. The letter designations refer to the color of the compounds under ultraviolet light after extraction from the grain and separation by chromatography. The B_1 and B_2 compounds fluoresce with a blue color, and the G_1 and G_2 appear green.

Figure 43.7   Aflatoxin Analysis. Aflatoxins, if present in foods, can be detected after extraction of the foods with solvents and thin-layer chromatography. A completed chromatographic plate used to test for aflatoxins in foods is shown. The chromatographic plate has been photographed with its bottom down so that the fastest moving components are at the top of the figure. The two left lanes contain the aflatoxin standards B_1, B_2, G_1, and G_2 (B_1 moved the farthest, and G_2 the least). The remaining 15 samples contained varying amounts of the four different aflatoxins, as well as other components.

Figure 43.8   Fumonisin Structure. The basic structure of fumonisins produced by *Fusarium moniliforme,* a fungal contaminant that can grow in improperly stored corn. A total of six different fumonisins have been isolated that have different "R" groups. These are strongly polar compounds that cause diseases in domestic animals and also in humans.

recent discovery, the "no-effect" levels in animals and humans have not been established. These results make it extremely important to store corn and corn products under dry conditions where these fungi cannot develop.

Mutagens and mutagenic mechanisms (pp. 261–68).

Meat and dairy products, with their high nutritional value and the presence of easily usable carbohydrates, fats, and proteins, are ideal environments for spoilage by microorganisms (figure 43.9). Proteolysis and putrefaction are typical results of microbial spoilage of such high-protein materials. Unpasteurized milk undergoes a predictable four-step succession during spoilage: Acid production by *Lactococcus lactis* subsp. *lactis* (formerly *Streptococcus lactis*) is followed by additional acid production associated with the growth of more acid tolerant organisms such as *Lactobacillus*. At this point yeasts and molds become dominant and degrade the accumulated lactic acid, and the acidity gradually decreases. Eventually protein-digesting bacteria become active, resulting in a putrid odor and bitter flavor. The milk, originally opaque, can eventually become clear (figure 43.9c).

In comparison, most fruits and vegetables have a much lower protein and fat content and undergo a different kind of spoilage. Readily degradable carbohydrates favor vegetable spoilage by bacteria, especially bacteria that cause soft rots, such as *Erwinia carotovora,* which produces hydrolytic enzymes. The high oxidation-reduction potential and lack of reduced conditions permits aerobes and facultative anaerobes to contribute to the decomposition processes (figure 43.10). Bacteria do not seem important in the initial spoilage of whole fruits; instead such spoilage often is initiated by molds. These organisms have enzymes that contribute to the weakening and penetration of the protective outer skin.

Food spoilage problems occur with minimally processed, concentrated frozen citrus products. These are prepared with little or no heat treatment, and major spoilage can be caused by *Lactobacillus* and *Leuconostoc* spp., which produce diacetyl-butter flavors. *Saccharomyces* and *Candida* can also spoil juices. Concentrated juice has a decreased water activity (a_w = 0.8 to 0.83), and when kept frozen at about –9°C, the juices can be stored for long periods. However, when concentrated juices are diluted with water that contains spoilage organisms, or if the juice is stored in improperly washed containers, problems can occur. Also, microorganisms in the frozen concentrated juices can begin the spoilage process after addition of water. Ready-to-serve (RTS) juices present other problems as the a_w values are sufficiently high to allow microbial growth. This is especially true with extended storage at refrigeration temperatures. Although pasteurization can be used, most consumers are sensitive to the loss of flavor that this process entails.

(a)

(b)

(c)

Figure 43.9   Spoilage of Meat and Dairy Products. (*a*) Fresh and spoiled (moldy) meat. (*b*) Moldy cheese. (*c*) Fresh and curdled milk. The curdled milk has undergone a natural sequence of spoilage organism activity, resulting in separated curds and whey.

(a)

(b)

Figure 43.10 **Spoilage of Fruits.** (*a*) *Penicillium* mold on oranges. (*b*) Mold growing on grapes.

Despite efforts to eliminate spoilage microorganisms during canning, sometimes canned foods are spoiled (figure 43.11). This may be due to spoilage before canning, underprocessing during canning, and leakage of contaminated water through can seams during cooling. Spoiled food can be altered in such characteristics as color, texture, odor, and taste. Fermentation acids, sulfides, and gases (particularly CO_2 and H_2S) may be produced. In flat sour spoilage no gas is generated and the can does not swell, but its contents are rendered sour by the presence of fermentation acids. If spoilage microorganisms produce gas, both ends of the can will bulge outward to give a swell. Sometimes the swollen ends can be moved by thumb pressure (soft swells); in other cases the gas pressure is so great that the ends cannot be

Figure 43.11 **Food Preservation by Canning.** This technique is widely used and very effective. Improper canning can occasionally occur, as shown by this bent and leaking can.

dented by hand (hard swells). It should be noted that swelling is not always due to microbial spoilage. Acid in high-acid foods may react with the iron of the can to release hydrogen and generate a hydrogen swell. Hydrogen sulfide production by *Desulfotomaculum* can cause "sulfur stinkers."

1. What fungal genus produces ergot alkaloids? What conditions are required for the synthesis of these substances?
2. Aflatoxins are produced by which fungal genus? How do they damage animals that eat the contaminated food?
3. What microbial genus produces fumonisins and why are these compounds of concern? If improperly stored, what are the major foods and feeds in which these chemicals might be found?
4. Describe in general how food spoilage occurs. What factors influence the nature of the spoilage organisms responsible?
5. Why do concentrated citrus juices present such interesting spoilage problems?
6. How may canned food become spoiled and what changes can occur? Why is swelling not always an indication of microbial spoilage?

Microbiology of Fermented Foods

Over the last several thousand years, fermentation has been a major way of preserving food. Microbial growth, either of natural or inoculated populations, causes chemical and/or textural changes forming a product that can be stored for extended periods. The fermentation process also is used to create new, pleasing food flavors and odors.

TABLE 43.6   Major Fermented Milk Products and Fermenting Microorganisms

Fermented Product	Fermenting Microorganism	Description
Acidophilus milk	*Lactobacillus acidophilus*	Skim milk is sterilized and then inoculated with *L. acidophilus*
Cultured buttermilk	*Lactococcus lactis* subsp. *diacetilactis, Leuconostoc cremoris, Lactococcus cremoris*	Product is made with skimmed or low-fat pasteurized milk
Kefir	*Lactococcus lactis, Lactobacillus bulgaricus, Saccharomyces* spp.	Produced from a mixed lactic acid and alcoholic fermentation of cow's milk; 1% alcoholic content
Kumiss	*Lactobacillus bulgaricus, Lactobacillus leichmannii, Torula* spp.	Made with raw mare's milk; 2% alcoholic content
Sour cream	*Lactococcus* spp., *Leuconostoc* spp.	Cream is inoculated and incubated until acidity develops
Yogurt	*Streptococcus thermophilus* and *Lactobacillus bulgaricus*	Product is made from nonfat or low-fat milk to which stabilizers like gelatin are added
Butter	*Lactococcus lactis*	Cream is incubated until the desired acidity is achieved, followed by churning, washing, and salting

---

Box 43.4

Starter Cultures, Bacteriophage Infections, and Plasmids

Cultures of lactic acid bacteria, called **starter cultures,** are added to milk during the preparation of many dairy products. For example, *Streptococcus lactis* and *S. cremoris* are used in the production of cheese. One of the greatest problems for the dairy industry is the presence of bacteriophages that destroy these starter cultures. Lactic acid production by a heavily phage-infected starter culture can come to a halt within 30 minutes. The industry has tried to overcome this problem by practicing aseptic techniques in order to reduce phage contamination, and by selecting for phage-resistant bacterial cultures.

Most efforts at control have not been successful in the longer term. It has been found that the very aseptic techniques and phage-resistant pure cultures used to attempt to solve this problem actually were parts of the problem. The most stable and dependable cultures,

called P starter cultures, contain the bacteriophages in a pseudolysogenic state. When cultures are grown without phages or under aseptic conditions (L starters), they lose their phage resistance. The key to this riddle appears to be plasmids, which can block phage adsorption. The loss of the plasmids in a subpopulation of the bacteria allows the phage carrier state to be established. New phages can develop by acquiring restriction enzymes (*see chapter 15*) from plasmids. These modified phages again become lytic, establishing a new equilibrium in the population.

Other control approaches are being tested. Antisense RNA is now being used in an attempt to provide an agent against bacteriophage genes to help in the constant struggle between lactic acid bacteria and their phages.

Dairy Products

Fermented milk has distinct flavors and aromas, depending on the incubation conditions and the microbial inocula used (table 43.6). All fermented dairy products result from similar manufacturing techniques, in which acid produced through microbial activity causes protein denaturation. To carry out the process, one usually inoculates milk with the desired starter culture, incubates it at optimum temperature, and then stops microbial growth by cooling (Box 43.4). *Lactobacillus* spp. and *Lactococcus lactis* cultures are used for aroma and acid production. The organism *Lactococcus lactis* subsp. *diacetilactis* converts milk citrate to diacetyl, which gives a special buttery flavor to the

finished product. The use of these microorganisms with skim milk produces cultured buttermilk, and when cream is used, sour cream is the result.

The genus Lactobacillus (p. 449).
The genus Lactococcus (pp. 443–45).

Yogurt is produced by a special starter culture in which two major bacteria are present in a 1:1 ratio: *S. thermophilus* and *L. bulgaricus*. With these organisms growing in concert, acid is produced by *Streptococcus,* and aroma components are formed by the *Lactobacillus*. Freshly prepared yogurt contains 10^9 bacteria per gram. Acidophilus milk is produced by using *Lactobacillus acidophilus*.

Figure 43.12   Bifidobacterium. Cultured milks are increasing in popularity. A light micrograph of *Bifidobacterium,* a microorganism suggested to provide many health benefits.

Figure 43.13   Examples of Bifid-Amended Dairy Products. These are produced in many countries.

Fermented milks may have beneficial effects. *L. acidophilus* may modify the microbial flora in the lower intestine, thus improving general health, and it often is used as a dietary adjunct. Many microorganisms in fermented dairy products stabilize the bowel microflora, and some appear to have antimicrobial properties. The exact nature and extent of health benefits are still unclear but may involve minimizing lactose intolerance, lowering serum cholesterol, and possibly exhibiting anticancer activity. Several lactobacilli have antitumor compounds in their cell walls. Such findings suggest that diets including lactic acid bacteria, especially *L. acidophilus,* may contribute to the control of colon cancer.

Another interesting group are the bifidobacteria. The genus *Bifidobacterium* contains irregular, nonsporing, gram-positive rods that may be club-shaped or forked at the end (figure 43.12). Bifidobacteria are nonmotile, anaerobic, and ferment lactose and other sugars to acetic and lactic acids. They are typical residents of the human intestinal tract and were discovered in 1906. Many beneficial properties are attributed to them. Bifidobacteria are thought to help maintain the normal intestinal balance, while improving lactose tolerance; to possess antitumorigenic activity; and to reduce serum cholesterol levels. In addition, some believe that they promote calcium absorption and the synthesis of B-complex vitamins. Bifid-amended fermented milk products are now available in various parts of the world (figure 43.13).

Cheese is one of the oldest human foods and is thought to have been developed approximately 8,000 years ago. About 2,000 distinct varieties of cheese are produced throughout the world, representing approximately 20 general types (table 43.7 and figure 43.14). Often cheeses are classified based on texture or hardness as soft cheeses (cottage, cream, Brie), semi-soft cheeses (Muenster, Limburger, blue), hard cheeses (cheddar, Colby, Swiss), or very hard cheeses (Parmesan). All cheese results from a lactic acid fermentation of milk, which results in coagulation of milk proteins and formation of a curd. Rennin, an enzyme from calf stomachs, but now produced by genetically engineered microorganisms, can also be used to promote curd formation. After the curd is formed, it is heated and pressed to remove the watery part of the milk or whey, salted, and then usually ripened (figure 43.15). The cheese curd can be packaged for ripening with or without additional microorganisms. Cheese curd inoculation is used in the manufacture of Roquefort and blue cheese. In this case *Penicillium roqueforti* spores are added to the curds just before the final cheese processing. Sometimes the surface of an already formed cheese is inoculated at the start of ripening; for example, Camembert cheese is inoculated with spores of *Penicillium camemberti.* The final hardness of the cheese is partially a function of the length of ripening. Soft cheeses are ripened for only about 1 to 5 months, whereas hard cheeses need 3 to 12 months, and very hard cheeses like Parmesan require 12 to 16 months' ripening.

The ripening process also is critical for Swiss cheese. Gas production by *Propionibacterium* contributes to final flavor development and hole or eye formation in this cheese. Some cheeses are soaked in brine to stimulate the development of specific fungi and bacteria; Limburger is one such cheese.

Meat and Fish

Besides the fermentation of dairy products, a variety of meat products, especially sausage, can be fermented: country-cured hams, summer sausage, salami, cervelat, Lebanon bologna, fish sauces (processed by halophilic *Bacillus* species), izushi, and katsuobushi. *Pediococcus cerevisiae* and *Lactobacillus plantarum* are

TABLE 43.7   Major Types of Cheese and Microorganisms Used in Their Production

Cheese (Country of Origin)	Contributing Microorganisms[a]	
	Earlier Stages of Production	Later Stages of Production
Soft, unripened		
Cottage	*Lactococcus lactis*	*Leuconostoc cremoris*
Cream	*L. cremoris,* *L. diacetylactis,* *S. thermophilus,* *Lactobacillus bulgaricus*	
Mozzarella (Italy)	*S. thermophilus,* *Lactobacillus bulgaricus*	
Soft, ripened		
Brie (France)	*Lactococcus lactis,* *L. cremoris*	*Penicillium camemberti,* *P. candidum,* *Brevibacterium linens*
Camembert (France)	*L. lactis,* *L. cremoris*	*Penicillium camemberti,* *Brevibacterium linens*
Semisoft		
Blue (France)	*Lactococcus lactis,* *L. cremoris*	*Penicillium roqueforti*
Brick (United States)	*L. lactis,* *L. cremoris*	*Brevibacterium linens*
Limburger (Belgium)	*L. lactis,* *L. cremoris*	*Brevibacterium linens*
Monterey (United States)	*L. lactis,* *L. cremoris*	
Muenster (United States)	*L. lactis,* *L. cremoris*	*Brevibacterium linens*
Roquefort (France)	*L. lactis,* *L. cremoris*	*Penicillium roqueforti*
Hard, ripened		
Cheddar (Britain)	*Lactococcus lactis,* *L. cremoris,* *E. durans*	*Lactobacillus casei,* *L. plantarum*
Colby (United States)	*L. lactis,* *L. cremoris,* *E. durans*	*L. casei*
Edam (Netherlands)	*L. lactis,* *L. cremoris*	
Gouda (Netherlands)	*L. lactis,* *L. cremoris, L. diacetylactis*	
Swiss (Switzerland)	*L. lactis, L. helveticus,* *S. thermophilus*	*Propionibacterium shermanii,* *P. freudenreichii*
Very hard, ripened		
Parmesan (Italy)	*Lactococcus lactis,* *L. cremoris,* *S. thermophilus*	*Lactobacillus bulgaricus*

[a]*Lactococcus lactis* stands for *L. lactis* subsp. *lactis. Lactococcus cremoris* is *L. lactis* subsp. *cremoris,* and *Lactococcus diacetilactis* is *L. lactis* subsp. *diacetilactis.*

(a)

(b)

(c)

(d)

(e)

Figure 43.14 Cheese. A bewildering array of cheeses are produced around the world using microorganisms. (*a*) Gouda (top left) and cheddar cheese (lower right). Note the typical indentations on the surface of Gouda caused by the cheesecloth and red wax covering. (*b*) Roquefort cheese crumbled for use in salad dressing. The dark areas are the result of extensive *Penicillium* growth. (*c*) Swiss cheese. This hard, ripened cheese contains holes formed by carbon dioxide from a *Propionibacterium* fermentation. (*d*) Brie (left) and Limburger (right) cheeses are soft, ripened cheeses. Ripening results from the surface growth of microorganisms like *Penicillium camemberti* (Brie) and *Brevibacterium linens* (Limburger). (*e*) Cottage cheese and cream cheese (spread on crackers) are soft, unripened cheeses. They are sold immediately after production, and the curd is consumed without further modification by microorganisms.

Figure 43.15   Cheddar Cheese Production. Cheddar, a village in England, has given its name to a cheese made in many parts of the world. "Cheddaring" is the process of turning and piling the curd to express whey and develop desired cheese texture.

most often involved in sausage fermentations. Izushi is based on the fermentation of fresh fish, rice, and vegetables by *Lactobacillus* spp.; katsuobushi results from the fermentation of tuna by *Aspergillus glaucus*. Both meat fermentations originated in Japan.

Wine, Beer, and Other Fermented Alcoholic Beverages

Fermented alcoholic beverages are produced throughout the world from a variety of plant products that contain readily utilizable carbohydrates. To produce such fermentable substrates, called **musts,** it is often only necessary to allow natural fermentation to occur. The must also can be sterilized by pasteurization or the use of sulfur dioxide, and then the desired microbial culture is added.

In contrast, before cereals and other starchy materials can be used as substrates for the production of alcohol, their complex carbohydrates must be degraded. They are mixed with water and incubated in a process called **mashing.** The insoluble material is then removed to yield the **wort,** a clear liquid containing fermentable sugars and other simple molecules. Much of the art of beer and ale production involves the controlled hydrolysis of protein and carbohydrates to provide the desired body and flavor of the final product.

Wines and Champagnes

Wine production, or the science of **enology** [Greek *oinos,* wine, and *ology,* the science of], starts with the collection of grapes, continues with their crushing and the separation of the liquid (must) before fermentation, and concludes with a variety of storage and aging steps (figure 43.16). All grapes have white juices. To make a red wine from a red grape, the grape skins are allowed to remain in contact with the must before fermentation to release their skin-coloring components. Wines can be created by using the natural grape skin microorganisms. This natural mixture of bacteria and yeasts gives unpredictable fermentation results. To avoid such problems, one can treat the fresh must with a sulfur dioxide fumigant and add a desired strain of *Saccharomyces cerevisiae* or *S. ellipsoideus.* After inoculation the juice is fermented for 3 to 5 days at temperatures varying between 20 and 28°C. Depending on the alcohol tolerance of the yeast strain, the final product may contain 10 to 18% alcohol. Clearing and development of flavor occur during the aging process.

Yeasts (pp. 506; 511–14).

A critical part of wine making involves the choice of whether to produce a dry (no remaining free sugar) or a sweeter (varying amounts of free sugar) wine. This can be controlled by regulating the initial must sugar concentration. With higher levels of sugar, alcohol will accumulate and inhibit the fermentation before the sugar can be completely used, thus producing a sweeter wine. During final fermentation in the aging process, flavoring compounds accumulate and so influence the bouquet of the wine.

Microbial growth during the fermentation process produces sediments, which are removed during **racking.** Racking can be carried out at the time the fermented wine is transferred to bottles or casks for aging or even after the wine is placed in bottles.

Many processing variations can be used during wine production. The wine can be distilled to make a "burned wine" or brandy. *Acetobacter* and *Gluconobacter* can be allowed to oxidize the ethanol to acetic acid and form a **wine vinegar.** In the past an acetic acid generator was used to recirculate the wine over a bed of wood chips, where the desired microorganisms developed as a surface growth. Today the process is carried out in large aerobic submerged cultures under much more controlled conditions.

Natural champagnes are produced by continuing the fermentation in bottles to produce a naturally sparkling wine. Sediments that remain are collected in the necks of inverted champagne bottles after the bottles have been carefully turned. The necks of the bottles are then frozen and the corks removed to disgorge the accumulated sediments. The bottles are refilled with clear champagne from another disgorged bottle, and the product is ready for final packaging and labeling.

Beers and Ales

Beer and ale production uses cereal grains such as barley, wheat, and rice. The complex starches and proteins in these grains must be changed to a more readily usable mixture of simpler carbohydrates and amino acids. This process, shown in figure 43.17, involves germination of the barley grains and activation of their enzymes to produce a **malt.** The malt is then mixed with water and the desired grains, and the mixture is transferred to the mash tun or cask in order to hydrolyze the starch to usable carbohydrates. Once this process is completed, the **mash** is heated with hops (dried flowers of the female vine *Humulus lupulis*), which were originally added to the mash to inhibit spoilage microorganisms. The hops also provide flavor and assist in clarification of the wort. In this heating step the hydrolytic enzymes are inactivated and the wort can be **pitched**—inoculated—with the desired yeast.

Most beers are fermented with **bottom yeasts,** related to *Saccharomyces carlsbergensis,* which settle at the bottom of the fermentation vat. The beer flavor also is influenced by the production of small amounts of glycerol and acetic acid. Bottom yeasts produce beer with a pH of 4.1 to 4.2 and requiring 7 to 12 days of fermentation (figure 43.18). With a top yeast, such as *Saccharomyces cerevisiae,* the pH is lowered to 3.8 to produce ales. Freshly fermented (green) beers are aged or **lagered,** and when they are bottled, CO_2 is usually added. Beer can be pasteurized at 140°F or higher or sterilized by passage through membrane filters to minimize flavor changes.

Distilled Spirits

Distilled spirits are produced by an extension of beer production processes. The fermented liquid is boiled, and the volatile components are condensed to yield a product with a higher alcohol content than beer. Rye and bourbon are examples of whiskeys. Rye whiskey must contain at least 51% rye grain, and bourbon must contain at least 51% corn. Scotch whiskey is made primarily of barley. Usually a **sour mash** is used; the mash is inoculated with a homolactic (lactic acid is the major fermentation product) bacterium such as *Lactobacillus delbrueckii,* which can lower the mash pH to around 3.8 in 6 to 10 hours. This limits the development of undesirable organisms. Vodka and grain alcohols are also produced by distillation. Gin is vodka to which resinous flavoring agents—often juniper berries—have been added to provide a unique aroma and flavor.

Bread and Other Fermented Plant Products

Bread is one of the most ancient of human foods, and the use of yeasts to leaven bread is carefully depicted in paintings

Processing step	Biological change
Grape pressing	
Sterilization Yeast addition	Elimination of contaminants Addition of desired organisms
Fermentation of must	Alcohol production from sugars
	Excess yeast
Settling vat	Malolactic fermentation
	Excess yeast
Aging	Development of final wine bouquet
Bottling	

(a)

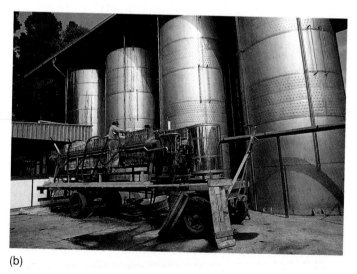

(b)

Figure 43.16   Wine Making. Once grapes are pressed, the sugars in the juice (the must) can be immediately fermented to produce wine. (*a*) Must preparation, fermentation, and aging are critical steps. (*b*) A photograph of wine fermentation vats.

Processing step   Biological change

Barley moistening
and germination

Enzymatic
release of
soluble
carbohydrates

Malting floor

Drying and
crushing

Mashing

Further enzymatic
activity—release
of maltose, dextrins,
and proteins

Mash tun

Add hops

Heat in brew
kettle

Spoilage organism inhibition
Enzyme inactivation
Flavoring from hops
Clarification

Brew kettle

Add yeast   Remove hops

Fermentation   Alcoholic fermentation

Storage (lagering)

Final flavor
development

Packaging

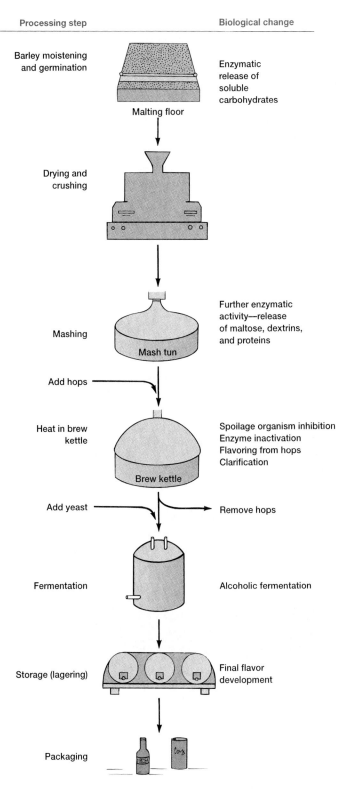

Figure 43.17   Producing Beer. To make beer, the complex carbohydrates in the grain must first be transformed into a fermentable substrate. Beer production thus requires the important steps of malting, and the use of hops and boiling for clarification, flavor development, and inactivation of malting enzymes. Only after completion of these steps can the actual fermentation be carried out.

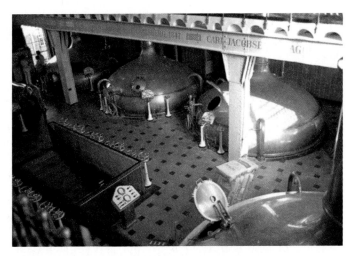

Figure 43.18   Beer Fermentation. Beer making can be carried out on a large scale or in the kitchen. In large-scale processes, copper brewing vats are often used, as shown here in the Carlsberg Brewery, Copenhagen, Denmark.

from ancient Egypt. Samples of bread from 2100 B.C. are on display in the British Museum. In breadmaking, yeast growth is carried out under aerobic conditions. This results in increased CO_2 production and minimum alcohol accumulation. The fermentation of bread involves several steps: alpha- and beta-amylases present in the moistened dough release maltose and sucrose from starch. Then a baker's strain of the yeast *Saccharomyces cerevisiae,* which contains maltase, invertase, and zymase enzymes, is added. The CO_2 produced by the yeast results in the light texture of many breads, and traces of fermentation products contribute to the final flavor. Usually bakers add sufficient yeast to allow the bread to rise within 2 hours—the longer the rising time, the more additional growth by contaminating bacteria and fungi can occur, making the product less desirable.

By using more complex assemblages of microorganisms, bakers can produce special breads such as sour doughs. The yeast *Saccharomyces exiguus,* with a *Lactobacillus* species, produces the characteristic acidic flavor and aroma of such breads.

Bread products can be spoiled by *Bacillus* species that produce ropiness. If the dough is baked after these organisms have grown, stringy and ropy bread will result, leading to decreased consumer acceptance.

Many other plant products can be fermented, as summarized in table 43.8. These include sufu, which is produced by the fermentation of tofu, a chemically coagulated soybean milk product. To carry out the fermentation, the tofu curd is cut into small chunks and dipped into a solution of salt and citric acid. After the cubes are heated to pasteurize their surfaces, *Actinimucor elegans* and other *Mucor* species are added. When a white mycelium develops, the cubes, now called pehtze, are aged in salted rice wine. This product has

TABLE **43.8**   **Fermented Foods Produced from Fruits, Vegetables, Beans, and Related Substrates**

Foods	Raw Ingredients	Fermenting Microorganisms	Area
Coffee	Coffee beans	*Erwinia dissolvens, Saccharomyces* spp.	Brazil, Congo, Hawaii, India
Gari	Cassava	*Corynebacterium manihot, Geotrichum* spp.	West Africa
Kenkey	Corn	*Aspergillus* spp., *Penicillium* spp., Lactobacilli, yeasts	Ghana, Nigeria
Kimchi	Cabbage and other vegetables	Lactic acid bacteria	Korea
Miso	Soybeans	*Aspergillus oryzae, Saccharomyces rouxii*	Japan
Ogi	Corn	*Lactobacillus plantarum, Lactococcus lactis, Saccharomyces rouxii*	Nigeria
Olives	Green olives	*Leuconostoc mesenteroides, Lactobacillus plantarum*	Worldwide
Ontjom	Peanut presscake	*Neurospora sitophila*	Indonesia
Peujeum	Cassava	Molds	Indonesia
Pickles	Cucumbers	*Pediococcus cerevisiae, L. plantarum*	Worldwide
Poi	Taro roots	Lactic acid bacteria	Hawaii
Sauerkraut	Cabbage	*L. mesenteroides, L. plantarum*	Worldwide
Soy sauce	Soybeans	*Aspergillus oryzae* or *A. soyae, S. rouxii, Lactobacillus delbrueckii*	Japan
Sufu	Soybeans	*Mucor* spp.	China
Tao-si	Soybeans	*A. oryzae*	Philippines
Tempeh	Soybeans	*Rhizopus oligosporus, R. oryzae*	Indonesia, New Guinea, Surinam

From James M. Jay, *Modern Food Microbiology*, 4th edition. Reprinted by permission of Chapman & Hall, New York, NY.

achieved the status of a delicacy in many parts of the Western world. Another popular product is tempeh, a soybean mash fermented by *Rhizopus* (figure 43.19).

Sauerkraut or sour cabbage, is produced from wilted, shredded cabbage, as shown in figure 43.20. Usually the mixed microbial community of the cabbage is used. A concentration of 2.2 to 2.8% sodium chloride restricts the growth of gram-negative bacteria while favoring the development of the lactic acid bacteria. The primary microorganisms contributing to this product are *Leuconostoc mesenteroides* and *Lactobacillus plantarum*. A predictable microbial succession occurs in sauerkraut's development. The activities of the lactic acid-producing cocci usually cease when the acid content reaches 0.7 to 1.0%. At this point *Lactobacillus plantarum* and *Lactobacillus brevis* continue to function. The final acidity is generally 1.6 to 1.8, with lactic acid comprising 1.0 to 1.3% of the total acid in a satisfactory product.

Pickles are produced by placing cucumbers and such components as dill seeds in casks filled with a brine. The sodium chloride concentration begins at 5% and rises to about 16% in 6 to 9 weeks. The salt not only inhibits the growth of undesirable bacteria but also extracts water and water-soluble constituents from the cucumbers. These soluble carbohydrates are converted to lactic acid. The fermentation, which can require 10 to 12 days, involves the development of *Leuconostoc mesenteroides, Enterococcus faecalis, Pediococcus cerevisiae, Lactobacillus brevis,* and *L. plantarum. L. plantarum* plays the dominant role in this fermentation process. Sometimes, to achieve more uniform pickle quality, natural microorganisms are first destroyed and the cucumbers are fermented using pure cultures of *P. cerevisiae* and *L. plantarum*.

Figure 43.19   Fermented Soybean Products. Microorganisms also can be used to process soybean products. Tempeh is produced from soybeans and *Rhizopus*. Both the raw cake and the fried food are shown.

Grass, chopped corn, and other fresh animal feeds, if stored under moist anaerobic conditions, will undergo a lactic-type mixed fermentation that produces pleasant-smelling **silages.** Trenches or more traditional vertical steel or concrete silos are used to store the silage. The accumulation of organic acids in silage can cause rapid deterioration of these silos. Older wooden stave silos, if not properly maintained, allow the outer portions of the silage to become aerobic, resulting in spoilage of a large portion of the plant material.

Processing step   Biological change

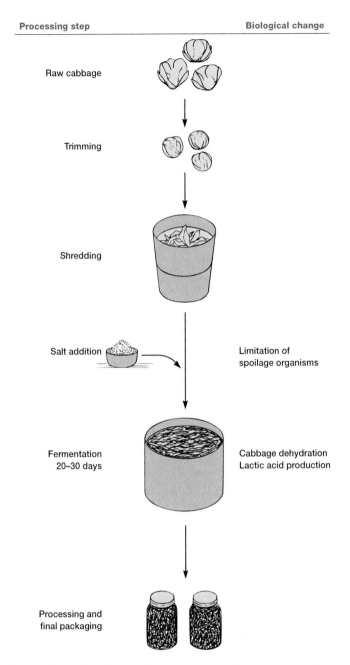

Raw cabbage

Trimming

Shredding

Salt addition   Limitation of
                                   spoilage organisms

Fermentation   Cabbage dehydration
20–30 days   Lactic acid production

Processing and
final packaging

Figure 43.20   Sauerkraut. Sauerkraut production employs a lactic acid fermentation. The basic process involves fermentation of shredded cabbage in the presence of 2.25–2.5% by weight of salt to inhibit spoilage organisms.

Figure 43.21   Mushroom Farming. Growing mushrooms requires careful preparation of the growth medium and control of environmental conditions. The mushroom bed is a carefully developed compost, which can be steam sterilized to improve mushroom growth.

4. Describe and contrast the processes of wine and beer production. How are red wines produced when the juice of all grapes is white? How are champagnes made?
5. Describe how the brewing process is continued to produce distilled spirits like whiskey.
6. How are bread, sauerkraut, and pickles produced? What microorganisms are most important in these fermentations?

Microorganisms as Sources of Food

Besides microorganisms' actions in fermentation as agents of physical and biological change, they themselves can be used as a food source. A variety of bacteria, yeasts, and other fungi have been used as animal and human food sources. Mushrooms (*Agaricus bisporus*) are one of the most important fungi used directly as a food source. Large caves are used to maintain optimal conditions for the production of this delicacy (figure 43.21). Microorganisms can be used directly as a food source or as a supplement to other foods and are then called single-cell protein. One of the more popular microbial food supplements is the cyanobacterium *Spirulina* (*see chapter 22*). It is used as a food source in Africa and is now being sold in United States health food stores as a dried cake or powdered product.

1. What gives fermented milk products their flavor?
2. What major steps are used to produce cheese? How is the cheese curd formed in this process? What is whey? How does Swiss cheese get its holes?
3. At what point in the production of Roquefort or blue cheese is a fungus added to the cheese? What fungal genus has been used in the production of these cheeses?

1. What conditions are needed to have most efficient production of edible mushrooms? What is the scientific name of the most important fungus used for this purpose?
2. A cyanobacterium is widely used as a food supplement. What is this genus, and in what part of the world was it first used as a significant food source?

Summary

1. Most foods, especially when raw, provide an excellent environment for microbial growth. This growth can lead to spoilage or preservation, depending on the microorganisms present and environmental conditions.

2. The course of microbial development in a food is influenced by the characteristics of the food itself—pH, salt content, substrates present, water presence and availability—and the environment, including temperature, relative humidity, and atmospheric composition.

3. Foods can be preserved in a variety of physical and chemical ways, including filtration, alteration of temperature (cooling, pasteurization, sterilization), drying, the addition of chemicals, radiation, and fermentation.

4. Microorganisms in improperly prepared or handled food can produce diseases. These can be either food-borne infections or food intoxications. Usually transmission is by the fecal-oral route with food acting as an intermediary agent.

5. DNA- and RNA-based methods allow rapid detection of disease-causing microorganisms in foods.

6. Fungi, if they can grow in foods, especially cereals and grains, can produce important disease-causing chemicals, including aflatoxins and fumonisins (carcinogens), and ergot alkaloids (mind-altering drugs).

7. Microorganisms can spoil meat, dairy products, fruits, vegetables, and canned goods in several ways. Spices, with their antimicrobial compounds, sometimes protect foods.

8. Dairy products can be fermented to yield cultured liquid products and cheeses. *Lactobacillus acidophilus* and *Bifidobacterium* can modify the intestinal microflora and may contribute to the control of colon cancer.

9. Wines are produced from pressed grapes and can be dry or sweet, depending on the level of free sugar that remains at the end of the alcoholic fermentation. Champagne is produced when the fermentation, resulting in CO_2 formation, is continued in the bottle.

10. Beer and ale are produced from cereals and grains. The starches in these substrates are hydrolyzed, in the processes of malting and mashing, to produce a fermentable wort. *Saccharomyces cerevisiae* is a major yeast used in the production of beer and ale.

11. Many plant products can be fermented with bacteria, yeasts, and fungi. Important products are breads, soy sauce, sufu, and tempeh. Sauerkraut and pickles are produced in a fermentation process in which natural populations of lactobacilli play a major role.

12. Microorganisms themselves can serve as an important food source. Mushrooms (*Agaricus bisporus*) are one of the most important fungi used as a food source. *Spirulina*, a cyanobacterium, also is a popular food source sold in specialty stores.

Key Terms

aflatoxin *889*

bottom yeast *898*

canning *883*

enology *897*

ergotism *889*

food-borne infection *886*

food intoxication *887*

lagered *898*

malt *898*

mash *898*

mashing *897*

must *897*

osmophilic microorganisms *882*

pasteurization *883*

pitched *898*

probe *888*

putrefaction *881*

racking *897*

radappertization *885*

silages *900*

sour mash *898*

starter culture *893*

wine vinegar *897*

wort *897*

xerophilic microorganisms *882*

Questions for Thought and Review

1. You are going through a salad line in a cafeteria at the end of the day. Which type of foods would you tend to avoid, and why?

2. Why does the law only allow foods to be pasteurized a single time? What could be the consequences of repeated pasteurization?

3. Why is it recommended that frozen foods, once thawed, not be frozen again? Suggest some of the reasons for this important principle of food handling.

4. Hamburger is notorious for having relatively high populations of microorganisms. How can these populations be controlled in preparation and storage?

5. Coffee cream is now packaged in small-portion servings that can be held at room temperature for extended periods. How do you think this product is produced and packaged?

6. What factors might limit the application of DNA- and RNA-based molecular techniques to the detection of pathogens in foods? How might these problems be overcome?

7. Why were aflatoxins not discovered before the 1960s? Do you think this was the first time they had grown in a food product to cause disease?

8. What is the best means of controlling the possible development of fungi that might produce fumonisins in corn and corn products?

9. Why has gamma irradiation of foods not become a more common means of controlling spoilage of foods?

10. Because of the possible beneficial effects of microorganisms on the intestinal microflora, scientists are becoming increasingly interested in their use as dietary supplements. How might modern biotechnology be used to improve the effectiveness of these microorganisms?

Additional Reading

General

Banwart, G. J. 1989. *Basic food microbiology,* 2d ed. New York: Van Nostrand Reinhold Co.

Frazer, W. C., and Westhoff, D. C. 1988. *Food microbiology,* 4th ed. New York: McGraw-Hill.

Hurst, A., and Collins-Thompson, D. L. 1977. Food as a bacterial habitat. *Adv. Microbiol. Ecol.* 3:79–134.

Jay, J. M. 1992. *Modern food microbiology,* 4th ed. New York: Van Nostrand Reinhold Co.

Macrae, R.; Robinson, R. K.; and Sadler, M. J. 1993. Encyclopedia of food science, food technology, and nutrition. San Diego: Academic Press.

Marshall, R. T., ed. 1992. *Standard methods for the examination of dairy products.* 16th ed. Washington, D.C.: American Public Health Association.

McMeekin, T. A., and Thomas, C. J. 1992. *Food microbiology,* New York: Harwood Academic.

Patel, P. 1994. *Rapid analysis techniques in food microbiology.* Glasgow: Blackie Academic and Professional.

Pierson, M. D., and Corlett, D. A., Jr. 1992. *HACCP. Principles and Applications.* New York: Chapman and Hall.

Sofos, J. N. 1993. Current microbiological considerations in food preservation. *Int. J. Food Microbiol.* 19:87–108.

Stewart, G. S. A., and Williams, P. 1993. Shedding new light on food microbiology. *ASM News* 59(5):241–46.

Vandersant, C., and Splittstoesser, D. F. 1992. *Compendium of methods for the microbiological examination of foods,* 3d ed. Washington, D.C.: APHA.

Food Spoilage and Preservation Processes

Becker, H.; Schaller, G.; von Wiese, W.; and Terplan, G. 1994. *Bacillus cereus* in infant foods and dried milk products. *Int. J. Food Microbiol.* 23:1–15.

Bhatnagar, D.; Lillehoj, E. B.; and Arora, D. K. 1991. *Mycotoxins in ecological systems,* vol. 5, New York: Marcel Dekker, Inc.

Cullen, J. C. 1994. The miracle of Bolsena. *ASM News* 60(4):187–90.

Davidson, M. D., and Branen, A. L. 1993. *Antimicrobials in foods,* 2d ed. New York: Marcel Dekker, Inc.

Diker, K. S., and Hascelik, G. 1994. The bactericidal activity of tea against *Helicobacter pylori. Let. Appl. Microbiol.* 19:299–300.

Gonzalez-Fandos, E.; Garcia-Lopez, M. L.; Sierra, M. L.; and Otero, A. 1994. Staphylococcal growth and enterotoxins (A-D) and thermonuclease synthesis in the presence of dehydrated garlic. *J. Appl. Bacteriol.* 77:549–52.

Kraft, A. A. 1993. *Psychrotrophic bacteria in foods: Disease and spoilage.* Boca Raton, Fla., CRC Press.

Ray, B. 1993. Sublethal injury, bacteriocins, and food microbiology. *ASM News* 59(6):285–91.

Diseases and Foods

Abdul-Raouf, U. M.; Beuchat, L. R.; and Ammar, M. S. 1993. Survival and growth of *Escherichia coli* O157:H7 on salad vegetables. *Appl. Environ. Microbiol.* 59(7):1999–2006.

Baird-Parker, A. C. 1994. Foods and microbiological risks. *Microbiology* 140:687–95.

Brackett, R. E. 1992. Microbiological safety of chilled foods: current issues. *Trends Food Sci. & Technol.* 3(4):81–86.

Bunning, V. K. 1994. Immunopathogenic aspects of foodborne microbial disease. *Food Microbiol.* 11:89–95.

Cliver, D. O. 1990. *Food-borne diseases.* San Diego: Academic Press, Inc.

Doyle, M. P. 1989. *Foodborne bacterial pathogens.* New York: Marcel Dekker, Inc.

Doyle, M. P. 1991. *Escherichia coli* O157:H7 and its significance in foods. *Int. J. Food Microbiol.* 12:289–302.

Lee, A. B., and Cooper, T. A. 1995. Improved direct PCR screen for bacterial colonies: Wooden toothpicks inhibit PCR amplification. *BioTechniques* 18(2):225–26.

Nachamkin, I.; Blaser, M. J.; and Tompkins, L. S. 1992. Campylobacter jejuni: *Current status and future Trends.* Washington, D.C.: American Society for Microbiology.

Niederhauser, C.; Candrian, U.; Hofelein, C.; Jermini, M.; Buhler, H. P.; and Luthy, J. 1992. Use of polymerase chain reaction for detection of *Listeria monocytogenes* in food. *Appl. Environ. Microbiol.* 58(5):1564–68.

Patel, P. 1994. Microbiological applications of immunomagnetic techniques. In *Rapid analysis techniques in food microbiology.* P. Patel, editor, 104–30, Glasgow: Blackie Academic and Professional.

Food Spoilage

Pohland, A. E.; Dowell, V. R., Jr.; and Richard, J. L. 1990. *Microbial toxins in foods and feeds.* New York: Plenum Publishing.

Scott, P. M. 1993. Fumonisins. *Int. J. Food Microbiol.* 18:257–70.

Microbiology of Fermented Foods

Bernet, M. F.; Brassart, D.; Neeser, J. R.; and Servin, A. L. 1993. Adhesion of human bifidobacterial strains to cultured human intestinal epithelial cells and inhibition of enteropathogen-cell interactions. *Appl. Environ. Microbiol.* 59(12):4121–28.

Fleet, G. H. 1993. *Wine: Microbiology and biotechnology.* London: Harwood Academic Publishers.

Glatz, B. A. 1992. The classical propionibacteria: Their past, present, and future as industrial organisms. *ASM News* 58(4):197–201.

Hoier, E. 1992. The use of probiotic starter cultures in dairy products. *Food Australia* 44(9):418–21.

Kim, J. H.; Kim, S. G.; Chung, D. K.; Gor, Y. C.; and Batt, C. A. 1992. Use of antisense RNA to confer bacteriophage resistance in dairy starter cultures, *J. Ind. Microbiol.* 10:71–78.

Kosikowski, F. V. 1985. Cheese. *Sci. Am.* 252(5):88–99.

Lodics, T. A., and Steenson, L. R. 1993. Phage-host interactions in commercial mixed-strain dairy starter cultures: Practical significance—a review. *J. Dairy Sci.* 76(8):2380–91.

Robinson, R. K. 1991. *Therapeutic properties of fermented milk.* New York: Elsevier.

Rose, A. H. 1981. The microbiological production of food and drink. *Sci. Am.* 245(3):126–39.

Webb, A. D. 1984. The science of making wine. *American Scientist* 72:360–67.

Microorganisms as Sources of Food

Chang, S. T.; Buswell, J. A.; and Miles, P. G. 1993. *Genetics and breeding of edible mushrooms.* New York: Gordon and Breach.

Finkelstein, D. B., and Ball, C. 1991. *Biotechnology of filamentous fungi—technology and products.* Stoneham, Mass.: Butterworth-Heinemann.

CHAPTER 44
Industrial Microbiology and Biotechnology

Nothing is more agreeable to those devoted to a scientific career than to increase the number of discoveries, but when the results of these observations are demonstrated by practical utility, their joy is complete.

—Louis Pasteur

Biotechnology. The scientific manipulation of living organisms, especially at the molecular genetic level, to produce useful products.

—D. A. Micklos and
G. A. Freyer

Concepts

1. The use of microorganisms in industrial microbiology and as a part of the new biotechnology involves the management of microbial growth processes. Factors limiting microbial growth must also be understood and controlled.

2. In modern biotechnology, microorganisms with specific genetic characteristics can be constructed to meet desired objectives. Earlier, mutation and screening (selection) were the major means of improving cultures from the available gene pool.

3. The media and growth conditions used in industrial processes often are created to provide specific limitations on microbial growth and functioning.

4. Different types of products are produced during and after the completion of microbial growth. Stopping growth at a specific point often is a necessary part of an industrial process.

5. Through recombinant DNA technology, the fields of protein engineering and metabolic engineering have been created.

6. In many situations it is important to inhibit microbial degradation of materials, or biodegradation. In other situations it is desirable to facilitate these degradation processes through bioenhancement or bioremediation.

7. Microbial cells and their products can be used in immobilized forms. Such biocatalysts have many uses, including the removal of metals from process streams.

8. Microorganisms are being linked with electronics to produce biosensors. Biosensors have many medical, industrial, and environmental applications.

When microorganisms are used in industrial microbiology and biotechnology, the particular microorganism or microbial community often can be chosen. The microbial growth environment also can be defined.

Environmental conditions can be changed or held constant over time, depending on the goals for the particular process. Modern gene-cloning techniques also allow a considerable range of possibilities for manipulation of microorganisms and the use of plants and animals as factories for the expression of recombinant DNA.

Industrial microbiology and biotechnology have become important parts of the field of microbiology. **Industrial microbiology** involves the commercial and industrial use of microorganisms to produce organic chemicals, antibiotics, other pharmaceuticals, and food supplements (*see Box 9.1*). Microorganisms also are used for the control of insects and other pests, for the recovery of metals, and for the improvement and maintenance of environmental quality. At first, selection and mutation were the major means of improving cultures for use in industrial microbiology. In biotechnology it is now possible with recombinant DNA techniques to manipulate genetic information and design products such as proteins, or to modify microbial gene expression. In addition, genetic information can be transferred between markedly different groups of organisms and between bacteria and plants, as an example. This chapter describes the accomplishments, present activities, and the future potential of microbial uses in these important areas.

Industrial Microbiology and the New Biotechnology

It is now popular to discuss the area of biotechnology as a new development. The term **biotechnology,** however, is not new. It is a product of the 1930s, a time when efforts were made to use agricultural surpluses to produce plastic and hydrocarbon substrates.

In the earlier development of what became known as industrial microbiology, the primary methods of selecting and improving microorganisms involved what Guido Pontecorvo of the University of Glasgow, in 1974, described as a prehistoric technique: mutation and screening (selection) from the available gene pool. This approach, used with success over many decades, is still important. Recently a major pharmaceutical company helped establish a 12,000 square mile genetic preserve in Central America where the search for unique plants and microorganisms can be continued without disturbance. In modern biotechnology these classic selection, mutation, and genetic exchange approaches are complemented by the use of recombinant DNA technology for industrial purposes. With these procedures new genetic information can be added intentionally to desired cells. It is now possible to combine DNA sequences from animal and plant cells, as well as from different groups of microorganisms.

Genetic engineering and recombinants (chapter 15).

TABLE 44.1	Fermentation: A Word with Many Meanings for the Microbiologist

1. Any process involving the mass culture of microorganisms, either aerobic or anaerobic
2. Any biological process that occurs in the absence of O_2
3. Food spoilage
4. The production of alcoholic beverages
5. Use of an organic substrate as the electron donor and acceptor
6. Use of an organic substrate as a reductant, and of the same partially degraded organic substrate as an oxidant (electron acceptor)
7. Growth dependent on substrate-level phosphorylation

Once a suitable microorganism is available, the microbiologist must still grow it in such a way as to ensure that the desired products are formed. Thus the use of microorganisms in modern biotechnology is still based on the principles of microbial mass culture developed over decades by industrial microbiologists.

Growth of microorganisms (chapter 6).

The term **fermentation,** used in a physiological sense in earlier sections of the book, is employed in a much more general way in relation to industrial microbiology and biotechnology. As noted in table 44.1, the term can have several meanings, including the mass culture of microorganisms (or even plant and animal cells). The development of industrial fermentations requires appropriate culture media and the large-scale screening of microorganisms. Often years are needed to achieve optimum product yields. Many isolates are tested for their ability to produce a new product in the desired quantity. Few are successful.

Fermentation as a physiological process (pp. 175–76).

1. In what different ways can the term fermentation be used?
2. How might industrial microbiology be different from biotechnology?

Microbial Growth Processes

Microbial Culture

Microorganisms for industrial applications are grown in culture tubes, shake flasks, and stirred fermenters or other mass culture systems. Stirred fermenters can range in size from 3 to 4 liters to 100,000 liters or larger, depending on production requirements (figure 44.1). A typical large-scale stirred fermentation unit is illustrated in figure 44.2. This unit requires a large capital investment and skilled operators. All required steps in the growth and harvesting of products must be carried out under aseptic conditions. Not only must the medium be sterilized but aeration, pH adjustment, sampling, and process monitoring must be carried out under rigorously controlled conditions.

Figure 44.1   Industrial Stirred Fermenters. Modern biotechnology is very capital and labor intensive. With recent advances in molecular biology, close coordination between basic research, process development, manufacturing, quality control, and marketing is required to remain competitive. This picture shows large fermenters used by a pharmaceutical company for the microbial production of antibiotics.

Figure 44.2   A Large-Scale Fermentation Unit. For the growth of microorganisms in industrial processes, specially designed stirred fermenters often are used. These can be run under aerobic or anaerobic conditions. Nutrient additions, sampling, and fermentation monitoring can be carried out under aseptic conditions. Biosensors and infrared (near IR or Fourier transform) monitoring can provide real-time information on the course of the fermentation. Specific substrates, metabolic intermediates, and final products can be detected.

When required, foam control agents must be added, especially with high-protein media. Computers are commonly used to monitor outputs from probes that determine microbial biomass, levels of critical metabolic products, pH, input and exhaust gas composition, and other parameters. Such information is needed for precise process and product control.

Besides the traditional stirred aerobic or anaerobic fermenter, other approaches can be used to grow microorganisms. These alternatives, illustrated in figure 44.3, include lift-tube fermenters (figure 44.3*a*), which eliminate the need for stirrers that can cause problems with filamentous fungi. Also available is solid-state fermentation (figure 44.3*b*), in which the substrate is not diluted in water. In various types of fixed- (figure 44.3*c*) and fluidized-bed reactors (figure 44.3*d*), the microorganisms are attached to surfaces and medium flows past the fixed or suspended particles.

Dialysis culture units also can be used (figure 44.3*e*). These units allow toxic waste metabolites or end products to diffuse away from the microbial culture and permit new substrates to diffuse through the membrane toward the culture. Continuous culture techniques (figure 44.3*f*) can markedly improve cell outputs and rates of substrate use because microorganisms can be maintained in a continuous logarithmic phase. However, continuous maintenance of an organism in an active growth phase is undesirable in many industrial processes.

Medium Development and Growth Conditions

The medium used to grow a microorganism is critical because it can influence the economic competitiveness of a particular process. Frequently, lower-cost crude materials are used as sources of carbon, nitrogen, and phosphorus (table 44.2). Crude plant hydrolysates often are used as complex sources of carbon, nitrogen, and growth factors. By-products from the brewing industry often are employed because of their lower cost and greater availability. Other useful carbon sources include molasses, and whey from cheese manufacture.

Microbial growth media (pp. 106–8).

The levels and balance of minerals (especially iron) and growth factors can be critical in medium formulation. For example, biotin and thiamine, by influencing biosynthetic reactions, control product accumulation in many fermentations. The medium also may be designed so that carbon, nitrogen, phosphorus, iron, or a specific growth factor will become limiting after a given time during the fermentation. In such cases the limitation often causes a shift from growth to production of desired metabolites.

(a) Lift-tube fermenter

Density difference of
gas bubbles entrained
in medium results in
fluid circulation.

← Air in

(b) Solid-state fermentation

Growth of culture
without presence of
added free water

Flow in →

(c) Fixed-bed reactor

Microorganisms on surfaces
of support material;
flow can be up or down.

Fixed
support
material

→ Flow out

(d) Fluidized-bed reactor

Microorganisms on surfaces
of particles suspended
in liquid or gas stream—
upward flow

→ Flow out

Suspended
support particles

Flow in →

(e) Dialysis culture unit

Waste products diffuse
away from the culture.
Substrate may diffuse
through membrane to
the culture.

Membrane

Culture   Medium
               or buffer

Medium in →

(f) Continuous culture unit

Medium in and excess
medium to waste with
wasted cells

Medium and
cells out

Figure 44.3   Alternate Methods for Mass Culture. In addition to stirred fermenters, other methods can be used to culture microorganisms in industrial processes. In many cases these alternate approaches will have lower operating costs and can provide specialized growth conditions needed for product synthesis.

TABLE 44.2   Major Components of Growth Media Used in Industrial Processes

Source	Raw Material	Source	Raw Material
Carbon and energy	Molasses Whey Grains	Vitamins	Crude preparations of plant and animal products
	Agricultural wastes (corncobs)	Iron, trace salts	Crude inorganic chemicals
		Buffers	Chalk or crude carbonates Fertilizer-grade phosphates
Nitrogen	Corn-steep liquor Soybean meal Stick liquor (slaughterhouse products) Ammonia and ammonium salts Nitrates Distiller's solubles	Antifoam agents	Higher alcohols Silicones Natural esters Lard and vegetable oils

Figure 44.4 Barriers to Oxygen Transfer. Many barriers exist that limit oxygen transfer from air to an enzyme within the microorganism. In spite of having a high oxygen concentration within the air bubble, the barriers to transfer into the liquid medium and finally across the cell membrane result in a low level of oxygen at the enzyme. Only by maintaining a high aeration rate is it possible to assure that oxygen will not limit microbial activity. The oxygen concentration is expressed on a logarithmic scale, and the oxygen present at the site of use can be less than one millionth of that occurring in the air bubble.

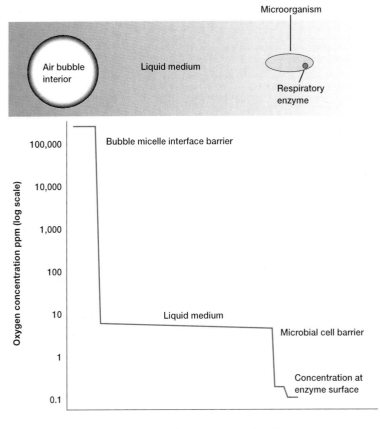

Frequently a critical component in the medium, often the carbon source, is added continuously—**continuous feed** or fed-batch mode—so that the microorganism will not have excess substrate available at any given time. An excess of substrate can cause undesirable metabolic waste products to accumulate.

Once a medium is developed, the physical environment for microbial functioning in the mass culture system must be defined. This often involves precise control of agitation, cooling, pH changes, and oxygenation. Phosphate buffers can be used to control pH while also functioning as a source of phosphorus. Oxygen limitations, especially, can be critical in aerobic growth processes. The O_2 concentration and flux rate must be sufficiently high to have O_2 in excess within the cells so that it is not limiting.

Factors limiting O_2 availability can be illustrated in terms of barriers to O_2 movement from a gas bubble to the dissolved O_2 at the site of respiratory enzymes within a microorganism (figure 44.4). Only with a sufficient O_2 concentration and flux rate from the bubble across each barrier will it be possible to avoid O_2 limitation. This is especially true when a dense microbial culture is growing. When filamentous fungi and actinomycetes are cultured, these physical aeration processes can be even further limited by filamentous growth (figure 44.5). Such filamentous growth results in a viscous, plastic medium, known as a **non-Newtonian broth,** which offers even more resistance to stirring and aeration. To minimize this problem, cultures can be grown as pellets or flocs or bound to artificial particles.

It is essential to assure that these physical factors are not limiting microbial growth. This is most critical during **scaleup,** where a successful procedure developed in a small shake flask is modified for use in a large fermenter. One must understand the microenvironment of the culture and maintain similar conditions near the individual cell despite increases in the culture volume. If a successful transition can be made from a process originally developed in a 250 ml Erlenmeyer flask to a 100,000 liter reactor, then the process of scaleup has been carried out properly.

Strain Selection, Improvement, and Preservation

Throughout most of the development of industrial microbiology, the major sources of microbial cultures were natural materials, such as soil samples, water, and spoiled bread and fruit. Cultures from all areas of the world were examined by industrial microbiologists in an attempt to identify isolates with improved characteristics. Once a promising culture was found, a variety of techniques were used for culture improvement, including the use of chemical mutagens and ultraviolet light (*see chapter 13*). As an example, the first cultures of *Penicillium notatum,* which could be grown only under static conditions, yielded low concentrations of penicillin. In 1943 a strain of *Penicillium chrysogenum* was isolated—strain NRRL 1951—which was further improved through screening (selection) and mutation. Today most penicillin is produced with cultures of *Penicillium chrysogenum,* which give 55-fold higher penicillin yields than the original culture when grown in aerobic stirred fermenters.

Genetic manipulations also have been used to identify microorganisms with new and desirable characteristics. The classical

Figure 44.5   Filamentous Growth During Fermentation. Filamentous fungi and actinomycetes can change their growth form during the course of a fermentation. The development of pelleted growth by fungi has major effects on oxygen transfer and energy required to agitate the culture. (*a*) Initial culture (time 0). (*b*) 14 hours. (*c*) 16 hours. (*d*) 18 hours.

(a)

(b)

(c)

(d)

methods of microbial genetics (*see chapter 14*) played a vital role in the earlier development of cultures for industrial microbiology. **Protoplast fusion** is now widely used with yeasts and molds. Most of these microorganisms are asexual or of a single mating type, which decreases the chance of random mutations that could lead to strain degeneration. To carry out genetic studies with these microorganisms, protoplasts are prepared by growing the cells in an isotonic solution while inhibiting cell wall growth (figure 44.6). The protoplasts are then mixed with polyethylene glycol (PEG), resulting in partial solubilization of the plasma membranes. If fusion occurs to form hybrids, desired recombinants are identified by means of selective plating techniques. After regeneration of the cell wall, the new protoplasm fusion product can be used in further studies.

A major advantage of the protoplast fusion technique is that protoplasts of different microbial species can be fused, even if they are not closely linked taxonomically. For example, protoplasts of *Penicillium roquefortii* have been fused with those of *P. chrysogenum*. Even yeast protoplasts and erythrocytes can be fused. Although many of these fusion products are unstable, useful applications continue to be found for them.

Newer molecular techniques continue to be discovered and applied to transfer genetic information and to construct microorganisms with new capabilities. The use of the gene gun or other means of biolistic delivery of DNA, discussed in chapter 15, and the use of artificial chromosomes are excellent examples of these new techniques.

Once a desired microorganism has been developed and is ready for use in a particular fermentation, strain stability and

preservation are of concern. A variety of culture preservation techniques may be used to maintain desired culture characteristics (table 44.3). **Lyophilization,** or freeze-drying, and storage in liquid nitrogen are frequently employed with microorganisms. Although lyophilization and liquid nitrogen storage are complicated and require expensive equipment, they do allow microbial cultures to be stored for years without loss of viability or an accumulation of mutations. Recently it has been found that many common bacteria can be maintained for extended periods in water agar and common growth media (Box 44.1).

1. What parameters can be monitored in a modern, large-scale industrial fermentation?

2. Besides the aerated, stirred fermenter, what other alternatives are available for the mass culture of microorganisms in industrial processes? What is the principle by which a dialysis culture system functions?

3. What are the major types of materials used as carbon and nitrogen sources in industrial fermentations?

4. Why is it often more difficult to aerate and agitate cultures of actinomycetes and molds in industrial fermentations?

5. Describe the barriers to O_2 transfer from the interior of an air bubble to the active respiratory site within a microorganism.

6. Describe the process of protoplast fusion.

7. What are the major advantages of lyophilization and liquid nitrogen for culture preservation, in comparison with other available techniques?

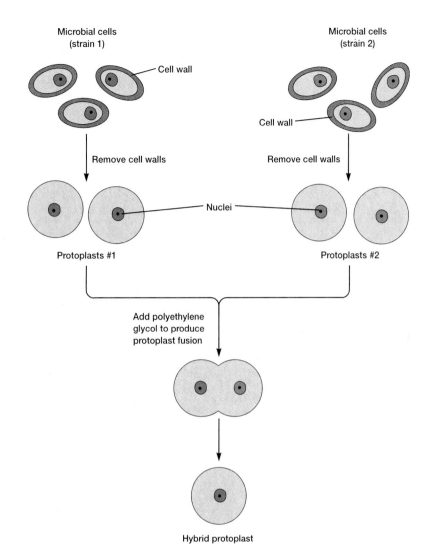

Figure 44.6   Protoplast Fusion and the Genetic Manipulation of Microorganisms. This technique is particularly useful for the genetic improvement of asexual or single-mating-type microorganisms used in many industrial processes.

TABLE 44.3	Methods Used to Preserve Cultures of Interest for Industrial Microbiology and Biotechnology
Method	**Comments**
Periodic transfer	Variables of periodic transfer to new media include transfer frequency, medium used, and holding temperature; this can lead to increased mutation rates and production of variants
Mineral oil slant	A stock culture is grown on a slant and covered with sterilized mineral oil; the slant can be stored at refrigerator temperature
Minimal medium, distilled water, or water agar	Washed cultures are stored under refrigeration; these cultures can be viable for 3 to 5 months or longer
Freezing in growth media	Not reliable; can result in damage to microbial structures; with some microorganisms, however, this can be a useful means of culture maintenance
Drying	Cultures are dried on sterile soil (soil stocks), on sterile filter paper disks, or in gelatin drops; these can be stored in a desiccator at refrigeration temperature, or frozen to improve viability
Freeze-drying (lyophilization)	Water is removed by sublimation, in the presence of a cryoprotective agent; sealing in an ampule can lead to long-term viability, with 30 years having been reported
Ultrafreezing	Liquid nitrogen at −196°C is used, and cultures of fastidious microorganisms have been preserved for more than 15 years

Major Products of Industrial Microbiology

Once a suitable microorganism is obtained using these selection and mutation processes, one can produce many compounds and transform a variety of substrates (table 44.4). The discipline responsible for such accomplishments is known as industrial microbiology. These products include pharmaceutical and medical compounds (antibiotics, hormones, transformed steroids), solvents, organic acids, chemical feedstocks, amino acids, and enzymes. Microorganisms also can be used to produce fuels and energy-related products. The economics of the production of these materials is constantly changing.

Microbial products often are classified as primary and secondary metabolites. As shown in figure 44.7, **primary metabolites** consist of compounds related to the synthesis of microbial cells and often are involved in the growth phase, or **trophophase.** They include amino acids, nucleotides, and fermentation end products such as ethanol and organic acids. In addition, industrially useful enzymes, either associated with the microbial cells or exoenzymes, often are synthesized by microorganisms during growth. These enzymes find many uses in food production and textile finishing.

Secondary metabolites usually accumulate during the period that follows the active growth phase, often called the **idiophase.** Compounds produced in the idiophase have no direct relationship to the synthesis of cell materials and normal growth. Most antibiotics and the mycotoxins fall into this category.

Microorganisms, if cultured under ideal conditions without environmental limitations, will attempt to maximize microbial biomass formation. Under such conditions of balanced growth,

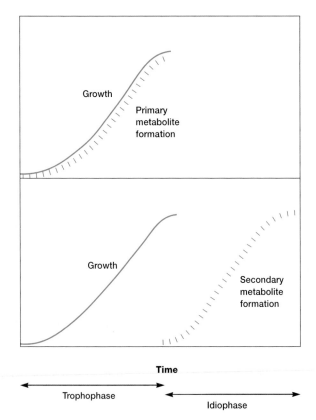

Figure 44.7   Primary and Secondary Metabolites. Depending on the particular organism, the desired product may be formed during or after growth. Primary metabolites are formed during the trophophase (active growth), whereas secondary metabolites are formed after growth is completed, the idiophase.

TABLE 44.4   Major Microbial Products and Processes of Interest in Industrial Microbiology and Biotechnology

Substances	Microorganisms
Industrial Products	
Ethanol (from glucose)	*Saccharomyces cerevisiae*
Ethanol (from lactose)	*Kluyveromyces fragilis*
Acetone and butanol	*Clostridium acetobutylicum*
2,3-butanediol	*Enterobacter, Serratia*
Enzymes	*Aspergillus, Bacillus, Mucor, Trichoderma*
Agricultural Products	
Gibberellins	*Gibberella fujikuroi*
Food Additives	
Amino acids (lysine, etc.)	*Corynebacterium glutamicum*
Organic acids (citric acid)	*Aspergillus niger*
Nucleotides	*Corynebacterium glutamicum*
Vitamins	*Ashbya, Eremothecium, Blakeslea*
Polysaccharides	*Xanthomonas*
Medical Products	
Antibiotics	*Penicillium, Streptomyces, Bacillus*
Alkaloids	*Claviceps purpurea*
Steroid transformations	*Rhizopus, Arthrobacter*
Insulin, human growth hormone, somatostatin, interferons	*Escherichia coli, Saccharomyces cerevisiae,* and others (recombinant DNA technology)
Biofuels	
Hydrogen	Photosynthetic microorganisms
Methane	*Methanobacterium*
Ethanol	*Zymomonas, Thermoanaerobacter*

the cells will minimize accumulation of any particular cellular building blocks in amounts beyond those required for growth (*see pp. 123–24*). Much of modern industrial microbiology and biochemistry involves "tricking" particular microorganisms into producing large excesses of desired compounds, usually by selecting mutants that have lost the ability to control synthesis of a particular end product. The following sections of this chapter describe examples of changes in genetic characteristics and environmental factors that induce the synthesis of excess amounts of microbial products.

Antibiotics

Many antibiotics are produced by microorganisms, mostly by actinomycetes—predominantly the genus *Streptomyces*—and filamentous fungi (table 44.5). Environmental control is crucial to the production of these important compounds.

Antibiotics in medicine (chapter 33).

Penicillin

Penicillin, produced by *Penicillium chrysogenum,* a fungus that can be grown in stirred fermenters (see figure 44.2), is an excellent example of a fermentation for which careful adjustment of

the medium composition is used to achieve maximum yields. Rapid production of cells, which can occur when high levels of glucose are used as a carbon source, does not lead to maximum antibiotic yields. Provision of the slowly hydrolyzed disaccharide lactose in combination with limited nitrogen availability, stimulates a greater accumulation of penicillin after growth has stopped (figure 44.8). The same result can be achieved by using a slow continuous feed of glucose. If a particular penicillin is needed, the specific precursor is added to the medium. For example, phenylacetic acid is added to maximize production of penicillin G, which has a benzyl side chain (*see figure 33.7*). The fermentation pH is maintained around neutrality by the addition of sterile alkali, which assures maximum stability of the newly synthesized penicillin. Once the fermentation is completed, normally in 6 to 7 days, the broth is separated from the fungal mycelium and processed by absorption, precipitation, and crystallization to yield the final product. This basic product can then be modified by chemical procedures to yield a variety of **semisynthetic penicillins.**

Streptomycin

Streptomycin is a secondary metabolite produced by *Streptomyces griseus,* for which changes in environmental conditions

TABLE 44.5   Major Antibiotics Produced by Actinomycetes, Other Bacteria, and Fungi, and the Microorganisms Affected

Microbial/Antibiotic Group	Produced by	Spectrum of Action
Actinomycetes		
Amphotericin-B	*Streptomyces nodosus*	Fungi
Carbomycin	*Streptomyces halstedii*	Gram-positive bacteria
Chlorotetracycline	*Streptomyces aureofaciens*	Broad spectrum
Chloramphenicol	*Streptomyces venezuelae*	Broad spectrum
Cycloheximide	*Streptomyces griseus*	Pathogenic yeasts
Erythromycin	*Streptomyces erythraeus*	Mostly gram-positive bacteria
Kanamycin	*Streptomyces kanamyceticus*	Gram-positive bacteria
Oleandomycin	*Streptomyces antibioticus*	Staphylococci
Oxytetracycline	*Streptomyces rimosus*	Broad spectrum
Neomycin-B	*Streptomyces fradiae*	Broad spectrum
Novobiocin	*Streptomyces niveus*	Gram-positive bacteria
Nystatin	*Streptomyces noursei*	Fungi
Streptomycin	*Streptomyces griseus*	Gram-negative bacteria, *Mycobacterium tuberculosis*
Other Bacteria		
Polymyxin-B	*Bacillus polymyxa*	Gram-negative bacteria
Bacitracin	*Bacillus licheniformis*	Gram-positive bacteria
Fungi		
Cephalosporin	*Cephalosporium acremonium*	Broad spectrum
Fumigillin	*Aspergillus fumigatis*	Amoebae
Griseofulvin	*Penicillium griseofulvum*	Fungi
Penicillin	*Penicillium chrysogenum*	Gram-positive bacteria

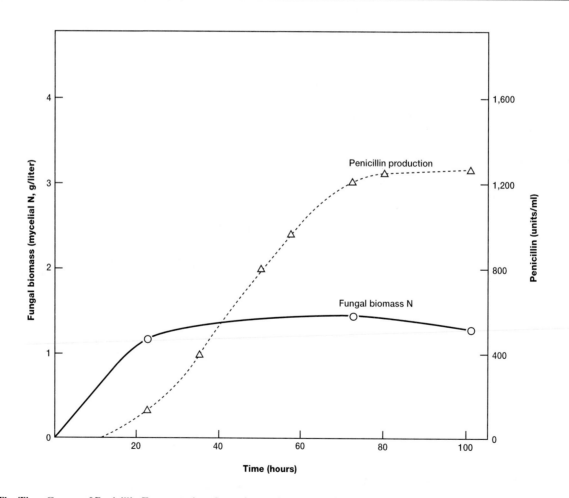

Figure 44.8   The Time Course of Penicillin Fermentation. Once nitrogen becomes limiting and growth stops, penicillin is produced at high levels.

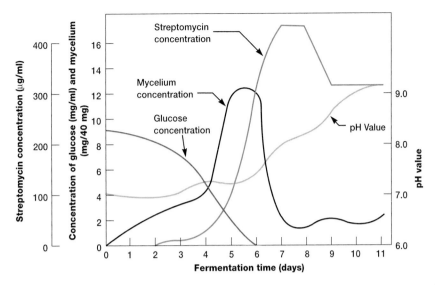

Figure 44.9   Streptomycin Production by *Streptomyces griseus*. Sequential use of different carbon sources leads to maximum antibiotic yields.

and substrate availability also influence final product accumulation. In this fermentation a soybean-based medium is used with glucose as a carbon source. The nitrogen source is thus in a combined form (soybean meal). After growth the antibiotic levels in the culture begin to increase (figure 44.9) under conditions of controlled nitrogen limitation.

Amino Acids

Amino acids such as lysine and glutamic acid are used in the food industry as nutritional supplements in bread products and as flavor-enhancing compounds such as monosodium glutamate (MSG).

Amino acid production is typically carried out by means of **regulatory mutants.** The normal microorganism avoids overproduction of biochemical intermediates by the careful regulation of cellular metabolism. Production of glutamic acid and several other amino acids in large quantities is now carried out using mutants of *Corynebacterium glutamicum* that lack, or have only a limited ability to process, the TCA cycle intermediate α-ketoglutarate (*see appendix II*) to succinyl-CoA as shown in figure 44.10. A controlled low biotin level and the addition of fatty acid derivatives results in increased membrane permeability and excretion of high concentrations of glutamic acid. The impaired bacteria use the glyoxylate pathway (*see chapter 10*) to meet their needs for essential biochemical intermediates, especially during the growth phase. After growth becomes limited because of changed nutrient availability, an almost complete molar conversion (or 81.7% weight conversion) of isocitrate to glutamate occurs.

Lysine, an essential amino acid used to supplement cereals and breads, was originally produced in a two-step microbial process. This has been replaced by a single-step fermentation in which the bacterium *Corynebacterium glutamicum,* blocked in the synthesis of homoserine, accumulates lysine. Over 44 g/liter can be produced in a 3 day fermentation.

Although not used extensively in the United States, microorganisms with related regulatory mutations have been employed to produce a series of 5′ purine nucleotides that serve as flavor enhancers for soups and meat products.

Organic Acids

Organic acid production by microorganisms is important in industrial microbiology and illustrates the effects of trace metal levels and balances on organic acid synthesis and excretion. Citric, acetic, lactic, fumaric, and gluconic acids are major products (table 44.6). Until microbial processes were developed, the major source of citric acid was citrus fruit from Italy. Today most citric acid is produced by microorganisms; 70% is used in the food and beverage industry, 20% in pharmaceutical, and the balance in other industrial applications.

The essence of citric acid fermentation involves limiting the amounts of trace metals such as manganese and iron to prevent growth of *Aspergillus niger* beyond a certain point. The medium often is treated with ion exchange resins to ensure low and controlled concentrations of available metals. Citric acid fermentation, which earlier was carried out by means of static surface growth, now takes place in aerobic stirred fermenters. Generally, high sugar concentrations (15 to 18%) are used, and copper has been found to counteract the inhibition of citric acid production by iron above 0.2 ppm. The success of this fermentation depends on the regulation and functioning of the glycolytic pathway and the tricarboxylic acid cycle (*see chapter 9*). During the idiophase, when the substrate level is high, citrate synthase activity increases and the activities of aconitase and isocitrate dehydrogenase decrease. This results in citric acid accumulation and excretion by the stressed microorganism.

In comparison, the production of gluconic acid involves a single microbial enzyme, glucose oxidase, found in *Aspergillus niger*. *A. niger* is grown under optimum conditions in a corn-steep liquor medium. Growth becomes limited by nitrogen, and the resting cells transform the remaining glucose

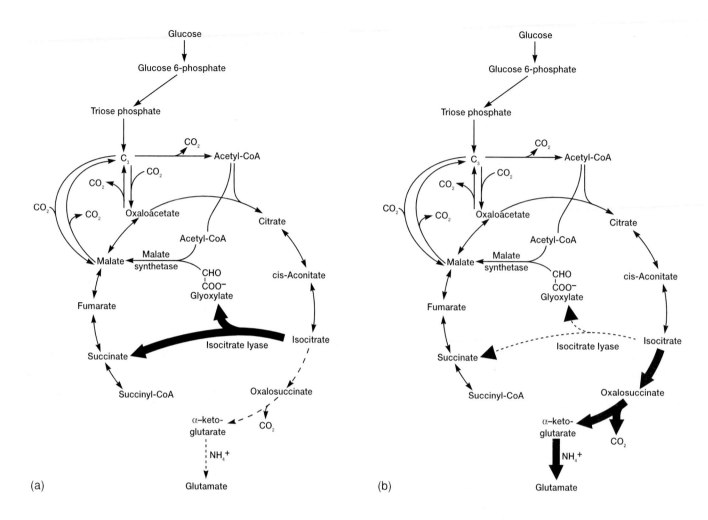

Figure 44.10 Glutamic Acid Production. The sequence of biosynthetic reactions leading from glucose to the accumulation of glutamate by *Corynebacterium glutamicum*. Major carbon flows are noted by bold arrows. (*a*) Growth with use of the glyoxylate bypass to provide critical intermediates in the TCA cycle. (*b*) After growth is completed, most of the substrate carbon is processed to glutamate (please note shifted bold arrows). The dashed lines indicate reactions that are being used to a lesser extent.

to gluconic acid in a single-step reaction (figure 44.11). Gluconic acid is used as a carrier for calcium and iron and as a component of detergents.

Bioconversion Processes

Various microorganisms can be used to carry out **bioconversions,** also known as **microbial transformations** or **biotransformations.** Such changes are usually specific, minor modifications of compounds that are not normally used for growth. Bioconversions have major advantages over chemical procedures. Enzymes carry out very specific reactions under mild conditions, and larger water-insoluble molecules can be transformed. Unicellular bacteria, actinomycetes, yeasts, and molds have been used in various bioconversions. The enzymes responsible for these conversions can be intracellular or extracellular.

Biocatalysts can be grown and processed in a variety of ways, as shown in table 44.7. Cells can be produced in batch or continuous culture and then dried for direct use, or they can be prepared in more specific ways to carry out desired bioconversions.

Biotransformations carried out by free enzymes or intact nongrowing cells do have limitations. Reactions that occur in the absence of active metabolism—without reducing power or ATP being available continually—are primarily exergonic reactions (*see chapter 8*). If ATP or reductants are required, an energy source such as glucose must be supplied under carefully controlled nongrowth conditions.

When freely suspended vegetative cells or spores are employed, the microbial biomass usually is used only once. At the end of the process, the cells are discarded. Cells often can be used repeatedly after attaching them to ion exchange resins by ionic interactions or immobilizing them in a polymeric matrix.

TABLE 44.6   Major Organic Acids Produced by Microbial Processes

Product	Microorganism Used	Representative Uses	Fermentation Conditions
Acetic acid	*Acetobacter* with ethanol solutions	Wide variety of food uses	Single-step oxidation, with 15% solutions produced; 95–99% yields
Citric acid	*Aspergillus niger* in molasses-based medium	Pharmaceuticals, as a food additive	High carbohydrate concentrations and controlled limitation of trace metals; 60–80% yields
Fumaric acid	*Rhizopus nigricans* in sugar-based medium	Resin manufacture, tanning, and sizing	Strongly aerobic fermentation; carbon-nitrogen ratio is critical; zinc should be limited; 60% yields
Gluconic acid	*Aspergillus niger* in glucose-mineral salts medium	A carrier for calcium and sodium	Uses agitation or stirred fermenters; 95% yields
Itaconic acid	*Aspergillus terreus* in molasses-salts medium	Esters can be polymerized to make plastics	Highly aerobic medium, below pH 2.2; 85% yields
Kojic acid	*Aspergillus flavus-oryzae* in carbohydrate-inorganic N medium	The manufacture of fungicides and insecticides when complexed with metals	Iron must be carefully controlled to avoid reaction with kojic acid after fermentation
Lactic acid	Homofermentative *Lactobacillus delbrueckii*	As a carrier for calcium and as an acidifier	Purified medium used to facilitate extraction

Figure 44.11   Gluconic Acid Production. This is an example of a transformation carried out by nitrogen-limited nongrowing cells. *Aspergillus niger* is used in this fermentation. The group that is altered is shown in color.

TABLE 44.7   Methods of Producing and Using Biocatalysts in Bioconversion Processes

Biocatalyst Production Procedure
  Continuous culture
  Batch culture

Biocatalyst Processing Alternatives
  Immobilized cells
  Dried cells
  Permeabilized cells
  Resting cells
  Cell extracts and enzyme purification
  Extracellular enzyme purification

Major types of **immobilization** techniques are shown in figure 44.12. Ionic, covalent, and physical entrapment approaches can be used to immobilize microbial cells, spores, and enzymes. Microorganisms also can be immobilized on the inner walls of fine tubes. The solution to be modified is then simply passed through the microorganism-lined tubing; this approach is finding applications in many industrial and environmental processes.

Immobilized microorganisms and their associated enzymes are used in bioconversions of steroids, degradation of phenol, and the production of a wide range of antibiotics, enzymes, organic acids, and metabolic intermediates (table 44.8). One application of cells as biocatalysts is the recovery of precious metals from dilute-process streams. As shown in figure 44.13, previously grown microbial cells—in this case an alga—are used to concentrate precious metals, including silver and gold.

1. What are the major differences between primary and secondary metabolites? Define trophophase and idiophase.

2. Why might the use of lactose and glucose in culture media result in different growth rates of a penicillin-producing microorganism?

3. Describe how mutations and the modification of culture conditions are used to stimulate glutamic acid synthesis.

4. What is the principal limitation created to stimulate citric acid accumulation by *Aspergillus niger?*

5. What types of bioconversion processes can be carried out by microbial cells?

6. Describe the major approaches used to immobilize microbial cells or cell components.

Figure 44.12 Immobilization Techniques. Immobilization of enzymes and whole microorganisms can lead to improved industrial processes. Alternatives for immobilizing (*a*) enzymes and (*b*) whole cells of microorganisms are shown.

TABLE 44.8   Immobilized Microbial Cells: Their Products and Reactions

Process Type	Product or Activity	Microorganisms Used
Antibiotics	Penicillin	*Penicillium chrysogenum*
	Bacitracin	*Bacillus* spp.
	Cephalosporins	*Streptomyces clavuligerus*
Amino acids	L-alanine	*Corynebacterium dismutans*
	L-glutamic acid	*Corynebacterium glutamicum*
	L-tryptophan	*Escherichia coli*
	L-lysine	*Microbacterium ammoniaphila*
Enzymes	Coenzyme A	*Brevibacterium ammoniagenes*
	Protease	*Streptomyces fradiae*
Vitamins	Pantothenic acid	*Escherichia coli*
Steroids	Prednisolone	*Curvularia lunata* and *Corynebacterium simplex*
Fermentation products	Ethanol	*Saccharomyces cerevisiae*
	Lactic acid	*Lactobacillus delbrueckii*
Environmental management	Degradation of paranitrophenol	*Pseudomonas* spp.
	Degradation of phenol	*Candida tropicalis*
	Denitrification	*Micrococcus* spp.
	Heavy-metal sorption (uranium, plutonium)	*Pseudomonas aeruginosa*

Figure 44.13   The Recovery of Precious Metals. Microorganisms can be used to remove metals from waters, either for water purification or metal recovery. In this electron micrograph gold has been recovered from a dilute-process stream by cells of *Chlorella vulgaris* (×26,400). Deposits of gold (black specks) are associated with the green algal cells. Bar = 1 μm.

Figure 44.14   Microbial DNA Purification. DNA isolated from microorganisms can be purified by centrifugation in a cesium chloride gradient. The DNA glows under ultraviolet light because it is complexed with the dye ethidium bromide. Recombinant DNA techniques are increasingly used in the pharmaceutical and agricultural industries.

Recombinant DNA Techniques in Biotechnology

An array of recombinant DNA techniques are now available for the selection and genetic manipulation of microorganisms (figure 44.14). In these approaches **biochemical markers,** such as nutrient requirements and antibiotic resistance patterns, are often used to identify mutant or recombinant organisms.

Several techniques can be used to construct desired genotypes, as discussed in chapter 15. These include use of restriction enzymes, preparation of probes and synthetic DNA, site-directed mutagenesis, PCR, and construction of specific cloning vectors. Newer areas of molecular applications in biotechnology include (1) modification of product expression (it is possible to have cells that contain 30% by weight of a specific desired product), (2) custom design of proteins and other products to give them more desirable characteristics, and (3) preparation of vectors to express genes in different organisms. This latter strategy allows the production of specific proteins and peptides without contamination by similar products that might be synthesized in the original organism. This approach can decrease the time and cost of recovering and purifying a product.

Recombinant DNA technology (pp. 309–23).

Modification of Gene Expression

With the ability to modify transcription and translation of genetic information, instead of merely selecting random overproducing mutants, it is now possible to make specific changes in control regions as noted in table 44.9. By modifying control mechanisms, enzymes that are normally expressed only in the absence of intermediates or end products can be produced on a constitutive basis. Recombinant plasminogen, for example, can be 20 to 40% of the soluble protein, a tenfold increase in production over the original strain.

Design of Proteins and Peptides

By the use of site-directed mutagenesis and the chemical synthesis of DNA, which can be inserted into the desired organism to produce a modified protein, the field of **protein engineering** has been created. This provides a new and exciting area of molecular biotechnology. Enzymes and bioactive peptides with markedly different characteristics (stability, kinetics, activities) can be created. The molecular basis for the functioning of these modified products also can be better understood.

As summarized in table 44.10, a variety of modified proteins and bioactive peptides has been created using bacteria. These products can only serve as examples because the field is expanding so rapidly. One of the most interesting areas is the design of enzyme-active sites to promote the modification of "unnatural substrates." This approach may lead to improved transformation of recalcitrant materials, or even the degradation of materials that have previously not been amenable to biological processing.

Yeasts are becoming even more important in the production of recombinant proteins that have particular biomedical importance. As noted in table 44.11, a variety of vaccines, diagnostics, and human therapeutic agents has been created using recombinant techniques in yeasts.

TABLE 44.9   Examples of Recombinant DNA Systems Used to Modify Gene Expression

Product	Microorganism	Change
Actinorhodin	*Streptomyces coelicolor*	Modification of gene transcription
Cellulase	*Clostridium* genes in *Bacillus*	Amplification of secretion through chromosomal DNA amplification
Recombinant protein albumin	*Saccharomyces cerevisiae*	Fusion to a high-production protein
Heterologous protein	*Saccharomyces cerevisiae*	Use of the inducible strong hybrid promoter $UAS_{gal}/CYCl$

TABLE 44.10   Examples of Modified Proteins Created through Recombinant DNA Technology

Characteristic Modified	Bacterium Used	Result	Mechanism
Glucose isomerase thermostability	*Actinoplanes*	Enhanced stability of soluble and immobilized forms	Amino acid substitution of arginine for lysine at position 253
Protein separation characteristics	*Escherichia coli*	96-fold difference in modified partition coefficient	Fusion with partitioning tetrapeptide
Protein structure and crystallinity	*Escherichia coli*	Unique macromolecular polyamide copolymers	Modification of leader region in message
Biopesticide activity	*Escherichia coli*	Modified moulting of fall armyworm	Insertion of *LacZ* gene and modification of hormone
Enzyme oxidation resistance	*Escherichia coli*	Alpha-antitrypsin more resistant to oxidation	Replacement of a critical oxygen-sensitive methionine with valine
Anticoagulant polypeptide alteration	*Escherichia coli*	Improve the 65 amino acid polypeptide hirudin	Substitution of asparagine residue 47 with lysine or arginine

TABLE 44.11   Recombinant Proteins Produced by *Saccharomyces cerevisiae*

Vaccines
Hepatitis B virus surface antigen
Malaria circumsporozoite protein
HIV-1 envelope protein

Diagnostics
Hepatitis C virus protein
HIV-1 antigens

Human Therapeutic Agents
Epidermal growth factor
Insulin
Insulinlike growth factor
Platelet-derived growth factor
Proinsulin
Fibroblast growth factor
Granulocyte-macrophage colony stimulating factor
α_1-antitrypsin
Blood coagulation factor XIIIa

From B. R. Glick and J. J. Pasternak, "Heterologous Protein Production in Eukaryotic Cells" in B. R. Glick and J. J. Pasternak (eds.), *Molecular Biotechnology: Principles and Applications of Recombinant DNA.* Copyright © 1994 ASM Press, Washington, D.C.

Progress has reached the point that an entirely new field has emerged: **metabolic engineering.** In this application of recombinant DNA technology, metabolic networks are restructured by the recruitment of proteins from different cells. The result is a change in pathway distribution and rate. Examples of altered small metabolites and protein-related end products are summarized in table 44.12. By combining the metabolic capabilities from two microorganisms, entirely new products and intermediates can be synthesized.

Many synthetic medical peptides have been developed through recombinant DNA technology (table 44.13). These include products to promote wound healing and blood coagulation, to treat cancer and AIDS, and to influence sexual dysfunctions. Other products include an intracellular adhesin molecule used to battle the common cold, nervous system growth factors, and compounds useful in hormone disorders. The market for polypeptide hormones is expected to expand especially rapidly.

A major advantage of peptide production with modern biotechnology is that only biologically active stereoisomers are produced. This specificity is required to avoid the harmful side effects of inactive stereoisomers, as occurred in the thalidomide disaster.

TABLE 44.12   Heterologous Activities Recruited to Alter Small Metabolite and Protein End Products[a]

Host Organism	Original Metabolite	Enzyme Added (source organism)	New Product
Erwinia herbicola	2,5-DKG[b]	2,5-DKG reductase (*Corynebacterium*)	2-KLG,[c] a vitamin C precursor in a single fermentation
Acremonium chrysogenum	Cephalosporin C	D-amino acid oxidase (*Fusarium solani*), cephalosporin acylase (*Pseudomonas diminuta*)	7ACA,[d] an antibiotic precursor by a single fermentation instead of two steps
Chinese hamster ovary cells	Terminal β-galactosyl residues in cell surface glycoproteins	β-galactoside 2,6-sialyltransferase (rat)	Increased β-galactosyl residues to better resemble human-derived forms

[a]The original metabolite is modified by the combined genetic information from two microorganisms to give a new product.
[b]2,5-DKG = 2,5-diketo-D-gluconic acid.
[c]2-KLG = 2-keto-L-gulonic acid.
[d]7ACA = 7-β-(5-carboxy-5-oxopentanamido)-cephalosporanic acid.
Reprinted with permission from J. E. Bailey, "Toward a Science of Metabolic Engineering" in *Science*, 252:1668. Copyright 1991 American Association for the Advancement of Science.

TABLE 44.13   New Therapeutic Synthetic Peptides Expected to Reach the Market in the Next Few Years

Peptide	Therapeutic Use	Comments
LHRH agonists	Prostate cancer	Luteinizing hormone-releasing hormone affects pituitary receptor, blocking release of luteinizing hormone
Octreotide	Diarrhea associated with intestinal tumors	Helps control diarrhea
Immune stimulator	Various cancers	Stimulates immune system to control intracellular recognition of self
RDG peptide	Wound healing	Synthetic tripeptide assists in cell anchoring and healing
HIV-1 blockers	AIDS	Peptides block virus replication inside cell
Pentigetide	Allergy-related problems	Interacts with active peptides to control allergies
Vasoactive intestinal peptide	Sexual dysfunction	Active in men and women
Hirudin	Anticoagulant	Product can compete with warfarin and heparin, two major currently used compounds

From *Genetic Engineering News*, May 1991. Copyright © 1991 by Mary Ann Liebert, Inc., New York, NY. Reprinted by permission.

Vectors for Product Expression

Genetic information from a wide range of biological forms can be inserted into microorganisms using vectors and recombinant DNA techniques. Vectors include artificial chromosomes, such as those for yeasts (YACs), bacteria (BACs), P1 bacteriophage-derived chromosomes (PACs), and mammalian artificial chromosomes (MACs). Information on some of these artificial chromosomes is given in chapter 15. YACs are especially valuable as large DNA sequences (>100 kb) can be maintained in the YAC as a separate chromosome in yeast cells.

A good example of vector use is provided by the foot-and-mouth disease of cattle and other livestock. Genetic information for a foot-and-mouth disease virus antigen can be incorporated into *E. coli*, followed by the expression of this genetic information and synthesis of the gene product for use in vaccine production (figure 44.15).

Recombinant DNA also can be expressed in markedly different organisms to improve production efficiency and minimize the purification steps required before the product is ready for use (table 44.14). For example, recombinant baculoviruses (*see p. 384*) can be replicated in insect larvae to achieve rapid large-scale production. Transgenic plants (*discussed on p. 321*) may be used to manufacture large quantities of a variety of metabolic products. A most imaginative way of incorporating new DNA into a plant is to simply shoot it in using DNA-coated microprojectiles and a gene gun, as discussed in chapter 15.

1. Why might one wish to express a gene in a foreign cell?
2. Define the terms protein engineering and metabolic engineering.
3. Why is it often important to produce peptides enzymatically rather than by chemical synthesis?
4. Why might it be as important to modify the genes for control of a product's synthesis as to modify its structural genes?
5. What are YACs, BACs, PACs, and MACs?

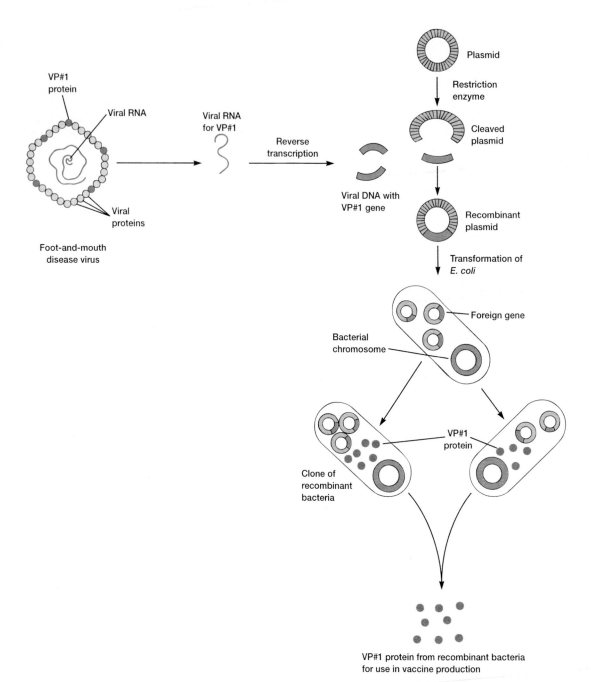

Figure 44.15   Recombinant Vaccine Production. Genes coding for desired products can be expressed in different organisms. By the use of recombinant DNA techniques, a foot-and-mouth disease vaccine is produced through cloning the vaccine genes into *Escherichia coli.*

TABLE 44.14   Examples of Expression of Recombinant DNA in Other Organisms to Improve Process Efficiency

Product	Donor	Recipient	Comment
Hemoglobin	Humans	Saccharomyces cerevisiae	Use of a hybrid promoter
Glucose oxidase	Aspergillus niger	Saccharomyces cerevisiae	Better temperature and pH stability, and no contamination
Interleukin-6	Humans	Escherichia coli	N-terminal methionine-free product without purification
Pectinases	Erwinia	Escherichia coli	Improvement of yield and purity
Monoclonal IgG antibody against Pseudomonas lipoprotein I	Mouse light and heavy chain cDNAs	Recombinant baculovirus vector grown in fall armyworms	High level of expression of recombinant monoclonal antibody genes
Alpha and beta portions of hemoglobin	Humans, with tobacco mosaic virus as vector	Tobacco	Large-scale production
Alpha-amylase	Bacillus licheniformis	Tobacco	Use of tobacco seeds directly in starch liquifaction, production of an industrial enzyme in plants

TABLE 44.15   The Use of Bacteria, Viruses, and Fungi As Bioinsecticides: An Older Technology with New Applications

Microbial Group	Major Organisms and Applications
Bacteria	Bacillus thuringiensis and Bacillus popilliae are the two major organisms of interest. Bacillus thuringiensis is used on a wide variety of vegetable and field crops, fruits, shade trees, and ornamentals. B. popilliae is used primarily against Japanese beetle larvae. Both bacteria are considered harmless to humans. Pseudomonas fluorescens, which contains the toxin-producing gene from B. thuringiensis, is used on maize to suppress black cutworms.
Viruses	Three major virus groups that do not appear to replicate in warm-blooded animals are used: nuclear polyhedrosis virus (NPV), granulosis virus (GV), and cytoplasmic polyhedrosis virus (CPV). These occluded viruses are more protected in the environment.
Fungi	Over 500 different fungi are associated with insects. Infection and disease occur primarily through the insect cuticle. Four major genera have been used. Beauveria bassiana and Metarhizium anisopliae are used for control of the Colorado potato beetle and the froghopper in sugarcane plantations, respectively. Verticillium lecanii and Entomophthora spp., have been associated with control of aphids in greenhouse and field environments.

Other Microbial Products

Microorganisms continue to find additional applications. For example, they are employed as insecticides and are used to produce biopolymers and surfactants.

Bioinsecticides

Many alternatives exist for the use of bacteria, viruses, and fungi as **bioinsecticides** (table 44.15). Bacterial agents include a variety of *Bacillus* species, primarily *B. thuringiensis* (*see* p. 448). This bacterium is only weakly toxic to insects as a vegetative cell, but during sporulation, it produces an intracellular protein toxin crystal, the parasporal body, that can act as a microbial insecticide.

Work on the molecular biology of *Bacillus thuringiensis* is progressing rapidly. The parasporal crystal, after exposure to alkaline conditions in the hindgut, fragments to release the protoxin. After this reacts with a protease enzyme, the active toxin is released (figure 44.16a). Six of the active toxin units integrate into the midgut membrane (figure 44.16b,c) to form a hexagonal-shaped pore through the midgut cell, as shown in figure 44.16d. This leads to the loss of osmotic balance and ATP, and finally to cell lysis.

The parasporal body will be expressed in *Bacillus thuringiensis* only if a specialized plasmid is present. Curing of the plasmid (*see plasmids and curing on p. 280*) results in a loss of the ability to synthesize the parasporal body, a discovery made in 1977. These parasporal bodies have been found to

(a)

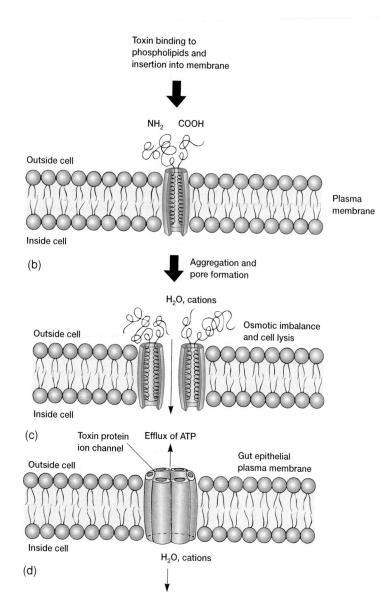

(b)

(c)

(d)

Figure 44.16   The Mode of Action of the *Bacillus thuringiensis* Toxin. (*a*) Release of the protoxin from the parasporal body and modification by proteases in the hindgut. (*b*) Insertion of the 68 kDa active toxin molecules into the membrane. (*c*) Aggregation and pore formation, showing a cross section of the pore. (*d*) Final creation of the hexagonal pore which causes an influx of water and cations as well as a loss of ATP, resulting in cell imbalance and lysis.

vary in their geometry, and it appears that the shape is related to host range and toxin effectiveness. The usual bipyramid-shaped parasporal bodies occur in *Lepidoptera*-active strains (*see figure 21.9*). In comparison, spherical crystals are produced by a strain that is effective against mosquitoes, whereas nontoxic strains produce irregular crystals. Thus there seems to be a relation between parasporal body morphology and activity.

The most recent advances in our understanding of *Bacillus thuringiensis* have involved the creation of pest-resistant plants. The first step in this work was to insert the toxin gene into *E. coli*. This work showed that the crystal protein could be expressed in another organism, and that the toxin was effective. This major scientific advance was followed in 1987 by the production of tomato plants that contained the toxin gene.

Additional types of transgenic plants have been produced in the last few years, and these later studies involved an intriguing use of modern molecular biology. A synthetic toxin gene was created in which natural AT-rich gene regions were replaced with more normal GC content sequences (this was done to minimize degradation of mRNA in the plant). After successful insertion of this modified artificial gene into the plant, over 0.2% of the transgenic plant weight was the active toxin protein. Efforts are under way to make such transgenic plant modifications more cost-effective for routine production of commercial crops. Questions of stability and development of resistance by the target pest are of concern. This research has been carried out with dicots such as soybean, oilseed rape, sugar beet, and sunflower. Work also is proceeding with monocots such as maize, rice, and wheat.

TABLE 44.16 Hosts for *Bacillus thuringiensis* Outside of the Order Lepidoptera

Order	Insects
Diptera	Mosquitoes
	Blackflies
	Gnats
	Midges
	Moth flies
	Black fungus gnats
	Craneflies
	Houseflies
	Blowbeetles
Coleoptera	Leaf beetles
Phthiraptera	Bird lice
	Mammalian lice
Acari	Animal ectoparasitic mites
	Phytophagous mites
Strongylida	Animal endoparasitic nematodes
Tylenchida	Phytophagous nematodes

Source: Data from J. Drummond and D. E. Pinnock, "Host Spectrum of *Bacillus Thuringiensis*" in *Ag. Ecosystems Env.* 49:15–19. 1994.

B. thuringiensis can be grown in fermenters. When the cells lyse, the spores and crystals are released into the medium. The medium is then centrifuged and made up as a dust or wettable powder for application to plants. It should be noted that *Bacillus thuringiensis* also has been found to be active against a wide range of target organisms in addition to the Lepidoptera, as noted in table 44.16.

A related organism, *Bacillus popilliae*, is used to combat the Japanese beetle. This bacterium, however, cannot be grown in fermenters, and inocula must be grown in the living host. The microorganism controls development of larvae, but destruction of the adult beetle requires chemical insecticides.

Viruses that are pathogenic for specific insects include nuclear polyhedrosis viruses (NPVs), granulosis viruses (GVs), and cytoplasmic polyhedrosis viruses (CPVs). Until 1980 only one of these viruses was registered as a control agent, a CPV of the pine caterpillar *Dendrolimus spectabilis*. Currently 125 types of NPVs are known, of which approximately 90% affect the Lepidoptera—butterflies and moths. Approximately 50 GVs are known, and they, too, primarily affect butterflies and moths. CPVs are the least host-specific viruses, affecting about 200 different types of insects. The first commercial viral pesticide was marketed under the trade name Elcar for control of the cotton bollworm *Heliothis zea*.

The development of biopesticides is progressing rapidly. One of the most exciting advances involves the use of baculoviruses that have been genetically modified to produce a potent scorpion toxin active against insect larvae. After ingestion by the larvae, viruses are dissolved in the midgut and are released. Because the recombinant baculovirus produces this insect-selective neurotoxin, it acts more rapidly than the parent virus, and the extent of leaf damage can be markedly decreased.

Characteristics of insect viruses (pp. 383–84).

Biopolymers

Biopolymers are produced by many microorganisms and have important uses in the pharmaceutical and food industries because of their ability to modify flow characteristics of liquids and to serve as gelling agents. The advantage of using microbial polysaccharides is that production is independent of climate, political events that can limit raw material supplies, and the depletion of natural resources. Production facilities can be located near sources of inexpensive substrates (e.g., near agricultural areas).

Bacterial exopolysaccharides (pp. 58–9).

At least 75% of all polysaccharides are used as stabilizers, for the dispersion of particulates, as film-forming agents, or to promote water retention in various products (table 44.17). Polysaccharides help maintain the texture of many frozen foods, such as ice cream, that are subject to drastic temperature changes. These polysaccharides must maintain their properties under the pH conditions in the particular food and be compatible with other polysaccharides. They should not lose their physical characteristics if heated.

Polysaccharides such as scleroglucan are used by the oil industry as drilling mud additives. Xanthan polymers enhance oil recovery by improving water flooding and the displacement of oil. This use of xanthan gum, produced by *Xanthomonas campestris*, represents a large potential market for this microbial product.

Biosurfactants

Many surfactants that have been used for commercial purposes are products of synthetic chemistry. At the present time there is an increasing interest in the use of biosurfactants. These are especially important for environmental applications where biodegradability is a major requirement. Biosurfactants are used for emulsification, increasing detergency, wetting and phase dispersion, as well as for solubilization. These properties are especially important in bioremediation, soil spill dispersion, and enhanced oil recovery (EOR).

The most widely used microbially produced biosurfactants are glycolipids, and important examples are shown in figure 44.17. These compounds have distinct hydrophilic and hydrophobic regions, and the final compound structure and characteristics depend on the particular growth conditions and the carbon source used. Good yields often are obtained with insoluble substrates. These biosurfactants are excellent

TABLE 44.17   Characteristics and Uses of Commercially Important Microbial Biopolymers

Name	State of Development	Organisms and Applications
Dextran	Production	*Klebsiella, Acetobacter,* and *Leuconostoc* are major genera; these alpha-linked glucans are used as blood expanders and absorbants; they can form a hydrophilic layer at burn surfaces
Erwinia exopolysaccharide	Production	Used as a paint component because of its compatibility with basic dyes
Xanthan	Production	Used in foods and in secondary oil recovery programs; produced by *Xanthomonas campestris*
Cellulose microfibrils	Production	Thickener for use in foods; forms films to create textures and barriers; a special *Acetobacter* strain is used
Pullulan	Development	Produced by *Aureobasidium pullulans,* can serve as biodegradable material for food coating, and may be able to replace starch in selected applications
Scleroglucan	Development	Produced by the fungus *Sclerotium,* and used in drilling muds; forms a viscous gel that exhibits pseudoplasticity
Microbial alginate	Development	*Azotobacter vinlandii* is a major organism; wide range of uses to replace algal alginates; use as a sizing agent and food stabilizer should provide an expanding market
Curdlan	Development	Produced by *Alcaligenes faecalis* var. *myxogenes;* swells in water and produces an irreversible gel; not digested in the intestinal tract
Polyesters	Development	*Pseudomonas oleovorans* produces optically active polyesters under nitrogen-limited conditions. Useful as a chemical feedstock, particularly for specialty plastics

From I. W. Sutherland and D. C. Ellwood, *Society for General Microbiology Symposium,* 29, p. 128, table 6. Published by Cambridge University Press (1979); data included from R. A. Kent, et al., *Food Technology,* June 1991.

(a)

(b)

(c)

Figure 44.17   Structures of Typical Microbially Produced Biosurfactants. These glycolipid-based compounds are among the most widely used products. (*a*) Rhamnolipid R3 from *Pseudomonas aeruginosa.* (*b*) Trehalose-2,3,4,2′-tetraester from *Rhodococcus erythropolis.* (*c*) Surfactin from *Bacillus subtilis.*

Figure 44.18   Microbial Products Can Aid in the Cleanup of Oil Spills. A surfactant glycolipid emulsifier (EM) from *P. aeruginosa* improves removal of crude oil from a solid surface (the interior of a beaker) in a laboratory experiment.

dispersing agents, as shown in figure 44.18, and these types of products have been used with the *Exxon Valdez* oil spill.

In many bioremediation programs the biosurfactant-producing microorganism can be added directly to the oil-containing geological structure, and the biosurfactant will be produced in situ. This may improve oil recovery, particularly from older oil fields.

---

1. What two major bacteria have been used as insecticides? What is a parasporal body? Briefly describe how the *B. thuringiensis* toxin kills insects. What other two microbial groups have been used as bioinsecticides?

2. Why might microbial biopolymers be more useful than similar polymers derived from plants and other natural sources? Give some examples of their uses.

3. Biosurfactants are finding many uses. What are possible reasons for the increased interest in these products?

Biodegradation

Biodegradation is a general term referring to the undesired microbial-mediated destruction of paper, paint, metals, textiles, concrete, and other materials. Particularly in humid climates, the effects of microbial growth on stored crops and foods have been estimated to cost over $50 billion per year. The problem is vast and extremely costly.

Biodegradation Control

The nature of biodegradation and the ways in which it can be limited vary with the substance being destroyed. The following section describes some examples of biodegradation and its control.

Jet Fuels

Some of the earliest difficulties with microbially produced changes in jet fuels resulted from microbial growth in the bottom of storage facilities and in airplane fuel tanks. Trace amounts of water accumulated in the fuel and created a water-hydrocarbon interface where microorganisms grew. Microorganisms in the fuel clogged critical pumps and orifices; the most common resident, *Cladosporium resinae,* has even been

called the kerosene fungus. A series of microbial inhibitors have been used to control growth at jet fuel/water interfaces. These inhibitors include organoborons and isothiazolone compounds. A methylchloro/methyl-isothiazolone mixture prevents growth of the fungus at 1 ppm. The problem has also been combated by cleaning storage and aircraft fuel tanks more frequently and using filters in fuel delivery and transfer lines of the aircraft.

Paper

Microbial degradation of stored softwoods used in paper manufacture can cause substantial economic losses. Microbial growth results in a decrease of paper strength, discoloration, and subsequent increased deterioration of the final paper product. Papermaking involves solubilization of lignin and hemicellulose by chemical processes, with the release of microbially usable wood sugars. These soluble waste products are removed from the residual cellulose by extensive washing. Microorganisms that grow in the wash solutions produce slimes and surface growths that lower process efficiency and final paper yield and quality. In the past mercury-containing organic and inorganic compounds were commonly used as **biocides** in pulp and paper manufacturing plants. The long-term effects of these mercury products on lakes and rivers are still being felt in many timber-producing areas of the world (*see chapter 40*). Today other compounds are used as biocides, including chlorine, phenols, and organosulfide compounds. Extensive washing with hot alkali often is used to maintain papermaking equipment in a slime-free condition.

Computer Chips

Microbial contamination also is a major concern in the computer chip and electronics industries. It can decrease transistor service life by impairing the adhesion of successive coatings on chips during their manufacture. Microbial growth at transistor junctions creates serious problems in critical electronic components. Thus ultrapure water, free of dissolved organic matter and microorganisms, is a necessity in modern electronics manufacturing. Because of the ability of microorganisms to grow while using low levels of organic compounds leached from plastic tubing or absorbed from the air, even nominally ultrapure water may support microbial growth. Despite careful and repeated membrane filtration, the presence of minicells can lead to microbial contaminants even in water that has been passed through 0.22 μm membrane filters.

Paints

A major challenge in the management of biodeterioration is the control of microbial growth in paint. Microorganisms can grow in the components of paint before paint formulation, in the container, and following application. It is especially necessary to use chemical inhibitors with modern water-soluble latex-based paints. Mercury compounds, primarily phenylmercuric acetate, were added to paint for many years until manufacturers and purchasers became aware of the environmental problems arising from the accumulation of mercury in the environment. The use of such inhibitor-containing paints produces a long-term, low

(a)

(b)

Figure 44.19   Microbial-Mediated Metal Corrosion. The microbiological corrosion of iron is a major problem. (*a*) The graphitization of iron under a rust bleb on the pipe surface allows microorganisms, including *Desulfovibrio,* to corrode the inner surface. (*b*) Evidence points to the importance of communities of microorganisms, as opposed to individual species acting alone, as a major factor in microbiologically influenced corrosion. This epifluorescence microscope view (×1,600) is of pipeline steel a few hours after colonization by sulfate-reducing and organic acid-producing bacteria such as species of *Enterobacter* and *Clostridium.*

level of volatile mercury in the atmosphere of painted rooms, possibly leading to mercury poisoning of the occupants. Mercury-containing inhibitors have been replaced by other compounds, including quaternary ammonium compounds, barium metaborate, and various chlorinated phenolic products.

Paints used on the hulls of marine vessels can contain high levels of organotin and copper compounds as biocides. These compounds accumulate in harbor muds, changing the metal resistance characteristics of the microorganisms in such environments.

Textiles and Leather

Degradation of cloth and leather materials, especially in tropical or humid environments, is of continuing economic concern. Fungal inhibitors, including a variety of phenolic-based compounds, may be added to natural materials such as cotton and wool. The use of copper compounds in many types of tanning processes leads to a greater resistance of leather products to microbiological degradation.

Metals

The microbial-mediated corrosion of metals is particularly critical where iron pipes are used in waterlogged anaerobic environments or in secondary petroleum recovery processes carried out at older oil fields. In these older fields water is pumped down a series of wells to force residual petroleum to a central collection point. If the water contains low levels of organic matter and sulfate, anaerobic microbial communities can develop in rust blebs or tubercles (figure 44.19), resulting in punctured iron pipe and loss of critical pumping pressure. Microorganisms that use elemental iron as an electron donor during the reduction of CO_2 in methanogenesis have recently been discovered (Box 44.2). Because of the wide range of interactions that occur between microorganisms and metals, the need to develop strategies to deal with corrosion problems is critical.

Concrete

Concrete can be dissolved by thiobacilli (*see chapter 22*). One member of the genus, formerly known as *Thiobacillus concretivorous,* now called *Thiobacillus thiooxidans,* causes major structural damage to highways and concrete sewage lines. In the

---

Box 44.2

Methanogens: A New Role for An Ancient Microbial Group

The methanogens, an important group of the archaeobacteria that can produce methane, are considered to be at least 3.5 billion years old. Despite intensive research, new discoveries are still being made concerning these microorganisms. Methanogenic bacteria have now been found to contribute to the anaerobic corrosion of soft iron. Previously the microbial group usually considered the major culprit in the anaerobic corrosion process was the genus *Desulfovibrio,* which can use sulfate as an oxidant and hydrogen produced in the corrosion process as a reductant. Methanogens can use elemental iron as an electron source in their metabolism. It appears that corrosion may occur even without the presence of sulfate, which is required for functioning of *Desulfovibrio.* Rates of iron removal by the methanogens are around 79 mg/1,000 cm^2 of surface area in a 24 hour period. This may not seem a high rate, but in relation to the planned service life of metal structures in muds and subsurface soils—possibly years and decades—such corrosion can become a major problem. Continuous efforts to improve protection of iron structures will be required in view of the diversity of iron-corroding bacteria.

TABLE 44.18 **Requirements for Successful Biostimulation of Hydrocarbon Degradation in a Marine Environment: Interactions of Microbial and Environmental Factors**

System Component	Requirement
Microorganism	Ability to degrade hydrocarbons
	Ability to be grown and stored prior to use in a biostimulation process
	Ability to survive and function in a marine environment with its low temperature (5°C or lower) and high salinity
	Ability to assist in bioemulsification
	Possible ability to fix nitrogen
	Ability to grow from low inoculum levels
	No secondary toxic effects
Environment	Accessory nutrients (N, P, K, Fe) available and in same physical location as microorganisms and hydrocarbon substrate
	Absence of toxic chemicals (phenolics, heavy metals, etc.) that can harm hydrocarbon-degrading microorganisms
	Adequate O_2 levels and transfer rates

presence of sufficient moisture and pyritic sulfur compounds of lower oxidation state than sulfate, this microorganism produces sulfuric acid, resulting in the dissolution of concrete. Thiobacilli can be controlled by the selection of cements that do not contain appreciable levels of oxidizable sulfur compounds, the avoidance of contamination of concrete with soil microorganisms, and the use of organic inhibitors.

Biodegradation Stimulation

Microbial activities can be used to restore or maintain environmental quality by **biodegradation stimulation.** This has led to the creation of an entire industry that specializes in the application of these technologies to restore the environment. The use of microorganisms to remove pollutants from the environment often is called **bioremediation.** Several examples of current bioremediation efforts will be discussed in this section.

Stimulation of Oil Spill Degradation

To reduce the damage caused by oil spills, the rate of petroleum degradation must be increased. The development of a mixture of suitable microorganisms to degrade the wide variety of aromatic, polycyclic, and aliphatic compounds in most petroleums has been a major goal of researchers.

The first patented bacterial life form—patented by A. M. Chakrabarty in 1974—was a *Pseudomonas* species containing hydrocarbon degradation plasmids. This bacterium is much better able to degrade mixed petroleum residues. The successful functioning of such a microorganism, if it is added to a habitat, depends on several factors (table 44.18). When working with dispersed hydrocarbons in the ocean, contact between the microorganism, the hydrocarbon substrate, and other essential nutrients must be maintained. To achieve this, pellets containing nutrients and an oleophilic (hydrocarbon soluble) preparation have been used. Such a nutritional supplement not only aids inoculated organisms but also stimulates the natural hydrocarbon-degrading microbial population. This technique has accelerated the degradation of different crude oil slicks by 30 to 40%, in comparison with control oil slicks where the additional nutrients were not available.

A unique challenge for this technology was the *Exxon Valdez* oil spill, which occurred in Alaska in March 1989. Several different approaches were used to increase biodegradation. These included nutrient additions, chemical dispersants, biosurfactant additions, and the use of high-pressure steam. The use of a microbially produced glycolipid emulsifier has proven helpful, as discussed earlier. High-temperature steam caused a major disruption in the microbial community, which has still not fully recovered.

Biodegradation also can be enhanced in surface soils and sediments. In one of the largest biodegradation enhancement projects carried out, 300,000 yd^3 of hydrocarbon-contaminated soil were treated in California, and a 94% reduction in hydrocarbons was achieved. A mixed population of pigmented microorganisms developed and contributed substantially to this successful project.

A unique two-stage degradation process is being used experimentally to degrade PCBs in sediments. The muds in the Hudson River area can carry out a partial dechlorination under anaerobic conditions. Following this first step, the muds are aerated to completely degrade the less-chlorinated anaerobically produced PCP residues (figure 44.20).

Biodegradation (pp. 873–74).

Interesting microorganisms with unique degradative capabilities continue to be isolated. One of the most famous of these is *Phanerochaete chrysosporium* (Box 44.3).

Subsurface Biodegradation Enhancement

The same principles can be used to stimulate the degradation of hydrocarbons and other chemical residues in contaminated subsurface environments. The major difference is that geological structures have limited permeability. Although subsurface regions in a pristine state often have O_2 concentrations approaching saturation, the penetration of small amounts of organic matter into these structures can quickly lead to O_2 depletion.

A major gasoline leak occurred in the central northeast coastal region of the United States following a break in a petroleum pipeline in the early 1970s. In this particular subsurface biodegradation enhancement effort, 58 tons of ammonium sulfate and 29 tons of monosodium and disodium phosphates were added to the subsurface and compressed air was pumped into the subsurface zone. An estimated 1,000 barrels of gasoline were degraded in this bioreclamation program. This technology is now being used in a wide range of surface—land farming—and subsurface environments, with and without the use of specific microbial inocula.

A typical approach that can be used to carry out such in situ bioremediation is shown in figure 44.21. Depending on

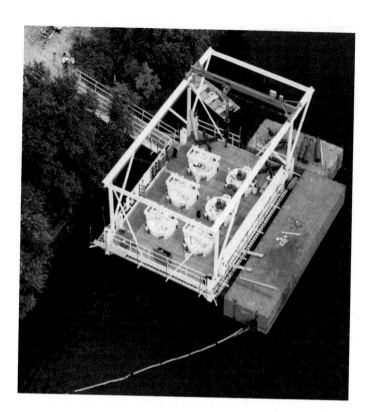

Figure 44.20 PCB Biodegradation. The environment can be managed to promote critical microbial bioremediation reactions. Hudson River (New York) muds are being aerated in cylindrical caissons to stimulate aerobic bacteria to complete the degradation of PCB molecules. Under previously anaerobic conditions the PCB molecules had been partially dechlorinated.

Figure 44.21   An in situ Bioremediation System Using Nutrient Additions and Bioventing. Monitoring and recovery wells are used to monitor the plume and its possible movement. Nutrients and oxygen (as peroxide or air) are added to the contaminated soil and groundwater. A bioventing well can be used to accelerate the removal of hydrocarbon vapors.

Box 44.3

Phanerochaete chrysosporium: A Wood-Degrading Fungus with a Voracious Appetite

The basidiomycete *Phanerochaete chrysosporium* (the scientific name means "visible hair, golden spore") is a fungus with unusual degradative capabilities. This organism is termed a "white rot fungus" because of its ability to degrade lignin, a randomly linked phenylpropane-based polymeric component of wood (*see chapter 40*). The cellulosic portion of wood is attacked to a lesser extent, resulting in the characteristic white color of the degraded wood. This organism also degrades a truly amazing range of xenobiotic compounds (nonbiological foreign chemicals) using both intracellular and extracellular enzymes.

As examples, the fungus degrades benzene, toluene, ethylbenzene, and xylenes (the so-called BTEX compounds), chlorinated compounds such as 2, 4, 5-trichloroethylene (TCE), and trichlorophenols. The latter are present as contaminants in wood preservatives and also are used as pesticides. In addition, other chlorinated benzenes can be degraded with or without toluenes being present. Even the insecticide Hydramethylnon is degraded!

How does this microorganism carry out such feats? Apparently most degradation of these xenobiotic compounds occurs after active growth, during the secondary metabolic lignin degradation phase. Degradation of some compounds involves important extracellular enzymes including lignin peroxidase, manganese-dependent peroxidase, and glyoxal oxidase. A critical enzyme is pyranose oxidase, which releases H_2O_2 for use by the manganese-dependent peroxidase enzyme. The H_2O_2 also is a precursor of the highly reactive hydroxyl radical, which participates in wood degradation. Apparently the pyranose oxidase enzyme is located in the interperiplasmic space of the fungal cell wall, where it can function either as a part of the fungus or be released from the fungus and penetrate into the wood substrate. It appears that the nonspecific enzymatic system that releases these oxidizing products degrades many cyclic, aromatic, and chlorinated compounds related to lignins.

We can expect to continue hearing of many new advances in work with this organism. Potentially valuable applications being studied include growth in bioreactors where intracellular and extracellular enzymes can be maintained in the bioreactor while liquid wastes flow past the immobilized fungi.

---

the petroleum contamination and the geological characteristics of the site, injection and monitoring wells can be installed. Nutrients and a source of oxygen (possibly compressed air or peroxide) also can be added. Often this process is combined with bioventing, the physical removal of vapors by a vacuum. Depending on the volume and the location of the contaminated soil, the process may require months or years to complete. In most cases the natural microbial communities are used in bioremediation. Only under very specialized circumstances will degradative microorganisms be added.

An interesting aspect of using microbial inocula in freshwater, marine, and soil environments is the potential attractiveness of laboratory-grown cultures as a food source. The possibility of their more rapid use as a preferred nutrient by protozoa may limit the effectiveness of such inocula in biodegradation enhancement.

Bioleaching of Metals

Another application of biodegradation enhancement is the recovery of metals from ores and mining tailings with metal levels too low for smelting (figure 44.22). Bioleaching, by means of natural populations of *Thiobacillus ferrooxidans* and related thiobacilli, for example, allows recovery of up to 70% of the copper in low-grade ores. Bioleaching, shown in figure 44.23, involves the biological oxidation of copper present in these ores to produce soluble copper sulfate. The copper sulfate can then be recovered by reacting the leaching solution, which contains up to 3.0 g/liter of soluble copper, with iron. The copper sulfate reacts with the elemental iron to form ferrosulfate, and

Figure 44.22   Bioleaching of Metals in Mining. Microorganisms can be used to recover metals from low-grade ores and tailings. In this aerial photograph of copper leach dumps at Santa Rita, New Mexico, the presence of iron in the ferrous form (green) and ferric form (orange) creates a mosaic pattern. About 10% of United States copper comes from leaching ore through the activity of bacteria such as *Thiobacillus* and *Leptospirillum.*

the copper is reduced to the elemental form, which precipitates out in a settling trench. The process is summarized in the following reaction:

$$CuSO_4 + Fe^0 \longrightarrow Cu^0 + FeSO_4$$

Bioleaching may require added phosphorus and nitrogen if these are limiting in the ore materials, and the same process can be used to solubilize uranium.

$$CuSO_4 + Fe^0 \longrightarrow Cu^0 + FeSO_4$$

Ore

$$2Fe_2(SO_4)_3 + CuFeS_2 + 2H_2O + 3O_2 \longrightarrow$$
$$\longrightarrow CuSO_4 + 5FeSO_4 + 2H_2SO_4$$

$Fe_2(SO_4)_3$

Air

Leached ore

$$CuSO_4 + Fe \longrightarrow FeSO_4 + Cu$$

$FeSO_4$

$FeSO_4 + CuSO_4$ Fe

Precipitation of copper

Pump

Thiobacillus

$$FeSO_4 \longrightarrow Fe_2(SO_4)_3$$

Figure 44.23   **Copper Leaching from Low-Grade Ores.** The chemistry and microbiology of copper ore leaching involve interesting complementary reactions. The microbial contribution is the oxidation of ferrous ion (2+) to ferric ion (3+). *Thiobacillus ferrooxidans* and related thiobacilli are very active in this oxidation. The ferric ion then reacts chemically to solubilize the copper. The soluble copper is recovered by a chemical reaction with elemental iron, which results in an elemental copper precipitate.

1. What microbial genus can grow in jet fuels? What problems does this cause and how can these be prevented?

2. What types of inhibitors have been used in the past to control microbial growth in paints?

3. Briefly describe the problems microorganisms pose for paper, computer chips, textiles, leather, metals, and concrete. How can biodegradation be avoided in each case?

4. What factors must one consider when attempting to stimulate the microbial degradation of a massive oil spill in a marine environment?

5. What might be additional applications of biosurfactants? Think of possible biomedical uses.

6. What is unique about *Phanerochaete chrysosporium?* What does its name mean?

7. What is the major microbial genus that contributes to bioleaching processes? How, in general terms, does the process work?

Molecule discriminating function

Receptacle substances (enzyme, antibiotic, antigen)

Physical and chemical change

Signal conversion

Electric signal

Substance to be measured

Transducer

Receptor

Figure 44.24   **Biosensor Design.** Biosensors are finding increasing applications in medicine, industrial microbiology, and environmental monitoring. In a biosensor a biomolecule or whole microorganism carries out a biological reaction, and the reaction products are used to produce an electrical signal.

Biosensors

A rapidly developing area of biotechnology, arousing intense international scientific interest, is that of **biosensor** production. In this new field of bioelectronics, living microorganisms (or their enzymes or organelles) are linked with electrodes, and biological reactions are converted into electrical currents by these biosensors (figure 44.24). Biosensors are being developed to measure specific components in beer, to monitor pollutants, and to detect flavor compounds in food. It is possible to measure the concentration of substances from many different environments (table 44.19). In the future biosensors will be able to measure ions and molecules, specific enzymatic activities, and the presence of bacteria. Applications include the detection of glucose, acetic acid, glutamic acid, ethanol, and biochemical oxygen demand. In addition, the application of biosensors to measure cephalosporin, nicotinic acid, and several B vitamins has been described. It appears that biosensors will also be able to detect phenols, methane, and carbon monoxide, which are important in environmental management.

TABLE 44.19   Biosensors: Potential Biomedical, Industrial, and Environmental Applications

Clinical diagnosis and biomedical monitoring

Agricultural, horticultural, veterinary analysis

Detection of pollution, and microbial contamination of water

Fermentation analysis and control

Monitoring of industrial gases and liquids

Measurement of toxic gas in mining industries

Direct biological measurement of flavors, essences, and pheromones

Figure 44.25   Glucose Monitoring with a Biosensor. A biosensor based on microbially derived enzymes for use in fermentation process monitoring or biomedical analyses.

One of the most interesting applications of this technology is to combine bacterial magnetite (*see p. 818 in chapter 40*) with a chromophoric system to measure glucose through a color change (figure 44.25). Rapid advances are being made in biosensor technology to improve the stability of these probes and allow real-time monitoring of glucose and other metabolites, a critical need in modern medicine.

1. What are biosensors and how do they detect substances?

2. What areas are biosensors being used in to assist in chemical and biological monitoring efforts?

Summary

1. Fermentation is a word with different meanings, ranging from a description of a microbial mass culture—aerobic or anaerobic—to rigorous physiological definitions.

2. Microorganisms can be grown in a variety of mass culture systems. Especially with filamentous microorganisms (actinomycetes, molds), it often is necessary to provide high agitation and aeration rates because of the occurrence of viscous plasticlike growth.

3. Fermentation media are usually made from crude materials, such as molasses, corn-steep liquor, and soybean hydrolysates, and often are designed to create specific limitations on microbial growth processes.

4. The mass culture of microorganisms involves transfers of oxygen, heat, and nutrients, requiring the establishment of driving forces or gradients. This is especially critical for O_2.

5. Scaleup is the critical process of carrying out a fermentation in progressively larger reactor vessels.

6. Besides selection, mutation, and screening processes, which have been used for decades in industrial microbiology, modern biotechnology uses molecular techniques and a wide range of vectors to create new DNA constructs for DNA expression.

7. Microbial products can be classified as primary or secondary metabolites, depending on whether or not they are produced during growth. Most antibiotics are secondary metabolites.

8. Most amino acids are produced by carefully selected microbial cultures that have regulatory mutations. This type of impairment, together with careful nutrient stresses, can increase the levels of amino acid excretion.

9. Bioconversions, carried out with nongrowing vegetative microorganisms or free enzymes, can cause minor but important modifications of complex molecules. Cells used in bioconversions can be immobilized.

10. Through recombinant DNA technology, gene expression and genes themselves can be modified. Genes also can be expressed in different microorganisms to improve process efficiency, yield, and purity.

11. Microbial applications of continuing interest include microbial insecticides, biopolymers, and biosensors.

12. *Bacillus thuringiensis* is finding increasing applications. Transgenic plants are now available that contain the toxic protein derived from the parasporal body.

13. Biodegradation management is a matter of continuing concern. Under special conditions biocides can be used to control microbial damage to certain materials.

14. Biodegradation stimulation involves the addition of nutrients, and microorganisms increase degradation rates and promote bioremediation. Bioleaching, a related process, can be used to solubilize metals from low-grade ores.

15. *Phanerochaete chrysosporium* is capable of degrading a wide range of cyclic, aromatic, and chlorinated compounds. This fungus will be useful in degrading complex mixtures of chemical wastes in bioremediation programs.

16. In situ bioremediation is finding widespread uses in biodegradation programs. This technology has resulted in the creation of an entire industry.

17. Biosensors promise to have expanding applications in industrial microbiology, medicine, and environmental monitoring.

Key Terms

biochemical markers *919*

biocide *927*

bioconversion *916*

biodegradation *927*

biodegradation stimulation *929*

bioinsecticides *923*

biopolymer *925*

bioremediation *929*

biosensor *932*

biotechnology *906*

biotransformation *916*

continuous feed *909*

fermentation *906*

idiophase *912*

immobilization *917*

industrial microbiology *906*

lyophilization *910*

metabolic engineering *920*

microbial transformation *916*

non-Newtonian broth *909*

primary metabolite *912*

protein engineering *919*

protoplast fusion *910*

regulatory mutant *915*

scaleup *909*

secondary metabolite *912*

semisynthetic penicillin *913*

trophophase *912*

Questions for Thought and Review

1. Although continuous culture has many advantages in terms of cell production rates, it is not widely used in industrial processes except for sewage treatment. What are some of the possible reasons for this?

2. Why has scaleup been such a problem in industrial microbiology? Consider mixing, aeration, and cooling processes in a 10 ml culture versus a 100,000 liter fermenter.

3. Most commercial antibiotics are produced by actinomycetes, and only a few are synthesized by fungi and other bacteria. From physiological and environmental viewpoints, how might you attempt to explain this observation?

4. We hear much about the beneficial uses of recombinant DNA technology. What are some of the problems and disadvantages that should be considered when using microorganisms for these applications?

5. As the president of a biotechnology company, you must decide whether a new process for production of a widely used organic acid should be kept secret or whether a patent should be obtained. What factors should you consider in making this decision?

6. Your company has developed a new biocide for use in a paper manufacturing plant. Before this new compound can be released for sale, what types of environmental questions should you ask and what tests should you complete?

7. Why might *Bacillus thuringiensis* bioinsecticides be of interest in other areas of biotechnology? Consider the molecular aspects of their mode of action.

8. What are some of the possible advantages of biosensors as opposed to more traditional physical and chemical measurement procedures?

Additional Reading

General

American Association for the Advancement of Science. 1991. Frontiers in biotechnology. *Science.* 252:1585–1756.

Crueger, W., and Crueger, A. 1990. *Biotechnology: A textbook of industrial microbiology.* 2d ed., ed. T. D. Brock. Sunderland, Mass.: Sinauer Associates.

Demain, A. L., and Solomon, N. A. 1986. Industrial microbiology. *Sci. Am.* 245(3):66–75.

Finkelstein, D. B., and Ball, C. 1992. *Biotechnology of filamentous fungi: Technology and products.* Stoneham, Mass.: Butterworth-Heinemann.

Glazer, A. N., and Nakaido, H. 1994. *Microbial biotechnology.* New York: W. H. Freeman and Co.

Glick, B. R., and Pasternak, J. J. 1994. *Molecular biotechnology: Principles and applications of recombinant DNA.* Washington, D.C.: ASM Press.

Leatham, G. 1992. *Frontiers in industrial mycology.* New York: Chapman and Hall.

Primrose, S. B. 1991. *Molecular Biotechnology,* 2d ed. Oxford: Blackwell Scientific Publishers.

Microbial Growth Processes

Dierkes, W.; Lohmeyer, M.; and Rehm, H.-J. 1993. Long-term production of ergot peptides by immobilized *Claviceps purpurea* in semicontinuous and continuous culture. *Appl. Environ. Microbiol.* 59(7):2029–33.

Hansen, J. N. 1993. Antibiotics synthesized by posttranslational modification. *Ann. Rev. Microbiol.* 47:535–64.

He, M.; Lorr, D.; and Wang, N. S. 1993. Microbial fermentation monitoring using near-infrared spectroscopy. *Am. Biotech. Lab.* July:56–57.

Scheper, T.-H., and Lammers, F. 1994. Fermentation monitoring and process control. *Curr. Opin. Biotech.* 5:187–91.

Stephanopoulos, G. 1994. Metabolic engineering. *Curr. Opin. Biotech.* 5:196–200.

Major Products of Industrial Microbiology

Aharonowitz, Y., and Cohen, G. 1981. The microbiological production of pharmaceuticals. *Sci. Am.* 245(9):140–52.

Eveleigh, D. E. 1981. The microbiological production of industrial chemicals. *Sci. Am.* 245(3):154–78.

Gallo, V. P. 1993. *The discovery of natural products with therapeutic potential.* New York: Butterworth-Heinemann.

Hopwood, D. A. 1981. The genetic programming of industrial microorganisms. *Sci. Am.* 245(3):90–102.

Hugo, W. B., and Russell, A. D. 1992. *Pharmaceutical microbiology,* 5th ed. New York: Blackwell Scientific Publications.

Ikeda, M., and Kasumata, R. 1992. Metabolic engineering to produce tyrosine or phenylalanine in a tryptophan-producing *Corynebacterium glutamicum* strain. *Appl. Environ. Microbiol.* 58:781–85.

King, L. A., and Possee, R. D. 1992. *The Baculovirus expression system.* New York: Chapman and Hall.

Phaff, H. J. 1981. Industrial microorganisms. *Sci. Am.* 245(3):76–89.

Sakurai, N.; Imai, Y.; Masuda, M.; Komatsubara, S.; and Tetsuya, T. 1993. Molecular breeding of a biotin-hyperproducing *Serratia marcescens* strain. *Appl. Environ. Microbiol.* 59:3225–32.

Stevenson, R. 1994. Extremozymes. *Am. Biotech. Lab.* 12(9):5–8.

Tanaka, A.; Tosa, T.; and Kobayashi, T. 1992. *Industrial application of immobilized biocatalysts.* New York: Marcel Dekker, Inc.

Recombinant DNA Techniques in Biotechnology

Blondelle, S. E.; Takahashi, E.; Dinh, K. T.; and Houghton, R. A. 1995. The antimicrobial activity of hexapeptides derived from synthetic combinatorial libraries. *J. Appl. Bacteriol.* 78:39–46.

McDaniel, R.; Ebert-Khosla, S.; Hopwood, D. A.; and Khosla, C. 1993. Engineered biosynthesis of novel polyketides. *Science* 262:1546–49.

Monaco, A. P., and Larin, Z. 1994. YAC's, BAC's, PAC's and MAC's: Artificial chromosomes as research tools. *Tibtech* 12:280–86.

Old, R. W., and Primrose, S. B. 1994. *Principles of gene manipulation,* 5th ed. Boston: Blackwell Science, Inc.

Toffaletti, D. L.; Rude, T. H.; Johnston, S. A.; Durack, D. T.; and Perfect, J. R. 1993. Gene transfer in *Cryptococcus neoformans* by use of biolistic delivery of DNA. *J. Bacteriol.* 175(5):1405–11.

Other Microbial Products

Abernethy, G. A., and Walker, J. R. L. 1993. Degradation of the insecticide Hydramethylnon by *Phanerochaete chrysosporium*. *Biodegradation* 4:131–39.

Chilcott, C. N., and Wigley, P. J. 1994. Opportunities for finding new *Bacillus thuringiensis* strains. *Agric. Ecosystems Environ.* 49:51–57.

Daniel, D.; Volc, J.; and Kubatova, E. 1994. Pyranose oxidase, a major source of H_2O_2 during wood degradation by *Phanerochaete chrysosporium, Trametes versicolor,* and *Oudemansiella mucida*. *Appl. Environ. Microbiol.* 60:2524–32.

Georgiou, G.; Lin, S.-C.; and Sharma, M. M. 1992. Surface-active compounds from microorganisms. *Bio/Technology* 10:60–65.

Gill, S. S.; Cowles, E. A.; and Pietrantonio, P. V. 1992. The mode of action of *Bacillus thuringiensis* endotoxins. *Ann. Rev. Entomol.* 37:615–36.

Hofte, H., and Whiteley, H. R. 1989. Insecticidal crystal proteins of *Bacillus thuringiensis*. *Microbiol. Rev.* 53(2):242–55.

Ignoffo, C. M., and Anderson, R. F. 1979. Bioinsecticides. In *Microbial Technology*. Peppler, H. J., and Perlman, D., editors, 1–28. New York: Academic Press.

Leathers, T. D.; Gupta, S. C.; and Alexander, N. J. 1993. Mycopesticides: Status, challenges and potential. *J. Ind. Microbiol.* 12:69–75.

Llewellyn, D.; Cousins, Y.; Mathews, A.; Hartweck, L.; and Lyon, B. 1994. Expression of *Bacillus thuringiensis* insecticidal protein genes in transgenic crop plants. *Agric. Ecosystems Environ.* 49:85–93.

Maramorosch, K. 1991. *Biotechnology for biological control of pests and vectors*. Boca Raton, Fla.: CRC Press.

Strobel, G. A. 1991. Biological control of weeds. *Sci. Am.* 265:72–78.

Wood, H. A., and Granados, R. R. 1991. Genetically engineered baculoviruses as agents for pest control. *Ann. Rev. Microbiol.* 45:69–87.

Biodegradation

Alexander, M. 1994. *Biodegradation and bioremediation*. New York: Academic Press.

Armenante, P. M.; Pal, N.; and Lewandowski, G. 1994. Role of mycelium and extracellular protein in the biodegradation of 2, 4, 6-trichlorophenol by *Phanerochaete chrysosporium*. *Appl. Environ. Microbiol.* 60(6):1711–18.

Atlas, R. M., and Cerniglia, C. E. 1995. Bioremediation of petroleum pollutants. *BioScience* 45(5):332–38.

Bragg, J. R.; Prince, R. C.; Harner, E. J.; and Atlas, R. M. 1994. Effectiveness of bioremediation for the Exxon Valdez oil spill. *Nature* 368:413–18.

Brierley, C. L. 1982. Microbiological mining. *Sci. Am.* 247(2):44–53.

Cookson, Jr., J. T. 1995. *Bioremediation engineering: Design and application*. New York: McGraw-Hill.

Holloway, M. 1991. Soiled shores. *Sci. Am.* 265(4):102–16.

Hutchins, S. R.; Davidson, M. S.; Brierley, J. A.; and Brierley, C. L. 1986. Microorganisms in reclamation of metals. *Ann. Rev. Microbiol.* 40:311–36.

Moffat, A. S. 1994. Microbial mining boosts the environment, bottom line. *Science* 264:778–79.

Sayler, G. S.; Fox, R.; and Blackburn, J. W. 1991. *Environmental biotechnology for waste treatment*. New York: Plenum Press.

Shannon, M. J. R., and Unterman, R. 1993. Evaluating bioremediation: Distinguishing fact from fiction. *Ann. Rev. Microbiol.* 47:715–38.

Warhurst, A. M., and Fewson, C. A. 1994. Biotransformations catalyzed by the genus *Rhodococcus*. *Crit. Rev. Biotech.* 14(1):29–73.

Biosensors

Henrysson, T., and Mattiasson, B. 1993. A microbial biosensor system for dihalomethanes. *Biodegradation* 4:101–105.

Mathewson, P. W., and Finley, J. W. 1992. *Biosensor design and application*. Washington, D.C.: American Chemical Society.

Scheper, T. 1992. Biosensors for process monitoring. *J. Ind. Microbiol.* 9:163–72.

Schultz, J. S. 1991. Biosensors. *Sci. Am.* 265(2):64–69.

Valcarcel, M., and de Castro, L. 1994. *Flow-through (bio) chemical sensors*. New York: Elsevier.

APPENDIX I
A Review of the Chemistry of Biological Molecules

A ppendix I provides a brief summary of the chemistry of organic molecules with particular emphasis on the molecules present in microbial cells. Only basic concepts and terminology are presented; introductory textbooks in biology and chemistry should be consulted for a more extensive treatment of these topics.

Atoms and Molecules

Matter is made of elements that are composed of atoms. An element contains only one kind of atom and cannot be broken down to simpler components by chemical reactions. An atom is the smallest unit characteristic of an element and can exist alone or in combination with other atoms. When atoms combine they form molecules. Molecules are the smallest particles of a substance. They have all the properties of the substance and are composed of two or more atoms.

Although atoms contain many subatomic particles, three directly influence their chemical behavior—protons, neutrons, and electrons. The atom's nucleus is located at its center and contains varying numbers of protons and neutrons (figure AI.1). Protons have a positive charge, and neutrons are uncharged. The mass of these particles and the atoms that they compose is given in terms of the atomic mass unit (AMU), which is 1/12 the mass of the most abundant carbon isotope. Often the term dalton (Da) is used to express the mass of molecules. It also is 1/12 the mass of an atom of ^{12}C or 1.661×10^{-24} grams. Both protons and neutrons have a mass of about one dalton. The atomic weight is the actual measured weight of an element and is almost identical to the mass number for the element, the total number of protons and neutrons in its nucleus. The mass number is indicated by a superscripted number preceding the element's symbol (e.g., ^{12}C, ^{16}O, and ^{14}N).

Negatively charged particles called electrons circle the atomic nucleus (figure AI.1). The number of electrons in a neutral atom equals the number of its protons and is given by the atomic number, the number of protons in an atomic nucleus. The atomic number is characteristic of a particular type of atom. For example, carbon has an atomic number of six, hydrogen's number is one, and oxygen's is eight (table AI.1).

The electrons move constantly within a volume of space surrounding the nucleus, even though their precise location in this volume cannot be determined accurately. This volume of space in which an electron is located is called its orbital. Each orbital can contain two electrons. Orbitals are grouped into shells of different energy that surround the nucleus. The first shell is closest to the nucleus and has the lowest energy; it contains only one orbital. The second shell contains four orbitals, one circular and three shaped like dumbbells (figure AI.2a). It can contain up to eight electrons. The third shell has even higher energy and holds more than eight electrons. Shells are filled beginning with the innermost and moving outward. For example, carbon has six electrons, two in its first shell and four in the second (figures AI.1 and AI.2b). The electrons in the outermost shell are the ones that participate in chemical reactions. The most stable condition is achieved when the outer shell is filled with electrons. Thus the number of bonds an element can form depends on the number of electrons required to fill the outer shell. Since carbon has four electrons in its outer shell and the shell is filled when it contains eight electrons, it can form four covalent bonds (table AI.1).

Chemical Bonds

Molecules are formed when two or more atoms associate through chemical bonding. Chemical bonds are attractive forces that hold together atoms, ions, or groups of atoms in a molecule or other substance. Many types of chemical bonds are present in organic molecules; three of the most important are covalent bonds, ionic bonds, and hydrogen bonds.

In covalent bonds, atoms are joined together by sharing pairs of electrons (figure AI.3). If the electrons are equally shared between identical atoms (e.g., in a carbon-carbon bond), the covalent bond is strong and nonpolar. When two different atoms such as carbon and oxygen share electrons, the covalent bond formed is polar since the electrons are pulled toward the more electronegative atom, the atom that more strongly attracts electrons. A single pair of electrons is shared in a single bond; a double bond is formed when two pairs of electrons are shared.

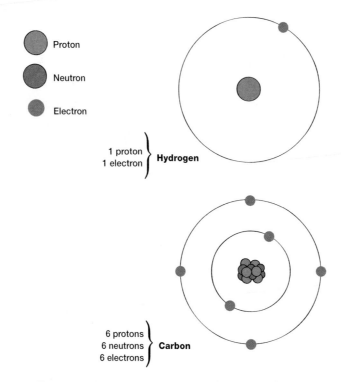

Figure AI.1   Diagrams of Hydrogen and Carbon Atoms. The electron orbitals are represented as concentric circles.

TABLE AI.1	Atoms Commonly Present in Organic Molecules			
Atom	**Symbol**	**Atomic Number**	**Atomic Weight**	**Number of Chemical Bonds**
Hydrogen	H	1	1.01	1
Carbon	C	6	12.01	4
Nitrogen	N	7	14.01	3
Oxygen	O	8	16.00	2
Phosphorus	P	15	30.97	5
Sulfur	S	16	32.06	2

Atoms often contain either more or fewer electrons than the number of protons in their nuclei. When this is the case, they carry a net negative or positive charge and are called ions. Cations carry positive charges and anions have a net negative charge. When a cation and an anion approach each other, they are attracted by their opposite charges. This ionic attraction that holds two groups together is called an ionic bond. Ionic bonds are much weaker than covalent bonds and are easily disrupted by a polar solvent such as water. For example, the Na^+ cation is strongly attracted to the Cl^- anion in a sodium chloride crystal, but sodium chloride dissociates into separate ions (ionizes) when dissolved in water. Ionic bonds are important in the structure and function of proteins and other biological molecules.

When a hydrogen atom is covalently bonded to a more electronegative atom such as oxygen or nitrogen, the electrons are unequally shared and the hydrogen atom carries a partial positive charge. It will be attracted to an electronegative atom such as oxygen or nitrogen, which carries an unshared pair of electrons; this attraction is called a hydrogen bond (figure AI.4). Although an individual hydrogen bond is weak, there are so many hydrogen bonds in proteins and nucleic acids that they play a major role in determining protein and nucleic acid structure.

Organic Molecules

Most molecules in cells are organic molecules, molecules that contain carbon. Since carbon has four electrons in its outer shell, it tends to form

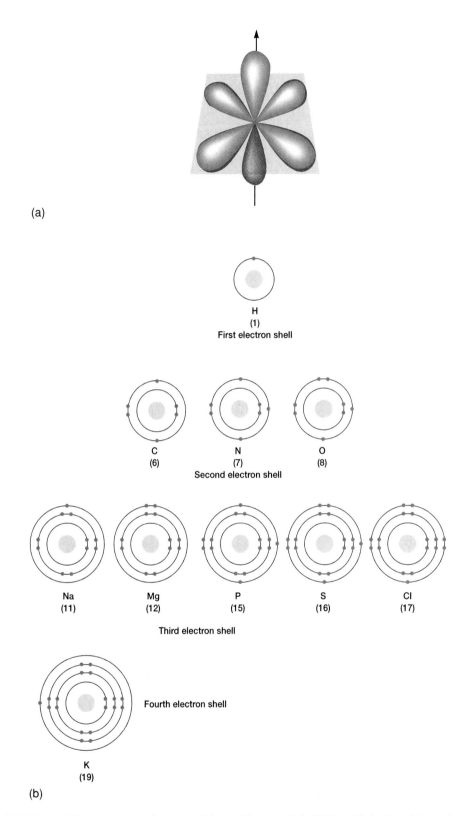

(a)

(b)

Figure AI.2   Electron Orbitals. (*a*) The three dumbbell-shaped orbitals of the second shell. The orbitals lie at right angles to each other. (*b*) The distribution of electrons in some common elements. Atomic numbers are given in parentheses.

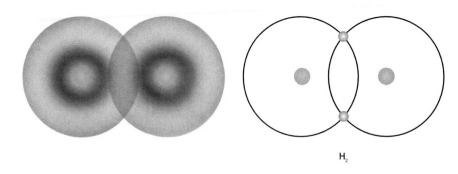

Figure AI.3   **The Covalent Bond.** A hydrogen molecule is formed when two hydrogen atoms share electrons.

H₂

four covalent bonds in order to fill its outer shell with eight electrons. This property makes it possible to form chains and rings of carbon atoms that also can bond with hydrogen and other atoms (figure AI.5). Although adjacent carbons usually are connected by single bonds, they may be joined by double or triple bonds. Rings that have alternating single and double bonds, like the benzene ring, are called aromatic rings. The hydrocarbon chain or ring provides a chemically inactive skeleton to which more reactive groups of atoms may be attached. These reactive groups with specific properties are known as functional groups. They usually contain atoms of oxygen, nitrogen, phosphorus, or sulfur (figure AI.6) and are largely responsible for most characteristic chemical properties of organic molecules.

Organic molecules are often divided into classes based on the nature of their functional groups. Ketones have a carbonyl group within the carbon chain, whereas alcohols have a hydroxyl on the chain. Organic acids have a carboxyl group, and amines have an amino group (figure AI.7).

Organic molecules may have the same chemical composition and yet differ in their molecular structure and properties. Such molecules are called isomers. One important class of isomers is the stereoisomers. Stereoisomers have the same atoms arranged in the same nucleus-to-nucleus sequence but differ in the spatial arrangement of their atoms. For example, an amino acid such as alanine can form stereoisomers (figure AI.8). L-Alanine and other L-amino acids are the stereoisomer forms normally present in proteins.

Figure AI.4   **Hydrogen Bonds.** Representative examples of hydrogen bonds present in biological molecules.

Carbohydrates

Carbohydrates are aldehyde or ketone derivatives of polyhydroxy alcohols. The smallest and least complex carbohydrates are the simple sugars or monosaccharides. The most common sugars have five or six carbons (figure AI.9). A sugar in its ring form has two isomeric structures, the α and β forms, which differ in the orientation of the hydroxyl on the aldehyde or ketone carbon, the anomeric or glycosidic carbon (figure AI.10). Microorganisms have many sugar derivatives in which a hydroxyl is replaced by an amino group or some other functional group (e.g., glucosamine).

Two monosaccharides can be joined by a bond between the anomeric carbon of one sugar and a hydroxyl or the anomeric carbon of the second (figure AI.11). The bond joining sugars is a glycosidic bond and may be either α or β depending on the orientation of the anomeric carbon. Two sugars linked in this way constitute a disaccharide. Some common disaccharides are maltose (two glucose molecules), lactose (glucose and galactose), and sucrose (glucose and fructose). If 10 or more sugars are linked together by glycosidic bonds, a polysaccharide is formed. For example, starch and glycogen are common polymers of glucose that are used as sources of carbon and energy (figure AI.12).

C₆H₁₄ (Hexane)

(a)

C₆H₁₂ (Cyclohexane)

(b)

C₆H₆ (Benzene)

(c)

Figure AI.5   **Hydrocarbons.** Examples of hydrocarbons that are (*a*) linear, (*b*) cyclic, and (*c*) aromatic.

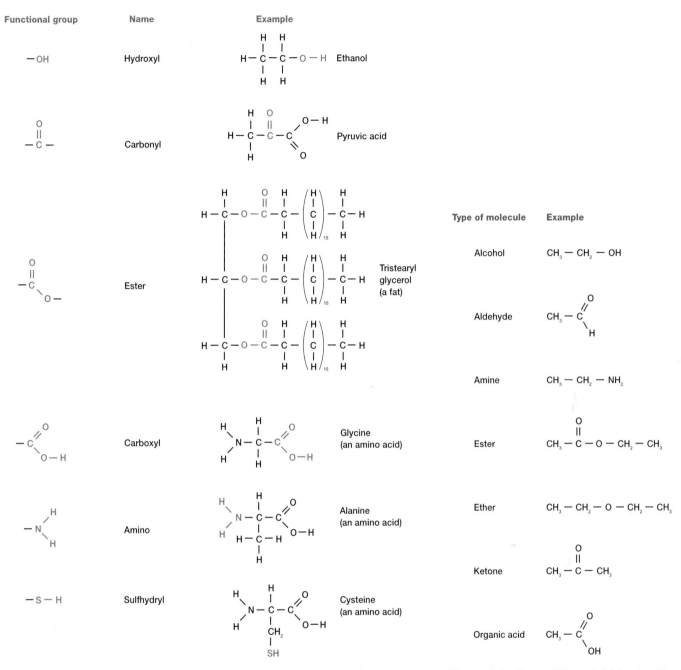

Figure AI.6 Functional Groups. Some common functional groups in organic molecules. The groups are shown in color.

Figure AI.7 Types of Organic Molecules. These are classified on the basis of their functional groups.

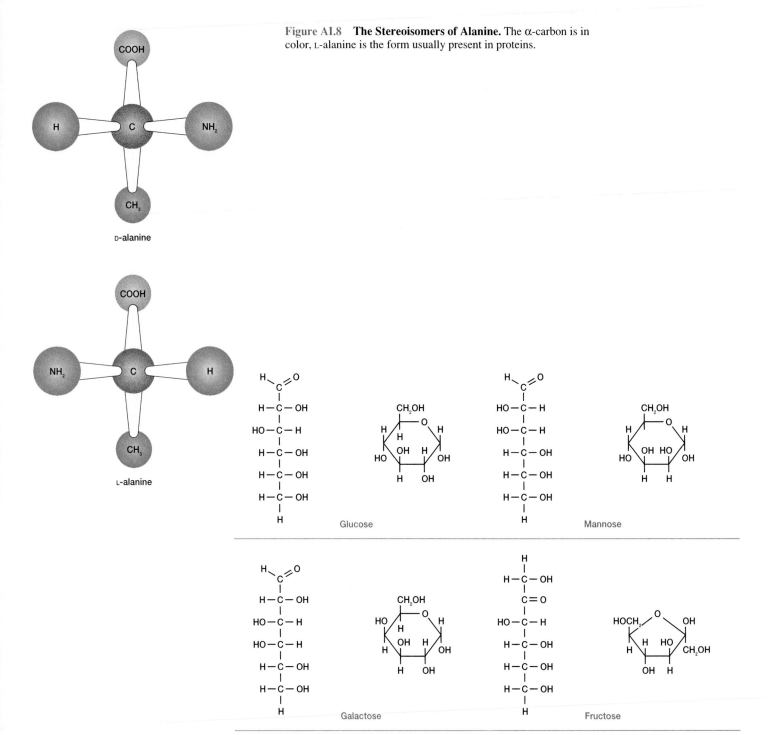

Figure AI.8 **The Stereoisomers of Alanine.** The α-carbon is in color, L-alanine is the form usually present in proteins.

Figure AI.9 **Common Monosaccharides.** Structural formulas for both the open chains and the ring forms are provided.

β-D-glucose   α-D-glucose

Figure AI.10   The Interconversion of Monosaccharide Structures. The open chain form of glucose and other sugars is in equilibrium with closed ring structures (depicted here with Haworth projections). Aldehyde sugars form cyclic hemiacetals, and keto sugars produce cyclic hemiketals. When the hydroxyl on carbon one of cyclic hemiacetals projects above the ring, the form is known as a β form. The α form has a hydroxyl that lies below the plane of the ring. The same convention is used in showing the α and β forms of hemiketals such as those formed by fructose.

(a)   Glucose   Glucose   Maltose

(b)   Sucrose

(c)   Lactose (α form)

Figure AI.11   Common Disaccharides. (*a*) The formation of maltose from two molecules of a α-glucose. The bond connecting the glucose extends between carbons one and four, and involves the α form of the anomeric carbon. Therefore, it is called an α (1 → 4) glycosidic bond. (*b*) Sucrose is composed of a glucose and a fructose joined to each other through their anomeric carbons, an αβ (1 → 2) bond. (*c*) The milk sugar lactose contains galactose and glucose joined by a β (1 → 4) glycosidic bond.

Figure AI.12   Glycogen and Starch Structure. (*a*) An overall view of the highly branched structure characteristic of glycogen and most starch. The circles represent glucose residues. (*b*) A close-up of a small part of the chain (shown in color in part *a*) revealing a branch point with its α (1 → 6) glycosidic bond.

Lipids

All cells contain a heterogeneous mixture of organic molecules that are relatively insoluble in water but very soluble in nonpolar solvents such as chloroform, ether, and benzene. These molecules are called lipids. Lipids vary greatly in structure and include triacylglycerols, phospholipids, steroids, carotenoids, and many other types. Among other functions, they serve as membrane components, storage forms for carbon and energy, precursors of other cell constituents, and protective barriers against water loss.

Most lipids contain fatty acids, monocarboxylic acids that often are straight chained but may be branched. Saturated fatty acids lack double bonds in their carbon chains, whereas unsaturated fatty acids have double bonds. The most common fatty acids are 16 or 18 carbons long.

Two good examples of common lipids are triacylglycerols and phospholipids. Triacylglycerols are composed of glycerol esterified to three fatty acids (figure AI.13*a*). They are used to store carbon and energy. Phospholipids are lipids that contain at least one phosphate group and often have a nitrogenous constituent as well. Phosphatidyl ethanolamine is an important phospholipid frequently present in bacterial membranes (figure AI.13*b*). It is composed of two fatty acids esterified to glycerol. The third glycerol hydroxyl is joined with a phosphate group, and ethanolamine is attached to the phosphate. The resulting lipid is very asymmetric with a hydrophobic nonpolar end contributed by the fatty acids and a polar, hydrophilic end. In cell membranes the hydrophobic end is buried in the interior of the membrane, while the polar-charged end is at the membrane surface and exposed to water.

O
||
CH₂ — O — C — R
| O
| ||
CH — O — C — R
| O
| ||
CH₂ — O — C — R

(a)

O
||
CH₂ — O — C — R
| O
| ||
CH — O — C — R
| O
| ||
CH₂ — O — C — O — CH₂ — CH₂ — NH₃
(b) | +
         O⁻

Figure AI.13   Examples of Common Lipids. (*a*) A triacylglycerol or neutral fat. (*b*) The phospholipid phosphatidyl ethanolamine. The R groups represent fatty acid side chains.

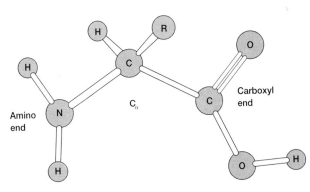

Figure AI.14   L-Amino Acid Structure. The uncharged form is shown.

Proteins

The basic building blocks of proteins are amino acids. An amino acid contains a carboxyl group and an amino group on its alpha carbon (figure AI.14). About 20 amino acids are normally found in proteins; they differ from each other with respect to their side chains (figure AI.15). In proteins, amino acids are linked together by peptide bonds between their carboxyls and α-amino groups to form linear polymers called peptides (figure AI.16). If a peptide contains more than 10 amino acids, it usually is called a polypeptide. Each protein is composed of one or more polypeptide chains and has a molecular weight greater than about 6,000 to 7,000.

Proteins have three or four levels of structural organization and complexity. The primary structure of a protein is the sequence of the amino acids in its polypeptide chain or chains. The structure of the polypeptide chain backbone is also considered part of the primary structure. Each different polypeptide has its own amino acid sequence that is a reflection of the nucleotide sequence in the gene that codes for its synthesis. The polypeptide chain can coil along one axis in space into various shapes like the α-helix (figure AI.17). This arrangement of the polypeptide in space around a single axis is called the secondary structure. Secondary structure is formed and stabilized by the interactions of amino acids that are fairly close to one another on the polypeptide chain. The polypeptide with its primary and secondary structure can be coiled or organized in space along three axes to form a more complex, three-dimensional shape (figure AI.18). This level of organization is the tertiary structure (figure AI.19). Amino acids more distant from one another on the polypeptide chain

contribute to tertiary structure. Secondary and tertiary structures are examples of conformation, molecular shape that can be changed by bond rotation and without breaking covalent bonds. When a protein contains more than one polypeptide chain, each chain with its own primary, secondary, and tertiary structure associates with the other chains to form the final molecule. The way in which polypeptides associate with each other in space to form the final protein is called the protein's quaternary structure (figure AI.20).

The final conformation of a protein is ultimately determined by the amino acid sequence of its polypeptide chains. Under proper conditions a completely unfolded polypeptide will fold into its normal final shape without assistance.

Protein secondary, tertiary, and quaternary structure is largely determined and stabilized by many weak noncovalent forces such as hydrogen bonds and ionic bonds. Because of this, protein shape often is very flexible and easily changed. This flexibility is very important in protein function and in the regulation of enzyme activity. Because of their flexibility, however, proteins readily lose their proper shape and activity when exposed to harsh conditions. The only covalent bond commonly involved in the secondary and tertiary structure of proteins is the disulfide bond. The disulfide bond is formed when two cysteines are linked through their sulfhydryl groups. Disulfide bonds generally strengthen or stabilize protein structure but are not especially important in directly determining protein conformation.

Nucleic Acids

The nucleic acids, deoxyribonucleic acid (DNA) and ribonucleic acid (RNA), are polymers of deoxyribonucleosides and ribonucleosides joined by phosphate groups. The nucleosides in DNA contain the purines adenine and guanine, and the pyrimidine bases thymine and cytosine. In RNA the pyrimidine uracil is substituted for thymine. Because of their importance for genetics and molecular biology, the chemistry of nucleic acids is introduced earlier in the text. The structure and synthesis of purines and pyrimidines are discussed in chapter 10 (*pp. 203–4*). The structures of DNA and RNA are described in chapter 11 (*pp. 212–14*).

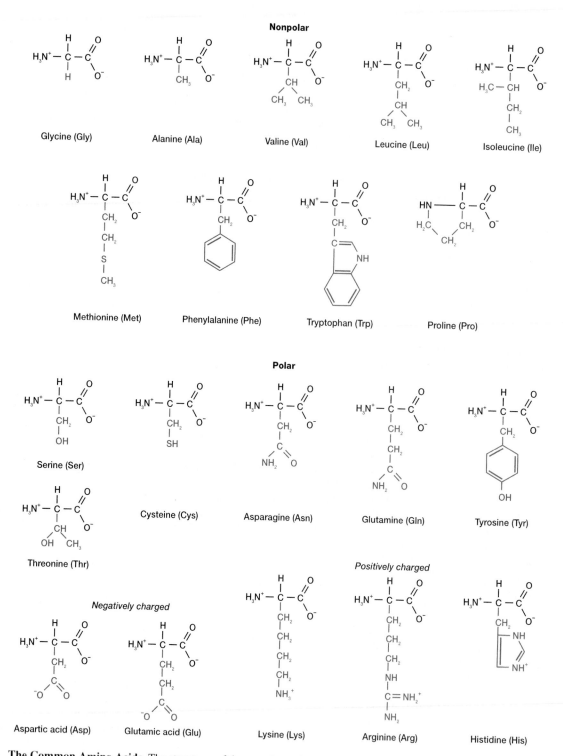

Figure AI.15   The Common Amino Acids. The structures of the α-amino acids normally found in proteins. Their side chains are shown in color, and they are grouped together based on the nature of their side chains—nonpolar, polar, negatively charged (acid), or positively charged (basic). Proline is actually an imino acid rather than an amino acid.

Figure AI.16   A Tetrapeptide Chain. The end of the chain with a free α-amino group is the amino or N terminal. The end with the free α-carboxyl is the carboxyl or C terminal. One peptide bond is shown in color.

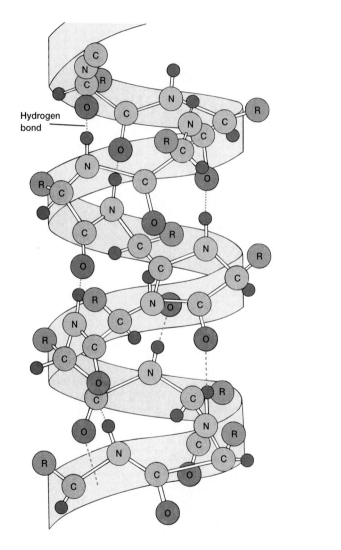

Figure AI.17   The α-Helix. A polypeptide twisted into one type of secondary structure, the α-helix. The helix is stabilized by hydrogen bonds joining peptide bonds that are separated by three amino acids.

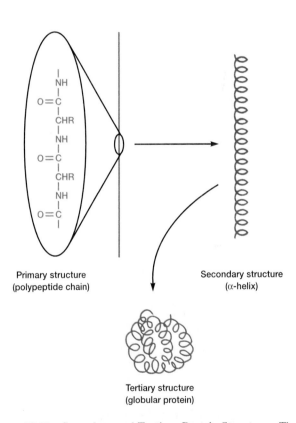

Figure AI.18   Secondary and Tertiary Protein Structures. The formation of secondary and tertiary protein structures by folding a polypeptide chain with its primary structure.

(a)

(b)

Figure AI.19   Lysozyme. The tertiary structure of the enzyme lysozyme. (*a*) A diagram of the protein's polypeptide backbone with the substrate hexasaccharide shown in color. The point of substrate cleavage is indicated. (*b*) A space-filling model of lysozyme. The left figure shows the empty active site with some of its more important amino acids indicated. On the right the enzyme has bound its substrate.

(a)

(b)

CTP site

(c)

Figure AI.20   An Example of Quaternary Structure. The enzyme aspartate carbamoyltransferase from *E. coli* has two types of subunits, catalytic and regulatory. The association between the two types of subunits is shown: (*a*) a top view, and (*b*) a side view of the enzyme. The catalytic (C) and regulatory (r) subunits are shown in different colors. (*c*) The peptide chains shown when viewed from the top as in (*a*). The active sites of the enzyme are located at the positions indicated by A. *(See pp. 238–40 for more details.)* (*a* and *b*) Adapted from Krause, et al., in *Proceedings of the National Academy of Sciences,* V. 82, 1985, as appeared in *Biochemistry,* 3rd edition by Lubert Stryer. Copyright © 1975, 1981, 1988. Reprinted with permission of W.H. Freeman and Company. (*c*) Adapted from Kantrowitz, et al., in *Trends in Biochemical Science*, V. 5, 1980, as appeared in *Biochemistry,* 3rd edition by Lubert Stryer. Copyright © 1975, 1981, 1988. Reprinted with permission of W.H. Freeman and Company.

APPENDIX II
Comparative Sizes of Molecules, Viruses, and Procaryotic and Eucaryotic Cells

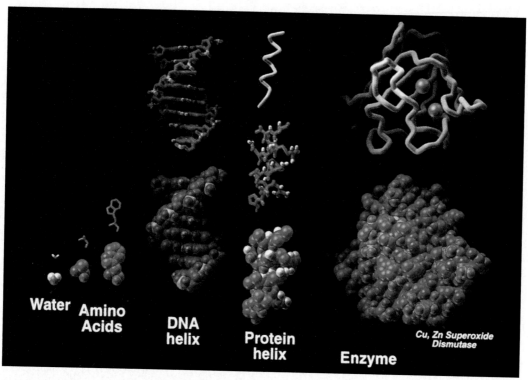

Figure AII.1 Water, Amino Acids, DNA, and Protein Structure. At the left is a water molecule with its single oxygen (red) and two hydrogen atoms (white). Water is a small molecule, a little less than 0.5 nanometer (nm) across. This molecule and the other molecules in this illustration are pictured in two ways. The upper illustrations show the chemical bonds of smaller molecules and the molecular backbones of larger molecules. The lower illustrations picture each structure as it appears when all atoms are drawn as space-filling spheres. Next to water are two common amino acids. Alanine is one of the smallest amino acids and has only 5 nonhydrogen atoms; tryptophan, with its 14 nonhydrogen atoms, is the largest common amino acid. The colors represent different atoms: green (carbon), blue (nitrogen), and red (oxygen). The next two models represent the DNA double helix (about 2 nm in diameter) and the α-helix, a type of secondary structure found in most proteins (approximately 1 nm across). A complete protein, superoxide dismutase, is included in this figure for reference and will be discussed in the next illustration.

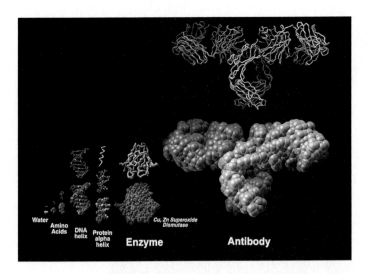

Figure AII.2   Proteins. This illustration shows two proteins, superoxide dismutase and immunoglobulin G, along with smaller molecules to provide perspective. Proteins are formed of one or more polypeptide chains, which already have primary and secondary structure as discussed in appendix I. Superoxide dismutase (SOD) is an example of a relatively small to middle-sized protein, about 6 nm across, which is made of a single polypeptide chain. It contains about 150 amino acids and two metal ions, zinc and copper, which are represented by the colored spheres in the backbone model. SOD catalyzes the destruction of the toxic superoxide radical and thus protects organisms from this toxic oxygen product. It also is involved in the destruction of pathogens by leukocytes (*see pp. 131 and 582*). The antibody immunoglobulin G is an excellent example of a large protein composed of more than one polypeptide chain. This molecule is approximately 16 nm across and contains four chains: two light chains (about 220 amino acids each) and two heavy chains (about 440 amino acids each). The light chains are shown in red or pink and the heavy chains are in blue and gray. Each sphere in the space-filling model represents an amino acid. Immunoglobulin structure is discussed in detail on pages 591–94.

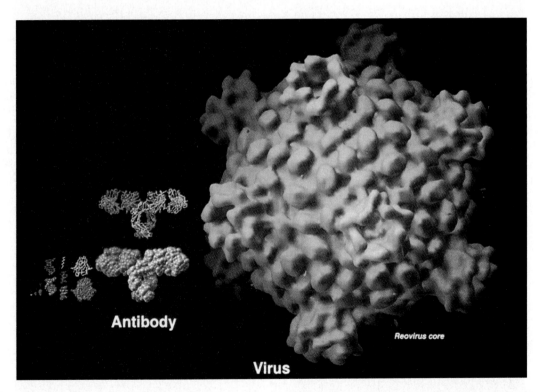

Figure AII.3   Proteins and Viruses. In this illustration the two proteins described in Figure AII.2 are compared with part of a virus capsid. Icosahedral virus capsids dwarf even the largest single protein and usually are composed of two different kinds of capsomers: hexamers and pentamers (*see pp. 341–43*). This illustration shows a reovirus core particle, an icosahedron about 60 nm or 0.06 μm across. Reoviruses are distinctive in containing 10–12 pieces of double-stranded RNA as their genetic material. The complete virus consists of two concentric icosahedral capsids surrounding a core particle, which contains the dsRNA. After the virus has entered the cell and its outer layers have been removed, newly transcribed mRNA is extruded into the cytoplasm through channels at the core particle's vertices. Sometimes orthoreoviruses will cause human respiratory and gastrointestinal tract infections.

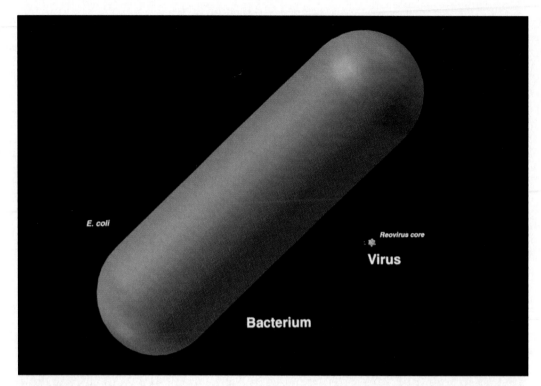

Figure AII.4   A Bacterial Cell. The bacterium *Escherichia coli,* a common inhabitant of the human gastrointestinal tract, is pictured here. *E. coli* is a gram-negative straight rod, 1.1–1.5 µm by 2.0–6.0 µm in size. Its length in this illustration is about 3,000 nm or 3 µm; by comparison, the reovirus core (60 nm) is barely visible to the right of the bacterium. The difference in size between bacteria and viruses is not always this great. The largest viruses are the poxviruses, which may reach about 400 × 240 × 200 nm in size. Mycoplasmas, probably the smallest independently living bacteria, can be as small as 0.3 µm or 300 nm in diameter.

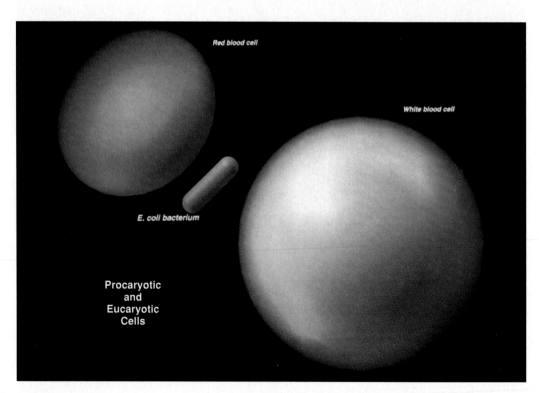

Figure AII.5   Eucaryotic Cells. Eucaryotic cells are almost always larger than procaryotic cells such as *E. coli.* In this illustration *E. coli* is surrounded by highly stylized representations of two typical human cells, a small white blood cell or leukocyte (about 10 µm or 10,000 nm in diameter) and an average size red blood cell or erythrocyte (7 µm in diameter). Three kinds of leukocytes are very important in the phagocytosis of pathogens: neutrophils (around 10–12 µm in diameter), monocytes (16–22 µm), and macrophages (15–25 µm). Clearly all three types are large enough to phagocytose and destroy an invading bacterium the size of *E. coli* (*see pp. 576–78, 582–83*).

APPENDIX III
Common Metabolic Pathways

This appendix contains a few of the more important pathways discussed in the text, particularly those involved in carbohydrate catabolism. Enzyme names and final end products are given in color. Consult the text for a description of each pathway and its roles.

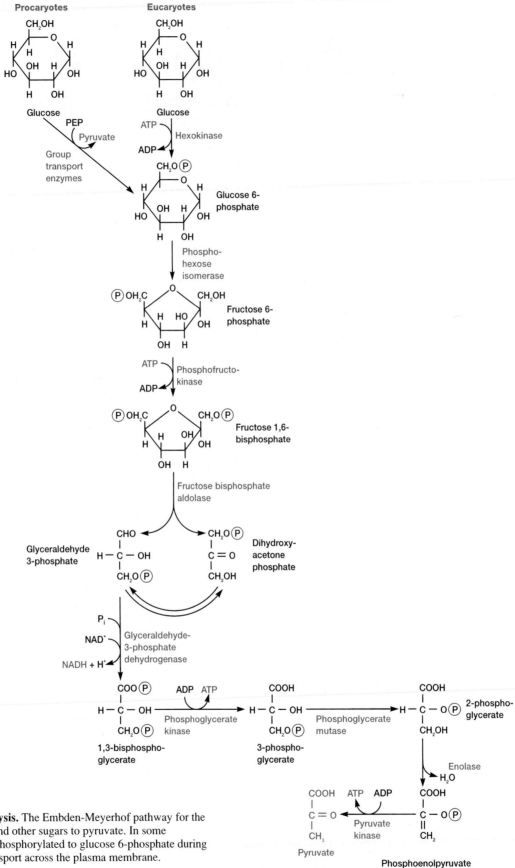

Figure AIII.1   Glycolysis. The Embden-Meyerhof pathway for the conversion of glucose and other sugars to pyruvate. In some procaryotes glucose is phosphorylated to glucose 6-phosphate during group translocation transport across the plasma membrane.

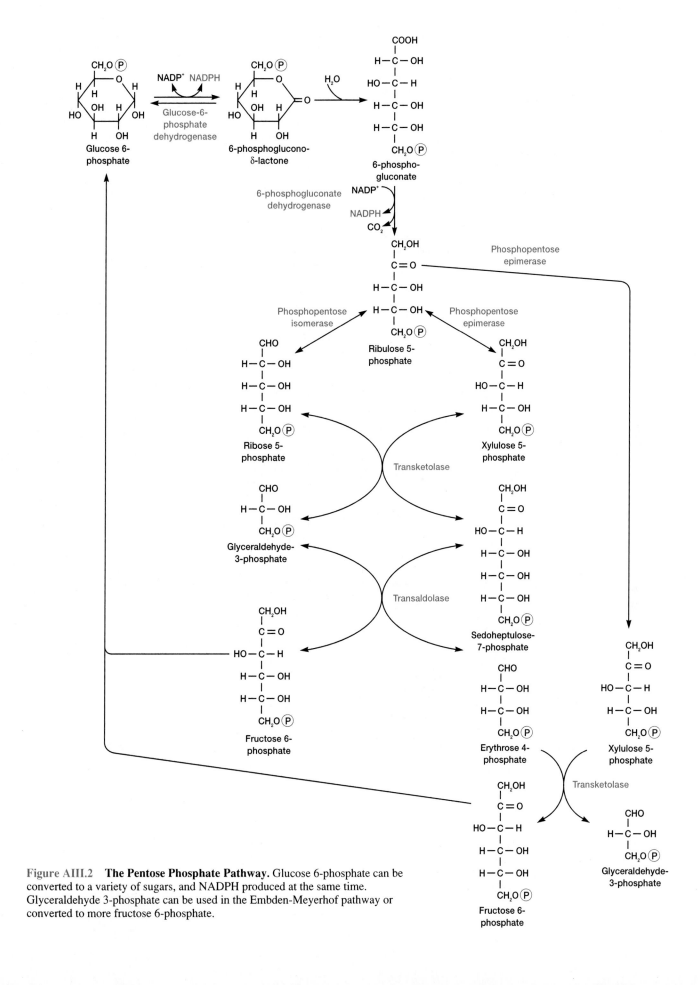

Figure AIII.2 The Pentose Phosphate Pathway. Glucose 6-phosphate can be converted to a variety of sugars, and NADPH produced at the same time. Glyceraldehyde 3-phosphate can be used in the Embden-Meyerhof pathway or converted to more fructose 6-phosphate.

Figure AIII.3   The Entner-Doudoroff Pathway.

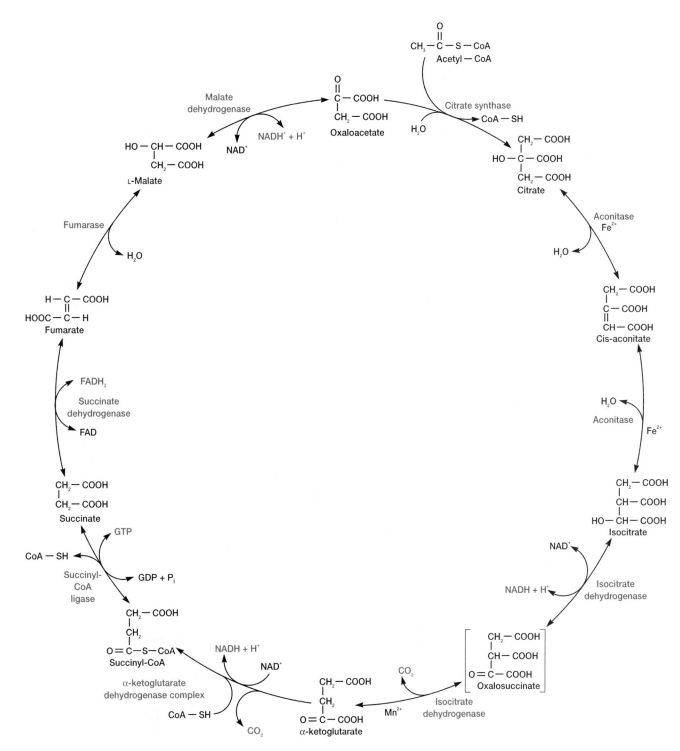

Figure AIII.4   The Tricarboxylic Acid Cycle. Cis-aconitate and oxalosuccinate remain bound to aconitase and isocitrate dehydrogenase. Oxalosuccinate has been placed in brackets since it is so unstable.

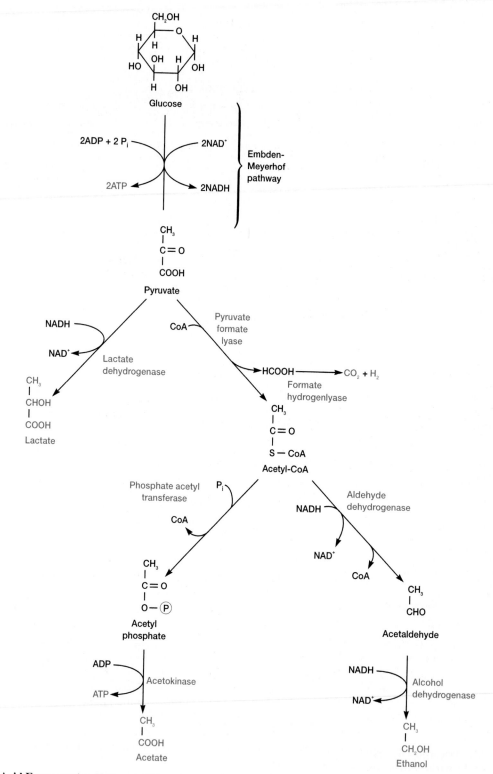

Figure AIII.5   The Mixed Acid Fermentation Pathway. This pathway is characteristic of many members of the *Enterobacteriaceae* such as *E. coli*.

Figure AIII.6   The Butanediol Fermentation Pathway. This pathway is characteristic of members of the *Enterobacteriaceae* such as *Enterobacter*. Other products may also be formed during butanediol fermentation.

Figure AIII.7 Lactic Acid Fermentations. (*a*) Homolactic fermentation pathway. (*b*) Heterolactic fermentation pathway.

(a)

(b)

Figure AIII.8   The Pathway for Purine Biosynthesis. Inosinic acid is the first purine end product. The purine skeleton is constructed while attached to a ribose phosphate.

APPENDIX IV
Classification of Bacteria According to *Bergey's Manual of Systematic Bacteriology*

Great progress has occurred in bacterial taxonomy since *Bergey's Manual of Systematic Bacteriology* was published between 1984 and 1989. Many completely new taxa have been added, and large, complex genera such as *Pseudomonas, Streptococcus,* and *Bacillus* have been subdivided to form new genera. Thus the classification presented here differs from that given in the 1994 edition of *Bergey's Manual of Determinative Bacteriology,* and even this more recent work is already out-of-date in some respects. Appendix IV summarizes the system presented in *Bergey's Manual of Systematic Bacteriology* and does not represent a completely current bacterial classification.

Volume I*

Section 1
The Spirochetes

Order I *Spirochaetales*
  Family I *Spirochaetaceae*
    Genus I *Spirochaeta*
    Genus II *Cristispira*
    Genus III *Treponema*
    Genus IV *Borrelia*
  Family II *Leptospiraceae*
    Genus I *Leptospira*
Other Organisms
  Hindgut Spirochetes of Termites and
    Cryptocercus punctulatus

Section 2
Aerobic/Microaerophilic, Motile, Helical/Vibrioid Gram-Negative Bacteria

  Genus *Aquaspirillum*
  Genus *Spirillum*
  Genus *Azospirillum*
  Genus *Oceanospirillum*
  Genus *Campylobacter*
  Genus *Bdellovibrio*
  Genus *Vampirovibrio*

Section 3
Nonmotile (or Rarely Motile), Gram-Negative Curved Bacteria

  Family I *Spirosomaceae*
    Genus I *Spirosoma*
    Genus II *Runella*
    Genus III *Flectobacillus*
  Other Genera
    Genus *Microcyclus*
    Genus *Meniscus*
    Genus *Brachyarcus*
    Genus *Pelosigma*

Section 4
Gram-Negative Aerobic Rods and Cocci

  Family I *Pseudomonadaceae*
    Genus I *Pseudomonas*
    Genus II *Xanthomonas*
    Genus III *Frateuria*
    Genus IV *Zoogloea*
  Family II *Azotobacteraceae*
    Genus I *Azotobacter*
    Genus II *Azomonas*
  Family III *Rhizobiaceae*
    Genus I *Rhizobium*
    Genus II *Bradyrhizobium*
    Genus III *Agrobacterium*
    Genus IV *Phyllobacterium*
  Family IV *Methylococcaceae*
    Genus I *Methylococcus*
    Genus II *Methylomonas*
  Family V *Halobacteriaceae*
    Genus I *Halobacterium*
    Genus II *Halococcus*
  Family VI *Acetobacteraceae*
    Genus I *Acetobacter*
    Genus II *Gluconobacter*
  Family VII *Legionellaceae*
    Genus I *Legionella*
  Family VIII *Neisseriaceae*
    Genus I *Neisseria*
    Genus II *Moraxella*
    Genus III *Acinetobacter*
    Genus IV *Kingella*
  Other Genera
    Genus *Beijerinckia*
    Genus *Derxia*
    Genus *Xanthobacter*
    Genus *Thermus*
    Genus *Thermomicrobium*
    Genus *Halomonas*
    Genus *Alteromonas*
  Genus *Flavobacterium*
  Genus *Alcaligenes*
  Genus *Serpens*
  Genus *Janthinobacterium*
  Genus *Brucella*
  Genus *Bordetella*
  Genus *Francisella*
  Genus *Paracoccus*
  Genus *Lampropedia*

Section 5
Facultatively Anaerobic Gram-Negative Rods

  Family I *Enterobacteriaceae*
    Genus I *Escherichia*
    Genus II *Shigella*
    Genus III *Salmonella*
    Genus IV *Citrobacter*
    Genus V *Klebsiella*
    Genus VI *Enterobacter*
    Genus VII *Erwinia*
    Genus VIII *Serratia*
    Genus IX *Hafnia*
    Genus X *Edwardsiella*
    Genus XI *Proteus*
    Genus XII *Providencia*
    Genus XIII *Morganella*
    Genus XIV *Yersina*
  Other Genera of the Family
    Enterobacteriaceae
    Genus *Obesumbacterium*
    Genus *Xenorhabdus*
    Genus *Kluyvera*
    Genus *Rahnella*
    Genus *Cedecea*
    Genus *Tatumella*
  Family II *Vibrionaceae*
    Genus I *Vibrio*
    Genus II *Photobacterium*
    Genus III *Aeromonas*
    Genus IV *Plesiomonas*

*Appendix IV is modified from *Bergey's Manual of Systematic Bacteriology,* edited by J.G. Holt and N.R. Krieg, 1984–1989. Copyright © Williams and Wilkins Company, Baltimore, MD. Reprinted by permission.

Family III *Pasteurellaceae*
 Genus I *Pasteurella*
 Genus II *Haemophilus*
 Genus III *Actinobacillus*
Other Genera
 Genus *Zymomonas*
 Genus *Chromobacterium*
 Genus *Cardiobacterium*
 Genus *Calymmatobacterium*
 Genus *Gardnerella*
 Genus *Eikenella*
 Genus *Streptobacillus*

Section 6
Anaerobic Gram-Negative Straight, Curved, and Helical Rods

Family I *Bacteroidaceae*
 Genus I *Bacteroides*
 Genus II *Fusobacterium*
 Genus III *Leptotrichia*
 Genus IV *Butyrivibrio*
 Genus V *Succinimonas*
 Genus VI *Succinivibrio*
 Genus VII *Anaerobiospirillum*
 Genus VIII *Wolinella*
 Genus IX *Selenomonas*
 Genus X *Anaerovibrio*
 Genus XI *Pectinatus*
 Genus XII *Acetivibrio*
 Genus XIII *Lachnospira*

Section 7
Dissimilatory Sulfate- or Sulfur-Reducing Bacteria

 Genus *Desulfuromonas*
 Genus *Desulfovibrio*
 Genus *Desulfomonas*
 Genus *Desulfococcus*
 Genus *Desulfobacter*
 Genus *Desulfobulbus*
 Genus *Desulfosarcina*

Section 8
Anaerobic Gram-Negative Cocci

Family I *Veillonellaceae*
 Genus I *Veillonella*
 Genus II *Acidaminococcus*
 Genus III *Megasphaera*

Section 9
The Rickettsias and Chlamydias

Order I *Rickettsiales*
 Family I *Rickettsiaceae*
 Tribe I *Rickettsieae*
 Genus I *Rickettsia*
 Genus II *Rochalimaea*
 Genus III *Coxiella*
 Tribe II *Ehrlichieae*
 Genus IV *Ehrlichia*
 Genus V *Cowdria*
 Genus VI *Neorickettsia*
 Tribe III *Wolbachieae*
 Genus VII *Wolbachia*
 Genus VIII *Rickettsiella*

Family II *Bartonellaceae*
 Genus I *Bartonella*
 Genus II *Grahamella*
 Family III *Anaplasmataceae*
 Genus I *Anaplasma*
 Genus II *Aegyptianella*
 Genus III *Haemobartonella*
 Genus IV *Eperythrozoon*
Order II *Chlamydiales*
 Family I *Chlamydiaceae*
 Genus I *Chlamydia*

Section 10
The Mycoplasmas
Division *Tenericutes*
Class I *Mollicutes*

Order I *Mycoplasmatales*
 Family I *Mycoplasmataceae*
 Genus I *Mycoplasma*
 Genus II *Ureaplasma*
 Family II *Acholeplasmataceae*
 Genus I *Acholeplasma*
 Family III *Spiroplasmataceae*
 Genus I *Spiroplasma*
 Other Genera
 Genus *Anaeroplasma*
 Genus *Thermoplasma*
 Mycoplasma-like Organisms of Plants
  and Invertebrates

Section 11
Endosymbionts

A. Endosymbionts of Protozoa
 Endosymbionts of ciliates
 Endosymbionts of flagellates
 Endosymbionts of amoebas
 Taxa of endosymbionts
 Genus I *Holospora*
 Genus II *Caedibacter*
 Genus III *Pseudocaedibacter*
 Genus IV *Lyticum*
 Genus V *Tectibacter*
B. Endosymbionts of Insects
 Bloodsucking insects
 Plant sap-sucking insects
 Cellulose and stored grain feeders
 Insects feeding on complex diets
 Taxon of endosymbionts:
 Genus *Blattabacterium*
C. Endosymbionts of Fungi and Invertebrates
  Other Than Arthropods
 Fungi
 Sponges
 Coelenterates
 Helminthes
 Annelids
 Marine worms and mollusks

Volume II

Section 12
Gram-Positive Cocci

Family I *Micrococcaceae*
 Genus I *Micrococcus*
 Genus II *Stomatococcus*

Genus III *Planococcus*
 Genus IV *Staphylococcus*
Family II *Deinococcaceae*
 Genus I *Deinococcus*
Other Genera
 Genus *Streptococcus*
 Pyogenic Hemolytic Streptococci
 Oral Streptococci
 Enterococci
 Lactic Acid Streptococci
 Anaerobic Streptococci
 Other Streptococci
 Genus *Leuconostoc*
 Genus *Pediococcus*
 Genus *Aerococcus*
 Genus *Gemella*
 Genus *Peptococcus*
 Genus *Peptostreptococcus*
 Genus *Ruminococcus*
 Genus *Coprococcus*
 Genus *Sarcina*

Section 13
Endospore-Forming Gram-Positive Rods and Cocci

 Genus *Bacillus*
 Genus *Sporolactobacillus*
 Genus *Clostridium*
 Genus *Desulfotomaculum*
 Genus *Sporosarcina*
 Genus *Oscillospira*

Section 14
Regular, Nonsporing, Gram-Positive Rods

 Genus *Lactobacillus*
 Genus *Listeria*
 Genus *Erysipelothrix*
 Genus *Brochothrix*
 Genus *Renibacterium*
 Genus *Kurthia*
 Genus *Caryophanon*

Section 15
Irregular, Nonsporing, Gram-Positive Rods

 Genus *Corynebacterium*
 Plant Pathogenic Species of
  Corynebacterium
 Genus *Gardnerella*
 Genus *Arcanobacterium*
 Genus *Arthrobacter*
 Genus *Brevibacterium*
 Genus *Curtobacterium*
 Genus *Caseobacter*
 Genus *Microbacterium*
 Genus *Aureobacterium*
 Genus *Cellulomonas*
 Genus *Agromyces*
 Genus *Arachnia*
 Genus *Rothia*
 Genus *Propionibacterium*
 Genus *Eubacterium*
 Genus *Acetobacterium*
 Genus *Lachnospira*
 Genus *Butyrivibrio*
 Genus *Thermoanaerobacter*

Genus *Actinomyces*
Genus *Bifidobacterium*

Section 16

The Mycobacteria

Family *Mycobacteriaceae*
Genus *Mycobacterium*

Section 17

Nocardioforms

Genus *Nocardia*
Genus *Rhodococcus*
Genus *Nocardioides*
Genus *Pseudonocardia*
Genus *Oerskovia*
Genus *Saccharopolyspora*
Genus *Micropolyspora*
Genus *Promicromonospora*
Genus *Intrasporangium*

Volume III

Section 18

Anoxygenic Phototrophic Bacteria

I. Purple Bacteria

Family I *Chromatiaceae*
Genus I *Chromatium*
Genus II *Thiocystis*
Genus III *Thiospirillum*
Genus IV *Thiocapsa*
Genus V *Lamprobacter*
Genus VI *Lamprocystis*
Genus VII *Thiodictyon*
Genus VIII *Amoebobacter*
Genus IX *Thiopedia*
Family II *Ectothiorhodospiraceae*
Genus *Ectothiorhodospira*
Purple Nonsulfur Bacteria
Genus *Rhodospirillum*
Genus *Rhodopila*
Genus *Rhodobacter*
Genus *Rhodopseudomonas*
Genus *Rhodomicrobium*
Genus *Rhodocyclus*

II. Green Bacteria

Green Sulfur Bacteria
Genus *Chlorobium*
Genus *Prosthecochloris*
Genus *Pelodictyon*
Genus *Ancalochloris*
Genus *Chloroherpeton*
Multicellular, Filamentous, Green Bacteria
Genus *Chloroflexus*
Genus *Heliothrix*
Genus *"Oscillochloris"*
Genus *Chloronema*

III. Genera Incertae Sedis

Genus *Heliobacterium*
Genus *Erythrobacter*

Section 19

Oxygenic Photosynthetic Bacteria

Group I Cyanobacteria

Subsection I Order *Chroococcales*
1. Genus I *Chamaesiphon*
2. Genus II *Gloeobacter*
3.   *Synechococcus*-group
4. Genus III *Gloeothece*
5.   *Cyanothece*-group
6.   *Gloeocapsa*-group
7.   *Synechocystis*-group
Subsection II Order *Pleurocapsales*
1. Genus I *Dermocarpa*
2. Genus II *Xenococcus*
3. Genus III *Dermocarpella*
4. Genus IV *Mxyosarcina*
5. Genus V *Chroococcidiopsis*
6.   *Pleurocapsa*-group
Subsection III Order *Oscillatoriales*
Genus I *Spirulina*
Genus II *Arthrospira*
Genus III *Oscillatoria*
Genus IV *Lyngbya*
Genus V *Pseudanabaena*
Genus VI *Starria*
Genus VII *Crinalium*
Genus VIII *Microcoleus*
Subsection IV Order *Nostocales*
Family I *Nostocaceae*
Genus I *Anabaena*
Genus II *Aphanizomenon*
Genus III *Nodularia*
Genus IV *Cylindrospermum*
Genus V *Nostoc*
Family II *Scytonemataceae*
Genus I *Scytonema*
Family III *Rivulariaceae*
Genus I *Calothrix*
Subsection V Order *Stigonematales*
Genus I *Chlorogloeopsis*
Genus II *Fischerella*
Genus III *Stigonema*
Genus IV *Geitleria*

Group II Order *Prochlorales*

Family I *Prochloraceae*
Genus *Prochloron*
Other taxa
Genus *"Prochlorothrix"*

Section 20

Aerobic Chemolithotrophic Bacteria and Associated Organisms

A. Nitrifying Bacteria
Family *Nitrobacteraceae*
Nitrite-oxidizing bacteria
Genus I *Nitrobacter*
Genus II *Nitrospina*
Genus III *Nitrococcus*
Genus IV *Nitrospira*
Ammonia-oxidizing bacteria
Genus V *Nitrosomonas*
Genus VI *Nitrosococcus*
Genus VII *Nitrosospira*
Genus VIII *Nitrosolobus*
Genus IX *"Nitrosovibrio"*
B. Colorless Sulfur Bacteria
Genus *Thiobacterium*

Genus *Macromonas*
Genus *Thiospira*
Genus *Thiovulum*
Genus *Thiobacillus*
Genus *Thiomicrospira*
Genus *Thiosphaera*
Genus *Acidiphilium*
Genus *Thermothrix*
C. Obligately Chemolithotrophic Hydrogen Bacteria
Genus *Hydrogenobacter*
D. Iron- and Manganese-Oxidizing and/or Depositing Bacteria
Family *"Siderocapsaceae"*
Genus I *"Siderocapsa"*
Genus II *"Naumanniella"*
Genus III *"Siderococcus"*
Genus IV *"Ochrobium"*
E. Magnetotactic Bacteria
Genus *Aquaspirillum (A. magnetotacticum)*
Genus *"Bilophococcus"*

Section 21

Budding and/or Appendaged Bacteria

I. Prosthecate Bacteria

A. Budding bacteria
1. Buds produced at tip of prostheca
Genus *Hyphomicrobium*
Genus *Hyphomonas*
Genus *Pedomicrobium*
2. Buds produced on cell surface
Genus *Ancalomicrobium*
Genus *Prosthecomicrobium*
Genus *Labrys*
Genus *Stella*
B. Bacteria that divide by binary transverse fission
Genus *Caulobacter*
Genus *Asticcacaulis*
Genus *Prosthecobacter*

II. Nonprosthecate Bacteria

A. Budding bacteria
1. Lack peptidoglycan
Genus *Planctomyces*
Genus *"Isosphaera"*
2. Contain peptidoglycan
Genus *Ensifer*
Genus *Blastobacter*
Genus *Angulomicrobium*
Genus *Gemmiger*
B. Nonbudding, stalked bacteria
Genus *Gallionella*
Genus *Nevskia*
C. Other bacteria
1. Nonspinate bacteria
Genus *Seliberia*
Genus *"Metallogenium"*
Genus *"Thiodendron"*
2. Spinate bacteria

Section 22

Sheathed Bacteria

Genus *Sphaerotilus*

Genus *Leptothrix*
Genus *Haliscomenobacter*
Genus "*Lieskeella*"
Genus "*Phragmidiothrix*"
Genus *Crenothrix*
Genus "*Clonothrix*"

Section 23
Nonphotosynthetic, Nonfruiting Gliding Bacteria

Order I *Cytophagales*
Family I *Cytophagaceae*
Genus I *Cytophaga*
Genus II *Capnocytophaga*
Genus III *Flexithrix*
Genus IV *Sporocytophaga*
Other genera
Genus *Flexibacter*
Genus *Microscilla*
Genus *Chitinophaga*
Genus *Saprospira*
Order II *Lysobacterales*
Family I *Lysobacteraceae*
Genus I *Lysobacter*
Order III *Beggiatoales*
Family I *Beggiatoaceae*
Genus I *Beggiatoa*
Genus II *Thiothrix*
Genus III *Thioploca*
Genus IV "*Thiospirillopsis*"
Other families and genera
Family *Simonsiellaceae*
Genus I *Simonsiella*
Genus II *Alysiella*
Family "*Pelonemataceae*"
Genus I "*Pelonema*"
Genus II "*Achroonema*"
Genus III "*Peloploca*"
Genus IV "*Desmanthos*"
Other genera
Genus *Toxothrix*
Genus *Leucothrix*
Genus *Vitreoscilla*
Genus *Desulfonema*
Genus *Achromatium*
Genus *Agitococcus*
Genus *Herpetosiphon*

Section 24
Fruiting Gliding Bacteria: The Myxobacteria

Order *Myxococcales*
Family I *Myxococcaceae*
Genus *Myxococcus*
Family II *Archangiaceae*
Genus *Archangium*
Family III *Cystobacteraceae*
Genus I *Cystobacter*
Genus II *Melittangium*
Genus III *Stigmatella*
Family IV *Polyangiaceae*
Genus I *Polyangium*
Genus II *Nannocystis*
Genus III *Chondromyces*

Section 25
Archaeobacteria
Group I Methanogenic Archaeobacteria

Order I *Methanobacteriales*
Family I *Methanobacteriaceae*
Genus I *Methanobacterium*
Genus II *Methanobrevibacter*
Family II *Methanothermaceae*
Genus *Methanothermus*
Order II *Methanococcales*
Family *Methanococcaceae*
Genus *Methanococcus*
Order III *Methanomicrobiales*
Family I *Methanomicrobiaceae*
Genus I *Methanomicrobium*
Genus II *Methanospirillum*
Genus III *Methanogenium*
Family II *Methanosarcinaceae*
Genus I *Methanosarcina*
Genus II *Methanolobus*
Genus III *Methanothrix*
Genus IV *Methanococcoides*
Other Taxa
Family *Methanoplanaceae*
Genus *Methanoplanus*
Other Genus *Methanosphaera*
Group II Archaeobacterial Sulfate Reducers
Order "*Archaeoglobales*"
Family "*Archaeoglobaceae*"
Genus *Archaeoglobus*
Group III Extremely Halophilic Archaeobacteria
Order *Halobacteriales*
Family *Halobacteriaceae*
Genus I *Halobacterium*
Genus II *Haloarcula*
Genus III *Haloferax*
Genus IV *Halococcus*
Genus V *Natronobacterium*
Genus VI *Natronococcus*
Group IV Cell Wall-less Archaeobacteria
Genus *Thermoplasma*
Group V Extremely Thermophilic S⁰-Metabolizers
Order I *Thermococcales*
Family *Thermococcaceae*
Genus I *Thermococcus*
Genus II *Pyrococcus*
Order II *Thermoproteales*
Family I *Thermoproteaceae*
Genus I *Thermoproteus*
Genus II *Thermofilum*
Family II *Desulfurococcaceae*
Genus *Desulfurococcus*
Other bacteria
Genus *Staphylothermus*
Genus *Pyrodictium*
Order III *Sulfolobales*
Family *Sulfolobaceae*
Genus I *Sulfolobus*
Genus II *Acidianus*

Volume IV

Section 26
Nocardioform Actinomycetes

Genus *Nocardia*
Genus *Rhodococcus*
Genus *Nocardioides*
Genus *Pseudonocardia*
Genus *Oerskovia*
Genus *Saccharopolyspora*
Genus *Faenia*
Genus *Promicromonospora*
Genus *Intrasporangium*
Genus *Actinopolyspora*
Genus *Saccharomonospora*

Section 27
Actinomycetes with Multilocular Sporangia

Genus *Geodermatophilus*
Genus *Dermatophilus*
Genus *Frankia*

Section 28
Actinoplanetes

Genus *Actinoplanes*
Genus *Ampullariella*
Genus *Pilimelia*
Genus *Dactylosporangium*
Genus *Micromonospora*

Section 29
Streptomycetes and Related Genera

Genus *Streptomyces*
Genus *Streptoverticillium*
Genus *Kineosporia*
Genus *Sporichthya*

Section 30
Maduromycetes

Genus *Actinomadura*
Genus *Microbispora*
Genus *Microtetraspora*
Genus *Planobispora*
Genus *Planomonospora*
Genus *Spirillospora*
Genus *Streptosporangium*

Section 31
Thermomonospora and Related Genera

Genus *Thermomonospora*
Genus *Actinosynnema*
Genus *Nocardiopsis*
Genus *Streptoalloteichus*

Section 32
Thermoactinomycetes

Genus *Thermoactinomyces*

Section 33
Other Genera

Genus *Glycomyces*
Genus *Kibdelosporangium*
Genus *Kitasatosporia*
Genus *Saccharothrix*
Genus *Pasteuria*

APPENDIX V
Classification of Viruses

I n this appendix selected groups of viruses are briefly described and a sketch of each is included. The illustrations are not to scale but should provide a general idea of the morphology of each group. The viruses are separated into six sections based on their host preferences. The following material has been adapted with permission from chapter 21 of *Introduction to Modern Virology,* 4th ed. by N. J. Dimmock and S. B. Primrose, copyright © 1994 by Blackwell Scientific Publications, Ltd.

Viruses Multiplying in Vertebrates and Other Hosts

Note that some or all of the *Reoviridae, Bunyaviridae, Rhabdoviridae,* and *Togaviridae* multiply in *both* vertebrates and other hosts. Other families include genera that multiply solely in vertebrates.

1. Family: *Iridoviridae*

Icosahedral particle (125 to 300 nm) consisting of a spherical nucleocapsid surrounded by lipid modified by protein subunits. Envelope may be present but is not required for infectivity. Double-stranded DNA of mol wt 100 to 250 × 10^6, which is circularly permuted with direct terminal repeats. Contain several enzymes. Transcription and DNA synthesis are nuclear. mRNAs have no polyA tails. Cytoplasmic.

Selected genera:
Iridovirus (small 120 nm blue iridescent viruses of insects)
Chloriridovirus (large 180 nm iridescent viruses of insects)
Ranavirus (frog virus group)
Lymphocystivirus (viruses of fish)

2. Family: *Poxviridae*

Double-stranded DNA of mol wt 85 to 240 × 10^6 with inverted terminal repeats. Largest viruses 170 to 260 × 300 to 450 nm. Complex structure composed of several layers and includes lipid. Core contains all enzymes required for mRNA synthesis. Cytoplasmic multiplication.

Selected genera:
Orthopoxvirus (vaccinia and related viruses)
Molluscipoxvirus (*Molluscum contagiosum* virus of humans)
Avipoxvirus (fowlpox and related viruses)
Parapoxvirus (milker's node and related viruses)

Poxviruses of vertebrates

Entomopoxvirus — Poxviruses of insects

3. Family: *Parvoviridae*

Single-stranded DNA of mol wt 1.5 to 2.0 × 10^6. Particle is an 18 to 22 nm icosahedron. Contains no enzymes. Multiplication has a nuclear stage.

Genera:
Parvovirus—viruses of vertebrates including humans. Virions mostly –DNA.
Dependovirus—adeno-associated virus. Infects vertebrates. Particles contain either +DNA or –DNA, which forms a double strand upon extraction. Require helper adenovirus or herpesvirus for efficient replication.
Densovirus—viruses of insects. Virions –DNA or +DNA. Helper not required.

4. Family: *Reoviridae*

Ten to twelve segments of double-stranded RNA of total mol wt 12 to 20×10^6. Particle is a 60 to 80 nm icosahedron. Has an isometric nucleocapsid with transcriptase activity. Cytoplasmic multiplication.

Selected genera:

   Orthoreovirus—of vertebrates
   Orbivirus—of vertebrates, but also multiply in insects
   Coltivirus—of vertebrates and ticks
   Rotavirus—of vertebrates
   Cypovirus—cytoplasmic polyhedrosis viruses of insects
   Phytoreovirus—wound tumor virus

5. Family: *Picornaviridae*

Single-stranded, positive-sense RNA of mol wt 2.5×10^6. Particle is 30 nm and icosahedral. Multiplication is cytoplasmic.

Genera:

   Enterovirus (acid-resistant, primarily viruses of gastrointestinal tract)
   Hepatovirus (hepatitis A virus group)
   Rhinovirus (acid-labile, mainly viruses of upper respiratory tract)
   Aphthovirus (foot-and-mouth disease virus)
   Cardiovirus (EMC virus of mice)
   Also various viruses of insects

6. Family: *Togaviridae*

Single-stranded, positive-sense RNA of mol wt 4×10^6. Enveloped particles 60 to 70 nm diameter contain an icosahedral nucleocapsid. Hemagglutinate. Cytoplasmic, budding from plasma membrane. Have a subgenomic mRNA.

Genera:

   Alphavirus (arboviruses, e.g., Semliki Forest and Sindbis viruses)
   Rubivirus (rubella virus) ⎫ Not
   Arterivirus (equine arteritis virus) ⎭ arboviruses

7. Family: *Flaviviridae*

Single-stranded RNA of mol wt $\sim4 \times 10^6$. Enveloped particles 40 to 60 nm diameter. Differ from *Alphaviridae* by presence of a matrix protein, the lack of intracellular subgenomic mRNAs, and budding from the endoplasmic reticulum. Hemagglutinate. Cytoplasmic.

Genera:

   Flavivirus (arboviruses, e.g., yellow fever virus)
   Pestivirus (e.g., hog cholera)
   Hepatitis C virus group

8. Family: *Rhabdoviridae*

Single-stranded, negative-sense RNA of mol wt 3.5 to 4.6×10^6 complementary to mRNA. The bullet-shaped or bacilliform (100 to 430×70 nm) particle is enveloped with 5 to 10 nm spikes. Inside is a helical nucleocapsid with transcriptase activity. Cytoplasmic, budding from plasma membrane. Some are arboviruses.

Genera:

   Vesiculovirus (vesicular stomatitis virus group) viruses of vertebrates and insects
   Lyssavirus (viruses of vertebrates, e.g., rabies, and of insects, e.g., sigma)
   Plant viruses (e.g., lettuce necrotic yellows, potato yellow dwarf, and many others)

9. Family: *Bunyaviridae*

Three segments (large, medium, and small) of single-stranded RNA of total mol wt 6×10^6. Enveloped 100 nm particles with spikes and three internal ribonucleoprotein filaments 2 nm wide. Cytoplasmic, budding from the Golgi apparatus. Arthropod-transmitted except *Hantavirus* genus.

Selected genera:

   Bunyavirus (Bunyamwera and 150 or so related viruses)
   Hantavirus (Korean hemorrhagic fever or Hantaan virus) not arboviruses
   Tospovirus (tomato spotted wilt group)

Viruses Multiplying Only in Vertebrates

1. Family: *Herpesviridae*

Double-stranded DNA of mol wt 80 to 150×10^6. Particle is a 100 to 110 nm icosahedron enclosed in a lipid envelope (120 to 200 nm virion diameter). Buds from nuclear membrane. Latency for the lifetime of the host is common.

Subfamily: *Alphaherpesvirinae*
Genera:

   Simplexvirus—human (alpha) herpesvirus 1 and 2 (herpes simplex virus types 1 and 2)
   Varicellovirus—human (alpha) herpesvirus 3 (varicella-zoster virus)

Subfamily: *Betaherpesvirinae* (cytomegaloviruses)
Genera:

   Cytomegalovirus—human cytomegalovirus group, e.g., human (beta) herpesvirus 5 (human cytomegalovirus)
   Muromegalovirus—mouse cytomegalovirus group

Subfamily: *Gammaherpesvirinae* (lymphoproliferative virus group)
Genera:

   Lymphocryptovirus—e.g., human (gamma) herpesvirus 4 (Epstein-Barr virus)

2. Family: *Adenoviridae*

Double-stranded DNA of mol wt 20 to 30×10^6. Particle is a 70 to 90 nm icosahedron, which replicates and is assembled in the nucleus.

Genera:

Mastadenovirus (adenoviruses of mammals)
Aviadenovirus (adenoviruses of birds)

3. Family: *Papovaviridae*

Double-stranded circular DNA. Particles have 72 capsomeres in a skew arrangement and are assembled in the nucleus. Hemagglutinate. Oncogenic.

Genera:

Papillomavirus (producing papillomas in several mammalian species) 50 to 55 nm particle; DNA 5×10^6 mol wt
Polyomavirus (found in rodents, humans, and other primates) 40 to 45 nm particle; DNA 3×10^6 mol wt. Includes SV40 and polyoma virus itself.

4. Family: *Hepadnaviridae*

One complete DNA minus strand of mol wt 1×10^6 with a 5′ terminal protein. DNA is circularized by an incomplete plus strand of variable length (50 to 100%), which overlaps the 3′ and 5′ termini of DNA minus. There is a 42 nm enveloped particle containing a core with DNA polymerase and protein kinase activities. Includes hepatitis B of humans, Pekin duck hepatitis, beechy ground squirrel hepatitis, and woodchuck hepatitis viruses. HBV is strongly associated with liver cancer.

Genus:

Orthohepadnavirus—hepatitis B virus

5. Family: *Caliciviridae*

Single-stranded, positive-sense RNA of mol wt 2.7×10^6. Icosahedral 37 nm particle with calixlike (cup-shaped) surface depressions.

Genus:

Calicivirus (vesicular exanthema of swine virus, Norwalk virus, possibly hepatitis E virus)

6. Family: *Arenaviridae*

Two (a large and a small) segments of single-stranded RNA of mol wt 3×10^6 and 1.3×10^6. Latter RNA is ambisense. Enveloped 50 to 300 nm particles with spikes. Contain ribosomes which have no known function. Cytoplasmic multiplication; buds from plasma membrane.

Genus:

Arenavirus (lymphocytic choriomeningitis virus and related viruses—e.g., Lassa, Junin, and Macupo viruses)

7. Family: *Paramyxoviridae*

Single-stranded RNA of mol wt 5 to 7×10^6. Enveloped 150 nm particles have spikes and contain a helical nucleocapsid 12 to 17 nm in diameter with transcriptase activity. Filamentous forms common. Cytoplasmic, budding from plasma membrane. Airborne transmission.

Subfamily: *Paramyxovirinae*
Genera:

Paramyxovirus (Newcastle disease virus, parainfluenza virus) Only this genus has a neuraminidase activity which is on the same protein (HN) as the hemagglutination activity.
Morbillivirus (measles virus, canine distemper) Hemagglutinate

Subfamily: *Pneumovirinae*

Pneumovirus (respiratory syncytial virus group)

8. Family: *Orthomyxoviridae*

Eight segments of single-stranded RNA of total mol wt 4×10^6. Enveloped 100 nm particles have spikes and contain a helical nucleocapsid 9 nm in diameter with transcriptase activity. Only A and B virions have separate hemagglutinin and neuraminidase proteins. Multiplication requires the nucleus. RNA segments in a mixed infection readily assort to form genetically stable hybrids within a virus. Buds from plasma membrane.

Genera:

Influenza virus A and B
Influenza C virus. Has seven RNA segments and a receptor-destroying activity (a sialic acid—O-acetyl esterase), which is on the hemagglutinin protein.

9. Family: *Filoviridae*

Long filamentous particles 800 to 900 (sometimes 14,000) × 80 nm with helical nucleocapsid of 50 nm diameter. Linear, single-stranded, negative-sense RNA of mol wt 4.2×10^6. Buds from plasma membrane. Contains Marburg and Ebola viruses, which are highly pathogenic for humans. Transmitted by contact.

Genus:

Filovirus—Marburg and Ebola viruses

10. Family: *Retroviridae*

Single-stranded "diploid" RNA with unique sequence having a mol wt 1 to 3×10^6. Enveloped 100 nm particles with a core containing a helical nucleoprotein. Contains RNA-dependent DNA polymerase. The DNA provirus is nuclear.

Selected genera:

MLV-related viruses (mammalian type C retrovirus group)—murine leukemia virus
Spumavirus—human foamy virus
HTLV-BLV group—human T-cell lymphotrophic viruses types 1 and 2
Lentivirus—human immunodeficiency viruses types 1 and 2; simian immunodeficiency virus, visna virus

Viruses Multiplying Only in Invertebrates

Viruses occur not only in insects, crustacea, and molluscs but probably in all groups of invertebrates. The *Poxviridae, Reoviridae, Parvoviridae, Rhabdoviridae*, and *Togaviridae* (see earlier) have representatives that multiply in invertebrates. Some plant viruses are transmitted by, but do not multiply in these vectors.

1. Family: *Baculoviridae*

Double-stranded circular DNA of mol wt 60 to 110×10^6. Bacilliform particles 30 to 60 nm $\times$ 250 to 300 nm with an outer membrane. May be occluded in a protein inclusion body containing usually one particle (granulosis viruses, upper illustration) or in a polyhedra containing many particles (polyhedrosis viruses, lower illustration).

Selected genera:
Nuclear polyhedrosis virus
Granulosis viruses
Nonoccluded baculoviruses

2. Family: *Tetraviridae*

Single-stranded RNA of mol wt 1.8×10^6 in a 35 nm particle. T = 4 (whereas in *Picornaviridae* T = 1). All isolated from Lepidoptera. No infection of cultured cells. *Nudaurelia β* virus.

Viruses Multiplying Only in Plants

Knowledge of virus multiplication in plants is relatively rudimentary since the cell culture systems are less manageable than the animal cell cultures. Work has concentrated on physical properties and disease characteristics. Designation into families or genera has not yet been decided. The *Reoviridae* and *Rhabdoviridae* have members that multiply in both plants and invertebrates. The plant viruses listed below are not known to multiply in their invertebrate vector. Differences in virus proteins and translation strategy are important classification criteria.

1. *Caulimovirus*—cauliflower mosaic virus group

Open circular dsDNA with single-strand discontinuities like the hepadnavirus. Mol wt 4 to 5×10^6. Isometric 50 nm particles. Aphid vectors.

2. Geminivirus

Circular single-stranded DNA of mol wt 0.7 to 0.8×10^6. Two incomplete icosahedral particles joined as a pair, ~18 × 30 nm, and usually found in the nucleus. One or two molecules of DNA per pair of particles. Persistent in whitefly or leafhopper vectors.

3. *Luteovirus*—barley yellow dwarf virus group

Single-stranded, positive-sense RNA of mol wt 2×10^6. Isometric 25 to 30 nm particle. Persistent retention by aphid vectors.

4. Maize chlorotic dwarf virus group

Single-stranded, positive-sense RNA of mol wt 3.2×10^6 in 30 nm particle. Only infects grasses and transmitted by leafhoppers.

5. *Necrovirus*—tobacco necrosis virus group

Single-stranded, linear, positive-sense RNA of mol wt 1.5×10^6. Isometric 28 nm particle. Cytoplasmic. Fungal vector.

6. *Tombusvirus*—tomato bushy stunt virus group

Single-stranded, linear, positive-sense RNA of mol wt 1.5×10^6. Not polyadenylated. Particle 30 nm in diameter. Cytoplasmic, nuclear, and sometimes mitochondrial location. Transmitted through the soil.

7. *Bromovirus*—brome mosaic virus group

Four single-stranded, linear, positive-sense RNAs of mol wt 1.1, 1.0, and 0.8, and 0.3×10^6. Particles 26 nm diameter. Infectivity requires the three largest RNAs. Each is in a different particle. Assembly in cytoplasm. Some with a beetle vector.

8. *Cucumovirus*—cucumber mosaic virus group

Single-stranded, linear, positive-sense RNAs of mol wt 1.3, 1.1, 0.8, and 0.3×10^6. Particles 29 nm in size. Infectivity requires the three largest RNAs, each of which is in a separate particle. Cytoplasmic. Nonpersistent in aphid vector.

9. Alfalfa mosaic virus group

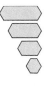

Four bacilliform particles 18 × 56, 18 × 43, 18 × 35, and 18 × 30 nm. Three largest contain single-stranded, linear, positive-sense RNA of mol wt 1.1, 0.8, 0.7, and 0.3×10^6, all needed for infectivity. Smallest contains two molecules of the coat protein mRNA. Cytoplasmic. Nonpersistent in aphid vector.

10. *Comovirus*—cowpea mosaic virus group

Two 28 nm particles containing single-stranded, linear, positive-sense RNA of mol wt 2.4 or 1.4×10^6. Two coat polypeptides encoded by the smaller RNA. Both RNAs needed for infectivity. Cytoplasmic. Transmitted by beetles or seed.

11. *Nepovirus*—tobacco ringspot virus group

Two 28 nm particles containing single-stranded, linear, positive-sense RNA of mol wt 2.4 to 2.8 or 1.3 to 2.4×10^6. Both RNAs needed for infectivity. One coat protein encoded on larger RNA. Some nepoviruses have a third particle containing two molecules of the second RNA. Cytoplasmic. Nematode vectors.

12. *Tobamovirus*—tobacco mosaic virus group

Single-stranded, linear, positive-sense RNA of mol wt 2×10^6. Rigid cylindrical particles 300×18 nm. Transmitted mechanically or by seed.

13. *Potyvirus*—potato virus Y group

Single-stranded, linear, positive-sense RNA of mol wt 3.0 to 3.5×10^6. Particle is a flexuous rod with helical symmetry, 680 to 900×11 nm. Cytoplasmic but some have nuclear inclusions. Nonpersistent in aphid vectors.

Viruses Multiplying Only in Fungi

1. Family: *Totiviridae*

Isometric 40 to 43 nm particles with a genome consisting of a single molecule of double-stranded RNA of mol wt 3.3 to 4.2×10^6. One major capsid protein. Single shell. Transcriptase present. Cytoplasmic.

Genus:

Totivirus (Saccharomyces cerevisiae virus LA)

2. Family: *Phycodnaviridae*

Large polyhedral 130 to 200 nm diameter particles containing linear, double-stranded DNA of mol wt 150 to 210 $\times 10^6$. Nonenveloped. Contain lipid in shell. Infect *Paramecium* and *Chlorella* sp.

Viruses Multiplying Only in Bacteria

Surprisingly little is known of the comparative biology of bacterial viruses as only a few representatives have been studied in detail.

1. Family: *Myoviridae* (Phages with contractile tails)

Linear double-stranded DNA of mol wt 120 $\times 10^6$. Head isometric or elongated, 110×80 nm; complex contractile tail 113×16 nm. Tail has collar, base plate, spikes, and fibers. Includes T2, T4, T6, PBS1, SP8, SP50, P2, Mu.

2. Family: *Siphoviridae* (Phages with long, noncontractile tails)

Linear double-stranded DNA of mol wt 33×10^6. Head 60 nm diameter; long noncontractile tail up to 570 nm. No breakdown of host DNA. Includes λ, χ (chi), and φ80.

3. Family: *Podoviridae* (Phages with short tails)

Linear double-stranded DNA of mol wt 25×10^6. Head 60 nm diameter. Short (17×8 nm) tail with 6 short fibers. Includes T7 and P22.

4. Family: *Tectiviridae* (Phages with double capsids)

Linear double-stranded DNA of mol wt 10×10^6 in 63 nm particles. Contains internal lipid. Double capsid with a rigid outer shell and flexible inner coat. After injection of DNA a tail structure of about 60 nm appears. PRD1, Bam35

5. Family: *Plasmaviridae* (Pleomorphic phages)

Circular double-stranded DNA of 8×10^6 mol wt in lipid-containing envelope. 50 to 125 nm particle with small dense core. Slightly pleomorphic. Formed by budding. Infects mycoplasmas.

6. Family: *Corticoviridae* (PM2 phage group)

Circular double-stranded DNA of mol wt 6×10^6, isometric 60 nm particle, lipid between protein shells, no envelope, no tail. Spikes at vertices. Infects *Pseudomonas*.

7. Family: *Microviridae* (Isometric phages with ssDNA)

Circular single-stranded DNA of mol wt 1.6 to 1.7×10^6. Icosahedron (25 to 27 nm) with knobs on 12 vertices. No envelope. Includes φX174 and G4.

8. Family: *Inoviridae* (Rod-shaped phages)

Helical, rod-shaped phages with a circular, single-stranded, positive-sense DNA. No host lysis.

Genera:

Inovirus

DNA of 1.9 to 2.7×10^6, long flexible filamentous particle 760 to $1,950 \times 6$ to 8 nm. Host bacteria not lysed. Includes M13 and fd phages.

Plectovirus

Mycoplasma virus type 1 phages. DNA 2.5 to 5.2×10^6 mol wt. Short rods of 85 to 280×14 nm.

9. Family: *Cystoviridae* (φ6 phage group)

Three molecules of linear, double-stranded RNA of mol wt 2.3, 3.1, and 5.0×10^6. Isometric 75 nm particle with lipid envelope. Infects *Pseudomonas*.

10. Family: *Leviviridae* (single-stranded RNA phages)

Linear, single-stranded, positive-sense RNA of mol wt 1.2×10^6. Icosahedral capsid is 24 nm.

Genera:

Levivirus—coliphage MS2 group
Allolevivirus—coliphage Qβ group

GLOSSARY

Pronunciation Guide

Many of the boldface terms in this glossary are followed by a phonetic spelling in parentheses. These pronunciation aids usually come from *Dorland's Illustrated Medical Dictionary*. The following rules are taken from this dictionary and will help in using its phonetic spelling system.

1. An unmarked vowel ending a syllable (an open syllable) is long; thus *ma* represents the pronunciation of *may; ne*, that of *knee; ri*, of *wry; so*, of *sew; too*, of *two;* and *vu*, of *view.*

2. An unmarked vowel in a syllable ending with a consonant (a closed syllable) is short; thus *kat* represents *cat; bed, bed; hit, hit; not, knot; foot, foot;* and *kusp, cusp.*

3. A long vowel in a closed syllable is indicated by a macron; thus *māt* stands for *mate; sēd*, for *seed; bīl*, for *bile; mōl*, for *mole; fūm*, for *fume;* and *fo͞ol*, for *fool.*

4. A short vowel that ends or itself constitutes a syllable is indicated by a breve; thus *ĕ-fekt´* for *effect, ĭ-mūn´* for *immune,* and *ŏ-klo͞od´* for *occlude.*

Primary (´) and secondary (″) accents are shown in polysyllabic words. Unstressed syllables are followed by hyphens.

Some common vowels are pronounced as indicated below.

ə	sof*a*	ē	m*e*t	ŏ	g*o*t
ā	m*a*te	ī	b*i*te	ū	f*ue*l
ă	b*a*t	ĭ	b*i*t	ŭ	b*u*t
ē	b*ea*m	ō	h*o*me		

From Dorland's Illustrated Medical Dictionary. *Copyright © 1988 W. B. Saunders, Philadelphia, Pa. Reprinted by permission.*

A

AB model   The model that describes the structure and activity of many exotoxins. The B portion of the toxin is responsible for toxin binding to a cell but does not directly harm it. The A portion enters the cell and disrupts its function.   *567*

ABO blood grouping   The major human blood typing system that reflects the presence or absence of the A and B glycoprotein antigens in/on erythrocyte plasma membranes.   *628*

accessory pigments   Photosynthetic pigments such as carotenoids and phycobiliproteins that aid chlorophyll in trapping light energy.   *183*

acetyl coenzyme A (acetyl-CoA)   A combination of acetic acid and coenzyme A that is energy rich; it is produced by many catabolic pathways and is the substrate for the tricarboxylic acid cycle, fatty acid biosynthesis, and other pathways.   *170*

acid dyes   Dyes that are anionic or have negatively charged groups such as carboxyls.   *26*

acid fast   Refers to bacteria like the mycobacteria that cannot be easily decolorized with acid alcohol after being stained with dyes such as basic fuchsin.   *452*

acid-fast staining   A staining procedure that differentiates between bacteria based on their ability to retain a dye when washed with an acid alcohol solution.   *26*

acidophile (as´id-o-fīl″)   A microorganism that has its growth optimum between about pH 0 and 5.5.   *126*

acne vulgaris   A chronic inflammatory disease of the sebaceous glands, the lesions occurring most frequently on the face, chest, and back. The inflamed glands may form small pink papules, which sometimes surround comedones so as to have black centers, or form pustules.   *554*

acquired enamel pellicle   A membranous layer on the tooth enamel surface formed by selectively adsorbing glycoproteins (mucins) from saliva. This pellicle confers a net negative charge to the tooth surface.   *782*

acquired immune deficiency syndrome (AIDS)   An infectious disease syndrome caused by the human immunodeficiency virus and is characterized by the loss of a normal immune response, followed by increased susceptibility to opportunistic infections and an increased risk of some cancers.   *722*

acquired immune tolerance (tolerance induction)   The ability to produce antibodies against nonself antigens while "tolerating" (not producing antibodies against) self-antigens.   *618*

acquired immunity   Refers to the type of specific immunity that develops after exposure to a suitable antigen or is produced after antibodies are transferred from one individual to another.   *588*

actinomycete (ak″tĭ-no-mi´sēt)   An aerobic, gram-positive bacterium that forms branching filaments (hyphae) and asexual spores.   *492*

actinorhizae   Associations between actinomycetes and plant roots.   *870*

activation energy   The energy required to bring reacting molecules together to reach the transition state in a chemical reaction.   *160*

active carrier   An individual who has an overt clinical case of a disease and who can transmit the infection to others.   *705*

active site   The part of an enzyme that binds the substrate to form an enzyme-substrate complex and catalyze the reaction. Also called the catalytic site.   *160*

active transport   The transport of solute molecules across a membrane against an electrochemical gradient; it requires a carrier protein and the input of energy.   *103*

acute carrier   *See* casual carrier.   *706*

acute infections   Virus infections with a fairly rapid onset that last for a relatively short time.   *379*

acute-phase reactant   Refers to a substance that participates in the qualitative and quantitative changes which occur in the host's blood plasma during the early stages of an acute infection. These substances are an important part of the host's general or nonspecific defense and are not immunoglobulins.   *573*

acute viral gastroenteritis   An inflammation of the stomach and intestines, normally caused by Norwalk and Norwalklike viruses, other caliciviruses, rotaviruses, and astroviruses.   *737*

acyclovir (a-si´klo-vir)   A synthetic purine nucleoside derivative with antiviral activity against herpes simplex virus.   *672*

adenine (ad´e-nēn)   A purine derivative, 6-aminopurine, found in nucleosides, nucleotides, coenzymes, and nucleic acids.   *204*

adenosine diphosphate (ADP; ah-den´o-sēn)   The nucleoside diphosphate usually formed upon the breakdown of ATP when it provides energy for work.   *153*

adenosine 5´-triphosphate (ATP)   The triphosphate of the nucleoside adenosine, which is a high energy molecule or has high phosphate group transfer potential and serves as the cell's major form of energy currency.   *153*

adjuvant (aj´ə-vant)   Material added to an antigen to increase its immunogenicity. Common examples are alum, killed *Bordetella pertussis*, and an oil emulsion of the antigen, either alone (Freund's incomplete adjuvant) or with killed mycobacteria (Freund's complete adjuvant).   *600*

adult T-cell leukemia   A type of white blood cell cancer caused by the HTLV-1 virus.   *733*

aerobe (a´er-ōb)   An organism that grows in the presence of atmospheric oxygen.   *130*

aerobic respiration (res″pĭ-ra´shun)   A metabolic process in which molecules, often organic, are oxidized with oxygen as the final electron acceptor. It is one type of respiration: an energy-yielding process in which an electron donor is oxidized using an inorganic electron acceptor.   *153*

aerotolerant anaerobes   Microorganisms that grow equally well whether or not oxygen is present.   *130*

aflatoxin (af″lah-tok´sin)   Fungal metabolite that can cause cancer.   *889*

agar (ahg´ar)   A complex sulfated polysaccharide, usually extracted from red algae, that is used as a solidifying agent in the preparation of culture media.   *107*

agglutinates The visible aggregates or clumps formed by an agglutination reaction. *641*

agglutination reaction (ah-gloo″tĭ-na′shun) The formation of an insoluble immune complex by the cross-linking of cells or particles. *640*

agglutinin (ah-gloo″tĭ-nin) The antibody responsible for an agglutination reaction. *640*

AIDS *See* acquired immune deficiency syndrome. *722*

AIDS-related complex (ARC) A collection of symptoms such as lymphadenopathy (swollen lymph glands), fever, malaise, fatigue, loss of appetite, and weight loss. It results from an HIV infection and may progress to frank AIDS. *724*

airborne transmission The type of infectious organism transmission in which the pathogen is truly suspended in the air and travels over a meter or more from the source to the host. *707*

akinetes Specialized, nonmotile, dormant, thick-walled resting cells formed by some cyanobacteria. *461*

alcoholic fermentation A fermentation process that produces ethanol and CO_2 from sugars. *175*

alga (al′gah) A common term for a series of unrelated groups of photosynthetic eucaryotic microorganisms lacking multicellular sex organs (except for the charophytes) and conducting vessels. *521*

algicide (al′jĭ-sīd) An agent that kills algae. *138*

algology (al-gol′o-je) The scientific study of algae. *521*

alkalophile A microorganism that grows best at pHs around 8.5 to 11.5. *126*

allergen (al′er-jen) A substance capable of inducing allergy or specific susceptibility. *621*

allergic contact dermatitis An allergic reaction caused by haptens that combine with proteins in the skin to form the allergen that produces the immune response. *623*

allergy (al′er-je) *See* hypersensitivity. *621*

allochthonous Microorganisms or substances that are not normally present in a particular habitat and that have been introduced from the outside. *822*

allograft (al′o-graft) A transplant between genetically different individuals of the same species. *626*

allosteric enzyme (al″o-ster′ik) An enzyme whose activity is altered by the binding of a small effector or modulator molecule at a regulatory site separate from the catalytic site; effector binding causes a conformational change in the enzyme and its catalytic site, which leads to enzyme activation or inhibition. *238*

allotype Allelic variants of antigenic determinant(s) found on antibody chains of some, but not all, members of a species, which are inherited as simple Mendelian traits. *592*

alpha hemolysis A greenish zone of partial clearing around a bacterial colony growing on blood agar. *572*

alternative pathway An antibody-independent pathway of complement activation that includes the C3–C9 components of the classical pathway and several other serum protein factors (e.g., factor B and properdin). *634*

amantadine (ah-man′tah-den) An antiviral compound used to prevent type A influenza infections. *672*

amebiasis (amebic dysentery) (am″e-bi′ah-sis) An infection with amoebae, often resulting in dysentery; usually it refers to an infection by *Entamoeba histolytica.* *799*

American trypanosomiasis (Chagas' disease) *See* trypanosomiasis. *805*

Ames test A test that uses a special *Salmonella* strain to test chemicals for mutagenicity and potential carcinogenicity. *270*

amino acid activation The initial stage of protein synthesis in which amino acids are attached to transfer RNA molecules. *226*

aminoacyl or **acceptor site (A site)** The site on the ribosome that contains an aminoacyl-tRNA at the beginning of the elongation cycle during protein synthesis; the growing peptide chain is transferred to the aminoacyl-tRNA and lengthens by an amino acid. *229*

aminoglycoside antibiotics (am″ĭ-no-gli′ko-sīd) A group of antibiotics synthesized by *Streptomyces* and *Micromonospora,* which contain a cyclohexane ring and amino sugars; all aminoglycoside antibiotics bind to the small ribosomal subunit and inhibit protein synthesis. *668*

amnesic shellfish poisoning (am-nes′tik) The disease arising in humans and animals that eat seafood such as mussels contaminated with domoic acid from diatoms. The disease produces short-term memory loss in its victims. *529*

amoeboid movement Moving by means of cytoplasmic flow and the formation of pseudopodia (temporary cytoplasmic protrusions of the cytoplasm). *536*

amphibolic pathways (am″fe-bol′ik) Metabolic pathways that function both catabolically and anabolically. *166*

amphitrichous (am-fit′rĕ-kus) A cell with a single flagellum at each end. *60*

amphotericin B (am″fo-ter′i-sin) An antibiotic from a strain of *Streptomyces nodosus* that is used to treat systemic fungal infections; it also is used topically to treat candidiasis. *671*

anabolism (ah-nab′o-lizm″) The synthesis of complex molecules from simpler molecules with the input of energy. *165*

anaerobe (an-a′er-ōb) An organism that grows in the absence of free oxygen. *130*

anaerobic digestion (an″a-er-o′bik) The microbiological treatment of sewage wastes under anaerobic conditions to produce methane. *845*

anaerobic respiration (an″a-er-o′bik) An energy-yielding process in which the electron transport chain acceptor is an inorganic molecule other than oxygen. *176*

anamnestic response (an″am-nes′tik) The recall, or the remembering, by the immune system of a prior response to a given antigen. *601*

anaphylaxis (an″ah-fĭ-lak′sis) An immediate (type I) hypersensitivity reaction following exposure of a sensitized individual to the appropriate antigen. Mediated by reagin antibodies, chiefly IgE. *621*

anaplerotic reactions (an″ah-plĕ-rot′ik) Reactions that replenish depleted tricarboxylic acid cycle intermediates. *202*

anogenital condylomata (venereal warts) (kon″dĭ-lo″ mah-tah) Warts that are sexually transmitted and caused by types 6, 11, and 42 human papilloma virus. Usually occur around the cervix, vulva, perineum, anus, anal canal, urethra, or glans penis. *739*

anoxygenic photosynthesis Photosynthesis that does not oxidize water to produce oxygen; the form of photosynthesis characteristic of purple and green photosynthetic bacteria. *187, 455*

antheridium (an″ther-id′e-um; pl., **antheridia**) A male gamete-producing organ, which may be unicellular or multicellular. *511, 522*

anthrax (an′thraks) An infectious disease of animals caused by ingesting *Bacillus anthracis* spores. Can also occur in humans and is sometimes called woolsorter's disease. *755*

antibiotic (an″tĭ-bi-ot′ik) A microbial product or its derivative that kills susceptible microorganisms or inhibits their growth. *657*

antibody (an′tĭ-bod″e) A glycoprotein produced in response to the introduction of an antigen; it has the ability to combine with the antigen that stimulated its production. Also known as an immunoglobulin (Ig). *591*

antibody-dependent cell-mediated cytotoxicity An antibody provides the bridging mechanism for the intimate association between the effector and target cell. The effector cell then releases its cytotoxic mediators or directly causes cytolysis—either of which kills the target cell. *637*

antibody-mediated immunity *See* humoral immunity. *589*

anticodon triplet The base triplet on a tRNA that is complementary to the triplet codon on mRNA. *226*

antigen (an′tĭ-jen) A foreign (nonself) substance (such as a protein, nucleoprotein, polysaccharide, or sometimes a glycolipid) to which lymphocytes respond; also known as an immunogen because it induces the immune response. *590*

antigen-binding fragment (Fab) The top of the Y portion of an antibody molecule; contains the sites that bind with compatible antigenic determinants. *592*

antigenic determinant site *See* epitope. *590*

antigenic drift A small change in the antigenic character of an organism that allows it to avoid attack by the immune system. *705*

antigenic shift A major change in the antigenic character of an organism that alters it to an antigenic strain unrecognized by host immune mechanisms. *703*

antigen-presenting cells Antigen-presenting cells (APCs) are cells that take in protein antigens, process them, and present antigen fragments to B cells and T cells in conjunction with class II MHC molecules so that the cells are activated. Macrophages, B cells, dendritic cells, and Langerhans' cells may act as APCs. *613*

antimetabolite (an″tĭ-mĕ-tab′o-līt) A compound that blocks metabolic pathway function by competitively inhibiting a key enzyme's use of a metabolite because it closely resembles the normal enzyme substrate. *663*

antimicrobial agent An agent that kills microorganisms or inhibits their growth. *139*

antisense RNA A single-stranded RNA with a base sequence complementary to a segment of another RNA molecule that can specifically bind to the target RNA and inhibit its activity. *248*

antisepsis (an″tĭ-sep′sis) The prevention of infection or sepsis. *138*

antiseptic (an″tĭ-sep′tik) Chemical agents applied to tissue to prevent infection by killing or inhibiting pathogens. *138*

antiserum (an″tĭ-se′rum) Serum containing induced antibodies. *601*

antitoxin (an″tĭ-tok′sin) An antibody to a microbial toxin, usually a bacterial exotoxin, that combines specifically with the toxin, in vivo and in vitro, neutralizing the toxin. *636*

apical complex (ap′ĭ-kal) A set of organelles characteristic of members of the phylum *Apicomplexa*: polar rings, subpellicular microtubules, conoid, rhoptries, and micronemes. *540*

apicomplexan (a′pĭ-kom-plek′san) A sporozoan protist that lacks special locomotor organelles but has an apical complex and a spore-forming stage. It is either an intra- or extracellular parasite of animals; a member of the phylum *Apicomplexa*. *540*

aplanospore (a′plan-o-spor) A nonflagellated, nonmotile spore that is involved in asexual reproduction. *522*

apoenzyme (ap″o-en′zīm) The protein part of an enzyme that also has a nonprotein component. *159*

apoptosis (ap″o-to′sis) Programmed cell death. The fragmentation of a cell into membrane-bound particles that are eliminated by phagocytosis. Apoptosis is a physiological suicide mechanism that preserves homeostasis and occurs during normal tissue turnover. It is responsible for cell death in pathological circumstances, such as exposure to low concentrations of xenobiotics and infections by HIV and various other viruses. Apoptotic cells display profound structural changes such as plasma membrane blebbing and nuclear collapse. DNA is cleaved into short oligonucleosomal length DNA fragments. Apoptosis usually occurs after the activation of a calcium-dependent endogenous endonuclease. *728*

arbuscules Branched, treelike structures formed in cells of plant roots colonized by endotrophic mycorrhizal fungi. *870*

archaea The domain or empire that contains procaryotes with isoprenoid glycerol diether or diglycerol tetraether lipids in their membranes and archaeobacterial rRNA (among many differences). *412*

archaeobacteria or **archaebacteria** (ar″keo-bak-tē′re-ah) Bacteria that lack muramic acid in their cell walls, have membrane lipids with ether-linked branched chain fatty acids, and differ in many other ways from eubacteria. *402, 478*

arthroconidium (ar′thro-ko-nid′e-um; pl., **arthroconidia**) A thallic conidium released by the fragmentation or lysis of hypha. It is not notably larger than the parental hypha, and separation occurs at a septum. *508*

arthrospore (ar′thro-spōr) A spore resulting from the fragmentation of a hypha. *508*

artificially acquired active immunity The type of immunity that results from immunizing an animal with a vaccine. The immunized animal now produces its own antibodies and activated lymphocytes. *588*

artificially acquired passive immunity The type of immunity that results from introducing into an animal antibodies that have been produced either in another animal or by in vitro methods. Immunity is only temporary. *588*

ascocarp (as′ko-karp) A multicellular structure in ascomycetes lined with specialized cells called asci in which nuclear fusion and meiosis produce ascospores. An ascocarp can be open or closed and may be referred to as a fruiting body. *511*

ascogenous hypha A specialized hypha that gives rise to one or more asci. *511*

ascogonium (as″ko-go′ne-um; pl., **ascogonia**) The receiving (female) organ in ascomycetous fungi which, after fertilization, gives rise to ascogenous hyphae and later to asci and ascospores. *511*

ascomycetes (as″ko-mi-se′tēz) A division of fungi that form ascospores. *510*

ascospore (as′ko-spor) A spore contained or produced in an ascus. *509*

ascus (as′kus) A specialized cell, characteristic of the ascomycetes, in which two haploid nuclei fuse to produce a zygote, which immediately divides by meiosis; at maturity an ascus will contain ascospores. *511*

aseptic meningitis syndrome *See* meningitis. *745*

aspergillosis (as″per-jil-o′sis) A fungal disease caused by species of *Aspergillus*. *797*

assimilatory reduction The reduction of an inorganic molecule to incorporate it into organic material. No energy is made available during this process. *197, 198, 816*

associative nitrogen fixation Nitrogen fixation by bacteria in the plant root zone (rhizosphere). *868*

athlete's foot *See* tinea pedis. *793*

attenuation (ah-ten″u-a′shun) 1. A mechanism for the regulation of transcription of some bacterial operons by aminoacyl-tRNAs. 2. A procedure that reduces or abolishes the virulence of a pathogen without altering its immunogenicity. *248*

attenuator A rho-independent termination site in the leader sequence that is involved in attenuation. *247*

autochthonous (aw-tok′tho-nus) Pertaining to substances or microorganisms indigenous to a particular environment. *862*

autoclave (aw′to-klāv) An apparatus for sterilizing objects by the use of steam under pressure. Its development tremendously stimulated the growth of microbiology. *140*

autogenous infection (aw-toj′e-nus) An infection that results from a patient's own microbiota, regardless of whether the infecting organism became part of the patient's microbiota subsequent to admission to a clinical care facility. *711*

autoimmune disease (aw″to-ĭ-mūn′) A disease produced by the immune system attacking self-antigens. Autoimmune disease results from the activation of self-reactive T and B cells that damage tissues after stimulation by genetic or environmental triggers. *626*

autoimmunity (aw″to-ĭ-mun′ĭ-te) Autoimmunity is a condition characterized by the presence of serum autoantibodies and self-reactive lymphocytes. It may be benign or pathogenic. Autoimmunity is a normal consequence of aging; is readily inducible by infectious agents, organisms, or drugs; and is potentially reversible in that it disappears when the offending "agent" is removed or eradicated. *626*

autolysins (aw-tol′ĭ-sins) Enzymes that partially digest peptidoglycan in growing bacteria so that the peptidoglycan can be enlarged. *209*

autotroph (aw′to-trōf) An organism that uses CO_2 as its sole or principal source of carbon. *99*

auxotroph (awk′so-trōf) A mutated prototroph that lacks the ability to synthesize an essential nutrient and therefore must obtain it or a precursor from its surroundings. *100, 262*

axial filament The organ of motility in spirochetes. It is made of axial fibrils or periplasmic flagella that extend from each end of the protoplasmic cylinder and overlap in the middle of the cell. The outer sheath lies outside the axial filament. *63, 417*

B

bacillus (bah-sil′lus) A rod-shaped bacterium. *39*

bacteremia (bak″ter-e′me-ah) The presence of bacteria in the blood. *785*

bacteria (bak-te′re-ə) The domain or empire that contains procaryotic cells with primarily diacyl glycerol diesters in their membranes and with eubacterial rRNA. Also a general term for organisms that are composed of procaryotic cells and are not multicellular. *412*

bacterial (septic) meningitis *See* meningitis. *745*

bacterial species A collection of strains that share many stable properties and differ significantly from other groups of strains. *392*

bacterial vaginosis (bak-te′re-əl vaj′′ĭ-no′sis) Bacterial vaginosis is a sexually transmitted disease caused by *Gardnerella vaginalis*, *Mobiluncus* spp., *Mycoplasma hominis*, and various anaerobic bacteria. Although a mild disease it is a risk factor for obstetric infections and pelvic inflammatory disease. *755*

bactericide (bak-tēr′ĭ-sid) An agent that kills bacteria. *138*

bacteriochlorophyll (bak-te″re-o-klo′ro-fil) A modified chlorophyll that serves as the primary light-trapping pigment in purple and green photosynthetic bacteria. *187*

bacteriocin (bak-te′re-o-sin)   A protein produced by a bacterial strain that kills other closely related strains.   *576*

bacteriophage (bak-te′re-o-fāj″) A virus that uses bacteria as its host; often called a phage.   *354*

bacteriophage (phage) typing   A technique in which strains of bacteria are identified based on their susceptibility to a variety of bacteriophages.   *693*

bacteriostasis   (bak-te″re-os′tah-sis)   *See* bacteriostatic.   *814*

bacteriostatic   (bak-te″re-o-stat′ik)   Inhibiting the growth and reproduction of bacteria.   *138*

bacteroid   (bak′tĕ-roid)   A modified, often pleomorphic, bacterial cell within the root nodule cells of legumes; after transformation into a symbiosome it carries out nitrogen fixation.   *865*

baeocytes   Small, spherical, reproductive cells produced by pleurocapsalean cyanobacteria through multiple fission.   *462*

balanced growth   Microbial growth in which all cellular constituents are synthesized at constant rates relative to each other.   *123*

balanitis (bal″ah-ni′tis)   Inflammation of the glans penis usually associated with *Candida* fungi; a sexually transmitted disease.   *799*

barophilic (bar″o-fil′ik) or **barophile**   Organisms that prefer or require high pressures for growth and reproduction.   *132, 836*

barotolerant   Organisms that can grow and reproduce at high pressures but do not require them.   *132*

basal body   The cylindrical structure at the base of procaryotic and eucaryotic flagella that attaches them to the cell.   *60, 91*

base analogs   Molecules that resemble normal DNA nucleotides and can substitute for them during DNA replication, leading to mutations.   *264*

basic dyes   Dyes that are cationic, or have positively charged groups, and bind to negatively charged cell structures. Usually sold as chloride salts.   *26*

basidiocarp (bah-sid′e-o-karp″)   The fruiting body of a basidiomycete that contains the basidia.   *514*

basidiomycetes (bah-sid″e-o-mi-se′tēz)   A division of fungi in which the spores are born on club-shaped organs called basidia.   *514*

basidiospore (bah-sid′e-o-spōr)   A spore born on the outside of a basidium following karyogamy and meiosis.   *509*

basidium (bah-sid′e-um; pl., **basidia**)   A structure that bears on its surface a definite number of basidiospores (typically four) that are formed following karyogamy and meiosis. Basidia are found in the basidiomycetes and are usually club-shaped.   *514*

batch culture   A culture of microorganisms produced by inoculating a closed culture vessel containing a single batch of medium.   *115*

B cell, also known as a B lymphocyte   A type of lymphocyte derived from bone marrow stem cells that matures into an immunologically competent cell under the influence of the bursa of Fabricius in the chicken and bone marrow in nonavian species. Following interaction with antigen, it becomes a plasma cell, which synthesizes and secretes antibody molecules involved in humoral immunity.   *589, 627*

benthic (ben′thic)   Pertaining to the bottom of the sea or another body of water.   *521*

beta hemolysis   A zone of complete clearing around a bacterial colony growing on blood agar. The zone does not change significantly in color.   *572*

beta-lysin (ba″tah-li′sin)   A polypeptide released from blood platelets that can kill some gram-positive bacteria by disrupting their plasma membranes.   *576*

binal symmetry   The symmetry of some virus capsids (e.g., those of complex phages) that is a combination of icosahedral and helical symmetry.   *347*

binary fission   Asexual reproduction in which a cell or an organism separates into two cells.   *466, 522, 534*

binomial system   The nomenclature system in which an organism is given two names; the first is the capitalized generic name, and the second is the uncapitalized specific epithet.   *392*

biochemical markers   Biochemical characteristics of microorganisms that can be monitored to detect genetic exchange (e.g., traits such as antibiotic resistance and changes in nutrient requirements).   *919*

biochemical oxygen demand (BOD)   The amount of oxygen used by organisms in water under certain standard conditions; it provides an index of the amount of microbially oxidizable organic matter present.   *826*

biocide   A substance that can kill organisms.   *927*

bioconversion   The use of living organisms to modify substances that are not normally used for growth. Also known as biotransformation or microbial transformation.   *916*

biodegradation   (bi″o-deg″rah-da′shun)   The breakdown of a complex chemical through biological processes that can result in minor loss of functional groups, fragmentation into larger constitutents, or complete breakdown to carbon dioxide and minerals. Often the term refers to the undesired microbial-mediated destruction of materials such as paper, paint, and textiles.   *927*

biodegradation stimulation   The modification of environmental conditions, including inoculation with microorganisms, to enhance degradation of particular compounds.   *929*

biofilms   Organized microbial systems consisting of layers of microbial cells associated with surfaces, often with complex structural and functional characteristics. Biofilms have physical/chemical gradients that influence microbial metabolic processes. They can form on inanimate devices (catheters, medical prosthetic devices) and also cause fouling (e.g., of ships' hauls, water pipes, cooling towers).   *813, 833*

biogeochemical cycling   The oxidation and reduction of substances carried out by living organisms and/or abiotic processes that results in the cycling of elements within and between different parts of the ecosystem (the soil, aquatic environment, and atmosphere).   *814*

bioinsecticide   A pathogen that is used to kill or disable unwanted insect pests. Bacteria, fungi, or viruses are used, either directly or after manipulation, to control insect populations.   *923*

biologic transmission   A type of vector-borne transmission in which a pathogen goes through some morphological or physiological change within the vector.   *709*

bioluminescence (bi′o-loo″mĭ-nes′ɔns)   The production of light by living cells, often through the oxidation of molecules by the enzyme luciferase.   *427*

biomagnification   The increase in concentration of a substance in higher level consumer organisms.   *819*

biopolymer (bi′o-pol′ĭ-mer)   A polymer produced by microorganisms or other organisms; used in food and industrial microbiology.   *925*

bioremediation   The use of biologically mediated processes to remove or degrade pollutants from specific environments. Bioremediation can be carried out by modification of the environment to accelerate biological processes, either with or without the addition of specific microorganisms.   *854, 929*

biosensor   The coupling of a biological process with production of an electrical signal or light to detect the presence of particular substances.   *932*

biosynthesis   *See* anabolism.   *165*

biotransformation   *See* bioconversion.   *916*

black peidra (pe-a′drah)   A fungal infection caused by *Piedraia hortae* that forms hard black nodules on the hairs of the scalp.   *791*

blastomycosis (blas″to-mi-ko′sis)   A systemic fungal infection caused by *Blastomyces dermatitidis* and marked by suppurating tumors in the skin or by lesions in the lungs.   *795*

blastospore (blas′to-spōr)   A spore formed by budding from a hypha.   *509*

blood type   One of the classes into which blood can be separated based on the presence or absence of specific erythrocyte glycoprotein antigens.   *627*

B lymphocyte   *See* B cell.   *589, 627*

bottom yeast   Yeast used in beer production that tends to settle to the bottom of the fermentation vessel.   *898*

botulism (boch′oo-lizm)   A form of food poisoning caused by a neurotoxin (botulin) produced by *Clostridium botulinum* serotypes A–G; sometimes found in improperly canned or preserved food.   *767*

bright-field microscope   A microscope that illuminates the specimen directly with bright light and forms a dark image on a brighter background.   *18*

Bright's disease   *See* glomerulonephiritis.   *748*

broad-spectrum drugs   Chemotherapeutic agents that are effective against many different kinds of pathogens.   *658*

bronchial asthma   An example of an atopic allergy involving the lower respiratory tract.   *622*

bubo (bu′bo)   A tender, inflamed, enlarged lymph node that results from a variety of infections.   *754*

bubonic plague   *See* plague.   *754*

budding   A vegetative outgrowth of yeast and some bacteria as a means of asexual reproduction; the daughter cell is smaller than the parent.   *466*

bulking sludge   Sludges produced in sewage treatment that do not settle properly, usually due to the development of filamentous microorganisms.   *845*

burst *See* rise period. *355*

burst size The number of phages released by a host cell during the lytic life cycle. *355*

butanediol fermentation A type of fermentation most often found in the family *Enterobacteriaceae* in which 2,3-butanediol is a major product; acetoin is an intermediate in the pathway and may be detected by the Voges-Proskauer test. *175*

C

Calvin cycle The main pathway for the fixation (or reduction and incorporation) of CO_2 into organic material by photoautotrophs during photosynthesis; it also is found in chemolithoautotrophs. *194*

cancer (kan'ser) A malignant tumor that expands locally by invasion of surrounding tissues, and systemically by metastasis. *379*

candidal vaginitis Vaginitis caused by *Candida*. *799*

candidiasis (kan″dĭ-di'ah-sis) An infection caused by *Candida* species of dimorphic fungi, commonly involving the skin. *797*

canning A method of food preservation in which the food is placed in containers, hermetically sealed, and heated to destroy pathogens and spoilage microorganisms. *883*

capsid (kap'sid) The protein coat or shell that surrounds a virion's nucleic acid. *340*

capsomer (kap'so-mer) The ring-shaped morphological unit of which icosahedral capsids are constructed. *341*

capsule A layer of well-organized material, not easily washed off, lying outside the bacterial cell wall. *58*

carboxysomes Polyhedral inclusion bodies that contain the CO_2 fixation enzyme ribulose 1,5-bisphosphate carboxylase; found in cyanobacteria, nitrifying bacteria, and thiobacilli. *45, 194*

caries (kar'e-ēz) Bone or tooth decay. *784*

carotenoids (kah-rot'e-noids) Pigment molecules, usually yellowish in color, that are often used to aid chlorophyll in trapping light energy during photosynthesis. *183*

carrier An infected individual who is a potential source of infection for others and plays an important role in the epidemiology of a disease. *705*

caseous lesion (ka'se-us) A lesion resembling cheese or curd; cheesy. Most caseous lesions are caused by *M. tuberculosis*. *750*

casual carrier An individual who harbors an infectious organism for only a short period. *706*

catabolism (kah-tab'o-lizm) That part of metabolism in which larger, more complex molecules are broken down into smaller, simpler molecules with the release of energy. *165*

catabolite repression (kah-tab'o-līt) Inhibition of the synthesis of several catabolic enzymes by a metabolite such as glucose. *247*

catalyst (kat'ah-list) A substance that accelerates a reaction without being permanently changed itself. *158*

catalytic antibody An antibody that catalyzes a specific chemical reaction by lowering the free energy of activation. *603*

catalytic site *See* active site. *160*

catheter (kath'ě-ter) A tubular surgical instrument for withdrawing fluids from a cavity of the body, especially one for introduction into the bladder through the urethra for the withdrawal of urine. *677*

cat-scratch disease (CSD) A loosely defined syndrome caused by either of the following gram-negative bacilli: *Bartonella* (*Rochalimaea*) *henselae* or *Afipia felis*. The typical case of CSD is self-limiting, with abatement of symptoms over a period of days to weeks. *755*

cell-associated differentiation antigens (CDs) Functional cell surface proteins or receptors that can be measured in situ from peripheral blood, biopsy samples, or other body fluids. They also are called cluster of differentiation antigens. They can be used to identify leukocyte subpopulations. Some examples include interleukin-2 receptor (IL-2R), CD4, CD8, CD25, and intercellular adhesion molecule-1 (ICAM-1). *628*

cell cycle The sequence of events in a cell's growth-division cycle between the end of one division and the end of the next. In eucaryotic cells, it is composed of the G_1 period, the S period in which DNA and histones are synthesized, the G_2 period, and the M period (mitosis). *86, 250*

cell-mediated immunity The type of immunity that results from T cells coming into close contact with foreign cells or infected cells to destroy them; it can be transferred to a nonimmune individual by the transfer of cells. *589*

cellular slime molds Slime molds with a vegetative phase consisting of amoeboid cells that aggregate to form a multicellular pseudoplasmodium; they belong to the division *Acrasiomycota*. *515*

cellulitis (sel″u-li'tis) A diffuse spreading infection of subcutaneous skin tissue caused by streptococci, staphylococci, or other organisms. The tissue is inflamed with edema, redness, pain, and interference with function. *747*

cell wall The strong layer or structure that lies outside the plasma membrane; it supports and protects the membrane and gives the cell shape. *88*

cephalosporin (sef″ah-lo-spōr'in) A group of β-lactam antibiotics derived from the fungus *Cephalosporium*, which share the 7-aminocephalosporanic acid nucleus. *667*

chancre (shang'ker) The primary lesion of syphilis, occurring at the site of entry of the infection. *762*

chancroid (shang'kroid) A sexually transmitted disease caused by the gram-negative bacterium *Haemophilus ducreyi*. Worldwide, chancroid is an important cofactor in the transmission of the AIDS virus. Also known as genital ulcer disease due to the painful circumscribed ulcers that form on the penis or entrance to the vagina. *756*

chemical oxygen demand (COD) The amount of chemical oxidation required to convert organic matter in water and wastewater to CO_2. *826*

chemiosmotic hypothesis (kem″e-o-os-mot'ik) The hypothesis that a proton gradient and an electrochemical gradient are generated by electron transport and then used to drive ATP synthesis by oxidative phosphorylation. *172*

chemoautotroph (ke″mo-aw'to-trōf″) *See* chemolithotrophic autotrophs. *101*

chemoheterotroph (ke″mo-het'er-o-trōf″) *See* chemoorganotrophic heterotrophs. *100*

chemokine (ke″mo-kīn) A low molecular weight proinflammatory cytokine that functions as a paracrine (a hormone that influences nearby cells). Chemokines play important roles in acute and chronic inflammation and have immunoregulatory functions. They stimulate leukocytes. Examples include interleukin-8, melanoma growth stimulation factor, monocyte chemoattractant protein-1, and macrophage inflammatory protein. *609*

chemolithotroph (ke″mo-lith'o-trōf) *See* chemolithotrophic autotrophs. *180*

chemolithotrophic autotrophs Microorganisms that oxidize reduced inorganic compounds to derive both energy and electrons; CO_2 is their carbon source. *101*

chemoorganotrophic heterotrophs Organisms that use organic compounds as sources of energy, hydrogen, electrons, and carbon for biosynthesis. *100*

chemoreceptors Special protein receptors in the plasma membrane or periplasmic space that bind chemicals and trigger the appropriate chemotaxic response. *63*

chemostat (ke′mo-stat) A continuous culture apparatus that feeds medium into the culture vessel at the same rate as medium containing microorganisms is removed; the medium in a chemostat contains one essential nutrient in limiting quantities. *122*

chemotaxis (ke″mo-tak'sis) The pattern of microbial behavior in which the microorganism moves toward chemical attractants and away from repellents. *63*

chemotherapeutic agents (ke″mo-ther-ah-pu'tik) Compounds used in the treatment of disease that destroy pathogens or inhibit their growth at concentrations low enough to avoid doing undesirable damage to the host. *657*

chemotrophs (ke′mo-trōfs) Organisms that obtain energy from the oxidation of chemical compounds. *100*

chickenpox (varicella; chik'en-poks) A highly contagious skin disease, usually affecting 2 to 7-year-old children; it is caused by the varicella-zoster virus, which is acquired by droplet inhalation into the respiratory system. *716*

chimera (ki-me'rah) A recombinant plasmid containing foreign DNA, which is used as a cloning vector in genetic engineering. *320*

chitin (ki'tin) A tough, resistant, nitrogen-containing polysaccharide forming the walls of certain fungi, the exoskeleton of arthropods, and the epidermal cuticle of other surface structures of certain protists and animals. *504*

chlamydial pneumonia (klə-mid'e-əl noo-mo'ne-ə) A pneumonia caused by *Chlamydia pneumoniae*. Clinically, infections are mild and 50% of adults have antibody to the chlamydiae. *777*

chlamydospore (klam'ĭ-do-spōr″) An asexually produced, thick-walled resting spore formed by some fungi. *508*

chloramphenicol (klo″ram-fen′ĭ-kol) A broad-spectrum antibiotic that is produced by *Streptomyces venezuelae* or synthetically; it binds to the large ribosomal subunit and inhibits the peptidyl transferase reaction. 668

chlorophyll (klor′o-fil) The green photosynthetic pigment that consists of a large tetrapyrrole ring with a magnesium atom in the center. 182

chloroplast (klo′ra-plast) A eucaryotic plastid that contains chlorophyll and is the site of photosynthesis. 82

cholera (kol′er-ah) An acute infectious enteritis endemic and epidemic in Asia and periodically spreading to the Middle East, Africa, Southern Europe, and South America; caused by *Vibrio cholerae*. 767

choleragen (kol′er-ah-gen) The cholera toxin; an extremely potent protein molecule elaborated by strains of *Vibrio cholerae* in the small intestine after ingestion of feces-contaminated water or food. It acts on epithelial cells to cause hypersecretion of chloride and bicarbonate and an outpouring of large quantities of fluid from the mucosal surface. 767

chromatin (kro′mah-tin) The DNA-containing portion of the eucaryotic nucleus; the DNA is almost always complexed with histones. It can be very condensed (heterochromatin) or more loosely organized and genetically active (euchromatin). 84

chromogen (kro′me-jen) A colorless substrate that is acted on by an enzyme to produce a colored end product. 646

chromoblastomycosis (kro″mo-blas″to-mi-ko′sis) A chronic fungal infection of the skin, producing wartlike nodules that may ulcerate. It is caused by the black molds *Phialophora verrucosa* or *Fonsecaea pedrosoi*. 794

chromophore group (kro′mo-fōr) A chemical group with double bonds that absorbs visible light and gives a dye its color. 25

chromosomes (kro′mo-somz) The bodies that have most or all of the cell's DNA and contain most of its genetic information (mitochondria and chloroplasts also contain DNA and genes). 84

chronic carrier An individual who harbors a pathogen for a long time. 706

chrysolaminarin The polysaccharide storage product of the chrysophytes and diatoms. 526

ciguatera (se″gwah-ta′rah) A form of ichthyosarcotoxism, marked by gastrointestinal and neurological symptoms from ingestion of marine fish (e.g., grouper and snapper) that store the toxin in their tissues; it occurs in tropical and subtropical coastal areas. 529

cilia (sil′e-ah) Threadlike appendages extending from the surface of some protozoa that beat rhythmically to propel them; cilia are membrane-bound cylinders with a complex internal array of microtubules, usually in a 9 + 2 pattern. 90

citric acid cycle *See* tricarboxylic acid (TCA) cycle. 170

classical pathway The antibody-dependent pathway of complement activation; it leads to the lysis of pathogens and stimulates phagocytosis and other host defenses. 634

classification The arrangement of organisms into groups based on mutual similarity or evolutionary relatedness. 391

clinical microbiologist A person who isolates and identifies microorganisms from clinical specimens. 677

clonal anergy (klōn′əl an′ər-je) The mechanism that turns off self-reactive lymphocytes as a consequence of the way in which "self"-molecules are presented to them. 619

clonal selection theory The theory that clones of effector B and T cells arise from single cells (or very small clones) that are stimulated to reproduce by antigen binding to their receptors. 600

clone (klōn) A group of genetically identical cells or organisms derived by asexual reproduction from a single parent. 255, 600

clostridial myonecrosis (klo-strid′e-al mi″o-ne-kro′sis) Death of individual muscle cells caused by clostridia. Also called gas gangrene. 756

clue cell A sloughed-off vaginal epithelial cell covered with bacteria—mostly *Gardnerella vaginalis*. 755

coaggregation The collection of a variety of bacteria on a surface such as a tooth surface because of cell-to-cell recognition of genetically distinct bacterial types. Many of these interactions appear to be mediated by a lectin on one bacterium that interacts with a complementary carbohydrate receptor on another bacterium. 782

coagulase (ko-ag′u-las) An enzyme that induces blood clotting; it is characteristically produced by pathogenic staphylococci. 442

coagulation (ko-ag″u-la′shun) 1. The process of blood clot formation. 2. In municipal water purification, it is the process that precipitates microorganisms and organic material out of water by the addition of chemicals. 570, 848

coccidioidomycosis (kok-sid″e-oi″do-mi-ko′sis) A fungal disease caused by *Coccidioides immitis* that exists in dry, highly alkaline soils. Also known as valley fever, San Joaquin fever, or desert rheumatism. 795

coccus (kok′us, pl. **cocci**, kok′si) A roughly spherical bacterial cell. 38

code degeneracy The genetic code is organized in such a way that often there is more than one codon for each amino acid. 257

codon (ko′don) A sequence of three nucleotides in mRNA that directs the incorporation of an amino acid during protein synthesis or signals the start or stop of translation. 257

coenocytic (se″no-sit′ik) Refers to a multinucleate cell or hypha formed by repeated nuclear divisions not accompanied by cell divisions. 115, 506

coenzyme (ko-en′zīm) A loosely bound cofactor that often dissociates from the enzyme active site after product has been formed. 157

cofactor The nonprotein component of an enzyme; it is required for catalytic activity. 159

cold sore A lesion caused by the herpes simplex virus; usually occurs on the border of the lips or nares. Also known as a fever blister or herpes labialis. 730

colicin (kol′ĭ-sin) A plasmid-encoded protein that is produced by enteric bacteria and binds to specific receptors on the cell envelope of sensitive target bacteria, where it may cause lysis or attack specific intracellular sites such as ribosomes. 576

coliform (ko′lĭ-form) A gram-negative, nonsporing, facultative rod that ferments lactose with gas formation within 48 hours at 35°C. 850

colony A cluster or assemblage of microorganisms growing on a solid surface such as the surface of an agar culture medium; the assemblage often is directly visible, but may be seen only microscopically. 108

colony forming units (**CFU**) The number of microorganisms that can form colonies when cultured using spread plates or pour plates, an indication of the number of viable microorganisms in a sample. 119

colony-stimulating factors (**CSFs**) Cytokines that stimulate the production of specific cell lines (mononuclear phagocytes, granulocytes) and modulate secretory capacity, chemotaxis, and phagocyte killing. 607

Colorado tick fever A disease that occurs in the mountainous regions of western states. It is caused by an RNA virus of the genus *Coltivirus* that is spread from ground squirrels, rabbits, and deer to humans by the tick, *Dermacentor andersoni*. Complications are rare. 722

colostrum (kō-los′trum) The thin milky fluid secreted by the mammary gland a few days before or after parturition. 588

comedo (kom′ĕ-do; pl., **comedones**) A plug of dried sebum in an excretory duct of the skin. 556

cometabolism The modification of a compound not used for growth by a microorganism, which occurs in the presence of another organic material that serves as a carbon and energy source. 873

commensal (kŏ-men′sal) Living on or within another organism without injuring or benefiting the other organism. 547

commensalism (kŏ-men′sal-izm″) A type of symbiosis in which one individual gains from the association and the other is neither harmed nor benefited. 547

common cold An acute, self-limiting, and highly contagious virus infection of the upper respiratory tract that produces inflammation, profuse discharge, and other symptoms. 730

common-source epidemic An epidemic that is characterized by a sharp rise to a peak and then a rapid, but not pronounced, decline in the number of individuals infected; it usually involves a single contaminated source from which individuals are infected. 703

common vehicle transmission The transmission of a pathogen to a host by means of an inanimate medium or vehicle. 709

communicable disease A disease associated with a pathogen that can be transmitted from one host to another. 705

community An assemblage of different types of organisms or a mixture of different microbial populations. 813

competent A bacterial cell that can take up free DNA fragments and incorporate them into its genome during transformation. 292

competitive inhibitor A molecule that inhibits enzyme activity by competing with the substrate at the enzyme's active site. 161

complementary DNA (cDNA) A DNA copy of an RNA molecule (e.g., a DNA copy of an mRNA). *214*

complement fixation The activation of the complement system with the binding of the components to the activator substance and to each other, with their consequent removal from serum. *634*

complement system A group of plasma proteins that plays a major role in an animal's defensive immune response. *633*

complex medium Culture medium that contains some ingredients of unknown chemical composition. *107*

complex viruses Viruses with capsids having a complex symmetry that is neither icosahedral nor helical. *340*

compromised host A host with lowered resistance to infection and disease for any of several reasons. The host may be seriously debilitated (due to malnutrition, cancer, diabetes, leukemia, or another infectious disease), traumatized (from surgery or injury), immunosuppressed, or have an altered microbiota due to prolonged use of antibiotics. *797*

concatemer A long DNA molecule consisting of several genomes linked together in a row. *358*

conditional mutations Mutations that are expressed only under certain environmental conditions. *262*

confocal scanning laser microscopy Microscopy with monochromatic laser-derived light allowing precise control of the depth of light collection and depth of focus, which are "cofocused" in the specimen at the same level or are "confocal." This technique allows horizontal and vertical sectioning at less than 1 μm resolution. *833*

conformational change hypothesis The hypothesis that the energy from electron transport is used to induce changes in the shape of the enzyme that synthesizes ATP and thereby drives ATP synthesis. *172*

congenital (neonatal) herpes An infection of a newborn caused by transmission of the herpesvirus during vaginal delivery. *732*

congenital rubella syndrome A wide array of congenital defects affecting the heart, eyes, and ears of a fetus during the first trimester of pregnancy and caused by the rubella virus. *720*

congenital syphilis Syphilis that is acquired in utero from the mother. *762*

conidiospore (ko-nid′e-o-spōr) An asexual, thin-walled spore borne on hyphae and not contained within a sporangium; it may be produced singly or in chains. *509*

conidium (ko-nid′e-um; pl., **conidia**) *See* conidiospore. *492*

conjugants (kon′joo-gants) Complementary mating types that participate in a form of protozoan sexual reproduction called conjugation. *535*

conjugation (kon″ju-ga′shun) 1. The form of gene transfer and recombination in bacteria that requires direct cell-to-cell contact. 2. A complex form of sexual reproduction commonly employed by protozoa. *288, 535*

conjugative plasmid A plasmid that carries the genes for sex pili and can transfer copies of itself to other bacteria during conjugation. *280*

conjunctivitis of the newborn *See* ophthalmia neonatorum. *757*

conoid (ko′noid) A hollow cone of spirally coiled filaments in the anterior tip of certain apicomplexan protozoa. *540*

constant region (C$_L$ and C$_H$) The part of an antibody molecule that does not vary greatly in amino acid sequence among molecules of the same class, subclass, or type. *592*

constitutive mutant A strain that produces an inducible enzyme continually, regardless of need, because of a mutation in either the operator or regulator gene. *243*

constructed wetlands Intentional creation of marshland plant communities and their associated microorganisms to purify water by the removal of bacteria, organic matter, and chemicals as the water passes through the aquatic plant communities. *846*

consumer An organism that feeds directly on living or dead animals, by ingestion or by phagocytosis. *811*

contact transmission Transmission of the pathogen by contact of the source or reservoir of the pathogen with the host. *709*

continuous culture system A culture system with constant environmental conditions maintained through continual provision of nutrients and removal of wastes. *122*

continuous feed The continuous addition of a substance to a culture during growth. *909*

contractile vacuole (vak′u-ōl) In protists and some animals, a clear fluid-filled cell vacuole that takes up water from within the cell and then contracts, releasing it to the outside through a pore in a cyclical manner. Contractile vacuoles function primarily in osmoregulation and excretion. *533*

convalescent carrier (kon″vah-les′ent) An individual who has recovered from an infectious disease but continues to harbor large numbers of the pathogen. *705*

corepressor (ko″re-pre′sor) A small molecule that inhibits the synthesis of a repressible enzyme. *243*

cortex (kor′teks) The layer of a bacterial endospore that is thought to be particularly important in conferring heat resistance on the endospore. *67*

coryza (kŏ-ri′zah) *See* common cold. *730*

cosmid (koz′mid) A plasmid vector with lambda phage cos sites that can be packaged in a phage capsid; it is useful for cloning large DNA fragments. *321*

cristae (kris′te) Infoldings of the inner mitochondrial membrane. *82*

crossing-over A process in which segments of two adjacent DNA strands are exchanged; breaks occur in both strands, and the exposed ends of each strand join to those of the opposite segment on the other strand. *278*

crustose Lichens that are compact and appressed to a substratum. *548*

cryptococcosis (krip″to-kok-o′sis) An infection caused by the basidiomycete, *Cryptococcus neoformans*, which may involve the skin, lungs, brain, or meninges. *514, 795*

crystallizable fragment (Fc) The stem of the Y portion of an antibody molecule. Cells such as macrophages bind to the Fc region, and it also is involved in complement activation. *592*

cutaneous anthrax (ku-ta′ne-us an′thraks) A form of anthrax involving the skin. *755*

cutaneous diphtheria (ku-ta′ne-us dif-the′re-ah) A skin disease caused by *Corynebacterium diphtheriae* that infects wound or skin lesions, causing a slow-healing ulceration. *744*

cyanobacteria (si″ah-no-bak-te′re-ah) A large group of photosynthetic bacteria with oxygenic photosynthesis and a photosynthetic system like that present in eucaryotic photosynthetic organisms. *459*

cyclic photophosphorylation (fo″to-fos″for-ĭ-la′shun) The formation of ATP when light energy is used to move electrons cyclically through an electron transport chain during photosynthesis; only photosystem I participates. *184*

cyst (sist) A general term used for a specialized microbial cell enclosed in a wall. Cysts are formed by protozoa and a few bacteria. They may be dormant, resistant structures formed in response to adverse conditions or reproductive cysts that are a normal stage in the life cycle. *534*

cytochromes (si′to-krōms) Heme proteins that carry electrons, usually as members of electron transport chains. *157*

cytokine (si′to-kīn) A general term for nonantibody proteins, released by a cell in response to inducing stimuli, which are mediators that influence other cells. Are produced by lymphocytes, monocytes, macrophages, and other cells. *589, 607*

cytokinesis (si″to-ki-ne′sis) The division of a cell's cytoplasm into two halves after nuclear division. *88*

cytomegalovirus inclusion disease (si″to-meg″ah-lo-vi′rus) An infection caused by the cytomegalovirus and marked by nuclear inclusion bodies in enlarged infected cells. *731*

cytopathic effect (si″to-path′ik) The observable change that occurs in cells as a result of viral replication. Examples include ballooning, binding together, clustering, or even death of the cultured cells. *335, 683*

cytoplasmic matrix (si″to-plaz′mik) The protoplasm of a cell that lies within the plasma membrane and outside any other organelles. In bacteria it is the substance between the cell membrane and the nucleoid. *45, 76*

cytoproct (si′to-prokt) Site on a protozoan where undigestible matter is expelled. *541*

cytosine (si′to-sēn) A pyrimidine 2-oxy-4-aminopyrimidine that is found in nucleosides, nucleotides, and nucleic acids. *204*

cytoskeleton (si″to-skel′ĕ-ton) A network of microfilaments, microtubules, intermediate filaments, and other components in the cytoplasm of eucaryotic cells that helps give them shape. *77*

cytostome (si′to-stōm) A permanent site in the protozoan ciliate body through which food is ingested. *534*

cytotoxic T (T_C) cell (si″to-tok′sik) A cell that is capable of recognizing virus-infected cells through the major histocompatability molecules and destroying the infected cell. Also called CD8 cell. *619*

cytotoxin (si′to-tok′sin) A toxin or antibody that has a specific toxic action upon cells; cytotoxins are named according to the cell for which they are specific (e.g., nephrotoxin). *567*

D

Dane particle A 42 nm spherical particle that is one of three that are seen in hepatitis B virus infections. The Dane particle is the complete virion. *735*

dark-field microscopy Microscopy in which the specimen is brightly illuminated while the background is dark. *21*

dark reactivation The excision and replacement of thymine dimers in DNA that occurs in the absence of light. *133*

deamination (de-am″i-na′shun) The removal of amino groups from amino acids. *179*

death phase The decrease in viable microorganisms that occurs after the completion of growth in a batch culture. *116*

decimal reduction time (D or D value) The time required to kill 90% of the microorganisms or spores in a sample at a specified temperature. *140*

decomposer An organism that breaks down complex materials into simpler ones, including the release of simple inorganic products. *811*

decomposer-reducer An organism, such as an insect or earthworm, that carries out decomposition and physically reduces the size of substrate particles. *862*

defensin (de-fens′sin) Specific peptides produced by neutrophils that permeabilize the outer and inner membranes of certain microorganisms, thus killing them. *583*

defined medium Culture medium made with components of known composition. *106*

delayed-type hypersensitivity T (T_{DTH}) cell A cell that is responsible for initiating type IV delayed-type hypersensitivity reactions through secretion of cytokines. *619*

Delta agent A defective RNA virus that is transmitted as an infectious agent, but cannot cause disease unless the individual is also infected with the hepatitis B virus. *See* hepatitis D. *737*

denaturation (de-na″chur-a′shun) A change in the shape of an enzyme that destroys its activity; the term is also applied to changes in nucleic-acid shape. *161*

dendritic cell (den-drit′ik) An antigen-presenting cell found in lymph nodes, spleen, and mucous membranes that possesses long processes that interdigitate among lymphoid cells. *613*

dendrogram A treelike diagram that is used to graphically summarize mutual similarities and relationships between organisms. *398*

denitrification (de-ni″tri-fi-ka′shon) The reduction of nitrate to gas products, primarily nitrogen gas, during anaerobic respiration. *177, 816*

dental plaque (plak) A thin film on the surface of teeth consisting of bacteria embedded in a matrix of bacterial polysaccharides, salivary glycoproteins, and other substances. *782*

deoxyribonucleic acid (DNA; de-ok″se-ri″bo-nu-kle′ik) The nucleic acid that constitutes the genetic material of all cellular organisms. It is a polynucleotide composed of deoxyribonucleotides connected by phosphodiester bonds. *49, 212*

dermatophyte (der′mah-to-fīt″) A fungus parasitic on the skin. *792*

dermatophytosis (der″mah-to-fi-to′sis) A fungal infection of the skin; the term is a general term that comprises the various forms of tinea, and it is sometimes used to specifically refer to athlete's foot (tinea pedis). *792*

desensitization (de-sen″si-ti-za′shun) To make a sensitized or hypersensitive individual insensitive or nonreactive to a sensitizing agent. *622*

detergent (de-ter′jent) An organic molecule, other than a soap, that serves as a wetting agent and emulsifier; it is normally used as cleanser, but some may be used as antimicrobial agents. *148*

diatomaceous earth (di″ə-to-ma′shəs) Oceanic sediments formed by accumulation of the silica frustules of diatoms. *527*

diatoms (di′ah-toms) Algal protists with siliceous cell walls called frustules. They constitute a substantial subfraction of the phytoplankton. *526*

diauxic growth (di-awk′sik) A biphasic growth pattern or response in which a microorganism, when exposed to two nutrients, initially uses one of them for growth and then alters its metabolism to make use of the second. *247*

dictyosomes (dik′te-o-sōms) Individual stacks of cisternae in a Golgi apparatus. *79*

differential media (dif″er-en′shal) Culture media that distinguish between groups of microorganisms based on differences in their biological characteristics. *108*

differential staining procedures Staining procedures that divide bacteria into separate groups based on staining properties. *26*

diffusely adhering *E. coli* (DAEC) DAEC strains of *E. coli* adhere over the entire surface of epithelial cells and usually cause diarrheal disease in immunologically naive and malnourished children. *771*

diffusion (dĭ-fu′zhun) *See* facilitated diffusion and passive diffusion. *102, 103*

dikaryotic stage (di-kar-e-ot′ik) In fungi, having pairs of nuclei within cells or compartments. Each cell contains two separate haploid nuclei, one from each parent. *509*

dinoflagellate (di″no-flaj′e-lāt) An algal protist characterized by two flagella used in swimming in a spinning pattern. Many are bioluminescent and an important part of marine phytoplankton. *528*

diphtheria (dif-the′re-ah) An acute, highly contagious childhood disease that generally affects the membranes of the throat and less frequently the nose. It is caused by *Corynebacterium diphtheriae*. *744*

dipicolinic acid A substance present in great quantities in the bacterial endospore. It is thought to contribute to the endospore's heat resistance. *67*

diplococcus (dip′lo-kok′us) A pair of cocci. *38*

diploid (dip′loid) Having double the basic chromosome number. *88*

disease (di-zez) A deviation or interruption of the normal structure or function of any part of the body that is manifested by a characteristic set of symptoms and signs. *700*

disease syndrome (sin′drŏm) A set of signs and symptoms that are characteristic of the disease. *702*

disinfectant (dis″in-fek′tant) An agent, usually chemical, that disinfects; normally, it is employed only with inanimate objects. *137*

disinfection (dis″in-fek′shun) The killing, inhibition, or removal of microorganisms that may cause disease. *137*

dissimilatory nitrate reduction The process in which some bacteria use nitrate as the electron acceptor at the end of their electron transport chain to produce ATP. The nitrate is reduced to nitrite or nitrogen gas. *177*

dissimilatory reduction The use of a substance as an electron acceptor in energy generation. The acceptor (e.g., sulfate or nitrate) is reduced but not incorporated into organic matter during biosynthetic processes. *197, 816*

diurnal oxygen shifts (di-er′nal) The changes in oxygen levels that occur in waters when algae produce and use oxygen on a cyclic basis during day and night. *842*

DNA ligase An enzyme that joins two DNA fragments together through the formation of a new phosphodiester bond. *219*

DNA polymerase (pol-im′er-ās) An enzyme that synthesizes new DNA using a parental DNA strand as a template. *218*

Döderlein's bacillus (ded′er-līnz) Large gram-positive bacteria (*Lactobacillus acidophilus*) found in the vagina that break down the glycogen produced by the vaginal epithelium and form lactic acid. *558*

domain (do-mān′) Each loop, along with approximately 25 amino acids on each side of the loop, of an antibody chain. It forms a compact, globular section. *592*

double diffusion agar assay (Ouchterlony technique) An immunodiffusion reaction in which both antibody and antigen diffuse through agar to form stable immune complexes, which can be observed visually. *649*

doubling time *See* generation time. *116*

DPT (diphtheria-pertussis-tetanus) vaccine A vaccine containing three antigens that is used to immunize people against diphtheria, pertussis or whooping cough, and tetanus. *744*

droplet nuclei Small particles (1 to 4 μm in diameter) that represent what is left from the evaporation of larger particles (10 μm or more in diameter) called droplets. *707*

D value *See* decimal reduction time. *140*

E

early mRNA Messenger RNA produced early in a virus infection that codes for proteins needed to take over the host cell and manufacture viral nucleic acids. *357*

Ebola virus hemorrhagic fever (a′bo-lə) An acute infection caused by a virus that produces varying degrees of hemorrhage, shock, and sometimes death. *721*

eclipse period (e-klips′) The initial part of the latent period in which infected host bacteria do not contain any complete virions. *356*

ecosystem (ek″o-sis′tem)   The total community of living organisms and their associated physical and chemical environment.   *811*

ectomycorrhizal   Referring to a mutualistic association between fungi and plant roots in which the fungus surrounds the root tip with a sheath.   *869*

ectoparasite (ek″to-par′ah-sīt)   A parasite that lives on the surface of its host.   *562*

ectoplasm (ek′to-plazm)   The outer stiffer portion or region of the cytoplasm in a protozoan, which may be differentiated in texture from the inner portion or endoplasm.   *533*

ectosymbiosis   A type of symbiosis in which one organism remains outside of the other organism.   *547*

effacing lesion (le′zhən)   The type of lesion caused by enteropathogenic strains of *E. coli* (EPEC) when the bacteria attach to and destroy the brush border of intestinal epithelial cells. The term AE *E. coli* is now used to designate true EPEC strains that are an important cause of diarrhea in children from developing countries and in traveler's diarrhea.   *770*

effector T cell   The general term for T cells that directly attack specific target cells.   *619*

ehrlichiosis (ar-lik″e-o′sis)   A tick-borne (*Dermacentor andersoni, Amblyomma americanum*) rickettsial disease caused by *Ehrlichia chaffeensis.* Once inside leukocytes, a nonspecific illness develops that resembles Rocky Mountain spotted fever.   *780*

electron transport chain   A series of electron carriers that operate together to transfer electrons from donors such as NADH and FADH$_2$ to acceptors such as oxygen.   *171*

electrophoresis (e-lek″tro-fo-re′sis)   A technique that separates substances through differences in their migration rate in an electrical field due to variations in the number and kinds of charged groups they have.   *313*

electroporation (e-lek″tro-pə-ra′shən)   The application of an electric field to create temporary pores in the plasma membrane in order to insert DNA into the cell and transform it.   *321*

elementary body   A small, dormant body that serves as the agent of transmission between host cells in the chlamydial life cycle.   *432*

ELISA   *See* enzyme-linked immunosorbent assay.   *644*

elongation cycle   The cycle in protein synthesis that results in the addition of an amino acid to the growing end of a peptide chain.   *229*

Embden-Meyerhof pathway (em′den mi′er-hof)   A pathway that degrades glucose to pyruvate; the six-carbon stage converts glucose to fructose 1,6-bisphosphate, and the three-carbon stage produces ATP while changing glyceraldehyde 3-phosphate to pyruvate.   *166*

encystation (en-sis-′ta′shen)   The formation of a cyst.   *534*

endemic disease (en-dem′ik)   A disease that is commonly or constantly present in a population, usually at a relatively steady low frequency.   *700*

endemic (murine) typhus (mu′rin ti′fus)   A form of typhus fever caused by the rickettsia *Rickettsia typhi* that occurs sporadically in individuals who come into contact with rats and their fleas.   *780*

endergonic reaction (end″er-gon′ik)   A reaction that does not spontaneously go to completion as written; the standard free energy change is positive, and the equilibrium constant is less than one.   *155*

endocytosis (en″do-si-to′sis)   The process in which a cell takes up solutes or particles by enclosing them in vesicles pinched off from its plasma membrane.   *79*

endogenote (en″do-je′nōt)   The recipient bacterial cell's own genetic material into which the donor DNA can integrate.   *279*

endogenous infection (en-doj′ĕ-nus in-fek′shun)   An infection by a member of an individual's own normal body microbiota.   *749*

endogenous pyrogen (en-doj′e-nus pi′ro-jen)   A substance such as the lymphokine interleukin-1, which is produced by host cells and induces a fever response in the host. It also is called simply a pyrogen.   *572*

endogenous respiration (en-doj′ĕ-nus res″pi-ra′shun)   The use of nutrients, usually internally stored reserves, to maintain an organism without growth.   *845*

endomycorrhizal   Referring to a mutualistic association of fungi and plant roots in which the fungus penetrates into the root cells and arbuscules and vesicles are formed.   *869*

endoparasite (en″do-par′ah-sīt)   A parasite that lives inside the body of its host.   *562*

endophyte (en′do-fīt)   A microorganism living within a plant, but not necessarily parasitic on it.   *871*

endoplasm (en′do-plazm)   The central portion of the cytoplasm in a protozoan.   *533*

endoplasmic reticulum (en″do-plas′mik rĕ-tik′u-lum)   A system of membranous tubules and flattened sacs (cisternae) in the cytoplasmic matrix of eucaryotic cells. Rough or granular endoplasmic reticulum (RER or GER) bears ribosomes on its surface; smooth or agranular endoplasmic reticulum (SER or AER) lacks them.   *77*

endosome (en′do-sōm)   A membranous vesicle formed by endocytosis.   *79*

endospore (en′do-spōr)   An extremely heat- and chemical-resistant, dormant, thick-walled spore that develops within bacteria.   *67*

endosymbiont (en″do-sim′be-ont)   An organism that lives within the body of another organism in a symbiotic association.   *550*

endosymbiosis (en″do-sim″bi-o′sis)   A type of symbiosis in which one organism is found within another organism.   *547*

endosymbiotic theory or **hypothesis**   The theory that eucaryotic organelles such as mitochondria and chloroplasts arose when bacteria established an endosymbiotic relationship with the eucaryotic ancestor and then evolved into eucaryotic organelles.   *84, 410*

endotoxin (en″do-tox′sin)   The heat-stable lipopolysaccharide in the outer membrane of the cell wall of gram-negative bacteria that is released when the bacterium lyses, or sometimes during growth, and is toxic to the host.   *569*

endotoxin units (E.U.)   The units in which the concentration of an endotoxin is measured.   *570*

end product inhibition   *See* feedback inhibition.   *241*

energy   The capacity to do work or cause particular changes.   *153*

enteric bacteria (enterobacteria; en-ter′ik)   Members of the family *Enterobacteriaceae* (gram-negative, peritrichous or nonmotile, facultatively anaerobic, straight rods with simple nutritional requirements); also used for bacteria that live in the intestinal tract.   *425*

enteroaggregative *E. coli* (EAggEC)   EAggEC strains of *E. coli* adhere to epithelial cells in localized regions of the small intestine forming clumps of bacteria with a "stacked brick" appearance. Conventional extracellular toxins have not been detected in EAggEC even though it causes diarrhea.   *771*

enterohemorrhagic *E. coli* (EHEC) (en′tər-o-hem″ə-raj′ik)   EHEC strains of *E. coli* (0157:H7) produce several cytotoxins that provoke fluid secretion in traveler's diarrhea; however, their mode of action is unknown.   *771*

enteroinvasive *E. coli* (EIEC) (en′tər-o-in-va′siv)   EIEC strains of *E. coli* cause traveler's diarrhea by penetrating and binding to the intestinal epithelial cells. EIEC may also produce a cytotoxin and enterotoxin.   *770*

enteropathogenic *E. coli* (EPEC) (en′tər-o-path-o-jen′ik)   EPEC strains of *E. coli* attach to the brush border of intestinal epithelial cells and cause a specific type of cell damage called effacing lesions that lead to traveler's diarrhea.   *770*

enterotoxigenic *E. coli* (ETEC) (en′tər-o-tok″sĭ-jen′ik)   ETEC strains of *E. coli* produce two plasmid-encoded enterotoxins (which are responsible for traveler's diarrhea) and are distinguished by their heat stability: heat-stable enterotoxin (ST) and heat-labile enterotoxin (LT).   *770*

enterotoxin (en″ter-o-tok′sin)   A toxin specifically affecting the cells of the intestinal mucosa, causing vomiting and diarrhea.   *567, 766*

enthalpy (en′thal-pe)   The heat content of a system; in a biological system it is essentially equivalent to the total energy of the system.   *155*

Entner-Doudoroff pathway   A pathway that converts glucose to pyruvate and glyceraldehyde 3-phosphate by producing 6-phosphogluconate and then dehydrating it.   *169*

entropy (en′tro-pe)   A measure of the randomness or disorder of a system; a measure of that part of the total energy in a system that is unavailable for useful work.   *154*

envelope (en′vĕ-lōp)   1. All the structures outside the plasma membrane in bacterial cells. 2. In virology it is an outer membranous layer that surrounds the nucleocapsid in some viruses.   *51, 340*

enzootic (en″zo-ot′ik)   The moderate prevalence of a disease in a given animal population.   *701*

enzyme (en′zīm)   A protein catalyst with specificity for both the reaction catalyzed and its substrates.   *158*

enzyme-linked immunosorbent assay (ELISA)   A technique used for detecting and quantifying specific antibodies and antigens.   *644*

epidemic (ep″ĭ-dem′ik)   A disease that suddenly increases in occurrence above the normal level in a given population.   *700*

epidemic (louse-borne) typhus (ep″i-dem′ik ti′fus)   A disease caused by *Rickettsia prowazekii* that is transmitted from person to person by the body louse.   *780*

epidemiologist (ep″ĭ-de″me-ol′o-jist)   A person who specializes in epidemiology.   *700*

epidemiology (epi″-de″me-ol′o-je)   The study of the factors determining and influencing the frequency and distribution of disease, injury, and other health-related events and their causes in defined human populations.   *700*

epilimnion (ep″e-lim′ne-on″)   The warmer aerobic layer at the surface of a thermally stratified lake. It lies above the thermocline.   *840*

episome (ep′ĭ-sōm)   A plasmid that can exist either independently of the host cell's chromosome or be integrated into it.   *280*

epitheca (ep″ĭ-the′kah)   The larger of two halves of a diatom frustule (shell).   *526*

epitope (ep′i-tōp)   An area of the antigen molecule that stimulates the production of, and combines with, specific antibodies; also known as antigenic determinant site.   *590*

epizootic (ep″ĭ-zo-ot′ik)   A sudden outbreak of a disease in an animal population.   *701*

epizootiology (ep″i-zo-ot″e-ol′o-je)   The field of science that deals with factors determining the frequency and distribution of a disease within an animal population.   *701*

equilibrium (e″kwĭ-lib′re-um)   The state of a system in which no net change is occurring and free energy is at a minimum; in a chemical reaction at equilibrium, the rates in the forward and reverse directions exactly balance each other out.   *155*

equilibrium constant   The constant that is characteristic of a chemical reaction at equilibrium and is based on the relative equilibrium concentrations or activities of products and reactants.   *155*

ergot (er′got)   The dried sclerotium of *Claviceps purpurea*. Also, an ascomycete that parasitizes rye and other higher plants causing the disease called ergot.   *510*

ergotism (er′got-izm)   The disease or toxic condition caused by eating grain infected with ergot; it is often accompanied by gangrene, psychotic delusions, nervous spasms, abortion, and convulsions in humans and in animals.   *510, 889*

eructation (ĕ-ruk-ta′shun)   To belch forth; what ruminants do in releasing CO₂, methane, and other gases.   *552*

erysipelas (er″ĭ-sip′ĕ-las)   An acute inflammation of the dermal layer of the skin, occurring primarily in infants and persons over 30 years of age with a history of streptococcal sore throat.   *747*

erythema infectiosum (er″ə-the′-mə)   A disease in children caused by the parvovirus B19. This disease is common in children between 4 and 11 years of age and is sometimes called fifth disease, since it was the fifth of six erythematous rash diseases in children in an older classification.   *733*

erythromycin (ĕ-rith′ro-mi′sin)   An intermediate spectrum macrolide antibiotic produced by *Streptomyces erythreus*.   *668*

eschar (es′kar)   A slough produced on the skin by a thermal burn, gangrene, or the anthrax bacillus.   *755*

eubacteria (u″bak-te′re-ah)   The large majority of bacteria that have cell wall peptidoglycan containing muramic acid (or are related to such bacteria) and membrane lipids with ester-linked straight chained fatty acids.   *402*

eucarya   The domain or empire that contains organisms composed of eucaryotic cells with primarily glycerol fatty acyl diesters in their membranes and eucaryotic rRNA.   *412*

eucaryotic cells (u″kar-e-ot′ik)   Cells that have a membrane-delimited nucleus and differ in many other ways from procaryotic cells; protists, algae, fungi, plants, and animals are all eucaryotic.   *13, 91*

euglenoids (u-gle′noids)   A group of algae (the division *Euglenophyta*) or protozoa (order *Euglenida*) that normally have chloroplasts containing chlorophyll *a* and *b*. They usually have a stigma and one or two flagella emerging from an anterior gullet.   *524*

eumycotic mycetoma (mi″se-to′mah)   *See* maduromycosis.   *794*

eurythermal (u″re-ther′mal)   Organisms that grow well in a wide range of temperatures.   *128*

eutrophic (u-trof′ik)   A nutrient-enriched environment.   *840*

eutrophication (u″tro-fĭ-ka′shun)   The enrichment of an aquatic environment with nutrients.   *841*

exanthem subitum (eg-zan′thəm sub″a-tum)   A skin eruption or rash in infants caused by the human herpesvirus 6 (HHV-6, type A and type B). It was originally called roseola infantum and then termed sixth disease to differentiate it from other exanthems and roseolas. This disease is characterized by a high fever of 3 to 4 days' duration, after which the temperature suddenly drops to normal, a macular rash appears on the trunk, and then spreads to other areas of the body.   *732*

excystation (ek″sis-ta′shun)   The escape of one or more cells or organisms from a cyst.   *534*

exergonic reaction (ek″ser-gon′ik)   A reaction that spontaneously goes to completion as written; the standard free energy change is negative, and the equilibrium constant is greater than one.   *155*

exfoliative toxin (eks-fo′le-a″tiv) or **exfoliatin** (eks-fō″le-a′tin)   An exotoxin produced by *Staphylococcus aureus* that causes the separation of epidermal layers and the loss of skin surface layers. It produces the symptoms of the scaled skin syndrome.   *761*

exit site (E site)   The location on a ribosome to which an empty tRNA moves from the P site before it finally leaves the ribosome during protein synthesis.   *232*

exoenzymes (ek″so-en′zīms)   Enzymes that are secreted by cells.   *52*

exogenote (eks″o-je′nōt)   The piece of donor DNA that enters a bacterial cell during gene exchange and recombination.   *279*

exon (eks′on)   The region in a split or interrupted gene that codes for RNA which ends up in the final product (e.g., mRNA).   *224*

exotoxin (ek″so-tok′sin)   A heat-labile, toxic protein produced by a bacterium as a result of its normal metabolism or because of the acquisition of a plasmid or prophage that redirects its metabolism. It is usually released into the bacterium's surroundings.   *567*

exponential phase (eks″po-nen′shul)   The phase of the growth curve during which the microbial population is growing at a constant and maximum rate, dividing and doubling at regular intervals.   *115*

expression vector   A special cloning vector used to express recombinant genes in host cells; the recombinant gene is transcribed and its protein synthesized.   *322*

extended aeration   The aeration of sewage to allow self-digestion of microorganisms that grew during use of a soluble waste.   *845*

extracutaneous sporotrichosis (spo″ro-tri-ko′sis)   An infection by the fungus *Sporothrix schenckii* that spreads throughout the body.   *794*

exudation (eks″u-da′shun)   The release of soluble substances by plant roots and other living organisms.   *864*

eyespot   *See* stigma.   *523*

F

F factor   The fertility factor, a plasmid that carries the genes for bacterial conjugation and makes its *E. coli* host cell the gene donor during conjugation.   *280*

F′ plasmid   An F plasmid that carries some bacterial genes and transmits them to recipient cells when the F′ cell carries out conjugation; the transfer of bacterial genes in this way is often called sexduction.   *291*

F₁ particle   Particle on the inner mitochondrial membrane, which is the site of ATP synthesis by oxidative phosphorylation.   *82*

facilitated diffusion   Diffusion across the plasma membrane that is aided by a carrier.   *103*

facultative anaerobes (fak′ul-ta″tiv an-a′er-ōbs)   Microorganisms that do not require oxygen for growth, but do grow better in its presence.   *130*

facultative psychrophile (fak′ul-ta″tiv si′kro-fīl)   *See* psychrotroph.   *130*

FAD   *See* flavin adenine dinucleotide.   *157*

fatty acid   A monocarboxylic acid that may be unbranched or branched, and saturated or unsaturated; the fatty acids most common in lipids are 16 or 18 carbons long.   *205*

fatty acid synthetase (sin′thĕ-tās)   The multienzyme complex that makes fatty acids; the product usually is palmitic acid.   *205*

fecal coliform test (fe′kal ko′lĭ-form)   Test for coliforms that are able to grow at 44.5°C, whose normal habitat is the intestinal tract.   *850*

fecal enterococci (fe′kal en″ter-o-kok′si)   Enterococci found in the intestine of humans and other warm-blooded animals. They are used as indicators of the fecal pollution of water.   *853*

feedback inhibition   A negative feedback mechanism in which a pathway end product inhibits the activity of an enzyme in the sequence leading to its formation; when the end product accumulates in excess, it inhibits its own synthesis.   *241*

fermentation (fer″men-ta′shun)   An energy-yielding process in which organic molecules serve as both electron donors and acceptors.   *175, 906*

fever blister   *See* cold sore.   *730*

fibronectin (fi″bro-nek′tin) A high molecular weight adhesive glycoprotein. It can interact with bacteria and mediate their nonspecific clearance from the body. Fibronectin is also involved in cell–cell interactions. *576*

fimbria (fim′bre-ah; pl., **fimbriae**) A fine, hairlike protein appendage on some gram-negative bacteria that helps attach them to surfaces. *59*

final host The host on/in which a parasitic organism either attains sexual maturity or reproduces. *562*

Firmicutes (fer-mik′u-tēz) A division in the kingdom *Procaryotae* containing bacteria with gram-positive type cell walls. *400*

first law of thermodynamics Energy can be neither created nor destroyed (even though it can be changed in form or redistributed). *154*

fixation (fik-sa′shun) The process in which the internal and external structures of cells and organisms are preserved and fixed in position. *25*

flagellin (flaj′ĕ-lin) The protein used to construct the filament of a bacterial flagellum. *60*

flagellum (flah-jel′um; pl., **flagella**) A thin, threadlike appendage on many procaryotic and eucaryotic cells that is responsible for their motility. *60, 90*

flat or plane warts Small, smooth, slightly raised warts. *739*

flavin adenine dinucleotide (**FAD;** fla′vin ad′ĕ-nēn) An electron carrying cofactor often involved in energy production (for example, in the tricarboxylic acid cycle and the β-oxidation pathway). *157*

fluid mosaic model The currently accepted model of cell membranes in which the membrane is a lipid bilayer with integral proteins buried in the lipid, and peripheral proteins more loosely attached to the membrane surface. *43*

fluorescent light (floo″o-res′ent) The light emitted by a substance when it is irradiated with light of a different wavelength. *24*

fluorescence microscope A microscope that exposes a specimen to light of a specific wavelength and then forms an image from the fluorescent light produced. Usually the specimen is stained with a fluorescent dye or fluorochrome. *24*

foliose Having a leaflike appearance, especially of lichens. *548*

fomite (fo′mīt; pl., **fomites**) An object that is not in itself harmful but is able to harbor and transmit pathogenic organisms. Also called fomes. *563, 709*

food-borne infection Gastrointestinal illness caused by ingestion of microorganisms, followed by their growth within the host. Symptoms arise from tissue invasion and/or toxin production. *766, 886*

food chain The flow of energy and matter in living organisms through a producer-consumer sequence (*See also* food web). *533*

food intoxication Food poisoning caused by microbial toxins produced in a food prior to consumption. The presence of living bacteria is not required. *766, 887*

food poisoning A general term usually referring to a gastrointestinal disease caused by the ingestion of food contaminated by pathogens or their toxins. *766*

food web A network of many interlinked food chains, encompassing primary producers, consumers, decomposers, and detritivores. *533, 813*

fractional steam sterilization *See* tyndallization. *142*

fragmentation (frag″men-ta′shun) A type of asexual reproduction in which a thallus breaks into two or more parts, each of which forms a new thallus. *522*

frameshift mutations Mutations arising from the loss or gain of a base or DNA segment, leading to a change in the codon reading frame and thus a change in the amino acids incorporated into protein. *268*

free energy change The total energy change in a system that is available to do useful work as the system goes from its initial state to its final state at constant temperature and pressure. *155*

French polio *See* Guillain-Barré syndrome. *718*

fruiting body A specialized structure that holds sexually or asexually produced spores; found in fungi and in some bacteria. *474*

frustule (frus′tūl) A silicified cell wall in the diatoms. *526*

fruticose Having a shrubby shape, especially of lichens. *548*

fungicide (fun′jĭ-sīd) An agent that kills fungi. *138*

fungistasis (fun-jĭ-sta′sis) *See* fungistatic. *814*

fungistatic (fun″jĭ-stat′ik) Inhibiting the growth and reproduction of fungi. *138*

fungus (fung′gus; pl., **fungi**) Achlorophyllous, heterotrophic, spore-bearing eucaryotes with absorptive nutrition; usually, they have a walled thallus. *504*

***F* value** The time in minutes at a specific temperature (usually 250°F) needed to kill a population of cells or spores. *140*

G

GALT (**gut associated lymphoid tissue**) *See* Peyer's patches. *596*

gametangium (gam-ĕ-tan′je-um; pl., **gametangia**) A structure that contains gametes or in which gametes are formed. *509*

gas gangrene (gang′grēn) A type of gangrene that arises from dirty, lacerated wounds infected by anaerobic bacteria, especially species of *Clostridium.* As the bacteria grow, they release toxins and ferment carbohydrates to produce carbon dioxide and hydrogen gas. *756*

gas vacuole A gas-filled vacuole found in cyanobacteria and some other aquatic bacteria that helps its possessor float. It is composed of gas vesicles, which are made of protein. *45*

gastritis (gas-tri′tis) Inflammation of the stomach. *758*

gastroenteritis (gas″tro-en-ter-i′tis) An acute inflammation of the lining of the stomach and intestines, characterized by anorexia, nausea, diarrhea, abdominal pain, and weakness. It has various causes including food poisoning due to such organisms as *E. coli, S. aureus, Campylobacter,* and *Salmonella* species; consumption of irritating food or drink; or psychological factors such as anger, stress, and fear. Also called enterogastritis. *768*

gastrointestinal anthrax *See* anthrax. *755*

gene (jēn) A DNA segment or sequence that codes for a polypeptide, rRNA, or tRNA. *257*

gene gun A device that uses a .22-caliber blank shell or other propellant to shoot a spray of DNA-coated microprojectiles into cells and transform them. Sometimes it is called a biolistic device. *321*

generalized transduction The transfer of any part of a bacterial genome when the DNA fragment is packaged within a phage capsid by mistake. *294*

general recombination Recombination involving a reciprocal exchange of a pair of homologous DNA sequences; it can occur any place on the chromosome. *278*

generation time The time required for a microbial population to double in number. *116*

genetic engineering The deliberate modification of an organism's genetic information by directly changing its nucleic acid genome. *306*

genital herpes (her′pēz) A sexually transmitted disease caused by the herpes simplex virus type 2. *731*

genital ulcer disease *See* chancroid. *756*

genome (je′nōm) The full set of genes present in a cell or virus; all the genetic material in an organism; a haploid set of genes in a cell. *255*

genus (je′nəs) A well-defined group of one or more species that is clearly separate from other genera. *392*

geosmin A group of compounds produced by actinomycetes and cyanobacteria that cause the characteristic odor of soils and waters. *498, 862*

German measles *See* rubella. *720*

germicide (jer′mĭ-sīd) An agent that kills pathogens and many nonpathogens but not necessarily bacterial endospores. *138*

germination (jer″mĭ-na′shun) The stage following bacterial endospore activation in which the endospore breaks its dormant state. Germination is followed by outgrowth. *69*

Ghon complex (gon) The initial focus of parenchymal infection in primary pulmonary tuberculosis. *750*

giardiasis (je″ar-di′ah-sis) A common intestinal disease caused by the parasitic protozoan *Giardia lamblia.* *801*

gingivitis (jin-jĭ-vi′tis) Inflammation of the gingival tissue. *785*

gliding motility A type of motility in which a microbial cell glides along when in contact with a solid surface. *63, 470*

glomerulonephritis (glo-mer″u-lo-nĕ-fri′tis) An inflammatory disease of the renal glomeruli. *748*

glucans Polysaccharides composed of glucose units held together by glycosidic linkages. Some types of glucans have $\alpha(1\rightarrow3)$ and $\alpha(1\rightarrow6)$ linkages and bind bacterial cells together on teeth forming a plaque ecosystem. *784*

gluconeogenesis (gloo″ko-ne″o-jen′e-sis) The synthesis of glucose from noncarbohydrate precursors such as lactate and amino acids. *196*

glycocalyx (gli″ko-kal′iks) A network of polysaccharides extending from the surface of bacteria and other cells. *58*

glycogen (gli′ko-jen) A highly branched polysaccharide containing glucose, which is used to store carbon and energy. *45*

glycolysis (gli-kol′ĭ-sis) The anaerobic conversion of glucose to lactic acid by use of the Embden-Meyerhof pathway. *166*

glycolytic pathway (gli″ko-lit′ik) *See* Embden-Meyerhof pathway. *166*

glyoxylate cycle (gli-ok′sĭ-lăt) A modified tricarboxylic acid cycle in which the decarboxylation reactions are bypassed by the enzymes isocitrate lyase and malate synthase; it is used to convert acetyl-CoA to succinate and other metabolites. *203*

gnotobiotic (no″to-bi-ot′ik) Animals that are germfree (microorganism free) or live in association with one or more known microorganisms. *553*

Golgi apparatus (gol′je) A membranous eucaryotic organelle composed of stacks of flattened sacs (cisternae), which is involved in packaging and modifying materials for secretion and many other processes. *78*

gonococci (gon′o-kok′si) Bacteria of the species *Neisseria gonorrhoeae*—the organism causing gonorrhea. *756*

gonorrhea (gon″o-re′ah) An acute infectious sexually transmitted disease of the mucous membranes of the genitourinary tract, eye, rectum, and throat. It is caused by *Neisseria gonorrhoeae*. *756*

Gracilicutes (gras″ĭ-lik′u-tez) A division in the kingdom *Procaryotae* containing bacteria with gram-negative type cell walls. *400*

gradient (gra′dĕ-ent) The change in concentration of a substance over distance. *813*

Gram stain A differential staining procedure that divides bacteria into gram-positive and gram-negative groups based on their ability to retain crystal violet when decolorized with an organic solvent such as ethanol. *26*

grana (gra′nah) A stack of thylakoids in the chloroplast stroma. *84*

greenhouse gases Gases released from the earth's surface through chemical and biological processes that interact with the chemicals in the stratosphere to decrease the release of radiation from the earth. It is believed that this leads to global warming. *874*

griseofulvin (gris″e-o-ful′vin) An antibiotic from *Penicillium griseofulvum* given orally to treat chronic dermatophyte infections of skin and nails. *671*

groundwater Water below the soil surface in the saturated zone; it supplies wells and springs. *854*

group translocation A transport process in which a molecule is moved across a membrane by carrier proteins while being chemically altered at the same time. *105*

growth An increase in cellular constituents. *115*

growth factors Organic compounds that must be supplied in the diet for growth because they are essential cell components or precursors of such components and cannot be synthesized by the organism. *101*

growth yield (Y) A quantitative measure of the amount of microbial mass produced from a nutrient. *122*

guanine (gwan′in) A purine derivative, 2-amino-6-oxypurine, found in nucleosides, nucleotides, and nucleic acids. *204*

Guillain-Barré syndrome (ge-yan″bar-ra′) A relatively rare disease affecting the peripheral nervous system, especially the spinal nerves, but also the cranial nerves. The cause is unknown, but it most often occurs after an influenza infection or flu vaccination. Also called French Polio. *718*

gumma (gum′ah) A soft, gummy tumor occurring in tertiary syphilis. *764*

H

halomethanes Chlorinated organic compounds formed during the process of water chlorination. *848*

halophile (hal′o-fĭl) A microorganism that requires high levels of sodium chloride for growth. *125*

Hansen's disease *See* leprosy. *757*

hantavirus pulmonary syndrome The disease in humans caused by the pulmonary syndrome hantavirus. Deer mice shed the virus in their feces, humans inhale the virus and first develop ordinary flulike aches and pains. Within a few days the hantavirus causes lung damage and capillary leakage. After about a week the infected person enters a crisis phase and may die. *721*

haploid (hap′loid) Having half the number of chromosomes present in a somatic cell or a single set of chromosomes. *88*

hapten (hap′ten) A molecule not immunogenic by itself but that, when coupled to a macromolecular carrier, can elicit antibodies directed against itself. *590*

harborage transmission The mode of transmission in which an infectious organism does not undergo morphological or physiological changes within the vector. *709*

hay fever Allergic rhinitis; a type of atopic allergy involving the upper respiratory tract. *621*

health (helth) A state of optimal physical, mental, and social well-being, and not merely the absence of disease and infirmity. *700*

healthy carrier An individual who harbors a pathogen, but is not ill. *706*

heat-shock proteins Proteins produced when cells are exposed to high temperatures or other stressful conditions. They protect the cells from damage and often aid in the proper folding of proteins. *49*

helical (hel′ĭ-kal) In virology this refers to a virus with a helical capsid surrounding its nucleic acid. *340*

helicases Enzymes that use ATP energy to unwind DNA ahead of the replication fork. *218*

hemadsorption (hem″ad-sorp′shun) The adherence of red blood cells to the surface of something, such as another cell or a virus. *683*

hemagglutination (hem″ah-gloo″tĭ-na′shun) The agglutination of red blood cells by antibodies. *640*

hemagglutinin (hem″ah-gloo′tĭ-nin) The antibody responsible for a hemagglutination reaction. *640*

hemoflagellate (he″mo-flaj′ĕ-lāt) A flagellated protozoan parasite that is found in the bloodstream. *804*

hemolysin (he-mol′ĭ-sin) A substance that causes hemolysis (the lysis of red blood cells). At least some hemolysins are enzymes that destroy the phospholipids in erythrocyte plasma membranes. *572*

hemolysis (he-mol′ĭ-sis) The disruption of red blood cells and release of their hemoglobin. There are several types of hemolysis when bacteria such as streptococci and staphylococci grow on blood agar. In α-hemolysis, a narrow greenish zone of incomplete hemolysis forms around the colony. A clear zone of complete hemolysis without any obvious color change is formed during β-hemolysis. *443*

hemolytic disease of the newborn A hemolytic anemia of a newborn child that results from the destruction of the infant's red blood cells by antibodies produced by the mother; usually the antibodies are due to Rh blood type incompatibility. Also called erythroblastosis fetalis. *628*

hemorrhagic fever A fever usually caused by a specific virus that may lead to hemorrhage, shock, and sometimes death. *721*

hepatitis (hep″ah-ti′tis) Any infection that results in inflammation of the liver. Also refers to liver inflammation as such. *734*

hepatitis A (formerly infectious hepatitis; hep″ah-ti′tis) A type of hepatitis that is transmitted by fecal-oral contamination; it primarily affects children and young adults, especially in environments where there is poor sanitation and overcrowding. It is caused by the hepatitis A virus, a single-stranded RNA virus. *738*

hepatitis B (formerly serum hepatitis; hep″ah-ti′tis) This form of hepatitis is caused by a double-stranded DNA virus (HBV) formerly called the "Dane particle." The virus is transmitted by body fluids. *735*

hepatitis C About 90% of all cases of viral hepatitis can be traced to either HAV or HBV. The remaining 10% is believed to be caused by one and possibly several other types of viruses. At least one of these is hepatitis C (formerly non-A, non-B). *736*

hepatitis D (formerly delta hepatitis) The liver diseases caused by the hepatitis D virus in those individuals already infected with the hepatitis B virus. *737*

hepatitis E (formerly enteric-transmitted NANB hepatitis) The liver disease caused by the hepatitis E virus. Usually, a subclinical, acute infection results; however, there is a high mortality in women in their last trimester of pregnancy. *738*

hepatitis F Also known as fulminant, post-transfusion hepatitis. *737*

hepatitis G Also known as syncytial giant-cell hepatitis. *737*

herd immunity The resistance of a population to infection and spread of an infectious organism due to the immunity of a high percentage of the population. *703*

herpes labialis *See* cold sore. *730*

herpetic keratitis (her-pet′ik ker″ah-ti′tis) An inflammation of the cornea and conjunctiva of the eye resulting from a herpes simplex virus infection. *730*

heterocysts Specialized cells produced by cyanobacteria that are the sites of nitrogen fixation. *461*

heteroduplex DNA A double-stranded stretch of DNA formed by two slightly different strands that are not completely complementary. *278*

heterogeneous nuclear RNA (hnRNA) The RNA transcript of DNA made by RNA polymerase II; it is then processed to form mRNA. *223*

heterolactic fermenters (het″er-o-lak′tik) Microorganisms that ferment sugars to form lactate, and also other products such as ethanol and CO_2. *175*

heterologous antigen *See* heterophile. *590*

heterophile (het″er-o-fīl)   An antigen that cross-reacts with antibodies produced in response to a different antigen; also called a heterologous antigen. Usually the term refers to antigens from unrelated species that cross-react because they possess identical or similar epitopes.   *590*

heterotroph (het′er-o-trōf″)   An organism that uses reduced, preformed organic molecules as its principal carbon source.   *99*

heterotrophic nitrification   Nitrification carried out by chemoheterotrophic microorganisms.   *816*

hexon or **hexamer**   A capsomer composed of six protomers.   *341*

hexose monophosphate pathway (hek′sōs mon″o-fos′fāt)   *See* pentose phosphate pathway.   *167*

Hfr strain   A bacterial strain that donates its genes with high frequency to a recipient cell during conjugation because the F factor is integrated into the bacterial chromosome.   *289*

high-energy molecule   A molecule whose hydrolysis under standard conditions makes available a large amount of free energy (the standard free energy change is more negative than about –7 kcal/mole); a high-energy molecule readily decomposes and transfers groups such as phosphate to acceptors.   *155*

histone (his′tōn)   A small basic protein with large amounts of lysine and arginine that is associated with eucaryotic DNA in chromatin.   *215*

histoplasmosis (his″to-plaz-mo′sis)   A systemic fungal infection caused by *Histoplasma capsulatum* var *capsulatum*.   *796*

hives (hīvz)   An eruption of the skin.   *622*

holdfast   A structure produced by some bacteria and algae that attaches the cell to a solid object.   *467*

holoenzyme   A complete enzyme consisting of the apoenzyme plus a cofactor.   *159*

holozoic nutrition (hol″o-zo′ik)   In this type of nutrition, nutrients (such as bacteria) are acquired by phagocytosis and the subsequent formation of a food vacuole or phagosome.   *534*

homolactic fermenters (ho″mo-lak′tik)   Organisms that ferment sugars almost completely to lactic acid.   *175*

hormogonia   Small motile fragments produced by fragmentation of filamentous cyanobacteria; used for asexual reproduction and dispersal.   *461*

host (hōst)   The body of an organism that harbors another organism. It can be viewed as a microenvironment that shelters and supports the growth and multiplication of a parasitic organism.   *562*

host restriction   The degradation of foreign genetic material by nucleases after the genetic material enters a host cell.   *280*

human herpesvirus 6 (HHV-6, type A and B) (hər′pēz)   HHV-6 was discovered in 1986 and was initially called the human B-lymphotropic virus. The virus was later shown to have a marked tropism for CD4⁺ T cells and was renamed HHV-6. HHV-6 is genetically similar to cytomegalovirus. HHV-6 causes exanthem subitum (roseola infantum or sixth disease) in infants and has been suspected of involvement in many conditions, including opportunistic infections in immunocompromised patients, hepatitis, lymphoproliferative diseases, synergistic interactions with HIV, lymphadenitis, and chronic fatigue syndrome.   *732*

human immunodeficiency virus (HIV)   A lentivirus of the family *Retroviridae* that is associated with the onset of AIDS.   *722*

human leukocyte antigen (HLA)   An antigen on the surface of cells of human tissues and organs that is recognized by the immune system cells and therefore is important in graft rejection and regulation of the immune response. This is the same as MHC class II.   *614*

human parvovirus B19 (pahr′vo-vi″rəs)   The human parvovirus B19 belongs to the family *Parvoviridae*, genus *Parvovirus*. B19 virions are uniform, unenveloped particles approximately 23 nm in diameter with a single-stranded DNA genome. There is a spectrum of disease caused by this virus ranging from mild symptoms in normal persons to more serious diseases in compromised individuals.   *733*

humoral immunity (hu′mor-al)   The type of immunity that results from the presence of soluble antibodies in blood and lymph; also known as antibody-mediated immunity.   *589*

humus (hu′mus)   The complex organic residues that remain in the soil after microbial degradation of plant and animal matter.   *860*

hybridoma (hi″brĭ-do′mah)   A fast-growing cell line produced by fusing a cancer cell (myeloma) to another cell, such as an antibody-producing cell.   *602*

hydrogenosome (hi-dro-jen′osom)   A microbodylike organelle that contains a unique electron transfer pathway in which hydrogenase transfers electrons to protons (which act as the terminal electron acceptors) and molecular hydrogen is formed.   *534*

hydrophilic (hi″dro-fil′ik)   A polar substance that has a strong affinity for water (or is readily soluble in water).   *42*

hydrophobic (hi″dro-fo′bik)   A nonpolar substance lacking affinity for water (or which is not readily soluble in water).   *42*

hydrothermal vent community (hi-dro-thər′məl)   The community of organisms that exists several thousand meters below the sea surface where the earth's crustal plates are spreading apart. The seawater surrounding these vents is hot, anoxic, and contains high concentrations of hydrogen sulfide. The dominant animal species are supported primarily by endosymbiotic chemolithoautotrophic bacteria. The main energy of these communities is the oxidation of geologically produced reduced substances.   *549*

hyperendemic disease (hi″per-en-dem′ik)   A disease that has a gradual increase in occurrence beyond the endemic level, but not at the epidemic level, in a given population; also may refer to a disease that is equally endemic in all age groups.   *700*

hyperferremia (hi″per-fer-re′me-ah)   An excess of iron in the blood.   *573*

hypersensitivity (hi″per-sen′si-tiv″i-te)   A condition of increased immune sensitivity in which the body reacts to an antigen with an exaggerated immune response that usually harms the individual. Also termed an allergy.   *621*

hyperthermophile (hi″per-ther′mo-fīl)   A bacterium that has its growth optimum between 80°C and about 110°C. Hyperthermophiles usually do not grow well below 55°C.   *130*

hypha (hi′fah; pl., **hyphae**)   The unit of structure of most fungi and some bacteria; a tubular filament.   *506*

hypoferremia (hi″po-fĕ-re′me-ah)   Deficiency of iron in the blood.   *573*

hypolimnion   The part of a lake lying below the thermocline (a layer of rapid temperature change) that usually has a relatively uniform, cold temperature and a low oxygen level.   *840*

hypotheca (hi-po-theca)   The smaller half of a diatom frustule.   *526*

hypothesis   A tentative assumption or educated guess developed to explain a set of observations.   *11*

I

icosahedral   In virology this term refers to a virus with an icosahedral capsid, which has the shape of a regular polyhedron having 20 equilateral triangular faces and 12 corners.   *340*

identification (i-den″tĭ-fĭ-ka′shun)   The process of determining that a particular isolate or organism belongs to a recognized taxon.   *391*

idiophase   The period in a batch culture in which secondary metabolites are synthesized in preference to primary metabolites; it generally corresponds to the stationary phase and the end of the log phase.   *912*

idiotype (id′e-o-tīp′)   A set of one or more unique epitopes in the variable region of an immunoglobulin that distinguishes it from immunoglobulins produced by different plasma cells.   *592*

IgA   Immunoglobulin A; the class of immunoglobulins that is present in dimeric form in many body secretions (e.g., saliva, tears, and bronchial and intestinal secretions) and protects body surfaces. IgA also is present in serum.   *595*

IgD   Immunoglobulin D; the class of immunoglobulins found on the surface of many B lymphocytes; thought to serve as an antigen receptor in the stimulation of antibody synthesis.   *595*

IgE   Immunoglobulin E; the immunoglobulin class that binds to mast cells and basophils, and is responsible for type I or anaphylactic hypersensitivity reactions such as hay fever and asthma. IgE is also involved in resistance to helminth parasites.   *597*

IgG   Immunoglobulin G; the predominant immunoglobulin class in serum. Has functions such as neutralizing toxins, opsonizing bacteria, activating complement, and crossing the placenta to protect the fetus and neonate.   *594*

IgM   Immunoglobulin M; the class of serum antibody first produced during an infection. It is a large, pentameric molecule that is active in agglutinating pathogens and activating complement. The monomeric form is present on the surface of some B lymphocytes.   *595*

immobilization (im-mo″bil-i-za′shun)   The incorporation of a simple, soluble substance into the body of an organism, making it unavailable for use by other organisms.   *917*

immune complex (ĭ-mūn′kom′pleks)   The product of an antigen-antibody reaction, which may also contain components of the complement system.   *640*

latent virus infections   Virus infections in which the virus quits reproducing and remains dormant for a period before becoming active again.   *379*

leader sequence   A nontranslated sequence at the 5′ end of mRNA that lies between the operator and the initiation codon; it aids in the initiation and regulation of transcription.   *222, 260*

legionellosis (le″jə-nel-o′sis)   *See* Legionnaires′ disease.   *744*

Legionnaires′ disease (legionellosis)   A pulmonary form of legionellosis, resulting from infection with *Legionella pneumophila.*   *744*

leishmanias (lēsh″ma′ne-ăs)   Zooflagellates, members of the genus *Leishmania,* that cause the disease leishmaniasis.   *805*

leishmaniasis (lēsh″mah-ni′ah-sis)   The disease caused by the protozoa called leishmanias.   *805*

lepromatous (progressive) leprosy (lep-ro′mah-tus lep′ro-se)   A relentless, progressive form of leprosy in which large numbers of *Mycobacterium leprae* develop in skin cells, killing the skin cells and resulting in the loss of features. Disfiguring nodules form all over the body.   *758*

leprosy (lep′ro-se) or **Hansen's disease**   A severe disfiguring skin disease caused by *Mycobacterium leprae.*   *757*

lethal dose 50 (LD$_{50}$)   Refers to the dose or number of organisms that will kill 50% of an experimental group of hosts within a specified time period.   *339, 563*

leukemia (loo-ke′me-ah)   A progressive, malignant disease of blood-forming organs, marked by distorted proliferation and development of leukocytes and their precursors in the blood and bone marrow. Certain leukemias are caused by viruses (HTLV-1, HTLV-2).   *733*

leukocidin (loo″ko-si′din)   A microbial toxin that can damage or kill leukocytes.   *572*

L forms   Pleomorphic bacterial cells formed by the complete or partial loss of their cell walls. Cell wall loss may be reversible or permanent.   *58*

lichen (li′ken)   An organism composed of a fungus and either green algae or cyanobacteria in a symbiotic association.   *548*

Liebig's law of the minimum (le′bigz)   Living organisms and populations will grow until lack of a resource begins to limit further growth.   *813*

lipopolysaccharide (lip″o-pol″e-sak′ah-rīd)   A molecule containing both lipid and polysaccharide, which is important in the outer membrane of the gram-negative cell wall.   *56*

liposome (lip′o-som)   A spherical particle formed by a lipid bilayer enclosing an aqueous solution. It may be used to administer chemotherapeutic agents or in diagnostic testing.   *693*

lithotroph (lith′o-trōf)   An organism that uses reduced inorganic compounds as its electron source.   *100*

log phase   *See* exponential phase.   *115*

lophotrichous (lo-fot′ri-kus)   A cell with a cluster of flagella at one or both ends.   *60*

LPS-binding protein   A special plasma protein that binds bacterial lipopolysaccharides and then attaches to receptors on monocytes, macrophages, and other cells. This triggers the release of IL-1 and other cytokines that stimulate the development of fever and additional endotoxin effects.   *572*

Lyme disease (LD, Lyme borreliosis; līm)   A tick-borne disease caused by the spirochete *Borrelia burgdorferi.*   *752*

lymphocyte (lim′fo-sīt)   A nonphagocytic, mononuclear leukocyte (white blood cell) that is an immunologically competent cell, or its precursor. Lymphocytes are present in the blood, lymph, and lymphoid tissues. *See* B cell and T cell.   *589*

lymphogranuloma venereum (LGV; lim″fo-gran″u-lo′mah)   A sexually transmitted disease caused by *Chlamydia trachomatis* serotypes L$_1$–L$_3$, which affect the lymph organs in the genital area.   *777*

lymphokine (lim′fo-kin)   A biologically active glycoprotein (e.g., IL-1) secreted by activated lymphocytes, especially sensitized T cells. It acts as an intercellular mediator of the immune response and transmits growth, differentiation, and behavioral signals.   *607*

lyophilization (li-of″ĭ-li-za′shun)   A method of freeze-drying used to preserve microbial cultures.   *910*

lysis (li′sis)   The rupture or physical disintegration of a cell.   *57*

lysogenic (li-so-jen′ik)   *See* lysogens.   *294, 362*

lysogens (li′so-jens)   Bacteria that are carrying a viral prophage and have the potential of producing bacteriophages under the proper conditions.   *294, 362*

lysogeny (li-soj′e-ne)   The state in which a phage genome remains within the bacterial host cell after infection and reproduces along with it rather than taking control of the host and destroying it.   *294, 362*

lysosome (li′so-sōm)   A spherical membranous eucaryotic organelle that contains hydrolytic enzymes and is responsible for the intracellular digestion of substances.   *79*

lysozyme (li′sō-zīm)   An enzyme that degrades peptidoglycan by hydrolyzing the $\beta(1 \rightarrow 4)$ bond that joins N-acetylmuramic acid and N-acetylglucosamine.   *57*

lytic cycle (lit′ik)   A virus life cycle that results in the lysis of the host cell.   *354*

M

macrolide antibiotic (mak′ro-līd)   An antibiotic containing a macrolide ring, a large lactone ring with multiple keto and hydroxyl groups, linked to one or more sugars.   *668*

macromolecule (mak″ro-mol′ĕ-kūl)   A large molecule that is a polymer of smaller units joined together.   *192*

macronucleus (mak″ro-nu′kle-us)   The larger of the two nuclei in ciliate protozoa. It is normally polyploid and directs the routine activities of the cell.   *533*

macrophage (mak′ro-fāj)   The name for a large mononuclear phagocytic cell, present in blood, lymph, and other tissues. Macrophages are derived from monocytes. They phagocytose and destroy pathogens; some macrophages also activate B cells and T cells.   *613*

maduromycosis (mah-du″ro-mi-ko′sis)   A subcutaneous fungal infection caused by *Madurella mycetoma;* also termed an eumycotic mycetoma.   *794*

madurose   The sugar derivative 3-O-methyl-D-galactose, which is characteristic of several actinomycete genera that are collectively called maduromycetes.   *500*

magnetosomes   Magnetite particles in magnetotactic bacteria that are tiny magnets and allow the bacteria to orient themselves in magnetic fields.   *64*

magnetotactic bacteria   Bacteria that can orient themselves in the earth's magnetic field.   *64*

maintenance energy   The energy a cell requires simply to maintain itself or remain alive and functioning properly. It does not include the energy needed for either growth or reproduction.   *122*

major histocompatibility complex (MHC)   A large set of cell surface molecules in each individual, encoded by a family of genes, that serves as a unique biochemical marker of individual identity. It can trigger T-cell responses that may lead to rejection of transplanted tissues and organs. MHC molecules are also involved in the regulation of the immune response and the interactions between immune cells.   *614*

malaria (mah-la′re-ah)   A serious infectious illness caused by the parasitic protozoan *Plasmodium.* Malaria is characterized by bouts of high chills and fever that occur at regular intervals.   *802*

malt (mawlt)   Grain soaked in water to soften it, induce germination, and activate its enzymes. The malt is then used in brewing and distilling.   *898*

Marburg viral hemorrhagic fever   An acute infection caused by a virus that produces varying degrees of hemorrhage, shock, and sometimes death.   *721*

mash   The soluble materials released from germinated grains and prepared as a microbial growth medium.   *898*

mashing   The process in which cereals are mixed with water and incubated in order to degrade their complex carbohydrates (e.g., starch) to more readily usable forms such as simple sugars.   *897*

mean growth rate constant (k)   The rate of microbial population growth expressed in terms of the number of generations per unit time.   *117*

measles (rubeola; me′zelz)   A highly contagious skin disease that is endemic throughout the world. It is caused by a morbilli virus in the family *Paramyxoviridae,* which enters the body through the respiratory tract or through the conjunctiva.   *718*

medical mycology (mi-kol′o-je)   The discipline that deals with the fungi that cause human disease.   *791*

meiosis (mi-o′sis)   The process in which a diploid cell divides and forms two haploid cells.   *88*

melting temperature (T_m) of DNA   The temperature at which double-stranded DNA separates into individual strands; it is dependent on the G + C content of the DNA and is used to compare genetic material in microbial taxonomy.   *394*

membrane attack complex (MAC)   The complex complement components (C5b–C9) that create a pore in the plasma membrane of a target cell and leads to cell lysis. C9 probably forms most of the actual pore.   *634*

membrane filter   A thin filter made of cellulose or synthetic polymers that has a carefully controlled pore size and is used to filter out microorganisms.   *142*

membrane filter technique The use of a thin porous filter made from cellulose acetate or some other polymer to collect microorganisms from water, air, and food. *119, 850*

memory B cell A lymphocyte capable of initiating the antibody-mediated immune response upon detection of a specific antigen molecule for which it is genetically programmed. It circulates freely in the blood and lymph and may live for years. *600*

Mendosicutes (men″do-sik′utēz) A division in the kingdom *Procaryotae* composed of bacteria that usually have a wall, but lack peptidoglycan with muramic acid. It contains the archaeobacteria. *400*

meningitis (men″in-ji′tis) A condition that refers to inflammation of the brain or spinal cord meninges (membranes). The disease can be divided into bacterial (septic) meningitis (caused by bacteria) and aseptic meningitis syndrome (caused by nonbacterial sources). *745*

mesophile (mes′o-fīl) A microorganism with a growth optimum around 20 to 45°C, a minimum of 15 to 20°C, and a maximum about 45°C or lower. *130*

messenger RNA (mRNA) Single-stranded RNA synthesized from a DNA template during transcription that binds to ribosomes and directs the synthesis of protein. *216*

metabolic channeling (mĕt″ah-bol′ik) The localization of metabolites and enzymes in different parts of a cell. *237*

metabolic engineering The rational modification of metabolic pathways and products by the use of molecular biology. *920*

metabolism (me-tab′o-lizm) The total of all chemical reactions in the cell; almost all are enzyme catalyzed. *165*

metachromatic granules (met″ah-kro-mat′ik) Granules of polyphosphate or volutin granules in the cytoplasm of some bacteria that appear a different color when stained with a blue basic dye. They are storage reservoirs for phosphate. *47*

metastasis (mĕ-tas′tah-sis) The transfer of a disease like cancer from one organ to another not directly connected with it. *379*

methanogens (meth′ə-no-jens″) Strictly anaerobic bacteria that derive energy by converting CO_2, H_2, formate, acetate, and other compounds to either methane or methane and CO_2; they are archaeobacteria. *482*

methylotroph A bacterium that uses one-carbon compounds such as methane and methanol as its sole source of carbon and energy. *424, 467*

Michaelis constant (K_m; mĭ-ka′lis) A kinetic constant for an enzyme reaction that equals the substrate concentration required for the enzyme to operate at half maximal velocity. *161*

microaerophile (mi″kro-a′er-o-fīl) A microorganism that requires low levels of oxygen for growth, around 2 to 10%, but is damaged by normal atmospheric oxygen levels. *130*

microbial ecology The study of microorganisms in their natural environments, with a major emphasis on physical conditions, processes, and interactions that occur on the scale of individual microbial cells. *813*

microbial transformation (mi-kro′be-al) *See* bioconversion. *916*

microbiology (mi″kro-bi-ol′o-je) The study of organisms that are usually too small to be seen with the naked eye. Special techniques are required to isolate and grow them. *3*

microbiostasis The inhibition of microbial growth and reproduction. *814*

microbiota (also **indigenous microbial population, microflora, microbial flora**; mi″kro-bi-o′tah) The microorganisms normally associated with a particular tissue or structure. *553*

microbivory The use of microorganisms as a food source by organisms that can ingest or phagocytose them. *862*

microenvironment (mi″kro-en-vi′ron-ment) The immediate environment surrounding a microbial cell or other structure, such as a root. *813*

microfilaments (mi″kro-fil′ah-ments) Protein filaments, about 4 to 7 nm in diameter, that are present in the cytoplasmic matrix of eucaryotic cells and play a role in cell structure and motion. *76*

micronucleus (mi″kro-nu′kle-us) The smaller of the two nuclei in ciliate protozoa. Micronuclei are diploid and involved only in genetic recombination and the regeneration of macronuclei. *533*

microorganism (mi″kro-or′gan-izm) An organism that is too small to be seen clearly with the naked eye. *3*

microtubules (mi″kro-tu′buls) Small cylinders, about 25 nm in diameter, made of tubulin proteins and present in the cytoplasmic matrix and flagella of eucaryotic cells; they are involved in cell structure and movement. *76*

migration inhibition factor (MIF) A cytokine, probably produced by macrophages, that inhibits the movement of macrophages away from the MIF source, thus promoting macrophage accumulation in the area. *609*

miliary tuberculosis (mil′e-a-re) An acute form of tuberculosis in which small tubercles are formed in a number of organs of the body because of dissemination of *M. tuberculosis* throughout the body by the bloodstream. Also known as reactivation tuberculosis. *750*

mineralization process The microbial breakdown of organic materials to inorganic substances. *423, 813*

minimal inhibitory concentration (MIC) The lowest concentration of a drug that will prevent the growth of a particular microorganism. *660*

minimal lethal concentration (MLC) The lowest concentration of a drug that will kill a particular microorganism. *660*

minus, or negative, strand The virus nucleic acid strand that is complementary in base sequence to the viral mRNA. *346*

missense mutation A single base substitution in DNA that changes a codon for one amino acid into a codon for another. *266*

mitochondrion (mi″to-kon′dre-on) The eucaryotic organelle that is the site of electron transport, oxidative phosphorylation, and pathways such as the Krebs cycle; it provides most of a nonphotosynthetic cell's energy under aerobic conditions. It is constructed of an outer membrane and an inner membrane, which contains the electron transport chain. *81*

mitosis (mi-to′sis) A process that takes place in the nucleus of a eucaryotic cell and results in the formation of two new nuclei, each with the same number of chromosomes as the parent. *86*

mixed acid fermentation A type of fermentation carried out by members of the family *Enterobacteriaceae* in which ethanol and a complex mixture of organic acids are produced. *175*

mixotrophic (mik″so-trof′ik) Refers to microorganisms that combine autotrophic and heterotrophic metabolic processes (they use inorganic electron sources and organic carbon sources). *101*

mold Any of a large group of fungi that cause mold or moldiness and that exist as multicellular filamentous colonies; also the deposit or growth caused by such fungi. Molds typically do not produce macroscopic fruiting bodies. *506*

molecular chaperones Proteins that aid in the proper folding of unfolded polypeptides or partly denatured proteins and often also help transport proteins across membranes. *49*

Monera (mo-ne′rah) *See Procaryotae.* *408*

mono *See* infectious mononucleosis. *733*

monoclonal antibody (MAb; mon″o-klōn′al) An antibody of a single type that is produced by a population of genetically identical plasma cells (a clone); a monoclonal antibody is typically produced from a cell culture derived from the fusion product of a cancer cell and an antibody-producing cell (a hybridoma). *602*

monokine (mon′o-kīn) A generic term for a cytokine produced by mononuclear phagocytes (macrophages or monocytes). *607*

mononuclear phagocyte system (MPS; mon″o-nu′kle-ar) The body's collection of macrophages, monocytes, and their precursors. The system's major function is the removal of foreign material, including pathogens, and cell debris. *576*

monotrichous (mon-ot′ri-kus) Having a single flagellum. *60*

morbidity rate (mor-bid′i-te) Measures the number of individuals who become ill as a result of a particular disease within a susceptible population during a specific time period. *701*

mordant (mor′dant) A substance that helps fix dye on or in a cell. *26*

mortality rate (mor-tal′i-te) The ratio of the number of deaths from a given disease to the total number of cases of the disease. *701*

most probable number (MPN) The statistical estimation of the probable population in a liquid by diluting and determining end points for microbial growth. *850*

mucociliary blanket The layer of cilia and mucus that lines certain portions of the respiratory system; it traps microorganisms up to 10 μm in diameter and then transports them by ciliary action away from the lungs. *575*

multi-drug-resistant strains of tuberculosis (MDR-TB) A multi-drug-resistant strain is defined as *Mycobacterium tuberculosis* resistant to isoniazid and rifampin, with or without resistance to other drugs. *750*

mumps An acute generalized disease that occurs primarily in school-age children and is caused by a paramyxovirus that is transmitted in saliva and respiratory droplets. The principal manifestation is swelling of the parotid salivary glands. *718*

murein *See* peptidoglycan. *51*

must The juices of fruits, including grapes, that can be fermented for the production of alcohol. *897*

mutagen (mu′tah-jen) A chemical or physical agent that causes mutations. *262*

mutation (mu-ta′shun) A permanent, heritable change in the genetic material. *261*

mutualism (mu′tu-al-izm″) A type of symbiosis in which both partners gain from the association and are unable to survive without it. The mutualist and the host are metabolically dependent on each other. 547

mutualist (mu′tu-al-ist) An organism associated with another in a relationship that is beneficial to both (and often obligatory). 547

mycelium (mi-se′le-um) A mass of branching hyphae found in fungi and some bacteria. 39, 506

mycobiont The fungal partner in a lichen. 548

mycolic acids Complex 60 to 90 carbon fatty acids with a hydroxyl on the β-carbon and an aliphatic chain on the α-carbon; found in the cell walls of mycobacteria. 452

mycologist (mi-kol′o-jist) A person specializing in mycology; a student of mycology. 504

mycology (mi-kol′o-je) The science and study of fungi. 504

mycoplasma (mi″ko-plaz′mah) Bacteria that are members of the class *Mollicutes* and order *Mycoplasmatales;* they lack cell walls and cannot synthesize peptidoglycan precursors; most require sterols for growth; they are the smallest organisms capable of independent reproduction. 434

mycoplasmal pneumonia (mi″ko-plaz′mal nu-mo′ne-ah) A type of pneumonia caused by *Mycoplasma pneumoniae*. Spread involves airborne droplets and close contact. 779

mycosis (mi-ko′sis; pl., **mycoses**) Any disease caused by a fungus. 504, 791

mycotoxicology (mi-ko′tok″si-kol′o-je) The study of fungal toxins and their effects on various organisms. 504

myeloma cell (mi″e-lo′mah) A tumor cell that is similar to the cell type found in bone marrow. Also, a malignant, neoplastic plasma cell that produces large quantities of antibodies and can be readily cultivated. 602

myositis (mi″o-si′tis) Inflammation of a striated or voluntary muscle. 748

myxamoeba (mik-sah-me′bah; pl., **myxamoebae**) A free-living amoeboid cell that can aggregate with other myxamoeba to form a plasmodium or pseudoplasmodium. Found in cellular slime molds and the myxomycetes. 515

myxobacteria A group of gram-negative, aerobic soil bacteria characterized by gliding motility, a complex life cycle with the production of fruiting bodies, and the formation of myxospores. 473

myxospores (mik′so-spōrs) Special dormant spores formed by the myxobacteria. 474

N

napkin (diaper) candidiasis Typically found in infants whose diapers are not changed frequently and are therefore not kept dry. Caused by *Candida* species of fungi. 797

narrow-spectrum drugs Chemotherapeutic agents that are effective only against a limited variety of microorganisms. 658

natural classification A classification system that arranges organisms into groups whose members share many characteristics and reflect as much as possible the biological nature of organisms. 397

natural environment An environment that has not been modified by human activity. 811

natural killer (NK) cell A non-T, non-B lymphocyte present in nonimmunized individuals that exhibits MHC-independent cytolytic activity against tumor cells. 589, 619

naturally acquired active immunity The type of active immunity that develops when an individual's immunologic system comes into contact with an appropriate antigenic stimulus during the course of normal activities; it usually arises as the result of recovering from an infection and lasts a long time. 588

naturally acquired passive immunity The type of temporary immunity that involves the transfer of antibodies from one individual to another. 588

necrotizing fasciitis (nek′ro-tīz″ing fas″e-i′tis) A disease that results from a severe invasive group A streptococcus infection. Necrotizing fasciitis is an infection of the subcutaneous soft tissues, particularly of fibrous tissue, and is most common on the extremities. It begins with skin reddening, swelling, pain, and cellulitis, and proceeds to skin breakdown and gangrene in 3 to 5 days after onset. 748

needle aspiration Using proper aseptic technique, the withdrawal of fluid into a sterile tube that has been treated with an anticoagulant. 677

negative staining A staining procedure in which a dye is used to make the background dark while the specimen is unstained. 27

Negri bodies (na′gre) Masses of viruses or unassembled viral subunits found within the brain neurons of rabies-infected animals. 734

neurotoxin (nu″ro-tok′sin) A toxin that is poisonous to or destroys nerve tissue; especially the toxins secreted by *C. tetani, Corynebacterium diphtheriae,* and *Shigella dysenteriae.* 567

neustonic (nu′ston″ik) The microorganisms that live at the atmospheric interface of a water body. 521, 567

neutrophile (nu′tro-fīl″) Microorganisms that grow best at a neutral pH range between pH 5.5 and 8.0. 126

nicotinamide adenine dinucleotide (NAD⁺; nik″o-tin′ah-mīd) An electron-carrying coenzyme; it is particularly important in catabolic processes and usually donates its electrons to the electron transport chain under aerobic conditions. 156

nicotinamide adenine dinucleotide phosphate (NADP⁺; nik″o-tin′ah-mīd) An electron-carrying coenzyme that most often participates as an electron carrier in biosynthetic metabolism. 157

nitrification (ni″trĭ-fĭ-ka′shun) The oxidation of ammonia to nitrate. 181, 464, 816

nitrifying bacteria (ni′trĭ-fi′ing) Chemolithotrophic, gram-negative bacteria that are members of the family *Nitrobacteriaceae* and convert ammonia to nitrate and nitrite to nitrate. 181, 464

nitrogen fixation The metabolic process in which atmospheric molecular nitrogen is reduced to ammonia; carried out by cyanobacteria, *Rhizobium,* and other nitrogen-fixing bacteria. 199, 817

nitrogen oxygen demand (NOD) The demand for oxygen in sewage treatment, caused by nitrifying microorganisms. 842

nitrogenase (ni′tro-jen-ās) The enzyme that catalyzes biological nitrogen fixation. 199

nocardioforms Bacteria that resemble members of the genus *Nocardia;* they develop a substrate mycelium that readily breaks up into rods and coccoid elements. 495

Nod factors Chemical signals produced by *Rhizobium,* a legume nodule-forming bacterium, in response to plant-derived inducer molecules. These factors promote both the binding of specific strains of *Rhizobium* to root hairs and root hair curling and thus help initiate nodulation. 865

nomenclature (no′men-kla″tūr) The branch of taxonomy concerned with the assignment of names to taxonomic groups in agreement with published rules. 391

noncompetitive inhibitor Inhibition of enzyme activity that results from the inhibitor binding at a site other than the active site and altering the enzyme's shape to make it less active. 162

noncyclic photophosphorylation (fo″to-fos″for-i-la′shun) The process in which light energy is used to make ATP when electrons are moved from water to NADP⁺ during photosynthesis; both photosystem I and photosystem II are involved. 184

nongonococcal urethritis (NGU) (u″rə-thri′tis) Any inflammation of the urethra not caused by *Neisseria gonorrhoeae.* 778

non-Newtonian broth A liquid whose viscosity can vary with the rate of agitation; it usually has plasticlike characteristics. 909

nonsense codon A codon that does not code for an amino acid but is a signal to terminate protein synthesis. 232, 257

nonsense mutation A mutation that converts a sense codon to a nonsense or stop codon. 266

nonspecific resistance Refers to those general defense mechanisms that are inherited as part of the innate structure and function of each animal; also known as nonspecific immunity. 587

nosocomial infection (nos″o-ko′me-al) An infection that develops within a hospital (or other type of clinical care facility) and is produced by an infectious organism acquired during the stay of the patient. 711, 785

nuclear envelope (nu′kle-ar) The complex double-membrane structure forming the outer boundary of the eucaryotic nucleus. It is covered by pores through which substances enter and leave the nucleus. 85

nucleic acid hybridization (nu-kle′ik) The process of forming a hybrid double-stranded DNA molecule using a heated mixture of single-stranded DNAs from two different sources; if the sequences are fairly complementary, stable hybrids will form. 395

nucleocapsid (nu″kle-o-kap′sid) The nucleic acid and its surrounding protein coat or capsid; the basic unit of virion structure. 340

nucleoid (nu′kle-oid) An irregularly shaped region in the procaryotic cell that contains its genetic material. 50

nucleolus (nu-kle′o-lus) The organelle, located within the eucaryotic nucleus and not bounded by a membrane, that is the location of ribosomal RNA synthesis and the assembly of ribosomal subunits. 85

nucleoside (nu′kle-o-sīd″) A combination of ribose or deoxyribose with a purine or pyrimidine base. 204

nucleosome (nu′kle-o-sōm″) A complex of histones and DNA found in eucaryotic chromatin; the DNA is wrapped around the surface of the beadlike histone complex. *215*

nucleotide (nu′kle-o-tīd) A combination of ribose or deoxyribose with phosphate and a purine or pyrimidine base; a nucleoside plus one or more phosphates. *204*

nucleus (nu′kle-us) The eucaryotic organelle enclosed by a double-membrane envelope that contains the cell's chromosomes. *84*

null cell A lymphoid cell that does not have characteristics of either T or B cells; also known as a third population cell. *589*

numerical aperture The property of a microscope lens that determines how much light can enter it and how great a resolution it can provide. *20*

numerical taxonomy The grouping by numerical methods of taxonomic units into taxa based on their character states. *397*

nutrient (nu′tre-ent) A substance that supports growth and reproduction. *99*

nystatin (nis′tah-tin) A polyene antibiotic from *Streptomyces noursei* that is used in the treatment of *Candida* infections of the skin, vagina, and alimentary tract. *671*

O

O antigen A polysaccharide antigen extending from the outer membrane of some gram-negative bacterial cell walls; it is part of the lipopolysaccharide. *56*

obligate aerobes Organisms that grow only in the presence of oxygen. *130*

obligate anaerobes Microorganisms that cannot tolerate the presence of oxygen and die when exposed to it. *130*

odontopathogens Dental pathogens. *782*

Okazaki fragments Short stretches of polynucleotides produced during discontinuous DNA replication. *219*

oligotroph A microorganism that can survive and function in a low-nutrient environment. *862*

oligotrophic (ol″ĭ-go-trof′ik) Bodies of water containing low levels of nutrients, particularly nutrients that support microbial growth. *840*

oncogene (ong′ko-jēn) A gene whose activity is associated with the conversion of normal cells to cancer cells. *380*

one-step growth experiment An experiment used to study the reproduction of lytic phages in which one round of phage reproduction occurs and ends with the lysis of the host bacterial population. *354*

onychomycosis (on″i-ko-mi-ko′sis) A fungal infection of the nail plate producing nails that are opaque, white, thickened, friable, and brittle. Also called ringworm of the nails and tinea unguium. Caused by *Trichophyton* and other fungi such as *C. albicans.* *797*

oocyst (o′o-sist) Cyst formed around a zygote of malaria and related protozoa. *540*

oogonia (o″o-go′ne-a) Mitotically dividing female structures that produce primary oocytes and gametes. *522*

oomycetes (o″o-mi-se′tēz) A collective name for members of the division *Oomycota;* also known as the water molds. *515*

operator The segment of DNA to which the repressor protein binds; it controls the expression of the genes adjacent to it. *243*

operon (op′er-on) The sequence of bases in DNA that contains one or more structural genes together with the operator controlling their expression. *244*

ophthalmia neonatorum (of-thal′me-ah ne″o-nator-um) A gonorrheal eye infection in a newborn, which may lead to blindness. Also called conjunctivitis of the newborn. *757*

opportunistic organism An organism that is generally harmless but becomes pathogenic in a compromised host. *797*

opsonization (op″so-ni-za′shun) The action of opsonins in making bacteria and other cells more readily phagocytosed. Antibodies, complement (especially C3b), and fibronectin are potent opsonins. *592, 637*

oral candidiasis *See* thrush. *797*

organelle (or″gah-nel′) A structure within or on a cell that performs specific functions and is related to the cell in a way similar to that of an organ to the body. *74*

organotrophs Organisms that use reduced organic compounds as their electron source. *100*

ornithosis *See* psittacosis. *778*

osmophilic microorganisms (oz″mo-fil′ik) Microorganisms that grow best in or on media of high solute concentration. *882*

osmosis (oz-mo′sis) The movement of water across a selectively permeable membrane from a dilute solution (higher water concentration) to a more concentrated solution. *57*

osmotolerant Organisms that grow over a fairly wide range of water activity or solute concentration. *125*

Ouchterlony technique *See* double diffusion agar assay. *649*

outbreak The sudden, unexpected occurrence of a disease in a given population. *700*

outer membrane A special membrane located outside the peptidoglycan layer in the cell walls of gram-negative bacteria. *51*

β-oxidation pathway The major pathway of fatty acid oxidation to produce NADH, FADH$_2$, and acetyl coenzyme A. *179*

oxidation-reduction (redox) reactions Reactions involving electron transfers; the reductant donates electrons to an oxidant. *156*

oxidative phosphorylation (fos″for-ĭ-la′shun) The synthesis of ATP from ADP using energy made available during electron transport. *171*

oxidizing agent or **oxidant** (ok′sĭ-dant) The electron acceptor in an oxidation-reduction reaction. *156*

oxygenic photosynthesis Photosynthesis that oxidizes water to form oxygen; the form of photosynthesis characteristic of eucaryotic algae and cyanobacteria. *187, 455*

P

pacemaker enzyme The enzyme in a metabolic pathway that catalyzes the slowest or rate-limiting reaction; if its rate changes, the pathway's activity changes. *241*

pandemic (pan-dem′ik) An increase in the occurrence of a disease within a large and geographically widespread population (often refers to a worldwide epidemic). *700*

pannus (pan′us) A superficial vascularization of the cornea with infiltration of granulation tissue. *779*

panzootic (pan″zo-ot′ik) Th[...] of a disease in an animal popul[...]

paralytic shellfish poisoning (pаɪ ə-lɪt′ɪk) Dinoflagellates (*Gonyaulax* spp.) produce a powerful neurotoxin called saxitoxin. Shellfish accumulate saxitoxin and are poisonous to animals and humans who eat them. Saxitoxin paralyzes the striated respiratory muscles in many vertebrates by inhibiting sodium transport. Paralytic shellfish poisoning is characterized by numbness of the mouth, lips, face, and extremities. Its duration ranges from a few hours to a few days and usually is not fatal. *529*

parasite (par′ah-sīt) An organism that lives on or within another organism (the host) and benefits from the association while harming its host. Often the parasite obtains nutrients from the host. *562*

parasitic organism *See* parasite. *562*

parasitism (par′ah-si″tizm) A type of symbiosis in which one organism adversely affects the other (the host), but cannot live without it. *562*

parenteral route (pah-ren′ter-al) A route of drug administration that is nonoral (e.g., by injection). *663*

parfocal (par-fo′kal) A microscope that retains proper focus when the objectives are changed. *19*

paronychia (par″o-nik′e-ah) Inflammation involving the folds of tissue surrounding the nail; usually caused by *Candida albicans.* *797*

passive diffusion The process in which molecules move from a region of higher concentration to one of lower concentration as a result of random thermal agitation. *102*

Pasteur effect (pas-tur′) The decrease in the rate of sugar catabolism and change to aerobic respiration that occurs when microorganisms are switched from anaerobic to aerobic conditions. *175*

pasteurization (pas″ter-ĭ-za′shun) The process of heating milk and other liquids to destroy microorganisms that can cause spoilage or disease. *142, 883*

pathogen (path′o-jen) A disease-producing organism or material. *562*

pathogenic potential The degree that a pathogen causes morbid signs and symptoms. *563*

pathogenicity (path″o-je-nis′ĭ-te) The ability to cause disease. *562*

pébrine (pa-brēn′) An infectious disease of silkworms caused by the protozoan *Nosema bombycis.* *540*

ped A natural soil aggregate, formed partly through bacterial and fungal growth in the soil. *860*

pellicle (pel′ĭ-k′l)   A relatively rigid layer of proteinaceous elements just beneath the plasma membrane in many protozoa and algae. The plasma membrane is sometimes considered part of the pellicle.   *90, 524, 533*

pelvic inflammatory disease (PID)   A severe infection of the female reproductive organs. The disease that results when gonococci and chlamydiae infect the uterine tubes and surrounding tissue.   *757, 778*

penicillins (pen″ĭ-sil-ins)   A group of antibiotics containing a β-lactam ring, which are active against gram-positive bacteria.   *57, 666*

penton or **pentamer**   A capsomer composed of five protomers.   *341*

pentose phosphate pathway (pen′tōs)   The pathway that oxidizes glucose 6-phosphate to ribulose 5-phosphate and then converts it to a variety of three to seven carbon sugars; it forms several important products (NADPH for biosynthesis, pentoses, and other sugars) and also can be used to degrade glucose to CO_2.   *167*

peplomer or **spike** (pep′lo-mer)   A protein or protein complex that extends from the virus envelope and often is important in virion attachment to the host cell surface.   *346*

peptic ulcer disease   A gastritis caused by *Helicobacter pylori.*   *758*

peptide interbridge (pep′tīd)   A short peptide chain that connects the tetrapeptide chains in some peptidoglycans.   *52*

peptidoglycan (pep″tĭ-do-gli′kan)   A large polymer composed of long chains of alternating N-acetylglucosamine and N-acetylmuramic acid residues. The polysaccharide chains are linked to each other through connections between tetrapeptide chains attached to the N-acetylmuramic acids. It provides much of the strength and rigidity possessed by bacterial cell walls.   *51, 439*

peptidyl or **donor site (P site)**   The site on the ribosome that contains the peptidyl-tRNA at the beginning of the elongation cycle during protein synthesis.   *229*

peptidyl transferase   The enzyme that catalyzes the transpeptidation reaction in protein synthesis; in this reaction, an amino acid is added to the growing peptide chain.   *232*

peptones (pep′tōns)   Water-soluble digests or hydrolysates of proteins that are used in the preparation of culture media.   *107*

period of infectivity   Refers to the time during which the source of an infectious disease is infectious or is disseminating the pathogen.   *705*

periodontal disease (per″e-o-don′tal)   A disease located around the teeth or in the periodontium—the tissue investing and supporting the teeth, including the cementum, periodontal ligament, alveolar bone, and gingiva.   *784*

periodontitis (per″e-o-don-ti′tis)   An inflammation of the periodontium.   *785*

periodontium (per″e-o-don′she-um)   *See* periodontal disease.   *784*

periodontosis (per″e-o-don-to′sis)   A degenerative, noninflammatory condition of the periodontium, which is characterized by destruction of tissue.   *785*

periplasm (per′ĭ-plaz-əm)   The substance that fills the periplasmic space.   *51*

periplasmic flagella   The flagella that lie under the outer sheath and extend from both ends of the spirochete cell to overlap in the middle and form the axial filament. Also called axial fibrils and endoflagella.   *417*

periplasmic space (per″i-plas′mik) or **periplasm** (per′ĭ-plazm)   The space between the plasma membrane and the outer membrane in gram-negative bacteria, and between the plasma membrane and the cell wall in gram-positive bacteria.   *51*

peristalsis (per″ĭ-stal′sis)   The progressive wavelike or wormlike movement seen in tubes such as the small intestine. This movement propels the contents of the tube down its length.   *575*

peritrichous (pĕ-rit′rĭ-kus)   A cell with flagella evenly distributed over its surface.   *60*

permease (per′me-ās)   A membrane carrier protein.   *103*

pertussis (pər-tus′is)   An acute, highly contagious infection of the respiratory tract, most frequently affecting young children, usually caused by *Bordetella pertussis* or *B. parapertussis.* Consists of peculiar paroxysms of coughing, ending in a prolonged crowing or whooping respiration; hence the name whooping cough.   *746*

petri dish (pe′tre)   A shallow dish consisting of two round, overlapping halves that is used to grow microorganisms on solid culture medium; the top is larger than the bottom of the dish to prevent contamination of the culture.   *110*

Peyer's patches (pi′erz)   Lymphatic aggregates in the submucosa of the small intestine, which contain a full repertoire of lymphoreticular cells, T and B cells, and macrophages. Is also a part of the GALT system.   *596*

phage (fāj)   *See* bacteriophage.   *354*

phagocytic vacuole (fag″o-sit′ik vak′u-ol)   A membrane-delimited vacuole produced by cells carrying out phagocytosis. It is formed by the invagination of the plasma membrane and contains solid material.   *534*

phagocytosis (fag″o-si-to′sis)   The endocytotic process in which a cell encloses large particles in a membrane-delimited phagocytic vacuole or phagosome and engulfs them.   *79, 582*

phagolysosome (fag″o-li′so-sōm)   The vacuole that results from the fusion of a phagosome with a lysosome.   *582*

phagosome (fag′o-sōm)   *See* phagocytic vacuole.   *582*

phagovar (fag′o-var)   A specific phage type.   *695*

pharyngitis (far″in-ji′tis)   Inflammation of the pharynx, often due to a *S. pyogenes* infection.   *749*

phase-contrast microscope   A microscope that converts slight differences in refractive index and cell density into easily observed differences in light intensity.   *21*

phenetic system   A classification system that groups organisms together based on their mutual similarity.   *397*

phenol coefficient test   A test to measure the effectiveness of disinfectants by comparing their activity against test bacteria with that of phenol.   *148*

phosphatase (fos′fah-tās″)   An enzyme that catalyzes the hydrolytic removal of phosphate from molecules.   *197*

phosphate group transfer potential   A measure of the ability of a phosphorylated molecule such as ATP to transfer its phosphate to water and other acceptors. It is the negative of the $\Delta G^{o\prime}$ for the hydrolytic removal of phosphate.   *155*

photoautotroph (fo′to-aw′to-trōf)   *See* photolithotrophic autotrophs.   *100*

photolithotrophic autotrophs   Organisms that use light energy, an inorganic electron source (e.g., H_2O, H_2, H_2S), and CO_2 as a carbon source.   *100*

photoorganotrophic heterotrophs   Microorganisms that use light energy and organic electron donors, and also employ simple organic molecules rather than CO_2 as their carbon source.   *101*

photoreactivation (fo″to-re-ak″tĭ-va′shun)   The process in which blue light is used by a photoreactivating enzyme to repair thymine dimers in DNA by splitting them apart.   *133, 272*

photosynthesis (fo″to-sin′thĕ-sis)   The trapping of light energy and its conversion to chemical energy, which is then used to reduce CO_2 and incorporate it into organic form.   *153, 181*

photosystem I   The photosystem in eucaryotic cells that absorbs longer wavelength light, usually greater than about 680 nm, and transfers the energy to chlorophyll P700 during photosynthesis; it is involved in both cyclic photophosphorylation and noncyclic photophosphorylation.   *184*

photosystem II   The photosystem in eucaryotic cells that absorbs shorter wavelength light, usually less than 680 nm, and transfers the energy to chlorophyll P680 during photosynthesis; it participates in noncyclic photophosphorylation.   *184*

phototrophs   Organisms that use light as their energy source.   *100*

phycobiliproteins   Photosynthetic pigments that are composed of proteins with attached tetrapyrroles; they are often found in cyanobacteria and red algae.   *183*

phycobilisomes   Special particles on the membranes of cyanobacteria that contain photosynthetic pigments and electron transport chains.   *460*

phycobiont (fi″ko-bi′ont)   The algal or cyanobacterial partner in a lichen.   *548*

phycocyanin (fi″ko-si′an-in)   A blue phycobiliprotein pigment used to trap light energy during photosynthesis.   *183*

phycoerythrin (fi″ko-er′i-thrin)   A red photosynthetic phycobiliprotein pigment used to trap light energy.   *183*

phycology (fi-kol′o-je)   The study of algae; algology.   *521*

phylogenetic or **phyletic classification system** (fi″lo-jĕ-net′ik, fi-let′ik)   A classification system based on evolutionary relationships rather than the general similarity of contemporary characteristics.   *397*

phytoplankton (fi″to-plank′ton)   A community of floating photosynthetic organisms, largely composed of algae and cyanobacteria.   *521, 838*

piedra (pe-a′drah)   A fungal disease of the hair in which white or black nodules of fungi form on the shafts.   *791*

pinocytosis (pi″no-si-to′sis) The endocytotic process in which a cell encloses a small amount of the surrounding liquid and its solutes in tiny pinocytotic vesicles or pinosomes. *79*

pitched Pertaining to inoculation of a nutrient medium with yeast, for example, in beer brewing. *898*

plague (plāg) An acute febrile, infectious disease, caused by the bacillus *Yersinia pestis,* which has a high mortality rate; the two major types are bubonic plague and pneumonic plague. *753*

plankton (plank′ton) Free-floating, mostly microscopic microorganisms that can be found in almost all waters; a collective name. *521, 533*

planktonic (adj.) *See* plankton. *521*

plantar wart A wart that occurs on the sole of the foot. *739*

plaque (plak) 1. A clear area in a lawn of bacteria or a localized area of cell destruction in a layer of animal cells that results from the lysis of the bacteria by bacteriophages or the destruction of the animal cells by animal viruses. 2. The term also refers to dental plaque, a film of food debris, polysaccharides, and dead cells that cover the teeth. It provides a medium for the growth of bacteria (which may be considered a part of the plaque), leading to a microbial plaque ecosystem that can produce dental decay. *335*

plasma cell A mature, differentiated B lymphocyte chiefly occupied with antibody synthesis and secretion; a plasma cell lives for only 5 to 7 days. *589*

plasma membrane The selectively permeable membrane surrounding the cell's cytoplasm; also called the cell membrane, plasmalemma, or cytoplasmic membrane. *42*

plasmid (plaz′mid) A circular, double-stranded DNA molecule that can exist and replicate independently of the chromosome or may be integrated with it. A plasmid is stably inherited, but is not required for the host cell's growth and reproduction. *280, 669*

plasmid fingerprinting A technique used to identify microbial isolates as belonging to the same strain because they contain the same number of plasmids with the identical molecular weights and similar phenotypes. *696*

plasmodial slime mold (plaz-mo′de-al) A member of the division *Myxomycota* that exists as a thin, streaming, multinucleate mass of protoplasm, which creeps along in an amoeboid fashion. *515*

plasmodium (plaz-mo′de-um; pl., **plasmodia**) A stage in the life cycle of myxomycetes (plasmodial slime molds); a multinucleate mass of protoplasm surrounded by a membrane. Also, a parasite of the genus *Plasmodium.* *515*

plasmolysis (plaz-mol′ĭ-sis) The process in which water osmotically leaves a cell, which causes the cytoplasm to shrivel up and pull the plasma membrane away from the cell wall. *57*

plastid (plas′tid) A cytoplasmic organelle of algae and higher plants that contains pigments such as chlorophyll, stores food reserves, and often carries out processes such as photosynthesis. *82*

pleomorphic (ple″o-mor′fik) Refers to bacteria that are variable in shape and lack a single, characteristic form. *39*

plus strand or **positive strand** The virus nucleic-acid strand that is equivalent in base sequence to the viral mRNA. *343*

pneumocystis pneumonia, *Pneumocystis carinii pneumonia* **(PCP)**; (noo″mo-sis-tis) A type of pneumonia caused by the protist *Pneumocystis carinii.* *799*

pneumonic plague *See* plague. *754*

point mutation A mutation that affects only a single base pair in a specific location. *266*

polar flagellum A flagellum located at one end of an elongated cell. *60*

polio *See* poliomyelitis. *738*

poliomyelitis (po″le-o-mi″e-li′tis) An acute, contagious viral disease that attacks the central nervous system, injuring or destroying the nerve cells that control the muscles and sometimes causing paralysis; also called polio or infantile paralysis. *738*

poly-ß-hydroxybutyrate (hi-drok″se-bu′tĭ-rāt) A linear polymer of ß-hydroxybutyrate used as a reserve of carbon and energy by many bacteria. *45*

polymerase chain reaction (PCR) An in vitro technique used to synthesize large quantities of specific nucleotide sequences from small amounts of DNA. It employs oligonucleotide primers complementary to specific sequences in the target gene and special heat-stable DNA polymerases. *310*

polyribosome (pol″e-ri′bo-sōm) A complex of several ribosomes with a messenger RNA; each ribosome is translating the same message. *81, 226*

polysome (pol′e-sōm) *See* polyribosome. *81*

Pontiac fever A bacterial disease caused by *Legionella pneumophila* that resembles an allergic disease more than an infection. First described from Pontiac, Michigan. *See* Legionnaires' disease. *745*

population An assemblage of organisms of the same type. *813*

porin proteins Proteins that form channels across the outer membrane of gram-negative bacterial cell walls. Small molecules are transported through these channels. *56*

posttranscriptional modification The processing of the initial RNA transcript, heterogeneous nuclear RNA, to form mRNA. *223*

potable (po′tah-b′l) Refers to water suitable for drinking. *850*

pour plate A petri dish of solid culture medium with isolated microbial colonies growing both on its surface and within the medium, which has been prepared by mixing microorganisms with cooled, still liquid medium and then allowing the medium to harden. *109*

precipitation (or **precipitin**) **reaction** (pre-sip″ĭ-ta′shun) The reaction of an antibody with a soluble antigen to form an insoluble precipitate. *640*

precipitin (pre-sip′ĭ-tin) The antibody responsible for a precipitation reaction. *640*

presence-absence test (P-A test) A simplified test for the presence of coliforms in water, in which a single 100 ml sample is incubated in a lactose broth. *850*

prevalence rate Refers to the total number of individuals infected at any one time in a given population regardless of when the disease began. *701*

Pribnow box A special base sequence in the promoter that is recognized by the RNA polymerase and is the site of initial polymerase binding. *222, 260*

primary amebic meningoencephalitis An infection of the meninges of the brain by the free-living amoebae *Naegleria* or *Acanthamoeba.* *801*

primary metabolites Microbial metabolites produced during the growth phase of an organism. *912*

primary producer Photoautotrophic and chemoautotrophic organisms that incorporate carbon dioxide into organic carbon and thus provide new biomass for the ecosystem. *811*

primary production The incorporation of carbon dioxide into organic matter by photosynthetic organisms and chemoautotrophic bacteria. *811*

primary treatment The first step of sewage treatment, in which physical settling and screening are used to remove particulate materials. *843*

prion (pri′on) An infectious particle that is the cause of slow diseases like scrapie in sheep and goats; it has a protein component, but no nucleic acid has yet been detected. *385*

probe (prōb) A short, labeled nucleic acid segment complementary in base sequence to part of another nucleic acid, which is used to identify or isolate the particular nucleic acid from a mixture through its ability to bind specifically with the target nucleic acid. *888*

probiotics The oral administration of either living microorganisms or substances to promote the health and growth of an animal or human. *558*

Procaryotae (pro-kar″e-o′te) A kingdom comprising all the procaryotic microorganisms (those lacking true membrane-enclosed nuclei). *408*

procaryotic cells (pro″kar-e-ot′ik) Cells that lack a true, membrane-enclosed nucleus; bacteria are procaryotic and have their genetic material located in a nucleoid. *13, 94*

prodromal stage (pro-dro′mal) The period during the course of a disease in which there is the appearance of signs and symptoms, but they are not yet distinctive and characteristic enough to make an accurate diagnosis. *702*

product The substance formed in an enzyme reaction. *159*

progametangium (pro-gam-ĕ″tan′je-um; pl., **progametangia**) The cell that gives rise to a gametangium and a proximal suspensor during the early stages of sexual reproduction in zygomycetous fungi. *510*

proliferative kidney disease (pro-lif′er-a-tiv) A protozoan disease caused by an unclassified myxozoan in salmonids throughout the world. *541*

promoter The region on DNA at the start of a gene that the RNA polymerase binds to before beginning transcription. *222, 259*

propagated epidemic   An epidemic that is characterized by a relatively slow and prolonged rise and then a gradual decline in the number of individuals infected. It usually results from the introduction of an infected individual into a susceptible population, and the pathogen is transmitted from person to person.   *703*

prophage (pro'fāj)   The latent form of a temperate phage that remains within the lysogen, usually integrated into the host chromosome.   *294, 362*

prostheca (pros-the'kah)   An extension of a bacterial cell, including the plasma membrane and cell wall, that is narrower than the mature cell.   *466*

prosthetic group (pros-thet'ik)   A tightly bound cofactor that remains at the active site of an enzyme during its catalytic activity.   *159*

protease (pro'te-ās)   An enzyme that hydrolyzes proteins to their constituent amino acids. Also called a proteinase.   *179*

protein engineering (pro'tēn)   The rational design of proteins by constructing specific amino acid sequences through molecular techniques, with the objective of modifying protein characteristics.   *919*

Protista (pro-tis'tah)   A kingdom containing eucaryotes with unicellular organization, either in the form of solitary cells or colonies of cells lacking true tissues; members of the kingdom are called protists.   *408*

protomer   An individual subunit of a viral capsid; a capsomer is made of protomers.   *341*

protonmotive force (PMF)   The force arising from a gradient of protons and a membrane potential that is thought to power ATP synthesis and other processes.   *172*

protoplast (pro'to-plast)   A bacterial or fungal cell with its cell wall completely removed. It is spherical in shape and osmotically sensitive.   *45, 57*

protoplast fusion   The joining of cells that have had their walls weakened or completely removed.   *910*

protothecosis (pro'to-the-ko'sis; pl., **protothecoses**)   A disease of humans and animals produced by the green alga *Prototheca moriformis*.   *523*

prototroph (pro'to-trōf)   A microorganism that requires the same nutrients as the majority of naturally occurring members of its species.   *100, 262*

protozoan or **protozoon** (pro''to-zo'an, pl. **protozoa**)   A microorganism belonging to the *Protozoa* subkingdom. A unicellular or acellular eucaryotic protist whose organelles have the functional role of organs and tissues in more complex forms. Protozoa vary greatly in size, morphology, nutrition, and life cycle.   *533*

protozoology (pro''to-zo-ol'o-je)   The study of protozoa.   *533*

proviral DNA   Viral DNA that has been integrated into host cell DNA. In retroviruses it is the double-stranded DNA copy of the RNA genome.   *376*

pseudomurein   A modified peptidoglycan lacking D-amino acids and containing N-acetyltalosaminuronic acid instead of N-acetylmuramic acid; found in methanogenic bacteria.   *478*

pseudoplasmodium (soo''do-plaz-mo'de-um; pl., **pseudoplasmodia**)   A sausage-shaped amoeboid structure consisting of many myxamoebae and behaving as a unit; the result of myxamoebal aggregation in the cellular slime molds; also called a slug.   *515*

pseudopodium or **pseudopod** (soo''do-po'de-um)   A nonpermanent cytoplasmic extension of the cell body by which amoebae and amoeboid organisms move and feed.   *534*

psittacosis (ornithosis; sit''ah-ko'sis)   A disease due to a strain of *Chlamydia psittaci*, first seen in parrots and later found in other birds and domestic fowl (in which it is called ornithosis). It is transmissible to humans.   *778*

psychrophile (si'kro-fīl)   A microorganism that grows well at 0°C and has an optimum growth temperature of 15°C or lower and a temperature maximum around 20°C.   *128*

psychrotroph   A microorganism that grows at 0°C, but has a growth optimum between 20 and 30°C, and a maximum of about 35°C.   *130*

puerperal fever (pu-er'per-al)   An acute, febrile condition following childbirth; it is characterized by infection of the uterus and/or adjacent regions and is caused by streptococci.   *708*

pulmonary anthrax (pul'mo-ner''e)   A form of anthrax involving the lungs. Also known as woolsorter's disease.   *755*

pulmonary syndrome hantavirus   *See* hantavirus pulmonary syndrome.   *721*

pure culture   A population of cells that are identical because they arise from a single cell.   *108*

purine (pu'rin)   A basic, heterocyclic, nitrogen-containing molecule with two joined rings that occurs in nucleic acids and other cell constituents; most purines are oxy or amino derivatives of the purine skeleton. The most important purines are adenine and guanine.   *204*

purple membrane   An area of the plasma membrane of *Halobacterium* that contains bacteriorhodopsin and is active in photosynthetic light energy trapping.   *486*

putrefaction (pu''trě-fak'shun)   The microbial decomposition of organic matter, especially the anaerobic breakdown of proteins, with the production of foul-smelling compounds such as hydrogen sulfide and amines.   *881*

pyrenoid (pi'rě-noid)   The differentiated region of the chloroplast that is a center of starch formation in green algae and stoneworts.   *84, 522*

pyrimidine (pi-rim'i-dēn)   A basic, heterocyclic, nitrogen-containing molecule with one ring that occurs in nucleic acids and other cell constituents; pyrimidines are oxy or amino derivatives of the pyrimidine skeleton. The most important pyrimidines are cytosine, thymine, and uracil.   *204*

Q

Q fever   An acute zoonotic disease caused by the rickettsia *Coxiella burnetii*.   *781*

Quellung reaction   The increase in visibility or the swelling of the capsule of a microorganism in the presence of antibodies against capsular antigens.   *652*

R

rabies (ra'bēz)   An acute infectious disease of the central nervous system, which affects all warm-blooded animals (including humans). It is caused by an ssRNA virus belonging to the genus *Lyssavirus* in the family *Rhabdoviridae*.   *734*

racking   The removal of sediments from wine bottles.   *897*

radappertization   The use of gamma rays from a cobalt source for control of microorganisms in foods.   *885*

radioimmunoassay (RIA; ra''de-o-im''u-no-as'a)   A very sensitive assay technique that uses a purified radioisotope-labeled antigen or antibody to compete for antibody or antigen with unlabeled standard and samples to determine the concentration of a substance in the samples.   *651*

rapid sand filter   A bed of sand through which water is passed at a rapid rate to trap physical impurities and purify the water.   *848*

reactivation tuberculosis   *See* miliary tuberculosis.   *750*

reagin (re'ah-jin)   Antibody that mediates immediate hypersensitivity reactions. IgE is the major reagin in humans.   *621*

recalcitrance   Stubbornness; the resistance of a substance to modification or degradation by microorganisms.   *873*

recombinant DNA technology   The techniques used in carrying out genetic engineering; they involve the identification and isolation of a specific gene, the insertion of the gene into a vector such as a plasmid to form a recombinant molecule, and the production of large quantities of the gene and its products.   *306*

recombination (re''kom-bĭ-na'shun)   The process in which a new recombinant chromosome is formed by combining genetic material from two organisms.   *278*

recombination repair   A DNA repair process that repairs damaged DNA when there is no remaining template; a piece of DNA from a sister molecule is used.   *273*

red tides   Red tides occur frequently in coastal areas and often are associated with population blooms of dinoflagellates. Dinoflagellate pigments are responsible for the red color of the water. Under these conditions, the dinoflagellates often produce saxitoxin, which can lead to paralytic shellfish poisoning.   *529*

reducing agent or **reductant** (re-duk'tant)   The electron donor in an oxidation-reduction reaction.   *156*

refraction (re-frak'shun)   The deflection of a light ray from a straight path as it passes from one medium (e.g., glass) to another (e.g., air).   *18*

refractive index (re-frak'tiv)   The ratio of the velocity of light in the first of two media to that in the second as it passes from the first to the second.   *18*

regulator T cell Regulator T cells control the development of effector T cells. Two types exist: T-helper cells (CD4⁺ cells) and T-suppressor cells. There are three subsets of T-helper cells: T_H1, T_H2, and T_HO. T_H1 cells produce IL-2, IFN-γ, and TNF-ß. They effect cell-mediated immunity and are responsible for delayed-type hypersensitivity reactions and macrophage activation. T_H2 cells produce IL-4, IL-5, IL-6, IL-10, IL-13. They are helpers for B-cell antibody responses and humoral immunity; they also support IgE responses and eosinophilia. T_HO cells exhibit an unrestricted cytokine profile. *616*

regulatory mutants Mutant organisms that have lost the ability to limit synthesis of a product, which normally occurs by regulation of activity of an earlier step in the biosynthetic pathway. *915*

replica plating A technique for isolating mutants from a population by plating cells from each colony growing on a nonselective agar medium onto plates with selective media or environmental conditions, such as the lack of a nutrient or the presence of an antibiotic or a phage; the location of mutants on the original plate can be determined from growth patterns on the replica plates. *268*

replication (rep″lĭ-ka′shun) The process in which an exact copy of parental DNA or RNA is made with the parental molecule serving as a template. *216*

replication fork The Y-shaped structure where DNA is replicated. The arms of the Y contain template strand and a newly synthesized DNA copy. *217*

replicative form A double-stranded form of nucleic acid that is formed from a single-stranded virus genome and used to synthesize new copies of the genome. *360, 376*

replicon (rep′lĭ-kon) A unit of the genome that contains an origin for the initiation of replication and in which DNA is replicated. *217, 280*

repressible enzyme An enzyme whose level drops in the presence of a small molecule, usually an end product of its metabolic pathway. *243*

repressor protein (re-pres′or) A protein coded for by a regulator gene that can bind to the operator and inhibit transcription; it may be active by itself or only when the corepressor is bound to it. *243*

reservoir (rez′er-vwar) A site, alternate host, or carrier that normally harbors pathogenic organisms and serves as a source from which other individuals can be infected. *705*

reservoir host An organism other than a human that is infected with a pathogen that can also infect humans. *562*

resolution (rez″o-lu′shun) The ability of a microscope to separate or distinguish between small objects that are close together. *20*

respiratory burst The respiratory burst occurs when an activated phagocytic cell increases its oxygen consumption to support the increased metabolic activity of phagocytosis. The burst generates highly toxic oxygen products such as singlet oxygen, superoxide radical, hydrogen peroxide, hydroxyl radical, and hypochlorite. *582*

respiratory syncytial virus (RSV; sin-sish′al) A member of the family *Paramyxoviridae* and genus *Pneumovirus*, it is a negative-sense ssRNA virus that causes respiratory infections in children. *719*

restricted transduction A transduction process in which only a specific set of bacterial genes are carried to another bacterium by a temperate phage; the bacterial genes are acquired because of a mistake in the excision of a prophage during the lysogenic life cycle. *295*

restriction enzymes Enzymes produced by host cells that cleave virus DNA at specific points and thus protect the cell from virus infection; they are used in carrying out genetic engineering. *306, 358*

reticulate body (RB) The form in the chlamydial life cycle whose role is growth and reproduction within the host cell. *432*

reticuloendothelial system (RES; re-tik″u-lo-en″do-the′le-al) The collection of fixed phagocytic cells (including macrophages, monocytes, and specialized endothelial cells) located in the liver, spleen, lymph nodes, and bone marrow. The RES is an important component of the host's general or nonspecific defense against pathogens. *576*

retroviruses (re″tro-vi′rus-es) A group of viruses with RNA genomes that carry the enzyme reverse transcriptase and form a DNA copy of their genome during their reproductive cycle. *376*

reverse transcriptase An RNA-dependent DNA polymerase that uses a viral RNA genome as a template to form a DNA copy; this is a reverse of the normal flow of genetic information, which proceeds from DNA to RNA. *376*

reversible covalent modification A mechanism of enzyme regulation in which the enzyme's activity is either increased or decreased by the reversible covalent addition of a group such as phosphate or AMP to the protein. *239*

Reye's syndrome An acute, potentially fatal disease of childhood that is characterized by severe edema of the brain and increased intracranial pressure, vomiting, hypoglycemia, and liver dysfunction. The cause is unknown but is almost always associated with a previous viral infection (e.g., influenza or varicella-zoster virus infections). *718*

R factors Plasmids bearing one or more drug resistant genes. *281*

rheumatic fever (roo-mat′ik) An autoimmune disease characterized by inflammatory lesions involving the heart valves, joints, subcutaneous tissues, and central nervous system. The disease is associated with hemolytic streptococci in the body. It is called rheumatic fever because two common symptoms are fever and pain in the joints similar to that of rheumatism. *748*

rhizosphere A region around the plant root where materials released from the root increase the microbial population and its activities. *864*

rho factor (ro) The protein that helps RNA polymerase dissociate from the terminator after it has stopped transcription. *223*

rhoptry Saclike, electron dense structure in the anterior portion of a zoite of a member of the phylum *Apicomplexa;* perhaps involved in the penetration of host cells. *540*

Rh system The classification of human blood types based on the presence of the Rh blood group antigens on erythrocytes. The RhD antigen is responsible for hemolytic disease of the newborn. *628*

ribonucleic acid (RNA; ri″bo-nu-kle′ik) A polynucleotide composed of ribonucleotides joined by phosphodiester bridges. *212*

ribosomal RNA (rRNA) The RNA present in ribosomes; ribosomes contain several sizes of single-stranded rRNA that contribute to ribosome structure and are also directly involved in the mechanism of protein synthesis. *86, 222*

ribosome (ri′bo-sōm) The organelle where protein synthesis occurs; the message encoded in mRNA is translated here. *48, 228*

ribotyping Ribotyping is the use of *E. coli* rRNA to probe chromosomal DNA in Southern blots for typing bacterial strains. This method is based on the fact that rRNA genes are scattered throughout the chromosome of most bacteria and therefore polymorphic restriction endonuclease patterns result when chromosomes are digested and probed with rRNA. *695*

ribulose-1,5-bisphosphate carboxylase (ri′bu-lōs) The enzyme that catalyzes the incorporation of CO_2 in the Calvin cycle. *194*

ringworm (ring′werm) The common name for a fungal infection of the skin, even though it is not caused by a worm and is not always ring-shaped in appearance. *792*

rise period or **burst** The period during the one-step growth experiment when host cells lyse and release phage particles. *355*

RNA polymerase The enzyme that catalyzes the synthesis of mRNA under the direction of a DNA template. *222*

Rocky Mountain spotted fever A disease caused by *Rickettsia rickettsii*. *782*

rolling-circle mechanism A mode of DNA replication in which the replication fork moves around a circular DNA molecule, displacing a strand to give a tail that is also copied to produce a new double-stranded DNA. *218*

root nodule Gall-like structures on roots that contain endosymbiotic nitrogen-fixing bacteria (e.g., *Rhizobium* or *Bradyrhizobium* is present in legume nodules). *865*

roseola infantum (ro-ze′o-lə) A skin eruption that produces a rose-colored rash in infants. Caused by the human herpesvirus 6. The disease is short-lived and characterized by a high fever of 3 to 4 days' duration. *See* exanthem subitum. *732*

rubella A moderately contagious skin disease that occurs primarily in children 5 to 9 years of age that is caused by the rubella virus, which is acquired by droplet inhalation into the respiratory system; German measles. *720*

rubeola *See* measles. *718*

rumen (roo-men) The expanded upper portion or first compartment of the stomach of ruminants. *551*

ruminant (roo′mĭ-nant) An herbivorous animal that has a stomach divided into four compartments and chews a cud consisting of regurgitated, partially digested food. *551*

run The straight line movement of a bacterium. *63*

S

salmonellosis (sal″mo-nel-o′sis) An infection with certain species of the genus *Salmonella,* usually caused by ingestion of food containing salmonellae or their products. Also known as *Salmonella* gastroenteritis or *Salmonella* food poisoning. *769*

sanitization (san″ĭ-ti-za′shun) Reduction of the microbial population on an inanimate object to levels judged safe by public health standards; usually, the object is cleaned. *137*

saprophyte (sap′ro-fit) An organism that takes up nonliving organic nutrients in dissolved form and usually grows on decomposing organic matter. *507*

saprozoic nutrition (sap″ro-zo′ik) Having the type of nutrition in which organic nutrients are taken up in dissolved form; normally refers to animals or animal-like organisms. *534*

scale (skāl) A platelike organic structure found on the surface of some cells (chrysophytes). *526*

scaleup The process of establishing conditions for growth of larger volumes of microorganisms, based on experience with smaller-scale cultures. *909*

scanning electron microscope An electron microscope that scans a beam of electrons over the surface of a specimen and forms an image of the surface from the electrons that are emitted by it. *32*

scanning probe microscope A microscope that studies surface features by moving a sharp probe over the object's surface (e.g., the scanning tunneling microscope). *33*

scarlatina (skahr″la-te′nah) *See* scarlet fever. *748*

scarlet fever (scarlatina; skar′let) A disease that results from infection with a strain of *Streptococcus pyogenes* that carries a lysogenic phage with the gene for erythrogenic (rash-inducing) toxin. The toxin causes shedding of the skin. This is a communicable disease spread by respiratory droplets. *748*

schizogony (skĭ-zog′o-ne) Multiple asexual fission. *540*

secondary metabolites Products of metabolism that are synthesized after growth has been completed. *912*

secondary treatment The biological degradation of dissolved organic matter in the process of sewage treatment; the organic material is either mineralized or changed to removable solids. *843*

second law of thermodynamics Physical and chemical processes proceed in such a way that the entropy of the universe (the system and its surroundings) increases to the maximum possible. *154*

secretory IgA (sIgA) The primary immunoglobulin of the secretory immune system. *See* IgA. *595*

secretory vacuole In protists and some animals, these organelles usually contain specific enzymes that perform various functions such as excystation. Their contents are released to the cell exterior during exocytosis. *534*

sedimentation basin A large basin often used in the initial stages of municipal water purification. Raw water is held in the basin until large particulate material settles out. *848*

segmented genome A virus genome that is divided into several parts or fragments, each probably coding for the synthesis of a single polypeptide; segmented genomes are very common among the RNA viruses. *346*

selectins (sə-lek′tins) A family of cell adhesion molecules that are displayed on activated endothelial cells; examples include P-selectin and E-selectin. Selectins mediate leukocyte binding to the vascular endothelium. *578*

selective media Culture media that favor the growth of specific microorganisms; this may be accomplished by inhibiting the growth of undesired microorganisms. *107*

selective toxicity The ability of a chemotherapeutic agent to kill or inhibit a microbial pathogen while damaging the host as little as possible. *658*

self-assembly The spontaneous formation of a complex structure from its component molecules without the aid of special enzymes or factors. *61, 193*

semisynthetic penicillins Penicillins that have been modified after biosynthesis, usually by chemical processes. *913*

sense strand The DNA strand that RNA polymerase copies to produce mRNA, rRNA, or tRNA. *222, 259*

sepsis (sep′sis) Systemic response to infection. This systemic response is manifested by two or more of the following conditions as a result of infection: temperature >38 or <36°C; heart rate >90 beats per min; respiratory rate >20 breaths per min, or pCO$_2$ <32 mm Hg; leukocyte count >12,000 cells per ml^3 or >10% immature (band) forms. Sepsis also has been defined as the presence of pathogens or their toxins in blood and other tissues. *771*

septate (sep′tāt) Divided by a septum or cross wall; also with more or less regular occurring cross walls. *506*

septic shock (sep′tik) Sepsis associated with severe hypotension despite adequate fluid resuscitation, along with the presence of perfusion abnormalities that may include, but are not limited to, lactic acidosis, oliguria, or an acute alteration in mental status. Gram-positive bacteria, fungi, and endotoxin-containing gram-negative bacteria can initiate the pathogenic cascade of sepsis leading to septic shock. *771*

septicemia (sep″tĭ-se′me-ah) A disease associated with the presence in the blood of pathogens or bacterial toxins. *420, 566*

septic tank (sep′tik) A tank used to process small quantities of domestic sewage. Solid material settles out and is partially degraded by bacteria as sewage slowly flows through the tank. The outflow is further treated or dispersed in the soil. *854*

septum (sep′tum; pl., **septa**) A wall or partition dividing a body space or cavity. *506*

serology (se-rol′o-je) The branch of immunology that is concerned with in vitro reactions involving one or more serum constituents (e.g., antibodies and complement). *641*

serotyping A technique or serological procedure that is used to differentiate between strains (serovars or serotypes) of microorganisms that have differences in the antigenic composition of a structure or product. *652*

serum (se′rum; pl., **serums** or **sera**) The clear, fluid portion of blood lacking both blood cells and fibrinogen. It is the fluid remaining after coagulation of plasma, the noncellular liquid faction of blood. *601*

settling basin A basin used during water purification to chemically precipitate out fine particles, microorganisms, and organic material by coagulation or flocculation. *848*

sewage treatment The use of physical and biological processes to remove particulate and dissolved material from sewage. *843*

sex pilus (pi′lus) A thin protein appendage required for bacterial mating or conjugation. The cell with sex pili donates DNA to recipient cells. *59, 289*

sheath (shēth) A hollow tubelike structure surrounding a chain of cells and present in several genera of bacteria. *468*

shigellosis (shĭ′gəl-o′sis) The diarrheal disease that arises from an infection with a member of the genus *Shigella.* Often called bacillary dysentery. *769*

Shine-Dalgarno sequence A segment in the leader of procaryotic mRNA that binds to a special sequence on the 16S rRNA of the small ribosomal subunit. This helps properly orient the mRNA on the ribosome. *261*

shingles (zoster; shing′g′lz) A reactivated form of chickenpox caused by the varicella-zoster virus. *716*

siderophore (sid′er-o-for″) A small molecule that complexes with ferric iron and supplies it to a cell by aiding in its transport across the plasma membrane. *106*

sigma factor A protein that helps the RNA polymerase core enzyme recognize the promoter at the start of a gene. *222*

sign An objective change in a diseased body that can be directly observed (e.g., a fever or rash). *702*

silent mutation A mutation that does not result in a change in the organism's proteins or phenotype even though the DNA base sequence has been changed. *266*

simple matching coefficient (S_{SM}) An association coefficient used in numerical taxonomy; the proportion of characters that match regardless of whether or not the attribute is present. *397*

single radial immunodiffusion (RID) assay An immunodiffusion technique that quantitates antigens by following their diffusion through a gel containing antibodies directed against the test antigens. *646*

site-specific recombination Recombination of nonhomologous genetic material with a chromosome at a specific site. *278*

sixth disease *See* exanthem subitum. *732*

S-layer A regularly structured layer composed of protein or glycoprotein that lies on the surface of many bacteria. It may protect the bacterium and help give it shape and rigidity. *59*

slime The viscous extracellular glycoproteins or glycolipids produced by staphylococci and *Pseudomonas aeruginosa* bacteria that allows them to adhere to smooth surfaces such as prosthetic medical devices and catheters. More generally, the term often refers to an easily removed, diffuse, unorganized layer of extracellular material that surrounds a bacterial cell. *58, 759*

slime layer A layer of diffuse, unorganized, easily removed material lying outside the bacterial cell wall. *58*

slime mold A common term for members of the divisions *Acrasiomycota* and *Myxomycota*. *515*

slow sand filter A bed of sand through which water slowly flows; the gelatinous microbial layer on the sand grain surface removes waterborne microorganisms, particularly *Giardia*, by adhesion to the gel. This type of filter is used in some water purification plants. *848*

slow virus disease A progressive, pathological process caused by a transmissible agent (virus or prion) that remains clinically silent during a prolonged incubation period of months to years after which progressive clinical disease becomes apparent. *379, 739*

sludge (sluj) A general term for the precipitated solid matter produced during water and sewage treatment; solid particles composed of organic matter and microorganisms that are involved in aerobic sewage treatment (activated sludge). *843*

smallpox (variola; smawl´poks) Once a highly contagious, often fatal disease caused by a poxvirus. Its most noticeable symptom was the appearance of blisters and pustules on the skin. Vaccination has eradicated smallpox throughout the world. *720*

snapping division A distinctive type of binary fission resulting in an angular or a palisade arrangement of cells, which is characteristic of the genera *Arthrobacter* and *Corynebacterium*. *451*

sorocarp The fructification of the *Acrasiomycetes*. *515*

SOS repair A complex, inducible repair process that is used to repair DNA when extensive damage has occurred. *274*

source The location or object from which a pathogen is immediately transmitted to the host, either directly or through an intermediate agent. *705*

sour mash A mash that has been inoculated with lactic-acid producing microorganisms to control undesirable growth and create a distinct final product taste. *898*

Southern blotting technique The procedure that isolates and identifies DNA fragments from a complex mixture. The isolated, denatured fragments are transferred from an agarose electrophoretic gel to a nitrocellulose filter and identified by hybridization with probes. *308*

specialized transduction *See* restricted transduction. *295*

species (spe´shēz) Species of higher organisms are groups of interbreeding or potentially interbreeding natural populations that are reproductively isolated. Bacterial species are collections of strains that have many stable properties in common and differ significantly from other groups of strains. *392*

spheroplast (sfēr´o-plast) A relatively spherical cell formed by the weakening or partial removal of the rigid cell wall component (e.g., by penicillin treatment of gram-negative bacteria). Spheroplasts are usually osmotically sensitive. *57*

spike *See* peplomer. *346*

spirillum (spi-ril´um) A rigid, spiral-shaped bacterium. *39*

spirochete (spi´ro-kēt) A flexible, spiral-shaped bacterium with periplasmic flagella. *39*

split or **interrupted gene** A structural gene with DNA sequences that code for the final RNA product (expressed sequences or exons) separated by regions coding for RNA absent from the mature RNA (intervening sequences or introns). *224*

spontaneous generation (spon-ta´ne-us) The hypothesis that living organisms can arise from nonliving matter. *3*

sporadic disease (spo-rad´ik) A disease that occurs occasionally and at random intervals in a population. *700*

sporangiospore (spo-ran´je-o-spōr) A spore born within a sporangium. *497*

sporangium (spo-ran´je-um; pl., **sporangia**) A saclike structure or cell, the contents of which are converted into an indefinite number of spores. It is borne on a special hypha called a sporangiophore. *67, 509*

spore (spōr) A differentiated, specialized form that can be used for dissemination, for survival of adverse conditions because of its heat and dessication resistance, and/or for reproduction. Spores are usually unicellular and may develop into vegetative organisms or gametes. They may be produced asexually or sexually and are of many types. *522*

sporogenesis (spor´o-jen´ĕ-sis) *See* sporulation. *69*

sporotrichosis (spo´ro-tri-ko´sis) A subcutaneous fungal infection caused by the dimorphic fungus *Sporothrix schenckii*. *794*

sporulation (spor´´u-la´shun) The process of spore formation. *69*

spread plate A petri dish of solid culture medium with isolated microbial colonies growing on its surface, which has been prepared by spreading a dilute microbial suspension evenly over the agar surface. *108*

sputum (spu´tum) The mucous secretion from the lungs, bronchi, and trachea that is ejected (expectorated) through the mouth. *677*

stalk (stawk) A nonliving bacterial appendage produced by the cell and extending from it. *466*

standard free energy change The free energy change of a reaction at 1 atmosphere pressure when all reactants and products are present in their standard states; usually the temperature is 25°C. *155*

standard reduction potential A measure of the tendency of a reductant to lose electrons in an oxidation-reduction (redox) reaction. The more negative the reduction potential of a compound, the better electron donor it is. *156*

staphylococcal food poisoning (staf´´i-lo-kok´al) A type of food poisoning caused by ingestion of improperly stored or cooked food in which *Staphylococcus aureus* has grown. The bacteria produce exotoxins that accumulate in the food. *770*

staphylococcal scalded skin syndrome (SSSS) A disease caused by staphylococci that produce exfoliative toxin. The skin becomes red (erythema) and sheets of epidermis may separate from the underlying tissue. *760*

starter culture An inoculum, consisting of a mixture of carefully selected microorganisms, used to start a commercial fermentation. *893*

stationary phase (sta´shun-er´´e) The phase of microbial growth in a batch culture when population growth ceases and the growth curve levels off. *115*

statistics (stah-tis´tiks) The mathematics of the collection, organization, and interpretation of numerical data. *701*

stenothermal (sten´´o-ther´mal) Organisms that can grow well only over a narrow temperature range. *128*

sterilization (ster´´ĭ-lĭ-za´shun) The process by which all living cells, viable spores, viruses, and viroids are either destroyed or removed from an object or habitat. *137*

stigma (stig´mah) A light-sensitive eyespot, which is found in some algae and photosynthetic protozoa; it is believed to be involved in phototaxis, at least in some cases. *523*

stoneworts A group of approximately 250 species of algae that have a complex growth pattern, with nodal regions from which whorls of branches arise; they are abundant in fresh to brackish waters. *523*

strain A population of organisms that descends from a single organism or pure culture isolate. *255, 392*

streak plate A petri dish of solid culture medium with isolated microbial colonies growing on its surface, which has been prepared by spreading a microbial mixture over the agar surface, using an inoculating loop. *108*

streptococcal sore throat (strep´´to-kok´al) One of the most common bacterial infections of humans. It is commonly referred to as "strep throat." The disease is spread by droplets of saliva or nasal secretions and is caused by *Streptococcus* spp. (particularly group A streptococci). *749*

streptococcal pneumonia An endogenous infection of the lungs caused by *Streptococcus pneumoniae* that occurs in predisposed individuals. *749*

streptolysin-O (SLO) (strep-tol´ĭ-sin) A specific hemolysin produced by *Streptococcus pyogenes* that is inactivated by oxygen (hence the "O" in its name). SLO causes beta-hemolysis of blood cells on agar plates incubated anaerobically. *572*

streptolysin-S (SLS) A product produced by *Streptococcus pyogenes* that is bound to the bacterial cell but may sometimes be released. SLS causes beta hemolysis on aerobically incubated blood-agar plates and can act as a leukocidin by killing leukocytes that phagocytose the bacterial cell to which it is bound. *572*

streptomycin (strep´to-mi´´sin) A bactericidal aminoglycoside antibiotic produced by *Streptomyces griseus*. *668*

strict anaerobes *See* obligate anaerobes.

stroma (stro´mah) The chloroplast matrix that is the location of the photosynthetic carbon dioxide fixation reactions. *84*

stromatolite (stro´´mah-to´lĭt) Dome-like microbial mat communities consisting of filamentous photosynthetic bacteria and occluded sediments (often calcareous or siliceous). They usually have a laminar structure. Many are fossilized, but some modern forms occur *833*

structural gene A gene that codes for the synthesis of a polypeptide or polynucleotide with a nonregulatory function. *244*

subgingival plaque (sub-jin´jĭ-val) The plaque that forms at the dentogingival margin and extends down into the gingival tissue. *785*

substrate (sub´strāt)   The substance an enzyme acts upon.   *159*

substrate-level phosphorylation   The synthesis of ATP from ADP by phosphorylation coupled with the exergonic breakdown of a high-energy organic substrate molecule.   *167*

sulfate reduction (sul´fat)   The process of sulfate use as an oxidizing agent, which results in the accumulation of reduced forms of sulfur such as sulfide, or incorporation of sulfur into organic molecules, usually as sulfhydryl groups.   *197, 816*

sulfonamide (sul-fon´ah-mīd)   A chemotherapeutic agent that has the SO_2-NH_2 group and is a derivative of sulfanilamide.   *663*

superantigen   Superantigens are bacterial proteins that stimulate the immune system much more extensively than do normal antigens. They stimulate T cells to proliferate nonspecifically through simultaneous interaction with class II MHC proteins on antigen-presenting cells and variable regions on the β chain of the T-cell receptor complex. Examples include streptococcal scarlet fever toxins, staphylococcal toxic shock syndrome toxin-1, and streptococcal M protein.   *620*

superinfection (soo˝per-in-fek´shun)   A new bacterial or fungal infection of a patient that is resistant to the drug(s) being used for treatment.   *670*

superoxide dismutase (soo˝per-ok´sīd dis-mu´tas)   An enzyme that protects many microorganisms by catalyzing the destruction of the toxic superoxide radical.   *131*

suppressor mutation   A mutation that overcomes the effect of another mutation and produces the normal phenotype.   *265*

Svedberg unit (sfed´berg)   The unit used in expressing the sedimentation coefficient; the greater a particle's Svedberg value, the faster it travels in a centrifuge.   *48*

swab (swahb)   A wad of absorbent material usually wound around one end of a small stick and used for applying medication or for removing material from an area; also, a dacron-tipped polystyrene applicator.   *677*

swarm cell   A flagellated cell; the term is usually applied to the motile cells of the *Myxomycota*.   *515*

symbiont (sim´bi-ont)   An organism that lives in a state of symbiosis; also known as a symbiote.   *547*

symbiosis (sim˝bi-o´sis)   The living together or close association of two dissimilar organisms, each of these organisms being known as a symbiont.   *547*

symbiosome   The final nitrogen-fixing form of *Rhizobium* that is active within root nodule cells.   *865*

symbiote (sim´bi-ōt)   *See* symbiont.   *547*

symptom (simp´təm)   A change during a disease that a person subjectively experiences (e.g., pain, bodily discomfort, fatigue, or loss of appetite). Sometimes the term symptom is used more broadly to include any observed signs.   *702*

syndrome   *See* disease syndrome.   *702*

synthetic medium   *See* defined medium.   *106*

syphilis (sif´ĭ-lis)   *See* venereal syphilis.   *762*

systematics (sis˝te-mat´iks)   The scientific study of organisms with the ultimate objective being to characterize and arrange them in an orderly manner; often considered synonymous with taxonomy.   *391*

systemic lupus erythematosus (loo´pus er˝ĭ-them-ah-to´sus)   An autoimmune, inflammatory disease that may affect every tissue of the body.   *623*

T

taxon (tak´son)   A group into which related organisms are classified.   *391*

taxonomy (tak-son´o-me)   The science of biological classification; it consists of three parts: classification, nomenclature, and identification.   *391*

TB skin test   Tuberculin hypersensitivity test for a previous or current infection with *Mycobacterium tuberculosis*.   *623*

T cell or **T lymphocyte**   A type of lymphocyte derived from bone marrow stem cells that matures into an immunologically competent cell under the influence of the thymus. T cells are involved in a variety of cell-mediated immune reactions.   *589*

T-dependent antigen   An antigen that effectively stimulates B-cell response only with the aid of T-helper cells that produce interleukin-2 and B-cell growth factor.   *611*

teichoic acids (ti-ko´ik)   Polymers of glycerol or ribitol joined by phosphates; they are found in the cell walls of gram-positive bacteria.   *54*

temperate phages   Bacteriophages that can infect bacteria and establish a lysogenic relationship rather than immediately lysing their hosts.   *294, 362*

template (tem´plat)   A strand of DNA or RNA that specifies the base sequence of a newly synthesized complementary strand of DNA or RNA.   *222*

Tenericutes (ten˝er-ik´u-tēz)   A division in the kingdom *Procaryotae* containing bacteria that lack an outer cell wall and do not synthesize peptidoglycan. It contains the mycoplasmas.   *400*

terminator   A sequence that marks the end of a gene and stops transcription.   *261*

tertiary treatment (ter´she-er-e)   The removal from sewage of inorganic nutrients, heavy metals, viruses, etc., by chemical and biological means after microorganisms have degraded dissolved organic material during secondary sewage treatment.   *843*

test   A loose-fitting shell of an amoeba.   *537*

tetanolysin (tet˝ah-nol´ĭ-sin)   A hemolysin that aids in tissue destruction and is produced by *Clostridium tetani*.   *765*

tetanospasmin (tet˝ah-no-spaz´min)   The neurotoxic component of the tetanus toxin, which causes the muscle spasms of tetanus. Tetanospasmin production is under the control of a plasmid gene.   *765*

tetanus (tet´ah-nus)   An often fatal disease caused by the anaerobic, spore-forming bacillus *Clostridium tetani*, and characterized by muscle spasms and convulsions.   *764*

tetracyclines (tet˝rah-si´klēns)   A family of antibiotics with a common four-ring structure, which are isolated from the genus *Streptomyces* or produced semisynthetically; all are related to chlortetracycline or oxytetracycline.   *668*

tetrapartite associations (tet˝rah-par´tīt)   A mutualistic association of the same plant with three different types of microorganisms.   *871*

T$_H$1 cell   *See* regulator T cell.   *616*

T$_H$2 cell   *See* regulator T cell.   *616*

T$_H$O cell   *See* regulator T cell.   *616*

thallus (thal´us)   A type of body that is devoid of root, stem, or leaf; characteristic of some algae, many fungi, and lichens.   *492, 504, 522*

T-helper (T$_H$) cell   A cell that is needed for T-cell–dependent antigens to be effectively presented to B cells. It also promotes cell-mediated immune responses.   *611*

theory   A set of principles and concepts that have survived rigorous testing and that provide a systematic account of some aspect of nature.   *11*

thermal death time (TDT)   The shortest period of time needed to kill all the organisms in a microbial population at a specified temperature and under defined conditions.   *140*

thermoacidophiles   A group of bacteria that grow best at acid pHs and high temperatures; they are members of the *Archaeobacteria*.   *488*

thermocline   The layer in a thermally stratified lake having the greatest rate of temperature change with depth; it separates the upper warmer epilimnion zone from the lower colder oxygen-poor hypolimnion.   *841*

thermodynamics (ther˝mo-di-nam´iks)   The scientific discipline that deals with heat and energy, and their interconversion.   *153*

thermophile (ther´mo-fīl)   A microorganism that can grow at temperatures of 55°C or higher; the minimum is usually around 45°C.   *130*

thrush (thrush)   Infection of the oral mucous membrane by the fungus *Candida albicans;* also known as oral candidiasis.   *797*

thylakoid (thi´lah-koid)   A flattened sac in the chloroplast stroma that contains photosynthetic pigments and the photosynthetic electron transport chain; light energy is trapped and used to form ATP and NAD(P)H in the thylakoid membrane.   *84*

thymine (thi´min)   The pyrimidine 5-methyluracil that is found in nucleosides, nucleotides, and DNA.   *204*

thymocyte (thi´mo-sīt)   A lymphocyte in the thymus.   *589*

T-independent antigen   An antigen that triggers a B cell into immunoglobulin production without T-cell cooperation.   *613*

tinea (tin´e-ah)   A name applied to many different kinds of superficial fungal infections of the skin, nails, and hair, the specific type (depending on characteristic appearance, etiologic agent, and site) usually designated by a modifying term.   *791*

tinea barbae   A fungal infection of beard hair caused by *Trichophyton mentagrophytes* or *T. verrucosum*.   *792*

tinea capitis   A fungal infection of scalp hair caused by species of *Trichophyton* or *Microsporum*.   *792*

tinea corporis   A fungal infection of the smooth parts of the skin caused by either *Trichophyton rubrum, T. mentagrophytes,* or *Microsporum canis*.   *792*

tinea cruris   A fungal infection of the groin caused by either *Epidermophyton floccosum, Trichophyton mentagrophytes,* or *T. rubrum;* also known as jock itch.   *792*

tinea manuum   A fungal infection of the hand caused by *Trichophyton rubrum, T. mentagrophytes,* or *E. floccosum*.   *793*

tinea pedis   A fungal infection of the foot caused by *Trichophyton rubrum, T. mentagrophytes,* or *E. floccosum;* also known as athlete's foot.   *792*

tinea unguium   A fungal infection of the nail bed caused by either *Trichophyton rubrum* or *T. mentagrophytes*.   *793*

tinea versicolor A fungal infection caused by the yeast, *Malassezia furfur*, that forms brownish-red scales on the skin of the trunk, neck, face, and arms. *791*

Ti plasmid A plasmid obtained from *Agrobacterium tumefaciens* that is used to insert genes into plant cells. *872*

titer (ti′ter) Reciprocal of the highest dilution of an antiserum that gives a positive reaction in the test being used. *601*

T lymphocyte *See* T cell. *589*

tonsillitis (ton′si-li′tis) Inflammation of the tonsils, especially the palatine tonsils often due to *S. pyogenes* infection. *749*

total organic carbon (TOC) The organic carbon in a biological sample, including both biologically utilizable and nonutilizable materials. *826, 843*

toxemia (tok-se′me-ah) The condition caused by toxins in the blood of the host. *567*

toxic shock-like syndrome (TSLS) A disease caused by an invasive group A streptococcus infection that is characterized by a rapid drop in blood pressure, failure of many organs, and a very high fever. It probably results from the release of one or more streptococcal pyrogenic exotoxins. *748*

toxic shock syndrome (tok′sik) A staphylococcal disease that most commonly affects females who use certain types of tampons during menstruation. It is associated with the production of toxic shock syndrome toxin by certain strains of *Staphylococcus aureus*. *760*

toxigenicity (tok″sĭ-jĕ-nis′i-tĕ) The capacity of an organism to produce a toxin. *563*

toxin (tok′sin) A microbial product or component that can injure another cell or organism at low concentrations. Often the term refers to a poisonous protein, but toxins may be lipids and other substances. *567*

toxin neutralization The inactivation of toxins by specific antibodies, called antitoxins, that react with them. *636*

toxoid (tok′soid) A bacterial exotoxin that has been modified so that it is no longer toxic but will still stimulate antitoxin formation when injected into a person or animal. *567*

toxoplasmosis (tok″so-plaz-mo′sis) A disease of animals and humans caused by the parasitic protozoan, *Toxoplasma gondii*. *806*

trachoma (trah-ko′mah) A chronic infectious disease of the conjunctiva and cornea, producing pain, inflammation and sometimes blindness. It is caused by *Chlamydia trachomatis* serotypes A–C. *778*

transamination (trans″am-i-na′shun) The removal of amino acid's amino group by transferring it to an α-keto acid acceptor. *179*

transcriptase (trans-krip′tās) An enzyme that catalyzes transcription; in viruses with RNA genomes, this enzyme is an RNA-dependent RNA polymerase that is used to make RNA copies of the RNA genomes. *376*

transcription (trans-krip′shun) The process in which single-stranded RNA with a base sequence complementary to the template strand of DNA or RNA is synthesized. *216*

transduction (trans-duk′shun) The transfer of genes between bacteria by bacteriophages. *294*

transfer host (trans′fer) A host that is not necessary for the completion of a parasite's life cycle, but is used as a vehicle for reaching a final host. *562*

transfer RNA (tRNA) A small RNA that binds an amino acid and delivers it to the ribosome for incorporation into a polypeptide chain during protein synthesis. *222*

transformation (trans″for-ma′shun) A mode of gene transfer in bacteria in which a piece of free DNA is taken up by a bacterial cell and integrated into the recipient genome. *255, 292*

transgenic animal or **plant** An animal or plant that has gained new genetic information from the insertion of foreign DNA. It may be produced by such techniques as injecting DNA into animal eggs, electroporation of mammalian cells and plant cell protoplasts, or shooting DNA into plant cells with a gene gun. *321*

transient carrier *See* casual carrier. *706*

transition mutations (tran-zish′un) Mutations that involve the substitution of a different purine base for the purine present at the site of the mutation or the substitution of a different pyrimidine for the normal pyrimidine. *263*

translation (trans-la′shun) Protein synthesis; the process by which the genetic message carried by mRNA directs the synthesis of polypeptides with the aid of ribosomes and other cell constituents. *216*

transmission electron microscope (trans-mish′un) A microscope that forms an image by passing an electron beam through a specimen and focusing the scattered electrons with magnetic lenses. *29*

transovarian passage (trans″o-va′re-an) The passage of a microorganism such as a rickettsia from generation to generation of hosts through tick eggs. (No humans or other mammals are needed as reservoirs for continued propagation of the rickettsias.) *782*

transpeptidation The reaction that forms the peptide cross-links during peptidoglycan synthesis. *208*

transposable elements *See* transposon. *283*

transposition (trans″po-zish′un) The movement of a piece of DNA around the chromosome. *283*

transposon (tranz-po′zon) A DNA segment that carries the genes required for transposition and moves about the chromosome; if it contains genes other than those required for transposition, it may be called a composite transposon. Often the name is reserved only for transposable elements that also contain genes unrelated to transposition. *283*

transversion mutations (trans-ver′zhun) Mutations that result from the substitution of a purine base for the normal pyrimidine or a pyrimidine for the normal purine. *263*

traveler's diarrhea A type of diarrhea resulting from an encounter with certain viruses, bacteria, or protozoa normally absent from the traveler's environment. One of the major pathogens is enterotoxigenic *Escherichia coli*. *770*

tricarboxylic acid cycle (TCA) The cycle that oxidizes acetyl coenzyme A to CO_2 and generates NADH and $FADH_2$ for oxidation in the electron transport chain; the cycle also supplies carbon skeletons for biosynthesis. *170*

trichome (tri′kōm) A row or filament of bacterial cells that are in close contact with one another over a large area. *461*

trichomoniasis (trik″o-mo-ni′ah-sis) A sexually transmitted disease caused by the parasitic protozoan *Trichomonas vaginalis*. *806*

tripartite associations (tri-par′tīt) A mutualistic association of the same plant with two types of microorganisms. *871*

trophophase The phase of active microbial growth in batch culture; primary metabolism (growth-directed metabolism) is dominant. *912*

trophozoite (trof″o-zo′īt) The active, motile feeding stage of a protozoan organism; in the malarial parasite, the stage of schizogony between the ring stage and the schizont. *534*

trypanosome (tri-pan′o-sōm) A protozoan of the genus *Trypanosoma*. Trypanosomes are parasitic flagellate protozoa that often live in the blood of humans and other vertebrates and are transmitted by insect bites. *536, 805*

trypanosomiasis (tri-pan″o-so-mi′ah-sis) An infection with trypanosomes that live in the blood and lymph of the infected host. *805*

T-suppressor (T_S) cell A cell that is involved in suppressing an immune response. *589*

tubercle (too′ber-k′l) A small, rounded nodular lesion produced by *Mycobacterium tuberculosis*. *617, 750*

tuberculoid (neural) leprosy (too-ber′ku-loid) A mild, nonprogressive form of leprosy that is associated with delayed-type hypersensitivity to antigens on the surface of *Mycobacterium leprae*. It is characterized by early nerve damage and regions of the skin that have lost sensation and are surrounded by a border of nodules. *758*

tuberculosis (too-ber″ku-lo′sis) An infectious disease of humans and other animals resulting from an infection by a species of *Mycobacterium* and characterized by the formation of tubercles and tissue necrosis, primarily as a result of host hypersensitivity and inflammation. Infection is usually by inhalation, and the disease commonly affects the lungs (pulmonary tuberculosis), although it may occur in any part of the body. *750*

tuberculous cavity (too-ber′ku-lus) An air-filled cavity that results from a tubercle lesion caused by *M. tuberculosis*. *750*

tularemia (too″lah-re′me-ah) A plaguelike disease of animals caused by the bacterium *Francisella tularensis* subsp. *tularensis* (Jellison type A), which may be transmitted to humans. *766*

tumble Random turning or tumbling movements made by bacteria when they stop moving in a straight line. *63*

tumor (too′mor) A growth of tissue resulting from abnormal new cell growth and reproduction (neoplasia). *379*

tumor necrosis factor-α (TNF-α) (too′mor nə-kro′sis) A cytokine produced by mononuclear phagocytes and some T cells. It promotes inflammation and is the major cytokine that mediates the host response to gram-negative bacteria, along with interferon-α. *607*

turbidostat A continuous culture system equipped with a photocell that adjusts the flow of medium through the culture vessel so as to maintain a constant cell density or turbidity. *122*

twiddle *See* tumble. *63*

tyndallization (fractional steam sterilization; tyn″dal-i-za′shun) A sterilization process in which a container is heated at 90 to 100°C for 30 minutes on three successive days and incubated at 37°C in between. *142*

type I (anaphylaxis) hypersensitivity A form of immediate hypersensitivity arising from the binding of antigen to IgE attached to mast cells, which then release anaphylaxis mediators such as histamine. Examples: hay fever, asthma, and food allergies. *621*

type II (cytotoxic) hypersensitivity A form of immediate hypersensitivity involving the binding of antibodies to antigens on cell surfaces followed by destruction of the target cells (e.g., through complement attack, phagocytosis, or agglutination). *622*

type III (immune complex) hypersensitivity A form of immediate hypersensitivity resulting from the exposure to excessive amounts of antigens in which antibodies bind to the antigens and produce antibody-antigen complexes. These activate complement and trigger an acute inflammatory response with subsequent tissue damage. Examples: poststreptococcal glomerulonephritis, serum sickness, and farmer's lung disease. *623*

type IV (cell-mediated) hypersensitivity A delayed hypersensitivity response (it appears 24 to 48 hours after antigen exposure). It results from the binding of antigen to activated T lymphocytes, which then release cytokines and trigger inflammation and macrophage attacks that damage tissue. Type IV hypersensitivity is seen in contact dermititis from poison ivy, leprosy, and tertiary syphilis. *623*

typhoid fever (ti-foid) A bacterial infection transmitted by contaminated food, water, milk, or shellfish. The causative organism is *Salmonella typhi*, which is present in human feces. *771*

U

ultraviolet (UV) radiation (ul″trah-vi′o-let) Radiation of fairly short wavelength, about 10 to 400 nm, and high energy. *133, 143*

uracil (u′rah-sil) The pyrimidine 2,4-dioxypyrimidine, which is found in nucleosides, nucleotides, and RNA. *204*

V

vaccine (vak′sēn) A preparation of either killed microorganisms; living, weakened (attenuated) microorganisms; or inactivated bacterial toxins (toxoids). It is administered to induce development of the immune response and protect the individual against a pathogen or a toxin. *588*

valence (va′lens) The number of antigenic determinant sites on the surface of an antigen or the number of antigen-binding sites possessed by an antibody molecule. *590*

VA mycorrhizae Endomycorrhizae, characterized by having vesicles and arbuscules present in the plant roots. *870*

variable region (V$_L$ and V$_H$) The region at the N-terminal end of immunoglobulin heavy and light chains whose amino acid sequence varies between antibodies of different specificity. Variable regions form the antigen binding site. *592*

vasculitis (vas″ku-li′tis) Inflammation of a blood vessel. *780*

vector (vek′tor) 1. In genetic engineering, another name for a cloning vector. A DNA molecule that can replicate (a replicon) and is used to transport a piece of inserted foreign DNA, such as a gene, into a recipient cell. It may be a plasmid, phage, or cosmid. 2. In epidemiology, it is a living organism, usually an arthropod or other animal, that transfers an infective agent from one host to another. *563, 706*

vector-borne transmission The transmission of an infectious pathogen between hosts by means of a vector. *709*

vehicle (ve′ĭ-k′l) An inanimate substance or medium involved in the transmission of a pathogen. *709*

venereal syphilis (ve-ne′re-al sif′ĭ-lis) A contagious, sexually transmitted disease caused by the spirochete *Treponema pallidum*. *762*

venereal warts *See* anogenital condylomata. *739*

verrucae vulgaris (v′ĕ-roo′se vul-ga′ris; s. **verruca vulgaris**) The common wart; a raised, epidermal lesion with horny surface caused by an infection with a human papillomavirus. *739*

vesicles (ves′i-k′ls) Small membrane-bound bodies. In endomycorrhizal plants infected with mycorrhizal fungi, swollen areas in the colonizing hyphae. *870*

vibrio (vib′re-o) A rod-shaped bacterial cell that is curved to form a comma or an incomplete spiral. *39*

viral hemagglutination (vi′ral hem″ah-gloo″tĭ-na′shun) The clumping or agglutination of red blood cells caused by some viruses. *641*

viral neutralization An antibody-mediated process in which IgA, IgM, and IgA antibodies bind to some viruses during their extracellular phase and inactivate or neutralize them. *637*

viricide (vir′i-sīd) An agent that inactivates viruses so that they cannot reproduce within host cells. *138*

virino *See* prion. *385*

virion (vi′re-on) A complete virus particle that represents the extracellular phase of the virus life cycle; at the simplest, it consists of a protein capsid surrounding a single nucleic acid molecule. *340*

viroid (vi′roid) An infectious agent of plants that is a single-stranded RNA not associated with any protein; the RNA does not code for any proteins and is not translated. *384*

virology (vi-rol′o-je) The branch of microbiology that is concerned with viruses and viral diseases. *333*

virulence (vir′u-lens) The degree or intensity of pathogenicity of an organism as indicated by case fatality rates and/or ability to invade host tissues and cause disease. *563*

virulent bacteriophages (vir′u-lent bak-te′re-o-fajs″) Bacteriophages that lyse their host cells during the reproductive cycle. *362*

virus (vi′rus) An infectious agent having a simple acellular organization with a protein coat and a single type of nucleic acid, lacking independent metabolism, and reproducing only within living host cells. *334*

vitamin (vi′tah-min) An organic compound required by organisms in minute quantities for growth and reproduction because it cannot be synthesized by the organism; vitamins often serve as enzyme cofactors or parts of cofactors. *101*

volutin granules (vo-lu′tin) *See* metachromatic granules. *47*

W

wart (wort) An epidermal tumor of viral origin. *739*

water mold A common term for a member of the division *Oomycota*. *515*

water activity (a$_w$) A quantitative measure of water availability in the habitat; the water activity of a solution is one-hundredth its relative humidity. *124*

Weil-Felix reaction A test for the diagnosis of typhus and certain other rickettsial diseases. In this test, the blood serum of a patient with suspected rickettsial disease is tested against certain strains of *Proteus vulgaris* (OX-2, OX-19, OX-K). The agglutination reactions, based on antigens common to both organisms, determine the presence and type of rickettsial infection. *781*

white piedra A fungal infection caused by the yeast *Trichosporon beigelii* that forms light-colored nodules on the beard and mustache. *791*

whole-genome shotgun sequencing An approach to genome sequencing in which the complete genome is broken into random fragments, which are then individually sequenced. Finally the fragments are placed in the proper order using sophisticated computer programs. *299*

whooping cough *See* pertussis. *746*

Widal test (ve-dahl′) A test involving agglutination of typhoid bacilli when they are mixed with serum containing typhoid antibodies from an individual having typhoid fever; used to detect the presence of *Salmonella typhi* and *S. paratyphi*. *641*

wine vinegar Vinegar produced by the oxidation of alcohol in wine by members of the genus *Acetobacter*. *897*

Winogradsky column A glass column with an anaerobic lower zone and an aerobic upper zone, which allows growth of microorganisms under conditions similar to those found in a nutrient-rich lake. *831*

woolsorter's disease *See* anthrax. *755*

wort The filtrate of malted grains used as the substrate for the production of beer and ale by fermentation. *897*

X

xenograft (zen′o-graft)   A tissue graft between animals of different species.   *626*

xerophilic microorganisms (ze″ro-fil′ik)   Microorganisms that grow best under low a_w conditions, and may not be able to grow at high a_w values.   *882*

Y

yeast (yēst)   A unicellular fungus that has a single nucleus and reproduces either asexually by budding or fission, or sexually through spore formation.   *506*

yellow fever   An acute infectious disease caused by a flavivirus, which is transmitted to humans by mosquitoes. The liver is affected and the skin turns yellow in this disease.   *722*

YM shift   The change in shape by dimorphic fungi when they shift from the yeast (Y) form in the animal body to the mold or mycelial form (M) in the environment.   *506*

Z

zooflagellates (zo″o-flăj′-e-lāts)   Flagellate protozoa that do not have chlorophyll and are either holozoic, saprozoic, or symbiotic.   *536*

zoonosis (zo″o-no′sis; pl. **zoonoses**)   A disease of animals that can be transmitted to humans.   *701*

zooplankton (zo″o-plank′ton)   A community of floating, aquatic, minute animals and nonphotosynthetic protists.   *521*

zoospore (zo′o-spōr)   A motile, flagellated spore.   *522*

z value   The increase in temperature required to reduce the decimal reduction time to one-tenth of its initial value.   *140*

zoster   *See* shingles.   *716*

zooxanthella (zo″o-zan-thel′ah)   A dinoflagellate found living symbiotically within cnidarians and other invertebrates.   *530, 549*

zygomycetes (zi″go-mi-se′tez)   A division of fungi that usually has a coenocytic mycelium with chitinous cell walls. Sexual reproduction normally involves the formation of zygospores. The group lacks motile spores.   *510*

zygospore (zi′go-spōr)   A thick-walled, sexual, resting spore characteristic of the zygomycetous fungi.   *509*

zygote (zi′gōt)   The diploid (2n) cell resulting from the fusion of male and female gametes.   *522*

zymogenous (zi-moj′ĕ-nus)   Fermentation producing. In environmental microbiology, the term refers to microorganisms, often transient or alien, that respond rapidly by enzyme production and growth when simple organic substrates become available.   *862*

Chapter 43

Chapter 44

Appendix I

Appendix II

Line Art

Chapter 2

Chapter 3

Chapter 7

Chapter 10

Chapter 11

Chapter 13

Chapter 14

Chapter 16

Chapter 17

Chapter 18

Chapter 19

Chapter 20

Chapter 22

Chapter 24

Chapter 27

Chapter 29

Chapter 30

Chapter 31

Chapter 32

Chapter 34

Chapter 36

Chapter 37

Chapter 38

Chapter 42

Appendix I

Illustrators

In this index, page numbers followed by a
t designate tables; page numbers followed
by an *f* refer to figures; page numbers set
in boldface refer to major discussions.

In this index, page numbers followed by a
t designate tables; page numbers followed
by an *f* refer to figures; page numbers set
in boldface refer to major discussions.

In this index, page numbers followed by a
t designate tables; page numbers followed
by an *f* refer to figures; page numbers set
in boldface refer to major discussions.

In this index, page numbers followed by a *t* designate tables; page numbers followed by an *f* refer to figures; page numbers set in boldface refer to major discussions.

In this index, page numbers followed by a *t* designate tables; page numbers followed by an *f* refer to figures; page numbers set in boldface refer to major discussions.

In this index, page numbers followed by a *t* designate tables; page numbers followed by an *f* refer to figures; page numbers set in boldface refer to major discussions.

Microorganism Pronunciation Guide

The pronunciation of each name is given in parentheses. The phonetic spelling system is explained at the beginning of the glossary (p. G1)

Bacteria

Acetobacter (ah-se″to-bak′ter)
Acinetobacter (as″ĭ-net″o-bak′ter)
Actinomyces (ak″tĭ-no-mi′sēz)
Agrobacterium (ag″ro-bak-te′re-um)
Alcaligenes (al″kah-lij′ĕ-nēz)
Anabaena (ah-nab′ē-nah)
Arthrobacter (ar″thro-bak′ter)
Bacillus (bah-sil′lus)
Bacteroides (bak″tĕ-roi′dēz)
Bdellovibrio (del″o-vib′re-o)
Beggiatoa (bej″je-ah-to′ah)
Beijerinckia (bi″jer-ink′e-ah)
Bifidobacterium (bi″fid-o-bak-te′re-um)
Bordetella (bor″dĕ-tel′lah)
Borrelia (bŏ-rel′e ah)
Brucella (broo-sel′lah)
Campylobacter (kam″pĭ-lo-bak′ter)
Caulobacter (kaw″lo-bak′ter)
Chlamydia (klah-mid′e-ah)
Chlorobium (klo-ro′be-um)
Chromatium (kro-ma′te-um)
Citrobacter (sit″ro-bak′ter)
Clostridium (klo-strid′e-um)
Corynebacterium (ko-ri″ne-bak-te′re-um)
Coxiella (kok″se-el′lah)
Cytophaga (si-tof′ah-gah)
Desulfovibrio (de-sul″fo-vib′re-o)
Enterobacter (en″ter-o-bak′ter)
Erwinia (er-win′e-ah)
Escherichia (esh″er-i′ke-ah)
Flexibacter (flek″sĭ-bak′ter)
Francisella (fran-si-sel′ah)
Frankia (frank′e-ah)
Gallionella (gal″le-o-nel′ah)
Haemophilus (he-mof′ĭ-lus)
Halobacterium (hal″o-bak-te′re-um)
Hydrogenomonas (hi-dro″jĕ-no-mo′nas)
Hyphomicrobium (hi″fo-mi-kro′be-um)
Klebsiella (kleb″se-el′lah)
Lactobacillus (lak″to-bah-sil′lus)
Legionella (le″jun-el′ah)
Leptospira (lep″to-spi′rah)
Leptothrix (lep′to-thriks)

Leuconostoc (loo″ko-nos′tok)
Listeria (lis-te′re-ah)
Methanobacterium (meth″ah-no-bak-te′re-um)
Methylococcus (meth″il-o-kok′-us)
Methylomonas (meth″il-o-mo′nas)
Micrococcus (mi″kro-kok′us)
Mycobacterium (mi″ko-bak-te′re-um)
Mycoplasma (mi″ko-plaz′mah)
Neisseria (nīs-se′re-ah)
Nitrobacter (ni″tro-bak′ter)
Nitrosomonas (ni-tro″so-mo′nas)
Nocardia (no-kar′de-ah)
Pasteurella (pas″tĕ-rel′ah)
Photobacterium (fo″to-bak-te′re-um)
Propionibacterium (pro″pe-on″e-bak-te′re-um)
Proteus (pro′te-us)
Pseudomonas (soo″do-mo′nas)
Rhizobium (ri-zo′be-um)
Rhodopseudomonas (ro″do-soo″do-mo′nas)
Rhodospirillum (ro″do-spi-ril′um)
Rickettsia (ri-ket′se-ah)
Salmonella (sal″mo-nel′ah)
Sarcina (sar′sĭ-nah)
Serratia (sĕ-ra′she-ah)
Shigella (shĭ-gel′ah)
Sphaerotilus (sfe-ro′tĭ-lus)
Spirillum (spi-ril′um)
Spirochaeta (spi″ro-ke′tah)
Spiroplasma (spi″ro-plaz′mah)
Staphylococcus (staf″ĭ-lo-kok′us)
Streptococcus (strep″to-kok′us)
Streptomyces (strep″to-mi′sēz)
Sulfolobus (sul″fo-lo′bus)
Thermoactinomyces (ther″mo-ak″tĭ-no-mi′sēz)
Thermoplasma (ther″mo-plaz′mah)
Thiobacillus (thi″o-bah-sil′us)
Thiothrix (thi′o-thriks)
Treponema (trep″o-ne′mah)
Ureaplasma (u-re′ah-plaz″ma)
Veillonella (va″yon-el′ah)
Vibrio (vib′re-o)
Xanthomonas (zan″tho-mo′nas)
Yersinia (yer-sin′e-ah)
Zoogloea (zo″o-gle′ah)